Computational Fluid Dynamics 2000

Springer
Berlin
Heidelberg
New York
Barcelona
Hong Kong
London
Milan
Paris
Tokyo

Nobuyuki Satofuka (Ed.)

Computational Fluid Dynamics 2000

Proceedings of the First International Conference on Computational Fluid Dynamics, ICCFD, Kyoto, Japan, 10–14 July 2000

Springer

Professor Nobuyuki Satofuka
Department of Mechanical and System Engineering
Kyoto Institute of Technology
Matsugasaki, Sakyo-ku
Kyoto, 606-8585, Japan

Library of Congress Cataloging-in-Publication Data
International Conference on Computational Fluid Dynamics (1st: 2000: Kyoto, Japan)
Computational fluid dynamics 2000: proceedings of the First International Conference on Computational Fluid Dynamics, ICCFD, held at Kyoto, Japan, 10–14 July, 2000 / Nobuyuki Satofuka (ed.)
p. cm.
Includes bibliographical references and index.
ISBN 3540414592 (alk. paper)
1. Fluid dynamics–Data processing–Congresses. I. Satofuka, N. (Nobuyuki) II. Title.
QA911 .I534 2000 535'.05–dc21 2001020771

ISBN 3-540-41459-2 Springer-Verlag Berlin Heidelberg New York

Springer-Verlag Berlin Heidelberg New York
a member of BertelsmannSpringer Science+Business Media GmbH

http://www.springer.de

Printed in Germany

Typesetting: Camera-ready by the authors/editors

Printed on acid-free paper SPIN 10793257 54/3141/xo – 5 4 3 2 1 0

Preface

This volume constitutes the Proceedings of the First International Conference on Computational Fluid Dynamics, held at the Kyoto Research Park, Kyoto, Japan on 10-14 July, 2000. The conference is the first one at which the International Conference on Numerical Methods in Fluid Dynamics (ICNMFD) and the International Symposium on Computational Fluid Dynamics (ISCFD) were merged. The purpose of the conference was to bring together scientists, mathematicians, and engineers to review and share recent advances in mathematical and computational techniques for modeling fluid dynamics.

The conference had the following format: Each day of the conference except Wednesday (July 12) started with a plenary session at which an invited lecture was delivered. During the rest of the day there were three sessions in parallel in which oral presentations were made. Poster presentations were also made on Monday, Tuesday and Thursday afternoons.

A total of 205 abstracts were submitted from all over the world, and were evaluated by five paper selection committees chaired by J.J. Chattot (USA), K.W. Morton (UK), M. Napolitano (Italy), K. Srinivas (Austraria) and myself. Out of 136 papers accepted for oral presentations, 17 were withdrawn and out of 46 papers accepted for poster presentation, 14 were cancelled.

In this volume the invited lectures grouped separately from the contributed papers appear first, in the order they were received. Then all the contributed papers accepted for oral presentation have been grouped under 21 topic headings. These are in the order that the topics first appeared in the conference programme. Finally, papers presented in poster sessions appear as Technical Notes.

The conference was attended by 183 people from 24 countries. I hope that all participants carried with them memories of a fruitful and pleasant meeting held in Kyoto, a city with both natural beauty and a rich historical legacy. Its success owes much to the effort of the Scientific Committee, to my associate K. Matsuno, K. Morinishi and H. Nishida of the Local Organizing Committee, and above all to Ms. F. Yanagawa, the conference secretary. Thanks are also due to M. Obata for preparing the Proceedings.

Kyoto, September 2000 *Nobuyuki Satofuka*

Preface

[illegible]

Sponsors

- Kyoto Institute of Technology
- Ministry of Education, Science, Sports and Culture
- National Aerospace Laboratory
- National Aeronautics and Space Administration, USA
- Commemorative Association for the Japan World Exposition(1970)
- Kyoto City
- Kyoto Prefecture
- Japan Society of Computational Fluid Dynamics
- Japan Society for Aeronautics and Space Sciences
- Japan Society of Fluid Dynamics
- Japan Society of Mechanical Engineers
- Japan Society for Computational Engineering and Sciences
- Hewlett-Packard Japan, Ltd.
- Ishikawajima-Harima Heavy Industries Co., Ltd.
- KGT Inc.
- NKK Corp.
- Toray Industries, Inc.
- Towa Corp.
- Toyota Motor Corp.

This conference is hosted by Kyoto Institute of Technology as a memorial event celebrating 100 years anniversary of establishment.

This conference has been executed with a grant from the Commemorative Association for the Japan World Exposition (1970).

Contents

Part X Flows with Free Surface

Part XI Numerical Techniques

Part XVI Adjoint Method

Part XVII Vortex Method

Part XVIII Two-phase Flows

Part XXI Moving Boundaries

Part XXII Level Set / VOF

Part XXIII Technical Notes

Part I

Invited Lectures

Progress in Unstructured-Grid CFD

Kazuhiro Nakahashi

Tohoku University, Aoba 01, Sendai 980-8579, JAPAN
email: naka@ad.mech.tohoku.ac.jp

1 Introduction

In the last 30 years, computational fluid dynamics (CFD) has achieved a significant progress in its algorithms and the applications. In these days CFD is considered to be very close to its maturing stage. It is certainly true that CFD is routinely applied to flows around airfoils and wings for the performance analysis and design.

However, CFD is still far from being a maturing subject for geometrically and physically complex flows. One of such problems is the computation of three-dimensional high-lift flows where complicated geometries, involving flaps, slats, and hinge fairings, must be faithfully reproduced in the computational space [1]. Rotorcraft aerodynamics is a more challenging subject for the current CFD, not only because of the geometrical complexity but also due to the complex flow physics ranging from the small vortices shedding from the rotor blades to the large scale flow around the airframe [2]. For these problems, the current CFD still needs painful efforts for the grid generation and the enormous CPU times to obtain the desired accuracy level. Currently, three approaches have been developed to treat such flow problems of complex geometries; overset structured grid, Cartesian grid, and unstructured grid.

Unstructured grids are becoming popular in the CFD community especially in the last decade because of their advantages over conventional structured grids for complex geometries. The unstructured grids are flexible in treating complex geometry and have an adaptive refinement/unrefinement capability that covers the various flow scales appeared in the complex flows. The computational efficiency of unstructured grid Euler and Navier-Stokes solvers has been significantly improved by recent developments in multigrid, implicit time integration, and other convergence acceleration methods (for example, [3], [4], [5], [6]). High Reynolds number viscous flows can be also accurately solved by employing an anisotropic stretched tetrahedral mesh or a hybrid grid near solid surfaces (for example, [7], [8], [9], [10]).

The present paper attempts to briefly review recent progresses in the CFD approaches for complex geometries, and describes several efforts performed by the author and his group to improve and extend the capability of the unstructured-grid CFD. It consists of four subjects; unstructured-grid CFD for geometrically complex flows, for physically complex flows, for high-Reynolds number viscous flows, and for moving bodies.

2 Unstructured grid for geometrically complex flows

As the performance of computers increase, the demand for easily applied pre-processing procedures for treating complicated configurations increases. The pre-processing includes the geometry definition, the surface meshing and the volume gridding. Among them, the process of volume grid generation has received considerable attention by researchers because of the need to efficiently and effectively distribute grid points to generate the most accurate flow solution possible. To date, very complicated configurations have been treated by the tetrahedral unstructured grid (for example [11], [12], [13], [14]). The techniques of the tetrahedral volume grid generation have almost matured.

However, as the complexity of the configuration increase, the burden for the surface gridding abruptly increases. Today, considerable progress has been made in the geometry definition by CAD/CAM softwares, but the surface meshing to translating the CAD output into data understandable by the flow solver is still not a routine process. The unstructured grid needs a 'waterproof' surface mesh and this fact is one of the critical disadvantages of the unstructured-grid CFD compared to the Cartesian grid method where the surface mesh is basically generated automatically during the process of the volume grid generation. The demand for a close-coupled procedure between the geometry definition and the surface meshing is becoming more and more important as the complexity of the geometry increases.

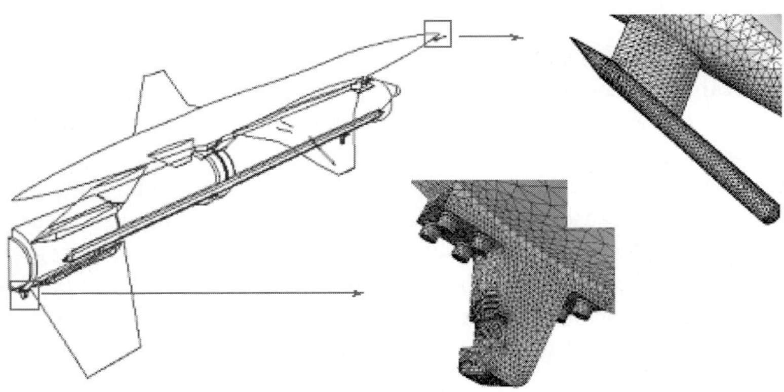

Fig. 1. Geometrical feature lines and the generated surface mesh [15]

We have developed a surface triangulation method [15] where the geometrical feature lines (Fig. 1) extracted from the CAD-defined data are utilized to simplify the procedure. The use of the geometrical feature lines is very useful to accurately retain the original configuration defined in CAD and to minimize the user interventions for the surface gridding. Figure 2 shows a result of the NAL experimental airplane piggybacked on a rocket booster obtained by solving the Euler equations.

Fig. 2. Computed pressure distributions on the NAL experimental airplane piggy-backed on a rocket booster at Mach 2 [15]

3 Unstructured grid for physically complex flows

Physically complex flows appear in high-speed problems where many shock waves, slip surfaces and shear layers complicatedly interact with each other, and in the separated flows which are characterized by shear layers and vortex lines. For these flows, solution adaptive grid methods have shown their effectiveness to improve the solution accuracy in structured, unstructured, and Cartesian grids.

There are two important key technologies in the solution adaptive grid method. One is the algorithm to efficiently and smoothly relocate/add mesh points in the specified regions. This moving/adding technique seems to be well developed already. However, another key technology, how to determine the clustering/refining regions from flow solutions, should be investigated more as the computational size increases. Most approaches specify a threshold value for some error indicators to identify regions where the grid density should be increased. The error indicators may be absolute values of the derivative or the second derivative of the computed density, for example in high speed flows. The threshold value controls the degree of the grid density and usually is determined by a try-and-error procedure to fit the required accuracy in a given computer environment. As the computational size increases, however, the cost and the user burden to determine the appropriate threshold increase.

One approach to remove or reduce the necessity to select the threshold value is to use some geometrical lines/surfaces, such as shock waves, shear layers and vortex lines, appeared in the flow solutions. These flow feature lines characterize the flow and are appropriate for the adaptive grid indicators. Although the identification of these flow feature lines may need some threshold value, it is less sensitive. Figure 3 shows the vortex centerlines and the reverse flow regions appeared in a flow around a high-alpha delta wing. By using these lines as the adaptive refinement indicator, the vortex breakdown was accurately predicted [16].

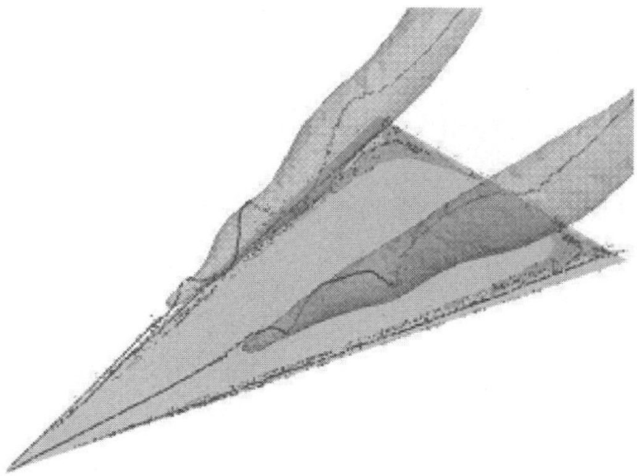

Fig. 3. Vortex centerlines used for the flow feature adaptation and the computed reverse flow regions on a delta wing at angle of attach of 40 degrees [16].

4 Unstructured grid for high-Reynolds number viscous flows

High-Reynolds number viscous flows are still tough problems for the current CFD, mainly because of the inadequate transition/turbulence models that required in the simulation using the Reynolds-Averaged Navier-Stokes (RANS) equations. In the unstructured- grid CFD, there are two more reasons for the difficulty. One of the reasons is the difficulty of generating a volume grid that accurately and efficiently resolves thin boundary layers. Another is a requirement of large computational time and memory to obtain the meaningful accuracy level for the complex three-dimensional problems.

If an isotropic tetrahedral mesh is used in the entire field of a high-Reynolds number viscous flow, the required number of mesh points to obtain the engineeringly meaningful solution easily becomes out of the current computer capacity. Therefore, there have been many discussions about the volume grid approaches for high-Reynolds number viscous flows: a combination of structured and unstructured grids [17], prismatic grid [18], anisotropic highly stretched tetrahedral grid [7], [9], and hybrid grids [8], [10]. At present, anisotropic and hybrid grids are often used for resolving wall boundary layers and wakes. Figure 4 is an example computed by a prismatic-tetrahedral hybrid grid [4]. However, the control of the grid anisotropy is still not straightforward for those unstructured grids. The difficulty in the viscous grid generation hinders the application of the unstructured- grid methods to high Reynolds number viscous flows. The reliable generation of high quality grids for RANS simulations is still one of the most important research subjects for the unstructured-grid CFD.

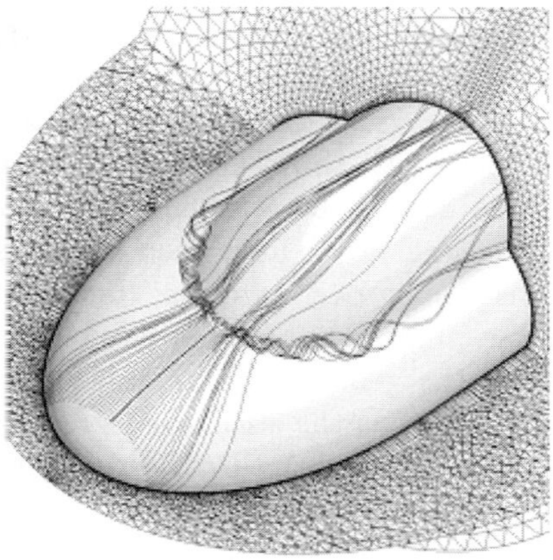

Fig. 4. Hypersonic viscous flow computation using hybrid grid [4].

Computational time is another problem for large-scale simulations using the unstructured grids. Flow solvers on unstructured grids suffer from the computational efficiency due to the indirect addressing. Moreover, very fine grids used for the thin boundary layers in high Reynolds number viscous flow computations causes a stiffness problem of the flow solver. At the early stage of unstructured-grid CFD development, explicit time integration schemes had been commonly used due to the difficulty of the implicit time integration on a randomly numbered unstructured mesh. Therefore, the three-dimensional RANS computations on unstructured grids were rarely seen in the past, though poor computer performance was another reason.

In the last five years, there has been a significant improvement in the computational efficiency of the unstructured-grid flow solvers as mentioned in the introduction. However we still need more efforts in improving the efficiency and solution accuracy of the unstructured-grid CFD especially for unsteady viscous flows around real configurations.

5 Unstructured grid for moving bodies

Even with the recent progress of the unstructured-grid CFD, difficulties still remain in its application to some real engineering problems. One such problem is the unsteady flow computation around multiple bodies in relative motion. The unstructured grid is capable of treating moving bodies as well. However, for computation of unsteady flows with moving bodies, a part of the computational grid or in some cases the whole grid has to be regenerated at every time step. This procedure may become quite complex and computationally expensive. The sophistication of the programs for moving/refinement/unrefinement and regeneration of the grid is a severe hurdle to overcome for most of the new researchers of unstructured grids. A simple and efficient approach should be developed to treat such a problem.

In structured grids, moving body problems have been successfully simulated by the overset grid approach [19]. In this method, valid Chimera-holes at first need to be cut in each grid in regions that overlap with solid bodies or any other non-flow regions which belong to the other grids of the overset grid system. Next, interpolation stencils have to be established for points that lie along the hole-boundary. Although several codes have been developed for constructing the intergrid communications among overset grids, they seem to require technical expertise and considerable man-hours. One of the reasons for this difficulty is that, as the complexity of the geometry increases, the number of grids which must be overlap increases so far as the structured grids are used.

It is a reasonable idea to use unstructured grids for the overset concept. By use of unstructured meshes, the number of submeshes required for covering the flow filed around complex geometries can be significantly reduced as compared with that needed in the overset structured grid. It can also extend the applicability of the unstructured-grid method to multiple moving-body problems without much need for code development. However, the overset unstructured-grid method is relatively unexplored, and various aspects such as automatic hole-cutting, efficient search, interpolations, etc., remain to be developed.

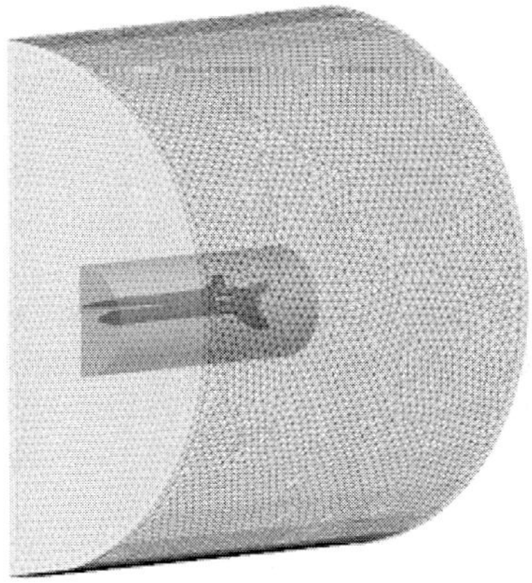

Fig. 5. Overset grids for supersonic airplane (outer cylindrical region) and rocket booster (inner cylindrical region) [21].

We proposed an efficient and reliable algorithm to automatically localize the intergrid boundaries for the overset unstructured-grid method [20]. The intergrid boundaries are localized using the distance to the wall. The search for donor cells is efficiently performed by the neighbor-to-neighbor jump search on a modified convex domain utilizing a byproduct of the Delaunay triangulation method. The method was applied to a separation simulation of a supersonic airplane and a rocket booster [21]. Figure 5 shows overset two unstructured grids and Fig. 6 show the intergrid boundaries during the separation process and the computed pressure distributions.

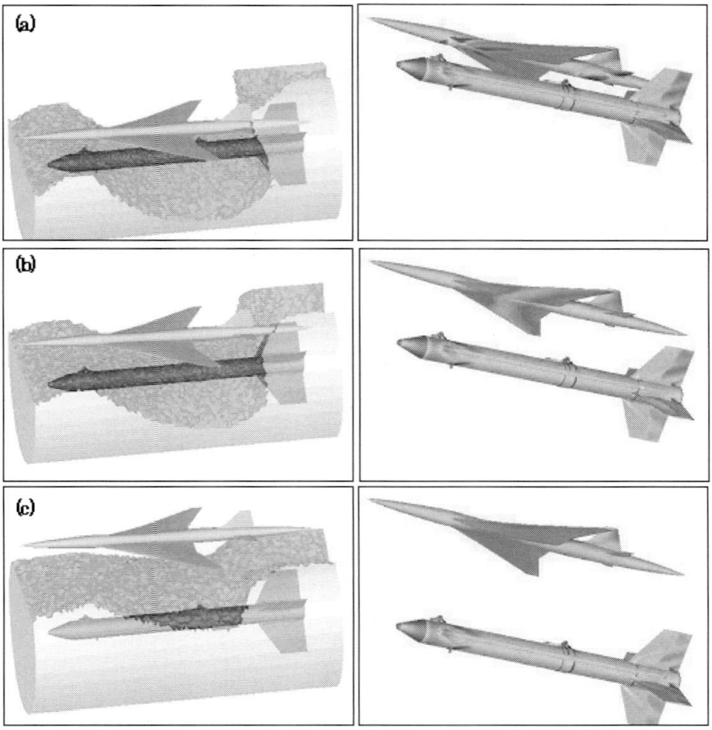

Fig. 6. Intergrid boundaries of the rocket booster subgrid (left) and the computed pressure distributions (right) of the airplane-booster separation simulation [21]

6 Challenges of the unstructured-grid CFD

Perhaps, one of the most challenging problems for the current CFD is a flow around an insect and a bird in flight as shown in Fig.7 and Fig.8. It includes not only the complexity of the geometry, but also the moving and deforming bodies, the fluid-structure coupling, and the complex flow physics due to periodically shedding vorticies from the flapping wings. Although it is still beyond our capability to accurately simulate it, such a challenge will advance the unstructured grid CFD.

Fig. 7. Surface and volume grids around a hornet in flight.

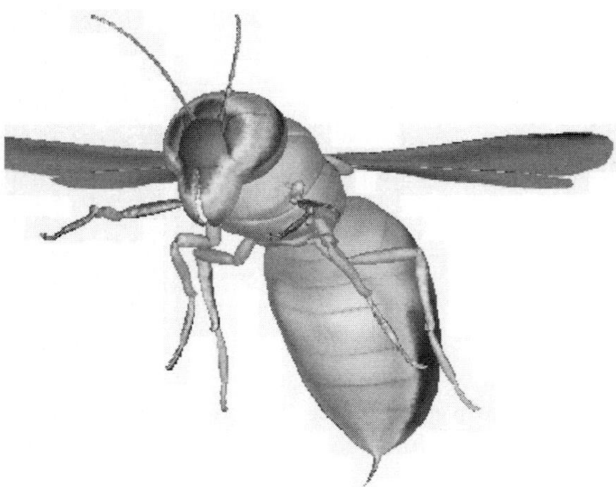

Fig. 8. Computed pressure distribution around a hornet (no-flapping, $Re = 10^4$.)

Another challenging subject, for example, is an accurate drag prediction of a golf ball. It is well known that the sudden decrease in the value of the drag coefficients occurs near $Re = 3.0 \times 10^5$ in the case of spheres. For a golf ball, this sudden decrease occurs about $Re = 4 \times 10^4$ due to the boundary layer transition caused by dimples. Already, beautiful flow features around a baseball ball were reproduced in the computational space by Himeno [22]. For a golf ball, however, unstructured grid will be more appropriate than the structured grid because of a number of dimples on the surface as shown in Fig.9. At present, the hybrid grid method can produce qualitatively reasonable flows around the golf ball (Fig.10). But the accurate flow simulation requires more dense grid, higher-order numerical scheme, and may need reliable transition and turbulence models.

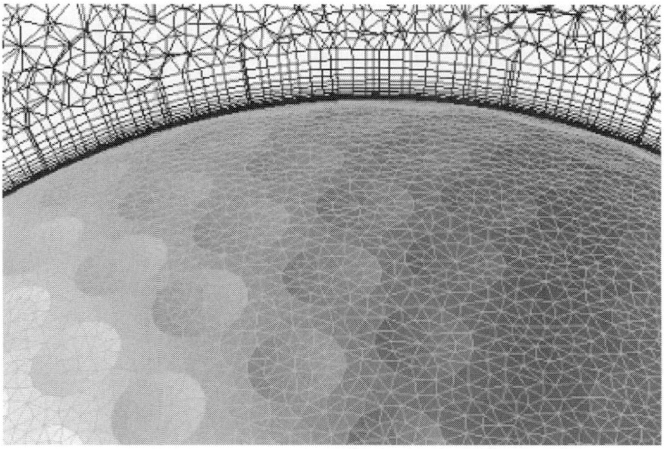

Fig. 9. Unstructured grid around a golf ball

In the last thirty years, computational fluid dynamics has been developed for the requirements mainly in the aerospace engineering. Recently, however, its application is enlarged in the various non-aerospace fields, such as environmental and biological flows as well as non-aerospace engineering problems. In these fields, the flowfield geometry is usually complex and the flow physics is more mysterious. Current CFD still has a severe limitation for these real world simulations that require very large-scale computations and physical modeling. However, it is the time for the computational fluid dynamics to open up a new world in the flow sciences. Beauty of flow physics, wonder of flying creatures and fun in the sports revealed by CFD will enrich our lives and attract young generation. This is the next role of the CFD as one of the computational sciences.

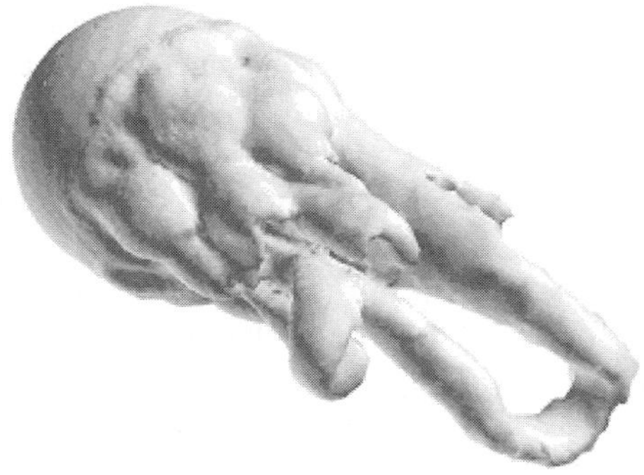

Fig. 10. Computed equi-volticity surface at $Re = 2 \times 10^5$

Acknowledgements

The author is grateful to his colleagues and students D. Sharov, T. Kato, Y. Ito, M. Murayama, F. Togashi and T. Fujita who contributed to various numerical developments and computations covered in the present paper.

References

1. S. E. Rogers, H. V. Cao, and T. Y. Su: 'Grid Generation for Complex High-Lift Configurations', AIAA Paper 98-3011, June (1998)
2. F. X. Caradonna: 'Developments and Challenges in Rotercraft Aerodynamics', AIAA Paper 2000-0109, January (2000)
3. D. J. Mavrpiplis: 'Three-Dimensional High-Lift Analysis Using a Parallel Unstructured Multigrid Solver', NASA/CR-1998-207682, ICASE Report No. 98-20, (1998)
4. D. Sharov and K. Nakahashi: 'Reordering of Hybrid Unstructured Grids for Lower-Upper Symmetric Gauss-Seidel Computations'. AIAA Journal, **36**, 3, pp.484–486, (1998)
5. H. Luo, J. Baum, and R. Löhner: 'A Fast, Matrix-free Implicit Method for Compressible Flows on Unstructured Grids', J. Comp. Phy., **146**, pp.664–690, (1998)
6. N. Frink: 'Tetrahedral Unstructured Navier-Stokes Method for Turbulent Flows', AIAA Journal, **36**, 11, pp.1975–1982, (1998).
7. S. Pirzadeh: 'Three-Dimensional Unstructured Viscous Grids by the Advancing Layers Method', AIAA Journal, **34**, 1, pp.43–49, (1996)
8. D. Sharov and K. Nakahashi: 'Hybrid Prismatic/Tetrahedral Grid Generation for Viscous Flow Applications', AIAA J., **36**, 2, pp.157–162, (1998).
9. R. Löhner: 'Generation of Unstructured Grids Suitable for RANS Calculations', AIAA Paper 99-0662, January (1999)
10. A. Khawaja and A. Kallinderis: 'Adaptive Hybrid Grid Generation for Turbomachinery and Aerospace Applications', AIAA Paper 99-0916, January (1999)
11. R. Löhner and P. Parikh: 'Three-Dimensional Grid Generation by the Advancing Front Method', Int. J. Num. Meth. Fluids, **8**, pp.1135–1149, (1988)

12. N. P. Weatherill and O. Hassan: 'Efficient Three-Dimensional Delaunay Triangulation with Automatic Point Creation and Imposed Boundary Constraints', Int. J., Num. Meth., Engng., **37**, pp.2005–2039, (1994)
13. D. Marcum and N. Weatherill: 'Unstructured Grid Generation Using Iterative Point Insertion and Local Reconnection', AIAA Journal, **33**, p. 1619, (1995)
14. D. Sharov and K. Nakahashi: 'A Boundary Recovery Algorithm for Delaunay Tetrahedral Meshing', 5th Int. Conf. on Numerical Grid Generation in Computational Field Simulations, pp.229–238, (1996)
15. Y. Ito and K. Nakahashi: 'Direct surface triangulation using stereolithography (STL) data', AIAA Paper 2000-0924, (2000)
16. M. Murayama, K. Nakahashi, and K. Sawada: 'Numerical Simulation of Vortex Breakdown Using Adaptive Grid Refinement with Vortex-Center Identification', AIAA Paper 2000-0806, (2000)
17. K. Nakahashi: 'FDM-FEM Zonal Approach for Computations of Compressible Viscous Flows', Proc. of Tenth Int. Conf. on Numerical Methods in Fluid Dynamics, Lecture Notes in Physics **264**, pp.494–498, (1986)
18. K. Nakahashi: 'A Finite-Element Method in Prismatic Elements for the Three-Dimensional Navier-Stokes Equations', Proc. of 11th Int. Conf. on Numerical Methods in Fluid Dynamics, Lecture Notes in Physics **323**, pp.434–438, (1988)
19. J. L. Steger, F. C. Dougherty, and J. A. Benek: 'A Chimera grid scheme', ASME Mini-Symposium on Advances in Grid Generation, (1982)
20. K. Nakahashi, F. Togashi, and D. Sharov: 'An Intergrid-Boundary Definition Method for Overset Unstructured Grid Approach', AIAA Paper 99-3304-CP, (1999) (to appear in AIAA J.)
21. F. Togashi, K. Nakahashi, Y. Ito, Y. Shinbo, T. Iwamiya: 'Flow Simulation of NAL Experimental Supersonic Airplane/Booster Separation', AIAA Paper 2000-1007, (2000)
22. R. Himeno: 'A Numerical Analysis of the Effect of Seam Line on Flows around a Rotating Baseball', Proceedings of the 32nd Fluid Dynamics Conference (in Japanese), pp. 255-258, (2000)

Gas-Kinetic Schemes for Fluid Simulations

Kun Xu

Mathematics Department, The Hong Kong University of Science & Technology, Clear Water Bay, Kowloon, Hong Kong

Abstract. In this paper, we are going to first present a gas-kinetic scheme based on the Bhatnagar-Gross-Krook (BGK) model for the Navier-Stokes flow simulations. Then, the relations among the BGK scheme, the central difference, Godunov, kinetic Flux Vector Splitting, and Lattice Boltzmann method will be discussed. A few comments on CFD algorithm development will also be presented.

1 Preliminaries

Based on the gas-kinetic theory, the Navier-Stokes equations can be derived from the Boltzmann equation using the Chapman-Enskog expansion. Therefore, a Navier-Stokes solver can be constructed by solving the Boltzmann equation, especially the simplified collision models [2]. In the gas-kinetic representation, all flow variables are moments of a particle distribution function. Since the gas distribution function is used to describe both equilibrium and nonequilibrium states, the inviscid and viscous fluxes are obtained simultaneously. Many kinetic schemes are based on the collisionless Boltzmann equation, $f_t + uf_x = 0$, as the governing equation to evaluate the numerical fluxes across a cell interface. The free transport mechanism implicitly sets the time step Δt as the particle collision time τ, which generates a numerical viscosity being proportional to it, i.e., $\mu_n \sim \Delta t$. Instead of solving the collisionless Boltzmann equation, a collisional BGK model is solved for the numerical fluxes in the BGK scheme, i.e., $f_t + uf_x = (g-f)/\tau$ [1]. As a result, the dissipation in the transport process is controlled by the collision time τ. This paper will present the BGK scheme for the Navier-Stokes simulations. It will be shown that the Navier-Stokes solutions can be obtained accurately in both $\tau > \Delta t$ and $\tau < \Delta t$ regions.

2 BGK Flow Solver

Following van Leer's MUSCL idea, a numerical scheme is composed of an initial reconstruction stage followed by a dynamical evolution stage. With the implementation of nonlinear limiter, the reconstructed distribution of conservative variables inside each cell becomes $\bar{w}_j(x) = w_j + L(s_+, s_-)(x - x_j)$, where w is the conservative varibales, such as the mass, momentum and energy, $s_+ = (w_{j+1} - w_j)/\Delta x$ and $s_- = (w_j - w_{j-1})/\Delta x$ are slopes, and $L(s_+, s_-)$ is the nonlinear limiter. After reconstruction, the flow distribution around a cell interface is shown in Fig.(1). The BGK scheme is basically to present a Navier-Stokes solution based on the kinetic equation for the above initial condition.

For a 2D flow, the BGK model in the x-direction is,

$$f_t + u f_x = \frac{g - f}{\tau}, \tag{1}$$

where f is the gas distribution function and g is the equilibrium state approached by f. The particle collision time τ is related to the viscosity and heat conduction coefficients. The equilibrium state is a Maxwellian distribution,

$$g = \rho\left(\frac{\lambda}{\pi}\right)^{\frac{K+2}{2}} e^{-\lambda((u-U)^2+(v-V)^2+\xi^2)},$$

where ρ is the density, U and V are the macroscopic velocities in the x and y directions, and λ is related to the gas temperature $m/2kT$. The total number of degrees of freedom K in ξ is equal to $(5-3\gamma)/(\gamma-1)+1$. In the equilibrium state, ξ^2 is equal to $\xi^2 = \xi_1^2 + \xi_2^2 + ... + \xi_K^2$. The relation between mass ρ, momentum $(m = \rho U, n = \rho V)$, and energy E densities with the distribution function f is

$$w = (\rho, m, n, E)^T = \int \psi_\alpha f d\Xi = \int \psi_\alpha g d\Xi, \qquad \alpha = 1,2,3,4, \tag{2}$$

where ψ_α is the vector of moments

$$\psi_\alpha = (1, u, v, \frac{1}{2}(u^2 + v^2 + \xi^2))^T,$$

and $d\Xi = dudvd\xi$ is the volume element in the phase space with $d\xi = d\xi_1 d\xi_2 ... d\xi_K$.

The BGK scheme is based on the solution f of the BGK model at a cell interface $x_{j+1/2}$,

$$f(x_{j+1/2}, t, u, v, \xi) = \frac{1}{\tau}\int_0^t g(x', t', u, v, \xi) e^{-(t-t')/\tau} dt' + e^{-t/\tau} f_0(x_{j+1/2} - ut), \tag{3}$$

where $x' = x_{j+1/2} - u(t-t')$ is the trajectory of a particle motion and f_0 is the initial gas distribution function f at the beginning of each time step $(t = 0)$. Two unknowns g and f_0 must be specified in Eq.(3) in order to obtain the solution f. In order to simplify the notation, $x_{j+1/2} = 0$ will be used in the following text.

In a general situation, the initial gas distribution function f_0 is assumed to have the form,

$$f_0 = \begin{cases} g^l\left(1 + a^l x - \tau(a^l u + A^l)\right), & x \leq 0 \\ g^r\left(1 + a^r x - \tau(a^r u + A^r)\right), & x \geq 0 \end{cases} \tag{4}$$

where a^l and a^r are coming from the spatial derivative of a Maxwellian and have a unique correspondence with the slopes of the conservative variables. The terms $-\tau(a^l u + A^l)g^l$ and $-\tau(a^r u + A^r)g^r$ account for the deviation of a gas distribution function away from a Maxwellian. The above nonequilibrium parts are obtained from the Chapman-Enskog expansion of the BGK model, and these parts have no direct contribution to the conservative variables, i.e.,

$$\int (a^l u + A^l)\psi_\alpha g^l d\Xi = 0 \quad , \quad \int (a^r u + A^r)\psi_\alpha g^l d\Xi = 0. \tag{5}$$

The equilibrium state g around $(x = 0, t = 0)$ is assumed to have the form,

$$g = g_0 \left(1 + (1 - \mathrm{H}[x])\bar{a}^l x + \mathrm{H}[x]\bar{a}^r x + \bar{A}t\right), \tag{6}$$

where $\mathrm{H}[x]$ is the Heaviside function. Here g_0 is a local Maxwellian distribution function located at $x = 0$. Even though, g is continuous at $x = 0$, but it has different slopes at $x < 0$ and $x > 0$ regions, see Fig.(2). The dependence of $a^l, a^r, ..., \bar{A}$ on the particle velocities can be obtained from a Taylor expansion of a Maxwellian and have the following form,

$$a^l = a_1^l + a_2^l u + a_3^l v + a_4^l \frac{1}{2}(u^2 + v^2 + \xi^2) = a_\alpha^l \psi_\alpha,$$

$$\dots$$

$$\bar{A} = \bar{A}_1 + \bar{A}_2 u + \bar{A}_3 v + \bar{A}_4 \frac{1}{2}(u^2 + \xi^2) = \bar{A}_\alpha \psi_\alpha,$$

where $\alpha = 1, 2, 3, 4$ and all coefficients $a_1^l, a_2^l, ..., \bar{A}_4$ are local constants, which depend on temporal and spatial derivatives of the local macroscopic flow variables.

In the reconstruction stage, we have obtained the distributions $\bar{\rho}_j(x)$, $\bar{m}_j(x)$, $\bar{n}_j(x)$, and $\bar{E}_j(x)$ inside each cell $x_{j-1/2} \leq x \leq x_{j+1/2}$. At the cell interface $x_{j+1/2}$, the left and right macroscopic states are $\bar{w}_j(x_{j+1/2})$ and $\bar{w}_{j+1}(x_{j+1/2})$. By using the relation between the gas distribution function f and the macroscopic variables (Eq.(2)), around $x_{j+1/2}$ we get

$$\int g^l \psi_\alpha d\Xi = \bar{w}_j(x_{j+1/2}) \; ; \; \int g^l a^l \psi_\alpha d\Xi = \frac{\bar{w}_j(x_{j+1/2}) - \bar{w}_j(x_j)}{\Delta x^-} \tag{7}$$

$$\int g^r \psi_\alpha d\Xi = \bar{w}_{j+1}(x_{j+1/2}) \; ; \; \int g^r a^r \psi_\alpha d\Xi = \frac{\bar{w}_{j+1}(x_{j+1}) - \bar{w}_{j+1}(x_{j+1/2})}{\Delta x^+} \tag{8}$$

where $\Delta x^- = x_{j+1/2} - x_j$ and $\Delta x^+ = x_{j+1} - x_{j+1/2}$. The above equations uniquely determine g^l, g^r, a^l and a^r. After having a^l and a^r, A^l and A^r terms in f_0 can be found from Eq.(5).

The equilibrium state g_0 is obtained from the compatibility condition at $(x_{j+1/2}, t = 0)$,

$$\int g_0 \psi_\alpha d\Xi = w_0 = \int_{u>0} \int g^l \psi_\alpha d\Xi + \int_{u<0} \int g^r \psi_\alpha d\Xi. \tag{9}$$

Then, $\bar{a}^l$ and $\bar{a}^r$ of g in Eq.(6) can be obtained through the relation of

$$\frac{\bar{w}_{j+1}(x_{j+1}) - w_0}{\rho_0 \Delta x^+} = \int g_0 \bar{a}^l \psi_\alpha d\Xi \quad , \quad \frac{w_0 - \bar{w}_j(x_j)}{\rho_0 \Delta x^-} == \int g_0 \bar{a}^r \psi_\alpha d\Xi. \tag{10}$$

After substituting Eq.(4) and Eq.(6) into Eq.(3), the gas distribution function f at a cell interface can be expressed as

$$\begin{aligned} f(x_{j+1/2},t) &= (1-e^{-t/\tau})g_0 + \tau(t/\tau - 1 + e^{-t/\tau})\bar{A}g_0 \\ &+ \left(\tau(-1+e^{-t/\tau}) + te^{-t/\tau}\right)\left(\bar{a}^l \mathrm{H}[u] + \bar{a}^r(1-\mathrm{H}[u])\right) ug_0 \\ &+ e^{-t/\tau}\left((1-u(t+\tau)a^l)\mathrm{H}[u]g^l + (1-u(t+\tau)a^r)(1-\mathrm{H}[u])g^r\right) \\ &+ e^{-t/\tau}\left(-\tau A^l \mathrm{H}[u]g^l - \tau A^r(1-\mathrm{H}[u])g^r\right). \end{aligned} \quad (11)$$

The only unknown term left in the above expression is $\bar{A}$. Since both f (Eq.(11)) and g (Eq.(6)) contain $\bar{A}$, the integration of the conservation constraint Eq.(2) at $x_{j+1/2}$ over the whole time step Δt determines $\bar{A}$ uniquely,

$$\int_0^{\Delta t} \int (g-f)\psi_\alpha dt d\Xi = 0.$$

Finally, the time-dependent numerical fluxes in the x-direction across the cell interface can be computed by

$$\begin{pmatrix} \mathcal{F}_\rho \\ \mathcal{F}_m \\ \mathcal{F}_n \\ \mathcal{F}_E \end{pmatrix}_{j+1/2} = \int u \begin{pmatrix} 1 \\ u \\ v \\ \frac{1}{2}(u^2+v^2+\xi^2) \end{pmatrix} f(x_{j+1/2},t,u,v,\xi) d\Xi, \quad (12)$$

where $f(x_{j+1/2},t,u,v,\xi)$ is given in Eq.(11). By integrating the above equation in a time step, we can get the total mass, momentum and energy transport.

3 Analysis

3.1 Navier-Stokes Solver

Due to the implementation of initial nonequilibrium state and the BGK gas evolution model, the BGK scheme presented in the last section gives an accurate Naver-Stokes solution in both smooth and discontinuity regions. Eq.(11) presents an explicit time-dependent gas distribution function f at the cell interface. In a resolved flow region, such as a resolved shock layer, the reconstructed conservative variables in Fig.(1) will be distributed approximately on a straight line. In such a case, the gas distribution function f at a cell interface becomes

$$f = g_0(1 - \tau(u\bar{a} + \bar{A}) + t\bar{A}), \quad (13)$$

where $-\tau(u\bar{a}+\bar{A})g_0$ is exactly the nonequilibrium state in the Chapman-Enskog expansion of the BGK model [23], and $g_0\bar{A}t$ is the temporal evolution part of the gas distribution function. The above equation (13) is actually the equation we used for the low Mach number viscous flow calculations [18]. Eq.(13) is basically a kinetic Lax-Wendroff type scheme for the Navier-Stokes equations. Therefore, in the smooth flow region the BGK method goes to a 2nd-order central difference scheme.

3.2 Collision time

In a well resolved dissipative region, such as the cell size Δx is smaller than the dissipative length scale determined by the physical viscosity, the collision time τ in the BGK scheme can be determined by the relation $\tau = \mu/p$, where μ is the dynamical viscosity coefficient and p is the pressure. Theoretically, any dissipative structure, such as the shock thickness, is determined by the physical viscosity, and the structure should be independent of the cell size. However, even though the Navier-Stokes equations are solved accurately by the BGK method, if the cell size is not fine enough to resolve the wave structure, the physical wave structure has to be replaced by a numerical one, such as the physical shock thickness being replaced by the numerical cell size. In such a situation, we cannot solve the Navier-Stokes equations with the original physical viscosity. The effective viscosity should be a combination of the physical and numerical ones. Different from many upwinding schemes, the BGK method cannot simply take the apology to say that the implicit numerical viscosity is included in the under-resolved flow region. Since the BGK method is so accurate in solving the Navier-Stokes equations even in the under-resolved discontinuity region, the required additional numerical viscosity has to be explicitly included. Because the jump in the flow variables at a cell interface, see Fig.(1), represents the under resolveness and it appears automatically in high gradient flow region, the collision time τ used in all simulations will take the jump into account,

$$\tau = \frac{\mu}{p} + \frac{|p^l - p^r|}{|p^l + p^r|}\Delta t, \tag{14}$$

where Δt is the CFL time step and the second term provides the numerical viscosity, which is a function of the pressure jump at the cell interface in the reconstructed initial data. The obvious advantage of the BGK method is that it solves a viscous equation with an explicit dissipative coefficient, which avoids the ambiguity of implicit dissipation in many upwinding schemes. Even though the shock jump can be captured nicely in the Godunov type schemes, but the dissipation to construct such a wide shock transition is mainly coming from numerics. It does not guarantee that the same numerical dissipation will not poison the physical viscous solution, or the dissipation will appear when necessary in multidimensional case.

Even the Navier-Stokes equations with an adaptive local viscous coefficient can be solved by the Godunov-type method, there are still difference between it and the BGK method. The BGK scheme gives a solution directly under the general initial condition (Fig.(1)), it is difficult to design an unsplitting Godunov method for the Navier-Stokes equations from the same initial condition. Even for the same mass, momentum, and energy distributions, see Fig.(1) again, the kinetic scheme uses a non-equilibrium state f_0 to describe it. The macroscopic description could only see an equilibrium state.

3.3 KFVS Limit

The BGK scheme is valid for the Navier-Stokes solution in both $\tau < \Delta t$ and $\tau > \Delta t$ region. In the limit of $\tau >> \Delta t$, the solution of the gas distribution function f in Eq.(11) reduces to

$$f = (1 - \tau(ua^l + A^l))\mathrm{H}[u]g^l + (1 - \tau(ua^r + A^r))(1 - \mathrm{H}[u])g^r. \tag{15}$$

This distribution is similar to the one used by Chou and Baganoff in their Kinetic Flux Vector Splitting (KFVS) Navier-Stokes (NS) solver [6]. In their approach, a direct implementation of the Chapman-Enskog distribution of the Boltzmann equation is used to split the flux. From the BGK scheme, we can clearly understand the limitation of the Chou-Baganoff's Kinetic KFVS NS method. Because Eq.(15) is the limiting case of Eq.(11) when $\tau >> \Delta t$, Eq.(15) is only valid for the Navier-Stokes solution under such a limiting condition. In other words, KFVS NS scheme approaches the Navier-Stokes solution accurately only in the case $\mu/p >> \Delta t$.

For the Euler solutions, Eq.(15) can be further simplified,

$$f = \mathrm{H}[u]g^l + (1 - \mathrm{H}[u])g^r, \tag{16}$$

where the nonequilibrium state is totally removed. This is precisely the KFVS scheme for the compressible Euler equations [15,11]. A earlier version of the above scheme is the beam scheme, where instead of Maxwellians the equilibrium states g^l and g^r are replaced by three Delta functions or particles [16]. As analyzed recently [19], the Steger-Warming method can be represented as a "beam scheme". Because of their slight difference in the particle representation, the Steger-Warming method is less robust than the beam scheme due to the lack of internal energy in its second "particle". Physically, the KFVS, beam, Steger-Warming and many other FVS schemes are equivalent. Therefore, it is easy to understand the poor performance of FVS schemes in the viscous boundary layer calculations [17,20,21], where the condition $\mu/p >> \Delta t$ is not satisfied once there are only a few grid points in the boundary layer.

Lattice Boltzmann method is very fortunate in this aspect. It does present an accurate NS solution in the incompressible limit [4]. The reason for this is that with a symmetric lattice, the free particle transport from one node to another node (no averaging) could generate anti-diffusive term which is consistent with the incompressible Navier-Stokes equations and the viscous coefficient is proportional to $-\frac{1}{2}\Delta t$. Therefore, the numerical dissipative term can be absorbed in the physical one [10]. As a result, the final viscosity coefficient in the Lattice BGK model is proportional to $(\tau - \Delta t/2)$, where $\Delta t = 1$ is used there. There is no a precise analogue between the KFVS scheme and the Lattice BGK method. Due to cell averaging process and the non-isotropic mesh, such as the lack of diagonal transport, the KFVS scheme has a much more complicated dissipative mechanism. But, the numerical viscosity coefficient ν_n can be still estimated using a simple shear flow model, which shows $\nu_n \sim \Delta t/2$. We believe that the development of a multidimensional scheme will depend not only on the wave *modeling*, but also on the *mesh construction*. It depends closely on wether a numerical mesh could preserve the isotropic property of the fluid equations. CFD community usually has less experience in this aspect. There is something we can learn from the Lattice Boltzmann method, where the symmetry, invariants, etc., are the main concerns in their algorithm developments. Recently, it has been realized by the CE/SE method that a triangular mesh is important to preserve the conservation in space and time and avoid the reconstruction or interpolation [5]. In some sense, a triangular mesh has more symmetry and isotropic properties than a rectangular one.

3.4 Slip Boundary Condition

Another important observation of the kinetic scheme is that it can introduce slip condition in the Navier-Stokes formulation naturally through the use of appropriate flux boundary condition. The basic formulation of kinetic slip boundary is based on the fact that with the introduction of gas distribution function, we can explicitly evaluate the amount of particles hitting the boundary, then according to the accommodation coefficients for the momentum and energy, or the temperature at the wall, we can re-emit the same amount of particles with a pre-described distribution function. As a result, the appearance of slip at the boundary is obtained naturally, and this step is consistent with the DSMC type boundary condition in the near continuum regime due to their common kinetic considerations. Therefore, the kinetic method is the one which can cover the whole spectrum from the Euler and Navier-Stokes to near continuum region, where a slip boundary can be used to match it to the DSMC method.

3.5 Prandtl Number Fix

It is well known that the BGK model corresponds to a unit Prandtl number. In order to change the above Prandtl number to any realistic value, many attempts have been proposed, such as the BGK-Ellipsoidal-Statistical (BGK-ES) collision operator [9]. The BGK model itself can always make one coefficient correct, the viscosity or heat conduction. From the BGK method, we have obtained explicitly the time dependent gas distribution function at the cell interface Eq.(11). Therefore, the heat flux can be precisely evaluated,

$$q = \frac{1}{2}\int (u-U)\left((u-U)^2 + (v-V)^2 + \xi^2\right) f d\Xi, \tag{17}$$

where the average velocities U and V are

$$U = \int u f d\Xi / \int f d\Xi \quad , \quad V = \int v f d\Xi / \int f d\Xi .$$

Then, the easiest way to fix the Prandtl number for the BGK scheme is to modify the energy flux by subtracting the above heat flux (17) and adding another amount with a correct Prandtl number,

$$F_E^{new} = F_E + (\frac{1}{\mathrm{Pr}} - 1) q, \tag{18}$$

where F_E is the energy flux in the original BGK scheme. A simple method to evaluate q is proposed in [3]. We believe that the above fix can be equally applied to the BGK Discrete Velocity Model (DVM), where the discrete distribution function f is known [12].

The Prandtl number fix (18) is a post-processing correction, which is basically a numerical fix. But, we believe that to the Navier-Stokes order, the above fix will work perfectly. Actually, the BGK-ES model is a numerical fix on the dynamical level. But, dynamically keeping an anisotropic Gaussian for the equilibrium distribution function seems no any physical basis. The real physical weakness of the BGK model is that the collision time is independent of particle velocity, this fact is different from the phenomena with an anisotropic temperature distribution, where the temperature is directionly dependent.

3.6 Artificial viscosity – Godunov – BGK method

In the CFD algorithm development, the two classical pioneering papers for the shock capturing schemes are by von Neumann and Richtmyer [22] and by Godunov [7]. Since any physical solution has to be described in the discretized space and time, the limitation of cell size and time step has to be considered. von Neumann and Richtmyer realized that the numerical shock thickness needs to be compatible with the cell size. So, the central idea in [22] is that a viscous governing equation with an enhanced viscosity coefficient has to be solved numerically.

The success in Godunov method is that he introduced a discontinuity in the flow representation. For the under resolved flow simulation, due to the large cell size, a discontinuity will naturally appear in the initial data. The implementation of a discontinuity is much more important than the introduction of the Riemann solver. The cell interface discontinuity gives a more realistic representation about the physical solution and the numerical dissipation involved in the discontinuity can hardly be recovered by a delicate viscosity coefficient. But, as we go to a second order scheme and include the non-linear limited slopes, the relability of the numerical dissipation inside the initial data for the capturing discontinuity solution becomes questionable. From our experience, we believe that if a *Generalized Riemann Solver* is used to solve the *Euler equations* with the inclusion of initial slopes in the gas evolution stage, see Fig.(1), a 2nd-order (in both space and time) scheme cannot properly capture the numerical shock waves. Even with the discontinuity at a cell interface, additional numerical dissipation is still needed.

The methodology of the BGK scheme is in somehow to combine the two important issues rised in the above two methods, 1. a viscous governing equation with an enhanced viscosity coefficient (14) is solved. 2. Account the discontinuity and slopes of the initial data in the gas evolution stage. Both factors are important for the development of a robust scheme for the fluid simulation. In the smooth region, the additional numerical viscosity and the discontinuity at the cell interface disappear. The BGK scheme goes back to the traditional Lax-Wendroff type central schemes for NS equations.

4 Numerical Experiments

Case(1) Couette Flow with a Temperature Gradient Couette flow with a temperature gradient provides a good test for the BGK scheme to describe the viscous heat conducting flow. With the bottom wall fixed, the top boundary is moving at a speed U. The temperatures at the bottom and top are fixed with values T_0 and T_1. The analytic steady state temperature distribution is

$$\frac{T - T_0}{T_1 - T_0} = \frac{y}{H} + \frac{\mathrm{PrEc}}{2}\frac{y}{H}(1 - \frac{y}{H}), \tag{19}$$

where H is the height of the channel, Pr is the Prandtl number, Ec is the Eckert number $\mathrm{Ec} = U^2/C_p(T_1 - T_0)$, and C_p is the specific heat ratio at constant pressure. The BGK solutions with 20 grid points are shown in Fig.(3). The Prandtl number fix does modify the heat conduction term correctly.
Case(2) Navier-Stokes Shock Structure This case is to show the performance of the BGK scheme from the shock capturing to shock structure calculation. Initially, a stationary shock with Mach number M = 1.5 is located at $x = 0$. The physical viscosity coefficient for the BGK scheme takes a value $\nu = 0.00025$, which corresponds to a shock thickness, $l_s \sim 1/300$. The density distributions with different cell sizes are shown in Fig.(4).
Case(3) Mach 3 Step Problem The computation is carried out on a uniform mesh with 120×40 cells, and the cell size used is $\Delta x = \Delta y = 1/40$. In order to test the viscous effect on the flow structure, we have used different Reynolds number $\mathrm{Re} = UL/\nu = 10^5$ and 50. The adiabatic slip Euler condition is imposed at the boundaries in order to avoid the formation of viscous boundary layer. The density and pressure distributions at different Reynolds number are presented in Fig.(5) and (6). The shock structure in (6) is well resolved.
Case(4) Laminar Boundary Layer Case A laminar boundary layer with Mach number $M = 0.15$ and $\mathrm{Re} = 10^5$ is tested over a flat plate. A rectangular mesh with 120×30 grid points clustering above the flat plate is used. The U and V velocity distributions at two locations are plotted in Fig.(7). Due to the high Reynolds number in this case, the physical collision time τ determined by the viscosity coefficient is much smaller than the time step Δt, i.e., $\tau << \Delta t$. There will have difficulties for the KFVS NS method to calculate this case.
Case(5) Shock Boundary Layer Interaction This test is about the interaction of an oblique shock at an angle 32.6^0 with a boundary layer. The Mach number of the shock wave is $M = 2.0$ and the Reynolds number for the upstream flow is $\mathrm{Re} = 2.96 \times 10^5$. The dynamical viscosity μ used here is the Sutherland's law, and the Prandtl number is equal to 0.72. A mesh clustering around the flat plate with 110×60 grid points are constructed. The skin friction and pressure distributions at the surface of the plate is shown in Fig.(8).
Case(6) A M_s=1.22 shock wave in air hits a Helium cylindrical bubble The BGK scheme has been recently extended to two component flow simulation. In the multicomponent case, each component has its individual BGK model and the momentum and energy coupling are achieved through the collision terms. The application to the shock-bubble interaction is shown in Fig.(9).

5 Conclusion

A gas-kinetic BGK scheme for the fluid simulations is presented in this paper. Due to the implemetation of non-equilibrium state at the beginning of each time step and the BGK gas evolution model, the BGK method could give accurate NS solution in both smooth and discontinuity regions. The KFVS method and the Lax-Wendroff type schemes are the limiting cases of the current kinetic BGK method. The comprehensive numerical results presented in this paper and the physical and numerical analysis about the scheme indicate the level of maturity achieved by the gas-kinetic method.

References

1. P.L. BHATNAGAR, E.P. GROSS, AND M. KROOK, Phys. Rev., **94** (1954), pp. 511.
2. G.A. BIRD, Orford Science Publications (1994).
3. D. CHAE, C. KIM, AND O. RHO, J. Comput. phys., **158** (2000), pp. 1-27.
4. S. CHEN AND G. DOOLEN, Ann. Rev. Fluid Mech., **30** (1998), pp. 329-364.
5. S.C.CHANG, ET. AL., CFD Review 1998, M. HAFEZ AND K. OSHIMA eds..
6. S.Y. CHOU AND D. BAGANOFF, J. Comput. Phys., **130** (1997), pp. 217-230.
7. S.K. GODUNOV, Math. Sbornik, **47** (1959), pp. 271.
8. R.J. HAKKINEN, ET. AL., NASA Memo. 2-18-59W (1959).
9. L.H. HOLWAY, Rarefied Gas Dynamics, Vol. 1 (1966), pp. 193-215.
10. M. JUNK AND S.V.R. RAO, J. Comput. Phys., **155** (1999), pp. 178-198.
11. J.C. MANDAL AND S.M. DESHPANDE, Computers & Fluids, **23** (1994), pp. 447.
12. L. MIEUSSENS, to appear in J. Comput. Phys. (2000).
13. G.N. PATTERSON, *Molecular Flow of Gases*, Wiley, New York (1956).
14. K.H. PRENDERGAST AND K. XU, J. Comput. Phys., **109** (1993), pp. 53.
15. D.I PULLIN, J. Comput. Phys., **34** (1980), pp. 231-244.
16. R.H. SANDERS AND K.H. PRENDERGAST, Astrophy. Journal, **188** (1974).
17. J. L. STEGER AND R. F. WARMING, J. Comput. Phys., **40** (1981), pp. 263-293.
18. M.D. SU, K. XU, M. GHIDAOUI, J. Comput. Phys., **150** (1999), pp. 17-39.
19. H.Z. TANG AND K. XU, accepted by ZAMP (2000).
20. B. VAN. LEER, ICASE report, NO.82-30 (1982).
21. B. VAN. LEER, J.L. THOMAS, P.L. ROE, R.W. NEWSOME, AIAA 87-1104.
22. J. VON NEUMANN AND R.D. RICHTMYER, J. Appl. Phys., **21** (1950), pp. 232-237.
23. K. XU, VKI Lecture Series 1998-03 for Computational Fluid Dynamics.
24. K. XU, L. MARTINELLI, AND A. JAMESON, J. Comput. Phys., **120** (1995), pp. 48.

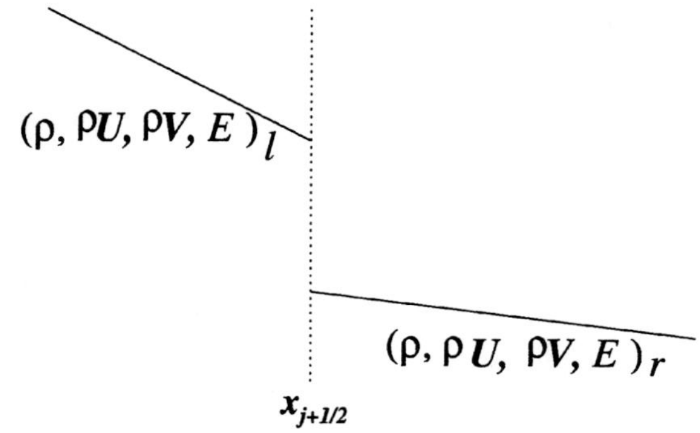

Fig. 1. Reconstructed initial conservative variables around a cell interface, from which the nonequilibrium state f_0 is constructed.

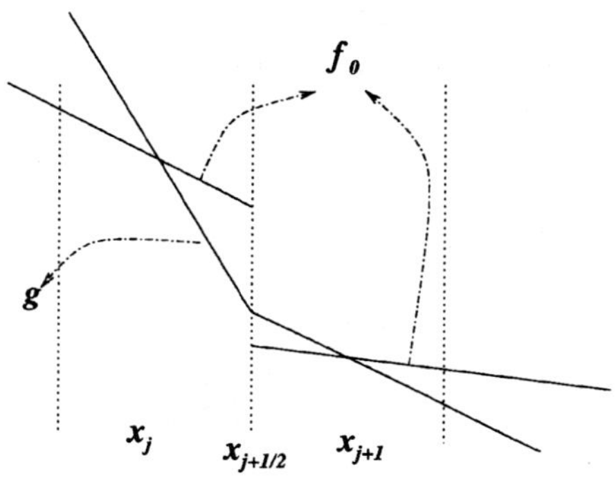

Fig. 2. The spatial distribution of the initial state f_0 and the equilibrium state g at $t = 0$.

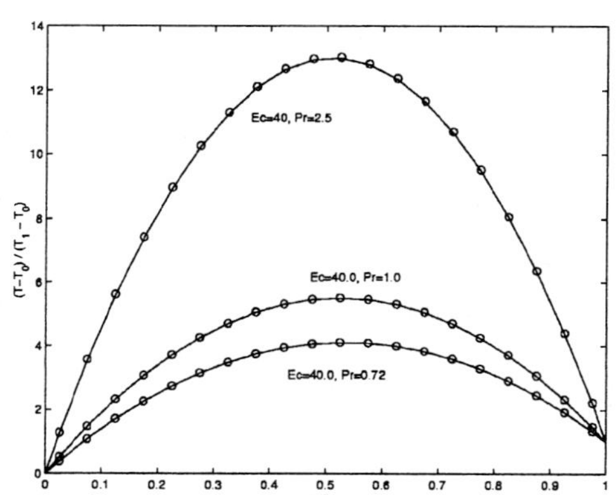

Fig. 3. Temperature ratio $(T - T_0)/(T_1 - T_0)$ in Couette flow. The solid line is the analytic solution and the circles are the numerical ones from the BGK scheme.

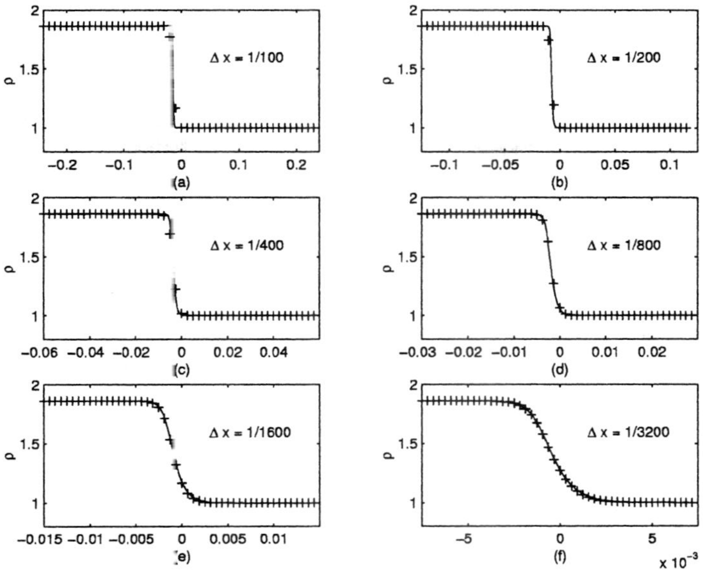

Fig. 4. Density distribution of a stationary shock wave with $M = 1.5$. The numerical solution (+ sign) is obtained from the BGK scheme with different cell sizes. The solid lines are the exact Navier-Stokes solution.

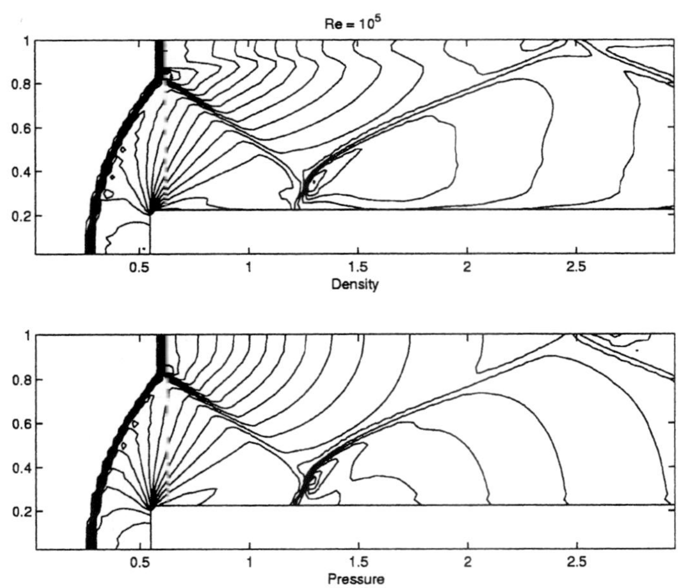

Fig. 5. Density and pressure contours in Mach 3 step problem on a mesh with 120×40 grid points. The Reynolds number used in this case is Re=10^5 w.r.t. the upstream velocity $U = 3.0$ and the channel height $L = 1.0$.

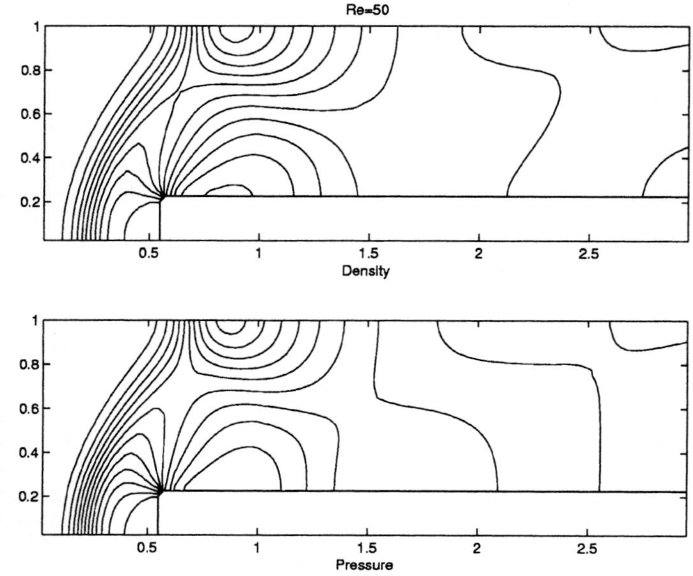

Fig. 6. Density and pressure contours. The Reynolds number is Re=50.

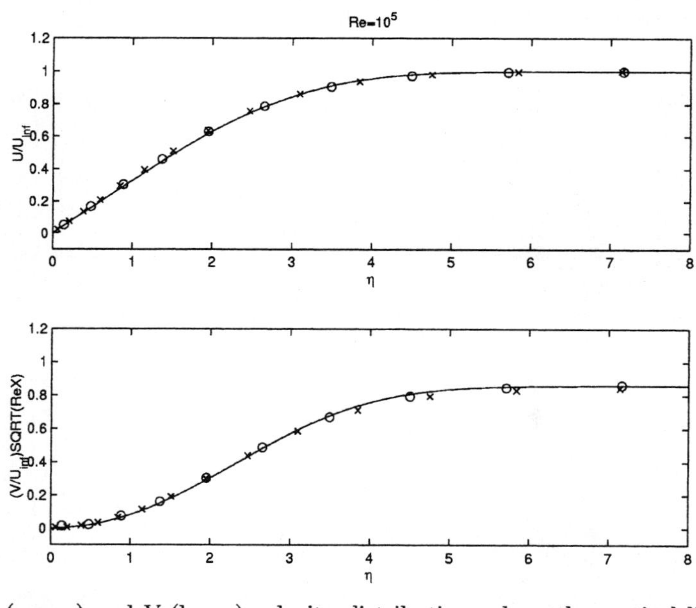

Fig. 7. U (upper) and V (lower) velocity distributions along the vertical lines at two different location of the boundary. The solid lines are the exact Blasius solutions.

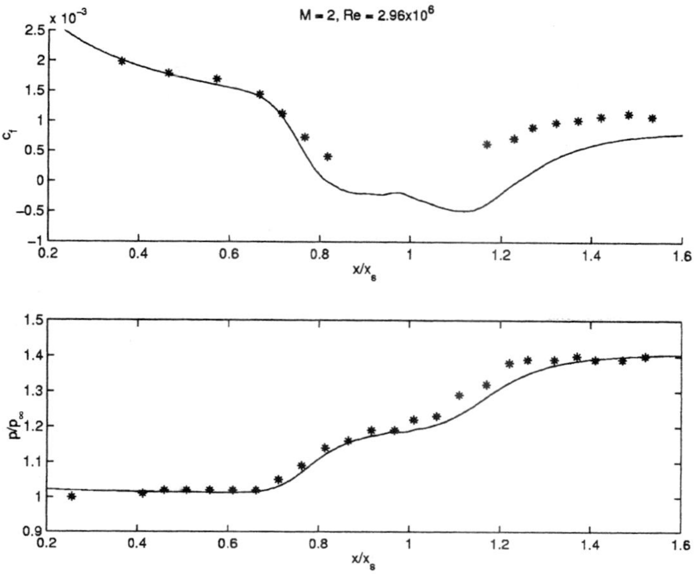

Fig. 8. Shock-boundary interaction case. (upper): skin friction, (lower): pressure distribution, along the flat plate. The * is the experimental data [8]. Solid line: numerical solution. The computation is done on a mesh with 110×60 grid points.

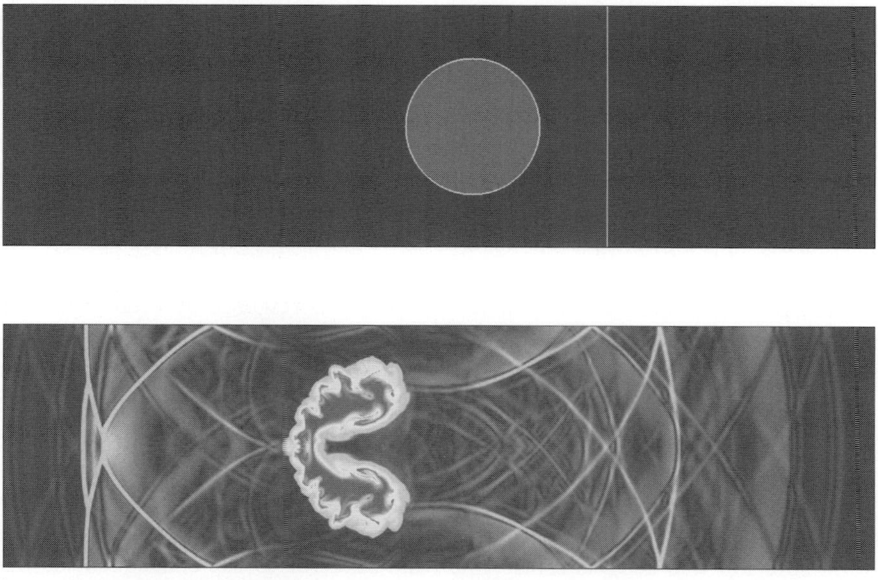

Fig. 9. Numerical Schlieren images of the interaction between a M_s=1.22 shock wave in the air and a Helium cylindrical bubble. The shock is moving from right to left. The second image describe the fields of the density gradient distribution after collision.

Defect and Adjoint Error Correction

Michael B. Giles

Oxford University Computing Laboratory, Oxford OX1 3QD, U.K.

Abstract. Motivated by applications in aero-acoustics and electromagnetics, this paper discusses the combined use of defect correction to improve the order of accuracy of numerical solutions, and adjoint error correction to improve the order of accuracy of derived output functionals such as far-field boundary integrals. Numerical results for the 1D Helmholtz equation on an irregular grid show fourth order accuracy for the numerical solution, and sixth order accuracy for the boundary value.

1 Introduction

The primary motivation for the work in this paper is the need for high order accuracy for aeroacoustic and electromagnetics calculations. In steady CFD calculations, grid adaptation can be used to provide high grid resolution in the limited areas which require it. However, using standard second order accurate methods, the wave-like nature of aeroacoustic and electromagnetic solutions would lead to grid refinement throughout the computational domain in order to reduce the wave dispersion and dissipation to acceptable levels. The preferable alternative is to use higher order methods, allowing one to use fewer points per wavelength, which can lead to a very substantial reduction in the total number of grid points for 3D calculations. The difficulty with this is that one often wants to use unstructured grids because of their geometric flexibility, and the construction of higher order approximations on unstructured grids is complicated and computationally expensive.

The current research also followed from previous research by Pierce and Giles on the use of adjoint error correction to obtain improved values for output functionals [6]. The relevance of this to aero-acoustic and electromagnetics is that one is often interested in the value of a far-field boundary integral giving the radiated acoustic energy in aeroacoustics, or the radar cross-section in electromagnetics [5]. Pierce and Giles achieved superconvergent results by using a reconstruction process to formulate a smooth approximate numerical solution. The residual error in approximating the original p.d.e. was then evaluated, and an approximate adjoint solution was used to relate this residual error to the consequential error in the output functional of interest. Removing this estimate of the error gave a doubling of the order of accuracy of the functional in a number of test cases, including the 2D Laplace and quasi-1D Euler equations [3]. An alternative use for the reconstruction and residual error evaluation would have been to use it to improve the whole solution through the well-known established of defect correction (e.g. [1,4,7,8]). However, defect correction and adjoint error correction are not mutually exclusive; the best accuracy is to be achieved through the simultaneous use of both techniques.

Accordingly, in this paper we examine the use of both to improve the accuracy in approximating the scalar Helmholtz equation on an irregular 1D grid. The first section describes the model problem and the simple second-order accurate, piecewise linear, Galerkin finite element method which is used as the basic approximation. The second section describes the defect correction in which a smooth solution is reconstructed by cubic spline interpolation. The residual error then produces the source term in a calculation of a correction using the Galerkin solver; this step is repeated if necessary. The third section very briefly recaps the adjoint error correction procedure. The final section presents the numerical results, showing global fourth order accuracy for the solution obtained with defect correction. For the output functional, which in this case in the solution at one end of the domain, fourth order accuracy is achieved using either defect or adjoint error correction on their own, but sixth order accuracy is obtained when using both.

2 Problem description and Galerkin method

The model problem to be solved is the 1D Helmholtz equation

$$u'' + \pi^2 u = 0, \quad 0 < x < 10,$$

subject to the Dirichlet boundary condition $u = 1$ at $x = 0$ and the radiation boundary condition $u' - i\pi u = 0$ at $x = 10$. The analytic solution is $u = \exp(i\pi x)$ and the domain contains precisely five wavelengths. The output functional of interest is the value $u(10)$ at the right hand boundary. This can be viewed as a model of a far-field boundary integral giving the radiated acoustic energy in aeroacoustics, or the radar cross-section in electromagnetics [5].

Integrating by parts, the weak form of the inhomogeneous equation

$$u'' + \pi^2 u = f, \quad 0 < x < 10,$$

subject to the same boundary conditions is

$$-(w', u') + \pi^2 (w, u) + i\pi w^*(10)\, u(10) = (w, f),$$

for any differentiable $w(x)$ with $w(0) = 0$. Here the inner product (w, u) is defined as

$$(w, u) \equiv \int_0^{10} w^* u \, \mathrm{d}x,$$

with w^* denoting the complex conjugate of w.

The Galerkin solution on the irregular grid $x_j, j = 0, 1, 2, \ldots, N$, is defined as

$$U(x) = \sum_{j=0}^{N} U_j \phi_j(x)$$

where the $\phi_j(x)$ are the usual piecewise linear 'hat' functions for which $\phi_j(x_i) = \delta_{ij}$. The value U_0 is given by the Dirichlet boundary condition. The values of the other coefficients U_j for $j>0$ are obtained from the equations

$$-(\phi_i', U') + \pi^2(\phi_i, U) + \mathrm{i}\pi\phi_i(10)\, U(10) = 0, \quad i = 1, 2, \ldots, N.$$

It is well established that this discretisation is second order accurate, producing dispersion but no dissipation on a uniform grid.

3 Defect correction

The first step in the defect correction is to define a new approximate solution $u_h(x)$ by cubic spline interpolation of the nodal values U_j. The choice of end conditions for the cubic spline is very important. A natural cubic spline would have $u_h''=0$ at both ends, but this would introduce small errors at each end since $u''\neq 0$ for the analytic solution. Instead, at $x=10$ we require the splined solution to satisfy the analytic boundary condition by imposing $u_h' - i\pi u_h = 0$. At $x=0$, the analytic boundary condition is already imposed through having the correct value for the end point $U(0)$. Therefore, here we require that $u_h'' + \pi^2 u_h = 0$ so the splined solution satisfies the o.d.e. at the boundary.

The solution error, $e = u(x) - u_h(x)$ satisfies the inhomogeneous Helmholtz equation

$$e'' + \pi^2 e = -(u_h'' + \pi^2 u_h), \quad 0 < x < 10,$$

the right-hand-side of which is the residual error of the approximation $u_h(x)$. Given the homogeneous Dirichlet boundary condition at $x = 0$, and the same radiation boundary condition at $x=10$, the Galerkin approximation to the error is given by the equations

$$-(\phi_i', E') + \pi^2(\phi_i, E) + \mathrm{i}\pi\phi_i(10)\, E(10) = -(\phi_i, u_h'' + \pi^2 u_h), \quad i = 1, 2, \ldots, N.$$

Adding the nodal corrections E_j to the original nodal values U_j gives a corrected solution. The whole procedure can then be repeated to improve the accuracy. This follows the procedure described by Barrett *et al* who also showed that it converges to a solution of an appropriately defined Petrov-Galerkin discretisation, with the trial space being the space of cubic splines, while the test space is the space of piecewise linear functions [1].

4 Adjoint error correction

We begin with a presentation of the linear theory for adjoint error correction in applications with inhomogeneous boundary conditions and boundary functionals; for the nonlinear theory, see [3].

Let u be the solution of the linear differential equation

$$Lu = f,$$

in the domain Ω, subject to the linear boundary conditions

$$Bu = e.$$

on the boundary $\partial\Omega$. In general, the operator B may be different on different parts of the boundary, and in some applications (e.g. inflow and outflow sections for the convection p.d.e.) even its dimension may differ.

The output functional of interest is taken to be

$$J = (g, u) + (h, Cu)_{\partial\Omega},$$

where $(.,.)$ represents an integral inner product over the domain Ω and $(.,.)_{\partial\Omega}$ represents an integral inner product over the boundary $\partial\Omega$. In [3], the theory was presented for real variables, but here we are considering complex variables and so for the general case of vector variables u, v the inner product (v, u) is defined as

$$(v, u) \equiv \int_{\Omega} v^H u \,\mathrm{d}V$$

with v^H being the Hermitian (complex conjugate transpose) of v.

The boundary operator C may be algebraic (e.g. $Cu \equiv u$) or differential (e.g. $Cu \equiv \frac{\partial u}{\partial n}$), but must have the same dimension as the adjoint boundary condition operator B^* to be defined shortly. Note that the components of h may be set to zero if the functional does not have a boundary integral contribution.

The corresponding linear adjoint problem is

$$L^* v = g,$$

in Ω, subject to the boundary conditions

$$B^* v = h,$$

on the boundary $\partial\Omega$. The fundamental identity defining L^*, B^* and the boundary operator C^* is

$$(L^* v, u) + (B^* v, Cu)_{\partial\Omega} = (v, Lu) + (C^* v, Bu)_{\partial\Omega},$$

for all u, v. This identity is obtained by integration by parts, and in a previous paper we describe the construction of the appropriate adjoint operators for the linearized Euler and Navier-Stokes equations [2]. We will follow the same process later to construct the adjoint boundary operators for the Helmholtz equation.

Given approximate solutions u_h, v_h we define e_h, f_h, g_h, h_h by

$$Lu_h = f_h, \qquad L^* v_h = g_h,$$
$$Bu_h = e_h, \qquad B^* v_h = h_h,$$

and hence obtain

$$
\begin{aligned}
(g,u) + (h,Cu)_{\partial\Omega} &= (g,u_h) + (h,Cu_h)_{\partial\Omega} \\
&\quad - (g_h, u_h - u) - (h_h, C(u_h - u))_{\partial\Omega} \\
&\quad + (g_h - g, u_h - u) + (h_h - h, C(u_h - u))_{\partial\Omega} \\
&= (g,u_h) + (h,Cu_h)_{\partial\Omega} \\
&\quad - (L^* v_h, u_h - u) - (B^* v_h, C(u_h - u))_{\partial\Omega} \\
&\quad + (g_h - g, u_h - u) + (h_h - h, C(u_h - u))_{\partial\Omega} \\
&= (g,u_h) + (h,Cu_h)_{\partial\Omega} \\
&\quad - (v_h, L(u_h - u)) - (C^* v_h, B(u_h - u))_{\partial\Omega} \\
&\quad + (g_h - g, u_h - u) + (h_h - h, C(u_h - u))_{\partial\Omega} \\
&= (g,u_h) + (h,Cu_h)_{\partial\Omega} \\
&\quad - (v_h, f_h - f) - (C^* v_h, e_h - e)_{\partial\Omega} \\
&\quad + (g_h - g, u_h - u) + (h_h - h, C(u_h - u))_{\partial\Omega}.
\end{aligned}
$$

In the final result, the first line is the functional based on the approximate solution u_h. The second line is the adjoint correction term which now includes a term related to the extent to which the primal solution does not correctly satisfy the boundary conditions. The third line is the remaining error for which an *a posteriori* error bound can be found, in principle [6].

To apply this theory to the Helmholtz problem, the first step is to construct the appropriate adjoint problem. Integration by parts reveals that the Helmholtz equation is self-adjoint, so

$$L^* v \equiv v'' + \pi^2 v,$$

and

$$(v,\, Lu) - (L^* v,\, u) = \left[\boldsymbol{v}^H \boldsymbol{A} \boldsymbol{u}\right]_0^{10},$$

where

$$\boldsymbol{u} = \begin{pmatrix} u \\ \frac{du}{dx} \end{pmatrix}, \quad \boldsymbol{v} = \begin{pmatrix} v \\ \frac{dv}{dx} \end{pmatrix},$$

and

$$\boldsymbol{A} = \begin{pmatrix} 0 & 1 \\ -1 & 0 \end{pmatrix}.$$

At $x = 10$ we have

$$Bu \equiv u' - \mathrm{i}\pi u \equiv \boldsymbol{B}\boldsymbol{u}, \quad \boldsymbol{B} = (-\mathrm{i}\pi \;\; 1),$$

and

$$Cu \equiv u \equiv \boldsymbol{C}\boldsymbol{u}, \quad \boldsymbol{C} = (1 \;\; 0).$$

To satisfy the adjoint identity [2], we require $\boldsymbol{B}^*$ and $\boldsymbol{C}^*$ such that

$$\boldsymbol{A} = \begin{pmatrix} -\boldsymbol{C}^* \\ \hline \boldsymbol{B}^* \end{pmatrix}^H \begin{pmatrix} \boldsymbol{B} \\ \hline \boldsymbol{C} \end{pmatrix}.$$

Solving this gives

$$\left(\frac{-C^*}{B^*}\right) = \left(\frac{B}{C}\right)^{-H} A^H = \begin{pmatrix} 1 & 0 \\ -i\pi & -1 \end{pmatrix}$$

and hence $B^* v \equiv -v' - i\pi v$ and $C^* v \equiv -v$. Similarly, at $x=0$, we obtain $B^* v = v$ and $C^* v = v'$.

Now, noting that in our application $f=g=0$, and h has value 0 at $x=0$ and 1 at $x=10$, then the full specification of the adjoint problem is

$$v'' + \pi^2 v = 0, \quad 0 < x < 10,$$

with $v=0$ at $x=0$ and $-v' - i\pi v = 1$ at $x=10$.

Let v_h be an approximate solution of this problem, obtained by the same Galerkin and cubic spline reconstruction approach as u_h, with or without defect correction. Noting that the cubic spline reconstruction ensures that the boundary conditions are satisfied exactly, the corrected approximation to the value $u(10)$ is

$$u_h(10) - (v_h, u_h'' + \pi^2 u_h).$$

The theory gives the error in this corrected functional as being

$$(v_h - v, u_h'' + \pi^2 u_h).$$

In the absence of defect correction, both terms in this inner product are second order in the average grid spacing and so the error is fourth order. With defect correction, the first term is fourth order while the second term remains second order. Therefore, the error remaining after the adjoint error correction is sixth order.

5 Numerical results

Numerical results have been obtained for grids with 4, 8, 16, 32, 64 and 128 points per wavelength. To test the ability to cope with irregular grids, the coordinates for the grid with N intervals are defined as

$$x_0 = 0, \quad x_N = 10, \quad x_j = \frac{10}{N}(j + \sigma_j), \; 0<j<N,$$

where σ_j is a uniformly distributed random variable in the range $[-0.3, 0.3]$.

Figure 1 shows the L_2 norm of the error in the reconstructed cubic spline solution before and after defect correction. Without defect correction, the error is second order, while with defect correction it is fourth order. Note that a second application of defect correction makes a significant reduction in the error even though it remains fourth order. This is because one application of the defect correction procedure gives a correction which is second order in magnitude, with a corresponding error which is second order in relative magnitude and therefore

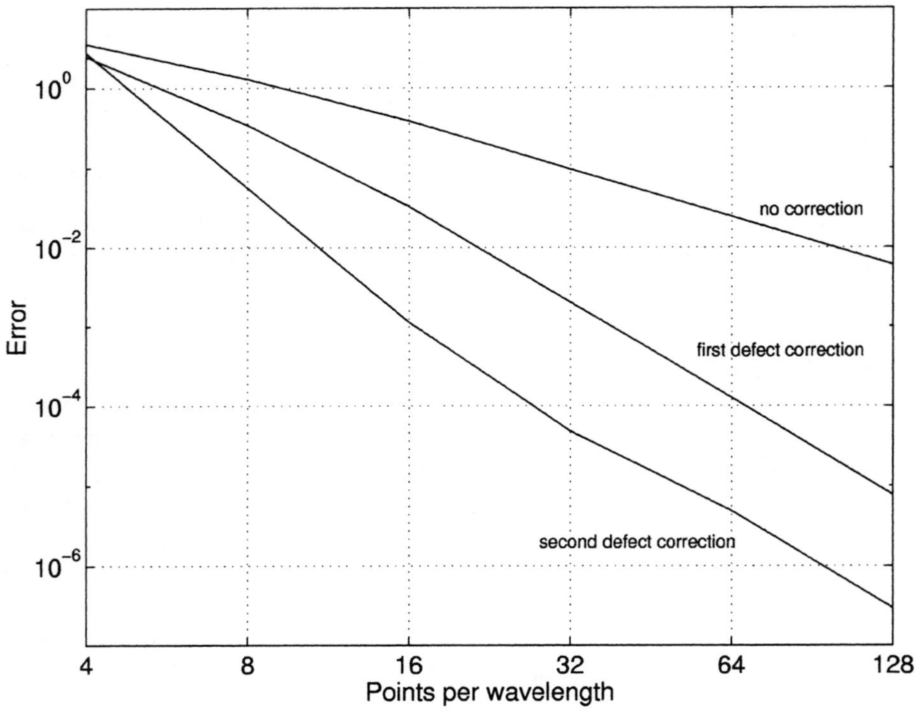

Fig. 1. L_2 error in the numerical approximation to $u(x)$

fourth order in absolute magnitude. It is this error which is corrected by a second application of the defect correction procedure.

Figure 2 shows the error in the numerical value for the output functional $u(10)$. Without any correction, the error is second order. Using either defect correction or adjoint error correction on their own increases the order of accuracy to fourth order, but using them both increases the accuracy to sixth order. Note that the calculation with 8 points per wavelength plus both defect and adjoint error correction gives an error which is approximately 2×10^{-3}. This is more accurate than the calculation with 128 points per wavelength and no corrections, and comparable to the results using 14 points and defect correction, or 30 points with adjoint error correction.

In 3D, the computational cost is proportional to the cube of the number of points per wavelength, so this indicates the potentially huge savings offered by the combination of defect and adjoint error correction. The cost of computing the corrections is five times the cost of the original calculation, due to the additional two calculations for the defect correction, and the one adjoint calculation plus its two defect corrections. In practice, the second defect correction for the primal and adjoint calculations make negligible difference to the value obtained after

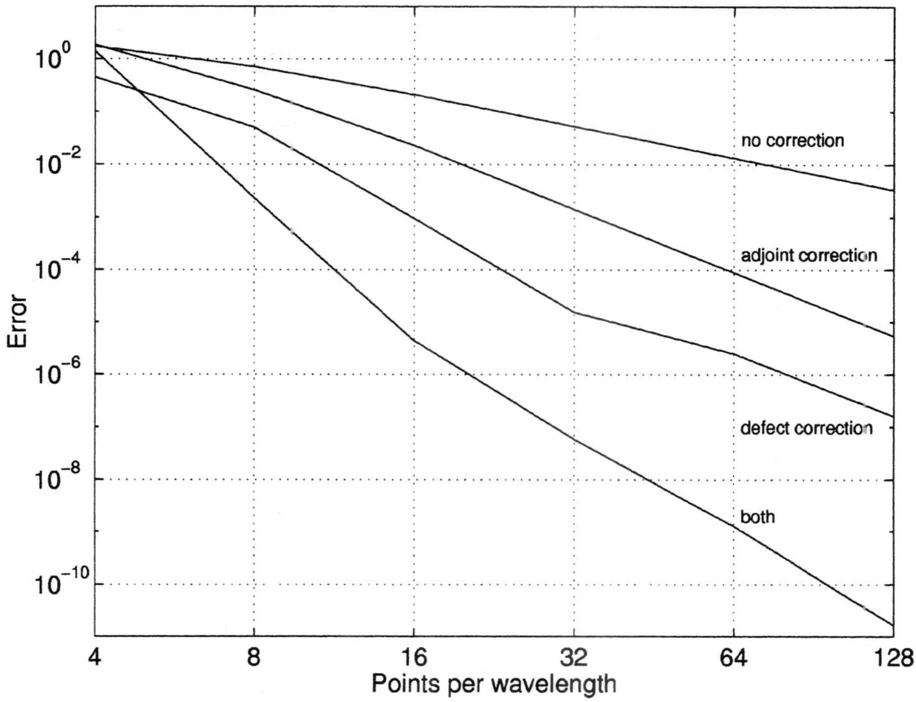

Fig. 2. Error in the numerical approximation to $u(10)$

the adjoint error correction, so these can be omitted, reducing the cost of the corrections to just three times the cost of the original calculation.

6 Concluding remarks

The numerical results which have been presented show the potential offered by defect correction and adjoint error correction, but there is much work to be done to achieve this potential for multi-dimensional applications. There will be some problems in the representation and approximation of curved boundaries and boundary integrals, but the key issue is likely to be the smooth reconstruction of a numerical solution from nodal data. On a structured grid, cubic spline interpolation can be used in each direction, but on an unstructured grid one would need a suitable generalisation of cubic spline interpolation to produce a reconstructed solution of sufficient smoothness. This will be the main challenge in trying to reproduce similar improvements in accuracy for aeroacoustic and electromagnetic calculations on 3D unstructured grids.

Acknowledgements

This research was supported by EPSRC grants GR/K91149 and GR/L95700, and by Rolls-Royce plc.

References

1. J.W. Barrett, G. Moore, and K.W. Morton. Optimal recovery in the finite element method, part 2: Defect correction for ordinary differential equations. *IMA J. Numer. Anal.*, 8:527-540, 1988.
2. M.B. Giles and N.A. Pierce. Adjoint equations in CFD: duality, boundary conditions and solution behaviour. AIAA Paper 97-1850, 1997.
3. M.B. Giles and N.A. Pierce. Improved lift and drag estimates using adjoint Euler equations. AIAA Paper 99-3293, 1999.
4. B. Koren. Defect correction and multigrid for an efficient and accurate computation of airfoil flows. *J. Comput. Phys.*, 76, 1988.
5. P. Monk and E. Süli. The adaptive computation of far field patterns by a posteriori error estimates of linear functionals. *SIAM J. Numer. Anal.*, 36(1):251-274, 1998.
6. N.A. Pierce and M.B. Giles. Adjoint recovery of superconvergent functionals from PDE approximations. *SIAM Rev.*, 42(2):247-264, 2000.
7. R.D. Skeel. A theoretical framework for proving accuracy results for deferred corrections. *SIAM J. Numer. Anal.*, 19:171-196, 1981.
8. H.J. Stetter. The defect correction principle and discretization methods. *Numer. Math.*, 29:425-443, 1978.

A Vertex-Centroid (V-C) Scheme for the Gas-Dynamics Equations

Antony Jameson[1*] and John C. Vassberg[2**]

[1] Department of Aeronautics & Astronautics, Stanford University, Stanford, CA 94305, USA

[2] Aerodynamics, Phantom Works, The Boeing Company, Long Beach, CA 90807, USA

1 Introduction

While the majority of aerodynamic simulations continue to be performed on structured meshes, and the associated algorithms have realized a high level of maturity, mesh generation continues to be a bottleneck in the treatment of really complex configurations. The generation of a structured mesh can span one to two months of elapsed time, depending on the complexity of the geometry. The use of unstructured meshes offers the prospect of substantially reducing the time required to generate a grid about a complex shape. It is difficult, however, to achieve a comparable level of accuracy on an unstructured mesh, especially for Reynolds-Averaged Navier-Stokes (RANS) simulations. Further, the computational costs associated with the indirect-addressing algorithms of an unstructured-mesh solver are substantially higher.

Following classical finite element theory, a number of vertex based schemes have been developed [1,2]. The upwind schemes which have proved so successful on structured meshes are typically formulated as cell-centered, finite-volume schemes, in which the integral form of the equations is used to update the cell-averaged values of the conserved flow variables. This approach has also been quite widely adopted on unstructured meshes [3–6]. In order to obtain accurate values at the center of each cell interface, it requires a reconstruction procedure to estimate the gradients in each cell.

In the past, the present authors have developed cell-vertex schemes related to the Galerkin formulation [7–9]. Both scalar diffusion and upwind biased schemes have been devised in this context [10,11]. One reason for preferring a cell-vertex scheme over a cell-centered one, is that the number of cells in a tetrahedral mesh is about six times larger than the number of vertices, and in the past, the available computer memory was a significant constraint on the size of the calculations which could be attempted. On modern machines this is no longer a dominating consideration. In this paper we propose a vertex-centroid (V-C) scheme which combines some of the features of cell-vertex and cell-centered schemes. This leads to a neat reconstruction procedure. It also simplifies the inter-mesh transfers of

* T. V. Jones Professor of Engineering

** Associate Technical Fellow

an unstructured-mesh multigrid scheme, in both the fine-to-coarse and coarse-to-fine directions.

2 Description of the Algorithms

2.1 Finite-Volume Scheme

We consider the two dimensional Euler equations for inviscid, compressible flow. Let p, ρ, u, v, E and H denote the pressure, density, Cartesian velocity components, total energy and total enthalphy. For a perfect gas,

$$\begin{aligned} p &= (\gamma-1)\rho(E-\frac{1}{2}(u^2+v^2)), \\ H &= \frac{p}{\rho}, \\ c^2 &= \gamma\frac{p}{\rho} \end{aligned}$$

where c is the speed of sound, and γ is the ratio of specific heats. The Euler equations can be written in integral form

$$\frac{\partial}{\partial t}\int_D w ds + \int_B f dy - g dx = 0 \tag{1}$$

for a domain D with boundary B, where the state and flux vectors are

$$\begin{aligned} w &= (\rho, \rho u, \rho v, \rho E)^T, \\ f &= (\rho u, \rho u^2 + p, \rho uv, \rho uH)^T, \\ g &= (\rho v, \rho uv, \rho v^2 + p, \rho vH)^T. \end{aligned} \tag{2}$$

We assume the domain to be subdivided into small triangles. Equation (1) is applied directly to each triangle to follow the evolution of the cell-averaged values of the state vector w. This leads to the semi-discrete form

$$\frac{dw_j}{dt} + R_j = 0$$

where R_j is the residual for cell j. This is evaluated as

$$R_j = \sum_{sides} h_{jk}$$

where h_{jk} is a numerical estimate of the flux $(f\Delta y - g\Delta x)$ across the edge separating cell j and a neighboring cell k. This is illustrated in Figure 1, which shows a cell numbered 0, and its three neighbors.

2.2 Reconstruction

The accuracy of the scheme depends on the accuracy of the estimate of the fluxes h_{jk}. Current upwind schemes [3,12] reconstruct values of the state vector w at the center of each edge from the cell-averaged values and the gradients in the cells on the left and right side of each edge. These values, which may be denoted as w_L and w_R, are then used to construct the numerical flux taking account of the convection and wave propogation across the edge. In order to estimate the gradients in each cell, a least squares fit is constructed from the values in a set of neighboring cells.

In the present work, we adopt an alternative reconstruction procedure, in which the left and right values are constructed in two steps. First we derive estimates of the state vectors at each vertex, by considering the aggregate conserved values of mass, momentum and energy in the polygon surrounding each vertex.

$$w_{vertex} = \frac{\sum_{cells} w_k A_k}{\sum_{cells} A_k}$$

where A_k is the area of cell k.

This recovery formula does not allow for the fact that the centroid of the polygon does not coincide with the vertex. We have found it necessary to use an alternative procedure to estimate the values at boundary verticies, where the discrepancy is particularly large. An improved recovery procedure which is exact for linearly varying fields will be presented in a forthcoming paper [13]. We anticipate that this will significantly improve the accuracy on irregular meshes.

In the second step, we simply use linear extrapolation along the lines connecting the verticies through the centroid of each cell to the center of the opposing edge, to form the left and right states across each edge. This is illustrated in Figure 2, where the left state across edge 12 is extrapolated from vertex 3 through centroid $C1$, and the right state is extrapolated from vertex 4 through centroid $C2$.

2.3 Limiters

In order to preserve monotonicity in the solution, the left and right values w_L and w_R must be limited so that they do not overshoot the values in the neighboring

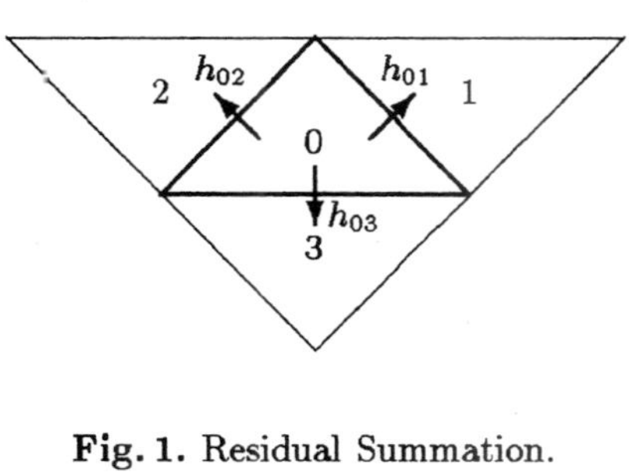

Fig. 1. Residual Summation.

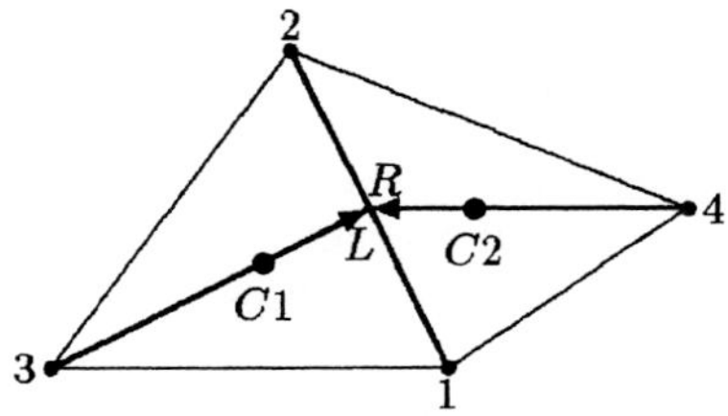

Fig. 2. Left-Right Extrapolation.

cells [3]. The results presented in this paper were obtained using a symmetric limiter similar to that proposed by Jameson [10]. Following the notation of Figure 2, define the left and right differences

$$\begin{aligned} \Delta w_L &= w_{C1} - w_3, \\ \Delta w_R &= w_4 - w_{C2}. \end{aligned}$$

Then set

$$\begin{aligned} w_L &= w_{C1} + \frac{1}{2}\mathcal{L}(\Delta w_L, \Delta w_R), \\ w_R &= w_{C2} - \frac{1}{2}\mathcal{L}(\Delta w_L, \Delta w_R) \end{aligned}$$

where the factor $\frac{1}{2}$ accounts for the relative distances along the directions of extrapolation, and $\mathcal{L}$ is a limited average satisfying the following four properties:

$$\begin{aligned} \mathcal{L}(u,v) &= \mathcal{L}(v,u) && (P1) \\ \mathcal{L}(\alpha u, \alpha v) &= \alpha \mathcal{L}(u,v) && (P2) \\ \mathcal{L}(u,u) &= u && (P3) \\ \mathcal{L}(u,v) &= 0 \ \ if \ uv < 0. && (P4) \end{aligned}$$

The limited average used in this work is

$$\mathcal{L}(u,v) = \frac{1}{2}(1 - \mathcal{R}(u,v))(u+v)$$

where

$$\mathcal{R}(u,v) = \left| \frac{u-v}{max((|u|+|v|), tol)} \right|^q$$

and tol is a threshold proportional to the mesh width to the $\frac{3}{2}$ power. Increasing the power q reduces the severity of the limiter. The value $q = 1$ corresponds to the minmod limiter. The calculations presented here were performed with $q = 3$. Alternative one-sided limiters have also been investigated.

2.4 Numerical Flux

The numerical flux is constructed with upwind biasing to take account of wave propogation. The flux across an edge has the form

$$h = f\Delta y - g\Delta x.$$

The Jacobian $\frac{\partial h}{\partial w}$ has the eigenvalues $Q, Q, Q \pm cs$, where s is the edge length, c is the speed of sound, and Q is the rate of convection across the edge,

$$Q = u\Delta y - v\Delta x.$$

Following Roe [14], we can introduce a matrix A_{LR} which exactly satisfies the relation

$$h_R - h_L = A_{LR}(w_R - w_L)$$

where

$$h_R = h(w_R), \quad h_L = h(w_L).$$

Then one can take the eigenvalues of A_{LR} to represent the eigenvalues of $\frac{\partial h}{\partial w}$ across the interface. Following Jameson [11] three families of numerical flux of increasing complexity can be identified:

(1) Scalar diffusion:

$$h_{LR} = \frac{1}{2}(h_R + h_L) - \frac{1}{2}\alpha(w_R - w_L) \tag{3}$$

where the diffusion coefficient should be no smaller that the largest wave speed

$$\alpha \geq Q + cs.$$

This is similar to the diffusion used in the Jameson-Schmidt-Turkel scheme [15] and has also been advocated by Tadmor [16].

(2) Schemes forming diffusion from differences of both flux and state vectors:

$$h_{LR} = \frac{1}{2}(h_R + h_L) - \frac{1}{2}\alpha(w_R - w_L) - \frac{1}{2}\beta(h_R - h_L). \tag{4}$$

In order to produce full upwinding in supersonic flow, we have: $\alpha = 0$ and $\beta = sign(M)$. In a subsonic flow, a blend is required to contain the appropriate domain of dependence.

(3) Characteristic diffusion:

$$h_{LR} = \frac{1}{2}(h_R + h_L) - \frac{1}{2}|A_{RL}|(w_R - w_L) \tag{5}$$

where the diffusion matrix $|A_{RL}|$ is formed by decomposing A_{RL} in terms of its eigenvectors and eigenvalues, and replacing the eigenvalues with their absolute values.

In this work, we have tested both scalar diffusion and the H-CUSP scheme, which defines coefficients in Equation 4 such that in one-dimensional flow the numerical shock structure contains a single interior point [11]. Results are presented for the H-CUSP scheme.

2.5 Discretization of Navier-Stokes Viscous Terms

Following the same philosophy of an interplay between nodal and cell-centered quantities, the V-C scheme can be readily extended to treat the Navier-Stokes equations for viscous flow calculations. First the velocity derivatives, $\frac{\partial u_l}{\partial x_m}$, are evaluated at the grid vertices by applying the Gauss theorem

$$\int \frac{\partial u}{\partial x} dA = \oint u dy$$

to a control volume around each interior vertex, as illustrated in Figure 3. The control volume is defined by lines through the centroids of the cells, drawn parallel to the outer edge of each cell, with a length equal to $\frac{2}{3}$ the length of the outer edge, and containing $\frac{4}{9}$ the area of the cell. Using a rectangle rule of integration, the derivative is estimated as

$$(D_x u)_V = \frac{\frac{2}{3}\sum u_k \Delta y_k}{\frac{4}{9}\sum A_k} = \frac{3}{2}\frac{\sum u_k \Delta y_k}{\sum A_k}$$

where u_k is the value of the velocity at the centroid, A_k is the triangle area, Δy_k is the outer edge length, and the sum is over the triangles surrounding the vertex.

Modified control volumes are used at boundary vertices, closed by edges along the boundary. The rate of strain tensor and energy flux are then calculated at each vertex. Finally the viscous fluxes for each triangle are evaluated by a trapezoidal rule of integration using the vertex values.

$$\sum_{edges} \left(\sigma_{xx}\Delta y - \sigma_{xy}\Delta x\right).$$

2.6 Time Stepping Scheme

Following earlier work by the present authors, a hybrid multi-stage time-stepping scheme is used in which the convective and diffusive terms are treated differently in order to expand the stability region. The residual is split as

$$R(w) = Q(w) + D(w)$$

where $Q(w)$ is the convective part and $D(w)$ is the diffusive part. Then the hybrid multi-stage time-stepping scheme is formulated as

$$\begin{aligned}
w^{(1)} &= w^{(0)} - \alpha_1 \Delta t(Q^{(0)} + D^{(0)}),\\
w^{(2)} &= w^{(0)} - \alpha_2 \Delta t(Q^{(1)} + D^{(1)}),\\
&\dots\\
w^{(m)} &= w^{(0)} - \alpha_m \Delta t(Q^{(m-1)} + D^{(m-1)})
\end{aligned}$$

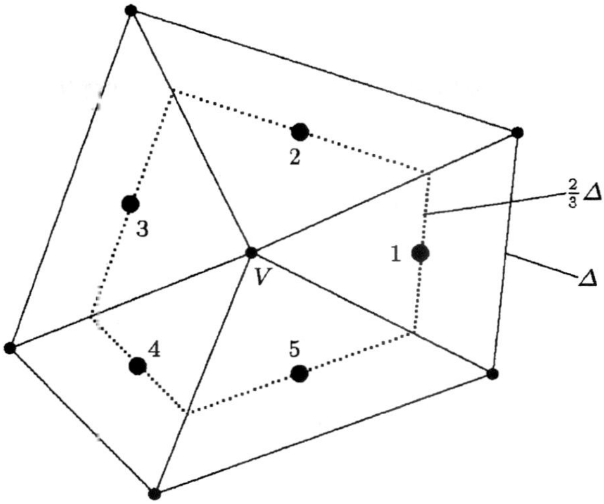

Fig. 3. Gradient Polygon about Vertex V.

where $w^{(0)}$ is the value of the state vector at the beginning of the time step, and for an m-stage scheme, $\alpha_m = 1$. Here

$$Q^{(0)} = Q(w^{(0)}), \quad D^{(0)} = D(w^{(0)})$$

and

$$Q^{(k)} = Q(w^{(k)}), \quad D^{(k)} = \beta_k D(w^{(k)}) + (1 - \beta_k) D(w^{(k-1)}).$$

The results presented herein have been calculated with a 5-stage scheme with the following coefficients:

$$\begin{array}{l} \alpha_1 = \frac{1}{4}, \alpha_2 = \frac{1}{6}, \alpha_3 = \frac{3}{8}, \quad \alpha_4 = \frac{1}{2}, \alpha_5 = 1 \\ \beta_1 = 1, \ \beta_2 = 0, \ \beta_3 = 0.56, \beta_4 = 0, \ \beta_5 = 0.44. \end{array}$$

2.7 Multigrid Scheme

It is well established that multigrid acceleration can drastically reduce the computational costs of flow simulations. With a structured mesh, a sequence of successively coarser grids can easily be generated by eliminating alternate points in each grid direction. With unstructured meshes it is not so easy to generate a suitable sequence of meshes, and in general, the cells of any pair of meshes in the sequence may overlap each other in an arbitrary manner. This complicates the construction of appropriate mesh transfer operators, and is one motivation for the introduction of automated grid-coarsening schemes such as that proposed by Vassberg [17], or the popular agglomeration techniques [18,19]. Here we simplify the construction of mesh transfer operators by transferring the flow variables from the cell centers to the vertices in an intermediate step.

The multigrid time stepping scheme is similar to that we have previously used [20]. Suppose the grids are numbered consecutively from the finest to the coarsest grid. Then on any grid the following procedure is used. First the solution vector on grid k must be initialized as

$$w_k^{(0)} = T_{k,k-1} w_{k-1},$$

where w_{k-1} is the current state vector on grid $k-1$, and $T_{k,k-1}$ is a transfer operator. Next it is necessary to transfer a residual forcing function such that the solution on grid k is driven by the residuals of grid $k-1$. This can be accomplished by setting

$$P_k = Q_{k,k-1} R_{k-1}(w_{k-1}) - R_k\left(w_k^{(0)}\right),$$

where $Q_{k,k-1}$ is another transfer operator. Then $R_k(w_k)$ is replaced with $R_k(w_k) + P_k$ in the time-stepping scheme. Thus the multi-stage scheme is reformulated as

$$w_k^{(q)} = w_k^{(0)} - \alpha_q \Delta t_k \left(R_k^{(q-1)} + P_k\right).$$

The result $w_k^{(m)}$ then provides initial data for grid $k+1$. Finally, the accumulated correction on grid k has to be transferred back to grid $k-1$ with the aid of an interpolation operator $I_{k-1,k}$.

The fine-to-coarse transfer operators are defined next. The flow variables are first distributed to the vertices with area weighting, as in the reconstruction procedure, and the sum of the residuals in the surrounding triangles is accumulated at each vertex. Next the flow variables are transferred from the vertices of the fine mesh to the vertices of the coarse mesh, by finding the fine-mesh triangle which contains the coarse-mesh vertex, and interpolating the fine-mesh values with area weighting, as shown in Figure 4(a).

$$w_{V_c} = \frac{A_1 w_1 + A_2 w_2 + A_3 w_3}{A_1 + A_2 + A_3}.$$

Also the residuals are transferred by finding the coarse-mesh triangle which contains a given fine-mesh vertex. The residual at the fine-mesh vertex is distributed to the three coarse-mesh vertices with the area weighting of Figure 4(b).

$$R_k = R_k + \frac{A_k R_{V_f}}{A_1 + A_2 + A_3}.$$

Finally the flow variables at the coarse-mesh centroids are formed as the arithematic average of the values at the vertices of each cell

$$w_c = \frac{1}{3}(w_1 + w_2 + w_3),$$

while the residuals are distributed from a vertex to the surrounding cells with the area weighting illustrated in Figure 5. Triangle 1 receives $(A_1 R_{V_c})/(\sum_k A_k)$

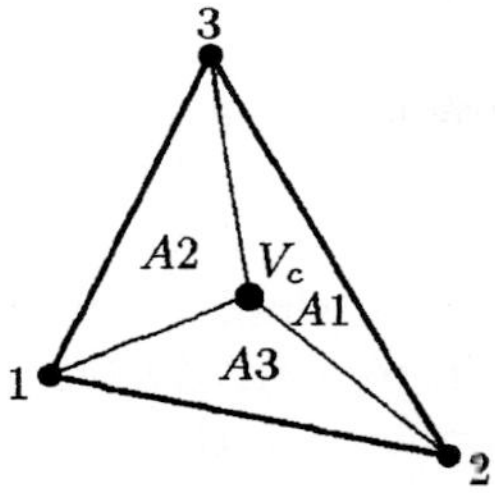

(a) Coarse-Mesh Vertex in Fine-Mesh Cell

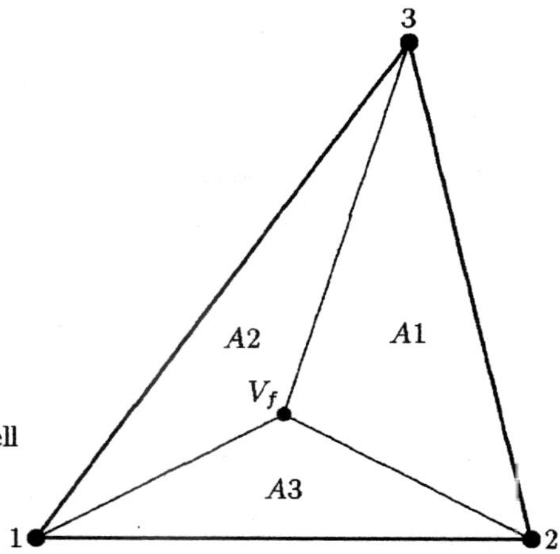

(b) Fine-Mesh Vertex in Coarse-Mesh Cell

Fig. 4. Area Stencils for Transfer Operators.

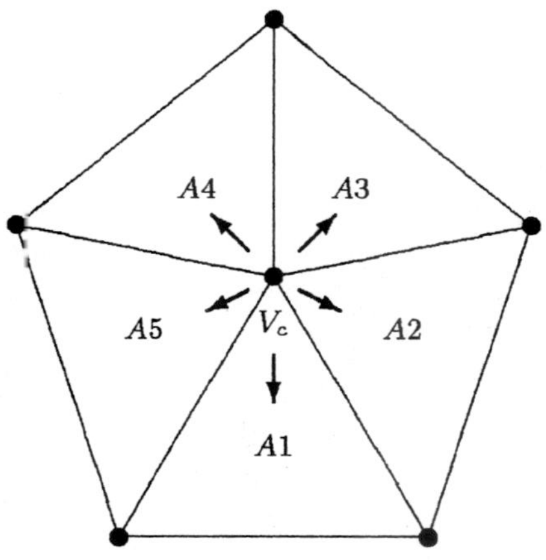

Fig. 5. Area Stencil for Residual Re-Distribution.

from V_c, with similar contributions from its other two vertices. This procedure conserves the sum of the residuals between the coarse and fine meshes.

The coarse-to-fine mesh interpolation operator reverses the process. The coarse mesh corrections are first transferred to the coarse-mesh vertices with area weighting, then the correction at each fine-mesh vertex is interpolated between the three vertices of the coarse-mesh triangle in which it is contained, with area weighting as in Figure 4(b). Finally, the correction in each fine-mesh cell is taken as the arithematic average of the corrections at the three vertices of that cell.

3 Results

In order to evaluate the accuracy and efficiency of the vertex-centroid (V-C) scheme, we have performed calculations for a variety of airfoils in subsonic and transonic flow. In this section we present results for the standard test case of a RAE-2822 airfoil at a Mach number of $M = 0.75$, and an angle of attack $\alpha = 3.0°$.

In order to assess the dependence of the accuracy on the mesh size, each result is provided for a series of increasingly fine meshes, with respectively (40×9), (80×17), (160×33) and (320×65) nodes; the first number is in the chordwise direction, and the second in the normal direction. In addition, the multigrid scheme with a W-cycle was used to accelerate the calculation on each of these meshes, so a total of 3 grids were used to obtain the result on the (40×9) mesh, while 4, 5 and 6 grids were used, respectively, for the calculations on the successively finer meshes. In every case, the coarsest grid used in the multigrid sequence contained (8×3) nodes.

For all the grids, the mesh nodes were generated via a conformal mapping of the airfoil section to a near circle. Nodes are evenly distributed in the mapped plane, so that they are bunched near the leading and trailing edges in the physical plane. Alternate rows of points were shifted a half-mesh interval in the mapped plane to produce a distribution of points which permits a very regular triangulation. Finally, a constrained Delaunay triangulation algorithm [21,22] was used to connect the mesh points while preserving the integrity of the profile. Figure 6 shows a close up of the trailing edge region for the four solution grids.

The results of the calculations are presented in the following figures. For each calculation, we show the solution on each of the four meshes. The (40×9) mesh in the top left, the (80×17) mesh in the top right, the (160×33) mesh in the bottom left, and the (320×65) mesh in the bottom right. To represent the solution, we display a plot of the pressure coefficient, $C_p = (p - p_\infty)/(\frac{1}{2}p_\infty q_\infty^2)$, with the axis upward in the negative direction. We also show a porcupine plot of the computed value of a measure of the entropy, $\frac{p}{p_\infty}/\left(\frac{\rho}{\rho_\infty}\right)^\gamma$, along the surface of the profile, with positive values in the outward normal direction. This is a useful measure of the numerical error, as entropy should be generated only through the occurrence of shock waves in the inviscid flow.

As a basis for comparison, we first show in Figure 7 results for the RAE-2822 airfoil computed with the cell-vertex scheme which we have previously used to perform calculations for complete aircraft [7–9]. This scheme may be viewed as a modified Galerkin method, stabilized by artificial diffusion adapted for unstructured meshes from the Jameson-Schmidt-Turkel (JST) scheme [15]. It is very robust, but exhibits a high level of numerical diffusion, as can be seen from the plots of the entropy along the surface. The multigrid acceleration scheme produces quite a high rate of convergence, dropping the residuals 6 orders of magnitude in 100 cycles.

Next in Figure 8 we show the results of introducing the CUSP numerical flux into the cell-vertex scheme. It can be seen both that the shock waves are captured more crisply, and that the levels of spurious entropy generated ahead of the shock waves are greatly reduced. The convergence rates are slightly slower, but still quite fast, dropping 4 orders in 100 cycles. Since 1993 it has been our normal practice to use the CUSP scheme in preference to the JST scheme for calculations on both structured and unstructured meshes.

In Figure 9 we present results computed with the new vertex-centroid (V-C) scheme using the CUSP numerical flux. It can be seen that the shocks are captured extremely crisply with only one interior point and without any overshoot in pressures. The production of spurious numerical entropy is significantly less than that of the cell-vertex scheme. On the finest mesh, there is no visible signs of entropy production ahead of the shock. However, the convergence rate is somewhat slower, taking almost 500 cycles to drop the residuals by 4 orders of magnitude.

4 Conclusion

We consider the results we have obtained with the V-C scheme to be extremely promising. In fact, the level of spurious numerical entropy production is lower than we have ever previously observed with any scheme on either a structured or an unstructured mesh. Our tests have also shown that the V-C scheme exhibits low numerical entropy production and remarkably crisp shocks when it is stabilized with a simple scalar artificial diffusion.

These results have benefited from the use of very regular triangulations: we believe it is essential to first validate the accuracy of the method with meshes of high quality. Our efforts are now focused on improving the reconstruction method to maintain the highest possible accuracy on less regular meshes [13]. We are also testing the scheme for viscous flows, and for three-dimensional calculations.

Acknowledgements

This work has benefited from the generous support of the Air Force Office of Scientific Research under Grant No. AF F49620-98-1-002, as well as from Internal Research and Development support from The Boeing Company.

References

1. T.J.R. Hughes, L.P. Franca, and M. Mallet. A new finite element formulation for computational fluid dynamics,I, Symmetric forms of the compressible Euler and Navier-Stokes equations and the second law of thermodynamics. *Comp. Meth. Appl. Mech. and Eng.*, 59:223–231, 1986.
2. R. Lohner, K. Morgan, and J. Peraire. Improved adaptive refinement strategies for the finite element aerodynamic configurations. AIAA Paper 86-0499, AIAA 24th Aerospace Sciences Meeting, Reno, NV, January 1986.
3. T.J. Barth and D.C. Jespersen. The design and application of upwind schemes on unstructured meshes. *AIAA paper 89-0366*, AIAA 27th Aerospace Sciences Meeting, Reno, NV, January 1989.

4. W.K. Anderson and D.L. Bonhaus. An implicit upwind algorithm for computing turbulent flows on unstructured grids. *Computers and Fluids*, 23:1–21, 1994.
5. N.T. Frink. Upwind scheme for solving the Euler equations on unstructured tetrahedral meshes. *AIAA Journal*, 30(1):70–77, 1992.
6. W.A. Smith. Multigrid solution of transonic flow on unstructured grids. Technical report, ASME WAM, Dallas, TX, November 1990.
7. A. Jameson, T.J. Baker, and N.P. Weatherill. Calculation of inviscid transonic flow over a complete aircraft. *AIAA paper 86-0103*, AIAA 24th Aerospace Sciences Meeting, Reno, Nevada, January 1986.
8. J.C. Vassberg and K.B. Dailey. AIRPLANE: Experiences, Benchmarks & Improvements. *AIAA paper 90-2998*, AIAA Applied Aerodynamics Conference, Portland, OR, August 1990.
9. J.C. Vassberg, K.B. Dailey, and D.M. Friedman. AIRPLANE: Unstructured-Mesh Applications. *SAE paper 901857*, SAE International, Long Beach, CA, October 1990.
10. A. Jameson. Analysis and design of numerical schemes for gas dynamics 1, artificial diffusion, upwind biasing, limiters and their effect on multigrid convergence. *Int. J. of Comp. Fluid Dyn.*, 4:171–218, 1995.
11. A. Jameson. Analysis and design of numerical schemes for gas dynamics 2, artificial diffusion and discrete shock structure. *Int. J. of Comp. Fluid Dyn.*, 5:1–38, 1995.
12. V. Venkatakrishnan. A perspective on unstructured grid flow solvers. *Report* 95-3, ICASE, Hampton, VA, 1995.
13. J.C. Vassberg and A. Jameson. Accurate reconstruction procedure for upwind algorithms on unstructured meshes. *In Preparation*, The Boeing Company, Long Beach, CA, August 2000.
14. P.L. Roe. Approximate Riemann solvers, parameter vectors, and difference schemes. *Journal of Computational Physics*, 43:357–372, 1981.
15. A. Jameson, W. Schmidt, and E. Turkel. Numerical solutions of the Euler equations by finite volume methods with Runge-Kutta time stepping schemes. *AIAA paper 81-1259*, AIAA 24th Aerospace Sciences Meeting, Reno, NV, January 1981.
16. A. Kurganov and E. Tadmor. New high-resolution central schemes for nonlinear conservation laws and convective-diffusion equations. *J. Comp. Phys.*, 160:241–282, 2000.
17. J.C. Vassberg. A fast, implicit unstructured-mesh Euler method. *AIAA paper 92-2693*, 10th AIAA Applied Aerodynamics Conference, Palo Alto, CA, June 1992.
18. D.J. Mavriplis and V. Venkatakrishnan. A 3D agglomeration multigrid solver for the Reynolds-averaged Navier-Stokes equations on unstructured meshes. *Int. J. Numerical Methods in Fluids*, 23:1–18, 1996.
19. M.H. Lallemand, H. Steve, and A. Dervieux. Unstructured multigridding by volume aggregation: Current status. *Computers and Fluids*, 21:397–433, 1992.
20. A. Jameson. Solution of the Euler equations for two dimensional transonic flow by a multigrid method. *Applied Mathematics and Computations*, 13:327–356, 1983.
21. B. Delaunay. Sur la sphere vide. *Bull. Acad. Science USSR VII: Class Scil, Mat. Nat.*, pages 793–800, 1934.
22. A. Jameson. Notes on constrained Delaunay triangulation for two-dimensional unstructured meshes. *Unpublished Notes*, Princeton University, Princeton, NJ, 1985.

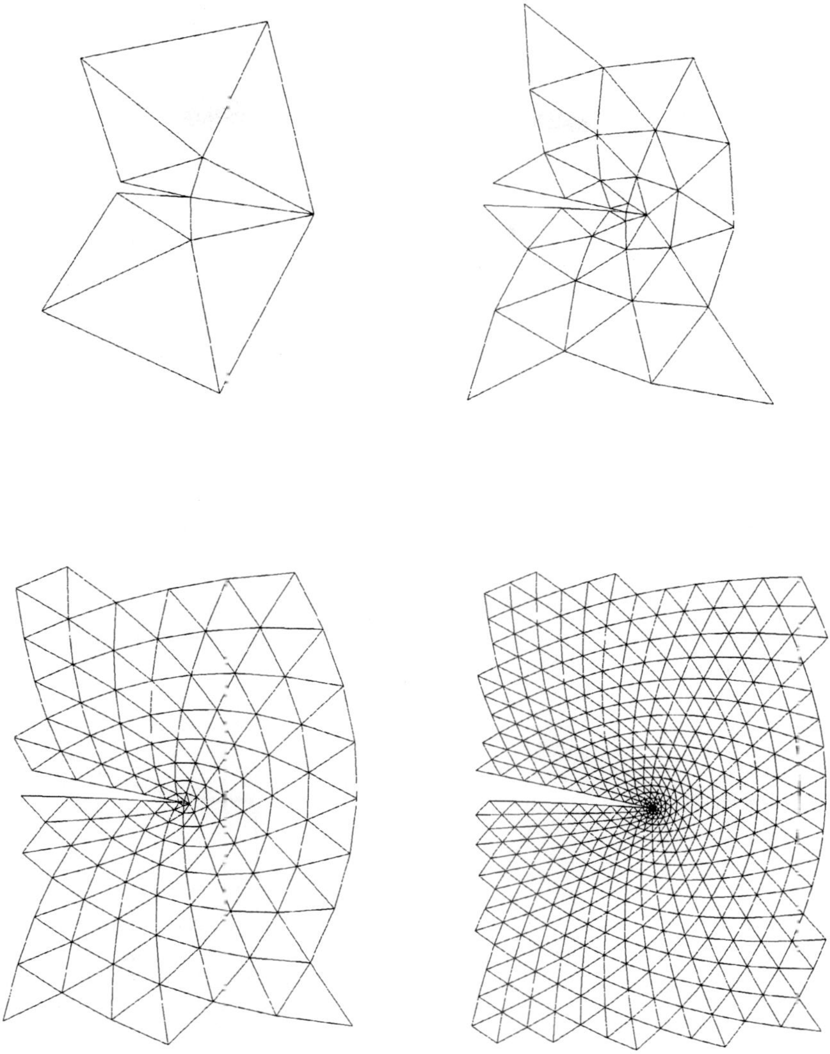

Fig. 6. RAE-2822 Grid Series, Close-Up of Trailing Edge Region.

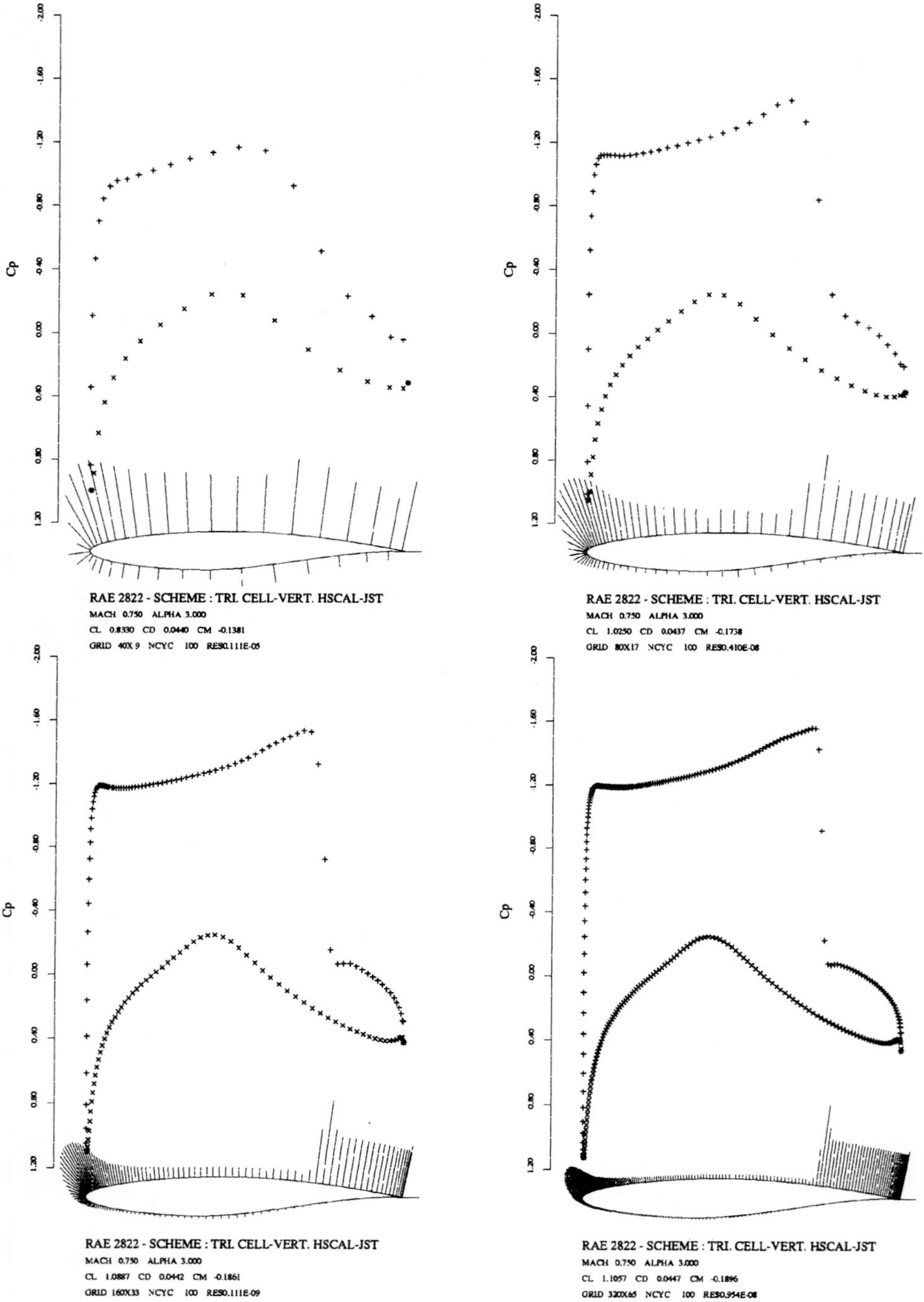

Fig. 7. RAE-2822 Solution Series, Cell-Vertex HSCAL-JST Scheme.

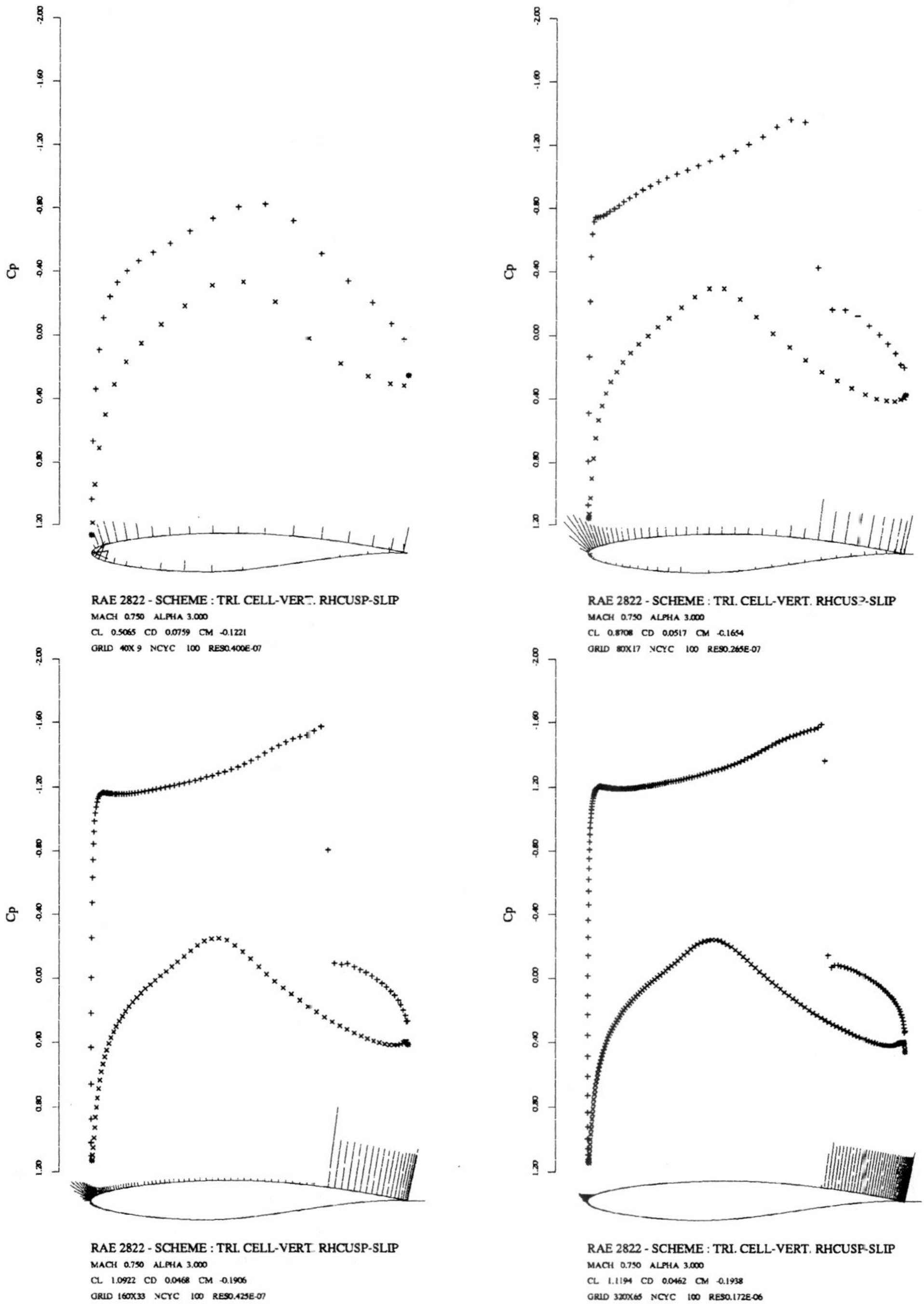

Fig. 8. RAE-2822 Solution Series, Cell-Vertex RHCUSP-SLIP Scheme.

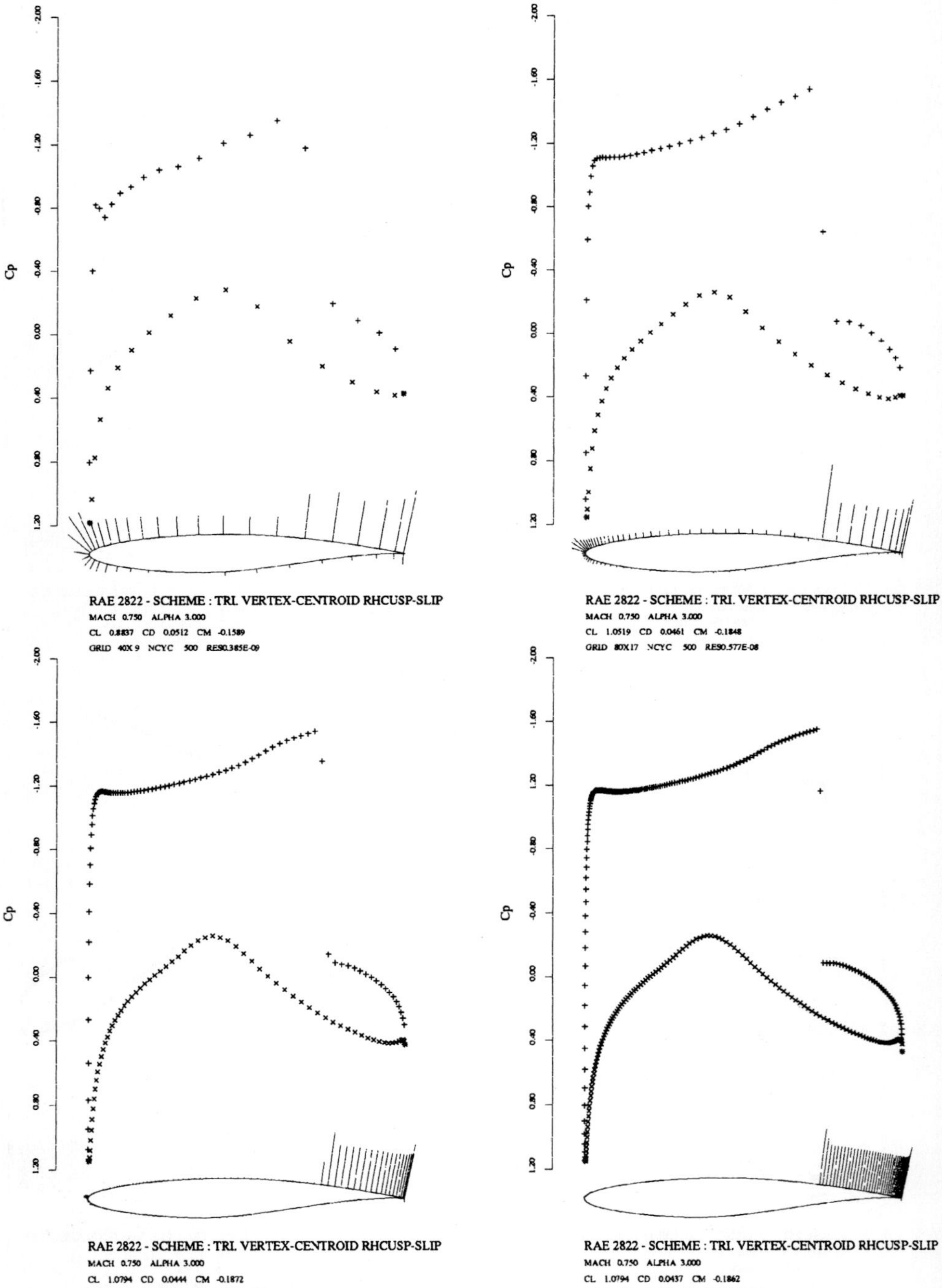

Fig. 9. RAE-2822 Solution Series, Vertex-Centroid RHCUSP-SLIP Scheme.

Part II

Adaptive Mesh

A Solution-Adaptive Technique Using Unstructured Hexahedral Grids

M. Sun and K. Takayama

Shock Wave Research Center, Institute of Fluid Science,
Tohoku University, Katahira 2-1-1, Aoba, Sendai 980, Japan

Abstract. This paper summarizes our work on the development of a solution-adaptive flow solver using unstructured hexahedral meshes for 3-D flows. A cell-face data structure is proposed to avoid data dependency that is faced when applying unstructured data for high performance computation. Two flow solvers are constructed using two different schemes. One solver uses the MUSCL-Hancock scheme, a second-order upwind scheme, another employs a second-order central-difference scheme with artificial viscosity. The parallel performance of the solvers on three supercomputers is reported. A result of shock motion from a square tube to a circular tube is also presented.

1 Introduction

Unstructured meshes have been widely used in computation fluid dynamics [8]. In order to improve the accuracy of numerical solutions with less computer time, various grid adaptation methods (*e.g.* [3,9]) have been developed and applied to the simulation of flows that contain disparate length scales, such as flows with shock waves. Most schemes were proposed for triangles and tetrahedra. Since a triangular mesh contains more edges then the quadrilateral counterpart, it consumes more storage and flux evaluations. Aftosmis et al [1] compared a variety of schemes on both meshes, and concluded that on regular and stretched meshes the additional edges do not lead to apparent accuracy advantage. For three dimensional meshes, Biswas and Strawn [2] also concluded that hexahedral meshes utilize computer resources more efficiently than tetrahedral meshes for the same level of accuracy by comparing the solutions of the Euler equations. In addition, unlike the hybrid methods that combine structured quadrilaterals/prisms around body with unstructured triangles/tetrahedrals outside in the computation of viscous flows, a technique using unstructured quadrilateral or hexahedral meshes will neatly generate a layer of body-fitted mesh by repeatedly refining the cells in the boundary layer. It is a real bonus that hexahedral grid generation becomes rather easy especially due to rapid progress on the automation of unstructured hexahedral meshing.

In the last symposium held in Bremen, we presented a 2-D vectorized adaptive solver (VAS2D) using a grid containing uniform quadrilateral cells [6], and have applied it to solve a variety of problems. The present work aims to develop an algorithm for three-dimensional flows using hexahedral cells. The paper is organized as follows. Section 2 is devoted to the basic ideas of the present data structure and grid adaptation. Section 3 reviews the main points of the present flow solver. Section 4 reports numerical results.

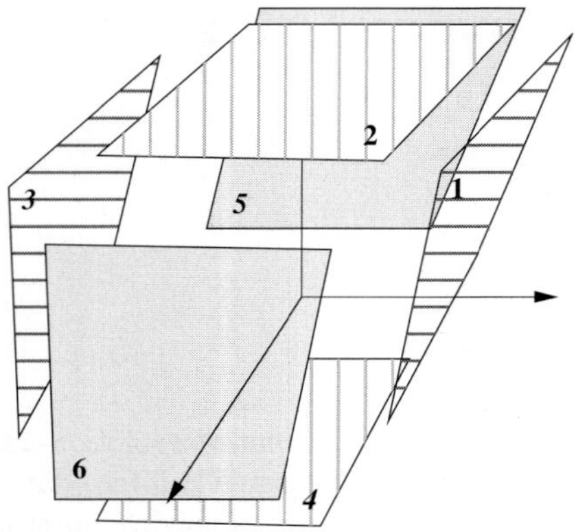

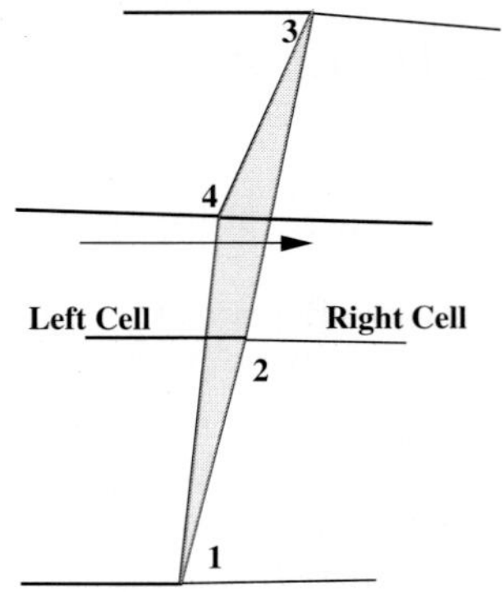

Fig. 1. Cell-face data structure: **a.** every cell stores its location and points to six faces; **b.** every face stores the locations of its four vertices and points to two neighboring cells.

2 Data structure and grid adaptation

The data structure used for unstructured quadrilateral and hexahedral meshes is optimized for the finite volume method, in order to enhance computational efficiency and to minimize memory requirement. The finite volume method evaluates the change of the conservative values of all control volumes by integrating their interface fluxes which consists of two steps, calculating fluxes at every interface and gathering interface fluxes for every control volume. The data structure is designed as having two primary arrays, one for control volumes, another for faces, with a bi-directional reference between them,

$$\text{control volume} \Longleftrightarrow \text{face}.$$

Thus the data structure is called a *cell-face* one. The data structure actually degenerates to a *cell-edge* one for 2-D quadrilateral grids. There are no reconcurrence appearing in the bi-directional reference, so that it naturally avoids sorting or coloring for vector/parallel supercomputing. Furthermore, because of the similarity of the data structure for two dimensions and three dimensions, it is rather convenient for programming.

A basic nature of the data structure for three dimensions is that every cell points to its six faces, and every face points to its two neighboring cells, as shown in Fig. 1. Physical variables are stored at cells, therefore it is a cell-centered data structure. Every face is set to have a direction, so that we may label two neibhouring cells, *left* and *right*, according to the direction. The six faces of each cell is ordered following these rules: two opposite faces must have the same direction, and three pairs of opposite faces must constitute a local coordinate system (the right-hand rule). These strict definitions in the data structure reduce the complexity of logic in the unstructured adaptive solver without losing generality. The existence of the ordering method for any unstructured quadrilateral and hexahedral grids has been proven in [4,6]. Note that the data structure does not

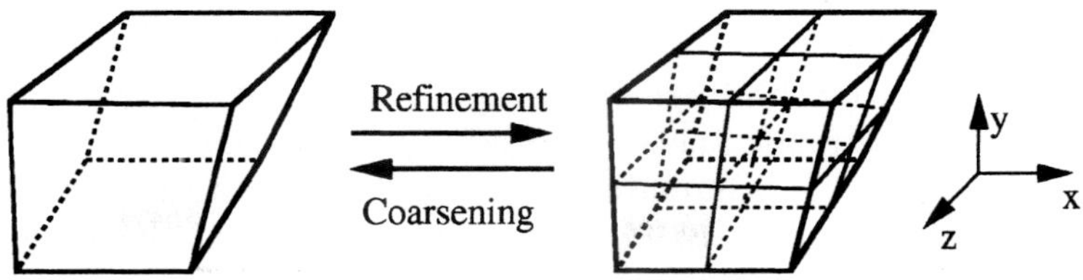

Fig. 2. Refinement strategy: a father cell is divided into eight sons

require additional memory to store the definitions, but stores the neighboring information following the rules.

A cell to be refined is divided into eight subcells or *sons* as shown in Fig. 2. To avoid unlimitedly adding fine cells around shock waves, either the maximum level of refinement or the minimum cell volume or both should be prescribed. When either the level or the volume of cell reaches the given limit, further refinement is prohibited. This strategy is similar to many other h-adaptive schemes. The criterion used for grid adaptation is based on the truncation error of the Taylor series expansion of density. The truncation error indicator is given by the ratio of the second-order derivative term to the first-order one in the Taylor series. The indicator is consistent with the limiting procedure in flow solvers so that the cells that turn on the artificial viscosity or the slope limiter must be refined.

Refinement and coarsening procedures are handled separately. Both procedures have similar steps for vector/parallel processing:

1. handling the inside of cells which are flagged to be refined or coarsened,
2. handling the faces of the flagged cells,
3. arranging memory.

Step 1 is conducted based on cells, and updates all information inside, such as face deletion and adding of finer cells, which is independent of other cells. Step 2, based on faces, renews every face of refined or coarsened cells and its two neighboring cells, which may be done without influencing other faces. In this way the adaptation procedure has no data dependence and can be naturally vectorized. Step 1 and 2 change the status of some cells and faces, for example from sons to fathers. Step 3 fills the "holls" that appear due to the deletion of cells, and make the memory easily to be accessed by other subroutines such as flow solvers.

Either the conservative or the primitive values at newly appeared cells in the adaptation procedure have to be imposed by those on the old mesh. In the refinement procedure the conservative variables of new sons are linearly interpolated from those of their father, while in coarsening procedure flow variables of a coarsened father cell are the volume-weighted average of these of its "dead" sons. The interpolation and the weighted average for new cells preserve conservation [4].

3 Flow solver

The conservation laws are solved by the finite volume method. The method solves the conservation laws by directly applying them to every non-overlapping discrete volume the summation of which covers the whole computational domain. The conservation laws written for a discrete control volume are, for a second-order scheme in both time and space,

$$\mathbf{U}_i^{n+1} = \mathbf{U}_i - \frac{\Delta t}{\Delta \Omega_i} \sum_{k=1}^{\text{faces}} \hat{\mathbf{F}}_k^{n+1/2}, \tag{1}$$

where $\hat{\mathbf{F}}_k^{n+1/2}$ are fluxes evolved by half a time step, and locate at the center point of the interface k. Since the values are unknown at the center point, they are interpolated from the centroid (cell-centered data are considered here). Difference between all conservative schemes is just how they calculate the numerical fluxes. Two schemes that belongs to the central-difference scheme and the upwind scheme respectively are used to calculate the fluxes.

For the solver using the central-difference scheme, a TVD smoothing scheme is used [4]. For the solver using the upwind scheme, the MUSCL-Hancock scheme (see, [7] for formulations on structured grids) is applied to unstructured grids. Primitive variables are first reconstructed by the least square method at cells, and other procedures are all conducted at interfaces. In slope limiting procedure which aims to maintain TVD conditions around sharp gradients, only gradients in the direction along two neighboring cells beside an interface are modified by limiters. The minmod limiter is used in most applications. The limited slopes are then used to evolve solutions at the two cells respectively by half a time step. A HLL approximate Riemann solver (see, [7]) is chosen to determine the flux at interface because of its simplicity and generality.

4 Numerical results

The data structure proposed contains no data dependency, so that it allows easy vector/parallel processing without recourse to any sorting or coloring procedure that most unstructured data structures require. The parallel performance on three different machines of our 2-D and 3-D solvers is summarized in Fig 3. On the shared-memory machines, SX-5 and Cray C90, the data are collected by testing an unsteady computation in which grid adaptation is performed at every iteration. For eight processors the speedup is approximately six, which is not bad for the two solvers are automatically parallelized by the compliers coming with the machines. On the distributed-memory machines, only steady flow computation is tested, and very few adaptations are conducted. It seems that the speedup achieved is not as high as well-tuned parallel codes using domain decomposition method. However, if the overheads of conducting domain decomposition is taken into account, the speedup is rather satisfactory because the data structure allows grid adaptation that will decrease computer time dramatically (but not the rate of speedup).

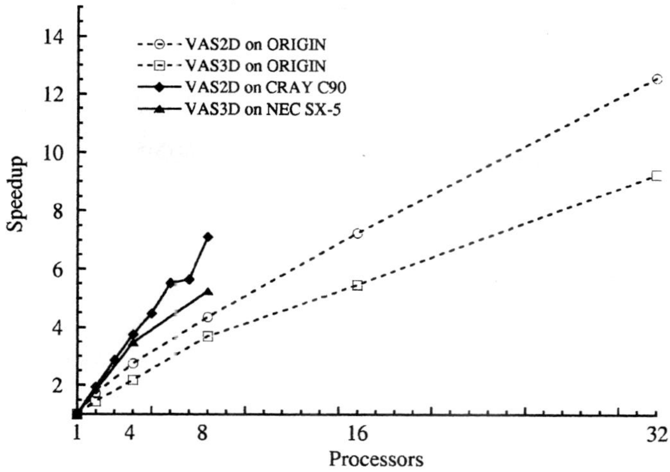

Fig. 3. Performance of the 2-D and the 3-D solvers on machines, Cray C90, Origin 2000, and SX-5

Numerical results of a 3-D case are shown in Fig. 4, in which an incident shock wave with M_s=2.0 moves from a square cross-section shock tube into a larger cylindrical tube. Initial mesh contains 299 cells. Notice that unstructured cells are used in the cylindrical cross-section. In this unsteady computation, because of the existence of the moving shock wave and complicated vortex formation whose locations are unknown in advance, the mesh adaptation is done at every time step. An adaptive mesh and the density distribution on the surface at some instant after shock wave entering the circular tube are shown in Fig. 4bc; the total number of cells is increased to about 0.78 million. It takes about three minutes to run this computation on SX-5 using one processor. It is seen that basic flow features, such as incident shock and vortices, are well reproduced. Hexahedral cells on a cut plane inside the tube are shown in Fig. 4d; fine cells are clearly distributed around shocks and vortex regions.

5 Summary

A solution-adaptive technique for hexahedral meshes has been invented using a vectorizable data structure, and it is successfully applied to the solution of the unsteady Euler equations. It seems that unstructured hexahedral meshes may be a viable alternative to tetrahedral meshes for the simulation of three dimensional flows.

References

1. Aftosmis MJ, Gaitonde D, Sean Tavares T (1994) On the accuracy, stability, and monotonicity of various reconstruction algorithms of unstructured meshes, AIAA paper 94-0415.

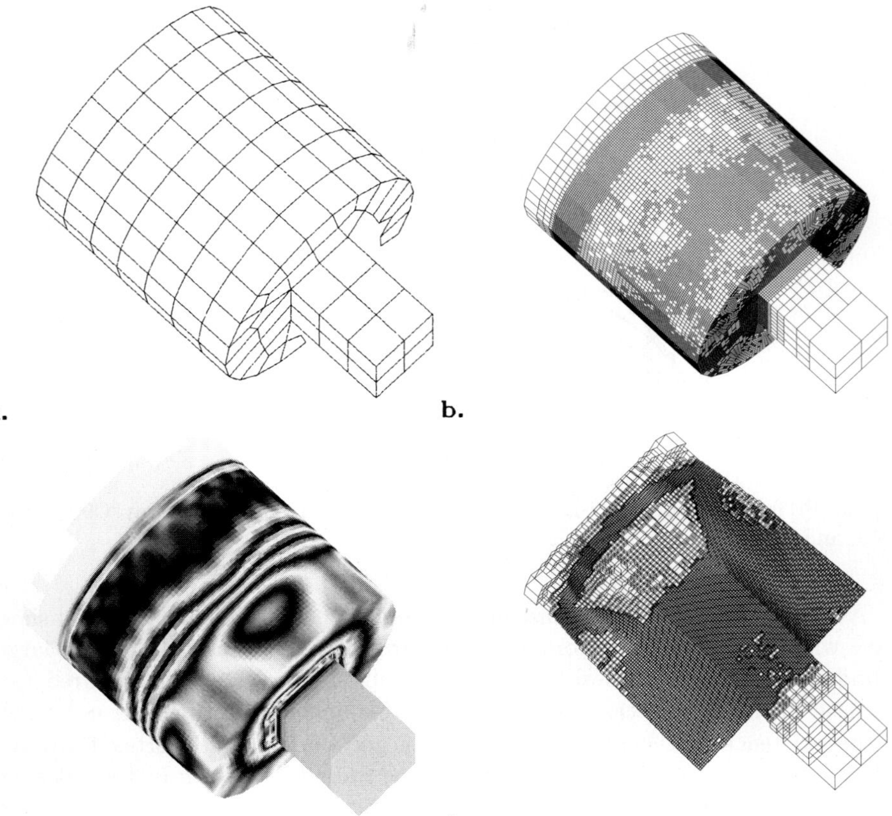

a. b. c. d.

Fig. 4. Shock wave entering a cylindrical tube from a smaller square tube: **a.** initial unstructured mesh, 299 cells; **b.** adaptive surface meshes at $t = 1.4$, totally 776,550 cells and 2,412,340 faces; **c.** corresponding density contours; **d.** adaptive cells on a cut plane.

2. Biswas R, Strawn RC (1997) Tetrahedral and hexahedral mesh adaptation for CFD problems, NAS-97-007, NASA Ames Research Center.
3. Löhner R (1987), *Comput. Meths. Appl. Meh. Engrg.* V61, pp.323-338.
4. Sun M (1998) Numerical and experimental studies of shock wave interaction with bodies, Ph.D. Thesis, Tohoku University, Japan. *http://ceres.ifs.tohoku.ac.jp/~sun/thesis.html*
5. Sun M, Takayama K (1999) *J. Comp. Phys*, V150 : 143-180.
6. Sun M, Takayama K (1999) H-adaption on unstructured quadrilaterals, 8th Intl. Symp. on Comput. Fluid Dynamics, Bremen, Germany.
7. Toro EF (1999) Riemann solvers and numerical methods for fluid dynamics, 2nd edition, Springer.
8. Venkatakrishnan V (1995) AIAA paper 95-0667. *see also AIAA J.* V34 : 533-547, 1996.
9. Voinovich P, Timofeev E, Takayama K, Saito T, Galyukov A (1998), AIAA paper 98-0540.

Local Mesh Refinement in the Space-Time CE/SE Method

S.C. Chang[1], Y.H. Wu[2], X.Y. Wang[3], V. Yang[4]

[1]NASA Glenn Research Center, Cleveland, Ohio 44135, USA
[2,4]Department of Mechanical Eng., Pennsylvania State University, University Park, PA 16802, USA
[3]Taitech Inc., NASA Glenn Research Center, Cleveland, Ohio 44135, USA
email: *sin-chung.chang@grc.nasa.gov*, *yxw119@psu.edu*, *wangxy@turbot.grc.nasa.gov*, *vigor@psu.edu*

Abstract. A local mesh refinement procedure for the CE/SE method which does not use an iterative procedure in the treatments of grid-to-grid communications is described. It is shown that a refinement ratio in the order of 10 can be applied successfully across a single coarse grid/fine grid interface.

1 Introduction

To accurately resolve flow properties in high-gradient regions without applying a very fine and costly uniform grid over the entire computational domain, simulation of practical unsteady CFD problems may require implementation of local mesh refinement (LMR) in both spatial and temporal directions. In this paper, we report a new LMR procedure for the CE/SE method [1-3].

It is a must for a LMR procedure to enforce space-time flux conservation across a coarse grid/fine grid interface so that numerical solutions will not be contaminated by spurious reflections originating there [4]. Fortunately, this requirement can be met easily in the CE/SE setting. In particular, an iterative procedure is not needed in the treatments of grid-to-grid communications. Furthermore, it will be shown that a refinement ratio in the order of 10 can be applied successfully across a single interface.

2 Preliminaries

To simplify the description of the LMR procedure, consider the PDE

$$\partial u/\partial t + a\partial u/\partial x = 0 \tag{1}$$

and its conservation form

$$\oint_{S(V)} \boldsymbol{h} \cdot d\boldsymbol{s} = 0 \tag{2}$$

where (i) $a > 0$ is a constant; (ii) $S(V)$ is the boundary of any region V in space-time; (iii) $\boldsymbol{h} = (au, u)$; and (iv) $d\boldsymbol{s} = d\sigma\,\boldsymbol{n}$ with $d\sigma$ and $\boldsymbol{n}$, respectively, being the area and the outward unit normal of a surface element on $S(V)$. Note that $\boldsymbol{h} \cdot d\boldsymbol{s}$ is the flux of $\boldsymbol{h}$ leaving V through the surface element $d\boldsymbol{s}$.

Let Ω_1 denote the set of all mesh points (j,n) (dots in Fig. 1) with $j+n$ being odd integers. For each $(j,n) \in \Omega_1$, let the solution element SE(j,n) be the *interior* of the *space-time* region bounded by a dashed curve depicted in Fig. 2). For any $(x,t) \in$ SE(j,n), let $u(x,t)$ and $\boldsymbol{h}(x,t)$, respectively, be approximated by $u^*(x,t\,;j,n)$ and $\boldsymbol{h}^*(x,t\,;j,n)$, where [1]

$$u^*(x,t\,;j,n) = u_j^n + (u_x)_j^n(x-x_j) + (u_t)_j^n(t-t^n) \tag{3}$$

$$\boldsymbol{h}^*(x,t\,;j,n) = \big(au^*(x,t\,;j,n),\, u^*(x,t\,;j,n)\big) \tag{4}$$

It is required that $u = u^*(x,t\,;j,n)$ satisfy Eq. (1). Thus

$$(u_t)_j^n = -a\,(u_x)_j^n \tag{5}$$

Hereafter, $\boldsymbol{q}_j^n$ denotes the column matrix formed by u_j^n and $(u_x)_j^n$, the independent marching variables at $(j,n) \in \Omega_1$.

Let space-time be divided into conservation elements (CEs). Two CEs, i.e., CE$_-(j,n)$ and CE$_+(j,n)$ (Figs. 3–4) are assigned to each $(j,n) \in \Omega_1$. Because the boundary of CE$_-(j,n)$ (CE$_+(j,n)$) lies within the union of SE(j,n) and SE$(j-1,n-1)$ (SE$(j+1,n-1)$), $\boldsymbol{q}_j^n$ can be solved for in terms of $\boldsymbol{q}_{j\pm1}^{n-1}$ by substituting $V = CE_\pm(j,n)$ and $\boldsymbol{h} = \boldsymbol{h}^*$ in Eq. (2). The resulting scheme is the *a* scheme which has the property that the flux of $\boldsymbol{h}^*$ is conserved over the union of any combination of CEs [1].

For the a-ϵ scheme [1], which reduces to the a scheme if $\epsilon = 0$, again $\boldsymbol{q}_j^n$ is a function of $\boldsymbol{q}_{j\pm1}^{n-1}$. However, the flux of $\boldsymbol{h}^*$ generally is not conserved over the CEs referred to above. Rather, for each $(j,n) \in \Omega_1$, it is conserved over CE(j,n) (the union of CE$_\pm(j,n)$—see Fig. 5). To proceed further, in the following, the concept of modified flux will be introduced.

By definition, the modified flux at each of the line segments $\overline{AB}$, $\overline{BC}$, $\overline{CD}$ $\overline{DE}$, $\overline{EF}$, and $\overline{FA}$, which form the boundary of CE(j,n), is the corresponding flux of $\boldsymbol{h}^*$. On the other hand, the modified flux at $\overline{AD}$, i.e., the interface that divides CE$_\pm(j,n)$, is defined such that the total modified flux leaving the boundary of CE$_-(j,n)$ vanishes. Note that the modified flux at $\overline{AD}$, except in the case of the a scheme, is not equal to the corresponding flux of $\boldsymbol{h}^*$. Because (i) the total modified flux leaving the boundary of CE(j,n) vanishes; and (ii) the modified flux leaving CE$_-(j,n)$ through $\overline{AD}$ is the negative of that leaving CE$_+(j,n)$ through $\overline{AD}$, one concludes that the total modified flux leaving the boundary of CE$_+(j,n)$ also vanishes. In other words, the modified flux is conserved over CE$_\pm(j,n)$ at each $(j,n) \in \Omega_1$.

The a-ϵ scheme obviously can also be defined over Ω_2, the set of (j,n) (not marked in Fig. 1) with $j+n$ being even integers. Moreover, two schemes defined over Ω_1 and Ω_2 can be fused into a "dual" scheme which is defined over $\Omega \stackrel{\text{def}}{=} \Omega_1 \cup \Omega_2$. Note that, in practical applications, a solution of a dual scheme may not be decoupled into two independent solutions.

Let $(j,n) \in \Omega_1$, then $(j+1,n) \in \Omega_2$. By definition, $\mathrm{CE}_+(j,n)$ coincides with $\mathrm{CE}_-(j+1,n)$. However, two different flux conservation conditions are defined over their co-boundary. In fact, two different modified fluxes tied to, respectively, these two coincident CEs, are defined at any boundary line segment. Obviously, the same complication arises for any CE. As a result, in case that the solution of a dual scheme becomes coupled, the concept of global flux conservation is not well defined in terms of these individual modified fluxes. However, *if the generalized flux at any line segment that forms the boundary of a CE is defined as the average of the two local modified fluxes, then the generalized flux is conserved over the union of any combination of CEs.* The proof follows from (i) the generalized flux is conserved over each CE—a fact that can be established by summing over the two local modified flux conservation conditions tied to this CE; and (ii) the generalized flux leaving one CE through an interface is the negative of that leaving the neighboring CE through the same interface.

3 Time Marching

As an example, a step-by-step description of the marching from the 0th to 1st time level (Fig. 6) will be described. Note that (i) the simple grid structure used here does not limit the generality of the following discussions; and (ii) *two* $\boldsymbol{q}$*s and thus two symbols (e.g.,* A_3^1 *and* B_1^1*) are assigned to any common mesh point of a coarse grid and a fine grid.*

I. Because $a > 0$, $\boldsymbol{q}(A_1^1)$ is specified as a part of the left boundary condition while $\boldsymbol{q}(C_3^1)$ is extrapolated from $\boldsymbol{q}(C_2^0)$, i.e., $\boldsymbol{q}(C_3^1) = \boldsymbol{q}(C_2^0)$.

II. $\boldsymbol{q}(A_2^1)$ is determined in terms of $\boldsymbol{q}(A_1^0)$ and $\boldsymbol{q}(A_3^0)$ using the dual a-ϵ scheme with $\epsilon = 0.5$ [1]. Similarly, $\boldsymbol{q}(B_2^{1/2})$ ($\boldsymbol{q}(C_2^1)$) is determined in terms of $\boldsymbol{q}(B_1^0)$ and $\boldsymbol{q}(B_3^0)$ ($\boldsymbol{q}(C_1^0)$ and $\boldsymbol{q}(C_3^0)$).

III. $\boldsymbol{q}(B_1^{1/2})$ is determined in terms of $\boldsymbol{q}(A_2^0)$, $\boldsymbol{q}(A_3^0)$ and $\boldsymbol{q}(B_2^0)$ assuming

$$u_x(B_1^{1/2}) = \left[u(B_1^{1/2}) - \left(u(A_2^0) + (\Delta t/2)u_t(A_2^0)\right)\right] / \Delta x \tag{6}$$

$$F(\overrightarrow{A_3^0 B_1^{1/2}}) + F(\overrightarrow{B_1^{1/2} B_2^{1/2}}) + F(\overrightarrow{B_2^0 B_2^{1/2}}) + F(\overrightarrow{B_2^0 B_1^0}) = 0 \tag{7}$$

Here (i) $u_t(A_2^0) = -a\, u_x(A_2^0)$ (see Eq. (5)); and (ii) $F(\overrightarrow{A_3^0 B_1^{1/2}})$, as an example, denotes the flux of $\boldsymbol{h}^*$ leaving the rectangle $A_3^0 B_1^{1/2} B_2^{1/2} B_2^0$ through $\overline{A_3^0 B_1^{1/2}}$ as evaluated using $\boldsymbol{q}(A_3^0)$. Similarly, $\boldsymbol{q}(B_3^{1/2})$ is determined in terms of $\boldsymbol{q}(B_2^0)$, $\boldsymbol{q}(C_1^0)$ and $\boldsymbol{q}(C_2^0)$. $\boldsymbol{q}(B_2^1)$ is then determined in terms of $\boldsymbol{q}(B_1^{1/2})$ and $\boldsymbol{q}(B_3^{1/2})$ using the special dual a-ϵ scheme referred to above.

IV. $\boldsymbol{q}(B_1^1)$ is determined in terms of $\boldsymbol{q}(B_2^{1/2})$, $\boldsymbol{q}(A_3^0)$ and $\boldsymbol{q}(A_2^0)$ assuming

$$u_x(B_1^1) = \left[u(B_1^1) - \left(u(A_2^0) + \Delta t\, u_t(A_2^0)\right)\right] / \Delta x \tag{8}$$

$$F(A_3^0, \overline{B_1^{1/2} B_1^1}) + F(\overrightarrow{B_1^1 B_2^1}) + F(\overrightarrow{B_2^{1/2} B_2^1}) + F(\overrightarrow{B_2^{1/2} B_1^{1/2}}) = 0 \tag{9}$$

Here $F(A_3^0, \overline{B_1^{1/2}B_1^1})$ denotes the flux of $\boldsymbol{h}^*$ at $\overline{B_1^{1/2}B_1^1}$ evaluated using $\boldsymbol{q}(A_3^0)$. Similarly, $\boldsymbol{q}(B_3^1)$ is determined in terms of $\boldsymbol{q}(B_2^{1/2})$, $\boldsymbol{q}(C_1^0)$ and $\boldsymbol{q}(C_2^0)$.

V. $\boldsymbol{q}(A_3^1)$ is determined in terms of $\boldsymbol{q}(B_1^0)$, $\boldsymbol{q}(B_1^{1/2})$, $\boldsymbol{q}(B_2^{1/2})$ and $\boldsymbol{q}(A_2^0)$ assuming

$$u_x(A_3^1) = \left[\left(u(B_2^{1/2}) + (\Delta t/2)u_t(B_2^{1/2})\right) - u(A_3^1)\right] / (\Delta x/2) \tag{10}$$

$$F(\overrightarrow{B_1^0B_1^{1/2}}) + F(\overrightarrow{B_1^{1/2}B_1^1}) + F(\overrightarrow{A_3^1A_2^1}) + F(\overrightarrow{A_2^0A_2^1}) + F(\overrightarrow{A_2^0A_3^0}) = 0 \tag{11}$$

Similarly, $\boldsymbol{q}(C_1^1)$ is determined in terms of $\boldsymbol{q}(B_3^0)$, $\boldsymbol{q}(B_3^{1/2})$, $\boldsymbol{q}(B_2^{1/2})$ and $\boldsymbol{q}(C_2^0)$.

Note that, *at each line segment on a coarse grid/fine grid interface, two fluxes of $\boldsymbol{h}^*$ are evaluated using a coarse-grid $\boldsymbol{q}$ and a fine-grid $\boldsymbol{q}$, respectively. Moreover, each of these fluxes is involved in a coarse-grid conservation condition and a fine-grid conservation condition. As a result, for the current marching procedure, it can be shown that the generalized flux is conserved over the union of any combination of interior CEs.*

4 Mesh Reconstruction

For time-dependent problems, spatial mesh reconstruction may be implemented at any designated time level such that high-gradient regions are always covered by fine grids. In the following, a step-by-step description of a reconstruction procedure that preserves generalized-flux conservation will be explained using Fig. 7 (Note: $C_{3/2}$ is the midpoint of $\overline{C_1C_2}$).

I. No reconstruction is required for the mesh points to the left of A_2 or to the right of C_2. Thus $\boldsymbol{q}(A_1') = \boldsymbol{q}(A_1)$ and $\boldsymbol{q}(C_2') = \boldsymbol{q}(C_3)$.

II. $\boldsymbol{q}(A_2')$ is determined in term of $\boldsymbol{q}(A_2)$, $\boldsymbol{q}(B_1)$ and $\boldsymbol{q}(B_2)$ assuming

$$F(\overrightarrow{A_2'A_1'}) = F(\overrightarrow{A_2A_1}), \quad F(\overrightarrow{A_2'A_3'}) = F(\overrightarrow{B_1B_2}) + F(\overrightarrow{B_2B_3}) \tag{12}$$

III. $\boldsymbol{q}(B_2')$ is determined in terms of $\boldsymbol{q}(C_1)$ and $\boldsymbol{q}(C_2)$ assuming

$$F(\overrightarrow{B_2'B_1'}) = F(C_2, \overline{C_{3/2}C_1}), \quad F(\overrightarrow{B_2'B_3'}) = F(C_1, \overline{C_{3/2}C_2}) \tag{13}$$

IV. $\boldsymbol{q}(A_3')$ is determined in terms of $\boldsymbol{q}(B_2)$, $\boldsymbol{q}(B_3)$ and $\boldsymbol{q}(B_2')$ assuming

$$u_x(A_3') = (u(B_2') - u(A_3'))/(\Delta x/2), \quad F(\overrightarrow{A_3'A_2'}) = F(\overrightarrow{B_3B_2}) + F(\overrightarrow{B_2B_1}) \tag{14}$$

V. $\boldsymbol{q}(B_1')$ is determined in terms of $\boldsymbol{q}(C_1)$ and $\boldsymbol{q}(A_2')$ assuming

$$u_x(B_1') = (u(B_1') - u(A_2'))/\Delta x, \quad F(\overrightarrow{B_1'B_2'}) = F(\overrightarrow{C_1C_{3/2}}) \tag{15}$$

VI. $\boldsymbol{q}(B_3')$ is determined in terms of $\boldsymbol{q}(C_2)$ and $\boldsymbol{q}(C_2')$ assuming

$$u_x(B_3') = (u(C_2') - u(B_3'))/\Delta x, \quad F(\overrightarrow{B_3'B_2'}) = F(\overrightarrow{C_2C_{3/2}}) \tag{16}$$

VII. $\boldsymbol{q}(C_1')$ is determined in terms of $\boldsymbol{q}(C_2)$ and $\boldsymbol{q}(B_2')$ assuming

$$u_x(C_1') = (u(C_1') - u(B_2'))/(\Delta x/2), \quad F(\overrightarrow{C_1'C_2'}) = F(\overrightarrow{C_2C_3}) \tag{17}$$

5 Numerical Results

Assuming $a = 1$, $\Delta x = 0.01$ and $\Delta t = 0.009$, an exact solution to Eq. (1), i.e.,

$$u = (10/\sqrt{\pi}) \exp\left[-100(x - t - 0.3)^2\right] \tag{18}$$

is solved numerically using the current LMR procedure with a grid refinement ratio $R = 10$ in both space and time. The initial condition ($t = 0$) and mesh distribution are shown in Fig. 8 while the exact solution (a solid line) and numerical solution (dots) along with the mesh distribution at $t = 0.9$ are depicted in Fig. 9 (Note: due to a consideration related to generalized flux, here each dot represents a midpoint solution, i.e., the average mesh value of two neighboring mesh points). It is seen that the numerical solution match very smoothly across the two coarse grid/fine grid interfaces.

Assuming $\Delta x = 4 \times 10^{-3}$, $\Delta t = 1.5 \times 10^{-3}$ and $R = 4$, the Sod's shock tube problem [1] is also solved using a similar Euler LMR procedure. The results at $t = 0.195$ are shown in Fig. 10(a) while a close-up view covering only the neighborhoods of grid interfaces is depicted in Fig. 10(b). Note that a simulation using $R = 64$ also has been carried out successfully.

6 Conclusions

A simple CE/SE LMR procedure has been described. Its robustness is demonstrated by numerical examples with grid refinement in the order of 10.

References

[1] S.C. Chang: The method of space-time conservation element and solution element – A new approach for solving the Navier-Stokes and Euler equations. *J. Computational Physics*, **119**, 1995, pp. 295–324.

[2] S.C. Chang, X.Y. Wang, and C.Y. Chow: The space-time conservation element and solution element method – A new high-resolution and genuinely multidimensional paradigm for solving conservation laws. *J. Computational Physics*, **156**, 1999, pp. 89-136.

[3] Other CE/SE references: www.grc.nasa.gov/www/microbus.

[4] M.J. Berger: On conservation at grid interfaces. *SIAM J. Numer. Anal.*, **24(5)**, 1986, pp. 967-984.

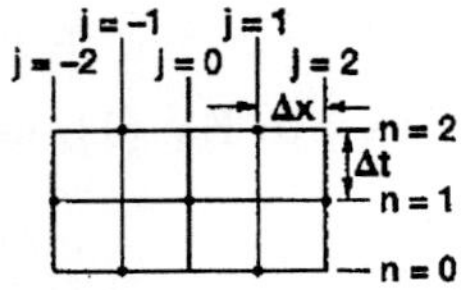

Figure 1.—A staggered space-time mesh.

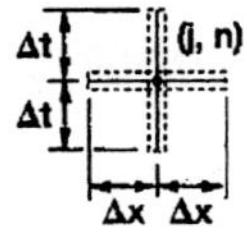

Figure 2.—SE (j, n).

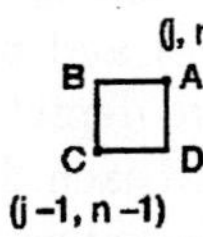

Figure 3.—CE_(j, n).

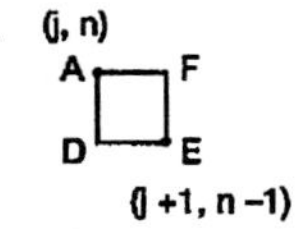

Figure 4.—CE+ (j, n).

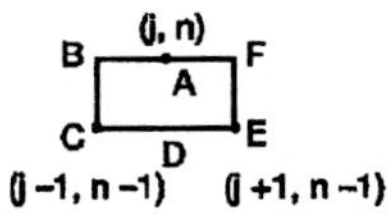

Figure 5.—CE (j, n).

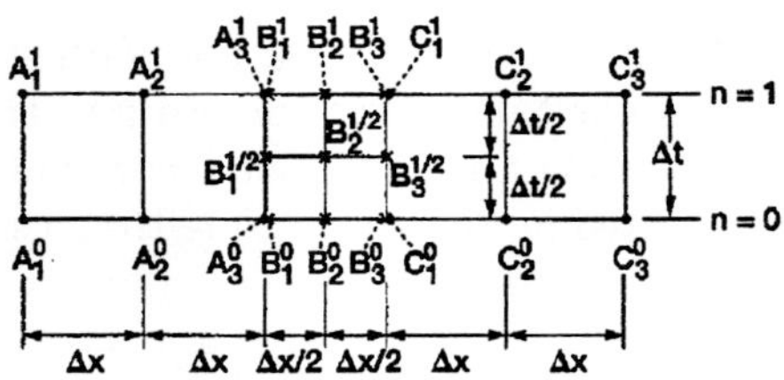

Figure 6.—A coarse-grid/fine-grid space-time mesh.

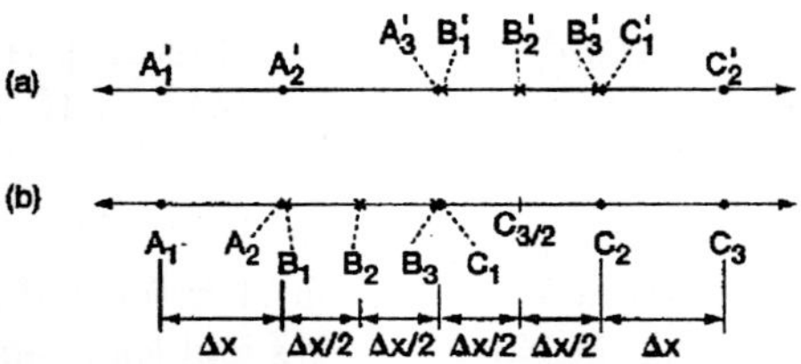

Figure 7.—Mesh reconstruction. (a) Reconstructed mesh. (b) Original mesh.

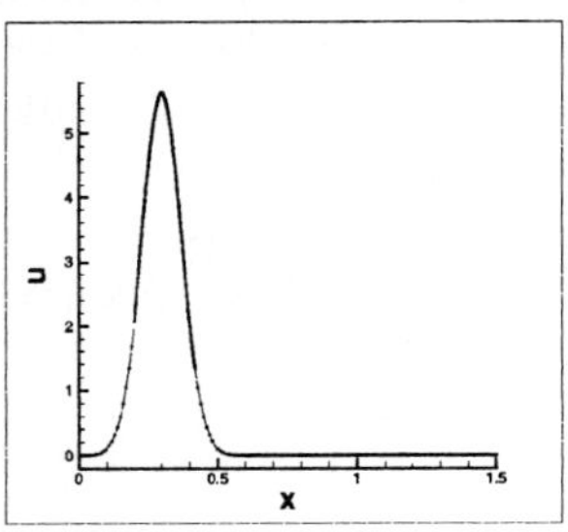

Fig. 8. Initial condition (Eq. (18)).

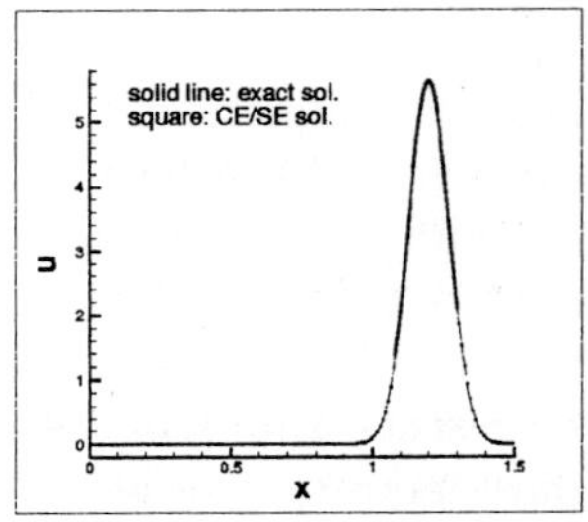

Fig. 9. Exact and numerical solutions at $t = 0.9$.

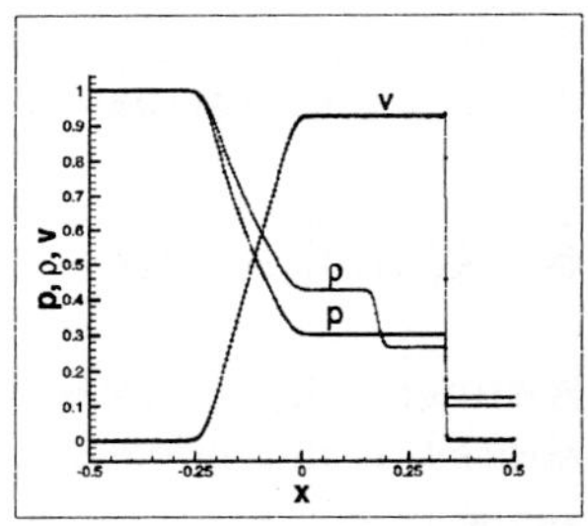

(a) Overall solution.

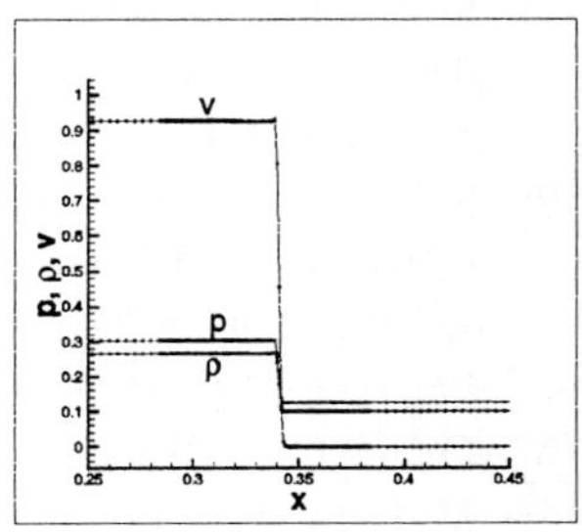

(b) Close-up view.

Fig. 10. Solution for Sod's problem at $t = 0.195$.

3-D Grid Adaptation Using a Sensor Based on Directed Divergence Between Maxwellians

Raghavendra N.V.[1], Mohan Varma[1], Biju Uthup[2], and Deshpande S. M.[1]

[1] Centre of Excellence for Aerospace CFD, Department of Aerospace Engineering, Indian Institute of Science, Bangalore, India.
[2] CFD Group, Aeronautical Development Agency, Bangalore, India

Abstract. A new *posteriori* sensor for Euler equations based on directed divergence between two Maxwellian distributions is presented in this paper. This sensor is used to adapt structured meshes in conjunction with a node movement method based on weighted averaging. Adaptation using this sensor is tested for flows over wings which are obtained by an in-house developed 3-D Euler code. The adapted meshes are shown to improve the solution accuracy considerably.

1 Introduction

Solution adaptive mesh generation procedures start with defining a sensor which indicates regions of poor solution accuracy. All these sensors can be classified into two groups. The first group contains most frequently used sensors which use functions of the gradient and curvature of a flow variable such as Mach number, pressure, density etc. These methods are based on well-founded reasoning, that one needs to have good resolution in regions of high gradients such as shocks, boundary layers, free shear layers. Alternately a boolean sum of gradient and curvature of different flow parameters [1] can be used. The second category involves sensors which indicate the local accuracy of a solution using some measure of the error in the vicinity. One of such sensors is the truncation error obtained from the Taylor series expansion [2]. Another such sensor [3] uses spurious entropy generated in the solution of the equations.

In this paper, we discuss a new sensor for Euler equations based on the theory of multivariate statistical analysis. This sensor is based on directed divergence between two Maxwellian distributions as proposed in [4] and applied to 2D grid adaptation [4,5]. This sensor includes all the primitive variables, and is expected to resolve all the flow features.

2 Sensor based on Directed Divergence

In the Euler limit of Boltzmann equation, the velocity distribution functions of two pseudo-particles assumed at node locations P_i and P_j can be represented by Maxwellians F_i and F_j. For a 3-d case, the Maxwellian velocity distribution function of the pseudo-particle at node i is given by

$$F_i^n = F\left(t_n, \boldsymbol{x}_i : \boldsymbol{v}, I\right) = \frac{\rho_i}{I_{0_i}} \left(\frac{1}{2\pi R T_i}\right)^{\frac{3}{2}} \exp\left[-\frac{(\boldsymbol{v}-\boldsymbol{u}_i)^2}{2RT_i} - \frac{I}{I_{0_i}}\right] \quad (1)$$

where t is the time, $\boldsymbol{v}$ is the molecular velocity, $\boldsymbol{u}$ is the fluid velocity, ρ is the fluid density, I is the internal energy due to non-translational degrees of freedom, $I_0 = \frac{2RT(3-\gamma)}{4(\gamma-1)}$, T is the temperature, R is Gas constant per unit mass, and γ is the ratio of specific heats. Here $(\boldsymbol{v}-\boldsymbol{u}_i)^2 = (v_1-u_{1_i})^2+(v_2-u_{2_i})^2+(v_3-u_{3_i})^2$.

Symmetric relative entropy or directed divergence proposed by Mahalanobis (see Kullback [6]) gives a measure of the difference between two bivariate normal populations. This is also known as Mahalanobis distance or D^2 distance and is widely used in hypothesis testing. The D^2 distance between two Maxwellians is given by

$$D^2(F_i,F_j) = \int_0^\infty \int_{-\infty}^\infty \int_{-\infty}^\infty \int_{-\infty}^\infty (F_j - F_i)\ln\left(\frac{F_j}{F_i}\right) d\boldsymbol{v}dI \tag{2}$$

where $\boldsymbol{v}$ is the molecular velocity and I is the internal energy due to non-translational degrees of freedom. On integration, we have

$$D^2\left(F_i,F_j\right) = (\rho_j-\rho_i)\ln\left[\frac{\rho_j}{\rho_i}\left(\frac{T_i}{T_j}\right)^{\frac{5}{2}}\right] + \frac{\rho_i}{2RT_i}\left[\frac{\rho_j}{\rho_i}+\frac{T_i}{T_j}\right]\left(\boldsymbol{u_i}-\boldsymbol{u_j}\right)^2 + \frac{5}{2}\left[\frac{\rho_i(T_i-T_j)}{T_j}+\frac{\rho_j(T_j-T_i)}{T_i}\right] \tag{3}$$

clearly $D^2(F_i,F_j) \geq 0$ and the equality holds iff $F_i = F_j$. This is true for both the continuous (2) and discrete forms (3) of Mahalanobis distance between two Maxwellians [7].

It is also interesting to note that, D^2 distance appears as an entropy production term in the BGK model for the collision operator in Boltzmann equation. Hence, this distance is used as a measure of the error in the local vicinity in the solution. This sensor is first tested for 2-D Euler solutions over structured meshes by Ghosh [4] followed by Euler solutions over 2-D unstructured meshes by Mathur [5]. Navier-Stokes solutions can be adapted using a sensor based on directed divergence between two Chapman-Enskog distributions.

3 Node Movement Strategy

This section will briefly describe the grid adaptation technique proposed in [8] and applied in both two and three dimensions [9–11]. The grid adaptation technique adopted here is based on grid point redistribution and is carried out by moving each grid point towards the high gradient regions by using weighted averaging in an iterative manner . In each iteration of the adaptation cycle, new coordinates of each interior node are calculated by weighted averaging over the coordinates of its neighbors and itself. The weights for D^2 distance based adaptation are obtained using the following sensor function in the vicinity of the grid point and its neighbors.

$$\omega_i = \sqrt{1+\alpha^2\left(\frac{D^2(F_0,F_i)}{\Delta s(P_0,P_i)}\right)^2}, \quad i = 1 \,..\, N \tag{4}$$

Here α is a control parameter, and $\Delta s(P_0, P_i)$ is the distance between a grid node P_0 and its neighbor node P_i. Weights for Density based adaptation can be obtained by replacing $D^2(F_0, F_i)$ with density difference between grid nodes P_0 and P_i. To adapt the surface grid, the surface is approximated by using bi-cubic splines. The grid points on the surface are then adapted in the parametric plane. This ensures that the adapted surface grid points always lie on the surface. In order to avoid wiggles in the adapted grid, smoothing with unit weights had been carried out at regular intervals.

4 Results

The above defined sensor function in conjunction with the adaptation process described in (3) is used to adapt flows over an ONERA-M6 wing. Figure (1) shows the initial coarse grid of an ONERA-M6 wing and pressure contours obtained using the 3-D Euler code HYENA [12]. This grid is adapted using both density based and D^2 distance based sensors and the results are shown in figures (2,3). The difference of flow features of solutions obtained on these two adapted grids are not very clear from the figures, except that the grid obtained by adapted using D^2 distance based sensor mimics the shock better. Figure (4) shows the zoomed-up views of the parametric planes of the initial and adapted wing surfaces near the L.E. The density-based adaptation deteriorates the clustering of grid lines at the L.E. while D^2 distance based approach improves it. Figure (5(a)) compares the C_p distributions obtained on adapted crude grids against the C_p distribution with unadapted fine grid. D^2 distance-based adaptation picks up the L.E. suction peak better and gives C_p values closer to those obtained on the unadapted fine grid. Figure (5(b)) shows C_p distribution on adapted fine grid and it indicates that D^2 distance-based adaptation performs better over density-based adaptation.

The adaptation procedure is tested for flow over a generic fighter aircraft wing also. Figure (6) shows the initial grid over it and pressure contours obtained on it. The adapted grid and pressure contours obtained adapted grid are shown in Figure (7). The shocks are crisper and the flow features are more evident in the solution from the adapted grid.

Acknowledgement

The authors would like to thank Dr. T. G. Pai, Project Director, Technology Development directorate and Dr. K. P. Singh, Group Director, CFD for the support given to carry out this work. They would also like to thank Dr. J. S. Mathur for his valuable suggestions and Dr. S. P. Koruthu for allowing them to use the HYENA code.

References

1. B. K. Soni, N. P. Weatherill, J. F. Thompson: 'Grid adaptive strategies in CFD'. In: *Advances in Hydro-Science & Engineering.* ed. by S.S.Y. Wang, (University of Mississippi Press, June 1993) pp. 201-208
2. M.M. Rai, D.A. Anderson: 'Application of adaptive grids to fluid flow problems with asymptotic solutions' AIAA Journal, V. 20, pp. 496–502, 1981
3. J. G. Andrews, K.W. Morton: 'Spurious entropy generation as a mesh quality indicator'. In: *Fourteenth International Conference on Numerical Methods in Fluid Dynamics, Bangalore, 1994.* ed. by Deshpande et. al. (Springer-Verlag, Berlin) pp. 122-126
4. A. K. Ghosh: Robust Least Squares Kinetic Upwind Method for inviscid compressible flows. Ph.D. Thesis, Indian Institute of Science, Bangalore, 1996
5. J. S. Mathur, S.M. Deshpande: 'Studies on CFD Algorithms', Report No. 99FM3, Indian Institute of Science, Bangalore, 1999.
6. S. Kullback: 'Information Theory and Statistics', (Dover publications, Inc., New York 1967)
7. N. V. Raghavendra: D^2-distance based 3-D Grid Adaptation for a Generic Fighter Aircraft Wing, M. E. Report, Department of Aerospace Engineering, Indian Institute of Science, January 2000
8. Biju Uthup: 'A novel grid adaptation technique by weighted algebraic smoothing', ADA/TD/CFD/TR/060, Aeronautical Development Agency, Bangalore, May 1997
9. Mohan Varma: Grid adaptation by weighted averaging, M. E. Report, Department of Aerospace Engineering, Indian Institute of Science, January 1998
10. Biju Uthup, Mohan Varma: 'Novel grid adaptation for accurate shock capturing'. In: *Computational Fluid Dynamics 98, Proceedings of the Fourteenth European CFD Conference, Athens, 1998.* ed. by Papailiou et. al. (John Wiley & Sons, September 1998) vol 1, pp. 681
11. Mohan Varma, Biju Uthup, S. M. Deshpande: 'Grid Adaptation by simple Weighted Averaging for 3D Transonic Flows' *Proceedings of the Third Asian Computational Fluid Dynamics Conference Bangalore, 1998*
12. S. P. Koruthu: 'Preliminary validation of HYENA (Version 1.0) Code', ADA/TD/CFD/TR/058, Aeronautical Development Agency, Bangalore, 1997

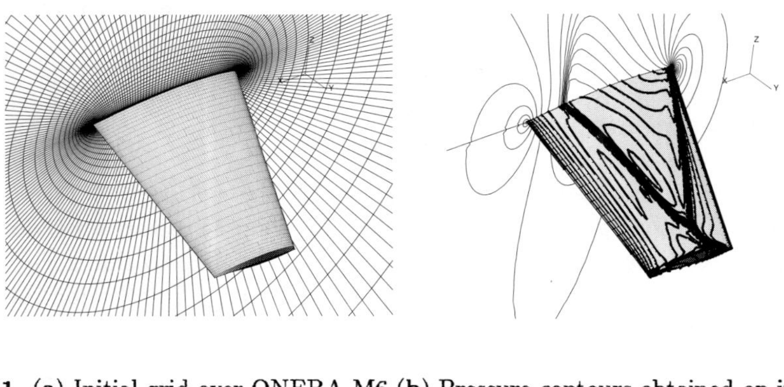

Fig. 1. (a) Initial grid over ONERA-M6 (b) Pressure contours obtained on initial grid

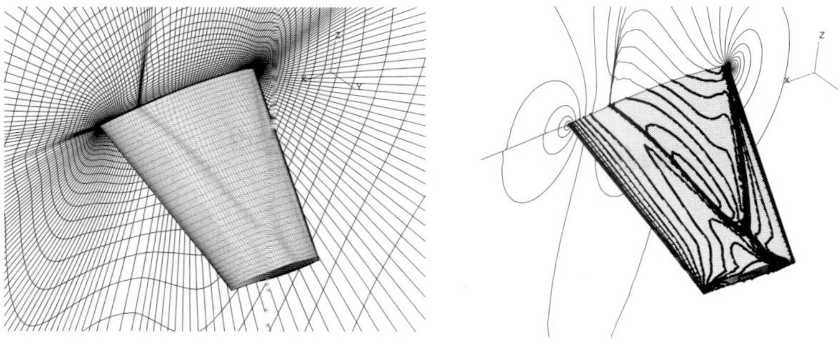

Fig. 2. Density based adaptation: (**a**) adapted grid (**b**) Pressure contours

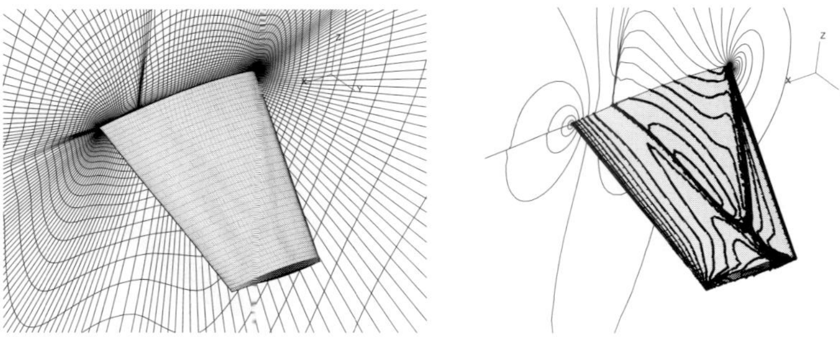

Fig. 3. D^2 distance based adaptation: (**a**) adapted grid (**b**) Pressure contours

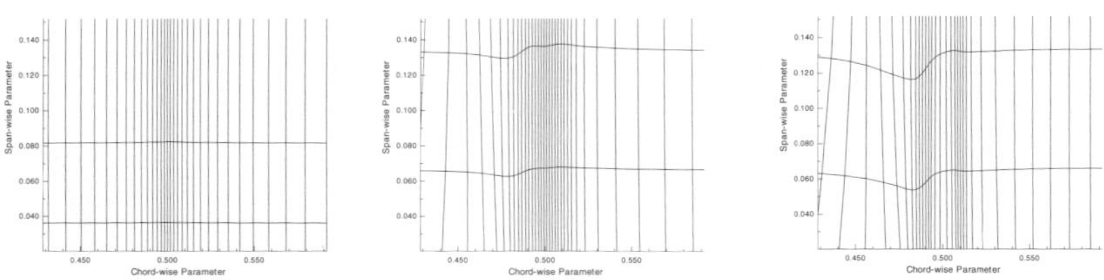

Fig. 4. Zoomed view of parametric plane near L.E. of (**a**) initial grid (**b**) density-based adapted grid (**c**) D^2 distance-based adapted grid

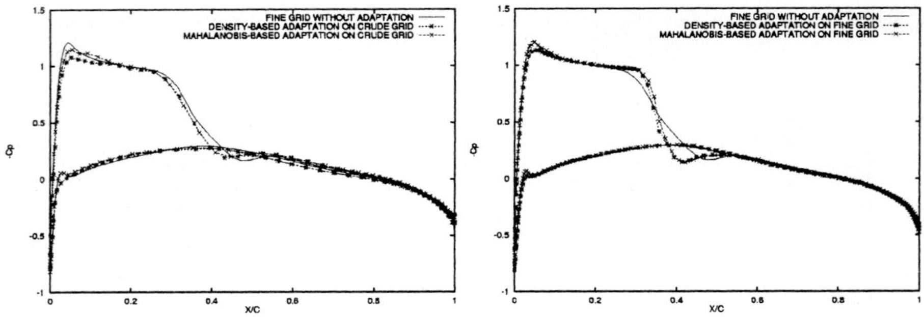

Fig. 5. Comparison of C_p distribution at 80% chord length for (**a**)fine unadapted and coarse adapted grids (**b**)fine unadapted and fine adapted grids

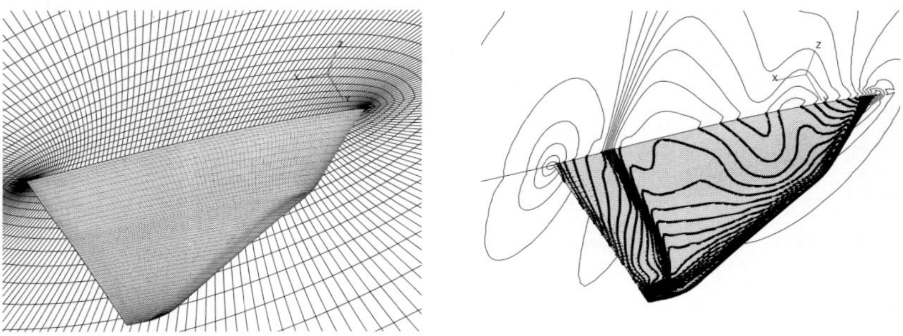

Fig. 6. (**a**)Initial grid over a generic aircraft wing (**b**)Pressure contours

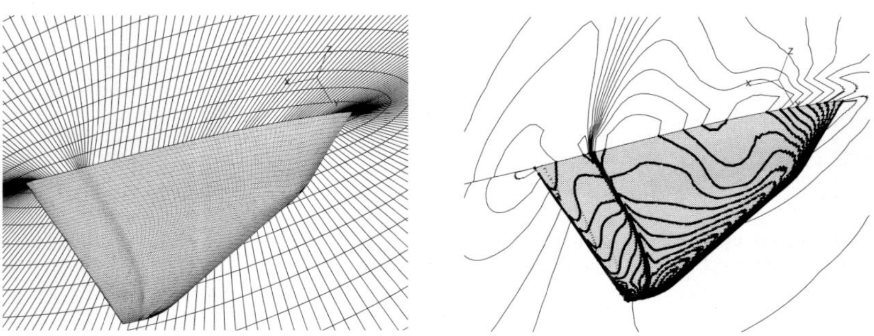

Fig. 7. D^2 distance based adaptation: (**a**)adapted grid (**b**)pressure contours

Anisotropic Cartesian Grid Adaptation

Paulus R. Lahur[1] and Yoshiaki Nakamura[2]

[1] Graduate Student
[2] Professor, Dept. of Aerospace Eng., Nagoya University, Furo-cho, Chikusa-ku, Nagoya 464-8603, Japan

Abstract. A three-dimensional anisotropic Cartesian grid adaptation method has been developed to improve computational efficiency by reducing the number of cells. The method can produce cells with unlimited aspect ratio while keeping the grid reasonably smooth. A test case of 2D supersonic flow around a cylinder shows that the number of cells is reduced to one to two orders of magnitude less than that of the corresponding isotropic grid, because all cells can extend in the spanwise direction. It is also found that the present method is not sensitive to variation in initial grid. Furthermore, a test case of ONERA M6 wing in transonic flow shows good agreement with experimental data, where the anisotropic grid needs only about half of the total number of cells in an equivalent isotropic grid.

1 Introduction

Cartesian grid is becoming one of popular choices in simulating the flow around a complex geometry due to their advantages of automatic and fast grid generation. Of particular advantage is ease in performing solution adaptive computation, due to its flexibility as to simple local grid clustering, where cells are divided isotropically into eight sub-cells of equal size in 3D cases and four in 2D. An efficient tree data structure such as octree or quadtree in 3D or 2D, respectively, is normally employed.

However, isotropic refinement generally results in an excessive number of cells, due to unnecessary refinement in the direction where flow gradient is low. To avoid this, anisotropic grid refinement should be employed instead.

There are at least two approaches in anisotropic Cartesian grid adaptation. The first approach starts with relatively coarse grid and refines the grid anisotropically [1–3]. In this approach a cell is divided into two, four, or eight sub-cells. Unstructured approach is used in [1]. A generalization of tree data structure that comprises two, four, or eight sub-cells has also been proposed, which is referred to as omni-tree [2], or 2N tree [3].

The second approach is the reverse of the first, where a fine grid is coarsened anisotropically by merging cells in the direction of low flow gradient [4]. The fine grid itself is obtained by using isotropic refinement.

The present study employs a combination of the first and the second method, where both grid refinement and coarsening are carried out anisotropically. Furthermore, no limitation on cell aspect ratio is enforced. These features compel us to use an unstructured approach. The 3D grid adaptation method proposed here is then assessed for 2D and 3D test cases.

2 Anisotropic Grid Adaptation Method

The procedure starts with a rather rough isotropic grid. Then flow computation is carried out up to a certain degree of convergence, and the result is used to improve the grid quality at the next cycle. The grid adaptation consists of three processes: grid coarsening, refinement, and smoothing. The following subsections will address the main aspects of the method.

2.1 Adaptation Parameter

A modified second finite difference is used to compute adaptation parameter G, as shown in (1), where only x direction is shown for simplicity.

$$(\mathrm{G_i})_\mathrm{x} = (\Delta \mathrm{x_i})^\mathrm{L} \left| (\mathrm{Q'_{ij}})_\mathrm{x,xmax} - (\mathrm{Q'_{ij}})_\mathrm{x,xmin} \right| \Delta \mathrm{x_i} \ . \tag{1}$$

where $\mathrm{Q'_{ij}}$ is the first difference of flow solution between cell i and its neighbor j at their common cell interface, and $\Delta \mathrm{x_i}$ is the cell size. L is a length scale to determine the balance of refinement between small and large cells, whose value is somewhere between 2 and 3, which gives satisfactory results for the test cases presented here. A larger value of L will result in more refinement on large cells, and a smaller value will result in more refinement on small cells. Furthermore, to encourage more refinement in the case of a Cartesian grid cell with many neighbors, the maximum difference is used in (1).

Having computed adaptation parameter G for all cells, the threshold values for grid coarsening and refinement: $\mathrm{T_{low}}$ and $\mathrm{T_{high}}$ are determined using the mean value, mean(G), the standard deviation, sdev(G), and user-defined parameters $\mathrm{R_{low}}$ and $\mathrm{R_{high}}$.

$$\mathrm{T_{low}} = \max\left[0, (\mathrm{mean(G)} - \mathrm{R_{low}} \times \mathrm{sdev(G)}\right] \ . \tag{2}$$

$$\mathrm{T_{high}} = \mathrm{mean(G)} + \mathrm{R_{high}} \times \mathrm{sdev(G)} \ . \tag{3}$$

2.2 Grid Modification

Grid modification consists of three consecutive processes: grid coarsening, refinement, and smoothing. First, the grid is coarsened by removing the interface between a cell and its neighbor if for both cells $\mathrm{G} < \mathrm{T_{low}}$ in the direction connecting the two cells.

In the next step, the grid is refined by dividing cells into two equal parts in *any* of the x, y, and z direction where $\mathrm{G} > \mathrm{T_{high}}$. Thus, a cell can be refined into 2, 4 or 8 sub-cells. For practical reasons, limitations are imposed on the minimum cell size and the maximum number of cells.

Finally, the grid is smoothed out to limit the number of neighbors of a cell so that solution accuracy is not degraded. Since this process takes time, excessive coarsening and refinement processes are basically avoided by setting $\mathrm{R_{low}} = \mathrm{R_{high}} = 0.5$ in (2) and (3). Smoothing is carried out by dividing a cell into two in the direction where there are more than two neighbors on its face. A smooth grid is produced when a cell is divided such that the division line coincides with its neighbors' boundaries. The combined approach proposed here seems to produce a reasonably smooth grid with a relatively small effort.

3 Grid Generation and Flow Solver

The Cartesian grid generation employed here is similar to the authors' previous research, except that the previous data structure was of octree type [5]. The grid generator can handle non-regular geometry such as thin and sharp body by using cell splitting, where intersected body surface is approximated as planar.

The flow solver is based on the Euler equations. Cell-centered finite volume approach is employed, where numerical flux is computed using Hännel's flux-vector-splitting scheme. Spatial accuracy is second order using the MUSCL method. Time integration is carried out using a 3-stage Runge-Kutta method, where local time stepping is employed to accelerate the convergence.

4 Test Cases

4.1 Cylinder

The first test case is a supersonic flow around a circular cylinder of unit diameter at $M_\infty = 3.0$. The 2D flow is computed using the 3D method outlined above on two different initial grids: (a) uniform and (b) multi-level (Fig. 1). For the uniform initial grid, the computational domain is divided into 64x1x48 in x, y and z direction, respectively, resulting in 3,072 cells. For the multi-level initial grid, the domain is divided into 8x1x6, and refined isotropically 5 times around the cylinder, resulting in 46,480 isotropic cells. The cells are then merged to produce a single cell in spanwise direction, resulting in 1,643 anisotropic cells.

The flow is solved on these grids until its maximum residue decreases two orders of magnitude. Then the grids are adapted using density and Mach number as the adaptation parameters, with $L = 0.3$ in (1). The final grids are shown in Fig. 2. The number of cells converge to around 14,000 for both cases. The distribution of Mach number can be found in Fig. 3, where it is observed that the shocks, shear layers, and smooth flow regions are captured successfully.

It is also observed that the final grids and solutions for the two cases are quite similar. This shows that the adaptation method presented here is not sensitive to variation in initial grid, which is a highly desirable property. Main differences exist only in the undisturbed flow region upstream the main shock, due to coarsening into very large cells, which has no significant effect on flow solution. Some differences are also present in the region of interaction between shear layer and shock due to flow unsteadiness in this region.

Another way to assess the performance of the adaptation method is to see the change in the mean value and variation of adaptation parameter G (Fig. 4). The mean value of G and the range between T_{high} and T_{low} decrease with adaptation cycle and converge to similar values for both cases of initial grid.

The advantage of the anisotropic approach is clear when the number of cells is considered. For the case of uniform initial grid, the number of equivalent isotropic cells in the final grid reaches over 100,000 cells, which is about 8 times as many as the anisotropic one, while for the case of multi-level initial grid it even reaches almost one million cells, which is about 67 times as many.

4.2 ONERA M6 Wing

The second test is a transonic flow over ONERA M6 wing. The wing has an ONERA D profile with an aspect ratio of 3.8 and a taper ratio of 0.562. The flow conditions are $M_\infty = 0.84$ and $\alpha = 3.06°$.

A multi-level grid generated with an isotropic grid refinement approach is used as the initial grid. The finest cells are of level 7, with the dimension of 1/64 of the wing chord at its root. The computational boundary is placed at a distance 30 times the root chord from the wing in all directions.

Density is chosen to compute the adaptation parameter, with $L = 2.0$ in (1). The adaptation is carried out twice, beyond which the flow on the wing surface changes little. The final grids are shown in Fig. 5.

Figure 6 shows comparison of pressure distribution along the wing surface at some spanwise sections between computed results and experimental data [6]. The final solution shows good agreement with the experiment, except for differences in the position and sharpness of the shocks due to the inviscid characteristics of the Euler equations.

Although not as dramatic as in the case of cylinder, here the saving in the number of cells is also significant. The final anisotropic grid contains 171,753 cells, which is 2.5 times fewer than the equivalent isotropic grid.

5 Concluding Remarks

An anisotropic Cartesian grid adaptation method has been developed in this study. It consists of three processes: grid coarsening, refinement, and smoothing, where no limitation is imposed on the cell aspect ratio.

The present 3D method was validated for 2D and 3D test cases: a cylinder in supersonic flow and ONERA M6 wing in transonic flow, respectively. For both cases, the method successfully captured the flow features and improved their solutions with significantly fewer grid cells, as compared to the corresponding isotropic grid. Good agreement with experimental data has been obtained.

The 2D flow around a cylinder was computed with high efficiency, since a single cell can span the whole computational domain in the spanwise direction. As a result, the number of anisotropic grid cells becomes comparable to that in 2D computation. In the final grids, the number of anisotropic cells are about 8 and 67 times less than that of the isotropic grid for the case of uniform initial and multi-level initial grid, respectively.

For the ONERA M6 test case, the number of anisotropic cells is 2.5 times less than that of the isotropic grid.

The adaptation method developed in this study does not seem to be sensitive to variation in initial grid. Very similar final grids are produced from the cases of uniform initial grid and multi-level initial grid.

References

1. M.J. Aftosmis: 'Solution Adaptive Cartesian Grid Methods for Aerodynamic Flows with Complex Geometries,' von Karman Institute for Fluid Dynamics, Lecture Series 1997-02 (1997)
2. N.D. Domel and S.L. Karman: 'Splitflow: Progress in 3D CFD with Cartesian Omni-tree Grids for Complex Geometries,' AIAA 2000-1006, 2000
3. Z.J. Wang and R.F. Chen: 'Anisotropic Cartesian Grid Method for Viscous Turbulent Flow,' AIAA 2000-0395, 2000
4. F. Deister, D. Rocher, E.H. Hirschel, and F. Monnoyer: 'Three-dimensional Adaptively Refined Cartesian Grid Generation and Euler Flow Solutions for Arbitrary Geometries,' Proc. of the 4th European CFD Conference, Vol. 1, Part 1 (1998), pp. 96-101
5. P.R. Lahur and Y. Nakamura: 'A New Method for Thin Body Problem in Cartesian Grid Generation,' AIAA 99-0919, 1999
6. V. Schmitt and F. Charpin: 'Pressure Distributions on The ONERA-M6-Wing at Transonic Mach Numbers,' Experiment Data Base for Computer Program Assessment, AGARD AR-138 (1979)

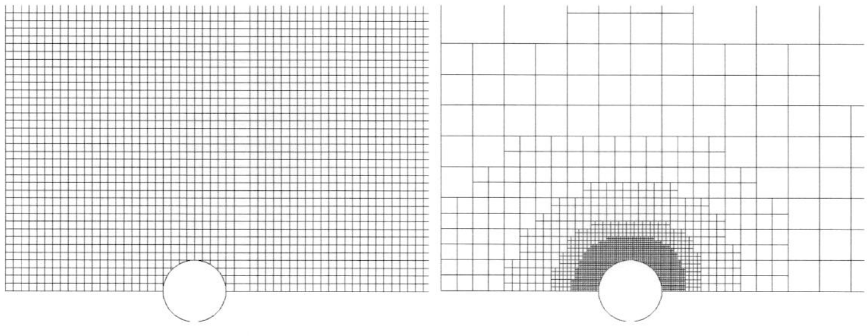

Fig. 1. Initial grids around cylinder. (a) Uniform. (b) Multi level

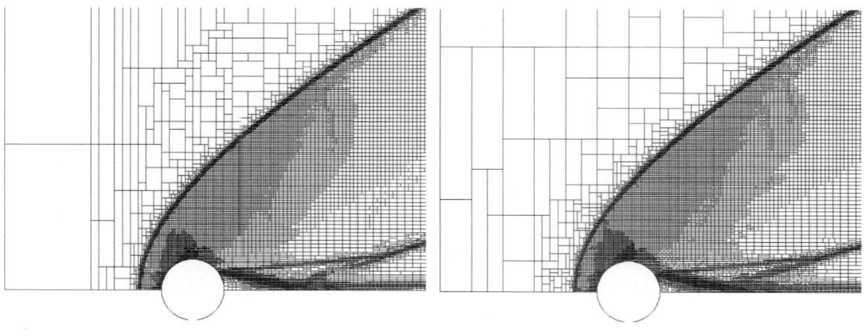

Fig. 2. Final grids around cylinder. (a) Uniform. (b) Multi level

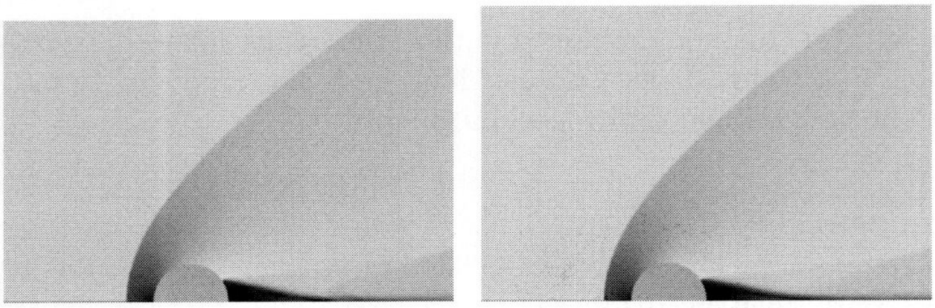

Fig. 3. Mach number around cylinder on final grids. (**a**) Uniform. (**b**) Multi level

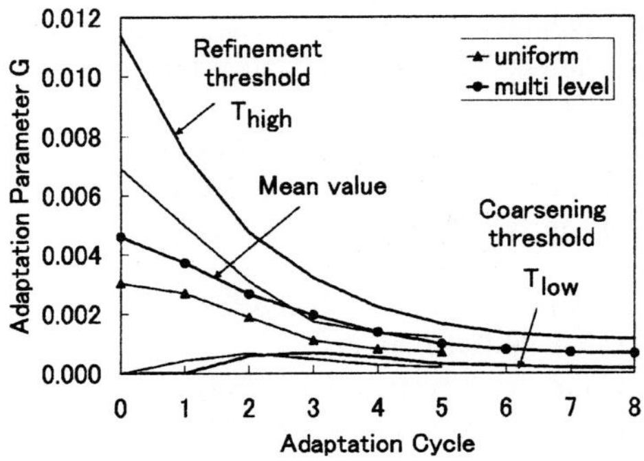

Fig. 4. Adaptation parameter history for the cylinder case

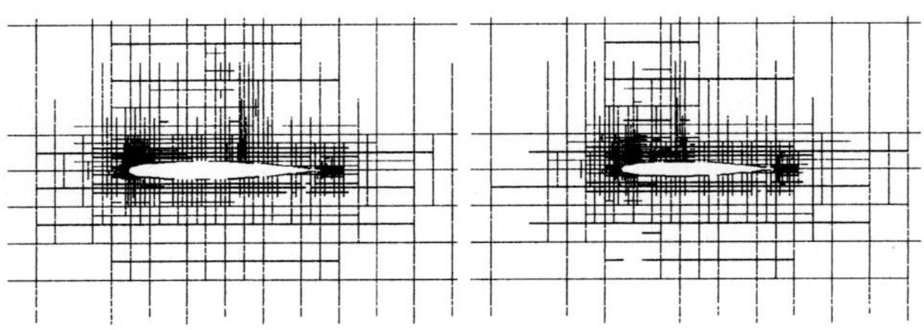

Fig. 5. Final grids around ONERA M6 wing. (**a**) 44% span. (**b**) 80% span

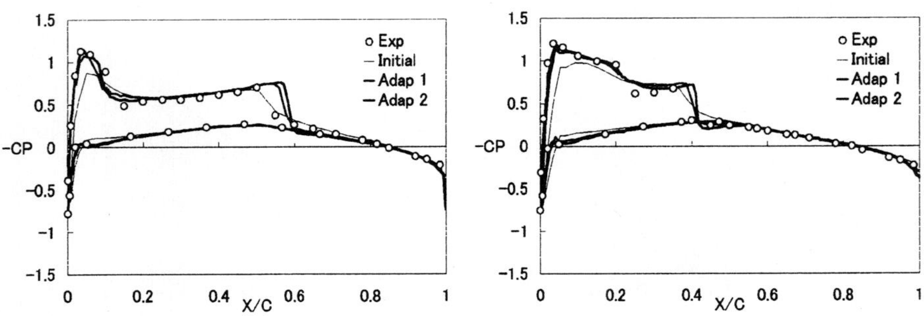

Fig. 6. Pressure distribution around ONERA M6 wing. (**a**) 44% span. (**b**) 80% span

Part III

Unstructured Grid

Directional Refinement for Viscous Flows Using Adaptive Unstructured Quadrilaterals

Kazuyuki Yada, Mingyu Sun, and Kazuyoshi Takayama

Shock Wave Research Center, Institute of Fluid Science,
Tohoku University, Katahira 2-1-1, Aoba, Sendai, 980, Japan

Abstract. Directional refinement of unstructured quadrilateral meshes is simple but improve the solution accuracy, memory storage and computational time. A compressible steady flow over a flat plate is chosen for demonstrating this performance. An application for the practical flow problem is also presented such as unsteady nozzle starting process in this paper.

1 Introduction

Computational fluid dynamics has been developed remarkably and now is able to simulate three-dimensional complex flow patterns. Recently unstructured adaptive mesh refinement methods, which automatically adjust meshes to improve the rigions that require, has been used to enhance the resolution at relatively low computational cost. Typical adaptive unstructured methods consist of triangular meshes some of which are repeatedly divided until fine enough to describe flow features. This approach is known to be more effective than any other methods, which do not use the adaptation for high-resolution calculations. The triangular meshes are approximately considered as the consequence of dividing quadrilaterals by joining their diagonal vertices. However, these additional computation does not lead to obvious improvement of solution accuracy, at least in the calculation of the Euler equations. Triangular meshes, therefore, require more memory storage and CPU time[3]. It is found that numerical solutions using quadrilateral meshes in 2-D and hexahedral meshes in 3-D are sometimes even more accurate than triangles and tetrahedra[2,3].

In order to solve this problem, a triangle-quadrilateral hybrid method is proposed by some authors (for instance, [4]). This strategy is, in 2-D, quadrilateral girds are used in the boundary layer, and triangle meshes are used outside of the boundary layer. Because physical phenomena in the boundary layer strongly depends on normal direction rather than tangential direction above the body surface. In the case of quadrilateral meshes, the directions of the numerical flux are normal to the boundary. Even though this hybrid approach manages to achieve high accuracy in the boundary layer, its algorithm becomes complex and larger memory storage is required. Therefore, the present work uses effectively a uniformly unstructured quadrilateral grid adaptation.

2 Numerical methods

In this paper, VAS2D (Two-dimensional Vectorizable Adaptive Solver)[1] is utilized. In VAS2D, unstructured adaptive refinement of quadrilateral meshes are taken into account and this is the finite volume method which is optimized for vector and parallel computatoin. Current fundamental equations are steady/unsteady Navier-Stokes equations and are solved as follows,

$$\mathbf{U}^{n+1} = N(t/2)E(t)N(t/2)\mathbf{U}^{n} \tag{1}$$

here, $\mathbf{U}$ represents conservative variables, $E(t)$ represents Eulerian terms step and $N(t/2)$ means Navier-Stokes specific terms step. This method is called Operator-Splitting method. In eq.(1), if each step indicates second-order accuracy, overall step (left hand side of eq.(1)) is second-order accuracy.

This scheme consists of predictor-corrector method. In predictor step, it is not necessary to preserve conservative systems, hence quasi-linear Euler system equations are solved by Lax-Friedrichs scheme in order to save computational time.

There are two solvers considered in corrector step. One is central-difference scheme, a TVD smoothing scheme. The other is the second-order upwind scheme, MUSCL type scheme proposed by Hancock[6]. In the former case, TVD condition is satisfied by adding artificial viscosity and this scheme will be first-order accuracy if there are a large gradient like a region of shock waves. In other region, the accuracy of this scheme will be second-order.

In the later case, HLL solver is utilized as an approximate Riemann solver. Furthermore, in Navier-Stokes specific terms, $N(t/2)$ step, FTCS (Forward in Time and Central difference in Space) scheme is used for the calculations. The indicator of the refinment and coarsening of adaptive meshes are expressed by the ratio of the second-order derivatives to the first-order one in the Taylor series expansion of density. Since this indicator is not sensitive for the strength of shock waves, it is not necessary to adjust for various flow simulations.

2.1 Directional refinement

In VAS2D, simple but effective strategy is employed, namely the directional refinement of adaptive unstructured quadrilateral meshes. This is designed for increasing the solution accuracy in the boundary layer. Figure 1 shows schematic of the directional refinement of quadrilateral meshes in the boundary layer. By changing a new vertex form the middle point of one edge to a point closer to the boundary, it is possible to finely capture the boundary layer phenomena.[1] Parameter α_{bl} is defined as the ratio of the segment, for example, α_{bl} = HD/AH or FC/BF in Fig.1a.

3 Steady compressible flow over a flat plate

In this section, we see the performance of the directional refinement on a flow over a flat plate as an example. TVD smoothing scheme was chosen in this section.

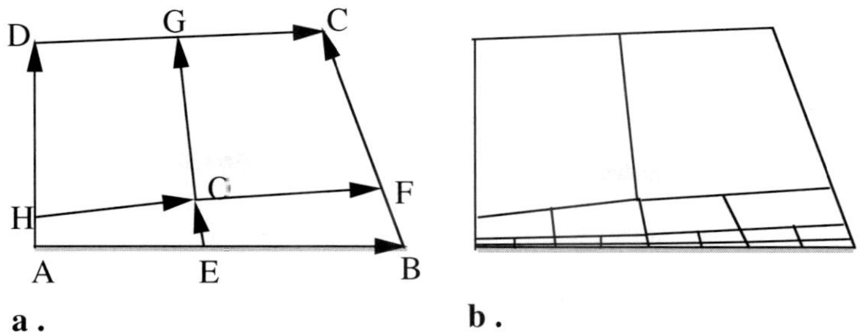

Fig. 1. Directional refinement inside the boundary layer. A highly stretched mesh can be generated by repeatedly refining cells along the viscous boundary

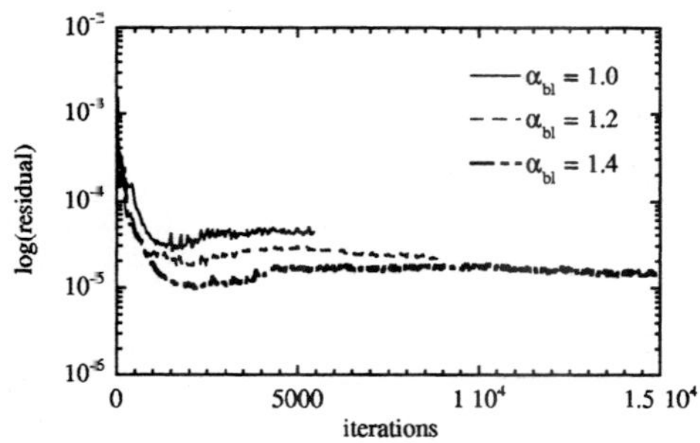

Fig. 2. Dependence of the convergence histories, refinement level 6. Density was chosen as a parameter

Before discussing the efficiency, it is necessary to analyze enough to converge the steady solutions. Fig. 2 shows the dependence of the convergence histories,

$$e_i = |\varrho_i^{n-1} - \varrho_i^{n-2}| \qquad L_p = \left(\frac{\sum e_i^p}{N}\right)^{1/p} \tag{2}$$

where e_i denotes error of density for the ith cells and n time steps, and L_p is the residual error[11]. In this case, $p = 2$ was chosen as a geometric mean. Every 20 time steps, these formula are solved and mesh adaptations are done during computations. The region of the estimation of e_i are $0.5 \leq x \leq 3.0$. Because in the region close to the tip of the wall, the variations normal to the wall become the same order as those along the wall, the boundary layer assumption is not valid there. From Fig.2 all of three cases are convergent, though iteration steps are different, nondimensional computation time is the same, $\tilde{t} = 10$.

Flow conditions over a flat plate are as follows, Mach number $M_\infty = 1.0$, wall temperature $T_w = T_\infty$, Reynolds number $Re = 10^4$, initial thickness of nodes

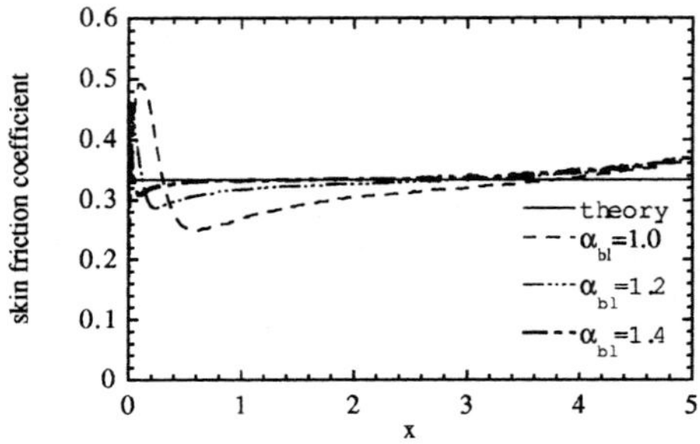

Fig. 3. Steady boundary layer flow over a flat plate by computing different directional refinement. Normal axis indicates nondimensional skin friction force $\tau_w/\sqrt{u^3/xRe}$. From boundary theory, its value is 0.332[10]

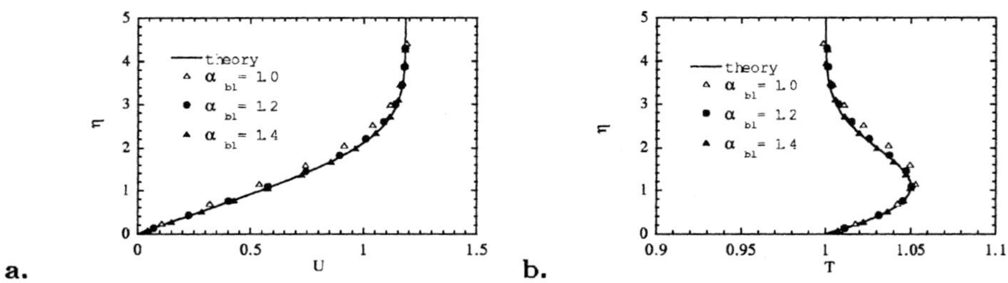

Fig. 4. Quantitative comparison between theoretical and numerical solutions. The point of a flat plate is x=1.0; **a.** Velocity profiles (x direction), **b.** Temperature profiles.

$\Delta x_{init} = 0.4$, Prandtl number $Pr = 1.0$, and viscous coefficient $\mu = T$ (power law)[7].

Fig.3 shows skin friction coefficient on the flat plate surface. Black solid line represents theoretical solution of skin friction coefficient. From this figure, discrepancies between theory and numerical solutions become smaller if the value of α_{bl} is larger. Fig.4a, 4b are the velocity and temperature profile in the laminar boundary layer. Again, black solid profiles are theoretical solutions. Theoretical equations are based on Blasius equations[7,8]. Vertical axis η is transformed suitable for comparison numerical solutions with theoretical one. From these figure, it is obvious that the larger α_{bl} is, the closer the values between theory and simulation are. Hence the usage of the directional refinement is effective way to solve viscous flow problems.

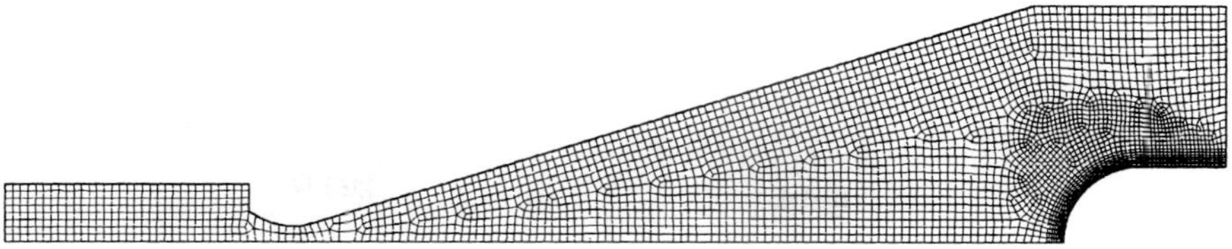

Fig. 5. Geometry of the nozzle and a spherical model; axisymmetric grids

4 Unsteady nozzle flows

In this section, unsteady nozzle flow and consequence of establishment of flows around a spherical model are considered. MUSCL-Hancock scheme is used for the solver in this section. Fig.5 shows grid geometry of the nozzle and a spherical model. The existence of a diaphragm between the shock tube and the nozzle, is not considered here. Incident shock Mach number in the shock tube is 6.1 and the initial pressure is 60 kPa [9]. Reynolds number is 1.3×10^7 per meter.

Fig.6 are shown as unsteady flows of the nozzle starting process. Shock wave propageting in the shock tube is reflected shock wave from the end of the shock tube in Fig. 6a. Fig. 6b and 6c are in the same time at 260 μs from nozzle starting. Fig. 6b presents density profile in the nozzle. Reflected shock wave is observed on a spherical model. Velocity contour is shown in Fig.6c and there are vortices arised from the flow separation. Secondary shock wave is also seen in the middle of the nozzle.

5 Summary

Adaptive unstructured quadrilateral meshes have been employed. It is found that by quantitative comparison of the solutions of the boundary layer over a flat plate, the accuracy is improved by using the directional refinement. Additionally, it is known that it is capable of applying to complex flow problems.

References

1. M.Sun: Numerical and Experimental Studies of Shock Wave Interaction with Bodies. Ph.D. Thesis, Tohoku University (1998)
2. R.Viswas, R.C.Strawn: Tetrahedral and hexahedral mesh adaptation for CFD problems, NAS-97-007.
3. M.Aftosmis, D.Gaitonde, T.S.Tavares: On the Accuracy, Stability and Monotonicity of Various Reconstruction Algorithms for Unstructured Meshes. AIAA 94-0415 (1994)
4. K.A.Sörensen, O.Hansen, K.Morgan and N.P.Weatherill: An Agglomerated Unstructured Hybrid Mesh Method for Turburent Compressible Flows, 9th ISCFD, pp. 604-611 (1999)

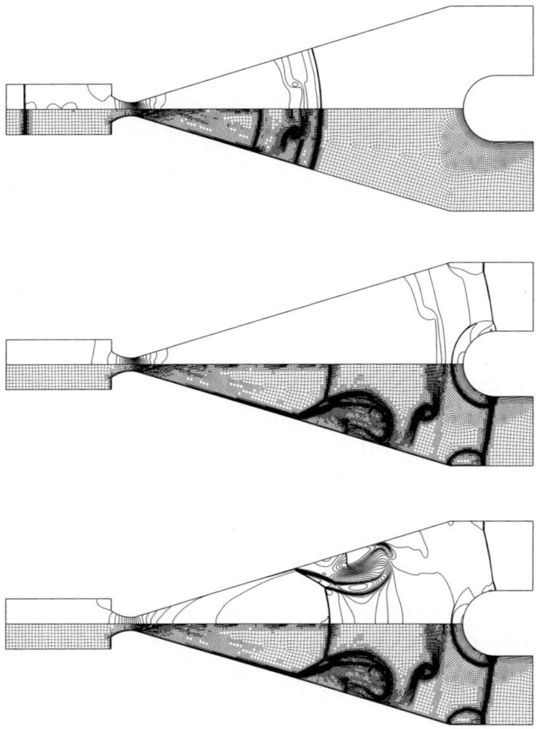

a.

b.

c.

Fig. 6. contours of the unsteady nozzle starting process; **a.** Density contour at 100μs, **b.** Density contour at 260 μs, **c.** Velocity contour at 260 μs; notice the vortex generations

5. M.Sun and K.Takayama: Conservative smoothing on an adaptive quadlirateral grid, *J.Comp. Phys,* **150**: pp. 143-180 (1999)
6. EF.Toro: *Riemann solvers and numerical methods for fluid dynamics* 2nd edition, Springer (1999)
7. F.M.White: *viscous fluid flow*, McGraw-Hill (1974)
8. B.K.Shivamoggi: *Theoretical fluid dynamics*, Matrinus nijhoff publishers (1985)
9. K.Koremoto: Optimization of the Characteristics of a Free Piston Shock Tunnel, Ph.D. Thesis, Tohoku University (1999)
10. L.G.Loitsyanskii: *Mechanics of Liquids and Gases*, Pergamon Press (1966)
11. W.J.Coirier, K.G.Powell: An Accuracy Assessment of Cartesian-Mesh Approaches for the Euler Equations, *J.Comp. Phys,* **117**: pp. 121-131 (1995)
12. K.Fujii: *Numerical Methods for Computational Fluid Dynamics*, University of Tokyo Press, (1994)

All Hexahedra Unstructured Mesh Adaptative Solver for Turbulent Flow Simulations

Benoit Leonard[1], Alpesh Patel[2], Michel Delanaye[2], and Charles Hirsch[1]

[1] Vrije Universiteit Brussel, Department of Fluid Mechanics, Pleinlaan 2, B-1050 Brussels, Belgium
[2] Numeca International, Avenue Franklin Roosevelt 5, B-1050 Brussels, Belgium

1 Introduction

The accurate and fast simulations of high-Reynolds number viscous flows around complex geometries is of paramount importance in the design of aerodynamic vehicles. The geometry complexity mixed with the presence of very different length scales phenomena pose a great burden to classical numerical techniques based on structured grids. In this context, unstructured grid technology is a promising approach since it provides the ability to automatically cluster grid points where required by geometry particularities and/or flow physics, hence optimizing computational cost with respect to accuracy. Most unstructured codes use tetrahedral meshes [1], essentially because of the availibility of algorithms to automatically generate meshes for complicated geometries. However for high-Reynolds number viscous flow simulations, the number of tetrahedra required for the accurate resolution of boundary layers significantly grows outbound. Furthermore the use of tetrahedra degrades the accuracy of classical numerical methods. This problem has been partly resolved by resorting to hybrid grids [2]. The present paper is based on an approach using hexadra only [3], [4]. The starting mesh is created using an octree-based hexahedral mesh generator [5], [6]. A finite volume discretization using a central scheme with artificial dissipation is combined with several two-equation turbulence models. Time integration is based on an explicit Runge-Kutta scheme accelerated by a hierarchical multigrid strategy. Mesh adaptation algorithms - anisotropic refinement and hierarchical coarsening - have been integrated into the flow solver. Adaptation criteria, based on flow physics, are used to govern adaptation. The generation of coarse levels for the multigrid scheme is tightly coupled with adaptation through the hierarchical data structure involving parent-child connectivities. Several simulations of internal and external steady flows demonstrate the capabilities of our approach.

2 Mesh adaptation

Mesh adaptation is governed by criteria based on flow physics called sensors. The sensor parameters [7] are density, velocity, pressure, vorticity, ... Threshold values are calculated based on a statistical approach depending on a user-defined parameter α. Sensors can be combined together to improve the detection of flow phenomena. The mesh anisotropic refinement procedure has the following features: cells can have 2, 4 or 8 children and a balancing function performs further refinement in order to avoid mesh incompatibilities. A hierarchical mesh coarsening technique has also been integrated. To remove the children of a cell, a minimum ratio of 75 percent of them have to be flagged for coarsening. Then, the parent cell becomes a leaf cell by the removal of all its children, including the non-flagged ones.

3 Spatial Discretization Schemes

The discretized Navier-Stokes equations associated with the conservative variables matrix U_J on a control volume Ω_J can be written:

$$\frac{\partial}{\partial t}(U_J \Omega_J) + \sum_{faces} (\mathcal{F}.\boldsymbol{n})\, \Delta S = 0 \tag{1}$$

where $\mathcal{F}$ is the numerical flux. The following expression, encompassing both upwind and central formulation, is used for $\mathcal{F}$ [8].

$$\mathcal{F}_{i+\frac{1}{2}}.\boldsymbol{n} = \frac{1}{2}\left[(\boldsymbol{F} - \boldsymbol{V})_i + (\boldsymbol{F} - \boldsymbol{V})_{i+1}\right].\boldsymbol{n} - d_{i+\frac{1}{2}} \tag{2}$$

where $\boldsymbol{F}$ is the convective flux vector and $\boldsymbol{V}$ comprises the diffusion, heat conduction and shear stresses. The central part is derived by flux averaging and the second term on the right hand side is a general dissipation term, which, depending on its formulation, leads to a central scheme with Jameson type dissipation or upwind type schemes. In the present unstructured solver, only the central scheme is implemented. Turbulence is modelled using the low-Reynolds-Number k-ϵ turbulence model of Yang-Shih [9]. A linear wall function model based on DNS data [10], derived from the previous one is also available. The determination of the viscous fluxes requires the knowledge of the gradients on the faces. In order to avoid odd-even decoupling problems, a staggered control volume consisting of two pyramids, each based on the cell face and the cell centre of the adjacent cells is used.

4 Time Integration and Multigrid

The time integration is based on explicit Runge-Kutta schemes. In order to increase the efficiency, local time stepping and multigrid acceleration are used. The current multigrid procedure is derived from the well proven approach followed in the multiblock/multigrid EURANUS code [8].

In order to avoid the problems of conservation along the interface between refined and unrefined regions, the so-called zonal coarse approach [11], has been considered. In this strategy, the finest grid corresponds to the composite one, made up of leaf (i.e. unrefined) cells only, whereas on coarser levels only the cells which have children on next level are kept. This approach is illustrated in Fig.1.

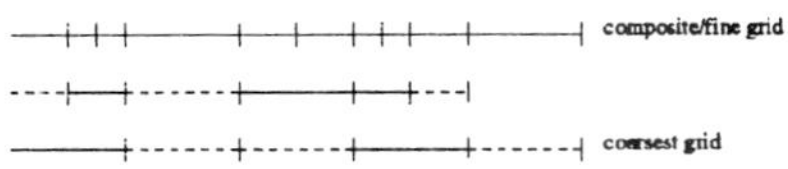

Fig. 1. Zonal coarse approach

The coarser grids generated do not cover the computational domain, so that usually, their boundaries do not correspond to the flow domain ones and require special boundary conditions [3]. The latter are implemented by introducing virtual or dummy (dotted line in Fig. 1) cells, where the solution on the corresponding physical cells (lying on another grid level) is stored, at the expense of an increased complexity in the data structure. This complexity has been removed by allowing a cell to belong to different grid levels. This strategy avoids the problems of conservation at the interface between refined and non-refined regions as well as the presence of holes and islands on coarser grids. Figure 2 illustrates the strategy currently used in the solver.

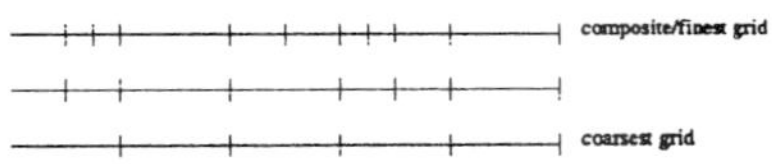

Fig. 2. Derived zonal coarse approach

The restriction of the residual is the sum of residuals of the children, whereas for the restriction of the solution, a weighted averaging is used:

$$I_f^c U_f = \frac{\sum_f U_f \Omega_f}{\Omega c} \tag{3}$$

where the indeces c and f respectively represent coarse and fine (children) cells. Ω represents the cell volume. When a cell belongs to two successive grid levels, the restricted solution (resp. residual) is the solution (resp. residual). The prolongation operator interpolates the solution correction from a coarse level to a fine level. A zero-order prolongation consists in setting a fine cell correction equal to the correction in its parent:

$$I_c^f \partial U_c = \partial U_c = U_c - I_f^c U_f \tag{4}$$

For cells belonging to different grid levels, the correction is the difference between the new solution on the current grid level and the restricted solution on the same level. The zero-order prolongation was not robust, so that a first-order prolongation was required in many cases. The implementation of the first order prolongation [3] can be summarized as follows: using the node-cell connectivity, the correction in the cells of the coarse grid are interpolated in the nodes of the fine grid. In a second phase, the correction in the nodes of the fine grid are distributed to the cell centres of the fine grid. The first order prolongation, combined with the above-mentioned strategy proved to be very robust.

5 Results

The first testcase deals with a 3D, 90 degree bend square duct. This geometry was studied experimentally and numerically by Humphrey *et al.* [12]. The velocity distribution at the inlet of the bend is a fully developed laminar flow with a bulk velocity of $100\,m/s$. The Reynolds number based on hydraulic diameter is 790. The walls of the duct are assumed to be adiabatic and the outlet static pressure is assumed to be constant equal to $10^5 Pa$. Since the numerical results of Humphrey *et al.* show only reasonable agreement with the experimental data, our results will be compared with the ones obtained with the EURANUS code using a structured grid of 129x65x65 cells, and with the results of Lopez *et al.* [13] computed on a 121x61x61 grid. The initial mesh used for this computation consists of 15x5x4 cells. The final mesh consists of 110000 nodes and is obtained after two isotropic refinements and one local refinement using vorticity and the z component of velocity as sensors with $\alpha = -0.1$ for both of them.

In Fig. 3.a the radial variation of the streamwise velocity is compared with experimental data and numerical results along the plane halfway between the symmetry plane and the outer wall at the 60 degree cross section. Figure 3.b shows the variation in the symmetry plane at the same cross section.

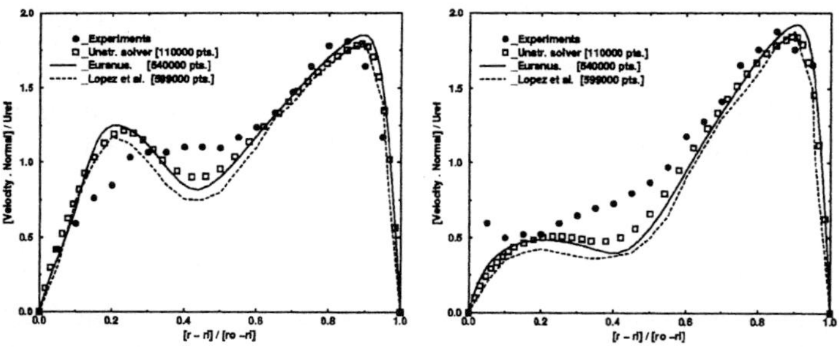

a] halfway between symmetry plane and outer wall

b] symmetry plane

Fig. 3. radial variation of the streamwise velocity

Figure 3 shows that the results obtained with the unstructured solver are in fair agreement with experimental data, but significant differences are nevertheless identified. The present results are in good agreement with previous computational results of Lopez and Oberkampf [13] and with results obtained with EURANUS. Good agreement between different codes lends significant credence to the accuracy of these predictions.

In order to validate the turbulence models and the adaptive capabilities of the code, the simulation of the flow over a flat plate is performed at a Reynolds number based on the plate length of $4.2\,10^6$, and a Mach number of 0.35. The linear low Reynolds k-ϵ of Yang-Shih is used. The starting mesh consists of two cells and the final mesh obtained after 15 local refinements contains 3565 cells.

The sensor chosen for performing mesh adaptation is the x-component of the velocity. After each local refinement, the number of grids levels used in the V-cycle multigrid strategy is increased with a maximum of five levels. Figure 4 shows the convergence and the solution in the boundary layer at $Re_x = 4.2\,10^6$. Good agreement is found with the Musker profile [14].

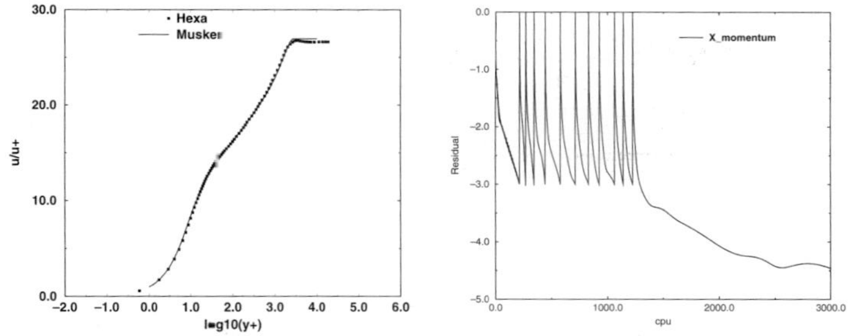

Fig. 4. logarithmic plot of u^+ vs y^+ at $Re_x = 4.2\,10^6$ and convergence history

Finally, an inviscid unsteady flow simulation has been performed using the dual time stepping method. This simulation is only for qualitative information since the flow solver cannot handle incompressible fluid for the moment; a flow of air is assumed at a relatively high speed for such a configuration $100m/s$. The propeller speed of rotation is chosen equal to 33 rpm. Figure 5 shows the final mesh and flow particles tracing is displayed for the flow around the propeller rotated by an angle of 41 degrees.

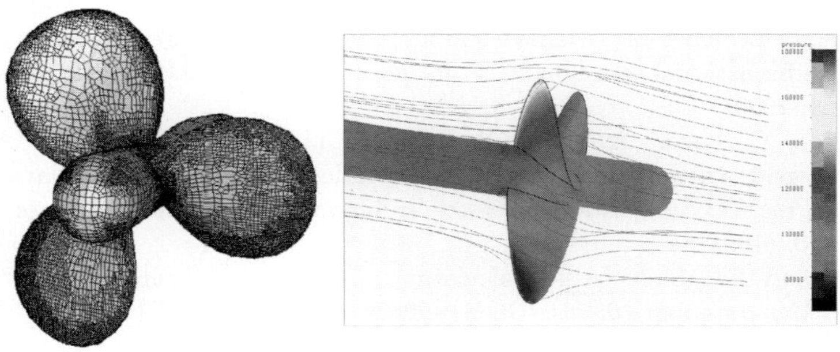

Fig. 5. Final mesh and pressure distribution

6 Conclusions

An unstructured hexahedral adaptive solver has been presented. The code uses central scheme with Jameson type artificial dissipation for the inviscid terms and a compact discretization of the viscous fluxes based on staggered control volumes. Several k-ϵ turbulence models are implemented. Convergence acceleration is obtained with an efficient multigrid strategy. The mesh adaptation module performs anisotropic grid refinement and hierarchical coarsening, allowing to optimize the computational cost with respect to accuracy. The efficiency of the adaptive solver has been demonstrated on several testcases.

7 Acknowledgment

This research is funded by ESTEC/ESA, Noordwijk, the Nederlands, under Contract no. 12110/96/NL/FG. This support is gratefully acknowledged.

References

[1] D.L. Marcum: 'Generation of unstructured grids for viscous flow applications'. *AIAA paper 95-0212*, (1995)

[2] A. Khawaja, H. Mc Morris and Y. Kallinderis: 'Hybrid grids for viscous flows around complex 3D geometries including multiple bodies'. *AIAA paper 95-1685* (1995)

[3] C. Lacor , C. Hirsch, B. Leonard and B. Lessani: 'A 3D Navier-Stokes solver using unstructured, hexahedral meshes with adaptation'. *AIAA paper 99-3364*, (1999)

[4] B. Leonard, A. Patel and C. Hirsch: 'Multigrid acceleration in a 3-D Navier-Stokes solver using unstructured hexahedra meshes'. In: *Proc. 4th Conference on Multigrid.* (Gent, 1999)

[5] K.-F. Tchon, C. Hirsch and R. Schneiders: 'Octree-based hexahedral mesh generation for viscous flow simulations'. *AIAA paper 97-1980.* (1997)

[6] M. Delanaye, K.-F. Tchon, A. Patel and C. Hirsch: ' All hexahedra unstructured grid generation'. In: *Proc. of the ECCOMAS 2000 Conference.* (Barcelona, 2000)

[7] A. Patel, B. Leonard, M. Delanaye and C. Hirsch: 'Unstructured unsteady adaptive simulations for external aerodynamics'. In: *Proc. of the ECCOMAS 2000 Conference.* (Barcelona, 2000)

[8] C. Hirsch, C. Lacor, A. Rizzi, P. Eliasson, I. Linblad, J. Hauser: 'A multiblock/multigrid software for the efficient solution of complex 3-D Navier-Stokes flows'. In: *Proc. 1st European Symposium on Aerothermodynamics for Space Vehicles.* pp. 415–420. (1991)

[9] Z. Yang and T.H, Shih: 'A k-ϵ calculation of transitional boundary layer', *NASA Technicel Memorandum 105604 ICOMP-92-08, CMOTT-92-05.* (1992)

[10] N. Hakimi: 'Preconditioning methods for time dependent Navier-Stokes equations / Application to environmental and low speed flows'. *PhD Thesis.* (Vrije Universiteit Brussel, 1997)

[11] D. Mavriplis: 'Multigrid techniques for unstructured meshes'. *ICASE Report 95-27.* (1995)

[12] J. Humphrey, A. Taylor J. Whitelaw: 'Laminar flow in a square duct of strong curvature'. *Journal of Fluid Mechanics, Vol. 83.* pp. 509–527 (1977)

[13] A.R. Lopez, W.L. Oberkampf: 'Numerical simulation of laminar flow in a curved duct. *AIAA paper 95-0865* (1995)

[14] A.J. Musker: 'Explicit expression for the smooth wall velocity distribution in a turbulent boundary layer'. *AIAA Journal 17, No 6.* (1979)

A Class of Matrix-Free Implicit Methods for Compressible Flows on Unstructured Grids

Hong Luo[1], Dmitri Sharov[1], Joseph D. Baum[1], and Rainald Löhner[2]

[1] Science Applications International Corporation, McLean VA 22030, USA
[2] George Mason University, Fairfax, VA 22030, USA

1 Introduction

In recent years, significant progress has been made in developing numerical algorithms for the solution of the compressible Euler and Navier-Stokes equations on unstructured grids. Early efforts in the development of temporal discretization methods using unstructured grids, focused on the explicit schemes. In general, explicit temporal discretization methods such as multi-stage Runge-Kutta schemes and their boundary conditions are easy to implement, vectorize and parallelize, and require only limited memory storage. However, for large-scale problems and especially for the solution of the Navier-Stokes equations, the rate of convergence slows down dramatically, resulting in inefficient solution techniques. In order to speed up convergence, a multigrid strategy or an implicit temporal discretization is required. In general, implicit methods can outperform their explicit counterparts by a factor of 10 or even more. Unfortunately, this performance is usually achieved at the expense of memory increment, as some efficient implicit methods might need up to an order of magnitude more storage than the explicit methods. This is mainly due to the fact that implicit methods usually require the computation of the left-hand-side Jacobian matrix in addition to construction of the right-hand-side residual. The storage of the Jacobian matrix consumes an enormous amount of memory, which may be prohibitive for large scale problems.

The objective of this paper is to develop and compare a class of matrix-free implicit methods for compressible flow problems around complex, realistic aerodynamic configurations on unstructured grids. Typically, hundreds of thousands of mesh points are necessary to represent such engineering type configurations accurately. Any implicit methods requiring the storage of the Jacobian matrix would be impractical, if not impossible, to use to solve such large scale problems, where the storage requirement can easily exceed the memory limitation of present computers. In the present study, an approximate Newton linearization based on the scalar dissipation is used to derive the left-hand-side matrix. The class of implicit methods considered to solve the resulting system of linear equations includes Symmetric Gauss-Seidel iterations (SGS), Lower-Upper Symmetric Gauss-Seidel method (LU-SGS)[1], and Generalized Minimum RESidual algorithm with a Lower-Upper Symmetric Gauss-Seidel preconditioner (GMRES+LU-SGS)[2]. A remarkable feature of these implicit methods is that the storage of the lower and upper parts of left-hand-side Jacobian matrix can be completely eliminated by approximating the Jacobian with numerical fluxes, resulting in matrix-free implicit methods. Furthermore, both lower and upper sweeps involved in these methods can be vectorized by appropriately reordering the grid points, resulting in a very efficient algorithm. The developed methods have been used to compute both inviscid and turbulent flows around complex aerodynamic configurations for a wide range of flow conditions. The numerical results obtained indicate that both iterative methods GMRES+LU-SGS and SGS provide substantial improvement over the approximate factorization method LU-SGS, and GMRES+LU-SGS method outperforms both LU-SGS and SGS methods at the expense of small memory increment.

2 Numerical Methods

The governing Navier-Stokes equations are discretized using a hybrid finite volume and finite element formulation, where the finite volume approximation based on a containment dual control volume rather than the more popular median-dual control volume is used to discretize the inviscid fluxes, and the finite element approximation is used to evaluate the viscous flux terms. In the present study, the numerical flux functions for inviscid fluxes at the dual mesh cell interface are computed using AUSM+[3] (Advection Upwind Splitting Method) scheme. A MUSCL[4] approach is used to achieve high-order accuracy. The Van Albada limiter based on primitive variables is used to suppress the spurious oscillation in the vicinity of the discontinuities. The Spalart and Allmaras one equation turbulence model has been used to simulate the effects of fluid turbulence at high Reynolds numbers. The mean-flow and turbulence-model equations are decoupled in the time integration in order to facilitate the incorporation of different turbulence models and reduce memory requirements.

Using Euler implicit time-integration, the spatially discretized Navier-Stokes equations can be linearized in time and written as

$$A\Delta\mathbf{U} = \mathbf{R}, \tag{1}$$

where

$$A = \frac{V}{\Delta t}\mathbf{I} - \frac{\partial \mathbf{R}^n}{\partial \mathbf{U}}, \tag{2}$$

where $\mathbf{R}$ is the right-hand-side residual and equals to zero for a steady state solution.

Equation (1) represents a large system of linear simultaneous algebraic equations and needs to be solved at each time step. The three methods considered to solve the linear system (1) in this study include SGS, LU-SGS, and GMRES+LU-SGS methods. For the linear system (1),The SGS method can be written as

0. Initialization:

$$\Delta\mathbf{U}^0 = 0, \tag{3}$$

1. Forward Gauss-Seidel iteration:

$$(D + L)\Delta\mathbf{U}^{n+1/2} + U\Delta\mathbf{U}^n = \mathbf{R}, \tag{4}$$

2. Backward Gauss-Seidel iteration:

$$(D+U)\Delta\mathbf{U}^{n+1} + L\Delta\mathbf{U}^{n+1/2} = \mathbf{R}, \tag{5}$$

where U, L, and D represent strict upper matrix, strict lower matrix, and diagonal matrix, respectively, and n is the number of iterations for SGS method. The main advantage of SGS method is that it does not require any additional storage beyond that of the matrix itself. Note that if only one iteration is used in the SGS method and the initial guess $\Delta\mathbf{U}^0 = 0$, the resulting method is nothing but so called LU-SGS method, which can be written as

Lower (forward) sweep:

$$(D+L)\Delta\mathbf{U}^{\star} = \mathbf{R} \tag{6}$$

Upper (backward) sweep:

$$(D+U)\Delta\mathbf{U} = D\Delta\mathbf{U}^{\star}. \tag{7}$$

The third method used to solve equation 1 is GMRES+LU-SGS method[2]. The idea behind this method is to combine the efficiency of the iterative method GMRES and low memory requirement of approximate factorization method LU-SGS in an effort to develop a fast, low storage implicit method. A clear advantage of the LU-SGS preconditioner is that it does not require any additional storage beyond that of the matrix itself, as compared with a more popular ILU preconditioner, and consequently does not require any additional memory storage and computational effort to store and compute the preconditioner matrix. However, GMRES method itself does need some additional memory to store the vectors of search directions. The GMRES+LU-SGS method has proven to be very reliable, robust, and computationally efficient to solve both steady and unsteady flow problems.

The common feature of the above three methods is that their operation involves primarily the Jacobian matrix-solution incremental vector product. Such Jacobian matrix-solution incremental vector can then be approximately replaced by computing increments of the flux vector. This is achieved by using scalar dissipation to derive the left-hand-side matrix. The detailed matrix-free approach can be found in the references 1, 2, and 5.

3 Numerical Results

The developed implicit methods have been used to compute a variety of compressible flow problems around complex aerodynamic configurations for a wide range of flow conditions, from subsonic to supersonic, for both inviscid and viscous flows, in both 2D and 3D. Due to space limitation, only two examples are presented here to illustrate the effectiveness of all three methods. The efficiency of the methods is measured in terms of the computing time required to obtain a steady-state solution for these problems. These computations were intentionally performed on an Intel Pentium II 400 MHz PC using Linux operating system to demonstrate that a large scale problems can be solved efficiently on a PC platform using the present methods. All computations were started with uniform flow. The relative L_2 norm of the density residual is taken as a criterion to test convergence history. The solution tolerance for GMRES is set to 0.1 with 10 search directions and 20 iterations.

The first, well-documented case is the inviscid transonic flow over a ONERA M6 wing configuration. The flow solutions are presented at a Mach number of 0.84 and an angle of attack of 3.06°. The computation has been performed on two different size grids to study the behavior of convergence history as the grid is refined. The coarse mesh contains 386,179 elements, 75,162 grid points, and 20,659 boundary points, and the fine one consists of 2,166,336 elements, 382,397 grid points, and 26,785 boundary points. The coarse mesh, and the computed pressure contours in the flow field are displayed in Fig. 1. Figure 2 shows a comparison of convergence histories among these implicit methods versus time steps and CPU time, respectively, for the coarse mesh, while the same comparison for the fine mesh is displayed in figure 3. Note that 10 iterations were used for the SGS method. It is clear that both SGS and GMRES+LU-SGS methods are far superior to LU-SGS method. The GMRES+LU-SGS method is faster than SGS method, especially on the fine grid.

The second test case involves the modeling of turbulent subsonic flow past a Douglas three-element airfoil in a landing configuration at a Mach number of 0.2, an angle of attack of 16°, and a Reynolds number of 9 million. The computational mesh, consisting of 66,530 elements, 33,618 points, and 710 boundary points and the computed Mach number contours in the flow field are displayed in Fig. 4. Figure 5 shows a comparison of convergence histories among the explicit, matrix-free LU-SGS, SGS, and GMRES+LU-SGS methods versus time steps and CPU time, respectively. The explicit method used a four-stage Runge-Kutta time-stepping scheme with local time stepping and implicit residual smoothing and a CFL number of 0.6. A CFL number of 200 was used by all implicit methods in the computation. Note that it is necessary to use 20 iterations for the SGS method

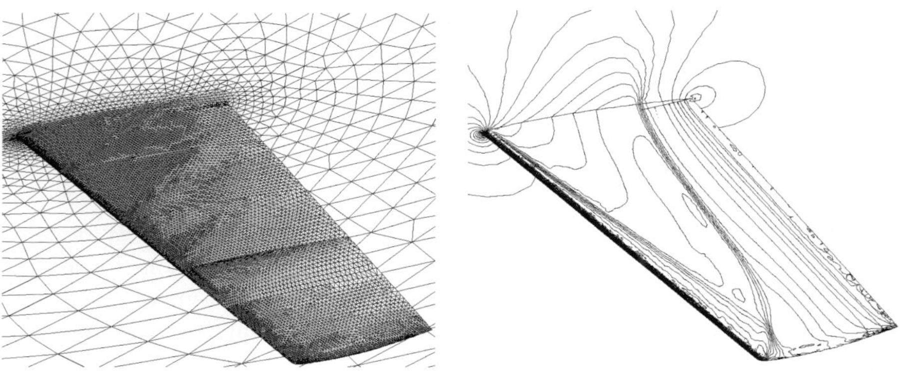

Fig. 1. The coarse mesh and computed pressure contours for inviscid transonic flow over M6 wing (M_∞=0.84, α=3.06°)

for this case. The effectiveness of the GMRES+LU-SGS method, compared to both LU-SGS and SGS methods, is especially apparent for this test case. Note that both iterative methods offered substantial saving over the approximate factorization method.

4 Conclusions

A class of matrix-free implicit methods has been developed to solve the compressible flow problems on unstructured grids. The developed methods are compared for a number of test cases for a wide range of flow conditions. The numerical experience indicates that both iterative methods, SGS and GMRES+LU-SGS lead to a significant increase in performance over the approximate factorization method, LU-SGS. The GMRES+LU-SGS provides the best convergence especially for large size and stiff problems at the expense of small memory increment.

References

1. Sharov, D. and Nakahashi, K.: 'Reordering of 3-D Hybrid Unstructured Grids for Vectorized LU-SGS Navier-Stokes Computations', AIAA Paper 97-2102, 1997.
2. Luo H., Baum, J. D., and Löhner, R.: 'A Fast, Matrix-free Implicit Method for Compressible Flows on Unstructured Grids', *Journal of Computational Physics*, Vol. 146, pp. 664-690, 1998.
3. Liou, M. S.: 'Progress towards an Improved CFD Method: AUSM+', AIAA paper-95-1701, Jun. 1995.
4. Van Leer, B.: 'Towards the Ultimate Conservative Difference Scheme, II. Monotonicity and Conservation Combined in a Second Order Scheme', *Journal of Computational Physics*, 14, pp. 361-370, 1974.
5. Men'shov, I., and Nakamura, Y.: 'An Implicit Advection Upwind Splitting Scheme for Hypersonic Air Flows in Thermochemical Nonequilibrium', *6th Int. Symp. on CFD*, 1995.

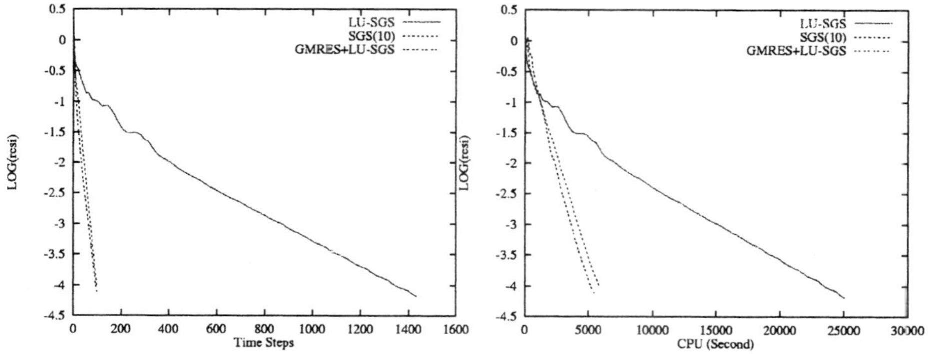

Fig. 2. Convergence history of the residual for coarse mesh for inviscid transonic flow by matrix-free LU-SGS, SGS, and GMRES+LU-SGS methods.

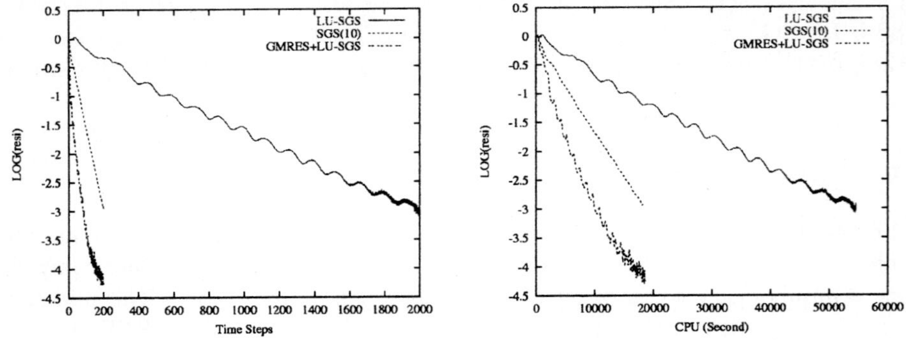

Fig. 3. Convergence history of the residual for fine mesh for inviscid transonic flow by matrix-free LU-SGS, SGS, and GMRES+LU-SGS methods.

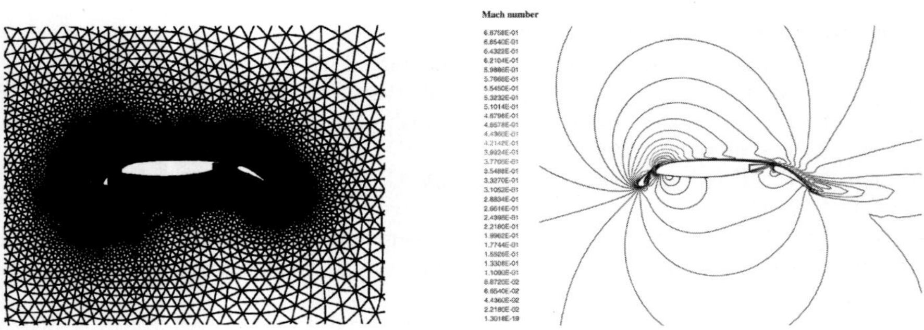

Fig. 4. Computational mesh and computed Mach number contours for turbulent flow over Douglas three-element airfoil (M_∞=0.2, Re=9 million, α=16°)

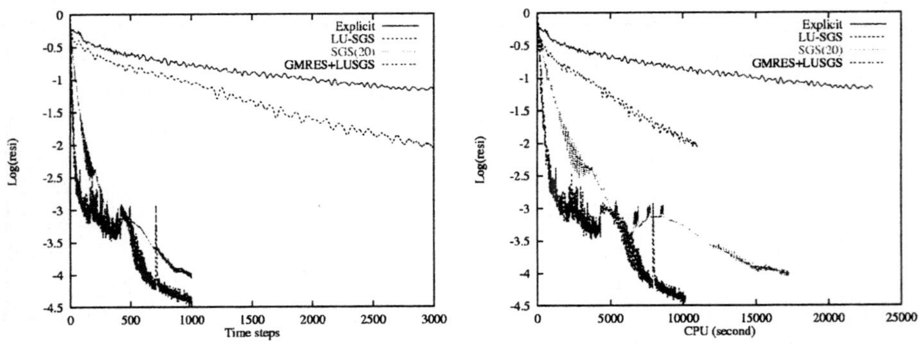

Fig. 5. Residual convergence history versus time steps and CPU time for turbulent flow over a Douglas 3-element airfoil using different methods: explicit, matrix-free LU-SGS, matrix-free SGS and matrix-free GMRES+LU-SGS.

Large-Scale Parallel Unstructured Mesh Computations Using a Multigrid Approach

Dimitri. J. Mavriplis[1]

ICASE MS 132C, NASA Langley Research Center, Hampton, VA 23681, USA

1 Introduction

The goal of this work is to demonstrate increasingly high resolution unstructured grid analyses of aerodynamic flows on current day supercomputers. This is particularly challenging because, while unstructured grid strategies greatly simplify the grid generation process for complex industrial geometries, they also require substantial additional memory and CPU time compared to block-structured or overset grid methods for a given mesh size.

In order to tackle large problems, the unstructured grid solver must be efficient in terms of numerical convergence and well adapted to current supercomputer architectures. Rapid convergence to steady state is obtained through the use of a preconditioned agglomeration multigrid algorithm. The solver is parallelized following the domain decomposition approach, using a hybrid two-level MPI and OpenMP communication strategy suited for distributed memory machines, shared memory machines, and clusters of shared memory multiprocessors.

2 Base Solver

The Reynolds averaged Navier-Stokes equations are discretized by a finite-volume technique on meshes of mixed element types which may include tetrahedra, pyramids, prisms, and hexahedra. All elements of the grid are handled by a single unifying edge-based data-structure in the flow solver [1].

The basic time-stepping scheme is a three-stage explicit multistage scheme with stage coefficients optimized for high frequency damping properties. An agglomeration multigrid algorithm [1] is used to further enhance convergence to steady-state. In this approach, coarse levels are constructed by fusing together neighboring fine grid control volumes to form a smaller number of larger and more complex control volumes on the coarse grid. A multigrid cycle consists of performing a time-step on the fine grid of the sequence, transferring the flow solution and residuals to the coarser level, performing a time-step on the coarser level, and then interpolating the corrections back from the coarse level to update the fine grid solution. The process is applied recursively to the coarser grids of the sequence. A directional smoothing technique [2] is employed to overcome stiffness due to the highly stretched grids required in the boundary layer regions. Directional smoothing is achieved by constructing lines in the unstructured mesh along the direction of strong coupling (i.e. normal to the boundary layer) and solving the implicit system along these lines using a tridiagonal line solver, during the smoothing process on each level of the multigrid algorithm. For additional convergence acceleration, the multigrid algorithm can be used as a preconditioner to a GMRES Krylov method, rather than as a solver itself. This approach is easily implemented but requires additional memory due to the need to store the Krylov vectors in the GMRES algorithm.

3 Parallel Implementation

The unstructured multigrid solver is parallelized by partitioning the domain using a standard graph partitioner [3], and allocating one or more partitions to each processor of a parallel computer or cluster of interconnected machines.

For each partition, the edges of the mesh which straddle two adjacent partitions are assigned to one of the processors, and a "ghost vertex" is constructed in this processor, which corresponds to the vertex originally accessed by the edge in the adjacent processor. During a residual evaluation, the fluxes are computed along edges and accumulated to the vertices. The flux contributions accumulated at the ghost vertices must then be added to the flux contributions at their corresponding physical vertex locations in order to obtain the complete residual at these points. This phase incurs interprocessor communication. For distributed memory architectures, this communication is implemented using the MPI message-passing library. For shared memory architectures, a potentially more efficient communication strategy is to simply copy (or copy-add) the values from the individual ghost points into the locations which correspond to their real images, since the memory on different partitions is addressable from any other partition. Additionally, the OpenMP standard provides a simple strategy for parallelizing shared memory programs through the use of compiler directives. The current implementation results in a code which can be run in a purely MPI mode, suitable for distributed memory architectures, a purely OpenMP mode, suitable for shared memory architectures, or a two-level hybrid MPI-OpenMP mode, suitable for clusters of shared-memory processors. In the hybrid mode, the MPI communication is handled by a single master OpenMP thread, which sends/receives a composite message gathered-from/scattered-to other threads [4].

4 Results

The high-lift flow over a complete aircraft configuration including slat, flaps, pylon and nacelle, has been computed on a grid of approximately 3 million points. This grid, which is depicted in Figure 1, was generated directly on a workstation using the VGRID grid generation software [5]. Figure 2 illustrates the computed steady-state solution as a set of surface pressure contours. This test case illustrates the flexibility of the unstructured mesh approach in handling complex geometries. More extensive computations have been performed on the same configuration with the nacelle and pylon removed, since the experimental data for this simpler configuration was more readily available for comparison. For this case, the freestream Mach number is 0.2, and the Reynolds number is 1.6 million based on the wing reference chord, and the experimental flow incidence varies over a range of -4 degrees up to 24 degrees.

Two grids of different resolution have been employed: the coarse grid, which contains 3.1 million vertices was generated using VGRID [5] on a workstation. A finer grid was obtained by uniform refinement of this grid resulting in a grid of 24.7 million vertices. The refinement operation was performed sequentially on a single processor of an SGI ORIGIN 2000, and required approximately 10 Gbytes of memory and 30 minutes of CPU time. Both grids contain prismatic elements in the boundary layer regions, and tetrahedral elements elsewhere. The multigrid convergence histories for the coarse and fine grids (at 10 degrees incidence) are depicted in Figure 3. Convergence to steady-state is nearly identical for both grids, resulting in a 4 order of magnitude reduction in the residual in 500 multigrid cycles. Additionally, substantial convergence acceleration is obtained in the coarse grid case by the incorporation of the GMRES technique, using 20 search directions. However, the additional memory required by this Krylov technique precluded its use in the fine grid case.

Figure 4 illustrates the comparison between then experimental wind tunnel drag polar and computed results for both the coarse and fine grids for the full range of incidences. Agreement is generally good in the fully attached regions of the drag polar, with the fine grid demonstrating increased accuracy over the coarse grid in this region. The maximum lift location and value is slightly overpredicted on both grids, with the overprediction being more pronounced for the fine grid.

Figures 5 through 9 illustrate the parallel scalability of the unstructured mesh solver on various computer architectures ranging from some of the most exotic and expensive custom-built machines to a relatively inexpensive PC cluster, using MPI communication exclusively in all cases. The overall best performance was obtained on a T3E-1200E where excellent scalability up to 1450 processors was observed for the 25 million point case. On this machine, the fine grid could be solved in about 1 hour of wall clock time. At the other end of the spectrum, Figure 9 illustrates the scalability of the coarse grid case on an in-house PC cluster. On this system, the coarse grid case could be solved in about 5.5 hours. Figure 6 illustrates the scalability for the 3 million point and the 25 million point case running on a collection of interconnected 128 processor SGI Origin 2000 machines at Los Alamos National Laboratory, in New Mexico, USA. The scalability of the coarse grid case is seen to degrade upwards of 256 processors. However, the fine grid case produces reasonable scalability up to about 1024 processors. Observed scalabilities for the coarser 3 million point grid an Intel-based machine at Sandia National Laboratory in New Mexico USA, and on an IBM-based machine at Lawrence Livermore National Laboratory in California, USA, are depicted in Figures 7 and 8. The Intel machine is seen to deliver good scalability up to 2048 processors, while the IBM machine provides reasonable scalability up to 512 cpus. The finer grid case can be expected to scale better on both of these machines. In these plots, the scalability of the multigrid grid algorithm is seen to be somewhat inferior to that of the single grid algorithm. This illustrates the increased communication induced by the coarse multigrid levels. In practice, however, the multigrid algorithm provides much faster numerical convergence and thus shorter overall solution time.

Figure 10 provides a comparison of the performance obtained using MPI exclusively, OpenMP exclusively, and the two-level hybrid MPI-OpenMP communication strategy, on an SGI Origin 2000. The scalability for exclusive MPI and exclusive OpenMP is very similar, right up to the maximum 128 processors of this shared memory machine. Scalability of the hybrid MPI-OpenMP communication strategy degrades when large numbers of OpenMP threads are employed, as for example is the case when using 64 OpenMP threads, with 2 MPI processes. This is attributed to the OpenMP sequentialization which occurs when only the master OpenMP thread is assigned the task of performing the MPI send-receive operations. However, this implementation produces a smaller number of larger MPI messages, which may in the future prove beneficial on machines or clusters with large latency networks.

5 Conclusion

The above cases are intended to demonstrate the numerical and computational efficiency of a parallel unstructured mesh multigrid solver for large problems on complex geometries. Currently, the simulation capability is paced by the requirements of the pre-processing operations, including grid partitioning and multigrid coarse level agglomeration, which must be performed sequentially. The parallel implementation of these tasks is currently under development. This work was partly performed under the auspices of the U.S. Department of Energy under subcontract B347882 from Lawrence Livermore National Laboratory and was also partially supported by the Air Force Office of Scientific Research (AFOSR) under grant AFOSR-PO-99-0005 monitored by L. Sakell.

References

1. D. J. Mavriplis and V. Venkatakrishnan. A unified multigrid solver for the Navier-Stokes equations on mixed element meshes. *International Journal for Computational Fluid Dynamics*, 8:247–263, 1997.
2. D. J. Mavriplis. Directional agglomeration multigrid techniques for high-Reynolds number viscous flows. AIAA Paper 98-0612, January 1998.
3. G. Karypis and V. Kumar. A fast and high quality multilevel scheme for partitioning irregular graphs. Technical Report Technical Report 95-035, University of Minnesota, 1995. A short version appears in Intl. Conf. on Parallel Processing, 1995.
4. D. J. Mavriplis. Parallel performance investigations of an unstructured mesh navier-stokes solver. ICASE Report 2000-13, NASA CR-2000-210088, March 2000.

5. S. Pirzadeh. Three-dimensional unstructured viscous grids by the advancing-layers method. *AIAA Journal*, 34(1):43–49, 1996.

6 Figures

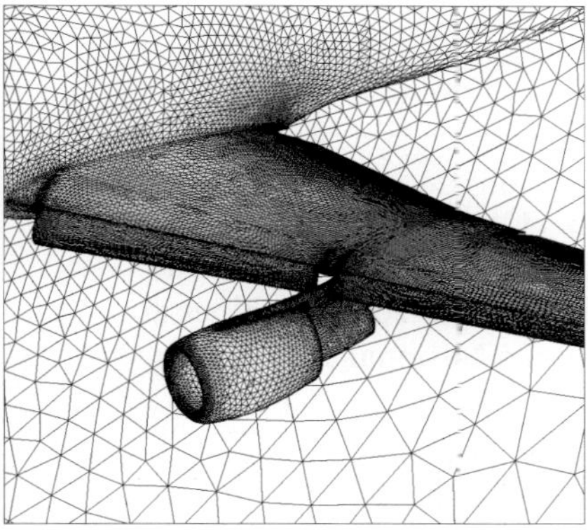

Fig. 1. Unstructured grid for complete wing-body-nacelle-pylon geometry.

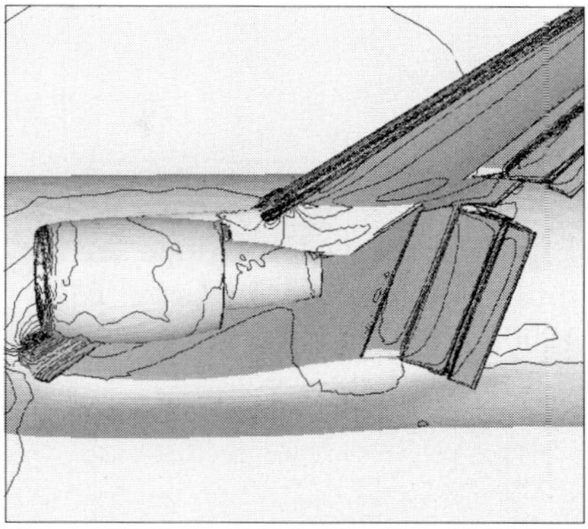

Fig. 2. Computed pressure contours for flow over complete wing-body-nacelle-pylon high-lift configuration.

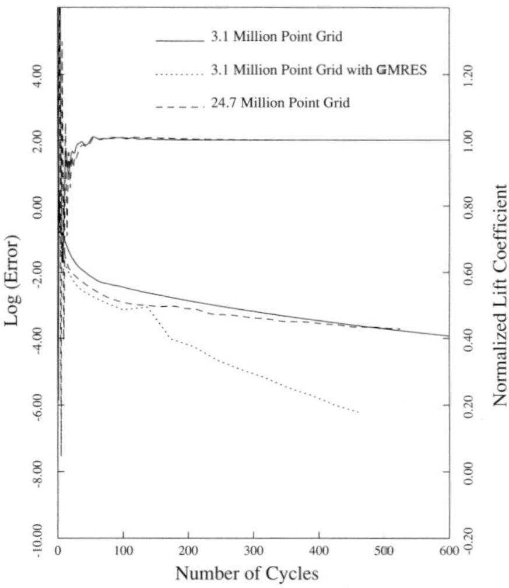

Fig. 3. Multigrid Convergence rates for coarse grid (3.1 million points, with and without GMRES acceleration) and fine grid (25 million points) on wing-body (without nacelle) high-lift configuration.

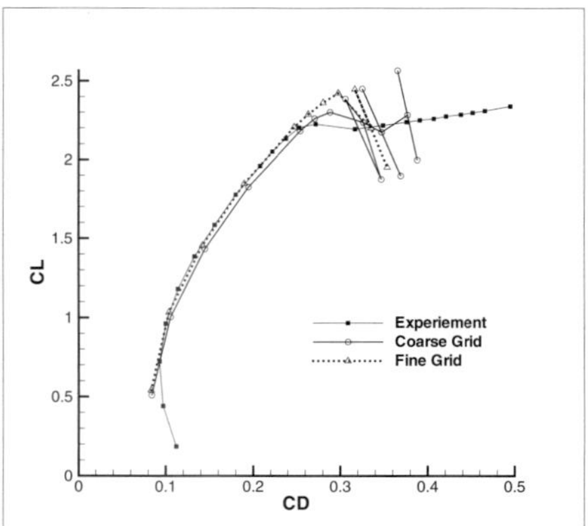

Fig. 4. Comparison of experimental drag polar with coarse and fine grid computed values.

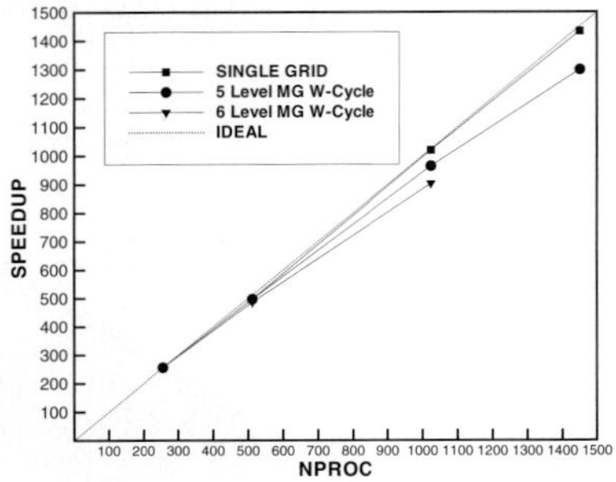

Fig. 5. Observed speedups for 25 million point case on Cray T3E-1200E

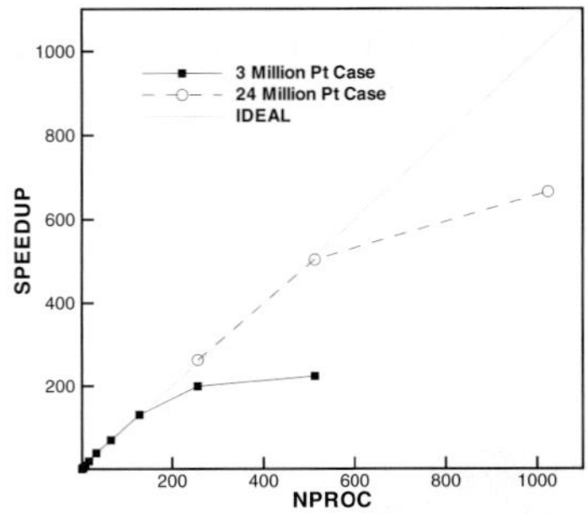

Fig. 6. Observed speedups for coarse and fine grid cases on cluster of SGI Origin 2000 machines at Los Alamos National Laboratory.

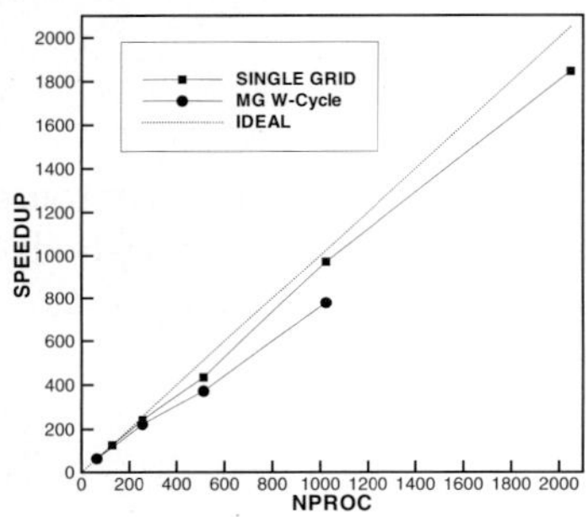

Fig. 7. Observed speedups (3.1 million pt mesh) on Intel-Based Machine at Sandia National Laboratory.

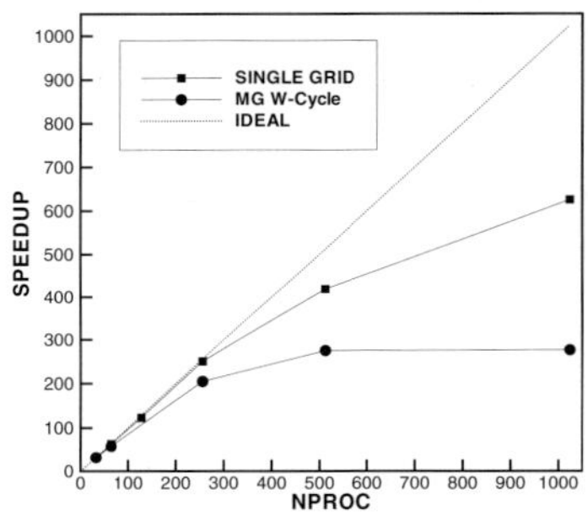

Fig. 8. Observed speedups (3.1 million pt mesh) on Intel-Based Machine at Sandia National Laboratory.

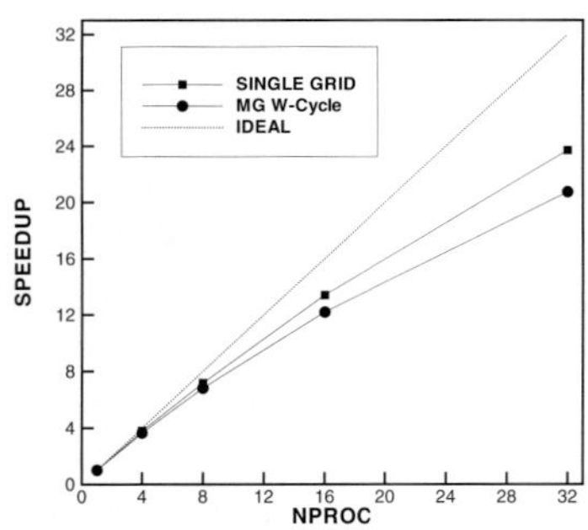

Fig. 9. Observed Speedup for 3.1 million point multigrid case on ICASE 400Mhz Pentium PC cluster.

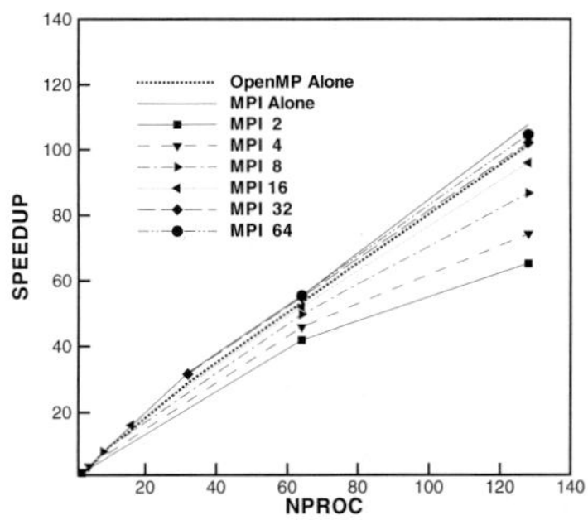

Fig. 10. Observed speedups for two-level hybrid MPI-OpenMP implementation on SGI Origin 2000.

An Adaptive Unstructured Explicit/Implicit Numerical Technique for Unsteady Viscous Flows with Shock Waves

E.V. Timofeev[1,3], P.A. Voinovich[1,2], K. Takayama[1], and Yu.P. Golovachov[3]

[1] Shock Wave Research Center, Institute of Fluid Science, Tohoku University, 2-1-1 Katahira, Aoba-ku, Sendai 980-8577, Japan
[2] Advanced Technology Center, P.O. Box 29, St. Petersburg 194156, Russia
[3] Ioffe Physico-Technical Institute, Russian Academy of Sciences, Polytekhnicheskaya 26, St. Petersburg 194021, Russia

Abstract. The paper describes further improvement and generalization of a numerical approach proposed earlier by the authors for simulations of essentially unsteady shock wave interactions with bodies. The mathematical model is based on the Navier-Stokes equations. The most important ingredients of the method are specially tailored adaptive unstructured grids and an efficient combination of explicit and implicit time integration modes which are chosen to be best suited for underlying flow physics.

1 Introduction

In the present paper we continue our quest for efficient and accurate numerical simulation of *unsteady compressible flows exhibiting significant influence of viscosity and heat conductivity* (our previous development and applications on the subject may be found in [1–4]). The class of the flows is very wide. It is hardly possible to design a universal numerical method covering the whole range of possible flow parameters. We restrict our consideration in favor of *high speed flows*, i.e. transonic and supersonic flows, being, at the same time, *essentially unsteady ones* with flow patterns undergoing fast and qualitative transformations. Reynolds numbers Re are typically $\geq 10^3$-10^4.

Flows of this type are of frequent occurrence in shock tube experiments when a shock wave and the flow behind it interact with a body. From practical point of view the respective numerical code may be applied, for instance, to study unsteady flows in experimental facilities aiming at improvements in their operational characteristics as well as to investigate starting process of nozzles and supersonic inlets serving as a numerical test tunnel.

It is generally recognized that a good numerical technique should essentially take into account physical features of the flow under simulation. The flows mentioned above typically have two areas being very different from the gasdynamical point of view. These are: 1) basically inviscid flow away from a body or solid walls with multiple discontinuities (shock waves, contact discontinuities, slip surfaces etc) interacting with each other and varying qualitatively in time; 2) viscous boundary layers on solid surfaces having essentially different characteristic scales along and normal to the surface (their ratio is proportional to $1/\sqrt{\mathrm{Re}}$). Below we develop a new numerical technique trying to employ optimal options from the point of view of being best suitable for the above flow physics and providing efficient codes.

2 Numerical Technique

2.1 Grid System

Near solid surfaces we use a structured-like grid clustered towards the boundaries, however (unlike [1–4]) do not require the same number of grid nodes along grid lines being normal to the body surface (Fig. 1). The grid can be then interpreted as a set of *stations* placed along solid surfaces. The nodes on the neighboring stations are in correspondence (like on structured meshes), however, each station may be terminated inside computational domain at a point being different from neighboring ones. The grid based on the stations is treated as *one-dimensional unstructured grid* (with the data structure similar to that used for the triangular unstructured mesh covering the rest of the domain) and is subject to transient reversible adaptation deleting or inserting *a station as a whole*. This allows to keep finer meshes in the vicinity of separation/reattachment points and other areas of special interest.

The rest of computational domain is covered by triangular unstructured grid subject to a conventional h-refinement/derefinement in the course of flow development, thus, allowing accurate and efficient resolution of all discontinuities and areas of special importance.

The thickness of the 'station layer' is not strictly defined. Since the full Navier-Stokes equations are to be solved everywhere, a heuristic rule is only used: the thickness should be chosen so that the size of control volumes near the stations/unstructured grid interface would be nearly the same; the boundary layers must be submerged into the station layer.

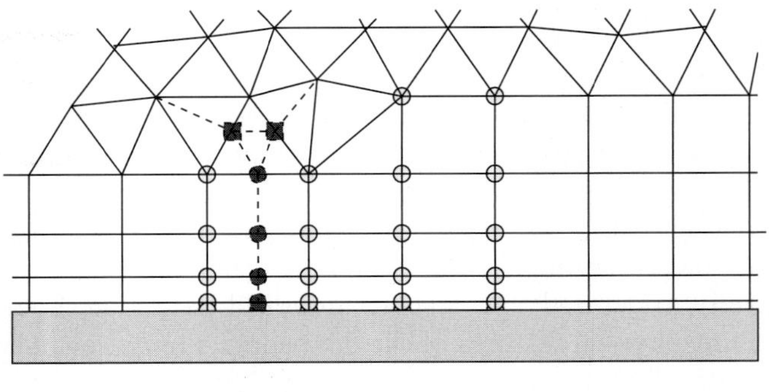

Fig. 1. A typical pattern of the near-wall adaptive grid. Four grid stations are highlighted with gray circles. A newly inserted station is represented by black circles. Its creation is accompanied by proper refinement of external triangular unstructured mesh (new nodes are shown as black squares)

2.2 Governing Equations

The mathematical model is based on the Navier-Stokes equations written in an integral form with respective boundary and initial conditions:

$$\frac{\partial}{\partial t}\int_{\text{Vol}} \mathbf{U}\,\mathrm{d}V + \oint_{\sigma} (\mathbf{F}-\mathbf{G})\cdot \mathrm{d}\mathbf{s} = 0, \tag{1}$$

where $\mathbf{U}$ is the vector of conserved variables, $\mathbf{F}$ and $\mathbf{G}$ are convective and viscous fluxes, respectively. Vol represent a gas volume bounded by the closed surface σ.

Nonoverlapping control volumes based on centroids of grid elements and middle points of edges are established for all grid nodes. The numerical method is formulated in terms of fluxes through control volume interfaces.

2.3 Numerical Integration

As to the spatial integration, we routinely employ central differences to approximate viscous terms while a second-order Godunov-type scheme with an exact Riemann solver is applied to convective terms. In the latter case, a linear reconstruction $\overline{\nabla \mathbf{V}}$ of primitive gasdynamic variables $\mathbf{V} = (\rho; u_x; u_y; p)$ with a TVD limiter is used for each control volume.

The key point is the time integration. Second order in time is mandatory for unsteady flows under study. Besides, due to the essential unsteadiness and relatively insignificant influence of viscosity away from solid surfaces, an explicit time integration represents a reasonable choice for the external unstructured triangular grid. For this purpose we employ here (similar to [1–4]) a generalization for unstructured grids and viscous flows of the explicit Rodionov scheme [5] which is very close but not fully identical to the MUSCL-Hancock approach ([6]). The second order in time is achieved via a predictor-corrector procedure with the first-order in time predictor step performed for the full time step Δt:

$$\frac{\mathbf{U}_i^{n+1/2} - \mathbf{U}_i^n}{\Delta t}\mathrm{Vol}_i + \sum_{j(i)} \mathbf{F}(\mathbf{V}_{ij}^n)\cdot \mathbf{S}_{ij} - \sum_{j(i)} \mathbf{G}(\mathbf{V}^n)\cdot \mathbf{S}_{ij} = 0 \ , \tag{2}$$

where $\mathbf{V}_{ij}^n = \mathbf{V}_i^n + 0.5\overline{\nabla \mathbf{V}_i}^n \cdot \boldsymbol{ij}$; Vol_i is the value of the control volume i; $\mathbf{S}_{ij}$ is the area vector of the control volume face between the nodes i and j; the index j denotes neighbors of the node i. The predictor values $\mathbf{U}^{n+1/2}$ (transformed into the primitive variables $\mathbf{V}^{n+1/2}$) are then used at the corrector step to obtain the second order accuracy in time.

However, in some cases, not to be discussed here due to lack of space, an implicit procedure may be locally needed even on the external unstructured grid. Undoubtedly, an implicit approximation is mandatory on the grid stations (at least, near walls) due to very small size of control volumes close to solid surfaces and the influence of viscous terms. A smooth transition between approximation modes would be a good asset too. To achieve all these ends, we propose here the following hybrid explicit/implicit approach.

Let's formally replace the predictor values $\mathbf{V}^{n+1/2}$ by a weighted average $\mathbf{V}_i^\tau$ with the fully implicit value $\mathbf{V}^{n+1}$: $\mathbf{V}_i^\tau = (1-\tau_i)\mathbf{V}_i^{n+1} + \tau_i \mathbf{V}_i^{n+1/2}$, where τ_i is a coefficient of 'implicitness' being different for *each* grid node and $0 \le \tau_i \le 1$. Thus, we would arrive to the following expression for the corrector step:

$$\frac{\mathbf{U}_i^{n+1} - \mathbf{U}_i^n}{\Delta t}\mathrm{Vol}_i + \sum_{j(i)} \mathbf{F}(\mathbf{W}_{ij}^{n+1}) \cdot \mathbf{S}_{ij} - \sum_{j(i)} \mathbf{G}(\tilde{\mathbf{V}}^{n+1}) \cdot \mathbf{S}_{ij} = 0 \ . \quad (3)$$

The viscous fluxes are computed at the state

$$\tilde{\mathbf{V}}^{n+1} = 0.5[\mathbf{V}^n + (1-\tau)\mathbf{V}^{n+1} + \tau \mathbf{V}^{n+1/2}]. \quad (4)$$

$\mathbf{W}_{ij}^{n+1}$ is the exact solution of the Riemann problem with the following 'left' and 'right' values:

$$\mathbf{V}_{ij}^l = 0.5[\mathbf{V}_i^n + (1-\tau_i)\mathbf{V}_i^{n+1} + \tau_i \mathbf{V}_i^{n+1/2} + \overline{\nabla \mathbf{V}_i^n} \cdot \boldsymbol{ij}]; \quad (5)$$

$$\mathbf{V}_{ij}^r = 0.5[\mathbf{V}_j^n + (1-\tau_j)\mathbf{V}_j^{n+1} + \tau_j \mathbf{V}_j^{n+1/2} + \overline{\nabla \mathbf{V}_j^n} \cdot \boldsymbol{ji}] \ . \quad (6)$$

Under $\tau = 1$ the scheme becomes the second-order predictor-corrector explicit approximation (the generalized Rodionov scheme) while with $\tau = 0$ we obtain fully implicit second-order in space and time formulation. Intermediate values provide a smooth transition between the two modes, thus allowing to flexibly maintain 'implicit pockets' in a flowfield. Somewhat similar schemes being first-order in time (in the implicit mode) may be found in [7–9].

To solve the above set of equations with the nonlinear fluxes $\mathbf{F}$ and $\mathbf{G}$ it should be linearized and internal iterations to account for the nonlinearity should be imposed at each time step. The resulting set of linear equations is solved via the LUSGS technique with the LU-decomposition by Jameson and Turkel which for the first time has been considered for arbitrary grids in [10,11] and those ideas were recently presented in more details in [12]. In [10–12] it has been shown that the algorithm does not require not only any actual matrix inversions but also any Jacobian computations at all. The two-step algorithm for one iteration may be written as follows:

$$\delta \mathbf{U}_i^{s,*} = -\frac{\Delta t}{\mathrm{Vol}_i + d_i}\{\mathrm{RHS}_i^s + 0.5 \sum_{j \in L(i)} \frac{1-\tau_j}{2}(\delta \mathbf{F}_j^{s,*} - r_j^s \delta \mathbf{U}_j^{s,*})\}, \quad (7)$$

$$\delta \mathbf{U}_i^s = \delta \mathbf{U}_i^{s,*} - \frac{0.5\Delta t}{\mathrm{Vol}_i + d_i} \sum_{j \in U(i)} \frac{1-\tau_j}{2}(\delta \mathbf{F}_j^s - r_j^s \delta \mathbf{U}_j^s), \quad (8)$$

where the right-hand-side RHS_i^s is given by

$$\mathrm{RHS}_i^s = \frac{\mathbf{U}_i^{n+1,s} - \mathbf{U}_i^n}{\Delta t}\mathrm{Vol}_i + \sum_{j(i)} \mathbf{F}(\mathbf{W}_{ij}^{n+1,s}) \cdot \mathbf{S}_{ij} - \sum_{j(i)} \mathbf{G}(\tilde{\mathbf{V}}^{n+1,s}) \cdot \mathbf{S}_{ij} \ . \quad (9)$$

Since we consider unsteady flows, the time derivative is included to the right-hand side. Other notations are: $d_i = 0.5\Delta t[(1-\tau_i)/2]\sum_{j(i)} r_i^s$; $r^s = (|\mathbf{u} \cdot \mathbf{n}_{ij}| +$

$c + 2\mu/\rho/|\boldsymbol{ij} \cdot \boldsymbol{n}_{ij}|)|\boldsymbol{S}_{ij}|$; $\delta \boldsymbol{U}^s = \boldsymbol{U}^{n+1,s+1} - \boldsymbol{U}^{n+1,s}$ – the iterative residual; $\delta \boldsymbol{F}^{s,*} = \boldsymbol{F}(\boldsymbol{U}^{n+1,s} + \delta \boldsymbol{U}^{s,*}) - \boldsymbol{F}(\boldsymbol{U}^{n+1,s})$; $\delta \boldsymbol{F}^s = \boldsymbol{F}(\boldsymbol{U}^{n+1,s} + \delta \boldsymbol{U}^s) - \boldsymbol{F}(\boldsymbol{U}^{n+1,s})$; $\boldsymbol{u} = (u_x; u_y)$, c, μ, and $\boldsymbol{n}_{ij}$ are the velocity vector, the speed of sound, the coefficient of dynamic viscosity and the normal vector of a control volume face, respectively; s is the superscript for iteration number. The distribution of the neighboring nodes $j(i)$ into lower $L(i)$ and upper $U(i)$ subgroups may be carried out in a number of ways taking into account specific grid structures and other considerations.

The above technique is very general and may be applied in a variety of ways. In the context of the present paper it is used with $\tau = 1$ (fully explicit mode) on the external triangular unstructured mesh while the hybrid explicit/implicit procedure with consecutive 'up' and 'down' sweeps is performed along the near-wall grid stations (in this case, the subdivisions $L(i)$ and $U(i)$ are obviously formed). The procedure ensures a smooth, gradual transition from fully explicit approximation at the outer end of the station to almost fully implicit approximation at the solid surface. It is important to note that we treat implicitly only fluxes along grid stations. All other fluxes are approximated explicitly (i.e. they appear only in the right-hand-side). Thus, the numerical procedure on the grid stations is a unidirectional (see [13] for more details on unidirectional techniques for viscous flows) explicit/implicit technique. In contrast to our previous fully-split approach on near-wall structured meshes [1–4], it is unsplit and provides a 'seamless' coupling with the numerical integration on the fully unstructured mesh away from solid surfaces.

It may be suggested from linear analysis to evaluate τ_i as $(\Delta t_i)^n_{\mathrm{CFL}}/\Delta t$ if the ratio is less than 1, and as $\tau_i = 1$ otherwise, where $(\Delta t_i)^n_{\mathrm{CFL}}$ is the time step base on the CFL condition for the node i. For a similar (but first-order in time in the implicit mode) scheme [7] it is noted that non-linear instability may develop. In [9] it is suggested, as a remedy, to evaluate $(\Delta t_i)_{\mathrm{CFL}}$ at the weighted level τ: $\boldsymbol{U}^\tau = \tau \boldsymbol{U}^n + (1 - \tau)\boldsymbol{U}^{n+1}$. The issue will be elaborated further in subsequent publications.

3 Concluding Remarks

An advantage of the approach is a possibility of simple generalization for 3D flows. A triangular unstructured grid covering solid surfaces in 3D provides a base on which grid stations could be grown up for each node, thus forming a prismatic near-wall mesh. Such a grid can be adapted in a natural way by inserting/deleting nodes on the quasi-2D unstructured surface mesh and the respective grid stations. The rest of the computational domain can be discretized using adaptive tetrahedral grid. The space and time integration can be done exactly in the same way as discussed in the previous section: using the explicit predictor-corrector or hybrid explicit/implicit LUSGS scheme for tetrahedral grid and the unidirectional explicit/implicit LUSGS procedure along near-wall grid stations.

The proposed approach leads to an efficient, flexible, easy-to-handle code suitable for a wide range of problems. Preliminary results show good agreement with analytical solutions, experimental data and other computations.

Acknowledgement

The work is partially supported by the Russian Foundation for Basic Research (grant 99-01-00735).

References

1. M.K. Beryozkina et al.: 'A Combined Adaptive Structured/Unstructured Technique for Essentially Unsteady Shock-Obstacle Interactions at High Reynolds Numbers'. In: *Proceedings of the 5th International Conference on Hyperbolic Problems: Theory, Numerics, Applications, Stony Brook, New York, USA, June 13–17, 1994*, ed. by J. Glimm et al. (World Scientific Publishing, Singapore 1996) pp. 464–470
2. D. Ofengeim et al.: 'A Locally Adaptive Structured/Unstructured 2D/3D Navier-Stokes Finite-Volume Solvers for Steady and Unsteady Compressible Flows.' In: *Computational Fluid Dynamics'96, Proceedings of the 3rd European Conference on CFD, Paris, France, September 9-13, 1996*, ed. by J.A. Desideri et al. (John Wiley & Sons, Chichester 1996) pp. 187–192
3. D. Drikakis, D. Ofengeim, E. Timofeev, P. Voinovich: J. Fluid & Structures **11**, 665 (1997)
4. D.Kh. Ofengeim, E.V. Timofeev, P.A. Voinovich, Yu.P. Golovachov: 'Numerical Investigations of Unsteady Viscous-Inviscid Interactions Induced by Shock Wave Diffraction Over Bodies'. In: *Proceedings of the 22nd International Symposium on Shock Waves, London, UK, July 18-23, 1999*, ed. by G.J. Ball et al. (University of Southhampton 2000) vol. 2, pp. 1053–1058
5. A.V. Rodionov: Zh. Vychisl. Mat. i Mat. Fiziki (J. Comp. Math. and Math. Phys.) **27**, 1853 (1987) (in Russian)
6. E.F. Toro: *Riemann Solvers and Numerical Methods for Fluid Dynamics*, 2nd edn. (Springer, Berlin 1999)
7. J.P. Collins et al.: J. Comp. Phys. **116**, 195 (1995)
8. P.J. O'Rourke, M.S. Sahota: J. Comp. Phys. **143**, 312 (1998)
9. I. Men'shov, Y. Nakamura: 'A Hybrid Explicit/Implicit Unconditionally Stable Method for Viscous Compressible Flows'. In: *CD-ROM Proc. of 13th National CFD Symp., Tokyo, Japan, December 21–23, 1999*, (Chuo University) paper Nr. F05-3. In: *Special Publication of National Areospace Laboratory, SP-44, Dec. 1999*, pp. 315–321
10. I. Men'shov, Y. Nakamura: 'An Implicit Advection Upwind Splitting Scheme for Hypersonic Air Flows in Thermochemical Nonequilibrium'. In: *Collection of Techinical Papers of the 6th International Symposium on CFD, Lake Tahoe, USA, September 4-8, 1995*, vol. I, pp. 815–820
11. I. Men'shov, Y. Nakamura: 'Implementation of the LU-SGS Method for an Arbitrary Finite Volume Discretization'. In: *Proceedings of 9th National CFD Symposium, Tokyo, Japan, December 21–23, 1995*, (Chuo University) pp. 123–124
12. I. Men'shov, Y. Nakamura: Computers & Fluids **29**, 595 (2000)
13. Yu.P. Golovachov: *Numerical Simulation of viscous Shock Layer Flows*, (Kluwer Academic Publishers, Dordrecht 1995)

Part IV

Mesh Generation

A Three-Dimensional Hybrid Grid: DRAGON Grid

Meng-Sing Liou[1] and Yao Zheng[2]

[1] NASA Glenn Research Center at Lewis Field, Cleveland, OH 44135, USA.
[2] Taitech, Inc., NASA Glenn Research Center at Lewis Field, Cleveland, OH 44135, USA.

1 Introduction

For a practical three-dimensional geometry, the time spent in generating a grid represents a serious bottleneck in the entire CFD analysis cycle. Hence, grid generation is an area where significant payoff can be realized. Furthermore, high quality grids for encompassing viscous regions are essential to yielding an accurate and efficient solution.

Overset and unstructured grid methods schemes currently are two main-stream approaches for dealing with complex geometry.

The overset grid method can yield structured grids with good quality, such as orthogonality and smoothness, and with ease to control grid spacing. However, the nonconservative interpolations in the overlapped region can give rise to spurious solution, especially through regions of sharp gradients.

The unstructured grid method is also flexible to generate grids around complex geometries. However, the unstructured grid method has been shown to be memory and computation intensive [3]. Also, it has been known that prismatic meshes are necessary to accurately resolve viscous layers around bodies.

The DRAGON grid method [8,5] will result in structured grids in the major portion of the computation domain and only small regions filled with unstructured grids. It has three important advantages: (1) preserving strengths of the Chimera grid, (2) eliminating difficulties encountered in the Chimera scheme, and (3) enabling grid communication in a fully conservative and consistent manner. Consequently, the DRAGON grid method can deliver a high quality grid for a complex multi-components geometry in an efficient and robust manner.

2 DRAGON Grid

In [8,5], the DRAGON grid method was first proposed and demonstrated for 2D problems. It adapted the thinking the Chimera grid method. However, it is more general and robust than the Chimera grid method since it does not require overlapping of grids and it completely avoids the issues about orphan points and bad quality points, which usually are the most time consuming part of the Chimera method. Figure 1 shows the connection between the Chimera and the DRAGON grids. In the overlapped region of the Chimera grid, a hole is created using the PEGSUS code [9] and subsequently replaced with triangular meshes in the DRAGON grid.

Figure 2 shows an interface connecting both structured and unstructured grids. The numerical fluxes, evaluated at the cell interface, are based on the conditions of neighboring cells (denoted as L and R cells, respectively). Consequently, the interface fluxes which have been evaluated in the unstructured grid code can now be directly applied in computing the cell volume residuals for the structured grid. Thus the DRAGON grid method provides fluxes on the same basis for both the structured and unstructured grids, completely avoids interpolations, and guarantees conservation of fluxes locally (element by element), and hence globally.

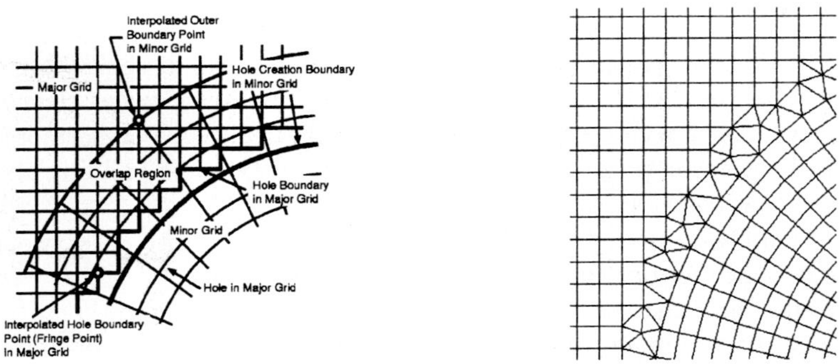

Fig. 1. Comparison of the Chimera grid (Left) and the DRAGON grid (Right).

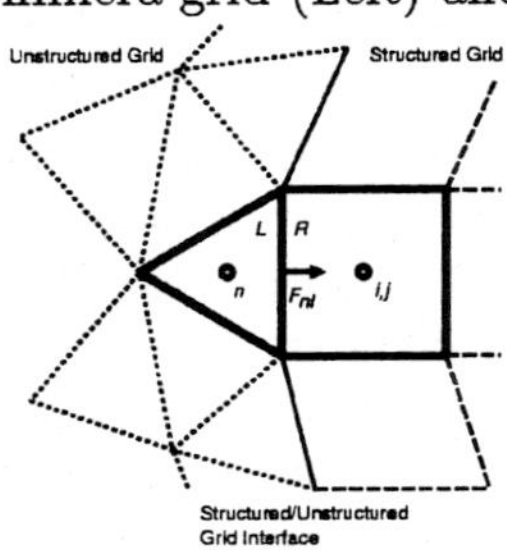

Fig. 2. Fluxes at the cell face connecting the structured and unstructured grids.

3 3D DRAGON Grid Generation

The extension of the concept of DRAGON grid [8,5] to that in three-dimensional space is straightforward. However, more difficulties are anticipated in actual **implementation**. Unique challenges for the case of three dimensions exist in various aspects, such as unstructured grid generation algorithms and code implementation. The development of the three-dimensional DRAGON gridding methodology is emphasized in this paper.

The faces of the structured grid in general do not necessarily match with the faces of the unstructured grid in the same location for three-dimensional cases, as illustrated in Fig. 3. This restriction may arise from the grid generation stage or the requirement to adapt to the physical phenomena. In Fig. 3, Points 1, 2, 3, 4, 5 and 6 on the structured grid are made coincident with the corresponding ones on the unstructured grid. Therefore, a scheme can be easily developed to meet flux conservation at common faces without resorting to interpolation of fluxes.

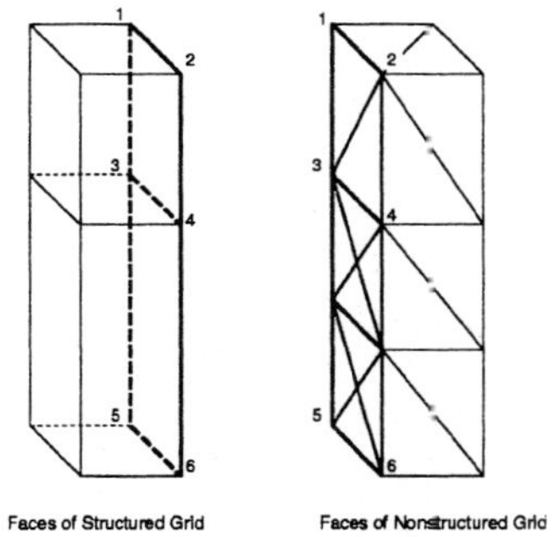

Fig. 3. Interface between structured and nonstructured grids in a three-dimensional DRAGON grid.

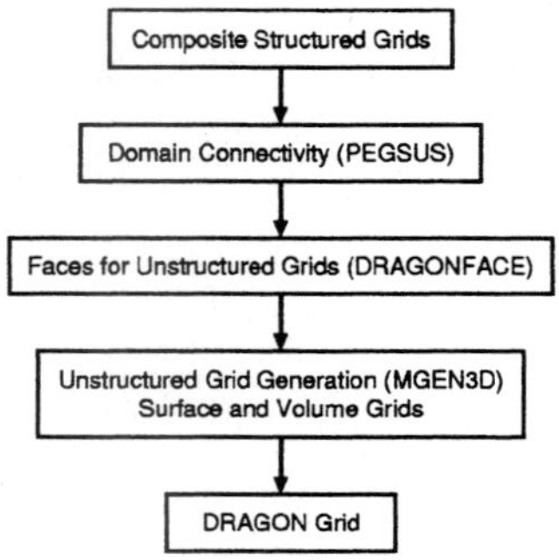

Fig. 4. Program modules involved in the three-dimensional DRAGON grid generation.

Fig. 4 shows the relationships among the modules involved, where two main modules, called DRAGONFACE and MGEN3D, are created in the three-dimensional DRAGON grid generation.

DRAGONFACE, a FORTRAN code, reads in composite structured grids with IBLANK values assigned by the PEGSUS code and writes out topology information of the faces as a surface description for the unstructured grid. Primary steps involved in the DRAGONFACE procedure include: identification of new faces, creation of isolated edges of the surface, and filling openings of the surface to ensure closedness of the surface.

MGEN3D is a C code. It inputs the surface description from the DRAGONFACE procedure and generates volumetric unstructured meshes and provides topology information of the corresponding surface grids, giving data communication at the interface between structured and unstructured grids. Main steps involved in the MGEN3D procedure include: spacing-source creation, three-dimensional surface triangulation, face reorientation, domain classification, and generation of volumetric grids, see [10,6].

To demonstrate the capability of the DRAGON grid code, we consider a typical film-cooled turbine vane, it includes the vane, coolant plena, and 33 holes inside of the vane. Figure 5 depicts the resulted DRAGON grid, where the connecting regions between the 33 holes and the flow domain are filled with unstructured grids. It gives a deep insight into this DRAGON grid by means of cutting through the grid, where attention has been paid to the leading edge region. The 33 individual structured grids for the holes have been created without trying to topologically join them to the grids representing the external flow domain. In this regard, the DRAGON grid scheme could be considered as a more flexible and easier approach.

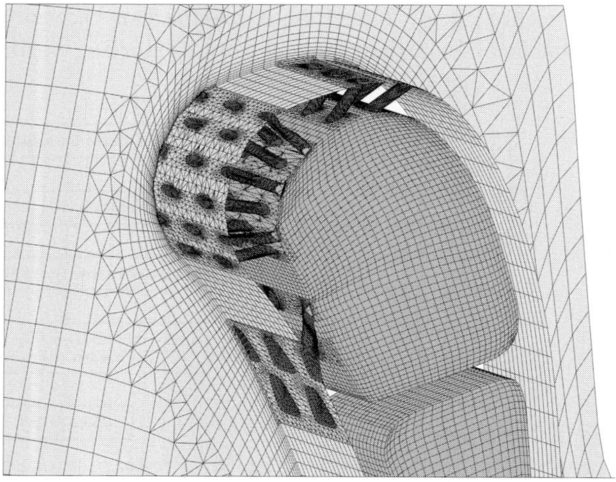

Fig. 5. Cut-through view of the DRAGON grid in the leading edge region of the film-cooled turbine vane.

4 Flow Solver

The flow code, termed DRAGONFLOW, is made up of two well-validated NASA codes, OVERFLOW [1] and USM3D [2], which are respectively structured- and unstructured-grid codes. A significant effort has been spent to add an interface and modify both codes, more so on the OVERFLOW code. The results presented in this paper also represents part of the validation process of the DRAGONFLOW code. A spatial-factored implicit algorithm is applied in the OVERFLOW code, while a point implicit one in the USM3D code. Since we are only interested in seeking steady-state solutions, the differences of convergence rate due to the incompatibility between the structured and unstructured regions can be tolerated. As for the spatial discretization, we use the AUSM$^+$ flux scheme [7], to express the inviscid flux at the cell faces in both codes. Also, the viscous fluxes are approximated, as usual, by a centered scheme.

5 Test Cases

5.1 Supersonic Flow in a Converging Duct

First we consider a supersonic flow of Mach 3 through a convergent channel, where the top and front side walls are bent by a wedge angle of 10 degrees respectively, thus creating two wedge shocks of equal strength and subsequent interaction between them. A strip of unstructured-meshed region is placed in the mid-section of the channel (not shown). The two wedge shocks intersect and generate a corner shock region manifested by the slip lines emanating from the triple point. As seen in Fig. 6, this region progressively becomes larger as the wedge shocks sweep towards the opposite walls. Eventually these two wedge shocks reflect and interact with the flow previously generated by the corner shocks, making the flowfield even more complicated.

The result on a single structured grid is also included for comparison. We see that both sets of results are essentially the same, except in the core region characterized by the slip lines.

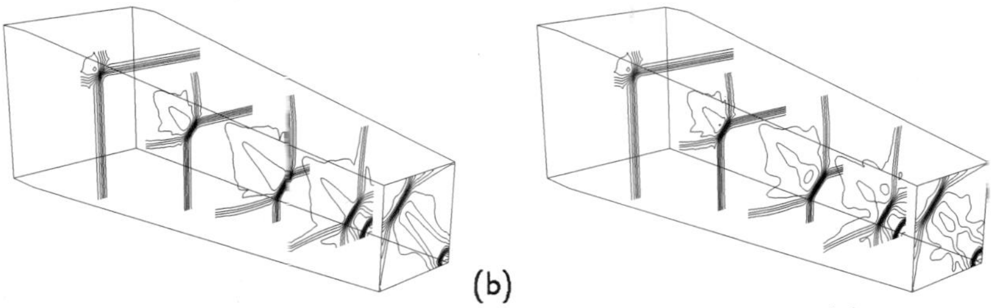

(a) (b)

Fig. 6. Mach number distribution of the flow at various stations: (a) Single grid; (b) DRAGON grid.

5.2 Viscous Flow through Annular Cascades

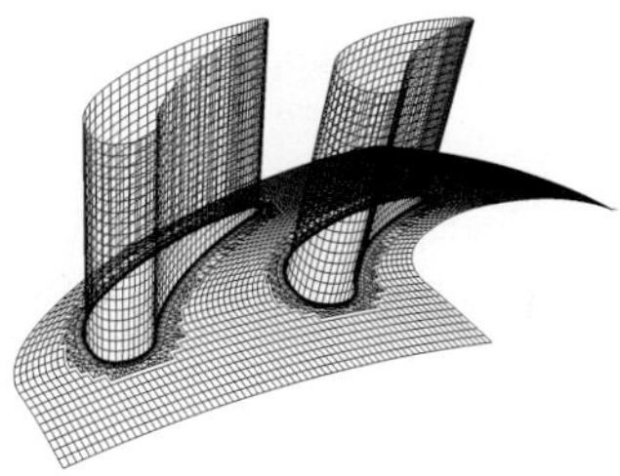

Fig. 7. DRAGON grid of the annular turbine cascade.

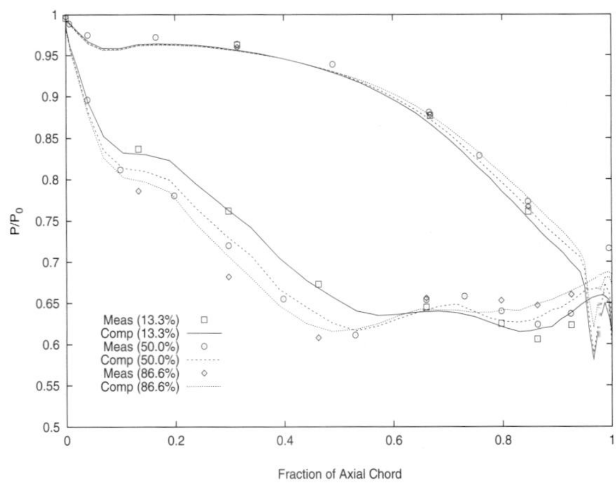

Fig. 8. Computed and measured static pressure distributions at 13.3%, 50% and 86.6% of span from hub.

Now, we give the results of a flow over turbine stator vanes [4]. Figure 7 shows the DRAGON grid, which consists of a background H-type grid, an O-type viscous grid around the vane, and an unstructured grid region between them. The static pressure distributions of the present DRAGON grid solution are plotted in Fig. 8 and they are in very good agreement with the data [4] measured at three spanwise locations, 13.3%, 50%, and 86.6% respectively.

6 Concluding Remarks

We have presented the extension of the DRAGON grid method into three-dimensional space. This method preserves the advantageous features of both the structured and unstructured grids, while eliminating or minimizing their respective shortcomings. As a result, it is amenable to **quickly** creating **quality** viscous grids for various individual components with **complex** shapes found in an engineering system.

Furthermore, the flow solutions confirm conservation of fluxes through the interfaces of structured-unstructured regions, and the results are in very good agreement with the measured data, thus demonstrating the reliability of the method.

Acknowledgment

This work is supported under the Turbomachinery and Combustion Technology, managed by Robert Corrigan, NASA Glenn Research Center.

References

1. P. G. Buning et. al.: 'OVERFLOW user's manual, version 1.8f', Unpublished NASA report (1998)
2. N. T. Frink: AIAA J. **36**, 1975 (1998)
3. F. Ghaffari: AIAA Paper 94-0163 (1994)
4. L. J. Goldman, R. G. Seasholtz: NASA Technical Paper 2018 (1982)
5. K.-H. Kao, M.-S. Liou: AIAA J. **33**, 1809 (1995)
6. R. W. Lewis, Y. Zheng, D. T. Gethin: Computer Methods in Appl. Mech. and Eng. **134**, 285 (1996)
7. M.-S. Liou: J. Comput. Phys. **129**, 364 (1996)
8. M.-S. Liou, K.-H. Kao: NASA TM 106709 (1994)
9. N. E. Suhs, R. W. Tramel: 'PEGSUS 4.0 user's manual', AEDC-TR-91-8 (1991)
10. Y. Zheng, R. W. Lewis, D. T. Gethin: Computer Methods in Appl. Mech. and Eng. **134**, 269 (1996)

Grids and Solutions from Residual Minimisation

Hiroaki Nishikawa, Mani Rad, and Philip Roe

Department of Aerospace Engineering,
University of Michigan, Ann Arbor, MI 48109, USA

Abstract. In this paper, a least-squares method that incorporates node movement is presented. For nonlinear hyperbolic problems with bilinear flux functions, a class of quadrature formulae for fluctuation computation is proposed which allows a stable structure of discrete shocks, yet preventing nonphysical shocks. Computational results for Burgers' equation are shown to demonstrate its sharp shock capturing ability.

1 Introduction

One way to construct solutions of partial differential equations on triangular grids is to minimise residuals in a suitable norm with respect to nodal positions as well as solution values. It has been shown that such a method is capable of producing exact solutions for linear hyperbolic equations, automatically adjusting the mesh into a characteristic configuration [1–4]. It is an advantage of the method that a shock can be captured with many fewer nodes than the usual curvature adaptation [5]. However, its extension to nonlinear problems is not yet well-developed because of the difficulty in forming a characteristic mesh at nonlinear shocks. In this paper, we propose a way to accomplish this, i.e. a way to capture exact nonlinear shocks.

2 Quadrature Formulae

We consider sets of two-dimensional conservation laws of the form

$$\partial_x \mathbf{F}(\mathbf{w}) + \partial_y \mathbf{G}(\mathbf{w}) = 0 \tag{1}$$

where $\mathbf{F}$, $\mathbf{G}$, $\mathbf{w} \in \mathcal{R}^m$ in which each component of the fluxes is a bilinear function of the components of $\mathbf{w}$.

$$\mathbf{F}(\mathbf{w}) = \mathbf{w}^t \mathbf{C} \mathbf{w}, \qquad \mathbf{G}(\mathbf{w}) = \mathbf{w}^t \mathbf{D} \mathbf{w} \tag{2}$$

where $\mathbf{C}$ and $\mathbf{D}$ are constant symmetric third-order tensors. This structure includes the Euler equations of compressible inviscid flow if $\mathbf{w}$ is taken to be Roe's parameter vector. We are interested in solving (1) by a least-squares method on triangular grids in which the first step is to compute the fluctuation.

$$\mathbf{\Phi}_{123} = \int\!\!\int_{123} \left[\partial_x \mathbf{F}(\mathbf{w}) + \partial_y \mathbf{G}(\mathbf{w})\right] dx dy = \oint_{123} \left[\mathbf{F}(\mathbf{w}) dy + \mathbf{G}(\mathbf{w}) dx\right] \tag{3}$$

where the data is available at the vertices 1,2,3. The second step is then to distribute some fraction of the fluctuation to each vertex. In this least-squares method, as will be explained in Section 5, these fractions sum to zero over any given element, so that the method will be conservative however the fluctuations are evaluated. Therefore we focus first on different properties of various evaluations. Consider a single term, arising from integrating just one component of the flux vector over on edge of the element.

$$\mathbf{F}_{12} = \int_1^2 \mathbf{F}(\mathbf{w})\, dy = \int_1^2 \mathbf{w}^t \mathbf{C} \mathbf{w}\, dy \tag{4}$$

To evaluate this integral there is a class of simple formulae;

$$\mathbf{F}_{12}(\alpha) = \frac{(y_2 - y_1)}{2} \left[(\mathbf{w}_1^t \mathbf{C} \mathbf{w}_2 + \mathbf{w}_2^t \mathbf{C} \mathbf{w}_1) + \alpha(\mathbf{w}_1 - \mathbf{w}_2) \mathbf{C} (\mathbf{w}_1 - \mathbf{w}_2) \right] \tag{5}$$

where α is a parameter that has only a second-order relative effect on the accuracy of the estimate as clearly seen.

3 Formulae for the Fluctuation

The formula for the fluctuation is obtained by summing up the contributions from three sides. Introducing the notation,

$$\mathbf{w}_{ij}(\alpha) = \frac{\alpha}{2}(\mathbf{w}_i + \mathbf{w}_j) + (1 - \alpha)\mathbf{w}_k \tag{6}$$

where i, j, k are cyclically permuted for $1, 2, 3$, and rearranging terms, we obtain

$$\begin{aligned} \boldsymbol{\Phi}_{123}(\alpha) = {} & \mathbf{w}_{31}^t(\alpha) \left\{ \Delta y_3 \mathbf{C} - \Delta x_3 \mathbf{D} \right\} (\mathbf{w}_3 - \mathbf{w}_1) + \\ & \mathbf{w}_{12}^t(\alpha) \left\{ \Delta y_2 \mathbf{C} - \Delta x_2 \mathbf{D} \right\} (\mathbf{w}_2 - \mathbf{w}_1) \end{aligned} \tag{7}$$

where Δy_i is the difference of y taken anticlockwise along the edge opposite to the node i, similarly for Δx_i. Note that this formula is not symmetric with respect to the vertices. There are two other rearrangements possible, and all three will give the same numerical value. This particular arrangement can be written

$$\begin{aligned} \boldsymbol{\Phi}_{123}(\alpha) = {} & \left\{ \frac{\partial \mathbf{F}}{\partial \mathbf{w}} \Delta y_3 - \frac{\partial \mathbf{G}}{\partial \mathbf{w}} \Delta x_3 \right\} (\mathbf{w}_3 - \mathbf{w}_1) + \\ & \left\{ \frac{\partial \mathbf{F}}{\partial \mathbf{w}} \Delta y_2 - \frac{\partial \mathbf{G}}{\partial \mathbf{w}} \Delta x_2 \right\} (\mathbf{w}_2 - \mathbf{w}_1) \end{aligned} \tag{8}$$

where the Jacobian matrices in the first term are evaluated at the state $\mathbf{w}_{31}(\alpha)$ and those in the second terms at the state $\mathbf{w}_{12}(\alpha)$, which shows that the parameter α acts on the Jacobian matrices. For example, taking $\alpha = 2/3$ we obtain the so-called conservative linearisation which has been utilised in the fluctuation splitting method [6].

3.1 Shock Recognition: $\alpha = 1$

Suppose that two of the states happen to be equal, $\mathbf{w}_1 = \mathbf{w}_2 = \mathbf{w}_c$. Then the second line of (7) vanishes and the first line becomes

$$\boldsymbol{\Phi}_{123}(\alpha) = \mathbf{w}_{31}^t(\alpha)\left\{\Delta y_3\mathbf{C} - \Delta x_3\mathbf{D}\right\}(\mathbf{w}_3 - \mathbf{w}_c). \tag{9}$$

For the special choice $\alpha = 1$, which gives $\mathbf{w}_{31}(\alpha) = \overline{\mathbf{w}}_{31} = (\mathbf{w}_3 + \mathbf{w}_c)/2$, this corresponds precisely with the Hugoniot condition,

$$\overline{\mathbf{w}}^t(S\mathbf{C} - \mathbf{D})\Delta\mathbf{w} = 0 \tag{10}$$

along the edge 31, provided the edge 12 is aligned in shock of the speed $dy/dx = S$. Note that this is independent of the position of the other node. However, unfortunately, this admits nonphysical shocks as well.

3.2 Special Properties: $\alpha = 0$

For scalar problems, the choice $\alpha = 0$ has a special property. With $\alpha = 0$, the fluctuation becomes

$$\begin{aligned}\boldsymbol{\Phi}_{123}(0) = {} & \mathbf{w}_2^t\left\{\Delta y_3\mathbf{C} - \Delta x_3\mathbf{D}\right\}(\mathbf{w}_3 - \mathbf{w}_1) + \\ & \mathbf{w}_3^t\left\{\Delta y_2\mathbf{C} - \Delta x_2\mathbf{D}\right\}(\mathbf{w}_2 - \mathbf{w}_1)\end{aligned} \tag{11}$$

where all quantities are thought of as scalar. Now suppose the edge 12 is aligned with the characteristic whose speed evaluated at the state 2, i.e. *on the edge itself*. Then the first term vanishes. And the second approximates the characteristic equation $d\mathbf{w} = 0$ exactly, again for any position of the third node.

Another important property associated with this choice is the ability to compute physical rarefaction. To see this, consider Burgers' equation $\partial_t u + \partial_x f = 0$ where $f = u^2/2$. Discretising the spatial derivative term by the quadrature formula (5), rearranging the terms, we find

$$\partial_t u_j + \frac{f_{j+1} - f_{j-1}}{2\Delta x} = (1-\alpha)\Delta x^2 \frac{u_{j+1} - u_{j-1}}{2\Delta x}\frac{u_{j+1} - 2u_j + u_{j-1}}{\Delta x^2} \tag{12}$$

which is a second order discretisation of

$$\partial_t u + \partial_x f = (1-\alpha)\Delta x^2\,\partial_x u\,\partial_x^2 u. \tag{13}$$

This shows that for diverging characteristics, $\partial_x u > 0$, taking $\alpha < 1$ gives positive dissipation. Experimentally we find that α needs to be well below unity, and that zero is a good choice.

4 Detecting Compression/Expansion

One possible way to take advantage of the formulae for the fluctuation is to use $\alpha = 1$ for elements in shocks and $\alpha = 0$ for others, thus capturing shocks exactly at the same time avoiding possible rarefaction shocks. The decision can be made based on the rate of change of the triangle area due to the virtual vertex motion caused by the characteristic speeds. Imagine that a triangle is convected in a characteristic field. If the characteristics are diverging, implying expansion, the triangle area would increase. On the other hand, if the characteristics are converging, implying compression, the triangle area would decrease. It is easy to show that the rate of change of the area is given by

$$\frac{dS_T}{dt} = \frac{1}{2} \sum_{i=1,2,3} \boldsymbol{\lambda}_i \cdot \mathbf{n}_i \tag{14}$$

where $\mathbf{n}_i$ is the scaled inward normal vector opposite to the vertex i, and $\boldsymbol{\lambda}_i$ is a characteristic speed vector at vertex i. For system of equations, we compute this for each wave-like component.

5 Least-Squares Formulation

The solutions are sought that minimise the norm

$$\mathcal{F} = \sum_{T \in \{T\}} F_T = \frac{1}{2} \sum_{T \in \{T\}} \frac{\boldsymbol{\Phi}_T^t Q_T \boldsymbol{\Phi}_T}{S_T} \tag{15}$$

over a set $\{T\}$ of triangular elements that divides the domain of interest. Q_T is a positive definite symmetric matrix that assigns relative weight to the different equations [1]. The change made to each vertex $\{j\}$ is the sum of the contribution from the surrounding triangles $\{T_j\}$, and can be written in the fluctuation splitting format as follows.

$$\delta \mathbf{w}_j = -\omega_u \sum_{T \in \{T_j\}} \mathcal{A}_j^T \boldsymbol{\Phi}_T \tag{16}$$

$$\delta \mathbf{x}_j = -\omega_x \sum_{T \in \{T_j\}} \left\{ \mathcal{B}_j^T \boldsymbol{\Phi}_T - \frac{F_T}{2S_T} \mathbf{n}_j \right\} \tag{17}$$

where $\mathbf{w}_j$ is a solution vector at the node, $\mathbf{x}_j$ is the nodal position vector, and ω_u and ω_x are small constants. The second term in (17) comes from differentiating the weight $1/S_T$ in the norm. Note that the scheme is equivalent to the steepest descent method. $\mathcal{A}_j^T$ and $\mathcal{B}_j^T$ are the distribution matrices given by

$$\mathcal{A}_j^T = \frac{1}{S_T} \left(Q_T \frac{\partial \boldsymbol{\Phi}_T}{\partial \mathbf{w}_j} \right)^t \tag{18}$$

$$\mathcal{B}_j^T = \frac{1}{S_T} \left(Q_T \frac{\partial \boldsymbol{\Phi}_T}{\partial \mathbf{x}_j} \right)^t . \tag{19}$$

Here the computation of the derivative in $\mathcal{A}_j^T$ requires careful consideration. We have observed that taking the derivative straightforwardly can result in completely wrong solutions. As suggested in [7], it is important to linearise the equations first and then apply the least-squares method. This means that we take the derivatives assuming $\mathbf{w}_{ij}(\alpha)$ are constant in (7) because these quantities act on the Jacobian matrix as mentioned before. Yet this raises another important point of consideration. Recall that the fluctuation has two other rearrangements other than (7). The derivative computed as above now depends on this arrangement. Then it would be reasonable to take the arithmetic average of the three possible forms and take the derivative, which yields

$$\frac{\partial \Phi_T}{\partial \mathbf{w}_1} = \frac{\mathbf{w}_{31}^T(\alpha)}{3}\left[(\Delta y_1 - \Delta y_3)\mathbf{C} - (\Delta x_1 - \Delta x_3)\mathbf{D}\right] + \frac{\mathbf{w}_{12}^T(\alpha)}{3}\left[(\Delta y_1 - \Delta y_2)\mathbf{C} - (\Delta x_1 - \Delta x_2)\mathbf{D}\right]. \tag{20}$$

6 Results

Preliminary results are available for Burgers' equation $\partial_y u + u\,\partial_x u = 0$, for which Q_T was taken to be unity. Figures 1 and 2 show the final grid and solution for a right-moving curved shock for the boundary conditions; $u = 3$ for $x \leq -0.8$ and $\frac{30}{17}(x-1)$ for $x \geq -0.7$ where the data is interpolated linearly between $x = -0.8$ and $x = -0.7$, modeling an initial discontinuity. The grid and the solution were obtained by repeating the following cycle.

1. Converge the solution on a fixed grid
2. Assign α to each element, depending the sign of (14)
3. Remove undesirable nodes
4. Update solutions and coordinates, with edge swappings interleaved (2000 iterations maximum)

and the method terminates when the changes to solutions and coordinates are both small in the step 4 or the maximum number of cycles is reached, say 20. The edge swapping is based on the norm reduction, which attempts to create a characteristic mesh as clearly seen in the results. Also, because two edges sometimes try to represent the same characteristic we implemented a scheme for removing redundant nodes. The method converged at 10 cycles. And the final values of α are 1 for the elements forming a shock, and 0 elsewhere.

A grid aligned expansion wave was computed with $u = -0.8$ on the left and $u = 0.8$ on the right, to demonstrate the ability of the method to compute entropy-satisfying solutions. The values of α are automatically set to be zero everywhere in the step 2, and as can be seen the final solution is the correct smooth expansion fan. The method converges at just 1 cycle. Increasing the value of α, we found that a rarefaction shock appeared in the middle of the rarefaction and it became finally a perfectly-resolved rarefaction shock when $\alpha = 1$. This is consistent with the observation in section 3.2.

7 Conclusions

A least-squares method that incorporates node movement and a class of integration rules has been presented. The preliminary results demonstrated its ability

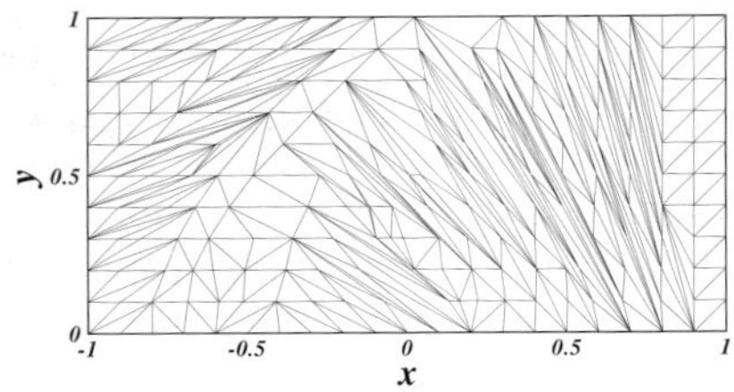

Fig. 1. The final grid for the curved shock.

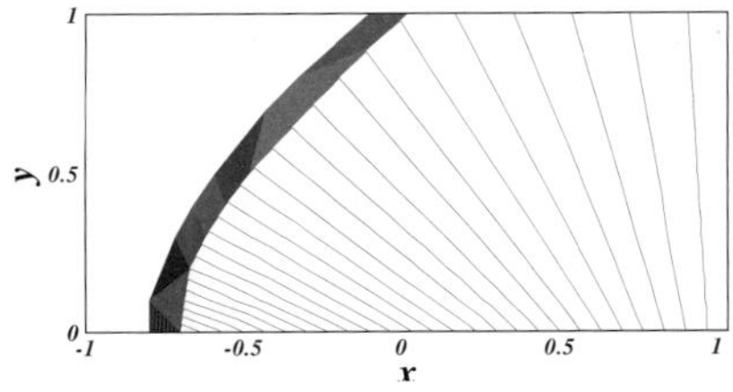

Fig. 2. Solution contours for the shock.

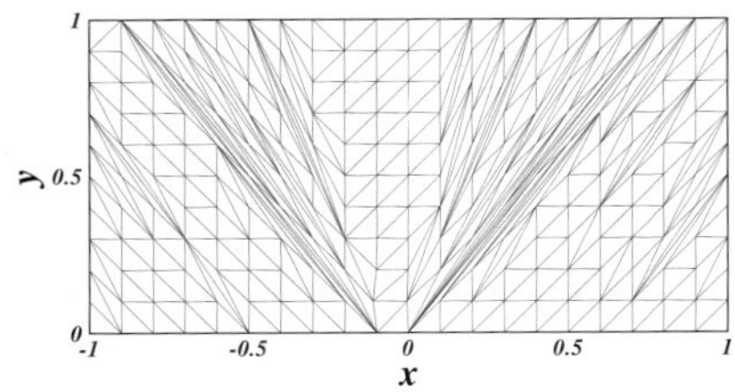

Fig. 3. The final grid for the expansion.

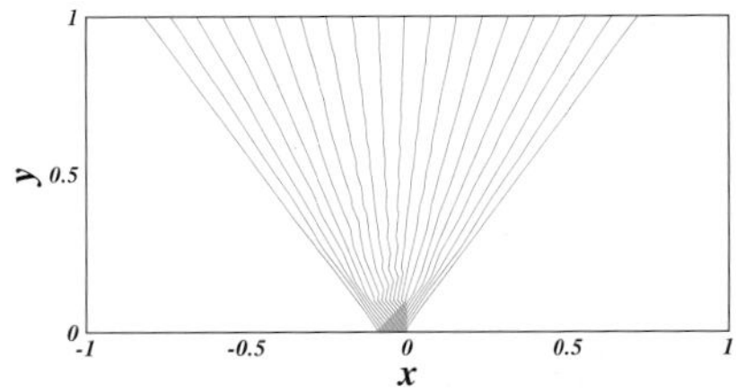

Fig. 4. Solution contours for the expansion.

to compute shocks extremely sharply as well as to avoid nonphysical shocks. In future work we will focus on improving the efficiency of the method, especially the mechanism for removing redundant nodes, and on the extension to systems of equations. In this latter task we expect the proper choice of Q_T in (15), which defines the minimization norm, to be important.

References

1. P. L. Roe: 'Compounded of Many Simples'. In: *Barriers and Challenges in Computational Fluid Dynamics*, ed. by V. Venkatakrishnan, M. D. Salas, Sukumar R. Chakravarthy (Kluwer Academic Publishers 1998) pp. 241-158
2. P. L. Roe: 'Fluctuation Splitting Scheme on Optimal Grids'. In: *AIAA 13th CFD Conference*, June 1997, Paper 97-2034
3. M. J. Baines, S. J. Leary, M. E. Hubbard: 'A Finite Volume Method for Steady Hyperbolic Equations'. In: *Finite Volumes for Complex Applications II, problems and perspectives, 2nd International Symposium at Duisburg, Germany, July 19–22, 1999*, ed. by R. Vilsmeier, F. Benkhaldoun, D. Hanel (Hermes Science Publications, Paris 1999) pp. 787–794
4. W. A. Wood, W. L. Kleb: 'On Multi-dimensional Unstructured Mesh Adaptation'. In: *AIAA 14th CFD Conference*, June 1999, Paper 99-3254
5. W. G. Habashi, J. Dompierre, Y. Bourgault, D. Ait-Ali-Yahia, M. Fortin, M. Vallet: In: Int. J. for Num. Meth. Fluids **32**, 725-744 (2000)
6. H. Deconinck, P. L. Roe, R. J. Struijs: In: Computers and Fluids **22**, No. 2/3, 215-222, (1993)
7. B. Jiang: *The Least-Squares Finite Element Method: Theory and Applications in Computational Fluid Dynamics and Electromagnetics* (Springer-Verlag New York 1998)

Part V

Compressible Flows

Calculations of Low-Mach-Number Viscous Flows without Preconditioning by the Space-Time CE/SE Method

Z.-C. Zhang[1], S.T. John Yu[1], S.-C. Chang[2], and P.C.E. Jorgenson[2]

[1] Mechanical Engineering Department, Wayne State University, Detroit, MI 48202
[2] NASA Glenn Research Center, Cleveland, OH 44135

Abstract. In this paper, we report the calculations of low-Mach-number viscous flows without preconditioning by the space-time CE/SE method. It is shown that the CE/SE method can sustain highly resolved solution for long-term calculations with large numbers of time steps. Several numerical examples are calculated to demonstrate the efficiency and accuracy of the CE/SE method in calculating low-speed flows.

1 Introduction

Many low-Mach-number flows contain both compressibility effects and low-speed regions, and the compressible equations must be used in the modling. For example, low speed flows with significant temperature variations are compressible due to density variation induced by heat addition. Although flow speeds are slow, one must employ the compressible flow equations to model these flow fields. However, it is well known that the conventional CFD methods, which can handle high-speed compressible flow easily, fail miserably when applied to low-Mach-number flows. In the past, due to wide applications of low-speed compressible flows, extensive studies have been conducted to construct viable numerical methods. Conventional approaches can be divided into two categories: the pressure-based methods and the density-based methods. The pressure-based methods are extension of the projection method for incompressible flows. The SIMPLE schemes are typical examples. The density-based methods are extension of methods for compressible flows with two modifications: (1) adding a pseudo compressibility term to the continuity equation, and (2) preconditioning the Jacobian matrix of the convection terms such that the eigenvalues are artificially brought closer to each other. Combination of the two special treatments enables one to fabricate an artificial hyperbolic system, which is amenable to be solved by conventional CFD methods in a limited number of time steps.

To date, the newly developed space-time CE/SE method has been used to calculate high-speed flows with shocks and aero acoustic flows. In this paper, we apply the CE/SE method to low-Mach-number compressible flows as well as incompressible by using the same flow solver for high speed flows. We show that numerical difficulties encountered when using the conventional CFD methods to calculate low-Mach-number flows do not exist if the space-time CE/SE method was used. Demonstrated by numerical examples, we show that the same CFD code, which has been used for shock capturing in high-speed flows, can be directly applied to low-Mach-number flows without any modification. We remark that the size of the time step employed in the present calculations is still restricted by the CFL number and the von Neumann number. Thus a large number of time steps are needed to march the flow solution to its full development. Note that when conventional methods without preconditioning were used, false solutions would be obtained due to excessive artificial damping and accumulation of errors. The preconditioning procedure will artificially enhance numerical wave propagation and thus reduce the number of time steps required for calculations. As such, the error accumulation by the numerical schemes is avoided. Therefore, the key issue here is *whether the numerical scheme employed can sustain highly accurate solution for a large number of time steps.* This issue is not restricted to low-speed flows only. In simulating a propagating normal shock over a long period of time, all upwind schemes show the so-called carbuncle effect, leading to numerical overflow.

The surprising finding reported here shows that the CE/SE method maintains the integrity of the flow solution when the simulation is carried out in a large number of time steps. In other words, the accumulation of errors is minimum, and the local and global flux conservation in a space-time sense has been truly maintained. In this paper, we first briefly summarize the CE/SE method for Navier Stokes equation. Then three numerical examples are included to demonstrate capabilities of the newly developed Navier-Stokes solver for solving these low-speed flows.

2 The CE/SE Method for the Navier Stokes Equations

Consider the following two-dimensional Navier-Stokes Equations,

$$\frac{\partial U_m}{\partial t} + \frac{\partial f_{im}}{\partial x} + \frac{\partial g_{im}}{\partial y} - \frac{\partial f_{vm}}{\partial x} - \frac{\partial g_{vm}}{\partial y} = 0 \tag{1}$$

where $m = 1, 2, 3, 4$ for the continuity, x and y momentum and the energy equations, and $\boldsymbol{U} = (\rho, \rho\mathrm{u}, \rho\mathrm{v}, \mathrm{E}_t)$. Here f_{im} and g_{im} are the inviscid fluxes, and f_{vm} and g_{vm} are viscous ones. Let $x_1 = x, x_2 = y$ and $x_3 = t$ be the coordinates of a three-dimensional Euclidean space E_3. The integral counterpart of Eq. (1) is

$$\oint_{S(V)} \boldsymbol{h}_m \cdot d\boldsymbol{s} = 0, \quad m = 1, 2, 3, 4 \tag{2}$$

Here $S(V)$ is the boundary surface of a space-time region V in E_3, and $\boldsymbol{h}_m = (f_{im}\text{-}f_{vm},\ g_{im}\text{-}g_{vm},\ U_m)$, and it can be decomposed into the following inviscid and viscous parts:

$$\boldsymbol{h}_m = \boldsymbol{h}_{im} - \boldsymbol{h}_{vm} = (f_{im}, g_{im}, U_m) - (f_{vm}, g_{vm}, 0) \tag{3}$$

To perform the space-time integration in two spatial dimensions, triangular spatial mesh is used. Refer to Fig. 1. The grid points are distributed at the center of each triangle. In connection with its three neighbors, we define three conservation elements (CEs) and one solution element (SE) associated with each mesh node. At point G', three CE$^{(l)}$ ($l = 1, 2, 3$) are the cylinders $ABGFA'B'G'F'$, $BCDGB'C'D'G'$ and $DEFGD'E'F'G'$. The SE is the union of four planes $A'B'C'D'E'F'$, $GG''F''F$, $GG''B''B$, $GG''D''D$ and their immediate neighborhood.

Inside each SE(j,n), the flow variables are assumed continuous, and the first-order Taylor series expansions is used to approximate $U_m(x,y,t)$, $f_{im}(x,y,t)$ and $g_{im}(x,y,t)$. To calculate the viscose fluxes, the midpoint rule is used. From Eq. (3), the third component of h_{vm} is a null. Thus, in calculating the viscous fluxes, we only need to calculate integrals over lateral surfaces in the space-time domain. For example, in CE$^{(1)}$, the quadrilateral cylinder $ABGFA'B'G'F'$, we only need to integrate the viscous terms over four lateral surfaces $ABA'B'$, $AFA'F'$, $GBG'B'$ and $GFG'F'$. Consider the integral of $\boldsymbol{h}_{vm}$ on the surface $AFA'F'$. Let $\Delta \boldsymbol{S} = (S_x, S_y, 0)$ be the unit outward normal vector of $AFA'F'$ multiplied by its area. Using the midpoint rule, we have

$$\int_{AFA'F'} \boldsymbol{h}_{vm} \cdot d\boldsymbol{s} \approx S_x f_{vm}\left((u_m)_Q, (u_{mx})_Q, (u_{my})_Q\right) + S_y g_{vm}\left((u_m)_Q, (u_{mx})_Q, (u_{my})_Q\right) \tag{4}$$

where Q is the centroid of surface $AFA'F'$. Because the surface $AFA'F'$ belong to the SE of point A, we can use the parameters at point A to approximate U_m, U_{mx} and U_{my} at point Q. Similar procedure is performed for surfaces $ABA'B'$, $GBG'B'$ and $GFG'F'$.

The space-time flux conservation over the three CE's provides three algebraic equations for the three unknowns, U, U_x and U_y at each mesh node. Since the viscous terms are included, the addition of the artificial damping as that in the a-ε scheme is optional. For flow with shocks, a re-weighting procedure is used for the jump condition, i.e., the a-α scheme [1-2]. This concludes the brief description the CE/SE method for solving the two-dimensional Navier-Stokes equations.

3 Numerical Results

Three low-Mach-number flows are calculated using the viscous CE/SE scheme. On the wall boundaries, a newly developed unified solid boundary treatment [6] is used. On the outlet and/or open boundaries, the non-reflective boundary condition is used.

3.1 Buoyancy-driven gas flows

The first example is a buoyancy-driven gas flow in a square box. The configuration consists of two insulated horizontal walls and two lateral walls with constant temperatures of T_h(left) and T_c(right). For a small temperature difference between these two vertical walls, this problem has been extensively studied based on the incompressible flow equations with Boussinesq model for buoyancy force. For a large temperature difference, the compressible formulation must be employed.

The flow features of this buoyancy-driven cavity flow depend on Raylaigh number R_a, Frounde number F_r, the aspect ratio of the cavity, and the temperature difference parameter ε. Here only one case is shown in Fig.2 with Raylaigh number $R_a = 10^5$ and temperature difference parameter $\varepsilon = 0.6$, which represents $T_h/T_c = 4$. The Frounde number and the aspect ratio are unity. 7200 triangles (cells) are used. The calculation reaches the steady-state solution after about 10,000 iterations. Fig.2 shows the velocity vectors and x-direction velocity distribution along the vertical centerline. The solution agrees well with previously reported data [7].

3.2 Driven cavity flow

This is a benchmark problem for incompressible viscous flow calculations. Here the full compressible Navier-Stokes equations are solved to demonstrate the capabilities of the CE/SE scheme at the incompressible limit. 7,688 triangles are used. Figure 3 shows velocity vectors, and x-velocity distribution along the vertical centerline at Re = 5000. This solution agrees well with Ghia's data [8].

3.3 Flows over a circular cylinder

The third example is an external flow over a circular cylinder. First, we consider $R_e = 40$, and a steady-state solution is shown in Fig. 3. Again, the full compressible equations are solved by the CE/SE method without preconditioning. The computational domain is [-5,15]×[-5, 5], and 10,092 triangles are used. Figure 4(a) shows the unstructured mesh near the circular cylinder. Figures 4(b)-(c) show the velocity vectors and Mach number contours of the flow solution. The location of the boundary layer separation on the cylinder surface and the size of the recirculating region ($L/d \approx 2.0$) compare favorably with the experiment data and previously reported results [9]. If the Reynolds number is increased to 200, the flow becomes unsteady, and Fig. 5 shows vorticity contours at two different times. The oscillating frequency compares well with the experimental data.

4 Concluding Remarks

In this paper, we report the applications of the CE/SE method to low-speed flows without preconditioning. Numerical results show that the present CE/SE method for Navier Stokes equations can be used for flows at all speeds without preconditioning. This surprising finding, mainly achieved by low numerical dissipation in calculating space-time flux conservation, is in contrast to the traditional wisdom of treating low-Mach number flows. It shows that the present

CE/SE solver can maintain solution integrity for calculation with a large number of time steps and error accumulation has been avoided by space-time flux conservation and a robust non-reflective boundary condition.

References

1. Chang, S.C., *J. Comp. Phys.*, **119** , pp. 295-324, 1995.
2. Chang, S.C., Wang, X.Y. and Chow, C.Y., *J. Comp. Phys.*, **156**, pp.89-136, 1999.
3. Zhang, Z.C., Yu, S.T., Chang, S.C., Himansu, A. and Jorgenson, P., AIAA Paper 99-3277, June. 1999.
4. Choi, D. and Merkle, C.L., *AIAA J.*, **23**, p.1518, 1985.
5. Choi, Y.H., Ph.D. Thesis, The Pennsylvania State University, 1989.
6. Chang, S.C., Zhang, Z.C., Yu, S.T. and Jorgenson, P., Proceeding of ICCFD, Kyoto, Japan 2000.
7. Yu, S.T., Jian, B.N., Wu, J. and Liu, N.S., *Comput. Methods Appl. Mech. Engrg.*, **137**, pp. 59-88, 1996.
8. Guia, U., Guia, K.N. and Shin, C.T., *J. Comp. Phys.*, **48**, pp. 387-411, 1982.
9. Dennis, S.C.R. and Chang, G.Z., *J. Fluid Mechs.*, **42**, 471-489, 1970.

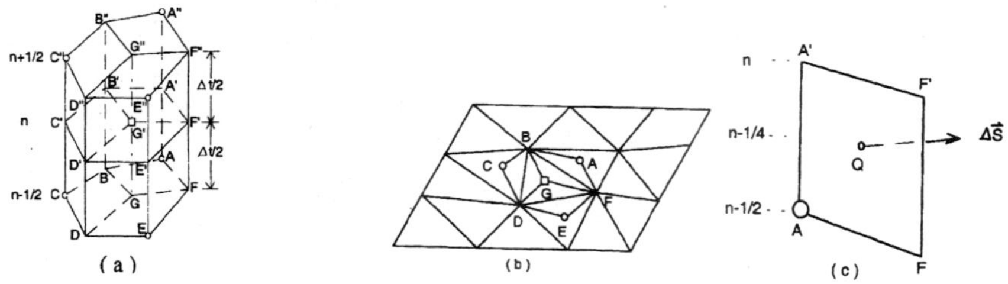

Fig. 1. A schematic of the CE/SE scheme: (a) a space-time mesh; (b) a triangular spatial mesh; (c) calculation of viscous fluxes.

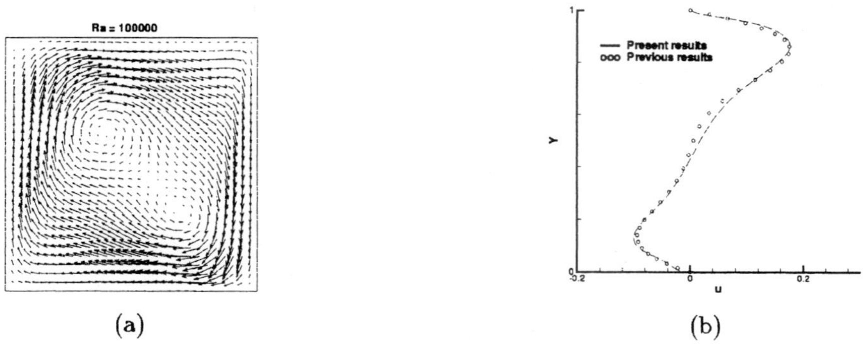

Fig. 2. Solution of the buoyancy-driven flow, (a) velocity vectors, (b) u along the vertical centerline.

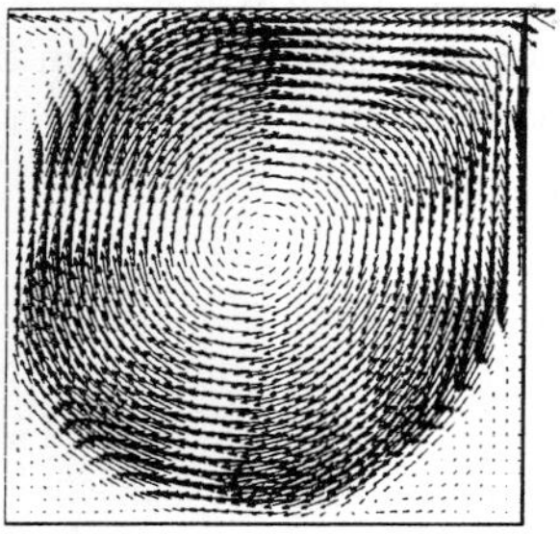

(a) velocity vectors.

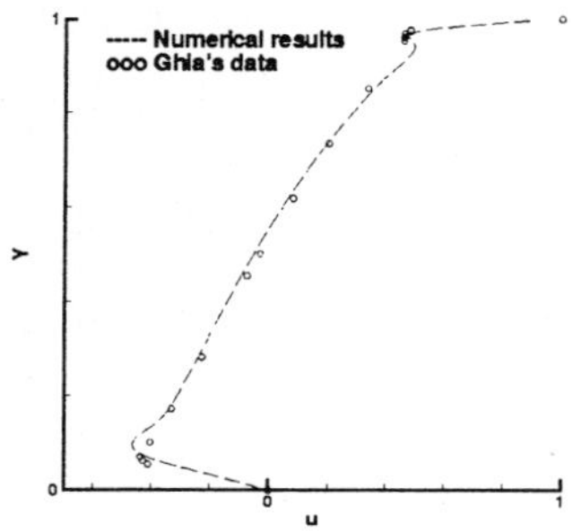

(b) u along the vertical centerline.

Fig. 3. Solution of a driven cavity flow.

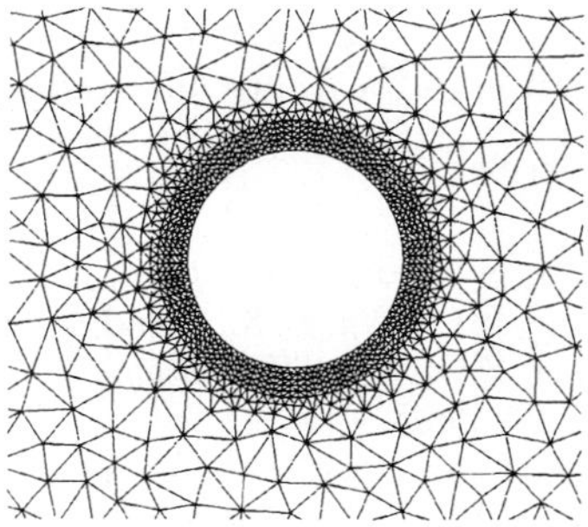

(a) mesh around the cylinder.

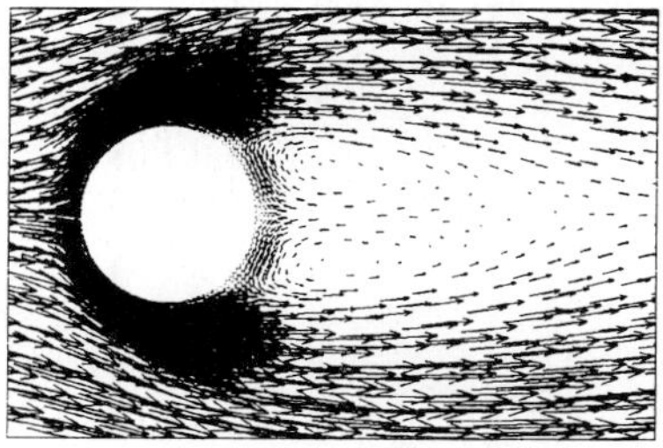

(b) velocity vectors.

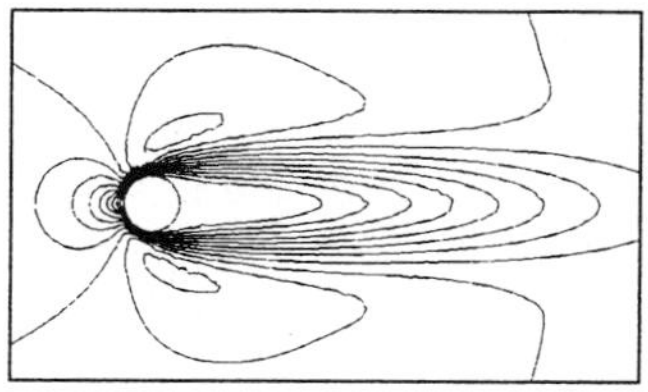

(c) Mach number contours.

Fig. 4. A flow over a cylinder ($R_e = 40$).

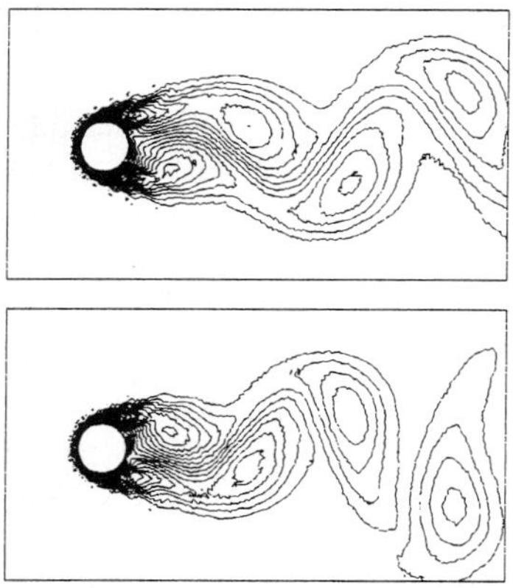

Fig. 5. Vorticity contours of the flow over a cylinder ($R_e = 200$) at two different time levels.

Time-marching Method for Complicated Compressible Flow Equations with Source Term

Satoru Yamamoto

Dept. of Aeronautics and Space Engineering, Tohoku University, Sendai 980-8579, Japan

1 Introduction

Time-marching methods for simultaneously solving compressible flow equations and additional fundamental equations of complicated physics, such as thermochemical nonequilibrium, nonequilibrium condensation, and magnet-plasma dynamics(MPD), are presented. Hypersonic thermochemical nonequilibrium flows can be simulated solving continuity equations of chemical species, momentum equations, a total energy equation, and a vibrational energy equation simultaneously. Chemical reactions and the relaxation of vibrational energy are modeled in the source term of those equations. Our research group proposed a high-resolution method for solving hypersonic thermochemical nonequilibrium flows[1]. Unsteady Type IV shock interference flows with thermochemical nonequilibrium effects were successfully calculated, too[2]. Transonic flows with nonequilibrium condensation have been calculated solving flow equations for gas and liquid phases. In most of them, the Euler equations are solved using the time-marching method, while the nonequilibrium wet-steam gas is approximated integrating the growth of droplet along each streamline. Therefore, it is difficult to solve all equations simultaneously using the time-marching method. Axisymmetric MPD flows have been generally calculated solving plasma flow equations and the equation of magnetic induction derived from the Maxwell's equations and the Ohm's law. In most of existing methods, the equation of magnetic induction was solved separately by a relaxation method as an elliptic equation. The reason may be that the characteristic time of magnetic field is much shorter than that of flow field.

In this study, fundamental equations for the above three flow problems are derived as a time-marching form. All equations are solved simultaneously using the numerical method based on the fourth-order compact MUSCL scheme[3] and the maximum second-order LU-SGS scheme coupled with the diagonal point-implicit method[4].

2 Fundamental Equations

The fundamental equations in this study can be written by the following vector form in curvilinear coordinates.

$$\frac{\partial Q}{\partial t} + \mathcal{F}(Q) = \frac{\partial Q}{\partial t} + \frac{\partial F_i}{\partial x_i} + S_P + S_{NP} + H_{NP} = 0 \tag{1}$$

$$Q = J \begin{bmatrix} q_1 \\ \vdots \\ q_s \\ \rho u_1 \\ \rho u_2 \\ E \end{bmatrix}, \quad F_i = J \begin{bmatrix} q_1 U_i \\ \vdots \\ q_s U_i \\ \rho u_1 U_i + \partial\xi_i/\partial x_1 p \\ \rho u_2 U_i + \partial\xi_i/\partial x_2 p \\ (E+p) U_i \end{bmatrix}$$

$$S_P = -J \frac{\partial \xi_i}{\partial x_j} \frac{\partial}{\partial \xi_i} \begin{bmatrix} 0 \\ \vdots \\ 0 \\ \tau_{1j} \\ \tau_{2j} \\ \tau_{kj} u_k + \kappa \partial T/\partial x_j \end{bmatrix}$$

Q is the vector of unknown variables. F_i and S_P are the flux vector and the diffusion term for compressible Navier-Stokes(N-S) equations. S_{NP} and H_{NP} are the diffusion term and the source term corresponding to additional equations. The vector Q contains not only unknown variables for the N-S equations but, for example, densities of chemical species and a vibrational energy for thermochemical nonequilibrium flows. Equations for thermochemical nonequilibrium flow(THC), flow with nonequilibrium condensation(CON), and axisymmetric MPD flow(MPD) derived or used by our research group are described as follows.

The q_s in Q is actually defined by

THC: $s = 6$, $(q_1, q_2, q_3, q_4, q_5, q_6) = (\rho_{N_2}, \rho_{O_2}, \rho_{NO}, \rho_N, \rho_O, E_v)$
CON: $s = 4$, $(q_1, q_2, q_3, q_4) = (\rho, \rho_v, \rho\beta, \rho n)$
MPD: $s = 2$, $(q_1, q_2) = (\rho, \Psi)$

ρ_{N_2} to ρ_O are the density of chemical species for N_2 to O. E_v is the vibrational energy. ρ_v is the density of vapor. β is the mass fraction of liquid. n is the number density of droplet. Ψ is the stream function of magnetic field. S_{NP} and H_{NP} are also defined by

THC:

$$S_{NP} = -J \frac{\partial \xi_i}{\partial x_j} \frac{\partial}{\partial \xi_i} \begin{bmatrix} q_1 v_{1j} \\ \vdots \\ q_5 v_{5j} \\ q_{vj} + q_k e_{vk} v_{kj} \\ 0 \\ 0 \\ q_k v_{kj} h_k \end{bmatrix}, \quad H_{NP} = -J \begin{bmatrix} w_1 \\ \vdots \\ w_5 \\ W_v \\ 0 \\ 0 \\ 0 \end{bmatrix}$$

where, v_{sj} is the diffusion velocity of species s. q_j and q_{vj} are the translational-rotational and the vibrational heat fluxes. e_{vj} is the vibrational energy per unit mass. h_j is the enthalpy per unit mass. w_s and W_v are the sources due to

chemical reactions and vibrational excitations. This is a well-known and standard set of equations for solving thermochemical nonequilibrium flows. The detail of chemical reaction models is not explained herein.
CON:

$$H_{NP} = -J \begin{bmatrix} 0 \\ -\Gamma \\ \Gamma \\ \rho I \\ 0 \\ 0 \\ 0 \end{bmatrix}$$

where, Γ is the mass generation rate due to condensation. I is the homogeneous nucleation rate. S_{NP} can be neglected assuming a homogeneous flow. This is an unique set of equations derived by our research group[5]. In most of the existing approaches, flow equations of air are solved using the time-marching method and the condensation is evaluated integrating the growth of the droplet along each streamline. Therefore, equations for air and liquid must be solved separately. The present equations involve the conservation laws of vapor, liquid, and the number density of droplet. They can be solved simultaneously assuming that the number density of droplet is conservative and the radius is very small.
MPD:

$$S_{NP} = -J \frac{1}{R_m} \begin{bmatrix} 0 \\ \frac{\partial \xi_i}{\partial x_k} \frac{\varepsilon}{\partial \xi_i} \frac{\partial \Psi}{\partial x_k} - \frac{1}{x_2} \frac{\partial \Psi}{\partial x_2} \\ 0 \\ 0 \\ 0 \end{bmatrix}, \qquad H_{NP} = H_{MPD} + H_{Axi}$$

where, R_m is the magnetic Reynolds number. $(x_1, x_2) = (z, r)$. z and r are axial and radial coordinates. H_{MPD} contains Lorentz force and Joule's heat. H_{Axi} includes derived terms due to the axisymmetric assumption. The equation of magnetic induction is additionally introduced in the set of equations assuming axisymmetric flows and coupling with the Maxwell's equations and Ohm's law. It has been solved separately like an elliptic equation because of very large difference of time-scale between particle motions and electric motions. While in this study, this equation and other equations are solved simultaneously by the time-marching method based on the LU-SGS method coupled with the diagonal point-implicit method.

3 Numerical Method

Fundamental equations for THC, CON and MPD are solved simultaneously using the time-marching method. The numerical flux defined at the interface between the control volume ℓ and $\ell+1$ in each coordinate $i (i = 1, 2)$ is written by

$(F_i)_{\ell+1/2}$. The flux-vector splitting approach can divide it into $(F_i^{\pm})_{\ell+1/2}$. Since F_i is a homogeneous function of degree one in Q, $F_i = A_i Q$ and $\delta F_i = A_i \delta Q$, where A_i is the Jacobian matrix in each coordinate. Then, the numerical flux $(F_i^{\pm})_{\ell+1/2}$ and $(\delta F_i^{\pm})_{\ell+1/2}$ can be written using $(A_i^{\pm})_{\ell+1/2}$ and Q^M. It is noted that the Q^M can act as not only Q but δQ extrapolated by the MUSCL approach. Finally, $(A_i^{\pm})_{\ell+1/2}Q^M$ can be rewritten using a set of sub-vector forms by

$$(A_i^{\pm})_{\ell+1/2}Q^M = \lambda_{i1}^{\pm}Q^M + \frac{\lambda_{ia}^{\pm}}{c\sqrt{g_{ii}}}Q_{ia} + \frac{\lambda_{ib}^{\pm}}{c^2}Q_{ib} \tag{2}$$

$$\lambda_{i1} = U_i, \quad \lambda_{i3} = U_i + c\sqrt{g_{ii}}, \quad \lambda_{i4} = U_i - c\sqrt{g_{ii}}$$

$$\lambda_{ij}^{\pm} = (\lambda_{ij} \pm \mid \lambda_{ij} \mid)/2, \quad \lambda_{ia}^{\pm} = (\lambda_{i3}^{\pm} - \lambda_{i4}^{\pm})/2$$

$$\lambda_{ib}^{\pm} = (\lambda_{i3}^{\pm} + \lambda_{i4}^{\pm})/2 - \lambda_{i1}^{\pm}$$

$$Q_{ia} = \bar{p}Q_{ic} + \Delta\bar{m}_i Q_d, \quad Q_{ib} = \Delta\bar{m}_i Q_{ic} + \bar{p}Q_d$$

$$\bar{p} = Q_p \cdot Q^M, \quad \Delta\bar{m}_i = Q_{im} \cdot Q^M$$

where, $g_{ii} = \nabla\xi_i \cdot \nabla\xi_i$. The sub-vectors Q_{ic},Q_p,Q_{im}, and Q_d are derived for THC, CON, and MPD as follows.
THC:

$$Q_{ic} = [0 \ \cdots 0 \ \ 0 \ \ \partial\xi_i/\partial x_1 \ \ \partial\xi_i/\partial x_2 \ \ U_i]^T$$

$$Q_p = [\phi_{N_2}^2 \ \cdots \phi_O^2 \ \ \tilde{\gamma} \ \ -\tilde{\gamma}u_1 \ \ -\tilde{\gamma}u_2 \ \ \tilde{\gamma}]^T$$

$$Q_{im} = [-U_i \ \cdots - U_i \ \ 0 \ \ \partial\xi_i/\partial x_1 \ \ \partial\xi_i/\partial x_2 \ \ 0]^T$$

$$Q_d = [\rho_{N_2}/\rho_0 \ \cdots \rho_O/\rho_0 \ \ E_v/\rho_0 \ \ u_1 \ \ u_2 \ \ (\chi^2 + c^2)/\tilde{\gamma}]^T$$

$\tilde{\gamma} = \gamma - 1$. γ is the specific heat. ρ_0 is the total density. The detail of ϕ_s^2 and χ^2 for THC is written in Refs.[1][6].
CON:

$$Q_{ic} = [0 \ \ 0 \ \ 0 \ \ 0 \ \ \partial\xi_i/\partial x_1 \ \ \partial\xi_i/\partial x_2 \ \ U_i]^T$$

$$Q_p = [\phi^2 \ \ 0 \ \ 0 \ \ 0 \ \ -\tilde{\gamma}u_1 \ \ -\tilde{\gamma}u_2 \ \ \tilde{\gamma}]^T$$

$$Q_{im} = [-U_i \ \ 0 \ \ 0 \ \ 0 \ \ \partial\xi_i/\partial x_1 \ \ \partial\xi_i/\partial x_2 \ \ 0]^T$$

$$Q_d = [1 \ \ \rho_v/\rho \ \ \beta \ \ n \ \ u_1 \ \ u_2 \ \ (\chi^2 + c^2)/\tilde{\gamma}]^T$$

$\phi^2 = \tilde{\gamma}u_j u_j/2 - h_{0m}$. h_{0m} is the heat of formation. $(\chi^2 + c^2)/\tilde{\gamma} = (E + p)/\rho$.
MPD:

$$Q_{ic} = [0 \ \ 0 \ \ \partial\xi_i/\partial x_1 \ \ \partial\xi_i/\partial x_2 \ \ U_i]^T$$

$$Q_p = [\phi^2 \ \ 0 \ \ -\tilde{\gamma}u_1 \ \ -\tilde{\gamma}u_2 \ \ \tilde{\gamma}]^T$$

$$Q_{im} = [-U_i \ \ 0 \ \ \partial\xi_i/\partial x_1 \ \ \partial\xi_i/\partial x_2 \ \ 0]^T$$

$$Q_d = [1 \ \ \Psi/\rho \ \ u_1 \ \ u_2 \ \ (\chi^2 + c^2)/\tilde{\gamma}]^T$$

$\phi^2 = \tilde{\gamma}u_j u_j/2$. $(\chi^2 + c^2)/\tilde{\gamma} = (E + p)/\rho$.
The advantage of doing these derivations may be that you don't need to know

all elements in the Jacobian matrix. You have only to know these set of sub-vectors. In the present study, the maximum second-order LU-SGS method coupled with the Newton iteration, the Crank-Nicholson method and the diagonal point-implicit method is applied to all cases, THC, CON, and MPD. This method can be written by

$$\begin{aligned}
\Delta Q^{*m}_{i,j} &= D^{-1}\{(RHS)^m_{i,j} + \theta_L \Delta t[(A_1^+ \Delta Q^*)^m_{i-1,j} + (A_2^+ \Delta Q^*)^m_{i,j-1}]\} \\
\Delta Q^m_{i,j} &= \Delta Q^{*m}_{i,j} - D^{-1}\theta_L \Delta t[(A_1^- \Delta Q^*)^m_{i+1,j} + (A_2^- \Delta Q^*)^m_{i,j+1}] \\
D &= I + \theta_L \Delta t\{\mathrm{diag}[(A_1)_{i,j}] + \mathrm{diag}[(A_2)_{i,j}] + \mathrm{diag}[viscosity] + \mathrm{diag}[source]\} \\
(RHS)^m_{i,j} &= -(Q^m_{i,j} - Q^n_{i,j}) - \Delta t[\mathcal{F}(Q^m)_{i,j} + \mathcal{F}(Q^n)_{i,j}]/2
\end{aligned}$$

m is the newton iteration number. n is the number of time step. $\theta_L = 1$ herein. The numerical fluxes $(A_1^\pm \Delta Q)_{i\pm 1,j}$ and $(A_2^\pm \Delta Q)_{i,j\pm 1}$ can be calculated using Eq.(2).

4 Results

We developed computational codes of 2-D and 3-D shock/shock interference flows with and without thermochemical nonequilibrium effects, those of 2-D steady, 2-D unsteady and 3-D steady transonic viscous flows in atmospheric wind tunnel conditions considering the relative humidity, and a code of axisymmetric MPD thruster viscous flows. In this paper, convergences of solution for the 2-D transonic viscous flow with nonequilibrium condensation and the axisymmetric MPD viscous flow are only discussed.

The result in the case of nonequilibrium condensation is first shown. The flow around the RAE2822 airfoil is calculated. The flow conditions are specified that the uniform mach number is 0.734, the uniform temperature is 293.15[K], the uniform pressure is 1.24×10^5[Pa], the Reynolds number(dry air) is 6.5×10^6 and the angle of attack is 2.54°. The C-type computational grid of 221×93 grid points algebraically generated are used. The flow at 90% relative humidity is calculated. Figure 1 shows the convergence history of solution(L2-norm). The histories obtained using the present LU-SGS method are compared with that of the two-step Runge-Kutta method. Consequently, similar convergence is obtained among three cases of the present method using the locally uniform CFL number at 0.5 and 5.0 and of the Runge-Kutta method at 0.5. The convergence is not always fast under the residue about at 10^{-2} because of probably a very humid condition. However, 37% CPU time per iteration is successfully saved by the present method compared with the Runge-Kutta method using single CPU of NEC SX-4.

Next, the convergence history of solution for the flow through the axisymmetric MPD thruster is shown. The configuration is not shown because of the limit of pages. Flow conditions are that the inlet Mach number is 1.2, the inlet temperature is 10000[K], the inter-electrode current is 5[kA], and the magnetic Reynolds number is 10. In this case, no convergence is obtained by the Runge-Kutta method. Figure 2 shows the convergence history of solution(L2-norm).

The parameter ε indicates the empirical parameter of the diagonal point-implicit term derived from the diffusion term S_{NP} in the equation of magnetic induction. According to the change of ε, the convergence is changed. Good convergences are obtained by $\varepsilon = 0.5, 1.0$. The convergences at $\varepsilon = 2.0, 4.0$ are adequate, and no solution is obtained at $\varepsilon = 0.0$. This result indicates that an optimum finite value of ε would exist and the diagonal point-implicit method is at least necessary.

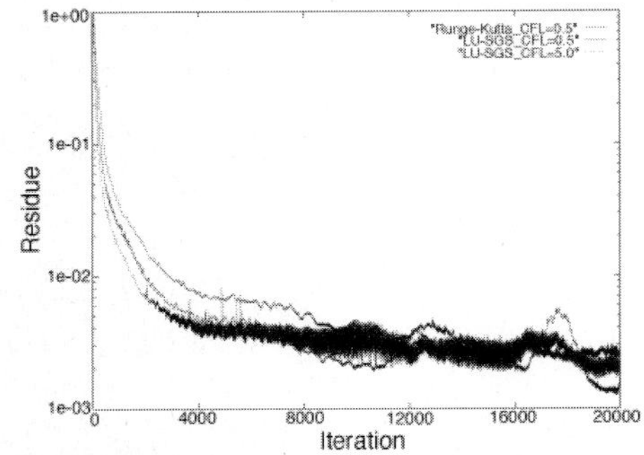

Fig.1 The convergence history of solution(RAE2822)

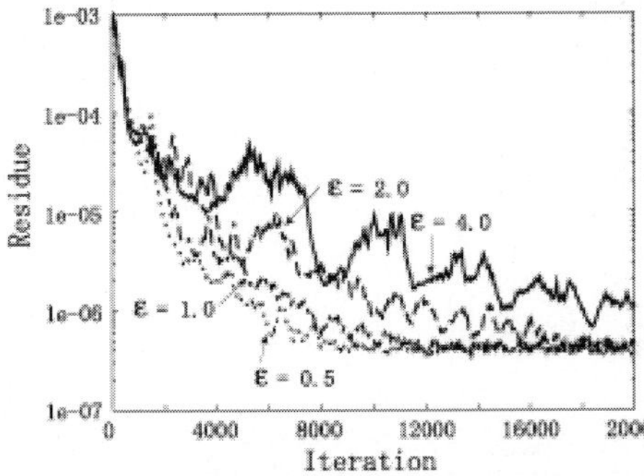

Fig.2 The convergence history of solution(MPD)

5 Conclusions

Complicated compressible flow equations with source term for thermochemical nonequilibrium flows, flows with nonequilibrium condensation, and axisymmetric MPD flows were derived as a time-marching form. An unified flux-vector splitting form for these equations was also derived and applied to the present LU-SGS method. The LU-SGS method worked very well especially for the MPD flow calculations. These three sets of equations can be also solved simultaneously for more complicated flow problems involving these flow problems at a same time.

References

1. S.Yamamoto, H. Nagatomo, and H. Daiguji, Proc. of 14th ICNMFD, Lecture Notes in Phys., Vol. 453, 1994, pp.314-319.
2. S. Yamamoto, N. Takasu, and H. Nagatomo, J. of Spacecraft and Rockets, Vol. 36, No.2, 1999, pp.240-246.
3. S. Yamamoto and H. Daiguji, Computers and Fluids, Vol.22, No.2/3, 1993, pp.259-270.
4. S. Yamamoto and S. Kano, Computers and Fluids, Vol. 27, No.5/6, 1998, pp.571-580.
5. S. Yamamoto, H. Hagari and M. Murayama, Trans. of Japan Society for Aeronautics and Space Sciences, Vol.42, No.138, 2000, pp.182-189.
6. S. Yamamoto and H. Daiguji, Proc. of Int. Symp. on Aerospace and Fluid Science, Vol.1, 1993, pp.156-163.

Numerical Prediction of Shear Flows by Upwind Methods

Maurizio Pandolfi and Domenic D'Ambrosio

Dipartimento di Ingegneria Aeronautica e Spaziale - Politecnico di Torino
e-mail: pandolfi@polito.it - domenic@athena.polito.it

1 Introduction

Flow fields in high speed regimes exhibit strong shock waves and sharp boundary layers (BL). Their numerical prediction is generally obtained with some of the several upwind methods that have been proposed in the literature over the past twenty years. Though shock waves are computed with a satisfactory numerical capturing, the description of shear flows, such as boundary layers, can suffer of severe shortcomings. These problems have widely experimented more than a decade ago for certain upwind methods and, more recently, for others. In the following we discuss this matter and focus attention on the motivations for such numerical deficiencies.

2 A Significant Problem and Some Numerical Experiments

We consider a simple but significant problem: *the high speed flow over a flat plate in the laminar regime*. Downstrean of the front region of strong interaction, just behind the leading edge of the plate, the flow field structure is characterized by the viscous region of the BL, closely to the wall. An outer oblique shock wave, induced by the BL obstruction, propagates higher over the plate. Inside the BL, the velocity presents, along the normal to the plate, severe variations, as well as do the density or the temperature. On the contrary, the pressure is very uniform, since the streamlines run almost parallel each other. The diffusive fluxes (typical of the N-S eqs.) play a dominant role inside the BL, by generating the correct physical dissipation and diffusion. On the contrary, the convective fluxes (typical of the Euler eqs.) should be almost inactive and must avoid any spurious injection of numerical (therefore artificial) dissipation, even in the presence of strong variations of the density and the velocity. However,we find that not all the upwind methods give the correct estimate of the convective fluxes and, depending on the upwind method, the thickness of the BL can result anomalously thicker because of the additional numerical dissipation or the pressure distribution inside the BL can show questionable oscillations.
In order to investigate the effects due to the different upwind methods, we have computed the flat plate flow field with a simple code based on the time-dependent

integration of the laws of conservation (laminar N-S), a finite volumes discretization on a properly stretched structured grid and the plain first order scheme. The diffusive fluxes at the interface between two adjacent volumes are evaluated with a centered approximation and the convective fluxes are estimated with several upwind methods that appear in the literature; the first order scheme is adopted to emphasize the features of each upwind method.
We classify in groups the upwind methods that we consider in the following numerical experiments: (**a**) flux-difference splitting methods (FDS), with the approximate solvers shown in [15] (FDSROE) and [11] (FDSPAN, a mirror image of the solver proposed in [10]), (**b**) flux-vector splitting methods (FVS) in the version reported in [16] (FVSSW), [19] (FVSVL) and [12] (FVSEFM), (**c**) the method HLL (or HLLE) proposed in [5] and [3], (**d**) the AUSM methods presented in [7] (AUSM-VEL), [8] (AUSM-M) and [9] (AUSM+), and (**e**) methods, FDS mimes, initially different from the FDS ones, that tend to mimic them through some implementations and are discussed in [1] (HUS, based on FVS),[3] (HLLEM, based on HLL), [17] (HLLC, also based on HLL), [20] [21] (AUSMD and AUSMV).
We have performed numerical experiments with the above methods by predictiong the flow field of the flat plate problem characterized by the following data: length of the plate $L = 1.0m$ and upstream conditions $M_\infty = 5.0$, $Re_{\infty L} = 62,000m^{-1}$, $T_\infty = 72.2K$, $T_w = 300K$ and $Pr = 0.72$. The results, presented as pressure (p/p_∞) distribution on the normal to the plate, at the distance $x = 0,89m$ from the leading edge, are compared with a reference solution obtained with a code that is based on FDSPAN, accurate ENO scheme and a much finer grid, and has been checked against test-cases reported in the literature.
The FDS predictions by FDSROE and FDSPAN, almost coincident, appear superimposed in Fig.1. The shock wave and the expansion fan are not so sharp as they should be, due to the poor accuracy of the first order scheme and the rough grid. Nonetheless, the overall prediction is good. The shock location and the pressure value at the wall are correct: these are indications that the numerical dissipation has been kept low.
The results obtained with the FVS methods are different. Those from FVSSW and FVSVL are shown in Fig.2 and those from FVSEFM and HLL appear in Fig.3. The sharpness of the captured oblique shock is very good for FVSVL and HLL, not so much for FVSSW and FVSEFM, but the most important fact is that its location is too far from the wall and the pressure level there is definitely overestimated. These results are due to a remarkable numerical dissipation injected in the BL.
The results given by the three version of AUSM are reported in Fig.4. The versions, AUSM-VEL and AUSM-M, almost coincident except at the wall, give a sharp capturing of the oblique shock, whereas AUSM+ shows a thicker capturing zone. The pressure level at the wall seems to have been well predicted, a sign that the numerical dissipation is minimal. However, a questionable result is represented by the pressure oscillation inside the BL, in the proximity of the

wall. Such an oscillation shows up clearly for AUSM-VEL, but also, though with smaller amplitude, in AUSM-M and AUSM+, as it could be seen in enlarged scale of the abscissa (Fig.5) or by increasing the Reynolds number.
The results obtained with FDS mimes methods are very close to the FDS ones. We do not report any result from them, but we remark that these methods fully reproduce the FDS features.
The remarkable discrepancies among the numerical predictions in a trivial fluid-dynamic case require a critical interpretation of the algorithms that are dictated by the different upwind methods.

3 A Critical Analysis on Upwind Methods

We borrow here the *odd-even decoupling problem* suggested in [13] in the investigation of numerical instabilities typical of some upwind methods in flows with strong shocks, and we extend that analysis to a more general case and to the many considered upwind methods.
The present analysis starts from the Euler eqs. (upwinding refers to the convective fluxes) in the 2D (x, y) problem, written in the linearized form about the uniform parallel flow characterized by $\rho_0 = 1.0$, $u_0 \neq 0$, $v_0 = 0$, $p_0 = 1.0$ (u and v are the velocity components along x and y). For each upwind method we develop the algorithm within the frame of the above linearization and we estimate the flux on any interface between two adjacent volumes. The flow is assumed uniform along x and variations can only occur on the transversal direction y. Initial conditions are imposed with alternate perturbations along y: $\rho^0 = 1 \pm \hat{\rho}^0$, $u^0 = u_0 \pm \hat{u}^0$, $v^0 = 0$, $p^0 = 1 \pm \hat{p}^0$, where the signs $+$ and $-$ hold alternatively in the cells in the y direction. Then, the first order scheme is constructed and the integration in time is carried out on the basis of the initial conditions (ρ^0, u^0, v^0, p^0). Finally, it is possible to obtain simple recursive formulas that determine the flow properties at any integration step $K+1$, starting at the step K. These formulas are reported in Tab.1. The label MOC refers to the FDS approach as originally proposed in [4] and founded on the method of characteristics. The formulas provide the evolution in time of the perturbations of density ($\hat{\rho}$), velocity ($\hat{u}$) and pressure ($\hat{p}$).
In order to interpret the results shown in the Figs.1-4, we assume initial perturbations of density and velocity ($\hat{\rho}^0 \neq 0$, $\hat{u}^0 \neq 0$) and uniform pressure ($\hat{p}^0 = 0$), a situation somewhat similar to a BL. The correct solution of the Euler eqs. requires the preservation of the initial conditions. It is now interesting to see how the different upwind schemes react in this problem.
The FDS methods (MOC, FDSROE, FDSPAN) give the correct answer: at any K (and for $K \to \infty$), we have $\hat{\rho}^\infty = \hat{\rho}^K = \hat{\rho}^0$ and $\hat{u}^\infty = \hat{u}^K = \hat{u}^0$. Therefore, we expect that, in the flat plate problem, the FDS methods do not react to the density and velocity profiles generated by the only diffusive terms of the N-S equations and that the convective terms do not add any incorrect dissipation. The recursive formulas for MOC, the approximate solvers FDSROE and

FDSPAN and the FDS mime HLLC are coincident. Similar results are obtained also for the other FDS mimes (HUS, AUSMD, AUSMV), since the only slight variation appears in the coefficient of the pressure perturbation that remains null in this analysis ($\hat{p}^{\infty} = \hat{p}^{K} = \hat{p}^{0} = 0$), but the structure of the formulas remains the same.

On the contrary, the reaction of the FVS methods is very different. Even if the coefficients in the formulas are not equal, the structure of the formulas results the same for FVSSW, FVSVL and FVSEFM. The initial perturbations , $\hat{\rho}^0$ and $\hat{u}^0$ decrease continuously by denoting the generation of spurious numerical dissipation. Also HLL presents the same incorrect behavior, even if the coefficient are different. It is now clear why the thickness of the BL is overestimated in Figs. 2 and 3, as well as the wall pressure: the inviscid fluxes of FVS and HLL methods, inside regions of finite gradients of density and velocity, tend to flatten these gradients by injecting numerical dissipation.

The AUSM methods react differently. If we look at AUSM-VEL, we recognize that the velocity perturbation is preserved correctly, but the density perturbation is amplified and triggers a stable pressure perturbation initially absent: $\hat{\rho}^{\infty} = \hat{p}^{\infty} = \frac{\gamma}{\gamma - 1}\hat{\rho}^0$. This very anomalous behavior is confirmed, in computations carried out for the flat plate problem, by the generation of steady pressure oscillations in the BL that are induced by the density gradient, just as shown in Figs. 4 and 5. The recursive formulas for AUSM-M and AUSM+ are identical each other and very simple: any initial perturbation is preserved. This is correct for density and velocity. However, the pressure should change through travelling waves as for FDS, whereas the recursive formulas indicate that they will be maintained. Therefore we shall be not surprised in detecting small but persistent pressure oscillations even with these two methods. We remind that the generation of more or less remarkable pressure oscillations with these methods has been already experimented in computations of practical interest and reported in the literature [14] [18] [6].

References

1. Coquel,F. and Liou.M.S., AIAA Paper 93-3302-CP, 1993.
2. Einfeldt,B.,SIAM J.Numer. Anal., Vol.25, N.2, pp.294,318., 1988.
3. Einfeldt,B., Munz,C.D., Roe,P.L. and Sjögreen,B.,*J.C.P.*, Vol.92,1991, pp.273-295.
4. Godunov,S.K.,*Mat.Sb.*, Vol.47, pp.271,290, 1959.
5. Harten,A., Lax,P.D. and Van Leer,B.,*SIAM Review*, Vol.25, N.1, pp.35,61, 1983.
6. Kim,K.H. et al., *Computers and Fluids*, Vol. 29, No.3,pp.311,346, 1998.
7. Liou.M.S.,*Lecture Notes in Physics*, Vol.414, Springer-Verlag, 1993.
8. Liou,M.S. and Steffen,C.J., *J.Comput.Phys.*, Vol.107,, pp.23,39, 1993.
9. Liou,M.S., *J.Comput.Phys.*, Vol. 129, pp.364,382, 1996.
10. Osher,S. and Solomon,F.,*Mathematics of Computations*, Vol.38, 1982, pp.339,377.
11. Pandolfi,M., *AIAA J.*, Vol.22, pp.602,610, 1984.
12. Pullin, *J.Comput.Phys.*, Vol. 34, pp.231,244, 1980.
13. Quirk,J.J., *ICASE* Report 92-64, 1992.
14. Radespiel,R. and Kroll,N., *J.Comput.Phys.*, Vol.121, pp.66,78, 1991.

15. Roe,P.L., *J.Comput.Phys.*, Vol. 43, pp.357,372, 1981.
16. Steger,J.L. and Warming,R.F., *J.Comput.Phys.*, Vol. 40, pp.263,293, 1981.
17. Toro,E.F., Spruce,M., and Speares,W., *Shock Waves*, Vol.4, pp.25,34, 1994.
18. van Keuk,J.et al., *AIAA 8th Int.Hyp.Syst. and Techn.Conf.*, Norfolk, VA, 1998.
19. Van Leer,B., *Lectures Notes in Physics*, Vol 170, Springer-Verlag, 1982.
20. Wada,Y. and Liou,M.S., *NASA T.M. 106452*, 1994.
21. Wada,Y. and Liou,M.S., *SIAM J.Sci.Comput.*, Vol.18, N.3, pp.633,657, 1997.

Table 1. Integration with the linearized form of the different methods.

MOC	$\hat{\rho}^{K+1} =$	$\hat{\rho}^K$		$+$	$(-\frac{2\nu}{\gamma})\hat{p}^K$
FDSROE	$\hat{u}^{K+1} =$		$\hat{u}^K$		
FDSPAN	$\hat{p}^{K+1} =$				$(1-2\nu)\hat{p}^K$
HLLC					
FVSSW	$\hat{\rho}^{K+1} =$	$(1-\frac{\nu}{\gamma})\hat{\rho}^K$		$+$	$(-\frac{\nu}{\gamma})\hat{p}^K$
	$\hat{u}^{K+1} =$		$(1-\frac{2\nu}{\gamma})\hat{u}^K$		
	$\hat{p}^{K+1} =$	$(\nu)\hat{\rho}^K$			$(1-3\nu)\hat{p}^K$
FVSVL	$\hat{\rho}^{K+1} =$	$(1-\frac{\nu}{2})\hat{\rho}^K$		$+$	$(-\frac{\nu}{2})\hat{p}^K$
	$\hat{u}^{K+1} =$		$(1-2\nu)\hat{u}^K$		
	$\hat{p}^{K+1} =$	$(\nu\frac{\gamma}{\gamma+1})\hat{\rho}^K$			$(1-3\nu\frac{\gamma}{\gamma+1})\hat{p}^K$
FVSEFM	$\hat{\rho}^{K+1} =$	$(1-\frac{\sqrt{2}\nu}{\sqrt{\pi\gamma}})\hat{\rho}^K$		$+$	$(-\frac{\sqrt{2}\nu}{\sqrt{\pi\gamma}})\hat{p}^K$
	$\hat{u}^{K+1} =$		$(1-2\frac{\sqrt{2}\nu}{\sqrt{\pi\gamma}})\hat{u}^K$		
	$\hat{p}^{K+1} =$	$(\frac{(\gamma+1)\nu}{\sqrt{2}\sqrt{\pi\gamma}})\hat{\rho}^K$			$(1-3\frac{(\gamma+1)\nu}{\sqrt{2}\sqrt{\pi\gamma}})\hat{p}^K$
HUS	$\hat{\rho}^{K+1} =$	$\hat{\rho}^K$		$+$	$(-\frac{\nu}{2}\frac{\gamma+1}{\gamma})\hat{p}^K$
	$\hat{u}^{K+1} =$		$\hat{u}^K$		
	$\hat{p}^{K+1} =$				$(1-\nu\frac{3\gamma-1}{\gamma+1})\hat{p}^K$
HLL	$\hat{\rho}^{K+1} =$	$(1-2\nu)\hat{\rho}^K$			
	$\hat{u}^{K+1} =$		$(1-2\nu)\hat{u}^K$		
	$\hat{p}^{K+1} =$				$(1-2\nu)\hat{p}^K$
AUSM-VEL	$\hat{\rho}^{K+1} =$	$(1+\frac{\nu}{2})\hat{\rho}^K$		$+$	$(-\frac{\nu}{2})\hat{p}^K$
	$\hat{u}^{K+1} =$		$\hat{u}^K$		
	$\hat{p}^{K+1} =$	$(\frac{\nu\gamma}{2})\hat{\rho}^K$			$(1-\frac{\nu\gamma}{2})\hat{p}^K$
AUSM-M	$\hat{\rho}^{K+1} =$	$\hat{\rho}^K$			
AUSM+	$\hat{u}^{K+1} =$		$\hat{u}^K$		
	$\hat{p}^{K+1} =$				$\hat{p}^K$
AUSMD	$\hat{\rho}^{K+1} =$	$\hat{\rho}^K$		$+$	$(-\nu)\hat{p}^K$
AUSMV	$\hat{u}^{K+1} =$		$\hat{u}^K$		
	$\hat{p}^{K+1} =$				$(1-\nu\gamma)\hat{p}^K$

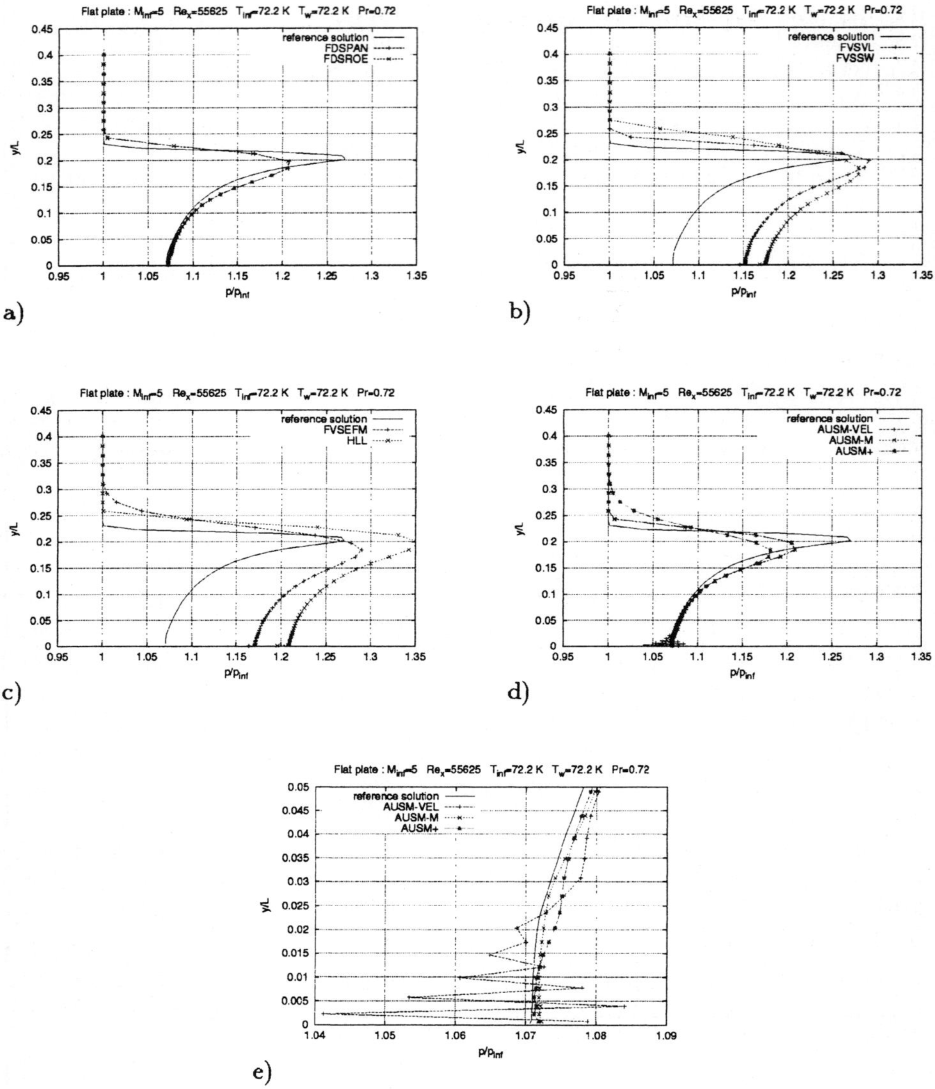

Fig. 1. Pressure plot at $x = 0.89$ m; a)FDSPAN and FDSROE; b)FVSVL and FVSSW; c)FVSEFM and HLL; d)AUSM-VEL, AUSM-M and AUSM+; e)enlarged view of AUSM-VEL, AUSM-M and AUSM+ behavior close to the wall.

Numerical Modeling of Discontinuous Gas Dynamic and MHD Astrophysical Flows

Nikolai V. Pogorelov

Institute for Problems in Mechanics, Russian Academy of Sciences
101-1 Vernadskii Avenue, Moscow 117526, Russia

Abstract. The application of computational fluid dynamics methods to complicated astrophysical problems is discussed. Special attention is paid to methodological aspects of numerical modeling of gas dynamic and MHD flows occurring in the outer space. The main emphasis is made on the correlation between numerical and hydrodynamic instabilities and the structural instability of solutions involving nonevolutionary shocks.

1 Introduction

Development of CFD provided a variety efficient numerical tools for resolving multi-shocked high-speed gas dynamic flows. Similar methods are expected to be developed and applied to much more complicated magnetohydrodynamic (MHD) flows. A number of astrophysical phenomena can be modeled using a continuum mechanics approach. However, they possess to an extreme extent the difficulties which one can encounter in engineering gas dynamics: a large variation of parameters in the computational region, an enormous difference in the characteristic wave propagation speeds, an interaction domain with boundary conditions to be stated at infinity, and a variety of interacting discontinuities. In solution of the mentioned problems we use newly developed numerical methods applicable both to purely gasdynamic and multidimensional MHD flows. Among them are high-resolution numerical schemes [1]–[3], irregular adaptive and overlapping grids [4], and efficient absorbing (nonreflecting) boundary conditions [5]. It becomes widely known that even application of well-validated codes for insufficiently understood problems can result in wrong results, excessive slow convergence, or even nonconvergent solutions. This is supported by the fact that high-resolution schemes are not strictly monotone if the amount of numerical dissipation is below a certain limit [6]. On the contrary, they produce high-amplitude oscillations of the difference derivatives which can, in principle, launch instabilities [7]. The correlation of these instabilities with genuinely hydrodynamic ones represents a challenging problem of applied mathematics.

The solar wind interaction with the interstellar medium from the mechanical viewpoint represents a fundamental problem of the interaction of the supersonic free stream with the supersonic source flow. This problem, in fact, can be treated as a combination of blunt-body and jet flows. The solution in this case consists of at least two shocks and a tangential discontinuity separating the winds. The last surface can fairly well be convectively unstable [8]. On the other hand, standard numerical meshes which are quite sufficient for the resolution of all discontinuities can be accompanied by numerical dissipation capable of damping hydrodynamic instabilities. In [9], the above problem was solved in the assumption of certain space-periodic nonuniformity of the interstellar medium and hydrodynamic instability of the tangential discontinuity (heliopause) manifested itself even with the application of fairly rough meshes.

As far as magnetohydrodynamic (MHD) flows are concerned, of great importance becomes the question of structural instability of discontinuous solutions. This effect takes its origin from the fact that not all entropy increasing shocks are admissible in MHD. Moreover, in the approximation of ideal plasma, which is very well suitable for problems of cosmic gas dynamics, only evolutionary shocks can exist. The shock is evolutionary if the problem of its interaction of small disturbances is well posed. In this case it can adjust itself to perturbations of the boundary conditions. The peculiarity of the problem is that some nonevolutionary shocks are structurally unstable only against rotational disturbances. Thus, if the space dimension of the problem is reduced and we solve a plane or an axisymmetric problem, out-of-plane rotations are not allowed and we can obtain a solution involving nonevolutionary discontinuities [1]. As will be shown later, this sometimes can occur even in the problems possessing a symmetry plane.

2 Solar wind interaction with the magnetized interstellar medium

In [10], the problem was considered of the solar wind interaction with the magnetized interstellar medium in which the magnetic field and velocity vectors, $\boldsymbol{B}_\infty$ and $\boldsymbol{V}_\infty$, are parallel. It turned out that in a certain range of parameters the solution very much differs from those obtained for smaller magnetic fields [5]. As a result, various additional discontinuities originate between the bow shock and the tangential discontinuity. In Fig. 1 we show the distribution of density (below the symmetry axis) and thermal pressure isolines. The Sun is located at the origin and the interstellar medium moves from the right to the left. Since the problem is obviously axisymmetric, there is no mechanism for the magnetic field vector to go outside the chosen plane of calculation. Thus, the reversals of the magnetic field component tangent to the shock normal can occur only on nonevolutionary shocks. Besides, the parameters of the external wind are characterized by the Mach number $M_\infty = 2$ and the Alfvén $A_\infty = \sqrt{2}$, where the former and the latter are the ratios of the velocity value to the thermal speed of sound and the Alfvén velocity, respectively. For the specific heat ratio γ equal to 5/3, these values of parameters imply that the hydrodynamic shock cannot exist on the symmetry axis, since it is nonevolutionary [11]. In the field-aligned case by a hydrodynamic shock we mean the shock with zero tangential components both ahead and behind the front. On the other hand, switch-on shocks (the ones switching on the tangential component of the magnetic field) are admissible. The geometry of the flow, however, does not allow any switching, for we have an infinite number of switching directions at the axis. As a result, most of obtained shocks are nonevolutionary. Since we know that such shocks are unstable to ro-

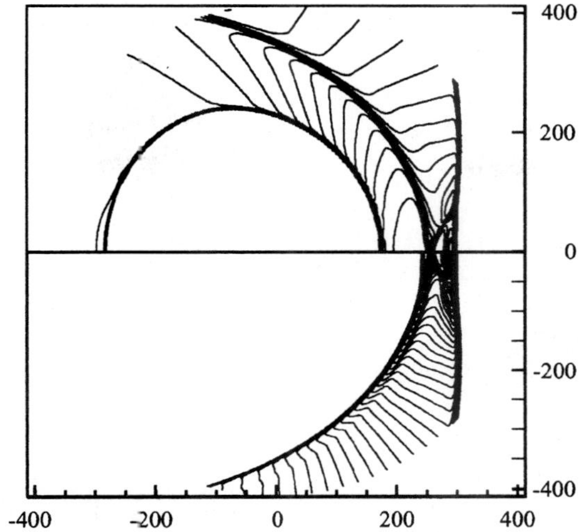

Fig. 1. Density (below the symmetry axis) and thermal pressure isolines: unperturbed wind

tational perturbations, we add small out-of-plane components to the magnetic field and velocity vectors, still remaining in the axisymmetric statement of the problem. The obtained streamlines, together with the bow shock and the isoline $A = 1$ are shown in Fig. 2.

Since nonevolutionary shocks are always trans-Alfvénic, we see that the bow shock, except for a small vicinity of the symmetry axis became super-Alfvénic and therefore admissible. The secondary shock coincide now with the $A = 1$ isoline and is an evolutionary switch-off shock. As far as the mentioned trans-Alfvénic zone on the bow shock is concerned, the analysis shows that here we have two merged shocks there. The first of them is a switch-on and the second one is a switch-off shock. This combination is conventionally considered structurally unstable. We cannot destroy it, however, since in our perturbed statement the out-of-plane component of the magnetic field vector must vanish on the symmetry axis. In addition, due to the necessity to resolve the complicated flow structure as a whole, we may loose in the resolution of the questionable zone. To look deeper inside the structural instability of the obtained nonevolutionary shocks, we shall consider a problem of the MHD flow over an infinitely conducting cylinder.

3 MHD flow over an infinitely conducting cylinder

In [12] the controversial problem was considered of the two-dimensional, supersonic and super-Alfvénic MHD flow over an infinitely conducting cylinder in the parameter range corresponding to the nonevolutionarity described in the previous section. The obtained solution exhibited a number of shocks which are not

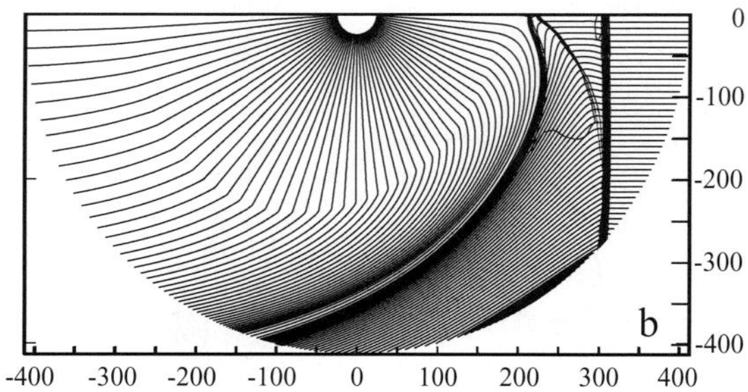

Fig. 2. Streamlines and $A = 1$ lines: perturbed wind

common in traditional blunt-body problems. To clear out the point, we obtained the solution of this field-aligned ideal plasma flow for dimensionless parameters of the uniform stream $M_\infty = \sqrt{13.5}$, $A_\infty = 1.5$, and $\gamma = 5/3$.

The steady-state pattern is shown in Fig. 3a, where the isolines of density logarithms are presented. In Fig. 4 we show the enlarged picture of the zone of our interest. In addition, the magnetic field lines are presented in the same figure. The analysis shows that if we move upward from the point of the bow shock intersection with the symmetry axis, this shock is fast and evolutionary. Note that the bow shock is concave outward, in agreement with our solution of the SW–LISM interaction problem and [12]. We see that further on this shock becomes convex outward, and we must have an inflection point on its profile. From this point the shock becomes nonevolutionary. This lasts until we reach the extremum convex point where the nonevolutionary shock transforms into the switch-on shock. Further upward it becomes fast and evolutionary again. The shocks lying between the bow shock and the body are all nonevolutionary. One can also see several tangential discontinuities in the shock layer. Their presence becomes clear if we inspect the streamline behaviour shown in Fig. 4. Note that in our case the streamlines are everywhere parallel to the magnetic field lines.

As noted in [12], the obtained solution is metastable if we introduce a small angle between $\boldsymbol{V}_\infty$ and $\boldsymbol{B}_\infty$. It is important that there is no necessity to pass to the three-dimensional formulation of the problem to destroy the obtained solution. It suffices to introduce a small magnetic field and velocity component parallel to the cylinder axis. In this way, we let the magnetic field vector go out of the initial plane. Since in this case we distinguish one of the directions, it becomes the direction for switching on the magnetic field in the forward part of the bow shock and no controversial irregular configurations are necessary to account for the flow symmetry. The steady-state solution is shown in Fig. 3b. The bow shock is everywhere fast, all additional discontinuities disappeared, and the solution in the plane perpendicular to the cylinder axis looks very much alike those occurring in the regular flow. Note that by introducing the disturbance we in no way destroyed the invariability of the flow along the cylinder axis. Of course, the velocity and magnetic field components along this axis acquire fairly large values behind the bow shock.

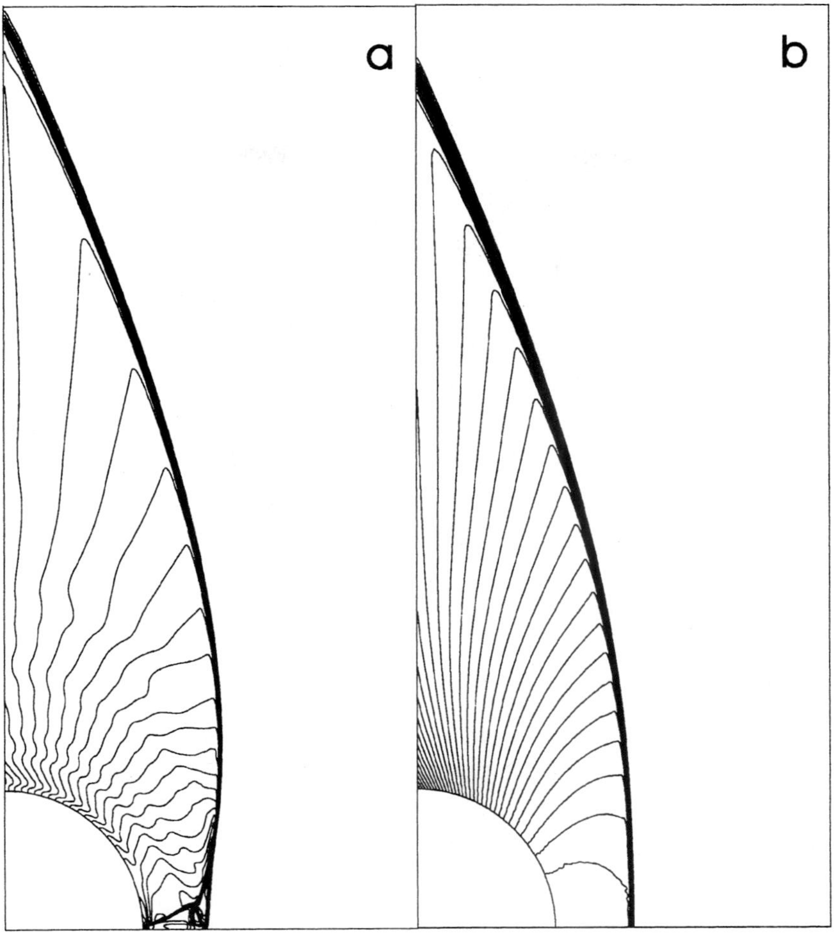

Fig. 3. Density isolines for the purely planar (a) and perturbed (b) flows of ideal plasma over an infinitely conducting cylinder.

It is claimed in [12] that various types of nonevolutionary shocks are inevitable if the angle between $\mathbf{B}_\infty$ and $\mathbf{V}_\infty$ is not zero and the problem becomes three-dimensional. This may happen in the vicinity of the bow shock point where the magnetic field vector is parallel to the shock normal. The reason of this statement is in the fact that one can always choose a plane of symmetry containing both the velocity and magnetic field vectors. It is apparent, however, that if we would like to destroy the nonevolutionary solution, it is necessary to perturb the symmetry plane. This means that if nonevolutionary shocks are expected to exist in the symmetric statement, the problem must be solved by disregarding the presence of this plane and out-of-plane disturbances of $\mathbf{B}_\infty$ and $\mathbf{V}_\infty$ must be introduced.

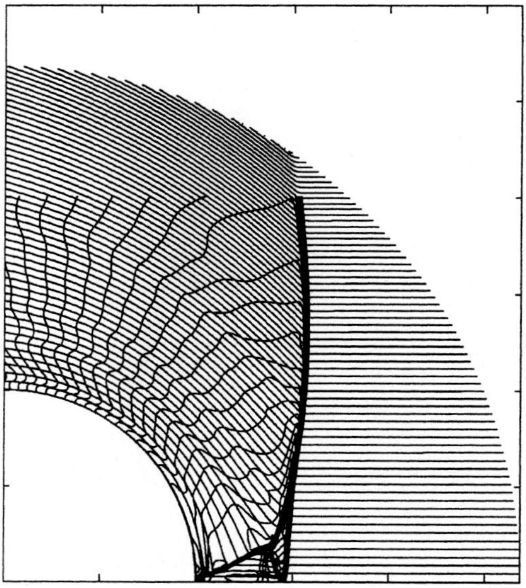

Fig. 4. Density isolines and magnetic field lines for the purely planar cylinder.

Acknowledgement

This work was supported, in part, by the Russian Foundation for Basic Research under grant 98-01-00352. The author is grateful to A.G. Kulikovskii and T. Matsuda for useful discussions.

References

1. A.A. Barmin, A.G. Kulikovskii, N.V. Pogorelov: J. Comput. Phys. **126**, 77 (1996)
2. N.V. Pogorelov, A.Yu. Semenov: Comput. Maths Math. Phys. **36**, 320 (1997)
3. N.V. Pogorelov, T. Matsuda: J. Geophys. Res. **103**, 237 (1998)
4. I.E. Ivanov, I.A. Kryukov, N.V. Pogorelov: 'Grid Adaptation for Gas Dynamic and Astrophysical Flows'. In: *Proc. 8th Int. Meshing Roundtable, Lake Tahoe, Oct. 10–13, 1999* (Sandia Report 99-2288, Albuquerque, NM) pp. 331–319
5. N.V. Pogorelov, A.Yu. Semenov: Astron. Astrophys. **321**, 330 (1997)
6. V.V. Ostapenko: Comput. Maths Math Phys. **37**, 1119 (1998)
7. N.V. Pogorelov, Y. Ohsugi, T. Matsuda: Month. Not. Royal Astron. Soc. **313**, 198 (2000)
8. N.A. Belov, A.V. Myasnikov: Izvestiya AN. Fluid Dynamics **34**, 379 (1999)
9. N.V. Pogorelov: Astrophys. Space Sci., to appear (2000)
10. N.V. Pogorelov, T. Matsuda: Astron. Astrophys. **354**, 697 (2000)
11. L.D. Landau, E.M. Lifshitz: *Electrodynamics of Continuous Media* (Pergamon, London 1960)
12. H. De Sterk: *Numerical Simulation and Aanalysis of Magnetically Dominated MHD Bow Shock Flows with Applications in Space Physics* (Ph.D. Thesis, Katholiele Universiteit Leuven)

A Computational Study of the Wake Structure of a Plunging Airfoil

J. Young, J.C.S. Lai

School of Mechanical and Aerospace Engineering, Australian Defence Force Academy, Canberra, ACT, 2600, Australia

1 Introduction

Oscillating airfoils have been extensively studied in a variety of contexts. Experimental and numerical studies have explored the dynamics of the wake and vortex roll-up, the dynamic stall phenomenon (e.g. Ekaterinaris and Platzer [1]), details of fish and marine mammal propulsion (e.g. Lighthill [2]), and the use of oscillating airfoils in flow control and flow reattachment (e.g. Tuncer *et al* [3]). The use of flapping wings for generation of lift and propulsion has also gained recent attention with application to Micro Air Vehicles (e.g. Shyy *et al* [4]).

Early theoretical analyses and a number of experimental studies (e.g. Koochesfahani [5], Dohring *et al* [6]) have shown that oscillating airfoils generate thrust at certain combinations of plunge frequency and amplitude. Garrick [7] determined the propulsive efficiency of a plunging and pitching flat-plate airfoil as a function of the oscillation frequency, for potential flow and small amplitude oscillation.

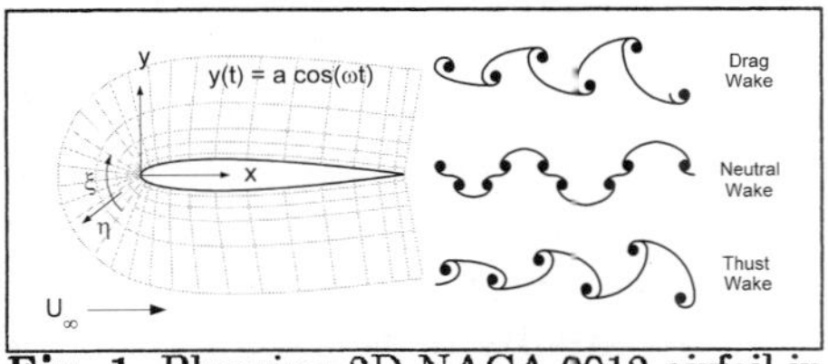

Fig. 1. Plunging 2D NACA 0012 airfoil in steady free stream flow

Pure plunging airfoil motion has received relatively less attention than pitching motion. Freymuth [8] and Lai and Platzer [9], among others, show that the wakes of plunging airfoils can be characterized as drag-producing, neutral, or thrust-producing depending on the plunge frequency and amplitude. Drag-producing wakes show momentum-deficit time-averaged velocity profiles, and either Karman vortex street or more complicated patterns with multiple vortices shed from the trailing edge per half-cycle of airfoil motion. Vortex pairs form 'mushrooms' that are tilted upstream. Thrust-producing wakes show a momentum-surfeit, or jet, superimposed on the momentum-deficit time-averaged velocity profile, such that the thrust of the jet is greater than the inherent drag of the airfoil. The wake configuration is typically a reverse Karman vortex street, so that vortex pairs form downstream-tilted 'mushrooms'. Neutral wakes, where the thrust due to plunging balances the inherent drag, may show multiple vortex shedding per half-cycle, and vortex pairs are not tilted. The vortex shedding process that leads to the wake patterns described above is not well understood.

The objective of this paper is to illustrate the use of a compressible, laminar/turbulent (Reynolds averaged) Navier Stokes solver in combination with a scalar particle tracing method to simulate the wake of a plunging airfoil for a range of oscillation frequencies and amplitudes. Comparisons are made between the numerical simulations and experimental flow visualization results of Lai and Platzer [9].

2 Analysis

The geometry under consideration is a sinusoidally plunging 2D NACA 0012 airfoil in a steady free stream, as shown in Fig. 1. In establishing the parameters important to the shape of the wake, it is instructive to examine the classical approach based on linearized potential flow theory. Garrick [7] provides the expression $C_{T_{LIN}} = 4\pi (kh)^2 (F^2 + G^2)$ for the time-averaged thrust coefficient per unit span $C_{T_{LIN}}$ as a function of the non-dimensional plunge amplitude $h = a/c$ and reduced frequency $k = \omega c/2U_\infty$ for a flat plate, where $F = F(k)$ and $G = G(k)$ are the real and imaginary components of the Theodorsen Function. Lines of constant $C_{T_{LIN}}$ in Fig. 2 asymptote to lines of constant kh (equivalent to constant Strouhal number $St = 2kh/\pi$) for $k > 3.0$, in which case $C_{T_{LIN}}$ depends only on the single parameter kh.

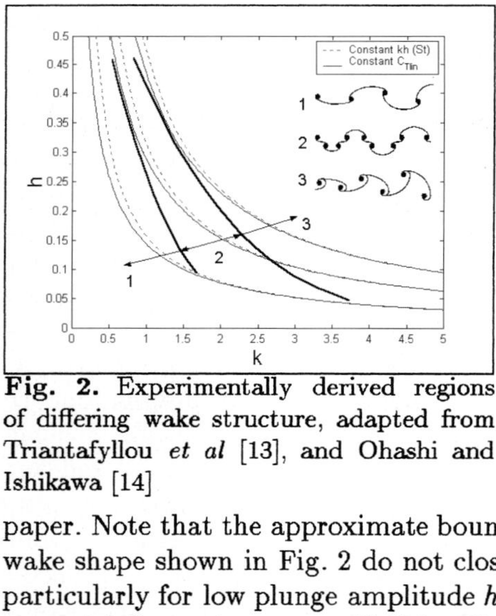

Fig. 2. Experimentally derived regions of differing wake structure, adapted from Triantafyllou *et al* [13], and Ohashi and Ishikawa [14]

Garrick assumes the wake consists of a small-width reverse Karman vortex street, which is valid only for small h and non-linearities due to self-induced evolution of the wake are not taken into account. This is in contrast to the real flow where the thrust developed and the shape of the vortex street in the wake are closely related. A third assumption of the theoretical approach by Garrick is that the Kutta condition is satisfied at the trailing edge. The limited validity of this assumption is discussed by McCroskey [10] and illustrated by results presented later in this paper. Note that the approximate boundaries separating the regions of different wake shape shown in Fig. 2 do not closely follow lines of constant kh or $C_{T_{LIN}}$ particularly for low plunge amplitude h where the thin-wake assumption of the potential theory is most applicable. The linear theory would imply that for $k >$ 3.0, two flows with the same kh but differing values of k and h should have the same vortex structure in the wake, and develop the same average thrust coefficient. This is not the case in Fig. 2 and is contradicted by experimental and numerical results presented in this paper, suggesting that the linearized theory is insufficient to explain the observed phenomena. Furthermore both the reduced frequency of plunging k and the amplitude of motion h must be considered as separate parameters in determining the structure of the wake.

3 Numerical Method

The unsteady, viscous, compressible 2D Reynolds averaged Navier Stokes equations are solved in generalized coordinates on a 537 × 81 element structured C-grid (187 points on the airfoil), which is deformed around the airfoil to introduce the plunging motion. The thin-layer approximation [11] is implemented, as is the Baldwin-Lomax turbulence model [11] for fully turbulent flow calculations. The flow solver uses an implicit Osher upwind-biased flux difference splitting scheme, second-order-accurate in space and time, to solve the strong conservation law form of the equations. Details of the method may be found in Tuncer *et al* [12].

Scalar particle tracing is performed by releasing particles at locations fixed relative to the airfoil trailing edge, and particle paths are integrated in ξ–η grid coordinates. Physical velocities (u,v) at the corners of the current cell (determined by the particle ξ–η grid coordinates) are converted to equivalent velocities in the x–y domain via the expression:

$$\dot{\xi} = \frac{\partial \xi}{\partial x}\left(u - \frac{\partial x}{\partial t}\right) + \frac{\partial \xi}{\partial y}\left(v - \frac{\partial y}{\partial t}\right),\ \dot{\eta} = \frac{\partial \eta}{\partial x}\left(u - \frac{\partial x}{\partial t}\right) + \frac{\partial \eta}{\partial y}\left(v - \frac{\partial y}{\partial t}\right) \quad (1)$$

where $\partial x/\partial t$ and $\partial y/\partial t$ are the grid deformation rates and $\partial \xi/\partial x$ etc. are grid metrics taken to be uniform across a cell. The particle equivalent ξ–η velocity is calculated by bilinear interpolation of the cell corner values, based on the current ξ–η coordinates. The new ξ–η particle location is calculated with a fourth-order predictor-corrector method. If adjacent cells are not identical (ie. the grid is non-uniform), differing grid metrics in the cells will introduce an error as a particle crosses the cell boundary, unless care is taken to recalculate $\dot{\xi}$ and $\dot{\eta}$ based on the new cell grid metrics for the remainder of that time step. The new x–y particle location is determined by bilinear interpolation of the position values at the corners of the new cell, based on the new ξ–η grid coordinates.

The method was verified for steady potential flow around a cylinder, where it was found to accurately reproduce the calculated streamlines. It was also compared to integration of particle paths in the x–y domain for unsteady flow, and found to produce indistinguishable results for the grid size and time steps used in the current work. For the cases run here, ξ–η domain tracing added approximately 20% overhead to the fluid solver calculations, whereas x–y domain tracing added up to 300% overhead due to the computationally intensive requirement to locate the grid cell containing the particle being integrated, for every particle at each time step.

4 Results

Numerical results were generated at a Reynolds number Re = 30,000 and Mach number M_∞ = 0.2, with both fully laminar and fully turbulent flow assumptions. Experimental results are reproduced from Lai and Platzer [9], generated in a water tunnel at Re = 20,000. Local instantaneous Mach numbers in the wake in the numerical results rise as high as 0.45 for the highest-kh case presented ($k = 4.0, h = 0.05$), so that compressibility effects are expected to become more apparent as kh is increased further. Also laminar numerical calculations are expected to be more sensitive to the differences in Reynolds number than turbulent calculations, due to larger predicted separation regions at the trailing edge. Consequently this paper seeks only qualitative comparison with the experimental results, and the full effects of Mach and Reynolds numbers remain to be explored.

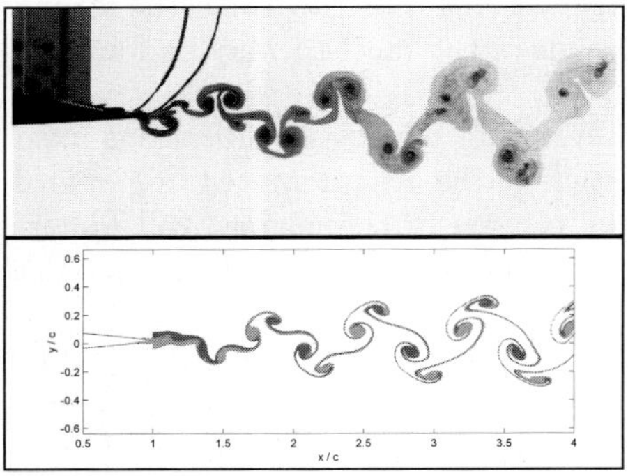

Fig. 3. **(a)** Experimental $k = 3.93$, $h = 0.0125$, **(b)** Numerical $k = 4.0$, $h = 0.0125$

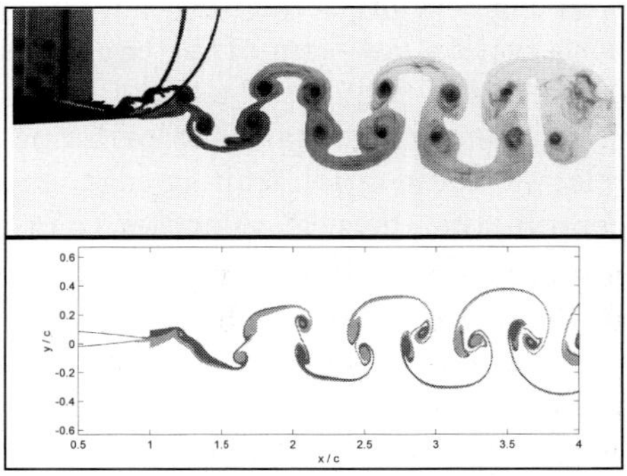

Fig. 4. **(a)** Experimental $k = 3.93$, $h = 0.025$, **(b)** Numerical $k = 4.0$, $h = 0.02$

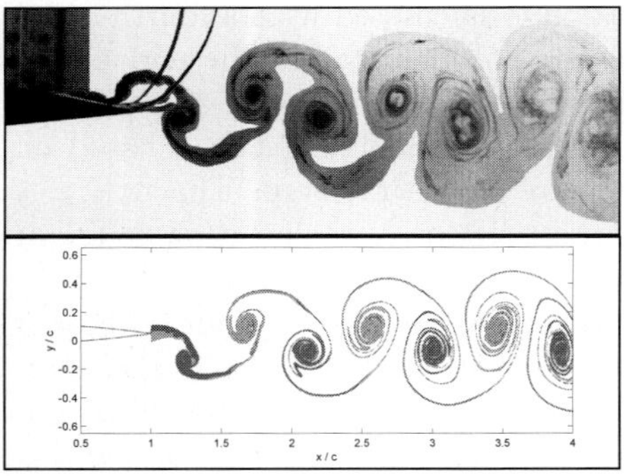

Fig. 5. **(a)** Experimental $k = 3.93$, $h = 0.05$, **(b)** Numerical $k = 4.0$, $h = 0.05$

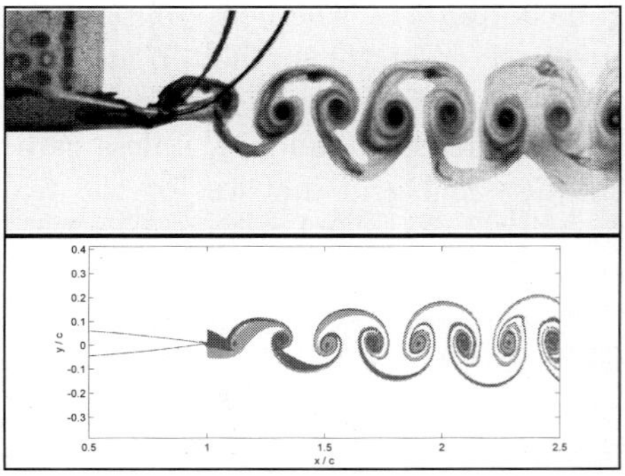

Fig. 6. **(a)** Experimental $k = 7.85$, $h = 0.0125$, **(b)** Numerical $k = 8.0$, $h = 0.0125$

Figures 3 to 5 compare experimental and laminar numerical results at $k \approx 4.0$ and $h = 0.0125$ to 0.05, representing drag-producing, neutral and thrust-producing wakes. Note that the experimental results have been processed to highlight the structures in the wake. The experimental behavior is reproduced qualitatively by the laminar numerical results in each case, although the neutral wake is predicted at a lower plunge amplitude ($h = 0.02$) than in the experiment ($h = 0.025$). Turbulent numerical results (not shown) were qualitatively different for $h = 0.0125$ and 0.02, appearing as conventional Karman vortex streets, but very similar for $h = 0.05$ albeit with less prominent Helmholtz instability vortices between the large shed vortices.

Figure 6 compares experimental and numerical results at $k \approx 8.0$, $h = 0.0125$. Contrasting these figures to the experimental results in Fig. 4, produced at the same value of kh, we see a marked difference in the wake structure. Similarly Fig. 7 compares three numerical laminar wakes produced at $kh = 0.05$, but with varying k and h. The wakes are shown at the same equivalent scale (normalized against plunge amplitude h). We see a straightforward Karman vortex street at $k = 8.0$, $h = 0.00625$, but multiple-vortex-per-half-cycle type shedding at $k = 4.0$, $h = 0.0125$ and $k = 2.0$, $h = 0.025$. Both Figs. 6 and 7 support the assertion that kh is not the sole determinant of wake structure. The good agreement between experimental and laminar numerical results for the range of k and h presented in this paper increases confidence that the features observed in Fig. 7 are physically realistic.

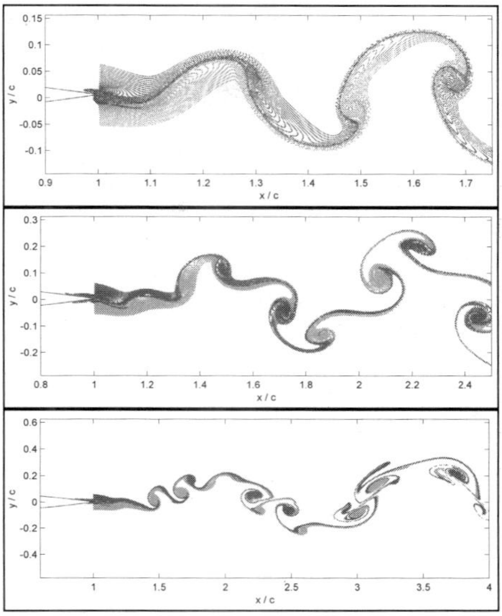

Fig. 7. Numerical **(a)** $k = 8.0$, $h = 0.00625$, **(b)** $k = 4.0$, $h = 0.0125$, **(c)** $k = 2.0$, $h = 0.025$

With this in mind, the method presented here may be used to examine the details of the vortex shedding process immediately downstream of the trailing edge. As an example, an instantaneous streamline plot for $k = 4.0$, $h = 0.0125$ (Fig. 8) shows a pair of counter-rotating vortices attached to the trailing edge, around which the remainder of the fluid flows, creating an effectively blunt-edged body. The degree of bluntness of the trailing edge varies throughout the plunge cycle, as these vortices separate and re-form on the opposite surface of the airfoil. The multiple-vortex-per-half-cycle structures apparent in Figs. 3, 4 and 7 may be the result of bluff-body shedding, and the apparent invalidity of the Kutta condition, whereby the trailing edge does not act as a fixed separation point, suggests shedding frequency modal competition similar to that found for tranversely vibrating circular cylinders [15].

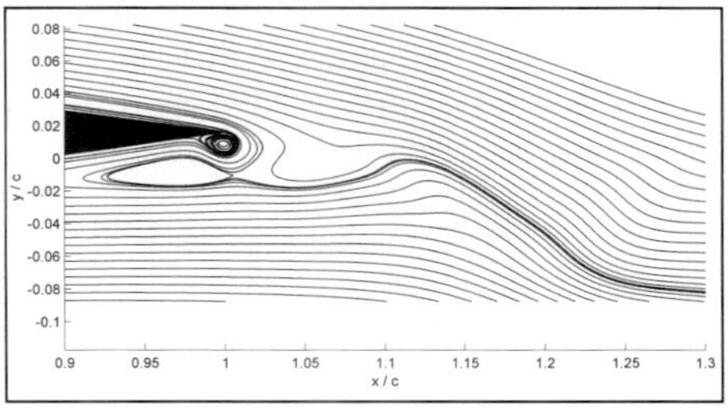

Fig. 8. Instantaneous numerical streamlines, velocity field relative to the airfoil, $k = 4.0$, $h = 0.0125$

5 Concluding Remarks

The form of the wake is shown to depend on both plunge amplitude h and plunge frequency k and to be sensitive to whether the flow is laminar or turbulent. Laminar flow calculations reproduce the prominent features observed by dye flow visualization better than turbulent flow calculations, for the range of k and h considered. The laminar results closely reproduced the experimental results of Lai and Platzer [9], as well as the qualitative aspects of experimental results in the literature [13,14] (reproduced at Fig. 2), for drag-producing, neutral and thrust-producing conditions. The method described here will be used to further explore the variation of the vortex shedding mechanism of a plunging airfoil with changes in k and h.

References

1. J.A. Ekaterinaris, M.F. Platzer: Computational Prediction of Airfoil Dynamic Stall. Progress in Aerospace Sciences **33** pp.759-846 (1997)
2. M.J. Lighthill: A Note on the Swimming of Slender Fish. J. Fluid Mechanics **9**, pp.305- 317 (1960)
3. I.H. Tuncer, J.C.S. Lai, M.F. Platzer: A Computational Study of Flow Reattachment Over a Stationary/Flapping Airfoil Combination. AIAA Paper 98-0109 (1998)
4. W. Shyy, M. Berg, D. Ljungqvist: Flapping and Flexible Wings for Biological and Micro Air Vehicles. Progress in Aerospace Sciences **35** pp.455-506 (1999)
5. M.M. Koochesfahani: Vortical Patterns in the Wake of an Oscillating Airfoil. AIAA J. **27**(9) pp.1200-1205 (1989)
6. C.M. Dohring, M.F. Platzer, K.D. Jones, I.H. Tuncer: Computational and Experimental Investigation of the Wakes Shed from Flapping Airfoils and their Wake Interference/Impingement Characteristics. In: *Characterisation and Modification of Wakes from Lifting Vehicles in Fluids CP-584*, AGARD, pp.33-1 to 33-9 (1996)
7. I.E. Garrick: Propulsion of a Flapping and Oscillating Airfoil. NACA Report No-567 (1937)
8. P. Freymuth: Propulsive Vortical Signature of Plunging and Pitching Airfoils. AIAA J. **26**(7) pp.881-883 (1988)
9. J.C.S. Lai, M.F. Platzer: The Jet Characteristics of a Plunging Airfoil. AIAA J. **37**(12) pp.1529-1537 (1999)
10. W.J. McCroskey: Unsteady Airfoils. Ann. Rev. Fluid Mech. **14** pp.285-311 (1982)
11. B.S. Baldwin, H. Lomax: Thin Layer Approximation and Algebraic Model for Separated Turbulent Flows. AIAA Paper 78-257 (1978)
12. I.H. Tuncer, J.C.S. Lai, M.A. Ortiz, M.F. Platzer: Unsteady Aerodynamics of Stationary / Flapping Airfoil Combination in Tandem. AIAA Paper 97-0659 (1997)
13. G.S. Triantafyllou, M.S. Triantafyllou, M.A. Grosenbaugh: Optimal Thrust Development in Oscillating Foils with Application to Fish Propulsion. J. Fluids and Structures **7** pp.205-224 (1993)
14. H. Ohashi, N. Ishikawa: Visualization Study of a Flow Near the Trailing Edge of an Oscillating Airfoil. Bulletin of the Japanese Society of Mechanical Engineers **15** pp.840-845 (1972)
15. B.S.V. Patnaik and P.A.A. Narayana: Numerical Simulation of Laminar Flow Past a Transversely Vibrating Circular Cylinder. J. Sound and Vibration **228**(3) pp.459-475 (1999)

Numerical Study of Transonic Shock/Boundary Layer Interactions on an Oscillating Airfoil Using a Third-Order Scheme and Nonlinear Turbulence Models

Paola Cinnella

SINUMEF Laboratory, Ecole Nationale Supérieure d'Arts et Métiers,
151 bd. de l'Hôpital, 75013 Paris, France
e-mail: Paola.Cinnella@paris.ensam.fr

1 Introduction

The understanding and the prediction of unsteady shock/turbulent boundary layer interactions is of great importance for many external and internal aerodynamic applications, notably when the shock is strong enough to produce massive boundary layer separation. This kind of phenomena represents a really challenging CFD problem, both for turbulence models and numerical methods. The modeling difficulties are related to the simultaneous occurrence of compressibility effects, strong pressure gradients, boundary layer separation, streamline curvature, all of which give rise to unequal normal Reynolds stresses. In such cases, standard eddy viscosity models generally show all their deficiencies, inherent to the linear description chosen for the Reynolds stress tensor. Additional difficulties are due to flow unsteadiness: as pointed out in [1], standard transport equation models, by their nature, are not completely suitable for time-dependent calculations; on the other hand more sofisticated approaches are still not reliable enough and too much expensive to be used in such extremely complex unsteady flow fields. It is then necessary to look for some compromise solution: an "optimal" turbulence model for transonic separated unsteady flows should account for more flow physics than standard Boussinesq models and, at the same time, have good numerical robustness and moderate computational cost. ¿From this point of view, the new generation of algebraic Reynolds stress closures seems to be quite promising.

In the present work, the shock stall of an oscillating NACA64A010 airfoil for which detailed experimental data are available is studied. In order to compute the shock/boundary layer interactions and the separated zones without a significant influence of the numerical errors, a high-order accurate numerical scheme [5],[6],[7] is used to solve the compressible Reynolds Averaged Navier–Stokes equations and several linear and nonlinear turbulence models are assessed against experimental data, both for the flow past the fixed and the oscillating airfoil.

2 Governing Equations and Turbulence Modeling

The compressible, unsteady, Reynolds–Averaged Navier–Stokes equations are written in conservative form in a curvilinear, body fitted coordinate system. Several linear (Boussinesq) and nonlinear turbulence models are considered. The nonlinear models are constructed using the EASM (Explicit Algebraic Stress Model) closure of Gatski and Speziale [8], with coefficients derived from the Launder–Reece–Rodi [9] pressure–strain correlation model. These nonlinear constitutive relations can be easily implemented to upgrade existing codes containing standard k-based transport-equation models; moreover, they do not introduce any additional differential equations with respect to the linear models. More exactly, the turbulence models considered for the present study are:

- the linear Spalart–Allmaras (S–A) one-equation model [10] and the low-Reynolds linear Jones-Launder k-ε [11] and Smith k-l two-equation models [12];
- EASM models constructed using the above nonlinear constitutive relations combined with transport equations for k and ε or k and l respectively.

The algebraic model of Baldwin and Lomax (B–L) was also considered, as a reference in terms of computational cost and numerical robustness.

3 Numerical Method

The RANS equations are discretised using a high-order accurate cell-centered finite volume scheme [5],[6]. The centered, five points, third-order accurate spatial approximation is constructed by correcting the dispersive error of the classical second-order Jameson scheme for the Euler equations. Particular care is taken to properly extend the scheme to multidimensional curvilinear meshes while insuring truly third-order accuracy. The numerical flux is constructed using suitably weighted discretisation formulae, which take into account mesh deformations [6]. The resulting scheme is third-order accurate on mildly deformed grids and second-order accurate on strongly deformed grids. The numerical method is then extended to the Navier–Stokes equations using a second-order discretisation for the viscous terms. If steady problems are considered, the system of governing equations is advanced in time using a four-stage Runge–Kutta scheme [13] with implicit residual smoothing [14]. Local time stepping and multigrid techniques are used to speed up convergence to the steady state solution. For unsteady flow computations, a fully implicit third-order accurate time-discretisation is used [7], which is iteratively solved at each time-step by a dual time-stepping technique [15] The above methodology has been implemented within the CANARI code of ONERA [16] and validated through systematic comparisons with the Jameson scheme, implemented within the same code.

4 Results and Discussion

The CANARI code originally contained only linear turbulence models. In the present study, the EASM closure of Gatski and Speziale was also implemented in. Numerous steady and unsteady calculations have been carried out using the formulation just described. Here, only a subset of results are presented.

NACA64A010 airfoil at fixed incidence. The turbulent flow past a NACA64A010 airfoil was simulated at Mach 0.8, 6.2^0 angle of attack and $Re = 2 \times 10^6$. For this problem, we dispose of the detailed experimental results of Johnson and Bachalo [2]. The flow is characterized by a strong shock wave at the upper surface of the airfoil and an extended recirculating region downstream the shock (see Fig. 1). After some tests, a (relatively coarse) C-grid of 311×75 points ($y^+ \approx 1.5$) was selected for the present calculations. Figure 2 represents the pressure coefficient for the turbulence models under study. As it could be expected, the B–L model performs quite poorly. On the contrary, the S–A model gives results in good agreement with the experiments. This model is moreover numerically robust and low CPU-demanding (it requires about 1.2 more CPU time per iteration and per point than B–L). The linear $k-l$ and $k-\varepsilon$ models perform better than B–L, but tend to locate the shock too much downstream with respect to S–A and the experiments. The EASMs allow to improve the results and both shock location and intensity are predicted with good accuracy. The EASM $k-l$ also allows to improve pressure distribution at the lower surface. Figures 3 and 4 show a typical velocity and shear stress profile at the airfoil upper surface. The EASMs give noticeably better results than B–L and S–A models for the shear stress and improve the $k-l$ and $k-\varepsilon$ solution. Less significant differences are observed for the velocity profiles. From the numerical cost and robustness point of view, a clear superiority of the EASM $k-l$ over the EASM $k-\epsilon$ (and of the linear $k-l$ over $k-\epsilon$) has been observed. Unfortunately, the nonlinear models are less robust than their linear versions which, in their turn, are more expensive and less robust than S–A and B–L. Since EASM $k-\varepsilon$ is less accurate and less robust than EASM $k-l$, it has not been retained for the following unsteady calculations.

NACA64A010 in sinusoidal pitching. The unsteady flow past a NACA64A010 airfoil oscillating at high angle of attack (AGARD SSC test case) has been calculated with Mach 0.789 and $Re = 12 \times 10^6$. The airfoil is pitched of $\pm 1.01^0$ about a mean angle of attack of 4^0 at a reduced frequency $k = 0.204$. The iso-Mach lines obtained using the EASM $k-l$ are presented in Fig. 5 to qualitatively illustrate the flow physics: the flow is attached or slightly separated during the most part of the upstroke (Fig. 5a); then, at an incidence of about 4.5^0 in the upstroke, shock stall occurs and during the downstroke, the flow is massively separated (Fig. 5b). Figure 6 compares the solution of the considered turbulence models with the experimental data available for the pressure distribution at the airfoil upper surface. The EASM $k-l$ solution, and in particular shock location, even if not perfect, is in rather good agreement with the experience, especially when the flow is mostly attached. Moreover, the solution compares well with that of the S–A model, which is considered as one of the best models for aerodynamic flows, and is significantly more accurate than the B–L or the $k-l$ solution. Contrary to linear models, EASM $k-l$ seems to predict the formation of a lambda-shock during the downstroke (see Fig. 5b and Fig. 6, for $\alpha = 4.28^0$), but no trace of such a structure appears in the experimental data. As in the previous steady case, EASM $k-l$ improves the solution accuracy but is less robust than linear models. This shortcoming, together with its higher numerical cost, can make its application to strongly separated unsteady transonic flows rather difficult.

5 Concluding Remarks

A numerical investigation on the use of EASM closures to predict unsteady transonic shock/boundary layer interactions has been carried out. From the accuracy point of view, the comparison of the EASM solution with experimental data and with some well-known linear turbulence models is rather positive. On the other hand the EASMs, at least in the present implementation, are numerically less robust than linear models. This could create difficulties for severe transonic unsteady flow computations. Furher testing of the EASMs on complex unsteady flows is currently in progress.

References

1. C.G. Speziale: AIAA J. **36**, pp. 173–184 (1998)
2. D.A. Johnson, W.D. Bachalo: AIAA J. **18**, pp. 16–24 (1980)
3. S.S. Davis: 'NACA64A010 Oscillatory Pitching'. AGARD R-702 (1982)
4. P. Cinnella, A. Lerat: CFD J. **9**, 1 (2000)
5. Y. Huang, P. Cinnella, A. Lerat: 'A Third-Order Accurate Centered Scheme for Turbulent Compressible Flow Calculations in Aerodynamics'. In: *Numerical Meth. Fluid Dynam. VI*, (Will Print 1998) pp. 355–361
6. A. Rezgui, P. Cinnella, A. Lerat: 'Third-Order Finite Volume Scheme for Euler Computations on Curvilinear Meshes', to appear (2000)
7. P. Cinnella, A. Lerat: 'A Fully Implicit Third-Order Scheme in Time and Space for Unsteady Turbulent Compressible Flow Simulations'. To be published in: *EC-COMAS 2000, Barcelona, Spain, September 2000*
8. T.B Gatski, C.G. Speziale: J. Fluid Mech. **254**, pp. 59–78 (1993)
9. B.E. Launder, G.J. Reece, W. Rodi: J. Fluid Mech. **72**, pp. 331–340 (1975)
10. P.R. Spalart, S.R. Allmaras: 'A One-Equation Turbulence Model for Aerodynamic Flows'. AIAA Paper 92-0439
11. W.P. Jones, B.E. Launder: J. Heat Mass Trans. **16**, pp. 1119–1130 (1973)
12. B.R. Smith: 'A Near Wall Model for the k-l Two-Equation Turbulence Model'. AIAA Paper 84-2386
13. A. Jameson, W. Schmidt, E. Turkel: 'Numerical Solutions of the Euler Equations by Finite Volume Methods Using Runge–Kutta Time Stepping'. AIAA Paper 81-1259
14. A. Lerat, J. Sidès, V. Daru: 'An Implicit Finite-Volume Method for Solving the Euler Equations'. In: *Lecture Notes in Physics 170* (Springer, Berlin, Heidelberg 1982) pp. 343–349
15. A. Jameson: 'Time-Dependent Calculations Using Multigrid, with Applications to Unsteady Flows Past Airfoils and Wings'. AIAA Paper 91-1596
16. A.M. Vuillot, V. Couailler: '3D Turbomachinery Euler and Navier–Stokes Calculations with a Multidomain Cell-Centered Approach'. AIAA Paper 93-2576

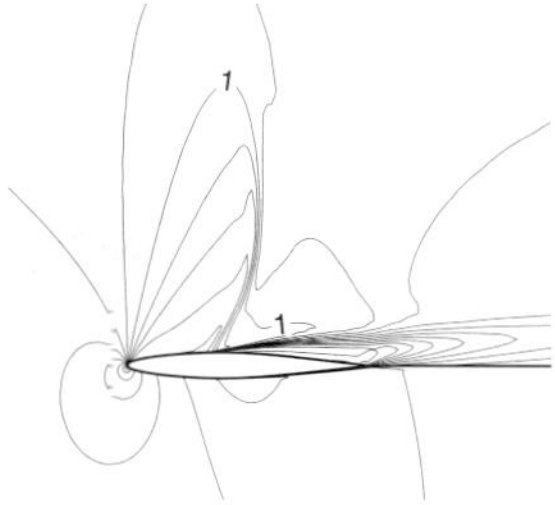

FIG. 1: Steady flow past a NACA64A010 airfoil. Iso-Mach lines (EASM $k-l$, $\Delta M = 0.1$)

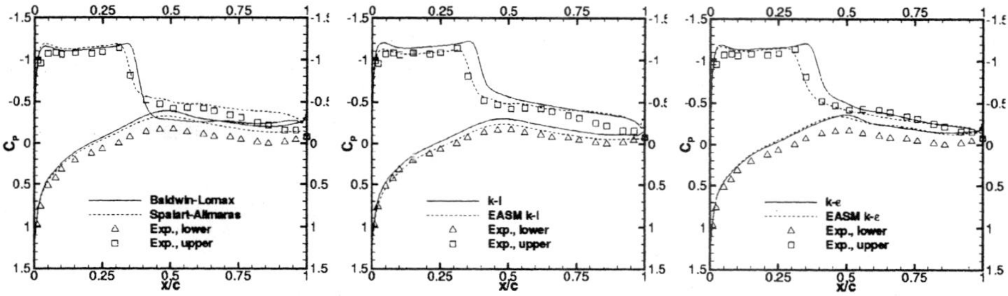

FIG. 2: Steady flow past a NACA64A010 airfoil. Pressure coefficient

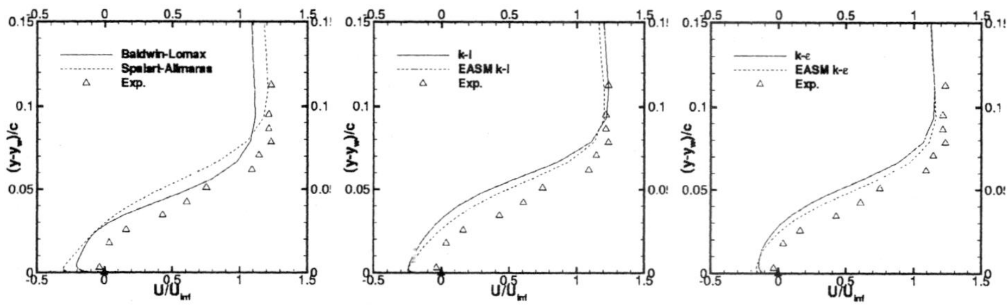

FIG. 3: Steady flow past a NACA64A010 airfoil. Velocity profiles at $x/c = 0.67$

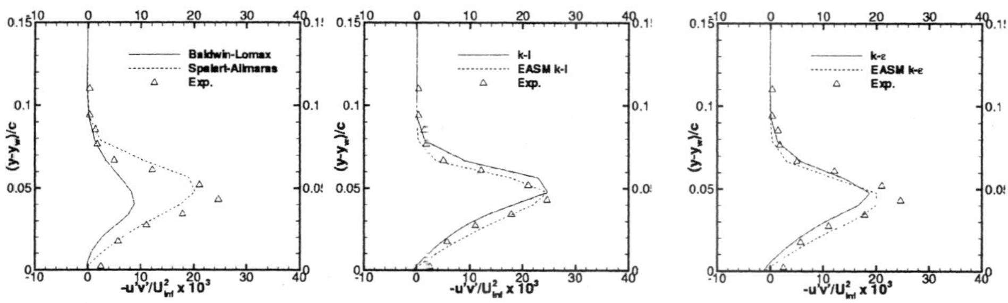

FIG. 4: Steady flow past a NACA64A010 airfoil. Shear stress profiles at $x/c = 0.67$

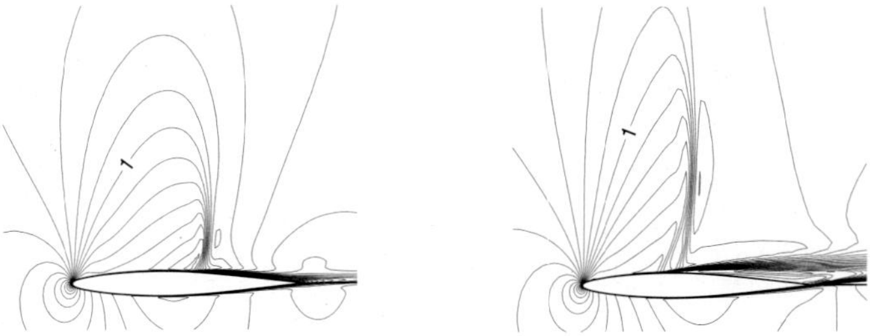

(a) $\alpha = 4.41^0$ (upstroke) (b) $\alpha = 4.28^0$ (downstroke)

FIG. 5: Unsteady flow past a NACA64A010 airfoil. Iso-Mach line (EASM $k-l$, $\Delta M = 0.05$)

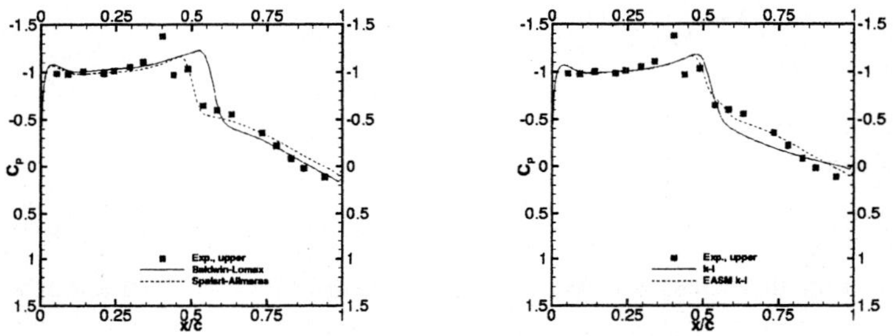

$\alpha = 4.87^0$ (upstroke)

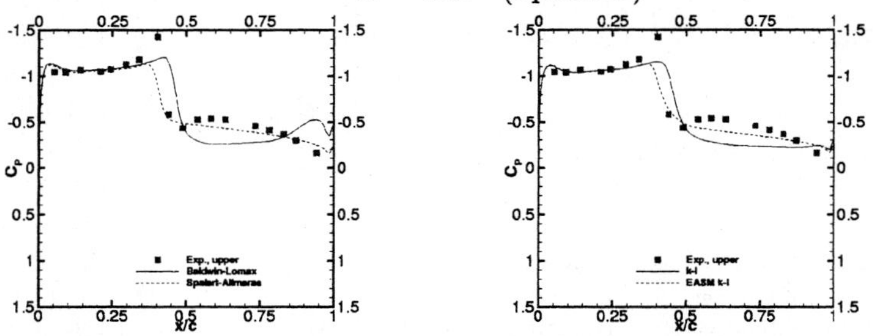

$\alpha = 4.87^0$ (downstroke)

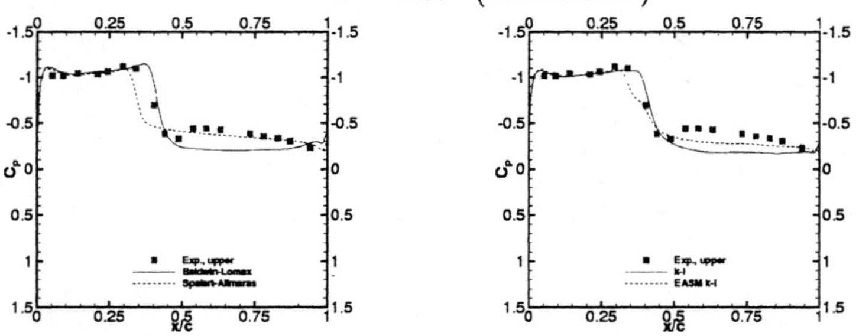

$\alpha = 4.28^0$ (downstroke)

FIG. 6: Unsteady flow past a NACA64A010 airfoil. Pressure coefficient

Numerical Study on Oscillation Modes of Supersonic Multijets

Nobuyoshi Fujimatsu and Kojiro Suzuki

The University of Tokyo, Japan

Abstract. Oscillation modes of circular single jet, two jets, and four jets configurations were numerically investigated and the details of the flowfield structures are discussed. The supersonic jet periodically oscillates behind the third or fourth shock cell and the oscillating shock cell generates the vortex structures. The flowfield structure and the acoustical characteristics of the oscillation modes for supersonic jet are in good agreement with the experimental results.

1 Introduction

In a supersonic free jet, there are several shock cell, and vortices produced by the instability in the shear layer interact with the shock cells. Interaction between the vortices and the shock waves produces very strong acoustic pulses. The resulting sound waves propagate upstream and excite the disturbunces in the thin shear layer near the nozzle lip. In this manner, the feedback loop instability is created [1]. In the measured spectra of sound emitted from a supersonic jet, a narrow-band frequency noise component, called as a screech tone, can be observed. This feedback loop instability is responsible for the self-excited oscillation of the jet.

The propulsion systems of many aircraft and rockets use clusters of axisymmetric nozzles. When multiple supersonic jets are closely spaced, the sound level of the screech may be greatly amplified due to the resonance between the jets, and the sonic fatigue of external panels can be caused. Such damage is believed to be caused by the components of screech radiating upstream. Thus, the instability of multiple supersonic jets has gathered much interest in recent years[2,3,7].

However, the dynamics of multi-jets configurations, such as two and four jets, has remained less explored. In this paper, supersonic multi-jets are numerically simulated, and the fundamental characteristics of the unsteady phenomena of multi-jets are discussed.

2 Computational Method

Self-excited oscillating supersonic free jets are numerically simulated by the finite volume method. The governing equations are the unsteady three dimensional Euler equations. The numerical fluxes for the convective terms are evaluated by the SHUS scheme (Simple High-resolution Upwind Scheme)[4], and is extended to third-order by MUSCL interpolation based on the primitive variables. The

SHUS scheme is a kind of AUSM-type scheme that was developed by Shima et al[4]. The two stage Runge-Kutta method is used for time integration.

The nozzle exit diameter D is 0.01[m]. The nozzle exit Mach number is 1.0. The pressure ratio PR, that is defined as the ratio of the reservoir pressure to the ambient pressure, is varied from 3.03 to 5.31. The sonic jet flow condition which is isentropically expanded from reservoir condition is imposed on the nozzle exit. The slip wall condition and the characteristic non-reflecting boundary condition[5] are used on the nozzle lip and on the outer boundary, respectively. The initial condition for the ambient air is 1.01325×10^5[Pa] static pressure, 293[K] static temperature and zero flow velocity. The numerical time step is 0.1 μ [sec].

Typical layout of the computational mesh used to simulate the case of four jets are shown in Fig. 1. The zone A is created around the nozzle and is connected to the zone B for far field. The computational domain for four jets consists of the eight zones. All the zones are solved simultaneously.

The mesh system with $42 \times 31 \times 120$ points and $21 \times 30 \times 120$ points are used for zone A and B, respectively. Mesh size at the nozzle exit is 1/30 times of the nozzle exit diameter. Mesh size in the downstream is 1/10 times of the nozzle exit diameter. The angular spacing in the nozzle is 9.0[deg]. The center-to-center distance of the nozzles for multi-jets simulation is set as $2.8 \times$ D for comparison with experiments[7].

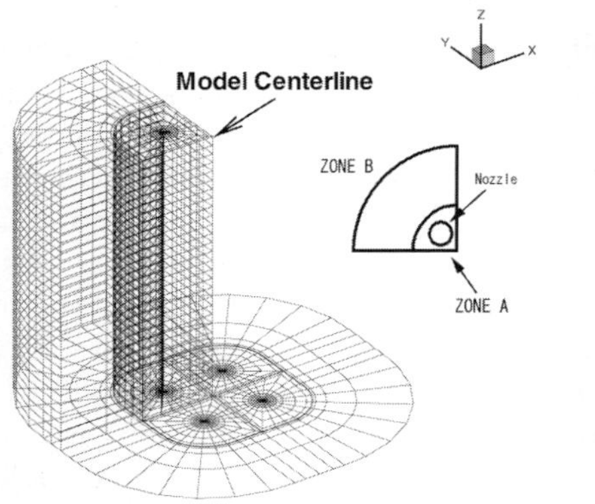

Fig. 1. Mesh system for the case of four jets

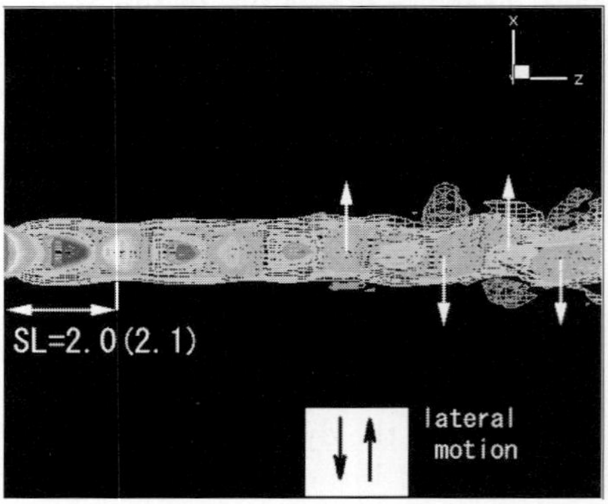

Fig. 2. Density contours of symmetric mode in the case of single jet(PR=4.87)

3 Results and Discussion

The general characteristics for supersonic free jet are discribed as follows. The sonic jet plume is accelerated to a supersonic jet by the expansion waves at the nozzle lip. The shock cell structure and the shear layer is created by the supersonic jet plume. The shock cell length becomes larger with increasing the pressure ratio. The length of the first shock cell(SL) in the numerical result is shown in Fig. 2 and the experimental result is shown in the parenthesis. These numerical results are in good agreement with the experimental results in all the cases. The present numerical simulations sucessfully capture the qualitative characteristics of the supersonic jet. The frequency of the oscillation in the case of single jet is about 10 [kHz] and is almost the same as those in the cases of two and four jets.

3.1 Single jet

Flowfield structure and oscillation mode Figure 2 shows the sideview of the instantaneous density contours for the single jet at PR=4.87. The first and second shock cells are clearly observed, but in further downstream the shock cell structure is not clear. The flowfield is dominated by large-scale disturbunces after the fourth shock cell. The supersonic jet periodically oscillates in the lateral direction behind the third shock cell.

Acoustical characteristics In the experimental study, the oscillation mode is discriminated by measuring the phase difference of the screech tone. The oscillation modes in the case of single jet are also investigated in the numerical study from a view point of acoustical characteristics.

The sound pressure level(SPL) is calculated from the pressure histories near the nozzle lip. SPL is defined as Eq. (1). The phase difference ψ is calculated from Eq. (2).

$$SPL(f) = 20\log\frac{\sqrt{P(f)/T}}{P_{ref}}[dB] \quad (1)$$

$$\psi(f) = tan^{-1}\frac{I_m(X(f)Y^*(f))}{R_e(X(f)Y^*(f))} \quad (2)$$

f : *frequency*
$P(f)$: *powerspectrum*
P_{ref} : $20\mu Pa$
$*$: *complex conjugate*

$X(f)$ and $Y(f)$ are the Fourier transforms calculated from the pressure histories at the reference point and at the point with circumferential angle θ from the reference point, respectively. The pressure histories are recorded after the flowfield has reached almost periodic state. The total sampling time T is 0.01 (sec) in the present calculation.

Figure 3 represents the relationship between SPL and the frequency calculated from the pressure histories near the nozzle lip at PR=4.87 in the case of single jet. In this simulation, two distinct peaks are observed, and the features of the screech tone of the oscillating supersonic free jet are captured. The frequency of the first peak is 8496Hz. The experimental result is 9125Hz.

The screech frequency is related to the feedback loop length between the nozzle lip and the shock cell point at which the sound wave is generated. These flowfields are simulated by solving the non-viscous equations, and the viscous effect is not included in interaction of the shock cell and the vortex. In real situation, the shock cell length becomes shorter after the first shock cell than predicted by inviscid analysis. Accordingly, the shock cell length in the present numerical simulation is longer than the viscous case, and the screech frequency is underestimated.

The relationship between the phase difference ψ and the separation angle θ in the case of single jet at PR=4.87 is shown in Fig. 4. At PR=4.87, the phase difference of 0 and 180(deg) is separated on the plane of symmetry (θ=180(deg)). This feature indicates that the present oscillation mode is the lateral mode and is in good agreement with the experimental results [7].

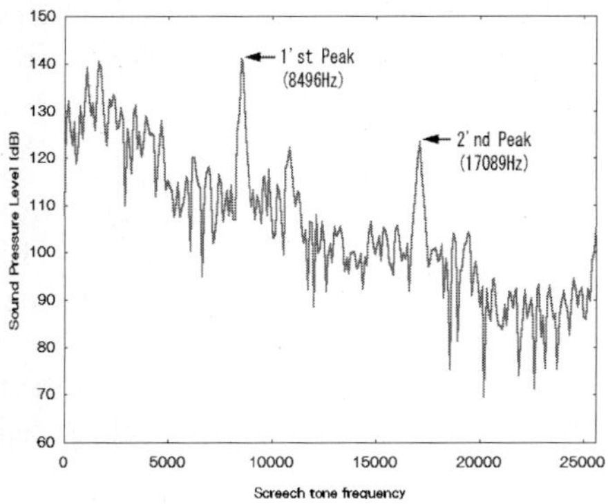

Fig. 3. Screech characteristics of the single jet at PR=4.87

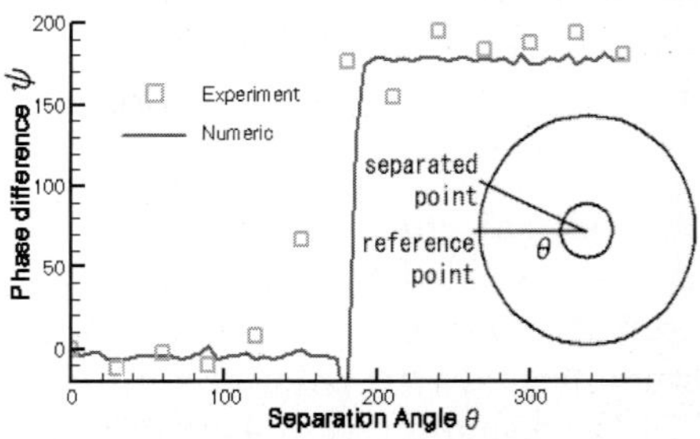

Fig. 4. Relationship between phase difference ψ and separation angle θ o the single jet at PR=4.87

3.2 Two jets

Figure 5 shows the sideview of the instantaneous density contours for the case of two jets at PR=3.71. The density contours clearly show how the large-scale vortices grow. The shear layer begins to roll up at the third shock cell, and the flow behind the third shock cell oscillates helically.

3.3 Four jets

Figure 6 shows the instantaneous density contours at PR=3.03. In this case, the quadrupole-like oscillation mode appears. The vortex structure oscillates diagonally after the third shock cell.

The helical mode appears at PR=3.90 as shown in Fig. 7. The shear layer begins to roll up at the third shock cell and the helical deformation occurs. After the fourth shock cell, the helical deformation has been completely grown up into the periodical vortex shedding.

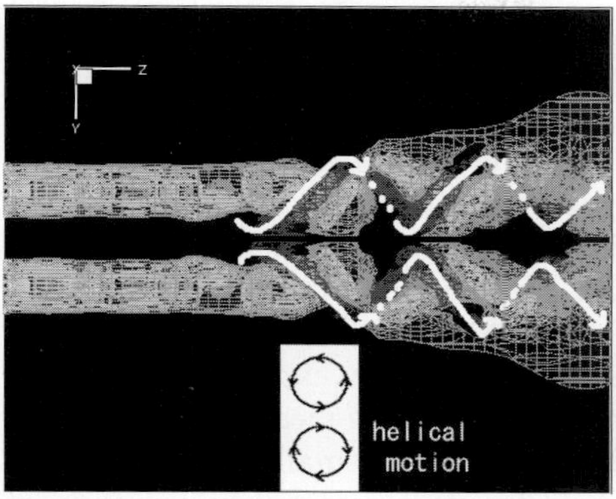

Fig. 5. Density contours of helical mode in the case of two jets(PR=3.71)

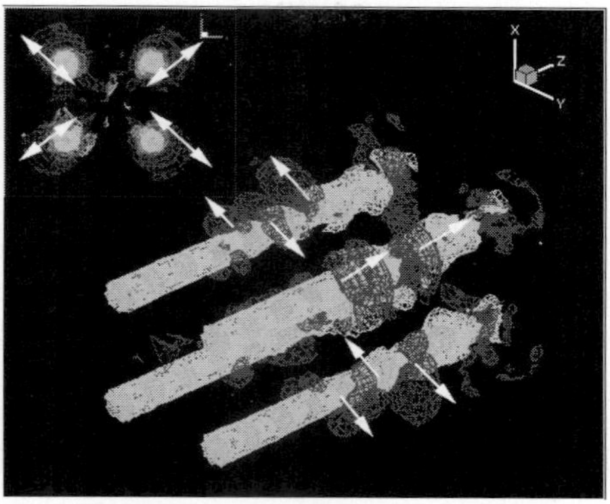

Fig. 6. Density contours of symmetric mode in the case of four jets(PR=3.03)

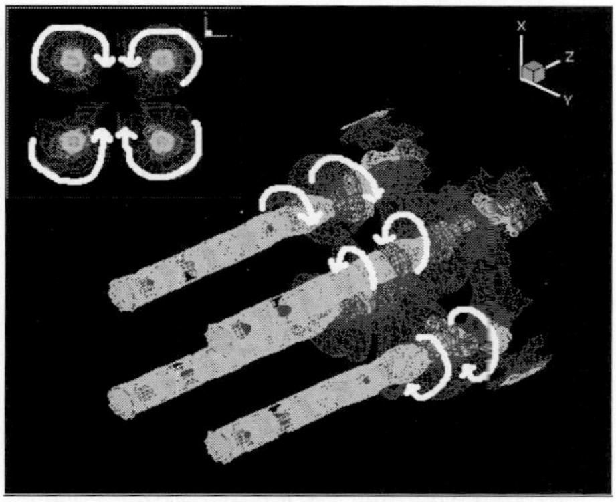

Fig. 7. Density contours of symmetric mode in the case of four jets(PR=3.90)

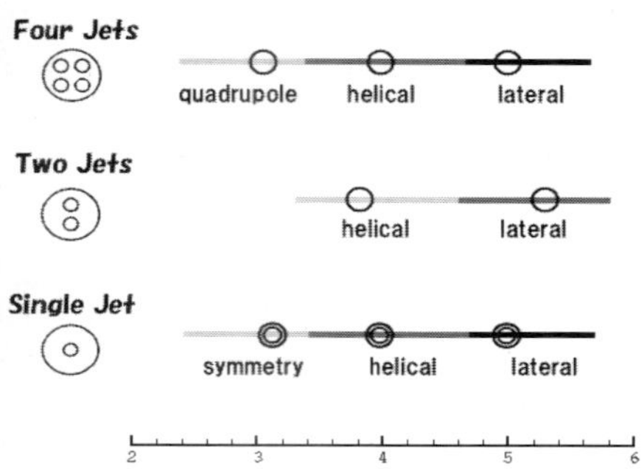

Fig. 8. Oscillation modes to pressure ratio

3.4 Comarison with experimental study

The oscillation modes of single jet, two jets, and four jets are reported in the experimental studies [6,7]. The range of the pressure ratios in each oscillation modes are shown in Fig. 8. In transition region, one oscillation mode gradually changes to another oscillation mode [6]. In Fig. 8, the co-encentric and the circular symbol represent the condition in which the computation is done. It is confirmed that the oscillation modes of the supersonic jet are coincide with the experimental results by using the computational animation. In the case of single jet, the acoustical characteristics are also in good agreement with the experimental results.

4 Conclusion

The oscillation modes of supersonic multi-jets are numerically simulated by solving the three-dimensional Euler equations. In the case of single jet, the flowfield structure and the acoustical characteristics of the oscillation modes are in good agreement with the experimental results.

It is important that the oscillation modes observed in experiments are successfully captured by the inviscid flow computation. This means that the effect of the viscous force on the overall flowfield structure is small. As future work, the details of the three dimensional flowfield for multi-jets will be investigated more precisely to clarify the reason why the oscillation modes change at their critical pressure ratios.

Acknowledgement

This work is supported in part by Grant-in-Aid for Scientific Research (B)(2) No.11450376 at the Ministry of Education, Science, Sports and Culture, Japan and is carried out under the collaborating research program at Intelligent Modeling Laboratory in the University of Tokyo.

References

1. Powell, A., "On the Noise Emanating from a Two-Dimensional Jet above the Critical Pressure," The Aeronautical Quarterly, Vol. IV, 1953, pp. 103-122.
2. Hay, J. A. & Rose, E. G., "In-flight shock cell noise", Journal of Sound and Vibration, 11, 411-420, 1970.
3. Berndt, D. E., "Dynamic Pressure fluctuations in the internozzle region of a twin-jet nacelle", SAE Paper 84-1540., SAE Aerospace Cong. & Expo., Long Beach, CA.
4. Shima, E., Jounouchi, T., "Role of CFD in Aeronautical Engineering (No.14) - AUSM Type Upwind Schemes-", NAL-SP30, Proceedings of 14th NAL symposium on Aircraft Computational Aerodynamics, 1996, pp.7-12.
5. Kevin. W. Thompson, "Time Dependent Boundary Condition for Hyperbolic Systems", Journal of Computational Physics, Vol.68, 1987, pp.1-24.
6. Powell, A., Y. Umeda and R. Ishii., "Observations of the oscillation modes of chocked circular jets", Journal of Acoustics Society America, Vol.92(5), November 1992.
7. Y. Umeda and R.Ishii, "Oscillation modes of supersonic multi-jets", J. Acoust. Soc. Am, **101** (6), June 1997.

Part VI

Hypersonic and Reactive Flows

Computation of Görtler Vortices in Separated Hypersonic Flows

Heinrich Lüdeke

DLR, Institute of Design Aerodynamics,Lilienthalplatz 7,
D-38108 Braunschweig, Germany, email: heinrich.luedeke@dlr.de

Abstract. In concave boundary layers as well as in separated flows counter-rotating streamwise vortices appear if certain flow conditions are satisfied. These socalled Görtler vortices lead to significant transversal heat flux fluctuations in hypersonic flows. Until now numerical simulations of such instabilities have been carried out only for simplified geometries. Nevertheless, fully three-dimensional computations are of special interest for hypersonic re-entry vehicles especially on deflected control surfaces (flaps). For the present investigation a lifting body vehicle is chosen to show the capability of Navier-Stokes codes to simulate the influences of Görtler vortices at Mach 6 and $Re = 3.1 \cdot 10^6$. Resolution of laminar vortices is possible by generation of local grids in the flap region with extensive spanwise grid refinement.

1 Introduction

As shown in different theoretical and experimental investigations [1] the flow in reattaching hypersonic boundary layers shows counter-rotating streamwise vortices if certain flow conditions are satisfied. It is well known, that these transitional 3D structures can increase thermal fluxes at the surface of hypersonic re-entry vehicles beyond the level obtained with fully turbulent flow.
The vortices are induced by the radial increase of centrifugal forces in boundary layers on concave surfaces, and they are able to distort the temperature layer significantly. The theoretical description of the vortices was given for the first time by H. Görtler (1941) with the approach of parallel instabilities of the Taylor type. The resulting local eigenvalue problem leads to the well known stability diagram. P. Hall was the first one who questioned the possibility of applying a quasi parallel eigenvalue analysis and derived a parabolic set of partial differential equations which do not allow a separation of variables. Furthermore he proved the dependence of the neutral curve on the position at which the flow is initiated and on the perturbations imposed at this position. So the approach of a unique neutral stability curve seemed to be refuted.
However, it was shown later on, that the agreement between Hall's theory and the local approach is quite good far downstream for appropriate choices of the amplification factor and initial conditions [2]. Recent investigations by Bottaro and Luchini take a look at the cases where local eigenvalue analysis is possible and in which cases parabolic equations are needed. Furthermore, they solved the receptivity problem of the instability by a Greens function approach marching backwards in time [3].

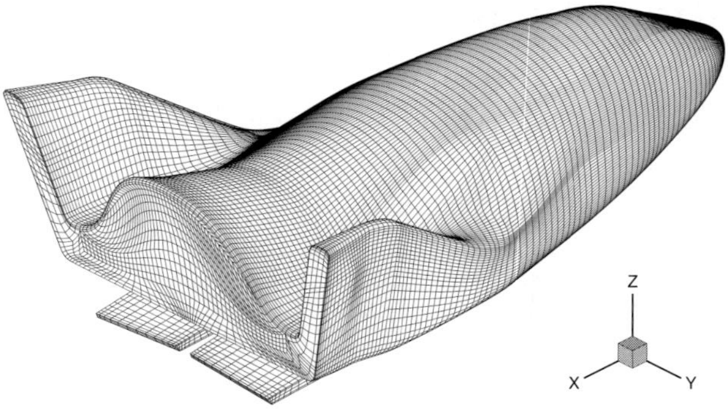

Fig. 1. surface mesh of the investigated re-entry configuration, 20° flap deflection angle

The effect of Görtler vortices is of special interest with respect to the control surfaces of re-entry vehicles since the topology of separated flow at the hinge lines of deflected control surfaces displays concave curved boundary layers at the reattachment line on the flap. The resulting Görtler vortices lead to significant spanwise heat flux peaks in this region.
For complicated 3D cases, as treated in the present study, the only efficient way of Görtler vortex prediction and investigation of the influences on the surface heat flux is a combined CFD/wind tunnel campaign to combine the advantages of both techniques. In the present paper we report the results of laminar computations that are carried out with especially designed computational grids around the body flaps of a realistic lifting body (Fig. 1) that represents the demonstrator re-entry vehicle X-38. The results show the appearance of counter-rotating vortices in the separation region and after reattachment as well as theyr influence on the heat flux distribution.

2 Numerical simulation

The main part of the numerical investigation has been carried out using the DLR CFD code CEVCATS-N, which was validated in a broad range of sub-, trans- and hypersonic flow cases for a variety of flow conditions and configurations. Examples are found in [4]. CEVCATS-N is a multiblock finite volume flow solver for the stationary Euler and Navier-Stokes equations in the integral form. The discretization is done by means of a second-order accurate finite-volume scheme with flow variables stored in the cell vertices. The numerical scheme consists of a hybrid upwind flux-vector splitting: the van Leer scheme for regions with strong shocks and the AUSM scheme according to Liou and Steffen for smooth flow regions. Second-order accuracy is obtained by MUSCL extrapolation, in order to allow the capturing of strong shocks and contact discontinuities. A five-stage Runge-Kutta scheme is used to advance the solutions in time. For the acceleration of the convergence process local time stepping, implicit residual averaging and full multigrid are possible options. On a NEC SX-4 super computer the cpu time per grid point and multigrid cycle is 20 μsec for the present computations. For turbulent calculations different validated zero- and one equation models are possible. The algebraic Baldwin-Lomax (BL) model and derivations have been tested previously within the X-38 project. For attached flows this algebraic model has shown good performance, in separation regions the one-equation Spalart-Allmaras (SA) model which is being validated for fully turbulent as well as for transitional cases. The transition locations have to be specified. A semi-empirical transition correlation has recently been implemented with very promising results for hypersonic flow cases.

3 Grid generation

For the construction of all meshes the grid generation system MegaCads [5] has been used to derive the 3D grid from the original CAD surface file. The Navier-Stokes grid contains about 10^6 nodes and includes the bow shock and the wake behind the vehicle (Fig. 1). For extensive grid studies different partial grids have been extracted from the blocks around the flaps and refined in all directions (Fig.2). Refinement in spanwise direction was needed for resolution of streamwise vortices with a short wavelength. For this purpose, the local meshes were further subdivided six times in spanwise direction ($1.1 \cdot 10^6$ nodes). Further improvements were achieved by redistribution of the boundary layer cells in order to capture vortex structures in the outer boundary layer and by refinement of the grid in the separation region.

4 CFD results

4.1 Flow conditions

For the present study flow conditions adapted to those in the Ludwieg tube facility in Göttingen are chosen. These are $M = 6$, $Re = 3.1 \cdot 10^6$, 40° angle of attack and 20° flap deflection angle. As a first step laminar computations are chosen although measurements suggest transitional behaviour after reattachment [5].

4.2 Flow over deflected body flap

Grid convergence results with respect to pressure and heat flux have been shown in previous investigations [5]. For laminar computations maximum changes due to grid refinement were found in the separation region at the flap hinge line where secondary separation occurred (Figs.2,3). This flow structure as well as the computed C_p distribution reproduces experimental observed data rather well, especially for the refined grids. However, experimental investigations of the heat

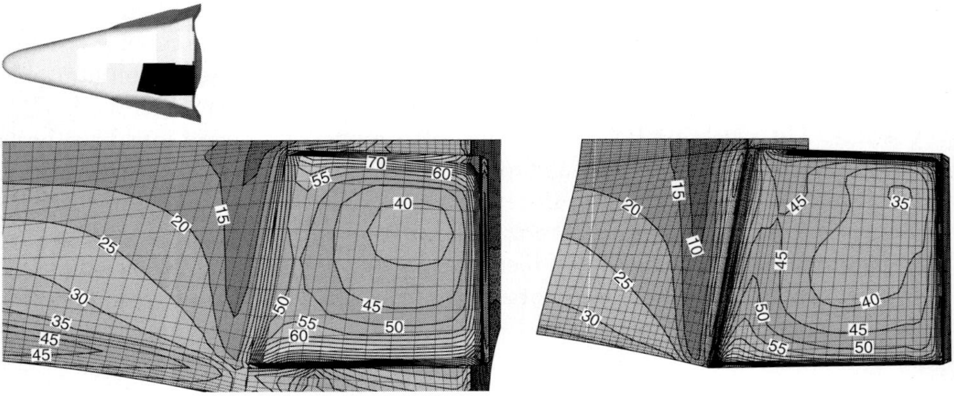

Fig. 2. Laminar heat flux distributions on 3- and 4-level grid at $M = 6$ $Re = 3.1 \cdot 10^6$. 3-level grid of the whole vehicle (*left*) . Reduced 4-level grid (*right*)

flux distribution indicate transition at the reattachment line which has only little effect on the pressure distribution.

Streamwise vortices are then resolved on extensively spanwise refined grids with adjusted point resolutions in streamwise and wall normal direction. All results from these grids have shown similar behaviour, resulting in a flow which is not completely stationary after vortices occur. The transient behaviour is studied by time-accurate test computations, showing an oscillation of the vortex structures which is still under study [6]. Visualisation of the perturbation flow, generated by subtracting the unperturbed basic solution from the perturbed one, shows the generation of the Görtler vortices in the separation region (Fig.5). The influence of the vortices on the wall heat flux is shown in Fig.4 with significant spanwise heat flux fluctuations. Heat flux peaks up to 25 % above the average value are generated by the perturbation. The perturbations in the velocity as well as in the heat flux are prolongated downstream the reattachment line. A direct

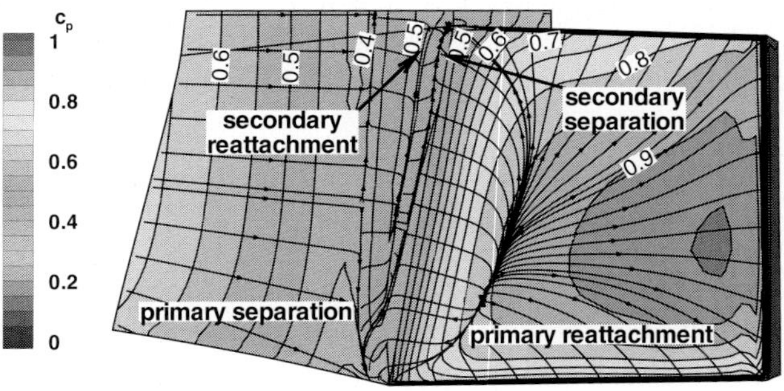

Fig. 3. Laminar flow topology on partial 4-level grid at $M = 6$ and $Re = 3.1 \cdot 10^6$

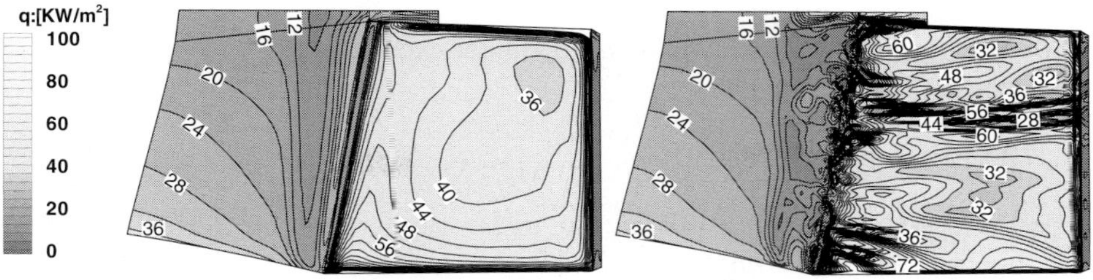

Fig. 4. Heat flux distribution in the flap region. Undisturbed basic solution (*left*). Influence of the vortices (*right*)

comparison of the CFD results with the wind tunnel data was not possible, since the measurements indicate turbulent flow conditions on the flap.

5 Conclusions and outlook

For the first time Görtler vortices on a realistic vehicle, as the X-38, including secondary separations and crossflow influences, are investigated by full Navier-Stokes computations.
Using extensively spanwise refined local grids in the flap region, resolution of streamwise vortices and their influence on the surface heat flux is possible. Preliminary grid refinement studies have shown very good results for the unperturbed flow. Vortex computation on grids with different streamwise and wall normal point distributions show similar behaviour of the perturbation flow and the heat flux. The suggestion of spanwise oscillations of the surface structures in time is supported by the results of time accurate computations [6]. Next steps will be the investigation of turbulent and transitional vortices on the X-38 flap and their transient behaviour. Furthermore even more refined grids have to be

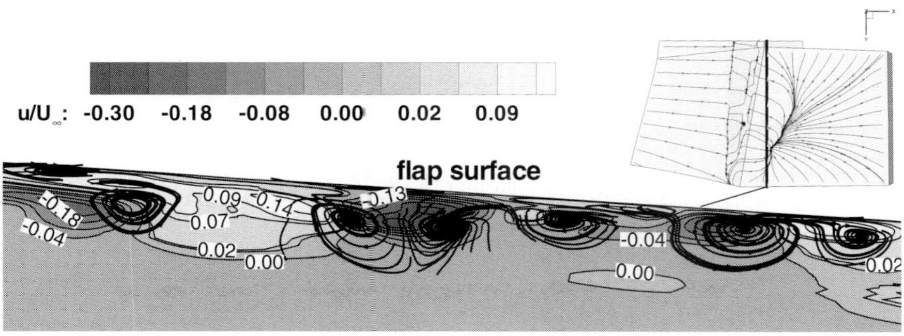

Fig. 5. Isolies of the perturbation flow in a perpendicular cut through the boundary layer on the body flap

generated to test the grid influence on the vortex diameter, the heat flux, and the discrepancies to the results of wind tunnel experiments.

6 Acknowledgements

This investigation is being carried out under support of the German national program on re-entry technologies, TETRA.

References

1. A.F. Kreins, A. Henkels, F. Maurer: 'Experimental studies of hypersonic shock induced boundary layer separation'. ZfW **20**, (1996) 80–88
2. P. Goulpie et.al.: 'Görtler vortices in boundary layers with streamwise pressure gradient, linear theory'. Phys. Fluids **8**, (1996) 451–459
3. P. Luchini, A. Bottaro: 'Görtler vortices: a backward in time approach to the receptivity problem'. J. Fluid. Mech. **363** (1998) 1–23
4. S. Brück, R. Radespiel, J.M.A. Longo: 'Comparison of Nonequilibrium Flows past a simplified Space-Shuttle Configuration'. AIAA Paper **97-0275** (1997).
5. H. Lüdeke, P. Krogmann: 'Investigation of Three-Dimensional Hypersonic Boundary Layer Structures'. Paper of the ISCFD Bremen, (1999)
6. H. Lüdeke,P. Krogmann: 'Numerical and Experimental Investigations of Laminar/Turbulent Boundary Layer Transition'. Paper of the ECCOMAS 2000 Barcellona, (2000)
7. J.J. Ginoux: 'Streamwise vortices in Reattaching High-Speed Flows: A Suggested Approach'. AIAA Journal **9** (1971) 759-760
8. G.R. Inger: 'Three-Dimensional Heat- and Mass-Transfer Effects across High-Speed Reattaching Flows'. AIAA Journal **15** (1977) 383–389

Numerical Simulations of Shock/Vortex Interaction in Non-Circular Jets

Zonglin Jiang[1] and Kazuyoshi Takayama[2]

[1] LHD, Institute of Mechanics, Chinese Academy of Sciences, Beijing 100080, China
[2] Shock Wave Research Center, Institute of Fluid Science, Tohoku University
2-1-1 Katahira, Aoba-ku, Sendai 980-8577, Japan

Abstract. In this paper, a numerical study of self-induced flow motions of non-circular jets is described to investigate the underlying mechanism of shock/vortex interaction, and axis-switching of the primary vortex loop in its initial stage. Numerical results were obtained by solving three-dimensional Euler equations with a dispersion-controlled scheme and displayed in the form of iso-vorticity surfaces. From the results, it was concluded that secondary flows, secondary shock waves and axis-switching play an important role in the jet instability development, and these are also the primary driving forces leading to the generation of hairpin vortices and complex coherence structures.

1 Introduction

Jets are widely applied in various industrial devices, e.g. rocket launching, fuel/air mixing and thrust producing in aircraft engines. Extensive investigations have been made on both planar and axisymmetric jets to enhance mixing process and thrust production or to reduce screech tones in nozzle flows. It is well known that non-circular jets are more unstable than two-dimensional ones (Krothapalli et al. 1981). This is not only because the non-circular jets have large-scale coherent structures and streamwise vortices, but also because of its special features, such as three-dimensional shock/vortex interactions, asymmetrical vorticity distributions and axis-switching (Husain and Hussain 1993, Koshigoe et al. 1989, and Tam and Thies 1993). Recently, there has been a growing interest in the use of non-circular jets in high-speed propulsive systems (Raman and Taghavi 1998). In such the systems, the residence time of air in hypersonic airplane engines is on order of 1 *ms* for typical flight conditions. Within this short time, the fuel must be injected, and penetrate deeply into air-stream, and be fully mixed for high quality combustion. Non-circular jets can offer possibilities of mixing process enhancement. The relevant research of axis switching and shock-enhanced mixing process have been reported by Krothapalli et al.1981 and Lee et al. 1998, but their research is mainly limited to either experimental measurements of average jet properties or theoretical analysis of jet instability models. The objective of this paper is to numerically investigate into the initial deformation of the primary vortex loop and the development of non-circular jet flows to explore the origin resulting in axis-switching and jet instability.

One way to study non-circular jets in laboratory is discharging a transmitting shock wave from the open end of a rectangular shock tube into ambient air

(Onodera et al. 1998). The primary vortex loop which develops near the open end of the shock tube behind the transmitting shock wave takes the shape of the shock tube cross section and is planar, but deforms quickly into a three-dimensional configuration. Three-dimensional secondary shock waves will also develop later for supersonic jets. This shock/vortex interaction will result in additional flow instability on the shear layer.

In order to remove uncertain factors and focus our attention on shock/vortex interaction and axis-switching, jet flows in the present study are assumed to be symmetrical and viscosity effects on the motion of the primary vortex loop at its early stage are negligible. Numerical simulations were carried out by solving the three-dimensional Euler equations with a dispersion-controlled scheme (Jiang et al. 1995). Four cases presented are of similar geometries with square or rectangular cross-sections, but with different aspect ratio and Mach number. The flow spreading, the secondary flow motion, the secondary shock wave generation, axis-switching, and aspect ratio effects on axis-switching are discussed in detail.

2 Governing Equations

Under the assumption mentioned in the above section, a hyperbolic system of three-dimensional conservation law for a perfect gas in Cartesian coordinates can be written as

$$\frac{\partial \mathbf{U}}{\partial t} + \frac{\partial \mathbf{F}}{\partial x} + \frac{\partial \mathbf{G}}{\partial y} + \frac{\partial \mathbf{E}}{\partial z} = 0, \tag{1}$$

where **U**, **F**, **G** and **E** denote the state variable and flux, respectively, given by

$$\mathbf{U} = \begin{pmatrix} \rho \\ \rho u \\ \rho v \\ \rho w \\ e \end{pmatrix}, \mathbf{F} = \begin{pmatrix} \rho u \\ \rho u^2 + p \\ \rho u v \\ \rho u w \\ (e+p)u \end{pmatrix}, \mathbf{G} = \begin{pmatrix} \rho v \\ \rho v u \\ \rho v^2 + p \\ \rho v w \\ (e+p)v \end{pmatrix}, \mathbf{E} = \begin{pmatrix} \rho w \\ \rho w u \\ \rho w v \\ \rho w^2 + p \\ (e+p)w \end{pmatrix}, \tag{2}$$

where primitive variables in the unknown **U** are density ρ, velocity components u, v and w, respectively, total energy per unit volume e, and fluid pressure p. The total energy e is related to the equation of state by

$$e = \frac{p}{\gamma - 1} + \frac{\rho}{2}(u^2 + v^2 + w^2), \tag{3}$$

where γ, the specific heat ratio, is taken as 1.4 in the present computation.

3 Numerical Methods

The explicit difference equations of (1) discretized in space by using the dispersion-controlled scheme (Jiang et al. 1995) are given as

$$\left(\frac{\partial \mathbf{U}}{\partial t}\right)^n_{i,j,k} = -\frac{1}{\Delta x}(\mathbf{H}^n_{i+\frac{1}{2},j,k} - \mathbf{H}^n_{i-\frac{1}{2},j,k}) - \frac{1}{\Delta y}(\mathbf{P}^n_{i,j+\frac{1}{2},k} - \mathbf{P}^n_{i,j-\frac{1}{2},k})$$

$$-\frac{1}{\Delta z}(\mathbf{Q}^n_{i,j,k+\frac{1}{2}} - \mathbf{Q}^n_{i,j,k-\frac{1}{2}}) \tag{4}$$

with

$$\begin{cases} \mathbf{H}^n_{i+\frac{1}{2},j,k} = \mathbf{F}^+_{i+\frac{1}{2}L,j,k} + \mathbf{F}^-_{i+\frac{1}{2}R,j,k} \\ \mathbf{P}^n_{i,j+\frac{1}{2},k} = \mathbf{G}^+_{i,j+\frac{1}{2}L,k} + \mathbf{G}^-_{i,j+\frac{1}{2}R,k} \\ \mathbf{Q}^n_{i,j,k+\frac{1}{2}} = \mathbf{E}^+_{i,j,k+\frac{1}{2}L} + \mathbf{E}^-_{i,j,k+\frac{1}{2}R} \end{cases}, \tag{5}$$

where the $(\cdot)^+$ or $(\cdot)^-$ superscript denotes flux splitting according to the Steger and Warming method, and more detailed introduction to the scheme was described by Jiang et al. 1995. Numerical solutions were marched in time by using the Runge-Kutta algorithm of second-order accuracy. Reflecting boundary conditions were specified both on solid walls and the planes of symmetry. Non-reflecting boundary conditions were applied at inflow and outflow boundaries. Computations were carried out in one quarter of the domain as shown in Fig. 1 because of the symmetric assumption. The equally-spaced grid of $600 \times 150 \times 150$ mesh points was used in case 1,2 and 3, and 50×50 points were distributed along the cross section of the shock tube. In case 3, the grid is composed of $400 \times 180 \times 120$ mesh points and 60×40 points were in the cross-section. Several view directions for observations are also defined in Fig. 1, which are the side view, the corner view; the back view, the front view, the diagonal plane and the mid-wall symmetrical plane.

4 Results and Discussion

Four test cases are carried out by discharging shock waves from the open end of shock tubes into ambient air for different Mach number measured before the shock waves move out of the shock tubes. In the discussion below, four main physical processes are emphasized, which are the secondary flow motion; secondary shock wave generation; axis-switching and the role of aspect ratio in axis-switching. These processes are self-induced flow motions closely linked to non-circular jet instability and have no relation to any disturbance in background flows.

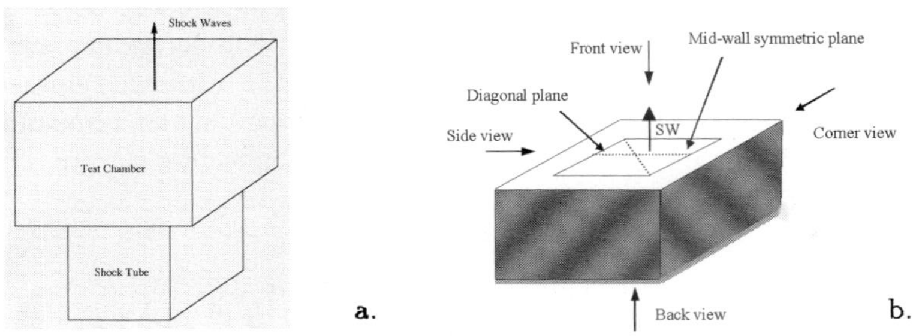

Fig. 1. Schematic of computational domain and view directions for flow visualization

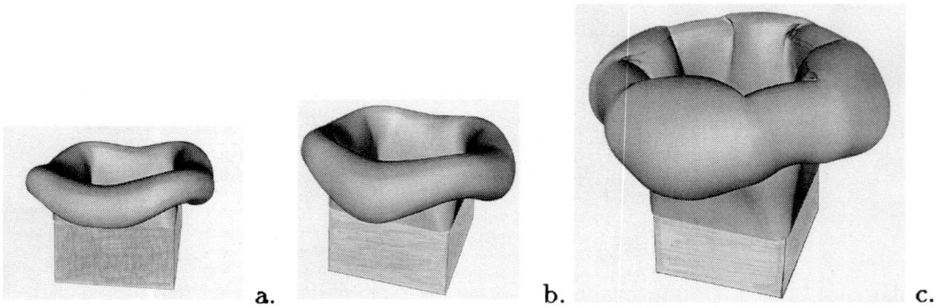

Fig. 2. Distortion of the primary vortex loop due to secondary flow shown with the iso-vorticity surface for $M_s = 1.29$: a. $t = 200\ \mu s$; b. $t = 300\ \mu s$; c. $t = 500\ \mu s$

4.1 Secondary Flow Motion

Numerical results of case 1 is calculated for $M_s = 1.29$ in a square cross-section shock tube. Three sequential three-dimensional displays of iso-vorticity surface are presented. Figure 2a shows the starting of the secondary flow motion. The flow fronts can be observed in the primary vortex loop that undergoes a diameter change. In this stage, straight parts of the primary vortex loop, not affected by the secondary flows, are still observable nearby the mid-wall symmetrical plane, but are shortened with an increase in loop diameter. The larger diameter of the primary vortex loop is observable in Fig. 2b, which indicates that the secondary flows imploded in the mid-wall symmetrical plane and will move back again to corners. Generally speaking, once the transmitting shock wave is discharged from the open end of a shock tube, the secondary flows will develop at shock tube corners due to the non-uniform expansion, move to the mid-wall symmetric plane, and finally result in jet instability as shown in Fig. 2c.

4.2 Secondary Shock Waves

The second case is similar to the first case but for a higher Mach number of 1.5. A special feature observable is some discontinuities in the iso-vorticity surface, as shown in Fig. 3b. These iso-vorticity surfaces appears very smooth in its earlier stages as shown in Fig. 3a. The discontinuities are induced by secondary shock waves due to locally-developed supersonic flows. The secondary shock waves appear earlier near the corner plane and then do in the mid-wall symmetric plane somewhere downstream. Therefore, these shock waves are of three-dimensional characters and can impose three-dimensional disturbances on non-circular jet flows as shown in Fig. 3c.

4.3 Axis-Switching of the Primary Vortex Loop

Figure 4 shows results of case 3. The primary vortex loop from the corners of the shock tube exit moves further downstream whereas other parts of the primary vortex loop in the mid-wall symmetrical plane do so more in the radial

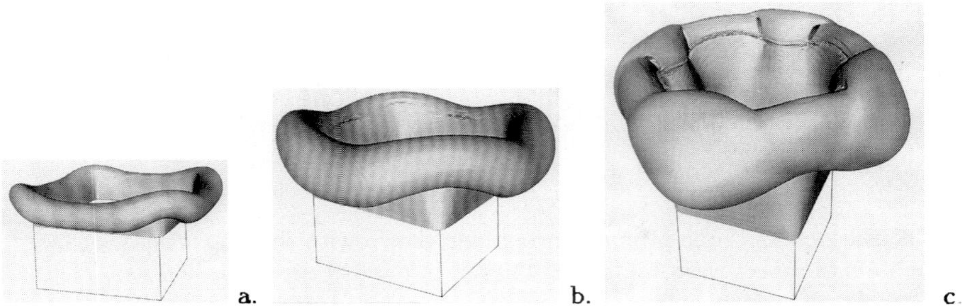

Fig. 3. Disturbances due to secondary shock waves shown with the iso-vorticity surface for $M_s = 1.5$: **a.** $t = 146\ \mu s$; **b.** $t = 386\ \mu s$; **c.** $t = 536\ \mu s$

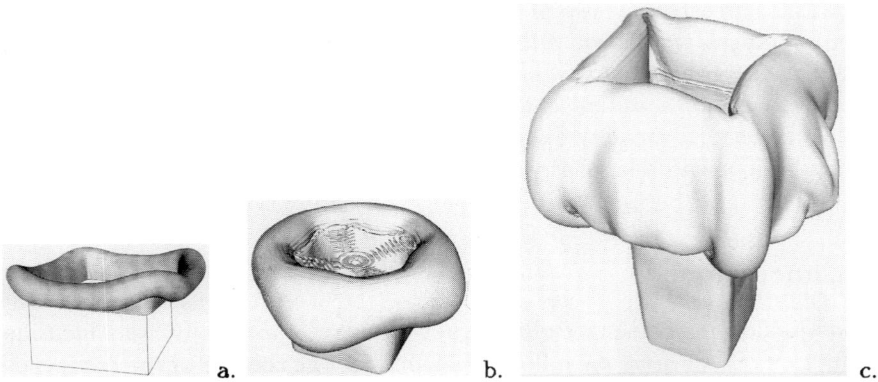

Fig. 4. Axis-switching of non-circular jets shown with the iso-vorticity surface for $M_s = 1.5$: **a.** $t = 146\ \mu s$; **b.** $t = 450\ \mu s$; **c.** $t = 737\ \mu s$

direction, as shown in Fig. 4a. This is the so-called flow spreading. The flow expansion at the corners is stronger than that in the mid-wall symmetric plane due to diffraction of the transmitting shock wave. Actually, the expansion process in the diagonal plane appears similar to axis-symmetric cases whereas that in the mid-wall symmetrical plane has a somewhat two-dimensional characteristics. Therefore, the primary vortex loop is generated with non-uniform vorticity distribution, which is considered to be self-induced and an intrinsic character of non-circular jets. The larger diameter of the primary vortex loop is observable in Fig. 4b, and the first 45° axis-switching is completed. Figure 4c shows the further deformation of the primary vortex loop, and the second phase of the axis-switching is finished. The first phase of axis-switching can be explained with flow expanding but the second phase cannot be. Therefore, the secondary flow motion may be the driving force for the second phase of axis-switching.

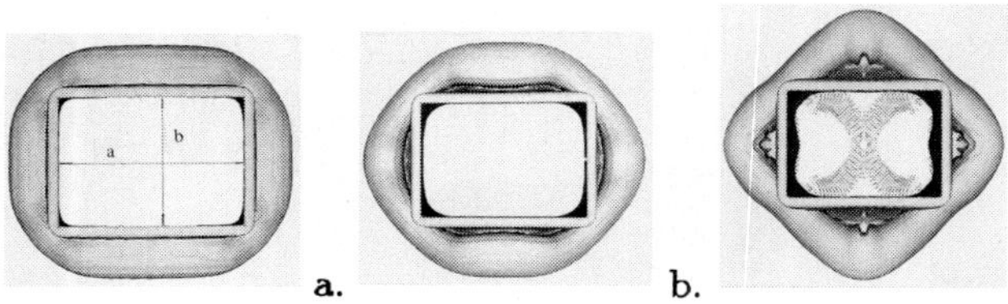

a. b. c.

Fig. 5. Back views of the primary vortex loop shown with the iso-vorticity surface for $M_s = 1.5$ and aspect ratio 1.5: **a.** $t = 140\ \mu s$; **b.** $t = 310\ \mu s$; **c.** $t = 466\ \mu s$

4.4 Role of Aspect Ratio

The last case is similar to case 3, but the shock tube is rectangular not square. The aspect ratio of the shock tube is 1.5 and Mach number was taken to be $M_s = 1.5$. Three back views of the primary vortex loop deformation at three sequential stages are shown with the iso-vorticity surface in Fig. 5. From this figure it was observed that within the first 45° axis-switching, the primary vortex loop develops into a square one. If comparing with the results in case 3 the more asymmetrical vorticity distribution is predicted here. This may imply that rectangular jets with a large aspect ratio will develop more easily into instability.

5 Conclusions

Two major flow phenomena leading to non-circular jet instability were identified. The first one is the secondary flow developing at the corners of the open end of shock tubes and moving down along the primary vortex loop. This motion results from asymmetrical pressure and vorticity distributions, and plays an important role in the axis-switching of the primary vortex loop. It is also a key driving force that leads to the deformation of the primary vortex loop and the development of hairpin vortices and coherence structures. The other is the secondary shock wave that is generated due to locally-developed supersonic flows and induces three-dimensional disturbances. These two physical phenomena are self-induced and considered to be the intrinsic characters of non-circular jets.

References

1. H.S. Husain and F. Hussain: J. Fluid Mech, **248**, (1993)
2. S.H. Lee, I.S. Jeung and Y. Yoon: AIAA Journal, **36**, (1998)
3. G. Raman: Physics Fluids, **11**, (1998)
4. G.Raman and R. Taghavi: J. Fluid Mech, **354**, (1998)
5. S. Koshigoe, E. Gutmark and K.C. Schadow: AIAA Journal, 27,(1989)
6. A. Krothapalli, D. Baganoff and K. Karamcheti: J. Fluid Mech, **170**,(1981)
7. Z. Jiang, K. Takayama and Y.S. Chen: Comp Fluid Dynamics J, **2**, (1995)
8. O. Onodera, Z. Jiang and K. Takayama: JSME Int J **41**, (1998)
9. G. Raman: J. Fluid Mech, **330**(1997)
10. C.K.W. Tam and A.T. Thies: J. Fluid Mech, **248**, (1993)

Nonequilibrium Effects on Shock Wave/Bounday Layer Interaction in High Enthalpy Flow

Akira Matsumoto[1] and Shigeru Aso

Department of Aeronautics and Astronautics, Kyushu University, Fukuoka, JAPAN

1 Introduction

Developments of space transportation systems are in progress. In the development of those high speed vehicles aerodynamic heating phenomena are quite important problems for hypersonic vehicles. In the flight of those hypersonic planes shock wave/boundary layer interaction occurs in several portion of the body as shown in Fig. 1. It causes local peak heating and peak pressure. However, especailly at very high speed flight high temperature effects in shock wave/boundary layer interaction flow fields are not sufficiently understood.

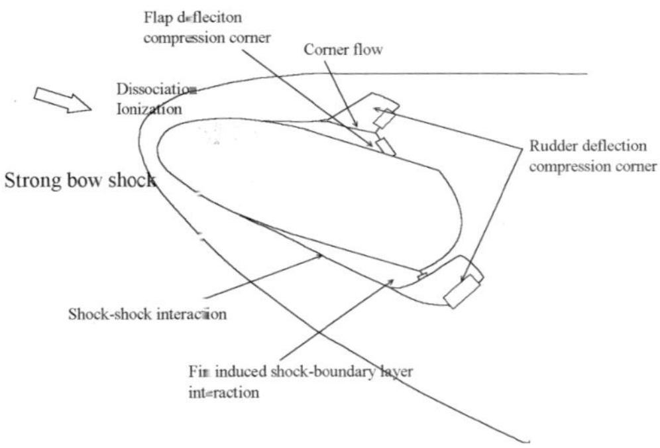

Fig. 1. Shock wave interference problems on a three-dimensional configuration vehicle

Many researches have been conducted on those problems, researches including high temperature effects are quite insufficient. Especially, dissociation and surface recombination play very important roles in a shock-wave/boundary-layer interaction flow field.

Schematics of shock-wave/boundary-layer interaction region at a two-dimensional compression corner is shown in Fig. 2. A strong normal shock wave impinges on a ramp surface and reflects by the surface. Then steady flow is induced behind a normal shock wave. The flow separates due to inverse pressure gradient at compression corner. Then shock-wave/boundary-layer interaction region is formed.

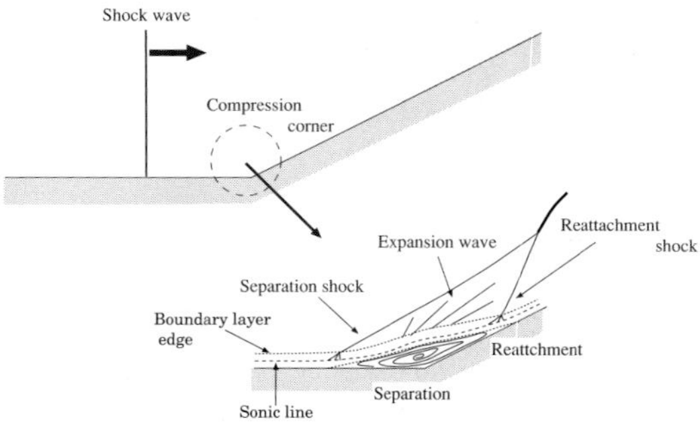

Fig. 2. Schematics of shock-wave/boundary-layer interaction near a two-dimensional compression corner

2 Numerical Method

Full Navier-Stokes equations extend to chemically reacting (thermal equilibrium) flows are employed as governing equations. In two-dimensional Cartesian coordinate, they are described in conservative, dimensional and tensor form shown as follows ([1], [4]):

$$\frac{\partial \boldsymbol{q}}{\partial t} + \frac{\partial \boldsymbol{F}_k}{\partial x_k} = \frac{\partial \boldsymbol{F}_{vk}}{\partial x_k} + \boldsymbol{S} \tag{1}$$

Components of each vectors are written as follows:

$$\boldsymbol{q} = \begin{bmatrix} \rho \\ \rho u_1 \\ \rho u_2 \\ E \\ \rho_1 \\ \vdots \\ \rho_{ns} \end{bmatrix}, \boldsymbol{F}_k = \begin{bmatrix} \rho u_k \\ \rho u_1 u_k + \delta_{1,k} p \\ \rho u_2 u_k + \delta_{2,k} p \\ (E+p) u_k \\ \rho_1 u_k \\ \vdots \\ \rho_{ns} u_k \end{bmatrix}, \boldsymbol{F}_{vk} = \begin{bmatrix} 0 \\ \tau_{k,1} \\ \tau_{k,2} \\ q_k + \tau_{k,1} u_1 + \tau_{k,2} u_2 \\ \rho D_1 \frac{\partial c_1}{\partial x_k} \\ \vdots \\ \rho D_{ns} \frac{\partial c_{ns}}{\partial x_k} \end{bmatrix}, \boldsymbol{S} = \begin{bmatrix} 0 \\ 0 \\ 0 \\ 0 \\ \dot{w}_1 \\ \vdots \\ \dot{w}_{ns} \end{bmatrix}$$

Governing equations are extended to generalized coordinate system, and solved by finite difference method. Conservative variables are defined at cell vertex. Harten-Yee's non-MUSCL modified flux type TVD scheme ([6],[8]) is applied to evaluate numerical flux for convective terms. Conventional Roe's average ([7]) is applied to determine the cell centered physical variables. For viscous terms a central difference scheme is applied. For chemically reacting source term familiar point implicit scheme is not applied, because stiffness due to chemically reacting source term is not so serious by the assumption of thermal equilibrium flow in the present calculations. chemically reacting source term can be explicitly treated. Time integration is performed by Strang-type fractional time step method to maintain second order time accuracy for unsteady problem.

Dissociated air without ionization is considered as test gas (N_2, O_2, N, O, NO is considered as dominant species). For the species, 17 reactions are considered as elementary reactions. Each reaction rates are evaluated by Arrhenius's law importing Blottner's data ([3], [5]).

Transport properties for single species are evaluated equations from Chapman-Enskog theory with Lennard-Jones potential ([1], [2], [5]). Viscosity for multicomponent gas is evaluated by Wilke's mixture formula. Thermal conductivities for molecules are corrected by Euken's relation, then Wilke's mixture formula is applied for multicomponent gas. Diffusivities for multicomponent gas are evaluated from binary diffusion coefficients with pressure and thermal diffusion is neglected.

Initial condition is obtained by solving steady inviscid one-dimensional conservative equations by Runge-Kutta method. For boundary conditions zero derivatives of physical variables are imposed at incoming, upper and exit boundary. Zero derivative of pressure, non-slip, constant wall temperature and fully catalytic or non catalytic wall conditions are imposed at wall boundary.

3 Results and discussion

Flow conditions are shown in Table 1. Difference between two cases is only

Table 1. Flow conditions

p_∞	T_∞	M_s	θ_w	Re
50 Torr	2,000 K	5	30°	5.8×10^4

downstream pressure. At downstream, species concentrations can be approximately set as $c_N = c_O = c_{NO} = 0$, $c_{N_2} = 0.788$, $c_{O_2} = 0.222$ (dissociation at downstream is neglected because of small dissociation degree). For each cases chemically frozen flow and chemical nonequilibrium flow (thermal equilibrium) are assumed. For chemical nonequilibrium flow two extreme conditions; fully catalytic (k_w=∞) or non catalytic ($k_w = 0$) wall conditions are imposed as species condition at a wall.

Typical shock reflection pattern through the computations is shown in Fig. 3. Shock reflection pattern shows double Mach reflection (DMR). All reflection patterns show double Mach reflection. A flow is separated near a compression corner and λ shape shock wave is formed. Slip line is formed from primary triple point and interacts to secondary Mach stem, then flow field is complicated.

The results of one-dimensional calculation as initial condition are shown in Fig. 4. From the calculation chemical equilibrium is established about 30 mm behind incident normal shock wave. Temperature contours around a compression corner are shown in Fig. 5. As shown in the figures, length of separation region, separation point and reattachment point are different form each other. Comparing frozen case with nonequilibrium case, reattachment point is far from a corner in nonequilibrium case.

Distributions of surface heat flux, shear stress and surface pressure are shown Fig. 6. The results show chemical reaction delays flow separation. Higher heating

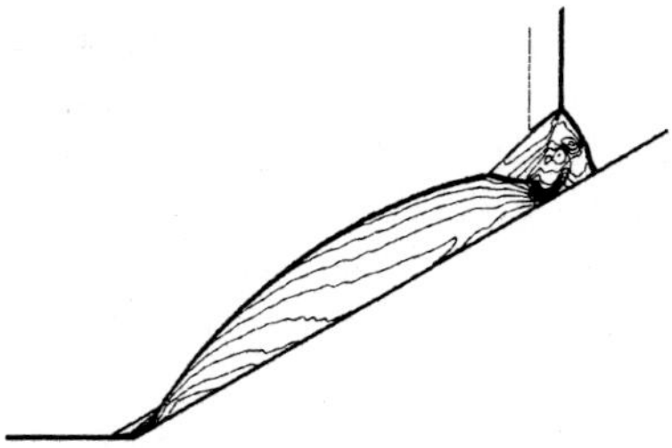

Fig. 3. Typical shock reflection pattern (Instantaneous density contours), Double Mach reflection, chemical nonequilibrium flow (case 1, FCW)

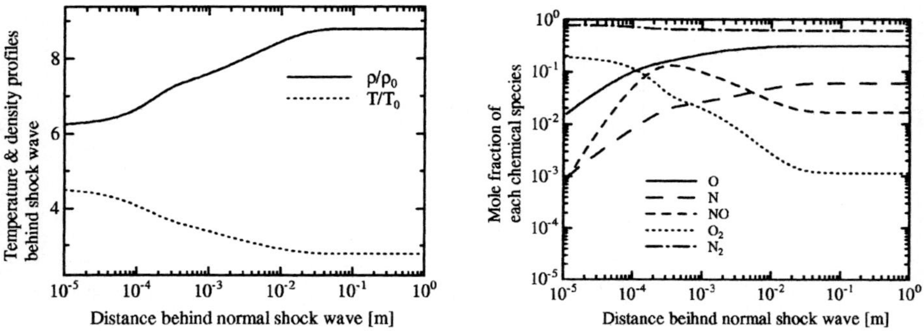

Fig. 4. Results from one dimensional calculation; Left : Density and temperature profiles behind normal shock wave, Right : Mole fraction profiles of each chemical species behind normal shock wave

is observed near reattachment point for FCW case. Comparing two nonequilibrium cases, difference of heat flux distributions are observed as effects of wall catalysis. Difference of between the two cases is quite large. Under fully catalytic wall condition, almost twice as large heat flux as that with non catalytic wall is observed. Also the results show peak heating and peak pressure locate a little after reattachment point especailly for??non-equilibrium cases. Separation point, reattachment point and separation length are measured from the figures. The results are shown in Table 2. The result shows that nonequilibrium effects make

Table 2. State of separation

	x_{sep}	x_{reat}	L_{sep}
Frozen	-2.59	2.59	5.18
Non-eq. (NCW)	-2.96	1.66	4.62
Non-eq. (FCW)	-1.85	1.29	3.10

unit : [mm]

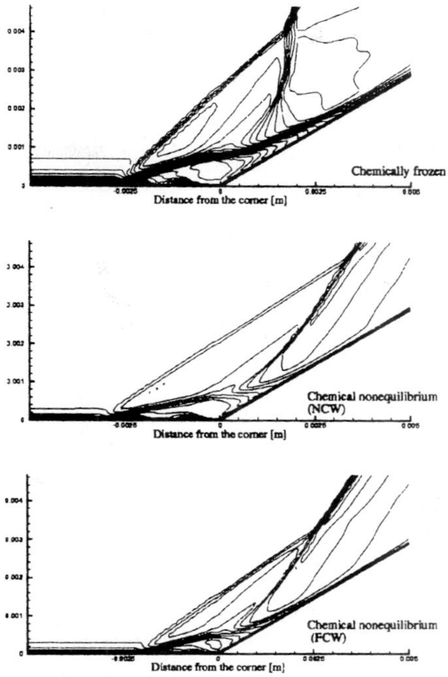

Fig. 5. Instantaneous temperature contours around a separation region

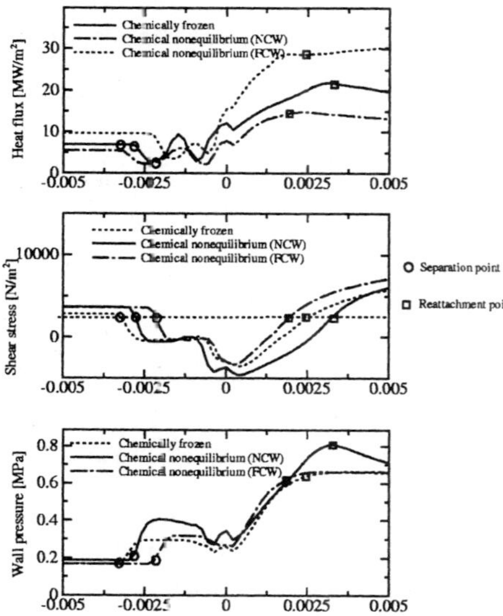

Fig. 6. Instantaneous distributions of surface heat flux, shear stress, and surface pressure

smaller angle of separation shock wave to the wall. Then a reattachment point becomes slightly near from the corner in nonequilibrium case. Also it is observed wall catalysis affects separation point due to its enhancement of surface recombination of atoms. Then FCW condition generates shorter separation length than NCW condition. Largest surface pressure in interaction region is observed under frozen flow case. In the pressure and heat flux distributions of nonequilibrium cases significant local peak pressure and local peak heating are not observed. This is quite different point compared wth that of frozen flow case.

In addition to this case the calculation of p_∞=20 Torr is also conducted. The results show that lower initial pressure causes longer separation length and overall above tendency is preserved. The maximum values of pressure near reattachment point of NCW and FCW at p_∞=50 Torr are almost same. However, at p_∞=20 Torr, the maximum values of pressure near reattachment point of NCW and FCW are quite different. The results show that effects of degree of non-equilibrium on aerodynamic heating and pressure including wall catalycity are different.

4 Conclusions

The conclusions of the present study are summarized as follows:

1. Chemical reaction delays flow separation.
2. Chemical reaction shortens separation length.
3. FCW condition generates shorter separation length than NCW condition.
4. Lower initial pressure causes longer separation length and overall above tendency is preserved.
5. Significant local peak heating and local peak pressure are not observed in chemically reacting flows for both of FCW and NCW conditions.
6. Peak heating and peak pressure locate a little after reattachment point.
7. Effects of degree of non-equilibrium on aerodynamic heating and pressure including wall catalycity are different.

References

1. Anderson Jr., JD : Hypersonic and High Temperature Gas Dynamics. (Mc Graw-Hill, Inc. 1989)
2. R. B. Bird, W. E. Stewart and E. N. Lightfoot : *Transport Phenomena* (John Wiley & Sons, Inc. 1960)
3. F. G. Blottner : AIAA J., Vol. 7, No. 12, pp. 2281–2288 (1969)
4. P. A. Gnoffo, R. N. Gupta, RN and J. L. Shinn : 'Conservation equations and physical models for hypersonic air flows in thermal and chemical nonequilibrium'. (NASA TP-2867. 1989)
5. R. N. Gupta, J. M. Yos,R. A. Thompson and K.-P. Lee : 'A review of reaction rates and thermodynamic and transport properties for 11-Species air model for chemical and thermal nonequilibrium calculation to 30,000 K'. (NASA RP-1232. 1989)
6. A. Harten : J. Comp. Phys., Vol. 49, pp. 357–393 (1983)
7. P. L. Roe : J. Comp. Phys., Vol. 43, pp. 357–372 (1981)
8. H. C. Yee and J. L. Shinn : AIAA J., Vol. 27, pp. 299–307 (1989)

Numerical Investigation on the Transition between Regular and Mach Reflections in Shock Waves

Domenic D'Ambrosio, Maurizio Pandolfi

Dipartimento di Ingegneria Aeronautica e Spaziale
Politecnico di Torino
Corso Duca degli Abruzzi, 24, 10129 Torino, Italy
e-mail: domenic@athena.polito.it - pandolfi@polito.it

1 Introduction

The transition between regular and irregular shock reflections, a challenging problem in fluid dynamics, has gained practical importance in the last years because of its implications in the design of super/hypersonic air intakes.

Regular and irregular shock wave reflections were first observed and reported by Mach about 120 years ago and Von Neumann, during World War II, theoretically demonstrated the existence of a *dual solution domain*, where both configuration are potentially possible. Despite this, the problem remained open (consult the survey by Hornung [1] for a detailed bibliograpy) untill the last decade, when Li and Ben-Dor [2] theoretically demonstrated that, opposite to the contemporary belief, both the regular reflection (RR) and the irregular reflection (now frequently called *Mach reflection* (MR)) are stable in the dual-solution domain. In the same period, the hysteresis was observed experimentally for the first time [3], and a tandem numerical study [4] confirmed such results. Other experimental and numerical investigations have been reported in the last years, confirming the presence of the hysteresis, but demonstrating that the experiments are particularly sensitive to the presence of disturbances in the wind tunnel [5].

In this paper, the authors present numerical results obtained in simulating shock reflections by solving the Euler equations in two dimensions. The attention is focused on the effect exerted on the transition by the grid refinement and on the Mach stem height by the geometrical parameters that characterize the geometrical configuration. In the following section, the shock reflection problem will be shortly reminded. Then, the numerical method used will be outlined. Finally the numerical results obtained will be presented and commented.

2 Regular and Mach Reflections

Two different configurations are possible when an incident shock wave is reflected on a plane surface: the *regular reflection configuration* (RR) and the *Mach reflection configuration* (MR) (figure 1). The *regular reflection configuration* is composed of two shock waves, the *incident shock* and the *reflected shock* (fig.1a).

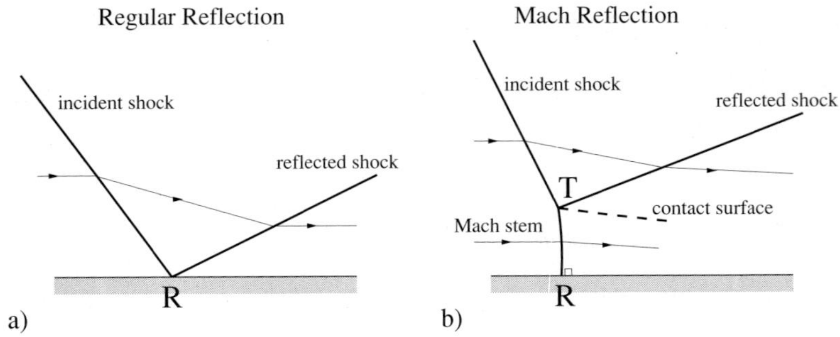

Fig. 1. Illustration of the two possible shock reflection solution: a)regular reflection and b)Mach reflection.

The flow field in the vicinity of the reflection point is divided in three parts: the undisturbed incoming flow, the region behind the incident shock and the region behind the reflected shock. In the *Mach reflection configuration*, three distinct shock waves are present: the *incident shock*, the slightly curved *Mach stem* and the *reflected shock* (fig.1b). The three shocks meet at the *triple point*, from which a *contact surface*, or *slipstream*, departs, separating the flow particles that experienced a compression through the Mach stem and those that were compressed through the incident and the reflected shocks at the same final pressure level and to the same final deflection angle, but with different entropy values. Thus, it is possible to distinguish four different flow regions in the vicinity of the triple point.

Theoretically, on the basis of the three-shocks theory developed by von Neumann, two different criteria for transition between *regular* and *Mach reflection* shock configurations can be defined. One is called the *mechanical equilibrium* or *von Neumann criterion*: it defines the minimum incident shock angle at which it is possible to obtain a MR configuration. The other is the *detachment criterion*, which defines the larger incident shock angle at which a regular reflection is still possible. In the range between the *von Neumann* incident shock angle and the *detachment* incident shock angle, both the regular and the Mach reflection are theoretically possible. The theoretical existence of a dual solution domain, demonstrated by von Neumann, implies other questions, such as the existence of an hysteresis phenomenon and the stability of the two different configurations within the domain.

Today, the existence of the hysteresis has been demonstrated both theoretically and experimentally (see, for instance [3] [4]), a theoretical study [2] has demonstrated that both the regular and the Mach reflection configurations are stable in the dual solution domain, and an analogy that provides the magnitude of the disturbance required to trigger the transition from one configuration to the other has been proposed [6].

3 Numerical Method

In the present numerical investigation, Euler computations have been performed using a time-marching technique, a finite volume discretization over structured grids and an upwind flux-difference splitting scheme of the Osher's family [7] for the evaluation of the convective fluxes at volumes surfaces. Second order accuracy in space and in time is achieved by means of ENO-like [8] linear interpolation of Riemann invariants in the whole flow field, using a *minmod* limiter. The physical domain is divided in different zones according to a multiblock strategy.

4 Numerical Results

4.1 The geometrical setup

We refer to an experimental geometrical setup [3] composed of two symmetric wedges whose slope with respect to the incoming flow is varied by rotation around the trailing edge, so that the ratio between the exit cross-sectional area at the trailing edge and the wedge length is constant. The symmetrical configuration, where the reflection occurs at the symmetry axis, is necessary to avoid the shock/boundary layer interaction that would take place in experiments and in Navier-Stokes computations if the reflecting surface were a physical wall. The actual values of the geometric parameters and the free-stream conditions are chosen in order to simulate the experimental campaigns, in particular those conducted in the SH2 tunnel in Meudon, France [3] and in the T-313 and T-326 tunnels in Novosibirsk, Russia [9]. Since we will assume the flow to be inviscid, the only free-stream parameter needed is the Mach number, which is equal to 4.96. The important geometrical parameter is the ratio between the exit cross-sectional area at the trailing edge of the wedge and the wedge length, H/w. This is set equal to 0.42 when focusing on the hysteresis curve, but will be also varied to investigate the dependence of the Mach stem height from it.

The numerical geometrical setup is composed of a wedge of length w followed by a flat plate. The actual experimental configuration is different, as the flat plate is not present, but, as it will be mentioned later, such a difference does not have effect on the shock reflection. Reflecting boundary conditions are imposed at the walls and at the symmetry line, supersonic inflow and outflow conditions at the inlet and at the outlet.

Provided the Mach number is sufficiently high and if the wedge angle is in the range between the *Von Neumann angle* α^N and the *detachment angle* α^D, both the RR and the MR configurations are possible. At Mach 4.96, their theoretical values are respectively $\alpha^N = 20.87^o$ and $\alpha^D = 27.71^o$. The results obtained for a wedge angle $\alpha = 25^o$ are shown in figure 2 in terms of Mach number and pressure contours for the Mach reflection configuration only.

It is possible to recognize the *incident shock*, the *Mach stem*, the *reflected shock* and the *slipstream* (*contact surface*). The dimension of the Mach stem is controlled by the expansion fan that emanates from the trailing edge of the wedge. When the expansion fan interacts with the reflected shock, the latter

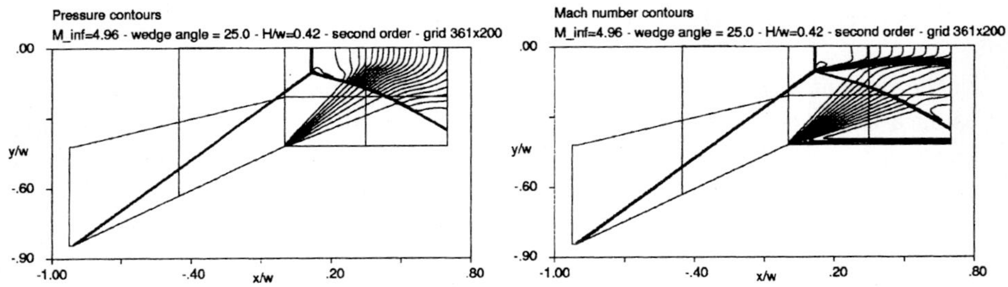

Fig. 2. Mach reflection configuration. $M_\infty = 4.96$, wedge angle $\delta = 25^o$, $H/w = 0.42$. Pressure contours (upper) and Mach number contours (lower).

is transformed in a transmitted-reflected shock, and the former in a transmitted expansion. An thick entropy layer is also generated because the transmitted/reflected shock is bended. The transmitted expansion fan reaches the slipstream and accelerates the subsonic flow behind the Mach stem until it reaches sonic conditions. A convergent two-dimensional channel is thus formed between the symmetry axis and the contact surface. After the sonic point, the channel diverges and the flow continues to accelerate in supersonic conditions. Between the Mach stem and the sonic throat, limited by the symmetry axis and the slipstream, a subsonic pocket exists in an otherwise supersonic flow. The dimensions of the subsonic pocket, and therefore the height of the Mach stem, is determined by the expansion fan, which brings the information about the geometrical configuration to the subsonic flow in the initial part of the channel. Such information, related to the ratio between the exit cross-sectional area at the trailing edge of the wedge and the wedge length H/w, propagates upstream to the Mach stem.

The presence of the flat plate behind the wedge does not have any influence on the Mach reflection, because the additional expansion waves that would emanate from the trailing edge of the wedge in case it would end abruptly, thus producing a stronger expansion, should hit the two-dimensional channel in a point where the flow is supersonic. Since the subsonic pocket is isolated from the supersonic region downstream the line composed by the sonic line at the throat and by the expansion/reflected expansion wave that hits the slipstream at the sonic point, whatever happens in such supersonic region will not affect the Mach stem.

In figure 3, the hysteresis curves obtained for Mach 4.96 and $H/w = 0.42$ using two different grids are shown. Both grids are equally spaced respectively in the direction of the symmetry axis and in the direction normal to it. The coarser grid is composed of 100 points along the wedge and 81 points along the flat plate, while 100 points discretize the domain in the direction normal to the symmetry axis. The grid refinement is doubled in the finer grid, which is composed of 200 points along the wedge, of 161 points along the flat plate and of 200 points in the normal direction. The hysteresis curves were obtained as follows. At first, computations were carried out with wedge angles ranging from values slightly lower than the *von Neumann angle* α^N to values higher than the *detachment angle* α^D. The step in the wedge angle variation was 0.5 degrees.

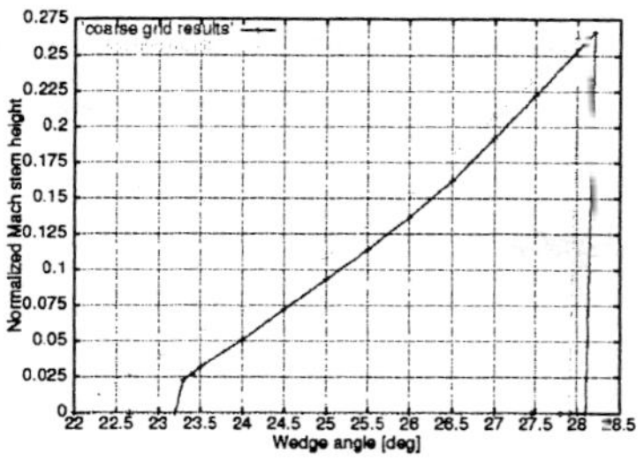

	theor.	num. fine grid	num. coarse grid
α^N	20.87°	22.20° - 22.30°	23.20° - 23.30°
α^D	27.71°	27.90° - 28.00°	28.10° - 28.20°

Fig. 3. Numerical hysteresis curves for Mach 4.96 and $H/w = 0.42$. The red line refers to the coarse grid (181x100), the green line to the fine grid (361x200). Theoretical and the numerical angles α^D and α^N for $M_\infty = 4.96$ and $H/w = 0.42$.

Starting from uniform flow conditions, the configuration which results stable in the dual-solution domain was the regular reflection. When the wedge angle was sufficiently high, however, only a Mach reflection was possible. Refining the study in the wedge angle interval where the RR $\rightarrow$ MR transition occurred, the *numerical detachment angle* α^D_{num} was determined. Then, starting from a Mach reflection configuration obtained with a wedge angle higher than α^D_{num}, the wedge angle was discontinuously decreased (with a step of 0.5 degrees) and, for each step, a time integration was performed until convergence to the steady state was achieved. In this case, the Mach reflection configuration was maintained until the *numerical von Neumann angle* α^N_{num} was reached. A further decrease in the incident shock angle resulted in a smooth transition from the MR to the RR configuration. The values for the theoretical and the numerical angles α^D and α^N are given in figure 3(right).

The grid refinement study shows that the height of the Mach stem is influenced by the grid refinement. Moreover, while the RR $\rightarrow$ MR transition angle varies slightly with the grid (about 0.2°, in figure 3), the MR $\rightarrow$ RR angle is strongly influenced by it (about 1°), because when the two-dimensional channel becomes smaller and smaller, a very fine grid should be provided to capture it. In some preliminary experiments, not shown here, a small but not negligible influence of the limiter used in the second order on the RR $\rightarrow$ MR angle has been found. This indicates that the numerical dissipation inherent of the numerical scheme plays a certain role in determining the RR $\rightarrow$ MR transition point.

Other experiments have been conducted to determine the dependence of the Mach stem height on the H/w parameter. Maintaining the Mach number equal to 4.96 and fixing the wedge angle to 25.00°, H/w was varied from 0.27 to 0.47, using always the same grid spacing. The related results, shown in figure 4, are in sufficiently good agreement with those obtained with the theoretical model recently proposed by Li and Ben-Dor [10] (the results shown in figure 15 of reference [10] have been extracted from the paper and are reported as line in figure 4). The effect of the ratio H/w on the fluid dynamics configuration is shown in figure 4(right) for values of H/w equal to 0.27.

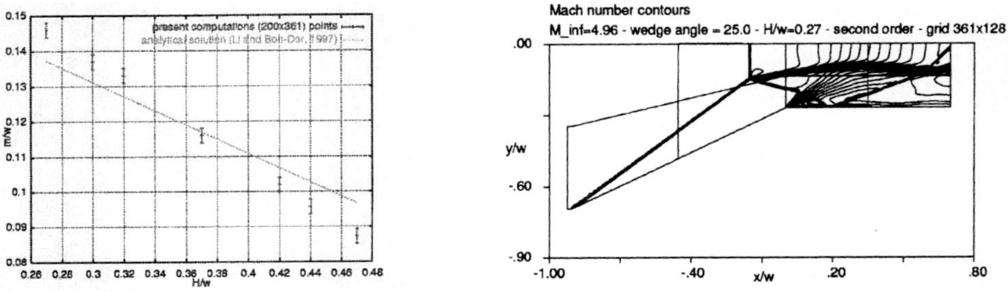

Fig. 4. Left: effect of H/w on the normalized Mach stem height m/w for Mach 4.96 and $\delta = 25^o$. Red bars: numerical results (the error bar represents the dimension of one cell in the direction normal to the symmetry axis). Green line: theoretical prediction (from figure 15 of [10]). Right: the Mach reflection configuration for $H/w = 0.27$

References

1. H.G. Hornung, "Regular and Mach Reflection of Shock Waves," *Ann. Rev. Fluid Mech.*, vol. 18, pp. 33–58, 1986.
2. H. Li and G. Ben-Dor, "Application of the principle of minimum entropy production to shock wave reflections. I. steady flows," *J. Appl. Phys.*, vol. 80, no. 4, pp. 2027–, 1996.
3. A. Chpoun, D. Passerel, H. Li, and G. Ben-Dor, "Reconsideration of oblique shock wave reflections in steady flows. part 1. experimental investigation," *J. Fluid Mech.*, vol. 301, pp. 19–35, 1995.
4. J. Vuillon, D. Zeitoun, and G. Ben-Dor, "Reconsideration of oblique shock wave reflections in steady flows. part 2. numerical investigation," *J. Fluid Mech.*, vol. 301, pp. 37–50, 1995.
5. A. Fomin, H. Hornung, M. Ivanov, A. Kharitonov, and G. Klemenkov, "The study of transition between regular and Mach reflection of shock waves in different wind tunnels," in *Proceedings of the 12st International Mach Reflection Symposium* (S. Africa, ed.), (Pilanesberg National Park), 1996.
6. N. Sudani and H. Hornung, "Stability and analogy of shock wave reflection in steady flow," *Shock Waves*, no. 8, pp. 367–374, 1998.
7. M. Pandolfi, "A contribution to the numerical prediction of unsteady flows," *AIAA Journal*, vol. 22, no. 5, 1983.
8. A. Harten, B. Engquist, S. Osher, and S. R. Chakravarthy, "Uniformly high order accurate Essentially Non-Oscillatory schemes, III," *Journal of Computational Physics*, vol. 71, 1987.
9. M. Ivanov, G. Klemenkov, A. Kudryavtsev, S. Nikiforov, A. Pavlov, A. Fomin, A. Kharitonov, D. Khotyanovsky, and H. Hornung, "Experimental and numerical study of the transition between regular and mach reflections of shock waves in steady flows," in *Proceedings of the 21st International Shock Waves Symposium* (A. Panther, Fyshwick, ed.), (Great Keppel Island, Australia), pp. 819–824, 1997.
10. H. Li and G. Ben-Dor, "A parametric study of Mach reflection in steady flows," *J. Fluid Mech.*, vol. 341, pp. 101–125, 1997.

Numerical Study on Flame Propagation along a Vortex Tube

Tatsuya Hasegawa, Shinnosuke Nishiki, and Ryosuke Nakamichi

Department of Environmental Technology and Urban Planning, Nagoya Institute of Technology, Gokiso-cho, Showa-ku, Nagoya 466-8555, JAPAN

Abstract. An unstretched straight vortex tube perpendicularly interacting with a premixed flame without any background turbulence is numerically simulated. The flame propagates along the vortex tube while producing precursor azimuthal vorticity in front of the flame. Baroclinic effect produces the azimuthal vorticity on the flame at the initial stage of propagation but convection and stretch effects produce it at the later stage. The flame propagation becomes unsteady for low density ratio and the peak of the azimuthal vorticity separates from the flame tip. Propagation velocity of the flame is generally proportional to the maximum circumferential velocity of the vortex core and depending on the power of the density ratio of the flame.

1 Introduction

Vortex tubes are recognized as sinews of turbulence. Concerning with turbulent premixed combustion, we focus on a simple configuration in which an unstretched straight vortex tube perpendicularly interacts with a premixed flame without any background turbulence. It is well known that a premixed flame propagates very fast with the velocity similar to the maximum circumferential velocity of the vortex core [1–3]. It is also shown that the premixed flame can propagate along the vortex tube when the maximum circumferential velocity is faster than the burning velocity and the core diameter is larger than the flame thickness [4]. However, the mechanism of the flame propagation is still not clear though several models have been proposed. In this study, the flame propagation along a fine vortex tube is numerically simulated and the mechanism that provokes the flame to propagate along the vortex tube is discussed in terms of the vorticity formation due to the interaction of the vortex tube and the flame.

2 Simulation Method

Three-dimensional governing equations of compressible reactive flows are solved with the following assumptions: (i) Chemical reaction is a one-step irreversible one with heat release, where the molecular weights of the reactants and products are the same. (ii) The bulk viscosity, the Soret and the Dufour effects and the pressure gradient diffusion are neglected. (iii) The specific heat at constant pressure and the specific heat ratio are constant. (iv) The equation of the state of the burned and unburned gases is that of an ideal gas.

A straight vortex tube is placed at the center of a volume of 2.5 mm × 1 mm × 1 mm as shown in Fig. 1. The side boundaries are periodic, but longitudinal boundaries are non-periodic and posed by non-reflecting boundary conditions [5]. The initial vortex tube is assumed to have a Gaussian vorticity distribution, therefore the circumferential velocity field is given by the following equation:

$$V(r) = \frac{\Gamma_0}{2\pi r}\left(1 - \exp\left(-\frac{r^2}{(\sigma_0/2)^2}\right)\right),$$

where

$$\Gamma_0 = \pi\sigma_0 V_{m0}.$$

The maximum velocity $0.63V_{m0}$ is achieved at the point $r = \sigma_0/2$. The initial pressure was assumed to be 0.5 MPa. The initial maximum circumferential velocity is V_{m0}=14.5 m/s and the initial core diameter is σ_0=0.156 mm.

A planar flame perpendicular to the vortex tube is placed simultaneously in the calculation domain. The heat of reaction is given as Q=2.446×10^6 J/kg, 1.223×10^6 J/kg and 0.6115×10^6 J/kg with the specific heats of $C_p = 1.25 \times 10^3$ J/kg·K and the thermal conductivity of λ=7.4 ×10^{-2} W/m·K. Thus the adiabatic flame temperature becomes 2260 K, 1280 K and 790 K which corresponds to the density ratio of the flame of 7.53, 4.27 and 2.63, respectively. The viscosity and the diffusion coefficient are given by the thermal conductivity assuming the Prandtl number of unity and the Lewis number of unity. The transport properties are assumed not to depend on the temperature change but depend on the density change. The frequency factor and the activation temperature of the one-step irreversible reaction are arranged to give the same burning velocity of 0.236 m/s and the same flame thickness of 0.0762 mm for different density ratios.

Fourier spectral collocation method is used for y, z directions and 6th-order central finite difference method is used for x direction to treat non-periodic boundary conditions. The number of grid points is 640 in x direction and 128 in y, z directions. The 3rd-order Runge-Kutta method is used in time integration and the time step is selected to satisfy the CFL condition. A vector-parallel computer (Fujitsu VPP 700) with 16 PEs is used for the direct numerical simulation. The total memory used is 10 GB.

3 Results and Discussions

Formation of azimuthal component of vorticity like a vortex ring is found to be the main reason that provokes the flame to propagate along a vortex tube [6,7]. Evolution of the azimuthal component of vorticity for the density ratio of 7.53 is shown in Fig. 2. The peak of the azimuthal vorticity exists on the flame at 31 μs but the peak separates from the flame and exists in front of the flame after 62 μs. Similar behaviors are observed for lower density ratios, but the position of the vorticity peak separates from the flame tip as shown in Fig. 3. Evolution of positions of the flame tip and the peak of the azimuthal vorticity for the density ratio 7.53 are shown in Fig. 4 (a). The peak of the azimuthal vorticity moves slowly at the initial stage but it abruptly moves down at 38.7 μs. It is found that the peak of the azimuthal vorticity moves with the flame tip after the positional transition. This means that the propagation of the peak of the azimuthal vorticity strongly relates to the flame propagation along the vortex tube. Evolution of positions of the flame tip and the peak of the azimuthal vorticity for the density ratio 2.63 are shown in Fig. 4 (b). In this case, the peak of the azimuthal vorticity moves faster than the flame tip after transition of the peak position at 54.2 μs, and as a result, the vorticity peak separates from the flame tip.

Next the formation of the azimuthal component of vorticity is analyzed by the transport equation of vorticity:

$$\frac{\partial \omega}{\partial t} = -(u \cdot \nabla)\omega + (\omega \cdot \nabla)u - \omega(\nabla \cdot u) - \nabla\left(\frac{1}{\rho}\right) \times \nabla p + \nabla\left(\frac{1}{\rho}\right) \times \nabla \cdot \tau + \frac{1}{\rho}\nabla \times \nabla \cdot \tau$$

The first term in the right hand side is the convection, the second term is the stretch, the third term is the dilatation, the fourth term is the baroclinic torque, the fifth and the sixth terms are the viscous diffusion and the dissipation. After integrating these terms in time, the produced azimuthal vorticity is evaluated term by term. Time evolution of the peak values of total azimuthal vorticity, the azimuthal vorticity caused by convection and stretch terms and that caused by baroclinic, dilatation and viscous terms are compared in Fig. 5. The baroclinic term produces azimuthal vorticity but the dilatation and the viscous terms reduce it. This figure clearly shows that the peak of the total azimuthal vorticity relates to that of the baroclinic term at the initial stage, but the peak strictly corresponds to that of the convection and stretch terms at the later stage. Thus it is concluded that the convection and stretch terms play important roles for flame propagation along a vortex tube though the baroclinic term plays a role at the initial stage [8].

Evolution of the propagation velocity for different density ratios of the flame is shown in Fig. 6. The velocity increases at the initial stage, but it decreases constantly at the later stage. As the decay of the maximum circumferential velocity of the vortex tube is the same for different density ratios, it is found that the rate of decay for the density ratio of 2.63 is smaller than the other cases. Fig. 7 shows evolution of the proportional factor of the propagation velocity to the maximum circumferential velocity of the vortex tube. The average proportional factor becomes 1.22, 1.05 and 0.82 for the density ratio of 7.53, 4.26 and 2.63, respectively. However, the proportional factor is still increasing for the density ratio of 2.63. This means the unsteady nature of propagation for the density ratio of 2.63. The relation between the proportional factor and the density ratio is shown in Fig. 8. By interpolating numerical data, we obtain the following empirical equation for the propagation velocity:

$$u_V = 0.725\left(\frac{\rho_u}{\rho_b} - 1\right)^{0.285} V_m$$

This relation should be compared with the theoretical one and discussed in future.

4 Conclusions

1) Flame propagation along a vortex tube relates to the formation of the azimuthal component of vorticity propagating in front of the flame.
2) Flame propagation becomes unsteady for lower density ratio and the peak of the azimuthal vorticity separates from the flame tip.
3) Convection and stretch effects provoke the flame to accelerate along the vortex tube at the later stage.
4) The empirical expression of propagation velocity of the flame shows proportionality to the maximum circumferential velocity of the vortex tube and to the power of the density ratio minus unity.

5 Acknowledgment

This work was partially supported by the Grant-in-Aid for Scientific Research (11650221) and by the Tanigawa Foundation.

References

1. J. Chomiak: 'Dissipation Fluctuations and the Structure and Propagation of Turbulent Flames in Premixed Gases at High Reynolds numbers'. In: *The Sixteenth Symposium (International) on Combustion*, (The Combustion Institute, Pittsburgh PA 1976) pp. 1665–1673
2. K. Asato, H. Wada, T. Hiruma and Y. Takeuchi: Combust. Flame **110**, 418 (1997)
3. S. Ishizuka, T. Murakami, T. Hamazaki, K. Koumura and R. Hasegawa: Combust. Flame. **113**, 542 (1998)
4. T. Hasegawa, K. Nishikado and J. Chomiak: Combust. Sci. Technol. **108**, 67 (1995)
5. T. J. Poinsot and S. K. Lele: J. Computat. Phys. **101**, 104 (1992)
6. T. Hasegawa and K. Nishikado: 'Effect of density ratio on flame propagation along a vortex tube'. In: *The Twenty-Sixth Symposium (International) on Combustion*, (The Combustion Institute, Pittsburgh PA 1996) pp. 291–297
7. T. Hasegawa, S. Nishiki and S. Michikami: *IUTAM Symposium on Geometry and Statistics of Turbulence*, (Kluwer Academic Publishers, 2000) to be published
8. W. T. Ashurst: Combust. Sci. Technol. **112**, 175 (1996)

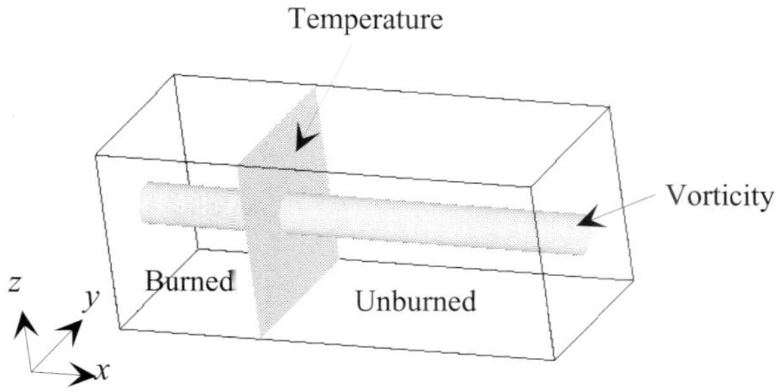

Fig. 1. Initial configuration of simulation.

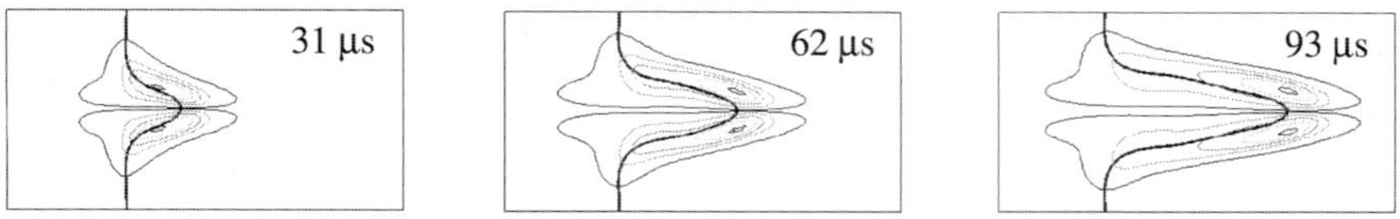

Fig. 2. Evolution of azimuthal component of vorticity (V_{m0} = 14.5 m/s, σ_0 = 0.156 mm, $\rho_u/\rho_b = 7.53$).

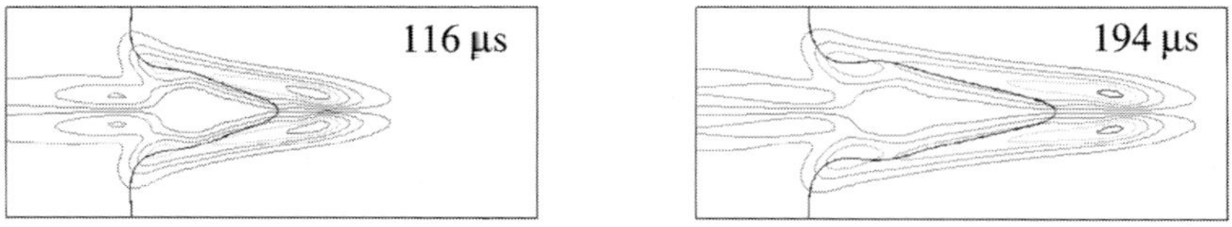

Fig. 3. Evolution of azimuthal component of vorticity (V_{m0} = 14.5 m/s, σ_0 = 0.156 mm, $\rho_u/\rho_b = 2.63$).

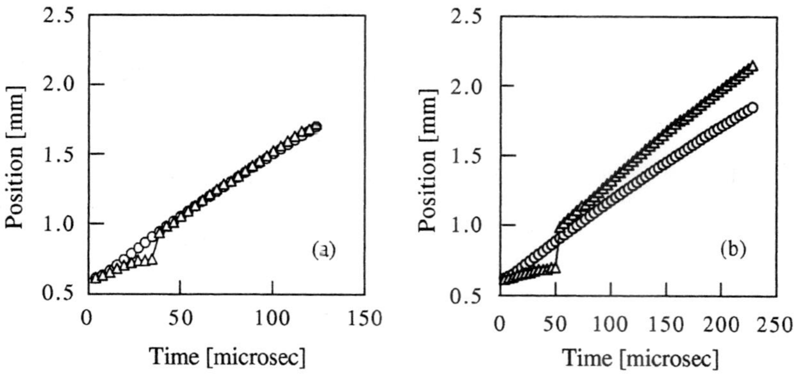

Fig. 4. Positional evolution of the flame tip (circle) and the peak of the azimuthal vorticity (triangle) for (a) $\rho_u/\rho_b = 7.53$, (b) $\rho_u/\rho_b = 2.63$.

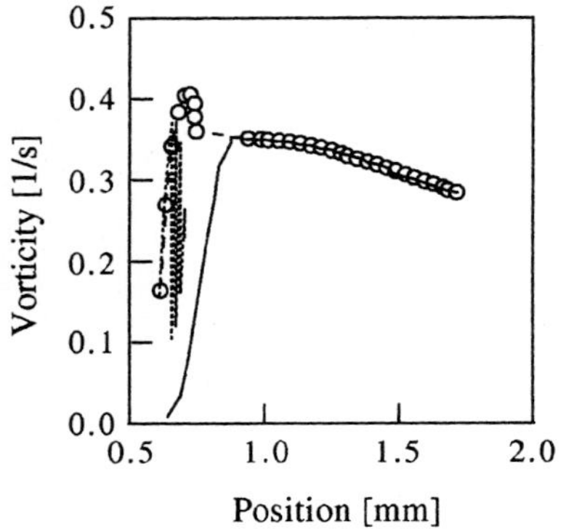

Fig. 5. Positional evolution of peaks of total azimuthal vorticity (circle), baroclinic, dilatation and viscous terms (dashed line) and convection and stretch terms (solid line) for $\rho_u/\rho_b = 7.53$.

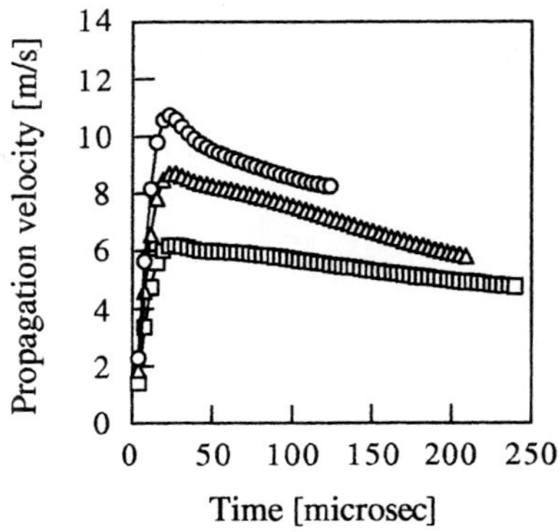

Fig. 6. Evolution of propagation velocity for different density ratios (circle: $\rho_u/\rho_b = 7.53$, triangle: $\rho_u/\rho_b = 4.26$, box: $\rho_u/\rho_b = 2.63$).

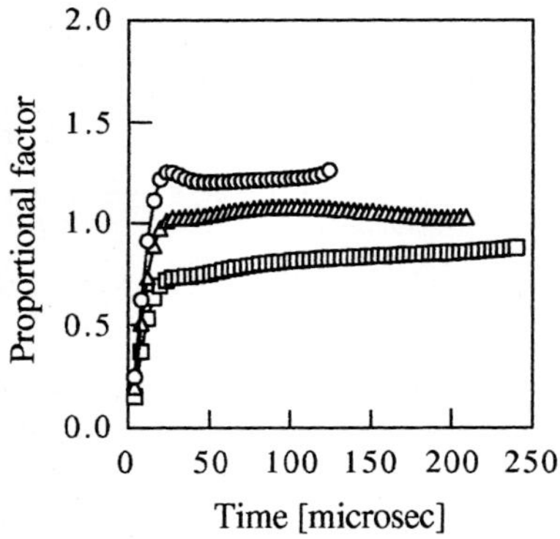

Fig. 7. Evolution of the proportional factor to the maximum circumferential velocity u_V/V_m for different density ratios (circle: $\rho_u/\rho_b = 7.53$, triangle: $\rho_u/\rho_b = 4.26$, box: $\rho_u/\rho_b = 2.63$).

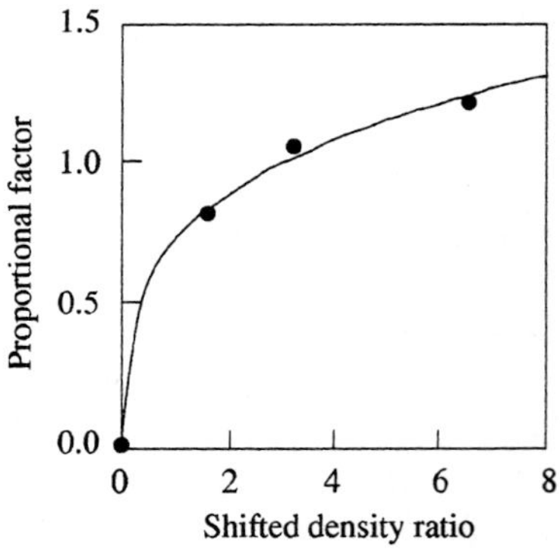

Fig. 8. Relation betwen the proportional factor u_V/V_m and the shifted density ratio $\rho_u/\rho_b - 1$ of the flame.

The Numerical Analysis of Cellular Premixed Flames Based on the Diffusive–Thermal and Navier–Stokes Equations

Satoshi Kadowaki and Shin-ichirow Goma

Department of Mechanical Engineering, Nagoya Institute of Technology
Gokiso-cho, Showa-ku, Nagoya 466-8555, Japan

Abstract. Unsteady calculations of two-dimensional (2-D) and three-dimensional (3-D) reactive flows are performed to investigate the hydrodynamic effect on the shape of cellular premixed flames at low Lewis numbers. The diffusive–thermal (D–T) and Navier–Stokes (N–S) equations, which contain a one-step irreversible chemical reaction, are employed. The D–T equation excludes the hydrodynamic effect caused by thermal expansion, since the constant-density approximation is used. The N–S equation, on the other hand, includes the hydrodynamic effect, since the compressibility of gases is taken into account. The hydrodynamic effect has a great influence on the flame instability and cellular-flame formation. The calculated growth rate based on the D–T equation is much smaller than that based on the N–S equation, and the obtained unstable range from the D–T equation is narrower. In addition, the calculated cell depth and flame length of a cellular flame based on the D–T equation are considerably smaller. Thus, it is necessary to include the hydrodynamic effect when studying the intrinsic instability of premixed flames and the shape of cellular-flame fronts.

1 Introduction

The hydrodynamic effect, together with the diffusive–thermal effect, is the significant factor in the intrinsic instability of premixed flames. The hydrodynamic effect caused by thermal expansion has a destabilizing influence on the flame front [5,11]. The diffusive–thermal effect, which is induced by the preferential diffusion of mass versus heat, has a destabilizing (stabilizing) influence when the Lewis number, defined as the ratio of the thermal diffusivity to the diffusion coefficient, is lower (higher) than unity [4,14,15]. Since diffusive–thermal instability appears at low Lewis numbers, most cellular flames are more likely to be formed under this condition.

The numerical calculations of unsteady reactive flows at low Lewis numbers based on the D–T equation were performed to study the temporal motion of flame fronts [1,6,12]. In the D–T equation the constant-density approximation is used, so that the hydrodynamic effect, which is indispensable to intrinsic instability, is excluded. Thus, the obtained results are valid for premixed flames with sufficiently small heat release. After that, based on the N–S equation, 2-D unsteady reactive flows were calculated [2,7,8,13]. The compressibility of gases is taken into account, and the hydrodynamic effect caused by thermal expansion is included. The hydrodynamic effect is of importance to the intrinsic instability of premixed flames and to the behavior of cellular-flame fronts. Moreover, 3-D flows were calculated to study the difference in shape between 2-D and 3-D cellular flames [9,10]. Now we can simulate unsteady behaviors of 2-D and 3-D premixed flames, by use of the D–T and N–S equations.

The hydrodynamic effect has a great influence on the flame instability and cellular-flame formation, so that we need to compare with the calculated results based on the D–T and N–S equations. In this work, we perform unsteady calculations of 2-D and 3-D reactive flows to investigate the hydrodynamic effect on the evolution of disturbed flame fronts and on the shape of cellular flames at low Lewis numbers.

2 Numerical Analysis

The single-reactant flame is treated, and the D–T and N–S equations, which contain a one-step irreversible chemical reaction, are employed in the present calculations. We take the direction tangential to the flame front as the yz-surface, with the gas velocity in the positive x-direction. The flow variables are nondimensionalized by the preheat zone thickness, the burning velocity, and the density and temperature of the unburned gas. The preheat zone thickness is defined as the thermal diffusivity divided by the burning velocity.

The physical parameters are given to simulate a gas mixture whose burning velocity is 0.83 m/s and flame temperature is 2086 K at atmospheric pressure and room temperature, where the nondimensionalized flame temperature is 7.0. The burning velocity is sufficiently small compared with the velocity of sound, so that the results nondimensionalized change very little, even though the burning velocity is changed. Thus, the obtained results are valid for slowly propagating flames with any burning velocity. The frequency factor and heating value are determined by the condition under which the calculated burning velocity and flame temperature of a plane flame are equivalent to the set values. To examine the diffusive–thermal effect, we take the Lewis number $\mathrm{Le} = 0.3$–1.0. Under this condition the diffusive–thermal effect has a destabilizing influence, and intrinsic instability becomes stronger as the Lewis number becomes lower.

The initial conditions are provided by the solution of a stationary plane flame, on which we superimpose a disturbance periodic in y- and z-directions. The boundary conditions are as follows: In the x-direction, except for the inlet velocity of the unburned gas, free-flow conditions are used upstream and downstream. The inlet velocity is set equal to the burning velocity. In y- and z-directions, spatially periodic conditions are used.

The explicit MacCormack scheme, which has second-order accuracy in both time and space, is adopted for calculations. The computational domain is 160 times the preheat zone thickness in the x-direction and one wavelength of a disturbance in y- and z-directions, which is resolved by a 341×65 variably spaced grid for the 2-D flow and by a $261 \times 31 \times 53$ variably spaced grid for the 3-D flow. The minimum grid size in the x-direction is one-fifth of the preheat zone thickness. These grids are fine enough to prevent numerical errors from contaminating the solutions. The time-step interval is set to satisfy the CFL condition.

When we obtain the relation between the growth rate ω and the wave number k, i.e., the dispersion relation, an infinitesimal disturbance is superimposed on a stationary plane flame. When we study the shape of a cellular flame, the wavelength of the superimposed disturbance is set equal to the peculiar wavelength, which corresponds to the maximum growth rate, since the spacing between cells in a cellular flame is equivalent to the peculiar wavelength.

3 Results and Discussion

We superimpose an infinitesimal sinusoidal disturbance on a stationary plane flame and calculate the evolution of a disturbed flame front to obtain the dispersion relation. Figure 1 shows the dispersion relations of 2-D and 3-D flames, based on the D–T and N–S equations. When the Lewis number is unity, the

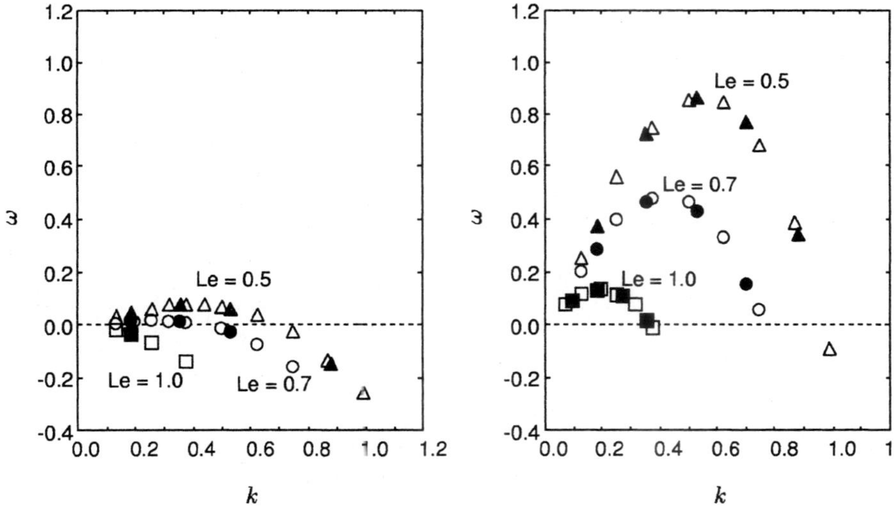

(a) Based on the D–T equation (b) Based on the N–S equation

Fig. 1. Dispersion relations of 2-D and 3-D flames at Le = 0.5, 0.7, and 1.0; open and closed symbols denote 2-D and 3-D flames, respectively

calculated growth rate based on the D–T (N–S) equation is negative (positive) at small wave numbers. The difference in growth rate between the D–T and N–S equations is due to the hydrodynamic effect. As the Lewis number becomes lower, the growth rate increases and the unstable range widens. This is due to the diffusive–thermal effect at low Lewis numbers. The calculated growth rate (unstable range) based on the D–T equation is much smaller (narrower) than

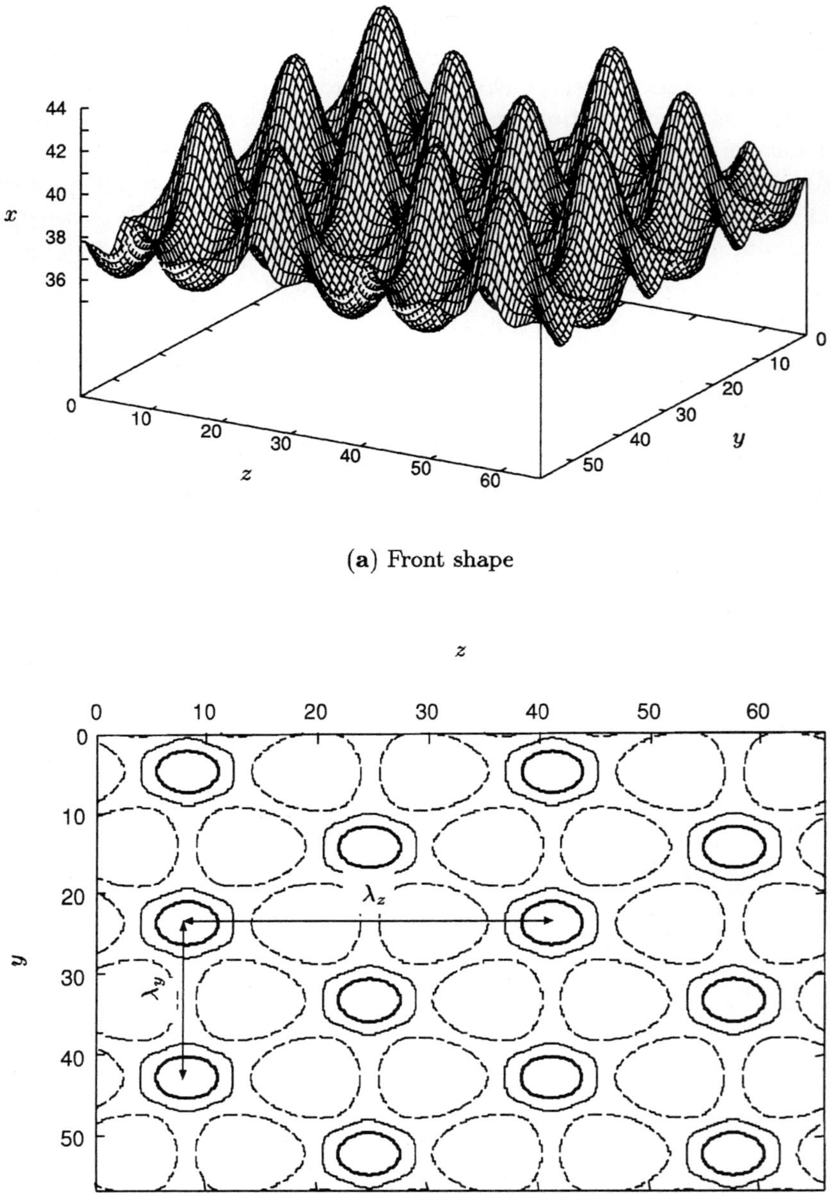

(**a**) Front shape

(**b**) Cell distribution

Fig. 2. Front shape and cell distribution of the 3-D flame at Le = 0.5, based on the D–T equation; the schematic domain is $3\lambda_y \times 2\lambda_z$

that based on the N–S equation. This means that the hydrodynamic effect has a great influence on intrinsic instability not only at the Lewis number unity but also at low Lewis numbers. In addition, the dispersion relation of the 2-D flame is nearly the same as that of the 3-D flame.

Next, we superimpose the disturbance with the peculiar wavelength to obtain a cellular flame. The front shape and cell distribution of the 3-D flame at Le = 0.5, based on the D–T equation, are illustrated in Fig. 2. The wavelength in y- and z-directions are $\lambda_y = 19.0$ and $\lambda_z = 32.9$, respectively, and the relation $\lambda_z/\lambda_y = \sqrt{3}$ is realized [3]. The unburned gas flows in from the bottom, and the burned gas flows out to the top. The cellular flame with regularly arranged hexagonal cells is obtained. The obtained front shape and cell distribution are in qualitative agreement with the calculated results based on the N–S equation, except for the cell depth and flame length.

The cell depths D_{cell} of 2-D and 3-D flames are shown in Fig. 3. At Lewis numbers higher than 0.8, no cellular flames are formed in calculations based on the D–T equation, since the growth rate is negative. The calculated cell depth

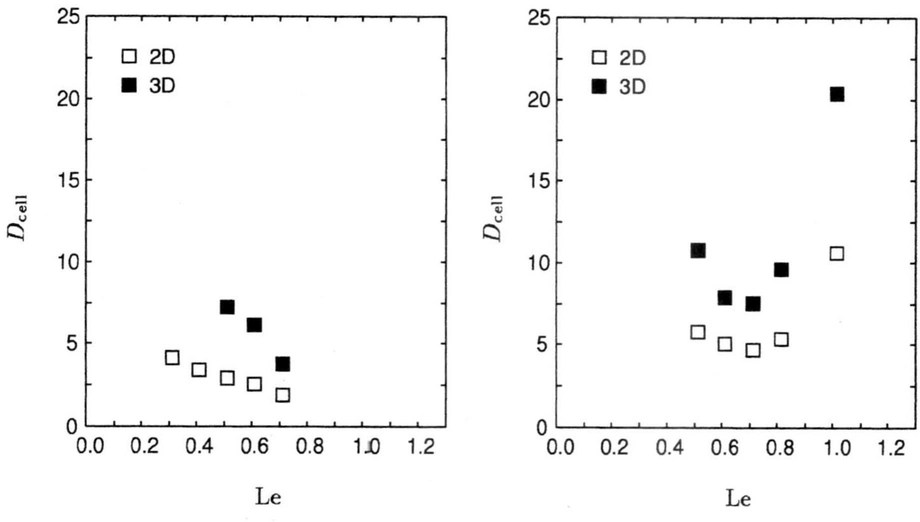

(a) Based on the D–T equation (b) Based on the N–S equation

Fig. 3. Cell depths of 2-D and 3-D flames at Le = 0.3–1.0

based on the D–T equation is considerably smaller than that based on the N–S equation, since the hydrodynamic effect is excluded in the D–T equation. In addition, the cell depth of 3-D flames is about twice that of 2-D flames. This is due to the difference in disposition of cells. The flame lengths L_{cf}, which are normalized by those of plane flames, are shown in Fig. 4. The calculated flame length based on the D–T equation is considerably smaller, just as the calculated

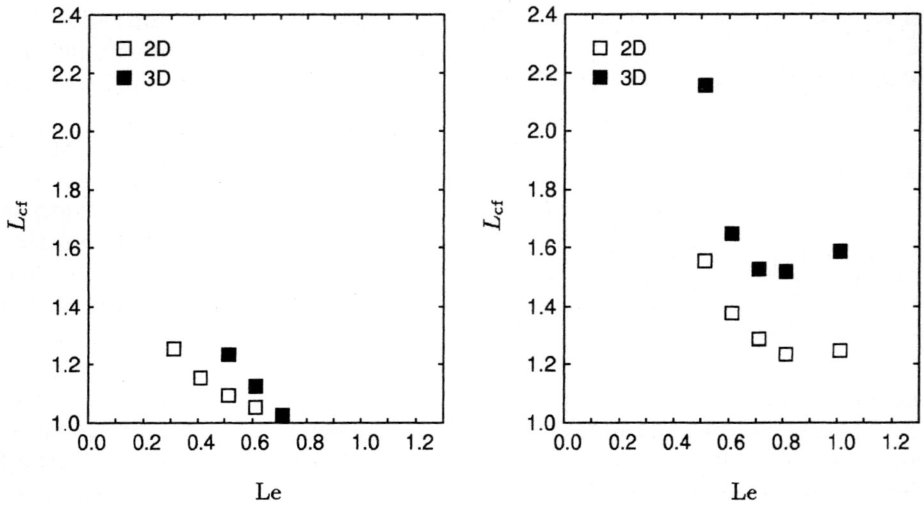

(**a**) Based on the D–T equation (**b**) Based on the N–S equation

Fig. 4. Flame lengths of 2-D and 3-D flames at Le = 0.3–1.0

cell depth was. The lower the Lewis number, the larger the flame length. Thus, the flame speed of a cellular flame increases by decreasing Le.

In conclusion, the hydrodynamic effect has a great influence, so that we need to include it when studying the intrinsic instability of premixed flames and the shape of cellular flames.

References

1. Bayliss, A., Matkowsky, B. J., and Riecke, H.: Physica D **74**, 1 (1994)
2. Bychkov, V. V., Golberg, S. M., Liberman, M. A., and Eriksson, L. E.: Phys. Rev. E **54**, 3713 (1996)
3. Christopherson, D. G.: Quart. J. Math. **11**, 63 (1940)
4. Clavin, P.: Prog. Energy Combust. Sci. **11**, 1 (1985)
5. Darrieus, G.: Unpublished Works Presented at La Technique Moderne (1938)
6. Denet, B. and Haldenwang, P.: Combust. Sci. Technol. **86**, 199 (1992)
7. Denet, B. and Haldenwang, P.: Combust. Sci. Technol. **104**, 143 (1995)
8. Kadowaki, S.: Phys. Rev. E **56**, 2966 (1997)
9. Kadowaki, S.: Int. J. Heat Fluid Flow **20**, 649 (1999)
10. Kadowaki, S.: Phys. Fluids **11**, 3426 (1999)
11. Landau, L. D.: Acta Phys. (USSR) **19**, 77 (1944)
12. Margolis, S. B. and Matkowsky, B. J.: Combust. Sci. Technol. **34**, 45 (1983)
13. Patnaik, G. and Kailasanath, K.: Combust. Flame **99**, 247 (1994)
14. Sivashinsky, G. I.: Phil. Trans. R. Soc. Lond. A **332**, 135 (1990)
15. Williams, F. A.: *Combustion Theory*, 2nd ed. (Addison-Wesley, Reading 1985) pp. 349–365

Space-Time Method for Chemically Reacting Flows with Detailed Kinetics

Yuhui Wu[1], Vigor Yang[1], and Sin-Chung Chang[2]

[1] Department of Mechanical Eng., Pennsylvania State University, University Park, PA 16802, USA
[2] NASA Glenn Research Center, Cleveland, Ohio 44135, USA
email: *yxw119@psu.edu, vigor@psu.edu, sin-chung.chang@grc.nasa.gov*

Abstract. The space-time CE/SE method originally developed for non-reacting flows was extended to accommodate finite-rate chemical kinetics for multi-component systems. The complete conservation equations of mass, momentum, energy, and species concentrations were directly tackled, and a sub-time step technique was established to perform the time integration of stiff chemical source terms in the formulation. In addition, a local grid refinement technique in the framework of the CE/SE method was incorporated to enhance the flow resolution in areas of interest. The algorithm was tested against several detonation problems with detailed chemical kinetics.

1 Introduction

Chemically reacting flows are much more difficult to simulate than their non-reacting counterparts for two reasons. First, the large number of species concentration equations are strongly coupled with the rest of the conservation laws, and consequently increase the computational burden. Second, perhaps more importantly, reacting flows usually involve a wide range of time and length scales. The resultant disparity of eigenvalues renders the solution scheme extremely stiff, and much effort is required to circumvent the problem.

Existing methods for reacting-flow simulations can be grouped into two categories. The first is the splitting method in which the original governing equations are divided into two sets. One represents pure hydrodynamics without chemical source terms so that traditional numerical schemes, such as TVD schemes, can be directly employed to solve it. The other treats motionless chemical reactions and can be handled by stiff ODE solvers. Since the two sets of equations are solved alternately at every time step, the conservation laws become difficult to be enforced, and non-physical waves may arise, as discussed by Ref.[1]. The second category attempts to directly solve the entire system of governing equations, e.g., Ref.[2]. Although much progress has been made, a number of fundamental issues still remain unresolved. Most of these schemes either evolve from their non-reacting counterparts or are established without careful treatment of steep gradients. For problems involving strong chemical reactions, some underlying assumptions such as the monotonicity property of TVD schemes may no longer be valid due to the presence of source terms, especially for flows containing discontinuities.

In view of the deficiencies of existing schemes, the present work attempts to develop an efficient algorithm for chemically reacting flows based on the space-time CE/SE method [3–5], which was originally developed for non-reacting flows. One of its unique features is the direct enforcement of integral form of the conservation equations in the space-time domain so that flow quantities are kept strictly conservative locally and globally. In an effort to extend the CE/SE method to treat flows involving strong source terms, Yu *et al.* [6] designed a new mesh system and successfully solved a ZND detonation problem by integration of the source term using a one-step, fully implicit treatment. Wu *et al.* [7] later developed a sub-time step technique to tackle strong sources in multi-component flows involving finite-rate chemical kinetics. Yu *et al.* [8] also tested the sub-time step integration method based on the trapezoidal rule and a linearization procedure. Guo and Hsu [9] integrated source terms using an adaptive multi-step procedure. The present work attempts to optimize the sub-time step integration technique described in Ref.[7] for treating chemical source terms within the context of the CE/SE method. In addition, local mesh refinement [10] is incorporated into the scheme to resolve steep gradients of flow properties arising from chemical reactions. The scheme is tested against several hydrogen/oxygen detonation problems with detailed chemical kinetics.

2 Discription of Numerical Scheme

2.1 CE/SE Method for Reacting Flow

Here we focus on one-dimensional chemically reacting flows described by the following vector form of the conservation equations of mass, momentum, energy, and species concentrations for multi-component systems.

$$\partial Q/\partial t + \partial E/\partial x = H \tag{1}$$

where Q is the dependent variable vector. The flux and source vectors are denoted by E and H, respectively. The model also accommodates finite-rate chemical kinetics and variable thermophysical properties. For reacting flows involving non-equilibrium chemistry and thermodynamics, the flux vector E becomes inhomogeneous with respect to Q.

Following the discretization procedure of the CE/SE method, we first define $\boldsymbol{h} = (E, Q)$ as the current density vector in a two-dimensional Euclidean subspace $E_2 = (x, t)$. By applying Gauss' divergence theorem in the E_2 space, Eq. (1) is shown to be the differential form of the following integral conservation law.

$$\oint_{S(V)} \boldsymbol{h} \cdot d\boldsymbol{S} = \int_V H \, dV \tag{2}$$

where $S(V)$ is the boundary of an arbitrary space-time region V in the E_2 space and $d\boldsymbol{S} = d\sigma\, \boldsymbol{n}$, with $d\sigma$ and $\boldsymbol{n}$, respectively, being the area and outward normal of a surface element on $S(V)$.

To proceed, we define the dotted area $\overline{ABCDEFG}$ in Fig. 1 as the conservation element $CE(j,n)$. The solution element $SE(j,n)$ is chosen to be coincident with $CE(j,n)$. Each $CE(j,n)$ is further divided into two sub-elements $CE_-(j,n)$ and $CE_+(j,n)$, whose domains are the enclosed areas of $\overline{ABCD}$ and $\overline{DEFA}$, respectively. Following the procedure described in [3], the vectors, $Q(x,t)$, $E(x,t)$, and $H(x,t)$ of the solution element $SE(j,n)$ are approximated by their respective first-order Taylor expansions about the point (x_j,t^n) . The integral form of the conservation laws, Eq. (2), is then enforced for the conservation sub-elements $CE_-(j,n)$ and $CE_+(j,n)$. After some manipulations, we obtain the following result.

$$Q_j^n - \frac{1}{\Delta x}\int_{CE(j,n)} H\,dV = RHS \tag{3}$$

where

$$RHS \equiv \frac{1}{2}[Q_{j-1/2}^{n-1/2} + Q_{j+1/2}^{n-1/2} + R_{j-1/2}^{n-1/2} + R_{j+1/2}^{n-1/2}]$$

$$R_j^n \equiv \frac{\Delta x}{4}(Q_x)_j^n + \frac{\Delta t}{\Delta x}E_j^n + \frac{\Delta t^2}{4\Delta x}(E_t)_j^n$$

Those terms on the right hand side of Eq. (3) take values at the last half-time step and thus are known quantities. Since the source vector H is only a function of Q, Eq. (3) can be solved to obtain Q_j^n. Finally, $(Q_x)_j^n$ is evaluated using an oscillation-suppressing procedure described in [3].

2.2 Treatment of Chemical Source Terms: Sub-Time Step Technique

In order to solve Eq. (3), integration of chemical source terms over the conservation element $CE(j,n)$ must be first performed. A simple treatment results in the following form.

$$\int_{CE(j,n)} H\,dV \cong \frac{\Delta x\,\Delta t}{2}H_j^n \tag{4}$$

Substitution into Eq. (3) leads to a nonlinear equation which can be solved using the Newton-Raphson iteration technique. Numerical experiments on reactive systems with finite-rate chemical kinetics, however, indicate that an extremely small time marching step Δt is required to ensure convergence of the iteration procedure. This is mainly attributed to the fact that the shortest chemical time scales are usually several orders smaller than the hydrodynamic scales and the source terms are extremely stiff. Numerical stability is achieved only when the time-marching step size is sufficiently small to resolve the smallest time scale of flow evolution. The resultant computational load may become unacceptably heavy. Moreover, the CFL number becomes very small, rendering large numerical dissipation.

In order to circumvent the above difficulty, an efficient integration scheme based on the sub-time step technique is developed as follows.

$$\int_{CE(j,n)} H\, dV \cong \Delta x \cdot \frac{\Delta t/2}{N_t} \left(\sum_{n_t=0}^{N_t-1} H_j^{n-n_t} \right) \tag{5}$$

where N_t is the number of sub-time steps within Δt. Physically, Eq. (5) introduces two time scales: $\Delta t/2$ and $(\Delta t/2)/N_t$, characterizing the flow evolution associated with hydrodynamics and chemical reactions, respectively. Δt is determined by the requirement for hydrodynamic flow resolution, and N_t is dictated by the requirement for numerical stability. It is obvious that N_t is dependent on the chemical kinetics involved. For flows in which chemical reactions are confined in a small region, N_t may be treated as a spatial function to save computational effort.

Substitution of Eq. (5) into Eq. (3) leads to the following equation.

$$Q_j^n = \frac{\Delta t/2}{N_t} \left(\sum_{n_t=0}^{N_t-1} H_j^{n-n_t} \right) + RHS \tag{6}$$

It can be solved using the procedure given below.

$$Q_j^0 = RHS\ , \tag{7a}$$

$$Q_j^1 = \frac{\Delta t/2}{N_t} H_j^1 + Q_j^0\ , \tag{7b}$$

$$\vdots$$

$$Q_j^{N_t} = \frac{\Delta t/2}{N_t} H_j^{N_t} + Q_j^{N_t-1}\ , \tag{7c}$$

$$Q_j^n = Q_j^{N_t}\ . \tag{7d}$$

Since chemical source terms are treated implicitly in the above equations, the Newton-Raphson iteration technique can be employed for each sub-time step.

3 Numerical Test

The scheme has been validated against several detonation problems for stoichiometric hydrogen/oxygen mixtures. The chemical kinetic mechanism involves eight species $(H_2, O_2, OH, H_2O, H, O, HO_2, H_2O_2)$ and twenty-four elementary reactions [11]. The computational domain includes a 1-D tube of 20 cm length with the head end closed and downstream end open. The solid-wall boundary condition was enforced at the head end and non-reflecting boundary condition was applied to the exit, according to the treatment described in [4] and [12]. The initial temperature and pressure of the mixture were 300 K and 1 atm, respectively. Detonation was initiated by a driver gas spanning a length of 0.2 cm from the head end. The pressure and temperature were 30 atm and 3000 K, respectively, and the species concentrations were determined from the chemical equilibrium calculation [13]. The mixture in the entire tube was initially at rest. In order to enhance the resolution of the detonation wave front, a local mesh refinement technique recently developed by the authors [10] in the framework of the CE/SE method was incorporated. Two meshes were used. The coarse mesh was uniformly distributed throughout the entire computational domain, and the fine mesh only covered a small region around the detonation wave front. The ratio of the coarse to the fine mesh size was R=8. Figures 2 and 3 show the close-up views of pressure, temperature, and species mass fractions near the detonation wave front at 65.3 ms. The calculated detonation wave speed of 2805 m/s and species concentrations at the Chapman-Jouguet point are in good agreement with the chemical equilibrium results [13].

4 Conclusions

A new scheme based on the CE/SE method has been developed for chemically reacting flows. The complete conservation equations were discretized using the CE/SE method and the time evolution of stiff source terms were calculated using a sub-time step technique. The scheme is capable of solving flows with intensive chemical reactions, as demonstrated by numerical experiments on several detonation problems. The local mesh refinement technique also proved to perform effectively in terms of numerical efficiency and accuracy.

References

1. LeVeque, R.J., and Yee, H.C., 'A Study of Numerical Methods for Hyperbolic Conservation Laws with Stiff Source Terms,' J. Comput. Phys., **86**, pp.187-210, 1990.
2. Hsieh, S.Y., Yang, V., 'A Preconditioned Flux-Differencing Scheme for Chemically Reacting Flows at all Mach Numbers,' Int'l J. Comput. Fluid Dynamics, **8**, pp.31-49, 1997.
3. Chang, S.C., 'The Method of Space-Time Conservation Element and Solution Element – A New Approach for Solving the Navier-Stokes and Euler Equations,' J. Comput. Phys., **119** , pp.295-324, 1995.
4. Chang, S.C., Wang, X.Y., and Chow, C.Y., 'The Space-Time Conservation Element and Solution Element: A New High Resolution and Genuinely Multidimensional Paradigm for Solving Conservation Laws,' J. Comput. Phys., **156**, pp.89-136, 1999.
5. Chang, S.C., Yu, S.T., Himansu, A., Wang, X.Y., Chow, C.Y. and Loh, C.Y., 'The Method of Space-Time Conservation Element and Solution Element– A New Paradigm for Numerical Solution of Conservation Laws,' in *Computational Fluid Dynamics Review 1998*, edited by M.M. Hafez and K. Oshima (World Scientific, Singapore, 1998), Vol.1.
6. Yu, S.T. and Chang, S.C., 'Treatment of Stiff Source Terms in Conservation Laws by the Method of Space-Time Conservation Element and Solution Element,' AIAA 97-0435.

7. Wu, Y.H., Yang, V., and Hsieh, K.C., 'Simulation of Unsteady Premixed Flames with Detailed Chemical Kinetics Using the Space-Time Method,' AIAA 99-1010.
8. Yu, S.T., Chang, S.C. and Jorgenson, Philip C.E., 'Direct Calculation of Detonation with Multi-Step Finite-Rate Chemistry by the Space-Time Conservation Element and Solution Element Method,' AIAA 99-3772.
9. Guo, Y., Hsu, A.T., 'Numerical Simulation of 2D Combustion Flows with the CE/SE Method,' AIAA 2000-0865.
10. Chang, S.J., Wu, Y., Wang, X.Y., and Yang, V., 'Local Mesh Refinement in the Space-Time CE/SE Method,' In *The First International Conference on Computational Fluid Dynamics (ICCFD), Kyoto, Japan, July 10-14, 2000.*
11. Franklach, M., Wang, H., Goldenburg, M., Smith, G.P., Golden, D.M., Bowman, C.T., Hanson, R.K., Gardiner, W.C., and Lissianski, V., 'GRI-Mech–An Optimized Detailed Chemical Reaction Mechanism for Methane Combustion,' Technical Report GRI-95/0058, Gas Research Institute, 1995 (http://www.me.berkeley.edu/gri_mech).
12. Chang, S.C., Loh, C.Y., Yu, S.T., Himansu, A., Wang, X.Y., Jorgenson, P., 'Robust and Simple Non-Reflecting Boundary Conditions for the Space-Time Consevation Element and Solution Element Method,' AIAA 97-2077.
13. McBride, B.J. and Gordon, S., 'Computer Program for Calculation of Complex Chemical Equilibrium Compositions and Applications,' NASA Reference Publication 1311, Jun. 1996.

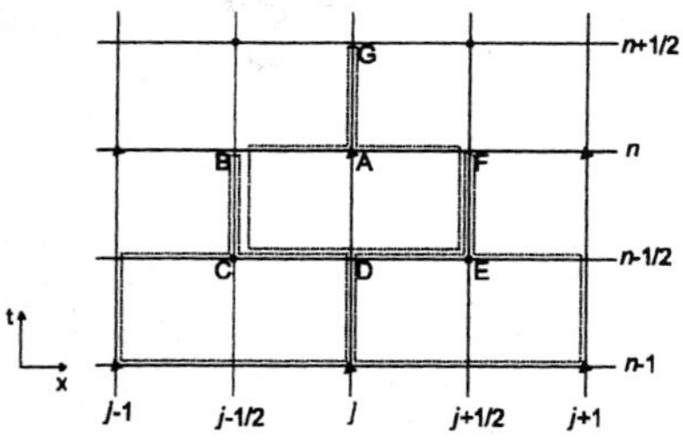

Fig. 1. Conservation element and solution element in CE/SE method

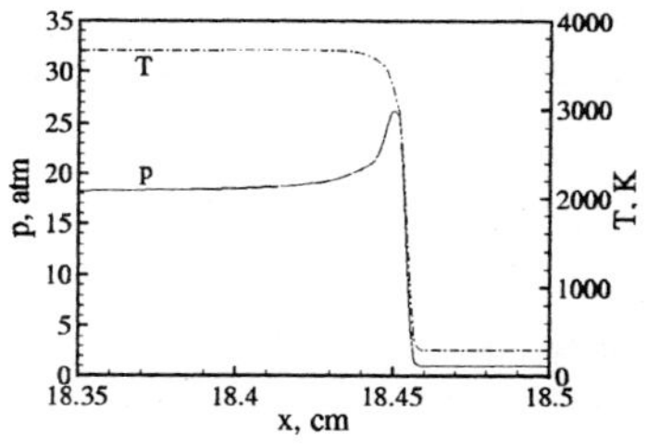

Fig. 2. Close-up view of pressure and temperature at 65.3 μs

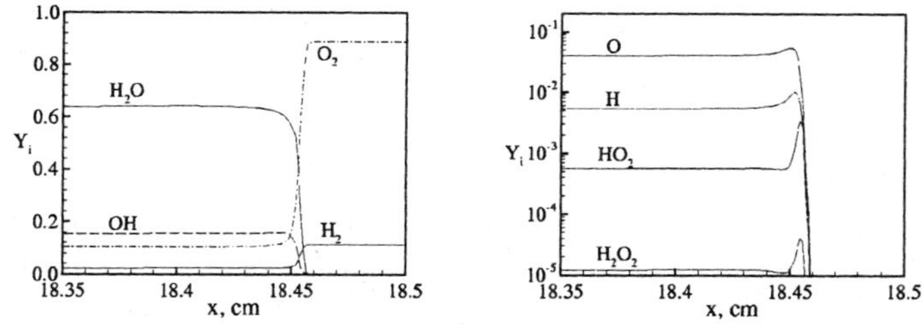

Fig. 3. Close-up view of species mass fractions at 65.3 μs

Condensed Burning Simulation in a Tube with Heated Wall Using Splitting and Pade Approximations

Andrei A. Markov

Institute for Problems in Mechanics, Russian Academy of Sciences
101-1 Vernadskii Avenue, Moscow 117526, Russia, e-mail: markov@ipmnet.ru

Abstract. Using simultaneous numerical calculation of the process for both the macroscopic and microscopic scales, the possibility is shown to analyze the chemical conversion, thermal process, transition conditions from heterogeneous condensation to homogeneous one, dispersion of the condensed particles. The new technique is developed in order to provide the stable algorithm for numerical simulation of a reacting twophase flow in a tube. As a result the processes whose scales differ from each other by the orders of magnitude are considered. The compact 5th-order convective term and 4th-order diffusion term approximations are applied. Highly accurate difference scheme enables to control the scales interaction and justify various flow structures. As an example, the model of burning of the acetylene–argon mixture is considered. The fields of temperature, mass fractions, and particle radius are presented for the various values of similarity parameters of the problem. A possibility of condensed phase forming on the reactor wall and in the flow is found. The model allows to consider various flow structures.

1 Introduction

The kinetics of nonequilibrium condensation can be studied in two stages:

1 The initial new phase formation in the form of condensation nuclei, called nucleation. 2. The subsequent growth of these nuclei into small drops (or crystals), called droplet growth. Small drops or crystals may be formed in presence of inert impurities (heterogeneous nucleation) or by statistical cluster formation due to thermal fluctuations (homogeneous nucleation). The first attempt on the mechanism of spontaneous formation of condensation nuclei in metastable vapors was taken by Volmer and Weber further improved by Farkas, Becker, and Doring, Zel'dovich and Frenkel (today this theory is known as the classical nucleation theory). Despite all, no universally accepted theory of nucleation exists yet. Processes of chemical condensation from gas mixtures have been so far described only at the stage of a primary nuclei formation or at final stages of a new phase generation. The kinetics of a new phase generation accompanied by the change in the supersaturation degree (the intermediate stage of gas-to-solid conversion) has not been considered as yet, except some particular cases concerning vapor condensation and mass crystallization. Nevertheless, the intermediate stage determines the condensed product dispersion. The distributed systems are the most complex for investigation. Present work deals with chemical-condensation model

[1] accounting of initial and intermediate stages by simulating numerically the direct flow CVD reactor modes.

2 Mathematical Descriptions

The process of chemical conversion in a flow of dusted gas mixture in a cylindrical tube with smooth walls at a constant temperature is considered using the kinetics [2]. The initial temperature of the walls is assumed to be equal to that of the gas mixture flowing. At some instant, the wall temperature is instantaneously increased to remain constant during the total process. The gas mixture consists of the reacting gas (A), the condensed vapor product (C), inert component, and the fine particles (P) uniformly distributed in the gas at the initial instant. Due to the heterogeneous chemical reaction, the size of condensed particles varies. The chemical conversion of gas-to-solid type occurs both in the gas volume (homogeneously) and on the reactor walls or the suspended particles (heterogeneously):

$$2A \xrightarrow[k_0]{Q_0} C$$

$$2A \underset{k_1\, k_{-1}}{\overset{Q_1}{\longleftrightarrow}} C \xrightarrow[k_2]{Q_2} P$$

Here k_1, k_2 and k_{-1} are the forward and the backward reaction rates, Q_0,Q_1, and Q_2 characterize the thermal effect as ($Q_0 < 0$, $Q_2 > Q_1 > 0$). Kinetic scheme is the simplest brute - scheme of thermal decay for light hydrocarbons only. We denote below reactions rate constants via K_1^0, K_{-1}^0, K_2^0. Let mass-transfer and heat-transfer coefficients be β_a, β_c, and α; mass fractions of A, C species, and temperature near a particle be a_{ex}, c_{ex}, T_{ex}. Particle size ,mass fractions of A, C species, and temperature in the flow as function of time, axial, and radial tube coordinate are denoted by $b(t, x, R), C_1, C_2, T$

2.1 Micro-level equations

Using the dimensionless form of balance equations of mass and energy, we write down micro-level equations on the surface of condensed particle and in the flow near the particle using the notations $\rho = \rho_G$, $d/dt = \partial/\partial t + \sum_i \nabla^i v^i$

At the particle surface:

$$\frac{da_s\rho}{dt} = \rho(-2k_1 a_s^2 \rho + k_{-1}c_s + \beta^a(a - a_s))b^2$$

$$\frac{dc_s\rho}{dt} = \rho(2k_1 a_s^2 \rho - k_{-1}c_s + \beta^c(c - c_s) - k_2(c_s - C_b^{sat}))b^2$$

$$\frac{d\,\rho T_s b^3}{dt} = \rho(Q_1(2k_1 a_s^2 \rho - k_{-1}c_s) + \alpha(T - T_s) + Q_2 k_2(c_s - C_b^{sat}))$$

The subscript s is referred to values at the particle surface.

The heat transfer α and mass transfer coefficients β^a and β^c are found from the solution of gas flow problem around a single particle in a region close to the particle.

2.2 Macro-level equations

On macro-level we use the continuum media approach accounting of the actual micro-flows $\overline{J}_a, \overline{J}_c$ and corresponding mass and thermal flows. The linear, surface, and volume scales of averaging are assumed to be much more than the scale of micro-heterogeneity (the particle sizes, the film thickness at the wall) as well as much less than the channel radius. We apply the averaging over a micro-volume as follows

$$\frac{1}{V_{micro}} \int \overline{J}_\alpha dv = \frac{1}{V_{micro}} N S_b J_\alpha^{get} + J_\alpha^{hom}$$

where J_α^{get} and J_α^{hom} are the averaged micro flows due to surface and bulk reactions respectively, S_b denotes the surface of a single particle.

The basic set of the equations is the Navier-Stokes equations supplemented by the terms resulting from microscopic analysis together with relaxation equations for the chemical species. The microscopic and macroscopic analysis in the general case was performed in [1].

The governing equations represent the total mass, particle mass, species mass fraction, $\alpha = 1, 2$, momentum , and energy, conservation in a reacting flow [3].

$$\frac{1}{Sh}\frac{\partial \varrho\sqrt{g}}{\partial t} + \frac{\partial \varrho u^j \sqrt{g}}{\partial x^j} + I_\varrho = 0$$

$$\frac{1}{Sh}\frac{\partial \rho_P \sqrt{g}}{\partial t} + \frac{\partial}{\partial x^j}(\rho_P\, u^j \sqrt{g}) + I_b = 0$$

$$\frac{1}{Sh}\frac{\partial \varrho C_\alpha \sqrt{g}}{\partial t} + \frac{\partial}{\partial x^j}(\sqrt{g}\varrho C_\alpha\, u^j) + I_\alpha - \frac{1}{Pe_\alpha}\frac{\partial}{\partial x^j}(g^{ij}\, D_\alpha \frac{\partial C_\alpha}{\partial x^i}\sqrt{g}) = 0,$$

The momentum conservation equations is as follows:

$$\frac{1}{Sh}\frac{\partial}{\partial t}(\varrho u^i \sqrt{g}) + \frac{\partial}{\partial x^j}(\varrho u^j u^i + p g^{ij})\sqrt{g} + D^i - \frac{1}{Re}\frac{\partial(\tau^{ij}\sqrt{g})}{\partial x^j} = 0$$

The energy conservation equation is presented as:

$$\frac{1}{Sh}\frac{\partial \varrho E\sqrt{g}}{\partial t} + \frac{\partial}{\partial x^j}(\sqrt{g}\varrho E u^j + \sqrt{g}\, p u^j) +$$
$$+ I_T = \tau_h^{-1} Q_0 C_1^2 \sqrt{g} + \frac{M_A^2}{Re}\frac{\partial R^j \sqrt{g}}{\partial x^j} + \frac{1}{Re}\frac{\partial}{\partial x^j}(\frac{\sqrt{g}}{\Pr} Q^j)$$

The equation of state reads

$$P_G = \varrho T(C_1 + \frac{C_2}{m_2} + \frac{1 - C_1 - C_2}{m_3})$$

The number of particles per unit volume obeys the equation

$$\frac{1}{Sh}\frac{\partial N\sqrt{g}}{\partial t} + \frac{\partial}{\partial x^j}(\sqrt{g}\, N\, u^j) - \frac{1}{Pe_N}\frac{\partial}{\partial x^j}(g^{ij}\, D_N\, \frac{\partial N}{\partial x^i}\sqrt{g}) = 0,$$

3 Numerical Method

The space-marching method [3] for subsonic and/or incompressible flow is combined with the time-marching for heat-mass transfer. This approach can accurately predict both the velocity and pressure field in ducts. The method is based on splitting of streamwise pressure gradient into three terms:

$$\frac{\partial p}{\partial x} = \chi \frac{\partial p}{\partial x}^{impl} - \chi \frac{\partial p}{\partial x}^{\exp l} + \frac{\partial p}{\partial x}^{glob}$$

The first and second terms are treated implicitly and explicitly in numerical scheme. The parameter χ provides the parabolization to the flow problem [3]. The last term is the physical streamwise pressure gradient term and is the one which contains all elliptic information of the flow field and is computed via splitting of the flow and heat-mass transfer equations by using the equation of state. The "elliptic information" is coming through the initial velocity profile as well as through the pressure gradient, computed at every time level according with the total density and temperature distributions in the channel.

The equations supplemented by the boundary conditions are numerically solved using the implicit finite–difference approximations and the splitting of the differential operator as a sum of two terms L_D and L_{KI}.

The splitting procedure is based on the algorithm

$$(\mathbf{A}^0_f)^n \frac{f^{n+1/2} - f^n}{\tau} + \mathbf{L}_{KI}(f^{n+1/2}) = 0; \quad f^{n+1/2}|_{t^n} = f^n$$

$$(\mathbf{A}^0_f)^n \frac{f^{n+1} - f^{n+1/2}}{\tau} + \mathbf{L}_D(f^{n+1}) = 0; \quad f^{n+1}|_{t^{n+1/2}} = f^{n+1/2}$$

Here $(\mathbf{A}^0_f)^n$ denotes the Jacobean matrix of the vector function $\mathbf{A}^0(\mathbf{f})$ with respect to the vector f. The superscript is referred to the current instant $t = t_n$. In this case $t_{n+1} = t_n + \tau$, where the time step τ is a function of n.

The $\mathbf{L}_D(f)$ is the transfer via advection and diffusion. New technique is developed in order to provide the stable algorithm for numerical simulation of a reacting two-phase flow in a tube. The approach uses the second order approximations on a nonuniform grid. As a result the processes whose scales differ from each other by the orders of magnitude are considered. The compact fifth order advection term and fourth order diffusion term approximations are applied using the Tolstykh finite difference scheme [4].

The approximation of advection terms using the fifth order compact scheme reads

$$\left(\frac{\partial A(f)}{\partial x^j}\right)_{x^j = x^j_l} = \frac{1}{2h}\{\Delta(s)(A + cf) + \Delta(-s)(A - cf)\}_l + \frac{s}{4h}w_l - \frac{s}{4h}v_l$$

Here c majors the Jacobian $c \geq ||\partial A/\partial f||$ and the following finite-difference operators are used

$$R(s)w_l = Q(s)(E + \frac{1}{12}\Delta_2)^{-1}\Delta_2(A + c\, f)_l$$

$$R(-s)v_l = Q(-s)(E + \frac{1}{12}\Delta_2)^{-1}\Delta_2(A - c\, f)_l$$

$R(s) = E + b_1\Delta_0 + b_2\Delta_2$, $Q(s) = E + a_1\Delta_0 + a_2\Delta_2$, $\Delta_0 f_l = f_{l+1} - f_{l-1}$; $\Delta_2 f_l = f_{l+1} + f_{l-1} - 2f_l$

See [4] in detail. Highly accurate difference scheme enables to control the scales interaction and justify various flow structures. The comparison of numerical results is presented. Using simultaneous numerical calculation of the process for both the macroscopic and microscopic scales [3], the possibility is shown to analyze the chemical conversion, thermal process, transition conditions from heterogeneous condensation to homogeneous one, and dispersion of the condensed particles.

4 Results of Computation

As an example, the model of burning of the acetylene–argon mixture is considered. The fields of temperature, mass fractions, and particle radius are presented for the various values of similarity parameters of the problem. The specific feature of the above investigation is an analysis of alternative physical-mechanical condensation accounting of a possibility of condensed phase forming not only on the reactor wall, but also in the flow. The model allows to consider various flow structures.

The axial velocity profiles for $\tau_h^{-1} = 5$, $\tau_g^{-1} = 6{\cdot}10^{-4}$ are given in Fig. 1. (left). Line 1 is the entrance velocity; line 2 refers to $x = x_5$ and $n = 50$; lines 3, 4 refer to $x = x_5$ and $n = 850, n = 1700$ $(t' = 45c$, $t' = 89c)$. Comparison of $T(t, x, R)$ for 2-nd and 5-th accurate scheme at $n = 1000$ $(t' = 42c)$ $x = x_5$ for $\tau_h^{-1} = 5 \cdot 10^5$, $\tau_g^{-1} = 6 \cdot 10^{-4}$ is shown in Fig. 1. (right).

Mass fractions and particle size against t for micro-problem are presented in Fig. 2. (left) for $K_1^0 = 10^7, K_{-1}^0 = 0.1$, $K_2^0 = 0.05$, $\beta_c = 250, \alpha = 5.1, a_{ex} = 0.4, c_{ex} = 0.05, T_{ex} = 4.1, \beta_a = 100$ (solid line), $\beta_a = 400$ (dash line). Particle size $b(t, x, R)$ against R, is shown in Fig. 2. (right) for $\tau_h^{-1} = 5{\cdot}10^2$, $\tau_g^{-1} = 6{\cdot}10^{-7}$ $0.9 < R < 1$ at $x = x_5$ and $n = 250, n = 1000$ ($t' = 53c$, $t' = 130c$).

The reagent mass fraction isolines are demonstrated in Fig. 3. (left). The "Spot" structure in the developed tube flow is presented via $b(t, x, R)$ isolines at $n = 10000$, $(t' = 520c)$, $\tau_g^{-1} = 6 \cdot 10^{-4}$ ($R = 1$ refers to the hot wall) in Fig. 3. (right).

References

1. Yu. M. Grigoryev and I. A. Filimonov: [in Russian], J. Chem. Phys.,**13**, 10 (1994)
2. Yu.M. Grigoryev, A.A. Markov and I.A. Filimonov: 'Modeling of the Condensed Burning in a Tube'. In: *Proc. Saint-Venant Symposium, Paris, Aug. 27–30, 1997* pp.221–228
3. A.A. Markov: *Gas-dynamic Investigation of tube flows with chemical condensation.* (The Institute for Problems in Mechanics RAS, Preprint No 641, Moscow 1999)
4. A.I. Tolstykh and D.A. Shirobokov: J. Comp. Maths. Math. Phys. **36**, 4 (1996)

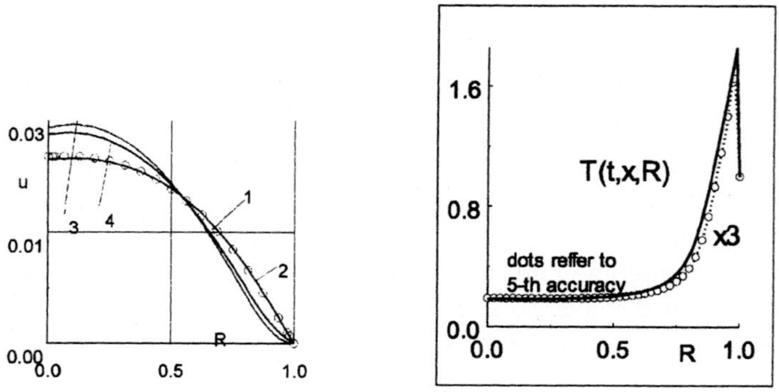

Fig. 1. The axial velocity profiles for $\tau_h^{-1} = 5$, $\tau_g^{-1} = 6 \cdot 10^{-4}$ (left); Comparison of $T(t, x, R)$ for 2-nd and 5-th accurate scheme (right)

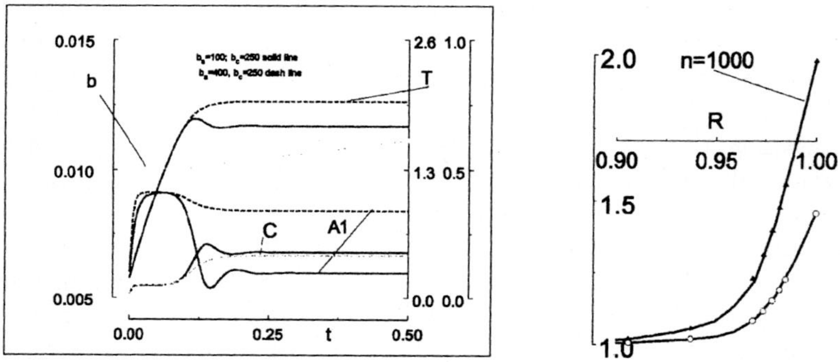

Fig. 2. Mass fractions and particle size against t for micro-problem (left); Particle size $b(t, x, R)$ against R (right)

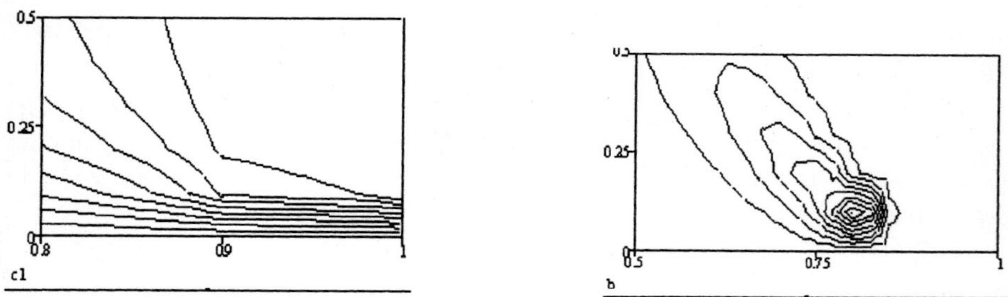

Fig. 3. The reagent massfraction isolines (left); "Spot" structure in the developed tubeflow (right)

Part VII

Unsteady Flows

A Visualization Technique to Identify the Flow Mechanism of Complicated Unsteady Flows

Susumu Teramoto and Kozo Fujii

The Institute of Space and Astronautical Science
3-1-1 Yoshinodai, Sagamihara, Kanagawa 229-8510,Japan

1 Introduction

Advances in computing technology have enabled us to carry out large-scale simulations, and unsteady simulations for complicated flowfields are carried out using tens of millions grid points within reasonable time. It has been pointed out that the pre-processing stage of such large-scale simulations occupies large part of the total analysis time. At the same time, the time required for the post-processing increases as the simulation becomes sophisticated.

The typical size of the flowfield data file for one time step with ten million grid points is 400 MB, and hundreds of such data sets are required to resolve temporal characteristics of unsteady phenomena. Hundreds giga bytes of data files have to be manipulated in the post-processing of such large-scale unsteady flow simulations, and handling of the data files becomes difficult and time consuming work. In addition, such flowfield data necessarily contain several spatial and temporal features, i.e. there may be several vortices and shock waves interacting each other, and the flowfield may oscillates with several frequencies. It is often the case that we are interested in only a part of these features, but it is difficult to distinguish characteristic features in interest, since all the features are overlaid in the raw data of the simulation results.

Several methods have been developed for the detection of spatial features of the computed flowfields, such as vortex core[1,2], shock wave[3], and separation and reattachment[4], out of complicated flowfields efficiently. These methods give clearer view of the spatial structure of the flowfields, and they help our understandings of the flowfield. However, they do not reduce the temporal complexity of the flowfields. In most cases, the temporal variation of the flowfield is discussed only by the the animations created from raw data of the simulation results, and there seems to be little effort to develop a method to detect temporal characteristics from the time series of the flow variables.

With increasing temporal complexities of the CFD problem, visualization techniques that detect the temporal characteristics of the flowfield would become important in the post-processing of unsteady flow simulations. In this paper, a simple but useful filtering technique is proposed for the better understandings of the temporal characteristics of the flowfield. The unsteady flow mechanism that is obscure in the animation of the raw data can be clearly visualized by the proposed approach.

2 Visualization of unsteady flowfields

Key temporal features that are discussed on the unsteady flowfields are

- amplitude and period of oscillation
- propagation of disturbance
- correlation between temporal variations of different flow structure (e.g. correlation between the shear layer and the vortex)

Therefore, major objective of the post-processing of unsteady flow simulation is to derive these informations from the raw data of the simulation results. Animations are commonly used for the visualization of unsteady flowfields. However, it is difficult to distinguish these temporal features from the animation, when the flowfield is complicated.

In such cases, we usually select several representative locations, extract time histories, analyze the temporal characteristics of the flow variables by for instance, the spectrum analysis, and then discuss the behavior of the whole flowfield. It is not easy, however, to discuss the whole flowfield only from few pointwise informations. Besides, large amount of data obtained in the simulation is not efficiently used by this approach.

In the present paper, temporal characteristic of the flow variables are extracted at all points, and secondary reconstructed data are visualized instead of the raw data. There may be many approaches to detect temporal characteristics. As the first step, the time histories of flow variables are pre-processed by frequency filters and visualized.

3 Numerical Method

Resultant data of unsteady flow simulations are post-processed using frequency filters as follows:

1. extract time history of a flow variable at a point that represents the flowfield
2. decompose the history into modes using Fourier analysis
3. select the mode to visualize
4. process the time history of the flow variables at all the points by the frequency filter and eliminate surplus frequency components.
5. visualize the filtered flow variables
6. create animation

Note that the size of the data files after the filtering are same as before, and the steps (5) and (6) are the same processes as those used in the conventional post-processing of unsteady simulations. However, the temporal complexity is already reduced by the filtering, and the filtered data will provide clearer view of the temporal features of the flowfield.

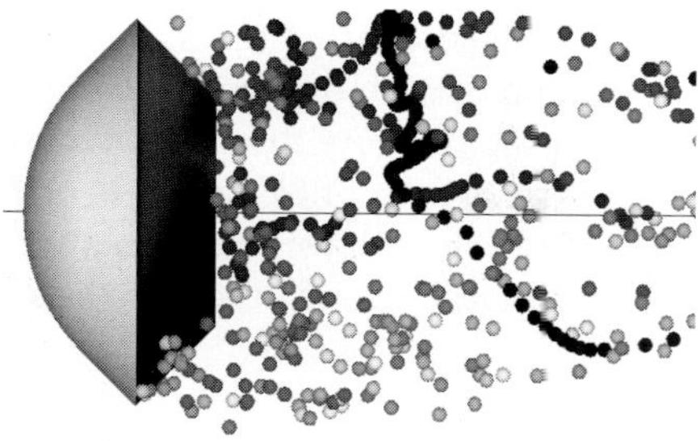

Fig. 1. Particle traces (raw data)

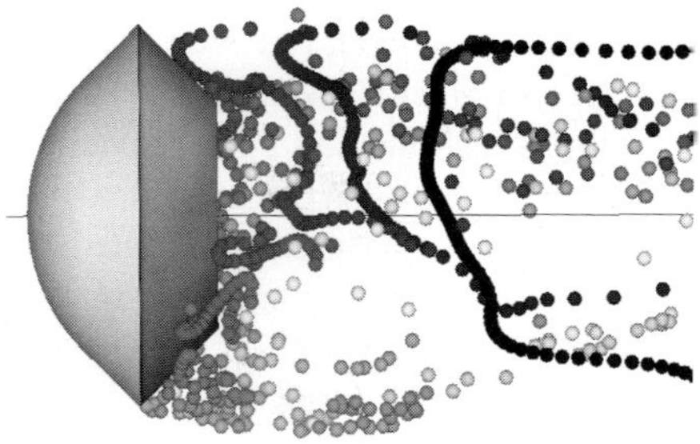

Fig. 2. Particle traces (Filtered, 150Hz, LPF)

4 Numerical Examples

4.1 Vortex structure behind a capsule in the pitching oscillation

Figure 1 shows the particle-trace plots behind a blunt capsule under the forced pitching oscillation at 20 Hz [5]. The free stream Mach number is 1.3 and the diameter of the capsule is 10 cm. Under these conditions, the wake oscillates at approximately 1kHz. The fluctuation behind the capsule is composed of the oscillation at frequencies both 20Hz and 1kHz, and unfortunately, the amplitude of the two frequency components are in the same order.

The particles are released at downstream of the capsule, from $X/D = 1.25$ in a line at a certain interval, and they convect with the local flow velocity. The flow structure behind the capsule is expected to be visualized by the deformation of the particle lines, but the particles are scattered by the high frequency disturbances as soon as released, and it is difficult to distinguish the large-scale flow structure from Figure 1.

Figure 2 shows the particle traces plots created from the flowfield data pre-processed by the low-pass filter (LPF). The LPF used in the pre-processing has cut-off frequency of 150Hz, and therefore high frequency disturbances are eliminated from the flowfield. The particles are released from the same positions at the same intervals. A large counterclockwise vortex, which is recognized as the region where the particle does not exist, is observed right behind the capsule. The vortex, which was not identified before the filtering (Fig 1), is clearly observed behind the capsule.

The animations of the particle-trace of Figs.1 and 2 clearly show the advantage of the filtered one, although it is impossible to present them in the manuscript.

4.2 Oscillation of the shear layer in the supersonic cavity flow

Next example is a supersonic flowfield around a cavity. This flowfields around cavities have been well studied, and it is well known that the pressure fluctuation in the cavity shows several modes with different frequencies. Each mode is considered to be correlated with particular oscillation patterns of the shear layer, but there are still discussions about the mechanism of each oscillation mode.

Figure 3 shows the power spectrum density of the pressure fluctuation at the rear wall of the cavity. Three peaks are observed at 1.4kHz, 4.6kHz and 9.1kHz. Each peak corresponds to the oscillation mode of the shear layer, but it is almost impossible to identify each mode in the animation created from the raw data of the simulation results.

Figures 4 and 5 show the contour surface of streamwise momentum fluctuation, pre-processed by the band-pass filter(BPF) of 3kHz–6kHz and 500Hz–2kHz respectively. As the streamwise momentum at the upper side of the shear layer is larger than that at the lower side, the fluctuation of the streamwise momentum is negative (colored dark) where the shear layer shifts upward , and it is positive (colored bright) where the shear layer shifts downward.

Figure 4 reveals that the most prevailing mode at 4.6kHz is the two-dimensional flapping fluctuation. The fluctuation originates from the leading edge of the cavity, and the wavelength of the fluctuation is equal to the length of the cavity. The first mode at 1.4kHz shown in Fig. 5 is quite different from the prevailing mode. The bright regions and the dark regions appears by turns laterally. The flow structure shows strong three-dimensionality, and it indicates the existence of the longitudinal disturbance.

The filtering process visualized the difference in the fluctuation of the shear layer at each frequencies. It would be difficult to recognize these differences from the pointwise information by the conventional approach.

5 Discussions

These examples clearly show the advantage the present temporal pre-processing in the post-processing of unsteady flow simulations. Next, the function and the limitation of the filtering are discussed in this section.

Figure 6 shows an example of oscillations of a physical property at two different points. The oscillations shown in Fig. 6 are composed of three components with different frequencies, and the oscillation corresponding to each component is shown in Figs. 7(a)–7(c). It is not difficult to understand the relationship (i.e. amplitude ratio and phase difference) between oscillations at two points for each component one by one. But once three components are superimposed into one (Fig. 6), it is almost impossible to distinguish each component.

The fluctuation of the realistic flowfield is often composed of several oscillation modes like Fig. 6. The frequency filter decompose the fluctuation into individual modes, and it enables us to observe and discuss the characteristics of each component separately.

The validity of applying the frequency filter is based on the assumption that the fluctuation of the flowfield can be linearly decomposed into individual modes. Unfortunately, the governing equations of fluid are non-linear, and the above assumption is not self-evident. The fluctuations for the different modes may

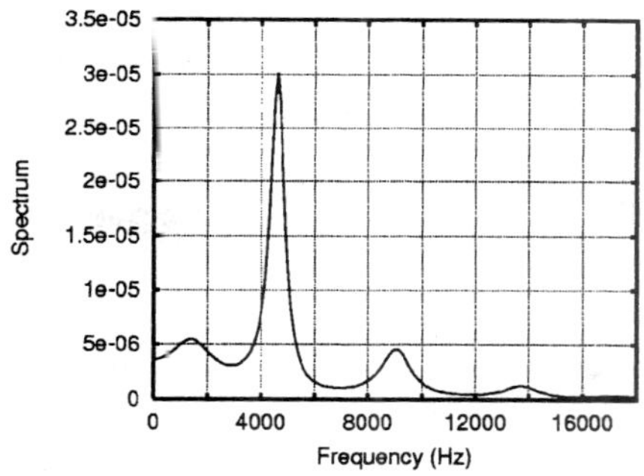

Fig. 3. Power spectrum density for rear-wall pressure

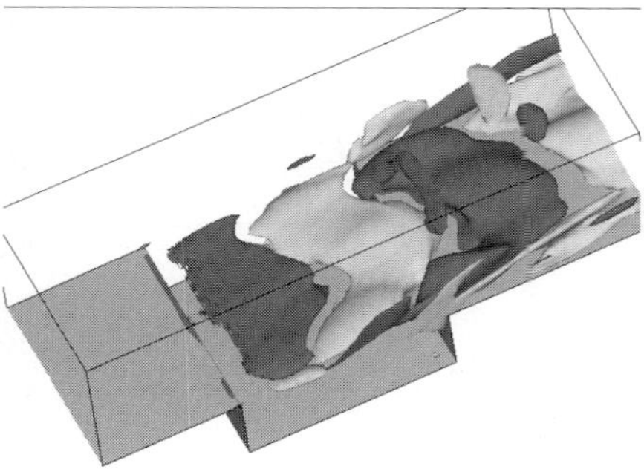

Fig. 4. Streamwise momentum (Filtered, 3000-6000Hz, BPF)

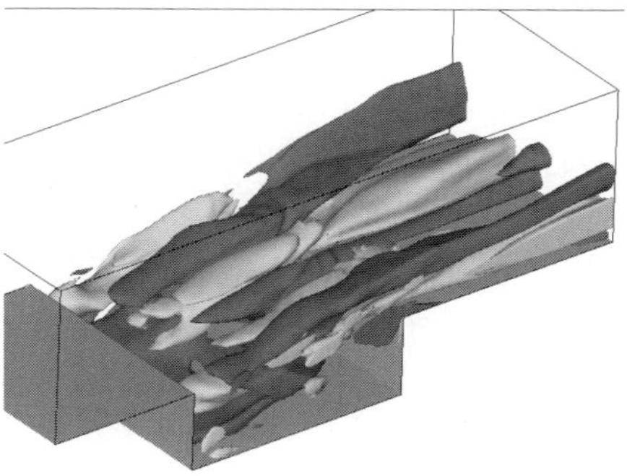

Fig. 5. Streamwise momentum (Filtered, 500-2000Hz, BPF)

influence each other, and the filtering may derive misleading results in some cases, especially when the frequencies are close to each other.

Therefore the present filtering technique should be complementally used with the conventional post-processing such as the raw animation and pointwise analysis. Nevertheless, the present idea gives intuitive view of the temporal variation of the whole flowfield, which is not given by the conventional methods, and this new idea would be an useful technique for the discussion of the results of unsteady flow simulation.

6 Conclusions

A simple idea that pre-processes raw outputs of simulation by frequency filters and separate the modes before the visualization, was proposed as a new technique for the post-processing of unsteady flow simulations. The idea was applied to several data obtained by unsteady flow simulations, and it was demonstrated that this simple idea clearly improves the understanding of the flow mechanism.

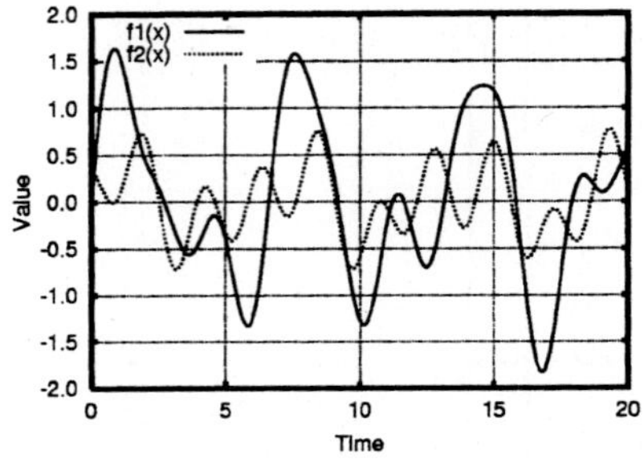

Fig. 6. Time histories of the physical property

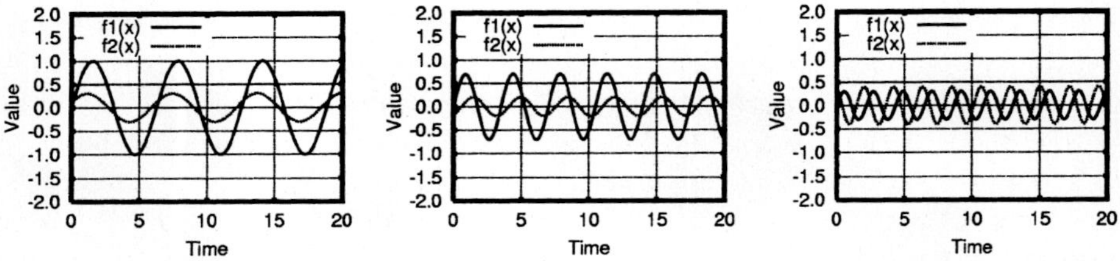

(a) $f = 6.28$ (b) $f = 11.31$ (c) $f = 18.22$

Fig. 7. Time histories of the each component

References

1. K. Sawada, "A convenient visualization method for identifying vortex cores," *Trans. Japan Soc. Aero. Space Sci.*, vol. 38, no. 120, pp. 102–120, 1995.
2. R. Haimes and D. Kenwright, "On the Velocity Gradient Tensor and Fluid Feature Extraction", AIAA Paper 99-3288, Jun. 1999
3. D. Lovely, "Shock Detection from Computational Fluid Dynamics Results", AIAA Paper 99-3285, Jun. 1999
4. D.L. Kao and H. Shen, "Automatic Surface Flow Feature Visualization", AIAA Paper 99-3287, Jun. 1999
5. S. Teramoto and K. Fujii, "Study on the mechanism of the instability of a re-entry capsule at transonic speeds", AIAA Paper 2000-2603, Jun. 2000
6. R. Kuc, *Introduction to Digital Signal Processing.* New York: McGraw-Hill, 1982.

Unsteady Three-Dimension Flow in Curved Distensible Tubes

Matthew Nelsenador[1], Angela Cheer[1], Harry Dwyer[2], and Thomas Rutaganira[1]

[1] Department of Mathematics and Institute of Theoretical Dynamics
[2] Department of Mechanical Engineering,
University of California, Davis, CA 95616, USA

1 Introduction

The main objective of our current study is to investigate the impact of wall distensibility and wall motion of the aortic arch on the unsteady flow field and the resulting shear on the wall. There are currently very few computational studies in the literature that incorporate unsteady three-dimensional flow solutions with fluid-wall interactions. A major motivation for the present work is its application to fluid flow in the aortic arch where early atherosclerotic lesions important in the development of cardiovascular disease are found. The inability of researchers to precisely and accurately characterize fluid flow in blood vessels, has impeded elucidation of the link between fluiddynamics and vascular disease. The present paper is intended to serve as a part of the foundation for future studies modelling the cardiovascular system which will include the realistic geometry and physiological conditions.

2 Equations and Methods

The three-dimensional, time-dependent and incompressible Navier Stokes equations in control volume form are:

$$\int\int_A \boldsymbol{V} \cdot d\boldsymbol{A} = 0, \qquad \text{(Continuity Equation)}$$

$$\int\int\int_V \rho \frac{D\boldsymbol{V}}{Dt} dV = -\int\int_A p d\boldsymbol{A} - \int\int_A \bar{\bar{\tau}} \cdot d\boldsymbol{A} \qquad \text{(Momentum)}$$

where $\boldsymbol{V}$ is the fluid velocity, p the dynamic pressure, $\bar{\bar{\tau}}$ the Newtonian stress tensor and ρ the fluid density. The equations are transformed into a generalized coordinate system and these equations are solved using a projection like method [1]. The pressure Poisson equation is solved with a matrix-free and pre-condition version of GMRES. To accelerate the convergence of the pressure Poisson equation, we calculate two predicted pressure correction terms, one resulting from the unsteady flow and one based on continuity [2].

One of the unique features of this study is the motion of the walls. The wall material is very complex with variable thickness, nonlinear material properties, non-homogeneous tethering and it is surrounded by fluid whose properties are not completely characterized. In short, there is not enough known about its properties so that it could be accurately modelled at the present time. However, based on experimental data, a distensible model of the following form $\frac{dA}{dp}\frac{1}{A} = \hat{k}$ where A is the cross-sectional area, p is the pressure and $\hat{k}$ is the distensibility constant, can be used [3]. In this model, the radius of the tube is determined by the fluid pressure in the tube. The value $\hat{k} = 0$ corresponds to a rigid pipe, and as the value of $\hat{k}$ increases, the pipe becomes less stiff. This parameter can be adjusted to fit experimental data obtained using MRI technology. The solution to this differential equation has the functional form

$$r(p) = r_0 e^{k(p(r)-p_0)}$$

where r_0 and p_0 are the radius and the pressure at rest conditions respectively. Here k and $\hat{k}$ differ by a constant. Analyzing such a distensible model is not straightforward because the radius is determined by pressure while the pressure distribution is, in turn, determined by the radius as well as the rate of change of the radius. In order to satisfy this distensibility relationship at n different points along the curved pipe would require solving a nonlinear system of n-equations in n-unknowns. An iterative method is used to obtain the solution at the discrete points and an interpolation procedure is used to obtain the values in between. This gives us the full three-dimensional motion of the curved pipe. We note here that the entrance and exit regions of the pipe are straight and the shape of the cross sections can be made irregular (Fig. 2). In this study, the entrance and exits are held fixed while the value of n is limited to 3.

3 Straight Pipe Calculations

To test the code, several simulations were done in a straight, rigid (non-moving) pipe. These included steady state simulations with constant inflow velocity and unsteady calculations with inflow velocity varying sinusoidally. Several simulations varied the Reynolds number (Re) which is the ratio of viscous to inertial forces from 150 to 750, and the Wormesly parameter (W_o) which is the ratio of the diffusion time scale to the cycle time from 2.2 to 15. For the distensible tube, the value of k in our model above was chosen to give a maximum deflection of 10%. These values represent a range of physiologic conditions.

The steady state simulations produced flow fields with a longitudinally and azimuthally symmetric parabolic profile, consistent with what is known for this problem. The unsteady simulations produced data which matched Wormersley's 1955 analysis of flow well [4]. For a distensible calculation with low Re and low W_o, the average shear is nearly constant along the length of the tube. The oscillating shear index (OSI) which is the average divided by the absolute value of the tangential component of shear, is also nearly constant and very close to 1 since there is very little flow reversal in this regime. This statistic estimates how extreme the oscillations in shear are. On the other hand for high Re and high W_o, the average shear goes down nearly linearly from 13 $\frac{dy}{cm^2}$ to almost 1 $\frac{dy}{cm^2}$.

The OSI also decreases very quickly to nearly zero indicating that the direction of shear is large and reversed for a substantial portion of the cycle. We also note that the peak shear occurs before the peak velocity. This indicates that shear changes, over time, more quickly than one might predict using a quasi-steady approximation. This in turn increases the rate of change of shear, a factor which may have implications for the pathogenesis of atherosclerosis.

4 Curved Pipe Calculations

In the curved pipe calculations, solution to both steady and unsteady flows through a curved distensible pipe were studied. The Reynolds number was varied from $Re = 150$ to $Re = 1800$ with a frequency parameter ranging from $W_o = 2.2$ to $W_o = 18$. These parameters were chosen to reflect the extreme physiologic conditions of rest and exercise of an average human. For steady flow, typical displacements of the radius of the curved pipe satisfying the distensibility condition for one point with linear interpolation inbetween is given in Fig. 1A. Figures 1B and 1C represents the radius displacement satisfying three points along the pipe with linear and spline interpolations inbetween the points respectively. As expected, the displacement is largest where the pressure is largest. Figures 3A and 3B compares the velocities at the same time of the cycle and at the cross-section located midway along the pipe for a low W_o and low Re flow against a high W_o and high Re calculation. The tangential velocities are plotted as contour lines, and the secondary flow as arrows. The maximum magnitude of the secondary flow is about 15-20% of the maximum tangential flow and this is a substantial contribution to the velocity field. For high W_o and high Re, the velocity profile rises very sharply from the wall and is nearly flat in the middle and the velocity field is skewed toward the inner wall for most of the cycle, whereas for low W_o and low Re, it is always skewed towards the outer wall. The secondary flow in the high W_o and high Re case is concentrated in the region near the wall.

For the unsteady flow calculations, two different pulsatile inflow were used and selected results for the low W_o and low Re calculations are presented in figure 4. Figure 4A, plots the radius versus time for the three point distensible model. The points are equally spaced along the pipe with the second point on the apex. Figure 4B plots shear versus time on the outer wall, middle wall and inner wall at the halfway point (top of the arch). We remark that the middle wall is on the wall of the tube, halfway between the outer and inner wall. Since absolute magnitude of shear, not just the tangential component is considered, this is a non-negative quantity. We see from this figure that the shear on the outer and middle walls are large and comparable, while the shear on the inner wall is smaller. The pressure midway along the pipe on the outer, middle and inner walls are plotted in Figure 4C.

Corresponding plots for the high W_o and high Re simulations are presented in Fig. 5 for comparison. The results for these two simulations are noticeably different. In the radius versus time plots, we see that the period in which the radius oscillates is longer for high W_o and high Re than for low W_o and low Re. The oscillations do not die out during the zero-velocity inflow phase in high W_o and high Re whereas it does in the low W_o and low Re case. Every distensibility simulation exhibited radius oscillations to some degree. These radius oscillations are not observed experimentally and occur in our calculations because our model does not include any damping.

Oscillations also appear in the plots of shear vs. time during the non-zero velocity phase, indicating that the rate of change of shear is affected by both the motion of the wall and the inflow. The time average shear for high W_o and high Re is larger on the inner wall than the outer wall. The larger negative shear on the inner wall contibutes to this since we average the absolute magnitudes. Examining the shear vs distance data (not shown) indicates that shear responds to changes in geometry where local extrema in the time averaged shear corresponds to local changes in radius of the vessel. When compared to the rigid wall calculations, the shear vs distance can change by as much as 40% in the distensibile simulations. The shear gradients change accordingly.

Shear values are generally higher in the curved pipe than they are for the straight pipe simulations when compared to the same W_o and Re; more than twice as high on the outer and middle walls. Unusual shear patterns appear at the transition form straight to curved, and curved to straight regions. The shear gradients are very large at these transiton regions and the shear on the inner wall is briefly higher than on the outer wall.

In the pressure plots for the high W_o and high Re case, the oscillations dominate the distensibile simulations. In this case, the pressure at the three points along the wall is almost equal as compared to the low W_o and low Re case where the pressure on the outer wall is larger than that on the inner wall. When compared to the steady calculations, we note that the peak pressure at the entrance is about the same but the peak pressure at the middle varies greatly with values nearly doubled.

5 Conclusions

Results indicate that distensibility can affect time average shear,increase shear gradients and increase rate of change of shear. All of these have been linked to the development of atherosclerotic plaque. This suggests that it is necessary to incorporate the effects of wall motion in order to understand the fluiddynamic implications in the pathogenesis of atherosclerosis.

References

1. D. S. Dandy, H. Dwyer, Journal of Fluid Mech. **216**, 381 (1990)
2. H. Dwyer, A.Y. Cheer, T. Rutaganira, Lecture Notes in Physics. **515**, 153 (1998)
3. S.A. Berger, Contemporary Mathematics Series. **141**, 479 (1993)
4. J.R. Womersley, Journal of Physiology. **127**, 553 (1955)

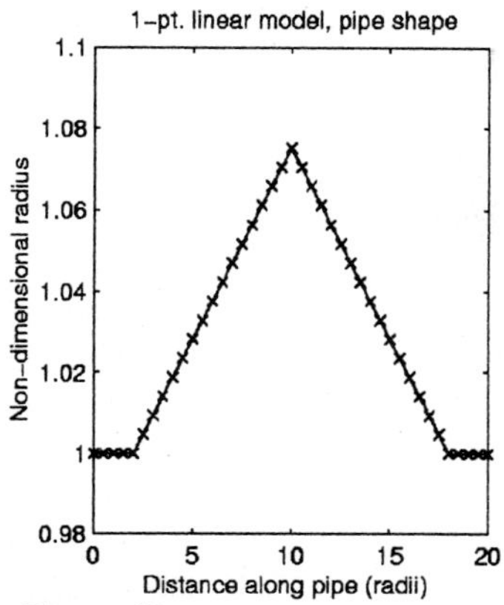

Figure 1A

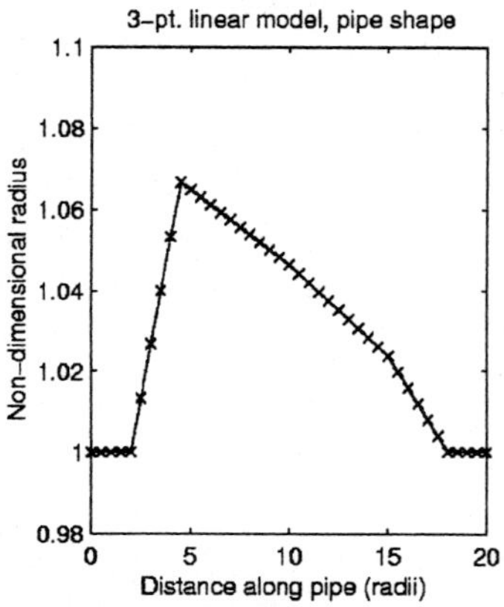

Figure 1B

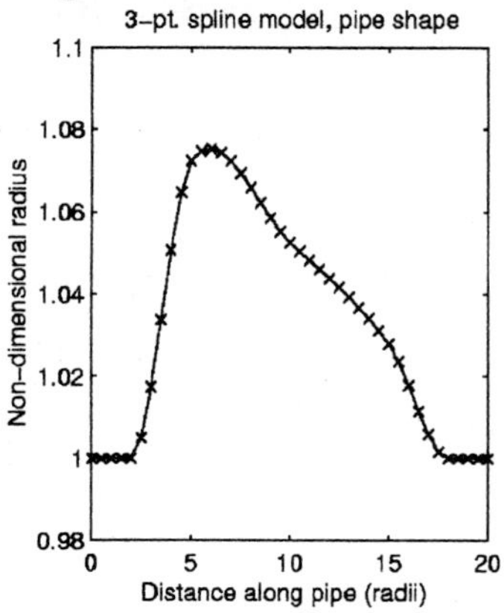

Figure 1: Displacement along the tube for various distensibility models

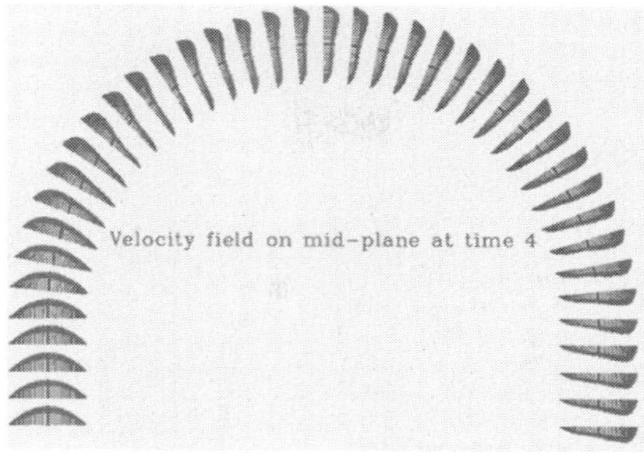

Figure 2: velocity and pressure on mid-plane for steady and Low Wo Low Re calculations

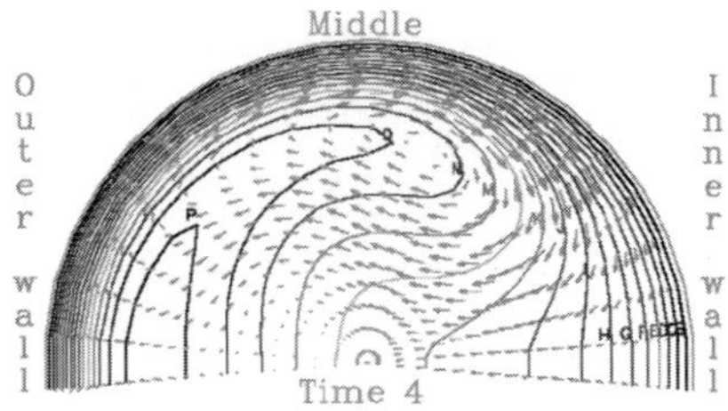

Figure 3A

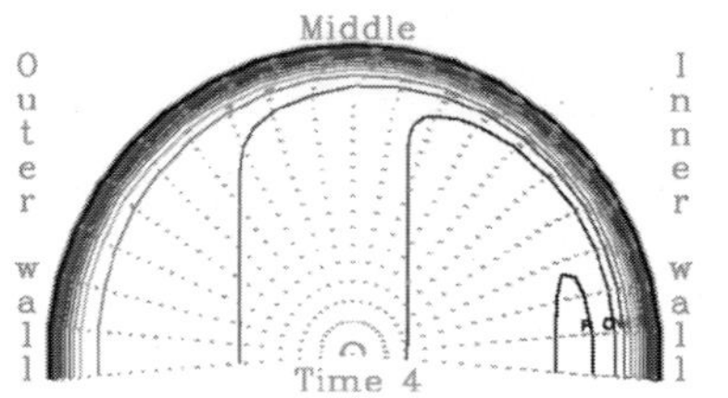

Figure 3B

Figure 3: Unsteady curved pipe calculations
A: Low Wo Low Re calculations
B: High Wo High Re

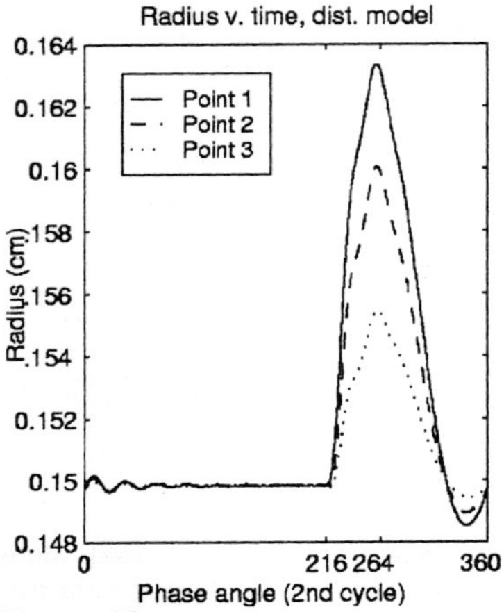

Figure 4A

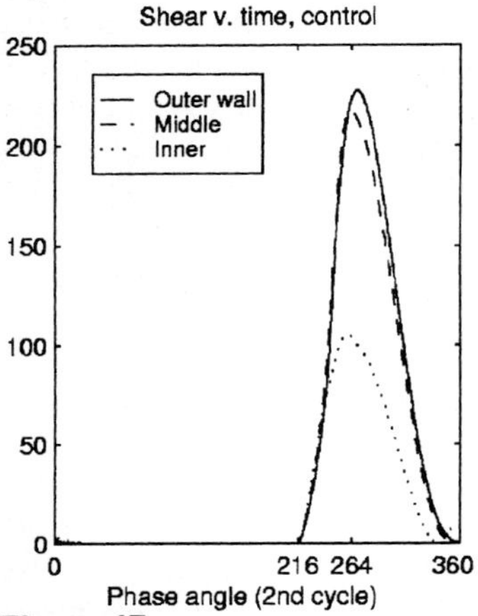

Figure 4B

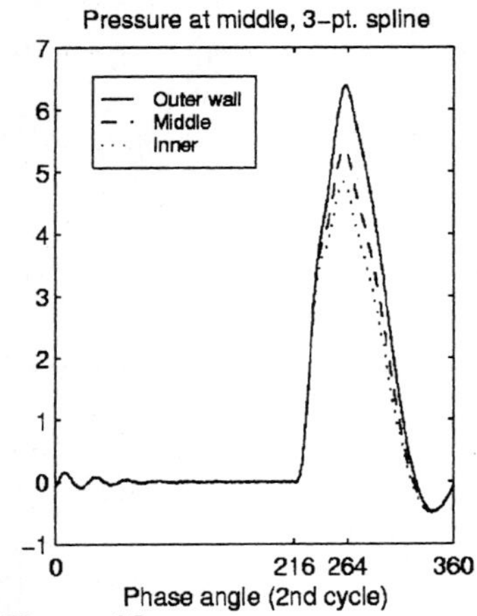

Figure 4C

Figure 4: Distensibility Calculation, Low W_O Low Re; Re=150 W_O=2.2

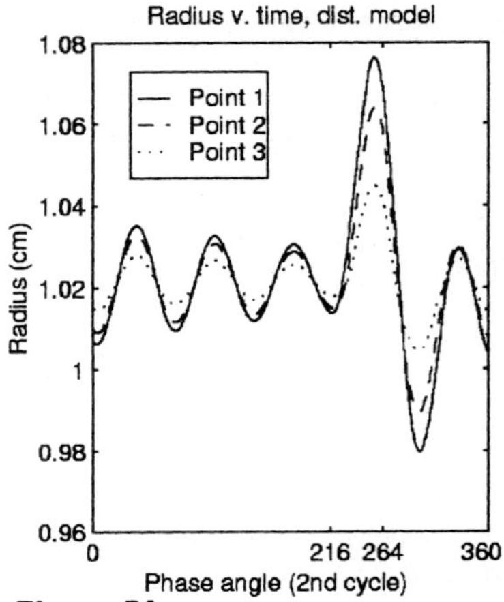

Figure 5A

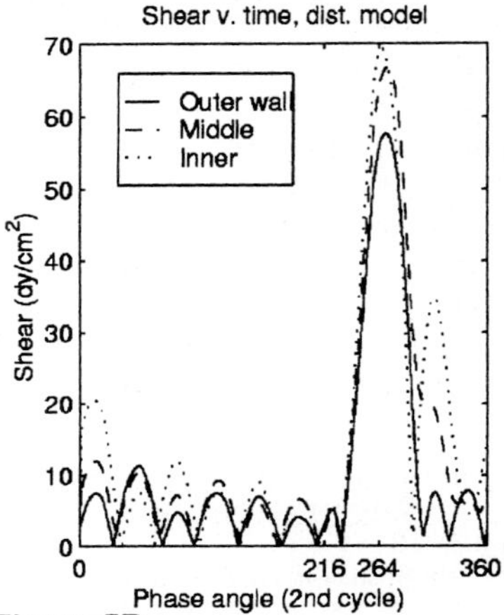

Figure 5B

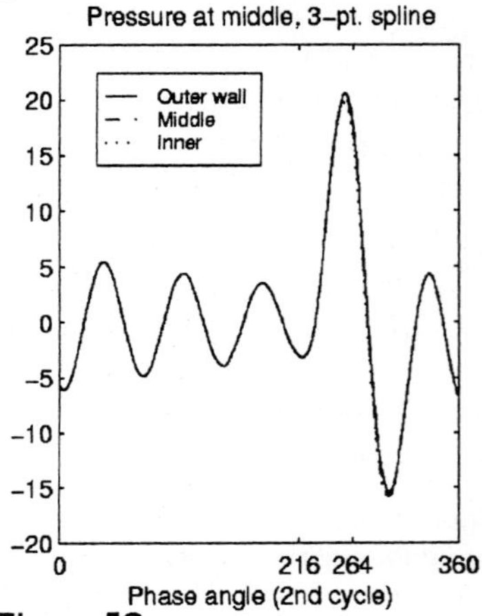

Figure 5C

Figure 5: Distensibility Calculation, High W_O High Re; W_O=15.0 Re=1500

Unsteady and Three-Dimensional Flow Simulations in the Human Aorta

Nader Shahcheraghi[1], Harry A. Dwyer[1], Angela Y. Cheer[2], Thomas Rutaganira[2], and Abdul Barakat[1]

[1] Mechanical and Aeronautical Engineering Department
[2] Institute for Theoretical Dynamics, University of California, Davis 95616, USA

Abstract. Three dimensional and pulsatile blood flow in the human aorta and its first three branches was considered. Reynolds number of 2500 and frequency parameter of 10 were used in the computational model. The flow field was solved using a generalized coordinate low Mach number projection like numerical scheme. A finite volume formulation of the Navier-Stokes equations along with overset grids were used. A particular feature of the method of solution includes prediction of the pressure field using fluid physics. These predictions, that apply at large frequency parameter range considered in this study, speed up the convergence of the pressure Poisson equations by an order of magnitude. The fluid mechanical stress distributions on the aorta wall and its branch walls are discussed. These stresses are of particular importance due to their suspected correlation with atherosclerosis, which develop at artery bifurcation and branches.

1 Introduction and Method of Approach

In general, incompressible unsteady internal flow in ducts and pipe bifurcation and branches have many applications. One particular application and a major motivation for the present work is simulation of fluid flow in the aortic arch, where early atherosclerotic lesions are found. These lesions are more often found in regions of arterial branching and curvature where the flow field is disturbed. This has motivated the notion that fluid mechanical stresses may act as a localizing factor for the disease. In the present research we have successfully simulated the unsteady and three-dimensional flow in a physiologically correct human aorta [1]. The human aorta and its three branches are shown in the fig. 1(a). Using Computerized Axial Tomography (CAT) scan data taken from the aorta section of a human subject, we have constructed the mesh shown in fig. 1(b). The time-dependent flow pulse has been based on measurements, and it involves both positive and negative flow in the aorta,[2], fig. 1(b). An important aspect of this study is to address the problem of slow convergence due to large aspect ratio computational cells in solving the Poisson equation, which provides the pressure field. The slow convergence has been accelerated by accurate predictions of the pressure at each time step using fluid physics.

In the present report we have used the overset grid approach to treat the aortic branches shown in fig. 1(a). Previously, we have used this approach to treat a steady flow through the celiac bifurcation, [3]. The principle advantage of the method is that it allows each branch to be meshed independently. The three dimensional data set from CAT scans is used by the overset grid generator part of the OVERTURE software from the Lawrence Livermore National Laboratory (LLNL) to generate the overset computational grids. The equations solved in this paper are the continuity, and time-dependent incompressible Navier-Stokes equations in control volume form,

$$\oint_S \boldsymbol{V} \cdot \boldsymbol{n} dA = 0$$

$$\frac{\alpha^2}{Re}\frac{\partial}{\partial t}\int_V \boldsymbol{V} dv + \int_V \boldsymbol{V} \cdot \nabla \boldsymbol{V} dv = -\oint_S P \cdot \boldsymbol{n} dA + \frac{1}{Re}\oint_S \boldsymbol{n} \cdot \bar{\tau} dA$$

$$\frac{\alpha^2}{Re}\frac{\partial}{\partial t}\int_V T dv + \int_V \boldsymbol{V} \cdot \nabla T dv = \frac{1}{RePr}\oint_S \nabla T \cdot \boldsymbol{n} dA$$

Here, T, is tracer concentration, which is governed by an unsteady convection diffusion equation and is used to illustrate the flow patterns. These equations are transformed into a generalized coordinate system to treat complex geometries. The parameter Reynolds number, Re, is set to 2500 at its peak value (maximum positive flow), and the frequency parameter, α, is set to 10. which are typical of a small human. The prandtl number, Pr, which is the ratio of kinematic viscosity to tracer diffusion is set to 1. Therefore, tracer concentration patterns represent the viscus flow features. These equations are solved using a projection like method, [4], and the pressure Poisson equation is solved with a matrix free and pre-conditioned version of GMRES. However, even with this level of sophistication, the accurate solution of the Poisson equation requires an excessive computational effort. In order to have the correct mass flow at each section of the pipe it is necessary to iterate excessively on the Poisson equation leading to a computationally expensive simulation. In this paper, a new approach to predicting the pressure field has been extended to bifurcating flows,[5]. Exploiting information about the unsteady flow field physics, this algorithm predicts two accurate pressure corrections based on a Bernoulli type flow, which is a reasonable assumption at high frequency parameter values. The two components of the pressure prediction are: 1.) A pressure correction based on the defect or excess of mass at a local pipe cross-section, P', which is given by

$$\nabla p' = -\rho\frac{\Delta \boldsymbol{U}}{\Delta t} \quad where \quad \Delta \boldsymbol{U} = \nabla\phi - \frac{\int_A \boldsymbol{V} \cdot \boldsymbol{n} dA}{A}, \quad and \quad \nabla^2\phi = 0$$

2.) The second pressure correction, p", is based on the inlet flow acceleration or deceleration:

$$\nabla p'' = -\rho\frac{\Delta \boldsymbol{U}_i}{\Delta t}$$

Therefore, the pressure at time n+1 is given by:

$$P^{n+1} = P^n + \alpha^{n+1} + p' + P''$$

, where α^{n+1} is the pressure correction from the Poisson solve. Previously, this approach had been successfully applied to flow in a single curved pipe, [3], however, in a bifurcating flow we do not know the mass at all pipe sections due to the diversion of mass into the mesh branches. In order to treat bifurcating flow problems, we have approximated mass flow at the pipe cross sections in the bifurcation regions based on known division of the mass flow. when properly applied, these approximations have lead to efficient predictions of the pressure field.

2 Results

The presentation of results will begin fig. 2. The top part of this figure illustrates the velocity vectors near the mid-plane and at three important times in the flow cycle, maximum, minimum, and zero flow. The size of the velocity vectors has been adjusted to give maximum visibility, and they do not scale relative to each other. At maximum flow, fig. 2(a), we see strong influences caused by the curvature of the aortic arch, and the flow into the arch branches is highly skewed towards the distal (downstream) walls of the branches. Because wall shear stress and fluid dynamic pressure are likely the most relevant parameters from the standpoint of the development of atherosclerosis, we present contours of these quantities in figs. 2(d) and 2(e). Maximum wall shear stress values, fig. 2(d), occur near the entrance of the aortic section and within the branches (particularly along the distal walls). On the other hand, maximum dynamic pressure values, fig. 2(e), occur at stagnation points at branch flow dividers, and in the aortic arch. The values of wall shear stress and pressure can be used as input to structural models of the walls yielding valuable information about the stresses to which the vascular wall is exposed. Retrograde flow in the aorta in the latter portion of the pulse likely provides blood flow to the coronary arteries, fig. 2(b). An important task is the three-dimensional presentation of the flow field, and an example is given in fig. 3, where we show the three-dimensional velocity field at various cross-sections of the main aorta and arch branches at maximum positive flow (peak systole). At the entrance to the aorta, sections A and B, the flow is skewed toward the inner wall. Gradually, artery curvature and the presence of the branches change the flow in sections B, C, and D. Far downstream of the branches, the flow is highly three-dimensional with little resemblance to classical parabolic profile for steady flow in a pipe. An important aspect of this work is the relative changes in flow that occur as a result of alterations in the shape of the aorta and its branches due to advanced atherosclerotic lesions. For example, the flows in the arch branches, sections G thru L, are highly skewed towards the distal walls, and there is evidence of reversed flow near the branch entrances (at peak systole). It is to be expected that atherosclerosis-induced area constrictions in the branches will have a strong influence on the flow patterns. Due to the highly unsteady and three-dimensional nature of blood flow in the large arteries, it is not an easy task to keep track of the paths of fluid particles within the flow field. A distinct advantage of scientific simulations is that detailed trajectories of fluids particles can be followed and studied. An effective way to investigate these fluid motions is to numerically inject a passive tracer into the flow, and we have carried out such an investigation in the present work. Shown in fig. 4 are the results of our tracer studies in which a tracer is introduced at the arterial wall. In the present paper, the presentation of the tracer studies is restricted to the time points within the cardiac cycle shown previously (fig. 2). The distribution of tracer concentration is shown in fig. 4; with the tracer concentration ranging from 1 to zero. At peak systole, fig. 4(a), the convection is quite strong, and the flow moves quickly with respect to wall diffusion. This dominance of convection causes thin tracer diffusion region as shown in fig. 4(a), except for isolated pockets of large and intermediate tracer concentration values in regions of low shear stress or reversed flow within the branches, where the velocity field is retarded. It should also be mentioned that diffusion processes proceed on a time scale that is proportional to the square of the local length scale, therefore, the wall tracer diffusion will be more effective in the smaller diameter branches. This is partially observed in fig. 4(a) and becomes more evident in figs. 4(b) and 4(c), where the flow rates are considerably smaller. For maximum negative flow (peak diastole), fig. 4(b), the flow in the branches strongly influence the distribution of the wall tracer while the influence within the main aortic section is less pronounced. At zero flow condition in the cycle, fig, 4(c), tracer concentration levels are in the intermediate to high range in the computational domain. This implies that all of the fluid at this time in the cycle has had time to experience the effect of the wall tracer.

3 Conclusions

The unsteady, pulsatile blood flow in a physiologically correct, rigid, geometric model of the human aorta and its three branches were successfully simulated. The method of solution employed the overset grid technique in a projection like formulation of the continuity and Navier-Stokes equations. A fluid physics model was used to provide two part predictions of the pressure field in order to accelerate the convergence of the Poisson pressure equation. The results show a strong three dimensional characteristics, which are due to the curved and twisted three dimensional geometry of the aorta and its branches, and the unsteady nature of the flow. The results show that the velocity profile is highly skewed in the aorta and contains strong recirculation regions near the entrance of the branches near maximum positive flow in the cycle. Near the maximum negative flow velocity profiles change shape with a different skewness. With zero inlet the fluid inertia effects set up complex flow patterns with many recirculation regions. The pressure field and the wall shear stress results indicate that these variables obtain their maximum values near peak systole. The maximum pressure occur near the branch intersections, where stagnation regions develop. The maximum wall shear stress also occurs in the same vicinity due to large acceleration and deceleration of the fluid. The use of physical pressure prediction strategies provided fast convergence of the Poisson equation.

References

1. Private communications and data- Dr. John Anderson, UC Davis Medical School
2. T. J. Pedley: *The Fluid Mechanics of Large Blood Vessels*, (Cambridge University Press, 1980)
3. A. Y. Cheer, H. A. Dwyer, et. al: Biorheology **35**:6 (1998), pp. 415-435
4. D. S. Dandy, and H. A. Dwyer, J. Fluid Mechanics **216** (1990), pp. 381-410
5. H. A. Dwyer, A. Y. Cheer, and T. Rutaganira, Lecture Notes in Physics, **515** (1998), Springer, pp. 153-158

4 Figures

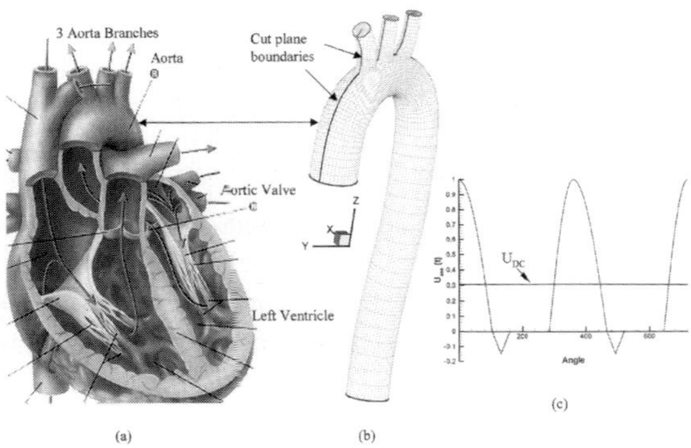

Fig. 1. (a)Blood flow in human heart and aorta, from Cardiovascular physiology; PERMA-CHART QUICK reference Guide,(b)Overset grid for the aorta and its branches.(c) Heart pulse (0 to 360 degrees is 1 cycle),normalized transient inlet (spatial-average) velocity, U_{ave}(t),and time-average (direct current) inlet velocity, U_{DC}

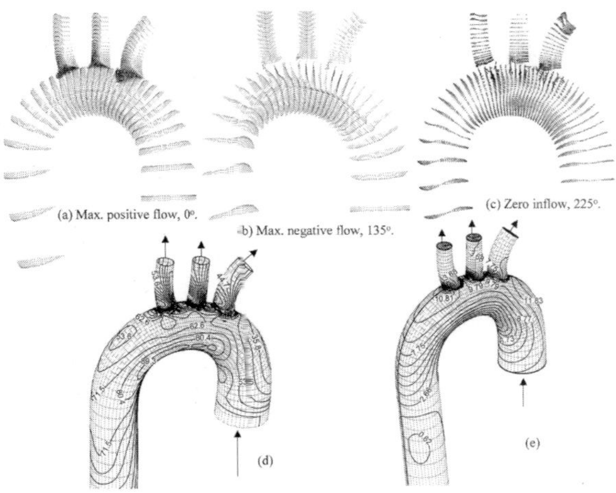

Fig. 2. (a), (b), (c) Velocity vectors in the center slice at various points of the cycle. (d) Wall shear stress contours at maximum positive flow. (e)Pressure contours at maximum positive flow.

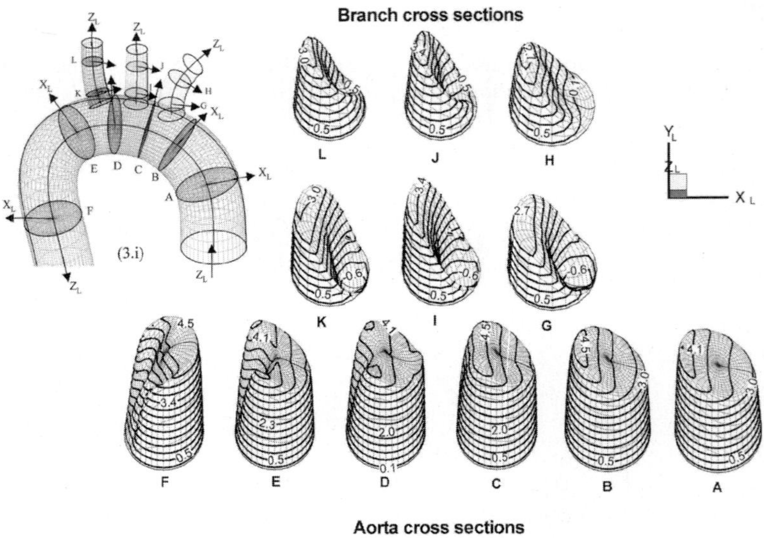

Fig. 3. Primary flow velocity profiles for maximum positive flow at selected cross sections (A to L). The local orientation of the velocity profiles is given by th elocal coordinate vectors X_L and Z_L in (3.i).

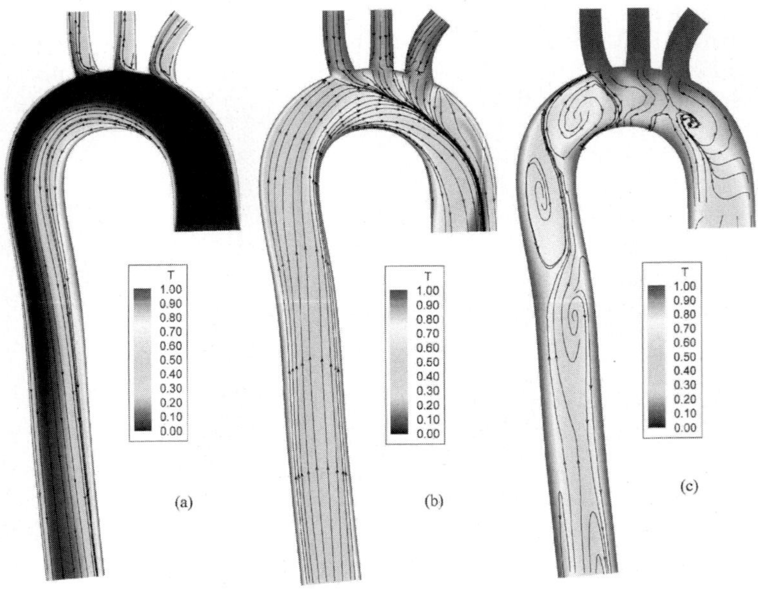

Fig. 4. Tracer concentration contours and stream lines in the aorta and its branches (a) maximum positive flow,0 degrees (b) maximum negative flow, 135 degrees, and (c) zero flow, 225 degrees.

Numerical Simulation of Unsteady Internal Flows using Dual Time Stepping Method

Olivier Rouzaud and Sylvie Plot

ONERA - CFD and Aeroacoustics Department, BP72, F-92322 Châtillon, France

1 Introduction

The Dual Time Stepping method has been applied to the numerical simulation of unsteady internal flows using a Reynolds-Averaged Navier-Stokes (RANS) solver with transport equation turbulence models. The first two test cases are purely numerical. Simulations have provided comparisons of computational performances between the DTS method and a time-explicit method. Furthermore, the turbulent computations have demonstrated the influence of the solution strategy for the RANS/turbulent decoupled systems in order to improve the efficiency. For the last test case – cavity flow –, the existence of a database has allowed to verify the relevancy of the DTS results. The numerical prediction is satisfactory since it provides a good description of such a flow.

2 Numerical Method

The governing equations correspond to the three-dimensional compressible RANS equations written in a cartesian frame of reference. The unknowns are the conservative variables, the static pressure being given through the equation of state for a perfect gas. For turbulent applications, the eddy viscosity is given by transport-equation turbulence models of Jones-Launder [1], Wilcox [2], Smith [3] and Spalart-Allmaras [4].

Computations have been performed with the codes CANARI [5–9] and elsA [10]. The former one is written in Fortran language. The latter one lays on Oriented-Object methodology and combines C++/Fortran languages. For the present work, both codes share the same numerical approach. They correspond to multidomain solvers for structured meshes with a cell-centred finite volume formulation.

2.1 Steady State Time Marching Solver

The steady state time marching solver stands on the numerical scheme of Jameson et al. [11]. Concerning the explicit stage, the spatial discretisation lays on second-order central differences. Such a choice involves the use of artificial dissipation terms of the first and third orders. Time integration is performed through a 4-stage Runge-Kutta algorithm. Besides, the steady state implementation contains convergence acceleration techniques such as local time stepping, implicit residual smoothing and FAS multigrid method. To maintain the modular feature of the solver, the RANS equations and the turbulent equations are decoupled. The multigrid method is only available for the RANS system. However, at each time step, the algorithm allows to perform several single grid iterations on the turbulent system for each multigrid iteration on the RANS system. This possibility may prove useful to improve the convergence. In the rest of the work, it will be referred to as the solution strategy while MF and TF will respectively stand for mean field and turbulent field number of iterations.

2.2 Time Explicit Solver (TES) for Unsteady Simulations

The previous solver allows the treatment of unsteady simulations as soon as global time stepping replaces local time stepping. By the way, the solver becomes time consistent, leading to a time-explicit solver. The implicit residual smoothing is always possible but the multigrid method is no longer available. The time accuracy can be of order one or two, depending on the algorithmic choice, but only first-order accurate computations are presented in this work. Note also that the choice of the time step value depends on the stability criterion of the scheme.

2.3 Dual Time Stepping (DTS) Method for Unsteady Simulations

The DTS method [12] is a time-implicit method. The choice of the physical time step lays no longer on stability of the numerical scheme but on accuracy. It ensures a second-order time accuracy. By introducing a fictitious time τ (i.e. dual time), the system of unsteady equations becomes a system of pseudo-unsteady equations relative to the dual time. At each physical time step, the unsteady solution is obtained using the previous steady-state time-marching algorithm, by performing subiterations in dual time. A tolerance criterion allows to manage the dual time convergence. Two definitions have been tested and are based, either on the squared L2 norm of density variation $\Delta\rho$ for the RANS system and of ρk variation between two consecutive dual time steps (first definition), or on the L2 norm of the density residual $\Delta\rho/\Delta\tau$ (second definition).

3 Numerical Results

For DTS computations, all the convergence acceleration techniques have been employed. Concerning the boundary treatment, the slip condition is ensured at the wall for inviscid computation while viscous tests use the adiabatic no-slip condition. At inlet boundaries, stagnation pressure, stagnation temperature and velocity direction are prescribed. At the outlet boundaries, only the static pressure is imposed. Finally, the boundaries between two domains are coincident. Computations have been performed on the ONERA NEC SX5 computer.

3.1 NASA Rotor 37 Test Case

The test case refers to NASA Rotor 37 [13] for which experimental data are available for steady configuration. It has allowed to define a numerical unsteady test case on a three-dimensional configuration in order to validate the DTS method relatively to a TES computation. Besides, only Euler computations have been considered. The row contains 36 blades. The rotation speed is equal to 17,188 rpm. The inlet stagnation pressure p_{i0} and temperature T_{i_0} are respectively equal to 101330 N/m^2 and 288.15 K. The unsteadiness is forced. It is due to a sinusoidal fluctuation of the donwstream static pressure at mid outlet boundary for which the period is equal to that of the rotation of the row: $p_s = 1.15 p_{i0}(1 + 0.2\sin\omega t)$. The static pressure at the downstream boundary

follows the simplified radial equilibrium law.
The computation is performed on only one blade-to-blade domain because of periodic symmetry. The grid is composed of two domains : an O grid around the blade and a H grid in the remaining space. It contains 185,000 points. The gap between the blade tip and the casing is not modeled.
Table 1 summarizes the computational performances. A first computation has been realised with TES method. The dimensionless time step Δt is equal to 0.01 and 12000 time steps are necessary to cover one cycle. For DTS method, several simulations have been performed by modifying the time step or the tolerance criterion (first definition). In particular, the method allows a great increase of the time step. For these computations, the CFL number is equal to 2 and the multigrid method uses two levels of grid. Table 1 also specifies the average number of subiterations for each physical time step. Both methods provide similar performances although better results could be expected from the time-implicit approach. Two facts explain this behavior. At first, the mesh coarseness alleviates the stability restriction for the TES method. On the other hand, the grid dimensions restrict the use of the multigrid algorithm to two grid levels, reducing DTS efficiency. Concerning the description of the unsteady phenomenon, identical solutions are obtained. Relative Mach number lines are plotted in Fig.1 at different times during the cycle for the first DTS computation. They show the motion of the supersonic-subsonic shock which moves upstream to interact with the supersonic-supersonic shock.

3.2 B Bump of Délery Test Case

The second test case involves a two-dimensional viscous transonic flow over a bump into a channel [14]. It provides a very simple numerical test case for unsteady simulations by prescribing a sinusoidal law in time to the downstream pressure $p_s = 0.65 p_{i0}(1 + 0.2 \sin \omega t)$. Such a test case is useful to compare the performances of the DTS method relative to the TES one, and to determine the influence of the solution strategy on the DTS method. Due to the symmetry of the flow, the computational domain covers only half of the channel. The grid contains 181 points in the streamwise direction and 65 in the normal direction. The Reynolds number based on the stagnation values and the channel height is equal to 2.078 millions. The dimensionless period T with respect to the channel height and the infinite stagnation speed of sound is equal to 4.
For TES computation, only result with Jones-Launder model is available. Besides, the choice of the time step value ($\Delta t = 10^{-4}$) stands on stability restriction. For DTS simulations, the time step value is much greater ($\Delta t = 10^{-1}$) and the tolerance is fixed to 10^{-4} (first definition). Four turbulence models have been used, two solution strategies being considered for each model. In the computations, the CFL number is equal to 2 and the multigrid algorithm uses two grid levels. Table 2 gives the computational time for one cycle for all the computations. In the case of DTS method, it also specifies the solution strategy and the average number of subiterations per physical time step. Performances with DTS algorithm are clearly better than with TES algorithm. The ratio of computational time is at least of 2.2. On the other hand, the results demonstrate

the interest of the solution strategy. The use of two single grid iterations on the turbulent field for each multigrid iteration on the mean field leads to a CPU time decrease. Figure 2 describes the Mach number contours during the sixth cycle of computation with Smith model. Due to the unsteadiness of the flow, a pressure wave appears and moves towards the shock before interacting with it. Thus, the shock position is modified together with the shock/turbulent boundary-layer interaction. All the computations yield similar qualitative results.

3.3 Cavity Test Case

The last test case concerns the simulation of an open subsonic cavity flow. Extensive experimental work has been done at ONERA [15]. Stagnation pressure and temperature are respectively equal to 98000 Pa and 300 K. The Reynolds number based on the momentum thickness just upstream of the separation point is worth to 7800. Using the cavity length and the infinite stagnation speed of sound as reference values, the dimensionless period T is equal to 3.85. The computational domain is divided into three subdomains(Fig. 3) which respectively contain 49x73, 97x145 and 49x73 grid points. The reduced height y_+ is close to 1 at the channel wall.

For the simulation, DTS method uses Jones-Launder model and all the steady-state acceleration techniques. The CFL number value is equal to 4 and two grid levels are employed with the multigrid method (MF=TF=1). The time step value Δt corresponds to $T/100$ and the tolerance criterion value is 0.025, using the second definition of the tolerance. With such choices, the computation of one cycle requires an average number of 8400 subiterations while the computational time is close to 4900 s. Figure 4 presents the frequency spectra of the sound pressure level at a point located on the upstream side of the cavity, 35 mm beneath the upstream corner of the cavity. The computation is able to capture several harmonics. Besides, the values of the first two peaks are in fair agreement with the experimental ones (1975 Hz, 3954 Hz). Figure 5 shows the vorticity contours evolution at some moments of the 100th period. The plots highlight the existence of three vortices, the first one being convected in the downstream part of the channel, the second one hitting the cavity sidewall before moving downwards and the third one breaking on the downstream corner. These behaviors are in agreement with the experimental observations.

4 Conclusion

The test cases demonstrate the capability of DTS method for the numerical simulation of unsteady internal flows using RANS solver with transport-equation turbulence model. Computational performances underline the efficiency of DTS method relatively to a time-explicit scheme and the influence of the solution strategy when decoupled systems of equations are solved. Further work is in progress for the more complex cavity flow.

References

1. W.P. Jones, B.E. Launder: Int. J. of Heat and Mass Transfer **15**, (1972)
2. D.C. Wilcox: AIAA Journal **26**, 11 (1988)

3. B.R. Smith: AIAA Paper 90-1483, (1990)
4. P. Spalart, S. Allmaras: AIAA Paper 92-0439, (1992)
5. A.M. Vuillot, V. Couaillier, N. Liamis: AIAA Paper 93-2576, (1993)
6. N. Liamis and V. Couaillier: 'Unsteady Euler and Navier-Stokes Flow Simulations with an Implicit Runge-Kutta Method'. In: *2nd ECCOMAS Conference, Stuttgart, Germany*, ed. by John Wiley & Sons (1994)
7. R. Collercandy: 'Multigrid Strategy for Euler and Navier-Stokes Computations for Flows around Complex Aerospace Configurations'. In: *ECCOMAS Conference, Athens, Greece*, ed. by John Wiley & Sons (1998)
8. O. Rouzaud, A. Plop: '2D Unsteady Navier-Stokes Computations at ONERA for Prediction of Dynamic Stall'. In: 24th *European Rotorcraft Forum, Marseille, France*, (1998)
9. V. Couaillier: AIAA Paper 99-0154, (1999)
10. L. Cambier: 'The elsA project'. In: *1st ONERA-DLR Aerospace Symposium, Paris, France* (1999)
11. A. Jameson, W. Schmidt, E. Turkel: AIAA Paper 81-1259, (1981)
12. A. Jameson: AIAA Paper 91-1259, (1991)
13. R.D. Moore, L. Reid: 'Performance of a single-Stage Axial-Flow Transonic Compressor with Rotor and Stator Aspect Ratios of 1.19 and 1.26, respectively, and with Design Pressure Ratio of 2.05'. *NASA TP 1659*, (1980)
14. J. Reisz: 'Analyse expérimentale d'une interaction choc-couche limite turbulente à Mach 1,45'. *ONERA RT N°44/7078 AY*, (1981)
15. N. Forestier, Ph. Geffroy, L. Jacquin: 'Flow over Cavities in Transonic Regime: a Test Case for Numerical Simulations'. In: *First International Symposium on Turbulence Shear Flow Phenomena, Santa-Barbara, CA*, (1999) pp. 933–938

Table 1. Computational results - NASA Rotor 37

Computation	subit./Δt	CPU/period
TES - Δt=0.01	/	3050 s
DTS - $(\Delta t, \varepsilon_{tol}) = (0.5; 10^{-3})$	27	3650 s
DTS - $(\Delta t, \varepsilon_{tol}) = (0.5; 10^{-2})$	20	2700 s
DTS - $(\Delta t, \varepsilon_{tol}) = (1.0; 10^{-3})$	46	3120 s

Table 2. Computational results - B bump of Délery

Computation	solution strategy (TF/MF)	subit./Δt	CPU/period
TES - $k-\varepsilon$	/	/	2100 s
DTS - $k-\varepsilon$	1/1	310	950 s
	2/1	230	765 s
DTS - SA	1/1	58	147 s
	2/1	55	130 s
DTS - $k-\omega$	1/1	115	350 s
	2/1	71	220 s
DTS - $k-l$	1/1	140	360 s
	2/1	84	256 s

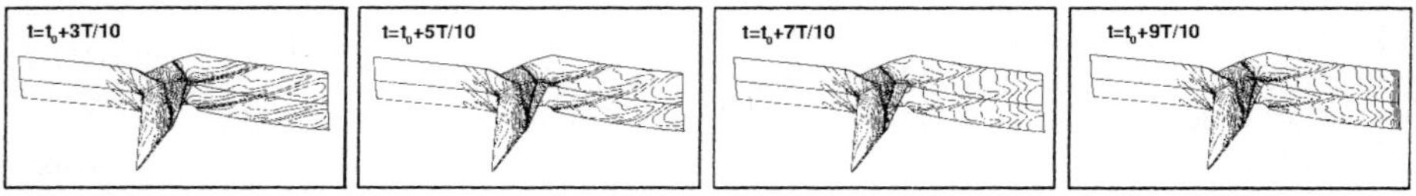

Fig. 1. Mach number contours - NASA Rotor 37

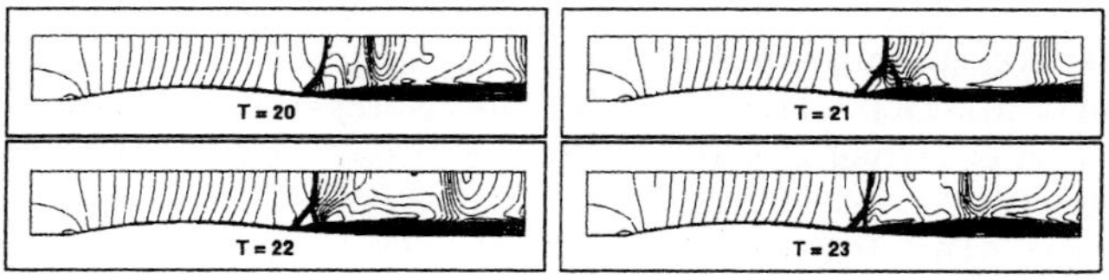

Fig. 2. Mach number contours - B bump of Délery

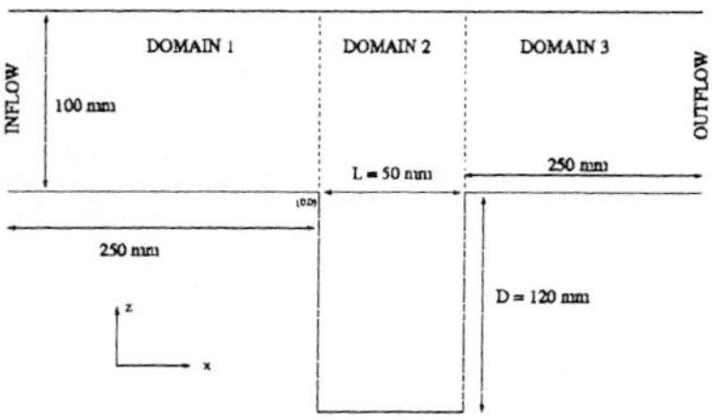

Fig. 3. Configuration sketch - Cavity

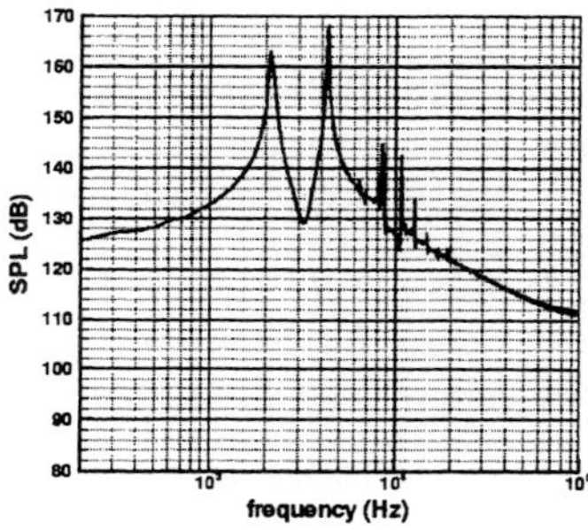

Fig. 4. Frequency spectra - Cavity

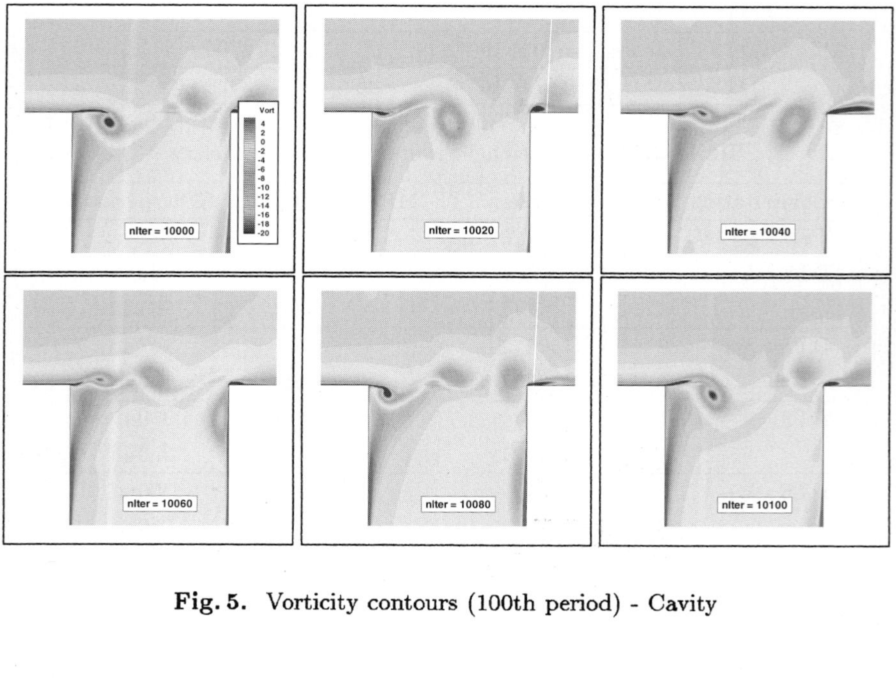

Fig. 5. Vorticity contours (100th period) - Cavity

Part VIII

Incompressible Flows

High-Reynolds Number Solutions of Incompressible Navier-Stokes Equations using Vectorial Operator Splitting

Rossitza Marinova[1], Christo Christov[2], and Tchavdar Marinov[3]

[1] Varna Free University, Varna 9007, Bulgaria
[2] University of Louisiana at Lafayette, Lafayette, LA 70504-1010, USA
[3] Saitama Institute of Technology, 1690, Fusaiji, Okabe, Saitama 369-0293, Japan

Abstract. The steady incompressible Navier-Stokes equations in primitive variables are solved by implicit vectorial operator-splitting. The method allows for complete coupling of the boundary conditions. Conservative approximations for the advective terms are employed on irregular staggered grids. The technique is used here for solving two benchmark problems. Numerical solutions for the flow in a lid-driven rectangular cavity with aspect ratio two (up to $Re = 6000$) and for the flow over backward-facing step in a channel (up to $Re = 1400$) on appropriate grids are presented.

1 Introduction

Consider the steady incompressible Navier-Stokes equations with equation for pressure instead the continuity equation (see [2,3] for details)

$$\frac{1}{Re}\Delta \boldsymbol{u} - \boldsymbol{\nabla} p - C[\boldsymbol{u}] = 0\ , \quad \frac{1}{Re}\Delta p - \boldsymbol{\nabla}\cdot\boldsymbol{u} + \frac{1}{Re}\boldsymbol{\nabla}\cdot C[\boldsymbol{u}] = 0 \tag{1}$$

in a closed domain Ω with a piece wise smooth boundary $\partial\Omega$. Here $\boldsymbol{x} \in \Omega$, $\boldsymbol{u} = \boldsymbol{u}(\boldsymbol{x})$ is the velocity vector, $p = p(\boldsymbol{x})$ – the pressure. The Reynolds number is defined as $Re = UL/\nu$, where U is the characteristic velocity, L – characteristic length, ν – kinematic coefficient of viscosity. The operator $C[\boldsymbol{u}]$ is a short-hand notation for the advective term. For this term we use the skew-symmetric form

$$C[\boldsymbol{u}] = \boldsymbol{\nabla}\cdot(\boldsymbol{u}\boldsymbol{u}) - \frac{1}{2}\boldsymbol{u}(\boldsymbol{\nabla}\cdot\boldsymbol{u})\ , \tag{2}$$

which follows from the continuity equation.

We assume that the velocity components are prescribed at the boundaries. The formulation with the equation for pressure is equivalent to the original system only if the continuity equation is satisfied also on the boundary [2,8]. Therefore we take

$$\boldsymbol{u}\big|_{\partial\Omega} = \boldsymbol{u}_b\ , \quad \boldsymbol{\nabla}\cdot\boldsymbol{u}\big|_{\partial\Omega} = 0\ . \tag{3}$$

For incompressible flows the pressure is defined up to an arbitrary function of time. For the sake of convenience we define this function as the average of the pressure at the specific time stage, i.e., we assume (for pressure uniqueness) that the following relation is satisfied $\int_\Omega p(\boldsymbol{x},t)\mathrm{d}\boldsymbol{x} = 0$, $\boldsymbol{x} \in \Omega$, $t > 0$.

In this work we consider two-dimensional case only. Upon introducing fictitious time t we arrive at the following vectorial parabolic system

$$\frac{\partial \boldsymbol{\theta}}{\partial t} = L[\boldsymbol{\theta}] + N[\boldsymbol{\theta}] + F[\boldsymbol{\theta}]\,, \quad \text{where} \quad \boldsymbol{\theta} = \begin{pmatrix} u \\ v \\ p \end{pmatrix},$$

$$L = \begin{pmatrix} \Delta/Re & -\nabla & 0 \\ 0 & \Delta Re & -\nabla \\ -\nabla & -\nabla & \Delta/Re \end{pmatrix}, N = -\begin{pmatrix} C & 0 & 0 \\ 0 & C & 0 \\ 0 & 0 & 0 \end{pmatrix}, F[\boldsymbol{\theta}] = \begin{pmatrix} 0 \\ 0 \\ (\boldsymbol{\nabla} \cdot C[\boldsymbol{u}])/Re \end{pmatrix}.$$

Solving this problem numerically is a formidable task in the case of high Reynolds numbers. The proposed vectorial version of the operator-splitting implicit scheme preserves the coupling between the sought functions at each fractional-time step.

2 Vectorial Operator-Splitting

We employ the second Douglas scheme [4] which is sometimes called "stabilizing correction". It is of first order in time, but has some advantages in stability, especially for non-commuting operators [10]. The order of approximation with respect to the fictitious time is not important here as far as we are looking for the stationary solution after the time stepping (iterations) converges. The two steps of the scheme of stabilizing correction are

$$\frac{\widehat{\boldsymbol{\theta}} - \boldsymbol{\theta}^n}{\tau} = A_1\widehat{\boldsymbol{\theta}} + A_2\boldsymbol{\theta}^n + G^n\,, \qquad \frac{\boldsymbol{\theta}^{n+1} - \widehat{\boldsymbol{\theta}}}{\tau} = A_2(\boldsymbol{\theta}^{n+1} - \boldsymbol{\theta}^n)\,, \tag{4}$$

τ is the increment of the fictitious time. The half-time-step variable $\widehat{\boldsymbol{\theta}}$ can be eliminated from (4) to get

$$(I + \tau^2 A_1 A_2)\,\frac{\boldsymbol{\theta}^{n+1} - \boldsymbol{\theta}^n}{\tau} = (A_1 + A_2)\boldsymbol{\theta}^{n+1} + G^n\,. \tag{5}$$

It is seen that upon convergence, i.e. $\|\boldsymbol{\theta}^{n+1} - \boldsymbol{\theta}^n\| \to 0$, the solution of the evolution problem satisfies the steady problem and the order of approximation does not depend on τ.

We treat implicitly the nonlinear operators, i.e., $A_1 = L_1 + N_1$, $A_2 = L_2 + N_2$. Here L_1 and L_2 are the operators of the derivatives with respect to x and y of the operator $L = L_1 + L_2$. Respectively N_1 and N_2 are the operators of the derivatives with respect to y and x of the operator $N = N_1 + N_2$.

We select a mesh which is non-uniform and staggered in x-direction for u and in y-direction for v. For boundary conditions involving derivatives this allows one to use central differences with second-order of approximation on two-point stencils. We employ difference approximations for the derivatives, which inherit the properties of the respective differential operators.

The gist of the new scheme is that it allows a special treatment of the algebraic systems. One of the systems for the respective velocity component is always conjugated to the system for pressure set function. For instance, on the first half-time stage the x operators are inverted along the line $y = y_j$ when u and p are conjugated. Then two linear algebraic systems are to be solved: one with tridiagonal matrix for v, and another (a conjugated one) – for the "composite" difference function obtained from the special arrangement of the two functions u and p. The latter has a pentadiagonal matrix twice the size of the tridiagonal one. In the same manner, the second half-time step requires to solve a tridiagonal and a pentadiagonal system.

The proposed difference scheme and algorithm are robust for large values of Reynolds number. To illustrate the advantages of the new method we conduct numerical tests for two different benchmark problems using uniform and/or non-uniform grids. Three grids are used in our computations. All computations are done with tolerance 10^{-10} for the criterion of convergence in terms of the normalized residual vector.

3 Lid-Driven Flow in Rectangular Cavity

Computational results for Reynolds number up to 11,000 for the flow in square cavity can be found in [3]. Here we consider rectangular cavity with aspect ratio 2. In studying this flow our aim is to verify that the scheme is able to determine the solutions in such a geometry for large enough Reynolds number. Steady solutions for this flow are presented in [1,6].

The streamlines of the solution obtained using Richardson extrapolation of the solutions with $h = 1/256$ and $h = 1/512$ for $Re = 1000$ and 6000 are plotted in Fig. 1. There are differences between our results and the results in [1] in the lower part of the cavity. Probably it is due to the presence of instabilities during iterations in [1]. The bottom-right and bottom-left secondary vortices are more intensive in our work comparing with [6]. The data concerning the location and

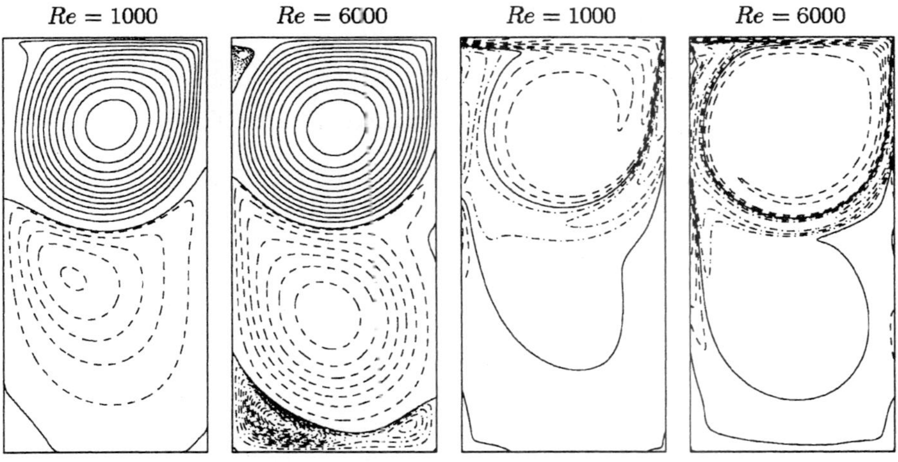

Fig. 1. Streamlines (left) and vorticity isolines (right) for $Re = 1000$ and $Re = 6000$

Table 1. Coordinates of the extrema of the stream function: Pr. – present results; R256 – Richardson extrapolation from solutions with $h = 1/256$ and $h = 1/512$

Re	Ref. data	$1/h$	Primary Top Vortex		Primary Bottom Vortex	
			ψ_{min}	(x_{min}, y_{min})	ψ_{max}	(x_{max}, y_{max})
1000	[1]	256	−0.1169	(0.5273, 1.5625)	0.0148	(0.3516, 0.7891)
	[6]	256	−0.1187	(0.5313, 1.5781)	0.0132	(0.3359, 0.8476)
	Pr.	R256	−0.117945	(0.5299, 1.5795)	0.013330	(0.3424, 0.8379)
6000	Pr.	R256	−0.119759	(0.5147, 1.5645)	0.021418	(0.4694, 0.6612)

strengths of the primary top vortex and the primary bottom vortex in the flow are presented in Table 1. Our results for $Re = 1000$ are compared with those in [1,6]. For $Re > 1000$ we were unable to find any steady state benchmark solution for comparisons. Therefore we present only our results.

4 Flow over a Backward-Facing Step in a Channel

Consider the flow over a backward-facing step, which is another well-studied test case. Fig. 2 shows the geometry of the flow and the defined flow parameters. Steady state numerical solutions are obtained up to $Re = 800$ in a fairly large number of works, see [5,7,9]. The Reynolds number is based on the mean inflow velocity U, the channel height H, and the viscosity ν. On the walls of the channel no-slip boundary conditions are used. At the inlet boundary ($x = 0,\ h \le y \le H$) a parabolic profile $u = 6U(H-y)(y-h)/(H-h)^2,\ v = 0$ is prescribed. The outflow velocity profile is also taken to be parabolic $u = 6U(H-y)y/H^2,\ v = 0$.

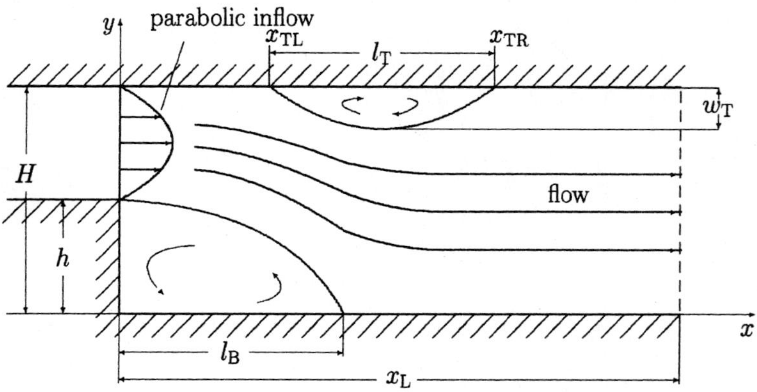

Fig. 2. Geometry of the flow over a backward-facing step in a channel ($h : H = 1 : 2$)

Table 2. Comparison of characteristic flow parameters for $Re = 800$, $H = 1$

Ref.	l_B	l_T	x_{TL}	x_{TR}	ψ_{min}	ψ_{max}	Nodes
[5]	6.10	5.63	4.85	10.48	−0.0342	0.5064	129681
[7]FD	6.082	5.6260	4.8388	10.4648	−0.034195	0.50661	245760
[9]	6.0964	5.6251	4.8534	10.4785	−0.03420	0.50653	3737
Pr.	6.0909	5.6505	4.8214	10.4719	−0.03421	0.50653	Extr.

Our tests have shown that for $Re \lesssim 1400$ it is safe to take $x_L \approx 100h$. The grid is non-uniform with numbers of the sells for the coarsest grid (Grid 1) in x and y direction $N_x = 256$ and $N_y = 64$, respectively. The grid in x-direction is denser near the walls, while near the outlet boundary lesser number of grid

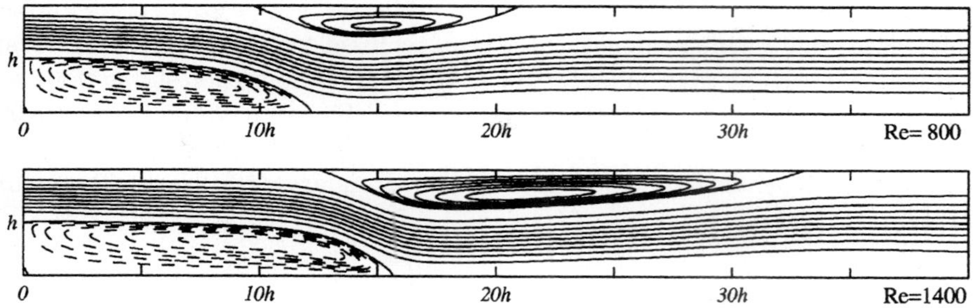

Fig. 3. Streamlines (expansion ratio 1:3)

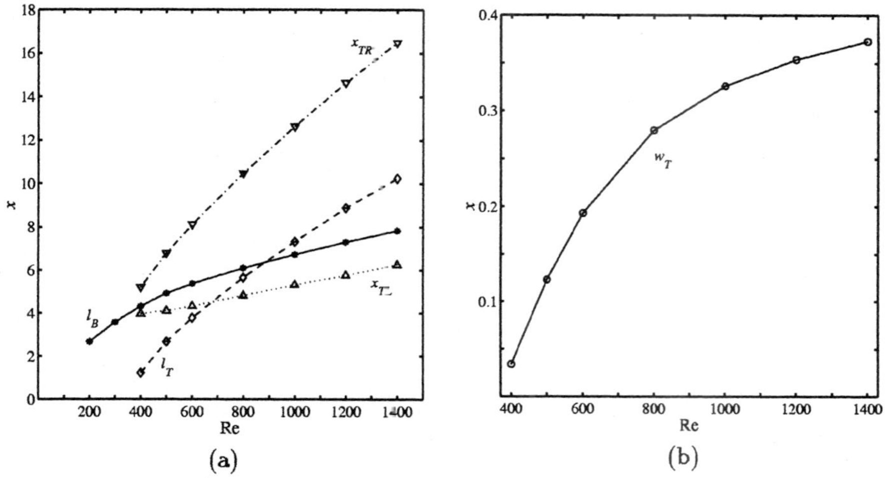

Fig. 4. Variation in different flow characteristics with Re: (a) reattachment and separation lengths ($* - l_B$; $\diamond - l_T$; $\triangle - x_{TL}$; $\nabla - x_{TR}$); (b) width w_T of the upper eddy

lines are used. In the interval $[0, x_c]$, where $x_c > x_{TR}$, a fine uniform grid is employed with number of grid cells $N_c = 7(N_y - 1)/8$. The finer grid (Grid 2) is constructed on the base of of the coarse Grid 1 through adding a middle point between two grid points. In a similar fashion, the finest grid (Grid 3) is constructed on the basis of Grid 2. Finally, using the obtained solutions on the last two grids, a Richardson extrapolation is performed and the accuracy of the solutions is increased to fourth order. We have obtained the steady solutions on all grids except for $Re = 1400$ on Grid 1. For the highest Reynolds number, convergence could not be reached on the course Grid 1.

Fig. 3 shows the streamline plots for $Re = 800$ and $Re = 1400$. In Table 2 the values of reattachment lengths l_B, l_T are presented as well as the longitudinal positions x_{TL}, x_{TR}, respectively of the start and end of the upper eddy. Our computations for $Re = 800$ are compared with benchmark computations of [5,7,9]. In Fig. 4, the obtained computational results of the most representative characteristics of the flow as function of the Reynolds number are shown. Our results for l_B are close to those in many other works for Reynolds number up to 800. Up to $Re = 1400$ our calculations show that the appearance of the upper eddy slows down the growth of the bottom eddy, see Fig. 4–(a). For Reynolds number $Re \geq 600$ the dependence of the reattachment length l_B on Re is almost linear with a less slope than those for $Re \leq 400$. Our computation shows that the length of the upper eddy l_T grows faster than l_B. For some value of $Re \leq 900$ the length l_T becomes longer than the length of the bottom eddy l_B. Fig. 4–(b) shows the dependence of the width of the secondary eddy w_T on Reynolds number.

Acknowledgements

The work of C.C. is supported in part by Grant LEQSF(1999-2002)-RD-A-49 from the Louisiana Board of Regents. The support for T.M. from the JSPS under Grant No P99745 is gratefully acknowledged.

References

1. C.-H. Bruneau, C. Jouron: J. Comput. Phys. **89**, pp. 389–413 (1990)
2. C. I. Christov, R. S. Marinova: 'Implicit vectorial operator splitting for incompressible Navier–Stokes equations in primitive variables'. Submitted to: Comput. Methods Appl. Mech. Engn.
3. C. I. Christov, R. S. Marinova: 'Implicit scheme for Navier–Stokes equations in primitive variables via vectorial operator splitting'. In: *Notes on Numer. Fluid Mech. 62*, ed. by M. Griebel et al. (Vieweg, Wiesbaden 1998) pp. 251–259
4. J. Douglas: SIAM Journal **3**, pp. 42–65 (1955)
5. D. Gartling: Int. J. Numer. Meth. Fluids **11**, pp. 953–967 (1990)
6. O. Goyon: Comput. Meth. Appl. Mech. Engrg. **130**, pp. 319–355 (1996)
7. P. M. Gresho et al: Int. J. Numer. Meth. Fluids **17**, pp. 501–541 (1993)
8. A. D. Henshaw: J. Comp. Phys. **113**, pp. 13–35 (1994)
9. Jayant Keskar, D. A. Lyn: Int. J. Numer. Meth. Fluids **29**, pp. 411–427 (1999)
10. N. N. Yanenko: *Method of fractional steps* (Gordon and Breach, New York, 1971)

Numerical Simulation of Incompressible Laminar Flow over 3-D Cavity

H Yao, R K Cooper, and S R Raghunathan

School of Aeronautical Engineering,
The Queen's University of Belfast, Belfast BT7 1NN, UK

Abstract. This paper investigates unsteady incompressible flow over cavities. Previous research in incompressible cavity flow has included flow inside and past a 2-D cavity, and flow inside a 3-D cavity, driven by a moving lid. The present research is focused on incompressible flow past a 3-D cavity. This involves the complex interaction between the external flow and the re-circulating flow within the cavity.

Computation has been performed on a 3-D rectangular cavity assuming laminar boundary layer as the inflow, with varying boundary layer thickness, at a Reynolds number of 10,000. A CFD approach, based on the unsteady Navier-Stokes (NS) equations for 3-D incompressible flow, was used in this study. These results indicate that, given the Reynolds number, the flow field is affected by the thickness of the inflow boundary layer. The flow becomes unstable and complex as the thickness of the boundary layer decreases.

1 Introduction

Cavities are present in many aerodynamic configurations as a direct result of manufacturing defects. The presence of cavity generates unsteady velocity and density, and unsteady pressure fluctuation [1]. Development of control techniques to alleviate the pressure fluctuation produced by a cavity is a topic of practical significance, which requires a fundamental understanding of aerodynamics of complex flow over such a geometry [2]. Simulation of flow with cavity is also a topic relevant to aeroacoustics (noise generation) and transition studies.

Previous studies were mainly focused on compressible cavity flows [3,4]. Although there have been some studies considering incompressible cavity flow, these are mainly focused on flow inside or past 2-D cavity [5,6], or flow *inside* 3-D cavity, known as lid-driven cavity [7,8]. Lid-driven cavity flow does not take into consideration the interaction between the external flow and the re-circulating flow within the cavity. This interaction can be observed for the flows passing over the cavity. A laminar boundary layer flow over a cavity may undergo transition in the shear layer to turbulence downstream of the cavity.

This paper presents results of computation of incompressible flow past a 3-D rectangular cavity, assuming a laminar boundary layer inflow with varying boundary layer thickness, at a Reynolds number of 10,000. This work is an extension of the authors' previous work [9] which considered uniform inflow and lower Reynolds numbers ($\leq$3,000).

2 Methodology

The flows to be modelled are assumed to be three dimensional, incompressible and unsteady. In addition, they are also assumed to be laminar.

2.1 Governing Equations

The governing equations are based on the 3-D, unsteady, incompressible NS equations. In Cartesian coordinates, these equations can be written, in a dimensionless form, as

$$\nabla \cdot \boldsymbol{u} = 0 \tag{1}$$

$$\frac{\partial \boldsymbol{u}}{\partial t} + \boldsymbol{u} \cdot \nabla \boldsymbol{u} = -\nabla p + \frac{1}{R_e} \nabla^2 \boldsymbol{u} \tag{2}$$

where

$\boldsymbol{u} = \boldsymbol{u}(\boldsymbol{x}, t)$ is the dimensionless velocity vector, with $\boldsymbol{x}$ and t representing the dimensionless vector of coordinates and dimensionless time, respectively;
$p = p(\boldsymbol{x}, t)$ is the dimensionless pressure;
$R_e = U_\infty L/\nu$ is the Reynolds number, with U_∞ being the free stream velocity, L being the length of the cavity and ν being the kinematic viscosity.

A pressure Poisson equation is used obtain the pressure field, required for solving Equations (1) and (2). This equation is obtained by taking the divergence of Equation (2), which gives

$$\nabla^2 p = \nabla \cdot (\frac{1}{R_e} \nabla \boldsymbol{u} - \boldsymbol{u} \cdot \nabla \boldsymbol{u}) - \nabla \cdot (\frac{\partial \boldsymbol{u}}{\partial t}) \tag{3}$$

Equations (1)-(3) constitute the mathematical model used in the computation.

2.2 Numerical method

The numerical scheme used was the second-order implicit Crank-Nicolson finite difference scheme, with second-order accuracy in both time and space. Fig.1 shows the geometric configuration of the rectangular cavity used in the computation, with a length L, width W and depth D. It is assumed that the flow is symmetric with respect to the longitudinal center plane, therefore only a half span of the cavity is chosen to be the computational domain.

An upstream laminar Blasius boundary layer is specified along the inflow of the computational domain. The outflow is specified by zero'th-order extrapolation from the computation zone. The entire flow field above the cavity is initialized with the inflow condition, and the flow field within the cavity is initialized to zero. The initial pressure is set to its free stream value. No-slip boundary condition is used on solid surfaces, where the pressure is obtained by an extrapolation

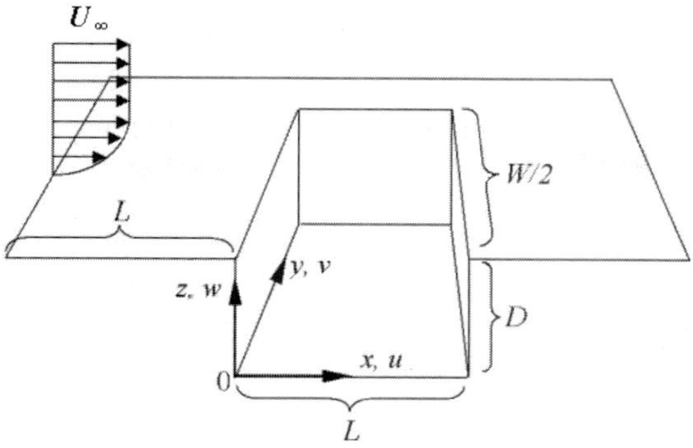

Fig. 1. Geometry of the rectangular half-span cavity used for computation

of the interior point values of the pressure in the direction normal to the wall. At the upstream boundary, the y (lateral) and z (normal) component velocities, v and w, are assumed to be zero. The flow variables at the downstream boundary are also obtained based on zero'th-order extrapolation.

Three-dimensional Cartesian non-uniform Cartesian grids are generated with clustering of nodes near walls and in the shear layer region. These clustered nodes account for greater gradients in velocity and pressure in these regions.

2.3 Code Validation

The code has been examined using both analytical and numerical examples given in [7]; the numerical examples used for the validation include both 2-D and 3-D cavity flow fields. A detailed description of these validation studies can be found in [9]. Both the analytical and numerical case studies have validated the code.

3 Typical Results and Discussion

This section describes the results of computation of incompressible flow over the 3-D rectangular cavity shown in Fig.1, with a Reynolds number of 10,000, and a laminar Blasius boundary layer inflow. The cavity has a depth-to-length ratio $D/L = 1$, and a width-to-length ratio $W/L = 3$. Different thickness of the inflow boundary layer were considered. In particular, Fig.2, Fig.3 and Fig.4 show the instantaneous flow fields with an inflow boundary layer thickness of 0.0, $0.1L$ and $0.2L$, respectively. Shown in these figures are the flows of the longitudinal (i.e. x) direction at the $y = 0$ (i.e. symmetry) plane, at the dimensionless time $t = 288$. Note that a boundary layer with zero thickness corresponds to a uniform inflow condition.

Fig.2, corresponding to a uniform inflow, indicates the Kelvin-Helmholtz instability in the shear layer, due to the interaction between the external flow and the re-circulating flow within the cavity. This is different from the flow structure found within the lid-driven cavity. Fig.2(b) also reveals a vortex shed from the shear layer near the rear wall of the cavity; the shear layer did not impinge on the downstream boundary. Fig.3 corresponds to the boundary layer inflow with a boundary thickness of $0.1L$. It clearly shows that the occurrence of the Kelvin-Helmholtz instability was moved closer to the rear wall of the cavity than it was found in Fig.2. Again, some interaction between the flow in the cavity and the shear layer was seen. But in Fig.3 there is no shed vortex, and the shear layer directly impinged on the downstream boundary of the cavity. Also, in Fig.2(a) two secondary eddies can be clearly seen at the lower corners of the cavity. But these became much weaker (almost unseen) at the same places in Fig.3(a). Compared to Fig.2 and Fig.3, Fig.4, corresponding to a boundary layer thickness of $0.2L$, shows a more stable flow field in terms of much smaller oscillations in the shear layer.

Fig.5 and Fig.6 show the instantaneous cross flow velocity vectors with the inflow boundary layer thickness being 0.0 and $0.1L$, respectively. Two lateral planes are indicated, i.e. x=0.6 and x=0.8, respectively (relative to the front wall of the cavity). Both figures indicate longitudinal vortices at the floor of the cavity. In Fig.5(b), with a uniform inflow, the longitudinal vortices were also found in the shear layer above the cavity. These, however, were not seen in Fig.6(b), with a boundary layer inflow (with a thickness of $0.1L$). These indicate that the cavity flow fields can be complex and can exhibit a strong 3-dimensional nature. Fig.7 shows the cross flow velocity vectors at the lateral plane x=0.6, with an inflow boundary layer thickness of $0.2L$. This figure shows that the cross flow became much weaker than those shown in Fig.5 and Fig.6 – actually, it looks like a 2-D cavity flow. As a conclusion, given the Reynolds number, the cavity flow structure is affected by the thickness of the inflow boundary layer; the flow becomes unstable and complex as the thickness of the inflow boundary layer decreases.

References

1. D. Rockwell, E. Naudascher: 'Review-Self- Sustaining Oscillations of Flow Past Cavities'. ASME Journal of Fluid Engineering, **100**, pp. 152–165 (1978) Suhs N E. Computions of
2. N. E. Suhs: 'Computions of Three-Dimensional Cavity Flow at Subsonic and Supersonic Mach Numbers'. AIAA Paper 87-1208 (1987)
3. O. Baysal, S. Srinivasan, and L. Stallings Jr: 'Unsteady Viscous Calculations of Supersonic Flows Past Deep and Shallow Three-Dimensional Cavities'. AIAA Paper 88-0101 (1988)
4. A. Morgenstern Jr, N. Chokani: 'Hypersonic Flow Past Cavities'. AIAA Journal, **32**, 12, pp. 2387–2393 (1994)
5. S. N. Sinha, A. K. Gupta, and M, M.Oberia: 'Laminar Separating Flow over Backsteps and Cavities Part II: Cavities'. AIAA Journal **20**, 3, pp. 370–375 (1982)
6. J. C. F. Pereira, J. M. M.Sousa: 'Experimental and Numerical Investigation of Flow Oscillations in a Rectangular Cavity'. ASME Journal of Fluid Engineering **117**, pp. 68–74, (1995)

7. T. P. Chiang, R. R. Hwang, and W. H.Sheu: 'On End-Wall Corner Vortices in a Lid-Driven Cavity'. ASME Journal of Fluid Engineering **119**, pp. 201–204 (1997)
8. C. J. Freitas, R. L. Street, A. N. Findikakis, and J. R. Koseff, R. L. Street: 'Numerical Simulation of Three-Dimensional Flow in a Cavity'. International Journal for Numerical Method Fluids **5**, pp. 561–575 (1985)
9. H.Yao, R. K. Cooper, and S. R. Raghunathan: 'Computation of Incompressible Flow over Three-Dimensional Cavities'. In: *Advances in Fluid Mechanics III, 3rd International Conference on Advances in Fluid Mechanics, Canada, May 24–26, 2000*, ed. by M. Rahman and C. A. Brebbia, pp. 125–133

(**a**) Velocity vectors (**b**) Vorticity contours of (a)

Fig. 2. Flow fields of x direction, uniform inflow, $y = 0.0$ plane, $R_e = 10,000$

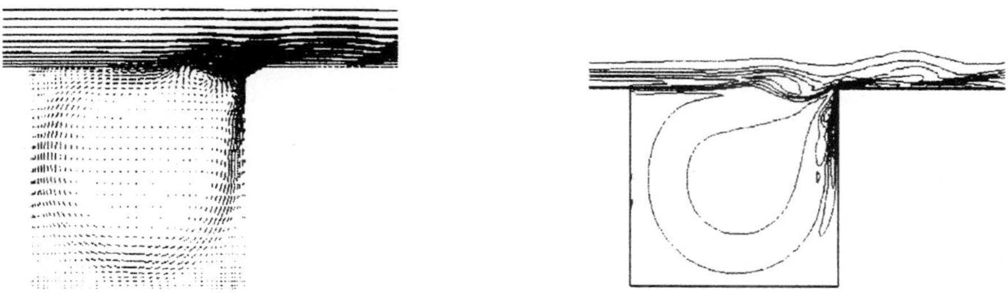

(**a**) Velocity vectors (**b**) Vorticity contours of (a)

Fig. 3. Flow fields of x direction, boundary layer inflow with thickness of $0.1L$, $y = 0.0$ plane, $R_e = 10,000$

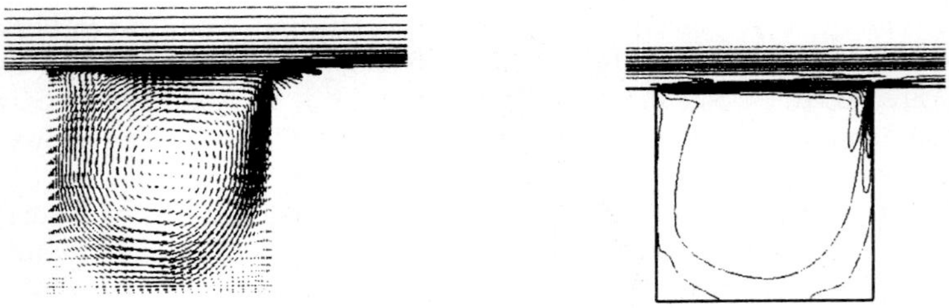

(a) Velocity vectors (b) Vorticity contours of (a)

Fig. 4. Flow fields of x direction, boundary layer inflow with thickness of $0.2L$, $y = 0.0$ plane, $R_e = 10,000$

(a) Velocity vectors at the lateral $x = 0.6$ plane (b) Velocity vectors at the lateral $x = 0.8$ plane

Fig. 5. Cross-flow fields, uniform inflow, $R_e = 10,000$

(a) Velocity vectors at the lateral $x = 0.6$ plane (b) Velocity vectors at the lateral $x = 0.8$ plane

Fig. 6. Cross-flow fields, boundary layer inflow with thickness of $0.1L$, $R_e = 10,000$

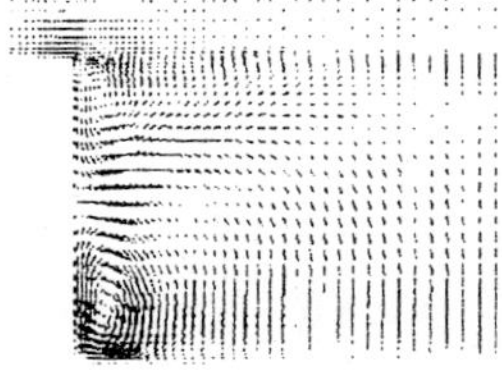

Fig. 7. Cross-flow fields, boundary layer inflow with thickness of $0.2L$, $x = 0.6$ plane, $R_e = 10,000$

Numerical Study on Aerodynamic Mechanism of Hovering Flight of Dragonfly

Koji Isogai and Yasuhisa Shinmoto

Department of Aeronautics and Astronautics
Kyushu University
6-10-1 Hakozaki, Higashiku, Fukuoka 812-8581
Japan

Abstract. In order to clarify the fundamental mechanism of the hovering flight of a dragonfly, the numerical simulation of unsteady viscous flow around a tandem airfoil configuration oscillating in still air has been conducted by using a two-dimensional Navier-Stokes code. It is shown that the mutual flow interactions between the fore- and hind-airfoils are playing the dominant role in generating the time mean aerodynamic force acting in the direction of the stroke plane, which is indispensable for the dragonfly to hover with the body axis horizontal. The total amounts of the lifting force and the necessary power are also estimated and shown to be very close to those estimated by other researchers.

1 Introduction

Doragonfly can hover with the body axis almost horizontal [11]. The mechanism of this has not yet been well understood, though many experimental [9,3,12] and theoretical [7,8,4,5,13] investigations have been conducted so far. These theoretical studies have been concentrated on quasi-steady or unsteady aerodynamics of a single airfoil, that is, the effects of the mutual interactions of the fore and hind wings have been completely neglected. According to the experimental flow visualization study by Somps and Luttges [9], however, it has been suggested that the mutual interactions between the vortices generated by the fore- and hind-wings might be playing the dominant role in generating the hovering forces. However, further details of the vortex interactions and its correlation with the lift genetation mechanism have remained to be investigated. As far as we are concerned with the hovering flight of a dragonfly, there has been no theoretical analysis nor numerical simulation which takes into account the effects of the aerodynamic interactions of the fore and hind wings.

2 Method of Analysis

In Fig. 1, the definitions of coordinate and forces are shown, and in Fig. 2, the definition of the airfoil motion in $X-Y$ plane is given (see [6] for further details). The fore- and hind-airfoils are oscillated in a coupled heaving and pitching mode (pitching motion advances about 90 deg phase angle ahead of the heaving), and the heaving oscillation of the hindairfoil advances some phase angle ahead of the foreairfoil. The 2D compressible Navier-Stokes Equations are nondimentionalized by the maximum velocity of the flapping motion of the foreairfoil, the semichord length of the foreairfoil and the far-field air density. The airfoils are assumed to be the flat-plate of zero thickness. Multi-block method is employed and the boundary conforming grid is generated at each time step by an algebraic method. The time differenced form of the GCL (Geometric Conservation Law) [10] is solved coupled with the conservation form of the Navier-Stokes equations. At the far-field boundaries located about 50 chord lengths away from the airfoils, the flow quantities are given by zeroth order extrapolation from the inner points. The total 280 × 160 grid points are employed for the whole flow field.

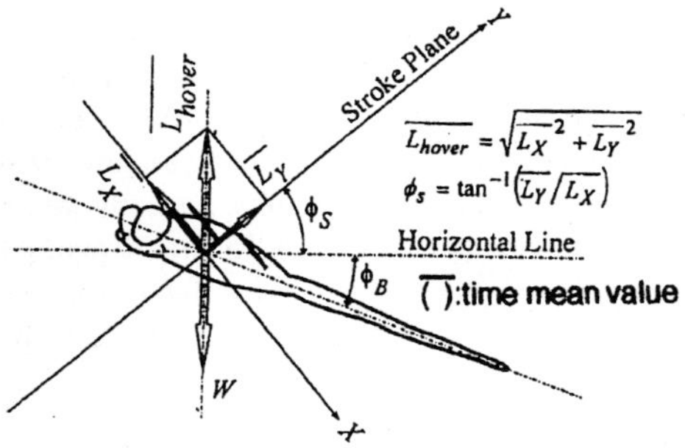

Fig. 1. Definition of coordinates and forces

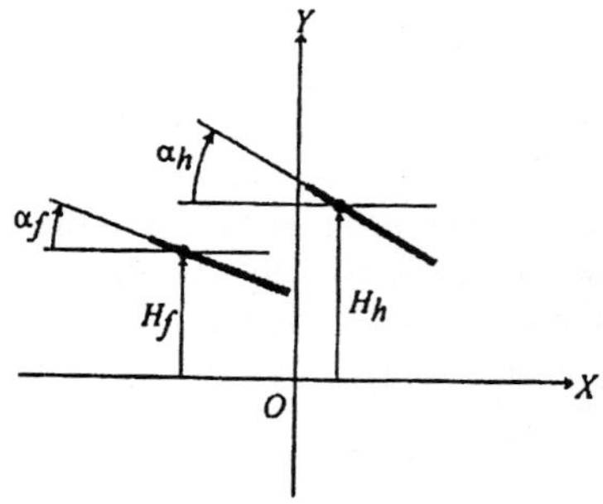

Fig. 2. Definition of airfoil motion

3 Dragonfly Model and Results of Simulation

In the present simulation, we take *Anax parthenope julius* as a typical example of a dragonfly. The flow around the 75 percent semispan station is simulated, where the amplitudes of heaving oscillations of the fore- and hind-airfoil are both $0.0241m$ and semichord lengths of each airfoil are $0.005m$ and $0.0057m$, respectively. The amplitude of pitching oscillation is 64 deg for upward motion and 40 deg for downward motion for the foreairfoil and 51 deg for upward motion and 24 deg for downward motion for hindairfoil. The frequency of oscillation is 28 Hz. The gap between the fore- and hind-airfoils is 96 percent of the chord length of the foreairfoil. The phase advance angle of the heaving oscillation of the hindairfoil ahead of that of the foreairfoil is about 40 deg. These data are taken from the paper by Azuma and Watanabe [1]. The numerical simulations using the Navier-Stokes code have been performed for the tandem airfoil configuration thus set. Since one of the purposes of the present study is to investigate the role of the flow interactions between the fore- and hind-airfoils, the flow simulations about the sigle airfoil configuration (the fore-airfoil or the hind-airfoil only) have also been performed for comparison with those for the tandem airfoil configuration.

The detailed values of the time mean aerodynamic forces and the mean rate of work, that are obtained by the present simulations are as follows:

A) Tandem airfoil configuration (with interaction)

Fore-airfoil: $\overline{L_{X,f}} = 0.0868 N/m$ $\overline{L_{Y,f}} = 0.0285 N/m$ $\overline{W_f} = 0.375 W/m$

Hind-airfoil: $\overline{L_{X,h}} = 0.0908 N/m$ $\overline{L_{Y,h}} = 0.0309 N/m$ $\overline{W_h} = 0.598 W/m$

Sum of fore- and hind-airfoils:

$\overline{L_X} = 0.178 N/m$ $\overline{L_Y} = 0.0594 N/m$ $\overline{L_{hover}} = 0.187 N/m$ $\phi_s = 18.5$ deg

$\overline{W_t} = 0.973 W/m$

B) Single airfoil configuration (the fore-airfoil only or the hind-airfoil only without interaction)

Fore-airfoil: $\overline{L_{X,f}} = 0.0686 N/m$ $\overline{L_{Y,f}} = 0.0092 N/m$ $\overline{W_f} = 0.279 W/m$

Hind-airfoil: $\overline{L_{X,h}} = 0.0783 N/m$ $\overline{L_{Y,h}} = 0.0074 N/m$ $\overline{W_h} = 0.477 W/m$

Sum of fore- and hind-airfoils:

$\overline{L_X} = 0.147 N/m$ $\overline{L_Y} = 0\ 0166 N/m$ $\overline{L_{hover}} = 0.148 N/m$ $\phi_s = 6.4$ deg

$\overline{W_t} = 0.756 W/m$

The increment of $\overline{L_{hover}}$ gained by aerodynamic interaction is about 27 percent of $\overline{L_{hover}}$ obtained for the single airfoil configuration (without interaction). Especially, the increase of $\overline{L_Y}$ gained by the flow interaction, which is indispensable for the dragonfly to hover with the body axis horizontal, is about 260 percent of that of the single airfoil configuration (without flow interaction).

In Fig. 3, the variation of L_Y and L_X acting on the fore- and hind-airfoils during one cycle of oscillation are plotted by the solid line for the tandem airfoil configuration and by the dotted line for the single airfoil configuration, respectively. It can be seen in the figure that both the fore- and hind-airfoils of the tandem configuration gain L_Y and L_X, (especially L_Y) during $kt = \pi/2 \sim 1.28\pi$ when both airfoils are in downward motion. The reason for this can be seen in Fig. 4 where iso-vorticity distributions around the tandem airfoil configuration at several phases of the oscillation are plotted. In the process of the downward motion, the strong vortices are generated on the upper surfaces of the fore- and hind-airfoils, that produce the strong suction forces. Although the vortices on the upper surfaces also appear for the single airfoil configuration in downward motion, the strength of the vortices seems to be enhanced for the tandem airfoil configuration by the flow interactions between the two airfoils.

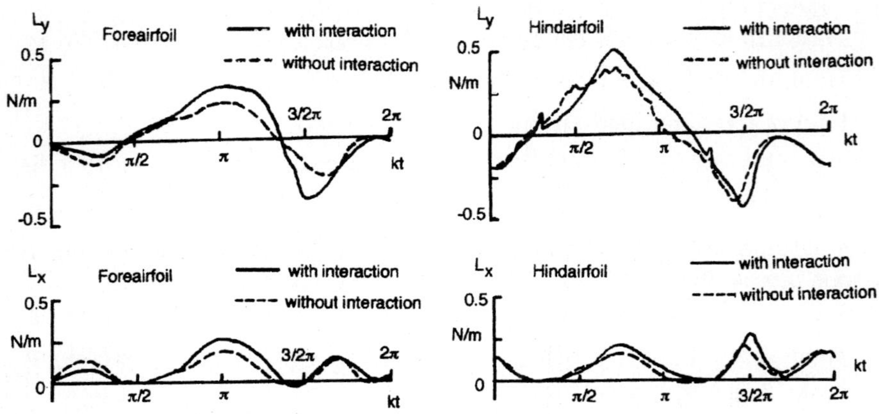

Fig. 3. Variations of L_Y and L_X during one cycle of oscillation

At this point, it is quite interesting to see whether the lifting force, $\overline{L_{hover}}$, predicted by the present numerical simulation can sustain the weight of the present dragonfly model ($7.74 \times 10^{-3}N$). In order to estimate the total hovering force acting on the dragonfly, the span-wise distribution of $\overline{L_{hover}}$ should be given. In addition to the simulation for 75 percent semi-span station, we also performed the simulation for 50 percent semi-span station, obtaining $\overline{L_{hover}} = 0.113N/m$. By using these data, we have estimated the total lifting force acting on the full span wings of the present tandem configuration as $12.0 \times 10^{-3}N$. Since the weight of the dragonfly is $7.74 \times 10^{-3}N$, the load factor n becomes $n = 1.55$, and this means that the dragonfly can sustain its weight. Similarly, the necessary power for the present hovering flight can also be estimated. The total necessary power we have obtained for the full span wings of the present tandem configuration is $P_n = 0.0593W$. Using these data, the necessary power for 1 g hover can be estimated as $P_n' = P_n/n^{1.5} = 0.0307W$ [6]. The available power P_a of the present dragonfly model is estimated [2] to be $P_a = 0.0514W$, which is higher than the necessary power of $0.0307W$ estimated by the present numerical simulation. It is also easy to estimate the necessary power for $1Kg$ of muscular mass, and it becomes $155W/Kg$, which is also quite reasonable value compared with $160W/Kg$ estimated by Wakeling and Ellington [13] for the dragonflies, *S. sanguineum and C. splendens.*

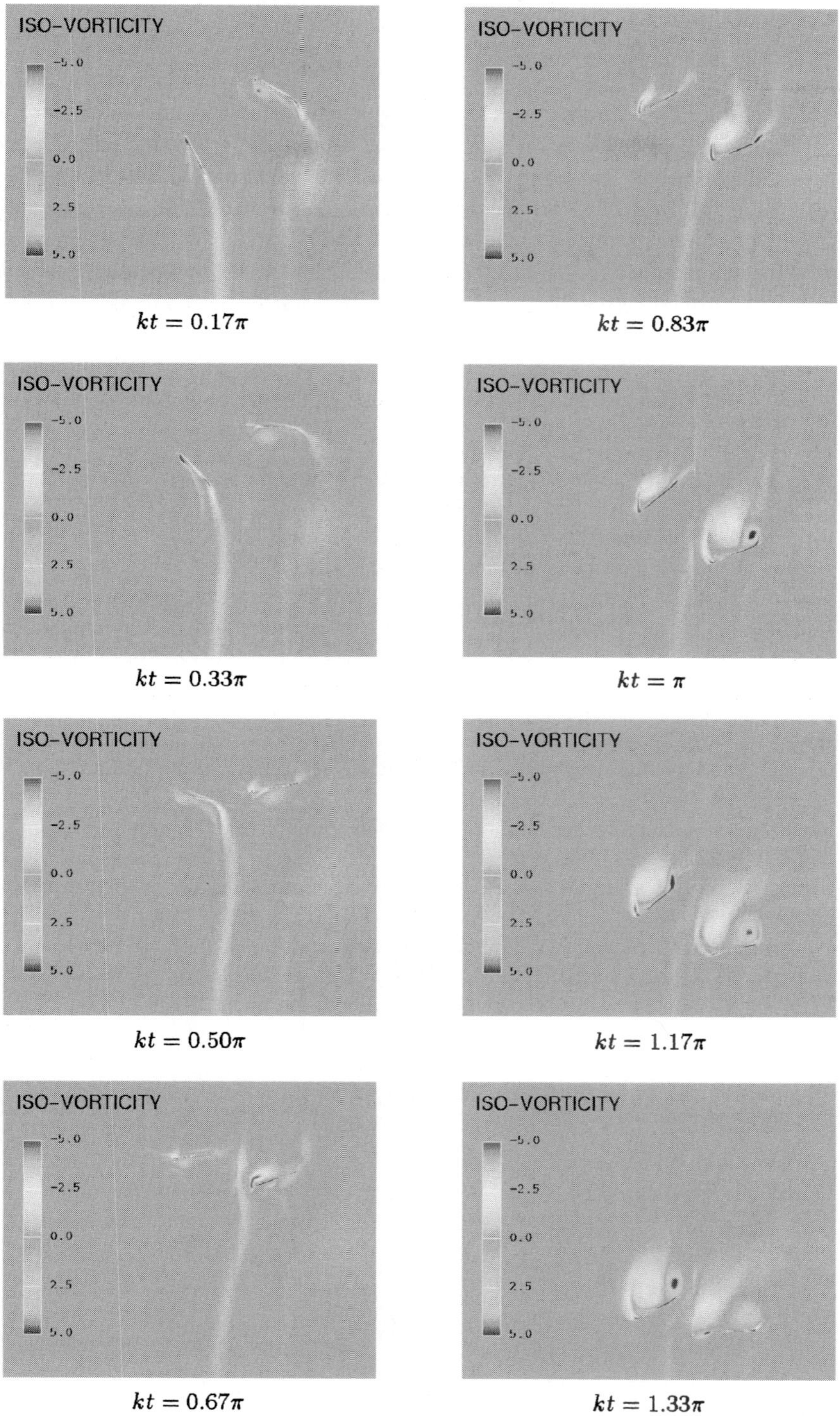

$kt = 0.17\pi$ $kt = 0.83\pi$

$kt = 0.33\pi$ $kt = \pi$

$kt = 0.50\pi$ $kt = 1.17\pi$

$kt = 0.67\pi$ $kt = 1.33\pi$

Fig. 4. Iso-vorticity distributicn around the tandem airfoil configuration

4 Conclusion

In order to clarify the fundamental mechanism of the hovering flight of a dragonfly, the numerical simulation of unsteady viscous flow around a tandem airfoil configuration oscillating in still air has been conducted by using a two-dimensional Navier-Stokes code. It is shown that the mutual flow interactions between the fore- and hind-airfoils are playing the dominant role in generating the time mean aerodynamic force acting in the direction of the stroke plane, which is indispensable for the dragonfly to hover with the body axis horizontal. The total amounts of the lifting force and the necessary power are also estimated and shown to be very close to those estimated by other researchers.

References

1. Azuma, A. and Watanabe, T.: Journal of Experimental Biology, 137, pp. 221-252 (1988)
2. Azuma, A.: The Biokinetics of Flying and Swimming, Springer-Verlag (1992)
3. Freymuth, P.: Experiments in Fluids 9, pp. 17-24 (1990)
4. Gustafson, K. and Leben, R.: Computer Physics Communications, 65, pp. 121-132 (1991a)
5. Gustafson, K., Jones, K., Leben, R. and McArthur, J.: Fourth International Symposium on Computational Fluid Dynamics, University of California, Davis, California, pp. 461-466 (1991b)
6. Isogai, K. and Shinmoto, Y.: Proceedings of the 2nd Pacific Symposium on Flow Visualization and Image Processing 1999, PSFVIP-2, Honolulu, Hawaii, 16-19 May (1999)
7. Norberg, R.A.: Swimming and Flying in Nature, Vol. 2, New York, Plenum, pp. 763-781 (1975)
8. Savage, S.B., Newman, B.G. and Wong, D.T.-B.: Journal of Experimental Biology, 83, pp. 59-77 (1979)
9. Somps, C., and Luttges, M.: Science, Vol. 228, 14, pp. 1326-1329 (1985)
10. Thomas, P.D. and Lombard, C.K.: AIAA J. Vol. 17, No.10, pp. 1030-1037 (1979)
11. Weis-Fogh, T.: Journal of Experimental Biology, 59, pp. 169-230 (1973)
12. Wakeling, J.M. and Ellington, C.P.: Journal of Experimental Biology, 200, pp. 557-582 (1997a)
13. Wakeling, J.M. and Ellington, C.P.: Journal of Experimental Biology, 200, pp. 583-600 (1997b)

Multigrid Solutions for the Three-Dimensional Incompressible Navier-Stokes Equations in Artificial Compressibility Formulation

Li Yuan

Institute of Computational Mathematics and LSEC, Academy of Mathematics and System Sciences, Chinese Academy of Sciences, Beijing 100080, P. R. CHINA

Abstract. The full-approximation-storage (FAS) multigrid algorithm is applied in conjunction with the artificial compressibility method to accelerate steady solutions of the 3D incompressible Navier-Stokes equations. The basic scheme used is the diagonalized Beam-Warming scheme, and the finite differences for inviscid fluxes include a MUSCL scheme and a symmetric TVD scheme for comparing purpose. Neumann boundary conditions on the coarse grids are implemented for the correction rather than for the flow variable itself. The performance of the present method is studied for the entry flow through a 90° bent square duct and the flow past an inclined prolate spheroid with an axes ratio of 4:1. It is found that the multigrid acceleration behaves well only by using the proposed treatment of Neumann boundary conditions on coarse grids, and that MUSCL scheme has better resolution than TVD scheme, due to the fact that TVD scheme with limiters degenerates to first-order accuracy at extrema.

1 Introduction

Most computational simulations of the 3D incompressible flows involve solving a Poisson equation for the pressure, and are thus computationally expensive on general 3D curvilinear coordinates. Nevertheless, for simulating engineering flow problems, it is desirable to use cost-effective numerical methods with reasonable accuracy. The artificial compressibility (AC) method [1] that avoids solution of the pressure Poisson equation belongs to such a method. Over the past thirty years there are numerous studies that applied well-established compressible flow solvers to the AC formulation for solving steady as well as unsteady incompressible flow problems, e.g. Beam-Warming scheme [2,3], line-relaxation scheme [4] and Runge-Kutta scheme [5]. Various spatial difference schemes were utilized, such as flux-difference splitting [4], MUSCL [6], TVD [7], and WENO schemes [8]. However, there remains space to improve the convergence rate of the AC method. One approach is to use multigrid (MG) methods. Although MG methods have been widely used in flow simulations, applications to the AC method are limited [9]. In addition, there are few papers that addressed the computational effects by using different spatial difference schemes.

In this work, a relatively simple multigrid method based on FAS algorithm and fixed V-cycle strategy is used in conjunction with the AC method for accelerating steady solutions of the 3D incompressible Navier-Stokes equations. The multigrid method is implemented with an ADI finite difference scheme. Appropriate treatment of Neumann boundary conditions on coarse grids is proposed.

2 Governing Equations and Basic Numerical Scheme

The starting point is the 3D incompressible Navier-Stokes equations written in conservative form in generalized coordinate system. Introduction of a pseudo-time derivative of pressure to the continuity equation gives the AC formulation

$$\partial_\tau \hat{\boldsymbol{Q}} + \partial_\xi(\hat{\boldsymbol{E}} - \hat{\boldsymbol{E}}_\nu) + \partial_\eta(\hat{\boldsymbol{F}} - \hat{\boldsymbol{F}}_\nu) + \partial_\zeta(\hat{\boldsymbol{G}} - \hat{\boldsymbol{G}}_\nu) = 0, \tag{1}$$

where $\hat{\boldsymbol{Q}} = \boldsymbol{Q}/J = (p,u,v,w)^{\mathrm{T}}/J$, J is the Jacobian of the coordinate transformation, τ is the time, and the detailed formulations of the inviscid flux vectors $\hat{\boldsymbol{E}}, \hat{\boldsymbol{F}}, \hat{\boldsymbol{G}}$ and viscous flux vectors $\hat{\boldsymbol{E}}_\nu, \hat{\boldsymbol{F}}_\nu, \hat{\boldsymbol{G}}_\nu$ can be found elsewhere [2,4].

By applying finite difference discretization and diagonalized Beam-Warming approximate factorization scheme to the above equations [3,7], one obtains

$$\begin{aligned}
&\mathbf{T}_\xi[\mathbf{I} + h \nabla \boldsymbol{\Lambda}_\xi^+ + h \triangle \boldsymbol{\Lambda}_\xi^- - h\delta_\xi \Gamma_1 \bar{\delta}_\xi]\mathbf{T}_\xi^{-1}\mathbf{T}_\eta[\mathbf{I} + h \nabla \boldsymbol{\Lambda}_\eta^+ + h \triangle \boldsymbol{\Lambda}_\eta^- - h\delta_\eta \Gamma_2 \bar{\delta}_\eta] \\
&\mathbf{T}_\zeta[\mathbf{I} + h \nabla \boldsymbol{\Lambda}_\zeta^+ + h \triangle \boldsymbol{\Lambda}_\zeta^- - h\delta_\zeta \Gamma_3 \bar{\delta}_\zeta]\mathbf{T}_\zeta^{-1}\Delta \boldsymbol{Q}^n = -h[(\tilde{\boldsymbol{E}}_{i+1/2} - \tilde{\boldsymbol{E}}_{i-1/2}) + \\
&(\tilde{\boldsymbol{F}}_{j+1/2} - \tilde{\boldsymbol{F}}_{j-1/2}) + (\tilde{\boldsymbol{G}}_{k+1/2} - \tilde{\boldsymbol{G}}_{k-1/2}) - \delta_\xi \hat{\boldsymbol{E}}_\nu - \delta_\eta \hat{\boldsymbol{F}}_\nu - \delta_\zeta \hat{\boldsymbol{G}}_\nu]^n = -hR^n. \quad (2)
\end{aligned}$$

Here $h = \Delta\tau J$, $\mathbf{I}$ is the identity matrix, $\triangle, \nabla$, and δ ($\bar{\delta}$) are forward, backward and central (midpoint central) differences respectively, Γ_1, Γ_2, and Γ_3 are the orthogonal part of viscous terms, $\mathbf{T}_\xi^{-1}$, $\mathbf{T}_\xi$ and $\boldsymbol{\Lambda}_\xi$ are the left and right eigenvector matrices and eigenvalues of the Jacobian $\mathbf{A} \equiv \partial\hat{\boldsymbol{E}}/\partial\boldsymbol{Q}$, $\boldsymbol{\Lambda}_\xi^\pm = \frac{1}{2}(\boldsymbol{\Lambda}_\xi \pm |\boldsymbol{\Lambda}_\xi|)$, cf. [3]. First-order flux-difference splitting scheme is used in the left-hand side to obtain tridiagonal matrix equations, while Roe's approximate Riemann solver together with Van Leer's MUSCL formulation [10,6], and Yee-Roe-Davis symmetric TVD scheme [11] are applied to the inviscid fluxes in the right-hand side. In the MUSCL scheme, the numerical flux at a grid interface is

$$\tilde{\boldsymbol{E}}_{i+1/2} = \hat{\boldsymbol{E}}_{i+1/2}(\boldsymbol{Q}^{\mathrm{L}}_{i+1/2}) + \bar{\mathbf{A}}^-_{i+1/2}(\bar{\boldsymbol{Q}}_{i+1/2}) \cdot (\boldsymbol{Q}^{\mathrm{R}}_{i+1/2} - \boldsymbol{Q}^{\mathrm{L}}_{i+1/2}), \tag{3}$$

where $\bar{\boldsymbol{Q}}_{i+1/2} = \frac{1}{2}(\boldsymbol{Q}^{\mathrm{R}}_{i+1/2} + \boldsymbol{Q}^{\mathrm{L}}_{i+1/2})$, and the flow variables to the right and the left side of a grid interface is obtained from the MUSCL formulation

$$\boldsymbol{Q}^{\mathrm{L}}_{i+1/2} = \boldsymbol{Q}_i + \left\{\frac{s}{4}\left[(1-\kappa s)\nabla + (1+\kappa s)\triangle\right]\boldsymbol{Q}\right\}_i, \tag{4}$$

$$\boldsymbol{Q}^{\mathrm{R}}_{i+1/2} = \boldsymbol{Q}_{i+1} - \left\{\frac{s}{4}\left[(1-\kappa s)\triangle + (1+\kappa s)\nabla\right]\boldsymbol{Q}\right\}_{i+1}, \tag{5}$$

where third-order upwind-biased scheme ($\kappa = 1/3$) is used and no limiter is used ($s = 1$). First-order extrapolation of $\nabla\boldsymbol{Q}$ or $\triangle\boldsymbol{Q}$ is used near boundaries.

In the symmetric TVD scheme used, the numerical flux in ξ direction is

$$\tilde{\boldsymbol{E}}_{i+1/2} = \frac{1}{2}\left[\hat{\boldsymbol{E}}_i + \hat{\boldsymbol{E}}_{i+1} + \frac{\mathbf{T}_{\xi_{i+1/2}}\boldsymbol{\Phi}_{i+1/2}}{J_{i+1/2}}\right], \tag{6}$$

and the l-th element of $\boldsymbol{\Phi}_{i+1/2}$ is

$$\phi^l_{i+\frac{1}{2}} = -\lambda(\lambda^l_{i+\frac{1}{2}})^2 s^l_{i+\frac{1}{2}} - \psi(\lambda^l_{i+\frac{1}{2}})(\alpha^l_{i+\frac{1}{2}} - s^l_{i+\frac{1}{2}}), \tag{7}$$

where $\lambda = \Delta\tau/\Delta\xi$, $\boldsymbol{\alpha}_{i+1/2} = \mathbf{T}^{-1}_{\xi_{i+1/2}}(\boldsymbol{Q}_{i+1} - \boldsymbol{Q}_i)$, $\psi(z)$ is the entropy fix function, taking to be $|z|$ here, and $s^l_{i-1/2}$ is the limiter function given by

$$s^l_{i+\frac{1}{2}} = \text{minmod}\left[2\alpha^l_{i-\frac{1}{2}}, 2\alpha^l_{i+\frac{1}{2}}, 2\alpha^l_{i+\frac{3}{2}}, 0.5(\alpha^l_{i-\frac{1}{2}} + \alpha^l_{i+\frac{3}{2}})\right]. \tag{8}$$

and is modified near boundaries via $s^l_{i-1/2} = s^l_{i+1/2}$ and $s^l_{i+1/2} = s^l_{i-1/2}$.

3 Multigrid Procedure

An efficient way to overcome the difficulty of slow convergence is to use MG methods. The concept of a MG method is simple, i.e. the fine grid serves to damp the high-frequency errors, while the coarser grid damp the low-frequency errors. But its implementation is diversified, e.g. full-coarsening, semi-coarsening, V, W and FMG cycling, and different restriction and prolongation operators may be used based on equation natures and flow regimes [9]. The MG algorithm adopted in this work is a full-coarsening FAS scheme with fixed V cycling, and trilinear interpolation for corrections, similar to the method of Jameson et al [12], except that there are some modifications to the restriction operator and the coarse grid boundary solutions, which is discussed below.

Firstly, the restriction operator for the solution is a simple injection due to the collocation of the solution variables on both fine and coarse grids. However, the restriction operator for the residual is a weighted average of the residuals on the fine grid multiplied by 8, as the residual expressed in R^n of (2) represents a control volume integral and the flux conservation must be met. The transferred residual must be underrelaxed to guarantee convergence in cases when the grid size is large. Secondly, the coarse-grid equations are iterated for all flow variables including those on Neumann boundaries. It is found that either explicit treatment of Neumann boundary conditions for the solution itself or simple retention of the boundary solution can not guarantee convergence. In the present work, the coarse grid solution on Neumann boundaries is updated by using

$$\boldsymbol{Q}^{s+1}_{\mathrm{b}} = \boldsymbol{Q}^s_{\mathrm{b}} + \Delta\boldsymbol{Q}^s_{\mathrm{b}}, \tag{9}$$

where s represents the iteration index, and $\Delta\boldsymbol{Q}^s_{\mathrm{b}}$ is obtained by linearization of Neumann boundary conditions that relate boundary values with inner values. As the test will show later, this treatment makes MG method converge better.

4 Results and Discussions

The present MG algorithm is applied to two laminar flow problems, i.e. the flow through a square duct with a 90° bend and the flow past a prolate spheroid with an axes ratio of 4:1. The pseudocompressibility factor is set to 1, and the time step is spatially varied based on a CFL number of 10~20, used on all grid levels. However, there is an upper bound of the time step used due to the factorization errors, nonlinearity and boundary condition problems, which is determined by numerical tests. Initial flow condition is the free stream or the analytic profile of a fully developed laminar square duct flow.

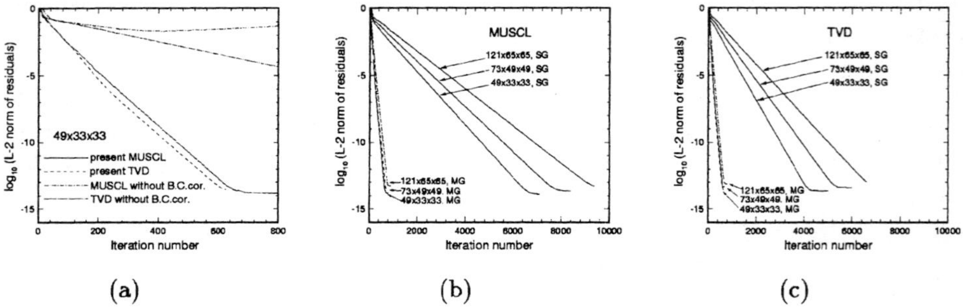

Fig. 1. Convergence histories by using (**a**) different Neumann B.C. treatments, (**b**) MUSCL scheme and (**c**) TVD scheme, for the flow through a 90° bending square duct

4.1 Flow Through a 90° Bent Square Duct

This case corresponds to the experiment of Humphrey *et al* [13]. The Reynolds number was 790 based on the side of the square cross section and the average inflow velocity. Three different grid systems with mesh sizes of 49 (streamwise) $\times$ 33 (normal) $\times$ 33 (spanwise), $73 \times 49 \times 49$ and $121 \times 65 \times 65$ were used to test grid convergence. As an example of how the coarse-grid boundary correction behaves, the convergence history using the coarsest grid is shown in Fig. 1(a). It can be seen that rapid convergence is obtained by using (9), while no renewal of boundary values leads to slow convergence or divergence. Fig. 1(b) and (c) show comparisons of the convergence history of the MG iteration with that of a single grid (SG) iteration. The convergence using MG iteration is very fast, and although it slows down with the increase of grid size, it is not so dramatically as the single grid iteration. Concerning the overall CPU time, the 3-level MG iteration, though costing nearly double CPU time per cycle, still takes only 1/6 or 1/5 of the overall CPU time of the single grid iteration in reducing the residuals by six order of magnitude, for MUSCL and TVD schemes, respectively. It is evident that MG method will save much CPU time in large-scale computations.

It is found that the grid size $73 \times 49 \times 49$ and above can give grid-independent results. The computed results of MUSCL and TVD schemes are compared to the experimental data [13] in Fig. 2. The streamwise velocity (V_θ) profile at

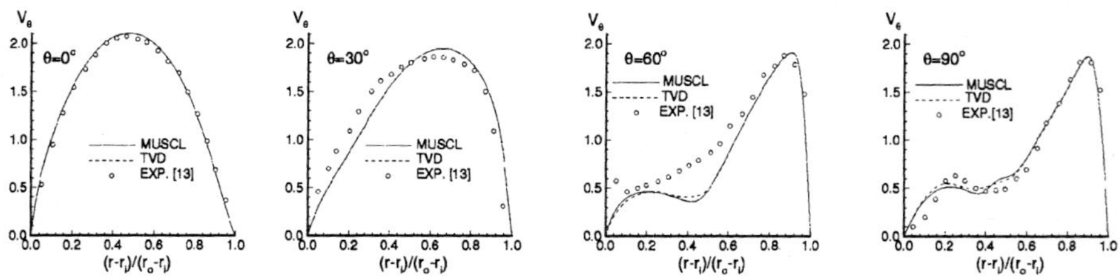

Fig. 2. Comparison of computed streamwise velocity profiles at different streamwise locations by using MUSCL and TVD schemes with the experimental results

the symmetry plane of the duct is shown for four different streamwise stations. The results are fairly well close to one another at the first station. However, some discrepancy is found between the numerical and experimental results and between MUSCL and TVD results at downstream stations. This deviation can also be found for other numerical calculations [4,8]. Nevertheless, the peak of streamwise velocity near the outside wall and the second peak near the inside wall at $\theta = 60°$ and $90°$ are well captured.

4.2 Flow Past a 4:1 Prolate Spheroid

Two cases, $\mathrm{Re_L} = 3000, \alpha = 30°$ and $\mathrm{Re_L} = 11000, \alpha = 10°$, have been selected. The grids are generated by using an ellipsoid coordinates. The first and the last ξ grid planes are located at half grid interval off the polar axes to avoid the singularity of the axes. The results presented below were obtained with 81 (streamwise) $\times$ 81 (circumferential) $\times$ 65 (normal) grid points.

The convergence histories of the multigrid and single grid iterations for the two cases are shown in Fig. 3 (a) and (b), respectively. From Fig. 3 (a), it can be seen that the overall CPU time is reduced by more than half in the MG iterations of both MUSCL and TVD schemes, while from Fig. 3(b), one can see that the MG iterations drive the residuals toward machine zero, but the residuals stall after initial drop for the single grid iterations.

For comparing the computational resolutions of the vortical flow field computed by using MUSCL and TVD schemes, the surface-streamline patterns are shown in Fig. 4 (a) and (b). It can be seen that the major difference between them seems to be the deeper penetration of the separated region into the upstream region in Fig. 4 (a). Furthermore, the result from MUSCL scheme reveals the very fine $S - N_a - N_s$ structure in the separated region while that of TVD scheme smears node N_a in this structure. Another difference is a false separation node N_s in the windward region in the TVD result. The reason may be the use of limiter, which causes degeneration of accuracy to first-order at extrema. To validate this point, MUSCL scheme was also added with a limiter in (4) and (5), and the results were similar to Fig. 4 (b). This shows better resolution of MUSCL scheme and inadequacy of TVD scheme in computing vortical flows.

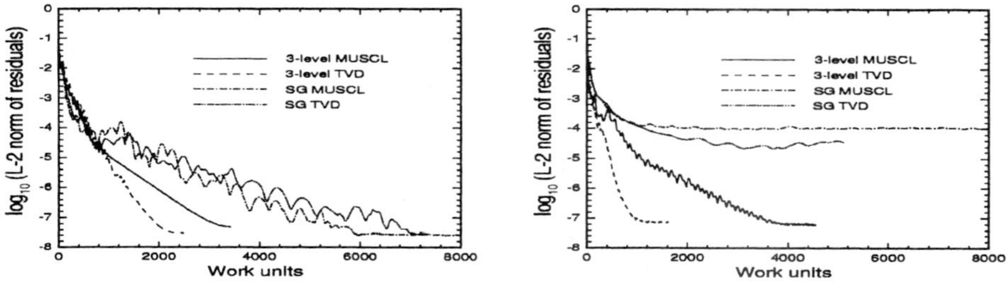

(a) $\mathrm{Re_L} = 3000, \alpha = 30°$ (b) $\mathrm{Re_L} = 11000, \alpha = 10°$

Fig. 3. The convergence history for the flow past the spheroid on $81 \times 81 \times 65$ grid

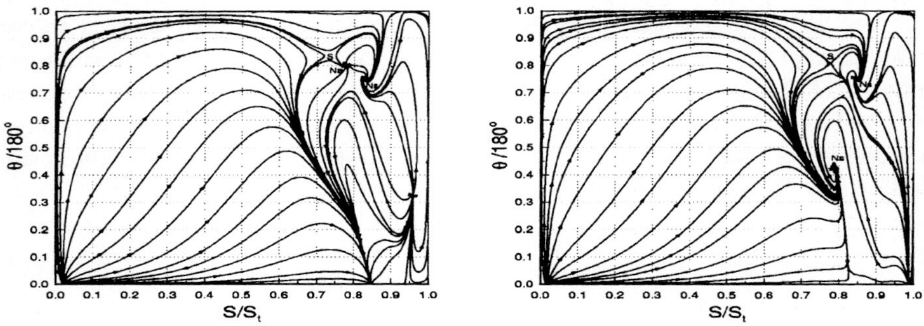

(a) MUSCL (b) TVD

Fig. 4. The surface streamlines on the unwrapped spheroid surface, $\mathrm{Re_L} = 11000, \alpha = 10°$, computed with $81 \times 81 \times 65$ grid points.

5 Conclusions

An implicit multigrid method has been developed that promotes convergence of steady solutions of the 3D incompressible Navier-Stokes equations. Suitable treatment of Neumann boundary conditions on coarse grids is suggested for obtaining good convergence. The results computed by using MUSCL and TVD schemes are compared and the former scheme is plausible to resolve vortical flows while the latter scheme is inadequate due to the use of limiter functions.

This work was supported by ≪ state key program for developing basic sciences≫ G1999032801 and National Laboratory of Computational Physics.

References

1. A. Chorin: J. Comput. Phys. **2**, 12 (1967)
2. D. Kwak *et al*: AIAA J. **24**, 390 (1986)
3. S. Rogers, J. Chang , D. Kwak: J. Comput. Phys. **73**, 364 (1987)
4. S. Roger, D. Kwak: AIAA J. **28**, 253 (1990) AIAA J. **29**, 603 (1991)
5. C. Merke, M. Athavale: 'Time-accurate unsteady incompressible flow algorithms based on artificial compressibility'. AIAA Paper 87-1137, 1987
6. W. Briley, S. Neerarambam, D. Whitfield: J. Comput. Phys. **128**, 32 (1996)
7. L. Yuan, and D. Xing: Chinese J. Aeronautics. **11**, 257 (1998)
8. J. Yang *et al* : J. Comput. Phys. **146**, 464 (1998)
9. D. Drikakis , O.Iliev, D. Vassileva: J. Comput. Phys. **146**, 301 (1998)
10. W. Anderson, L. Thomas, B. Van Leer: 'A comparison of finite volume flux vector splittings for the Euler equations'. AIAA Paper 85-0122, 1985
11. H. Yee: 'A class of high-resolution explicit and implicit shock-capturing methods'. NASA TM 101088, 1989
12. A. Jameson, S. Yoon: AIAA J. **24**, 1737 (1986)
13. J. Humphrey *et al* : J. Fluid Mech. **83**, 509 (1977)

Direct 3D Simulation for Falling Leaves

Takayuki Aoki[1]

Research Laboratory for Nuclear Reactors, Tokyo Institute of Technology 2-12-1 O-okayama, Meguro-ku, Tokyo 152-8550, Japan E-mail: taoki@nr.titech.ac.jp

1 Introduction

In autumn or winter, we often see dead leaves, falling down with complex trajectories, flattering, swaying, rotating, and twisting. The behaviors are determined by the sensitive interaction between leave's motion and the surrounding air flow. Many people recognize the difficulty to do the simulation, however we try to solve this problem as one of the grand challenge in CFD field.

There are several major difficulties caused by the small thickness of the leaves, the interaction between the structure and the air flow, the complex geometry of the leaves, the elastic deformation, and so on. In order to achieve the simulation for falling leaves, we introduce three key methods, 1) IDO Scheme, 2) Cut Cell technique and 3) Overlapping grid system. In this paper, it is assumed that the leaves do not deform the geometrical structure during the falling, as the first step simulation.

2 Key Method

2.1 IDO Scheme

For solving a wide variety of partial differential equations, we have developed a new numerical method "Interpolated Differential Operator (IDO) Scheme"[1]. A local interpolation function is used as an approximate solution of the given partial differential equation. For numerical stability, we choose a Hermite interpolation, which is constructed by the dependent variable and its spatial derivative given independently. The derivative terms in the equation are treated as an operator to take differentiation of the interpolation function instead of using a finite difference. We confirm that the IDO solver for Poisson equation by using 5-th order polynomial function has 4-th accuracy in space [2]. The use of the second mutual derivatives such as $\partial^2 f/\partial xy$ makes it easy to extend the multi-dimension [3]. The IDO scheme shows high accurate and stable results in many applications.

2.2 Cut-Cell Method

On the orthogonal grid, it is difficult to describe accurately the shape of complex geometry. Figure 1 shows that new grids (cut-cells) are introduced in between

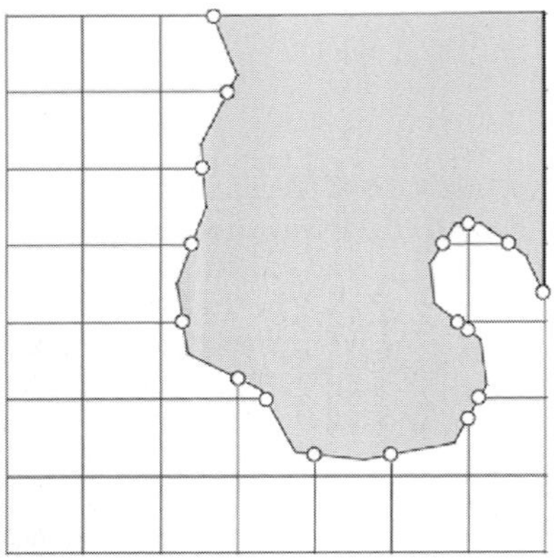

Fig. 1

Fig.2a

Fig.2b

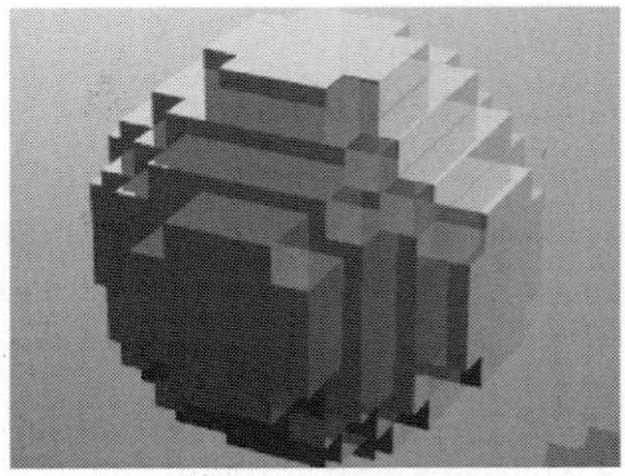

Fig.2c

Fig. 1. Complex Boundary described on orthogonal Structured Grid and cut-cells.
Fig.2a. The original shape of the object.
Fig.2b. The object surface described in a boxcell grid.
Fig. 2c. The object shape described by cut-cells.

the grids, and the boundary conditions are solved on the cut-cell grid. In this paper, the shape of the leaves does not change during the falling, and it is possible to apply BFC (Boundary Fitted Coordinate). However we develop the cut-cell method applicable to the elastic deformation of the leaves for the next step. Although the grid neighboring to the cut cell grid might have extremely non-equal grid intervals, it is possible to solve it with comparable accuracy to the case of the equal interval grid by IDO scheme. In the simulation for falling leaves, each cut-cell grid has two values at the same position representing the front and the rear side pressures, because the thickness of the leaves is negligible in comparison with the mesh size.

When the object has the shape shown in Fig.2a, the orthogonal boxcell grid describes it as shown in Fig.2b. In order to describe more accurately, a lot of grids are required. On the other hand, the object is expressed by the cut cell method in Fig.2c. In spite of the same grid number as Fig.2b, the original shape is treated much more accurately.

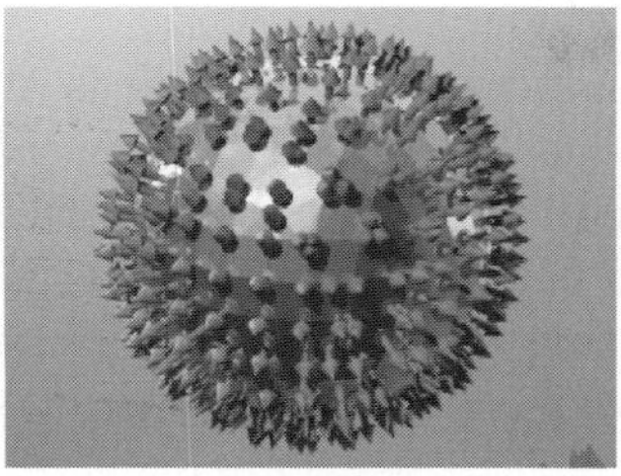

Fig.3

Fig.5

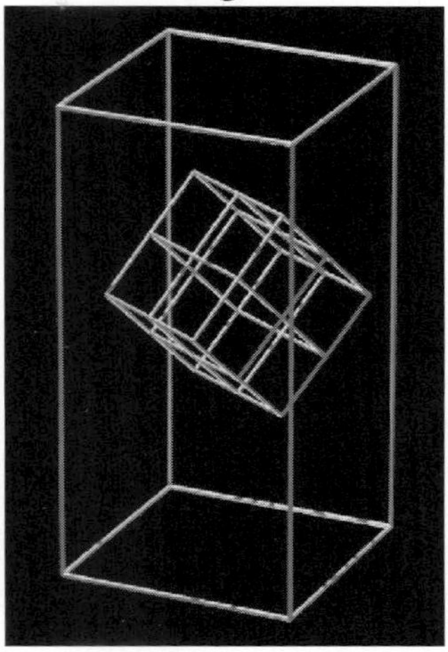

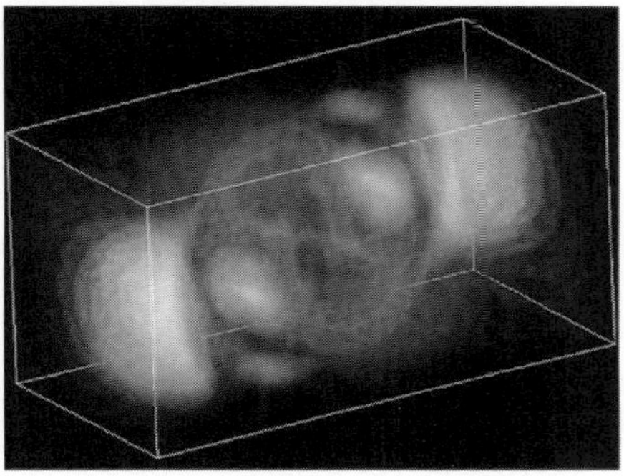

Fig.4

Fig. 3. The normal vectors of the object surface.
Fig. 4. A schematic view of the two over-lapping grid.
Fig.5. The pressure profile driven by the rotating fan.

The normal vector to the object surface is necessary to solve Navior Stokes equation on the cut-cell. We introduce Level-Set function ϕ of the distance from the surface, and propose to solve Laplace equation to determine the spatial derivatives of ϕ by IDO scheme. The gradient vector $\boldsymbol{n} = (\phi_x, \phi_y, \phi_z)$ becomes the normal vector and we can easily obtain the surface curvature by using only the local values. The arrows in Fig.3 indicate the normal vectors for the case of the surface in Fig. 2c.

2.3 Overlapping Grid

In order to study accurately a flow around the small thickness object such as a paper or a leaf, it is important to have enough resolution in the normal direction to the surface for the boundary layer. Because the object moves to arbitrary directions, we introduce a computational sub-domain motion with the fine resolution, moving in association with the object.

The computational mother domain is at rest in the background with an orthogonal coarse grid as shown in Fig.4. Due to the translation and the rotation of the grid coordinate, the virtual flow, the centrifugal and the Coriolis force appears in Navior Stokes equation of the sub-domain in spite of the orthogonal grid. The virtual acceleration is,

$$f = -a - \frac{d\omega}{dt} \times r - 2\omega \times u - \omega \times (\omega \times r),$$

where the notation a is the acceleration of the sub-domain system, r is the position vector, u is the fluid velocity, and w is angular velocity of the sub-grid system. The communication between the sub-domain and the mother domain is done by a cubic interpolation at the sub-domain boundary. It is well known that the attitude of the sub-domain is described by three independent rotational angles like Euler angle, but we introduce the following Euler's parameter $q = [q_1, q_2, q_3, q_4]$,

$$[q_1, q_2, q_3]^T = [\lambda_1, \lambda_2, \lambda_3]^T \sin(\theta/2),$$
$$q_4 = \cos(\theta/2)$$

to avoid the singularity, where l indicates the rotational axis and q is the rotational angle. The rotation matrix C to transform the coordinate has the following form [4],

$$C = \begin{pmatrix} 1 - 2(q_2^2 + q_3^2) & 2(q_1 q_2 + q_3 q_4) & 2(q_1 q_3 - q_2 q_4) \\ 2(q_2 q_1 - q_3 q_4) & 1 - 2(q_3^2 + q_1^2) & 2(q_2 q_3 + q_1 q_4) \\ 2(q_3 q_1 + q_2 q_4) & 2(q_3 q_2 - q_1 q_4) & 1 - 2(q_1^2 + q_2^2) \end{pmatrix}.$$

We set the priority to the computational domains in the overlapping case and do not calculate the shade area of the mother domain covered by the sub-domain of higher priority.

3 3D-Simulation for Falling Leaves

It is assumed that the shape of the leaf does not change during the falling such as elastic deformation for simplification. The simulation consists of two computational domains; the sub-domain including the leaf and the background mother domain. The incompressible Navior Stokes equation is solved by SMAC algorithm and the Poisson equation for the pressure is solved by IDO scheme using the center interpolation. We apply the up-wind interpolation to the advection term, and the diffusion term is solved by the implicit scheme [5].

As a preliminary simulation, we study the flow driven by the rotating fan (square paper) around the fixed central axis with a constant rotational velocity (in Fig.5). The vortices are generated near the fan and moves far away due to the centrifugal force, and we confirm that the overlapping grid works well.

Another preliminary simulation is the falling of a piece of square paper like a name card. We choose the z-axis in the gravity direction. Initially, the paper is located obliquely to the z-axis, that is, the paper is rotated around y-axis by 15 degree from the horizontal plane. The attack angle toward the z-direction generates the lift force in the plus x-direction and the paper moves with the transnational velocity. The drag force due to the pressure and the friction force decelerate the velocity in the x-direction gradually. The pressure at the top of the paper is higher than the that of the tail in the x-direction, and it acts on torque generation. The attack angle decreases to the negative value, and the lift force works on the push in the minus x-direction. This periodic sway motion is familiar with our daily life, and it also appears in the simulation result. The time sequence of the falling paper is shown in Fig.6.

Fig.6. The simulation result of the falling paper showing the periodic sway motion.

In order to describe a realistic shape for the leaf, we use the geometry data modeled by 200 polygons. It is able to convert it to DXF format, which is often used by CAD software, and our simulation code accepts the geometry data with DXF format. In the simulation for the falling leaf, the initial angle is quite sensitive to the falling trajectory. In a certain angle, the falling path becomes the similar swaying with the case of the paper shown in Fig.7, but falls irregularly in some angles.

4 Conclusion

The combination of the IDO scheme, the cut-cell method, and the overlapping grid make it possible to simulate the falling process such as a paper or a leaf.

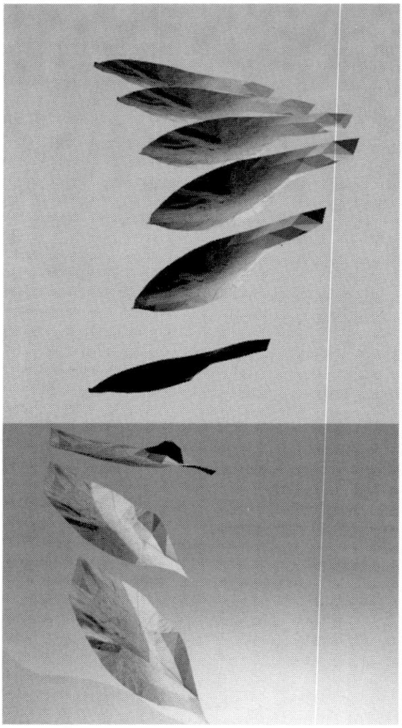

Fig. 7. An example of the time sequence of the falling leaf.

We can study the sensitive behavior of the interaction between the flow and the object with a small thickness, but we have to increase the grid number and the polygon number to compare with the real phenomena. The elastic deformation makes a strong effect on the falling motion, and movable cut-cells should be implemented in the nest stage.

Acknowledgement

The simulations in this paper were done by Silicon Graphics Systems Origin 2000 of the Computer Center at Tokyo Institute of Technology. This work was supported in part by Grant-in-Aid for Scientific Research of the Japanese Ministry of Education.

References

1. T. Aoki: Comp.Phys.Comm.,**102** (1996)
2. K. Sakurai and T. Aoki: Comp. Math. Appl. (submitted)
3. T. Aoki, J. Comp. Fluid Dynamics, **4**, **3** (1995)
4. P. C. Hughes, Spacecraft Attitude Dynamics, John Wiley & Sons, (1986)
5. K. Sakurai and T. Aoki, Comp. Fluid Dynamics Journal, **8**, 1 (1999)

Part IX

Stability of Boundary Layer

Multigrid Methods for Thin Liquid Film Spreading Flows

Neil Daniels[1], Pascal Ehret[1], Philip H. Gaskell[1], Harvey M. Thompson[1], and Michel Decré[2]

[1] Engineering Fluid Mechanics Research Group, School of Mechanical Engineering, The University of Leeds, Leeds, LS2 9JT, UK
[2] Philips Research Laboratories, Prof. Holstlaan 4 (WB31) 5656 AA Eindhoven, The Netherlands

1 Introduction

A range of industrial processes involve the spreading of thin layers of liquid or molten polymeric materials that subsequently solidify. Their function may be: to form a protective barrier; to give added value to a particular product (magnetic/optical storage media, etc.); the production of laminates, tapes, sheets ([5], [7], [8], [12]). Their quality depends ultimately on the homogeneity of the resultant polymer film or surface covering.

Recently, the most popular numerical approach for solving such time dependent, three-dimensional fluid problems has been to use a semi-implicit *time splitting* technique ([3], [4], [9], [10], [14]) where each time step is composed of two half time steps in which only one of the operators is implemented implicitly. Such methods combine some of the stability properties of implicit schemes with the cost efficiency of explicit ones. Here this approach is compared to an alternative method of solution which uses a fully implicit formulation, termed *multigrid* ([1]). Advantages of the latter include an order of magnitude increase in the rate of convergence, guaranteed accurate solutions and low memory requirements. Moreover, multigrid schemes improve robustness of the numerical process and their rate of convergence is independent of the resolution of the domain.

Sample solutions obtained using the multigrid solver are shown for flow over a heterogeneous surface and topography.

2 The mathematical model

The basic equation to model the spreading of liquid on a substrate inclined at an angle α (see fig. 1) is obtained by using a thin film approximation of the Navier-Stokes equations, and its dimensionless form is given by

$$\frac{\partial h}{\partial t} = \frac{1}{C}\nabla . \left[\frac{h^3}{3\mu}\left(\nabla p - \frac{B}{\epsilon}\sin\alpha \underline{i}\right) - M\nabla\sigma\frac{h^2}{2\mu}\right] - \frac{E}{C} \tag{1}$$

where $h = h(x, y, t)$ describes the height of the fluid surface over a topography $s = s(x, y)$, $p(x, y, t)$ is the pressure, ϵ is the ratio of drop height to substrate length, E is the evaporation rate, μ and σ are respectively the variable viscosity and surface tension of the liquid, and C, B and M are constant coefficients defined by

$$C = \frac{\sigma_0 \epsilon^3}{u_0 \nu_0}, \quad B = \frac{\rho g L^2 \epsilon^3}{u_0 \nu_0}, \quad M = \frac{\sigma_0 \epsilon}{u_0 \nu_0}.$$

These three constants represent surface tension, gravity and Marangoni effects respectively ([6]). If the surface is pre-wet, the pressure is then given by

$$\begin{aligned} p = & -C\nabla^2(h+s) + Bh\cos\alpha \\ & - \frac{M(n-1)(m-1)(1-\cos\theta_e)}{h*(n-m)} \left[\left(\frac{h*}{h}\right)^n - \left(\frac{h*}{h}\right)^m \right] \end{aligned} \tag{2}$$

where $h*$ is the pre-wet film thickness, n and m are positive constants (with $n > m > 1$) and θ_e is the contact angle.

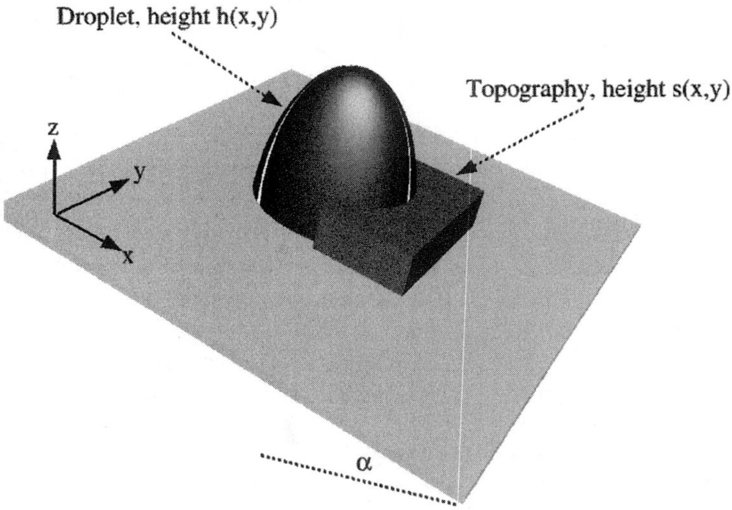

Fig. 1. Schematic of the flow of a droplet, over a topography, down an inclined plane.

3 Results

Good agreement was found between solutions obtained using both the time-splitting and the multigrid approaches for the same set of operating conditions. Table 1 shows a comparison between the CPU time needed for 50 iterations with a time step of 10^{-5} for grids with various numbers of internal points for the case of an initial paraboloid drop spreading with $B = M = \theta_e = 0.0$ and no evaporation or variation of viscosity and surface tension. The maximum possible time step on the 513 × 513 grid is 3×10^{-6} so the CPU time given is the time needed to complete the required number of time steps for a total time of 5×10^{-4}. The time-splitting technique deteriorated further at grid level 1025 × 1025, where the maximum time step was 6×10^{-7}. This table has 3 key features:

- The time-splitting technique is more efficient for the coarser grids.
- On the finer grids, which are required to examine more complicated flows, the multigrid approach is more efficient.
- The time-splitting approach is restricted to small time steps for fine grids (due to mass conservation problems and numerical instability), a problem not experienced by the multigrid approach.

Table 1. A comparison between CPU times for the Time-splitting and Multigrid approaches for various grid levels to reach a dimensionless time of 5×10^{-4}

Mesh size	Multigrid CPU time			Timesplitting
	1 V-cycle	3 V-cycles	5 V-cycles	CPU time
65×65	15.84	20.32	24.20	9.51
129×129	24.20	38.35	52.22	42.03
257×257	74.40	145.75	216.88	203.46
513×513	459.36	976.81	1500.08	3360.15

The effect of the number of multigrid V-cycles employed on the final residuals is examined via tables 1 and 2 from which it can be seen that increasing the number of V-cycles increases the accuracy (by reducing the residual size by a factor of 10 for each additional V-cycle). Note that the CPU time is not proportional to the number of unknowns for the multigrid technique. This result is unexpected but due to the fact that the convergence at the first time steps is dependent on the initial conditions and time step and so varies according to the discretization.

It should also be noted that the introduction of further physics/terms in the equation set does not necessitate reformulation of the multigrid solver and it is therefore more flexible and robust than the time-splitting technique. Further, being a fully implicit technique, it is inherently more stable than the semi-implicit time-splitting method thus allowing larger time-steps to be used.

3.1 Effect of the disjoining pressure model

Following [13], although there are various possibilities (see [9], [10]), (n, m) has been set to $(3, 2)$. Results show that drops spread more quickly with increasing $h*$ and smaller contact angles. Note that for small values of $h*$ a fine grid must

Table 2. A comparison between final residuals using the multigrid approach with different numbers of V-cycles and various grid levels

Mesh size	Number of V-cycles				
	1	2	3	4	5
65×65	5.47×10^{-7}	3.55×10^{-8}	1.46×10^{-9}	1.33×10^{-10}	7.20×10^{-12}
129×129	3.31×10^{-7}	2.00×10^{-8}	1.37×10^{-9}	1.16×10^{-10}	1.16×10^{-11}
257×257	3.56×10^{-7}	2.46×10^{-8}	2.28×10^{-9}	2.19×10^{-10}	2.14×10^{-11}
513×513	4.62×10^{-7}	4.52×10^{-8}	3.79×10^{-9}	4.06×10^{-10}	4.52×10^{-11}

be used and so it can again be seen that the multigrid approach is more efficient than time splitting.

Using this model, it was also possible to reproduce results in [10] for which a drop spread over a higher contact angle material cross, see figure 2. It is clear that the behaviour is qualitatively the same with the initial single drop splitting into a separate drop in each of the four quadrants. Note that a further disadvantage of a time splitting approach (the method used in [10]) is that the contact angle distribution must be differentiable. This means that a relatively large region must be incorporated at substrate material boundaries inside which a cubic polynomial smoothing function is applied.

3.2 Flow over topography

Following [7] and [11], an arc tangent function is used to describe the steepness of a topography (with variable breadth, length, height and location).

In figure 3, the flow of an initial drop over a ridge (of non-dimensional height 0.1) is examined via contours of h and $h + s$. Recall that h is the height of the liquid above the topography (s) and so to study how the liquid flows throughout the domain it is more practical to observe $h + s$. It is clear that as time increases from t_0 to t_2 the liquid spreads further off the ridge than across it. This leads to a lack of liquid spreading to the bottom left corner (in the diagram).

4 Conclusions

The main conclusions from this study are that although the time splitting approach is faster than the multigrid solver on coarser grids, it is significantly slower on the finer grids one needs to capture more complicated liquid flows e.g. flow over a fine topographical feature. Other benefits of the multigrid approach are that it allows additional physics to be implemented with ease without altering the quality of the multi-resolution solver; being fully implicit, the multigrid approach does not suffer from numerical instability caused by the choice of time

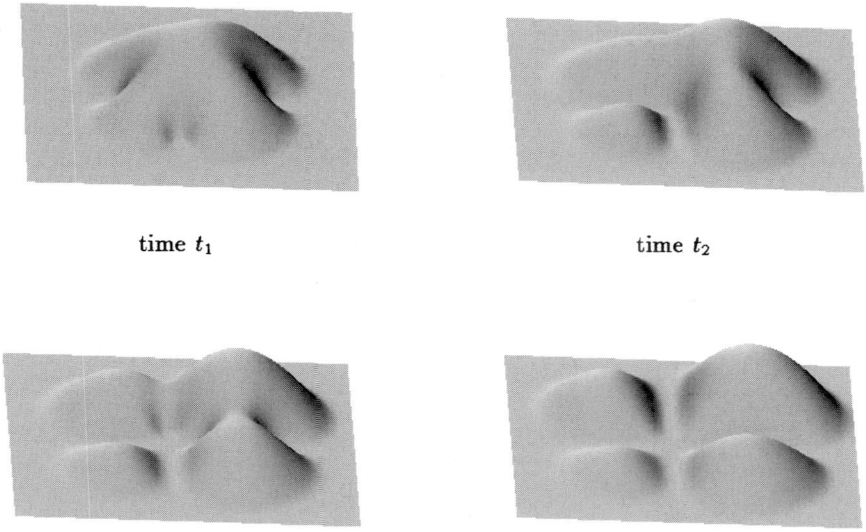

time t_1 time t_2

time t_3 time t_4

Fig. 2. Evolution of an initial parabaloid drop over a higher contact angle material cross at increasing times of t_1 to t_4

step or spatial discretisation and can use relatively large time increments. In contrast, our study of the time-splitting technique shows that the method is more prone to suffer from numerical instabilies which cause mass conservation problems, a factor which can restrict the maximum possible time-step.

References

1. A. Brandt: Mathematics of Computation **131**, 333 (1997)
2. F. Chou, P. Wu, S. Gong: Jpn. J. Appl. Phys. **37**, 4321 (1998)
3. C. I. Christov, J. Pontes, D. Walgraef, M. G. Velarde: Comput. Methods Appl. Mech. Engrg **148**, 209 (1997)
4. M. H. Eres, D. E. Weidner, L. W. Schwartz: Langmuir **15(5)**, 1859 (1999)
5. N. Fraysse, G. M. Homsy: Phys. Fluids **6(4)**, 1491 (1994)
6. T. G. Myers: Siam Rev. **40(3)**, 441 (1998)
7. L. M. Peurrung, D. B. Graves: IEEE Transactions on Semiconductor Manufacturing **6(1)**, 72 (1993)
8. D. Qin, Y. Xia, J. A. Rogers, R. J. Jackman, X. Zhao, G. M. Whitesides: Topics in Current Chemistry **194**, 1 (1998)
9. L. W. Schwartz: *Hysteretic effects in droplet motions on heterogeneous substrates: direct numerical simulation* (Dept. Mech. Engrg and Math. Studies, University of Delaware, Newark 1998)
10. L. W. Schwartz, R. R. Eley: J. Colloid and Interface Science **202**, 173 (1998)
11. L. E. Stillwagon, R. G. Larson: J. Appl. Phys. **63(11)**, 5251 (1988)
12. L. E. Stillwagon, R. G. Larson: Phys. Fluids A **2(11)**, 1937 (1990)
13. G. F. Teletzke, H. T. Davis, L. E. Scriven: Chem. Eng. Comm. **55**, 41 (1987)

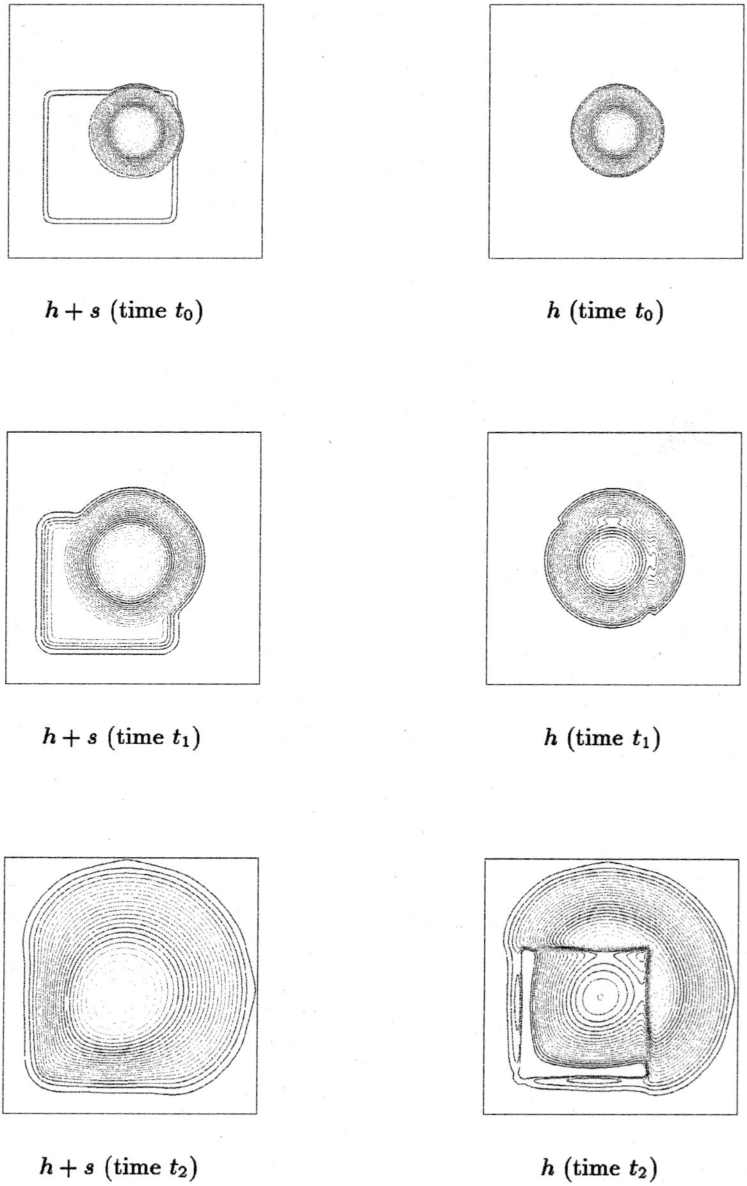

Fig. 3. Contours of $h + s$ and h illustrating the evolution of an initial parabaloid drop over a ridge at times $t_0 < t_1 < t_2$

14. W. E. Weidner, L. W. Schwartz, R. R. Eley: J. Colloid and Interface Science **179**, pp.66-75 [1996]

Instabilities of Dynamic Thermo- and Solutocapillary Liquid Layers

Gunar Schmidt, Hendrik C. Kuhlmann, and Hans J. Rath

Center of Applied Space Technology and Microgravity ZARM, University of Bremen, 28329 Bremen, Germany

Abstract. The variation of the surface tension with temperature and concentration constitutes a driving mechanism known as Marangoni convection. We study a coupled thermo- and solutocapillary surface-tension-driven shear flow in an unbounded plane layer. The basic flow can be calculated analytically. The stability of the flow is investigated by a linear analysis. Results are presented for some special contributions and for Pr = 1 and Sc = 100.

1 Model and Basic Flow

We consider an unbounded plane layer of a binary mixture of thickness d. On the non-deformable surface it is assumed that the temperature is a decreasing and the concentration an increasing linear function of the lateral coordinate.

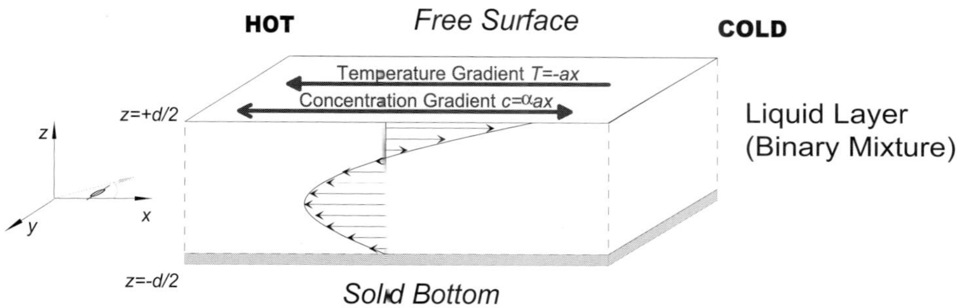

Fig. 1. Model a of soluto-thermocapillary liquid layer

As the surface tension σ can be approximated by a linear function of temperature T and concentration c, $\sigma(T, c) = \sigma_0 - \sigma_T(T - T_0) + \sigma_c(c - c_0)$, a shear flow is induced, which is an extension of the *return flow* of Smith and Davis [1]. The flow is governed by the Navier–Stokes, the energy, the diffusion and the continuity equation. Keeping the Soret effect (thermal diffusion) [2] we get after scaling (Table 1) in non-dimensionalized form

$$(\partial_t + \boldsymbol{v} \cdot \nabla)\boldsymbol{v} = -\nabla p + \Delta \boldsymbol{v}, \tag{1a}$$

$$(\partial_t + \boldsymbol{v} \cdot \nabla)c = \frac{1}{\mathrm{Sc}}(\Delta c + \Delta T), \tag{1b}$$

$$(\partial_t + \boldsymbol{v} \cdot \nabla)T = \frac{1}{\mathrm{Pr}}\Delta T, \tag{1c}$$

$$\nabla \cdot \boldsymbol{v} = 0. \tag{1d}$$

Table 1. Scaling

$\boldsymbol{x}$	$\boldsymbol{v}$	t	$T - T_0$	$c - c_0$	p
$\frac{d}{2}$	$\frac{2\nu}{d}$	$\frac{d^2}{4\nu}$	$\frac{ad}{2}$	$\frac{ad}{2} \cdot \frac{k_T}{T_0}$	$\frac{4\nu^2 \varrho}{d^2}$

On the bottom at $z = -1$ we impose no-slip, no mass transport and thermal insulation ($\boldsymbol{v} = \partial_z T = \partial_z c = 0$). The boundary conditions at $z = +1$ are $w = 0$ (non-deformable surface), $T = -x$, $c = \alpha x$, (prescribed temperature and concentration gradient), $0 = \boldsymbol{e_z} \times (\partial_z \boldsymbol{v} + \nabla(\mathrm{Re_T} T + \mathrm{Re_S} c))$ (surface-tension-driven flow) and $\partial_z T = \partial_z c = 0$ (insulating and impermeable surface). The characteristic numbers (thermal, solutal, and effective Reynolds number $\mathrm{Re_T}$, $\mathrm{Re_S}$, $\mathrm{Re_\Delta}$ respectively, the Prandtl, and the Schmidt number, Pr and Sc, respectively, are defined in Table 2. Here ϱ, ν, D, κ and k_T are, respectively, density, kinematic viscosity, mass diffusivity, thermal diffusivity and thermodiffusion ratio.

Table 2. Characteristic numbers

$\mathrm{Re_T}$	$\mathrm{Re_S}$	$\mathrm{Re_\Delta}$	Pr	Sc
$\frac{ad^2\sigma_T}{4\varrho\nu^2}$	$\frac{ad^2\sigma_c k_T}{4\varrho\nu^2 T_0}$	$\mathrm{Re_T} - \alpha \mathrm{Re_S}$	$\frac{\nu}{\kappa}$	$\frac{D}{\kappa}$

The steady unidirectional basic flow $\partial_t = 0$, $\boldsymbol{v} = (u(z), 0, 0)$, $T = -x + \theta(z)$, and $c = \alpha x + \gamma(z)$ with vanishing mean flow through the y-z-plane, $\int_{-1}^{1} u(z)dz = 0$, is given by

$$u(z) = -g''(z), \tag{2a}$$

$$T(x, z) = -x + \mathrm{Pr}\, g(z), \tag{2b}$$

$$c(x, z) = \alpha x - (\mathrm{Pr} + \alpha \mathrm{Sc}) g(z), \tag{2c}$$

where

$$g(z) = \frac{1}{8}\mathrm{Re_\Delta}\left(-\frac{11}{12} + z + \frac{1}{2}z^2 - \frac{1}{3}z^3 - \frac{1}{4}z^4\right).$$

2 Linear Stability Analysis

The stability of the flow is investigated by a linear analysis. The normal modes are plane waves at an angle ϕ with respect to the x-axis (see Fig. 1). For simplification we rotate the coordinate system such that the perturbation wave propagates in the x-direction and is invariant in y-direction. Expanding the solution of the stability problem in Chebyshev polynomials $\tau_n(z) = \cos(n \arccos z)$, we get

$$\{\psi, v, T, c\} = \sum_{n=0}^{N} \{\psi_n, v_n, T_n, c_n\} e^{(\sigma - i\omega)t} e^{ikx} + c.c. \tag{3}$$

Here σ is the growth rate, ω the frequency, k the wave number of the perturbation, and ψ the 2D-stream function along the propagating wave. The equations are invariant with respect to the transformations

$$\begin{pmatrix} k \\ \omega \end{pmatrix} \to \begin{pmatrix} -k \\ -\omega \end{pmatrix}, \begin{pmatrix} \mathrm{Re_T} \\ \mathrm{Re_S} \\ \omega \end{pmatrix} \to \begin{pmatrix} -\mathrm{Re_T} \\ -\mathrm{Re_S} \\ -\omega \end{pmatrix}, \begin{pmatrix} \omega \\ \phi \end{pmatrix} \to \begin{pmatrix} -\omega \\ \pi - \phi \end{pmatrix}, \phi \to -\phi. \tag{4}$$

Therefore we consider only waves for $\phi \in [0, \pi/2]$ and $k > 0$.

Solving the perturbed equations with help of Galerkin's method, we get

$$\mathbf{A}(\mathrm{Re_T}, \mathrm{Re_S}, \mathrm{Pr}, \mathrm{Sc}, k, \phi, \omega, \sigma) \cdot (\psi_{0 \ldots N}, v_{0 \ldots N}, T_{0 \ldots N}, c_{0 \ldots N})^T = 0. \tag{5}$$

Truncation on $N = 12$ leads to a sufficient accuracy.

The neutral hyper-surfaces are given by non-trivial solutions of (5) for $\sigma = 0$, while the critical surface is the surface curve of all neutral surfaces. Beyond the critical surface the basic flow is unstable.

3 Results

At first we want to investigate some special configurations. The pure thermocapillary flow ($\mathrm{Re_S} = 0$) was investigated by Smith and Davis [1]. We used the results for the validation of the numerical code. The Schmidt number of the binary mixture has in case of $\mathrm{Re_S} = 0$ no influence on the stability of the basic flow.

Due to the symmetry of the equations and the boundary conditions, the pure soluto-capillary flow has the same stability limits as the pure thermo-capillary flow (with $\mathrm{Re_S}$ instead of $\mathrm{Re_T}$ and Sc instead of Pr), but we have to assume $\mathrm{Pr} = 0$ additionally, otherwise the Soret-effect has influence on the flow (see (1b)). Figure 2 shows the critical solutal Reynolds number as function of the Schmidt number for different Prandtl numbers and $\mathrm{Re_T} = 0$. Obviously the influence of the Prandtl number can be neglected for very high Schmidt numbers.

The critical solutal Marangoni number $\mathrm{Ma_S} = \mathrm{Re_S}\,\mathrm{Sc}$ tends to 100 for high Schmidt numbers, furthermore we get $\lim_{\mathrm{Sc}\to\infty} \omega \mathrm{Sc} \approx 15.32$ (or $\mathrm{Re_S} \approx 6.53\omega$), $\lim_{\mathrm{Sc}\to\infty} k \approx 1.23$, and $\lim_{\mathrm{Sc}\to\infty} \phi \approx 7.75°$. Nevertheless, these limits are very

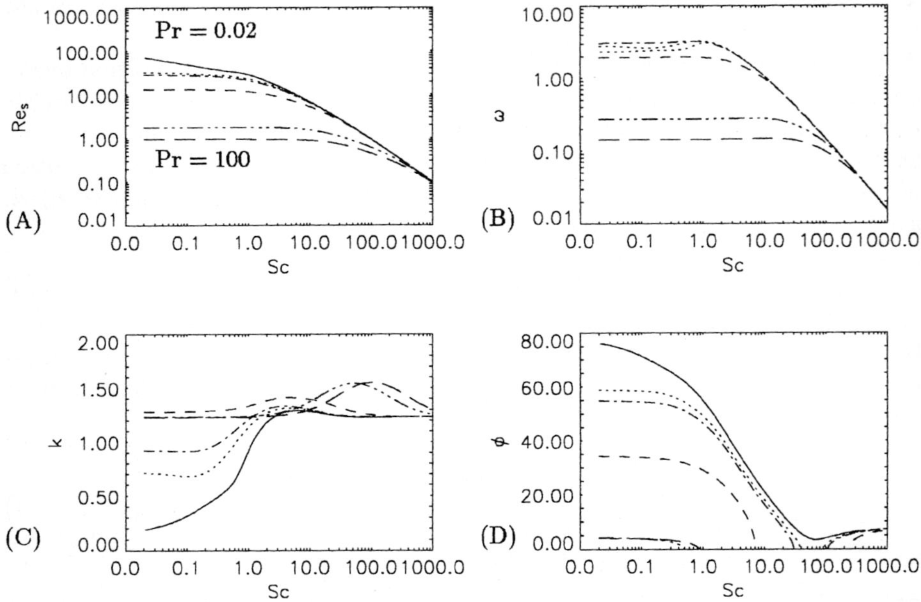

Fig.2. (A) Critical solutal Reynolds number, and corresponding (B) frequency, (C) wave number and (D) direction of perturbation waves for $Re_T = 0.0$ and $Pr = 0.02, 0.7, 1, 4, 50, 100$.

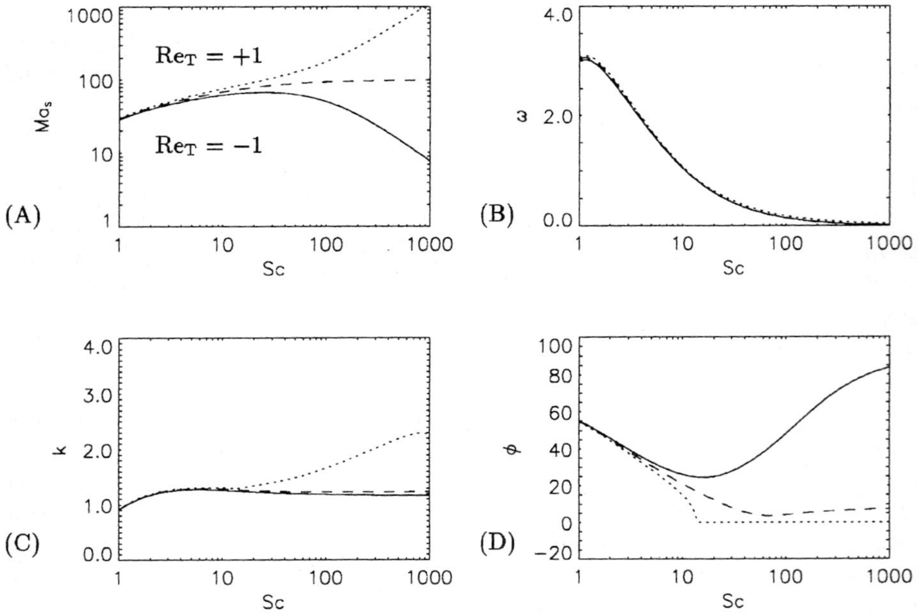

Fig.3. (A) Critical solutal Marangoni number, and corresponding (B) frequency, (C) wave number and (D) direction of perturbation waves for $Re_T = -1.0$ (solid), 0.0 (dashed) and $+1.0$ (dotted) and $Pr = 0.0$.

sensitive to the influence of the thermocapillary effect. Figure 3 shows the critical solutal Reynolds number for $\mathrm{Pr} = 0$ in case of $\mathrm{Re_T} = -1, 0$ and 1.

In case of $\mathrm{Re_T Re_S} < 0$, i.e. the thermocapillary and the solutcapillary surface tension drive act in the same direction, the direction of the perturbation wave is changing to nearly orthogonal direction (Fig. 3(D)) with respect to the basic flow. On the other hand, in case of $\mathrm{Re_T Re_S} > 0$, the direction of the perturbation wave changes from $\phi = 7°$ to $\phi = 0°$. The wave propagates in opposite direction to the basic flow on the free surface ($\omega < 0$). More remarkable is the increasing of the wave number in case of $\mathrm{Re_T} > 0$(Fig. 3(C)).

Figure 4 shows the streamlines and the concentration field of the critical mode in case of $\mathrm{Pr} = 0$, $\mathrm{Sc} = 1000$ and $\mathrm{Re_T} = -1, 0, +1$.

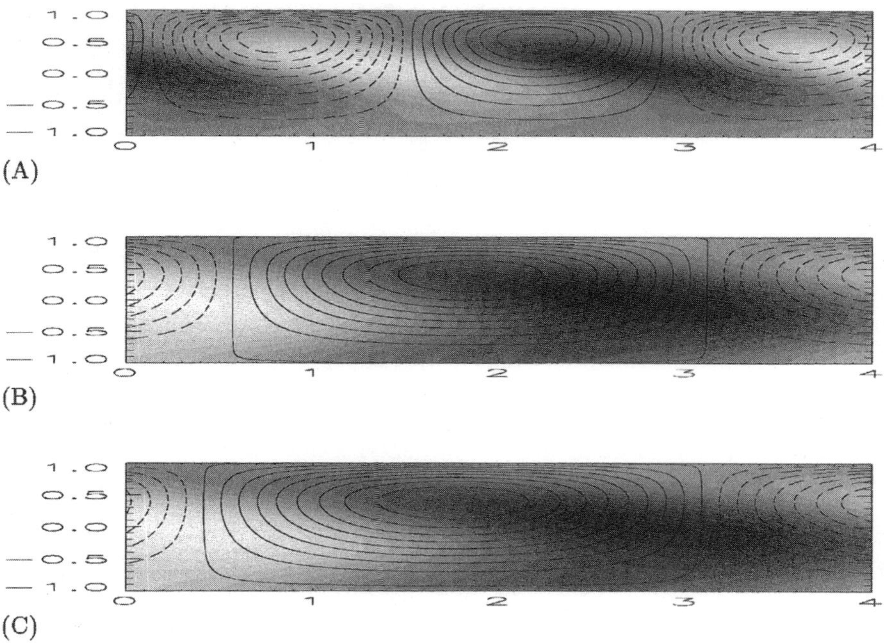

Fig.4. Streamlines and concentration field (shaded) for $\mathrm{Pr} = 0.0$, $\mathrm{Sc} = 100$ and $\mathrm{Re_T} = -1$ (A), $\mathrm{Re_T} = 0$ (B), $\mathrm{Re_T} = +1$(C)

Finally, we want to show as an example the stability diagram for $\mathrm{Pr} = 1$ and $\mathrm{Sc} = 100$. Starting from the known critical points $(\mathrm{Re_T}, \mathrm{Re_S}) = (0, \pm 0.93)$ and $(29.00, 0)$ we find within the stable region an unstable region for $\mathrm{Re_T Re_S} < 0$. Because of numerical problems due to ambiguous solutions of (5) and problems in finding the global minimum the diagram is not complete.

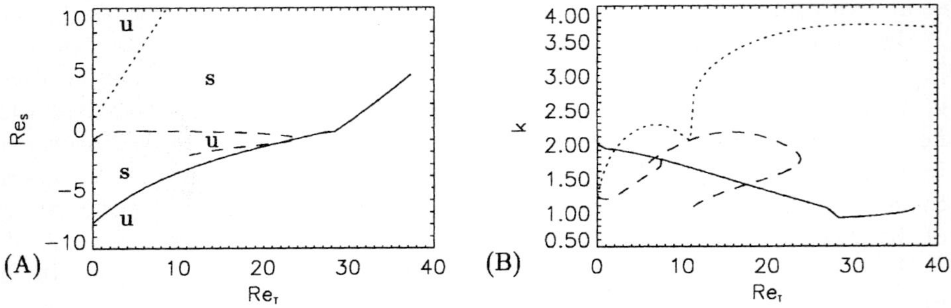

Fig.5. (A) Stable (s) and unstable (u) regions for Pr = 1.0 and Sc = 100, (B) corresponding wave number

4 Concluding remarks

A linear stability analysis is performed for a binary liquid layer with thermo- and solutocapillary driving forces. For solving we use the Chebyshev-Galerkin-method, that leads for 12th order to results with satisfying accuracy. Numerical problems can occur in case of ambiguous solutions or in finding the global minimum.

For high Schmidt numbers and neglected thermo-capillary effect the critical parameters are converging to limit values, which are independent of Prandtl number. Including the thermocapillary driving force, we found out, that depending on the direction of the temperature gradient the wave number or the propagation direction changes.

In the future it is planned to investigate the mechanisms which are leading to the instability. Therefore an a-posteriori energy analysis of the perturbations [3] shall be performed

References

1. M. K. Smith and S. H. Davis: J. Fluid Mech., **132** (1983)
2. L. D. Landau and E. M. Lifshitz: Fluid mechanics. Pergamon Press, New York (1959)
3. H. C. Kuhlmann: Thermocapillary Convection in Models of Crystal Growth. Vol. 152 of Springer Tracts in Modern Physics, Springer, Berlin, Heidelberg (1999)

Part X

Flows with Free Surface

Numerical Approach to Free Surface Generation using Cut Cell Method

Kouichi Sakurai[1] and Takayuki Aoki[2]

[1] Dept. of Energy Science, Tokyo Inst. of Tech., 2-12-1 O-okayama, Tokyo, Japan
[2] Research Lab. of Nuclear Reactors, Tokyo Inst. of Tech., 2-12-1 O-okayama, Tokyo, Japan

1 Introduction

In numerical simulation there are many problems including the free surface generation with time advance such as crack growth in material or vaporization in fluid dynamics. An ordinary approach to such situations based on coarse models has some problems. For example, in order to simulate the vapolization, such a distribution function defined as the rate of liquid (γ_l) and gas (γ_g); ($\gamma_l + \gamma_g = 1$) may be used. In the cell where liquid and gas are mixtured $(0 < \gamma_s < 1)_{s=l,g}$, liquid and gas are solved individualy, however, we can not split them spatially. When the Lagrangean approach is taken, of course, different materials can be distinguished easily, however, there is some possibility to require rezoning in multi-dimensional simulation. There is no doubtful that rezoning decreases the accuracy of calculation. In addition to this problem, for generating a free surface, we should prepare extra grids *initially* where the free surface will appear.

In order to overcome those problems, we developed a "Movable Cut Cell method" on a rectangular grid for separating different matters. For the reason of the irregularity in the grid interval with Cut Cell, we need to use high-order accurate scheme. In this paper Interpolated Differential Operator (IDO) scheme[1,2] is chosen.

2 Formalization

2.1 Cut Cell Method

By using the Cut Cell method [3] on a Cartesian grid, the discontinuity of physical values can be described exactly as shown in Fig. 1. On a Cut Cell grid, each physical values of different materials are defined and calculated individually only using the information of the corresponding material belonged to.

The Cut Cell Method is usually used as a fixed boundary condition, but we introduce it dynamically when a free surface or a discontinuity appears on simulating. When the area which has obviously different phase from the other is distinguished with a propaty such as the boiling point in vapolization, Cut Cells are generated at the interface. When the Cut Cell method is coupled with discrimination method such as Level Set Method and with higher-order interpolation method, the position of surface and physical values on it can be calculated accurately.

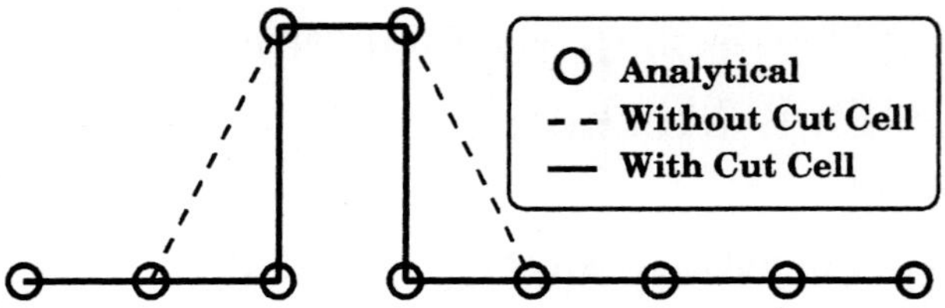

Fig. 1. The concept of Cut Cell method;
The Cut Cell can describe correctly the discontinuity on a Cartesian grid.

2.2 Interpolated Differential Operator (IDO) Scheme

In this paper, we use the Interpolated Differential Operator (IDO) scheme based on Euler scheme. For time advance, the IDO scheme uses Taylor expantion (Eq. (1)) respect to time.

$$f^{n+1} = f^n + \frac{\partial f}{\partial t}\Delta t + \frac{1}{2}\frac{\partial^2 f}{\partial t^2}\Delta t^2 + \frac{1}{6}\frac{\partial^3 f}{\partial t^3}\Delta t^3 \cdots \tag{1}$$

Time derivatives included in Eq. (1) are replaced with spatial derivatives by using Partial Differential Equation. This results in the generation of higher order spatial derivatives. These terms are estimated by differentiating a Hermite interpolation constructed with physical values and its first order spatial derivatives which are defined at each grid point. As an interpolation, we use 3rd-order and 5th-order polynomial for solving "Advection term" and "Non-Advection term". The former is used in CIP scheme[4], which is known as hyperboric equation solver, and the latter shows 4th-order spatial accuracy [1].

2.3 1-Dimensional Movable Cut Cell Method

For the simplicity of explanation, we first show the procedure required in constructing 1-dimensional Movable Cut Cell. For the computational simulation, we required the following four procedures;

1. Calculation of physical value on Euler grid.
2. Consider the numerical stability with CFL (Courant-Friedrich-Levy) condition.
3. Calculation of discontinuity position,
4. Calculation of physical value on discontinuity

For estimating the physical value at next time step on Euler grid, we choose the IDO scheme. In order to estimate the higher order spatial derivatives required in time advance, we construct interpolation between the Euler grids or Euler grid and Cut Cell as shown in Fig. 2. From this reason, we use interpolation for irregular interval grid. A problem arisen from CFL condition should be considered for calculation of Euler grid. When the Cut Cell grid takes over some Euler grids within Δt, the time advance based on Taylor expansion on Euler grid will

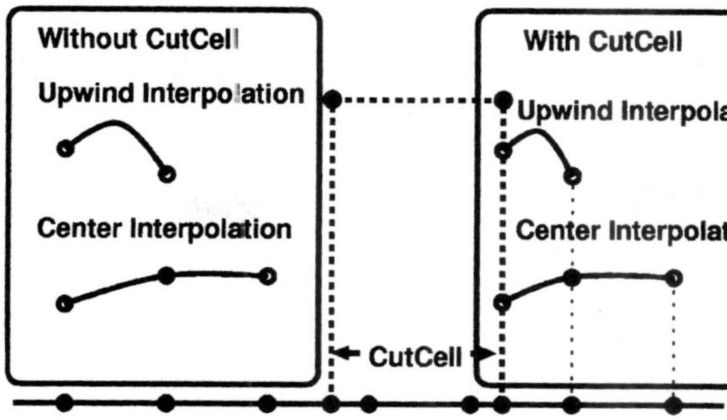

Fig. 2. Interpolation with the Cut Cell.
When the Cut Cell comes close to Euler grid, the interpolation will be constructed on irregular interval grid.

produce wrong value. This reason is quite simple that the estimated value within the outside of profile will be the value of outside, however, the region belonged to at $t = t^n$ and at $t = t^{n+1}$ is different on such Euler grids. For these grids, we use 3rd-order interpolation after time integrating on Euler grids and Cut Cell grids.

The position of discontinuity (Cut Cell) is estimated by integrating the velocity on it as Lagrange scheme;

$$x^{n+1} = x^n + \int u dt. \tag{2}$$

The Partial Differential Equation describing the phenomena on discontinuity will be rewritten in Lagrange scheme.

3 Numerical Examples

For a simple example which has discontinuities, we solve the 1-dimensional conservation equation ($f_t + uf_x + u_x f = 0$). By solving this prognostic partial differential equation, physical value calculation on the Movable Cut Cell is checked the accuracy. The advection velocity is set as $u(x) = 1/(1+0.5\sin(2\pi x))$ on the calculation space $0 \le x \le 2, \Delta x = 2/100$. An initial profile $f = 1 (0.4 \le x \le 0.6)$ and $f = 0$ (other points) is calculated by $t = 0.28$. The numerical result solved by the IDO scheme shown in Fig. 3(a) shows some numerical diffusion. On the contrary, the numerical result solved with Movable Cut Cell agrees with the analytical solution shown in Fig. 3(b). From this result, we can conclude that the Movable Cut Cell method shows good effect for 1-dimensional discontinuity.

From the view point of free surface generation, it is very useful to solve different matters which has completely different velocity on the same grid(Fig. 4). The calculation conditions are set as $\Delta x = 2/100 (0 \le x \le 2), \Delta t = 0.1 \times \Delta x$. In the left region velocity is given as $u(x) = -1/(1 + 0.5\sin(2\pi x))$, and in the

(a)

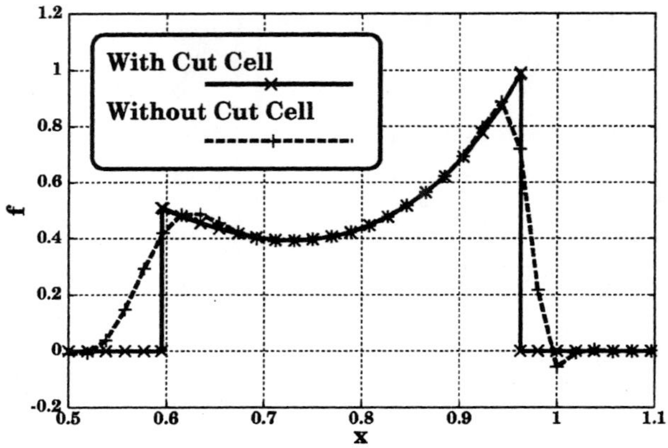

(b)

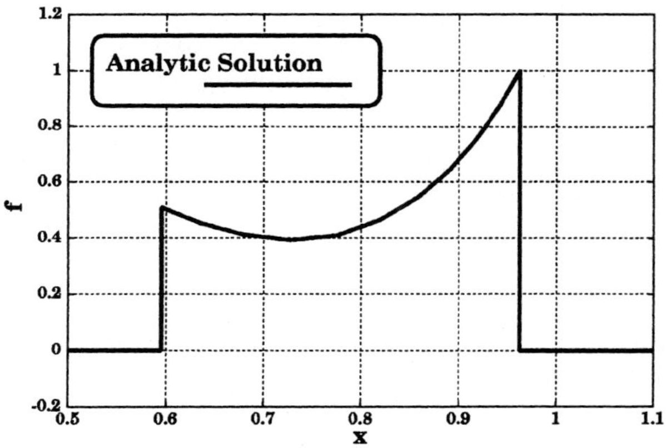

Fig. 3. Numerical result of 1-dimensional conservation equation:
On the calculation space $0 \leq x \leq 2, \Delta x = 2/100$, the advection velocity u is defined as $u(x) = 1/(1+0.5\sin(2\pi x))$. An initial profile ($f = 1(0.4 \leq x \leq 0.6)$ and $f = 0(otherpoints)$) moves and shows some diffusive error without Movable Cut Cell. On the other hand, the profile calculated with Movable Cut Cell solves the discontinuity accurately.

right $u(x) = 1/(1+0.5\cos(2\pi x))$. The initial profile is given as solid line in Fig. 4. As shown in Fig. 4, we can estimate the discontinuity accurately even if the distance of the interface is less than cell size (Δx).

This method is extended straightforwardly to 2-Dimensional case, and solves 2-dimensional advection equation and 2-dimensional conservation equation. For each problem, the caulclation area; $-1 \leq x \leq 1, -1 \leq y \leq 1, \Delta x = 2/100$, $\Delta y = 2/100$ is used. As an advection velocity, $u(x,y) = y, v(x,y) = -x$ and $u(x,y) = \frac{y}{1-0.2\cos\theta}, v(x,y) = -\frac{x}{1-0.2\cos\theta}$ are adopted for solving advection

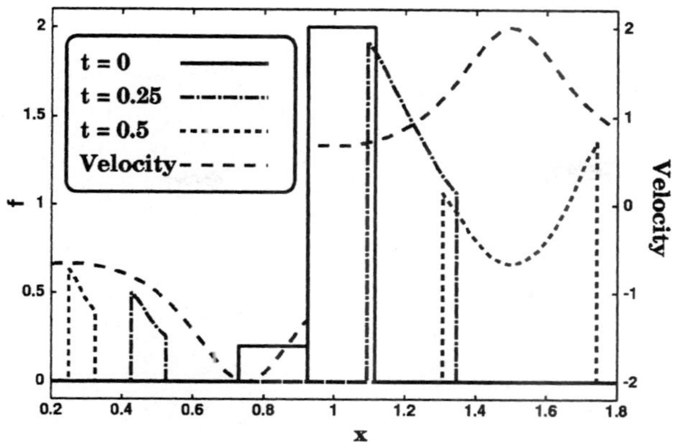

Fig. 4. By using the Cut Cell method two matters which have different velocity space can be solved accurately.

equation and conservation equation respectively, where $\sin(\theta) = x/\sqrt{x^2+y^2}$. The initial profile shown in Fig. 5(a) is set as; f = 1 inside of the circle centerd (0, 0.5) and its radius 0.3 (= $15\Delta x$), and f = 0 outside of it.

The numerical result calculated by one complete revolution of initial profile given from the 2-Dimensional advection equation is shown in Fig. 5(b). The position of the surface has some error in the order of numerical diffusion concerning the advection of Level Set function, however, we can treat the discontinuity correctly.

The numerical results at $t = 1.6, 3.52, 6.272$ with $\Delta t = 0.02\Delta x$ are shown in Fig. 5(c). In this case, the surface of the profile is deformed while calculation is continued and traced accurately. The numerical result still has some error, however, the direction for multi-dimensional case may be cleared. We can say that straightforwarding the 2-dimensional approach, multi-dimensional one can be constructed easily.

Acknowledgement

The simulations in this paper were done by Silicon Graphics Systems Origin 2000 of the Computer Center at Tokyo Institute of Technology. This work was supported in part by Grant-in-Aid for Scientific Research of the Japanese Ministry of Education.

References

1. T. Aoki: Comp.Phys.Comm.,**102** (1996)
2. K. Sakurai and T. Aoki: Comp. Fluids Dynamics Journal, **8**, 1 (1999)
3. J.J.Quirk: Comp. Fluids, **23**, 1 (1994)
4. T. Aoki: Comp. Fluids Dynamics Journal, **4**, 3 (1995)

(a)

(b)

(c)

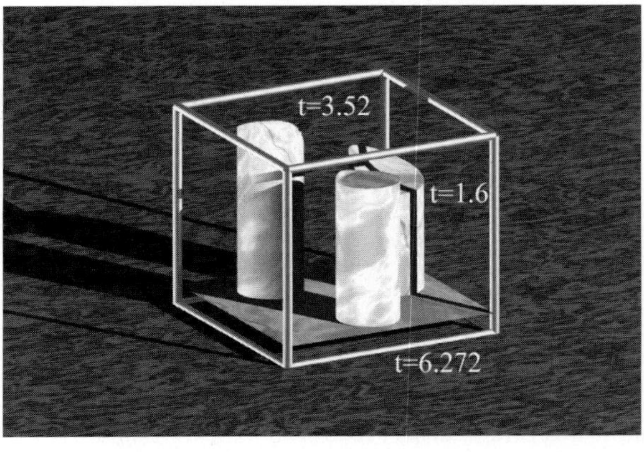

Fig. 5. Numerical results solved by using CutCell Method for describing the discontinuous surface in 2-Dimensional Case.

Numerical Simulations of Curtain Coating with a Moving Contact Line

M.C.T. Wilson[1], J.L. Summers[1], P.H. Gaskell[1], Y.D. Shikhmurzaev[2], T.D. Blake[3], and A. Clarke[3]

[1] The Engineering Fluid Mechanics Research Group, School of Mechanical Engineering, University of Leeds, Leeds, LS2 9JT, UK
[2] School of Mathematics and Statistics, University of Birmingham, Edgbaston, Birmingham B15 2TT, UK
[3] Kodak European Research, Kodak Limited, Headstone Drive, Harrow, HA1 4TY, UK

1 Introduction

Curtain coating is a high-speed manufacturing process for the simultaneous deposition of one or more thin liquid films onto a variety of surfaces. As its name implies, a thin sheet of liquid falls under gravity and impinges on the dry moving (solid) substrate being coated. The speed of the process (and hence productivity) is, however, limited by the onset of air entrainment at the point of contact between the liquid and the solid. In studying the conditions under which this disruptive instability occurs, Blake *et al.* [1] discovered that changing the flow rate or curtain height can increase the value of the substrate speed at which air is entrained. An additional result, confirmed subsequently [2], was that the dynamic contact angle, θ_d, formed between the liquid-gas and liquid-solid interfaces also depends on these parameters.

These observations raise an interesting question regarding CFD simulations of the flow, since they usually rely on a value for θ_d being specified to close the problem — see for example [3]. The work presented here attempts to explain the experimental results in [2] by *(i)* generating CFD (Galerkin Finite Element) simulations of the flow and thus establishing the motion of the fluid in the heel region of the curtain coater, see figure 1, and *(ii)* incorporating a more predictive wetting line model. For the latter, there exist many models in the open literature (e.g. [4],[5],[6]), however in this paper a small capillary number asymptotic solution to the theory of hydrodynamic wetting proposed by Shikhmurzaev [7] is applied. This model enables the value of θ_d to vary as a function of the capillary number, $Ca = \frac{\mu U_{sub}}{\sigma}$ (where U_{sub} is the speed of the substrate, μ and σ are respectively the viscosity and surface tension of the liquid) and other properties that characterise the liquid, but this approach cannot explain how the contact angle varies with flow rate, Q. If however, as is shown below, the free surface fluid velocity in the vicinity of the wetting line is coupled to Shikhmurzaev's asymptotic solution, information from the bulk flow is able to interact with the asymptotics, thus causing the dynamic contact angle to become a function of the flow rate or even the curtain height.

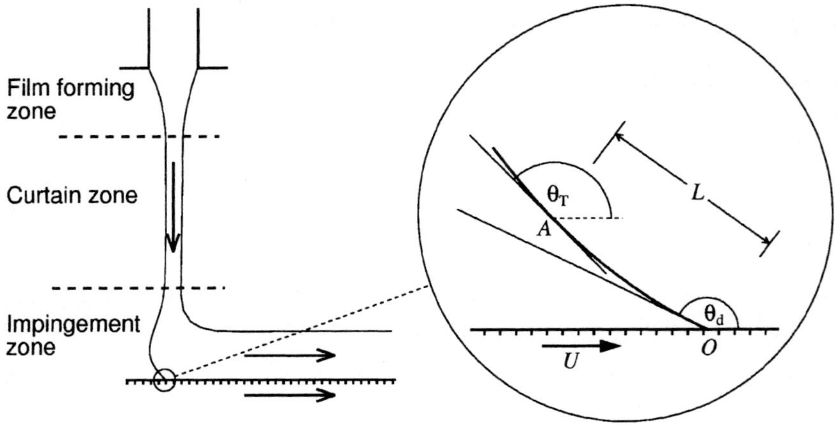

Fig. 1. Full solution domain

2 Solution Method

The steady Navier-Stokes equation governs the motion of the fluid (of density ρ) and its pressure within the curtain coater, and is written in nondimensional form as:

$$Re(\mathbf{u}.\nabla)\mathbf{u} = \nabla.\underline{\underline{\sigma}} + St\mathbf{f} \tag{1a}$$

$$\nabla.\mathbf{u} = 0 \tag{1b}$$

where $\underline{\underline{\sigma}} = -p\underline{\underline{I}} + (\nabla\mathbf{u}) + (\nabla\mathbf{u})^{\mathbf{T}}$ is the stress tensor, Re is the Reynolds number $(\frac{\rho Q}{\mu})$, St is the Stokes number $(\frac{\rho g Q^2}{\mu U_{sub}^3})$ and g is the acceleration due to gravity. The solution to (1a) and (1b) is subject to the following free surface boundary conditions, written as:

$$\mathbf{u}.\hat{\mathbf{n}} = 0 \tag{2a}$$

$$\underline{\underline{\sigma}}.\hat{\mathbf{n}} = \frac{\kappa}{Ca}\hat{\mathbf{n}} \tag{2b}$$

where κ is the free surface curvature. The Navier condition is applied as the substrate boundary condition and is written as:

$$\sigma_{nt} = \bar{\beta}(\mathbf{u}.\hat{\mathbf{t}} - 1) \tag{3}$$

where $\bar{\beta}$ is the non-dimensional coefficient of sliding friction defined by $\bar{\beta} = \frac{\beta Q}{\mu U_{sub}}$. Equation (3) is applied along the entire length of the substrate boundary and, apart from close to the wetting line, corresponds to the classical no-slip condition. Inflow and outflow boundaries have their velocity prescribed respectively by a Poiseuille and fully developed profile.

Finally, two conditions are required to determine the location of the dynamic wetting line and its associated contact angle. The definition of the dynamic contact angle yields one of the two equations, namely $\hat{\mathbf{t}}_{\mathbf{fs}}.\hat{\mathbf{t}}_{\mathbf{sub}} + \cos\theta_{\mathbf{d}} = \mathbf{0}$, where $\hat{\mathbf{t}}_{\mathbf{fs}}$ and $\hat{\mathbf{t}}_{\mathbf{sub}}$ are respectively the tangents to the free surface and the substrate at the contact line. The second condition is based on the tangential projection of forces acting on the contact line by Young's equation, rewritten in terms of the asymptotic solution of the full wetting line model [7]. The equation takes the form:

$$1 + \rho_{1e}^{s}\mathbf{u}|_{r=Ca\bar{\tau}} + (1 - \rho_{1e}^{s})\cos\theta_s - \frac{v + \sqrt{v^2+1}}{2v}\cos\theta_d = 0 \tag{4}$$

where ρ_{1e}^{s} is the nondimensional surface density of the liquid-gas interface, $\bar{\tau} = \frac{\tau\sigma U_{sub}}{\mu Q}$ is the dimensionless relaxation time of the surface tension, θ_s the static contact angle and $v = Ca\sqrt{\bar{\tau}\beta(1-\rho_{1e}^{s})/12}$ the dimensionless contact line speed.

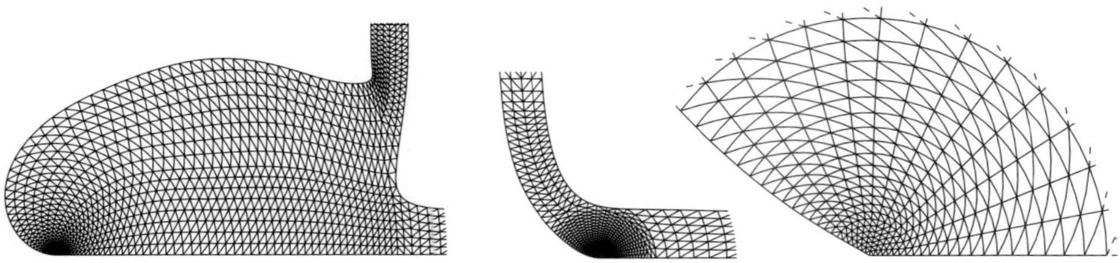

Fig. 2. Part of the finite element mesh.

The fluid domain of the curtain coater is bounded almost entirely by free surfaces, the positions of which are unknown *a priori* and governed to a great extent by the conditions at the contact line. The numerical method employed here is finite element analysis which has the topological flexibility to cope with the extreme situations of either the curtain being pulled by the substrate (starved flow condition) or the occurrence of a large heel (flooded flow condition) as indicated in figure 2.

The free-surface parametrisation is based on the spine method of Kistler and Scriven [8] and uses a combination of inter-dependent linear [9] and circular spines. The mesh in the heel region hinges on the *heel* spine that emanates from the wetting line and intersects the upper free surface at right-angles. The kinematic condition applied at the upper free surface together with the condition on the orientation are sufficient to establish both its length and angle to the horizontal. The radii of the circular spines are determined by distributing nodes exponentially along this heel spine, so that nodes are closer together near the wetting line. The arc length between the substrate and the free surface is then found for each spine by solving the associated kinematic condition.

3 Results

Experimental results by Clarke [10] showed that the liquid heel of the curtain coating process experiences large changes in its size when either the inlet flux (flow rate) or the substrate speed are changed. These changes have a profound effect on the flow within the heel; in particular, recirculation flow arises through flow bifurcations. In a sequence of experiments where hydrogen bubbles were generated in the flow by electrolysis, Clarke was able to demonstrate that for large inlet fluxes and low substrate speed, the heel becomes very large allowing for the generation of closed recirculating regions together with a separatrix in the flow enabling fluid to be transported into the wetting line region.

Examples of CFD simulations throughout *inlet flux, substrate speed* control space are shown in figure 3, where different flow regimes are identified. At high flow rates the liquid heel forms, usually with two recirculations, one bound to the free surface and the other within the flow. In this situation when the substrate speed is increased, the heel diminishes in size together with the internal recirculation while the other recirculation grows slightly, see top two flow patterns in figure 3. At the low flow rate end, the curtain is actually pulled on increasing the substrate speed. Under these conditions the curtain becomes unstable and a small reduction in flow rate will cause the curtain to break up [12].

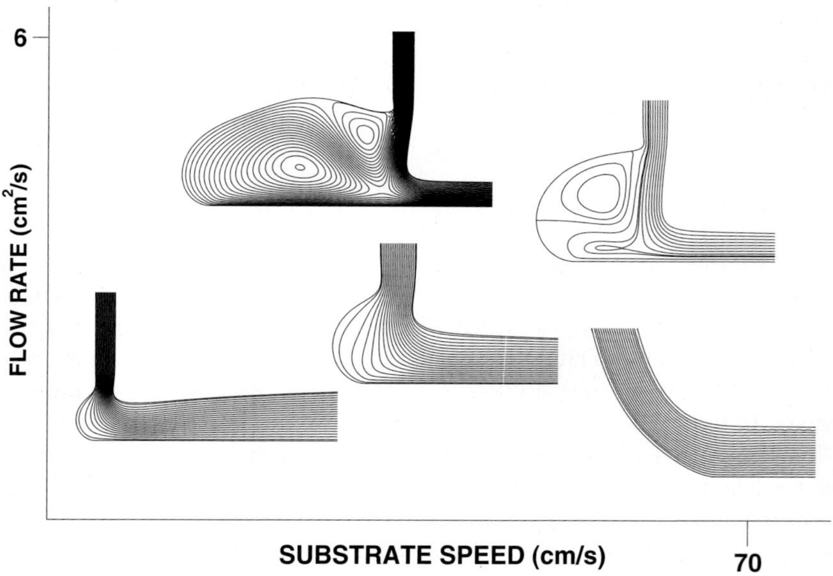

Fig. 3. Flow regimes of curtain coating. (solutions not shown to same scale)

Blake *et al.* [2], demonstrated that in curtain coating the dynamic contact angle varies not only with substrate speed but also with other (independent) parameters such as flow rate and curtain height. An attempt to explain this variation is made by accounting for the bending of the meniscus close to the contact line by measuring an 'apparent' contact angle at a certain resolution from the contact line – see [11] for more detail. Here the apparent contact angle, θ_T, is taken to be the tangent to the free surface at a distance L (set to the experimental resolution of $20\mu m$) from the contact line – see right diagram in figure 1. The four broken curves in figure 4 show the apparent contact angle variation with flow rate at a fixed substrate speed (i.e. fixed Ca) for four different values of the Navier slip coefficient, β, and in all cases the angle is matched with experiments at the high flow rate by varying the value of the imposed contact angle, θ_d. It is clear that even when β is large the bending of the meniscus alone is insufficient to explain the observed contact angle profile (shown in figure 4 as $+++$).

The coupling of the asymptotics with the numerical model involves a point at which the two are matched and in this case it is a point on the free surface a distance $Ca\bar{\tau}$ from the wetting line. Since the velocity at the matching point varies with flow rate, the value of θ_d will also vary accordingly. In this case the apparent contact angle is again measured at $20\mu m$ from the contact line, and matched to the experimental value at high flow rate by varying $\bar{\tau}$ and two other key parameters in the asymptotics, namely the interfacial density and thickness which in turn fixes the Navier slip coefficient. The results are shown in figure 4 as the solid line. The asymptotic solution to the theory proposed by Shikhmurzaev, when coupled to the numerical model for curtain coating, does offer a better explanation than just the local bending of the meniscus, but there is clearly still a discrepancy to be explained. Work is currently underway to incorporate the full model into the finite element simulations.

4 Conclusions

In curtain coating at high flow rates a heel appears together with recirculating flow close to the dynamic contact line. In this situation the recirculating zone tends to decouple the bulk hydrodynamics, such as flow rate, from the dynamic contact line region.

In modelling fluid flow problems that involve dynamic contact lines, it is insufficient to impose the contact angle arbitrarily. Additionally, it appears that wetting line models that depend only on Ca are also incomplete in that they cannot model the dynamic contact angle variations with other parameters, such as flow rate. While the small Ca coupled asymptotics applied to curtain coating in this paper do give an improved comparison with experiments, the large discrepancy may be because Ca ($= 0.27$) is not sufficiently small for the asymptotics to be accurate. The full model proposed by Shikhmurzaev is not restricted to small capillary number and may predict the observed experimental behaviour.

References

1. T.D. Blake, A. Clarke, and K.J. Ruschak: AIChE J. **40**, 229 (1994)
2. T.D. Blake, M. Bracke, and Y.D. Shikhmurzaev: Phys. Fluids **11**, 1995 (1999)
3. S.F. Kistler: The Fluid Mechanics of Curtain Coatings and Related Viscous Free Surface Flows with Contact Lines. PhD Thesis, University of Minnesota (1983)

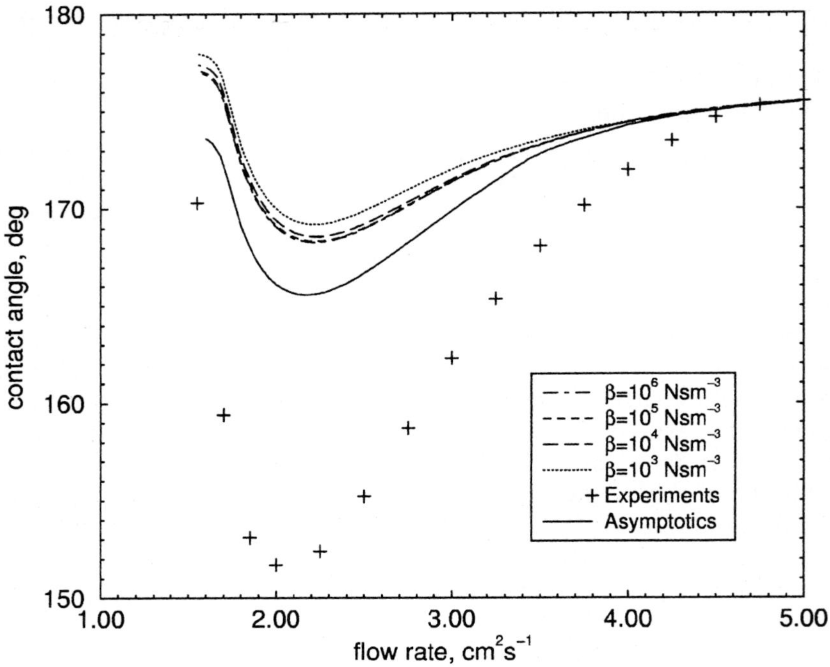

Fig. 4. Variation of apparent contact angle with flow rate for a fixed substrate speed of $70cms^{-1}$.

4. T.D. Blake: 'Dynamic contact angles and wetting kinetics'. In *Wettability.* ed. by J.C. Berg (Marcel Dekker Publ. 1983) pp. 252–309
5. E.B. Dussan V., S.H. Davis: J. Fluid Mech. **65**, 71 (1974)
6. L.M. Hocking: J. Fluid Mech. **76**, 801 (1976)
7. Y.D. Shikhmurzaev: Int. J. Multiphase Flow **19**, 589 (1993)
8. S.F. Kistler, L.E. Scriven: 'Coating flows'. In: *Computational Analysis of Polymer Processing.* ed. by J.R.A. Pearson, S.M. Richardson (Appl. Science Publishers, 1983) pp. 243–299
9. P.H. Gaskell, M.D. Savage, J.L. Summers, H.M. Thompson: 'Flow topology and transformation in fixed-gap symmetric forward roll coating systems'. In: *Proc. 9th Intl. Conf. Numerical Methods in Laminar and Turbulent Flows, Atlanta, July 1995,* ed. by C. Taylor, P. Durbetaki (Pineridge Press, 1995) pp. 984–995
10. A. Clarke: 'Recirculating flows in curtain coating'. In: *The mechanics of thin film coatings, Ist European Coating Symposium, Leeds, September 1995,* ed. by P.H. Gaskell, M.D. Savage, J.L. Summers (World Scientific, Singapore, 1996) pp. 32–41
11. M.C.T. Wilson, J.L. Summers, Y.D. Shikhmurzaev: 'Hydrodynamic assist of wetting: theoretical results'. In: *Advances in coating and drying of thin flims, 3rd European Coating Symposium, Erlangen, September 1999,* ed. by F. Durst, H. Raszillier (1999) pp. 75–80
12. K. Miyamoto, Y. Katagiri: 'Curtain coating'. In: *Liquid film coating.* ed. by S.F. Kistler, P.M. Schweizer (Chapman and Hall 1987) pp. 463–494

Numerical Solution of Steady Free-Surface Navier-Stokes Flow

E.H. van Brummelen[1], H.C. Raven[2], and B. Koren[1]

[1] CWI, P.O. Box 94079, NL-1098 GB Amsterdam, The Netherlands
[2] MARIN, P.O. Box 28, NL-6700 AA Wageningen, The Netherlands

Abstract. The usual time integration approach for solving steady viscous free-surface flow problems has several drawbacks. We propose an efficient iterative method, which relies on a so-called quasi free-surface condition. It is shown that the method displays asymptotically mesh-width independent convergence behavior. Numerical results for flow over an obstacle in a channel are presented. The results confirm mesh-width independence of the convergence behavior. Comparison of the numerical results with measurements shows good agreement.

1 Introduction

The numerical solution of flows which are partially bounded by a freely moving boundary is of great practical importance, e.g., in ship hydrodynamics. Whereas dedicated techniques have been developed for the solution of steady free surface potential flow [1,2], methods for steady Navier-Stokes flow simply continue the usual time integration process until a steady state is reached. In [3], several drawbacks of this approach are discussed. One can show that at sub-critical Froude numbers, the asymptotic temporal behavior of transient gravity waves in $\mathbb{R}^d$ is $O(t^{(1-d)/2})$. Moreover, due to separate treatment of the Navier-Stokes equations and the kinematic boundary condition the time-step is restricted by a CFL-condition. Hence, if the objective is to reduce the amplitude of transient waves to the order of spatial discretization errors, the efficiency of the time integration approach deteriorates rapidly with decreasing mesh-width. In practical computations, thousands of time steps are usually required, making the transient approach prohibitively expensive in actual design processes.

In the present work an efficient iterative solution method is proposed to reduce the computational cost of solving the steady free-surface flow problem. The method solves a sequence of steady Navier-Stokes sub-problems with a so-called quasi free-surface condition imposed at the free surface. This condition ensures that the disturbance induced by the subsequent displacement of the boundary is negligible. Each sub-problem evaluation then yields an improved approximation to the steady free surface position. It is shown that the convergence behavior of the method is asymptotically mesh-width independent.

Numerical experiments are performed and results are presented for flow over an obstacle in a channel. The convergence behavior of the iterative method is examined for different test-cases. Numerical results are compared with measurements.

2 Problem Statement

We consider an incompressible, viscous fluid flow, subject to a constant gravitational force on a domain $\mathcal{V} \subset \mathbb{R}^d$ $(d = 2, 3)$. The domain is bounded by a free boundary, $\mathcal{S}$, and a fixed boundary $\partial\mathcal{V} \setminus \mathcal{S}$. The distinguishing parameters of the problem are the Froude number, Fr, and the Reynolds number, Re.

The (non-dimensionalized) fluid velocity and pressure are identified by $\boldsymbol{v}(\boldsymbol{x})$ and $p(\boldsymbol{x})$, respectively. The objective is to find $\mathcal{S}, \boldsymbol{v}(\boldsymbol{x})$ and $p(\boldsymbol{x})$ such that the steady incompressible Navier-Stokes equations are satisfied on $\mathcal{V}$,

$$\operatorname{div} \boldsymbol{v}\boldsymbol{v} + \boldsymbol{\nabla} p - \operatorname{div} \boldsymbol{\tau}(\boldsymbol{v}) = -\mathrm{Fr}^{-2} \boldsymbol{e}_d \,, \qquad \boldsymbol{x} \in \mathcal{V} \,, \tag{1a}$$

$$\operatorname{div} \boldsymbol{v} = 0 \,, \qquad \boldsymbol{x} \in \mathcal{V} \,, \tag{1b}$$

the free-surface conditions hold at $\mathcal{S}$,

$$\boldsymbol{t}^\alpha \cdot \boldsymbol{\tau}(\boldsymbol{v}) \cdot \boldsymbol{n} = 0 \,, \qquad \boldsymbol{x} \in \mathcal{S} \,, \tag{2a}$$

$$p = 0 \,, \qquad \boldsymbol{x} \in \mathcal{S} \,, \tag{2b}$$

$$\boldsymbol{v} \cdot \boldsymbol{n} = 0 \,, \qquad \boldsymbol{x} \in \mathcal{S} \,, \tag{2c}$$

and appropriate boundary conditions apply at fixed boundaries, denoted by

$$\boldsymbol{B}(\boldsymbol{v}, p) = \boldsymbol{b}(\boldsymbol{x}) \,, \qquad \boldsymbol{x} \in \partial\mathcal{V} \setminus \mathcal{S} \,. \tag{3}$$

In (1) to (3), $\boldsymbol{\tau}(\boldsymbol{v}) = \mathrm{Re}^{-1} \left[(\boldsymbol{\nabla v}) + (\boldsymbol{\nabla v})^T\right]$ is the viscous stress tensor for a Newtonian incompressible fluid, $\boldsymbol{e}_d$ is the vertical unit-vector, $\boldsymbol{n}(\boldsymbol{x})$ is the unit normal vector to $\mathcal{S}$ and $\boldsymbol{t}^\alpha(\boldsymbol{x})$ $(\alpha = 1, \ldots, d-1)$ are orthogonal tangential unit vectors to $\mathcal{S}$. Condition (2c) is called the kinematic condition. Conditions (2a) and (2b) are referred to as the tangential- and normal dynamic conditions, respectively. In the derivation of the normal dynamic condition (2b), viscous effects have been ignored.

3 Iterative Solution Method

The number of free-surface conditions is one more than the number of boundary conditions required by the incompressible Navier-Stokes equations. Iterative methods for solving free-surface flow problems typically proceed in two steps: *i)* Solve the Navier-Stokes equations with all but one free-surface conditions imposed. *ii)* Use the solution to adjust the free boundary position, such that the remaining free-surface condition is satisfied as closely as possible.

A fundamental problem in solving free-surface flow problems is that the change in the boundary position induces a disturbance in the solution of *i)* which can, in turn, spoil the approximation of the free-surface position obtained from *ii)*. An efficient iterative method for solving free-surface flow problems must control this induced disturbance.

The induced disturbance can be controlled by a suitable choice of the boundary conditions at the free surface. These conditions will be derived next. Let

$\mathcal{O}$ denote the space of admissible domains for the free-surface flow problem. If $\mathcal{V} \in \mathcal{O}$ is sufficiently regular, then for a smooth function $\lambda(\boldsymbol{x})$ on $\mathcal{S}$,

$$\mathcal{S}_{\varepsilon\lambda} = \{\boldsymbol{x} + \varepsilon\lambda(\boldsymbol{x})\boldsymbol{e}_d : \boldsymbol{x} \in \mathcal{S}\} , \tag{4}$$

is the boundary of a nearby domain $\mathcal{V}_{\varepsilon\lambda} \in \mathcal{O}$. Suppose that $p(\boldsymbol{x})$ can be extended smoothly beyond the boundary $\mathcal{S}$, so that it is well-defined on $\mathcal{S}_{\varepsilon\lambda}$. If (2b) holds at $\mathcal{S}_{\varepsilon\lambda}$, Taylor-series expansion of $p(\boldsymbol{x})$ in the neighborhood of $\mathcal{S}$ can be used to estimate the position of $\mathcal{S}_{\varepsilon\lambda}$:

$$\varepsilon\lambda(\boldsymbol{x}) = \frac{-p(\boldsymbol{x})}{\boldsymbol{e}_d \cdot \boldsymbol{\nabla} p(\boldsymbol{x})} - O(\varepsilon^2) = \mathrm{Fr}^2 p(\boldsymbol{x}) + O(\varepsilon\delta, \varepsilon^2) , \qquad \boldsymbol{x} \in \mathcal{S} , \tag{5}$$

with $\delta \equiv \|1 + \mathrm{Fr}^2 \boldsymbol{e}_d \cdot \boldsymbol{\nabla} p\|_{\infty,\mathcal{S}}$ assumed to be $\ll 1$.

Subsequently, we obtain conditions at $\mathcal{S}$ if (2a) and (2c) hold at $\mathcal{S}_{\varepsilon\lambda}$, with $\varepsilon\lambda(\boldsymbol{x})$ by (5). Let $\boldsymbol{t}^{\alpha}_{\varepsilon\lambda}(\boldsymbol{x})$ and $\boldsymbol{n}_{\varepsilon\lambda}(\boldsymbol{x})$ denote the tangential and normal vectors to $\mathcal{S}_{\varepsilon\lambda}$, respectively. Assuming that $\mathcal{S}$ is a smooth approximation to $\mathcal{S}_{\varepsilon\lambda}$, so that $\boldsymbol{t}^{\alpha} \cdot \boldsymbol{n}_{\varepsilon\lambda} = O(\varepsilon)$, and ignoring terms $O(\varepsilon^2, \varepsilon\delta)$,

$$\boldsymbol{t}^{\alpha}_{\varepsilon\lambda}(\boldsymbol{x}\prime) \;/\!/\; \boldsymbol{t}^{\alpha}(\boldsymbol{x}) + \boldsymbol{t}^{\alpha}(\boldsymbol{x}) \cdot \boldsymbol{\nabla}\mathrm{Fr}^2 p(\boldsymbol{x})\, \boldsymbol{e}_d , \qquad \boldsymbol{x} \in \mathcal{S} , \tag{6a}$$

$$\boldsymbol{n}_{\varepsilon\lambda}(\boldsymbol{x}\prime) \;/\!/\; \boldsymbol{\nabla} p(\boldsymbol{x}) , \qquad \boldsymbol{x} \in \mathcal{S} , \tag{6b}$$

with $\boldsymbol{x}\prime(\boldsymbol{x}) \equiv \boldsymbol{x} + \mathrm{Fr}^2 p(\boldsymbol{x})\, \boldsymbol{e}_d$. Therefore, if $\boldsymbol{v}(\boldsymbol{x})$ satisfies (2c) at $\mathcal{S}_{\varepsilon\lambda}$,

$$\boldsymbol{v} \cdot \boldsymbol{\nabla} p = 0 + O(\varepsilon\sigma) , \qquad \boldsymbol{x} \in \mathcal{S} , \tag{7}$$

with $\sigma \equiv \max\{\|\boldsymbol{w}\cdot\boldsymbol{\nabla}\boldsymbol{v}\|_2/\|\boldsymbol{v}\|_2 : \|\boldsymbol{w}\|_2 = 1, \boldsymbol{x} \in \mathcal{S}\}$ assumed to be $\ll 1$. Moreover,

$$[\boldsymbol{t}^{\alpha}_{\varepsilon\lambda} \cdot \boldsymbol{\tau}(\boldsymbol{v}) \cdot \boldsymbol{n}_{\varepsilon\lambda}](\boldsymbol{x}\prime) = [\boldsymbol{t}^{\alpha} \cdot \boldsymbol{\tau}(\boldsymbol{v}) \cdot \boldsymbol{n}](\boldsymbol{x}) + O(\varepsilon\sigma) , \qquad \boldsymbol{x} \in \mathcal{S} . \tag{8}$$

Hence, (2a) and (7) specify the conditions at $\mathcal{S}$ to $O(\varepsilon^2, \varepsilon\delta, \varepsilon\sigma)$ if (2a) and (2c) are fulfilled at $\mathcal{S}_{\varepsilon\lambda}$.

The conditions (2a) and (7) can be used to set up an iterative solution method for the free-surface flow problem. Let $\boldsymbol{v}, p$ and $\boldsymbol{v}_{\varepsilon\lambda}, p_{\varepsilon\lambda}$ denote the solutions of the boundary value problem (1), (2a), (7), (3) on domains $\mathcal{V}$ and $\mathcal{V}_{\varepsilon\lambda}$, respectively. It holds that

$$p_{\varepsilon\lambda}(\boldsymbol{x}\prime) = p(\boldsymbol{x}) + \mathrm{Fr}^2 p(\boldsymbol{x})\, \boldsymbol{e}_d \cdot \boldsymbol{\nabla} p(\boldsymbol{x}) + p\prime_{\varepsilon\lambda}(\boldsymbol{x}) + O(\varepsilon^2) , \qquad \boldsymbol{x} \in \mathcal{S} , \tag{9}$$

with $p\prime_{\varepsilon\lambda}(\boldsymbol{x})$ the *induced disturbance* in $p(\boldsymbol{x})$ due to the displacement of the boundary, defined by

$$p\prime_{\varepsilon\lambda}(\boldsymbol{x}) = p_{\varepsilon\lambda}(\boldsymbol{x}) - p(\boldsymbol{x}) , \qquad \boldsymbol{x} \in \mathcal{V}_{\varepsilon\lambda} \cup \partial\mathcal{V}_{\varepsilon\lambda} \cup \mathcal{V} \cup \partial\mathcal{V} . \tag{10}$$

Introducing the *contraction number* $\zeta \equiv \|p_{\varepsilon\lambda}\|_{\mathcal{S}_{\varepsilon\lambda}}/\|p\|_{\mathcal{S}}$, it follows from (9) that

$$\zeta \leq \|1 + \mathrm{Fr}^2 \boldsymbol{e}_d \cdot \boldsymbol{\nabla} p\|_{\mathcal{S}} + \|p\prime_{\varepsilon\lambda}\|_{\mathcal{S}}/\|p\|_{\mathcal{S}} + O(\varepsilon) \leq \delta + \|p\prime_{\varepsilon\lambda}\|_{\mathcal{S}}/\|p\|_{\mathcal{S}} + O(\varepsilon) . \tag{11}$$

By (6b) to (8), $\boldsymbol{v}, p$ complies to $O(\varepsilon^2, \varepsilon\delta, \varepsilon\sigma)$ with (2a) and (7) at $\mathcal{S}_{\varepsilon\lambda}$. Therefore, the difference between $\boldsymbol{v}_{\varepsilon\lambda}, p_{\varepsilon\lambda}$ and $\boldsymbol{v}, p$ is only $O(\varepsilon^2, \varepsilon\delta, \varepsilon\sigma)$. The ratio

$||p'_{\varepsilon\lambda}||_{\mathcal{S}}/||p||_{\mathcal{S}}$ is then $O(\varepsilon, \delta, \sigma)$ and, by (11), $\zeta = O(\varepsilon, \delta, \sigma)$. Hence, modifying the boundary from $\mathcal{S}$ to $\mathcal{S}_{\varepsilon\lambda}$ effectively improves the approximation to the actual free-boundary position. The free-surface flow problem can thus be solved by iterating the following operations:

i) Solve the boundary value problem (1), (2a), (7), (3) for $\boldsymbol{v}(\boldsymbol{x}), p(\boldsymbol{x})$,
ii) Adjust the boundary position by (4), (5) with $p(\boldsymbol{x})$ from *i*).

This approach is comparable to methods for solving steady free-surface potential flow, e.g., [2].

Equation (11) provides an upper bound for the contraction number of the iteration. One may note that (11) depends on properties of the continuum solution only. Therefore, if the free-surface flow problem is solved numerically, the convergence behavior of the iteration is asymptotically mesh-width independent.

The iterative method relies on condition (7) to ensure that the induced disturbance is negligible. This condition is called a *quasi free-surface condition*, because the boundary value problem with (7) imposed, displays similar behavior as the free-boundary problem. In [4] it is shown that for flow in a channel the Navier-Stokes equations subject to (7) allow gravity wave solutions that satisfy the usual dispersion relation.

4 Numerical Experiments & Results

The algorithm is tested for sub-critical flow over an obstacle in a channel at $\mathrm{Fr} = 0.43$ and $\mathrm{Re} = 1.5 \times 10^5$. The geometry of the obstacle is described in [1]. We consider obstacles with height $E = 0.15$ and $E = 0.20$. The experiments are performed on grids with horizontal mesh-widths $h = 2^{-5}, 2^{-6}$. The number of grid cells in the vertical direction is 70 and exponential grid stretching is applied to resolve the boundary layer at the bottom. The (Reynolds averaged) Navier-Stokes equations and the corresponding boundary conditions are solved by the method described in [5]. After each evaluation, the grid is adapted using relative vertical stretching. An initial estimate of the solution on the adapted grid is generated by linear interpolation from the solution on the previous grid. Details of the discretization method and the setup of the numerical experiments can be found in [6].

Figure 1 displays the pressure defect at the free surface after consecutive iterations, defined by $||p^n||_{\mathcal{S}} = \sum_j h_j |p_j^n| / \sum_k h_k$, with $n = 0, 1, \ldots$ the iteration number. The results confirm convergence of the method. For $E = 0.15$, the average contraction number is $\zeta \approx 0.15$. The convergence behavior is indeed independent of h. For $E = 0.20$, the average contraction number is $\zeta \approx 0.45$ for $h = 2^{-5}$ and $\zeta \approx 0.52$ for $h = 2^{-6}$. As a result of strong non-linearity, the asymptotic mesh-width independence of the convergence behavior is in this case not yet apparent.

Figure 2 compares the computed wave elevation with measurements from [1]. In [1], a non-dimensionalized wavelength $\lambda = 1.10 \pm 10\%$ and amplitude $a = 4.5 \times 10^{-2} \pm 15\%$ are reported for the trailing wave. The trailing wave of the

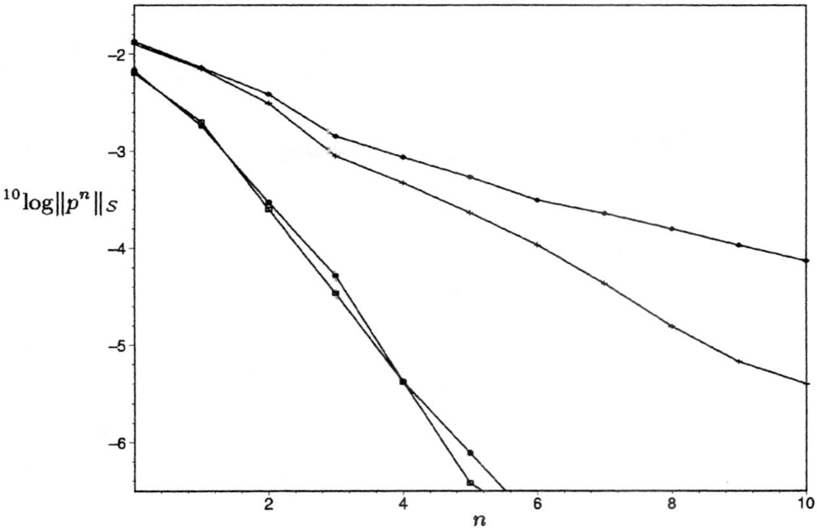

Fig. 1. Pressure defect at the free surface versus the iteration number for $E = 0.15$, $h = 2^{-5}$ (□), $h = 2^{-6}$ (○) and $E = 0.20$, $h = 2^{-5}$ (+), $h = 2^{-6}$ (◇).

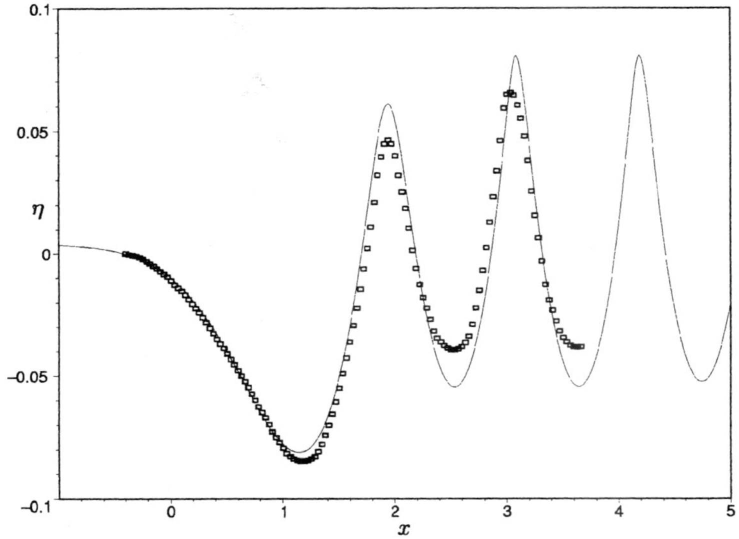

Fig. 2. Computed wave elevation for $h = 2^{-6}$ (*solid line*) and measurements from [1] (*markers only*), for $E = 0.20$. The obstacle is located in the interval $x \in [0, 2]$.

computed wave elevation on the grid with $h = 2^{-6}$ displays wavelength $\lambda = 1.11$ and amplitude $a = 6.5 \times 10^{-2}$. Clearly, the computed wavelength agrees well with the measurements. The amplitude is slightly overestimated.

5 Conclusions

An iterative method for the efficient numerical solution of steady viscous free-surface flow problems was presented. The method solves a sequence of steady Navier-Stokes sub-problems. After each sub-problem evaluation, the position of the free boundary is adjusted. The method relies on a quasi free-surface condition to control the disturbance induced by the displacement of the free boundary. An improved approximation to the free surface position is then obtained. It was shown that the convergence behavior of the method is asymptotically mesh-width independent.

Numerical results were presented for flow over an obstacle in a channel. The presented test-cases confirm mesh-width independent convergence behavior of the iterative method. The numerical results agree well with measurements. The results indicate that the method indeed permits efficient solution of steady free-surface Navier-Stokes flow problems.

References

1. J. Cahouet: Etude Numérique et Experimentale du Problème Bidimensionnel de la Résistance de Vagues Non-Linéaire. Ph.D. Thesis, ENSTA, Paris (1984), (In French)
2. H.C. Raven: A Solution Method for the Nonlinear Ship Wave Resistance Problem. Ph.D. Thesis, Delft University of Technology, Delft (1996)
3. H.C. Raven, E.H. van Brummelen: 'A New Approach to Computing Steady Free-Surface Viscous Flow Problems'. In: *1st MARNET-CFD Workshop, Spain, 1999* Available at `http://www.marin.nl/projects/cph_parnassos_720.html`
4. E.H. van Brummelen: Analysis of the Incompressible Navier-Stokes Equations with a Quasi Free-Surface Condition. Tech. Report MAS-R9922, CWI, Amsterdam (1999)
Available at `http://www.cwi.nl/ftp/CWIreports/MAS/MAS-R9922.ps.Z`
5. M. Hoekstra: Numerical Simulation of Ship Stern Flows with a Space-Marching Navier-Stokes Method. Ph.D. Thesis, Delft University of Technology, Delft (1999)
6. E.H. van Brummelen: Numerical Solution of Steady Free-Surface Navier-Stokes Flow. Tech. Report MAS-R0018, CWI, Amsterdam (2000)
Available at `http://www.cwi.nl/ftp/CWIreports/MAS/MAS-R0018.ps.Z`

Part XI

Numerical Techniques

A Navier-Stokes Solver Using Edge-Based Smoothing

J.J. Liu[1,2] and T.P. Hynes[1]

[1] Whittle Laboratory, University of Cambridge, Cambridge, UK
[2] **Current Address:** Institute of Engineering Thermophysics, Chinese Academy of Sciences, P.O. Box 2706, Beijing 100080, China

Abstract. A three dimensional Navier–Stokes solver based upon the one–step Lax–Wendroff scheme has been developed. The numerical smoothing term added to the basic scheme is derived by using the TVD approach and implemented in an edge-based manner in order to detect and eliminate numerical oscillations in multiple dimension. An algebraic turbulence model is applied to estimate turbulence viscosity. Multigrid algorithm is also incorporated to speed up the convergence to a steady state. Numerical results for subsonic flow over a flat plate and transonic flow in a 3D channel are presented.

1 Introduction

Numerical solution of the Navier–Stokes equations consists of discretization of the convection and diffusion terms. Second order central difference is commonly used for the diffusion terms and this is stable due to the dissipative nature of such terms. The boundary layer resolution of a Navier–Stokes solver mainly depends on the discretization method applied to the convection terms, or the way that numerical dissipation is introduced to stabilize the solution. There is a large variety of algorithms that can be used to discretize the convection terms, from simple central difference scheme to sophisticated high resolution TVD schemes. High resolution schemes with smart control on numerical dissipation are designed to cleanly capture flow discontinuities such as shock waves. Some of the schemes, like the one based on Roe's approximate Riemann solver[1], also resolve boundary layers well. In the other hand, central difference scheme is widely adopted owing to its simplicity in implementation, especially in multiple dimension. The artificial smoothing used in an ordinary central difference scheme, however, often gives spurious dissipation within boundary layers.

There is a connection between high resolution TVD schemes and central difference scheme. It has been shown by Roe[2] for a model equation that numerical flux of the TVD Lax–Wendroff scheme can be written as the flux of classic Lax–Wendroff scheme plus a numerical dissipation term, or equivalently the Lax–Wendroff scheme can be made TVD by adding a smoothing term derived by using the TVD approach. In the present work, this strategy has been extended to a three-dimensional Navier–Stokes solver using a cell–vertex scheme. As pointed out by Morton and Paisley[3], cell–vertex schemes have advantages over cell–center ones, such as less oscillation modes, higher accuracy at numerical boundaries and for skewed meshes. It is found in the present research that the smoothing for a cell–vertex scheme must apply in a manner of edge–based rather than cell–based, in order to allow for flux limiter effectively detecting and eliminating numerical oscillations. In the following sections the numerical method and typical results are described.

2 Numerical Method

The 3D Reynolds–averaged Navier–Stokes equations can be written in the following conservation form,

$$\frac{\partial U}{\partial t}+\frac{\partial F}{\partial x}+\frac{\partial G}{\partial y}+\frac{\partial H}{\partial z}=\frac{\partial F_v}{\partial x}+\frac{\partial G_v}{\partial y}+\frac{\partial H_v}{\partial z} \tag{1}$$

where t, x, y, z denote the time and the three Cartesian directions. U is the vector of conservative variables, F, G, H are the vectors of convection terms and F_v, G_v, H_v are the vectors of diffusion terms. With perfect gas assumption the static pressure is obtained. The detailed expressions of the NS equations can be found in literature.

The cell–vertex Lax–Wendroff scheme, with a smoothing term derived from the TVD approach and implemented by using Roe's wave–decomposition, is employed for the numerical solution of equation 1. Multigrid method[4] is also used to speed up the convergence to a steady state. The basic time–marching procedure on the fine mesh can be expressed as,

$$U^{n+1} = U^n + \delta U \tag{2}$$

$$\delta U = \delta U' + \delta U'' + \delta U_v' + \delta U^{TVD} \tag{3}$$

The first three terms on the right–hand side of equation 3 are the corrections calculated by using finite–volume discretization method. Among which, $\delta U'$ and $\delta U''$ are the first and second order corrections contributed by convection terms, $\delta U_v'$ is the first order correction contributed by diffusion terms. The 3D flow domain is discretized using a hexahedral mesh, i.e., each cell is formed by 8 vertices and 6 surfaces. The flow properties are stored at the cell vertices. The first order correction $\delta U'$ and the second order correction $\delta U''$ due to convection terms, and the first order correction $\delta U_v'$ due to diffusion terms are given by,

$$\Delta U_{cell} = -\frac{\Delta t_{cell}}{\Delta V_{cell}} \sum_{surf=1}^{6} \left(\widehat{F}S_x + \widehat{G}S_y + \widehat{H}S_z\right)_{surf} \tag{4}$$

$$\delta U_{node}' = \frac{\Delta t_{node}}{\Delta V_{node}'} \sum_{cell=1}^{N} \left(\frac{\Delta V_{cell}}{\Delta t_{cell}} \Delta U_{cell}\right)_{cell} \tag{5}$$

$$\delta U_{node}'' = -\frac{\Delta t_{node}}{\Delta V_{node}'} \sum_{cell=1}^{N} \left[\, \Delta F(S_x^i + S_x^j + S_x^k) + \Delta G(S_y^i + S_y^j + S_y^k) + \Delta H(S_z^i + S_z^j + S_z^k) \,\right]_{cell} \tag{6}$$

$$(\delta U'_v)_{node} = 2\frac{\Delta t_{node}}{\Delta V'_{node}} \sum_{cell=1}^{N} \left[F_v(S^i_x + S^j_x + S^k_x) + G_v(S^i_y + S^j_y + S^k_y) \right.$$
$$\left. + H_v(S^i_z + S^j_z + S^k_z)\right]_{cell} \tag{7}$$

Where ΔU_{cell}, Δt_{cell}, and ΔV_{cell} are first order change, time–step and volume of each cell, respectively. $\boldsymbol{S}$ is the outer area–vector normal to the cell surface. $\widehat{F}$, $\widehat{G}$, $\widehat{H}$ are the surface–averaged fluxes. N denotes the number of cells surrounding the given node. $\Delta t_{node} = min(\Delta t_{cell\ \ 1\ to\ N})$ is the time–step for the node. $\Delta V'_{node} = \sum_{cell=1}^{N}(\Delta V_{cell})_{cell}$ is the volume of primary control-volume formed by the cells surrounding the given node. $\boldsymbol{S}^i, \boldsymbol{S}^j, \boldsymbol{S}^k$ are the outer area–vector normal to the i–, j– and k–wise central surfaces of each cell. ΔF, ΔG, ΔH are the second order changes of each cell. ΔF, for example, is calculated by,

$$\Delta F = F(\bar{U} + \Delta U) - F(\bar{U}) \tag{8}$$

$\bar{U}$ is the cell-average conservative variables.

The last term on the right–hand side of equation 3, δU^{TVD}, is the numerical smoothing, which is obtained by collecting the contributions from each of the edges connecting to the given node,

$$\delta U^{TVD}_{ijk} = D_{i+\frac{1}{2}} - D_{i-\frac{1}{2}} + D_{j+\frac{1}{2}} - D_{j-\frac{1}{2}} + D_{k+\frac{1}{2}} - D_{k-\frac{1}{2}} \tag{9}$$

The amount of smoothing contributed from each edge is calculated by using Roe's wave–decomposition [1],

$$D_{edge} = \frac{1}{2}\frac{\Delta t}{\Delta V} S_{edge} \sum_{q=1}^{5} \boldsymbol{e}_q \mid \lambda_q \mid \alpha_q \{[1 - \Psi(r_q)](1 - \mid \nu_q \mid)\} \tag{10}$$

The edge interface is obtained by connecting the centers of the cells surrounding the edge. S_{edge} is the interface area; $\boldsymbol{e}_q$ and λ_q are the qth eigenvector and eigenvalue; α_q and ν_q are the wave strength and local Courant number corresponding to λ_q. The detailed expressions can be found in [5]. $\Psi(r_q)$ is a symmetric flux limiter[2], given by,

$$\Psi(r_q) = \max\left\{0,\ \min\left[2,\ 2r_q^-,\ 2r_q^+,\ 0.5(r_q^+ + r_q^-)\right]\right\} \tag{11}$$

Where r_q^+ and r_q^- are the ratios of wave strength,

$$r_q^+ = \frac{\alpha^q_{edge+1}}{\alpha^q_{edge}} \qquad r_q^- = \frac{\alpha^q_{edge-1}}{\alpha^q_{edge}} \tag{12}$$

To avoid the generation of expansion shock waves, and also to prevent from introducing extra diffusion into boundary layers, the eigenvalues corresponding to the pressure waves are modified[6] to make them differ from zero. The obtained numerical scheme, equation 2 to 12, is equivalent to symmetric TVD Lax–Wendroff Scheme.

The numerical smoothing term of equation 10 can be simplified by replacing the wave speed λ_q with a single representative wave speed $\tilde{\lambda}$, and the wave strength ratio r_q with a single representative wave strength ratio $\tilde{r}$. Thus, equation 10 becomes,

$$D_{edge} = \frac{1}{2}\frac{\Delta t}{\Delta V} S_{inter} \mid \tilde{\lambda} \mid [1 - \Psi(\tilde{r})](1 - \mid \tilde{\nu} \mid)\delta^+ U_{edge} \tag{13}$$

Where $\delta^+ U_{edge} = U_R - U_L$, L and R denote the left– and right–end of the edge. $\tilde{\nu}$ is the Courant number based on $\tilde{\lambda}$. One choice for $\tilde{\lambda}$ and $\tilde{r}$ is suggested by Davis[7],

$$\tilde{\lambda} = max(\mid \lambda_1 \mid \ \cdots, \ \mid \lambda_5 \mid) \qquad \tilde{r}^+ = \frac{\sum_{q=1}^5 \delta_+ U^q_{dege} \delta_+ U^q_{edge+1}}{\sum_{q=1}^5 \delta_+ U^q_{edge} \delta_+ U^q_{edge}} \tag{14}$$

Another choice is suggested by Roe[2],

$$\tilde{\lambda} = \sqrt{\frac{\sum_{q=1}^5 (\alpha_q^2 \lambda_q^2)_{edge}}{\sum_{q=1}^5 (\alpha_q^2)_{edge}}} \qquad \tilde{r}^+ = \frac{\sum_{q=1}^5 \alpha^q_{edge} \alpha^q_{edge+1}}{\sum_{q=1}^5 \alpha^q_{edge} \alpha^q_{edge}} \tag{15}$$

Numerical experiments showed that Davis' approach of equation 14 is computationally very efficient but too diffusive to be suitable for viscous flow calculations. Roe's simplification of equation 15 is slightly more diffusive than the original wave decomposition of equation 10.

By including the contributions from the multigrid method, the time marching procedure of equation 2 and 3 can be re-written as,

$$U^{n+1} = U^n + \delta U_h + \sum_{m=1}^{M} (T^h_{2mh} \delta U_{2mh}) \tag{16}$$

Where δU_h is the correction from the fine mesh. δU_{2mh} is the correction from the coarse meshes, T^h_{2mh} is interpolation operator from the coarse meshes to fine mesh. M refers to the total levels of coarse meshes applied.

3 Results

Laminar and Turbulent Flow Over a Flat Plate The length of the flat plate is one and a half units. The computational domain is extended a half unit towards the inflow boundary from the plate leading edge. The height of the channel is one unit. In order to resolve the high gradients near the plate leading edge and to resolve the boundary layer properly, the mesh is clustered at plate leading edge using an exponential function and stretched from the wall surface. Laminar Flow The free stream Mach number is 0.3 and the Reynolds number based on a unit length is 1×10^4. No–slip and adiabatic condition are imposed on the plate surface. Slip is allowed on the upper wall and the lower boundary before the plate leading edge. Molecular viscosity is calculated by Sutherland's law. A 3D mesh of $57 \times 25 \times 5$ is applied for this calculation. The computed skin-friction coefficient and the Blasius solution $C_f = 0.664 Re_x^{-0.5}$, the computed velocity profile at $x = 1.0$ and the corresponding Blasius solution, are given in Fig. 1. The comparison between the numerical and analytical solutions is very good. By using 2 levels of coarse mesh, the CPU time is reduced by 68% for this calculation. Other calculations are also carried out for three successive meshes, $57 \times 33 \times 5$, $57 \times 17 \times 5$ and $57 \times 9 \times 5$, in order to investigate the solution dependences on mesh and too on smoothing method. The computed velocity profiles at $x = 1.0$ and the Blasius solution are summarized in Fig. 2.

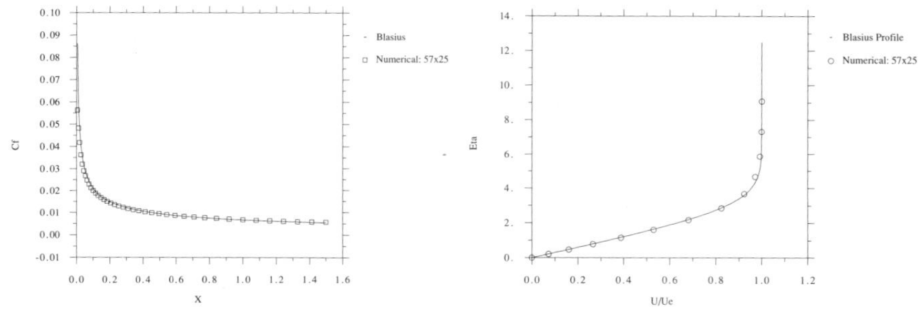

Fig. 1. Results for Laminar Flow Over a Flat Plate, Smoothing of Equation 13 and 15. **left**: Skin-Friction Coefficient; **right**: Velocity Profile

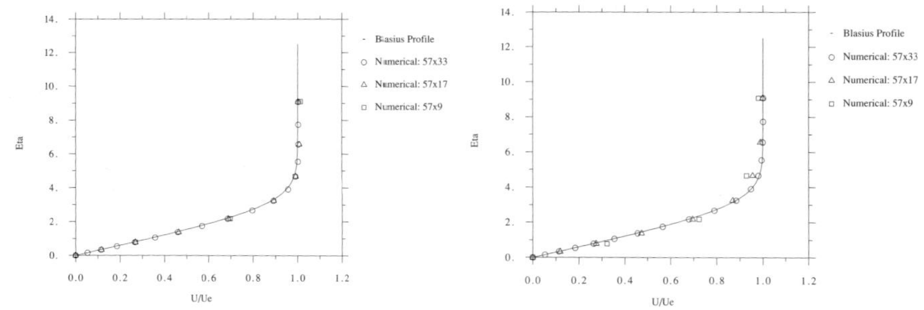

Fig. 2. Velocity Profile of Laminar Flow over a Flat Plate. **left**: Smoothing of Equation 10; **right**: Smoothing of Equation 13 and 15

Turbulent Flow The computational domain used for turbulent flow calculations is the same as that of the laminar flow. Calculations are performed at $M_\infty = 0.3$ and $Re_\infty = 4 \times 10^6$. The Baldwin–Lomax turbulence model is used to estimate eddy viscosity. Two meshes, $113 \times 41 \times 5$ and $113 \times 17 \times 5$, are respectively applied for the calculations with no-slip wall condition and with slip wall condition plus a wall function. The computed u^+ and y^+ at $x = 1.0$, together with the solution of Spalding's law of the wall are given in Fig. 3.

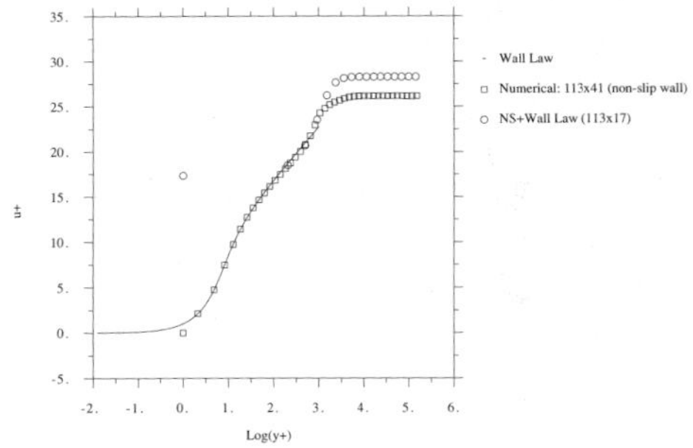

Fig. 3. Velocity Profile of Turbulent Flow over a Flat Plate, Smoothing of Equation 13 and 15

Turbulent Flow in a 3D Channel The geometry of the 3D channel has been reported in Cahen et.al.[8]. The Mesh dimension for the 3D channel is $113 \times 49 \times 49$ in the x, y and z directions, respectively. The inlet total temperature is 300K, total pressure 92000Pa, free stream Reynolds number 4.6×10^6. The static pressure at exit is 0.534 of the inlet total pressure. The Baldwin–Lomax model is employed to calculate eddy viscosity and a wall function is applied to avoid using extremely fine mesh near solid walls. A 2D cut of the 3D mesh at $Y = 10mm$ and computed Mach contours at $Y = 110mm$ are shown in Fig. 3.

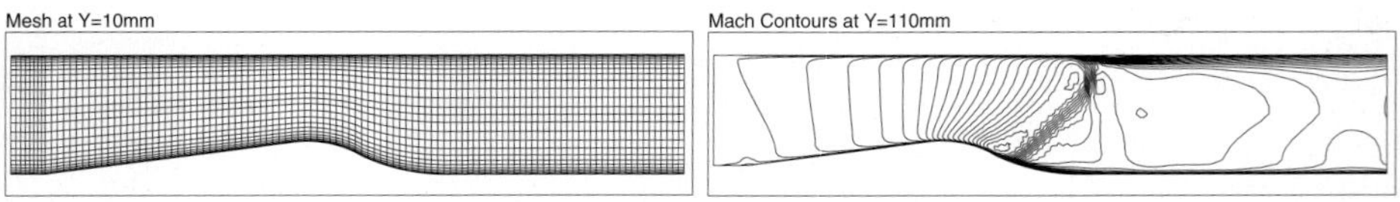

Fig. 4. Results for 3D Channel Flow, Smoothing of Equation 13 and 15. **left**: Mesh at Y=10mm; **right**: Mach contours at Y=110mm

References

1. P.L. Roe: J. Comp. Phys., Vol.43, pp.357–372, (1981)
2. P.L. Roe: ICASE Report, No.84-53, (1984)
3. K.W. Morton, M.F. Paisley: J. Comp. Phys., Vol.80, pp.168–203, (1989)
4. R.H. Ni: AIAA J., Vol.20, No.11, pp.1565-1571, (1982)
5. J.J. Liu: The Calculation of Asymmetric Flow in Turbine Exhaust Systems, PhD Thesis, Cambridge University, (1998)
6. H.C. Yee: J. Comp. Phys., Vol.68, pp.151–179, (1987)
7. S.F. Davis: ICASE Report No.84–20, (1984).
8. J. Cahen, V. Couaillier, J. Delery, T. Pot: AIAA J., Vol.33, No.4, pp671–679 (1995)

Iterative Space-Marching Method for Incompressible and Compressible Full Navier-Stokes Equations

Leonid I. Skurin

St. Petersburg University, Institut of Mathematics and Mechanics, St. Petersburg 198504, Russia

Abstract. The objective of this lecture is to show that global pressure relaxation technique can serve as a base for a computational method for fluid mechanics problems, which has the following characteristics: 1) it is quite universal - it can be applied for full Navier-Stokes equations for both compressible and incompressible fluids, for problems of any dimension, both steady and unsteady, and for flows without a main flow direction; 2) it allows for the solution of compressible and incompressible fluid problems based on a common algorithm; 3) its algorithm is simple because the method is based entirely on space-marching procedures (no splitting is used); 4) it is highly effective because it has good stability and convergence, it is applied for conservative form of equations with second or higher order of approximation with respect to all independent variables.

1 Introduction

Davis [1] was the first to solve a fluid mechanics problem using global iterations (GIs) for pressure. Later a number of authors used a similar technique for solving a limited range of problems on a base of reduced Navier-Stokes equations, namely, 2-D steady problems with a main direction of flow and only with the first order of the marching derivatives approximation.

Mathematical formalization of the method is as follows. In 2-D case one of the coordinate axes (let it be x) is declared a marching axis. The projection of the motion equation on this axis contains a transformation for the pressure p gradient of the following form: $\partial p/\partial x \Longrightarrow (1+\epsilon)\partial \overline{p}/\partial x - \epsilon \partial p/\partial x$, where ϵ is a given value, $\overline{p}$ - is a given function being refined through the GI process. The Navier-Stokes equations are presented in the form of finite-difference approximations and an iterational loop based on marching procedures is constructed. Function $\overline{p}$ for the s-th GI is calculated using the following relation: $\overline{p}^s = (1-\tau)\overline{p}^{s-1} + \tau p^{s-1}$, where τ is a parameter of the scheme. The objective of the GIs is to achieve the equality between $\overline{p}$ and p with a given level of accuracy: $|\overline{p}^s - p^s| < \epsilon_p << 1$. When this condition is met the transformation becomes trivial for all ϵ and thus solving the steady problem is completed, or in the case of an unsteady problem, one time step is completed.

3-D problems are solved by making multiple successive computations of a 2-D domain and within each 2-D domain the computation is performed using marching procedures [2, 3]. 1-D problems are solved by making multiple successive computations of grid functions at each point of sole space axis.

The first question that arises with an attempt to realize the above scheme is the existence of ϵ for which the transformed Euler equations are x-hyperbolic. A positive answer to this question for the case of a 2-D steady system for compressible fluid is provided by Vigneron, Rakich, Tannehill [4]. General results are presented in [2, 5]. In this study we show that for incompressible fluid it is necessary that $\epsilon > 0$, for compressible fluid the conditions for ϵ are different for steady and unsteady problems, and they depend on the problem dimension and on Mach number. It is obtained for any case such ϵ at which transformed system is x-hyperbolic. These results give hope that marching procedures for finite-difference analogs of the respective differential equation systems will be stable.

2 Finite-Difference Approximations

Let us determine indexes: σ - marching, j, k - transverse, m- time.

A. The derivatives with respect to marching coordinate.

For the first derivatives we use the following differences: 1) first order - upwind both central three-point and one-sided two-point (left-sided for pressure gradient) or 2) second order - upwind one-sided three-point (left-sided for pressure gradient). For the second derivatives:

$$\partial^2 f/\partial x^2 = [(f_{jkm}^{\sigma+1})^{s-1} - 2f_{jkm}^{\sigma} + f_{jkm}^{\sigma-1}]/\Delta x^2. \tag{1}$$

For $\overline{p}$: $\partial\overline{p}/\partial x = (\overline{p}_{jkm}^{\sigma+1} - \overline{p}_{jkm}^{\sigma-1})/(2\Delta x) + (\overline{p}_{jkm}^{\sigma-1} - p_{jkm}^{\sigma-1})n/(2\Delta x)$, $n = 1$ or 2.

B. The derivatives with respect to transverse coordinate: 1) central two-point differences second order; 2) one-sided three-point differences second order; 3) four- or six-point differences third or fifth order (using Lagrange formula).

C. The derivatives with respect to time: 1) two-point first order; 2) three point-second order.

3 The General Computational Scheme

An analytical investigation of the stability of marching procedures was performed in [2,3] using upwind first-order differences along the marching axes and two-point central differences in the transverse direction. This investigation shows that they are unconditionally stable for incompressible fluid. Another critical issue is the convergence of the GIs. Analytical investigation shows that the convergence is unconditional at $1 \leq \epsilon \leq 3$. These results are obtained in [6], using (1) and the same considerations that in [3]. Numerical solutions of various fluid mechanics problems support these results.

We refer to this method as the Iterative-Marching Method (IMM).

4 The Modified Computational Scheme

Consider now a computational scheme for an unsteady form of governing equations that follows from the IMM formulation for $\tau = 1$ and when the condition $|\overline{p}^s - p^s| < \epsilon_p << 1$ is not required to be met on each temporary layer. In this case one marching procedure (one GI) corresponds to each time step. Such scheme proved effective for 2-D and 3-D steady problems [7]. We refer to this scheme as "modified". We want to emphasize that this scheme is iterativeless. Actually, equations don't contain $\overline{p}$ because $\overline{p}^{\sigma+1}_{jkm} = p^{\sigma+1}_{jkm-1}$ in this case. An analytical examination of this scheme is performed in [6] for incompressible fluid and first order approximation of marching derivatives. It is shown that stability and time convergence of such scheme are unconditional. This scheme may be used for solving unsteady problems. In this case a time step has to satisfy a condition that time increments of unknown functions are much smaller than their space increments.

According to the calculations of the problems considered in [6, 11], the results mentioned here for the general and the modified schemes are also valid for compressible fluid. Also, good scheme stability and convergence take place for the case of the second order approximation of the derivatives with respect to the marching coordinate [3, 7].

5 Special Features of the IMM Realization

When the IMM is used there is no need in subiterations on nonlinearity. In each marching station we solve the algebraic system linearized using Newton method, while the convergence of the GIs results in a solution for a nonlinear equation system. To invert algebraic equations at each marching station sparse matrix solver is used.

A known drawback of the method when applied for solving steady problems is the increase in the number of GIs with an increase in the number of the grid points along the marching axis. In order to mitigate this problem, it is reasonable to take the initial approximation for the unknown functions from the problem solution on a rough grid. . This strategy gives good initial approximation for a fine grid and therefore significantly decreases the number of GIs and the computational time for a problem on the fine grid.

For unsteady problems good initial approximation at each time step always exists. It is the result of the previous time step solution. Therefore, only a few GIs are sufficient for each time step. We should also note the possibility and usefulness of an approximation of time derivatives using second-order three-point formula [8].

It is clear that problem of small Mach number does not exist for the IMM because the algorithm for both incompressible and compressible fluid is the same.

6 Computed Results

Unfortunately, space constraints don't allow us to show a number of results of IMM application. Therefore in this section we enumerate published solutions and mention problems that solved recently. The method was tested on various problems, specifically: 1) steady laminar and turbulent flows of incompressible fluid at swirl in variable cross section pipes [9-12] with large bubbles and recirculation regions and flow over a backward facing step [13]; 2) unsteady problems of evolution of uniform spot and vortex pair in stratified fluid with generating internal gravitational waves [14]; 3) 3-D fluid motions in a cubic cavity and 3-D flow development in a square duck [7]; 4) steady sub-, tran- and supersonic flows in ducks and Laval nozzles with bubbles and oblique shock waves [3, 15]; 5) unsteady 2-D flow in a Laval nozzle, when pressure at output is increasing with time, also 1-D termoacoustic waves and waves in a resonance tube [8]. Solutions of 1-D unsteady problems are obtained at $0 < \epsilon \leq 1, \tau = 1$. The rest of the problems solved at $\epsilon = 1$ (and most of the times $\tau = 1$). For steady problems modified scheme is more convenient. When it is used and finite-differences are determined the sole parameter of the scheme is time step. It depends on complexity of the problem, on grid type, and on the initial approximation for the grid functions.

Recently we obtained solutions of following 2-D steady problems for both compressible and incompressible fluid using modified scheme: 1) flow in vortex tube and jet - see Fig. 1a and 2)longitudinal flow over a body of limited size - see Fig. 1b. In case of computational domains that have big inner expansion or compression angles (for example, backward [13] or forward step) we use one-sided second or higher order differences (see section 2) for transverse derivatives to obtain smooth solutions. To compute these two problems we used differences of the same type.

Vortex tube has a unit radius and a length l. Boundary conditions: a) at $x < h$ fluid is injected in the tube: radial velocity $v_w = -1$, azimuth velocity $w_w = w_a$, axial velocity $u_w = 0$, temperature $T_w = 1$, b) at $r = 0$ - axis symmetry; c) at $x = l + L : p = const, \partial^2 u/\partial x^2 = \partial^2 T/\partial x^2 = \partial v/\partial x = \partial w/\partial x = 0$; d) at the rest of the boundary: $u = v = w = 0, T = 1$, normal pressure gradient is zero. The turbulence model is the simplest: Reynolds number in Navier-Stokes

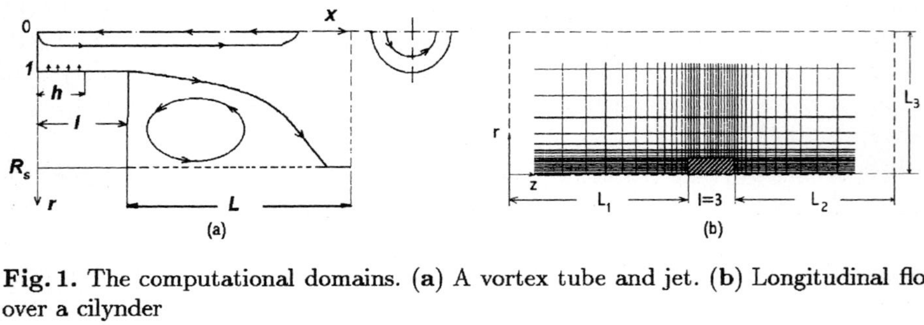

Fig. 1. The computational domains. **(a)** A vortex tube and jet. **(b)** Longitudinal flow over a cilynder

equations is calculated from formula $Re = 50h/w_a$ [16] . The calculation was run at $h = 2, l = 4, w_a$ varying from 10 to 100. The value of R_s varied from 3 to 8, the value of L - from 10 to 20. The typical marching direction step size is 0.1 and the transverse one is 0.05.

A characteristic peculiarity of this flow field is the presence of a long intensive axis recirculation region. According to the computations the absolute value of the velocity of the reversal flow may exceed the maximum value of the velocity of the attached flow. In case of compressible fluid the flow field has a similar structure (see Fig. 1a). The most interesting in this case is a temperature field. According to the computations the temperature at the axis is less than that at the entrance ($x < h, r = 1$). For example, when Mach number at entrance is 1.1, axis absolute temperature at $x = h$ is less than that at the entrance by approximately twenty percent. This is known as Runque effect.

The following example is a problem of flow over a body of limited length at sub- and transonic velocity. Solutions are obtained for axisymmetrical flows (see Fig. 1b) over a circular cylinder with a unit radius and a length of 3, at $0 \leq M \leq 0.95, 1 \leq Re \leq 300$. In Fig. 1b dashed lines are boundaries of the computational domain. At the right boundary $p = const$, soft conditions are used for u, v, T , at the other boundaries $u = T = 1, v = 0, \partial p/\partial n = 0$. Initial approximation is $u = T = 1, v = 0, p = const$ all over the domain. The values of L_1, L_2, L_3 vary from 10 to 50. We use non-uniform grids of 111×201 or 141×311 points. The grid points cluster near the body so that the spacing next to the surface is 0.05. The aspect ratio are 25-50 for marching axis and 5-10 for transverse axis. About 200 GIs are necessary to obtain an engineering solution in case of $M = 0.95$. There is a separation region next behind the body at $Re \geq 10$. In case of $M = 0.95$ there is a large local supersonic region above the body surface and in near wake. Maximum value of Mach number is 1.06, length of the separation region is 2.1 (at $Re = 100$).

7 Conclusion

Thus, IMM is characterized by algorithmic simplicity and it has a wide range of applications. It is highly effective because it requires a relatively small number of GIs for convergence, does not require any subiterations on nonlinearity, and has good stability and convergence. It allows: 1) the use of non-uniformity grids without significant loss of the convergence rate, 2) the solution of steady problems on the base of both steady and unsteady form equations, 3) the solution of gas dynamics problems in cases of small Mach (and Reynolds) numbers without a loss of the convergence rate and without changes in code, 4) the solution of problems for supersonic flows with embedded subsonic regions without change in code. Clearly it can be used for problems with non-homogeneous and multicomponent media motions.

ACKNOWLEGEMENTS. This work has been partially supported by the Russian Foundation for Basic Research (grants No. 96-01-00390 and No. 00-01-00633) and Ministry for Education (grant No.005.01.01.18).

References

1. Davis R.T. Numerical Solution of the Hypersonic Viscous Shock-Layer Equations. AIAA J. Vol. 8. 1970. N 5, pp. 843-851.
2. Skurin L.I. On Possibility Solution of an Unsteady Problems of Fluid and Gas Mechanics by Means of a Global Iteration Technique. St. Petersburg Univ. Mech. Bulletin. N 3. 1997, pp.78-84.
3. Skurin L.I. On Possibility of a Numerical Solution of Fluid and Gas Mechanics Problems on the Basis of Common Principle. St. Petersburg Univ. Mech. Bulletin. N 2. 1998, pp. 102-112.
4. Vigneron Y.C., Rakich J.V., Tannehill J.C. Calculation of Supersonic Viscous Flow Over Delta Wing With Sharp Subsonic Leading Edges. AIAA Paper 78-1137, 1978.
5. Skurin L.I. Iterative-Marching Method for Solving of Fluid and Gas Mechanics Problems. Siberian J. of Numer. Mathematics. RAS. Novosibirsk. N 2. 1998, pp. 171-181.
6. Skurin L.I. About effectiveness of Iterative-Marching Method for Solving of the Fluid and Gas Mechanics Problems. Vestnik St. Petersburg Univ. Mech. N 1. 2000, pp. 104-109.
7. Polyansky A.P., Skurin L.I. Numerical Investigation of 3-D Hydrodynamical Problems Using Iterative-Marching Method. St. Petersburg Univ. Mech. Bulletin. N 2. 1999, pp. 90-96.
8. Polyansky A.P., Skurin L.I. The Gas Dynamic Unsteady Problem Solutions Using Iterative-Marching method. Vestnik St. Petersburg Univ. Mech. N 2. 2000, pp. 92-97.
9. Arkhangelskaya L.A., Skurin L.I. The Use a Global Iteration Method for Solving Navier-Stokes Equations. St. Petersburg Univ. Mech. Bulletin. N 3. 1994, pp. 70-75.
10. Arkhangelskaya L.A., Skurin L.I. The Use a Global Iteration Method for Solving Equations System of Navier-Stokes. Vichislitelnie technologii. RAS. Novosibirsk. Vol 4. N 12. 1994, pp. 29-37.
11. Polyansky A.P., Skurin L.I. The Mathematical Modelling of Separated and Swirled Flows . St. Petersburg Univ. Mech. Bulletin. N 2. 1996, pp. 77- 82.
12. Polyansky A.P., Skurin L.I. Characteristics of a Vortex Sound Generator: a Calculation Procedure. Acoustical Physics. Vol. 43. N 6. 1997, pp. 728-732.
13. Polyansky A.P., Skurin L.I. Numerical Simulation of a vortex tube flows. Second Polyakhov's Reading. Selected papers. St. Petersburg. 2000, pp. 164-170.
14. Polyansky A.P., Skurin L.I. The Eddy Pair Evolution in Viscous Stratified Fluid. St. Petersburg Univ. Mech. Bulletin. N 4. 1998, pp. 125-131.
15. Polyansky A.P., Skurin L.I. Numerical Modelling Viscous Sub- and Supersonic Flows of Gas with Use of Iterative-Marching Method. St. Petersburg Univ. Mech. Bulletin. N 3. 1998, pp. 111-115.
16. Goltdshtik M.A. Variation Model of a Turbulent Swirling Flow. Izv. Akad. Nauk SSSR, Mekh. Zhidk. Gaza. N 9. 1985, pp. 22-32.

Effective Accuracy and Conservation Consistency of Gridless Type Solver

Koji Morinishi

Kyoto Institute of Technology, Kyoto 606-8585, Japan

Abstract. Gridless type solver can work on structured grids, on unstructured grids, and even on points arbitrarily distributed over computational domains. It is difficult, however, to demonstrate its accuracy order and conservation consistency theoretically except for a few special cases. In this paper, the effective accuracy and conservation consistency of the solver are numerically estimated by grid convergence studies. The effective second order of accuracy and conservation consistency are obtained in the numerical results for incompressible inviscid flows around a circular cylinder and a laminar boundary layer flow over a flat plate. The correct shock strength and position are obtained for a shock reflection problem.

1 Gridless Type Solver

In the gridless type solver, points are first distributed over the computational domain. For each point, its cloud is composed of the point itself and its neighbors. Numerical solutions of any partial differential equations are obtained at each point with the following manner.

1.1 Evaluation of Spatial Derivatives

At a point i shown in Fig. 1, the spatial derivatives of the solution function f can be evaluated using its values in the cloud $C(i)$ of the point i as:

$$\left.\frac{\partial f}{\partial x}\right|_i = \sum_{k \subset C(i)} a_{ik} f_{ik} , \quad \left.\frac{\partial f}{\partial y}\right|_i = \sum_{k \subset C(i)} b_{ik} f_{ik} \tag{1}$$

where the subscript k denotes the index of the point which belongs to the cloud $C(i)$ and f_{ik} is evaluated at the midpoint between the points i and k. The coefficients a_{ik} and b_{ik} can once be obtained with a least-squares curve fit and stored at the beginning of computations [1].

The second derivatives can be evaluated with following sequential manner:

$$\left.\frac{\partial^2 f}{\partial x^2}\right|_i = \sum_{k \subset C(i)} a_{ik} \left.\frac{\partial f}{\partial x}\right|_{ik} . \tag{2}$$

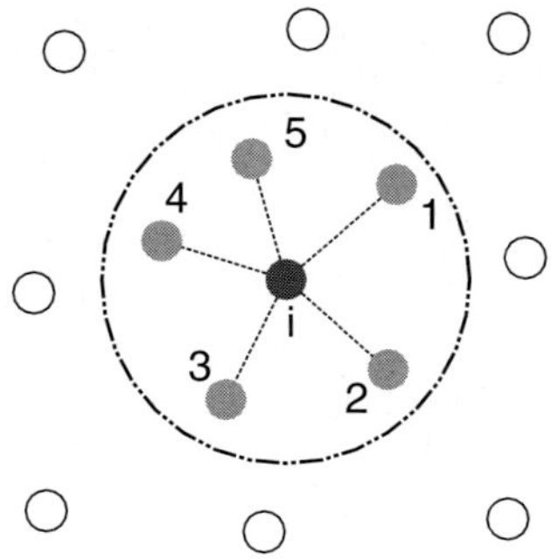

Fig. 1. Cloud C(i) for point i.

The first derivative at the midpoint is evaluated as follows:

$$\begin{aligned}\left.\frac{\partial f}{\partial x}\right|_{ik} &= \frac{\Delta x}{\Delta s^2}\left(f_k - f_i\right) \\ &+ \frac{1}{2}\frac{\Delta y}{\Delta s^2}\left[\Delta y\left(\left.\frac{\partial f}{\partial x}\right|_i + \left.\frac{\partial f}{\partial x}\right|_k\right)\right. \\ &\left. - \Delta x\left(\left.\frac{\partial f}{\partial y}\right|_i + \left.\frac{\partial f}{\partial y}\right|_k\right)\right]\end{aligned} \tag{3}$$

where

$$\Delta x = x_k - x_i \ , \quad \Delta y = y_k - y_i \ ,$$
$$\Delta s^2 = \Delta x^2 + \Delta y^2 \ .$$

A Laplace operator as well as the second derivatives can be directly obtained. For an example, the Laplace operator can be evaluated as:

$$\left(\frac{\partial^2 f}{\partial x^2} + \frac{\partial^2 f}{\partial y^2}\right)_i = \sum_{k \in C(i)} c_{ik} f_k \ . \tag{4}$$

The coefficients c_{ik} can be obtained as solutions of following linear equations using QR or singular value decompositions.

$$\sum_{k \in C(i)} c_{ik} f_k^{(m)} = d^{(m)} \tag{5}$$

Here $f^{(m)}$ and $d^{(m)}$ are respectively

$$f^{(m)} \in \left(1, x, y, x^2, xy, y^2, \cdots\right) \tag{6}$$

and

$$d^{(m)} \in (0, 0, 0, 2, 0, 2, \cdots) \ . \tag{7}$$

1.2 Evaluation of Inviscid Flux

The inviscid flux of the Navier-Stokes equations can be evaluated as:

$$\frac{\partial \mathbf{E}}{\partial x} + \frac{\partial \mathbf{F}}{\partial y} = \sum a_{ik}\mathbf{E}_{ik} + \sum b_{ik}\mathbf{F}_{ik} \\ = \sum \mathbf{G}_{ik} \; . \tag{8}$$

The flux term $\mathbf{G}$ at the midpoint are expressed as:

$$\mathbf{G} = \begin{pmatrix} \rho U \\ \rho u U + ap \\ \rho v U + bp \\ U(e+p) \end{pmatrix} \tag{9}$$

where U are defined like the so-called contravariant velocity as:

$$U = au + bv \; . \tag{10}$$

An upwind method may be obtained if the numerical fluxes on the midpoints are obtained using Roe's approximate Riemann solver as:

$$\mathbf{G}_{ik} = \frac{1}{2}\left(\mathbf{G}(\tilde{\mathbf{q}}_{ik}^{+}) + \mathbf{G}(\tilde{\mathbf{q}}_{ik}^{-}) - |\tilde{\mathbf{A}}|(\tilde{\mathbf{q}}_{ik}^{+} - \tilde{\mathbf{q}}_{ik}^{-})\right) \tag{11}$$

where $\tilde{\mathbf{q}}$ are the primitive variables and $\tilde{\mathbf{A}}$ are the flux Jacobian matrices.

The reconstructed primitive variables are obtained from:

$$\tilde{\mathbf{q}}_{ik}^{-} = \tilde{\mathbf{q}}_i + \frac{1}{2}\phi^{-}\nabla\tilde{\mathbf{q}}_i \cdot \mathbf{r}_{ik} \; , \quad \tilde{\mathbf{q}}_{ik}^{+} = \tilde{\mathbf{q}}_k - \frac{1}{2}\phi^{+}\nabla\tilde{\mathbf{q}}_k \cdot \mathbf{r}_{ik} \tag{12}$$

where $\mathbf{r}_{ik}$ are the vectors extending from the point i to the point k and $\phi^{\pm}$ are flux limiters.

The gradients of primitive variables $\nabla\tilde{\mathbf{q}}$ are obtained using (1) at each point. These gradients are also used for the evaluations of the viscous stresses in the Navier-Stokes equations with the sequential manner of (2).

1.3 Temporal Discretization

A linearized backward Euler method is used for the temporal discretization of the gridless type solver. We assume the following linearization.

$$\mathbf{G}_{ik}^{n+1} = \mathbf{G}_{ik}^{n} + \mathbf{A}_{ik}^{+}(\mathbf{q}_i)\Delta\mathbf{q}_i + \mathbf{A}_{ik}^{-}(\mathbf{q}_k)\Delta\mathbf{q}_k \tag{13}$$

Then the linearized backward Euler method is obtained as:

$$\left(\frac{1}{\Delta t_i}\mathbf{I} + \sum_{k\subset C(i)} \mathbf{A}_{ik}^{+}(\mathbf{q}_i)\right)\Delta\mathbf{q}_i \\ + \sum_{k\subset C(i)} \mathbf{A}_{ik}^{-}(\mathbf{q}_k)\Delta\mathbf{q}_k = \mathbf{W}(\mathbf{q}_i^{n}) \; . \tag{14}$$

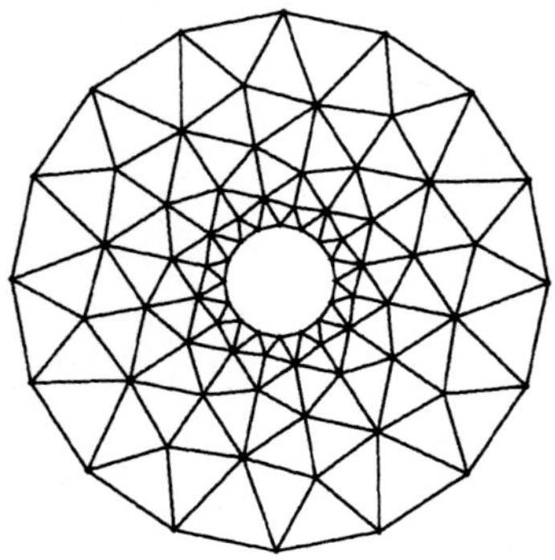

Fig. 2. Coarsest grid.

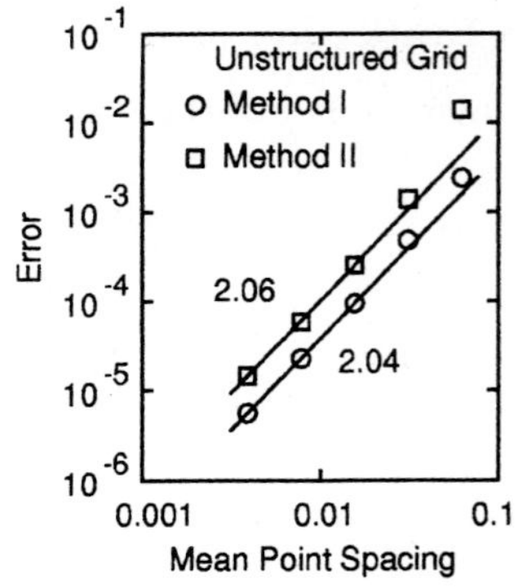

Fig. 3. L_2 error vs. mean point spacing.

Here $\mathbf{W}$ are residuals obtained with the gridless evaluations of the spatial derivatives in the Navier-Stoke equations and $\Delta\mathbf{q}$ are defined as:

$$\Delta\mathbf{q} = \mathbf{q}^{n+1} - \mathbf{q}^{n} \tag{15}$$

where the superscript n denotes the time index. The solution of this linear system of equations is obtained with LU-SGS procedure.

2 Numerical Experiments

The effective accuracy and conservation consistency of the gridless type solver are numerically estimated for four demonstrative test cases by means of grid convergence studies.

2.1 Experiment for a Laplace Equation

The first numerical experiment is carried out for the following Laplace equation:

$$\frac{\partial^2 \psi}{\partial x^2} + \frac{\partial^2 \psi}{\partial y^2} = 0 , \qquad 1.0 \le \sqrt{x^2 + y^2} \le 4.8 . \tag{16}$$

Dirichlet boundary conditions are given at the boundaries, so that following stream line function ψ of a irrotational flow about a circular cylinder is obtained.

$$\psi = U_\infty y \left(1 - \frac{1}{x^2 + y^2}\right) \tag{17}$$

Numerical solutions of the Laplace equations are obtained using GMRES method on five unstructured grids. Figure 2 shows the coarsest unstructured grid of this experiment. The other four grids are sequentially generated with systematical refinement. The L_2 error obtained on every grid is plotted as a function of mean point spacing in Fig. 3. Two series of numerical data are plotted.

Fig. 4. Iso-pressure contours.

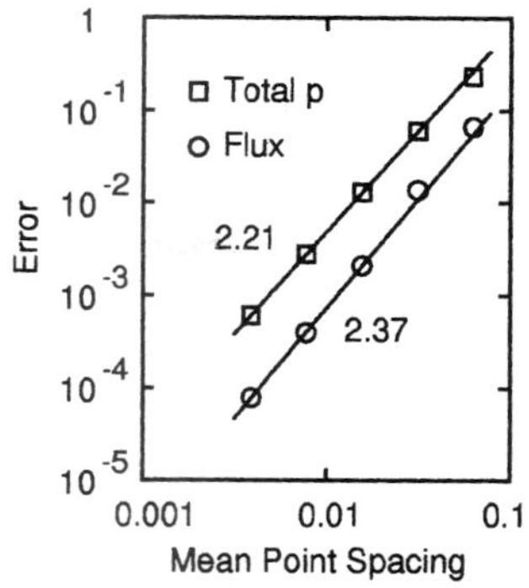

Fig. 5. L_2 errors for flux and total pressure versus mean point spacing.

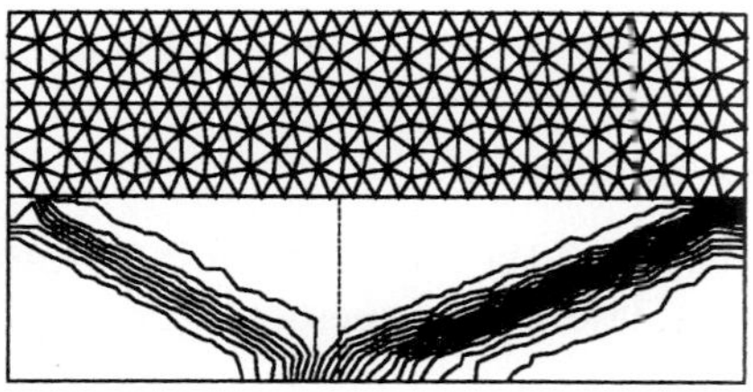

Fig. 6. Coarse grid and iso-pressure contours.

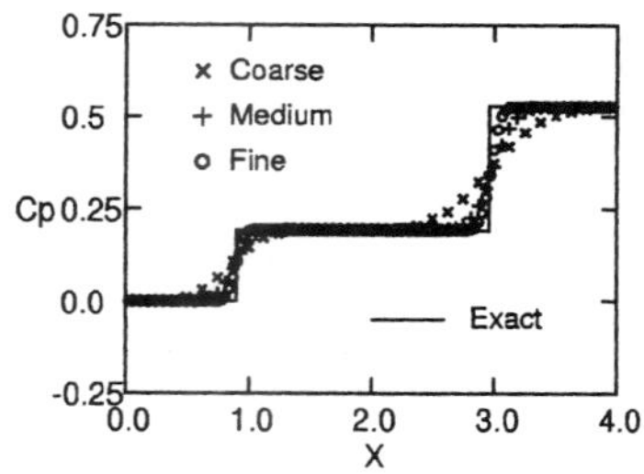

Fig. 7. Comparison of pressure distribution along y of 0.5.

The first is obtained with the sequential evaluation of the second derivatives (Method I) and the second with the direct evaluation of the Laplace operator (Method II). It is found that both methods are effectively second order accurate because both slopes of error curves are about 2.0.

2.2 Experiments for the Euler Equations

Figure 4 shows iso-pressure contours for an incompressible inviscid flow around a circular cylinder. The numerical results of the incompressible Euler equations are obtained for five structured O-grids which are generated with systematical refinement. The L_2 errors for the mass (flux) conservation and the energy (total pressure) conservation are plotted as a function of mean point spacing in Fig. 5. The expected slopes of the second order accuracy are again obtained.

Figures 6 and 7 show numerical results obtained with the compressible Euler equations for a shock reflection problem [2]. The numerical results are obtained for three unstructured grids which are generated with systematical refinement. The coarse grid and iso-pressure contours obtained for the grid are shown in Fig. 6. The pressure distributions along y of 0.5 are compared with the exact solution in Fig. 7. The location and strength of the shocks are correctly captured for all

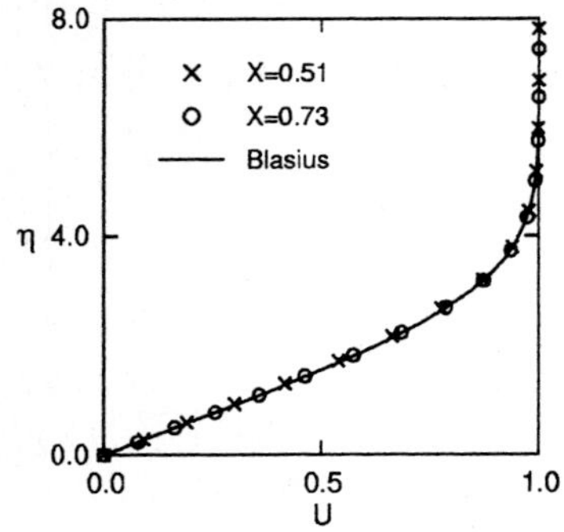

Fig. 8. Comparison of velocities.

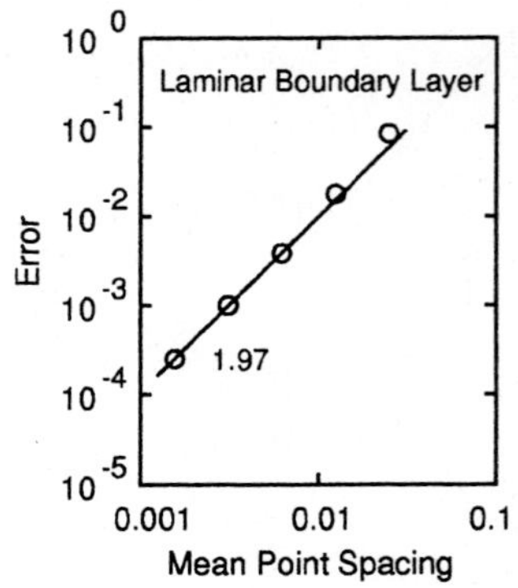

Fig. 9. L_2 error vs. mean point spacing.

grids, while the steepness is improved as the grid is refined. The conservation consistency of the gridless type solver may be responsible for this feature.

2.3 Experiment for the Navier-Stokes Equations

Figures 8 and 9 show numerical results obtained with the incompressible Navier-Stokes equations for a laminar flow over a flat plate. Five structured grids with systematical refinement are used. The coarsest grid has 17×9 points and the finest 257×129 points. The velocity distributions in the laminar boundary layer obtained with a medium grid (65×33) are compared with the similarity solution of Blasius in Fig. 8. The L_2 error for the velocity is plotted as a function of mean point spacing in Fig. 9. Instead of the similarity solution, the error is estimated using the grid-independent solution obtained with Richardson extrapolation. The slope of error curve is almost 2.0, which follows the expected slope of the second order accuracy.

3 Conclusions

The effective accuracy and conservation consistency of the gridless solver are numerically estimated by means of grid convergence studies. The effective second order of accuracy and conservation consistency are obtained for all test cases considered. The correct shock strength and position are obtained for a shock reflection problem.

This study was partially supported by the Research for the Future Program (97P01101) from Japan Society for the Promotion of Science.

References

1. K. Morinishi: CFD Journal **9**, (2000)
2. H.C. Yee, R.F. Warming and A. Harten: AIAA paper 83-1902 (1983)

An Artificial Boundary Condition for Subsonic Flows Governed by Compressible Navier–Stokes Equations

Charles-Henri Bruneau[1] and Emmanuel Creusé[1,2]

[1] Université Bordeaux I. 351, Cours de la Libération, 33405 Talence, FRANCE.
[2] Université de Valenciennes. Le Mont Houy, 59313 Valenciennes Cedex 9, FRANCE.

1 Introduction

In open flow simulations, the fluid evolves in an infinite domain and thus the computing domain must be obviously truncated. An artificial boundary Γ is then introduced, on which appropriate Boundary Conditions (BCs) have to be imposed. If the solution obtained on a given truncated domain Ω is equal to the restriction of the infinite solution on the domain Ω, then the artificial BC is perfect and called transparent BC. Unfortunately, such a configuration is only an ideal case. In this work, the compressible Navier–Stokes equations are approximated with a 2D Direct Numerical Simulation on an unstructured mesh composed of triangular cells. First, the governing equations are presented, and the amplitudes of the characteristic waves carefully defined. Then, the Non Reflecting Boundary Condition (NRBC) is described and its failures are pointed out. Finally, the Present Boundary Condition (PBC) is derived. Numerical simulation are performed to show the ability of the PBC to convey properly the phenomena without creation of strong acoustic waves.

2 Governing Equations and boundary conditions

2.1 Governing equations

The governing equations are the classical 2D compressible Navier–Stokes equations, given in their non-dimensionalized formulation, in an orthonormal system and with usual notations by :

$$\begin{cases} \dfrac{\partial U}{\partial t} + \dfrac{\partial F_x(U)}{\partial x} + \dfrac{\partial F_y(U)}{\partial y} = \dfrac{1}{Re}\left(\dfrac{\partial G_x(U,\nabla U)}{\partial x} + \dfrac{\partial G_y(U,\nabla U)}{\partial y}\right) \\ \\ U(t=0) = U_0 \end{cases}$$

with :

$$U = \begin{bmatrix} \rho \\ \rho u \\ \rho v \\ \rho E \end{bmatrix} \quad F_x(U) = \begin{bmatrix} \rho u \\ \rho u^2 + p \\ \rho uv \\ (\rho E + p)u \end{bmatrix} \quad F_y(U) = \begin{bmatrix} \rho v \\ \rho uv \\ \rho v^2 + p \\ (\rho E + p)v \end{bmatrix}$$

In order to close the system, the state equation $p = (\gamma - 1)\rho T$ has to be added, as well as the relation $\rho E = \dfrac{\rho(u^2 + v^2)}{2} + \dfrac{p}{\gamma - 1}$, where the Reynolds number of the flow, Re, is defined by $Re = \dfrac{u_c l_c}{\nu}$. u_c, l_c and ν are respectively a characteristic velocity, a characteristic length and the viscosity of the fluid.

2.2 Characteristic Waves

In order to simplify the expression, let us consider a node A, located on the artificial boundary Γ at $x = constant$, whose outward normal to the boundary is $\vec{n} = (1, 0)^T$. We reformulate the equations at node A to derive the amplitude of the characteristic waves normal to the boundary:

$$S^{-1}\frac{\partial V}{\partial t} + L + S^{-1}P^{-1}\frac{\partial F_y(U)}{\partial y} = \frac{S^{-1}P^{-1}}{Re}\left(\frac{\partial G_x(U, \nabla U)}{\partial x} + \frac{\partial G_y(U, \nabla U)}{\partial y}\right) \quad (1)$$

where $V = [\rho, u, v, p]^T$, $P = \dfrac{\partial U}{\partial V}$ and S is the transformation matrix whose columns are the $P^{-1}\dfrac{\partial F_x(U)}{\partial V}$ right eigenvectors. Then, by noting $c = \sqrt{\dfrac{\gamma p}{\rho}}$ the local sound speed, we get L whose components are the amplitudes of the characteristic waves :

$$L = \begin{bmatrix} L_1 \\ L_2 \\ L_3 \\ L_4 \end{bmatrix} = \begin{bmatrix} (u-c)\left(\dfrac{\partial p}{\partial x} - \rho c\dfrac{\partial u}{\partial x}\right) \\ u\left(c^2\dfrac{\partial \rho}{\partial x} - \dfrac{\partial p}{\partial x}\right) \\ u\left(\dfrac{\partial v}{\partial x}\right) \\ (u+c)\left(\dfrac{\partial p}{\partial x} + \rho c\dfrac{\partial u}{\partial x}\right) \end{bmatrix} = \begin{bmatrix} \lambda_1\left(\dfrac{\partial p}{\partial x} - \rho c\dfrac{\partial u}{\partial x}\right) \\ \lambda_2\left(c^2\dfrac{\partial \rho}{\partial x} - \dfrac{\partial p}{\partial x}\right) \\ \lambda_3\left(\dfrac{\partial v}{\partial x}\right) \\ \lambda_4\left(\dfrac{\partial p}{\partial x} + \rho c\dfrac{\partial u}{\partial x}\right) \end{bmatrix}. \quad (2)$$

According to the sign of the eigenvalues, the discretization of L_k at the boundary has to be done from inside ($\lambda_k > 0$) or outside ($\lambda_k < 0$) the domain.

2.3 Boundary conditions

Boundary conditions can be derived for each boundary with the use of the $L_k\,(1 \le k \le 4)$. The principle of the method as well as several examples are carefully described in [1]. The way to proceed is first to impose the physical BCs, and then to use (2) to derive the remaining unknown. When $\lambda_k > 0$, L_k is discretized easily from inside the domain, whereas if $\lambda_k < 0$, another way has to be found to evaluate L_k since the equation should be discretized from outside the domain. For the subsonic outflow boundary, no strongly BC is imposed. L_2, L_3 and L_4 are outgoing waves, and can be discretized from inside the computational domain. But L_1 is an entering wave. The Non-Reflecting Boundary Condition (NRBC) is then defined by $L_1 = 0$.

3 The present boundary condition

The NRBC above is not physically relevant in general. In this work, we prefer to find a way to discretize L_1 in order to better represent the actual flow. In order to compute U^{n+1}, the whole solution at time t_{n+1}, we need the value of the convective and diffusive residuals at time t_n, with $t_{n+1} = t_n + \Delta t$. We explain here how to determine the $(L_1)^n_{i,j}$ value on the outflow boundary, at node $A_{i\,j}$ on the Γ boundary.

3.1 The NRBC on two classical test cases

Let consider a boundary layer evolution on a flat plate. The computational domain is a rectangle, whose lower side corresponds to the wall (Fig. 1-(a)/plain line). For this test case, the solution is stationnary and the pressure is constant

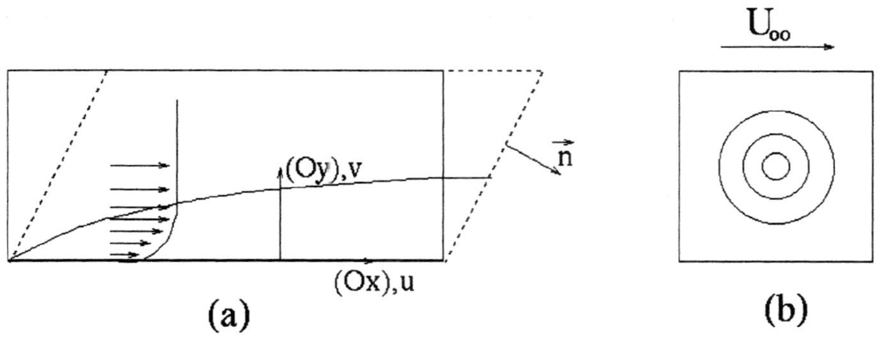

Fig. 1. Boundary layer and Vortex test cases

in the whole domain. The NRBC leads to $\left(\frac{\partial u}{\partial x}\right)^n_{i,j} = 0$. This approximation is probably close to the reality, especially if the Reynolds number is high enough. Let consider now the same test. This time, the computational domain is a parallelogram whose lower side corresponds to the wall (Fig. 1-(a)/dashed line). If $\vec{n}$ is the outward normal to this boundary, and u_n the velocity component according to $\vec{n}$, then the NRBC leads to $\left(\frac{\partial u_n}{\partial n}\right)^n_{i,j} = 0$. This time, this approximation is no more consistent. Numerical tests show that a recirculation area can arise without any physical meaning.

A second relevant test to understand the NRBC drawbacks is the simulation of an inviscid vortex leaving the computational domain (Fig. 1-(b)). The exact solution at each time of the simulation is equal to the initial datum translated of a given lenght depending on the convection velocity U_∞. As $L_1 \neq 0$ in the vortex, the NRBC is inevitably a rough approximation. As we shall see, this BC allows the hydrodynamic field to leave the computational domain, but generates significant acoustic waves, whereas the flow is inviscid and the outflow boundary perpendicular to the main direction of the flow.

Remark: The NRBC can be improved by taking the Partially Non Reflecting Boundary Condition (PNRBC) $(L_1)^n_{i,j} = K(p^n_{i,j} - p_\infty)$, where K is a local or global parameter, and p_∞ the static pressure at infinity. It is quite efficient for avoiding pressure drifts with a good choice of K, but is not enough for the test cases above.

3.2 The Present Boundary Condition

Let $A_{i+1,j}$ (resp. $A_{i+2,j}$) be a fictitious node, outside the domain, symmetric to $A_{i-1,j}$ (resp. $A_{i-2,j}$) with regards to Γ. If $U^n_{i+1,j}$ and $U^n_{i+2,j}$ were available, then we would write, with a second order scheme in space :

$$(L_1)^n_{i,j} = (u^n_{i,j} - c^n_{i,j})(\frac{-p^n_{i+2,j}+4p^n_{i+1,j}-3p^n_{i,j}}{2\Delta x} - \rho^n_{i,j}c^n_{i,j}\frac{-u^n_{i+2,j}+4u^n_{i+1,j}-3u^n_{i,j}}{2\Delta x}).$$

Nevertheless, as the $p^n_{i+2,j}$, $p^n_{i+1,j}$, $u^n_{i+2,j}$ and $u^n_{i+1,j}$ values are not known, we have to find a way to get them.

1. We first determine a global reference steady state $\overline{V}$, in the neighbourhood of $\Gamma : \overline{V} = [\overline{\rho}; \overline{u}; \overline{v}; \overline{p}]$. $\overline{V}$ depends on the flow under consideration, and can be obtained either analytically or numerically.
2. Then, $(V')^n$ is defined in the computational domain as a perturbation of $\overline{V}$: $(V')^n = V^n - \overline{V}$, and we write :

$$(L_1)^n_{i,j} = (\overline{L_1})^n_{i,j} + (L_1')^n_{i,j}$$

$$= (u^n_{i,j} - c^n_{i,j})(\frac{-\overline{p}_{i+2,j}+4\overline{p}_{i+1,j}-3\overline{p}_{i,j}}{2\Delta x} - \rho^n_{i,j}c^n_{i,j}\frac{-\overline{u}_{i+2,j}+4\overline{u}_{i+1,j}-3\overline{u}_{i,j}}{2\Delta x})$$

$$+ (u^n_{i,j} - c^n_{i,j})(\frac{-(p')^n_{i+2,j}+4(p')^n_{i+1,j}-(p')^n_{i,j}}{2\Delta x}$$
$$-\rho^n_{i,j}c^n_{i,j}\frac{-(u')^n_{i+2,j}+4(u')^n_{i+1,j}-3(u')^n_{i,j}}{2\Delta x}).$$

$(\overline{L_1})^n_{i,j}$ is easily computed with the knowledge of $\overline{V}$.
3. We assume that the perturbations arround $\overline{V}$ at node $A_{i,j}$ are convected with a convection velocity u_{conv}, which depends on the flow under consideration as we shall see later. Mainly, this convection velocity is the mean flow velocity. Then, we make the approximation :

$$(L_1')^n_{i,j} \approx (u^n_{i,j} - c^n_{i,j})(\frac{-(p')^{n-2\alpha}_{i,j}+4(p')^{n-\alpha}_{i,j}-3(p')^n_{i,j}}{2\Delta x}$$
$$-\rho^n_{i,j}c^n_{i,j}\frac{-(u')^{n-2\alpha}_{i,j}+4(u')^{n-\alpha}_{i,j}-3(u')^n_{i,j}}{2\Delta x}),$$

with $t_{n-\alpha} = t_n - \alpha\Delta t$ and $\alpha = \dfrac{\Delta x}{u_{conv}\Delta t}$. Since $\overline{V}$ is a steady state, we have:

$$(L_1')^n_{i,j} \approx (u^n_{i,j} - c^n_{i,j})(\frac{-p^{n-2\alpha}_{i,j}+4p^{n-\alpha}_{i,j}-3p^n_{i,j}}{2\Delta x} - \rho^n_{i,j}c^n_{i,j}\frac{-u^{n-2\alpha}_{i,j}+4u^{n-\alpha}_{i,j}-3u^n_{i,j}}{2\Delta x}).$$

The PBC can be used for a local subsonic inflow. In this particular case, $u^n_{i,j} \leq 0$, and $(L_2)^n_{i,j}$ and $(L_3)^n_{i,j}$ have also to be specified and can not be approximated from the solution inside the domain. In fact, the way to proceed is exactly the same than for $(L_1)^n_{i,j}$.

4 Numerical validation

The boundary layer evolution on a flat plate test case in then performed. The PBC is applied on the outflow boundary, with $\overline{V}$ equal to the Blasius profile and u_{conv} taken as the convection velocity very far from the plate u_∞. By noting η the classical reduced distance, (u_{ex}, v_{ex}) the velocity components of the exact solution, and Re_x the Reynolds number based on the distance to the leading edge, we get:

$$u_{ex} = u_\infty \, u_{analy}(\eta) \qquad \text{and} \qquad v_{ex} = \frac{u_\infty}{\sqrt{Re_x}} \, v_{analy}(\eta).$$

$v_{num}(\eta)$, obtained by the simulation, is the compared to $v_{analy}(\eta)$ on the middle transverse section and on the outflow transverse section (Fig. 2). The PBC gives better results. Then, the inviscid vortex test case is performed and the results

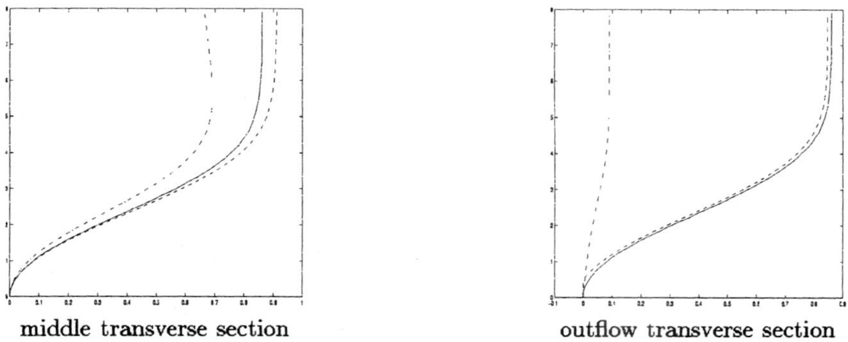

Fig. 2. $\eta = f(v_{num})$. Comparison of the NRBC solution (-.-.-.) and the PBC solution (- - -) to $\eta = f(v_{analy})$ (plain line).

are compared to a Reference Solution (RS). We plot the reflection coefficient $r(t) = \dfrac{\|div\,\vec{u}(t)\|_{L^\infty(\Omega)}}{\|\omega_0\|_{L^\infty(\Omega)}}$ as a function of the time (Fig. 3). Theoretically, this quantity has to decrease from the beginning to the end of the simulation, and to be equal to zero when the vortex is no more in the domain of simulation. This is the case when the PBC is used on the outflow boundary, with $\overline{V}$ equal to the state of the flow at infinity, and u_{conv} the velocity at infinity. But the NRBC exhibits a high increase of the graph, what is not physically relevant. Moreover, the PBC provides a correct pressure field distribution when the vortex center is located on the outflow boundary, whereas the NRBC destroys this distribution (Fig. 4). Up to now, we proved that the PBC leads to good results on simple well-known

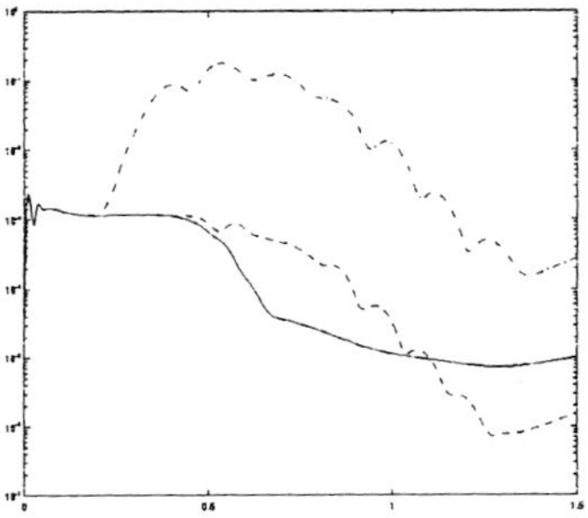

Fig. 3. $r(t)$. Comparison of the NRBC solution (-.-.-.) and the PBC solution (- - -) to the RS (plain line).

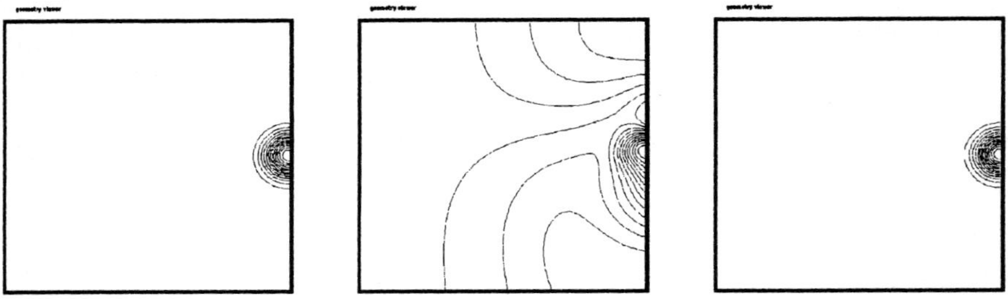

Fig. 4. Vortex center on the outflow boundary, pressure isolines. Comparison of the NRBC solution (middle) and the PBC solution (right) to RS (left).

test cases, for which the exact value of u_{conv} and $\overline{V}$ are available. Numerical tests have been performed on more complex configurations, for which the exact value of L_1 is not known, in order to actually evaluate the PBC efficiency in these more difficult simulations. $\overline{V}$ and u_{conv} are then obtained thanks to a reference flow given by a previous numerical simulation at a lower Reynolds number with the use of a buffer zone. Even if no analytical solution is available to interpret the results, we observed that the PBC provides a more physical behaviour of the solution than the NRBC does [2].

References

1. T.J. Poinsot, S.K. Lele : J. Comp. Phys. **101**, 104–129 (1992)
2. C.H. Bruneau, E. Creusé: 'Towards a transparent boundary conditions for the compressible Navier–Stokes Equations', submitted.

An Efficient Numerical Algorithm for the Tree-data Based Flow Solver

Takanobu Ogawa

Shimizu Corp., Institute of Technology, 3-4-17, Etchujima, Koto-ku, Tokyo, JAPAN
E-mail: ogawa@sit.shimz.co.jp

Abstract. An efficient numerical algorithm is developed for the tree data structure which have been widely used for the adaptive Cartesian mesh method. A simple neighbor cell finding with a "*pedigree*" and implementation of the implicit scheme and the Multigrid method are presented. It is found that the numerical schemes developed for the conventional flow solver can be easily implemented to the tree-data based flow solver.

1 Introduction

Mesh generation for complex geometry is still the most time-consuming task in CFD. Recently, the adaptive Cartesian mesh method has been successfully applied to practical complex flow fields [1],[2]. In this method, the Cartesian mesh is locally and recursively subdivided where high spatial resolution is needed, and mesh generation can be automated.

There are several data structures possible to organize the recursively subdivided Cartesian mesh. For example, the conventional unstructured mesh data structure is one alternative. Another approach is the tree data structure which is suitable for the hierarchical data such as the recursively subdivided mesh. The tree data structure is well known in the computer science field, however, it has not yet been a common data structure in CFD. This motivated the present author to examine the tree data structure for a flow solver.

In this paper, an efficient numerical algorithm for the tree-data based flow solver is developed. Although there are many topics about the adaptive Cartesian mesh method such as mesh generation around a curvilinear boundary, this paper mainly focuses on a fundamental algorithm for the tree data structure.

2 Numerical Algorithm

2.1 The tree data structure

The adaptive Cartesian mesh is generated from a single cell that covers an entire flow field and the cell is recursively subdivided into 8 identical child cells (4 child cells in 2D) as shown in Fig. 1 until sufficient spatial resolution is obtained.

Figure 2 shows a typical 2D cell distribution. The basic equations are dicretized with the cell-centered FVM. Although the artificial compressibility method [3] is used for demonstration in this study, the algorithm presented here can be applied to the other methods. The basic equations are discretized as,

$$\frac{\partial}{\partial t}(QV)_i + \sum_l (F - F_v) \cdot s_l = 0 \,. \tag{1}$$

The Roe's Flux Difference Splitting [4] is used for the convective terms and the flow variables at the both side of the cell interface are obtained by reconstruction. For the viscous terms, the velocity gradient is averaged at the cell interface. All these methods are based on the conventional unstructured mesh solver.

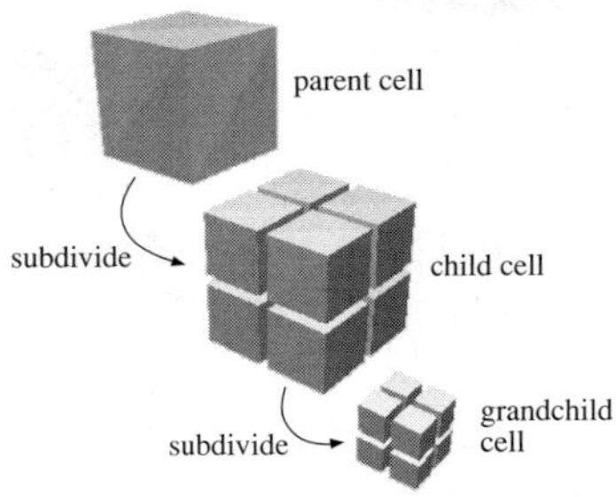

Fig. 1. Cell subdivision

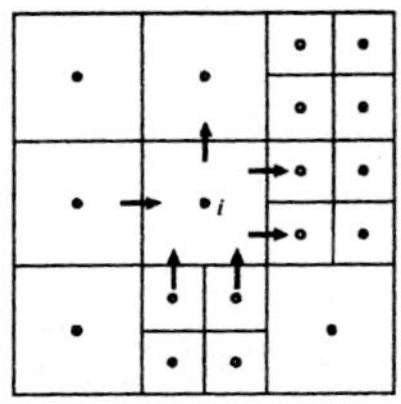

Fig. 2. The cell-centered FVM

Figure 3 shows a sample tree (octree) data structure. The cell at the top is called the root cell that is the largest cell. A parent cell has a pointer to its child cells. In this study, the generation of the cells is denoted as level and a cell at the lowest level is called a leaf cell. For mesh adaptation, cells are appended or removed only at the lowest level. The cell level difference between adjacent cells is restricted to 1.

Only the leaf cells are used for flow computation except for the case of the Multigrid method as is described later. In the tree data structure, the leaf cells can be tracked down from the root cell following the pointer to the child cells. This procedure can be carried out simply with the recursive subroutine call and the sample pseudo program code is shown in Fig. 4. `sub` is a subroutine for flow computation such as flux calculation which is done in "`procedures on the cell`" in the code.

2.2 Neighbor cell finding

The neighbor cell is needed to be found for flow computation, since it is not explicitly known in the tree data structure. In this study, the tree traversing is used for neighbor cell finding.

Figure 5 shows an example of the tree traversing in a 2D quadtree. To find the gray cell which is the neighbor of the black cell, one climbs the tree up to

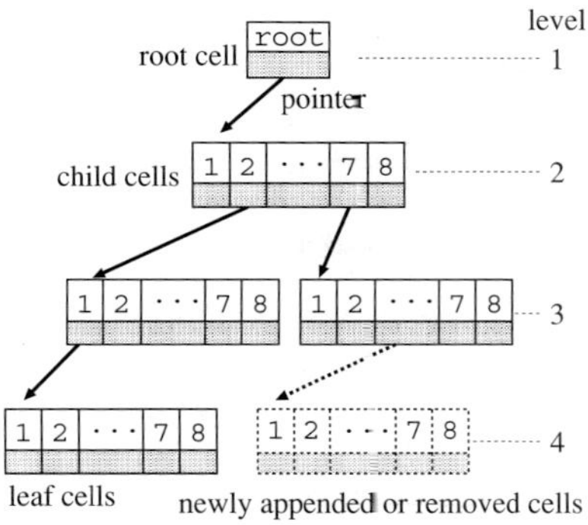

Fig. 3. A sample tree (octree) data structure

```
sub(cell pointer){
  if (the cell has children)
    for each child
      sub(child cell pointer);
  else
    procedures on the cell;
  endif
  return;
}
```

Fig. 4. The pseudo program code for the fundamental loop

the ancestor of the gray cell and then descends the tree to reach at the gray cell. This method can save memory since no pointers to the neighbor cells are stored, however, overhead for the tree traversing should be minimized.

In this study, the "*pedigree*" of a cell is utilized for fast and simple neighbor cell finding. Figure 6 shows the definition of the pedigree of the black cell for explanation. Cell position is denoted with 0 or 1 in each x and y direction according to the position in its parent cell. The pedigree of a cell is a table that contains all the cell positions of the family of the cell.

Figure 7 illustrates the meaning of the pedigree. Cell position in the root cell is expressed with the binary tree in each direction and the hatched area in the binary trees corresponds to the pedigree. It is found that the pedigree itself is the cell position in the root cell. Thus, the pedigree of the neighbor cell is obtained from the relative position to the target cell as shown in Fig. 8.

Using the pedigree, one can find a neighbor cell without climbing up the tree and the tree traversing becomes one-way. For a test case (3D octree, 200,000 cells at 5 level subdivision), the overhead for neighbor cell finding was about 12% to the total CPU time. Though the example shown here is two dimensional for simplicity, this algorithm can be extended to the three dimension without modification.

2.3 Implicit time integration scheme

The implicit time integration scheme is important for fast convergence particularly when extremely small cells appear after a number of subdivision. For the tree data structure, the implicit scheme is formulated as the conventional unstructured mesh method. For example, the LU-SGS method [5] is written as,

$$\begin{aligned} Forward: &\quad \Delta Q_i^* = D^{-1}(RHS^n - \textstyle\sum^{lower} sA^+\Delta Q), \\ Backward: &\quad \Delta Q_i = \Delta Q_i^* + D^{-1}\textstyle\sum^{upper} sA^-\Delta Q\,. \end{aligned} \tag{2}$$

D is the spectral radius of the flux Jacobian matrix $A(=\partial F/\partial Q)$. The summation is done for the upper and lower cells of the cell i.

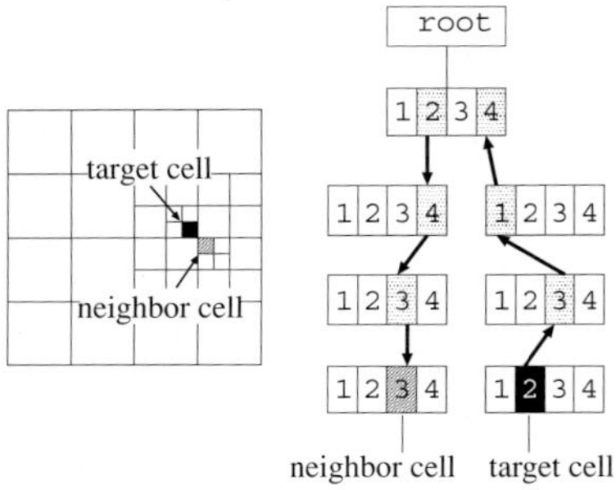

Fig. 5. Neighbor cell finding by tree traversing

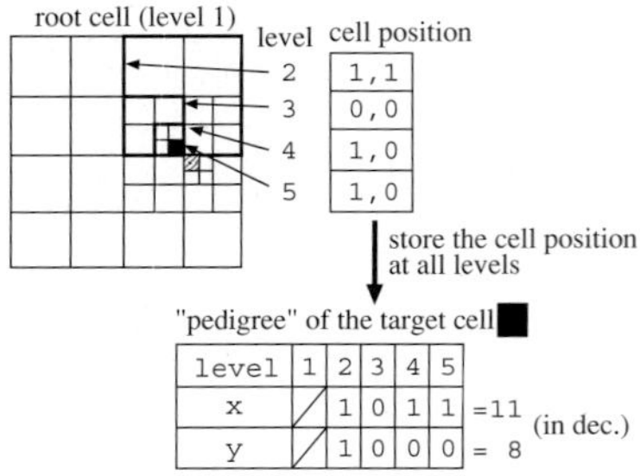

Fig. 6. Definition of the "pedigree"

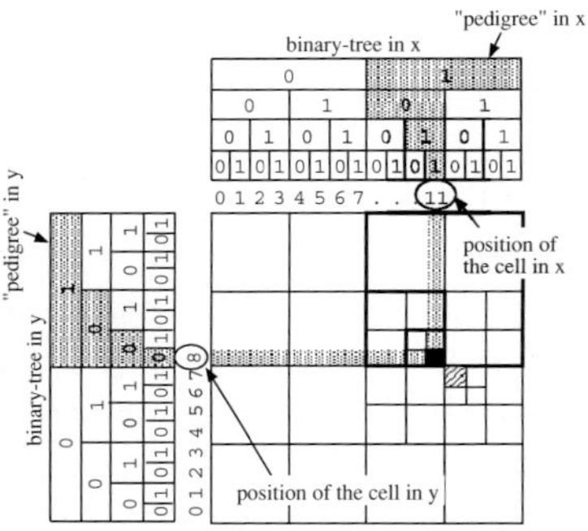

Fig. 7. Meaning of the "pedigree"

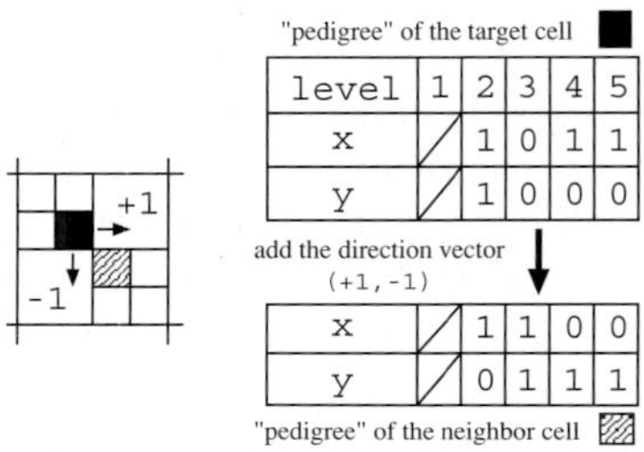

Fig. 8. The "pedigree" of the neighbor cell

Since the tree data structure has regularity in cell alignment, the lower and upper cells are naturally defined like the structured mesh. The sweep for the implicit schemes can be carried out by recursive subroutine call in appropriate order which can be controlled in "**for each child**" statement in Fig. 4. For example, Fig. 9 shows the order of subroutine call for the LU-SGS scheme. Even when a cell is newly subdivided or removed, the order does not need to be changed.

Since the type of the sweep can be controlled only by the order of subroutine call, not only the LU-SGS method but also the other methods, such as the ADI method, can be implemented in the similar manner.

2.4 Computational results

A simple flow field, a lid-driven cavity flow at $Re = 100$ is solved as a test case. Figure 10 (a) is the final cell distributions when the cells are divided up to 3 levels and Fig. 10 (b), (c) shows the pressure contour and the velocity distribution plots with the experimental data [6].

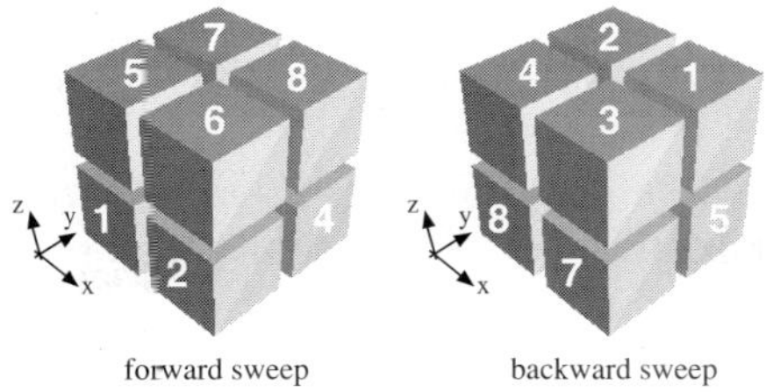

Fig. 9. Sweep order to call a subroutine for child cells in case of the LU-SGS scheme

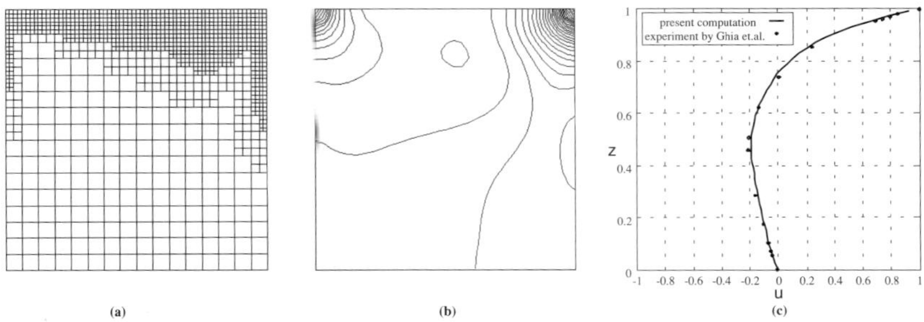

Fig. 10. The computational result of the lid-driven cavity at Re=100 (**a**) the final cell distributions, (**b**) the pressure contour plots, (**c**) the velocity distribution

2.5 The Multigrid method

In the tree data structure, a parent cell itself is a coarser mesh of its child cells as illustrated in Fig. 11. If not only the leaf cells but also their ancestor cells are used for computation, no additional data structure is required for the Multigrid method. Thus, the tree data structure is inherently suitable for the Multigrid method.

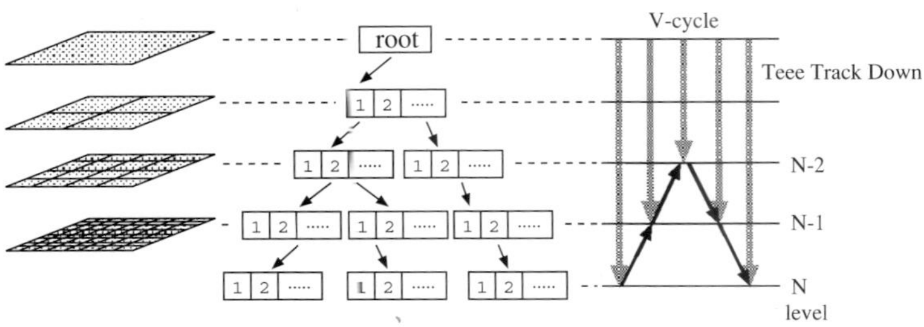

Fig. 11. The Multigrid method for the tree-data structure

The program code needs only minor modification. Fig. 12 shows the pseudo code of a subroutine inside the Multigrid cycle. Only one additional condition (level<levelMG) is appended to the original code (Fig. 4) so that the cells at the intermediate level (levelMG) are to be used for flow computation during the Multigrid cycle.

For the test case, the Full Approximate Storage [7] scheme is applied to the lid-driven cavity shown previously. The residual histories in Fig. 13 shows the convergence is accelerated for this simple test case.

```
sub(cell pointer){
  if ((the cell has children)
        and (level<levelMG))
    for each child
      sub(child cell pointer);
  else
    procedures on the cell;
  endif
  return;
}
```

Fig. 12. The pseudo program code for the Multigrid method

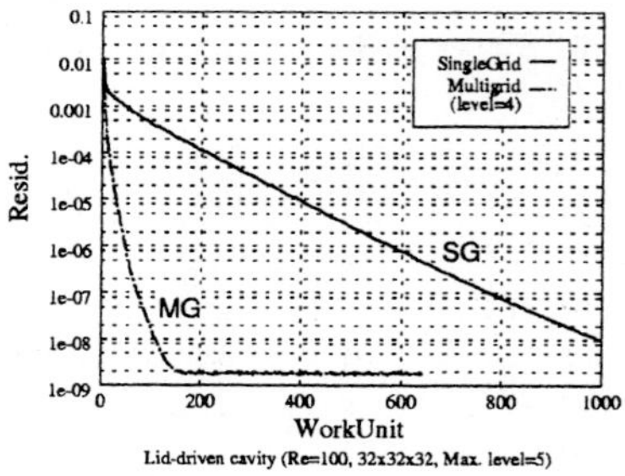

Fig. 13. The residual histories of the lid-driven cavity at Re=100

3 Conclusions

An efficient numerical algorithm is developed for the tree-data based flow solver. The one-way neighbor cell finding by a "*pedigree*" of a cell and implementation of the implicit scheme and the Multigrid method are presented with a very simple algorithm. It is found that the property of the tree data structure is between those of the conventional structured and unstructured mesh data structure owing to regularity of cell alignment and capability for mesh adaptation.

References

1. D.D. Zeeuw and K.G. Powell: J. of Comp. Phys., 104, pp.56-68, (1993)
2. S.L. Karman: AIAA paper 95-0343, (1995)
3. A.J. Chorin: J. of Comp. Physics, No.2, pp.12-26, (1967)
4. P.L. Roe: J. of Comp. Physics, No.43, pp.357-372, (1981)
5. S. Yoon and A. Jameson: AIAA J., 26, pp.1025-1026, (1988)
6. K.N. Ghia and J.S. Sokye: J. Fluids Eng., 99, pp.640-648, (1977)
7. P.W. Hemker and B. Koren: Lecture Notes, von Karman Institute for Fluid Dynamics, May, (1988)

Imposing Boundary Conditions with the Injection, the Projection and the Simultaneous Approximation Term Methods

Ken Mattsson

Department of Scientific Computing, Information Technology, Uppsala University, S-751 04 Uppsala, Sweden

Abstract. Four different methods of imposing boundary conditions for the linear advection-diffusion equation and a linear hyperbolic system are considered. The methods are analyzed using the energy method and the Laplace transform technique. Numerical calculations are done, considering in particular the case when the initial data and boundary data are inconsistent. As an application we consider the propagation of a two-dimensional vortex.

1 Introduction

In this paper we consider accurate and stable finite difference approximations of the linear advection-diffusion equation and of a linear hyperbolic system. There is a variety of different stability definitions [1]. We will consider strict stability. A strictly stable approximation has the same time-growth as the continuous problem. Finite difference operators that obey a summation by part property (SBP) [11] will result in a strictly stable approximation for a Cauchy problem. For an initial boundary value problem one must implement the boundary condition (BC) in a way that does not destroy the SBP property.

To impose the BC explicitly, i.e. to combine the difference operator and boundary operator into a modified operator might destroy the SBP property. This method is referred to as the injection method, and is often used in practical calculations. The basic idea behind the Simultaneous Approximation Term (SAT) method [7] and the projection method [3],[4] is to impose BC so that the SBP property is not destroyed. The method introduced in [10] is a "hybrid" between the injection method and the projection method. In this article differences between these methods are analyzed. A more detailed analysis can be found in [12].

2 Definitions

Consider the problem,

$$\begin{aligned} w_t + Qw &= F(x,t) \\ Lw(0,t) &= g(t) \\ w(x,0) &= f(x) \end{aligned} \tag{1}$$

where Q is the differential operator, L the boundary operator, F the forcing function, f the initial data and g the boundary data. Consider also the slightly perturbed version of (1), with solution v and initial data $f + \delta f$. The error $u = w - v$ satisfies (1) with $F = g = 0$ and initial data f.

If (1) is well posed [1], an energy estimate exists

$$||u|| \leq K_c e^{\alpha_c t} ||\delta f|| \ , \tag{2}$$

where K_c and α_c do not depend on t. For a strictly stable difference approximation of (1), a corresponding discrete energy estimate,

$$||u||_H \leq K_d e^{\alpha_d t} ||\delta f||_H \ , \tag{3}$$

where $\alpha_d \leq \alpha_c + \mathcal{O}(h)$ exist. This means that a strictly stable approximation has the same time-growth as the continuous problem.

3 The continuous scalar problem

Consider the model problem

$$\begin{aligned} & u_t + a u_x = \epsilon u_{xx} \qquad 0 \leq x \leq 1, \qquad t \geq 0 \\ & u(0,t) + \beta u_x(0,t) = 0 \quad , \ u_x(1,t) = 0 \\ & u(x,0) = f(x) \end{aligned} \tag{4}$$

where $a\,, \epsilon > 0$ and $a >> \epsilon$. The problem (4) has an energy estimate [12] if $-2\epsilon/a \leq \beta \leq 0$. Provided that the solution exist (can be shown by using Laplace-transform technique or via difference approximations: see [2] and [5]), we can conclude that (4) is well-posed.

The spectrum of the continuous problem (continuous spectrum) is calculated using the Laplace-transform technique. The result is shown in table 1. The utmost right eigenvalue in the continuous spectrum correspond to α_c, see (2). Roughly speaking, the continuous spectrum is obtained as the singular values of the solution to the Laplace-transformed version of (4). Details on how to compute the spectrum are given in [12].

4 The semi–discrete scalar problem

In this section we will discretize the problem (4), using second, fourth and sixth order accurate difference operators, satisfying the SBP property. These operators were presented in [6]. The BC are imposed using the injection method, the SAT method, the projection method and the modified projection method. A detailed analysis is given in [12].

In [12] the spectrum of the discrete problem (discrete spectrum) for the SAT, the projection, the modified projection and the injection methods is calculated.

Table 1. The two utmost right values of the continuous spectrum for different values of β; $\epsilon = 0.1$ and $a = 1$

β	spectrum (1)	spectrum (2)
$-\epsilon/2a$	-3.13023510	-5.15428651
$-\epsilon/a$	-3.02187288	-4.76698726
$-2\epsilon/a$	-2.67261695	-4.12696691

The semi-discrete equations are written as $v_t = A_{I,P,S,M}\, v$, where $A_{I,P,S,M}$ are the constant coefficient matrices of the semi-discrete problems. The discrete spectra are the eigenvalues of these matrices. The utmost right eigenvalue in the discrete spectrum converge to the utmost right eigenvalue in the continuous spectrum for all methods, except for the projection method.

The projection method has a zero eigenvalue in the spectrum. This means that any error introduced into the computation will remain in the solution. This is also verified in the numerical calculations.

We consider the case when the boundary and the initial data are inconsistent. The l_2 error as a function of time is presented in figure 1 . 50 grid points are used and the solutions are advanced to time $t = 10$, using the standard fourth-order Runge-Kutta method. The error is defined as the difference between the disturbed and the undisturbed solution.

Note that the error for the projection method remains in the solution due to the zero eigenvalue.

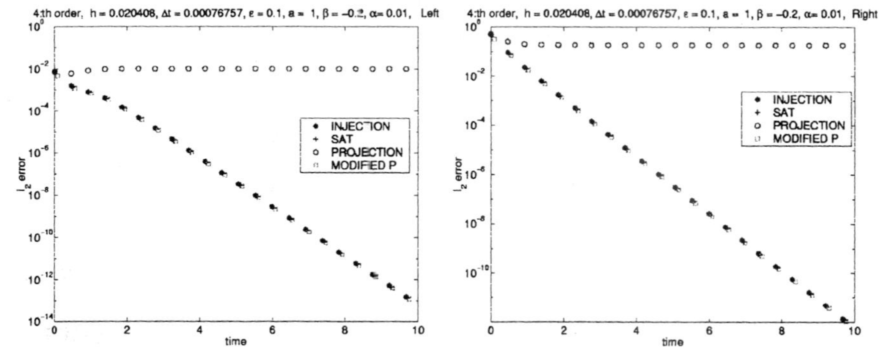

Fig. 1. The l_2 error as a function of time for the fourth order accurate case, inconsistency at the left and right boundary respectively

5 The hyperbolic systems

Consider the hyperbolic system

$$
\begin{aligned}
&u_t + u_x = 0 \qquad 0 \le x \le 1, \qquad t \ge 0\\
&v_t - v_x = 0 \qquad 0 \le x \le 1, \qquad t \ge 0\\
&u(0,t) = v(0,t) \quad , \; v(1,t) = u(1,t)\\
&u(x,0) = g(x) \quad , \; v(x,0) = h(x) \; .
\end{aligned}
\tag{5}
$$

The energy method on (5) leads to $\frac{d}{dt}\left(||u||^2 + ||v||^2\right) = 0$, i.e. the energy is constant, and the continuous spectrum consist of the single point, $s = 0$.

In [12] the discrete spectrum for the SAT, the projection, the modified projection and the injection methods are calculated. It is shown that the injection method have positive eigenvalues for the fourth and sixth order case, see table 2.

Table 2. The largest real part of the discrete spectra for the sixth order case

N	SAT	$Projection$	$Injection$	$ModifiedP$
50	0.000e+00	2.953e-07	1.112e+00	1.599e-14
100	-2.665e-15	6.388e-14	1.209e+00	2.331e-14
150	-3.553e-15	8.533e-07	1.249e+00	7.461e-14
200	1.277e-15	2.075e-12	1.278e+00	6.395e-14
250	-7.772e-16	5.690e-14	1.290e+00	8.527e-14

The exact solution to (5) is

$$u(x,t) = \sin 2\pi(x-t) \;,\; v(x,t) = -\sin 2\pi(x+t) \quad ,0 \le x \le 1,\; t \ge 0\,,$$

with initial data $u(x,0) = \sin 2\pi x$, $v(x,0) = -\sin 2\pi x$. To further investigate, numerically, if the the methods are strictly stable we compute the l_2-error for long time integrations, see figure 2.

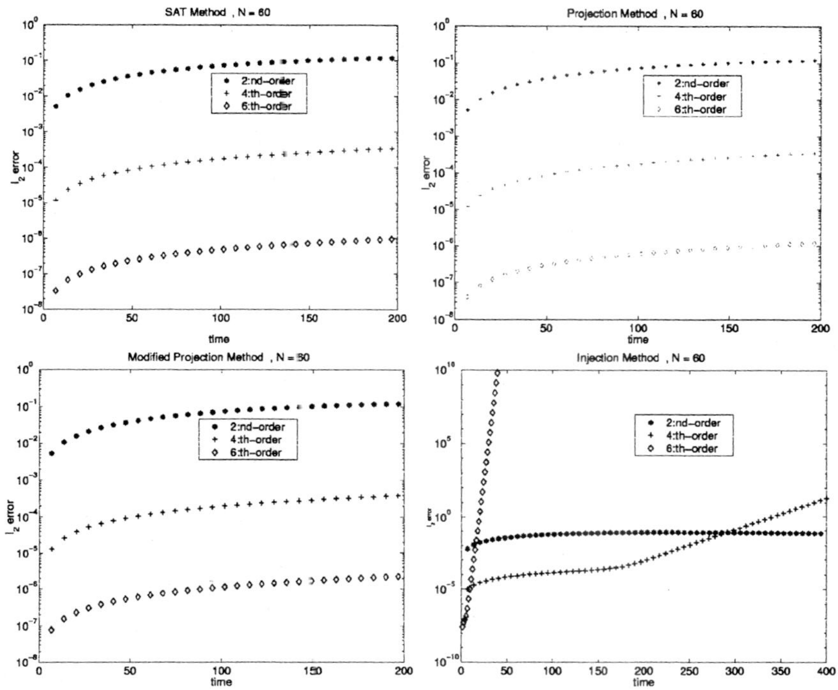

Fig. 2. l_2-error as a function of time, for 2:nd, 4:th and 6:th order case, $N = 60$

Obviously the SAT method, the projection method and the modified projection method have bounded error growth in time. The l_2-error for the injection method grows rapidly for the fourth and sixth order case, see figure 2, due to the eigenvalues with positive real part.

6 Application

As a typical application we consider the propagation of a vortex. Typical length and time scales are $L = 10$ and $t = 100$. The vortex model is presented in [8]. It satisfies the two-dimensional Euler equations, under the assumption of isentropy. In [9] it is shown that the solution is steady in the frame of reference moving with the freestream. The vortex is introduced into the computational domain by using the analytic solution as boundary data and initial data. The numerical calculations are done by using the SAT method, see figure 3.

The l_2-error at time $t = 100$ for the sixth order accurate case is of magnitude 10^{-5}, using 330×30 points. To produce the same order of accuracy using a second order accurate method would require $(330)^3 \times (30)^3$ points.

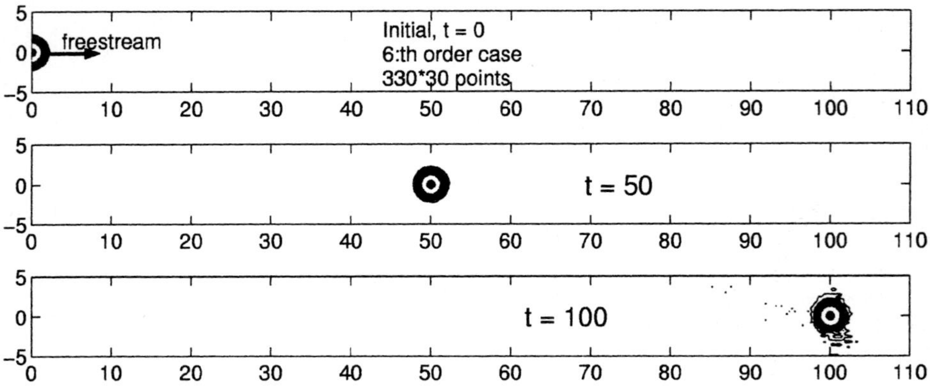

Fig. 3. Vortex calculation using a sixth order SAT method

References

1. B. Gustafsson, H.–O. Kreiss and J. Oliger: *Time dependent problems and difference methods* (Wiley, New York 1995)
2. H.–O. Kreiss and J. Oliger: *Initial Boundary Value Problems and the Navier–Stokes Equations* (Academic Press, New York 1989)
3. P. Olsson: Summation by parts, projections, and stability I. Math. Comp. **64**, 211 (1995)
4. P. Olsson: Summation by parts, projections, and stability II. Math. Comp. **64**, pp. 1473–1493 (1995)
5. J. Nordström: The Influence of Open Boundary Conditions on the Convergence to Steady State for the Navier–Stokes Equations. J. Comput. Phys. **85** (1989)
6. M. H. Carpenter, J. Nordström and D. Gottlieb: A Stable and Conservative Interface Treatment of Arbitrary Spatial Accuracy. J. Comput. Phys. **148** (1999)
7. M. H. Carpenter, D. Gottlieb and S. Abarbanel: The stability of numerical boundary treatments for compact high–order finite–difference schemes. J. Comput. Phys. **108**, 2 (1994)
8. G. Erlebacher, M. Y. Hussaini and C.–W. Shu: Interaction of a Shock with a Longitudinal Vortex. J. Fluid. Mech.. **337**, pp. 129–153 (1997)
9. E. Petrini, G. Efraimsson and J. Nordström: A Numerical Study of the Introduction and Propagation of a 2–D Vortex. Tech. Rep., The Aeronautical Research Institute of Sweden, Bromma, Sweden, FFA TN 1998–66
10. B. Gustafsson: On the implementation of boundary conditions for the methods of lines. BIT **38**, 2 (1998)
11. B. strand. High–Order difference approximations for hyperbolic initial boundary value problems. PhD thesis, Dept. of Information Technology, Uppsala Univ., Uppsala, Sweden (1996)
12. K. Mattsson. Imposing Boundary Conditions with the Injection, the Projection and the Simultaneous Approximation Term Methods. Tech. Rep., Dept. of Information Technology, Uppsala Univ., Uppsala, Sweden, 2000–016.

DNS of the Incompressible Flows Around a Sphere Using Cartesian Grid Approach with Virtual Boundary Method

Hidetoshi Nishida[1], Katsunori Horii[2], and Nobuyuki Satofuka[1]

[1] Department of Mechanical & System Engineering, Kyoto Institute of Technology, Matsugasaki, Sakyo-ku, Kyoto 606-8585, JAPAN
[2] Matsushita Refrigeration Company, Higashiosaka-city, Osaka 577-8501, JAPAN

Abstract. The DNSs of the incompressible flows around a sphere are performed by using the Cartesian grid approach with virtual boundary method. In the virtual boundary method, the body surface is expressed by a set of plural points, and the velocity components on virtual boundary are feedbacked to the momentum equations as the additional forcing terms. The DNSs are carried out in the range from $Re = 10$ to $Re = 1000$. In the steady flows up to $Re = 240$ in order to validate the present method, the flow characteristics are in good agreement with the other results. In the unsteady flows, it is clearly observed that the hairpin vortex ring is released and the flow transits from laminar to turbulence. Consequently, the present method is very promising for the DNSs of the incompressible complicated flow.

1 Introduction

In the direct numerical simulations (DNSs) around a body, the body fitted coordinates (BFC)[1] are usually used. This BFC approach has the high adaptability to the body configuration. However, in the BFC approach, it is necessary that the governing equations are transformed from the physical plane to the computational plane, so that the transformed governing equations have more terms than the original equations. Therefore, the computational effort is larger than the basic Cartesian coordinates. And it is difficult and takes more computational time that the body fitted coordinate system is generated for the complicated body geometries.

Another approach is to use the Cartesian grid, because it is more easy to generate the grid. But, it has a problem that the grid lines are not along the body geometries. Accordingly two techniques to solve the problem are considered, that is, the cut cell technique[2] and the virtual boundary method[3,4]. In the cut cell technique, the boundary points are defined by the intersections of the body surface and the Cartesian grid lines. And in the interior region of the boundary, it is not necessary to compute. On the other hand, the boundary points are defined by the body surface points independent of the Cartesian grid lines and it is necessary to compute the interior region, in the virtual boundary method. In this paper, we adopt the virtual boundary method and apply the method to the DNS of the incompressible flows around a sphere.

2 Cartesian Grid Approach with Virtual Boundary Method

2.1 Virtual Boundary Method

In the virtual boundary method, the body surface is expressed by a set of plural points. Figure 1 shows the concept of virtual boundary method. The computational procedure of virtual boundary method is explained by using Fig.1. For simplicity, we explain here in 2D. The extension to 3D is straightforward.

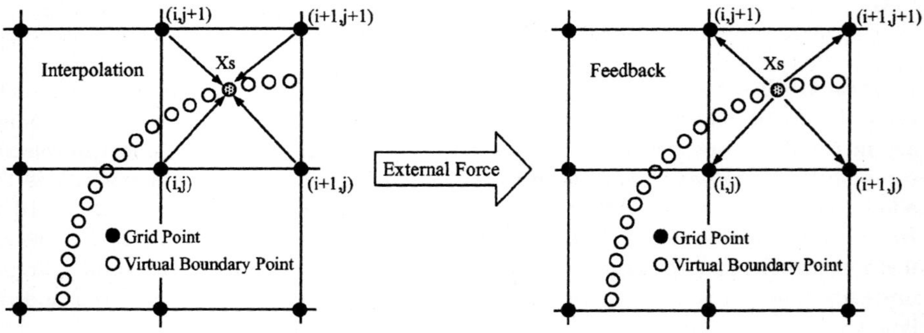

Fig. 1. Concept of virtual boundary method

First, the velocity components $\mathbf{U}(\mathbf{x}_s)$ on the virtual boundary point $\mathbf{x}_s = (x_s, y_s)$ are estimated by using the interpolation from four Cartesian grid points, i.e., (i,j), $(i+1,j)$, $(i,j+1)$, and $(i+1,j+1)$. Then, the velocity components $\mathbf{U}(\mathbf{x}_s)$ can be obtained by the relation,

$$\mathbf{U}(\mathbf{x}_s) = \sum_{i,j}^{i+1,j+1} W_{i,j}(\mathbf{x}_s)\mathbf{u}_{i,j}, \tag{1}$$

where $\mathbf{u}$ is the velocity on the Cartesian grid point and $W_{i,j}(\mathbf{x}_s)$ denotes the weighted coefficient defined by

$$W_{i,j}(\mathbf{x}_s) = w(x_s - x_i)w(y_s - y_j), \tag{2}$$

$$\begin{aligned} w(x_s - x_i) &= \frac{x_s - x_{i+1}}{x_i - x_{i+1}} \quad if \;\; x_i < x_s, \\ w(x_s - x_i) &= 1 \quad if \;\; x_i = x_s, \\ w(x_s - x_i) &= \frac{x_s - x_{i-1}}{x_i - x_{i-1}} \quad if \;\; x_i > x_s. \end{aligned} \tag{3}$$

Next, the additional forcing terms $\mathbf{G}(\mathbf{x}_s, t)$ on the virtual boundary point $\mathbf{x}_s$ is determined by

$$\mathbf{G}(\mathbf{x}_s, t) = \alpha \int_0^t (\mathbf{U}(\mathbf{x}_s) - \mathbf{v}(\mathbf{x}_s))dt + \beta(\mathbf{U}(\mathbf{x}_s) - \mathbf{v}(\mathbf{x}_s)), \tag{4}$$

where $\mathbf{G}$ and $\mathbf{v}$ denote the forcing terms and the specified velocity vector on the virtual boundary point. The parameters α and β are specified by $\alpha = -400000$, $\beta = -600$ [3]. Then, the forcing terms on the Cartesian grid point $x_{i,j}$, $\mathbf{G}_{i,j}$, can be obtained by

$$\mathbf{G}_{i,j} = \frac{1}{N_s} \sum_{n=1}^{N_s} W_{i,j}(\mathbf{x}_s)\mathbf{G}(\mathbf{x}_s, t). \tag{5}$$

N_s is the total virtual boundary points in the cell.

2.2 Governing Equations

The incompressible Navier-Stokes equations can be written in the nondimensional form by

$$\frac{\partial u_i}{\partial x_i} = 0, \tag{6}$$

$$\frac{\partial u_i}{\partial t} + u_j \frac{\partial u_i}{\partial x_j} = -\frac{\partial p}{\partial x_i} + \frac{1}{Re}\frac{\partial^2 u_i}{\partial {x_j}^2} + G_i, \tag{7}$$

where Re denotes the Reynolds number defined by $Re = UL/\nu$. The last term in Eq.(7), G_i, denotes the additional forcing term.

2.3 Numerical Method

The variable order method of lines[5] based on the fractional step approach[6] on the collocated grid is adopted to solve the Navier-Stokes equations (7). The spatial derivatives of the Navier-Stokes equations are discretized by using the variable order proper convective scheme[5] for convective terms and the modified differential quadrature method[5] for diffusion terms. The variable order method of lines can give the arbitrary order of spatial accuracy by changing only one parameter. In this paper, the spatial accuracy is fixed by the second order in all cases and the successive over relaxation method is adopted for pressure equation solver.

3 DNS of Incompressible Flows Around a Sphere

3.1 Computational Conditions

The computational domain is shown in Fig.2. The initial flow is fixed by the uniform flow. In the inflow boundary, the velocity components are fixed by the initial condition and the pressure is obtained by the Neumann condition. In the other boundaries, the pressure is estimeted by the Sommerfeld radiation condition and the velocity components are computed from the discrete continuity equation.

In this paper, the hierarchical Cartesian grids system shown in Fig.3 is introduced and the sphere surface is defined by the cubed sphere coordinates[7]. Table 1 shows the Cartesian grid system used in this work.

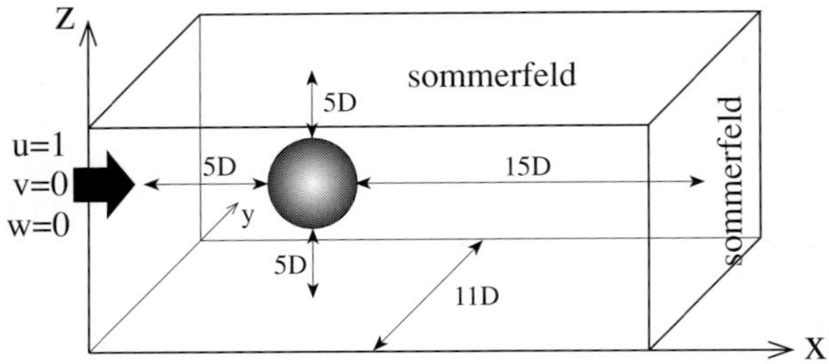

Fig. 2. Computational conditions

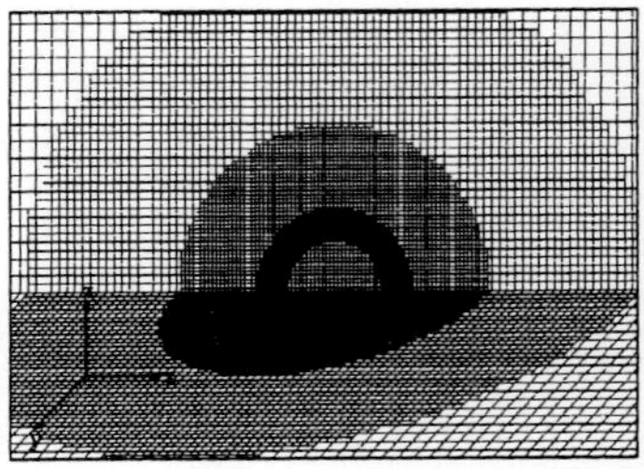

Fig. 3. Generated hierarchical Cartesian grid

Table 1. Hierarchical Cartesian grids system

	Reynolds number		
	10, 20, 40 , 100, 200, 240	300, 500	1000
Grid levels	5	5	7
Grid points	973,744	1,855,048	3,268,428
Virtual boundary points	153,600	153,600	614,400
Coarsest grid spacing	D/4	D/4	D/2

3.2 Numerical Results

For the incompressible flows past a sphere, it is repoted experimentally and computationally[8,9] that the flow is steady for $Re < 280$ and transits to turbulent flow for $Re > 800$. First we consider the steady state flows. In these cases, time step is set to be $\Delta t = 2.0 \times 10^{-3}$. The flow characteristics (Cd and vortex length L) are shown in Table 2. It is in excellent agreement with each other.

Table 2. Flow characteristics for steady flows

	Present		Shirayama[9]	
Re	Cd	L	Cd	L
10	4.201	0.000	4.255	0.000
20	2.610	0.000	2.622	0.029
40	1.733	0.266	1.741	0.289
100	1.099	0.872	1.104	0.844
200	0.752	1.333	0.774	1.310
240	0.711	1.543	0.723	—

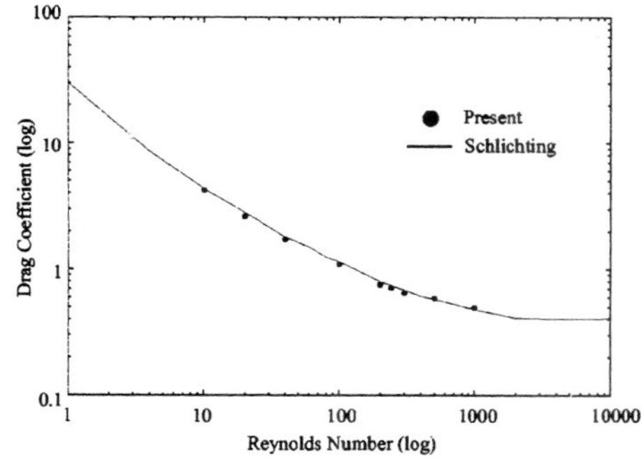

Fig. 4. Drag coefficients for a sphere

Next, we consider the unsteady flows up to Reynolds number $Re = 1000$. Figure 4 shows the drag coefficients in comparison with experiment[10]. The iso-surface of $\nabla^2 p$ with $Re = 300$ at nondimensional time $t = 121$, 124, 127, and 130 is shown in Fig.5. It is clearly observed that the hairpin vortex ring is released from sphere and flow field has the periodicity. Figure 6 shows the iso-surface with $Re = 1000$ and $\Delta t = 1.0 \times 10^{-3}$ similar to Fig.5. In this case, the released vortex ring from a sphere is broken down and generated many small eddies in the downstream. Then, the clear vortex ring is not observed in the wake region. These phenomena are repeated in every vortex ring release. Consequently the flow will be transited from the laminar to turbulent flow and it shows the same result as the experimental result[8].

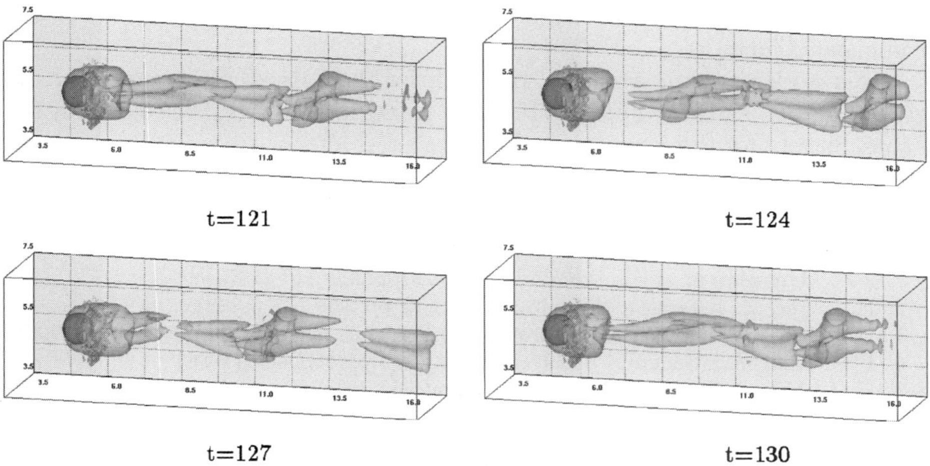

Fig. 5. Iso-surface of $\nabla^2 p$ ($Re = 300$)

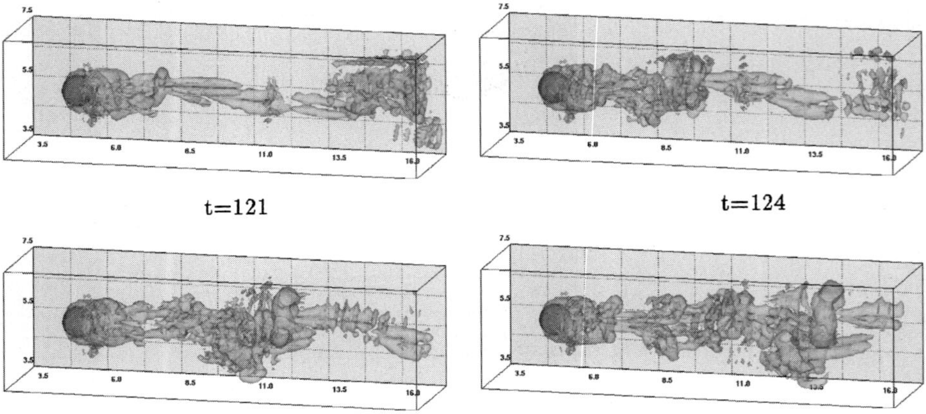

t=121 t=124

t=127 t=130

Fig. 6. Iso-surface of $\nabla^2 p$ ($Re = 1000$)

4 Concluding Remarks

In this paper, the Cartesian grid approach with virtual boundary method is developed for the DNS of incompressible flows around a sphere. The hierarchical grid system is introduced for ensuring the resolution. The DNSs are performed up to $Re = 1000$. The computational results show the good agreement with experimental and other DNS results. Especially, the transition from laminar to turbulent flow can be reappeared with $Re = 1000$. Then, it is concluded that the present method will be very promising for the DNSs of the incompressible complicated flows.

This work was supported by "Research for the Future" Program under the grant number JSPS-RFTF97P01101 and grant number No.12650165 from the Japan Society for the Promotion of Science.

References

1. J.F. Thompson et al.: J. Comput. Phys. **15** pp.299–319 (1974)
2. H. Forrer, R. Jeltsch: J. Comput. Phys. **140** pp.259–277 (1998)
3. D. Goldstein, R. Handler, L. Sirovich: J. Comput. Phys. **105** pp.354–366 (1993)
4. E.M. Saiki, S. Bringen: J. Comput. Phys. **123** pp.450–465 (1996)
5. H. Nishida: 'Higher order Direct Numerical SImulations of the Incompressible Navier-Stokes Equations Using a Variable Order Method of Lines'. In: *8th International Symposium on Computational Fluid Dynamics, Germany, September 5–10, 1999* (CD-ROM, 1999)
6. C.H. Rhie, W.L. Chow: AIAA J. **21-11** pp.1525–1532 (1983)
7. C. Ronchi et al.: J. Comput. Phys. **124** pp.93–114 (1996)
8. H. Sakamoto, H. Haniu: Transaction of the ASME **112** pp.386–392 (1990)
9. S. Shirayama: AIAA J. **30-2** pp.349–358 (1992)
10. H. Schlichting: *Boundary Layer Theory*, 7th edn. (McGraw-Hill, New York 1979)

Artificial Wind – a New Framework to Construct Simple and Efficient Upwind Shock-Capturing Schemes for Complex Hydrodynamic Problems

I.V. Sokolov[1], E.V. Timofeev[2], J. Sakai[1], and K. Takayama[2]

[1] Laboratory for Plasma Astrophysics, Faculty of Engineering, Toyama University, 3190 Gofuku, Toyama 930-8555, Japan
[2] Shock Wave Research Center, Institute of Fluid Science, Tohoku University, 2-1-1 Katahira, Aoba-ku, Sendai 980-8577, Japan

Abstract. Using a concept of Artificial Wind based on the Galilean invariance of the hydrodynamic equations, we outline a new framework to design upwind shock-capturing schemes. It allows to get simple, efficient and accurate schemes retaining their simplicity for a broad range of hydrodynamic problems. Test problems demonstrate that the newly derived schemes are quite competitive with traditional ones. For complex hydrodynamics systems (magnetohydrodynamics, relativistic hydrodynamics, complex equations-of-state etc.) they may provide at least one order of magnitude gains in efficiency while maintaining good accuracy.

1 Introduction

Many modern high-resolution numerical schemes for the hydrodynamic equations as well as for other hyperbolic systems of conservation laws employ to some extent an idea of upwinding. The numerical flux via a face between two adjacent cells should be decomposed into the sum of perturbations propagating across the face in one direction and those propagating in the opposite direction and numerical differences should be taken accordingly. This approach usually leads to high quality numerical results, but for nonlinear equations, such as the hydrodynamic ones, the practical construction of upwind numerical schemes is typically accompanied by a lot of technical difficulties to be overcome. For specific implementations of the upwinding approach it may be necessary to find, for instance, the Roe matrix with the total set of its eigenvalues and eigenvectors and/or calculate exact or approximate solutions of the Riemann problem. These tasks become more sophisticated for physical problems in which the effects of real fluids, MHD or special relativity should be taken into account. Each particular physical model may require new generalization of an upwind scheme and in some cases formulas for explicit calculations occupy literally several journal pages requiring tedious programming and reducing computational efficiency.

In the present paper we outline a new framework for constructing efficient upwind shock-capturing schemes for computational fluid dynamics. The basic idea for the first time proposed in [1] and subsequently elaborated in [2–5] is to solve the governing equations in different steadily moving frames of reference chosen in such a way that the flow would be supersonic there resulting in simple upwind formulas for fluxes across control volume faces. An additional velocity (*artificial wind*) is added to the velocity of the flow under simulation when the system of coordinates is changed. The approach, being very general, allows to derive new schemes as well as to get new interpretations or simplified versions of some already existing schemes ([3,4]).

2 The Concept of Artificial Wind

We pay attention to and use here the following fact: the 'technological complexity' of the upwinding problem is not Galilean invariant while the hydrodynamic equations are certainly Galilean invariant. That is why it is always possible to choose a frame of reference which moves with a velocity D with respect to the original frame of reference in such a way that all the flow under consideration becomes supersonic in the new frame. An additional velocity $-D$ is added to the velocity of the flow in the moving frame of reference making the flow to be supersonic. We refer to this velocity as an Artificial Wind (AW) velocity in order to emphasize that its value is a matter of our choice and that it is introduced to facilitate upwinding. The problem of upwinding becomes trivial in the moving frame of reference because in supersonic flows all perturbations propagate along one definite direction depending upon the direction of the AW velocity. The procedure for 'remapping' of numerical solution from the moving frame of reference to the original one may be arranged by several ways [3].

Instead of using a moving frame of reference, one may consider a spatial-temporal invariance of an arbitrary hyperbolic system $\partial\mathbf{U}/\partial t + \partial\mathbf{F}/\partial x = 0$, where $\mathbf{U}$ is the vector of conserved variables, $\mathbf{F}$ is the vector of their fluxes, t – time and x – a spatial coordinate. The equations are obviously invariant with respect to the transformation of time and space $t' = t$, $x' = x - Dt$ as long as the flux and the Jacobian matrix $\mathbf{A} = \partial\mathbf{F}/\partial\mathbf{U}$ are also transformed as follows:

$$\mathbf{F}' = \mathbf{F} - D\mathbf{U}, \quad \mathbf{A}' = \mathbf{A} - D\mathbf{E} \ . \tag{1}$$

Thus, all perturbations may be converted into those moving in the positive direction of x-axis (all eigenvalues of the Jacobian matrix become positive) or into those moving in the negative direction of x-axis (if all eigenvalues of the Jacobian matrix become negative) using solely the transformation (1) (the AW transformation).

3 Implementations of the Concept

It seems to be convenient to practically implement the idea considering motion of control volume faces rather than changing the frame of reference or transforming the governing equations. The integral form of conservation laws for any domain V in x-t space is: $\oint[\mathbf{U}dx - \mathbf{F}(\mathbf{U})dt] = 0$, the integration is performed along the boundary of the domain. The rectangle control volume $[x_{i-1/2}, x_{i+1/2}] \times [t^n, t^{n+1}]$ (Fig. 1) results in a standard finite volume formulation, but it is not imperative to perform the time integration along the *vertical lines* $x = x_{i-1/2}$ and $x = x_{i+1/2}$. Virtually *any path* C connecting points $(x_{i-1/2}, t^n)$ and $(x_{i-1/2}, t^{n+1})$ can be used for the integration. We choose the integration paths to ensure simplest possible upwinding. The path $C_{i-1/2}$ consists of two straight lines, $C^R_{i-1/2}$ and $C^L_{i-1/2}$, their slopes being AW velocities D^R and D^L. The integration path must arrive to the point $(x_{i-1/2}, t^{n+1})$, so that the total shift along t-axis must equal Δt while the total shift along x-axis must equal 0 i.e. $(\Delta t)^L + (\Delta t)^R = \Delta t$ and $D^R(\Delta t)^R + D^L(\Delta t)^L = 0$. The values of $(\Delta t)^R$ and $(\Delta t)^L$ can be easily found.

By a proper choice of D^R and D^L we may ensure that when moving along $C^R_{i-1/2}$ all disturbances come from the right side while when moving along the $C^L_{i-1/2}$ all disturbances come from the left side. Therefore, the task of upwinding becomes trivial and we eventually arrive at the flux formula

$$\mathbf{F}_{i-1/2} = [\xi^* \mathbf{F}(\mathbf{U}_i) + (1 - \xi^*)\mathbf{F}(\mathbf{U}_{i-1})] + \xi^* D^R(\mathbf{U}_{i-1} - \mathbf{U}_i), \tag{2}$$

where $\xi^* = D^L/(D^L - D^R)$.

The most straightforward way is to choose AW velocities D^L, D^R having the same values all over the computational domain (the *Global* AW). It turns out that the Lax-Friedrichs scheme is a particular case of the approach. In [4,5] we write out a new Global AW scheme which at least competitive with the Lax-Friedrichs one. The velocities D^L and D^R may be chosen separately for each face (the *Local* AW) based on the parameters in the adjacent cells. Among the schemes which may be obtained via the Local AW are the Rusanov scheme and HLL scheme family (HLL, HLLE, HLLR and others). Simple analysis and numerical tests ([3–5] and Fig. 2) show that greater AW velocities increase numerical diffusion.

In order to further decrease the numerical dissipation we should no longer limit AW velocities by the values based on the *averaged* conservative variables $\mathbf{U}_{i-1}$ and $\mathbf{U}_i$. We propose to assign them separately for each of the *intermediate states between* $\mathbf{U}_{i-1}$ and $\mathbf{U}_i$ which may be obtained using linear interpolation with the weight coefficient ξ $(0 \le \xi \le 1)$: $\overline{\mathbf{U}}(\xi, \mathbf{U}_{i-1}, \mathbf{U}_i) = (1-\xi)\cdot\mathbf{U}_{i-1} + \xi\cdot\mathbf{U}_i$.

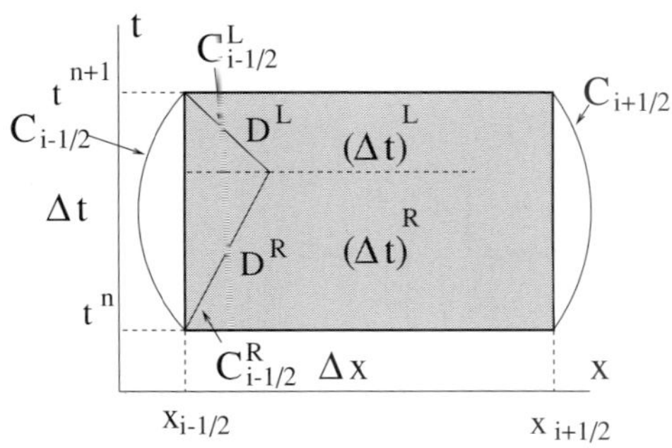

Fig. 1. A control volume on x-t plane

When integrating the Jacobian matrix $\mathbf{A}$ over the intermediate states we apply the AW transformation for the Jacobian under integration and obtain:

$$\mathbf{F}(\mathbf{U}_i) - \mathbf{F}(\mathbf{U}_{i-1}) = \int_0^1 (\mathbf{A}(\overline{\mathbf{U}}(\xi)) - D(\xi)\mathbf{E})d\xi \times (\mathbf{U}_i - \mathbf{U}_{i-1}) \tag{3}$$

as long as the following condition is satisfied: $\int_0^1 D(\xi)d\xi = 0$.

Now we choose the velocities of the *Differential* AW in the following way: $D(\xi) = D^R(\xi) \geq \max(0, v(\xi) + c(\xi))$, for $\xi \in (0, \xi^*)$ and $D(\xi) = D^L(\xi) \leq \min(0, v(\xi) - c(\xi))$ for $\xi \in (\xi^*, 1)$. As a result the integration in Eq.(3) between the limits 0 to ξ^* becomes a sum of perturbations propagating to the left, because $\mathbf{A}(\overline{\mathbf{U}}(\xi)) - D(\xi)\mathbf{E}$ has only negative eigenvalues. Then for the remaining part of the interval $\xi \in (\xi^*, 1)$ the integral becomes a sum of perturbations propagating to the right. The flux difference $\mathbf{F}(\mathbf{U}_i) - \mathbf{F}(\mathbf{U}_{i-1})$ is thus split and the problem of upwinding is solved. Constructing the numerical first order upwind flux in a usual manner and calculating the integral in Eq.(3) one can get a very simple flux formula:

$$\mathbf{F}_{i-1/2} = \mathbf{F}[\xi^*\mathbf{U}_i + (1-\xi^*)\mathbf{U}_{i-1}] + d \cdot (\mathbf{U}_{i-1} - \mathbf{U}_i) \tag{4}$$

where the diffusion coefficient is

$$d = \int_0^{\xi^*} D^R(\xi)d\xi = -\int_{\xi^*}^1 D^L(\xi)d\xi \; .$$

The velocities $D^R(\xi)$ and $D^L(\xi)$ may be taken to be equal to their limiting values (see the inequalities above). To get the value of ξ^* the above integrals may be taken or, which may be even more efficient, a simple iteration procedure may be involved [4,5].

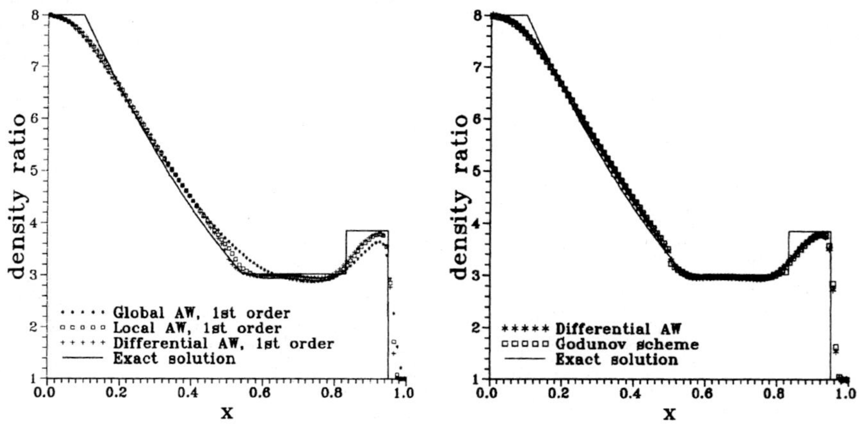

Fig. 2. Comparison of the first order results for the AW and Godunov schemes with the exact solution for '480:1' test problem

As to higher order extensions, a variety of ways can be employed including a standard one via the MUSCL predictor/corrector approach. The second-order numerical results shown below are based on the Rodionov scheme [6] which seems to be very close but not identical to the MUSCL-Hancock approach ([7]).

For the simplest example of a nonlinear scalar conservation equation with a convex flux and for the Differential AW velocities being equal to their limiting values, the numerical flux of the Differential AW scheme is identical to that of the Godunov scheme ([4,5]). Relationship of the AW schemes and Godunov-type schemes based on the exact Riemann solver are to be further explored in the forthcoming detailed paper [8]. Another interesting point to be presented there in detail is that a discontinuity (in particular, a contact discontinuity) may be incorporated into the integration path over intermediate states ξ. In some cases this may be beneficial in reducing numerical dissipation.

4 Numerical Results

The test results for the Riemann problem with the initial pressure ratio $p_2/p_1 = 480$, the density ratio $\rho_2/\rho_1 = 8$, zero initial velocities and $\gamma = 5/3$ are presented in Figs. 2,3a. The CFL number is 0.8. The initial position of the jump in parameters is at $x = 0.5$. The data are presented for the time instant $t = 0.04$. Fig. 3b shows some results for the Woodward & Colella test [9].

One can see that lower AW velocities leads to less dissipative results (Fig. 2) and the Differential AW scheme provides the solution of at least the same accuracy as that of the Godunov schemes while being incomparably simpler in

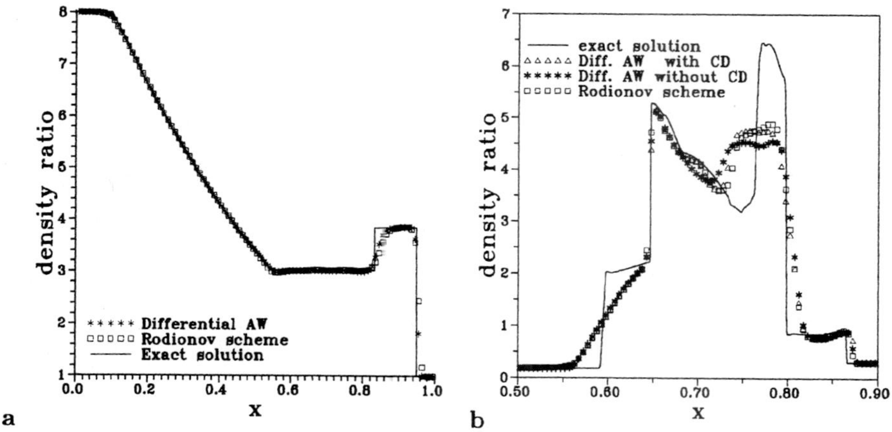

Fig. 3. Comparison of the second order results for the AW and Rodionov schemes with the exact solution for (a) '480:1' and (b) Woodward & Colella [9] test problems (the Rodionov scheme [6] is a second-order predictor-corrector Godunov scheme with the exact Riemann solver and a TVD limiter)

construction and, thus, more efficient. The positive influence of a contact discontinuity included into the integration path is clearly seen in Fig. 3b.

5 Conclusion

We have proposed a new general framework to construct upwind shock-capturing schemes. The framework can be viewed along with well-known general approaches, such as Riemann solvers and other techniques. It leads, in particular, to new simple, accurate and efficient upwind schemes. It is of extraordinary importance that the schemes retain the same and simple form for a wide range of hydrodynamical systems, such as MHD equations [10], relativistic hydrodynamics (considering the Lorenz invariance) [11], equations-of-state more complex than that for a perfect gas [12] and others. This results, in some cases, in at least one order of magnitude gains in efficiency while maintaining good accuracy. The above-cited applications to a variety of complex 2D and 3D problems convincingly prove the advantages of the new approach.

References

1. I.V. Sokolov, J. Sakai, E.V. Timofeev: 'On High-Resolution Shock-Capturing Schemes for Hydrodynamic and MHD Simulations via an Artificial Wind Concept'. In: *Proceedings of 12th National CFD Symposium, Tokyo, Japan, December 21–23, 1998*, (Chuo University) pp. 93–94
2. I.V. Sokolov, J. Sakai, E.V. Timofeev: 'Efficient and Accurate Shock-Capturing Schemes Using Artificial Wind: 1-D Test Results'. In: *Japan'99, Proceedings of Symposium on Shock Waves, Tokyo, Japan, March 19–21, 1999*, (Aoyama Gakuin University) pp. 425–428
3. I.V. Sokolov, E.V. Timofeev, J. Sakai, K. Takayama: Shock Waves **9**, 423 (1999)
4. I.V. Sokolov, E.V. Timofeev, J. Sakai, K. Takayama: 'Development and Applications of Artificial Wind Schemes for Hydrodynamics, MHD and Relativistic Hydrodynamics'. In: *CD-ROM Proceedings of 13th National CFD Symposium, Tokyo, Japan, December 21–23, 1999*, (Chuo University) paper Nr. E1-05, 6p.
5. E.V. Timofeev, I.V. Sokolov, J. Sakai, K. Takayama: Rep. Inst. Fluid Sci., Tohoku Univ. **12**, 71 (2000)
6. A.V. Rodionov: Zh. Vychisl. Mat. i Mat. Fiziki (J. Comp. Math. and Math. Phys.) **27**, 1853 (1987) (in Russian)
7. E.F. Toro: *Riemann Solvers and Numerical Methods for Fluid Dynamics*, 2nd edn. (Springer, Berlin 1999)
8. I.V. Sokolov, E.V. Timofeev, J. Sakai, K. Takayama: J. Comp. Phys. (in preparation)
9. P. Woodward, P. Colella: J. Comp. Phys. **54**, 115 (1984)
10. I.V. Sokolov, J. Sakai: J. Plasma Fusion Res. Series **2**, 482 (1998)
11. H.M. Zhang, I.V. Sokolov, K. Furusawa et al. Prog. Theor. Phys. Supp. **138**, 642 (2000)
12. E.V. Timofeev, P.A. Voinovich, M. Sun, K. Takayama: 'On the Long Distance Propagation of an Impact-Caused Shock Wave in the World Ocean. II. Underwater Shock Travelling over the Globe'. In: *Japan 2000, Proceedings of Symposium on Shock Waves, Tokyo, Japan, March 16–18, 2000*, (Tokyo University) pp. 601–604

Entropy Splitting for High Order Numerical Simulation of Compressible Turbulence

N.D. Sandham[1] and H.C. Yee[2]

[1] Aeronautics and Astronautics, School of Engineering Sciences, University of Southampton, Southampton SO17 1BJ, UK
[2] NASA Ames Research Center, Moffett Field, CA 94035, USA

1 Introduction

The requirements on a numerical method for high-speed turbulent flow simulation, including shock wave interactions with turbulence, are stringent. For turbulence the method must be capable of resolving accurately a wide range of length scales, whilst for shock waves the method must be stable and not generate excessive local oscillations. Classical methods are either too dissipative, incapable of handling complex geometries, or incapable of shock capturing. Higher-order ENO, WENO or hybrid schemes are too expensive for practical turbulent flow simulation. Previous work [1] developed efficient high order shock-capturing schemes which minimize the use of numerical dissipation away from shock waves. The objective of the present study is to add new techniques to these methods so that the use of numerical dissipation can further be minimized for shock-free compressible turbulent simulations. In order to achieve this, we believe it is important to condition the compressible Navier-Stokes equations so that they have certain nonlinear stability properties, including the physical boundaries. In addition, we need to construct high-order schemes that have a discrete analogue of the conditioned governing equations, including stable high order numerical boundary condition treatments. The theory described in [2] [3] for the perfect gas compressible Euler equations has led to a splitting of the inviscid flux derivatives of the governing equations, hereafter referred to as "Entropy Splitting". This splitting can improve nonlinear stability and minimize the use of numerical dissipation for both smooth flows and for problems containing complex shock-turbulence interactions [4]. Such stability properties are of much value in the context of turbulence simulation. In this contribution we consider application of the technique to direct numerical simulation of compressible turbulent channel flow.

2 Entropy splitting and summation-by-parts approach

We use an entropy splitting approach to transform the governing equations. For a perfect gas, suitable transformed variables W for the Euler equations with the conservative vector U can be found by defining $W = \partial\eta/\partial U$, where $\eta = \rho h(S)$ is an entropy function, $h(S)$ is an arbitrary but differentiable function of a dimensionless physical entropy S, and $S = \ln(p\rho^{-\gamma})$ is the physical entropy nondimensionalized by c_v and with a suitably defined datum. The choice of $h(S)$ is restricted by a homogeneity requirement and a positive definite condition on $U_W = \partial U/\partial W$. For the particular choice of $h(S)$ and a full derivation of this formulation see [5] [3]. The Euler equations, which would normally be written in the usual notation, as

$$U_t + F_x + G_y + H_z = 0 \tag{1}$$

are now rewritten in the entropy split form

$$U_t + \frac{\beta}{\beta+1}(F_x + G_y + H_z) + \frac{1}{\beta+1}(F_W W_x + G_W W_y + H_W W_z) = 0, \tag{2}$$

with entropy splitting parameter β. The complete formulation for a thermally perfect gas is given in [4].

In the proof of the generalised energy estimate for entropy splitting use is made of integration by parts. To obtain a similar stable estimate in numerical simulation for both the interior and boundary spatial schemes, we need difference operators that satisfy a discrete analogy of this, known as summation by parts (SBP) [7] [3]. In the current work we use the SBP schemes given in [6]

3 Navier-Stokes formulation

By itself, entropy splitting and compatible SBP boundary formulation might not be able to produce an ultimate robust method for the Navier-Stokes equations, and attention needs also to be focused on the viscous and heat conduction terms. A particular problem can be odd-even decoupling arising from succesive application of two first derivative operators. In order to remove this from the outset we rewrite the viscous terms in what we call a Laplacian form

$$\frac{\partial \tau_{ij}}{\partial x_j} = \mu \frac{\partial u_i}{\partial x_j \partial x_j} + \frac{\mu}{3}\frac{\partial}{\partial x_i}\left(\frac{\partial u_k}{\partial x_k}\right) + \frac{\partial \mu}{\partial x_j}\left(\frac{\partial u_i}{\partial x_j} + \frac{\partial u_j}{\partial x_i}\right) - \frac{2}{3}\frac{\partial \mu}{\partial x_i}\frac{\partial u_j}{\partial x_j}. \tag{3}$$

The first term is a Laplacian, corresponding to the incompressible formulation of the equations and is always treated with second-derivative operators.

The same procedure is applied to the energy equation leading to a first term containing the Laplacian of $\gamma e/\mathrm{Pr} + u_i u_i/2$, where Pr is the Prandtl number. This still leaves the continuity equation without a Laplacian on the right hand side, and the possibility of $2\widehat{h}$ waves appearing in the density and not being damped. Simulations showed that this could indeed happen. A remedy is to write the continuity equation in a new form, using a diffusion coefficient, which we take to be a constant $\nu^* = \mu/\rho$ at reference conditions,

$$\frac{\partial \rho}{\partial t} + \frac{\partial}{\partial x_i}\left(\rho u_i + \nu^* \frac{\partial \rho}{\partial x_i}\right) = \nu^* \frac{\partial^2 \rho}{\partial x_i \partial x_i}. \tag{4}$$

Since the second term in brackets on the left hand side of the left hand side balances the right hand side to the truncation error of the differentiation scheme,

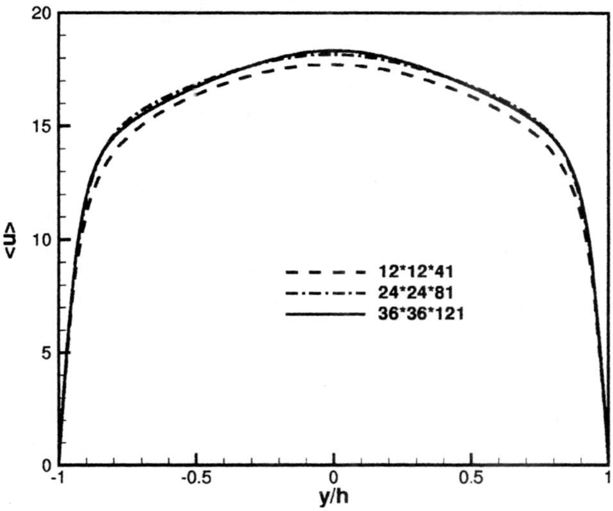

Fig. 1. Mean velocity profiles

when we are fully resolved there is no difference to the original formulation. However, with the new form any $2\widehat{h}$ waves ($\widehat{h}$ being the grid spacing) in density are automatically damped by the method. As we shall see this enables the calculation to proceed realistically even when resolution is very low.

4 Results for compressible channel flow

As a test case for application of the method we consider compressible turbulent channel flow with isothermal wall boundary conditions. The fluid mechanics of this problem have been studied in some detail by Coleman *et al.*[8]. They showed that the only compressibility effect at moderate Mach numbers comes from the variation of fluid properties with temperature. We choose a spatially-uniform forcing (equivalent to a constant pressure gradient) and solve the fixed pressure gradient problem rather than a constant mass flow. A Mach number of 0.1 is chosen, based on friction velocity and sound speed corresponding to the fixed wall temperature. Channel half width h, friction velocity u_τ, wall temperature and bulk (integrated) density are the normalising quantities for non-dimensionalisation with a Reynolds number chosen as 180. The splitting parameter was set to $\beta = 4$. The computational box has non-dimensional length 3 in the streamwise x-direction, 1.5 in the spanwise z-direction and 2 in the y-direction. The x- and z-directions have periodic boundary conditions with uniform grid spacing. In the y–direction, the grid is stretched according to $y/h = \tanh(c_\eta \eta)/\tanh c_\eta$ with η uniformly distributed on [-1,1] and $c_\eta = 1.7$. Incompressible flow simulations have guided the choice of box size (small, but not so

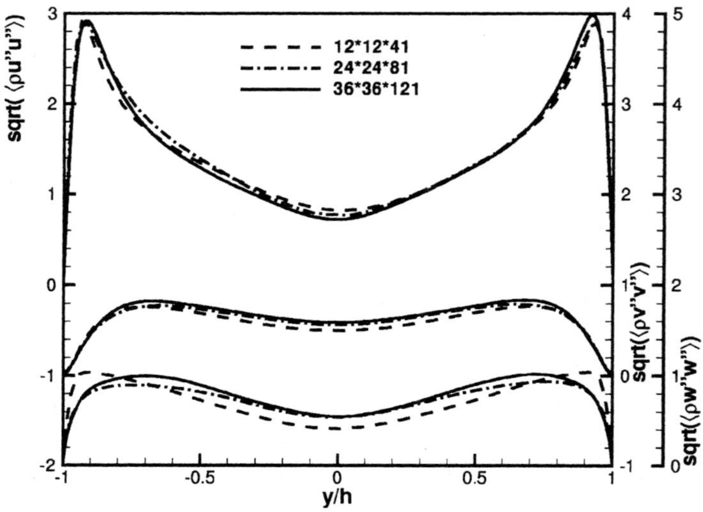

Fig. 2. Effect of grid refinement on normal stresses.

small as to cramp the turbulence excessively) and the ratio of grid points in each direction so that all directions have roughly the same degree of resolution of the relevant turbulence microscales. For these simulations, the classical fourth-order Runge-Kutta temporal discretization, and fourth-order non-compact (5-point) central spatial interior scheme have been used.

To demonstrate the effect of grid refinement on the simulated flow we compare three simulations with grids $12 \times 12 \times 41$, $24 \times 24 \times 81$ and $36 \times 36 \times 121$. The largest number in each case corresponds to the direction normal to the wall. Figure 1 shows the mean flow and Figure 2 the root-mean-square (rms) fluctuations across the channel. Angle brackets <> denote averages over the homogeneous spatial directions and time while in the usual notation double primes denote deviations from mass-weighted (Favre) averages. The convergence is not uniform across the channel but the change from medium to fine grid is much smaller than the change from coarse to medium grid.

The convergence of various global measures is shown on Table 1 for grids ranging from $6 \times 6 \times 21$ (on which a turbulent flow was sustained), up to $36 \times 36 \times 121$. For the pressure gradient and Reynolds number specified the velocity gradient at the wall should be 180, the difference away from this being an error of the simulation or an insufficient sample length. On Figure 3 we plot the absolute error, $|180 - Re_\tau|$ against number of points in one of the periodic directions. Here Re_τ is the Reynolds number based on u_τ, the mean density $< \rho_w >$ and the mean viscosity $< \mu_w >$ at the wall. The simulations demonstrate a robustness down to very coarse resolutions, comparable with the best incompressible turbulent

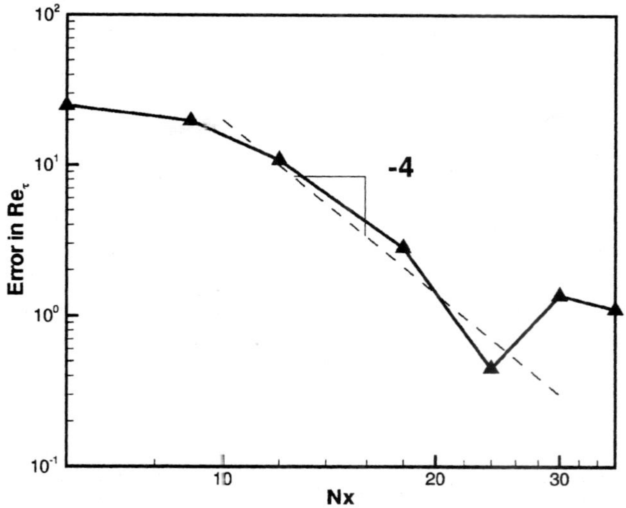

Fig. 3. Convergence of Re_τ compared with -4 slope expected of global fourth-order convergence

flow solvers. For a calculation on a grid $16 \times 16 \times 51$ the simulation was stable for $1.25 < \beta < 12$.

Grid	U_c	$\overline{U}$	$\left(\frac{d\overline{u}}{dy}\right)_w$	H
$6 \times 6 \times 21$	17.2	14.9	155.0	1.71
$12 \times 12 \times 41$	17.7	15.2	169.2	1.65
$24 \times 24 \times 81$	18.2	15.7	180.6	1.64
$36 \times 36 \times 121$	18.2	15.7	181.1	1.61
Incompressible	18.2	15.6	180.0	1.62

Table 1. Convergence of centreline velocity U_c, mean velocity $\overline{U}$, wall velocity gradient $\left(\frac{d\overline{u}}{dy}\right)_w$ and shape factor H.

5 Conclusions

A stable high order numerical scheme for direct numerical simulation (DNS) of shock-free compressible turbulence has been presented. The method contains no upwinding, artificial dissipation, or filtering. Instead the method relies on the stabilizing mechanisms of an appropriate conditioning of the governing equations and the use of compatible spatial difference operators. An entropy splitting approach splits the inviscid flux derivatives into conservative and non-conservative portions. The spatial difference operators satisfy a summation by parts condition leading to a stable scheme (combined interior and boundary schemes) for the initial boundary value problem using a generalized energy estimate. A Laplacian formulation of the viscous and heat conduction terms on the right hand side of the Navier-Stokes equations is used to ensure that any tendency to odd-even decoupling associated with central schemes can be countered by the fluid viscosity. A special formulation of the continuity equation is used, based on similar arguments. The methods were tested in a DNS of wall bounded compressible turbulent plane channel flow at a friction Mach number of 0.1. It has been demonstrated that the methods are robust in terms of grid resolution and that accurate turbulence statistics can be obtained with moderate grid sizes.

Acknowledgments

The authors would like to thank Q. Li of University of Southampton and M. Vinokur of NASA Ames Research Center for assistance in this research. This work has been supported in part by RIACS of NASA Ames Research Center. The first author would like to acknowledge support from EPSRC under grants GR/M 21546 and GR/M 84336. A full form of this paper appears as RIACS report 00.10, June 2000.

References

1. Yee, H.C., Sandham, N.D. and Djomehri, M.J. 1999 Low Dissipative High Order Shock-Capturing Methods Using Characteristic-Based Filters *J. comput. Phys.* **150**, 199-238.
2. Olsson, P. and Oliger, J. 1994 Energy and Maximum Norm Estimates for Nonlinear Conservation Laws. RIACS Technical Report 94-01, NASA Ames Research Center.
3. Gerritsen, M. and Olsson, P. 1996 Designing an Efficient Solution Strategy for Fluid Flows: I. A Stable High Order Finite Difference Scheme and Sharp Shock Resolution for the Euler Equations *J. Comput. Phys.* **129**, 245-262.
4. Yee, H.C., Vinokur, M. and Djomehri, M.J 1999 Entropy Splitting and Numerical Dissipation *NASA/TM-1999-208793*; J. Comput. Phys. **162**, 33-81, 2000.
5. Harten, A. 1983 On the Symmetric Form of Systems of Conservation Laws with Entropy. *J. Comput. Phys.* **49**, 151-164.
6. Carpenter,M.H., Nordstrom, J. and Gottlieb, D. 1998 A Stable and Conservative Interface Treatment of Arbitrary Spatial Accuracy. *NASA/CR-1998-206921 (also ICASE Report 98-12)*.
7. Strand, B. 1994 Summation by Parts for Finite Difference Approximations for d/dx *J. Comput. Phys.* **110**, 47-67.
8. Coleman, G.N., Kim, J. and Moser, R. 1995 A Numerical Study of Turbulent Supersonic Isothermal-Wall Channel Flow. *J. Fluid Mech.* **305**, 159-183.

An Implicit Fluctuation Splitting Scheme for Compressible Flows

A. Bonfiglioli[1], P. De Palma[2], G. Pascazio[3], and M. Napolitano[3]

[1] DIFA Università della Basilicata, 85100 Potenza, Italy, bonfiglioli@unibas.it
[2] DIM Università di Roma "Tor Vergata", 00133 Roma, Italy, depalma@uniroma2.it
[3] IME Politecnico di Bari, 70125 Bari, Italy, pascazio@poliba.it, napolita@poliba.it

1 Introduction

The authors are currently involved in a national research project on turbomachinery CFD. One major goal of this project is to develop an accurate, robust and efficient methodology for the three-dimensional compressible Navier–Stokes equations, based on a genuinely multidimensional upwind discretization of the Euler fluxes. In the last years, the first author has concentrated his efforts towards optimizing the efficiency of the solution procedure, using linear Fluctuation Splitting (FS) spacial discretizations of the Euler equations [1]. The method employs an implicit Euler time discretization and evaluates the Jacobian using either approximate analytical formulas (Picard linearisation) or two-point first-order-accurate differences. The resulting large sparse linear system is solved using a preconditioned GMRES strategy. The last three authors have developed and optimized a nonlinear hybrid formulation which, according to the local flow conditions, uses the most suitable FS spacial discretization among the matrix LDA and PSI schemes, which are optimal for subsonic and supersonic flow conditions, respectively, and the matrix N scheme, which is only first-order-accurate but can capture shocks monotonically [2]. Such a hybrid formulation employs a Runge–Kutta time integration procedure with multigrid and has been lately extended to the Reynolds averaged Navier–Stokes equations, using both an algebraic and a two-equation ($k - \omega$) turbulence model [3].

From the results obtained so far, it appears that the hybrid FS spacial discretization can solve the compressible Euler and Navier–Stokes equations more accurately than current state-of-the-art methods based on classical upwinding. On the other hand, implicit methods are needed to provide a solution procedure for the discrete equations being robust and efficient at the same time.

In the present work, the hybrid FS spacial discretization is combined with the implicit time integration procedure of Bonfiglioli [1] and used to solve two well known two-dimensional viscous flows as well as an inviscid three-dimensional flow.

2 Numerical Method

The Euler equations are discretized on a computational domain composed of linear finite elements (triangles/tetrahedra). The discrete conservative flux balance over each cell T, namely, the fluctuation, can be written in terms of appropriate fluxes through the sides of the cell (see, e.g., [4,5]) as:

$$\Phi_{U,T} = \bar{R}\Phi_T, \quad \Phi_T = -\sum_{j=1}^{d+1} \frac{l_j}{d} \bar{\boldsymbol{A}} \cdot \boldsymbol{n}_j Q_j = -\sum_{j=1}^{d+1} K_j Q_j. \tag{1}$$

In equation (1) $\bar{\boldsymbol{A}}$ is the Jacobian tensor with respect to the characteristic variables Q, $\bar{R}$ is the cell-averaged projection matrix from the variables Q to the conservative ones U, d is the number of space dimensions, l_j is the area of the side of the cell opposite to node j, $\boldsymbol{n}_j$ is the inward unit vector normal to l_j, and

$$K_j = \frac{l_j}{d} \bar{\boldsymbol{A}} \cdot \boldsymbol{n}_j. \tag{2}$$

Due to the hyperbolic nature of the system, K_j can be written as:

$$K_j = (\bar{R}_K \bar{\Lambda}_K \bar{L}_K)_j = (\bar{R}_K \bar{\Lambda}_K^+ \bar{L}_K)_j + (\bar{R}_K \bar{\Lambda}_K^- \bar{L}_K)_j = K_j^+ + K_j^-. \tag{3}$$

In equation (3), $\bar{R}_{K,j}$ and $\bar{L}_{K,j}$ are the right and left eigenvector matrices of K_j, whereas $\bar{\Lambda}_{K,j}^+$ and $\bar{\Lambda}_{K,j}^-$ are the corresponding positive and negative eigenvalue matrices. Introducing the following vectors,

$$Q_{\text{in}} = \left(\sum_{j=1}^{d+1} K_j^-\right)^{-1} \left(\sum_{j=1}^{d+1} K_j^- Q_j\right), \quad Q_{\text{out}} = \left(\sum_{j=1}^{d+1} K_j^+\right)^{-1} \left(\sum_{j=1}^{d+1} K_j^+ Q_j\right), \tag{4}$$

the linear matrix LDA scheme, which is linearity preserving [6], is obtained as:

$$\Phi_j = -K_j^+ \left[Q_{\text{out}} - Q_{\text{in}}\right]. \tag{5}$$

Such a scheme is very accurate for subsonic smooth flows and can be considered the optimal compact FS scheme for such conditions. For supersonic flows, a different set of characteristic variables, W, has been employed, which, for two-dimensional flows, allow to recast the Euler system into an equivalent set of four scalar advection equations [7]. These are then discretized by the nonlinear (scalar) PSI scheme [6], which can be written as,

$$\hat{\Phi}_j = -K_j^{nl+} \left(W_j - W_{\text{in}}^{nl}\right), \tag{6}$$

where K_j^{nl+} is computed using the nonlinear Jacobian $\overline{\hat{\boldsymbol{A}}}^{nl}$, as defined in [8]. For three-dimensional flows, the same scheme is used, although the Euler system does not decouple into five scalar equations. Finally, in order to compute transonic flows with strong shocks, it is necessary to use locally a monotone scheme, namely, the matrix N scheme of [5], given as:

$$\hat{\Phi}_j = -K_j^{l+} \left(W_j - W_{\text{in}}^{l}\right), \tag{7}$$

where K_j^l and W_{in}^l are computed using the Jacobian $\bar{A}_W$. In order to pinpoint the "shock cells", where such a lower-order scheme needs to be employed — disregarding those where the transition from supersonic flow conditions to subsonic ones is smooth enough to be properly handled by the LDA scheme — it is necessary to characterize them uniquely. By a careful analysis of the flow properties across a normal shock, it is concluded that "shock cells" are characterized by: i) average cell Mach number $M < 1$; ii) at least one supersonic node; iii) at least one subsonic node with its local Mach number lower than 0.9. In conclusion, at every step of the computational process, the present hybrid approach flags all elements of the computational domain and distributes the residuals for each of them using equation (5) at supersonic cells, equation (6) at subsonic ones, and equation (7) at "shock cells".

The residual of the Euler equations at each vertex j of the computational domain is then evaluated by collecting all contributions coming from the surrounding elements, as:

$$\left(\frac{\partial U}{\partial t}\right)_j = \frac{1}{S_j}\sum_T \Phi_{U,j}^T = R(U)_j, \tag{8}$$

S_j being the co-volume surrounding node j.

The numerical method is finally extended to the Navier–Stokes equations using a standard second-order-accurate Galerkin finite element scheme.

The time derivative in equation (8) is approximated using a two-point backward finite difference and the residual is evaluated implicitly. Linearizing $R(U)$ about time level n, the following scheme is obtained:

$$\left(\frac{S}{\Delta t}I - \left(\frac{\partial R}{\partial U}\right)\right)_j \left(U^{n+1} - U^n\right)_j = R(U)_j. \tag{9}$$

Equation (9) represents a large nonsymmetric sparse linear system to be solved at each time step. In the present computations the Jacobian, $\partial R/\partial U$, is evaluated numerically using first-order-accurate differences. The sparse matrix technology required to solve equation (9) is provided by the PETSc library [9]. The linear solver is based on an ILU(0) preconditioned GMRES algorithm, restarted every 30 linear iterations. The time step, Δt is increased, at each non-linear iteration, by the ratio between the initial and the current nonlinear residual, thus rapidly reaching very large values.

For turbulent flow calculations, the Spalart–Allmaras [10] one-equation turbulence model is employed and solved by the same implicit method, one nonlinear iteration being performed at the end of each nonlinear iteration of the flow equations. The convective and source terms are discretized in space by means of the PSI and LDA schemes, respectively, whereas the diffusive terms are discretized by the standard Galerkin scheme.

3 Results

The proposed methodology has been applied to solve three well documented test-cases. The first one is the laminar transonic flow past a NACA0012 airfoil with $\alpha = 10°$, $Re_\infty = 1000$ and $M_\infty = 2$. Two grids have been employed, with 21760 and 87040 cells, respectively. The values of the lift and drag coefficients are: $C_L = 0.3435$, $C_D = 0.2556$, and $C_L = 0.3392$, $C_D = 0.2576$, for the coarse and fine grids, respectively. The results agree quite well with the numerical data provided in [11]. The fine-grid Mach-number contours and the pressure-coefficient and skin-friction-coefficient distributions along the profile are given in figures 1, 2 and 3, respectively. The results are quite satisfactory; in particular, the shock is sharp and monotonic. Convergence to machine accuracy is obtained after about 300 iterations, which require 4 CPU hours on a single-processor (EV5.6-500 MHz) DIGITAL computer, whereas the memory requirement amounts to 200 Mbytes. The explicit scheme of [3] needs one order of magnitude more CPU time to converge but only 50 Mbytes of RAM.

The second test case is the turbulent flow past an RAE 2822 airfoil with $\alpha = 2.79°$, $Re_\infty = 6.5 \times 10^6$, and $M_\infty = 0.73$ (Case 9 in [12]). The computational domain is discretized with 47104 triangular elements. The computed values of the lift and drag coefficients, $C_L = 0.803$ and $C_D = 0.0176$, agree well with the available data. Figures 4 and 5 show the Mach-number contours and a comparison between the experimental and numerical pressure-coefficient distributions along the profile, respectively. The results are quite satisfactory; the minor discrepancy between the numerical and experimental results being due to the turbulence modeling, as already found by other investigators [13]. For such a very severe test case, a residual drop of three orders of magnitude is obtained after about 6 hours of CPU time starting from rest, with a memory requirement of about 100 Mbytes. Needless to say, the explicit approach of [3] would require weeks of CPU time to achieve a comparable convergence.

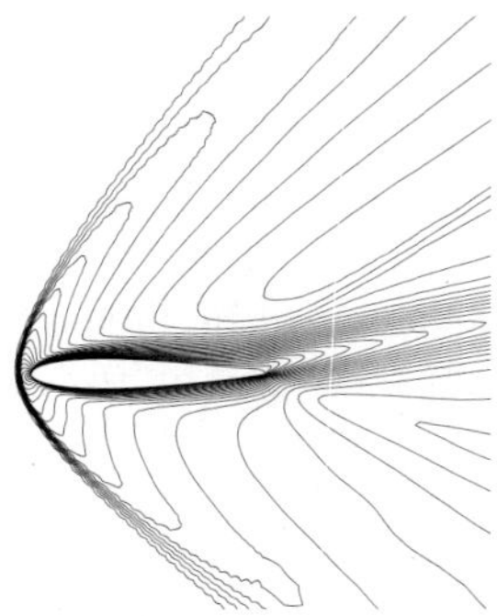

Fig. 1. NACA 0012: Mach-number contours ($\Delta M = 0.1$)

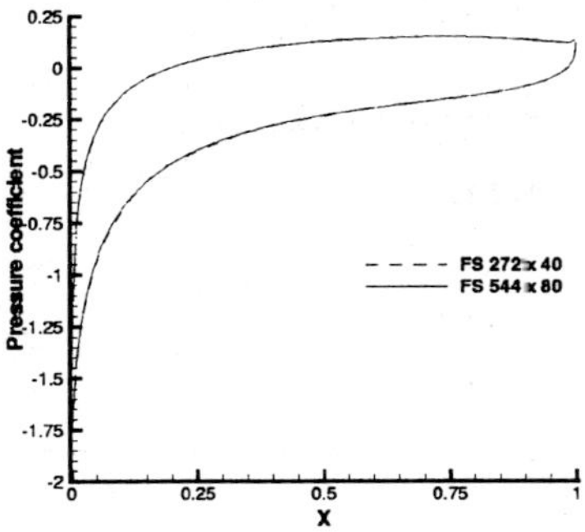

Fig. 2. NACA 0012: pressure-coefficient distribution

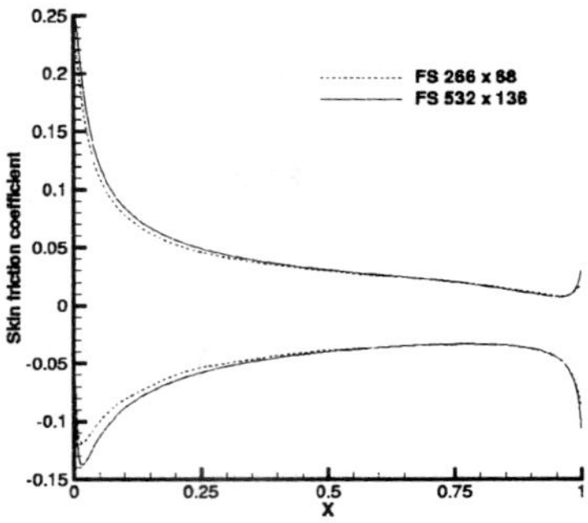

Fig. 3. NACA 0012: skin-friction coefficient distribution

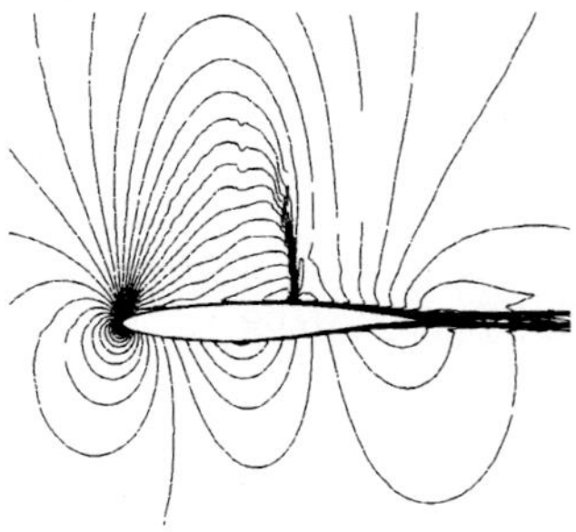

Fig. 4. RAE 2822: Mach-number contours ($\Delta M = 0.03$)

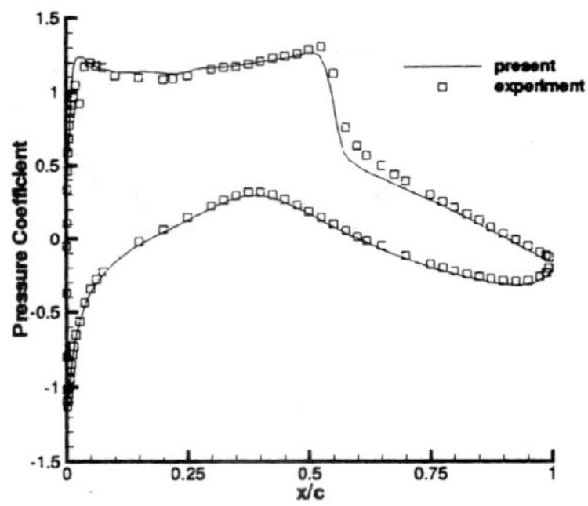

Fig. 5. RAE 2822: pressure-coefficient distribution

Supersonic inviscid flow (with the inlet Mach number equal to 1.4) through a three-dimensional channel with a circular bump having a linearly variable thickness (from 4.2% of the chord length to 0) has been finally computed using a grid with 98011 cells. The Mach-number contours on the two longitudinal endwalls are shown in figures 6 and 7, respectively. The very complex flow structure is well captured by the present results, which are in good agreement with the numerical solutions obtained with second-order accurate schemes available in the literature. A residual drop of five orders of magnitude is obtained after 200 nonlinear iterations, amounting to 5 hours of CPU time, with a memory requirement of 180 Mbytes. A comparison with the explicit scheme of [3] cannot be provided, since a three-dimensional code for such a method is not available.

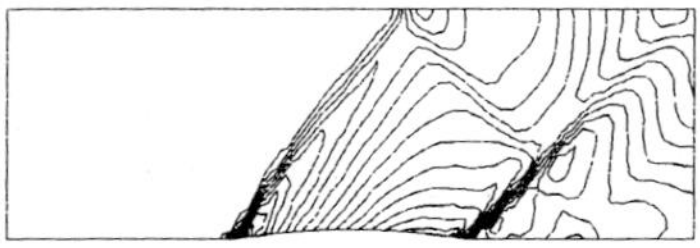

Fig. 6. Mach-number contours at the maximum-thickness endwall (ΔM=0.03)

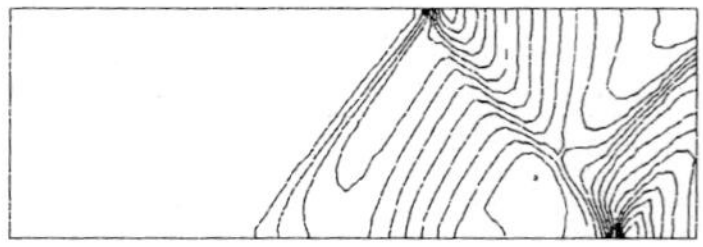

Fig. 7. Mach-number contours at the zero-thickness endwall ($\Delta M = 0.03$)

4 Conclusions

An implicit hybrid fluctuation splitting scheme has been developed for solving steady compressible flows in two and three dimensions. Results are presented for laminar and turbulent flows in two dimensions and an inviscid flow in three dimensions. The implicit GMRES integration procedure makes the scheme extremely robust and efficient, but its RAM requirements grows very rapidly with the problem size.

5 Acknowledgements

The present research has been supported by MURST/COFIN99.

References

1. A. Bonfiglioli: Lecture Notes in Physics **515**, pp. 254-259, Springer, 1998
2. P. De Palma, G. Pascazio, and M. Napolitano: *AIAA Paper 99-3513*, 1999
3. P. De Palma, G. Pascazio, and M. Napolitano: *8th International Symposium on Computational Fluid Dynamics*, Brema, 1999
4. E. van der Weide, H. Deconinck: *3rd ECCOMAS CFD Conference*, Paris, 1996
5. L. A. Catalano, P. De Palma, G. Pascazio, and M. Napolitano: Lecture Notes in Physics **490**, pp. 328-333, Springer, 1997
6. R. Struijs, H. Deconinck, and P. L. Roe: *VKI LS 1991-01, Computational Fluid Dynamics, von Karman Institute*, Belgium, 1991
7. B. van Leer, W. T. Lee, P. L. Roe: *AIAA Paper 91-1552*, 1991
8. P. De Palma, G. Pascazio, and M. Napolitano: *4th ECCOMAS CFD Conference*, Athens, 1998
9. S. Balay, W. D. Gropp, L. Curfman McInnes, and B. F. Smith: PETSc home page `http://www.mcs.anl.gov/petsc`, 1998
10. P. R. Spalart and S. R. Allmaras: La Recherche Aérospatiale **1**, pp. 5-21, 1994
11. M. O. Bristeau, R. Glowinski, J. Periaux, and H. Viviand (Eds.): Notes on Numerical Fluid Mechanics **18**, Vieweg, 1987
12. P. H. Cook, M. A. McDonald, and M. C. P. Firmin: *AGARD-AR-138*, 1979
13. R. Abid, C. Rumsey, and T. Gatski: AIAA Journal **33**, pp. 2026-2031, 1995

A Compact Residual-Based Scheme for Solving the Compressible Euler and Navier-Stokes Equations

Christophe Corre and Alain Lerat

SINUMEF Laboratory, ENSAM,
151 Bd de l'Hopital, 75013 Paris, France

Abstract. A dissipative compact scheme is presented for solving the steady compressible Euler and Navier-Stokes equations with a third-order accuracy. High-order accuracy is not obtained for each space derivative but for the whole residual, which avoids any linear algebra and provides compactness. Numerical dissipation is also constructed from derivatives of the residual only, providing simplicity and robustness. Applications of the scheme to model test-cases as well as compressible flows are presented.

1 Introduction

Accuracy of the numerical solution is a crucial issue in a number of CFD problems such as the simulation of turbulent flows, flow instabilities, vortex dynamics, aeroacoustics, etc. This has motivated the development of various high-order approximation methods. Among these methods, compact schemes are attractive for their narrow numerical dependence domain. In a classical compact scheme, a first derivative in a coordinate direction is approximated by a Pade formula using a small number of points in this direction. Here, we consider compact approximations that provide a high accuracy order not for each space-derivative treated apart, but for the whole residual r. This notion of compactness will be called *residual-based compactness*. Its main interest is to avoid the solution of the linear algebraic systems usually associated with compact approximations.
Dealing with the residual is also favourable for constructing the numerical dissipation necessary to a centered scheme. A *residual-based dissipation* contains derivatives of the residual only and thus is consistent with a partial differential operator vanishing at steady state. Such a dissipation can be of first-order in the transient phase and therefore can provide robustness in a simple way, without degrading the high accuracy at steady-state and the compactness.
The present paper describes the design of a third-order residual-based scheme for calculating steady solutions of multi-D hyperbolic systems of conservation laws on structured meshes and presents applications of the method to model test-cases and compressible flow problems.

2 Design principles

2.1 Residual-based compactness

Consider an initial-value problem for the hyperbolic system of conservation laws :

$$u_t + f_x + g_y = 0 \tag{1}$$

where t is the time, x and y are space coordinates, $w = w(x,y,t)$ is the state vector, $f = f(w)$ and $g = g(w)$ are flux components depending smoothly on w. Let $v_{j,k}$ be a mesh function defined on a uniform Cartesian mesh ($x_j = j\,\delta x, y_k = k\,\delta y$), with steps δx and δy of the same order, say $O(h)$. Let us define the difference and average operators : $(\delta_1 v)_{j+\frac{1}{2},k} = v_{j+1,k} - v_{j,k}$, $(\mu_1 v)_{j+\frac{1}{2},k} = \frac{1}{2}(v_{j+1,k} + v_{j,k})$ in the x-direction; δ_2, μ_2 are similarly defined in the y-direction. For smooth solutions, the classical centered discretization to f_x and g_y reads :

$$\frac{1}{\delta x}\delta_1\mu_1 f = f_x + \frac{\delta x^2}{6} f_{xxx} + O(h^4)\ , \quad \frac{1}{\delta y}\delta_2\mu_2 g = g_y + \frac{\delta y^2}{6} g_{yyy} + O(h^4) \qquad (2)$$

where subscripts j,k are omitted for concision. These 2nd-order approximations can be improved by correcting the leading error term. This can be achieved either in compact or non-compact formulation, using Pade fractions or 5 points per direction; in either case, the truncation error is corrected on each space derivative separately. We suggest instead the following approach : a 2nd-order term is added to the difference formulae (2) so that the *global* truncation error be expressed in terms of derivatives of the steady-state residual $r = f_x + g_y$ only :

$$\begin{aligned} \frac{1}{\delta x}\delta_1\mu_1(I + \frac{1}{6}\delta_2^2)f &= f_x + \frac{\delta x^2}{6} f_{xxx} + \frac{\delta y^2}{6} f_{xyy} + O(h^4) \\ \frac{1}{\delta y}\delta_2\mu_2(I + \frac{1}{6}\delta_1^2)g &= g_y + \frac{\delta y^2}{6} g_{yyy} + \frac{\delta x^2}{6} g_{xxy} + O(h^4) \end{aligned}$$

Using these centered difference formulae of second-order accuracy in Eq. (1), we obtain the scheme :

$$w_t + \frac{1}{\delta x}\delta_1\mu_1(I + \frac{1}{6}\delta_2^2)f + \frac{1}{\delta y}\delta_2\mu_2(I + \frac{1}{6}\delta_1^2)g = 0, \qquad (3)$$

which has a 3×3-point stencil and requires no linear algebra. Its truncation error is $\epsilon = w_t + r + \frac{\delta x^2}{6} r_{xx} + \frac{\delta y^2}{6} r_{yy} + O(h^4)$ so that, for an exact steady solution ($w_t = r = 0$), $\epsilon = O(h^4)$. The compact scheme (3) is thus fourth-order accurate at steady-state. Note that the same fourth-order spatial approximation was previously found in [1] using a different approach.

This type of approximation will be called *residual-based compactness*. It can be extended to any order of accuracy in 2-D or 3-D (see [2]-[3]) and can also be considered for unsteady problems : the physical time-derivative must simply be included in the residual and a dual time be introduced.

2.2 Residual-based dissipation

To be able to run the above method for compressible flows, some numerical dissipation is required. The present dissipation is also residual-based, *i.e.* constructed from derivatives of the residual r. This is advantageous because such a dissipation can be of order h in the transient phase and therefore can provide robustness in a simple way, while preserving a high accuracy at steady state. More precisely, the dissipative form of the scheme is :

$$w_t + \frac{1}{\delta x}\delta_1\mu_1(I + \frac{1}{6}\delta_2^2)f + \frac{1}{\delta y}\delta_2\mu_2(I + \frac{1}{6}\delta_1^2)g = \frac{\delta x}{2}(\Phi_1 r_1)_x + \frac{\delta y}{2}(\Phi_2 r_2)_y \quad (4)$$

where r_1 and r_2 are second-order centered approximations to r at $j+\frac{1}{2}, k$ and $j, k+\frac{1}{2}$ respectively. The coefficients Φ_1 and Φ_2 are matrices depending only on δx and δy and on the flux Jacobian matrices $A = df/dw$ and $B = dg/dw$. They have been determined to ensure an optimal multidimensional dissipation (see [4], [2]). For a scalar 2-D hyperbolic equation, they can be written as :

$$\Phi_1 = sgn(A)\min(1, \frac{1}{\alpha}), \quad \Phi_2 = sgn(B)\min(1, \alpha) \quad \text{where } \alpha = \frac{\delta x|B|}{\delta y|A|}.$$

For a 2-D hyperbolic system of conservation laws, Φ_1 and Φ_2 have rather simple expressions given in [2]. The truncation error of the dissipative scheme (4) is :

$$\epsilon = w_t + r + \frac{\delta x^2}{6}r_{xx} + \frac{\delta y^2}{6}r_{yy} - \frac{\delta x}{2}(\Phi_1 r)_x - \frac{\delta y}{2}(\Phi_2 r)_y + O(h^3), \quad (5)$$

so that, for an exact steady solution, we obtain $\epsilon = O(h^3)$, *i.e.* a third-order accuracy. Moreover, Scheme (4) is compact because its numerical dissipation has been discretized on the same 3×3 point stencil as its non-dissipative part.

3 Fully discrete scheme

For brevity's sake, the fully discrete form of the residual-based scheme will be presented for the following simplified system, modelling the full Navier-Stokes equations :

$$w_t + f_x + g_y = \nu\, w_{yy} \quad (6)$$

where $f = f(w)$ and $g = g(w)$ are the Euler fluxes and ν a positive constant coefficient. Residual-based compactness can be extended to (6) by adding the viscous terms in the residual, including the residual in the numerical dissipation. Being only interested here in steady solutions, the time derivative in (6) is simply approximated at first-order. The explicit version of the fully discrete scheme reads :

a) Residuals for numerical dissipation :

$$\begin{aligned} (\Pi_1)^n_{j+\frac{1}{2},k} &= \delta x\,(\frac{\delta_1 f}{\delta x} + \frac{\delta_2\mu_1\mu_2 g}{\delta y} - \nu\mu_1\frac{\delta_2^2 w}{\delta y^2})^n_{j+\frac{1}{2},k} \\ (\Pi_2)^n_{j,k+\frac{1}{2}} &= \delta y\,(\frac{\delta_1\mu_1\mu_2 f}{\delta x} + \frac{\delta_2 g}{\delta y} - \nu\mu_2\frac{\delta_2^2 w}{\delta y^2})^n_{j,k+\frac{1}{2}} \end{aligned} \quad (7)$$

b) Numerical fluxes :

$$\begin{aligned} F^n_{j+\frac{1}{2},k} &= [(I + \tfrac{1}{6}\delta_2^2)\mu_1 f]^n_{j+\frac{1}{2},k} - [\tfrac{1}{2}\Phi_1\Pi_1]^n_{j+\frac{1}{2},k} \\ G^n_{j,k+\frac{1}{2}} &= [(I + \tfrac{1}{6}\delta_1^2)(\mu_2 g - \nu\frac{\delta_2 w}{\delta y}) - \frac{1}{12}\nu\frac{\delta_2^3 w}{\delta y}]^n_{j,k+\frac{1}{2}} - [\frac{1}{2}\Phi_2\Pi_2]^n_{j,k+\frac{1}{2}} \end{aligned} \quad (8)$$

c) Explicit increment :

$$\Delta w_{j,k}^{expl} = -\Delta t \, (\frac{\delta_1 F}{\delta x} + \frac{\delta_2 G}{\delta y})_{j,k}^n \tag{9}$$

and $w_{j,k}^{n+1} = w_{j,k}^n + \Delta w_{j,k}^{expl}$, where $w_{j,k}^n$ denotes the numerical solution at time level $n\Delta t$ and location $(j\delta x, k\delta y)$. This scheme involves $(3 \times 3) + 2 = 11$ points (13 points for the complete Navier-Stokes equations) and (3×3) points when $\nu = 0$. For smooth solution of Eq. (6), the corrective terms added to basic 2nd-order centered approximations in the non-dissipative part of the fluxes (8) yield a truncation error given again by (5), with $r = f_x + g_y - \nu w_{yy}$. Thus, 3rd-order accuracy is obtained at steady state.
For an efficient solution of steady problems, an implicit version of scheme (7)-(9) is developed. A simplified linear implicit operator containing only derivatives in the x and y directions that can be discretized over 3 points is retained :

d) Implicit increment :

$$\begin{aligned} \Delta w_{j,k} &+ \frac{\Delta t}{\delta x}[\mu_1(A\delta_1 \Delta w^n)_{j,k} - \delta_1((\frac{1}{2}\Phi_1 A)\delta_1 \Delta w^n)_{j,k}] \\ &+ \frac{\Delta t}{\delta y}[\mu_2(B\delta_2 \Delta w^n)_{j,k} - \delta_2((\frac{1}{2}\Phi_2 B + \frac{\nu}{\delta y})\delta_2 \Delta w^n)_{j,k}] = \Delta w_{j,k}^{expl} \end{aligned} \tag{10}$$

with $w_{j,k}^{n+1} = w_{j,k}^n + \Delta w_{j,k}$ and A, B the Euler Jacobian matrices. The implicit scheme (7)-(10) is always linearly stable and offers good damping properties when used with large CFL numbers. In practice, the algebraic linear system (10) is solved by alternate-line relaxation of Jacobi or Gauss-Seidel type. Only a few (one or two) inner iterations of the relaxation process are necessary to ensure unconditional stability and a damping very close to that of (10).

4 Numerical results

4.1 Model problems

We consider first the linear advection with rotation around the point $(1,0)$ of a Gaussian profile over the square domain $[0,1] \times [0,1]$. The exact steady solution is $w = const$ on any circle of center $(1,0)$. The iso-lines of the numerical solution obtained on a 80×80 uniform Cartesian mesh using the residual-based scheme cannot be distinguished from those of the exact solution (see Fig. 1). The L_2-norm error between the exact and the numerical solution is computed for various mesh sizes, demonstrating the third-order accuracy of the scheme. The convergence history is monitored through the L_2-norm of $\frac{\Delta w}{\Delta t}$ on the grid. The present implicit scheme yields fast convergence to steady-state, with an even faster convergence of the error : though the scheme is only first-order in the transient phase, the third-order solution is obtained rather quickly, *i.e.* as the residual becomes lower than about 10^{-4}.
Next, we look for an approximate steady solution of $w_t + aw_x = \nu w_{yy}$, with

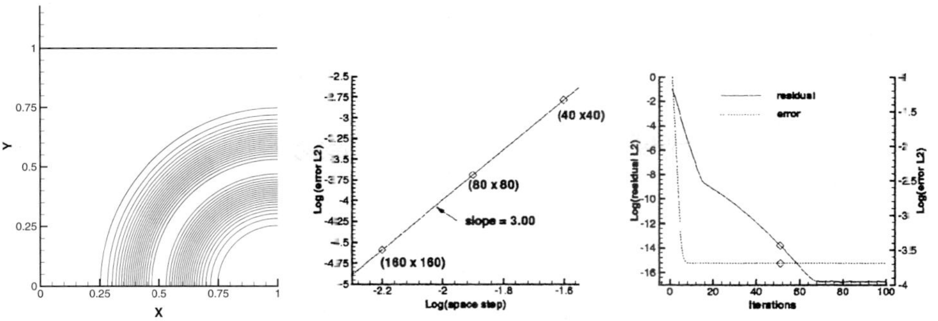

Fig. 1. Rotational advection. Left : Isovalues of $w(x,y)$ on a 80×80 grid. Center : computed accuracy order. Right : convergence and error history ($CFL = 100$).

boundary conditions such as to model a Poiseuille flow : an initial profile, prescribed at inflow $y = 0$, is advected and diffused between two solid walls on which its value is fixed to zero (see Fig. 2). The computed accuracy order of the scheme is in very good agreement with its theoretical value. Though the low cell Reynolds number, specific to this model problem, affects the convergence of the simplified implicit scheme, the slower decay of the residual has only a slight effect on the scheme error, which reaches anyway a steady-state after few iterations.

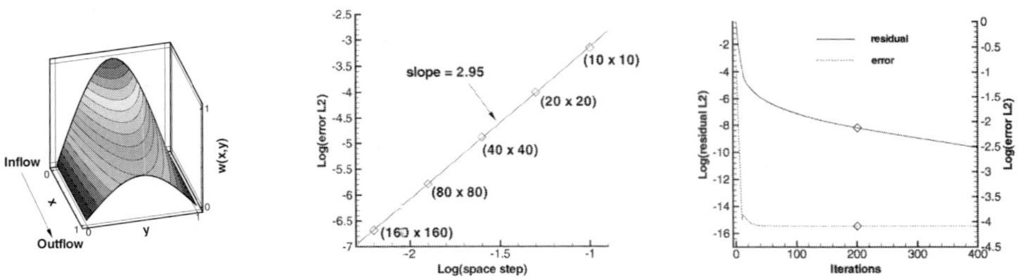

Fig. 2. Poiseuille model. Left : Isovalues of $w(x,y)$ ($Re_h = \frac{ah}{\nu} = \frac{1}{2}$) on a 20×20 grid. Center : computed accuracy order. Right : convergence and error history ($CFL = 100$).

4.2 Compressible flow problems

The classical inviscid flow over a NACA0012 airfoil at $M_\infty = 0.85$ and zero incidence is first computed, on a 125×34 grid, in order to demonstrate the good shock capturing properties of the scheme (see Fig. 3). Next, the laminar flow at $M_\infty = 0.5$ and $Re = 5000$ over a thermally insulated NACA0012 airfoil at zero incidence is computed using a series of 150×64 C meshes with a mesh spacing at the wall varying from 10^{-4} up to 10^{-2} chord. The 3rd-order scheme is compared with the 2nd-order one obtained when crossing out the corrections added to the basic centered terms in the non-dissipative part of the numerical fluxes (8). The low sensitivity of the 3rd-order scheme to grid coarsening is illustrated on Fig. 4 : the coarsening has only a weak influence on the 3rd-order skin-friction prediction outside the leading-edge region (the separation point shifts very slightly, even on very coarse grids), while the 2nd-order prediction is much more affected and eventually fails to converge.

Others applications of the residual-based scheme, such as turbulent flow over an airfoil or shock-boundary layer interaction, may be found in [3].

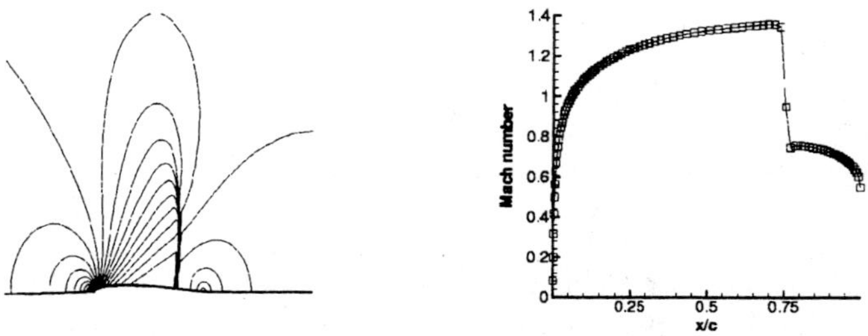

Fig. 3. Transonic inviscid flow. Left : pressure contours. Right : wall Mach distribution.

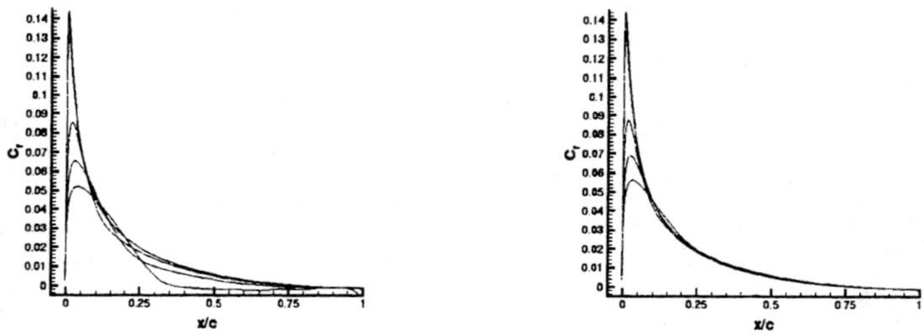

Fig. 4. Laminar flow. Effect of mesh coarsening on the computed skin-friction distribution. Left : second-order scheme. Right : third-order scheme.

References

1. Abarbanel S. and Kumar A., 'Compact high-order schemes for the Euler equations', *J. Scientific Computing*, **3**, 275 (1988).
2. Lerat A. and Corre C., 'Compact residual-based schemes for 2-D and 3-D hyperbolic systems of conservation laws', *Colloquium "State of the Art in CFD"*, Marseille, France, September 1999, to be published in *Comp. and Fluids.*
3. Lerat A. and Corre C., 'A compact 3rd-order accurate scheme using a 1st-order dissipation for the compressible Navier-Stokes equations', subm. to *J. Comp. Phys.*.
4. Huang Y. and Lerat A., 'Second-order upwinding through a characteristic time-step matrix for compressible flow calculations', *J. Comput. Phys.*, **142**, 445 (1998).

Some Properties of Residual Distribution Schemes for the Euler Equations

Mani Rad, Hiroaki Nishikawa, and Philip L. Roe

Department of Aerospace Engineering
University of Michigan, Ann Arbor MI 48105, USA

Abstract. A residual distribution technique is presented for the Euler equations, one which splits the updates into purely elliptic and hyperbolic contributions. The scheme combines fluctuation splitting methods ideal for scalar advection and a least squares minimisation appropriate for the elliptic subsystem. The minimisation is constrained to maximally decouple effects of the entropy, enthalpy and acoustic parts from each other. Results show that for subcritical flows where potential solutions should be obtained, the enthalpy is constant to machine zero, and the entropy is constant to an extremely high degree of accuracy. Still under investigation, fluctuation calculations with dissipative effects are presented. They are capable of breaking rarefaction shocks for fluctuation splitting schemes in grid-aligned cases.

1 Introduction

Fluctuation splitting (FS) schemes for Euler and other equations that share similar structure have become estabilished in the literature. As in past efforts, the present work takes a physical approach to strongly discontinuous flows while avoiding the defects of upwind methods applied to almost incompressible flow. Using the decomposed form of the Euler equations together with a residual distribution scheme ensures that the elliptic and hyperbolic parts of the problems are handled separately, with maximal decoupling. To illustrate the scheme's accuracy and test its properties, we choose a set of flow problems containing features that are difficult to capture. However, study reveals that FS is not immune to grid dependency and in combination with certain choices of meshes will produce non-physical solutions.

2 Uniquely Decomposed Form of the Euler Equations

Different properties of the Euler equations are emphasized by proper selection of the set of unknowns. For computing compressible flow, the most fundamental choice is the set of *conserved quantities* $\mathbf{u} = (\rho, \rho\mathbf{v}, \rho E)^t$ associated with the flux tensor $\underline{\underline{\mathbf{F}}} = (\rho\mathbf{v}, \rho\mathbf{v} \otimes \mathbf{v} + p\underline{\underline{\mathbf{I}}}, \mathbf{v}(E+p))^t$ since they satisfy the Rankine-Hugoniot conditions across discontinuities.

For computational convenience we use another set of unknowns, the *parameter vector* $\mathbf{z} = \sqrt{\rho}(1, \mathbf{v}, h)^t$, where h is the total specific enthalpy. This set allows the construction of local linearisations having conservation properties [1] [6]. Furthermore, of significance to the present approach, are the *natural variables* [3].

They are written as $\mathbf{x} = (S, h, p, \mathbf{v}/|\mathbf{v}|)^t$ and $\mathbf{x} = (S, h, W^+, W^-)^t$ respectively in subsonic and supersonic flow. In the supersonic regime, the acoustic Riemann invariants are $\partial W^+ = \beta\partial p + \rho q^2 \partial\theta$, $\partial W^- = \beta\partial p - \rho q^2\partial\theta$. It is uniquely in these variables that the Euler equations can be decoupled into maximally independent subsystems [3] [7] [9]. These variables enable the inherently different hyperbolic and elliptic behaviour to be computed independently with the minimum of "crosstalk". The decomposed form of the Euler equations can generically be written as $\hat{A}\mathbf{x}_s + \hat{B}\mathbf{x}_n = 0$ or more precisely as follows.

$$
\begin{array}{ll}
(1-M^2)\partial_s p - \rho q^2 \partial_n \theta = 0 & (\beta\partial_s + \partial_n)W^+ = 0 \\
\rho q^2 \partial_s \theta + \partial_n p = 0 & (\beta\partial_s - \partial_n)W^- = 0 \\
\partial_s S = 0 & \partial_s S = 0 \\
\partial_s h = 0 & \partial_s h = 0
\end{array}
\tag{1}
$$

3 Advection of Entropy, Enthalpy and Acoustic Invariants

Throughout this paper, the term 'update matrix' will reappear frequently because it is a convenient way to think about FS schemes. Let the twelve scalar quantities $\mathbf{W}^n = (\mathbf{w}_a, \mathbf{w}_b, \mathbf{w}_c)$ represent the initial state of a cell (in two dimensions) with vertices (a,b,c) and $\mathbf{w}$ be the quantity stored at the vertices. The update matrix is the constant matrix within the cell that defines the update $\mathbf{W}^{n+1} - \mathbf{W}^n = \omega U \mathbf{W}^n$. Among its nine blocks, $U_{i,j}$ would be the (i,j) block of the update matrix, signifying the effect of vertex i on j $(i, j = a, b, c)$.

As shown in (1), the Euler equations always exhibit a hyperbolic behaviour through entropy and enthalpy (S, h) advection. In supersonic regime, this behaviour is extended to the acoustic part. Through a rotation transformation, (1) becomes $\tilde{A}\mathbf{x}_x + \tilde{B}\mathbf{x}_y = 0$ where $\tilde{A} = \text{diag}(u, u, u\beta - v, v\beta + u)$ and $\tilde{B} = \text{diag}(v, v, u\beta + v, v\beta - u)$. The fluctuation then becomes

$$
\Phi = \frac{1}{A^T} \sum_{j=a,b,c} \mathcal{D}\left(\tilde{A}\Delta y_j - \tilde{B}\Delta x_j\right) \frac{\partial \mathbf{x}}{\partial \mathbf{z}} \mathbf{z_j}
$$

The acoustic part of the fluctuation is split apart from the (S, h) part using respectively the diagonal matrices $\mathcal{D} = \text{diag}(0, 0, 1, 1)$ and $\mathcal{D} = \text{diag}(1, 1, 0, 0)$. This splitting becomes important in the subsonic regime where the (S, h) part is treated identically as in the supersonic case. Meanwhile the acoustic part becomes an elliptic subsystem and is handled using a distribution scheme with no directional bias.

The fluctuation is then distributed by any favorite FS scheme having distribution coefficient matrix $\mathbf{\Psi}$. Assuming that we are storing the *parameter vector* $\mathbf{z}$ at the vertices, the (i,j) block of the update matrix becomes

$$
U_{i,j} = \frac{1}{A^T} \mathbf{\Psi} \left(\frac{\partial \mathbf{x}}{\partial \mathbf{z}}\right)^{-1} \mathcal{D}\left(\tilde{A}\Delta y_j - \tilde{B}\Delta x_j\right) \frac{\partial \mathbf{x}}{\partial \mathbf{z}}
$$

where

$$\frac{\partial \mathbf{x}}{\partial \mathbf{z}} = \begin{bmatrix} \frac{z_1 z_4(1-2\gamma)+\gamma(z_2^2+z_3^2)}{z_1} & -z_2 & -z_3 & z_1 \\ -\frac{z_4}{z_1} & 0 & 0 & 1 \\ \frac{\gamma-1}{\gamma}\beta z_4 & -\frac{\gamma-1}{\gamma}\beta z_2 - z_3 & -\frac{\gamma-1}{\gamma}\beta z_3 + z_2 & \frac{\gamma-1}{\gamma}\beta z_1 \\ \frac{\gamma-1}{\gamma}\beta z_4 & -\frac{\gamma-1}{\gamma}\beta z_2 + z_3 & -\frac{\gamma-1}{\gamma}\beta z_3 - z_2 & \frac{\gamma-1}{\gamma}\beta z_1 \end{bmatrix}$$

4 Treatment of the Elliptic Subsystem

This section briefly describes the solution method for the independant acoustic subsystem in (1). For a more detailed account, refer to [5]. If we choose to store the natural variables $\mathbf{x}$, the four residuals $\phi_{\mathbf{x}}$ in (1) can be written as a linear function of the vertex values

$$\phi_{\mathbf{x}} = \sum_{j=a,b,c} R_j \mathbf{x}_j$$

Since in fact we are storing the parameter vector $\mathbf{z}$ we must estimate the quantity to be minimised as

$$\phi_{\mathbf{z}} = \sum_{j=a,b,c} R_j \frac{\partial \mathbf{x}}{\partial \mathbf{z}} \mathbf{z_j}$$

Standard methods are used to evaluate R_j, the matrix of derivatives frozen during the update. The first two components of this vector comprise the elliptic part of the problem, if $M < 1$, so the norm to be minimized is

$$\mathcal{D} \sum_{j=a,b,c} R_j \frac{\partial \mathbf{x}}{\partial \mathbf{z}} \mathbf{z_j} \quad \text{where} \quad \mathcal{D} = \mathrm{diag}(1, \sqrt{1-M^2}, 0, 0)$$

where the relative weighting of the continuity and vorticity residuals follows from [9]. To converge towards steady-state, we choose a least squares minimisation using a steepest descent method. Steepest descent minimization is selected for illustration purposes only, one would revert to more efficient (mainly Newton-like) methods for any practical Euler code. After determining the gradient of the norm to be minimised, we conclude that the update matrix is given by

$$U = \left(\sum_{j=a,b,c} R_j \frac{\partial \mathbf{x}}{\partial \mathbf{z}} \right)^t \mathcal{D}^2 \left(\sum_{j=a,b,c} R_j \frac{\partial \mathbf{x}}{\partial \mathbf{z}} \right)$$

The update found in this manner has the disadvantage of changing the convected quantities, (S, h). Since we intend complete decoupling at the update level between the elliptic and hyperbolic parts, the elliptic part of the residual should only contribute to changes in pressure and flow angle and not to (S, h). A necessary step is to perform a constrained minimisation of the elliptic residual. This means that the updates from the elliptic residual should be in directions where only pressure and flow angle change. For the *natural variables*, the directions of vectors along which only one variable changes are $r_S = (\frac{1}{2}\frac{z_2^2+z_3^2}{z_4}, z_2, z_3, \frac{1}{2}\frac{z_2^2+z_3^2}{z_1})^t$, $r_h = (0, z_2, z_3, \frac{z_2^2+z_3^2}{z_1})^t$, $r_p = (\frac{M^2}{M^2-2}z_1, z_2, z_3, \frac{M^2}{M^2-2}z_4)^t$ and $r_\theta = (0, -z_3, z_2, 0)^t$.

5 Entropy Satisfying Solutions

In a study to test the robustness of FS schemes, we found some situations where they can produce non-physical solutions. As shown in Figure 4 (right), FS is not immune to grid dependency and in combination with certain choices of meshes will produce rarefaction shocks. Figure 4 (right) is a simulation of supersonic flow over a diamond shaped airfoil and a cut across segment AA' will show significant jumps in pressure and entropy. Where they are parallel to grid lines, expansion rays do not spread quickly enough. In search of a remedy, our investigation has led us to focus on the fluctuation calculation rather than a suitable distribution rule. More specifically, for the conservative form of the Euler equations, where the fluxes $\mathbf{F}$, $\mathbf{G}$ are bilinear functions of the parameter vector $\mathbf{z}$, the cell fluctuation can be written as

$$\Phi = \int\int \partial_x \mathbf{F}(\mathbf{z}) + \partial_y \mathbf{G}(\mathbf{z}) = \oint \mathbf{F}(\mathbf{z})dy - \mathbf{G}(\mathbf{z})dx = \oint \mathbf{z}^T \mathbf{C}\mathbf{z}dy - \mathbf{z}^T \mathbf{D}\mathbf{z}dx$$

This leads to an alternative expression for the fluctuation, a formula containing a parameter α which has only a second-order effect on the fluctuation's accuracy.

$$\begin{aligned}\Phi = &\left[\tfrac{\alpha}{2}(\mathbf{z}_1{}^T + \mathbf{z}_3{}^T) + (1-\alpha)\mathbf{z}_2{}^T\right]\{(y_2 - y_1)\mathbf{C} - (x_2 - x_1)\mathbf{D}\}(\mathbf{z}_1 - \mathbf{z}_3) + \\ &\left[\tfrac{\alpha}{2}(\mathbf{z}_1{}^T + \mathbf{z}_2{}^T) + (1-\alpha)\mathbf{z}_3{}^T\right]\{(y_3 - y_1)\mathbf{C} - (x_3 - x_1)\mathbf{D}\}(\mathbf{z}_1 - \mathbf{z}_2)\end{aligned}$$

Similarly, the above expression can be rearranged in two other ways by permuting the vertex indices. A family of fluctuations is obtained by varying the parameter α and at least three choices reveal interesting properties. For $\alpha = 1$, the fluctuation vanishes when two vertices are in the same state and the edge connecting them is aligned with the local characteristic. Work presented in [4] illustrates how this property is used to capture shocks exactly. For $\alpha = \frac{2}{3}$, one recovers the common method [1] for computing the fluctuation with $\mathbf{z}$ varying linearly along each edge and a single characteristic evaluated at the cell center. Finally, $\alpha = 0$ produces a fluctuation which has nice dissipative effects and breaks up rarefaction shocks. This is illustrated for a scalar example (Burgers equation) in Figure 1 where a solution profile free of discontinuities can be obtained for $\alpha = 0$. The extension from the scalar to a system (i.e. Euler) is currently under progress and is expected to produce entropy-satisfying solutions.

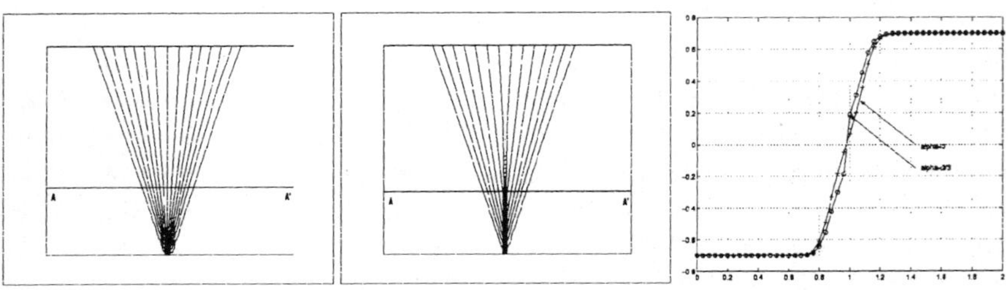

Fig. 1. Rarefaction solution to Burgers equation: $\alpha = 0$, $\alpha = \frac{2}{3}$, cut across AA'

6 Computational Examples and Concluding Remarks

Figure 2 shows computation of subcritical flow around a cylinder at $M_\infty = 0.38$ where constrained least-squares is used for the elliptic part and PSI for the hyperbolic. A potential flow solution should be obtained with perfect fore and aft symmetry. Note the satisfactory behaviour of the solver at stagnation points. Levels of entropy generation were low with $S_{min,max} = -8.5 \times 10^{-6}, 7.3 \times 10^{-6}$.

Figure 3 shows calculations around a supercritical cylinder at $M_\infty = 0.75$ with shock capturing. PSI is used for all advected quantities. The sonic transition is not treated in any special way. Even though the two schemes meeting at $M = 1$ are of very different nature, the solution there is still satisfactory. Note in Figure 3, a slight jump in pressure at the sonic line but the resulting changes of entropy are again extremely small.

To test the coupling between elliptic and hyperbolic parts, we consider incompressible flow with nonuniform enthalpy at inflow (Figure 4 (left)). In this problem it would be hard to exhibit the recirculation zones with any method that generates excessive entropy. Convergence tests revealed that our solution approached the exact solution [2] with second order accuracy.

We have illustrated the properties of residual distribution schemes when used in combination with the uniquely decomposed form of the Euler equations. Residual splitting proved appropriate for maximum decoupling between the hyperbolic and elliptic parts. Accurate results are obtained at sonic and stagnation regions, as measured by levels of spurious entropy generation. In rare circumstances, however, rarefaction shocks can appear, and their elimination is being studied.

References

[1] Deconinck, H., Roe, P. L., Struijs, R. J., 'A multidimensional generalisation of Roe's flux difference splitter for the Euler equations', *Computers and Fluids*, **22**, 1993.

[2] Fraenkel, L. E., 'On Corner Eddies in Plane Inviscid Shear Flow', Journal of Fluid Mechanics, Vol. 11, Part 3, November 1963.

[3] Hayes, W, Probstein, R. F., *Hypersonic Flow Theory*, Academic Press, 1963.

[4] Nishikawa, H., Rad, M., Roe, P. L., 'Grids and Solutions from Residual Minimization', ICCFD Proceedings, Kyoto. Springer-Verlag, 2000.

[5] Rad, M., Roe, P. L., 'An Euler Code that Can Compute Potential Flow', 2nd Inter. Symposium on FV Proceedings, Duisburg. Hermes, 1999

[6] Roe, P. L., 'Approximate Riemann solvers, parameter vectors and difference schemes', *J. Comput. Phys.*, **43**, pp357-372, 1981.

[7] Roe, P. L., Mesaros, L. M., 'Solving steady mixed conservation laws by E/H splitting', 13th Inter. Conf. on Num. Methods in Fl. Dyn., Monterey, July, 1996.

[8] Roe, P. L., Turkel E., 'The quest for diagonalization of differential systems', Workshop on Barriers and Challenges in CFD, NASA Langley 1996. Springer, 1997.

[9] Roe, P. L., 'Compounded of Many Simples, reflection on the role of model problems in CFD', Workshop on Barriers and Challenges in CFD, NASA Langley, August, 1996. Springer, 1997.

[10] Roe, P. L., Rad, M.,'A Form of the Euler Equations Conserving Potential Flow', AIAA CFD Meeting Proceedings, Norfolk, VA June, 1999.

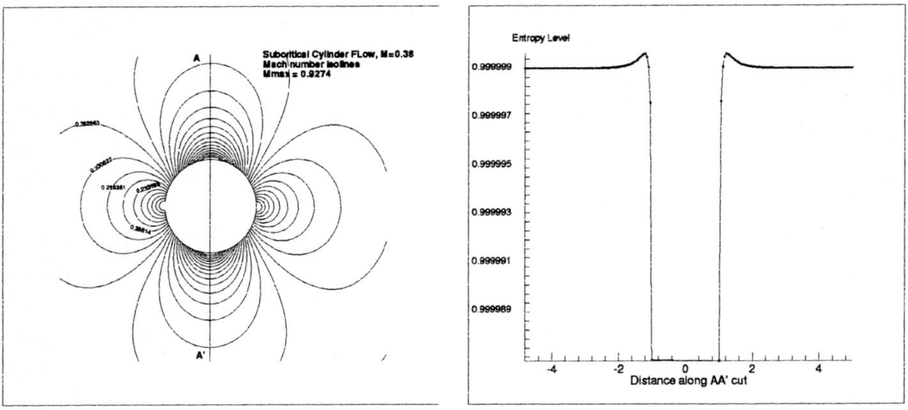

Fig. 2. Left: subcritical Cylinder Flow at $M_\infty = 0.38$, Mach contours. Right: entropy distributions along vertical cut through center of cylinder

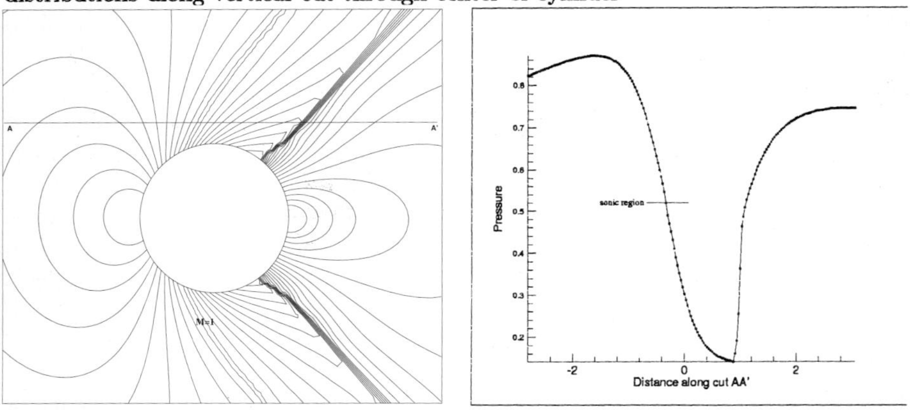

Fig. 3. Left: supercritical Cylinder Flow at $M_\infty = 0.75$, Mach contours. Right: pressure distributions across AA' cut. Even with no special treatment at the sonic transition, the two separate schemes match well at $M = 1$.

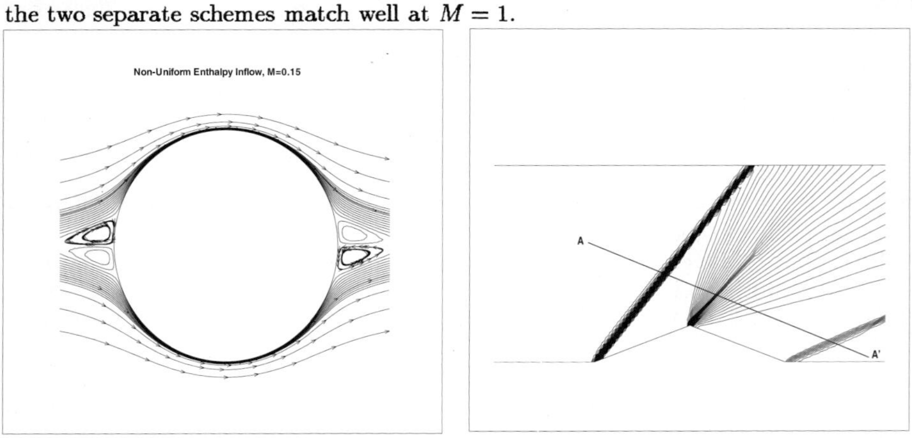

Fig. 4. Left: non-uniform enthalpy inflow at $M_\infty = 0.15$ showing symmetrical recirculations zones. Right: $M_\infty = 2.0$ flow over a diamond shaped airfoil, showing rarefaction shock.

Part XII

Domain Decomposition and Parallel Algorithms

Time-Accurate Implicit ALE Algorithm for Shared-Memory Parallel Computers

Dmitri Sharov[1], Hong Luo[1], Joseph D. Baum[1], and Rainald Löhner[2]

[1] Science Applications International Corporation, 1710 SAIC Drive, MS 2-6-9, McLean, VA 22102, USA
[2] Institute for Computational Sciences and Informatics, George Mason University, Fairfax, VA 22030, USA

Abstract. An unstructured grid matrix-free GMRES+LU-SGS scheme is used to simulate unsteady CFD problems with moving bodies. The method is applied on shared-memory, cache-based parallel machines. A special grid renumbering technique is used for the parallelization rather than the traditional method of partitioning the computational domain. Moving mesh with recurrent remeshing models the body motion.

1 Introduction

Accurate computations of fluid dynamics problems with moving bodies are required for such problems as pilot and canopy ejection, store or booster separation, etc. The characteristic time scale of the body movement is usually from tens to hundreds times larger than the characteristic time scale of the fluid motion. Thus, explicit computations with inherent small allowable time step are not efficient and implicit type solvers should be used.

Considerable progress has recently been made in the development of implicit schemes for unstructured grids. The GMRES+LU-SGS [1,2] can improve the efficiency of traditional explicit methods by one to more than two orders of magnitude. The scheme uses the Lower Upper-Symmetric Gauss-Seidel (LU-SGS) [3] scheme as a preconditioner for the Generalized Minimal Residual (GMRES) method [4] .

Another way to reduce turn-around time is to use massively parallel computers. With the advent of computers with over than 500 nodes, the exploitation of parallelism in solvers has become a major focus of attention. Most of the applications ported successfully to parallel machines to date have followed the Single Program Multiple Data (SPMD) paradigm. For grid-based solvers, a spatial subdomain was stored and updated in each processor. For obvious reasons, load balancing has been a major focus of activity.

Despite the striking successes reported to date, only the simplest of all solvers: explicit timestepping or implicit iterative schemes, perhaps with multigrid added on, have been ported without major changes and/or problems to massively parallel computers with distributed memory. Many code options that are essential for realistic simulations are not easy to parallelize on this type of machine. Among these, we mention local remeshing, repeated h-refinement, contact detection and force evaluation, some preconditioners, applications where particles, flow, and chemistry interact, and applications with rapidly varying load imbalances. Even if 99% of all operations required by these codes can be parallelized, the maximum achievable gain will be restricted to 1:100. If we accept as a fact that for most large-scale codes we may not be able to parallelize more than 99% of all operations, the shared-memory paradigm, discarded for a while as non-scalable, makes a comeback.

The objective of the present research effort is to implement the GMRES+LU-SGS scheme on shared memory parallel computers and apply it to moving body problems. Shared memory parallelization technique [5,6] is used. This method is based on extensive mesh renumbering which provides proper load balancing and avoids cache-misses and cache-line overwrite while allowing pipelining.

2 Governing Equations and their Discretisation

The Euler equations governing unsteady compressible inviscid flows can be expressed in the conservative form as

$$\frac{\partial \mathbf{Q}}{\partial t} + \frac{\partial \mathbf{F}^j}{\partial \mathbf{x}_j} = 0 \tag{1}$$

The governing equations are discretized by the finite volume method based on dual mesh cells associated with the nodes of the mesh, where the control volumes are nonoverlapping dual cells constructed by the median planes of the tetrahedra. In the present study the numerical flux functions for inviscid fluxes at the dual mesh cell interface are computed using the AUSM+ (Advection Upwind Splitting Method) scheme. Linear reconstruction of primitive variables is used with Van Albada limiter.

3 Implicit Time Integration

For the time-accurate implicit computation we use the method applied in Ref. [2]. The method is based on pseudo-timesteps. In this method, the discrete form of the governing equations Eq.(1) can be written as

$$\frac{V^{n+1}\mathbf{Q}^{n+1} - V^n\mathbf{Q}^n}{\Delta t} + \frac{\partial V\mathbf{Q}}{\partial \tau} = (1-\alpha)\mathbf{R}^{n+1} + \alpha\mathbf{R}^n \tag{2}$$

where Δt is the time increment; $\Delta \mathbf{Q}_n$ is the difference of an unknown vector between time levels n and $n+1$; V_i is the volume of the dual mesh cell; $\mathbf{R}_i$ is the right-hand side residual; τ is the pseudo time variable. If $\alpha = 1$, the scheme is the backward Euler method. If $\alpha = 0.5$, the resulting scheme known as Crank-Nicholson method is second-order in time. This yields the following system of linear equations

$$\left[\frac{V^{n+1}}{1-\alpha}\left(\frac{1}{\Delta t} + \frac{1}{\Delta \tau}\right)\mathbf{I} - \frac{\partial \mathbf{R}^m}{\partial \mathbf{Q}}\right]\Delta\mathbf{Q}^m = \mathbf{R}^m - \frac{V^{n+1}\mathbf{Q}^m}{(1-\alpha)\Delta t} + \frac{1}{1-\alpha}\left(\frac{V^n\mathbf{Q}^n}{\Delta t} + \alpha\mathbf{R}^n\right)$$

where m – pseudo-time levels.

It has been shown in Ref. [2] that the fully implicit scheme is more accurate than its linearized counterpart, since it requires several subiterations to achieve convergence on each time step. The linear system is solved by the matrix-free GMRES+LU-SGS method [1,2,6].

4 Shared Memory Parallelization Technique

The parallelization method is based on the combination of several renumbering and data regrouping techniques developed to avoid or considerably minimize cache-misses, cache-line overwrite, and memory contention [5,6].

Renumbering to minimize cache-misses. All unstructured CFD codes contain basic loops over nodes, edges, and elements. If a loop over the edges is considered and cache-misses are a concern, then the storage locations for the required point information should be as close as possible in memory when required by an edge. At the same time, as the loop progresses through the edges, the point information should be accessed as uniformly as possible. This may be achieved by first renumbering the points using a bandwidth-minimization technique such as the wavefront [7], or Peano-Hilbert type space-filling curves [8], and subsequently renumbering the edges according to the minimum point number on each edge. The same type of renumbering is done for all entities, which serve as basic loops in the code (e.g. elements, boundary faces, etc.). All of these algorithms are of complexity $O(N)$ or at most $O(NlogN)$, and well worth the effort.

Data and loop rearrangements to avoid memory contention. In order to achieve pipelining or vectorization, memory contention issues must be avoided. The memory contention can arise for instance in a loop over the edges while writing to corresponding points. To make sure that no point is accessed more than once, the loop is split into several contention-free loops over renumbered edges.

Data and loop rearrangements to avoid cache-line overwrite. An auto-parallelizing compiler can parallelize the inner loop. However, as has been mentioned in Ref. [5], such a parallelization is not efficient, because of both start-up penalties and cache-line overwrites. To keep vector-lengths short and enjoy small start-up cost, a special edge renumbering has been proposed [5].

Parallelization of the preconditioner. The parallelization of the LU-SGS preconditioner is performed by splitting the computational domain into several nonoverlapping regions according to the number of processors, and applying the LU-SGS method inside of each region without any special interprocessor boundary treatment [6]. This approach takes advantage of minimal parallelization overhead and good load balance.

For computational domain partitioning we use the fact that the grid nodes are already renumbered to minimize bandwidth, so we cut the entire array of nodes into equal-size pieces corresponding to the number of processors. This technique is very simple and provides perfect load balancing. Though the method doesn't provide good control over minimization of interprocessor boundary, this issue can be addressed by using alternative node renumbering techniques. In addition, since the shared-memory platforms are considered, the interprocessor communication overhead is not tightly connected to the area of the interprocessor boundaries.

5 Time Accurate Simulation of F18 Canopy/Pilot Ejection

The new method was used to compute an F/A18-C/D fighter canopy ejection. Single-pilot and two-pilot versions of the airplane were modeled. During the initial opening of the canopy, a number of topological changes occur in the geometry. The problem was computed with Mach number 0.76. The one-pilot computation was started at $t = 26$ ms, when the canopy just started to move, see Fig. 1, and ended at t=105ms when the canopy has moved to the tail part of the plane. Initially stage the canopy was hinged to the plane. After rotating 45 degrees (at $t = 40$ms), the canopy was released and allowed to move in response to the forces exerted on it. The average mesh size was 250,000 points and 1,300,000 elements. Mesh size varied little during the computation. Close-up of mesh near the canopy is shown in Fig. 2. A time step corresponding to CFL number of 50 was used in the computation. The simulation required approximately 4 CPU hours on 8 processors of SGI Origin 2000. A plot of CPU time vs. problem time is shown in Fig. 3a. The curve is not straight but rather has 8 steps attributed to global remeshing required by the algorithm. This computation is more than 10 times faster as compared to our explicit computation [9] . Only 184 time steps were required, compare with approximately 32,000 time steps with explicit scheme. This significant reduction in CPU requirements is attributed to the implicit GMRES scheme and parallel remeshing. Figure 3b shows the speedup of the time accurate computation. This speedup was computed by measuring the CPU time for a single timestep. The time accurate speedup result is very similar to those obtained for the steady-state cases. The results of the present implicit computation match very well with our explicit results. In fact, we have compared the canopy location at several time instances and observed no significant dislocation. At the same time, there is a good agreement between the results of the explicit computation and the experiment data [9].

The two-pilot plane has a larger canopy. The canopy was released from its hinges at $t = 65$ms. The weapons officer was ejected at $t = 170$ms, and the pilot was ejected at $t = 300$ms. Figure 4 shows their position at $t = 560$ms.

6 Conclusions

The parallel matrix-free GMRES+LU-SGS unstructured grid method is successfully applied to modeling moving body problems on shared-memory machine. The method requires no direct domain partitioning and can easily be combined

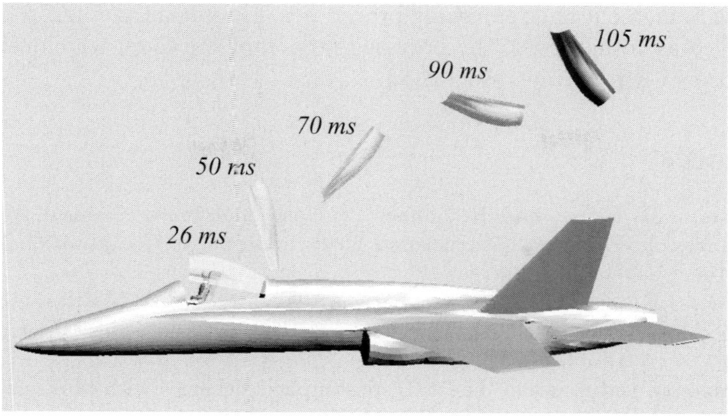

Fig. 1. F/A18-C/D fighter canopy ejection

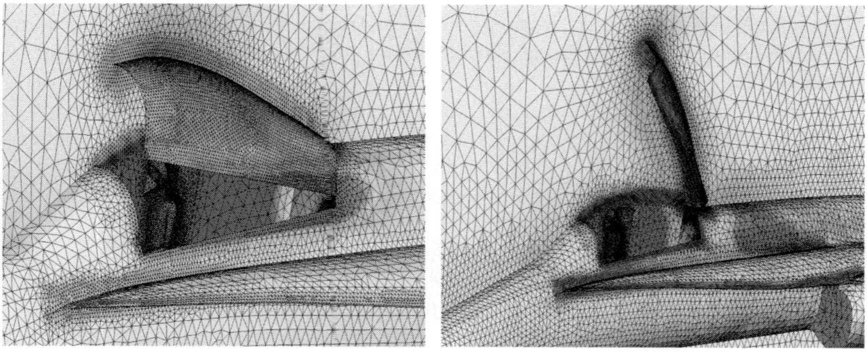

Fig. 2. F/A18-C/D fighter canopy ejection. Surface mesh

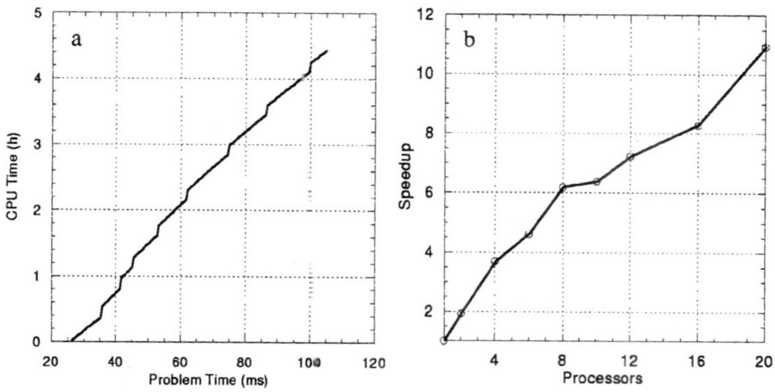

Fig. 3. Canopy ejection. (**a**) CPU time on 8 processors versus problem time. (**b**) Speedup for the canopy ejection case

with mesh refinement and remeshing procedures. Significant savings in CPU time are achieved as compared to the previous version of the code, which utilized the explicit Runge-Kutta time integration.

References

1. H. Luo, J.D. Baum, and R. Löhner: 'A Fast, matrix-free Implicit Method for Compressible Flows on Unstructured Grids,' Journal of Computational Physics, Vol. 146, pp.664-690, 1998
2. H. Luo, J.D. Baum, and R. Löhner: 'An Accurate, Fast Matrix-Free Implicit Method for Computing Unsteady Flows on Unstructured Grids,' AIAA 99-0937, 1999
3. A. Jameson, and S. Yoon: 'Lower-Upper Implicit Schemes with Multiple Grids for the Euler equations,' AIAA J., Vol. 25, No7, pp.929-935, 1987
4. Y. Saad, and M.H. Schultz: 'GMRES: a Generalized Minimal Residual Algorithm for Solving Nonsymmetric Linear systems,' SIAM J. Sci. Stat. Comp., Vol. 7, No 3 (1988), pp. 89-105
5. R. Löhner: 'Renumbering Strategies for Unstructured-Grid Solvers Operating on Shared-Memory, Cache-Based Parallel Machines,' AIAA 97-2045, 1997, pp.1015-1025
6. D. Sharov, H. Luo, J.D. Baum, R. Löhner: 'Implementation of Unstructured Grid GMRES+LU-SGS Method on Shared-Memory, Cache-Based Parallel Computers,' AIAA 2000-0927 (2000).
7. R. Löhner: 'Some Useful Renumbering Strategies for Unstructured Grids'; Int. J. Num. Meth. Eng., 36, pp.3259-3270, (1993)
8. H. Sagan: 'Space-Filling Curves,' Springer Verlag, New York, 1994
9. J.D. Baum, R. Löhner, T.J. Marquette, H. Luo: 'Numerical Simulation of Aircraft Canopy Trajectory,' AIAA-97-1885, (1997)

Fig. 4. F/A18-C/D fighter canopy and pilot ejection

Detailed Computations of Laminar Reactive Flows Using Multigrid Methods on Parallel Computers

Sébastien Paxion[1], Dominique Thévenin[2], Achim Gordner[1], Romain Baron[2], and Peter Bastian[1]

[1] Interdisziplinäreszentrum für Wissenschaftliches Rechnen (IWR), Universität Heidelberg, INF 368, D-69120, Germany
[2] Labo. EM2C, École Centrale Paris and CNRS, Grande Voie des Vignes, F-92295 Châtenay-Malabry, France

Accurate simulations of complex reactive flows remain difficult and costly, even today, although they are important from the scientific, economical and ecological point of view. For three years, we have been developing a computer code for steady laminar flame simulations in gaseous media, at low Mach numbers [1][2]. Detailed models are taken into account to describe reactive mechanisms, thermodynamic properties of the mixture and molecular diffusion of chemical species. A detailed thorough description of the model problem and the numerical methods, as well as a detailed presentation of the results and validation test cases can be found in [4].

1 Model Problem

The reactive mixture consists of a multicomponent gas of K different chemical species. The chemical activity is described by I elementary reactions involving no more than 3 reactants and 3 products. The molar production rate of each species k per unit volume is

$$\dot{\omega}_k = \sum_{i=1}^{I} (\nu''_{ki} - \nu'_{ki})\, q_i \quad \text{with} \quad q_i = k_{f_i} \prod_{k=1}^{K} [\mathcal{X}_k]^{\nu'_{ki}} - k_{r_i} \prod_{k=1}^{K} [\mathcal{X}_k]^{\nu''_{ki}}$$

where ν'_{ki} and ν''_{ki} are stœchiometric coefficients of species k in reaction I (typically 0, 1 or 2), $[\mathcal{X}_k]$ is the concentration of species k and the forward and backward rates of progress of each reaction are related to each other via the reaction equilibrium constant, and are modeled by an Arrhenius law:

$$k_{f_i} = A_i T^{\beta_i} \exp\left(\frac{-E_i}{RT}\right) \quad \text{and} \quad k_{r_i} = \frac{k_{f_i}}{\mathrm{Ke}_i}$$

the constant Ke_i being theoretically expressed from the standard enthalpy and entropy, and is mainly a function of temperature.

The thermodynamical and chemical description of the reactive mixture is complete, being given the set of chemical reactions, their related Arrhenius coefficients, and, for each species, the standard enthalpy and entropy, and the specific heats at constant pressure as a function of temperature. In practice, the latter is a fourth order polynomial fit of experimental measurements. Our computations are based on the JANAF thermodynamical database and use the CHEMKIN [5] processing library.

Detailed molecular transport mechanisms can only by formulated with the kinetic theory of multicomponent gases, based on the perturbation of the non-dimensionalized multi-component Boltzmann equation, around the Maxwellian velocity distribution. No explicit formulations can be obtained for diffusion coefficients of multicomponent gases. Linear systems of equations must be solved instead. This is performed with the EGLIB library developed by Ern and Giovangigli, for evaluating the transport coefficients [6][7][8]. Optimized iterative methods are used to solve to some accuracy these linear kinetic systems, based on a thorough mathematical analysis of their properties [9]. Molecular transport takes the form of diffusion velocities $\boldsymbol{V}_k$ proportional to the species molar fractions (ratio of the number of moles of species k to number of moles of the mixture, related to mass fractions via molar weights). Coupling between species diffusion and thermal diffusion, known as the Soret effect, has been neglected in our computations.

In flows at low Mach numbers M, acoustic waves travel much faster than entropy waves, with a velocity ratio of about M^{-1}, resulting in very ill-conditioned systems of equations when low speed flows are solved with compressible equations. To circumvent this problem, we write the Navier-Stokes equations in the low Mach number approximation, based on the observation that pressure variations through laminar flames at low Mach numbers stem for the very largest part from hydrodynamical and not from compressibility effects. The pressure is split into a bulk background uniform thermodynamic pressure p_u (the atmospheric pressure in our cases) and a hydrodynamic $\tilde{p}$ term. Giovangigli [10] shows from an asymptotic analysis of aerothermochemical structures of laminar flames that hydrodynamic pressure fluctuations lie within the Mach number squared of the overall thermodynamic pressure. A scale analysis then leads to the following so-called isobaric flame equations:

$$\frac{\partial \rho}{\partial t} + \boldsymbol{\nabla} \cdot (\rho \boldsymbol{v}) = 0 \tag{1}$$

$$\frac{\partial (\rho Y_k)}{\partial t} + \boldsymbol{\nabla} \cdot (\rho Y_k \boldsymbol{v}) = -\boldsymbol{\nabla} \cdot (\rho Y_k \boldsymbol{V}_k) + W_k \dot{\omega}_k \tag{2}$$

$$\frac{\partial (\rho \boldsymbol{v})}{\partial t} + \boldsymbol{\nabla} \cdot (\rho \boldsymbol{v} \boldsymbol{v}) = -\boldsymbol{\nabla} \tilde{p} + \boldsymbol{\nabla} \cdot \left\{ \mu \left(\boldsymbol{\nabla} \boldsymbol{v} + (\boldsymbol{\nabla} \boldsymbol{v})^T \right) \right\} \tag{3}$$

$$\frac{\partial (\rho T)}{\partial t} + \boldsymbol{\nabla} \cdot (\rho T \boldsymbol{v}) = \frac{1}{c_p} \frac{\mathrm{d} p_u}{\mathrm{d}t} - \frac{1}{c_p} \boldsymbol{\nabla} \cdot (-\lambda \boldsymbol{\nabla} T) - \frac{1}{c_p} \sum_{k=1}^{K} h_k W_k \dot{\omega}_k . \tag{4}$$

with the addition of the ideal gas law,

$$\rho = \frac{p_u W (Y_1, Y_2, \cdots, Y_K)}{RT} \tag{5}$$

This approach is now widely used in the low Mach combustion community [11] [12] [13], though it has to be pointed out that, to our knowledge, rigorous justifications of the isobaric flame equations with detailed chemical and diffusional models are still lacking. Majda and Sethian [14] proved the asymptotic consistency of the pressure decomposition and derived the equations from a rigorous asymptotic expansion but their analysis is restricted to premixed, simple diffusion and chemistry flames with adiabatic slip wall conditions. See [4] for a wider discussion.

2 Discretization and Implementation

The equations are discretised with a cell vertex finite volume method on conforming unstructured grids. All primitive variables ($\boldsymbol{v}$, $\tilde{p}$, T and the Y_k) are stored at the grid nodes. Stabilization is performed by the FIELDS method proposed by Schneider *et. al.*[15] initially for fully incompressible flows, which we have extended to the case of our incompressible dilatable flows. The density field is explicited with respect to the primitive variables. The Navier-Stokes equations (1) and (3) then appear implicitly decoupled from the thermo-reactive ones (2) and (4), and are marched in time separately but each sub-system with full implicit integration. Linearisation is achieved by fixed point iterations for the momentum convection term, and with an approximate-Newton method for non-linearities in the chemical and diffusional terms of the thermo-reactive subsystem.

The *UG-C* code is designed as an application of the *UG* library which is a modular, C-coded, numerical toolbox [16] aimed at both the development of simulation codes for a wide range of PDE problems. It is based on different levels of abstraction: the *UG library* itself, independent of the PDE problem to be solved; the *problem-class library*, concerned with discretization, using abstract grid elements description, independent of space dimensions (2D or 3D) and element types (triangles, quadrangles, tetrahedra...); finally the *application level* implements the boundary value problem (description of geometries and boundary conditions), the grid and any data files, and the definition of the solving strategy. The *UG-C* code automatically inherits all features of the *UG* library, among which the most important for our applications are: dynamic local refinement and coarsening of the multi-level grid; portability on distributed memory parallel machines with various spectral and geometrical load-balancing algorithms, specifically designed for multilevel-grid partitioning.

The linearized Navier-Stokes and thermo-reactive problems are solved by the Bi-CGSTAB method preconditioned by multigrid V-cycles with a Gauss-Seidel smoother. Special multigrid methods have been devised in *UG* for problems which require strong local mesh refinements [17]. Our general solving procedure consists in a rough flame initialization (from 1D flame database) and makes use of dynamic grid refinement and coarsening to track the flame front. Finally, the grid is further refined to achieve high spatial resolution at stationary convergence.

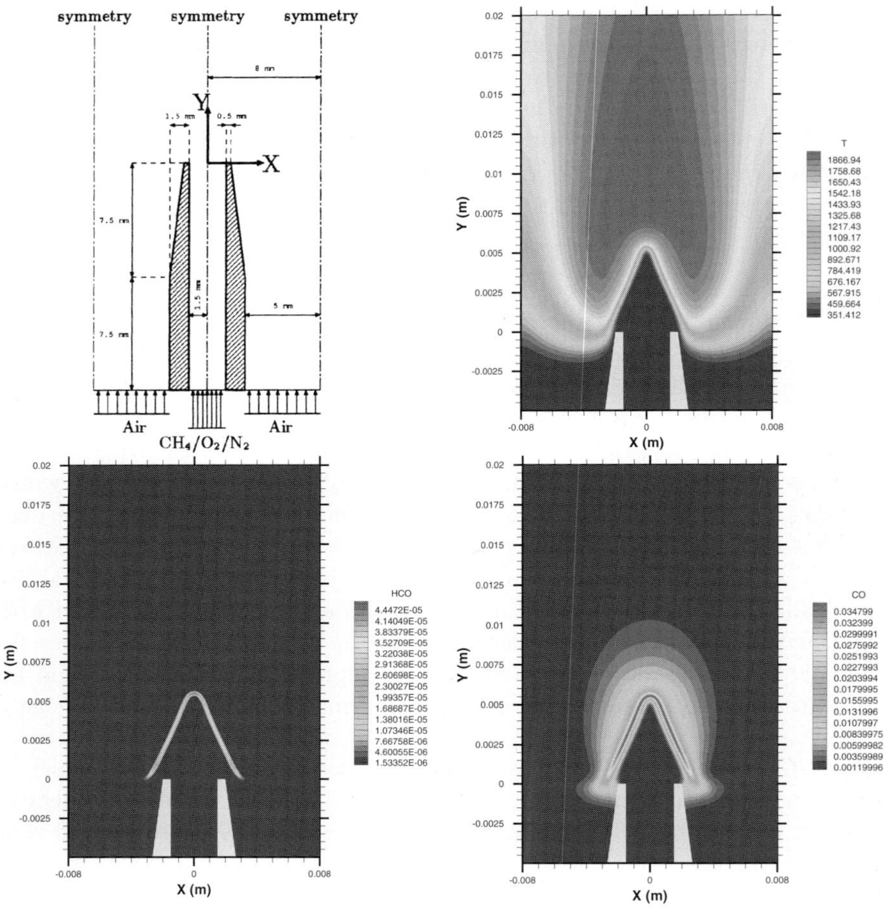

Fig. 1. Methane/air premixed flame on the TOPDEC burner: geometry, temperature and HCO, and COmass fractions

3 Results and Current Developments

The *UG-C* code has been validated for premixed and non-premixed laminar flame computations, on various configurations ranging from fundamental laboratory burners to industrial-like burners. We show in Fig. 1 computations for the TOPDEC burner which is a realistic configuration close to industrial combustion systems [18]. Methane is premixed with air in lean proportions (equivalence ratio of 0.83, with, at the injector inlet, $Y_{CH_4} = 0.046$ and $Y_{O_2} = 0.222$) and injected at 0.98 $m.s^{-1}$. A secondary air flow is injected between the injectors at 0.05 $m.s^{-1}$. The total computation from a rough initial solution takes about 20 hours on a single SUN Ultra-5 with less than 180 Mb and a final surface grid of about 15200 nodes. The same computation has been performed on 8 nodes of the 80-PC-cluster developed at IWR, Heidelberg. The partitioning of the 5-level grid

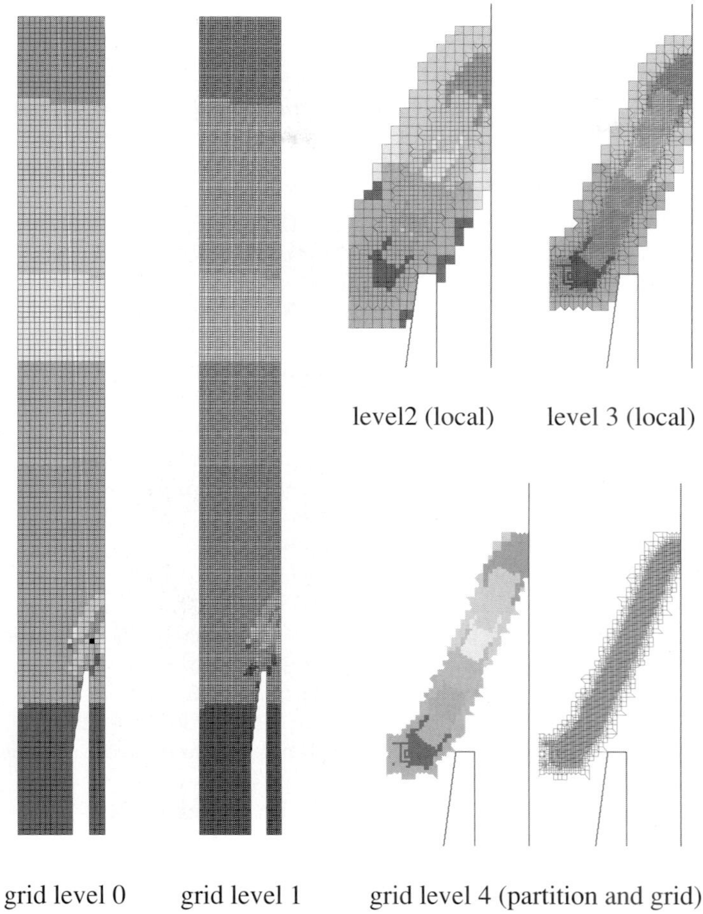

Fig. 2. Partitioning on 8 processors of the 5-level grid (whole domain for levels 0 and 1, zoom on the injector tip for levels 2, 3 and 4)

is achieved by the clustering method implemented in *UG* [19], see Fig. 2: father and son elements are clustered together up to a certain depth (here, one) and the resulting element clusters are distributed from fine to coarse levels among processors (a recursive inertial bisection performs very well for stretched refined grid along flames). This warranties minimum interprocessor data transfer due to vertical prolongations and restrictions involved in multigrid cycles. A global parallel efficiency of more than 80% is observed for this problem.

Current developments are concerned with realization of the first 3D laminar flame computations with detailed chemistry on our parallel PC-cluster. In parallel, the ILDM/FPI chemical kinetics reduction method [20] is being implemented, which should eventually decrease by an order of magnitude the global computation time.

References

1. S.C. Paxion, D. Thévenin, P. Bastian, and S. Candel: In C.-H. Bruneau, editor, *Sixteenth International Conference on Numerical Methods in Fluid Dynamics*, (Arcachon, France, 1998, Springer-Verlag) pp 536–541
2. S.C. Paxion, D. Thévenin, P. Bastian, and S. Candel: *7th International Conference on Numerical Combustion* (York, 1998)
3. S.C. Paxion, A. Gordner, N. Neuss, P. Bastian, and D. Thévenin: *8th International Symposium on Computational Fluid Dynamics* (Bremen, 1999)
4. S. Paxion: PhD thesis, École Centrale Paris (1999)
5. R.J. Kee, F.M. Rupley, and J.A. Miller: Technical Report SAND89-8009B, SANDIA (1991)
6. A. Ern and V. Giovangigli: *Journal of Computational Physics*, **120**, pp. 105–116 (1995)
7. A. Ern and V. Giovangigli: *Combust. Sci. Tech.*, **118**, pp. 387–395 (1996)
8. A. Ern and V. Giovangigli: Technical report, CERMICS (1997)
9. A. Ern and V. Giovangigli: *Multicomponent Transport Algorithms*, vol 24 of *Lecture Notes in Physics* (Springer-Verlag, Heidelberg, 1994)
10. V. Giovangigli: PhD thesis, École Centrale Paris (1988)
11. M. Braack: PhD thesis, Universität Heidelberg (1998)
12. M. Lai, J.B. Bell, and P. Colella: *AIAA Paper*, 93-3369, (1993).
13. R.B. Pember, L.H. Howell, J.B. Bell, P. Colella, W.Y. Crutchfield, W.A. Fiveland, and J.P. Jessee: *Combust. Sci. Tech.*, **140**, pp. 123–168 (1998)
14. A. Majda and J. Sethian: *Combust. Sci. Tech.*, **42**, pp. 185–205 (1985)
15. G.E. Schneider and M.J. Raw: *Num. Heat Transfer*, **11**, pp. 363–390 (1987)
16. P. Bastian, K. Birken, K. Johannsen, N. Neuss, H. Rentz-Reichert, and C. Wieners: *Comp. Vis. Sc.*, **1**, pp. 27–40 (1997)
17. P. Bastian: PhD thesis, Ruprecht-Karls-Universität Heidelberg (1994)
18. M. Perrin, A. Garnaud, F. Lasagni, S. Hasko, M. Fairweather, H. Levinsky, J.-C. Rolon, J.-P. Martin, J. Rolon, A. Soufiani, H. Volpp, T. Dreier, and R. Linstedt: *International Gas Research Conference* (San Diego, 1998)
19. P. Bastian: *SIAM J. Sci. Comput.*, **19**, pp. 1303–1321 (1998)
20. O. Gicquel, N. Darabiha, and D. Thévenin, 28th Symposium (Int.) on Combustion, (2000)

Massively Parallel Computing with Coupled Algorithm for Introduction of Large-Eddy Simulation

Chuichi Arakawa[1], Kenji Saito[2] and Kenjirou Shimano[3]

[1] Interfaculty Initiatives in Information Studies, University of Tokyo. Hongo 7-3-1, Bunkyo-ku, Tokyo, 113-8656 Japan,
[2] Department of Mechano-Informatics, Faculty of Engineering, University of Tokyo. Hongo 7-3-1, Bunkyo-ku, Tokyo, 113-8656 Japan
[3] Musashi Institute of Technology, Tamazutsumi 1-28-1, Setagaya-ku, Tokyo, 158-8557

Abstract. It was found that the coupled algorithm can keep comparatively larger efficiency of parallel computing for fine granularity in which the number of grid becomes very small. The authors extend it to LES simulation of 3-D circular cylinder in order to show its availability for the complex structure of flow. The parallel efficiency reaches about 30% even for fine granularity of 32 points for one processor with 1024 processors. The complex flow with Karman vortecies can be analyzed accurately and rapidly including turbulent quantities, which will be developed to the idea of real time simulation in the fluid dynamics.

1 Introduction

Massively parallel computer is expected to become a main platform in scientific numerical analysis, of course, in the CFD field. It has been often pointed out that efficiency of parallel computing becomes worse when the number of processors increases. The authors have focused on the coupled method based on SCGS method by Vanka[1] as an incompressible Navier-Stokes solver suitable for massively parallel computing. reported that parallel efficiency of calculating steady cavity flow by the coupled method was much higher than that by the MAC methods, and the authors[2] reported that parrallel efficiency of 2-D and 3-D unsteady flow was also high including Boundary Fitted Coordinate (B.F.C) calculations for circular cylinder.

The authors extend its algorithm to turbulent flows with LES on super computer Hitachi SR2201 of 1024 processors, and report the result of speed-up ratio and parallel efficiency in this paper. The complex flow of Karman vortices will be simulated accurately as well as instantaneously. The calculation speed will be the most important for the future world of real time simulation of CFD. Therefore the massively parallel computing with fine granularity and huge number of processors becomes to be a key technology. The algorithm proposed in this paper will take important role in the high speed simulation although the total grid number is limited. The minimum number of grid for one processor in this paper is 32, and total number of grid depends on the number of processors.

2 Calculation Method

2.1 Coupled Method

The standard methods for incompressible Navier-Stokes equations are SIMPLE and MAC family algorithms which solve momentum equations and continuity one separately. These decoupled methods are fundamentally so efficient for steady and unsteady flows that they have been widely used for the researches. On the contrary, the coupled method solves all of equations at the same time for each grid points.

The coupled algorithm has a main feature that they analyze equations of motion and equation of continuity simultaneously. Fig. 1 shows the arrangement of u, v, P in the grid for this research. The equations of motion are discretized using finite volume method in the same control volume. The equation of continuity uses a different control volume against momentum equations in order to avoid numerical oscillation of checker's board shown in Fig. 1.

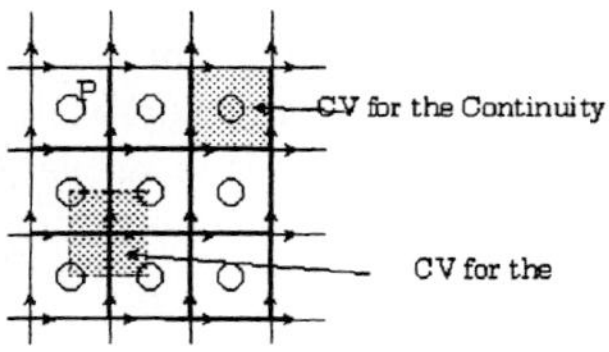

Fig. 1. Arrangement of variables and control volumes(CV) for 2-D B.F.C.

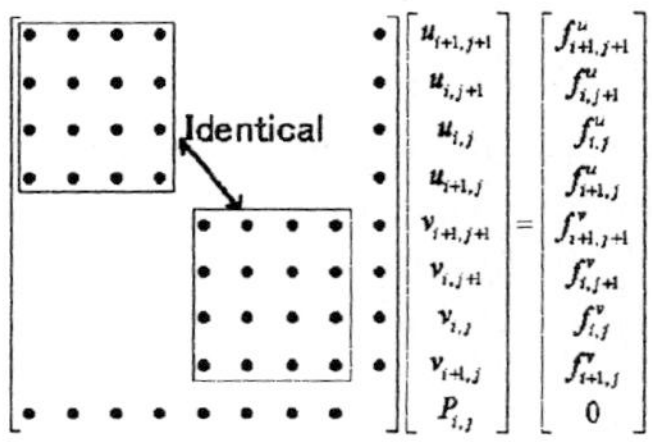

Fig. 2. Coefficient matrix of coupled method for 2-D B.F.C.

As for two dimensional analysis, one can derive one pressure and eight velocity components surrounding a control volume for the pressure at one time implicitly. In other words, 9×9 (in 3-D 25×25) matrix described in Fig. 2 in each cell is required to be solved. Because velocities of u and v are defined at the same grid point, the coefficients for u-velocity in the small matrix are coincident with the coefficient of v-velocity in the matrix. Finally one can require to inverse only smaller matrix of 5×5 (in 3-D 9×9).

2.2 Domain Decomposition for Parallel Algorithm

The domain decomposition is available for making the coupled method parallel. In order to equalize calculation time in each processor, the region is divided for distributing the same number of grids for one small domain. Each processor has the extra one outer line data from actual connecting boundary line, and communicates together with neighbor processors by sending data along the second outer line to neighbor's first outer line. While relaxing in each processor, variables along the first outer line are fixed as Dirichlet condition for the boundary.

2.3 Estimation of Parallel Algorithm

One can estimate parallel algorithm with speed-up rate S_N and efficiency ε_N as follows.

$$S_N = \frac{T_1}{T_N}, \qquad \varepsilon_N = \frac{S_N}{N} = \frac{T_1}{N \cdot T_N} = \frac{n_1}{n_N} \cdot \frac{T_1}{N \cdot t_N} = \varepsilon^{iter}{}_N \cdot \varepsilon^{ope}{}_N \tag{1}$$

Here T_1 is the total calculating time when the serial calculation is carried out in one processor. T_N is the total calculating time when the program runs in N processors. Furthermore n_1 and n_N mean iteration numbers to terminate iteration due to the same criterion for convergence. t_1 and t_N mean calculation time required for one loop, where the index shows the number of processors. $\varepsilon^{iter}{}_N$ is the parameter to show how better the convergence of iteration is for parallel processing. $\varepsilon^{ope}{}_N$ is the parameter to show how effectively the tasks are parallelized. $\varepsilon^{iter}{}_N$ and $\varepsilon^{ope}{}_N$ are called as convergence efficiency and parallelize efficiency respectively.

2.4 LES Simulation and Calculation Conditions

Large Eddy Simulation (LES) is a realistic approach for turbulent flows where the turbulence of small grid size are resolved with simple assumptions and larger structures of turbulence are directly derived without any modelling. Even recent computers are not available for exact and accurate LES computing due to the number of grid points and CPU time. Most of LES applications which has appeared in the journal and proceedings are ironically called as "Very Large Eddy Simulation (VLES)" because the grid is not so fine to resolve the turbulence of small scale where its structure is assumed to be isotropic. The high speed and large size of memory in the computer are strongly required for accurate use of LES, which means that the coupled method in massively parallel computing has a large possibility to accept this requirement in an early stage comparing with other methods. As a first step of LES, Smagorinsky model is employed for the assumption of sub-grid-scale turbulence. An objective flow in this paper is circular cylinder which provides the unsteady oscillating flow behind it, which becomes a typical example for LES evaluation.

The conditions for simulation of 3D circular cylinder with LES are described in the following.

- Reynolds number is 3900. The number of grids is $128\times64\times23$ (260 thousand). The minimum grid length is 0.005D
- O-grid system is employed. The calculation covers the area whose size is eight times as large as a diameter of circular in front of it, twenty four times behind it, and eight times in the vertical and spanwise directions.
- The finite volume method is employed for dicretize, QUICK method is installed for convection terms.
- Smagorinsky model is employed as LES model, and Smagorinsky constant is fixed to 0.15.

There are four simulations carried out in this research. It is summarized that case 1 is a simulation for 260k of grid points and $4.0*10^{-3}$ of time step. Case 2 is for 260k and $1.0*10^{-2}$, and case 3 is the largest simulation for 520k and $4.0*10^{-3}$. Finally no model is a simulation without LES for 260k and $1.0*10^{-2}$.

3 Calculation Results

At first the performance of the proposed algorithm is summarized using the basic examples of laminar flows such as 2-D square cylinder, 2-D and 3-D circular cylinders. Table 1 shows the efficiency of parallel algorithm and speed up ratio according to the change in the number of processors for the fixed grid number, which means the granularity changes from 256 to 32.

Generally speaking, the number of grid points for one processor is larger than 1,000 in the normal use for the application topics due to the limit of the number of processors. For example, the granularity is 1,000 for at most 1,000 processors in the simulation of 1,000,000 grid points. The efficiency in this case approaches nearly 1.0 if the program is appropriately designed for parallel computer. The target in this paper is different from this standard case, which means that the authors aim to develop the special program for fine granularity in the parallel computing.

It is found in the table 1 that the efficiency becomes worse as the granularity becomes to finer to 32. But it can keep the value of about 50% which will be acceptable for future use of massive parallel computing. The final speed-up ratio arrives at 500 with 1,024 processors in the total grid number of about 30,000. Of course, the efficiency in 3-D calculation is not so good as that in 2-D because the former case requires longer communication time for the six surfaces of contact region against four ones in the latter case. Finally it is found that the coupled method is very applicable for massive parallel computing with fine granularity for basic and laminar flows.

Table 1. Efficiency of parallel computation for laminar basic flow

Calculations	2D Square		2D Cylinder			3D Cicular Cylinder		
PE	256	512	256	512	1024	256	512	1024
Subdomain	16×16	32×16	16×16	32×16	32×32	$8 \times 8 \times 4$	$16 \times 8 \times 4$	$16 \times 8 \times 8$
Tocal CV	256×128		256×128			$64 \times 32 \times 16$		
CV per Proc.	128	64	128	64	32	128	64	32
Efficiency(%)	79.9	70.3	92.1	73.2	48.0	63.3	46.1	31.1
Speed-Up	204	360	236	375	491	162	236	319

Fig. 3 shows the iso-surfaces of vorticity around x-axis with 0.10 and -0.10 at non-dimensional time t=130 in the simulation of Reynolds number 3,900. It is found that it can represent the characters of flow very well such as Karman

Vortices, the rib-structure in the spanwise direction and the turbulence of vortex motion. It is interesting that the rib-structure becomes to have some disturbance due to turbulent phenomena comparing with the flow of Reynolds number 1,000. Fig. 4 describes the iso-surfaces of vorticity around z-axis with the same value as previous figure. It is found that the vortex flows out while twisting each other.

The following two figures from Fig. 5 to Fig. 6 show the detailed information of velocity field and turbulent quantities. Generally speaking, the case 3 is the best to predict the flow accurately in which the number of grid points is the largest and the time step is the smallest among all of simulations. The simulation without LES nor turbulence model reveals the large discrepancy in turbulent quantities.

In order to derive the basic information of parallel algorithm, the authors check the efficiency and speed up ratio with QUICK for convective terms, because they took Hybrid scheme for previous researches for laminar flows. It is found in the table 2 that the QUICK algorithm can keep similar efficiency as Hybrid one for laminar flow of Reynolds number 1,000 with $64 \times 32 \times 16(32{,}768)$ grids. For example,the efficiency in Hybrid scheme was reported to be 31.1%[2] in the same condition. The efficiency of 29.0% in QUICK is thought to be enough high for fine granularity of 32 grid points in one processor for 3-D calculations.

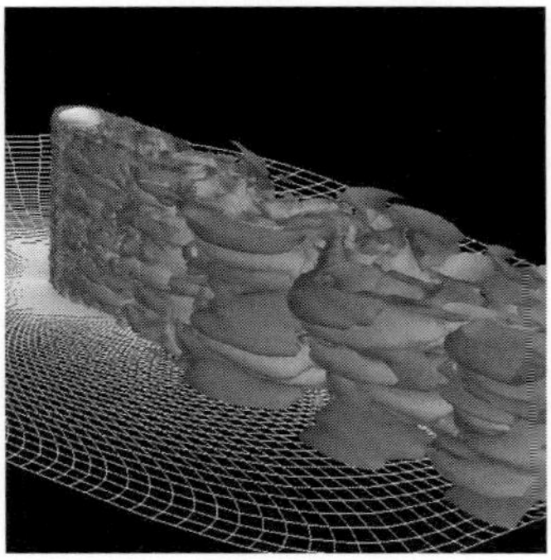

Fig. 3. Isosurfaces of vorticity strength around X-Axis

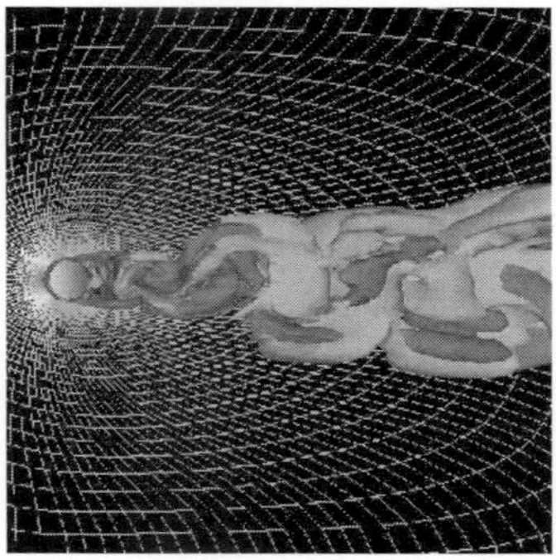

Fig. 4. Isosurfaces of vorticity strength around Z-Axis

4 Conclusion

The coupled method with fine granularity in the parallel computing is developed to LES calculation in body-fitted curvilinear coordinate system, and is found to be available for simulating three dimensional unsteady flow around cylinder on parallel computer of SR2201. The efficiency on 1024 processors (64 processors:standard) arrives at around 50% even if the granularity is fine, which results in the future development of massively parallel computing for real time

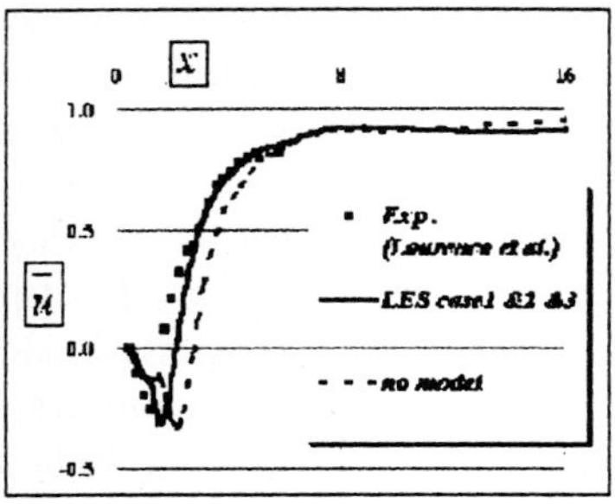

Fig. 5. Time averaged velocity $\overline{u}$ distribution along x-axis

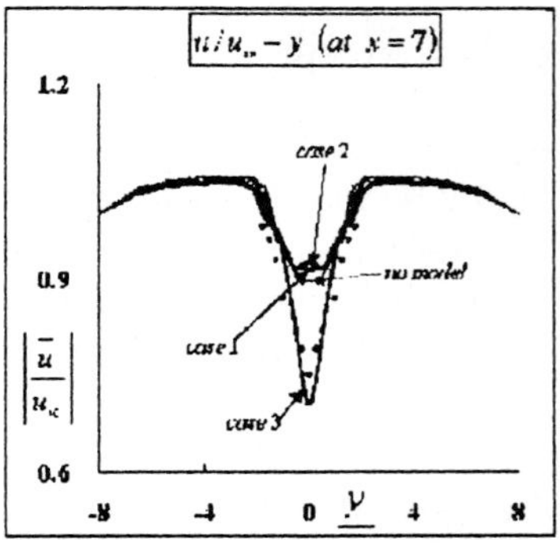

Fig. 6. Time averaged velocity $\overline{u}$ distribution in the y- direction at x=7

Table 2. Parallel computational results for laminar flows with QUICK

Calculation	QUICK(no model)		
PE	256	512	1024
Subdomain	$8 \times 8 \times 4$	$16 \times 8 \times 4$	$16 \times 8 \times 8$
Total CV	$64 \times 32 \times 16$		
CV per Proc.	128	64	32
$\varepsilon^{iter}{}_N$	49.0%	34.3%	22.4%
$\varepsilon^{ope}{}_N$	126%	128%	129%
ε	61.8%	44.0%	29.0%
Speed-Up	158	225	297

simulation with high speed. It is also confirmed that the complex flow with Karman vortex can be accurately simulated for turbulent flow of Reynolds number 3,900 with LES in the proposed algorithm.

References

1. Vanka, S.P., J.Comp.Phys., 65, 1986, pp138-158.
2. Hamajima Y, Arakawa C, Incompressible Navier-Stokes Solver on Massively Parallel Computer Adopting Coupled Method for B.F.C, Proc. of Parallel CFD '98, Elsevier (to be published)
3. Ong L. and Wallace J., The Velocity Field of the Turbulent Very Near Wake of Circular Cylinder, Exp. Fluids, 20, 1996, pp.441-453.

Numerical Simulation of Vacuum Plume Using Coupled-Parallel Computation

Cai Guobiao and Zhuang Fenggan

School of Astronautics, Beijing Univ. of Aero. and Astro., Beijing, 100083, China

Abstract. The numerical simulation of the attitude rocket engine vacuum plume is presented in this paper. A coupled computational method is developed, i.e., the nozzle flow is computed using Navier-Stokes equations and finite difference scheme and the plume using DSMC method and then coupled together. By this method, we obtained the physical variables of vacuum plume. The present work provides a good reference to the design of attitude controlling rocket engine nozzle, and is a good base to study vacuum plume characteristics and vacuum plume contamination.

1 Introduction

Gas dynamic force, aero-thermal load and contamination produced by vacuum plume are very important problems for the designers of spacecrafts. The region in vacuum plume includes continuum flow, transition flow and free molecular flow. Ground vacuum plume test is complex and costly. So using numerical simulation to study vacuum plume is certainly a reasonable approach. In order to numerically simulate vacuum plume successfully, we must set up a suitable mathematical model for different flow cases, and use relevant numerical methods. The DSMC method[1] may successfully simulate rarefied gas flow. It is known that DSMC method need the computation time step much less than the average collision time of molecules and computation mesh size much less than the mean free path of molecules, so it necessitates a very high computer performance with large memory. It is not suitable for continuum flow. Some successful simulation works have been reported[2,3,4]. To study nozzle flow and plume as an entirety, a coupled-parallel computation method is suggested in the paper.

2 Numerical Methods

When numerically simulating vacuum plume, according to the characteristics of attitude rocket engine, we adopt finite difference method for solving N-S equations to compute nozzle flow and use DSMC method for simulating the plume outside the nozzle, then couple them and calculate it repeatedly. By this way we expect to get the flow variables distribution of vacuum plume.

2.1 Nozzle flow simulation

Governing equations The governing equations we adopted here are Reynolds averaged axi-symmetric N-S equations that ignore the body force, chemical reaction and thermal radiation[5].

Turbulence model Turbulence model is a significant part in doing turbulent flow numerical computation. Its correctness will directly affect the accuracy of the whole numerical computation. This paper adopts the Baldwin-Lomax zero equation model(algebraic model)[6].

Difference method for solving N-S equations This paper adopts the MacCormack time marching predictor-corrector two-step scheme to solve the governing equations in computational plane[5].

2.2 DSMC method to simulate plume outside nozzle

The basic concept of DSMC is to use few simulating molecules instead of a large number of gas molecules in real flow field, and DSMC to simulate real physical processes. That is to say, tracing the simulating molecules, recording the variation of their position and velocity, calculating the collision of molecules with molecules and molecules with wall, then take statistical average value and obtaining the flow parameters we need[1]. The basic assumption of DSMC is that we need not consider the positions of the molecules when simulating the collision between molecules in the mesh. Although the relative place may affect the collision process, this effect can be neglected when mesh size much smaller than the local mean free path. Meanwhile, DSMC methods require the time interval of simulations much less than average molecular collision time. This is because there are always in the flow field some molecules colliding and others moving freely, hence we must decouple the free motion of molecules and molecular collision in a limited time interval. If the decoupled time is comparable to the average molecular collision time or even longer than it, there may happen many times collision for a molecule in this time interval and real simulation could not be made. Thus DSMC methods need simulation time step much shorter than average molecular collision time and this leads to a large amount of computational work. In turn a large computer memory is needed and it takes long calculation time.

Since the flow is axisymmetric, we only need to compute the flow parameters in a plane containing the axis and above it. Because the flow outside nozzle is a low density flow, in simulation, we make the following assumptions: a) only bimolecular collisions are considered; b) neglect internal freedom and chemical non-equilibrium effect; c) the overall gas flow is steady; d) the interaction forces between molecules are neglected.

2.3 Mesh treatment

When computing nozzle flow, we use the well-known TTM method to generate computational mesh [7]. In plume outside the nozzle, according to the characteristics that the lower plume density and pressure the farther away from nozzle exit, we adopt non-uniform meshes. In the region near the nozzle exit, the mesh is dense and the mesh is sparse in area far away from nozzle exit. Thus we can ensure each mesh unit have enough simulation molecules.

2.4 Boundary condition and coupling boundary data transfer

Boundary condition for nozzle flow computation Usual boundary conditions are applied here. The mention should be made for the exit boundary condition. If the flow is supersonic here, the flow parameters can be calculated with second order extrapolation; and if the flow is subsonic, the back pressure should be prescribed, the other parameters can be calculated with first order extrapolation.

Boundary condition for plume outside nozzle exit 1) The axisymmetric factor has been added in the mathematical model, the molecules will not move across the axis. 2) We consider the impinging wall as a completely diffuse reflection wall. 3) We eliminate the molecules that move out the simulation region during numerical simulation. 4) Molecules enter the simulation region continuously at the upstream boundary (nozzle exit). The number of molecules entering the simulation region per unit time and unit area with the angle between the incoming flow direction and the normal of surface element β is:

$$N_i = \frac{nv_m\{exp(-s^2cos^2\beta) + \sqrt{\pi}scos\beta[1 + erf(scos\beta)]\}}{2\sqrt{\pi}}$$

Here n is the number of molecules passing through unit area per unit time, v_m is the average molecular thermal velocity, $s = a_{cr}/v_m$, a_{cr} is the sonic velocity at nozzle exit and erf is the error function.

Transfer of coupled boundary data and the criteria of computation convergence We employ the coupled computation method to numerically simulate the attitude controlling rocket engine plume. First, we compute the nozzle flow using finite difference method to solve N-S equation with back pressure at the subsonic part of coupled boundary (nozzle exit boundary). After nozzle flow field computation is convergent, we put the nozzle exit parameters T,ρ, u and v as the entrance condition to start using DSMC method to compute plume outside the nozzle exit. If the computation is stable, we then do sampling. When the sampling number of molecules at each unit mesh is enough, we perform statistical average computation for each mesh unit molecular parameters, obtain the macro flow parameters at each mesh unit, and let the entrance pressure as the back pressure of nozzle flow simulating with difference method solving N-S equation next step. Repeat the foregoing steps, until the computational results of successive steps differ by only a very small amount.

2.5 Parallel Treatment

Parallel treatment is a necessity for large scal computation and especially for DSMC. It is relatively easy to make parallel treatment for the MacCormark time splitting scheme used to solve N-S equation in this research.

When using DSMC method to simulate flow field, because the molecular motion is random, no matter how the nodes of computational region are distributed, information transfer between every two computational nodes exist. When the number of computational node number is large, the type of information transfer seriously affects the efficiency of the parallel treatment. At the memory distribution type MIMD parallel computer, the realization of the communication between computational nodes in PVM parallel computational environment is through information packing in sending and unpacked in receiving. Considering the characteristic of DSMC methods and PVM parallel computational environment, this paper presents a relatively practical communication method — Twice Filtering Method (TFM, see Fig. 1). The meaning of Twice Filtering Method is that through twice filtering the molecular information of the pertinent computational nodes remains and the molecular information not belonging to them, filtered, then undergoing packing and sending to other computational nodes. After other computational nodes receive the information packet, through the second filtration the information belonging to it remains, thus complete information transfer.

3 Computational Results and Analysis

3.1 Numerical simulation of 5N engine plume and its contamination

Using coupled-parallel computation method we numerically computed 5N attitude controlling rocket engine vacuum plume and its contamination. For numerical calculation, we erect a board outside the nozzle exit lip. The initial conditions are chamber temperature being 1000K, pressure being 10 atm. and board temperature being 200K. Figures 2∼5 are the computational results of the flow field. Figures 2∼3 are pressure contoursand velocity vectors of nozzle flow. Figures 4∼5 are velocity vectors and pressure contours of plume outside nozzle exit. ¿From Fig.2∼5 we can find that the computational results appears reasonable. Flow field parameter contours distribute rationally. Outside the nozzle exit lip, due to the presence of the erect board, the normal velocity at plume wall boundary is zero. But the tangential velocity at plume wall boundary is not zero unlike nozzle wall boundary where we have a viscous boundary layer. This is because when we use DSMC method computing the plume we ignored the interaction force between molecules.

3.2 Numerical simulation of Satellite B-1 engine plume and its contamination

We used the numerical method introduced here to study the plume and its contamination about satellite B-1 (Fig.6). A and B are sensitive surface at the satellite. We want to know the contamination about A and B by the attitude controlling engine plume. The initial parameters are chamber temperature being 1000K, chamber pressure being 9 atm. and attitude controlling engine thrust force being

20N. The computation of the vacuum plume of satellite B-1 is now simplified as a axisymmetic problem. The computational results of vacuum plume are shown in Figs.7∼8.

Figure 7 shows the density contour of plume, and with the expansion of plume, the gas density reduced fast. At the place far away from the nozzle, plume density is very small. At the backflow of plume, because the plume backflow molecules are very few and the gas density is more less, we can judge that sensitive surface A and B not be contaminated by the attitude controlling engine plume. Figure 8 is the local velocity vector amplified picture above the nozzle with back flow. From these diagrams, we can see the free expansion and back flow of vacuum plume, this is in agreement with the real case.

4 Conclusion

Because real rocket engine nozzle flow density level is relatively high, numerical simulation of real rocket engine and vacuum plume flows with single DSMC method is not feasible. Based on the characteristics that in real rocket engine nozzle the flow is in the continuum regime and nozzle exit outside plume is a rarefied gas flow, we propose a coupled-parallel method with finite difference method for solving N-S equations to compute nozzle flow field and DSMC method to simulate plume outside nozzle exit. It not only overcomes the shortage of single method, but also ensures the exactness of numerical simulation by coupling the two methods in two regions. This paper provided a parallel treatment of the coupled codes for numerical simulation of real attitude controlling engine vacuum plume and obtained reasonable numerical results. The research work has an important direct effect to the design of attitude controlling engine, and provides a means to evaluate the plume contamination problem.

Acknowledgements

This work is supported by the National Natural Science Foundation of China.

References

1. G.A.Bird, *Molecular Gas Dynamics*, Clarendon Press, Oxford, 1976
2. I.D.Boyd, etc.: AIAA,1363(1991)
3. Jhon.J.Scialdone: AIAA 2592(1993)
4. Garrelie.F, Aubreton.J, etc, J. Appl. Phys., p5075(1998).
5. R.W.MacCormack, A.J.Paullay:AIAA Paper 72-154, (1972)
6. B.Baldwin, H.Lomax: AIAA Paper 78-257, (1978)
7. P.D.Thomas, J.F.Middlecoff:AIAA **18**, no. 6, (1980).

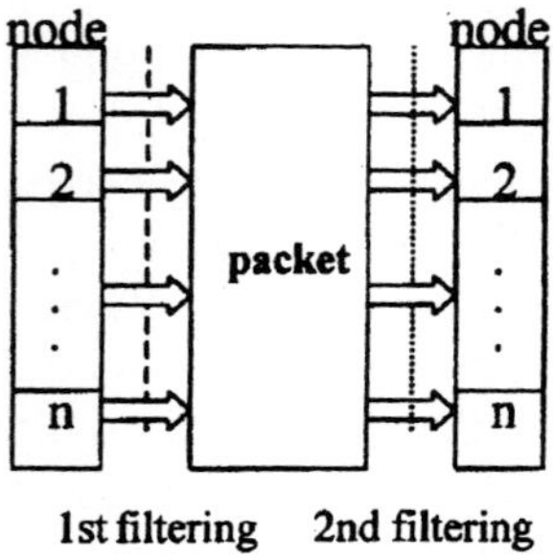

Fig.1 Illustration of TFM

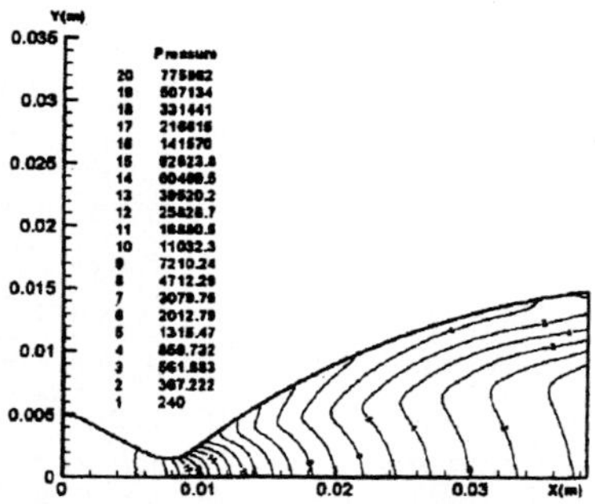

Fig.2 Pressure contour of nozzle flow

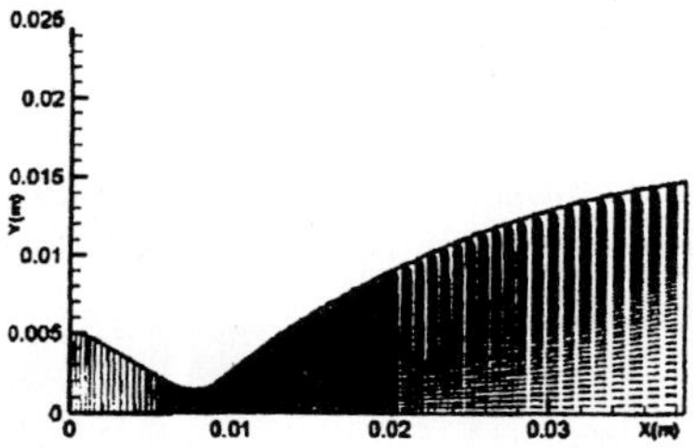

Fig. 3 Velocity vector of nozzle flow

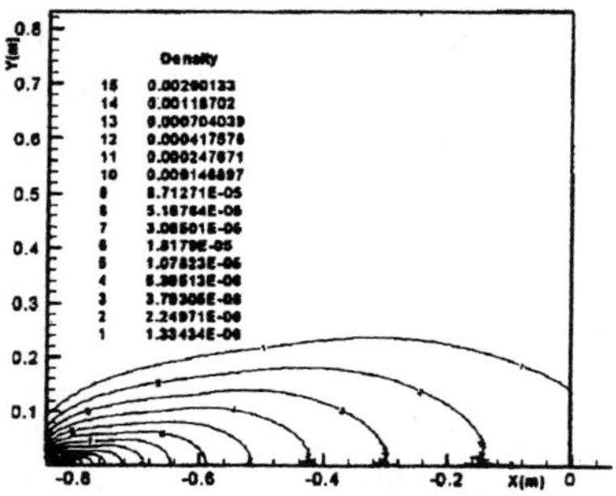

Fig.7 Density contour of plume

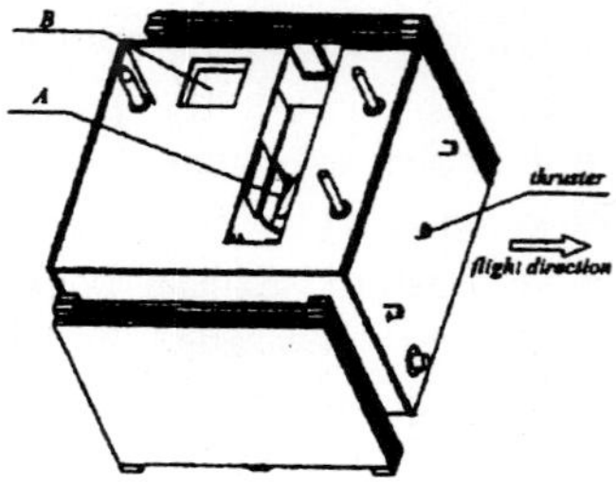

Fig.6 Satellite B-1

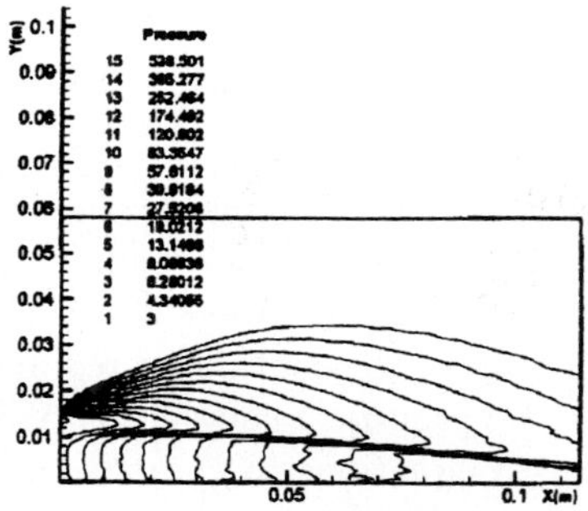

Fig.5 Pressure contour of plume

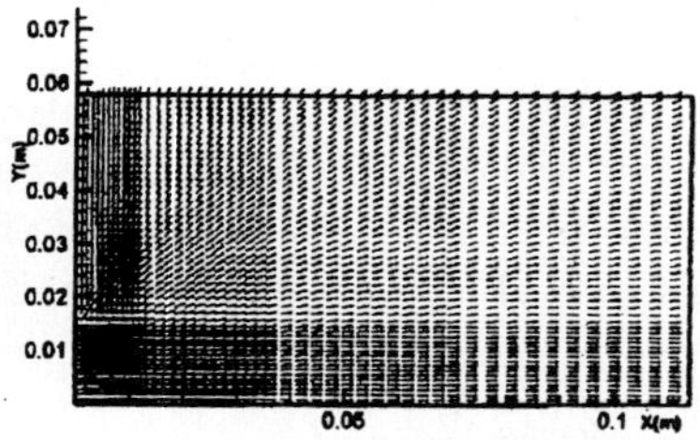

Fig.4 Velocity vector of plume

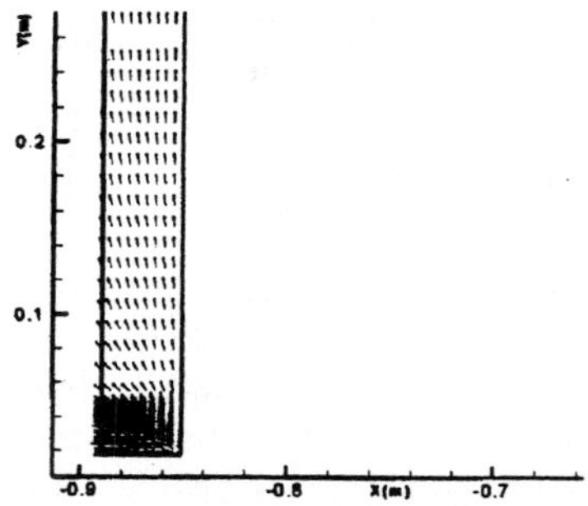

Fig. 8 Amplified local velocity vector above nozzle

Large-Scale Parallel Simulations in Computational Electromagnetics

Vijaya Shankar[1], Adour Kabakian[1], Chris Rowell[1], and Touraj Sahely[1]

HyPerComp, Inc., 31255, Cedar Valley Dr., Suite 327
Westlake Village, CA 91362, USA

1 Introduction

The ability to predict the electromagnetic response from complex structures with layered material media over a wide frequency range (100 MHz to 20 GHz) is a critical technology need for the development of stealth platforms. There are three basic approaches to numerical simulation of Maxwell's equations: 1) high frequency asymptotics, which treats scattering and diffraction as local phenomena; 2) solution of an integral equation for radiating sources on (or inside) the scattering body, which couples all parts of the body through a multiple scattering process; and 3) the direct integration of the differential form of Maxwell's equations in time. The integration of Maxwell's equations in time offers the most direct and general solution for broadband radar scattering and propagation of electromagnetic pulses in real materials. The challenge for time-domain methods has been to maintain global accuracy in the phase and amplitude of waves scattered by large, complex structures.

2 Maxwell's Equations

The sources of the electromagnetic fields of interest are typically currents in antenna structures and polarizations induced in target materials by incident radar pulses. The simplest case of plane-wave reflection by a metal aircraft can be treated as a problem in determining the currents induced on the aircraft surface. Other cases of interest, such as the interaction of high-power microwave pulses with shielding of various kinds, require tracking the buildup of field amplitudes inside the materials, which can trigger electrical breakdown or even changes in physical state. Such processes depend nonlinearly on the local field strength, and consequently they are not amenable to the frequency-domain description traditionally employed in radar scattering, which assumes that the target response is linear in the local fields.

In all cases, the evolution of the electric field $\mathbf{E}$ and magnetic induction field B is governed by the time-domain form of Maxwell's equations. Sources induced in target materials include currents $\mathbf{J}$, electrical polarization $\mathbf{P}$, and magnetic polarization $\mathbf{M}$, any of which may have nonlinear and nonlocal dependence on $\mathbf{E}$ and $\mathbf{B}$. To simplify the description of these interactions, one defines an electric displacement field $\mathbf{D} = (\epsilon_0 \mathbf{E} + \mathbf{P})$ and a magnetic field $\mathbf{H} = \mathbf{B}/(\epsilon_0 - \mathbf{M})$ in terms of which Maxwell's equations can be written as (in SI notation) as either local first-order partial differential equations or as integrals over an arbitrary volume V bounded by a regular surface ∂V:

Differential Form:

$$\frac{\partial \mathbf{B}}{\partial t} = -\nabla \times \mathbf{E} \qquad (1)$$
$$\frac{\partial \mathbf{D}}{\partial t} = \nabla \times \mathbf{H} - \mathbf{J}$$

Volume Integral Form:

$$\frac{d}{dt}\left[\int_V \mathbf{B} dV\right] = -\int_{\partial V} \mathbf{n} \times \mathbf{E} dS, \qquad (2)$$
$$\frac{d}{dt}\left[\int_V \mathbf{D} dV\right] = \int_{\partial V} \mathbf{n} \times \mathbf{H} dS - \int_V \mathbf{J} dV,$$

where $\mathbf{n}$ is the outward-pointing unit normal to ∂V . These equations are supplemented by the conditions $\nabla \cdot \mathbf{B} = 0$ and $\nabla \cdot \mathbf{D} = \rho$, where the electric charge density ρ obeys the basic conservation law $\partial\rho/\partial t = -\nabla \cdot \mathbf{J}$. These two relations can be regarded as initial conditions, since the curl equations guarantee that the time increments in $\mathbf{B}$ and $\mathbf{D}$ will also satisfy the conditions.
It is convenient to write for $\mathbf{B}$ and $\mathbf{D}$ a single "vector" Q of six field components, and to define a tensor flux $\mathbf{F}(\mathbf{E}, \mathbf{H})$ in terms of which Maxwell's equations can be written in standard conservation form:

Conservation Form

$$\begin{aligned} Q &= (\mathbf{B}, \mathbf{D}) \\ F_x &= (\mathbf{i} \times \mathbf{E}, -\mathbf{i} \times \mathbf{H}) \\ F_y &= (\mathbf{j} \times \mathbf{E}, -\mathbf{j} \times \mathbf{H}) \\ F_z &= (\mathbf{k} \times \mathbf{E}, -\mathbf{k} \times \mathbf{H}) \\ \frac{\partial Q}{\partial t} &+ \nabla \cdot F(Q) = 0 \end{aligned} \qquad (3)$$

In this way, the set of six equations is cast into the same mathematical structure as conservation of charge.
The differentiable solutions of Maxwell's equations satisfy continuity of tangential $\mathbf{E}$ and tangential $\mathbf{H}$ at an interface between two materials (or between a material region and free space). In the limit of a perfect electrical conductor (PEC), these magnetic field components terminate abruptly on a surface current $\mathbf{K}$, and the electric field components tend smoothly to zero. Conditions on

the normal field components also follow from the equations, but they are not usually required in solving scattering and propagation problems.

The first step in our basic approach to formulating numerical conditions that implement these physical requirements is to discretize space so as to fit the surface accurately. This simple measure avoids the problems associated with "floating" boundaries inside grid cells and deriving approximate homogenization conditions to high order. At the outer boundary of the computational domain, where the fields are those due to outgoing waves, the grid is chosen in such a way that its outward-pointing normals lie close to the dominant scattered-wave directions.

The second, and most important, step is to adopt the Godunov approach, regarding the field data within each grid cell as part of a global solution that is physically discontinuous across the cell boundary. The interface conditions to be satisfied by this weak solution to Maxwell's equations are determined by solving the Riemann problem exactly at each point on the cell boundary. That is, the discontinuities are resolved into waves propagating away from the cell face to either side, treating the discontinuous data as though they were true initial conditions at each time level.

The form of the solution at an interface between two isotropic materials in the weak-field (linear) limit is:

$$\mathbf{n} \times \mathbf{E}^* = \frac{\left([\mathbf{n} \times \epsilon_c \mathbf{E} - \mathbf{n} \times (\mathbf{n} \times \mathbf{H})]_L + [\mathbf{n} \times \epsilon_c \mathbf{E} - \mathbf{n} \times (\mathbf{n} \times \mathbf{H})]_R\right)}{(\epsilon_c)_L + (\epsilon_c)_R} \tag{4}$$

$$\mathbf{n} \times \mathbf{H}^* = \frac{\left([\mathbf{n} \times \mu_c \mathbf{H} - \mathbf{n} \times (\mathbf{n} \times \mathbf{E})]_L + [\mathbf{n} \times \mu_c \mathbf{H} - \mathbf{n} \times (\mathbf{n} \times \mathbf{E})]_R\right)}{(\mu_c)_L + (\mu_c)_R}$$

where L, R denote that quantities are evaluated approaching the left- and right-hand sides of the interface with respect to the unit normal n, which is outward-pointing for the left-hand cell; the material constants ϵ and μ may differ arbitrarily across the interface, and $c^2 = 1/\mu\epsilon$.

At the outer boundary of the computational domain, the Riemann solution reduces to the requirement that there be no scattered fields incoming along the normal to this boundary. This would be an exact condition if the outgoing fields from the cell interior were directed perfectly along the normal. Since in general they are not, there will be a small reflection of these scattered waves back into the computational domain. To effectively eliminate this numerically generated reflection, one could either implement a higher order absorbing boundary condition or design a perfectly-matched layer (PML) abutting the boundary to absorb the waves. Instead, we have taken advantage of the increased damping of Godunov schemes in large grid cells (where $c\Delta t$ is less than a typical cell dimension). By progressively expanding the average cell size in regions outside the convex hull of the target, we weaken the outgoing waves as they propagate away from the target, and absorb the weak reflections as they propagate inward from the computational boundary.

3 TEMPUS (Time-Domain EM Parallel Unstructured Simulator) Environment

Development of a time-domain EM computational environment that provides complete solutions to EM applications must consider all phases of the computational simulation. The following areas are critical for large-scale CEM simulations:

- Preprocessing: This stage begins with a CAD model of the geometry. Usually, a raw CAD model requires considerable repair to eliminate all CAD related geometry problems (missing patches, gaps in fairings, twisted pairs, overlapping patches,...). This is a bottleneck in the whole computational process and is manpower intensive. After the CAD geometry is cleaned up, one needs to set up the surface and volume gridding.
- Domain Decomposition: For parallel computations, the entire computational domain comprising of unstructured cells is split into as many subdomains as the number of processors nodes to be used. There are a number of techniques available to perform this domain decomposition phase.
- Solver: Appropriate space and time integration procedures are employed to solve the time-domain Maxwell's equations. Minimizing the phase and amplitude errors is critical to achieving reliable solutions. In addition, one needs to develop appropriate boundary condition procedures for material interface (solution to a Riemann problem), PEC boundaries, nonreflecting absorbing outer boundary conditions, impedance layers, and thin resistive cards.
- Physics Modeling: While Maxwell's equations are linear, they still pose a formidable challenge in being able to mathematically model different material media, such as dispersive, nonlinear, anisotropic, chiral, resistive and impedance layers, and special regions like gaps, cracks, thin wires, and sharp edges.
- Parallel Processing: In parallel processing, the goal is to achieve scalable performance by minimizing the overhead involved in message passing across nodal boundaries. Also, code portability across different parallel platforms (workstation clusters and fine-grain large-scale parallel machines) is a critical requirement.

The parallel, high-performance TEMPUS time-domain CEM environment is shown in Figure 1 [1,2].
Figure 2 shows the broadband RCS results, bistatic RCS at two different frequency cuts and the associated range profile for a sphere (radius of 40" or 1 meter) for frequencies up to 5 GHz were calculated using 5 million tetrahedra cells distributed on a 256-node IBM-SP. The RCS variation as a function of frequency is compared with a series solution. The range profile clearly exhibits the presence of a creeping wave. After the first range signal that corresponds to the pulse striking the nose of the sphere, a creeping wave signal shows at a time given by (diameter + π * radius)/c, where c is the velocity of light. In the range

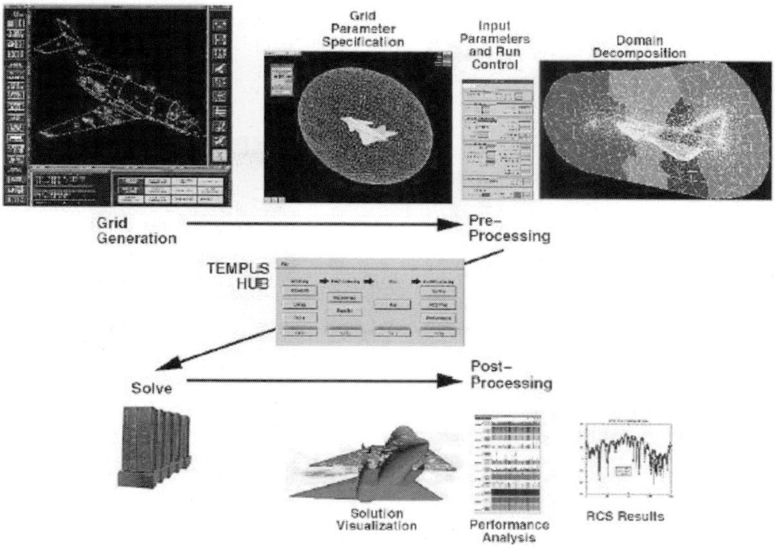

Fig. 1. TEMPUS environment for wide band CEM applications

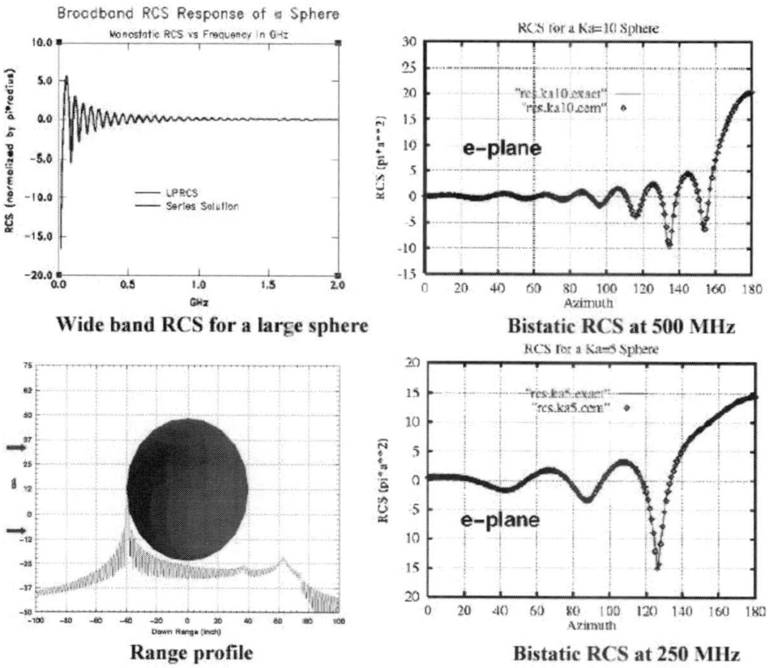

Fig. 2. Results demonstration for a sphere

profile (measured as half the distance traveled) this corresponds to about 62.8" from the center of the sphere.

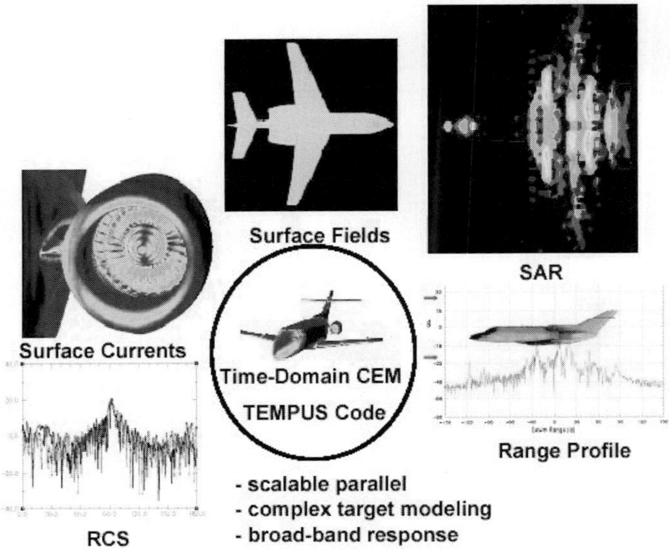

Fig. 3. DoD applications of TEMPUS

Figure 3 shows a typical surface current plot and the monostatic RCS at a given frequency for a complete aircraft geometry with engine details. It shows range profiles for nose-on and tail-on pulse incidence. To generate such range profile plots, TEMPUS processes the pulse transient response into many frequency cuts (usually 256 to 1024 frequency cuts). Ability to produce such plots is the real advantage of a wide-band CEM approach over frequency-domain formulations. Figure 4 also shows a SAR image for the C29 aircraft (such results require many frequency cuts (each frequency cut corresponding a pulse width of 1 GHz gives a downrange resolution of 6 inches) and many incident angles to generate the required data base for processing the SAR information). A typical aircraft EM calculation at frequencies of interest requires of the order of 100 million grid cells.

References

1. A. H. Mohammadian, V. Shankar and W. F. Hall: Computer Phys. Comm. **68**, 175 (1991)
2. V. Shankar, C. Rowell, A. Kabakian and W. F. Hall: 'A parallel time-domain CEM environment for large scale simulations'. Draft final report on Contract F33615-97-C-1000, May 2000

Simulation of Turbomachinery Flows by a Parallel Solver with Sub-iteration Recoupling

Vittorio Michelassi[1] and Paolo Giangiacomo[2]

[1] Dipartimento di Ingegneria Meccanica e Industriale, Università Roma Tre, Roma, Italy
[2] Dipartimento di Energetica 'Sergio Stecco', Università di Firenze, Firenze, Italy

1 Introduction

The increasing diffusion of gas turbine engines among energy conversion system pushes to increase the engine performance through a better understanding of the flow phenomena within the components. In this scope, a numerical simulation tool complements experimental practice and may even be alternative to it, at least in the preliminary phases of the design process [1][2][3]. The detailed numerical simulation of the complex turbomachinery flow fields may require very large grids with millions of nodes, if shear layers and shocks have to be accurately resolved. In this case, the computational time may become exceedingly long, and also the memory requirements of the code may overflow the capacity of most single processor computers. A considerable speed-up of the computations may come from parallel multiprocessor computing techniques, which split the computational task among several processors [4]. A simple parallel version of the time-marching implicit Navier-Stokes solver XFLOS [3] for distributed memory architectures has been developed, which takes advantage of the peculiar features of turbomachinery flows. The parallelised code has been used to simulate the flow in an axial transonic turbine stage.

2 Algorithm

The parallel version of the Navier-Stokes solver XFLOS solves the three-dimensional conservation equations in unsteady conservative form using the diagonal alternate direction implicit (DADI) algorithm. The transport equations are made nondimensional with respect to stage inlet stagnation temperature, pressure and inlet molecular viscosity. For rotor flows, the present set of computations adopts relative variables.

The equations are discretised in space by a centred finite difference scheme. A spectral-radius-weighed version of the non-linear second plus fourth order artificial damping terms is added on both the explicit and implicit side of the equations. The inversion of three scalar penta-diagonal systems along the three coordinates $\xi\eta\zeta$ completes the solution process.

3 Parallelisation Strategy and System Recoupling

For multi-processor runs, the simply connected grid is decomposed into non-overlapping blocks in the spanwise direction only (see Fig. 1). Thus, no interpolation is required to assemble fluxes at the interface, and each point of the overall grid belongs to one block only. This simple domain splitting was deemed particularly advantageous for turbomachinery flows with small spanwise convection, like for axial rows of high hub to tip radius ratio, and weak secondary flows. In fact, neither the predominant convective fluxes along the vane axis nor the transverse fluxes in the tangential direction are cut by the decomposition.

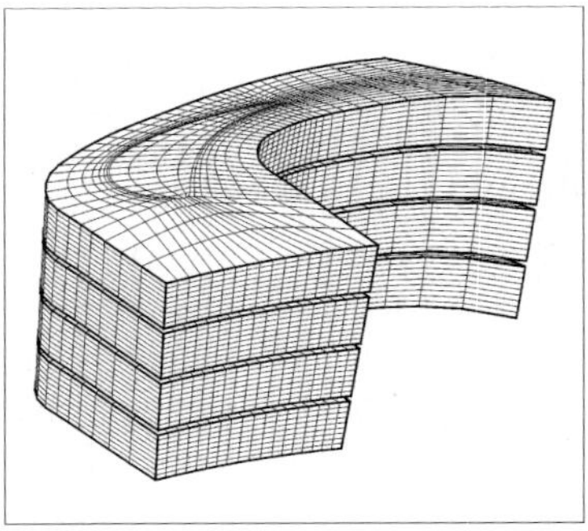

Fig. 1. Domain decomposition into non-overlapping blocks

To evaluate fluxes at the interfaces, the blocks are extended by two node layers in the spanwise direction. The equations are solved only in the unextended region, and, at each iteration, the solution in the extensions is simply exchanged between neighbouring blocks without interpolation. This procedure ensures the identity of single and multiprocessor solutions at convergence, apart from the different truncation errors. The complete RHS and the implicit operators in the axial and tangential directions are unaffected by the decomposition. Conversely, the implicit operator along the spanwise direction is uncoupled by the explicit evaluation of spanwise fluxes and artificial damping terms at interfaces between any pair of blocks n and n+1, as visible in the LHS of the following equation:

$$
\begin{array}{c} \mathrm{n} \left\{ \begin{array}{c} \\ \\ \\ \\ \end{array} \right. \\ \mathrm{n+1} \left\{ \begin{array}{c} \\ \\ \\ \\ \end{array} \right. \end{array}
\left[\begin{array}{ccccc|ccccc}
\ddots & \ddots & \ddots & \ddots & & & & & & \\
\ddots & C & A & B & D & & & & & \\
 & E & C & A & B & & & & & \\
 & & E & C & A & & & & & \\
\hline
 & & & & & A & B & D & & \\
 & & & & & C & A & B & D & \\
 & & & & & E & C & A & B & \ddots \\
 & & & & & & \ddots & \ddots & \ddots & \ddots
\end{array} \right]
\left[\begin{array}{c}
\vdots \\
\Delta Q^{(m)}_{n,K-2} \\
\Delta Q^{(m)}_{n,K-1} \\
\Delta Q^{(m)}_{n,K} \\
\hline
\Delta Q^{(m)}_{n+1,1} \\
\Delta Q^{(m)}_{n+1,2} \\
\Delta Q^{(m)}_{n+1,3} \\
\vdots
\end{array} \right]
=
\left[\begin{array}{l}
\vdots \\
RHS_{n,K-2} \\
RHS_{n,K-1} - D \cdot \Delta Q^{(m-1)}_{n+1,1} \\
RHS_{n,K} - B \cdot \Delta Q^{(m-1)}_{n+1,1} - D \cdot \Delta Q^{(m-1)}_{n+1,2} \\
\hline
RHS_{n+1,1} - C \cdot \Delta Q^{(m-1)}_{n,K} - E \cdot \Delta Q^{(m-1)}_{n,K-1} \\
RHS_{n+1,2} - E \cdot \Delta Q^{(m-1)}_{n,K} \\
RHS_{n+1,3} \\
\vdots
\end{array} \right]
$$

in which n,K refers to the last node K of the n-th block. This is known to reduce the implicitness of the algorithm. The neglected implicit terms are iteratively evaluated and reinserted as a correction to the RHS to restore a partial implicit link between the blocks. $\Delta Q^{(m)}$ is solved in function of $\Delta Q^{(m-1)}$ for a fixed number of sweeps, starting with $\Delta Q^{(0)} = 0$.

4 Stage Description

The parallel version of XFLOS has been applied to simulate the flow inside a research transonic axial turbine stage, with prismatic stator blades and fully three-dimensional rotor blades, in the frame of a joint research programme sponsored by EEC. Since the present tests are aimed mainly at testing the performances of the parallel code than at capturing the details of the flow fields, a simplified stage geometry is selected. Both stator and rotor grids have $76 \times 37 \times 82$ nodes in the axial, tangential and radial direction respectively. Grid points are equally spaced in the radial direction, and inviscid boundary conditions have been imposed at endwalls, although the computations are viscous and no-slip conditions are imposed on the blade. To balance the computational effort among processors, the blocks have been given the same number of nodes in the radial direction. The Authors [5] have shown the necessity to iterate on the stator exit boundary condition to find the working point of the full stage. For the purpose of this paper, only the stage solution at the design point has been considered.

5 Results

The computational time, the efficiency of the simple domain decomposition, and the effect of the sub-iterative recoupling are compared for variable domain decomposition choices. Details on the flow field may be found in [5] [6]. All the runs have been performed on a Cray T3E computer with MPI Fortran, and the execution times have been monitored by Cray's MPP Apprentice performance analysis tool, running on Cray MPP systems. For the stator, Fig. 2 shows the convergence rate for axial momentum with and without recoupling, the residuals for the other transport equations being very similar.

At least for these runs, with 2 and 4 processors the explicit flux assembling at interfaces does not affect the convergence rate, and with more processors the slowdown in convergence is small. The recoupling procedure successfully restores the convergence skills of single processor computations, in which radial fluxes are treated in a fully implicit manner. For 8 and 16 processors, respectively 2 and 3 sweeps suffice to recover the single processor residual history. For rotor runs, basically the same execution times as for the stator are found, due to the identity of grid dimensions between the two rows, and the single processor convergence rate is restored by the same number of sweeps.

Speed-up and efficiency for the most meaningful stator runs, at a fixed number of iterations, are summarised in Table 1. Surprisingly, without recoupling efficiencies larger than one are found. This is probably due to caching and/or

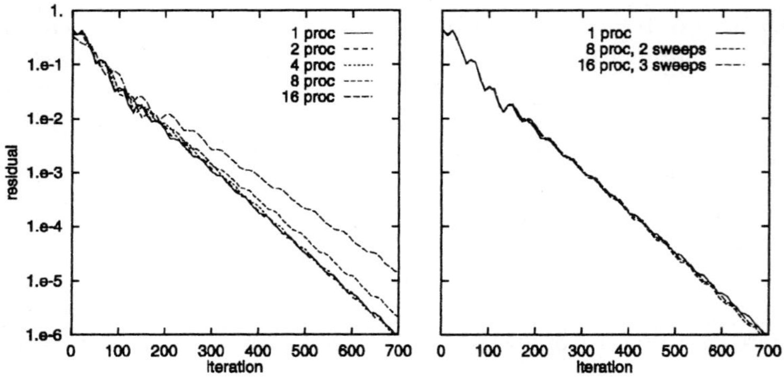

Fig. 2. Convergence rate for stator axial momentum .(a) Without recoupling. (b) With recoupling

striding, which overcompensate the communication time between processors. In fact, smaller blocks save memory access time.

Table 1. Speed-up and efficiency at 700 iterations

	speed-up	η
2 processors, 1 sweep	2.1	1.07
4 processors, 1 sweep	4.4	1.11
8 processors, 1 sweep	8.7	1.09
8 processors, 2 sweeps	7.7	0.96
16 processors, 1 sweep	16.8	1.05
16 processors, 3 sweeps	10.9	0.68

The sub-iterations considerably increase the execution time, and correspondingly reduce the speed-up. This is largely due to the incidence of MPI time over the total execution time. In fact, as is seen in Fig. 3a, without recoupling MPI procedures absorb no more than 5% of the total time. With 8 and 16 processors the first sub-iteration, which corresponds to 2 sweeps in the spanwise direction, causes a steep rise in MPI time, followed by a linear trend.

This sudden increase in MPI time is entirely due to overhead time of the sub-iteration procedure. In fact, parallel work time grows linearly with the number of sweeps (Fig. 3b). The same MPI overhead time is found for stator and rotor so that it can be concluded that this is a systematic feature of the code, the control of which through different sub-iteration or access to memory strategies will have to be optimised in the future. The comparison of execution time and speed-up factor at a fixed residual level, done in Table 2 for the stator runs,

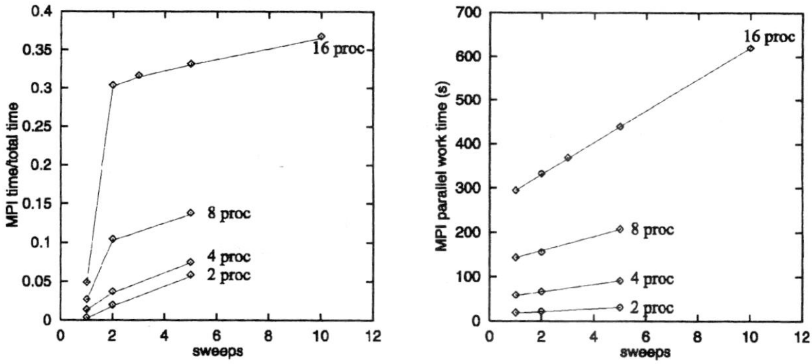

Fig. 3. Execution time for MPI subroutines .(a) MPI time over total time. (b) Parallel work time

shows the required number of time steps to decrease the average residual of four orders of magnitude.

Table 2. Stator speed-up and efficiency at 10^{-4} average residual level

	time steps	speed-up	η
1 processor	470		
2 processors, 1 sweep	480	2.1	1.05
4 processors, 1 sweep	460	4.5	1.13
8 processors, 1 sweep	500	8.2	1.02
8 processors, 2 sweeps	450	8.0	1.00
16 processors, 1 sweep	600	13.2	0.83
16 processors, 3 sweeps	480	10.7	0.67

The sub-iteration process allows the number of single processor time steps to be recovered. Still, the overall computational time is not necessarily reduced, at least in this stage of the code optimisation.

6 Conclusions

The proposed simple domain decomposition greatly eased code parallelisation, and proved to be both effective and efficient, thanks to the lack of recalculations and interpolations. The explicit evaluation of spanwise fluxes at interfaces reduces the convergence rate with increasing number of processors. However, with 16 processors the final residual level was only an order of magnitude larger than in single processor computations. A sub-iterative recoupling procedure has been developed, which successfully restores the single processor convergence rate by adding compensating terms into the RHS of the system. However, the MPI overhead time of the sub-iterations can reduce the attractiveness of the recoupling process. Without recoupling, at fixed number of iterations efficiencies larger than one are found, probably due to caching or striding, while at fixed residual levels the speed-up is about 13 with 16 processors. The parallelisation thus brings a large saving in the execution time, and makes the code a powerful tool for the simulation of turbomachinery flows.

7 Acknowledgements

The Authors wish to express their gratitude to Professor Francesco Martelli for the fruitful discussions and suggestions. The technical support and CPU time granted by Cineca, Italy, are gratefully acknowledged. This work has been sponsored by the European Community in the frame of the BRITE-EURAM project IMT Area 3, Contract No. TATEF BRPR-CT97-0519.

References

1. Emunds, R., Jennions, I.K., Bohn, D., Gier, J., 'The Computation of Adjacent Blade-Row Effects in a 1.5-Stage Axial Flow Turbine', ASME Journal of Turbomachinery, Vol. 121, pp. 1-10 (1999)
2. Michelassi, V., Martelli, F., Dnos, T. Arts, C.H. Sieverding, 'Unsteady Heat Transfer in Stator-Rotor Interaction by Two Equation Turbulence Model', ASME Journal of Turbomachinery, vol. 121, pp. 436-447 (1999)
3. Michelassi, V., Giangiacomo, P., Martelli F., 'Computer Simulation of a Three-Dimensional Transonic Turbine Stage', Proceedings XIV ISABE Conference, 5-10 September 1999, Florence, Italy (1999)
4. Schiano, P., Ecer, A., Periaux, J., Satofuka, N., (Ed.), 'Parallel Computational Fluid Dynamics - Algorithms and Results Using Advanced Computers', 1997, Proceedings of the Parallel CFD 96 Conference, Capri, Italy, May 20-23 1996
5. Giangiacomo, P., Michelassi, V., Martelli, F., 'Analysis of the Mixing Plane Interface Between Stator and Rotor of a Transonic Axial Turbine Stage', International Gas Turbines and Aeroengine Congress and Exhibition, 8-11 May 2000, Munich, Germany, ASME Paper 2000-GT-634 (2000)
6. Denos, R., Sieverding, C.H., Arts, T., Brouckaert, J.F., Paniagua, G., Michelassi, V., 'Experimental Investigation of the Unsteady Rotor Aerodynamics of a Transonic Turbine Stage', Third European Conference on Turbomachinery Fluid Dynamics and Thermodynamics, London 2-5 March 1999. Accepted for publication on IMECHE Journal

Parallel CE/SE Computations via Domain Decomposition

Ananda Himansu[1], Philip C.E. Jorgenson[2], Xiao-Yen Wang[1], and Sin-Chung Chang[2]

[1] Taitech, Inc., at NASA Glenn Research Center
[2] NASA Glenn Research Center, Cleveland OH 44135, USA

Abstract. This paper describes the parallelization strategy and achieved parallel efficiency of an explicit time-marching algorithm for solving conservation laws. The Space-Time Conservation Element and Solution Element (CE/SE) algorithm for solving the 2D and 3D Euler equations is parallelized with the aid of domain decomposition. The parallel efficiency of the resultant algorithm on a Silicon Graphics Origin 2000 parallel computer is checked.

1 Introduction

To accurately resolve complex three-dimensional turbulent flows, with possible acoustic waves or combustion phenomena, a time-dependent Euler or N-S solver must be used in Large Eddy Simulation or Direct Numerical Simulation mode. The need to resolve multiple space and time scales leads to very large computer memory and operation count requirements. The use of such computational resources is economically feasible only on modern parallel computing architectures. This paper presents the strategy and specific tools that were used to parallelize an existing serial code for solving the 3D Euler equations. While the concepts involved have been explored by other researchers over many years, the present authors hope that the description of a recent simple parallelization effort will be of value to many who are currently switching from serial supercomputers to more economical parallel architectures but who are not experts in parallel computing.

2 The CE/SE Euler Scheme

The Space-Time Conservation Element and Solution Element (CE/SE) method advanced by Chang [1] offers a simple numerical scheme that accurately resolves both shock waves and acoustic waves in the nearfield in a unified manner. The CE/SE 2D Euler solver is described in [2], while the CE/SE 3D Euler solver was presented in [3]. See the web site http://www.lerc.nasa.gov/WWW/microbus for details of the CE/SE method and a complete list of publications, including those detailing application of CE/SE schemes to computational aeroacoustics and combustion.

The success of the CE/SE technique for aeroacoustics, combustion, and other computations can be traced to the distinguishing features of the method, which are (i) an emphasis on solving a general space-time integral form of the conservation laws, (ii) the requirement of local and global conservation of space-time fluxes as a basis for the method, (iii) use of only the simplest discretization stencils on a staggered space-time mesh, and (iv) the treatment of flow property gradients as additional unknowns. The basic inviscid flow solvers are nondissipative (no inherent artificial viscosity) and neutrally stable. A specifiable amount of artificial viscosity is explicitly added to provide stabilizing damping and to minimize numerical oscillations in the vicinity of shock waves. Thus the CE/SE Method is well suited to aeroacoustics and combustion applications.

3 Code Parallelization via Domain Decomposition

The CE/SE Euler scheme is an explicit time-marching scheme, making it well-suited for parallelization via domain decomposition. If the solution is to be computed at a time level n, the computational stencil involves a single cell at the time level n, and involves only the cell and its immediate neighbors at the time level $n-1$. The algorithm to compute the solution is the same for every interior cell. Therefore, the scheme possesses a data parallelism. The only dependency in this data parallelism is that the solution and geometric information pertaining to immediate neighbors at the time level $n-1$ needs to be known by any cell before the solution at that cell can be computed at the level n. The natural technique for parallelization is by dividing the cells among the available processors. Given the low speed of communication between processors relative to the speed with which a processor accesses its local memory, the division should be such that communication between processors is minimized.

To enable efficient parallelization of the CE/SE codes, a suitable domain decomposition strategy for the numerical flow problem has been selected. This involves use of one of the METIS codes based on algorithms described in [4]. Specifically, the dual-mesh partitioning code named *partdmesh* is used. It is available at no cost, and can be downloaded from the website *http://www-users.cs.umn.edu/~karypis/metis/main.shtml*. Given an unstructured mesh as a map of cell numbers to node (cell vertex) numbers, the METIS mesh-partitioning code produces a k-way partition of the domain, where the input k is the desired number of subdomains. Generally, k is set equal to the number of computer processors used for the flow computation. The METIS code can partition both 2D meshes of triangular cells and 3D meshes of tetrahedral cells. The partitioning algorithm attempts to produce an optimal partition in three respects. The algorithm attempts to balance the load, i.e., to assign an equal number of cells to each subdomain. The algorithm also attempts to minimize the number of shared cell faces at the inter-subdomain boundaries, so as to lessen the amount of inter-processor communication required. Thirdly, the algorithm also attempts to minimize the maximum of the number of subdomains adjacent to any subdomain. The METIS code is based on a multilevel graph-partitioning algorithm, and is very rapid, producing a 32-way partition of a 96800-cell domain in about 2 seconds on a personal computer with a 450 MHz Pentium-II processor. In the current work, only the serial computer version of the METIS code is used, although a parallel computer version is also available. Other mesh-partitioning algorithms are available, but have not been evaluated for this work. In our experience, the METIS code produces partitions that are close to optimal, at very little computational cost.

The effort reported here consisted of "parallelizing" already existing 2D and 3D research codes that were written to run in serial fashion on a single processor. The parallelization was done in such a way as to introduce minimal modification of the existing flow solvers. The chief aim was to reduce the wallclock time of any given computation by using multiple processors. With regard to the actual flow computation, an additional benefit of the domain decomposition strategy is that the overall computer memory requirement is divided among the processors. However, little attention was paid to dividing among the processors the overall memory requirement of the domain decomposition and subdomain setup portions of the procedure. To enable the codes to be run on both parallel computers and workstation clusters, a message-passing approach was taken.

The CE/SE scheme is formulated on an unstructured mesh of triangles in the planar or axisymmetric case, and on an unstructured mesh of tetrahedra in the three-dimensional case. For simplicity of presentation, the case of a planar mesh of triangles will be discussed, though the parallelization procedure for volume meshes of tetrahedra is similar. The output typical of an unstructured mesh generator consists of the number of nodes, V, the number of cells, C, the number of boundary cell faces, B, the spatial coordinates X of each node, the map $C2V$ of cell number to the node numbers of its vertices (i.e., the cell definitions), a map $B2T$ of the boundary face number to a label denoting the type of boundary condition, the map $B2V$ of the boundary face number to the node numbers of its vertices, and the map $B2C$ of the boundary face number to the cell number of the cell to which the boundary face belongs. If the cell c has vertices with node numbers v_1, v_2 and v_3, then $C2V(c,i) = v_i$, $i = 1,2,3$. The map $C2V$ is supplied as input to the METIS mesh partitioning code together with the total number P of processors or subdomains desired. The output from the METIS code is a map $C2P$ of cell number to processor number, where the processor number to which the cell is assigned lies in the range 0 through $P-1$.

A simple example of a planar mesh is considered in Fig. 1. The mesh happens to be structured. However, structured meshes are treated in exactly the same way as a general unstructured mesh. The mesh of triangles covers the domain $ABCDEFGHA$. The total number of cells here is $C = 8$. The nodes are labeled A through I in the figure, to avoid confusion with the cell numbers. Suppose that the mesh is partitioned into two subdomains, with cells 1 through 4 being assigned to Processor 0, and cells 5 through 8 to Processor 1. The current practice in the CE/SE schemes is to implement boundary conditions by adjoining "ghost" cells of appropriate geometry outside the domain of the flow problem. These ghost cells, one at each boundary face, are numbered -1 through $-B$, see Fig. 1. A ghost cell at any boundary face that does not call for a periodicity type of boundary condition is assigned to the same processor as the interior cell with which it shares the boundary face. A ghost cell at a periodic boundary face is given the same geometric shape and assigned the same processor as the interior cell at the periodic image of the boundary face. For example, in Fig. 1, if the faces AB and GF are locations of periodicity of the solution, and if cell 2 is assigned to processor 0 while cell 5 is assigned to processor 1, then cell -6 is the periodic image of cell 2 and is assigned to processor 0, while cell -1 is the periodic image of cell 5 and is assigned to processor 1.

A face numbering is next generated, and the face-to-cell map is calculated. For very large meshes, this process may require too much memory to be done on a single processor, and a less memory-intensive approach would be needed to generate equivalent information. The cells marked for each processor p are given cell numbers "local" to the processor. Thus, the local numbering of the interior cells of processor p runs from 1 through C_p, and that of the ghost cells of processor p runs from -1 through $-B_p$. Local node numbering of the relevant nodes is also performed. The subdomain setup code then scans the face-to-cell map to find faces that are shared by processor p with a different processor. If there are D_p such faces, the cell at each face that is "foreign" to the processor p is assigned a local cell number in the range $C_p + 1$ through $C_p + D_p$, in the order in which the face is encountered in the face-to-cell map. The number of neighboring processors of any processor p is counted during this scan, and for each neighboring processor, the cell numbers (local to processor p) of both the "native" cells and the "foreign" cells are stored in the order in which the face is encountered in the face-to-cell map. An important property of this algorithm is that faces shared by two given processors p and p' are encountered in the same order when considering either p or p'. During the identification and local numbering of cells relevant to any processor p, maps from the local to global cell and node numberings are constructed, together with the inverse maps. Finally, for each processor p, mesh information is translated from global to local numbering and written to a disk file (with filename keyed to the processor number) about all the native interior and ghost cells, and about all the "neighboring" cells at the shared border of the subdomain. The "neighboring" cells appear merely as passive cells on processor p. They are used to store the flow data transferred from the neighboring processors. For each processor, the list of neighboring processors and the lists of native and foreign cells at the border with each processor are also written to disk files.

For a k-way partition of the domain, a CE/SE program executable is loaded into the node memory of each of k processors. Each of these processes then computes the solution in the corresponding subdomain. Only the cells assigned to each processor, together with passive border cells belonging to neighboring processors are stored, and the numbering is local. After each time step, the solution at the passive border cells is updated. This is done by message passing between processors, using Message Passing Interface (MPI) library calls. See [5] and the web site *http://www-unix.mcs.anl.gov/mpi* for details of the MPI standard. If necessary, the solution on the complete domain can be stitched together from the solutions on the individual subdomains by postprocessing.

4 Achieved Parallel Speedup

The parallel efficiency of the 3D Euler code when running on an SGI Origin 2000 computer architecture has been measured. The Origin 2000 uses a hypercube network topology to maximize the communication bandwidth. The parallel speedup achieved on this architecture is shown in Figs. 2 and 3. The test involved computing 100 time steps on a mesh of 172800 cells, using 1, 2, 4, 8, 16, 32 and 64 processors. The test problem was the reflection of an oblique shock wave from a wall, previously computed using the 2D Euler code in [2], but computed here using the 3D code. The parallel speedup of a run is defined as the inverse ratio of the wallclock time to the wallclock time of the single-processor case. The MPI_WTIME function call was used to reliably recover the wallclock time for each processor. The wallclock time of a run was taken to be the maximum time among the processors. The parallel speedup is seen to deviate visibly from ideal in the 32 processor case, indicating that for this implementation on this computer architecture, about 5400 mesh cells per processor is the limit below which the higher ratio of communication to computation will give poor parallel efficiency. Using 64 processors for this mesh size (about 2700 cells per processor) leads to a parallel speedup of only about 48. The same flow problem was run on a mesh four times as large, i.e. 691200 cells, with 64, 128 and 256 processors. This mesh was too large to be run on a single processor. Therefore, the 64 processor case was treated as having ideal speedup of 64, in order to examine the rate of falloff of the speedup, plotted in Fig. 3. Comparison of Figs. 2 and 3 shows that the rate of falloff (i.e., the slope) is approximately the same at 64 processors in Fig. 2 as it is at 256 processors in Fig. 3.

5 Summary and Conclusions

The CE/SE 2D and 3D Euler schemes were extended from serial computer versions to parallel computer versions. The parallel speedup of the 3D code was significantly less than ideal when an increased number of processors brought the number of mesh cells per processor below 5400. This indicates that tuning of the interprocessor communication will be beneficial when using a large number of processors on moderate sized meshes.

Acknowledgement

The first author gratefully acknowledges the support of this work by the HPCC Program at NASA Glenn Research Center, through Contract NAS3-97186.

References

1. S.C. Chang, "The Method of Space-Time Conservation Element and Solution Element – A New Approach for Solving the Navier-Stokes and Euler Equations", J. Comput. Phys., 119, pp. 295-324, (1995).

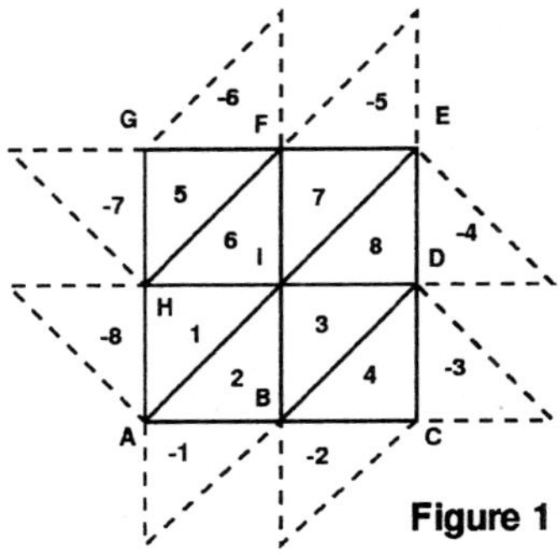

Figure 1

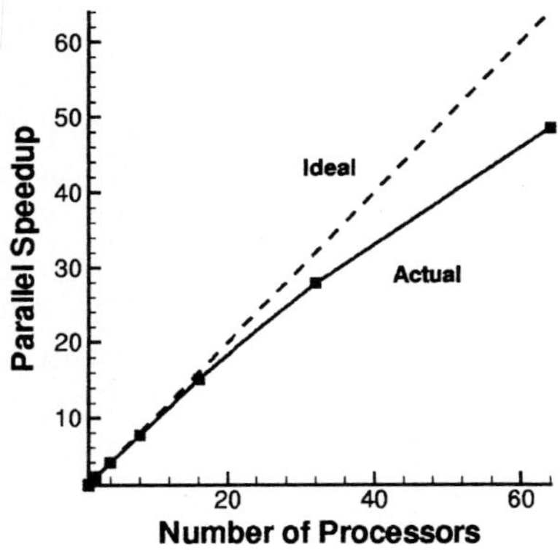

Figure 2

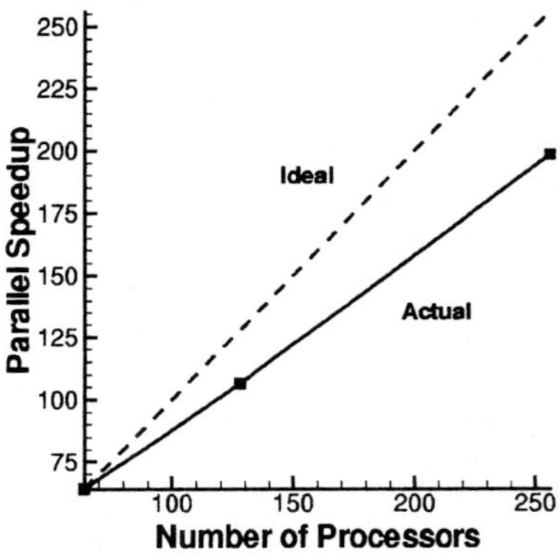

Figure 3

2. S.C. Chang, X.Y. Wang and C.Y. Chow, "The Space-Time Conservation Element and Solution Element Method : A New High-Resolution and Genuinely Multidimensional Paradigm for Solving Conservation Laws", J. Comput. Phys., 156 (1999), pp. 89-136.
3. X.Y. Wang and S.C. Chang, "A 3D Structured/Unstructured Euler Solver based on the Space-Time Conservation Element and Solution Element Method", 14th AIAA CFD Conference, June 1999, Norfolk, Virginia.
4. G. Karypis and V. Kumar, "Multilevel k-way Partitioning Scheme for Irregular Graphs", University of Minnesota, Dept. of Comp. Sc. / Army HPC Research Ctr., Technical Report 95-064, to appear in Journal of Parallel and Distributed Computing.
5. Message Passing Interface Forum, "MPI: A Message Passing Interface Standard", Intl. J. of Supercomputer Applications, vol. 8, no. 3/4, pp. 165–414, 1994. Available at http://www.mpi-forum.org.

Domain Decomposition Method and Fast Diagonalization Solver for Spectral Element Simulations

Dirk Wilhelm and Leonhard Kleiser

ETH Zürich, Institute of Fluid Dynamics, CH-8092 Zürich, Switzerland

1 Introduction

To perform direct numerical simulations of flows in more complex geometries such as the flow over an obstacle or over backward- and forward facing steps, numerical methods are required that combine high accuracy with geometric flexibility. In the present work, we employ the $P_N - P_{N-2}$ spectral element method (SEM) of Maday and Patera [6] for the discretization of the Navier-Stokes equations. In this approach, velocity and pressure are approximated by polynomials of order N and $N-2$, respectively. The efficiency of this method depends strongly on the solution algorithm for the algebraic systems associated with the Helmholtz equation for the velocity $\underline{u}$ and the pseudo-Laplacian equation for the pressure q. Since the pseudo-Laplacian equation can become ill-conditioned [4], the solution for q tends to be the computational bottleneck when iterative solvers are employed. Therefore, a considerable effort has been made on the development of efficient pressure preconditioners in the past [2–4]. On the other hand, the costly iterative procedure can by avoided if direct solvers are applied. Very efficient direct solvers exist that are based on fast diagonalization methods, which, however, can only be used for rectangular Cartesian domains [1,5]. For geometries that are more complex, but may still be decomposed into Cartesian subdomains, suitable domain decomposition methods are needed in order to apply fast diagonalization solvers. In the context of SEM, Couzy and Deville [3] employed a Schur complement method for the decomposition of the Helmholtz matrix in the momentum equation, while still solving the pressure equation iteratively. In the present work, we will extend the approach of Couzy and Deville [3] to obtain a direct solution of the pressure equation also, employing a suitable domain decomposition along with the influence-matrix technique of Schumann and Benner [9].

2 Numerical method

We consider the incompressible Navier-Stokes equations on a three-dimensional domain Ω given by

$$\frac{\partial \underline{u}}{\partial t} - \frac{1}{Re}\Delta \underline{u} + \nabla q = -\mathcal{C}(\underline{u})\,\underline{u} \tag{1}$$

$$\operatorname{div} \underline{u} = 0\,, \tag{2}$$

together with appropriate initial and boundary conditions. Here, $\underline{u}$ denotes the velocity, $\mathcal{C}(\underline{u})\,\underline{u}$ the convective term, and q the pressure. The Reynolds number $Re = UL/\nu$ is based on a characteristic velocity U, a characteristic length L, and the kinematic viscosity ν.

The spatial discretization of Eqs. (1) and (2) is performed by a spectral element method (SEM) in two spatial directions (x_1, x_2) and a Fourier spectral scheme in the third (spanwise) direction (x_3). For the discretization, we first employ a Fourier expansion of $\underline{u}$ and q

$$\underline{u}(x_1, x_2, x_3) = \sum_{m=-N_z/2}^{N_z/2-1} \hat{\underline{u}}^m(x_1, x_2)\, e^{i\,\alpha_z\, m\, x_3} \tag{3}$$

$$q(x_1, x_2, x_3) = \sum_{m=-N_z/2}^{N_z/2-1} \hat{q}^m(x_1, x_2)\, e^{i\,\alpha_z\, m\, x_3} \tag{4}$$

in x_3-direction, which results in a set of N_z equations for each Fourier mode m. Subsequently, these equations are discretized in the remaining two directions by a spectral element method. The SEM is based on the decomposition of the computational domain Ω into a number of K non-overlapping sub-domains (spectral elements). In each element, the solutions for velocity and pressure are expanded in tensor-product-based polynomials of order N (velocity) and $N-2$ (pressure), respectively. The velocities are discretized at the Gauss-Lobatto-Legendre (GL) points and the pressure at the Gauss-Lobatto (G) points. The basis functions for the velocity are continuous over element boundaries, while for the pressure such a requirement is not explicitly imposed. For a detailed presentation of the $P_N - P_{N-2}$ SEM see e.g. [4,6,11].

In primitive variables, the convective term $\mathcal{C}(\underline{u})\,\underline{u}$ can be written in different forms. In this work we employ the convective form, i.e. $\mathcal{C}(\underline{u})\,\underline{u} = (\underline{u}\cdot\nabla)\underline{u}$, since it was shown in [12] that in $P_N - P_{N-2}$ SEM the divergence form, the rotational form and the skew-symmetric form may lead to numerical instabilities and only the convective form is stable.

The spatial discretization leads to a semi-discrete system, which is then discretized in time. The linear terms are treated implicitly by a second-order backward-differentiation scheme, and the convective term explicitly by a second order extrapolation scheme (see [4,11]). The resulting system is decoupled by the fractional step method of Maday *et al.* [7]. In this approach, the system for velocity and pressure is solved in three successive steps employing a splitting velocity $[\hat{\underline{u}}^m]^*$

$$\left[\frac{1}{Re}\,\underline{\underline{E}} + \left(\frac{3}{2}\,\Delta t + \alpha_z^2\, m^2\right)\,\underline{\underline{B}}\right]\,[\hat{\underline{u}}^m]^* = \underline{\underline{B}}\,\hat{\underline{g}}^m + \underline{\underline{D}}^T\,[\hat{q}^m]^n \tag{5}$$

$$\frac{\Delta t}{\beta_0}\,\underbrace{\underline{\underline{D}}\,\underline{\underline{B}}^{-1}\,\underline{\underline{D}}^T}_{=:\underline{\underline{S}}_p}\,\delta\hat{q}^m = -\underline{\underline{D}}\,[\hat{\underline{u}}^m]^* \tag{6}$$

$$[\hat{\underline{u}}^m]^{n+1} = \frac{\Delta t}{\beta_0}\,\underline{\underline{B}}^{-1}\,\underline{\underline{D}}^T\,\delta\hat{q}^m + [\hat{\underline{u}}^m]^*\ , \tag{7}$$

where $[\underline{\hat{u}}^m]^n, [\hat{q}^m]^n$ are the mth Fourier modes at time $t^n := n\,\Delta t$ (Δt is the time step). In Eqs. (5)-(7) $\underline{\underline{E}}, \underline{\underline{D}}^T, \underline{\underline{D}}$ denote the discrete versions of the Laplacian, gradient, and divergence operator, respectively, $\underline{\underline{B}}$ is the diagonal mass matrix, and α_z is the wave number in the spanwise direction. The right-hand side $\underline{\hat{g}}^m$ contains the velocities and the convective operator at times t^n and t^{n-1}. In Eq. (6) we use the abbreviation $\delta\hat{q}^m := [\hat{q}^m]^{n+1} - [\hat{q}^m]^n$. Note that only one Helmholtz equation needs to be solved for the velocity, and one pseudo-Laplacian equation for the pressure. It can be shown [4] that the splitting error in the pressure equation is of second order in time.

3 Direct solution of the pressure equation

For virtually all flow problems of practical interest the pressure matrix is too large to be composed in full size. However, for a rectangular flow domain with a Cartesian grid, the pressure matrix can be decomposed into 1D operators. In this case, a direct solution using the fast diagonalization method of Lynch *et al.* [5] is feasible, where the 1D operators are inverted by an eigenvalue decomposition. Since the application of the fast diagonalization solver is restricted to rectangular domains, it is not directly applicable to flow problems in more complex domains like the flow over a forward-facing step. To deal with this problem, we decompose the flow domain into rectangular subdomains or 'macro-elements', each of them is again subdivided into a number of spectral elements (see Fig. 1). We denote the interfaces between the macro-elements by $\gamma_1 = M_1 \cap M_2, \gamma_2 = M_2 \cap M_3\ ..., \gamma_{d-1} = M_{d-1} \cap M_d$. Since in the $P_N - P_{N-2}$ method the pressure is defined only on interior points, no pressure points exist on γ_j, and thus straightforward domain-decomposition methods are not applicable. However, the pressure matrix $\underline{\underline{S}}_p$ of Eq. (6) contains operators acting on both the pressure and the velocity mesh. The operator $\underline{\underline{D}}^T$ evaluates the first derivative and interpolates the results from the pressure to the velocity nodes. The diagonal mass matrix $\underline{\underline{B}}$ is defined only on the velocity nodes and the operator $\underline{\underline{D}}$ evaluates the first derivative and interpolates back to the pressure nodes. Since the velocity mesh contains points on the internal boundaries γ_j, the matrix $\underline{\underline{S}}_p$ can be decomposed into submatrices (S_i) defined only on the inner points of the macro-elements M_i and submatrices (H_i, G_i) involving also points on γ_j. Thus, Eq. (6) can be rewritten in the form

$$S_i\, q_i + G_i\, g = r_i\ , \qquad\qquad i = 1, ..., d \tag{8}$$

$$g \qquad = \sum_{j=1}^{d} H_j\, q_j\ , \tag{9}$$

where q_i and r_i are the subvectors of q and r corresponding to the macro-element M_i. Here, we use the abbreviations $q := \delta\hat{q}^m, r := -\underline{\underline{D}}\,[\underline{\hat{u}}^m]^*$. In Eq. (8) and (9), g is an unknown vector defined on the internal boundaries γ_j. The matrices

G_i and H_i are defined by

$$G_i := \left[\underline{\underline{D}}\,\underline{\underline{B}}^{-1}\right]_i^b \ , \ H_i := \left[\underline{\underline{D}}^T\right]_i^b \ ,$$

where $[\ \cdot\ ]_i^b$ denotes operators that act on the internal boundaries γ_i. In Eq. (8), the term $G_i\, g$ is responsible for the coupling of the solution across the internal boundaries. The matrices S_i involve only inner points. The key feature of the present method is that the submatrices S_i on the macro-elements can be inverted very efficiently by the fast diagonalization solver [5].

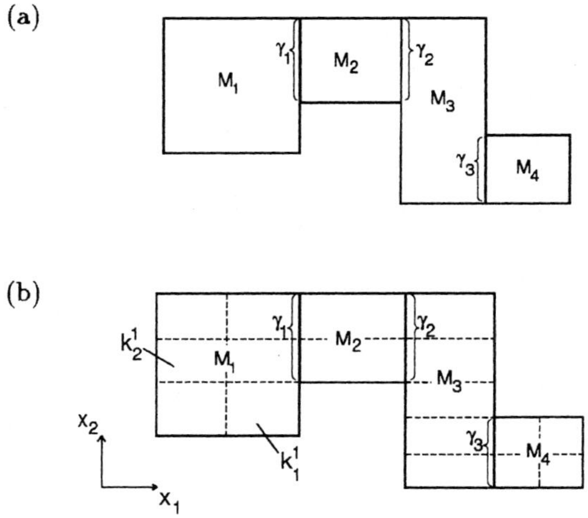

Fig. 1. Subdivision of the flow domain Ω into a number of macro-elements $M_1, ..., M_d$ (a), which are further subdivided into $(K_1^i \cdot K_2^i)$ spectral elements (b). The boundaries between the macro-elements are denoted by γ_j.

We emphasize here that the pressure matrix $\underline{\underline{S}}_p$ is singular due to the undetermined mean pressure level. Moreover, the d submatrices S_i are also singular. However, the complete system (8), (9) has only a rank defect of one. Thus, for solvability, additional consistency conditions of the form

$$\tilde{v}_i^T\,(r_i - G_i\, g) = 0 \ , \quad i = 1, ..., d \tag{10}$$

have to be satisfied, providing a proper coupling of the solution across the internal boundaries [9]. In Eq. (10), $\tilde{v}_i$ are the left eigenvectors of the zero eigenvalues of the matrices S_i. Employing such consistency conditions and fixing the mean pressure level to zero, a non-singular pressure system is obtained, which provides a unique solution of Eq. (6). The resulting system is then solved by the influence-matrix technique [9] (see also [11]).

Since for our applications the number d of macro-elements is small (typically $d < 10$), the number of points on the internal boundaries γ_j is also small, and therefore the largest amount of computing time is spent on the inversion of the matrices S_i in Eq. (8), which, however, is done efficiently by the fast diagonalization solver.

Table 1. Timing for disturbed 3D channel flow simulation on NEC SX4. Values for direct tensor-product solver and iterative solver (PCG). $N = 10$, K_m denotes the number of spectral elements in one macro-element in each spatial direction (here, $K_1 = K_2 = K_m$), t_{cpu} is the CPU-time for the pressure solution per time step (for one Fourier mode), and $MFLOP$ is the number of million floating point operations.

	direct		iterative	
K_m	t_{cpu}	$MFLOP$	t_{cpu}	$MFLOP$
6	0.011	6.89	0.94	363.8
8	0.020	16.98	1.08	462.2
10	0.034	34.15	1.28	764.2
20	0.198	262.39	4.09	2994.8

4 Results

In Table 1, the efficiency of the direct solver is illustrated by comparison with a standard conjugate gradient method (PCG) using a block preconditioner [2]. The results were obtained for the simulation of a 3D flow in a plane channel [11]. The flow domain was divided into two macro-elements and in each macro-element K_m spectral elements were used in every spatial direction. In Table 1, the number of iterations was limited to 600, which results in divergence errors of up to $7.3 \cdot 10^{-6}$. Note that for the direct solver the divergence error is zero within machine accuracy (about 10^{-15}) independent of the number of spectral elements. It is seen in the table that the direct solver is one to two orders of magnitude faster than the iterative solver. In Fig. 2, the number of million floating point operations ($MFLOP$) required for the complete pressure solution is given as a function of K_m. We remark that for the iterative solver the slope is 2.01 and for the direct solver it is 3.02. The expense of the iterative solution is $8\,\sigma N^3 K_m^2$ [11], where σ denotes the number of conjugate gradient iterations. In principle, σ can be reduced considerably if more sophisticated preconditioning techniques are employed (see Fischer [4]). However, for any iterative solver σ will remain problem-dependent and will in general increase strongly with mesh refinement. Clearly, this problem does not exist with direct solvers.

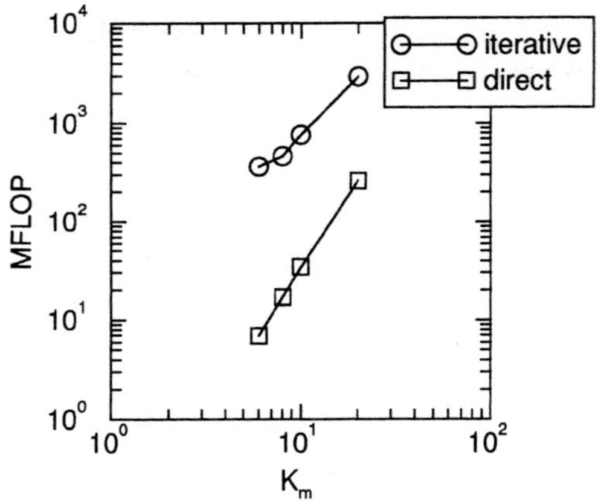

Fig. 2. Comparison of our novel direct solver with iterative solver (PCG). Million floating point operations as a function of the number of spectral elements (K_m) in one macro element in each spatial direction.

5 Conclusions

We developed a new direct domain-decomposition solver for the pseudo-Laplacian pressure equation in spectral element methods. The solver is based on a fast diagonalization strategy, which can be applied in parallel on each subdomain. The coupling between the elements is provided by the influence-matrix technique. We have shown that our direct solver is nearly two orders of magnitude faster than conventional preconditioned conjugate gradient solvers.

Acknowledgments: We would like to thank C. Härtel for helpful discussions.

References

1. C. Canuto, M. Y. Hussaini, A. Quarteroni, T. A. Zang: *Spectral Methods in Fluid Dynamics* (Springer, Heidelberg, 1988).
2. W. Couzy, M. O. Deville: J. Sci. Comput. **9**, 107 (1994).
3. W. Couzy, M. O. Deville: J. Comput. Phys. **116**, 135 (1995).
4. P. F. Fischer: J. Comput. Phys. **133**, 84 (1997).
5. R. E. Lynch, J. R. Rice, D. H. Thomas: Numer. Math. **6**, 185 (1964).
6. Y. Maday, A. T. Patera: 'Spectral element methods for the incompressible Navier-Stokes equations'. In *State-of-the-Art-Surveys in Computational Mechanics*, ed. by A. K. Noor (ASME, New York, 1989), pp. 71-143.
7. Y. Maday, A. T. Patera, E. M. Rønquist: J. Sci. Comput. **5**, 263 (1990).
8. E. Magère and M. O. Deville: Appl. Numer. Math. **33**, 241 (2000).
9. U. Schumann, J. Benner: J. Comput. Phys. **46**, 1 (1982).
10. H. Stüer, A. Gyr, W. Kinzelbach: Eur. J. Mech. B/Fluid **18**, 675 (1999).
11. D. Wilhelm: Numerical Investigation of Three-Dimensional Separation in a Forward-Facing Step Flow Using a Spectral Element Method, Ph.D. Thesis No. 13774, Eidgenössische Technische Hochschule Zürich (submitted).
12. D. Wilhelm, L. Kleiser: Appl. Numer. Math. **33**, 275 (2000).

Part XIII

Transition and Turbulence

Simulation of a Wake-Blade Interaction in Moving Grids

M. Opiela, M. Meinke, and W. Schröder

Aerodynamisches Institut, RWTH Aachen, Wüllnerstrasse zw. 5 u. 7
D-52062 Aachen, Germany

Abstract. A LES of a wake blade interaction carried out on block structured grids with a sliding interface. The wake is generated by a moving cylinder which interacts periodically with a T106 turbine blade. This test case was defined and the simulations were carried out in the framework of the BRITE-Euram project TURMUNSFLAT. The flow field is discussed for a Reynolds number of 160.000 based on the chord length of the blade. Time averaged velocity profiles are in acceptable agreement with experimental data, an accurate prediction of the transition location was not obtained yet.

1 Introduction

Increasing the performance of modern gas turbine engines requires the minimisation of the losses caused by the unsteady phenomena due to rotor-stator interaction. The stator blade loading and heat transfer as well as the boundary layer characteristics are influenced periodically by the wake flow generated by the moving rotor. To predict the unsteady flow accurately with numerical methods turbulence models are necessary which are capable to predict all phenomena like unsteady wakes, temporary separation and bypass transition.

Turbulence models based on the Reynolds-averaged Navier-Stokes equations must often be adapted for the different flow conditions and usually fail in regions of curved streamlines or swirling flow. Since a Direct Numerical Simulation (DNS) of such flow problems at the corresponding Reynolds numbers is currently not possible due to the high computational costs, Large-Eddy Simulations (LES) are carried out. Results of a model experiment of a wake-blade interaction in an axial turbine stage at Reynolds numbers of $O(10^5)$ are presented, which was defined in the BRITE-Euram project TURMUNSFLAT coordinated by the VKI Brussels in Belgium. Hereby the turbulent wake flow is generated by a moving cylinder to reduce the computational effort and to make experimental measurements easier.

2 Method of Solution

For the LES of the wake-blade interference a 13 block structured grid is used, which is partly shown in Fig. 2. A critical point in the simulation of the flow problem is the sufficient resolution of the wake generated by the cylinder. Due to the movement of the cylinder the wakes pass nearly through the whole flow field. Additionally, a nearly equidistant distribution of grid points is required across the sliding-interface, which is located between the grid blocks around the cylinder and those around the blade. Therefore, small spatial steps are required in the major part of the domain of integration which leads to a large number of grid points for the simulation of this flow problem. The first grid line normal to the cylinder wall in the generated grid is 0.01d, where d is the diameter of the cylinder, whereas normal to the blade wall the distance is 0.000025C, where C is the chord length of the blade. The total number of grid points is 1.86 million, of which 33 grid planes are used in span-wise direction.

The grid is generated by solving elliptic equations simultaneously for all grid blocks with a special treatment for non-regular cells so that block boundaries become part of the solution. Only with this procedure it was possible to generate a grid with sufficiently smooth and orthogonal grid lines, so that the flow simulation remained stable. The location of grid points is identical on the periodic lines in pitch-wise direction. A sliding interface is used for the connection of the moving and non-moving grid part to avoid interpolation and to ensure flux conservation across the interface.

For the inflow boundary a non-reflecting boundary condition with a fixed velocity profile and at the outflow boundary a non-reflecting boundary condition with pressure level relaxation is used [6]. The physical flow parameters used for the LES of the wake blade-interaction are defined in detail in [7]. They are chosen according to an experimentally investigated case by the University of Cambridge in the framework of the BRITE-Euram project TURMUNSFLAT. The Reynolds number is 160.000 based on the chord length of the blade and the Mach number of the incoming flow was set 0.1.

The numerical method applied is an explicit Runge-Kutta time stepping scheme of second-order accuracy. The convective terms are formulated fully conservative for moving and deforming grids with appropriate metric terms and are discretized with a AUSM scheme [3] which was slightly modified to reduce the numerical dissipation. This scheme was successfully applied to a various flow problems, results and a detailed description of the method used can be found e.g. in [4,5]. Results of one validation test case are presented in Fig. 1 for the turbulent wake behind a circular cylinder at a free stream Reynolds number of Re=3900 and a Mach number of Ma=0.3. Two 5-block grids were used, a coarse grid with 370.000 points (which has a similar resolution as was used for the cylinder in the wake-blade grid) and a fine grid with 1.4 million points. The obtained Strouhal number of Sr=0.211 and the pressure distribution shown in Fig. 1 are in good agreement with reference data. The velocity profile on the centerline deviates from the reference data due to the rather coarse mesh resolution in the wake. For the finer grid, the time averaging of the data was not yet sufficient to produce smooth results, but it can be seen that a much better agreement is obtained with the higher grid resolution.

The initial condition for the LES of the wake-blade interaction was obtained from a two-dimensional simulation, which has reached a fully periodical flow state, with only one grid plane in span-wise direction, but the same grid resolution in the other two space directions. The solution was then extended to 33 grid planes in span-wise direction and further integrated in time until a three-dimensional solution and an almost periodical flow state has been obtained. A comparison of the velocity profiles of the two different solutions in Fig. 4 for one location in stream-wise direction on the suction side of the blade and for different time levels during the passage of the wake shows that the two-dimensional simulation predicts totally different boundary layer profiles. Much larger velocity fluctuations are visible for the two-dimensional flow. It is also obvious that the two-dimensional solution predicts considerably smaller separated flow regions.

3 Results

In Fig. 3 the Mach number distribution, averaged over all grid planes in the span-wise direction, is shown for an instantaneous solution at a non-dimensional time level of T=0.66. The wake behind the cylinder is clearly visible. The convection of the wake through the blade generates a region of higher Mach number above the suction side surface. This flow structure is periodically generated at about 40% of chord length at a time of approx. T=0.15 and is convected downstream to the trailing edge and remains visible even in the wake of the blade. The wake flow behind the trailing edge of the turbine blade is periodically influenced by this structure. At the beginning of the formation of the higher Mach number region a regular vortex street behind the trailing edge is visible. At a time level of about T=0.5 the regular vortex shedding behind the trailing edge stops and a larger separation bubble is formed. At a time level of about T=0.66, which is shown in Fig. 3, the regular vortex shedding starts again. The surfaces of constant λ_2, which was proposed in [2] to visualise vortices, clearly indicate the main rollers and structures with a vortex axis in the stream-wise direction in the wake of the cylinder. The vortical structures of the wake behind the cylinder are stretched above the suction side after they have passed the leading edge of the turbine blade due to the acceleration caused by the pressure gradient in the stream-wise direction.

Velocity vectors averaged in span-wise direction are plotted in Fig. 5 for three different time levels during the passage of one cylinder around the trailing edge of the blade. They show two separation bubbles, which move downstream to the trailing edge and then vanish temporarily. The separated flow regions occur periodically and are directly connected to the passage of the wakes generated by the moving cylinders.

Time averaged velocity profiles at different locations on the suction side of the turbine blade obtained with the LES are plotted in Fig. 6 in comparison with experimental data. The results are compared to a case investigated by UNICAM at the same Reynolds number. The experimental data consists of time averaged velocity profiles. The velocity profiles from the numerical solution are averaged over all grid planes in span-wise direction and averaged over 8 different time levels during one passage of the cylinder. A good agreement between the LES and the experimental data can be observed concerning the shape of the profile and the boundary layer thickness, although the velocity profiles obtained with the LES are time-averaged only for one passage of the cylinder. The time and span-wise averaged suction side displacement (d1), the momentum thickness (d2) and the shape factor H12 obtained with the LES are plotted in Fig. 7 in comparison with time averaged experimental data. The displacement thickness from the simulation is first slightly smaller until approximately 85% of the surface arc-length are reached. Then the numerically predicted displacement thickness becomes larger. It has again to be noted that the numerical results are time averaged only over one cylinder passage resolved with eight different time levels, which might be a reason for the deviation. The predicted momentum thickness agrees with the experimental data until approximately 80% of the surface arc length. Then the slope of the numerical solution becomes smaller than that of the experimental data. Consequently the shape factor (H12) shown in Fig. 7 is first under- and for arc-lengths larger than about 80% over predicted. This deviation is caused by a different transition behaviour. In the experiment transition occurs at about 80%, while the numerical solution shows a tendency towards transition only at about 90% of arc length. Since no transition location is prescribed in the LES the overall agreement between the experimental data and the numerical solution is quite satisfactory.

4 Conclusions

A LES of a wake blade interaction has been presented on a moving grid with a sliding interface. An acceptable agreement with experimental data has been achieved for time averaged velocity profiles. The determination of turbulence statistics requires longer simulation time and has still to be carried out. The simulation of such flow cases for higher Reynolds numbers requires a more efficient algorithm, since the number of time steps required with the explicit scheme is extremely large. In addition an unsteady grid adaptive techniques is required for a proper resolution of the flow field which will be implemented in the near future.

References

1. P. Beaudan and P. Moin. Numerical experiments on the flow past a circular cylinder at sub-critical Reynolds numbers. Technical Report TF-62, Center Turb. Res., 1994.
2. J. Jeong and F. Hussain. On the identification of a vortex. *J. Fluid Mech.*, 285:69–94, 1995.
3. M. S. Liou and C. J. Steffen Jr. A new flux splitting scheme. *J. Comput. Phys.*, 107:23–39, 1993.
4. M. Meinke and E. Krause. Application of LES to jets and internal turbulent flows. In R. Peyret and E. Krause, editors, *Advanced Turbulent Flow Simulation*, pages 155–208. Springer, 1999.
5. M. Meinke, T. Rister, F. Rütten, and A. Schvorak. LES of free and wall bounded turbulent flows. In K. W. Morton, editor, *Lecture Notes in Physics, 16th International Conference on Numerical Methods in Fluid Dynamics, Arcachon, France*, volume 515, pages 201–206. Springer Verlag, July 1998.
6. T. J. Poinsot and S. K. Lele. Boundary conditions for direct simulations of compressible viscous flows. *J. Comput. Phys.*, 101:104–129, 1992.
7. C. Sieverding. Brite EURAM project TURMUNSFLAT, CT96-0143, 2000. Final Report.

5 Figures

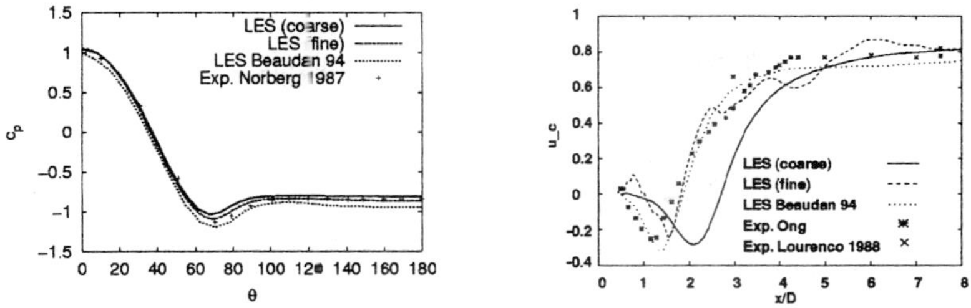

Fig. 1. Pressure coefficient on the surface of the cylinder (left) and velocity profile on the centerline (right) for a Reynolds number of 3900 in comparison to experimental data and LES of other authors [1]

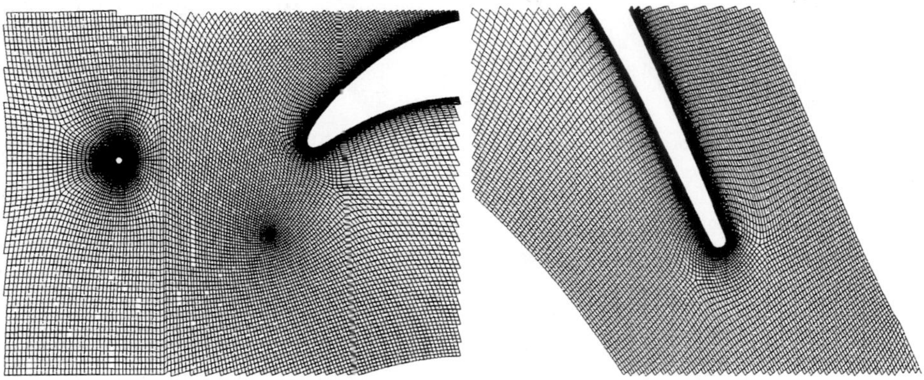

Fig. 2. Zoom of the leading- (left) and the trailing edge (right) of the block structured grid of cylinder and T106-blade region, 13 blocks with a total number of 1.86 million grid points

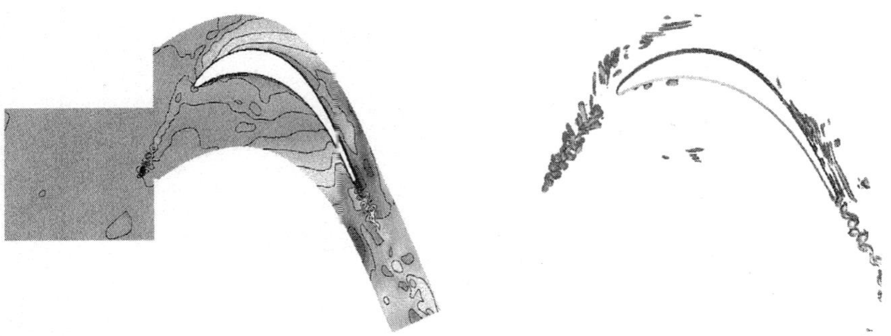

Fig. 3. Mach number distribution (left) and surfaces of λ_2 colour coded with the local Mach number for the wake-blade interaction (right) for a non-dimensional time level of 0.67. T=0 and T=1 correspond to the two consecutive time levels where the vertical positions of the cylinder and the leading edge of the blade are identical

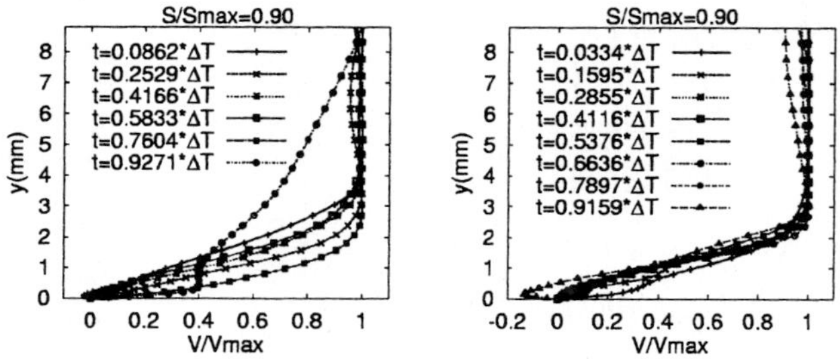

Fig. 4. Velocity profiles obtained with a two-dimensional solution (left) and the LES (right) for different time levels at S/Smax=0.9 (S/Smax is the curvilinear distance along the turbine blade, S=0 corresponds to the leading edge of the blade, Smax to the trailing edge)

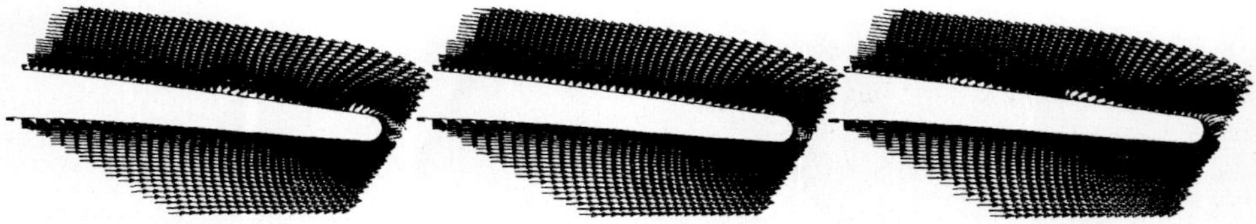

Fig. 5. Velocity vectors above the suction side of the blade near the trailing edge at different non dimensional time levels of T=0. (left), 0.29 (middle) and 0.92 (right)

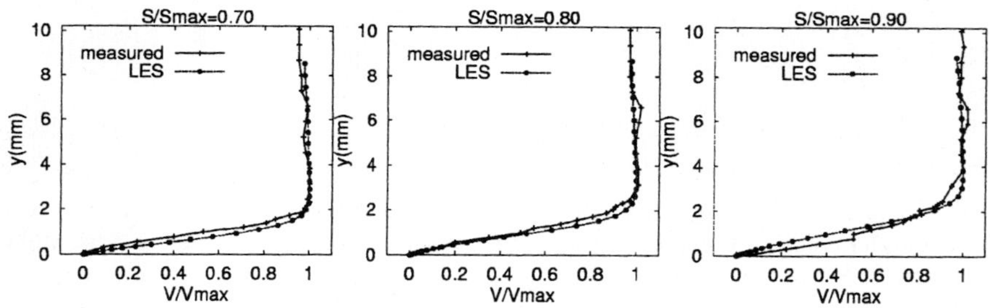

Fig. 6. Time and span-wise averaged velocity profiles obtained with the LES in comparison with measured time-mean velocity profiles from [7] at different locations along the suction side of the blade

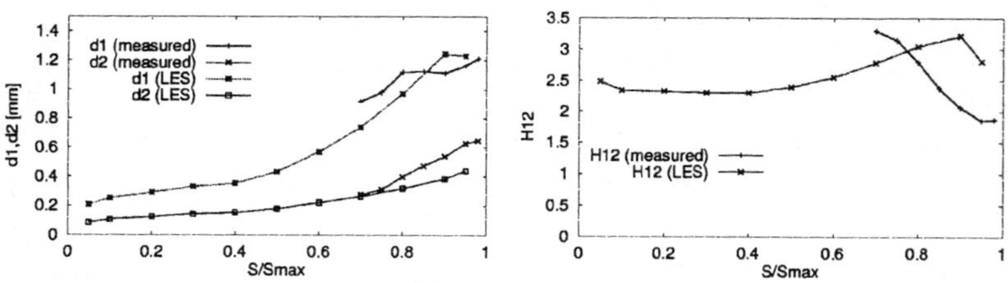

Fig. 7. Displacement- (d1) and momentum thickness (d2) (left) and shape factor H12 (right) obtained from the suction side boundary layer profiles (time and span-wise averaged) compared with time averaged experimental data from [7]

Large Eddy Simulation Analysis of Lobed Mixer Nozzle

Yoshinori Ooba[1], Hidekazu Kodama[1], Yoshiya Nakamura[1], Tsutomu Ooishi[1], Makoto Yamamoto[2], Chuichi Arakawa[3], Osamu Nozaki[4], Kazuo Kikuchi[4], Toshio Nishizawa[4], and Yuichi Matsuo[4]

[1] Ishikawajima-Harima Heavy Industries, 229, Tonogaya, Mizuho, Nishitama, Tokyo 190-12, JAPAN
[2] Science University of Tokyo, 1-3, Kagurazaka, Shinjuku, Tokyo, 162-8601, JAPAN
[3] University of Tokyo, 7-3-1, Hongo, Bunkyo, Tokyo, 113, JAPAN
[4] National Aerospace Laboratory, 7-44-1, Jindaijihigashi, Chofu, Tokyo, JAPAN

Abstract. Large Eddy Simulation (LES) was applied to predict the mixing characteristics of a lobed mixer nozzle analysis. LES was first validated by using experimental data of a convergent-divergent nozzle (CD nozzle) flow. Then, the LES results of the lobed mixer nozzle with a ejector duct were evaluated by comparing with measured data. LES analysis provided more reasonable agreement than Reynolds-averaged Navier Stokes (RANS) analysis.

1 Introduction

Intensity of jet noise of an aircraft engine is strongly influenced by flow structures downstream of the exhaust nozzle. Technology to reduce noise using a mixer nozzle requires understanding the detailed mixing mechanism, however complicated flow causes many difficulties in understanding the detailed phenomena occurring in a noise suppression system. The aerodynamic tests for a lobed mixer nozzle system with ejector duct were conducted by Nakamura, et al and the system performance was measured [1]. Concerning the detailed flow structure, traverse measurement was carried out only on the plane perpendicular to the nozzle center line at the outlet of the ejector duct. CFD simulations were conducted to understand the detailed flow structure in the ejector duct using two kinds of turbulence model, the Baldwin-Lomax (B-L model) and Lam-Bremhorst k-ϵ turbulence model (L-B model) [2]. From the results of the experimental tests, it was found that three high velocity core regions existed at the center of the ejector duct and at both upper and lower ends, whereas these simulations predicted quite different flow structures at the outlet of the ejector duct. This indicates RANS using simplistic turbulence models like the B-L model or the L-B model is not adequate for proper prediction of the complicated mixing phenomena. Choi et al (1999) applied LES to predict the jet mixing the flowfields of CD nozzle which was tested by Yamamoto et al [3][4]. Comparisons of profiles of the computed and measured axial velocity showed a good agreement. The importance of using 3D LES methods to predict the mixing characteristics was also demonstrated. In this paper, LES is applied to the more complicated exhaust nozzle system of a lobed mixer nozzle with a ejector duct, and the results are compared with the measured data to demonstrate the capability to predict the mixing characteristics.

2 Numerical Scheme

The governing equations are the Favre-filtered three-dimensional compressible Navier-Stokes equations. The convection terms are discretized using the 4th order TVD scheme developed by Chakravarthy and Osher [5]. Smagorinsky SGS model with Smagorinsky constant 0.15 is to determine the subgrid turbulence viscosity [6]. In the present calculation, multiblock method with parallel computation algorithm is used on a large computational domain with necessary grid refinement. Calculation for each sub-domain is executed on a different processing element of the Numerical Wind Tunnel (NWT), which is a parallel supercomputer developed by National Aerospace Laboratory (NAL) in Japan. In the calculation, each sub-domain exchanged boundary information with neighboring sub-domains.

3 Results and Discussion

3.1 LES analysis of convergent-divergent nozzle

LES is analysis conducted to predict the flowfield downstream of the CD nozzle. Table 1 shows the design parameters for the CD nozzle which was tested by Yamamoto et al [4]. Structured grids are arranged to fit the surfaces of the CD nozzle. Figure 1 shows comparison of the instantaneous Mach number contours between the LES analysis and RANS analysis using L-B model. The RANS result shows an almost axisymmetric mixing layer, while the LES result predicts the inherent three-dimensional structure of turbulence including a finite potential core.

Table 1. Design parameters for CD nozzle

Throat diameter (m)	0.1295
Exit diameter (m)	0.1370
Distance from throat to exit (m)	0.1403
Exit velocity (m/s)	734.3
Ambient velocity (m/s)	121.9
Pressure ratio	3.121

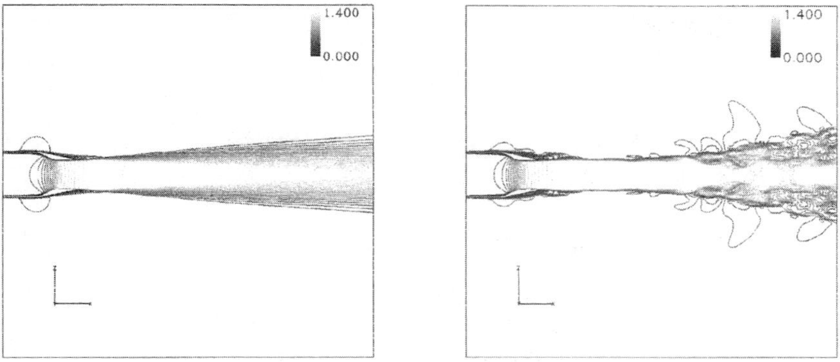

(a)RANS (L-B model) (b)LES

Fig. 1. Mach number contours of CD nozzle

Time-averaged axial velocity profiles of the LES results are compared with the measured data at three axial locations downstream of the CD nozzle (see Fig. 2). Here U denotes the axial velocity and Uj denotes the measured centerline axial velocity at the outlet of the CD nozzle. R and X denote the radial and axial location respectively. These parameters were normalized by the equivalent diameter of the jet flow, Deq. The agreement is not bad in general. RANS results using L-B model are also presented in Figure 2. The mixing layer predicted by RANS calulation is thicker than the measured (see Fig. 2 (b) and (c)). This indicates that RANS analysis tends to predict comparatively faster mixing.

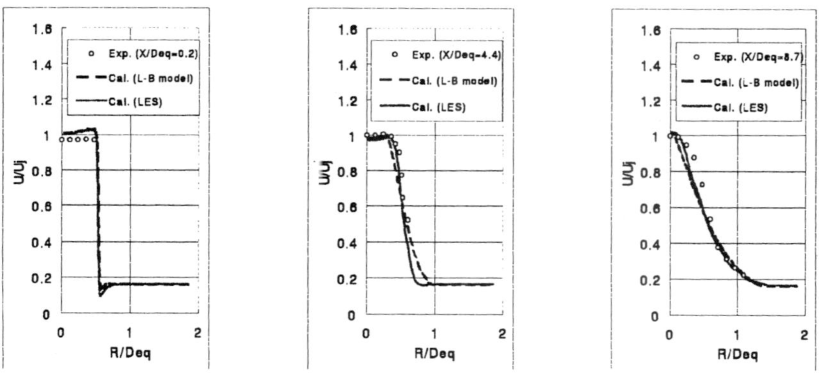

(a)X/Deq=0.5 (b)X/Deq=4.4 (c) X/Deq=8.7

Fig. 2. Comparison of axial velocity downstream of CD nozzle

3.2 LES analysis of lobed mixer nozzle with ejector duct

LES is applied to the prediction of flowfields downstream of the lobed mixer nozzle. Figure 3 (a) shows the cross section of the lobed mixer nozzle with the ejector duct used in the experiment [1]. The mixer nozzle has laterally eleven pairs of upper and lower lobes (see Fig. 3 (b)).

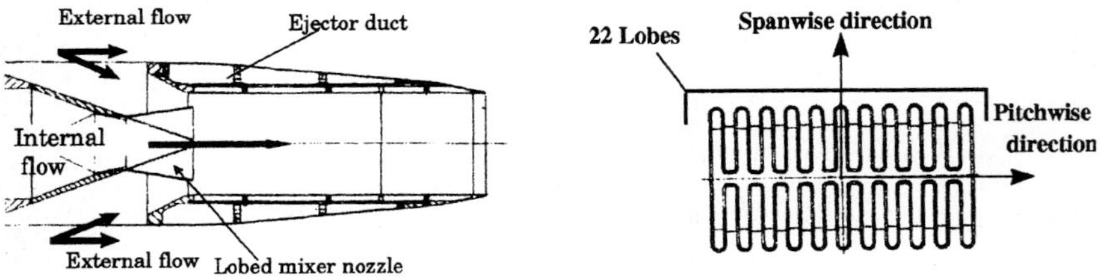

(a) Cross-section of mixer nozzle (b) Outlet shape of lobed mixer nozzle

Fig. 3. Schematic of lobed mixer nozzle with ejector duct

Figure 4 shows the calculated instantaneous Mach number contours of the lobed mixer nozzle with the ejector duct. In this calculation only one upper lobe was simulated using periodic boundary condition at pitchwise boundary ends and mirror boundary condition at the center of this nozzle. Here spanwise direction and pitchwise direction are defined as shown in Fig. 3 (b). In the LES result, high velocity core regions still exist at the outlet of the ejector duct. In contrast, the high velocity core regions disappear downstream of the ejector duct in the RANS result using L-B model.

Figure 5 (a) presents the comparison of time-averaged Mach number profiles at the outlet of the ejector duct with experimental data. Measurement plane was located at X/H=0.029 downstream of the outlet of the ejector duct (see Fig. 5 (b)). Here H denotes the height of the outlet of the ejector duct. X and Z denotes the axial and spanwise location respectively. LES results show a trend similar to the measured profile, while RANS calculation using L-B model predicted quite different profile which is the resultant faster mixing downstream of the lobed mixer nozzle. The LES analysis shows that the calculated spanwise location of the high velocity core near the upper end is higher than the measured, and the calculated velocity of the center high velocity core is considerably higher than the measured. One possible reason of this discrepancy may be that due to the periodic boundary condition imposed at pitchwise ends three-dimensional characteristics of turbulence was strongly affected in the calculation.

Therefore a LES analysis in which the number of lobe is increased to three to lessen the effect of periodic boundary condition is conducted. The grid in each lobe is same as those used in the one-lobe analysis. Periodic boundary condition

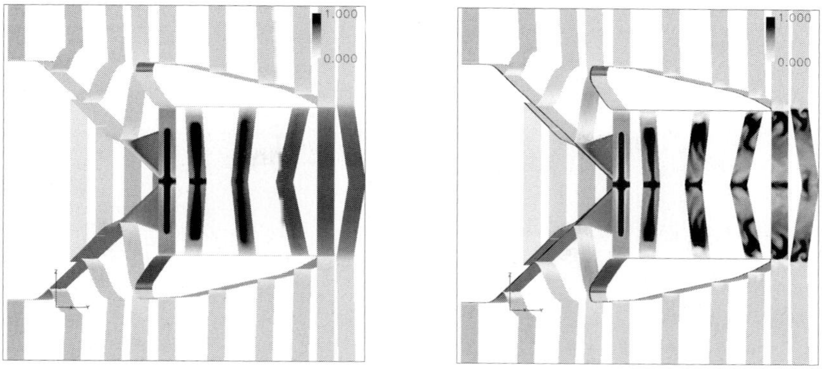

(a) RANS (L-B model) (b) LES

Fig. 4. Mach number contours downstream of mixer nozzle (one-lobe analysis)

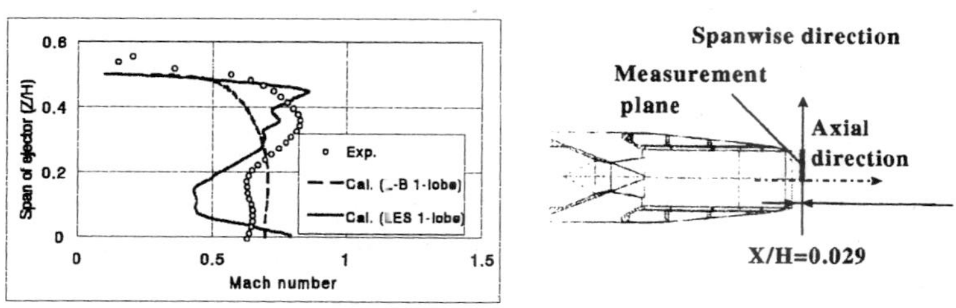

(a) Comparison of spanwise distributions (b) Schematic of measurement plane

Fig. 5. Distribution of Mach number at outlet of ejector duct (one-lobe analysis)

is used at right and left hand boundary. Figure 6 (a) shows the instantaneous contours of Mach number in the ejector duct. It is seen that the flow far downstream of the nozzle is not periodic in the pitchwise direction. The spanwise profiles of time-averaged Mach number are compared with the result of one-lobe analysis and the measured data at the outlet of the ejector duct in Fig. 6 (b). However the spanwise location of the upper high velocity core is still higher than the measured, the overall profile was improved. It is considered that for more accurate predictions, more realistic computational models are needed.

4 Conclusions

- LES analysis captured overall profile of jet velocity downstream of the lobed mixer nozzle with the ejector duct.
- The importance of more realistic computational model for more accurate prediction using LES was demonstrated.

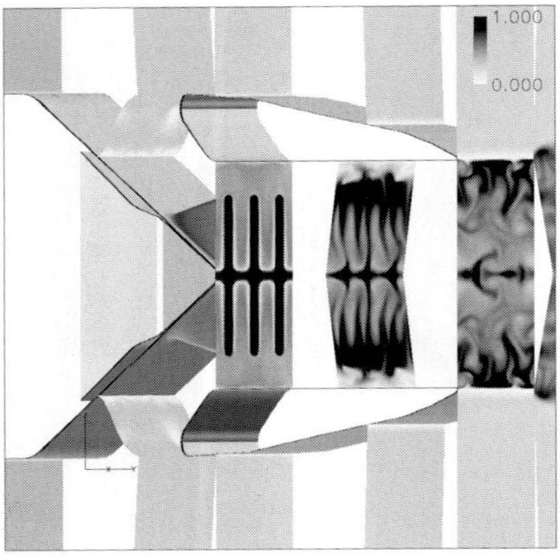

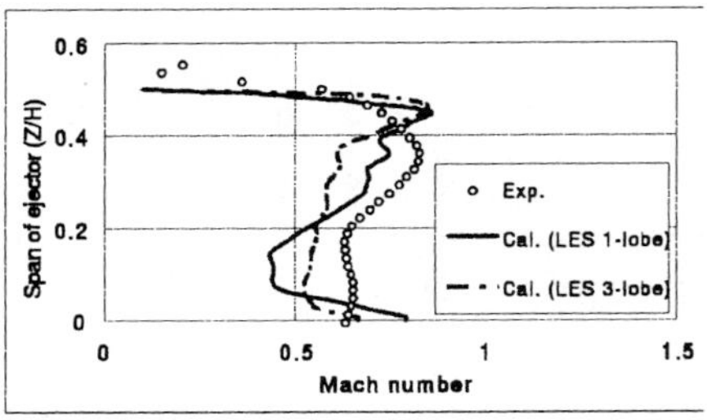

(a) Contours in ejector duct (b) Comparison of spanwise distribtions

Fig. 6. Distribution of Mach number (three-lobe analysis)

Acknowledgement

This study was conducted under the entrustment contact with the New Energy and Industrial Technology Development Organization (NEDO) as a part of the National Research and Technology Frontier Program of Agency of Industrial Science and Technology (AIST), Ministry of International Trade and Industry (MITI).

References

1. Nakamura,Y. and Oishi,T., 'Research and Development of Mixer-Ejector Nozzle', *Proceedings of the International Gas Turbines Congress 1999 Kobe*, 1999, pp219–224.
2. Lam,C.K.G. and Bremhorst, K.A., 'Modified Form k-ϵ model for Predicting Wall Turbulence', *ASME, J of Fluids Engineering*, Vol. 103, 1981, pp456–460.
3. Choi, D., et al, 'Large Eddy Simulation of High-Reynolds Number Jet Flows', *AIAA 99-0230*, 1999.
4. Yamamoto, K., et al, 'Experimental Investigation of Shock-Cell Noise Reduction for Single Stream Nozzles in Simulated Flight', *NASA CR 3845*, 1984.
5. Chakravarthy, S. R. and Osher, S., 'A New Class of High Accuracy TVD Schemes for Hyperbolic Conservation Laws', *AIAA 85-0363*, 1985.
6. Smagorinsky, J., 'General circulation experiments with the primitive equations, I The basic experiment', *Monthly Weather Review*, 91, 1963, pp99–164.

Numerical Simulation of Transitional Turbomachinery Flows

P. De Palma

Universitá degli Studi di Roma "Tor Vergata",
Dipartimento di Ingegneria Meccanica, Via di Tor Vergata, 110, 00133 Roma, Italy,
e-mail: depalma@uniroma2.it

1 Introduction

The numerical prediction of losses in turbomachinery flows can still be considered a challenge, due to the difficulty of modeling those flow regions which are strongly influenced by turbulence phenomena. Actually, a large part of numerical methods overestimate the magnitude of losses, whereas they can provide more reasonable loss-distribution profiles. In order to improve the loss-prediction capability of modern computational methods, improved turbulence models are needed, with particular attention to boundary-layer transition, which plays a fundamental role in turbomachinery flows. Commonly, such flows are characterized by transitional boundary layers, the transition being induced by: 1) the free-stream turbulence (bypass transition); 2) the formation of a separation bubble (separated-flow transition); 3) the wakes from upstream blades (wake-induced transition). Moreover, the curvature of the blade profile also influences the evolution of transition. Transitional boundary layers can be computed by introducing low-Reynolds-number modifications in the modeling of turbulence. Using the low-Reynolds-number $k - \omega$ model [1], the author has developed a methodology for the solution of the steady and unsteady Reynolds-averaged Navier–Stokes equations [2]. In the present work such a methodology is improved by using the Explicit Algebraic Stress Model (EASM) of [3], which accounts for mild nonequilibrium effects and turbulence anisotropy. Firstly, the subsonic flow through the T106 turbine cascade is considered as a suitable test case for validation in two space dimensions. In fact, such a test case is a formidable one due to the separated-flow transition which takes place in the rear part of the blade suction side. Then, the method is extended to three space dimensions and validated versus the experimental data available for the 3D flow through the T106 linear cascade. In particular, the loss-coefficient distribution downstream of the cascade is studied in order to demonstrate the effectiveness of the proposed approach.

2 Governing equations

The three-dimensional Favre–Reynolds-averaged Navier–Stokes equations are considered in the present work, closed by two differential equations for the turbulence kinetic energy, k, and the specific dissipation rate, ω:

$$\frac{\partial(\rho k)}{\partial t} + \frac{\partial(u_j k)}{\partial x_j} = \frac{\partial}{\partial x_j}\left((\mu + \sigma^\star \mu_t)\frac{\partial k}{\partial x_j}\right) + \tau_{ij}\frac{\partial u_i}{\partial x_j} - \beta^\star \rho \omega k, \quad (1)$$

$$\frac{\partial(\rho \omega)}{\partial t} + \frac{\partial(u_j \omega)}{\partial x_j} = \frac{\partial}{\partial x_j}\left((\mu + \sigma \mu_t)\frac{\partial \omega}{\partial x_j}\right) + \frac{\alpha \omega}{k}\tau_{ij}\frac{\partial u_i}{\partial x_j} - \beta \rho \omega^2. \quad (2)$$

The closure coefficients in the above equations are computed according to the low-Reynolds-number $k-\omega$ turbulence model proposed by Wilcox [1]. Moreover, the explicit nonlinear constitutive equation, obtained by Gatski and Speziale [3], has been considered for the Reynolds stress tensor:

$$\tau_{ij} = 2\mu_t[S_{ij} - \frac{1}{3}S_{kk}\delta_{ij} + \frac{\alpha_4}{\omega}(S_{ik}W_{kj} + S_{jk}W_{ki}) + \frac{\alpha_5}{\omega}(S_{ik}S_{kj} - \frac{1}{3}S_{kl}S_{kl})] - \frac{2}{3}\rho k \delta_{ij}, \quad (3)$$

where S_{ij} and W_{ij} are the mean-rate-of-strain tensor and the mean-vorticity tensor, respectively. In equation 3,

$$\mu_t = \rho \alpha^\star C_\mu k / (C_\mu^0 \omega), \quad (4)$$

where

$$C_\mu = \frac{3(1+\eta^2)\alpha_1}{3 + \eta^2 + 6\eta^2\xi^2 + 6\xi^2}, \; C_\mu^0 = 0.09, \; \eta^2 = \frac{\alpha_2}{\omega^2}(S_{ij}S_{ij}), \; \xi^2 = \frac{\alpha_3}{\omega^2}(W_{ij}W_{ij}). \quad (5)$$

The closure coefficients α_1-α_5 are evaluated as follows:

$$\alpha_1 = \left(\frac{4}{3} - C_2\right)\frac{g}{2}, \qquad \alpha_2 = (2 - C_3)^2\frac{g^2}{4}, \qquad \alpha_3 = (2 - C_4)\frac{g^2}{4}, \quad (6)$$

$$\alpha_4 = (2 - C_4)^2\frac{g}{2}, \qquad \alpha_5 = (2 - C_5)g, \qquad g = \frac{1}{0.5C_1 + C_5 - 1}, \quad (7)$$

with $C_1 = 3$, $C_2 = 0.8$, $C_3 = 1.75$, $C_4 = 1.31$, $C_5 = 2$ [4]. For high turbulence-Reynolds-number ($\alpha^\star \to 1$) the present formulation recovers the one proposed in [3,4].

Standard characteristic boundary conditions are imposed at inflow and outflow boundaries. The pressure at the wall is computed by linear extrapolation from the first two interior grid-points, whereas the temperature is obtained from zero normal temperature gradient (adiabatic wall). Concerning the value of ω at the wall, the following condition has been employed: $\omega_w = 60\nu/(\beta(\Delta y_w)^2)$, where ν is the kinematic viscosity coefficient and Δy_w is the distance between the wall and the first grid point away from the wall. Non-periodic C-grids are employed, namely, the two edges of the branch-cut have not the same number of cells in order to obtain a low distortion of the mesh in the blade channel [2]. For the present three-dimensional computations, the symmetry boundary condition is employed at midspan.

3 Numerical method

The Navier–Stokes and turbulence-model equations are solved coupled together as a unique set. An upwind modeling of the inviscid fluxes is employed, based on Roe's approximate Riemann solver with formally second-order-accurate upwind MUSCL extrapolation and minmod limiter. Heat transfer and viscous stress contributions are discretized using central differences. For the present steady-state computations, an explicit (four stage) Runge–Kutta time integration scheme is used, with multigrid acceleration, implicit residual smoothing and local time stepping, which allow the use of CFL number equal to 3 (see [2] for more details).

4 Numerical tests

The linear turbine cascade T106 has been considered as a suitable test case for 2D and 3D validation (see [5] for geometrical details). Such a cascade has been extensively investigated [5,6], detailed experimental data being available concerning the boundary-layer behavior and the secondary flows. The flow is subsonic with isentropic exit Mach number equal to 0.59 and Reynolds number, based on the chord length, c, and the exit conditions, equal to 5×10^5; the inlet flow angle is equal to 127.7°, whereas the inlet turbulence length scale is equal to $0.02c$.

Two-dimensional computations have been performed to study the boundary layer along the suction side of the blades at midspan, where 2D flow conditions exist. For an inlet turbulence intensity equal to $Tu_1 = 0.8\%$, experiments indicate that a separated-flow transition takes place [6].

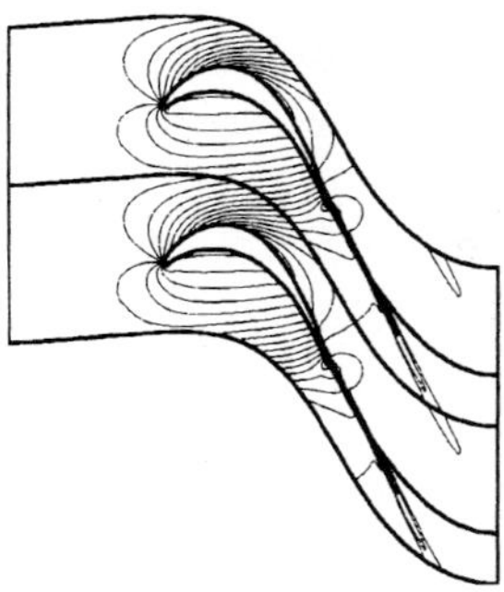

Fig. 1. Mach number ($\Delta M = 0.03$)

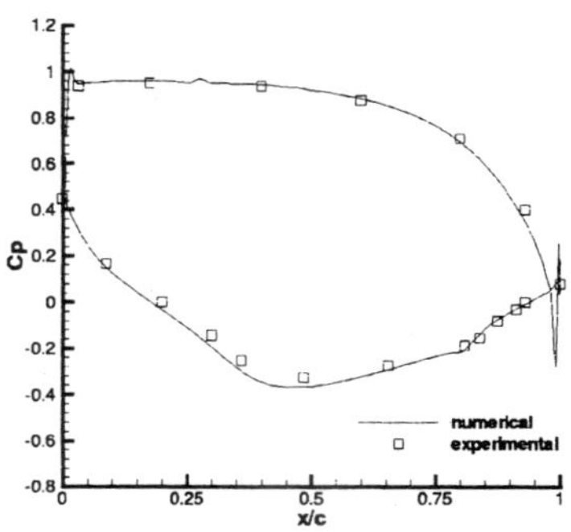

Fig. 2. Pressure coefficient

The computations have been performed using a 384×48 C-grid, the average non-dimensional distance of the first cell center from the wall being about $\overline{y}^+ = 1$.

Figure 1 shows the global view of the Mach number contours, whereas figure 2 provides the comparison between the computed and the experimental distributions of the pressure coefficient ($C_p = (p - p_2)/(p_{01} - p_2)$), which are very close each other. Figure 3 shows the comparison between numerical and experimental boundary layer profiles at five different locations along the suction side of the blade. Computations (solid line) show the presence of a separation bubble in good agreement with the experimental data. Before the trailing edge, the flow reattaches, as shown by the profiles at $x/c = 0.95$, but the predicted velocity profile recovers from separation too slowly. Finally, figure 4 provides a local view of the turbulent viscosity contours which clearly show that a large part of the suction-side boundary layer is laminar, in agreement with the experimental data of [6].

The three-dimensional flow through the linear turbine cascade T106, based on the previous profile, has been then considered in order to study the secondary flows and losses. Computations have been performed with inlet turbulence intensity equal to 5.8% to compare the results with the experimental data reported in [5]. Moreover, the measured [5] total-pressure spanwise profile has been imposed at inlet points. Two grids have been employed with $384 \times 32 \times 36$ cells and $384 \times 40 \times 48$ cells, respectively. For the coarse grid the average non-dimensional distance of the first cell center from the wall is about $\overline{y}^+ = 1$, along the blade surface, and $\overline{y}^+ = 3$, at the sidewall; for the fine grid, the corresponding values are 1 and 2, respectively. Also for this higher value of Tu_1, experimental data indicate that a separated-flow transition exists at midspan, in the rear part of the suction side, with a very small laminar separation bubble. However, for such a value of Tu_1, the computations do not show the separation. Figures 5-8 show the contours of the loss coefficient defined as $\zeta = (p_{t1} - p_t)/(p_{t1} - p_2)$, where p_t is the total pressure and the subscripts 1 and 2 indicate inlet and outlet conditions, respectively. All figures refer to a plane located at $x/c_{ax} = 1.5$. Figure 5 provides the experimental data of [5]: the distribution of the loss coefficient clearly shows a large two-dimensional flow-region around midspan. The loss-coefficient distribution is characterized by three loss cores which are typical of the flow through

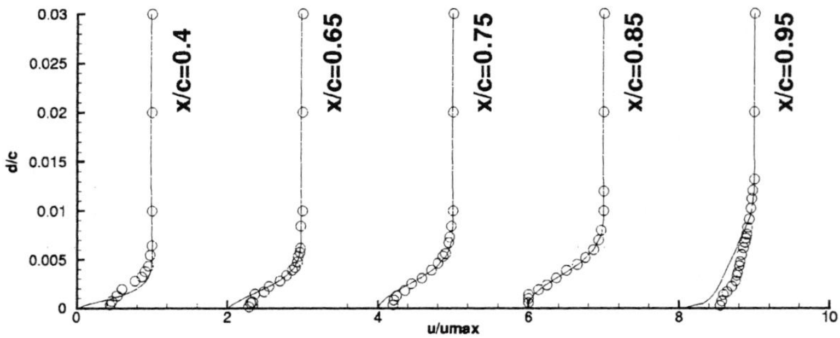

Fig. 3. Boundary layers profiles at the suction side

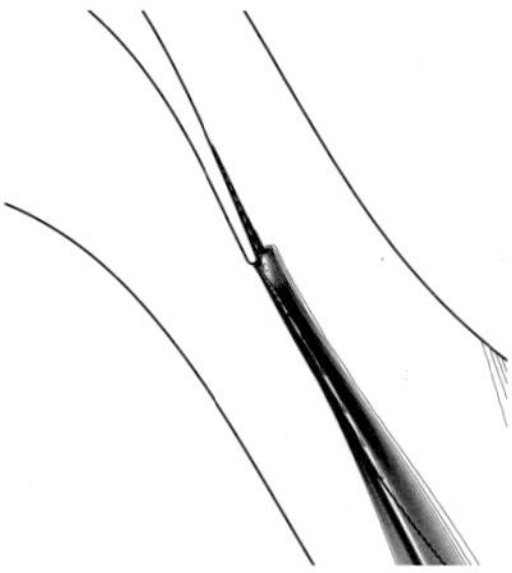

Fig. 4. Turbulent viscosity contours, μ_t/μ_{ref} ($\Delta = 2$, $(\mu_t/\mu_{ref})_{max} = 49$)

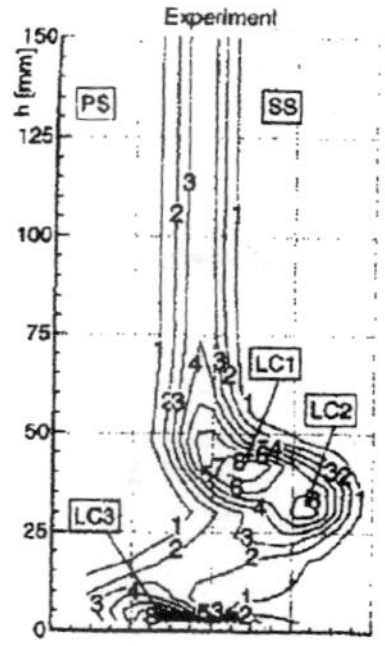

Fig. 5. Loss-coefficient contours ($\Delta\zeta = 0.03$)

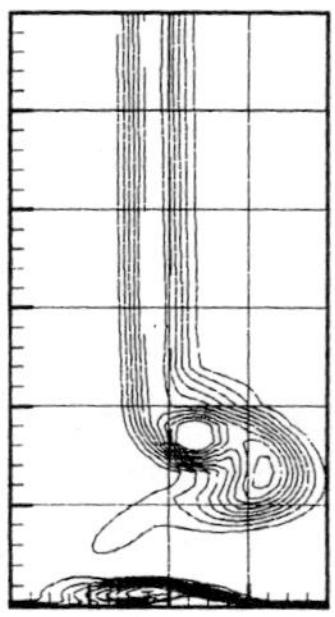

Fig. 6. EASM model: loss-coefficient contours on the coarse grid ($\Delta\zeta = 0.03$)

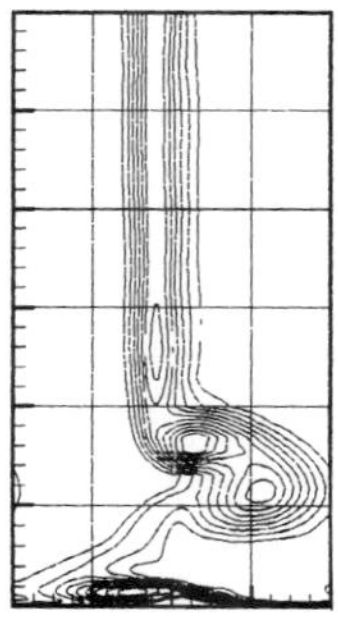

Fig. 7. EASM model: loss-coefficient contours on the fine grid ($\Delta\zeta = 0.03$)

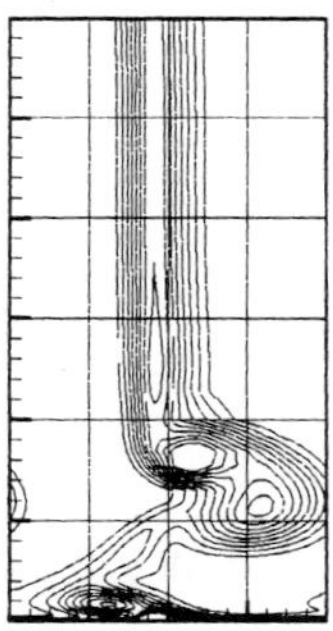

Fig. 8. Standard model: loss-coefficient contours on the fine grid ($\Delta\zeta = 0.03$)

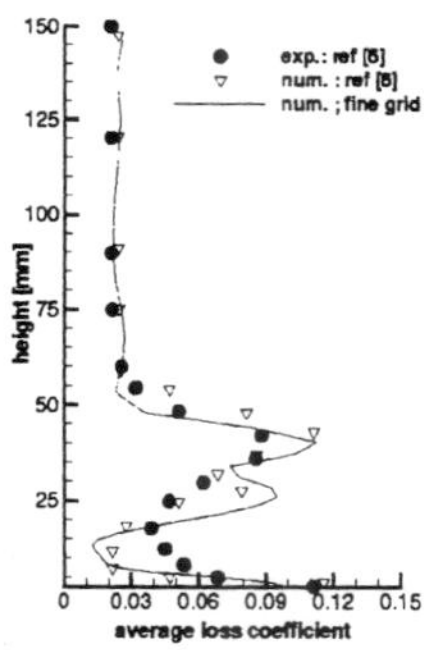

Fig. 9. Average loss-coefficient distributions

high-turning turbine cascades. The principal one (LC1) originates between the the suction side branch of the horseshoe vortex and the passage vortex; the loss core LC2 is directly related to the passage vortex, whereas the third loss core is due to the counter-rotating corner-vortex induced by the passage vortex in the corner between the sidewall and the blade suction side [5]. All these features are well captured by the computation performed using the coarse grid (see figure 6). In particular the loss cores LC1 and LC2 are well resolved and are located in the correct position. In figure 7, the numerical result obtained using the fine grid is shown. Unlike the previous result, the thickness of the wake agree quite well with the experimental data. The maximum value of the loss coefficient is overestimated, nevertheless such a value is lower than the one predicted using the standard low-Reynolds-number $k-\omega$ turbulence model (see figure 8). Finally, figure 9 provides the pitchwise-averaged loss coefficient distributions, defined as $\overline{\zeta} = (p_{t1} - \overline{p}_t)/(p_{t1} - p_2)$, where $\overline{p}$ is the mass-averaged total pressure, versus the blade height. The results obtained with the fine grid (solid line) agree reasonably well with the experimental data (circles) and are very close to the numerical results of [5] (triangles), obtained using a finer multi-block grid (1138000 points).

5 Conclusions

Two- and three-dimensional computations of the flow through the turbine cascade T106 have been performed using a methodology for solving the Favre–Reynolds-averaged Navier–Stokes equations. An Explicit Algebraic Stress Model (EASM), based on the low-Reynolds-number $k-\omega$ turbulence model of Wilcox, has been proposed. Although the prediction of losses in turbomachinery flows is still far from the the required accuracy ($\pm 0.5 \div 1\%$) two considerations may be drawn: i) the proposed model, at least for a low level of free-stream turbulence intensity, has been able to predict the separated-flow transition; ii) the maximum value of the loss coefficient and the thickness of the wake computed using the EASM are closer to the experimental data then the ones obtained employing the standard model. Such promising results, obtained at negligible additional cost, show the potential of this method, nevertheless further work is needed to validate the proposed methodology.

6 Acknowledgements

The author is grateful to prof. M. Napolitano for his precious suggestions and comments.

References

1. D. C. Wilcox: *Turbulence models for CFD*, 2nd edn., (DCW Industries, Inc., 1998)
2. P. De Palma: *8th ISCFD, Brema, Germany, September, 1999*
3. C. G. Speziale: AIAA Journal **36**, 2 (1998), pp. 173-184
4. R. Abid, C. Rumsey, T. Gatski: AIAA Journal **33**, 11 (1995), pp. 2026-2031
5. T. Hildebrandt, L. Fottner: ASME J. of Turbomachinery **121**, 4 (1999), pp. 709-716
6. H. Hoheisel: *AGARD-AR-275*, ed. by L. Fottner, (1990), pp. 112-123

Numerical Modeling of Internal Waves Generated by Turbulent Wakes Behind Self-Propelled and Towed Bodies in Stratified Media *

G.G. Chernykh[1], N.P. Moshkin[1,2], and O.F. Voropayeva[1]

[1] Institute of Computational Technologies, Siberian Division of Russian Academy of Sciences, 6 Lavrentjev Ave., Novosibirsk, 630090, Russia.
[2] Suranaree University of Technology, School of Mathematics, Institute of Science, Nakhon Ratchasima, 30000, Thailand.

1 Introduction

A flow that arises in a turbulent wake behind a body that moves in a stratified fluid is rather peculiar. With a relatively weak stratification a turbulent wake first develops essentially in the same way as in a homogeneous fluid and extends symmetrically. However, buoyancy forces oppose vertical turbulent diffusion. Therefore a wake has a flattened form at large distances from the body and, finally, ceases to extend in a vertical direction. Because of turbulent mixing the fluid density within the wake is distributed more uniformly than outside it. unperturbed state of a stable stratification. As a result, convective flows, which give rise to internal waves in an ambient fluid, arise in the plane perpendicular to the wake axis. Turbulent wakes behind bodies of revolution in stratified fluids have been considered in many publications [1-7]. Analyzing these works we note that the results of the numerical modeling of internal waves generated by turbulent wakes are incomplete. There are, in particular, no data on the numerical analysis of the characteristics of the internal waves generated by drag wake in stratified media, and there are no data on comparison of characteristics of the internal waves generated by the wakes behind the self-propelled and towed bodies. In the present work an attempt is made at filling the gaps available in the numerical modeling.

2 Problem formulation

To describe the flow in a far turbulent wake of a body of revolution in a stratified medium the following parabolized system of the averaged equations for the motion, continuity and incompressibility in the Oberbeck-Boussinesq approach

* This research was partially supported by Russian Foundation of Basic Research (98-01-00736) and the Siberian Division of the Russian Academy of Science (Integration grant No. 2000-1).

is used:

$$U_0 \frac{\partial U_d}{\partial x} + V \frac{\partial U_d}{\partial y} + W \frac{\partial U_d}{\partial z} = \frac{\partial}{\partial y} \langle u'v' \rangle + \frac{\partial}{\partial z} \langle u'w' \rangle, \tag{1}$$

$$U_0 \frac{\partial V}{\partial x} + V \frac{\partial V}{\partial y} + W \frac{\partial V}{\partial z} = -\frac{1}{\rho_0} \frac{\partial \langle p_1 \rangle}{\partial y} - \frac{\partial}{\partial y} \langle v'^2 \rangle - \frac{\partial}{\partial z} \langle v'w' \rangle, \tag{2}$$

$$U_0 \frac{\partial W}{\partial x} + V \frac{\partial W}{\partial y} + W \frac{\partial W}{\partial z} = -\frac{1}{\rho_0} \frac{\partial \langle p_1 \rangle}{\partial z} - \frac{\partial}{\partial y} \langle v'w' \rangle - \frac{\partial}{\partial z} \langle w'^2 \rangle - g \frac{\langle \rho_1 \rangle}{\rho_0}, \tag{3}$$

$$U_0 \frac{\partial \langle \rho_1 \rangle}{\partial x} + V \frac{\partial \langle \rho_1 \rangle}{\partial y} + W \frac{\partial \langle \rho_1 \rangle}{\partial z} + W \frac{d\rho_s}{dz} = -\frac{\partial}{\partial y} \langle v'\rho' \rangle - \frac{\partial}{\partial z} \langle w'\rho' \rangle, \tag{4}$$

$$\frac{\partial V}{\partial y} + \frac{\partial W}{\partial z} = \frac{\partial U_d}{\partial x} \tag{5}$$

In equations (1)–(5), U_0 is the free stream velocity; $U_d = U_0 - U$ is the defect of the mean streamwise velocity component; U, V, W are velocity components of the mean flow in the direction of the axes x, y, z; p_1 is the deviation of the pressure from the hydrostatic one conditioned by the stratification ρ_s; g is the gravity acceleration; $\langle \rho_1 \rangle$ is the mean density defect: $\rho_1 = \rho - \rho_s$; $\rho_s = \rho_s(z)$ is the undisturbed fluid density: $d\rho_s/dz \leq 0$ (stable stratification), $\rho_0 = \rho_s(0)$; the dash denotes the pulsation components; the symbol $\langle\ \rangle$ denotes the averaging. The coordinate system is related to the moving body in such a way that the velocity of its motion is equal to $-U_0$, and the z-axis is directed vertically upwards, in the counter-gravity direction. The fluid density is assumed to be a linear function of temperature; the stratification is assumed to be weak. Both small items involving the derivative with respect to the variable x and the factors in the form of a coefficient of laminar viscosity or diffusion have been omitted in the right hand sides of equations (1)–(4).

The system of equations (1)–(5) is nonclosed. The modified $e - \varepsilon$ model of turbulence is used for the system to be closed. In this model the unknown values of Reynolds stresses $\langle u_i'^2 \rangle$, $(i = 1,2,3)$, $\langle u'v' \rangle = \langle u_1' u_2' \rangle$, $\langle u'w' \rangle = \langle u_1' u_3' \rangle$, turbulent fluxes $\langle u_i' \rho' \rangle$ $(i = 1,2,3)$ are determined by the algebraic approximations (see Rodi [8]):

$$\frac{\langle u_i' u_j' \rangle}{e} = \frac{2}{3}\delta_{ij} + \frac{1 - c_2}{c_1} \left(\frac{P_{ij}}{\varepsilon} - \frac{2}{3}\delta_{ij}\frac{P}{\varepsilon} \right) + \frac{1 - c_3}{c_1} \left(\frac{G_{ij}}{\varepsilon} - \frac{2}{3}\delta_{ij}\frac{G}{\varepsilon} \right), \tag{6}$$

$$P_{ij} = -\left\{ \langle u_i' u_k' \rangle \frac{\partial U_j}{\partial x_k} + \langle u_j' u_k' \rangle \frac{\partial U_i}{\partial x_k} \right\}, \tag{7}$$

$$G_{ij} = \frac{1}{\rho_0} (\langle u_i' \rho' \rangle g_j + \langle u_j' \rho' \rangle g_i), \ i, j, k = 1, 2, 3;$$

$$\mathbf{g} = (g_1, g_2, g_3) = (0, 0, -g), \ 2P = P_{ii}, \ 2G = G_{ii}, \ U_1 = U, \ U_2 = V, \ U_3 = W,$$

$$-\langle u'\rho' \rangle = \frac{e}{c_{1T}\varepsilon} \left[\langle u'w' \rangle \frac{\partial \langle \rho \rangle}{\partial z} + (1 - c_{2T}) \langle w'\rho' \rangle \frac{\partial U}{\partial z} \right], \tag{8}$$

$$-\langle v'\rho' \rangle = \frac{\langle v'^2 \rangle}{c_{1T}} \frac{e}{\varepsilon} \frac{\partial \langle \rho \rangle}{\partial y}, \tag{9}$$

$$-\langle w'\rho'\rangle = \frac{e}{c_{1T}\varepsilon}\left[\langle w'^2\rangle\frac{\partial\langle\rho\rangle}{\partial z} + (1-c_{2T})\frac{g}{\rho_0}\langle\rho'^2\rangle\right], \tag{10}$$

$$\langle\rho'^2\rangle = -\frac{2}{c_T}\frac{e}{\varepsilon}\langle w'\rho'\rangle\frac{\partial\langle\rho\rangle}{\partial z}. \tag{11}$$

Here and below the summation is assumed over repeating indices. To determine the values of the turbulence energy e, the rate of dissipation ε and the shear Reynolds stress $\langle v'w'\rangle$ we make use of the differential equations:

$$U_0\frac{\partial e}{\partial x} + V\frac{\partial e}{\partial y} + W\frac{\partial e}{\partial z} = \frac{\partial}{\partial y}K_{ey}\frac{\partial e}{\partial y} + \frac{\partial}{\partial z}K_{ez}\frac{\partial e}{\partial z} + P + G - \varepsilon, \tag{12}$$

$$U_0\frac{\partial \varepsilon}{\partial x} + V\frac{\partial \varepsilon}{\partial y} + W\frac{\partial \varepsilon}{\partial z} = \frac{\partial}{\partial y}K_{\varepsilon y}\frac{\partial \varepsilon}{\partial y} + \frac{\partial}{\partial z}K_{\varepsilon z}\frac{\partial \varepsilon}{\partial z} + c_{\varepsilon 1}\frac{\varepsilon}{e}(P+G) - c_{\varepsilon 2}\frac{\varepsilon^2}{e}, \tag{13}$$

$$U_0\frac{\partial\langle v'w'\rangle}{\partial x} + V\frac{\partial\langle v'w'\rangle}{\partial y} + W\frac{\partial\langle v'w'\rangle}{\partial z} = \frac{\partial}{\partial y}K_{ey}\frac{\partial\langle v'w'\rangle}{\partial y} + \tag{14}$$

$$+\frac{\partial}{\partial z}K_{ez}\frac{\partial\langle v'w'\rangle}{\partial z} + (1-c_2)P_{23} + (1-c_3)G_{23} - c_1\frac{\varepsilon}{e}\langle v'w'\rangle.$$

We determine the coefficients of turbulent viscosity by simplified relations (6):

$$K_{ey} = \frac{1-c_2}{c_1}\frac{e\langle v'^2\rangle}{\varepsilon}, \quad K_{\varepsilon y} = K_{ey}/\sigma,$$

$$K_{ez} = \frac{\left[(1-c_2)e\langle w'^2\rangle - \dfrac{(1-c_3)(1-c_{2T})}{c_{1T}}\dfrac{g}{\rho_0}\dfrac{e^2}{\varepsilon}\langle w'\rho'\rangle\right]}{c_1\varepsilon\left(1 - \dfrac{(1-c_3)}{c_1c_{1T}}\dfrac{g}{\rho_0}\dfrac{e^2}{\varepsilon^2}\dfrac{\partial\langle\rho\rangle}{\partial z}\right)}, \quad K_{\varepsilon z} = K_{ez}/\sigma.$$

So that

$$-\langle u'v'\rangle = K_{ey}\frac{\partial U}{\partial y}, \quad -\langle u'w'\rangle = K_{ez}\frac{\partial U}{\partial z}.$$

The quantities c_1, c_2, c_3, c_{1T}, c_{2T}, c_T, $c_{\varepsilon 1}$, $c_{\varepsilon 2}$, σ are empirical constants. Their values are taken to be equal to 2.2, 0.55, 0.55, 3.2, 0.5, 1.25, 1.45, 1.90, 0.25, 1.3. The choice of this model of turbulence is due to the following: it is close to the standard $e-\varepsilon$ model of turbulence and we can take into account the anisotropy of the turbulence characteristics in the wakes in stratified fluid. Also a weak dependence of the internal wave characteristics on the applied mathematical model has been shown (see, for example [7]).

The variable x in equations (1)–(4), (12)–(14) plays the role of time. At a distance $x = x_0$ from the body the following initial conditions were specified:

$$U_d(x_0,y,z) = \Theta_1(r),\ e(x_0,y,z) = \Theta_2(r),\ \varepsilon(x_0,y,z) = \Theta_3(r),$$

$$r^2 = y^2 + z^2,\ 0 \le r < \infty;\ \langle\rho\rangle = \rho_s(z) = \rho_0(1-az),$$

$$V = W = \langle v'w'\rangle = 0,\ \langle u'^2\rangle = \langle v'^2\rangle = \langle w'^2\rangle = (2/3)e,$$

$$-\infty < y < \infty, \ -\infty < z < \infty, \ x = x_0; \ a = -(1/\rho_0)d\rho_s(0)/dz.$$

Here $\Theta_1(r)$, $\Theta_2(r)$, $\Theta_3(r)$ are the functions consistent with the experimental data of Lin and Pao [2,3] in the homogeneous fluid.

At $r \to \infty$ the free stream conditions were specified:

$$U_d = e = \varepsilon = \langle v'w' \rangle = \langle \rho_1 \rangle = V = W = 0, \ i = 1, 2, 3. \tag{15}$$

From the symmetry considerations the solution was determined only in the first quadrant of the (y, z) plane with the use of the boundary conditions

$$\frac{\partial \langle \rho_1 \rangle}{\partial y} = \frac{\partial e}{\partial y} = \frac{\partial \varepsilon}{\partial y} = V = \frac{\partial W}{\partial y} = \frac{\partial U_d}{\partial y} = \langle v'w' \rangle = 0, \ y = 0, \ z \geq 0;$$

$$\langle \rho_1 \rangle = \frac{\partial e}{\partial z} = \frac{\partial \varepsilon}{\partial z} = W = \frac{\partial V}{\partial z} = \frac{\partial U_d}{\partial z} = \langle v'w' \rangle = 0, z = 0, \ y \geq 0, \ i = 1, 2, 3.$$

In the numerical solution of the problem the boundary conditions (15) corresponding to $r \to \infty$ were translated to the boundaries of a sufficiently large rectangle $0 \leq y \leq y_*; \ 0 \leq z \leq z_*$.

The problem variables can be made dimensionless by using the characteristic length D, the body diameter, and the velocity scale U_0. The dimensionless variables are introduced as follows: $x^* = x/D, \ y^* = y/D, \ z^* = z/D; U_i^* = U_i/U_0; \ \langle u_i'u_j' \rangle^* = \langle u_i'u_j' \rangle / U_0^2, \ e^* = e/U_0^2; \ \varepsilon^* = \varepsilon D/U_0^3; \ \langle \rho_1 \rangle^* = \langle \rho_1 \rangle / (aD\rho_0);$ $\langle u_i'\rho' \rangle^* = \langle u_i'\rho' \rangle / (aD\rho_0 U_0), \ \ \langle \rho'^2 \rangle^* = \langle \rho'^2 \rangle / (aD\rho_0)^2$. As a result the quantity $4\pi^2/F_r^2$ will appear in the dimensionless equations instead of g, where F_r is the density Froude number defined as

$$F_r = \frac{U_0 T}{D}, \ T = \frac{2\pi}{\sqrt{ag}} = \frac{1}{N}.$$

T, N are the Brunt-Vaisala period and frequency. It is convenient to introduce for the interpretation of the computational results the time t related to the distance from the body:

$$t = \frac{x}{U_0}; \ t^* = \frac{t}{T} = \frac{xD}{U_0 DT} = \frac{x^*}{F_r}.$$

3 Algorithm of problem solution

For the construction of a finite difference algorithm new independent variables are introduced:

$$x = x', \ y = \varphi_1(\xi), \ z = \varphi_2(\eta)).$$

The functions φ_1 and φ_2 were specified in a tabular form; their choice enabled us to condense the grid nodes in the turbulent wake neighborhood. In the new coordinate system $(x', \ \xi, \ \eta)$ the nodes of the computing mesh in the $(\xi, \ \eta)$ plane were distributed uniformly: $\xi_i = i \cdot \Delta\xi, \ \eta_j = j \cdot \Delta\eta, \ i = 0, ..., M_1; \ j = 0, ..., M_2; \ \varphi_1(\xi_{M_1}) = y_*, \ \varphi_2(\eta_{M_2}) = z_*$. The size Δx of the mesh step in the direction of the x' axis was chosen to be variable. The algorithm of problem solution is based on the implicit splitting into space variables for equations (1), (4), (12)-(14) and on the application of an explicit splitting into physical processes to the system of equations (2),(3),(5). The equations of the mathematical model were integrated numerically term by term on each layer on x. The calculations are effected in the following sequential manner. First, velocity defect U_d was calculated by integration of equation (1). In the second place, the values $V, W, \langle p_1 \rangle$ were determined on the basis of equations (2), (3), (5). In the solution of these equations the terms $\langle v'^2 \rangle, \langle v'w' \rangle, \langle w'^2 \rangle, \langle \rho_1 \rangle$ appearing in the equations were set from the previous level on x. This stage is based on an explicit method of the splitting into the physical processes. At the last stage, the functions $\langle \rho_1 \rangle, e, \varepsilon, \langle v'w' \rangle$ were calculated from appropriate equations.

4 Computational results

In order to check the efficiency of the mathematical model we carried out a series of numerical experiments. In these experiments the calculation results are compared with experimental data of Lin and Pao [3] on the decay of momentumless and drag turbulent wakes in a linearly stratified medium. The calculations are carried out on a grid sequence. The computed axial values of velocity defect $U_d(x,0,0) = U_D(x)$, and the turbulence energy $e_0(x) = e(x,0,0)$ are in satisfactory agreement with the experimental data.

The pattern of the internal waves generated by turbulent wakes in a linearly stratified fluid is depicted in Fig. 1 where the dynamics of curves $\rho_0 - \langle \rho \rangle = \rho_0 - \rho_s(0.2D)$ is given for the moments of time $t/T = 1,2,3,4,5$; $F_r = 280$.

It will be noted that internal waves generated by the wake behind a towed body have essentially greater amplitude compare with waves behind a self-propelled body. Similar phenomenon have also seen in the case of nonlinearly stratified media. There exist some explanation of this effect. There is significant distinction between evolution of axisymmetric turbulent wakes past towed and self-propelled bodies. In the wakes behind towed bodies in a homogeneous fluid the production of turbulent energy due to the average flow has significant effect. In momentumless wakes the almost shearless flow is pointed out at the distance of several dozens of body diameters. In the wakes of self-propelled bodies the turbulence decays faster than in wakes past towed bodies. As a result, in the wakes behind towed bodies in the stratified fluid turbulence gives rise to mixing of a greater volume of fluid and action of gravity generates internal waves of greater amplitude. This phenomenon is illustrated in Fig. 2, where

$$E_t^*(t) = \iint_0^\infty e^* \, dy^* dz^*, \quad P_t^*(t) = \iint_0^\infty \left(\frac{U_d^{*2} + V^{*2} + W^{*2}}{2} + \frac{4\pi^2}{F_r^2} \langle \rho_1 \rangle^* z^* \right) dy^* dz^*$$

are dimensionless values of total turbulent energy and the energy of internal waves in the cross section of wake.

5 Conclusion

A numerical model of the internal waves generated by wakes in stratified fluids has been presented. The calculation results show that the turbulent wake behind a towed body generates waves of essentially greater amplitude than behind a self-propelled body.

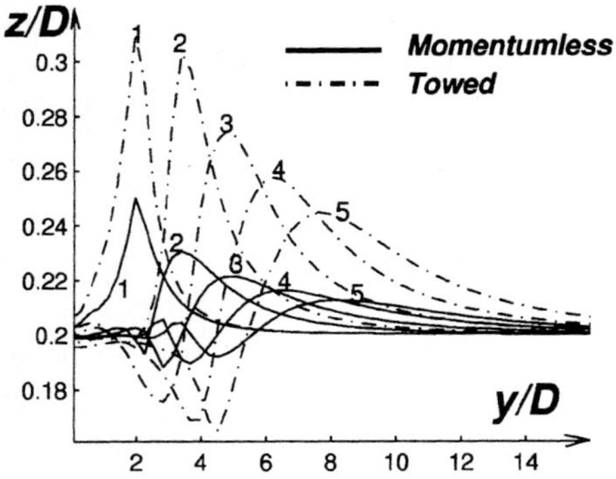

Fig. 1. Dynamics of the curve $\rho_0 - \langle\rho\rangle = \rho_0 - \rho_s(0.2D)$ for moments of time $t/T = 1, 2, 3, 4, 5$.

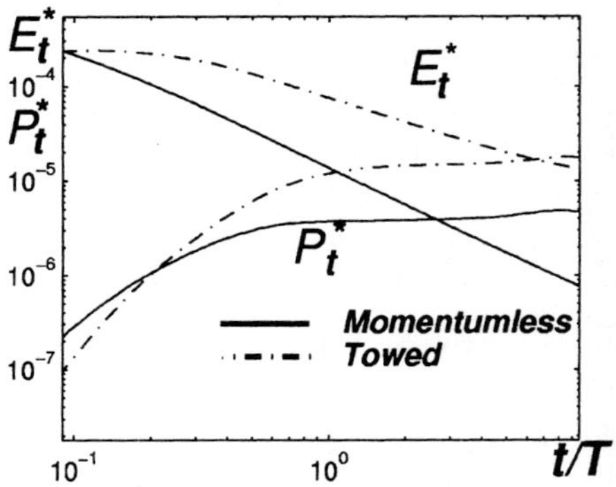

Fig. 2. Time variation of the values of the total turbulent energy $E_t^*(t)$ and the energy of internal waves $P_t^*(t)$

References

1. A.H. Schooley, R.W. Stewart: J. Fluid Mech. **15**, 83 (1963)
2. J.T. Lin, Y.H. Pao: Annu. Rev. Fluid Mech. **11**, 317 (1979)
3. S. Hassid: J. Hydronautics **14**, 25 (1980).
4. H.E. Gilreath, A. Brandt: AIAA Paper **83-1704** (1983).
5. Yu.D. Chashechkin: 'Internal Waves, Vortices and Turbulence in a Wake Past a Bluff Body in Continuously Stratified Fluid'. In: *Preprints of the Fourth Int. Symp. on Stratified Flows, Grenoble, France June 29-July 2, 1994*, ed. by E. Hopfinger, B. Voisin, G. Chavand (Grenoble Inst. of Mech., Grenoble, 1994, vol. 2, sess. B4) No. 29
6. G.G. Chernykh, A.G. Demenkov, N.P. Moshkin, O.F. Voropayeva: 'Numerical Models of Turbulent Wakes in Homogeneous and Stratified Fluids'. In: *Proc. of the Third ECCOMAS CFD Conf., Paris, Sept. 9–13, 1996*, ed. by J. –A. Desideri, C. Hirsch et al. (J. Wiley and Sons 1996) pp. 161-166
7. G.G. Chernykh, O.F. Voropayeva: Computers and Fluids **28**, 281 (1999)
8. W. Rodi: J. Geophys. Res., **92**, No.C5, 5305 (1987).

LES using a Parallel Multidimensional Upwind Solver

D. Caraeni and L. Fuchs

Lund Institute of Technology, Sweden

Abstract. Two second order solution algorithms based on the Residual Distribution approach are proposed for Large Eddy Simulation. Three different SGS models are tested for fully developed turbulent channel flow and the results compare well with the experimetal and DNS data. The parallel solution algorithm proves to be extremely efficient. The potential of extending the accuracy of the algorithm beyond the second order, by using a compact high order FEM approach to compute the cell residual, is highlighted.

1 Introduction

Large Eddy Simulation (LES) is a technique for computation of turbulent flows where the large scale of the flow are computed directly and the effect that the small scale field motions have on the large scales is modeled. Whereas most of the work in this field has been concentrated on structured meshes, in the context of finite-difference or spectral methods, the application of LES to engineering problems requires algorithms more suitable for complex geometries. We have developed a parallel Navier-Stokes solver on tetrahedral cells unstructured grids, for simulations of engineering turbulent compressible flows using LES. The flow-solver is based on a Residual Distribution Schemes algorithm. This approach proved capable of capturing better the real multidimensional physics than the classical approximate Riemann solvers. It borrows ideas both from finite-volume (FVM) and finite element (FEM) methods and has become popular due to its advantages, when compared with the classical FVM methods: more accurate, high order discretization using a compact stencil [9], extendible to higher order of accuracy for LES [2]. The compactness of these schemes makes them suitable for parallelization [3]. Extensions of these Residual Distribution schemes to accurately simulate time-dependent flows have been proposed in [4] and [11]. Here we propose our own approach based on a Jameson dual-time steps procedure. This, together with the use of a System Residual Distribution scheme [9] for the convective and real-time derivative term and a finite element Galerkin method [1] for the diffusive term of the Navier-Stokes equations, have been used to obtain second order accuracy both in space and in time. Three different Sub-Grid Scale (SGS) models have been tested for LES of fully developed turbulent channel flow. In this paper we report the numerical scheme and the parallel efficiency of the code. Finally some numerical results of LES simulations of fully developed turbulent channel flow are presented.

2 LES Equations

The compressible LES equations can be obtained by "top-hat" Favre filtering the time dependent Navier-Stokes equations:

$$\frac{\partial \overline{\rho}}{\partial t} + \frac{\partial \overline{\rho}\widetilde{u}_i}{\partial x_i} = 0 \tag{1}$$

$$\frac{\partial \overline{\rho}\widetilde{u}_i}{\partial t} + \frac{\partial [\overline{\rho}\widetilde{u}_i\widetilde{u}_j + \overline{p}\delta_{ij} - \overline{\sigma}_{ij} + \overline{\tau}_{ij}^{sgs}]}{\partial x_j} = 0 \tag{2}$$

$$\frac{\partial \overline{\rho}\widetilde{E}}{\partial t} + \frac{\partial [(\overline{\rho}\widetilde{E} + \overline{p})\widetilde{u}_j + \overline{q}_j - \widetilde{u}_i\overline{\sigma}_{ij} + H_j^{sgs} + G_j^{sgs}]}{\partial x_j} = 0 \tag{3}$$

The subgrid terms representing the subgrid stress tensor, $\overline{\tau}_{ij}^{sgs} = \overline{\rho}[\widetilde{u_i u_j} - \widetilde{u}_i\widetilde{u}_j]$, the subgrid heat flux, $H_j^{sgs} = \overline{\rho}[\widetilde{E u_j} - \widetilde{E}\widetilde{u}_j]$, and the subgrid viscous work, $G_j^{sgs} = \overline{\rho}[\widetilde{\tau_{ij} u_i} - \widetilde{\tau_{ij}}\widetilde{u}_i]$, all require modelling. In the present simulations, the $\overline{\tau}_{ij}^{sgs}$ was modeled by using a SGS model, as described below. All the other SGS terms presented above, were not explicitly modeled.

3 Numerical Algorithm

The solution algorithm uses the Jameson implicit scheme. This is a dual-time step scheme, with sub-iterations to converge the solution at every new real-time step. Multigrid iterations are performed to accelerate the convergence of the solution in these sub-iterations. The code works on an originally developed unstructured mesh, cell-vertex data structure. The discretization of the convective, diffusive and source terms for the Navier-Stokes equations is based on the Residual System-Distribution schemes approach [9]. Two different update schemes are proposed for time accurate simulations, which are both implicit of Jameson type:

A. The "upwind" approach:

$$U_i^{n+1,k+1} = U_i^{n+1,k} - \frac{\Delta\tau}{V_{\Omega_i}} \sum_{T,i\in T} [\mathbf{B}_i^{T,LDA}(\Phi_T^c + \Phi_T^{uns}) + (\frac{n_{ji}}{3})\overline{F}_j^V]^{n+1,k} \tag{4}$$

B. The "central" approach:

$$U_i^{n+1,k+1} = U_i^{n+1,k} - \frac{\Delta\tau}{V_{\Omega_i}} \sum_{T,i\in T} [\mathbf{B}_{i(\nu_{cell})}^{T,LW}.\Phi_T^c + \Phi_i^{uns} + (\frac{n_{ji}}{3})\overline{F}_j^V]^{n+1,k} \tag{5}$$

Here, $U_i^{n+1,k}$ is the approximation of U_i^{n+1} at the pseudo-time step k; $\mathbf{B}_i^{T,LDA}$ and $\mathbf{B}_{i(\nu_{cell})}^{T,LW}$ are the matrix-distribution coefficients for the convective term, for the LDA and the Lax-Wendroff distribution schemes, respectively; parameter ν_{cell} controls the artificial dissipation of the scheme, $\nu_{cell} = 0.05$; $\Phi_T^c =$

$\int\int\int_T F^c_{j,j}.d\Omega$ is the advective-residual over the cell T, $\Phi^{uns}_T = \int\int\int_T U_{,t}.dv$ is the unsteady residual over the tetrahedral cell T, and $\Phi^{uns}_i = \int\int\int_{T\cap\Omega_i} U_{,t}.dv$ is the unsteady residual over $T \cap \Omega_i$, where Ω_i is the dual volume around the node i. A central-Galerkin scheme has been used to discretize the viscous part. $\frac{1}{3}(\overline{F}^v_j n_{ji})$ represents the contribution of the viscous part of the Navier-Stokes equations to the nodal residual, where n_{ji} are the cell-face inward normal vectors. An implicit second order finite differences discretization in time is used for the real time derivative $U_{,t}$. All residual integrals above were computed with a second order accuracy. Because the schemes we use are Linearity Preserving (LP) schemes, the overall accuracy of the algorithm is second order. Extension to higher order accuracy is possible by using a FEM-like approach [2]. Details about the System Residual Distribution coefficients $\mathbf{B}^{T,LDA}_i$ and $\mathbf{B}^{T,LW}_{i(\nu_{cell})}$, and about the numerical algorithm can be found in [3].

4 SGS Models

Three different SGS models were implemented, which are suited for engineering applications: the selective Smagorinsky, the Dynamic Divergence (DDM) [5] and the dynamic Lilly's[8] models. The selective Smagorinsky model we used was proposed by Saugaut [10]. The model combines the classical Smagorinsky model with a selection function, f_s, that checks the structural properties of the flow field and turns off the SGS model when these properties do not correspond to the one expected from a fully turbulent field. Denote by θ the local angular fluctuation of the instantaneous vorticity. The selection function f_s turns on the SGS model only when θ is higher than a given threshold θ_0. Using the notations by Saugaut, the turbulent eddy viscosity is given by a modified Smagorinsky relation:

$$\mu_T = \overline{\rho} C' f_s(\theta, \theta_0) \overline{\Delta}^2 \sqrt{2\widetilde{S}_{kl}\widetilde{S}_{kl}} \tag{6}$$

where $C' = 0.0165$ and $\theta_0 = 20$. The model parameter(s) are computed dynamically and locally in the DDM and Lilly's dynamic models. Using the classical notations [8], the Lilly dynamic model defines the model coefficient C as:

$$C = \frac{m_{ij} L_{ij}}{m_{kl} m_{kl}} \tag{7}$$

where $m_{ij} = \alpha_{ij}(\widehat{\overline{\rho}}, \widehat{\widetilde{u}}, \widehat{\Delta}) - \left(\widehat{\beta_{ij}(\overline{\rho}, \widetilde{u}, \overline{\Delta})}\right)$. Thus, the subgrid stress tensor is modeled by using:

$$\tau^{SGS}_{ij} = 2\overline{\rho}\left(C\overline{\Delta}^2\sqrt{2\widetilde{S}_{kl}\widetilde{S}_{kl}}\right)\left(\widetilde{S}_{ij} - \frac{1}{3}\widetilde{S}_{kk}\delta_{ij}\right) \tag{8}$$

For the DDM model, the three model coefficients are independently determined using:

$$C^{(i)} = \frac{L_{ij,j}}{m_{ij,j}} \tag{9}$$

Here we use the Einstein notation. No summation is considered for index i. Then, the divergence of the SGS stress tensor is computed directly by using:

$$\tau_{ij,j}^{SGS} = C^{(i)} \beta_{ij,j} \tag{10}$$

No damping function has been used to modify the model parameter(s) close to the wall. These models proved to have the correct behavior near the wall and to self-adapt to the flow field characteristics, vanishing in laminar regions.

5 Results

The results for LES of fully developed channel flow at Reynolds number Re=5400 based on the bulk velocity and channel height are presented. Three different SGS models have been used. The results are compared with DNS [7] and experimental data [6]. The results for the friction velocity u_τ and for the centerline velocity, both normalized by the bulk velocity U_b, are summarized in the table below.

Simulation	Scheme	L_xxL_yxL_z	N_xxN_yxN_z	u_τ/U_b	U_c/U_b
LES-DDM	"central"	10 x 2 x 5	65 x 65 x 33	0.06386	1.176
LES-Lilly	"upwind"	10 x 2 x 5	65 x 65 x 33	0.06367	1.177
LES-Smag	"central"	10 x 2 x 5	65 x 65 x 33	0.0627	1.186
DNS[7]	—	2π x 1 x π	192 x 129 x 160	0.0643	1.162

Normalized friction and centerline velocities for channel flow computations

The friction velocity is in within 4% of the DNS and experimental value of 0.0643, for all LES simulations. The values of the centerline velocity are in within 2% of the experimental value of 1.162. Figure 1. shows the planar averages of the time averaged axial velocity, U_{mean}, normalized by the bulk velocity. Figure 2. presents mean axial velocity normalized by the friction velocity u_τ, $u^+ = \frac{U_{mean}}{u_\tau}$, in wall units, $y^+ = \frac{y.u_\tau}{\nu}$, compared with the velocity predicted by the linear law and with the experimental log law based on data at Re_c of 5600 and 8200. There is a good agreement between the LES results and experiments. Figures 3 to 5 compare the root-mean-square velocity fluctuations from our LES with the DNS and experimental results. The streamwise velocity fluctuations in the simulations using the DDM and LILLY's dynamic SGS model, in Figure 3, show good agreement with the experimental data, and are slightly higher than the DNS results. In the simulations using the selective Smagorinsky model the streamwise velocity fluctuations are overpredicted with as much as 10%. In Figures 4 and 5 one can see that the LES predictions for the wall-normal and spanwise velocity fluctuations are lower than the experiments (and DNS), although the position of the maximum for the wall-normal and spanwise fluctuations is in good agreement with the DNS data. However, the DNS results also show similar discrepancy with the experiments. The results obtained with both dynamic models show improvement when compared with the results obtained with the selective Smagorinsky model, in the level of these fluctuations. The results obtained with the DDM model, although comparable with the results obtained with Lilly's dynamic model, look to be slightly closer to the DNS results. This shows that the DDM model, which has three independently determined model parameters, is capable to deal better with the anisotropy of the channel flow. Figures 7 and 8 present instantaneous pictures of the velocity inside the channel.

6 Parallel results

The ParNAS3D code has been run on an SGI Origin 2000 platform on up to 32 processors. The results addressing the parallel speed-up for channel flow are presented in Figure 6.

References

1. T.J. Barth, "On unstructured grids and solvers". VKI CFD Lecture Series 1990-04, 1990.
2. D. Caraeni, L. Fuchs, "A new compact high order multidimensional upwind discretization", Presented at the 4th World CSCC conference, Vouliagmeni, July 10-15, Greece, 2000.
3. D. Caraeni, Ch. Bergström and L. Fuchs, "Parallel NAS3D: An Efficient Algorithm for Engineering Spray Simulation using LES". Proceedings of ParCFD'99 Williamsburg, pp. 155-164, Elsevier, 2000.
4. M.E. Hubbard, "Compact High Resolution Algorithms for Time-Dependent Advection on Unstructured Grids". To be published in the International Journal for Numerical Methods in Fluids, 2000.
5. J. Held, L. Fuchs, "Large Eddy Simulation of Separated Transonic Flows around a Wing Section", AIAA 98-0405, Reno, 1998.
6. H. Kreplin and H. Eckelmann. Behavior of the Three Fluctuating Velocity Components in the Wall Region of a Turbulent Channel Flow. Physics of Fluids, 22(7), pp. 1233-1239, 1979.
7. J. Kim, P. Moin and R. Moser. Turbulence Statistics in Fully Developed Channel Flow at Low Reynolds Number. In J. Fluid Mech., vol. 177, pp. 133-166, 1987.
8. D.K.Lilly, "A proposed modification of the Germano subgrid-scale closure method", Phys. Fluids A, Vol 4. , pp. 633-635, 1991.
9. H.Paillere, H.Deconinck, E. van der Weide, "Upwind Residual Distribution methods for compressible flows : An alternative for Finite Volume and Finite Element methods", VKI 28th CFD Lecture Series, March 1997.
10. P. Sagaut. "Subgrid-scale models for large-eddy simulation: presentation, comparison of performance". ECCOMAS 98 Proceedings, Eds. John Wiley& Sons, Ltd., 1998.
11. D. Sidilkover, "A New Time-Space Accurate Scheme for Hyperbolic Problems I: Quasi Explicit Case". ICASE report No. 98-25, 1998.

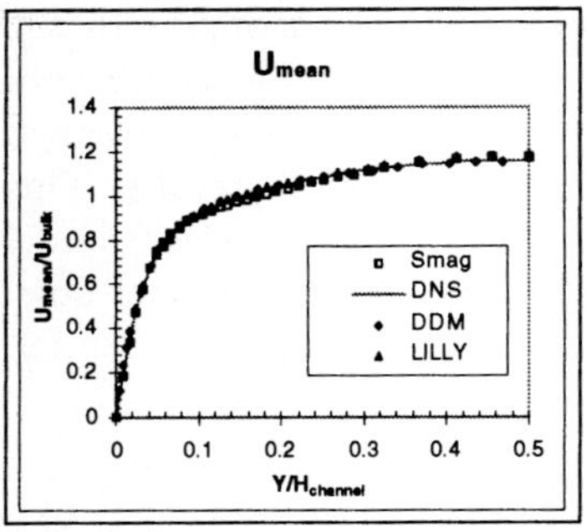

Figure 1. Mean axial velocity.

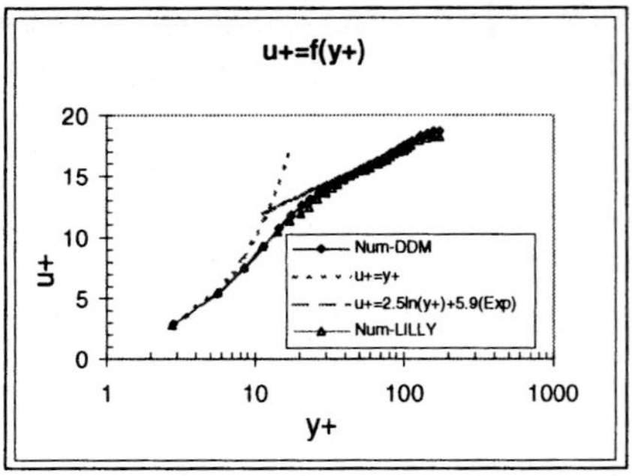

Figure 2. Mean axial velocity, in wall units.

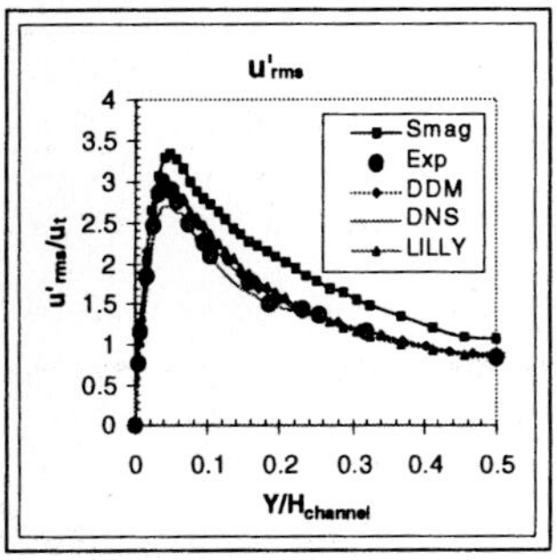

Figure 3. RMS of u′.

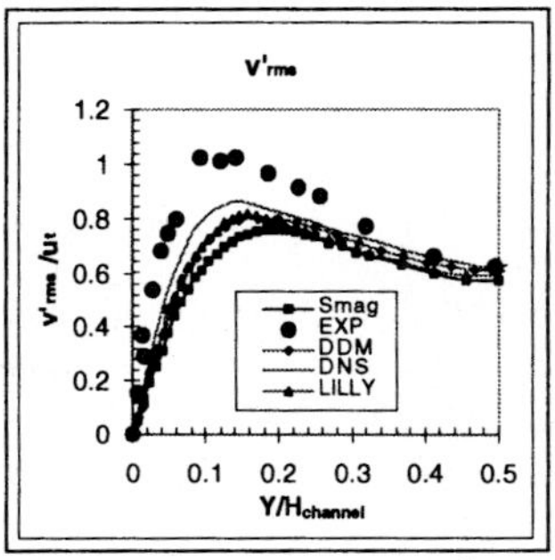

Figure 4. RMS of v′.

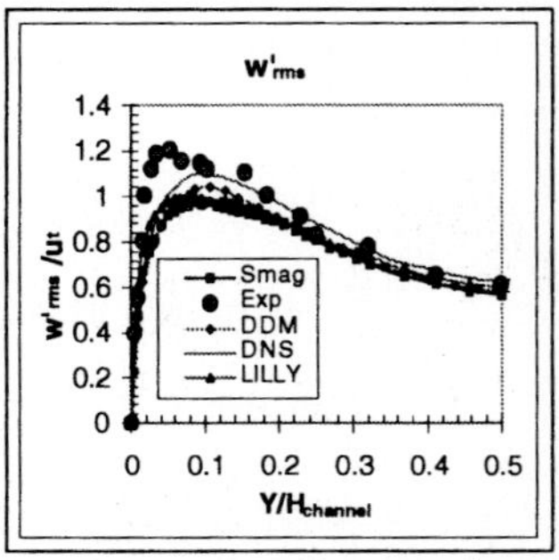

Figure 5. RMS of w′.

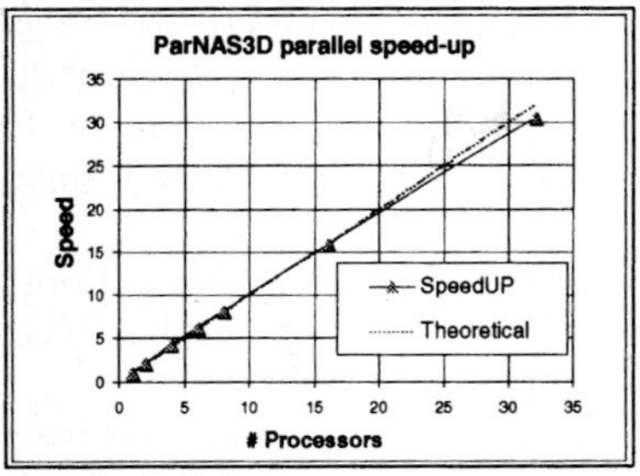

Figure 6. Parallel speed-up.

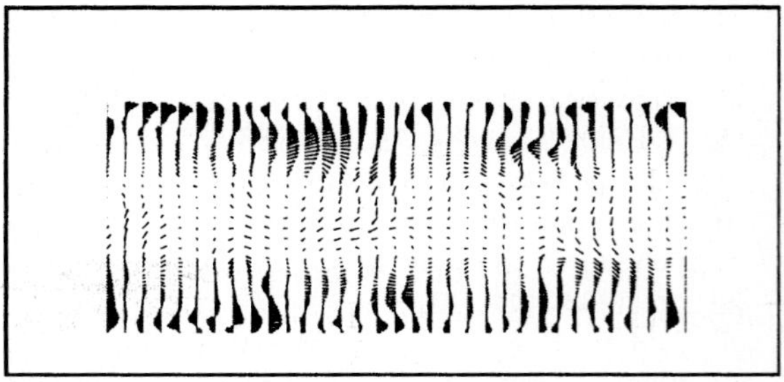

Figure 7. Velocity vectors in cross-section.

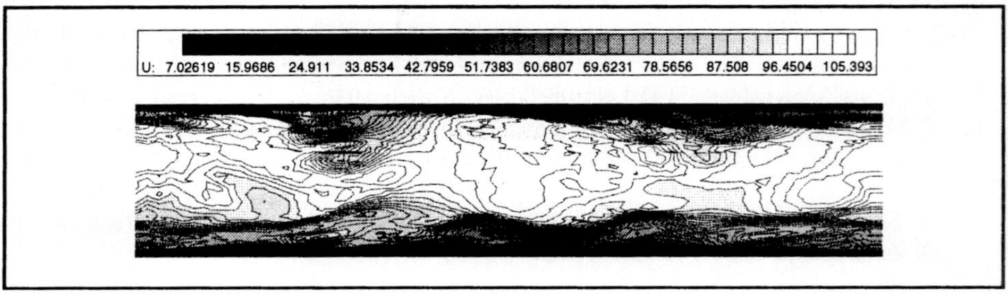

Figure 8. Instantaneous axial velocity distribution.

Large Eddy Simulation by Lagrangian Block Method

Wihel Altai and Vincent H. Chu

McGill University, Montreal, Quebec H3A 2K6, Canada

Abstract. The interrelation between numerical methods and Large Eddy Simulations of turbulent flow is examined in this paper using two very different computational procedures. Numerical simulations of the turbulent flow in a plane jet were conducted using the Lagrangian Block Method (LBM) and a conventional Central Difference Scheme (CDS). Compared with the LBM, the requirement for sub-grid viscosity in the simulation using CDS was significantly lower because the numerical diffusion in the scheme acted as the equivalent of sub-grid viscosity reducing the requirement for sub-grid viscosity. The results of the LES have suggested that the LBM is significantly less diffusive than the conventional CDS. The LBM is non-negative, mass conserving and unconditionally stable. The method is very efficient since accurate results can be obtained with the method using a Courant number close to the critical value of unity.

1 Introduction

Numerical simulations of the advective process have been a subject of continuous challenge to practitioners in the field of computational fluid dynamics. The advective term is nonlinear and has the tendency to produce unphysical oscillations. The negative undershot associated with the oscillations is most undesirable. Physical laws are violated and numerical instabilities are initiated when model parameters changes from positive to negative values.

One naturally positive (non-negative) scheme that stands out as a departure from the conventional approach, is the Lagrangian Block Method developed by Chu and Altai (1998). The LBM is unconditional stable, non-negative and mass conserving. These attributes of the LBM is superior compared with the conventional methods that are stable only when the Courant number is extremely small (less than 0.1). To demonstrate the capability of the LBM, Large Eddy Simulations (LES) of the turbulent flow in a plane jet are conducted using the method. The results are compared with experimental data and the results of a simulation obtained using a conventional Central Difference Scheme (CDS) with second order accuracy.

2 Lagrangian Block Method

The computational elements of the LBM are the cells and the blocks as shown in Figure 1. The cells are fixed in space and are defined in the figures by the dash lines. The mass in the cell are assumed to be contained in a block as shown in the shaded region. The block moves with the velocity $u(x_c, y_c, t)$ and $v(x_c, y_c, t)$. Over the period of one time step, from $n\Delta t$ to $(n+1)\Delta t$, the mass center of the block moves from (x_c^n, y_c^n) to (x_c^{n+1}, y_c^{n+1}). The displacements in the x- and y-directions over the period are given by the integrals

$$x_c^{n+1} - x_c^n = \int_{n\Delta t}^{(n+1)\Delta t} u(x_c, y_c, t)dt, \quad y_c^{n+1} - y_c^n = \int_{n\Delta t}^{(n+1)\Delta t} v(x_c, y_c, t)dt. \quad (1)$$

These integrals are determined numerically by iteration method. At the end of the time step, mass in the block is divided into 'portions' and then re-distributed onto the cells in the fixed Eulerian grid. The mass in the block is divided by the

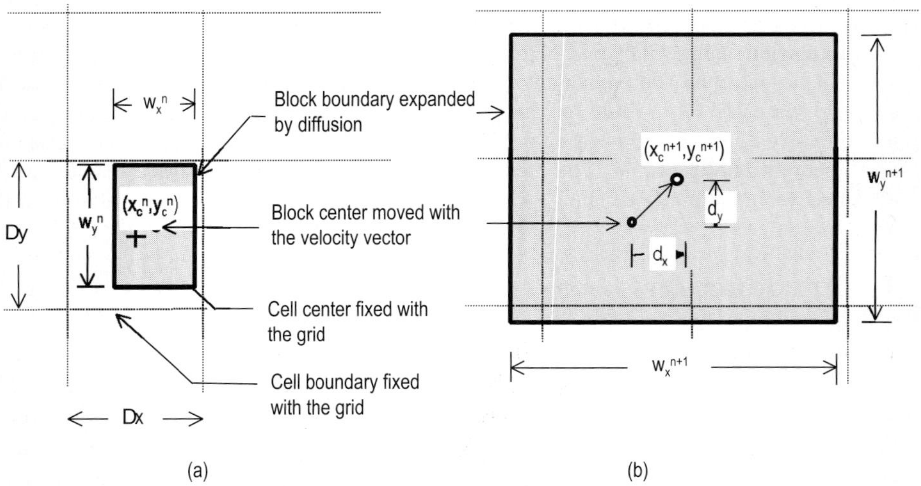

Fig. 1. Lagrangian Block Method. (a) The block in the cells at time $t = n\Delta t$. (b) The block at time $t = (n+1)\Delta t$, after the displacement and enlargement of the block by advection and diffusion. In general, the block is greater in size than the cells. At the end of the time step, the block is divided by the grid lines into 9 portions for re-distribution onto to the cells.

grid lines into 9 portions for re-distribution to its neighboring cells at the end of the time step. The rates at which the size of the block increases with time, are proportional to the diffusion coefficients, D_x and D_y as follows:

$$D_x = \frac{1}{2}\frac{d\sigma_x^2}{dt}, \quad D_y = \frac{1}{2}\frac{d\sigma_y^2}{dt} \quad (2)$$

in which (σ_x, σ_y) are the second moments that are related to the widths of the block, (w_x, w_y), by

$$\sigma_x^2 = \frac{1}{12}w_x^2, \quad \sigma_y^2 = \frac{1}{12}w_y^2. \quad (3)$$

In the diffusion problem of the random walk (Einstein, 1905; Taylor, 1921), D_x and D_y are the diffusion coefficients, and σ_x and σ_y are the root-mean-square displacements of the particle in the random walk. Integrating Equations 2 with time from t = $n\Delta t$ to t = $(n+1)\Delta t$, gives the widths

$$w_x^{n+1} = \sqrt{(w_x^n)^2 + 24 \int_t^{t+\Delta t} D_x(x, y, t)dt},$$

$$w_y^{n+1} = \sqrt{(w_y^n)^2 + 24 \int_t^{t+\Delta t} D_y(x, y, t)dt}. \tag{4}$$

By moving the block using the Lagrangian method according to Equation 1, and calculating the widths according to Equation 4, both the advective and the diffusive processes are now simulated in one single advection and diffusion step.

Aspects of the implementation details of the LBM have been reported in Chu and Altai (1998). An application of the method for turbulent flow simulation using a K-ϵ model was given by Altai, Zhang and Chu (1999). The method is unconditionally stable, non-negative and mass conserving. More computer memory is required for the implementation of the method. Compared with the conventional method, the LBM is more efficient because accurate results can be obtained with the LBM using a value of the Courant number very close to unity.

3 Large Eddy Simulation

The LES of the turbulent flow are conducted in this paper using the Smorgorinsky model for the sub-grid viscosity: $\nu_{SG} = C_s^2 \Delta^2 |S_{ij}|$ where $\Delta = \sqrt{(\Delta x)^2 + (\Delta y)^2}$ is the filter length and S_{ij} is the strain-rate tensor. This sub-grid viscosity ν_{SG} (equal to Δ^2 multiplying by a coefficient $C_s^2 = 0.056 \sim 0.014$) is quite small. Depending on computational method, the sub-grid viscosity can be smaller than the error of the numerical method. Successful LES of turbulent flows have been reported to be possible even without any sub-grid scale model (Kawamura *et al*, 1985; Nadaoka and Yagi, 1998) because the error in the finite difference approximation could act as a form of sub-grid viscosity to suppress the nonlinear instabilties (Ferziger, 1993; Rodi et al., 1998).

Since the effect of the numerical error is comparable to sub-grid viscosity, the simulation results are expected to depend on the numerical methods employed in the simulations. To investigate the dependency on the numerical methods, the turbulent flow of the plane jet were simulated using the LBM and the CDS. The LBM has been described in section 2 and in greater details by Chu and Altai (1998). The CDS is second order accurate as the time is advanced using a predictor and corrector procedure. The computation of the turbulent jet was conducted using a 96×96 grid covering an area 216 cm × 216 cm. The source dimension of the jet is 4 Δy. The velocity at the source is $U_o = 11.5$ cm/s. Hence, the momentum flux of the jet $m_o/\rho = 1190$ cm^3/s^2.

The simulation results using the LBM and CDS with different values of C_s are given in Figures 2 and 3. Figure 2 (a) shows the mean velocity and mean concentration profiles obtained from time average of the instantaneous values. The half-width of the velocity profile δ increase at a rate $d\delta/dx \simeq 0.095$. The width of the concentration profile is about 30% wider than the velocity profile. These are in exellent agreement with the experimental data. The Reynolds stresses $\overline{u'^2}/U_m^2$ and $\overline{u'v'}/U_m^2$ as shown in Figure 2 (b) and (c) are also in agreement with the experimental data. The mean flow data in the figure are normalized by the velocity scale $U_{\mathrm{sc}} = \sqrt{m_o/(\rho x)}$ and the concentration scale $C_{\mathrm{sc}} = \Gamma_o\sqrt{\rho/(m_o x)}$. where Γ_o is mass flux of the dye at the source.

The results of the LES were very sensitive to the Smagorinsky parameter, C_s, selected for the simulations. In order to produce the results that are in agreement with the experimental observation, the value of C_s must be precisely selected because small variation in the sub-grid viscosity can lead to huge difference in the results. The optimal value $C_s^2 = 0.0056$ was selected for the simulation using CDS but a considerably (2.6 times) higher value $C_s^2 = 0.0144$ has to be selected if the simulation results by LBM are to agree with the experimental data. Column 3 of Figure 3 shows how the turbulent jet would be like if the incorrect value of $C_s^2 = 0.0056$ were selected for the simulation by LBM. The jet would be correctly simulated by the LBM as shown in column 4 of Figure 3 only if the level of the sub-grid viscosity is increased 2.6 times, that is when $C_s^2 = 0.00144$ is selected for the simulation. The different required levels of sub-grid viscosity suggested that the LBM is numerically less diffusive compared with the CDS. The numerical diffusion in the CDS apparently acts as the equivalent of the sub-grid viscosity in the LES of the turbulent flow.

4 Conclusion

A semi-Lagrangian computational method was employed for the LES of the turbulent flow in a plane jet. The method is positive, unconditionally stable and mass conserving within the defined region of convergence. The positive experience obtained from the LES in this paper should encourage acceptance of the method in the field of computational fluid dynamic.

References

1. W. Altai, J. B. Zhang, and V. H. Chu: 'Shallow turbulent flow simulation using two-length-scale model', *J. Hydraulic Engineering*, **125**, 780-788 (1999).
2. V. H. Chu, and W. Altai: 'Generalized second-moment method for advection and diffusion processes,' *Proc. 2nd Intl Symp. on Envir. Hydraulics*, Hong Kong, Balkema Publisher, 357-362 (December, 1998).
3. A. Einstein: *Ann. d. Physik*, **17**, 549 (1905). This and other articles of Einstein is collected in a book entitled *Investigations on the Theory of the Brownian Movement*, edited by R. Füth, translated by A. D. Cowper (Dutton, New York).

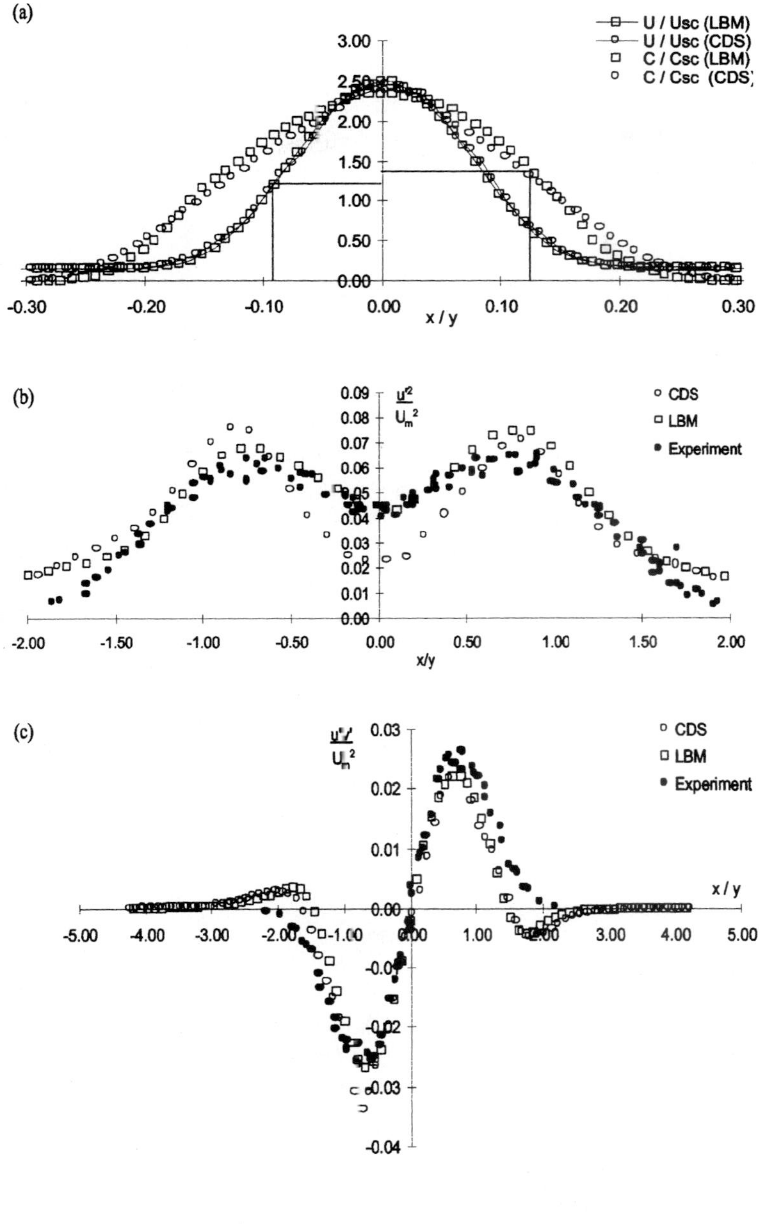

Fig. 2. (a) Mean velocity and concentration profiles obtained using the LBM with $C_s^2 = 0.0144$ and the CDS with $C_s^2 = 0.0056$. (b) $\overline{u'^2}/U_m^2$ profiles by LBM and CDS compared with experimental data. (c) $\overline{u'v'}/U_m^2$ profiles by LBM and CDS compared with experimental data.

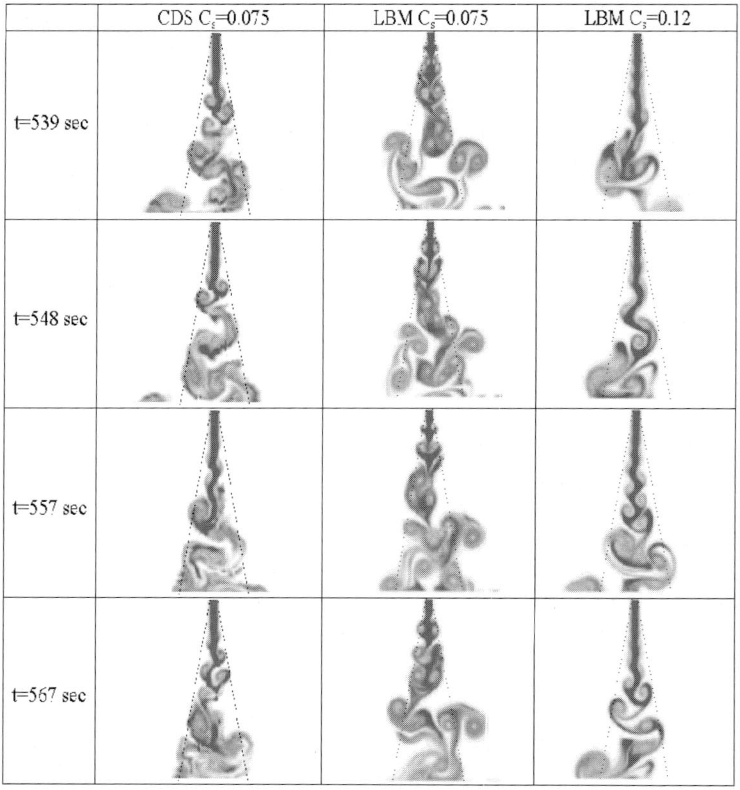

Fig. 3. Image sequence of the turbulent jet obtained by the LES.

4. J. H. Ferziger: 'Simulation of complex turbulent flows: recent advances and prospects in wind engineering,' *J. Wind Eng. and Ind. Aerodynamics,* **46,** . 195-212 (1993).
5. T. Kawamura, H. Takami and K. Kuwahara: 'A new higher order upwind scheme for incompressible Navier-Stokes equation', *Fluid Dyn. Res.,* **1,** 145162 (1985).
6. K. Nadaoka and H. Yagi: 'Shallow water turbulence modeling and horizontal large-eddy computation of river flow,' *J. Hydraulic Engineering,* **124,** 495-500 (1998).
7. W. Rodi, J. H. Ferziger, M. Breuer, and M. Pourquie: 'Status of Large Eddy Simulation: results of a workshop,' *J. Fluid Engineering,* ASME, **119,** 248-261 (1997).
8. G. I. Taylor: 'Diffusion by continuous movements', *Proc. Lond. Math. Soc.,* **20,** 2, 196-212 (1921).

Proper Orthogonal Decomposition of Turbulent Channel Flow

Giancarlo Alfonsi[1], Leonardo Primavera[2], Giuseppe Passoni[3], and Carlo Restano[3]

[1] Dipartimento di Difesa del Suolo, Università della Calabria
87040 Montalto Uffugo (Cosenza), Italy
[2] Dipartimento di Fisica, Università della Calabria
87036 Rende (Cosenza), Italy
[3] Dipartimento di Ingegneria Idraulica, Politecnico di Milano
Piazza L. da Vinci 32, 20133 Milano, Italy

Abstract. The technique of the Proper Orthogonal Decomposition (POD) is applied to the case of the turbulent channel flow at $Re = 40.000$ (based on channel half width). The turbulent flow database is obtained with the use of a finite volume computational code for the numerical integration of the Navier-Stokes equations; for turbulence modeling, the *LES* approach is followed and a *RNG* Sub Grid-Scale model is used. The three-dimensional, time-dependent velocity field in the channel is computed in more than 236.000 grid points and 100 non-dimensional time steps of the turbulent statistically steady state are considered. A reduced velocity field is reconstructed after decomposition by using three eigenfunctions containing up to the 96% of the turbulent kinetic energy of the originary field; results are presented in terms of vorticity, showing appropriate compositions of the most energetic modes of the flow in determinig the temporal dynamics of the field.

1 Introduction

In studying the fundamental physics of wall bounded turbulence, the case of the turbulent channel flow — the flow between two parallel plates driven by a uniform pressure gradient — has been often considered, both experimentally and numerically.

In the computational field, a possible approach to the numerical simulation of turbulence is the Direct Numerical Simulation (*DNS*); within this approach, the attitude of directly simulating all turbulent scales is followed by considering the Navier-Stokes equations with no modifications of any kind. The critical aspect of this method is the accuracy of the calculations, that in theory has to be as high as to resolve the Kolmogorov microscales in both space and time; in the framework of the *DNS* approach, results for the plane channel are reported, among others, in [1],[2],[3],[4],[5]. Due to the accuracy requirements and, as a consequence, the remarkable computational resources that are involved in *DNS* simulations, all *DNS* results are limited to relatively low values of the Reynolds number.

The other possible approach to turbulence calculation — through modeling — is the Large Eddy Simulation (*LES*); within this approach, the attitude of directly simulating only the larger scales of the flow and using a model for the smallest, is followed. A filter is applied to the Navier-Stokes equations for scale separation and a model is sought (the subgrid-scale model) for the term of the momentum equations that results not being a function of the filtered variables, the subgrid stress term. Several models have been devised, among others [6],[7],[8],[9],[10],[11]. In the framework of the *LES* approach, results for the plane channel are reported, among others, in [12]. Whether *DNS* or *LES*, modern techniques for the numerical integration of the Navier-Stokes equations have the potential of greatly increasing the amount of information gathered during a research of numerical nature; the continuous effort in studying turbulence in its full — three-dimensional and unsteady — complexity, has brought to manage very large amounts of data, the databases. A typical turbulent flow database includes all three components of the velocity in all points of a three-dimensional domain, gathered for an adequate number of time steps of the turbulent statistically steady state. Such a database contains many information about the physics of the turbulent flow, but in the formation of the value of each variable, all turbulent scales have contributed and the effect of each scale is nonlinearly combined with all other scales. It is also recognized that not all turbulent scales contribute to the same degree in determining the physical properties of a turbulent flow; methods have to be devised to extract the relevant information from a turbulent flow database, which has to be meant as to separate the effect of appropriately defined modes of the flow from the background flow, i.e. to extract the coherent structures of the flow.

There are several definitions of coherent structure ([13],[14]), of both qualitative and quantitative nature; in this work, the Proper Orthogonal Decomposition (*POD*) technique is used for the extraction of coherent turbulent motions in the channel flow case.

2 Proper Orthogonal Decomposition

The Proper Orthogonal Decomposition is an analytically founded statistical technique that can be applied for the extraction of coherent structures from a turbulent flow field; it allows the selection of a basis for a modal decomposition of an ensemble of signals by using the spectral theory of compact self-adjoint operators. The *POD*, also known as Karhunen-Loéve decomposition, was first introduced in turbulent flow analysis by Lumley [15]; the method is estensively presented in [16] and [17].

By considering an ensemble of temporal realizations of a non-homogeneous, square-integrable (finite kinetic energy), three-dimensional, real-valued velocity field $v_i(x_j, t)$ on a finite domain $(i, j = 1, 2, 3)$, one wants to find which is the most similar function to the elements of the ensemble on average, i.e. determine the best — the highest mean-square — correlated structure with all the elements of the ensemble. This corresponds to find a deterministic vector function $\phi_i(x_j)$ which maximizes the normalized inner product of the candidate structure with the field. A necessary condition for this problem is that $\phi_i(x_j)$ is an eigenfunction, solution of the eigenvalue problem and Fredholm integral equation ($i, j, l, k = 1, 2, 3$):

$$\int_D R_{ij}(x_l, x'_k)\ \phi_j(x'_k)\ dx'_k = \lambda\ \phi_i(x_l) \tag{1}$$

where R_{ij} is the two-point velocity correlation tensor; the maximum $\phi_i(x_l)$ corresponds to the largest eigenvalue λ of (1). When D is bounded, there exists a denumerable infinity of solutions of (1) (Hilbert-Schmidt theory), and these solutions are called empirical eigenfunctions $\phi_i^{(n)}(x_l)$ (normalized $||\phi_i^{(n)}|| = 1$); orthogonality (orthonormality in this case) implies, among others things, that structures of different order do not interact with each other in their contribution to the second order statistics. To each eigenfunction is associated a real positive eigenvalue $\lambda^{(n)}$ (R_{ij} is non-negative by construction) and the eigenfunctions form a complete set; every member of the ensemble can be reconstructed by means of a modal decomposition in the eigenfunctions themselves:

$$v_i(x_l, t) = \sum_n a_n(t)\ \phi_i^{(n)}(x_l) \tag{2}$$

which can be seen as a decomposition of the originary random field into deterministic structures $\phi_i^{(n)}(x_l)$ with random coefficients. The modal amplitudes, the time series of the coefficients $a_n(t)$ when related to velocity fields varying with time, are uncorrelated and their mean square values are the eigenvalues themselves. A diagonal decomposition of R_{ij} holds:

$$R_{ij}(x_l, x'_k) = \sum_n \lambda^{(n)}\ \phi_i^{(n)}(x_l)\phi_j^{(n)}(x'_k)$$

which in particular implies that the contribution of each different structure to the turbulent kinetic energy of the flow can be separately calculated; each eigenvalue represents the contribution of each correspondent structure to the total amount of kinetic energy in the domain. The *POD* technique is optimal for modeling or reconstructing a signal $v_i(x_l, t)$ in the sense that for a given number of modes, the projection on the subspace used for modeling will contain the most kinetic energy possible (on average) or, equivalently, the decay of the tail of the empirical eigenspectrum will be always faster (or at most as fast) than the tail of the spectrum based on any other possible basis, Fourier spectrum included.

The Proper Orthogonal Decomposition has been used in the case of the turbulent channel flow by some authors ([18],[19],[20]); more recently [21] reports a *POD* analysis of turbulent channel flow in the "minimal channel" computational domain [22] related to a *DNS* database at low Reynolds number. It is shown that the most energetic modes of the flow are streamwise rollers followed by outward tilted quasi-streamwise vortices, very similar to structures already observed in different scientific contests.

3 Results

The technique of the Proper Orthogonal Decomposition is applied to the case of the fully developed turbulent channel flow at Reynolds number *Re* = 40.000

(based on channel half width). The database of the three-dimensional time dependent flow field has been first produced; a computational code based on a finite volume algorithm has been used and, for turbulence modeling, the *LES* approach (Large Eddy Simulation) has been followed; a *RNG* based Sub Grid-Scale model [10] has been utilized. Overall, the three-dimensional time dependent (non-dimensional) velocity field in about 236.000 grid points (101x71x33 along x, y, z respectively) has been calculated and 100 non-dimensional time steps of the turbulent statistically steady state have been considered. Of the entire velocity field, mainly because of computational resources, a subdomain of 21x11x11 grid points (x, y, z directions respectively) has been selected in the central part of the channel.

The *POD* has been applied to the fluctuating portion of the velocity in the subdomain; problem (1) has been solved by means of a computational code developed by some of the authors and the three-dimensional empirical eigenfunctions with correspondent eigenvalues have been determined. One of the character of the computational code is that it computes the two-point velocity correlation tensor R_{ij} of (1) for all three components of the velocity in all three directions x, y, z; thus, the code can solve problem (1) for computational domains without particular symmetries. One of the remarkable aspects of the decomposition is that the first three eigenfunctions ϕ_1, ϕ_2, ϕ_3 (of a total of 7623) account for more than the 96% of the total turbulent kinetic energy of the originary field; a reduced velocity field, based on these first three eigenfunctions has been reconstructed with the use of the time dependent random coefficients (2).

The fluctuating energy $E(t)$ of the flow for the 100 non-dimensional time steps, related to the reduced velocity field, has been calculated; $E(t)$ exhibits

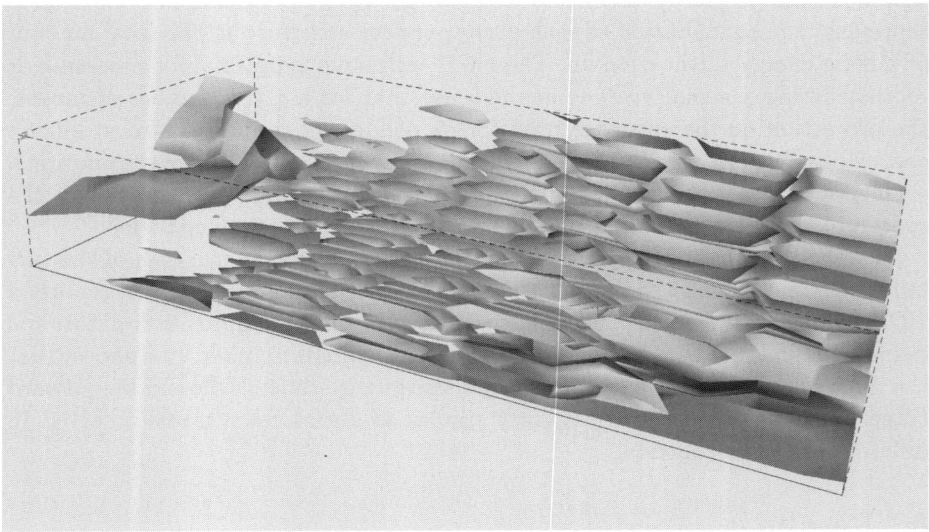

Fig. 1. Vorticity field in the subdomain at t=6

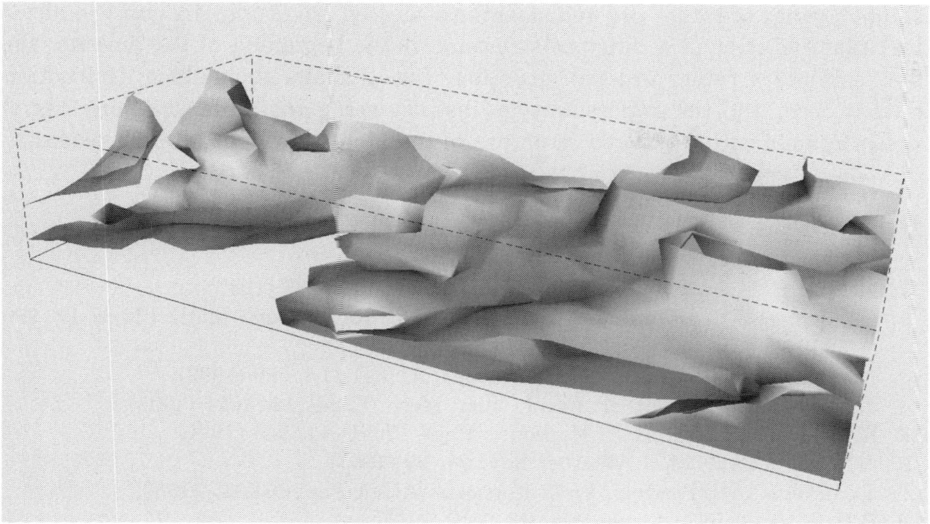

Fig. 2. Vorticity field in the subdomain at t=7

several sharp peaks at times $t = 7$, $t = 30$, $t = 68$, $t = 82$, besides a number of smaller spikes. The events occurring at these times indicate a sharp rise and subsequent fall of turbulent activity. The first event at $t = 7$ is examined in greater detail; in Figs. 1, 2 and 3 the vorticity field is shown at times $t = 6$ (immediately before), $t = 7$ (the event) and $t = 8$ (immediately after). Fig. 1

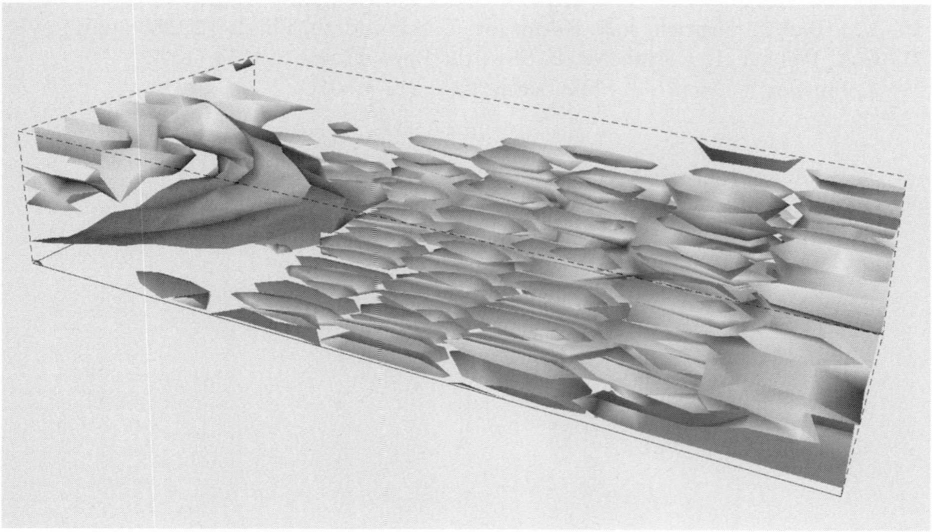

Fig. 3. Vorticity field in the subdomain at t=8

shows a series of rather ordered structures aligned longitudinally and the incipient manifestation of a different structure at the beginning of the domain; the flow exhibits a rather ordered structure. Fig. 2 shows the sudden destruction of that order and the impossibility of distinguishing any more structure. Fig. 3 shows again a rather ordered structure of the flow field, right after the event.

References

1. J. Kim, P. Moin, R.D. Moser : J. Fluid Mech. **177**, 133 (1987)
2. S.L. Lyons, T.J. Hanratty, J.B. Mc Laughlin : Int. J. Num. Meth. Fluids **13**, 999 (1991)
3. N. Kasagi, Y. Tomita, A. Kuroda: Trans. ASME **114**, 598 (1992)
4. J. Rutledge, C.A. Sleicher: Int. J. Num. Meth. Fluids **16**, 1051 (1993)
5. R.D. Moser, J. Kim, N.N. Mansour : Phys. Fluids **11**, 943 (1999)
6. J. Smagorinsky : Mon. Weather Rev. **91**, 99 (1963)
7. J. Bardina, J.H. Ferziger, W.C. Reynolds: AIAA Pap. 80-1357 (1980)
8. R.H. Kraichnan: J. Atmos. Sci. **33**, 1521 (1976)
9. O. Métais, M. Lesieur : J. Atmos. Sci. **43**, 857 (1986)
10. A. Yakhot, S.A. Orszag, V. Yakhot, M. Israeli : J. Sci. Comput. **4**, 139 (1989)
11. M. Germano : J. Fluid Mech. **238**, 325 (1992)
12. P. Moin, J. Kim : J. Fluid Mech. **118**, 341 (1982)
13. B.J. Cantwell : Ann. Rev. Fluid Mech. **13**, 457 (1981)
14. R.K. Robinson: Ann. Rev. Fluid Mech. **23**, 601 (1991)
15. J.L. Lumley: *Stochastic tools in turbulence* (Academic Press 1971)
16. L. Sirovich: Quart. Appl. Math. **45**, 561 (1987)
17. G. Berkooz, P. Holmes, J.L. Lumley: Ann. Rev. Fluid Mech. **25**, 539 (1993)
18. P. Moin, R.D. Moser: J. Fluid Mech. **200**, 471 (1989)
19. L. Sirovich, K.S. Ball, L.R. Keefe: Phys. Fluids A **2**, 2217 (1990)
20. K.S. Ball, L. Sirovich, L.R. Keefe: Int. J. Num. Meth. Fluids **12**, 585 (1991)
21. G.A. Webber, R.A. Handler, L. Sirovich: Phys. Fluids **9**, 1054 (1997)
22. J. Jiménez, P. Moin : J. Fluid Mech. **225**, 213 (1991)

A Finite Volume Formulation for Compact Schemes on Arbitrary Meshes with Applications to LES

Chris Lacor[1], Sergey Smirnov[1], and Martine Baelmans[2]

[1] Vrije Universiteit Brussel, Dept. Fluid Mechanics, Pleinlaan 2, 1050 Brussel, Belgium
[2] KULeuven, Department of Mechanical Engineering, Division of Applied Mechanics and Energy Conversion, Celestijnenlaan, 300 A, B-3001 Heverlee, Belgium

1 Introduction

In recent years there has been a great interest in compact (or Padé type) schemes [1], [2], [3]. On Cartesian uniform meshes these schemes offer a high order of accuracy with only a compact stencil. More importantly, the resolution of the shorter length scales of the solution of these schemes is better than for classical finite-difference methods [4], which makes these schemes especially attractive for applications such as DNS, LES and computational aeroacoustics [5], [6].

Originally, compact schemes were defined in the finite difference context [1]. Extension to curvilinear, non-uniform grids is then straightforward, e.g. [5], [7], but conservation may be lost.

A finite-volume formulation—inherently conservative—was proposed in [8] and further generalized in [9] for linear advection problems on Cartesian uniform meshes. However, no calculations on curvilinear meshes were considered. For applications on general meshes a compact 2nd order finite-volume scheme was proposed in [10], although this scheme was not of Padé type (i.e. this formulation did not contain an implicit operator).

In the current paper, a compact finite-volume approach for applications on general meshes is proposed. To test the obtained schemes a series of numerical experiments has been carried out for a 2D linear advection problem on different meshes.

2 The 1D linear case

Let us consider

$$\frac{\partial u}{\partial t} + \frac{\partial u}{\partial x} = 0 \tag{1}$$

In the finite-volume (FV) context equation (1) is discretized as

$$\int_{x_{i-1/2}}^{x_{i+1/2}} \frac{\partial u}{\partial t} dx = u_{i+1/2} - u_{i-1/2} \tag{2}$$

In the classical FV schemes the fluxes $u_{i+1/2}$ and $u_{i-1/2}$ are obtained by MUSCL extrapolation. A compact type scheme is obtained if, instead, one uses an implicit interpolation formula:

$$\alpha u_{i-1/2} + u_{i+1/2} + \alpha u_{i+3/2} = a\frac{u_{i+1} + u_i}{2} + b\frac{u_{i+2} + u_{i-1}}{2} \tag{3}$$

One can show that by tuning the coefficients α, a, b this formula can be made 6th order or, on a smaller stencil ($b = 0$), 4th order accurate. The values of the coefficients depend whether point-wise values (defined, for example, in the center of the cell) or cell-averaged ($u_i = \frac{1}{\Delta x}\int_{x_{i-1/2}}^{x_{i+1/2}} u(x)dx$) values are used in the rhs of (3). E.g. for the point-wise formulation the 4th order accurate interpolation formula will be given by $\alpha = 1/6,\ a = 4/3,\ b = 0$ [1], and if the cell-averaged values are used the coefficients will be $\alpha = 1/4,\ a = 3/2,\ b = 0$ [9].

In case of unsteady calculations, the choice between point-wise and cell-averaged values also influences the evaluation of the integral containing the time derivative in (2). For the cell-averaged formulation the calculation of the integral is straightforward and exact:

$$\int_{x_{i-1/2}}^{x_{i+1/2}} \frac{\partial u}{\partial t} dx = \frac{\partial u_i}{\partial t}\Delta x \tag{4}$$

while for the point-wise approach a more complex quadrature formula should be used to maintain a high accuracy in unsteady calculations.

The implicit interpolation approach is easily extendable to multi dimensions, but a high accuracy is only maintained on Cartesian uniform meshes unless the coefficients of interpolation formulae are tuned to account for the metrics of the grid.

3 The multidimensional, nonlinear case

Consider the 2D nonlinear equation (for clarity the 2D case is discussed; extension to 3D is straightforward).

$$\frac{\partial u}{\partial t} + \frac{\partial f(u)}{\partial x} + \frac{\partial g(u)}{\partial y} = 0 \tag{5}$$

By integrating this equation over the cell i, j (Fig. 1) one obtains the following exact integral relation

$$\int\int_{S_{i,j}} \frac{\partial u}{\partial t} dx dy + \oint_{ABB'A'A} (-f dx + g dy) = 0 \tag{6}$$

where $S_{i,j}$ is the area of the cell i, j.

Following the strategy of constructing the Padé type schemes, the idea is to use an implicit formula for evaluating the flux integrals over the cell faces.

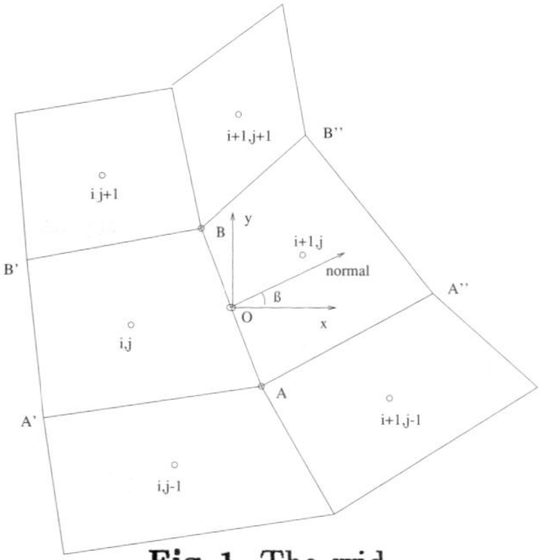

Fig. 1. The grid.

Consider, for instance, $\int_A^B f dy \equiv I_{i+1/2}$. The following expression for $I_{i+1/2}$ can be used:

$$\alpha I_{i-1/2} + I_{i+1/2} + \alpha I_{i+3/2} = a \frac{f_{i+1,j} + f_{i,j}}{2} (y_B - y_A) \tag{7}$$

where $I_{i-1/2} = \int_{A'}^{B'} f dy$, $I_{i+3/2} = \int_{A''}^{B''} f dy$, and $f_{i+1,j}$ and $f_{i,j}$ are the values of the flux f in the cells $i+1, j$ and i, j respectively. If the coefficients α and a are taken as specified in the previous section this approximation is 4th order accurate as long as the grid is Cartesian and uniform [9].

For calculations on arbitrary meshes a more general scheme, accounting for the metrics, is considered:

$$\beta I_{i-1/2} + I_{i+1/2} + \gamma I_{i+3/2} = \sum_{n=0}^{1} \sum_{m=-1}^{1} a_{n,m} f_{i+n,j+m} (y_B - y_A) \tag{8}$$

In the rhs of (8), the values of the flux in all six cells present in Fig 1 are used. The scheme is compact in the sense that the stencil for the calculation of the residual contains only 9 cells. Coefficients $a_{n,m}$, β, γ can be chosen so that (8) has the highest accuracy possible. To fit all the coefficients of the Taylor series of both rhs and lhs of (8) up to order 3 one has to satisfy 6 linear algebraic relations. Since the number of coefficients $a_{n,m}$, β, γ exceeds the number of equations (8 vs 6), the leading truncation error of order 3 can be reduced by means of the least square approach. This method allows to obtain a 3rd order accurate formula for the calculation of the flux integrals. On Cartesian uniform meshes this formula reduces to (7). Numerical tests, however, have shown that on very distorted grids this scheme has a strong oscillatory behaviour making the calculations unstable.

To overcome this problem, a scheme with formal accuracy of one order less was constructed. β and γ were set to $\frac{1}{4}$, their values in the formulation for a uniform Cartesian mesh, and the remaining 6 coefficients $a_{n,m}$ were chosen by imposing 2nd order accuracy of (8) and by reducing the leading truncation error by the least square approach.

4 Numerical tests

To test the present formulation of compact schemes, a series of numerical experiments has been carried out for a 2D linear advection problem in a domain $-0.5 < x < 0.5$, $-0.5 < y < 0.5$. A 2D Gaussian wave was used as an initial solution, and periodic boundary conditions were imposed in both x and y directions. For each test case the wave was propagated during 5 seconds with a speed given by $c_x = 0.8$, $c_y = 0.6$.

In all simulations a six stage low storage RK scheme was used to advance in time. The RK coefficients were taken as $1/6, 1/5, 1/4, 1/3, 1/2, 1$. This scheme is 6th order accurate in time for linear problems [11]. Since only the spatial accuracy was to be examined, the time step was taken very small to make sure that the errors occuring due to time integration are negligible as compared to those arising from spatial discretization.

In Fig 2 the results obtained on the Cartesian uniform mesh (80×80 cells) for the classical second order and 4th order compact finite-volume schemes are presented. Results with the compact scheme are in excellent agreement with the exact solution (circular isolines), while the second order scheme shows important errors both in terms of location of the peak of the solution and symmetry.

To show the importance of accounting for the non-uniformity of the mesh, a 1D test case was calculated on a non-uniform grid consisting of alternating bigger and smaller cells with the size ratio 11/9. In Fig 3 the results obtained with formulation (8) are compared with those of the compact scheme with constant coefficients (7). It is seen that, in spite of the relatively small non-uniformity, the latter scheme performs poorly.

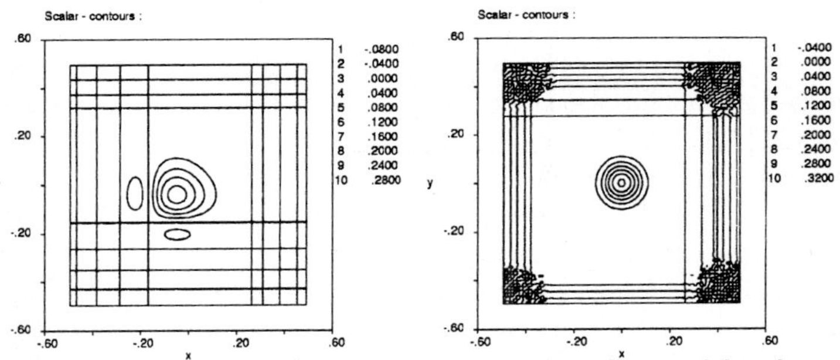

Fig. 2. Isolines of the solution (left picture: 2nd order scheme, right picture: 4th order scheme).

In Fig 4 the isolines of the solution on a curvilinear mesh (80×80 cells) (Fig 5) for both scheme (7) and scheme (8) are shown. The improvement in accuracy for the latter method, although less pronounced than on the clustered grid in the previous test case, is still visible from the isoline plot. An analysis of order of accuracy and absolute error on a sequence of meshes with different levels of resolution is provided in Fig 6. The measured order of accuracy for both schemes is 4. This can be explained by the smoothness of the grid, which, being refined,

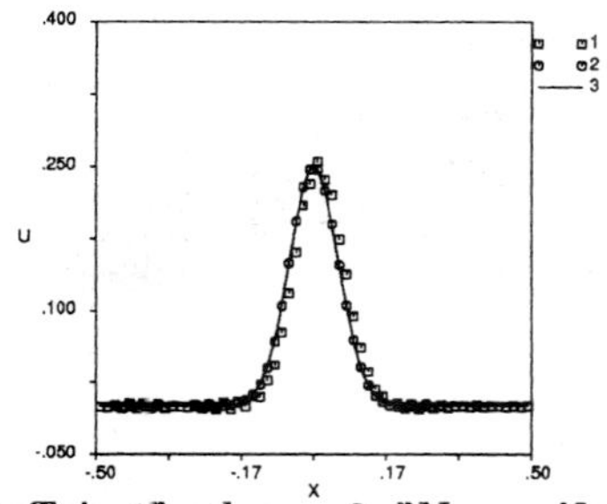

Fig. 3. 1: "Constant coefficient" scheme, 2: "Non-uniform" scheme, 3: exact solution.

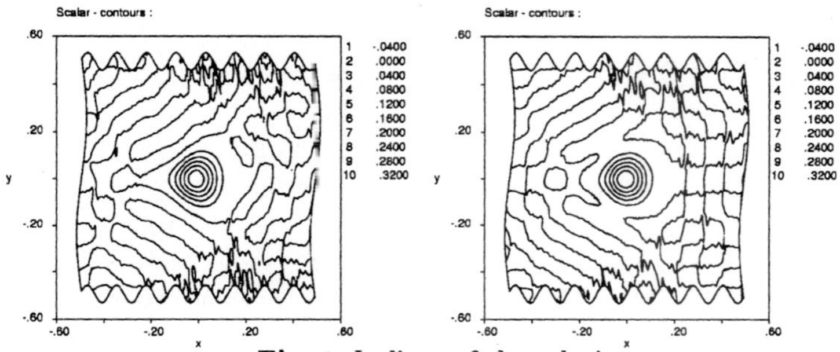

Fig. 4. Isolines of the solution.

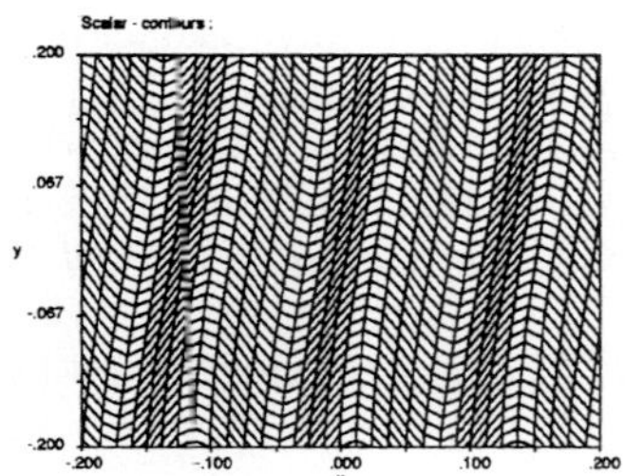

Fig. 5. Zoom of the mesh used in calculations

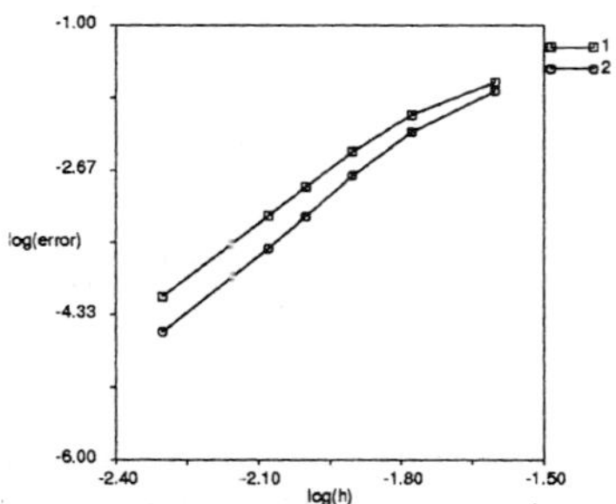

Fig. 6. Order of accuracy on the curvilinear mesh (1: "constant coefficient" scheme, 2: "non-uniform" scheme).

locally becomes straight and uniform, although non-orthonogal. Nevertheless, the absolute error is smaller (by approximately a factor of 2) for the scheme (8).

5 Conclusions

A Padé type FV scheme in multidimensions was proposed. The scheme is second order accurate in space on arbitrary meshes. The scheme was compared with compact schemes not accounting for the grid metrics on 1D and 2D convection problems. The tests show a significant improvement in the quality of the obtained solution with the new scheme.

6 Acknowledgement

This research work was partially funded by the Flemish Science Foundation (FWO). This support is gratefully acknowldged.

References

1. Lele Sanjiva K., "Compact Finite Difference Schemes with Spectral-like Resolution", *Journal of Computational Physics 103*, 16-42 (1992)
2. Kim J. W., Lee D. J., "Optimized Compact Finite Difference Schemes with Maximum Resolution", AIAA Journal, vol 34, No. 5, May 1996
3. Zigo Haras, Shlomo Ta'asan, "Finite difference Schemes for Long-Time Integration", Journal of Computational Physics 114, 265-279 (1994)
4. Tang L., Baeder J. D., "Uniformly Accurate Compact Difference Schemes", AIAA CFD Conference, June 29, 1997, Snowmass, CO
5. Gamet L., Ducros F., Nicoud F., Poinsot T., "Compact Finite Difference Schemes On Non-uniform Meshes. Application To Direct Numerical Simulations of Compressible Flows", *International Journal For Numerical Methods in Fluids*, 29, 159-191, (1999)
6. Povitsky A. and Morris P.J., "Parallel Compact Multi-Dimensional Numerical Algorithm With Application to Aeroacoustics", AIAA paper 99-3272, 1999, 14th AIAA CFD Meeting.
7. Visbal Miguel R., Gaitonde Datta V., "Higher-Order-Accurate Methods for Complex Unsteady Subsonic Flows", AIAA paper 98-0131, 36th Aerospace Science Meeting, Reno, NV, 1998
8. Gaitonde D., Shang J. S., "Optimized Compact-Difference-Based Finite-Volume Schemes for Linear Wave Phenomena", *Journal of Computational Physics 138*, 617-643 (1997)
9. Kobayashi Marcelo H., "On a Class of Padé Finite Volume Methods", Journal of Computational Physics 156, 137-180 (1999)
10. Ali Rezgui, "Schémas Numériques Compacts et non Compacts D'order Trois pour le Calcul D'écoulements Compressibles", Ph.D. Thesis, ENSAM Paris, 1997
11. Hu F. Q., Hussani M. Y., Manthey J. L., "Low-Dissipation and Low-Dispertion Runge-Kutta Schemes for Computational Acoustics", Journal of Computational Physics 124, 177-191 (1996)

A 3D Turbulent Boundary Layer Test for LES Models

Antonella Abbà[1], Carlo Cercignani[1], Gianluca Picarella[1], and Lorenzo Valdettaro[1]

Dipartimento di Matematica, Politecnico di Milano,
Piazza L. da Vinci 32, 20133 Milano, Italy

Abstract. In order to test turbulence models for Large Eddy Simulation (LES), a three-dimensional turbulent boundary layer flow [1] has been considered. First of all a LES simulation was performed over a grid with 66^3 points at a Reynolds number $Re = 767$ based on the freestream velocity. Second order central difference were used for approximation of spatial derivatives, and in time the equations are solved with a projection method combined with a 3 step explicit Runge–Kutta scheme. Thereafter an *a priori* test is made using the velocity field obtained by a Direct Numerical Simulation (DNS) of the 3D turbulent boundary layer with a mixed spectral–finite difference discretization [2].

1 Introduction

The LES technique introduces a spatial filter that removes unresolvable small scales of turbulence, usually called subgrid–scales (SGS), from the numerical solution of the Navier–Stokes equations, while for the unresolved scales, models are assumed. For this reason the LES is much cheaper than DNS. Moreover LES can be applied to high Reynolds 3D complex flows, where DNS is actually not achievable.

In LES the modelled subgrid scales are not, generally, isotropic, so the models based on the traditional turbulent viscosity concept are not adequate to represent the small turbulent scales in flows of practical interest, which are usually strongly inhomogeneous and anisotropic. Recently many SGS models have been developed to take these characteristics of unresolved scales into account.

In order to verify the quality of some of these models we have performed both a LES on a complex flow, and an *a priori* test.

The models are tested in an equilibrium three-dimensional turbulent boundary layer [1], which is a strongly anisotropic flow.

2 The Large Eddy Simulation

First of all a large-eddy simulation of an equilibrium three-dimensional turbulent boundary layer over an infinite plate was performed. The freestream velocity vector has magnitude U_0 and rotates with angular frequency f in planes parallel to the wall (Fig. 1) [1]. After a transient phase, the flow becomes statistically homogeneous in planes parallel to the wall. The mean flow is three-dimensional and the turbulence statistics, properly adimensionalized, in an appropriate reference frame rotating with the angular frequency f, depend only on the Reynolds number and on the distance from the wall. The flow is statistically homogeneous in parallel to the wall directions, in which peridical boundary conditions are imposed. Althoug its simplicity, this boundary layer flow is strongly anisotropic, making the present problem a very convenient test case for turbulence models.

Two different simulations have been performed, using the original dynamic model [3] [4] and the anisotropic one [3] [4]. The simulations has been performed at the Reynolds number $Re_{\delta_l} = \frac{U_0 \delta_l}{\nu} = 767$ over a grid with 66^3 collocation points. The equivalent Reynolds number based on the friction velocity is $Re_\tau = 914$. The grid spacing in wall unit is $\Delta_x+ = \Delta_z+ = 22$ in parallel to the wall direction. The parallel to the wall external velocity components are $U_{ext} = U_0 \cos(ft)$ and $W_{ext} = U_0 \sin(ft)$ where the nondimensional frequency is $f = 2/Re_{\delta_l}$.

2.1 Results

In Fig. 2 the resulting mean velocity profile in the reference frame aligned with the Reynolds shear stress vector at the wall, is compared with the values obtained by the LES [7] and by a DNS [1]. While isotropic model gives very bad results, the velocity values obtained with the anisotropic one is in better agreement with the DNS than the more expensive lagrangian model.

The same evaluation can be made observing the Fig. 3 representing the w velocity fluctuations profile. The superiority of the anisotropic model is clear mainly close to the wall.

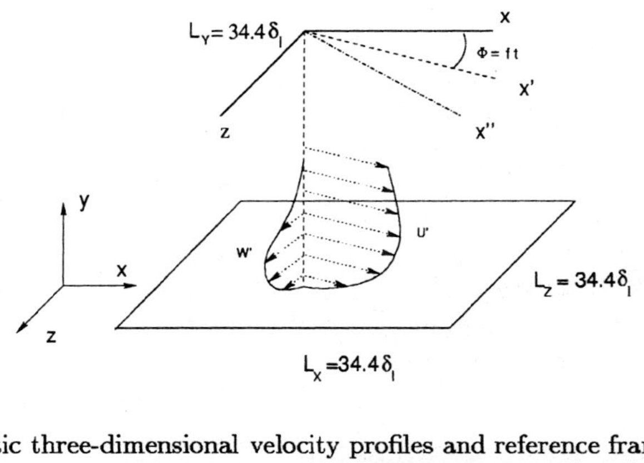

Fig. 1. Schematic three-dimensional velocity profiles and reference frames used for the present simulation: $x - z$ axes refer to the laboratory system; $x' - z'$ axes are aligned with the freestream velocity vector; $x'' - z''$ axes are aligned with Reynolds shear stress vector at wall [1]. L_x, L_y, L_z: dimensions of computational domain; $\delta_l = \sqrt{2\nu/f}$ is the viscous depth.

3 The *a priori* Test

In the *a priori* test, the quantities obtained filtering a field from a DNS are compared to those predicted by the models. The DNS was performed using a mixed finite difference–pseudospectral numerical code kindly supplied by [2].

4 The subgrid–scale Models

The *a priori* test has been performed for some of the recently developed models which take into account of the anisotropy of unresolved scales, making the SGS stress tensor not forced to be aligned with the strain rate tensor, as appens in traditional viscosity model. We will consider the original dynamic model [3] [4], the anisotropic model proposed by [5] [6], the dynamic mixed model by [8] and the anisotropic mixed model presented in the following.

4.1 The Anisotropic Mixed Model

We propose an anisotropic mixed model in which to the more dissipative term derived in the anisotropic model [5] [6], the modified Leonard tensor $L^m_{\mathrm{ij}} = \overline{\overline{u}_{\mathrm{i}}\overline{u}_{\mathrm{j}}} - \overline{\overline{u}}_{\mathrm{i}}\overline{\overline{u}}_{\mathrm{j}}$ [8] is added. So the anisotropic part of the SGS stresses τ_{ij} is modelled as

$$\tau_{\mathrm{ij}} - \frac{\delta_{\mathrm{ij}}}{3}\tau_{\mathrm{kk}} = -2C_{\alpha\beta}a_{\mathrm{i}\alpha}a_{\mathrm{j}\beta}a_{\mathrm{r}\alpha}a_{\mathrm{s}\beta}\overline{\Delta}^2|\overline{S}|\overline{S}_{\mathrm{rs}} + L^m_{\mathrm{ij}} \tag{1}$$

With the overbar are indicated the filtered quantities. $\overline{S}$ is the strain rate tensor obtained by the filtered velocity $\overline{\boldsymbol{u}}$, and $a_{\mathrm{i}\alpha}$ are the components of unit vectors

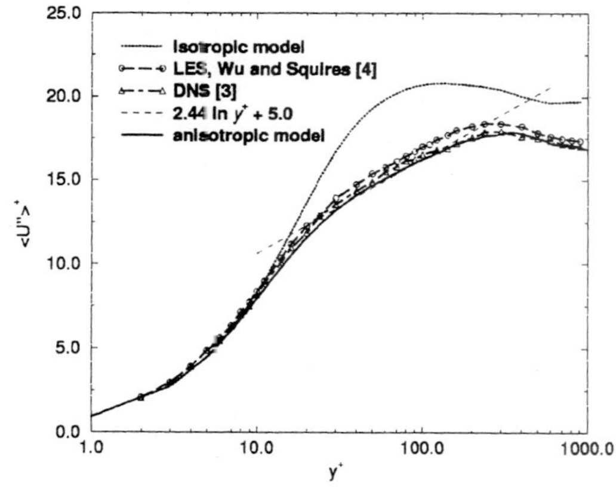

Fig. 2. Streamwise velocity component vs. distance from the wall. Superscript + denotes normalization by friction velocity.

$\boldsymbol{a}_\alpha(\alpha=1,2,3)$, to be chosen suitably. In this paper we have used

$$\boldsymbol{a}_1 = \frac{\boldsymbol{u}}{|\boldsymbol{u}|}, \quad \boldsymbol{a}_2 = \boldsymbol{a}_3 \times \boldsymbol{a}_1 \quad \boldsymbol{a}_3 = \frac{\nabla(|\boldsymbol{u}|^2) \times \boldsymbol{u}}{|\nabla(|\boldsymbol{u}|^2) \times \boldsymbol{u}|} \tag{2}$$

The components of the C matrix are determined with the dynamic procedure.

The formulation of this model derives from the observation of the results of the *a priori* test for the previous models. A combination of the advantages of both anisotropic and mixed models is espected.

4.2 Results

We present the analysis of models from the point of view of energetics. In Figg. 4, 5 the mean backward and forward scatter for the modelled stresses are compared with values computed on the filtered DNS field. The forward and backward scatter predicted by all the models are quite different from the exact ones, particularly for the latter. Among all, the anisotropic model and the anisotropic mixed one present the better behaviour.

The dynamic mixed model and the anisotropic mixed one, instead, give higher correlation coefficient between the modelled stresses and the exact ones (Fig. 6) than the other models.

5 Conclusion

LES and *a priori* test have been performed on a three–dimensional turbulent boundary layer flow. In the LES, the mean velocity profile and rms have demonstrate the good behaviour of the anisotropic model, especially near the wall. In

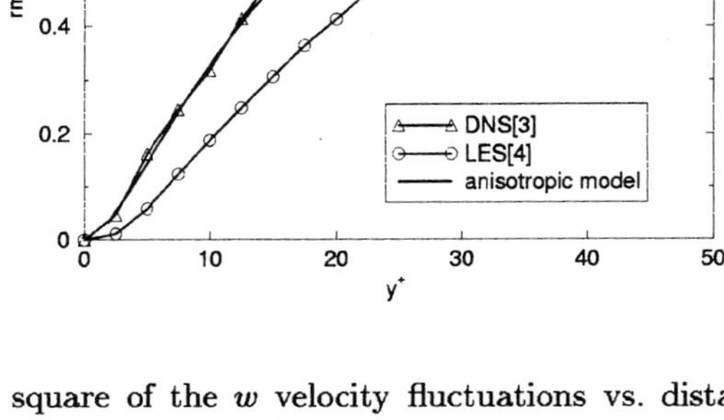

Fig. 3. Root mean square of the w velocity fluctuations vs. distance from the wall. Superscript + denotes normalization by friction velocity.

the *a priori* test the anisotropic model and its mixed version present a good capability, to reproduce the energetics of SGS, while better correlation coefficient are obtained for the dynamic mixed model and athe anisotropic mixed one.

In the future, it should be interesting to evaluate the correlation coefficient for the SGS stress divergence, which is really discretized in the Navier–Stokes equations. Moreover a LES using the anisotropic mixed model should be necessary to complete the analysis.

References

1. P.R. Spalart: J. Fluid Mech., 205 (1989)
2. G. Passoni, G. Alfonsi, G. Tula, U. Cardu: Parallel Computing, 25 (1999)
3. M. Germano, U. Piomelli, P. Moin, H. Cabot: Phys.Fluids A **3**, 7 (1991)
4. D.K. Lilly: Phys.Fluids A **4**, 3 (1992)
5. A. Abbà, C. Cercignani, L. Valdettaro: 'Anisotropy Effects in LES Simulation of Turbulent Flows'.In: *Annual GAMM Meeting 1997, Regensburg, Germany, March 1997.*
6. A. Abbà, C. Cercignani, L. Valdettaro: 'High Rayleigh Number Convection with LES Models'. In: *Computational Fluid Dynamics '98, ECCOMAS Conference, Athens, Greece, September 1998*, (Wiley, Chichester) pp. 280–285
7. X. Wu, K.D. Squires: AIAA Journal. **35**, 1 (1997)
8. M.V. Salvetti, S. Banerjee: Phys. Fluids **7**, 11 (1995)

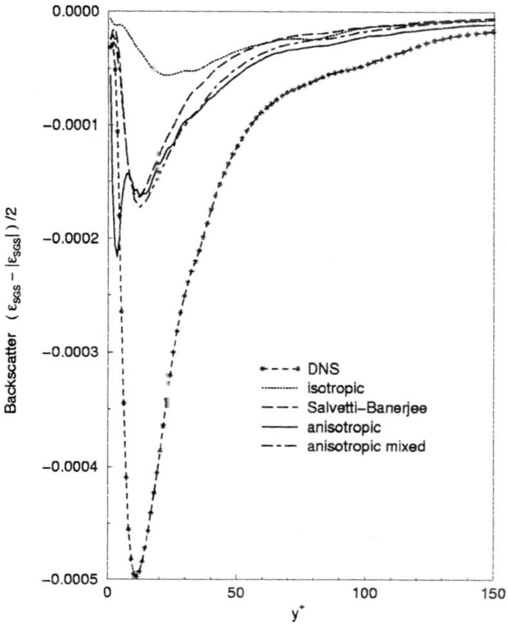

Fig. 4. Backward scatter vs. distance from the wall.

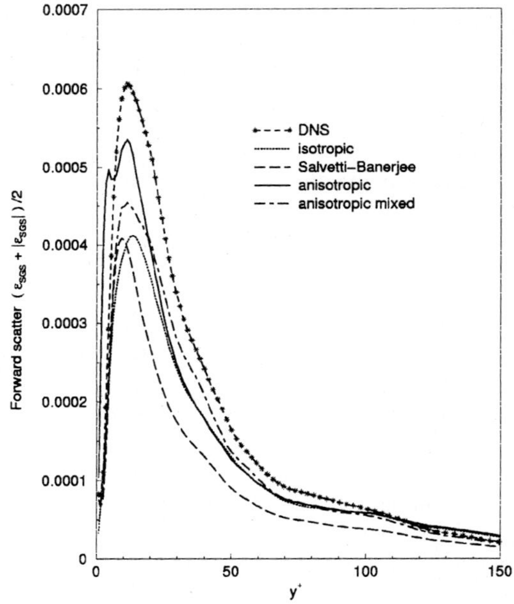

Fig. 5. Forward scatter vs. distance from the wall.

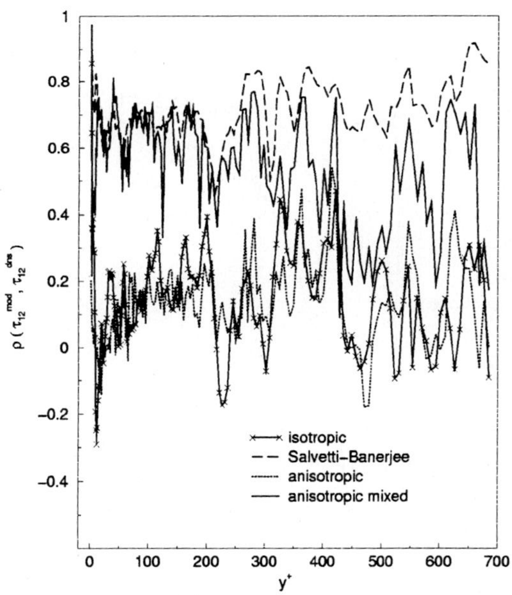

Fig. 6. Correlation coefficientvs. distance from the wall.

Part XIV

Lattice Boltzmann Method

Numerical Analysis of the Pressure Drop in Porous Media Flow using the Lattice Boltzmann Computational Technique

Jörg Bernsdorf[1], Gunther Brenner[2], Thomas Zeiser[2], Peter Lammers[2], and Franz Durst[2]

[1] C&C Research Laboratories, NEC Europe Ltd., Rathausallee 10, D – 53757 Sankt Augustin, Germany
[2] Institute of Fluid Mechanics (LSTM), University of Erlangen - Nuremberg, Cauerstr. 4, D – 91058 Erlangen, Germany

Abstract. In this paper, we describe detailed simulations of low mach number Newtonian flow through randomly generated porous media for a wide range of Reynolds numbers. The simulations were carried out using the lattice Boltzmann method, which has become an established tool for predicting flows in highly complex geometries. Several quantities were determined by an a posteriori analysis of the flow-fields: The pressure drop, the dissipation caused by shear and by elongation of the fluid, and the tortuosity. We were able to show that the pressure drop is mainly caused by the dissipation due to shear and elongation, and that the tortuosity is not related to the total dissipation, but to the relationship of dissipation and shear.

1 Introduction

For simulations of flows through highly complex geometries, the lattice Boltzmann method has several advantages when compared to Navier Stokes based methods. Due to its cellular automata based algorithm and the simple and efficient treatment of wall boundary conditions, simulations can be efficiently carried out on high performance vector-parallel computers with a very high discretization on simple equidistant orthogonal lattices. This simple lattice type allows one to implement arbitrary complex geometries by the *marker and cell* method, were geometry-covered cells of the lattice are marked as being occupied. The marker and cell geometry information can be automatically generated from CAD-data, three dimensional computer tomography or, as in the present paper, by computer programs.

In this paper, we used the lattice Boltzmann method for the prediction of flow through beds of randomly distributed cubes. The simulations were carried out for a wide range of Reynolds numbers, and the friction factor, which is a dimensionless quantity for the total dissipation, was determined from the pressure drop, the Reynolds number and the porosity of the porous media. A detailed a posteriori analysis of the velocity field gave the two parts adding up to the total dissipation of the fluid when passing through the porous media: the dissipation caused by sher and the dissipation caused by elongation. By integrating the path of the fluid inside the porous media, i.e., by measuring the length of the streamlines, the tortuosity, defined as the relationship of the length of the flow paths and the edge length of the porous media, was directly measured.

In the following section, we give a short introduction to the lattice Boltzmann method. The capillaric theory for predicting pressure losses in porous media flow will be reviewed in the third section and compared to other experimental and numerical results. In the fourth section we present the results of our simulations and address misleading assumptions within nthe capillaric theories.

2 The Lattice Boltzmann Method

The lattice Boltzmann (LBA) method treats the fluid on a statistical level, simulating the movement and interaction of ensemble-averaged particle density distributions by solving a velocity-discrete Boltzmann-type equation.

The lattice Boltzmann method prooved to be a very efficient tool for flow simulation in highly complex geometries (discretized by up to several million grid points) [4].

2.1 The Lattice – BGK – Approach

On every node $\boldsymbol{r}_*$, of an equidistant orthogonal lattice a set of i real numbers, the particle density distributions N_i, is stored. The updating of the lattice consists of basically two steps: a streaming process, where the particle densities are shifted in discrete time steps t_* through the lattice along the connection lines in direction $\boldsymbol{c}_i$ to their next neighboring nodes $\boldsymbol{r}_*+\boldsymbol{c}_i$, and a relaxation step, where locally a new particle distribution is computed by evaluating an equivalent to the Boltzmann collision integrals ($\triangle_i^{Boltz}$).

For every time step, all quantities appearing in the Navier-Stokes equations (velocity, density, pressure gradient and viscosity) can be locally computed in terms of moments of this density distribution and (for the viscosity) of the relaxation parameter ω.

For the present computations, a 3-D nineteen-speed (D3Q19) lattice Boltzmann automata with single time Bhatnager-Gross-Krook (BGK) relaxation collision operator $\triangle_i^{Boltz}$ proposed by Qian [5] is used

$$N_i(t_* + 1, \boldsymbol{r}_* + \boldsymbol{c}_i) = N_i(t_*, \boldsymbol{r}_*) + \triangle_i^{Boltz} \;, \tag{1}$$

$$\triangle_i^{Boltz} = \omega\left(N_i - N_i^{eq}\right) \;, \tag{2}$$

with a local equilibrium distribution function N_i^{eq}

$$N_i^{eq} = t_p\, \varrho \left\{ 1 + \frac{c_{i\alpha}\, u_\alpha}{c_s^2} + \frac{u_\alpha\, u_\beta}{2c_s^2} \left(\frac{c_{i\alpha}\, c_{i\beta}}{c_s^2} - \delta_{\alpha\beta} \right) \right\} \;. \tag{3}$$

This local equilibrium distribution function N_i^{eq} has to be computed every time step for every node from the components of the local flow velocity u_α, the fluid density ϱ, a lattice geometry weighting factor t_p and the speed of sound c_s. It is chosen to recover the incompressible time-dependent Navier-Stokes equations [5].

2.2 Introducing Complex Geometries

Single lattice nodes are either *occupied* by an elementary obstacle, or they are *free* (*marker and cell* approach) . Particle densities N_i, which are shifted to an occupied node owing to the streaming process, are simply bounced back to their original location during the next iteration, but with opposite velocity (indicated by the index $\bar{i}$). This results in the desired no-slip (zero velocity) wall boundary condition.

3 Pressure Drop in Porous Media

3.1 The Kozeny-Darcy Equation

The derivation of the Kozeny-Darcy equation is based on the assumption that porous media flow can be modeled as a flow through a bundle of channels without strongly changing cross-sections. For this geometry, the Navier Stokes equations can be integrated analytically. By introducing the hydraulic radius R_h as a fraction of fluid volume to wetted surface, this equation can be generalized for arbitrary geometries and rewritten in Ergun [3] coordinates:

$$f = \frac{\Lambda_{\text{th}}}{Re}. \tag{4}$$

with $\Lambda_{\text{th}} = 72$. The dimensionless quantities Reynolds number Re and friction factor f can, e.g., be defined for a sample of cubes with porosity ε and edge length l_c as:

$$f = \frac{\Delta P}{\Delta l} \frac{l_c}{\varrho U_0^2} \frac{\varepsilon^3}{(1-\varepsilon)}, \tag{5}$$

$$Re = \frac{U_0 l_c \varrho}{\mu(1-\varepsilon)}, \tag{6}$$

with a mean free stream velocity U_0, a viscosity μ, and a pressure drop ΔP over the length Δl.

Experimental and recent numerical investigations show the friction coefficient to be two or three times higher than this theoretical Λ_{th} [2]. Usually, this contradiction is explained by the longer flow paths compared to the edge length of the porous media, resulting in a higher pressure drop than predicted by this theory. With our numerical data we are able to show, that, in contrast to this widely accepted assumption, the tortuosity gives no significant contribution to the pressure drop.

3.2 Dissipation by Shear and Elongation

The total dissipation Φ of an incompressible flow passing through a porous media can be expressed by ([1]):

$$\Phi = -\mu \left(\frac{\partial U_j}{\partial x_i} + \frac{\partial U_i}{\partial x_j} \right) \frac{\partial U_j}{\partial x_i}$$

$$= -\mu \left\{ \overbrace{\left[\left(\frac{\partial u_1}{\partial x_1} \right)^2 + \left(\frac{\partial u_2}{\partial x_2} \right)^2 + \left(\frac{\partial u_3}{\partial x_3} \right)^2 \right]}^{\Phi_e \text{ (elongation)}} \right.$$

$$\left. + \underbrace{\left(\frac{\partial u_2}{\partial x_1} + \frac{\partial u_1}{\partial x_2} \right)^2 + \left(\frac{\partial u_3}{\partial x_2} + \frac{\partial u_2}{\partial x_3} \right)^2 + \left(\frac{\partial u_1}{\partial x_3} + \frac{\partial u_3}{\partial x_1} \right)^2}_{\Phi_s \text{ (shear)}} \right\} . \quad (7)$$

The total dissipation can thus be expressed as a sum of two parts: the dissipation caused by shear forces Φ_s and the dissipation caused by elongational strain Φ_e:

$$\Phi = \Phi_s + \Phi_e . \quad (8)$$

4 Numerical Simulations

4.1 Geometry and Boundary Conditions

Five samples with a porosity of $\epsilon = 0.75$ were generated by randomly distributing cubes of edge length $l_e = 10$ lattice units inside a $l_x \times l_y \times l_z = 100 \times 50 \times 50$ domain. Overlapping of the cubes was not allowed in order to control the shape and specific surface of this artificial porous media.

The whole set was centered inside a $l_x \times l_y \times l_z = 200 \times 50 \times 50$ sized channel, and periodic flow boundary conditions where applied orthogonal to the mean flow direction. At the inlet, a velocity block profile was applied, and at the outlet the pressure was fixed.

4.2 Simulation Parameters

For each of the five samples, six simulations with Reynolds Numbers in the range $0.01 \leq Re \leq 100$ were carried out on a SGI Enterprise 2000 with an 270 MHZ IP27 Processor. 180 Mbytes of memory were used for the 500000 lattice nodes, and for 20000 iterations 45000 s of CPU-time were consumed.

5 Simulation Results

The detailed flow field containing the flow velocity and pressure were stored after each simulation for further evaluation.

The friction factor can be derived from the flow field by the help of (4-6). It is known from experimental investigations, that the friction factor is a constant for $Re < 1$ and then increases linearly with the Reynolds number. Durst et. al. [2] measured a relationship for different beds of different shaped spheres of $\Lambda = 182 + 1.75 \times Re$. As can be seen in Fig. 1 (left), this prediction is approximated very well by the simulation results.

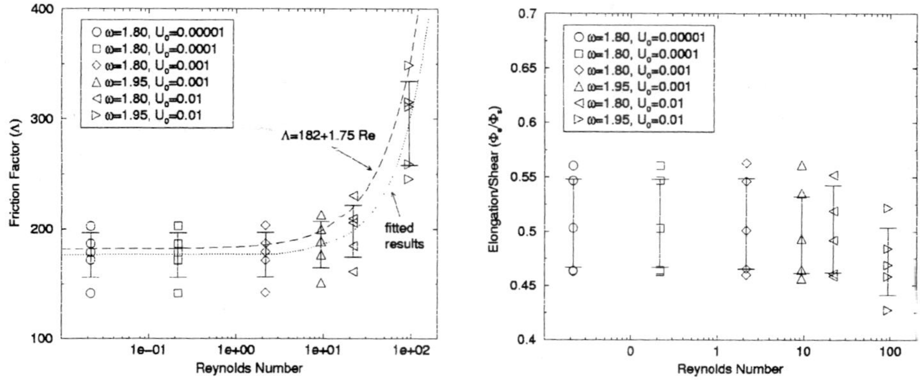

Fig. 1. Left: friction factor as a function of Reynolds number, **right:** relation of elongation and shear, both for five different geometries and six different Reynolds numbers.

The dissipation caused by shear and by elongation can be determined from the flow field by evaluating (7). As expressed by (8), these two parts sum up to the total dissipation and must therefore be directly related to the friction factor. This was investigated in detail for the five simulation results at $Re \approx 0.22$. As can be seen from the plot on the righthand side of Fig. 2, Φ is strongly related to Λ.

It can be concluded from these results, that a considerable error is being made if one neglects the dissipation due to the elongation of the fluid. This is normally done within the capillaric theories, since the Navier Stokes equations are only being solved for tubes, where no elongation takes place. By later generalizing the equations for arbitrary geometries a much too low friction factor is being predicted (4).

Comparing the tortuosity for the five different geometries (Fig. 3 left) with the total dissipation (Fig. 2 right), on can see that these two quantities are not related to each other, and that the average tortuosity ($\tau \approx 1.12$) is much too small to explain the gap between capillaric theory and numerical simulation. Comparing (Fig. 3 left) and (Fig. 3 right), there is a strong indication that the tortuosity is related to the relation of dissipation caused by elongation and the dissipation caused by shear. Thus, this relationship Φ_e/Φ_s is, as the tortuosity, directly related to the geometry for low Reynolds number flow. This is also

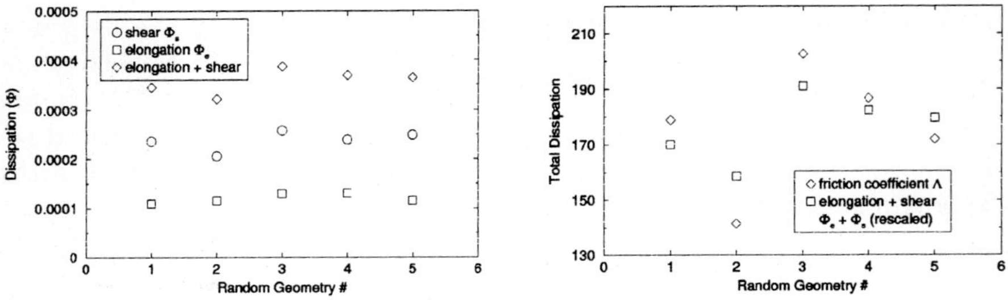

Fig. 2. Left: dissipation by shear, by elongation and sum, **right:** total dissipation.

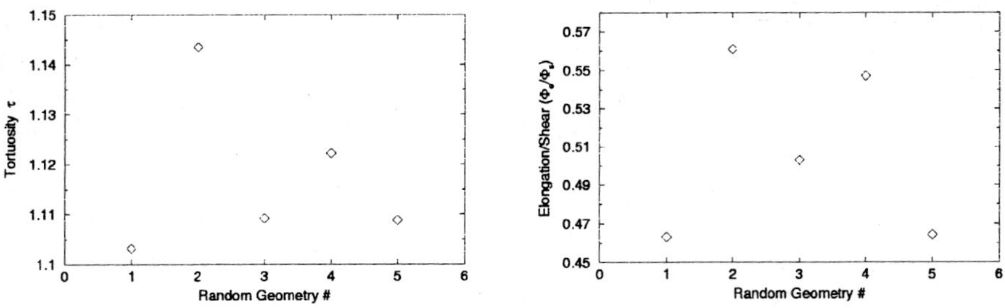

Fig. 3. Correlation between **left:** tortuosity, **right:** relationship of elongation and shear.

indicated by the almost constant Φ_e/Φ_s over six orders of magnitude for the Reynolds number (Fig. 1, right).

6 Conclusion

Applying the lattice Boltzmann technique for flow simulations through randomly generated geometries, we were able to determine the friction coefficient, the tortuosity and the dissipation caused by shear and elongation of the fluid. We demonstrated that the dissipation due to elongation of the fluid must not be neglected, and the tortuosity cannot explain the gap between the friction factor predicted by the capillaric theory and experimental or numerical results.

References

1. Bird R.B., Stewart W.E. and Lightfood E.N. (1960) Transport Phenomena, J. Wiley and Sons, New York
2. Durst F., Haas R. and Interthal W. (1987) Journal of Non–Newtonian Fluid Mechanics **22**: 169–189
3. Ergun S. (1952) Chem. Eng. Prog. **48(2)**: 84–89
4. Bernsdorf J. and Schäfer M. (1997) Notes on Numerical Fluid Mechanics, Volume **59**: 61–71
5. Qian Y.H. (1992) Europhys.Lett. **17(6)**: 479-484

Lattice Boltzmann Simulations of Drop Deformation and Breakup in a Simple Shear Flow

Takaji Inamuro, Takayuki Miyahara, and Fumimaru Ogino

Department of Chemical Engineering, Kyoto University, Kyoto 606-8501, Japan

Abstract. A new lattice Boltzmann method for multicomponent immiscible fluids with the same density is proposed. In the method, the Swift et al. model (1995) is used as an index function for the calculation of interface profiles, and another distribution function is introduced for the calculation of pressure and velocity of the fluids. The method is applied to simulations of drop deformation and breakup in a simple shear flow. Two types of drop breakup behavior can be simulated, and the results agree well with experimental results.

1 Introduction

The problems of drop deformation and breakup in viscous flows are very important in many science and engineering fields, such as mixing in multiphase viscous systems, blending of molten polymers, emulsion formation and rheology, deformation of biological cells, and so on [1]. The numerical approach to the problems has difficulties due to deformable boundaries of immiscible fluids, rupture of domain boundaries, and modeling of interfacial tension. Recently, the lattice Boltzmann method (LBM) has developed into an alternative and promising numerical scheme for simulating multicomponent fluid flows [2–5]. In this paper, a new lattice Boltzmann method for multicomponent immiscible fluids with the same density is proposed. In the method, by using the idea by He et al. [4], two distribution functions are introduced for the calculation of interface profiles and for the calculation of pressure and velocity of the fluids. The Swift et al. model [5] is used for the calculation of interface profiles. The method is applied to simulations of drop deformation and breakup in a simple shear flow.

2 Numerical Method

Non-dimensional variables, which are defined by using a characteristic length L, a characteristic particle speed c, a characteristic time scale $t_0 = L/U$ where U is a characteristic flow speed, and a reference density ρ_0, are used as in Ref. [6]. In the LBM, a modeled fluid, composed of identical particles whose velocities are restricted to a finite set of N vectors $\boldsymbol{c}_i (i = 1, 2, \cdots, N)$, is considered. The fifteen-velocity model ($N = 15$) [7] is used in the present paper. The velocity vectors of this model are given by

$$[\boldsymbol{c}_1,\boldsymbol{c}_2,\boldsymbol{c}_3,\boldsymbol{c}_4,\boldsymbol{c}_5,\boldsymbol{c}_6,\boldsymbol{c}_7,\boldsymbol{c}_8,\boldsymbol{c}_9,\boldsymbol{c}_{10},\boldsymbol{c}_{11},\boldsymbol{c}_{12},\boldsymbol{c}_{13},\boldsymbol{c}_{14},\boldsymbol{c}_{15}] = \begin{bmatrix} 0 & 1 & 0 & 0 & -1 & 0 & 0 & 1 & -1 & 1 & 1 & -1 & 1 & -1 & -1 \\ 0 & 0 & 1 & 0 & 0 & -1 & 0 & 1 & 1 & -1 & 1 & -1 & -1 & 1 & -1 \\ 0 & 0 & 0 & 1 & 0 & 0 & -1 & 1 & 1 & 1 & -1 & -1 & -1 & -1 & 1 \end{bmatrix}. \tag{1}$$

The physical space is divided into a cubic lattice, and the evolution of particle population at each lattice site is computed. Two particle distribution functions, f_i and g_i, are used. f_i is used as an index function for the calculation of interface profiles, and g_i is used for the calculation of pressure and velocity of the fluid. The evolution of the particle distribution functions $f_i(\boldsymbol{x},t)$ and $g_i(\boldsymbol{x},t)$ with velocity $\boldsymbol{c}_i$ at the point $\boldsymbol{x}$ and at time t is computed by the following equations:

$$f_i(\boldsymbol{x}+\boldsymbol{c}_i\Delta x, t+\Delta t) - f_i(\boldsymbol{x},t) = -\frac{1}{\tau_f}\left[f_i(\boldsymbol{x},t) - f_i^{\mathrm{eq}}(\boldsymbol{x},t)\right], \tag{2}$$

$$g_i(\boldsymbol{x}+\boldsymbol{c}_i\Delta x, t+\Delta t) - g_i(\boldsymbol{x},t) = -\frac{1}{\tau_g}\left[g_i(\boldsymbol{x},t) - g_i^{\mathrm{eq}}(\boldsymbol{x},t)\right], \tag{3}$$

where f_i^{eq} and g_i^{eq} are equilibrium distribution functions, τ_f and τ_g are dimensionless single relaxation times, Δx is a spacing of the cubic lattice, and Δt is a time step during which the particles travel the lattice spacing.

The index function ϕ representing an interface and the macroscopic variables of the multicomponent fluids (the pressure p and the velocity $\boldsymbol{u}$) are defined in terms of the two particle distribution functions as follows:

$$\phi = \sum_{i=1}^{15} f_i, \qquad p = \frac{1}{3}\sum_{i=1}^{15} g_i, \qquad \boldsymbol{u} = \sum_{i=1}^{15} g_i\boldsymbol{c}_i. \tag{4}$$

The equilibrium distribution functions are given by

$$\begin{aligned} f_i^{\mathrm{eq}} = {} & H_i\phi + F_i\left(p_0 - \kappa_f\phi\nabla^2\phi - \frac{\kappa_f}{6}|\nabla\phi|^2\right) \\ & + E_i\phi\left(3u_\alpha c_{i\alpha} - \frac{3}{2}u_\alpha u_\alpha + \frac{9}{2}u_\alpha u_\beta c_{i\alpha}c_{i\beta}\right) + E_i\kappa_f G_{\alpha\beta}c_{i\alpha}c_{i\beta}, \end{aligned} \tag{5}$$

$$g_i^{\mathrm{eq}} = E_i\left(3p + 3u_\alpha c_{i\alpha} - \frac{3}{2}u_\alpha u_\alpha + \frac{9}{2}u_\alpha u_\beta c_{i\alpha}c_{i\beta}\right) + E_i\kappa_g G_{\alpha\beta}c_{i\alpha}c_{i\beta}, \tag{6}$$

where

$$\left.\begin{array}{lll} E_1 = \frac{2}{9}, & E_2 = E_3 = \cdots = E_7 = \frac{1}{9}, & E_8 = E_9 = \cdots = E_{15} = \frac{1}{72}, \\ H_1 = 1, & H_2 = H_3 = \cdots = H_{15} = 0, & \\ F_1 = -\frac{7}{3}, & F_i = 3E_i \ \text{for} \ i = 2,3,\cdots,15, & \end{array}\right\} \tag{7}$$

and

$$G_{\alpha\beta} = \frac{9}{2}\frac{\partial\phi}{\partial x_\alpha}\frac{\partial\phi}{\partial x_\beta} - \frac{3}{2}\frac{\partial\phi}{\partial x_\gamma}\frac{\partial\phi}{\partial x_\gamma}\delta_{\alpha\beta}, \tag{8}$$

with $\alpha, \beta, \gamma = 1, 2, 3$ (subscripts α , β , and γ represent Cartesian coordinates and the summation convention is used). $\delta_{\alpha\beta}$ is the Kronecker delta. κ_f is a constant parameter determining the width of the interface, and κ_g is a constant parameter determining the strength of the interfacial tension. p_0 is given by

$$p_0 = \phi T \frac{1}{1 - b\phi} - a\phi^2, \tag{9}$$

where a, b, and T are free parameters determining the maximum and minimum values of the index function ϕ. It is noted that f_i is the same as that of the Swift et al. model [5], and the multicomponent fluids have the same density.

The viscosity μ of the fluid is related to the relaxation time τ_g and the lattice spacing Δx :

$$\mu = \frac{1}{3}\left(\tau_g - \frac{1}{2}\right)\Delta x. \tag{10}$$

One can change the viscosity of each component fluid through the value of τ_g. The interfacial tension σ is given by

$$\sigma = \kappa_g \int_{-\infty}^{\infty} \left(\frac{\partial \phi}{\partial \xi}\right)^2 d\xi, \tag{11}$$

with ξ being the coordinate normal to the interface [8].

3 Results and Discussion

The method is applied to simulations of drop deformation and breakup in a simple shear flow. A drop with radius R in a suspending fluid between two parallel walls with length H apart is considered (see Fig. 1). Both the fluids have the same density. The drop is brought to the equilibrium state at rest, and then the top and bottom walls begin to move with velocities u_w and $-u_w$, respectively, at $t = 0$. The no-slip boundary condition [9] is used on the top and bottom walls. A periodic boundary condition is used on the lateral sides of the domain. The two dimensionless parameters of the present problem are the capillary number

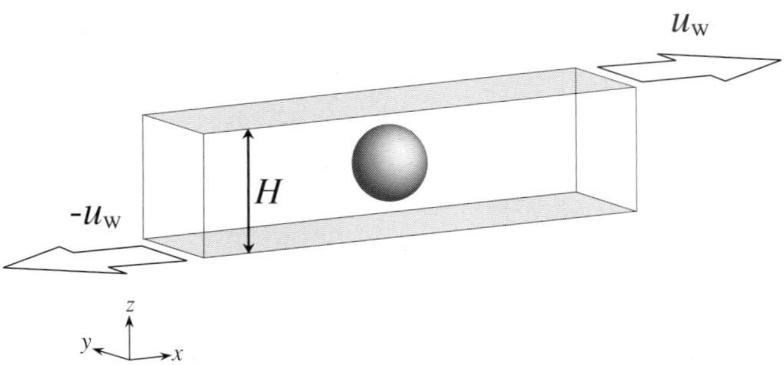

Fig. 1. A drop in a simple shear flow.

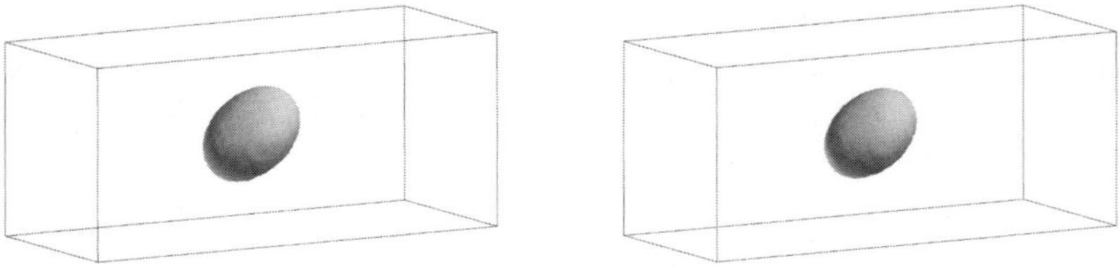

$t^* = 1.56$ $t^* = 20.3$

Fig. 2. Drop behavior for Ca = 0.0591 and $\eta = 1$ ($t^* = tu_w/H$). The contour surface represents the index function $\phi = (\phi_{max} + \phi_{min})/2$.

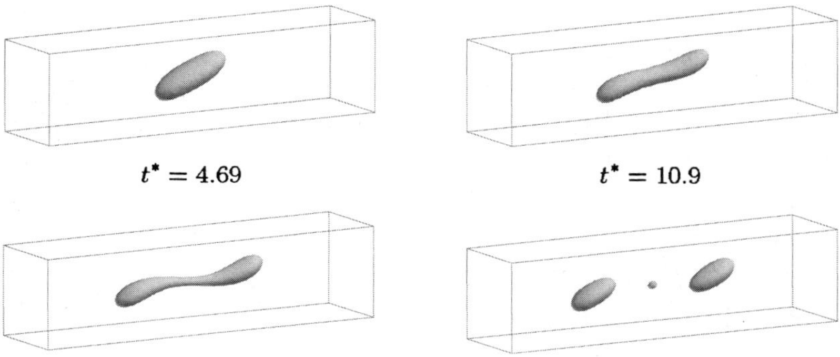

$t^* = 4.69$ $t^* = 10.9$

$t^* = 15.6$ $t^* = 18.8$

Fig. 3. Drop behavior for Ca = 0.331 and $\eta = 1$ ($t^* = tu_w/H$). The contour surface represents the index function $\phi = (\phi_{max} + \phi_{min})/2$.

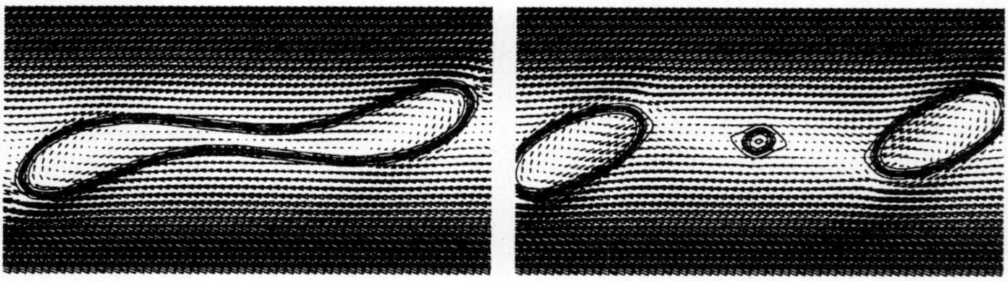

$t^* = 15.6$ $t^* = 18.8$

Fig. 4. Velocity vectors and contour lines of the index function on the x–z plane for Ca = 0.331 and $\eta = 1$ ($t^* = tu_w/H$).

Ca representing the ratio of flow forces to interfacial tension and the viscosity ratio η of the drop viscosity to that of the suspending fluid. The capillary number is defined by $\mathrm{Ca} = \mu_c \Gamma R/\sigma$ and the viscosity ratio is $\eta = \mu_d/\mu_c$, where μ_c is the viscosity of the suspending fluid, $\Gamma = 2u_w/H$ is the shear rate, σ is the

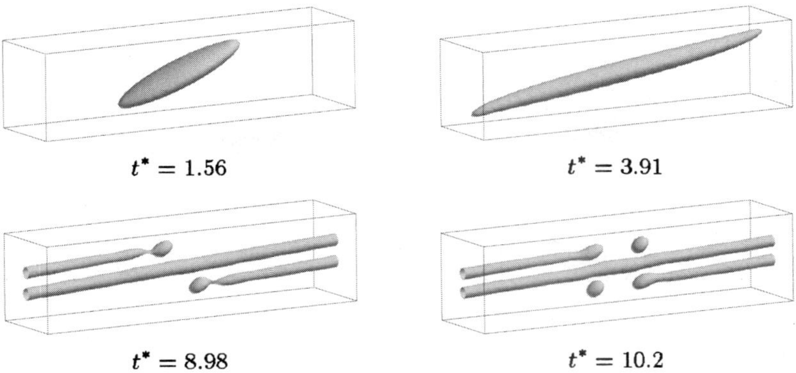

Fig. 5. Drop behavior for $\mathrm{Ca} = 0.381$ and $\eta = 0.1$ ($t^* = tu_\mathrm{w}/H$). The contour surface represents the index function $\phi = (\phi_\mathrm{max} + \phi_\mathrm{min})/2$.

interfacial tension, and μ_d is the drop viscosity. We chose $a = 9/49$, $b = 2/21$, and $T = 0.55$ in Eq. (9); it follows that the maximum and minimum values of the index function are $\phi_\mathrm{max} = 4.895$ and $\phi_\mathrm{min} = 2.211$. The viscosity in the interface is changed by $\mu = \mu_\mathrm{c} + (\mu_\mathrm{d} - \mu_\mathrm{c})(\phi - \phi_\mathrm{min})/(\phi_\mathrm{max} - \phi_\mathrm{min})$. The parameters are fixed at $\kappa_f = 0.01$, $\kappa_g = 0.02$, and $\tau_f = 0.7$. The other parameters, R, u_w, and τ_g, are changed so that $\mathrm{Ca} = 0.0591$, 0.331 and 0.381, and $\eta = 0.1$ and 1. The domain is divided into a $150 \times 64 \times 64$ cubic lattice or a $256 \times 64 \times 64$ cubic lattice.

Figure 2 shows the calculated results for the case of $\mathrm{Ca} = 0.0591$ and $\eta = 1$ ($R = 15\Delta x$, $u_\mathrm{w} = 0.05$, $\tau_g = 0.75$). Since Ca is small, the drop deforms into an elliptic shape and maintains the shape even if time goes on. Figure 3 shows the calculated results for the case of $\mathrm{Ca} = 0.331$ and $\eta = 1$ ($R = 15\Delta x$, $u_\mathrm{w} = 0.1$, $\tau_g = 1.2$). The drop is elongated into a dumb-bell shape and then fragments near the middle, forming two large droplets and a small droplet. This breakup behavior is called "fracture" [10]. The velocity vectors and the contour lines of the index function on the x–z plane are shown in Fig. 4. It is seen that the flows near the droplets are complicated, and circulatory flows are induced inside the deformed droplets. Figure 5 shows the calculated results for the case of $\mathrm{Ca} = 0.381$ and $\eta = 0.1$ ($R = 20\Delta x$, $u_\mathrm{w} = 0.1$, $\tau_g = 0.5625$ for the drop and 1.125 for the suspending fluid). The drop is elongated into a long sigmoidal shape, and then small droplets are ruptured off the tips of the drop. This breakup behavior is called "tipstreaming" [10] and has been experimentally observed only for low viscosity ratios. Finally, the above computational conditions are plotted in Fig. 6 which shows the experimental critical capillary number of drop breakup in simple shear flows by Torza et al. [11] and Grace [12]. The above results agree well with the experimental results. Further work remains in the computations for other conditions.

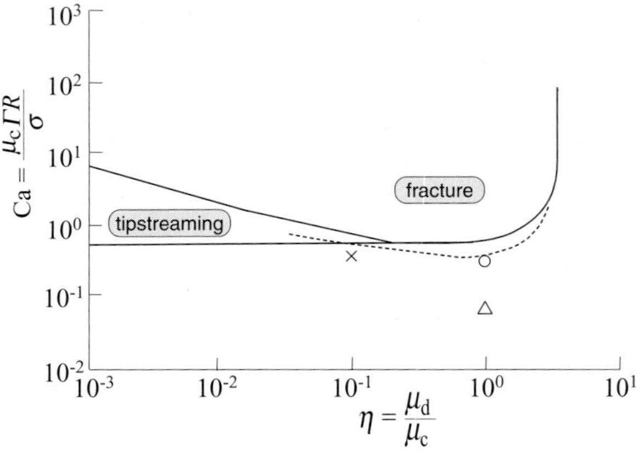

Fig. 6. Critical capillary number of drop breakup in simple shear flows. The solid curves are the experiments by Grace [12], and the dashed curve is the experiments by Torza et al. [11]. The symbols represent computational conditions: △, the case of $\mathrm{Ca} = 0.0591$ and $\eta = 1$ shown in Fig. 2; O, the case of $\mathrm{Ca} = 0.331$ and $\eta = 1$ shown in Fig. 3; ×, the case of $\mathrm{Ca} = 0.381$ and $\eta = 0.1$ shown in Fig. 5.

4 Conclusions

A new lattice Boltzmann method for multicomponent immiscible fluids with the same density is proposed. The method is applied to simulations of drop deformation and breakup in a simple shear flow. The method can simulate the two types of drop breakup behavior, "fracture" and "tipstreaming," and the calculated results agree well with experimental results.

References

1. H. A. Stone: Annu. Rev. Fluid Mech. **26**, 65 (1994)
2. A. K. Gunstensen, D. H. Rothman, S. Zaleski, and G. Zanetti: Phys. Rev. A **43**, 4320 (1991)
3. X. Shan and H. Chen: Phys. Rev. E **47**, 1815 (1993)
4. X. He, S. Chen, and R. Zhang: J. Comput. Phys. **152**, 642 (1999)
5. M. R. Swift, W. R. Osborn, and J. M. Yeomans: Phys. Rev. Lett. **75**, 830 (1995)
6. T. Inamuro, M. Yoshino, and F. Ogino: Phys. Fluids **9**, 3535 (1997)
7. Y. H. Qian, D. d'Humières, and P. Lallemand: Europhys. Lett. **17**, 479 (1992)
8. T. Inamuro, N. Konishi, and F. Ogino: Comput. Phys. Commun. **129**, 32 (2000)
9. T. Inamuro, M. Yoshino, and F. Ogino: Phys. Fluids **7**, 2928 (1995); Erratum: **8**, 1124 (1996)
10. R. A. de Bruijn: Chem. Eng. Sci. **48**, 277 (1993)
11. S. Torza, R. G. Cox, and S. G. Mason: J. Coll. Int. Sci. **38**, 395 (1972)
12. H. P. Grace: Chem. Eng. Commun. **14**, 225 (1982)

Comparison between Volume of Fluid Method and Lattice Boltzmann Method for Numerical Simulation of Droplet

Taro Imamura and Kojiro Suzuki

Univ. of Tokyo, Hongo 7-3-1 Bunkyo-ku, Tokyo 113-8656, Japan

Abstract. In this paper, the following two methods are used to solve two-phase flow problems. One is the Volume of Fluid (VOF) method, which solves the Navier-Stokes equation with a color function in order to distinguish the two phases. The Level-Set function is used as a color function to track the interface. The other is the Lattice Boltzmann (LB) method, which has recently become a powerful tool for the simulation of complex fluid flows such as multi-phase flow. The dynamics of static and oscillating droplet in two-dimension are chosen as test problems, and are solved by both methods. These results are compared with each other, and their accuracy is assessed by comparing with the analytical theory.

1 Introduction

A lot of study has been devoted to the dynamics of the droplet under micro-gravity environment in recent years because of its importance in space utilization, chemical engineering, and other related fields. However, performing micro-gravity experiments on the ground can be difficult. There are some experimental methods available to make micro-gravity environments (e.g., a drop tower or parabolic flight), but these methods have restrictions in time or in micro-gravity quality. So, it is very important to simulate numerically what will happen in such a situation. In this paper, two different numerical methods are used for the calculation of the two-dimensional static and oscillating droplet. One is Volume of Fluid (VOF) method, and the other is Lattice Boltzmann (LB) method. We will compare these two methods from the viewpoint of the accuracy of the oscillating period and the computational cost.

2 Volume of Fluid method (VOF method)

2.1 Governing Equations

The governing equations on the Cartesian coordinates including the surface tension and the viscosity, can be written as:

$$\frac{\partial \boldsymbol{f}}{\partial t} + (\boldsymbol{u} \cdot \nabla)\boldsymbol{f} = \boldsymbol{g}, \tag{1}$$

where $\boldsymbol{f} = \{\rho, \boldsymbol{u}, P, \phi\}$, $\boldsymbol{g} = \left\{-\rho(\nabla \cdot \boldsymbol{u}), \frac{1}{\rho}(-\nabla P + \boldsymbol{F}_s + \boldsymbol{F}_{vis}), -\gamma P(\nabla \cdot \boldsymbol{u}), 0\right\}$. The value ρ, $\boldsymbol{u}$, P are density, velocity and pressure, respectively. ϕ is an artificial function, which is called a color-function. The two phases are distinguished by the value of the color-function. The terms $\boldsymbol{F}_s$ and $\boldsymbol{F}_{vis}$ stand for the surface tension and the viscous force, respectively. The surface tension force is defined by $|\boldsymbol{F}_s| = \sigma\kappa$, where σ is the surface tension coefficient, and κ represents the local curvature of the interface.

The equations are divided into advection phase and non-advection phase. The advection phase is solved by the CIP(Cubic Interpolated Pseudo-Particle) method[1]. The CIP method has the third order accuracy in space. The pressure term of the non-advection phase is transformed into Poisson's equation:

$$\nabla \cdot \left(\frac{\nabla \cdot P^n}{\rho^n} \right) = \frac{\nabla \cdot \boldsymbol{u}^n}{\Delta t} + \nabla \cdot \left\{ \frac{1}{\rho^n} (\boldsymbol{F}_s + \boldsymbol{F}_{vis}) \right\} . \tag{2}$$

In this paper, ILDU-CGS method is used in solving (2). Since, the outer boundary is far enough from the droplet, computational results are not affected by the boundary conditions. Thus, we adopted pressure fixed boundary condition, in order to save the calculation time.

2.2 Level-Set Function and Heaviside Function

The Level-Set function[2] is introduced as a color function, in order to capture the interface between the two phases. The Level-Set function ϕ is initialized to be the signed normal distance from the interface. Thus, ϕ is set to zero at the interface of the droplet. The density ρ and the viscosity μ of the droplet and the surrounding fluid can be described in a unified way: $\boldsymbol{f}(\boldsymbol{x}) = \boldsymbol{f}_d \cdot (1 - H(\phi)) + \boldsymbol{f}_a \cdot H(\phi)$, where $\boldsymbol{f} = (\rho, \mu)$. The subscripts d and a stand for the droplet phase and the ambient fluid phase, respectively. The Heaviside function $H(\phi)$ is defined using the Level-Set function ϕ:

$$H(\phi) \equiv \begin{cases} 1 & \text{if } \phi > \alpha \quad \text{Ambient fluid} \\ 0 & \text{if } \phi < -\alpha \quad \text{Droplet} \\ \frac{1}{2}(1 + \frac{\phi}{\alpha} + \frac{1}{\pi}\sin(\pi\phi/\alpha)) & \text{otherwise} \quad \text{Transient region} . \end{cases} \tag{3}$$

In the actual case, density and pressure change discontinuously at the interface. For the stable calculation, we assume the transient region of the finite thickness, in which the flow properties gradually changes between the phases over several mesh points. From this point of view, it is important to calculate with sufficient fine grids to reduce the effects of this transient region. After some numerical experimentation, α is set to $1.5h$, where h is the size of the grid.

2.3 Continuum Surface Force Method

The Laplace's law determines the surface tension as a surface force which acts only on the interface. However, in order to include $\boldsymbol{F}_s$ into (1), $\boldsymbol{F}_s$ must be estimated as a volume force which acts in the transient region. The Continuum

Surface Force (CSF) method[3] was applied for this estimation. Using the Level-Set function and the Heaviside function that are defined in the previous section, the surface tension force can be estimated as follows:

$$\boldsymbol{F}_s(\boldsymbol{x}) = -\sigma\left\{\nabla\cdot\left(\frac{\nabla\phi(\boldsymbol{x})}{|\nabla\phi(\boldsymbol{x})|}\right)\right\}\cdot\nabla H(\boldsymbol{x}). \tag{4}$$

3 Lattice Boltzmann method (LB method)

The Lattice Boltzmann Equation(LBE) of k-th component for 2-dimension 9-velocity model[4], can be written as:

$$n_a^k(\boldsymbol{x}+\boldsymbol{e}_a, t+1) - n_a^k(\boldsymbol{x},t) = -\frac{1}{\tau_k}\left[n_a^k(\boldsymbol{x},t) - n_a^{k(eq)}(\boldsymbol{x},t)\right]; k=1,2, \tag{5}$$

where $n_a^k(\boldsymbol{x},t)$ is the distribution function of the k-th component which have the velocity $\boldsymbol{e}_a$ at lattice site $\boldsymbol{x}$ and time t. The collision interval of the k-th component is denoted as τ_k. Then, the viscosity of the pure k-th component is $(2\tau_k-1)/6$. $\{\boldsymbol{e}_a; a=0,\cdots,8\}$ is the set of vectors pointing from site $\boldsymbol{x}$ to its neighboring sites. The equilibrium distribution function is

$$n_a^{k(eq)}(n^k,\boldsymbol{u}) = E_a n^k[1+3\boldsymbol{e}_a\cdot\boldsymbol{u} + \frac{9}{2}(\boldsymbol{e}_a\cdot\boldsymbol{u})^2 - \frac{3}{2}\boldsymbol{u}\cdot\boldsymbol{u}] \tag{6}$$

for $i=0,\cdots,8$, where $E_0=4/9$, $E_1=\cdots=E_4=1/9$, and $E_5=\cdots=E_8=1/36$. The velocity of the k-th component is usually used for the estimation of the equilibrium function in (6). However, the surface tension model proposed by Shan and Chen (S-C model[5,6]) introduces the equilibrium velocity $\boldsymbol{u}_k^{eq}$ in place of $\boldsymbol{u}$ to consider the influence of interaction between the two phases. The equilibrium velocity is the parameters which will be calculated from the distribution functions and the interaction force:

$$\rho_k\boldsymbol{u}_k^{eq} = \rho_k\boldsymbol{u}' + \tau_k\boldsymbol{F}_k, \tag{7}$$

where

$$\boldsymbol{u}' = \sum_k\frac{\rho_k\boldsymbol{u}_k}{\tau_k}/\sum_k\frac{\rho_k}{\tau_k}, \tag{8}$$

$$\boldsymbol{F}_k(\boldsymbol{x}) = -\Psi_k(\boldsymbol{x})\sum_{\boldsymbol{x}'}\sum_{\bar{k}} G_{k\bar{k}}(\boldsymbol{x},\boldsymbol{x}')\Psi_{\bar{k}}(\boldsymbol{x}')(\boldsymbol{x}-\boldsymbol{x}'). \tag{9}$$

$G_{k\bar{k}}(\boldsymbol{x},\boldsymbol{x}')$ is a Green's function. However, if the collision is limited between the nearest neighbors, $G_{k\bar{k}}$ can be described in the following form.

$$G_{k\bar{k}}(\boldsymbol{x},\boldsymbol{x}') \equiv \begin{cases} 0, & |\boldsymbol{x}-\boldsymbol{x}'|>c \\ g_{k\bar{k}}, & |\boldsymbol{x}-\boldsymbol{x}'|\le c \end{cases} \tag{10}$$

The value c is the lattice spacing. $g_{k\bar{k}}$ is the interaction potential between component k and $\bar{k}$. $g_{k\bar{k}}$ is set to 0.20. The effective mass $\Psi_k(\boldsymbol{x})$ is taken as ρ_k in this study. The total density of the whole fluid is defined as $\rho=\sum_k\rho_k$, and the whole fluid velocity $\boldsymbol{u}$ is defined by $\rho\boldsymbol{u} = \sum_k\rho_k\boldsymbol{u}_k + \frac{1}{2}\sum_k\boldsymbol{F}_k$. Also, the pressure is written as $P=\frac{1}{3}\rho+\frac{3}{2}\sum g_{k\bar{k}}\Psi_k\Psi_{\bar{k}}$. At the boundary, periodic boundary condition is used.

4 Computational Results

4.1 Static Circular Droplet

To test the two calculation methods, the static circular droplet problem was investigated. The grid resolution for the VOF method and the lattice size for the LB method are 64×64. The droplet is located at the center of the computational field and the radius of the droplet is 12.8 in both cases. In the equilibrium state with no external forces, both inner and outer fluids should have constant density and pressure with no velocity fluctuation. The density ratio of ambient fluid to droplet is 0.97, since numerical instability occurs in the LB method for larger density ratio. In the VOF method, it is confirmed that computation can be done for the density ratio up to 800[7]. There is a pressure jump determined as $\Delta P = \sigma\kappa$ at the interface of the static droplet. This is called Laplace's law. In the VOF method, the surface tension coefficient can be specified arbitrarily. However, since only the value of the interaction potential $g_{k,\bar{k}}$ is given in the case of the LB method, σ is determined by Laplace's law using pressure difference ΔP at the interface and the curvature κ after the calculation. For the present calculation, σ is 0.28. The same value is used for the VOF method. Also, viscosity μ is set to 1/6 in both cases.

Figure 1 shows the velocity vectors and the shape of the interface calculated by the VOF and the LB methods, respectively. There are small velocity fluctuations around the interface, and the result calculated by the LB method shows larger fluctuation than that of the VOF method. However, these fluctuations do not cause the calculation to become unstable in either case. For the LB method. the shape is deformed in diagonal direction. The volume of the droplet is conserved through the computation in the VOF method. In the LB case, the volume increases for a while, but after 5000 steps, static state is obtained. Such rise in the volume of the droplet comes from the difficulty in setting the initial conditions. To keep numerical stability, the VOF method requires more than 5 grid points for the transient region. On the other hand, thickness of the transient region cannot be specified before the computation for the LB method (although the thickness of the interface is 5 lattice units after the droplet becomes stable).

4.2 Oscillation of the droplet

Another case is the oscillation of the droplet. Initial shape of the droplet is set as $r = a + A\cos n\theta$, where r is the distance from the center of the droplet. This disturbance was used to compare with the analytical solution by Lamb[8] The variable n is the oscillation mode, a and A are the radius of the droplet and the amplitude of the oscillation, respectively.

Figures 2 show the oscillation pattern of the droplet solved by the VOF and the LB methods. Because of the effect of the viscosity, the oscillation amplitude decreases in either cases.

Figure 3 shows the time history of the total kinetic energy of the calculation field ($K.E. = \sum \frac{1}{2}\rho \boldsymbol{u}^2$). For the LB case, velocity fluctuation is not negligible.

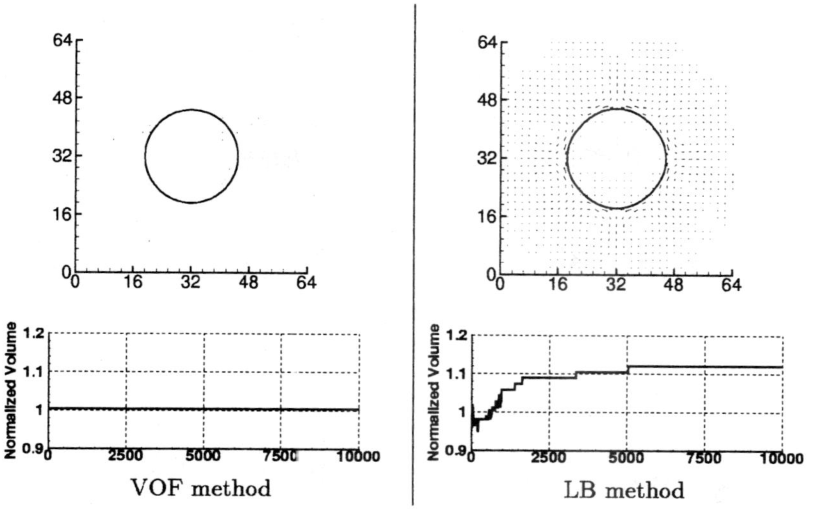

Fig. 1. The upper figures show the position of the interface and velocity fluctuations calculated by the VOF(left figure) and the LB methods(right figure). The lower figures show the time history of the normalized volume of the droplet.

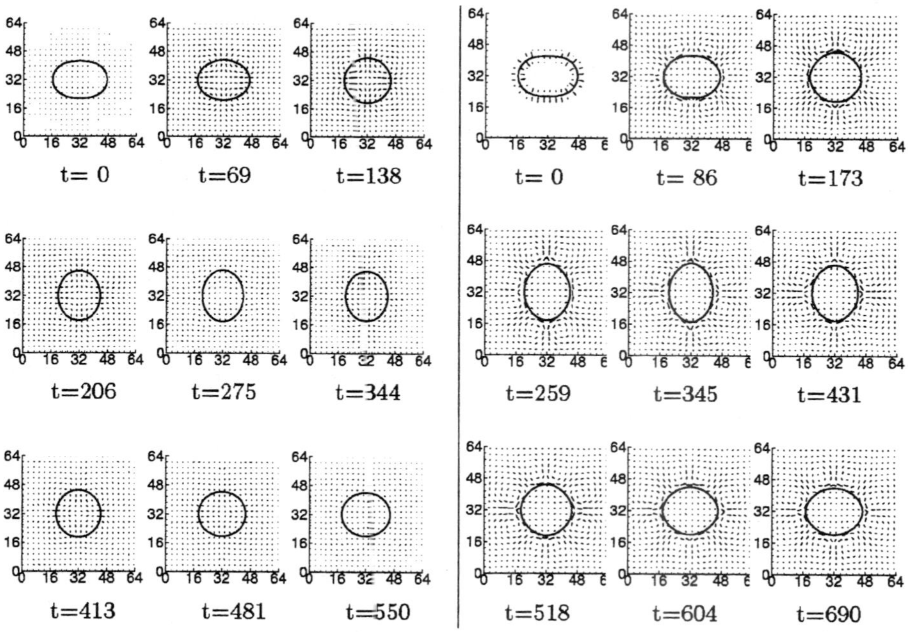

Fig. 2. The shape of the interface and the velocity vectors calculated by the VOF(left figures) and LB methods(right figures) : Oscillation Mode $n = 2$, Amplitude 20% of radius

Thus, total kinetic energy for static droplet is subtracted from the total kinetic energy of oscillating droplet. From these figures, the period is determined more clearly than from the time history of the droplet shape. The accuracy of our VOF method code has been already validated in our previous study[7], and for the present case, the period is 550. The period obtained by the LB method is 690; this period differs from the VOF method results by 25%. However, the calculation time is 40 times faster. Thus, using a 192×192 lattice gives a period of 595, which is within 10% of the VOF results. When a very fine lattice is used, the LB method has the potential to give the results with the same accuracy as the VOF method with less computational time.

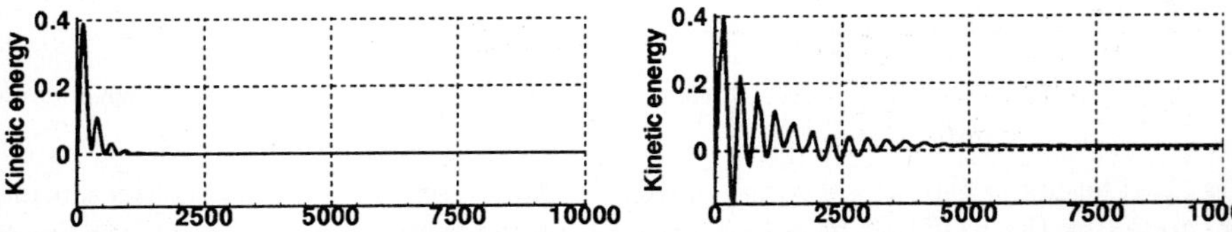

Fig. 3. The time history of the total kinetic energy of the calculation field calculated by the VOF(left figure) and the LB methods(right figure), relative to figure 2.

5 Conclusions

Using the Volume of Fluid method and the Lattice Boltzmann method, two-phase flow is simulated. From the static droplet calculation, the following results are obtained: 1)equilibrium state is obtained for both cases, 2)the thickness of the interface is same for both cases, and 3)the VOF method give results with less fluctuation around the interface as compared with the LB method. Comparing the period of the oscillating droplet, the LB results are similar to the VOF results with less computational time required. When a very fine lattice is used, the LB method has the potential to give results with the same accuracy as the VOF method.

References

1. T.Yabe, "Unified Solver CIP for Solid, Liquid, and Gas", Computational Fluid Dynamics Review 1997 (1997)
2. M.Sussman, P.Smereka and S.Osher: J.Comp.Phys.(1994) **114**, pp.146-159
3. J.U.Brackbill, D.B.Kothe and C.Zemach: J.Comp.Phys.(1992) **100**, pp.335-354
4. T. Inamuro, M. Yoshino, and F. Ogino, Phys. Fluids, 9(11), (1997)
5. S. Hou, X. Shan, et al, J. of Comp. Phys. 138, 695-713(1997)
6. X. Shan and G. Doolen, J. Stat. Phys. **81**, 379-393(1995)
7. T. Imamura and K. Suzuki, 22nd International Symposium on Space Technology and Science Morioka, Japan. (To be published)(2000)
8. H. Lamb, Hydrodynamics, 6th ed. (Cambridge University Press, Cambridge, 1932)

Lattice Boltzmann Simulation of Magnetohydrodynamic Flows

Ramesh K. Agarwal

Wichita State University, Wichita KS 67260, USA

Abstract. This paper presents the Lattice Boltzmann method for computation of low Mach number magnetohydrodynamic (MHD) flows. The equilibrium distribution function recently derived by Huba and Lyon for ideal MHD is used to obtain the particle distribution function for the Lattice BGK (LBGK) model for the MHD flow equations. In two-dimensions, a nine-velocity square lattice model is employed for solving the LBGK equations. The magnetic field equations are solved by a finite-difference method on the LBGK square lattice. Thus, the solution technique can be considered as hybrid–the MHD flow equations solved by the LBGK method and the magnetic field equations by a finite-difference technique. Computations are performed for fully developed Hartmann-Poisuelle flow in a channel and are compared with the analytical solution.

1 Introduction

In recent years, Lattice Boltzmann methods (LBM) for solving fluid flow problems have received considerable attention [1]. The Lattice Boltzmann algorithms are simple, fast and very suitable for parallel computing. It is also easy to incorporate complicated boundary conditions. The algorithms have been successfully employed to compute flows modeled by Euler and Navier-Stokes equations including reactive and multi-phase flows. Recently, a major simplification to the original LBM was achieved by Chen et al. [2] and Qian et al. [3] by employing the single relaxation time approximation due to Bhatnagar, Gross and Krook (BGK) to the collision operator in the Lattice Boltzmann equation. In this lattice BGK (LBGK) model, the local equilibrium particle distribution function is chosen to recover the Navier-Stokes equations. The LBGK model makes the computations more efficient and allows flexibility of the transport coefficients. In this paper, the LBGK method is formulated for the solution of viscous MHD equations at low Mach numbers and is demonstrated by computing the Hartmann-Poisuelle flow in a channel.

The attempts to solve the MHD equations by lattice gas methods date back to 1987. The solution has been attempted by the Lattice Gas Automata (LGA) and LBE methods using both hexagonal and square lattice in two-dimensions. A summary of the previous work in Lattice Boltzmann MHD is given in Martinez et al. [4].

2 Ideal MHD Equations and the Distribution Function

The 3D ideal MHD equations in conservation law form can be written as

$$\frac{\partial \rho}{\partial t} + \nabla \cdot (\rho \mathbf{V}) = 0 \tag{1}$$

$$\frac{\partial \rho \mathbf{V}}{\partial t} + \nabla \cdot \left(\rho \mathbf{V}\mathbf{V} + \left(P + \frac{B^2}{8\pi} \right) \underline{\mathrm{I}} - \frac{\mathbf{B}\mathbf{B}}{4\pi} \right) = 0 \tag{2}$$

$$\frac{\partial \varepsilon}{\partial t} + \nabla \left(\mathbf{V} \cdot \left(\varepsilon + P + \frac{B^2}{8\pi} \right) - \frac{\mathbf{B}}{4\pi} (\mathbf{V} \cdot \mathbf{B}) \right) = 0 \tag{3}$$

$$\frac{\partial \mathbf{B}}{\partial t} = \nabla \times (\mathbf{V} \times \mathbf{B}) \tag{4}$$

$$\varepsilon = \frac{1}{2} \rho V^2 + \frac{3P}{2} + \frac{B^2}{8\pi} \tag{5}$$

$$P/\rho^{\gamma} = \text{constant} \tag{6}$$

In (1) - (6), $\rho, P, \varepsilon, \mathbf{V} = (\mathrm{V}_x, \mathrm{V}_y, \mathrm{V}_z)$, $\mathbf{B} = (B_x, B_y, B_z)$are density, hydrodynamic pressure, total energy, velocity field components and magnetic field components respectively. Huba and Lyon [5]have shown that these equations can be derived from the generalized transport equation

$$\frac{\partial n \langle \chi \rangle}{\partial t} + \frac{\partial}{\partial x} \cdot n \langle \chi \mathbf{v} \rangle - n \left\langle \mathbf{a} \cdot \frac{\partial \chi}{\partial \mathbf{v}} \right\rangle = 0, \tag{7}$$

where $\quad \mathbf{a} = \dfrac{\nabla \mathrm{M}}{8\pi\rho}$ and $\mathrm{M} = B_x^2 \hat{\mathrm{e}}_x + B_y^2 \hat{\mathrm{e}}_y + B_z^2 \hat{\mathrm{e}}_z$

in the following manner. If the angular brackets in (7) are defined as

$$\langle G \rangle = \int d\mathbf{v} G(\mathbf{v}) F(\mathbf{v}) \text{ with } 1 = \int d\mathbf{v} F(\mathbf{v}),$$

the continuity equation is obtained by setting $\chi = m$, the momentum equations by setting $\chi = m\mathbf{v}$, and the energy equation by setting $\chi = mv^2/2$. A distribution function $F(\mathbf{v})$ can then be found such that, when it is substituted into the transport Equation (7), the ideal 3D MHD equations are recovered. Huba and Lyon [5] show that

$$F = F_1 + F_2, \tag{8}$$

where

$$F_1 = \frac{\exp\left(-u_1^2\right)}{\left(\pi v_{1x}^2\right)^{1/2}} \frac{\exp\left(-v_1^2\right)}{\left(\pi v_{1y}^2\right)^{1/2}} \frac{\exp\left(-w_1^2\right)}{\left(\pi v_{1z}^2\right)^{1/2}}, \tag{9}$$

$$F_2 = -\left(u_2 v_2 + u_2 w_2 + v_2 w_2\right) \frac{\exp\left(-u_2^2\right) \exp\left(-v_2^2\right) \exp\left(-w_2^2\right)}{\left(\pi v_{2x}^2\right)^{1/2} \left(\pi v_{2y}^2\right)^{1/2} \left(\pi v_{2z}^2\right)^{1/2}} \quad (10)$$

and

$$u_\alpha = \left(v_x V_x\right)/v_{\alpha x}, v_\alpha = \left(v_y V_y\right)/v_{\alpha y}, w_\alpha = \left(v_z V_z\right)/v_{\alpha z}, v_{1x}^2 = 2C_s^2/\gamma +$$

$$V_{Ay}^2 + V_{Az}^2, v_{1y}^2 = 2C_s^2/\gamma + V_{Ax}^2 + V_{Az}^2, v_{1z}^2 = 2C_s^2/\gamma + V_{Ax}^2 + V_{Ay}^2, v_{2x}^2 = 4V_{Ax}^2,$$

$$v_{2y}^2 = 4V_{Ay}^2, v_{2z}^2 = 4V_{Az}^2, C_s^2 = \gamma P/\rho, V_{A\alpha}^2 = B_\alpha^2/4\pi\rho.$$

3 LBGK Equation and Particle Distribution Function

3.1 Nine-Velocity Square Lattice Model

In 2D, as shown in Fig. 1, we consider a square lattice with unit spacing on which each node has eight nearest neighbors connected by eight links. Particles can only reside on the nodes and move to their nearest neighbors along the links in unit time. There are two types of moving particles: the particles that move along the axis with speed $|\mathbf{e}_i| = 1, i = 1,2,3,4$ and the particles that move along the diagonals with speed $|\mathbf{e}_i| = \sqrt{2}$, $i = 5,6,7,8$. Also, there are rest particles with speed zero at each node. The occupation of these three types of particles is described by the single particle distribution function $f_i(x,t)$ where the subscript i indicates the velocity direction. The distribution function f_i is the probability of finding a particle at node $\mathbf{x}$ at time t with velocity $\mathbf{e}_i$. The particle distribution function satisfies the LBGK equation

$$f_i\left(\mathbf{x} + \mathbf{e}_i,\ t+1\right) - f_i\left(x,t\right) = \Omega_i = -\frac{1}{\tau}\left(f_i - f_i^{(0)}\right), \quad (11)$$

where $f_i^{(0)}$ is the equilibrium particle distribution function for MHD equations and τ is the single relaxation time which controls the rate of approach to equilibrium. The density ρ and the macroscopic velocity $\mathbf{v}$ are given by

$$\rho = \sum_i f_i \text{ and } \rho\mathbf{v} = \sum f_i \mathbf{e}_i. \quad (12)$$

Now from (8)-(10), we derive the particle equilibrium distribution function as follows:

$$\text{Define } \xi_{1x} = \frac{\sqrt{2}v_x}{\sqrt{3}v_{1x}},\ \xi_{1y} = \frac{\sqrt{2}v_y}{\sqrt{3}v_{1y}},\ \xi_{1z} = \frac{\sqrt{2}v_z}{\sqrt{3}v_{1z}}$$

$$\text{and } q_{1x} = \frac{\sqrt{2}V_x}{\sqrt{3}v_{1x}},\ q_{1y} = \frac{\sqrt{2}V_y}{\sqrt{3}v_{1y}},\ q_{1z} = \frac{\sqrt{2}V_z}{\sqrt{3}v_{1z}}.$$

Then, for small $\mathbf{V} = (V_x, V_y, V_z)$, we obtain

$$F_1^p = \frac{e^{-\frac{3}{2}\xi^2}}{\left(\frac{2\pi}{3}\right)^{\frac{D}{2}}} \cdot \left[1 + 3(\xi \cdot \mathbf{q}) + \frac{9}{2}(\xi \cdot \mathbf{q})^2 - \frac{3}{2}q^2\right], \tag{13}$$

where $\xi = \xi_{1x}\hat{\imath} + \xi_{1y}\hat{\jmath} + \xi_{1z}\hat{k}$ and $\mathbf{q} = q_{1x}\hat{\imath} + q_{1y}\hat{\jmath} + q_{1z}\hat{k}$. D is the spatial dimension. Note that in the absence of magnetic field, $v_{1x} = 2/RT$, $\xi_{1x} = v_x/\sqrt{3RT}$ and $q_{1x} = V_x/\sqrt{3RT}$. Equation (13) is the particle equilibrium distribution function obtained by Qian et al. [3] among others in the absence of magnetic field.

Similarly define $\eta_{2x} = \frac{\sqrt{2}v_x}{\sqrt{3}v_{2x}}$, $\eta_{2y} = \frac{\sqrt{2}v_y}{\sqrt{3}v_{2y}}$, $\eta_{2z} = \frac{\sqrt{2}v_z}{\sqrt{3}v_{2z}}$ and $r_{2x} = \frac{\sqrt{2}V_x}{\sqrt{3}v_{2x}}$, $r_{2y} = \frac{\sqrt{2}V_y}{\sqrt{3}v_{2y}}$, $r_{2z} = \frac{\sqrt{2}V_z}{\sqrt{3}v_{2z}}$. Then for small values of $V = (V_x, V_y, V_z)$, obtain

$$F_2^p = e^{-\frac{3}{2}\eta^2}\left[(a-b)\{1 + 3(\eta \cdot \mathbf{r})\} + a\left\{\frac{9}{2}(\eta \cdot \mathbf{r})^2 - \frac{3}{2}r^2\right\} + c\right], \tag{14}$$

where

$$\begin{gathered}\eta = \eta_{2x}\hat{\imath} + \eta_{2y}\hat{\jmath} + \eta_{2z}\hat{k}, \quad \mathbf{r} = r_{2x}\hat{\imath} + r_{2y}\hat{\jmath} + r_{2z}\hat{k}, \\ a = \tfrac{3}{2}(\eta_{2x} \cdot \eta_{2y} + \eta_{2y} \cdot \eta_{2z} + \eta_{2z} \cdot \eta_{2x}), \\ b = \tfrac{3}{2}[r_{2x}(\eta_{2y} + \eta_{2z}) + r_{2y}(\eta_{2x} + \eta_{2z}) + r_{2z}(\eta_{2x} + \eta_{2y})], \\ \text{and } c = (r_{2x} \cdot r_{2y} + r_{2y} \cdot r_{2z} + r_{2z} \cdot r_{2x}).\end{gathered}$$

Note that in the absence of magnetic field, $F_2^p = 0$. From (13) and (14), the equilibrium particle distribution $f_i^{(0)}$ can be obtained as

$$\begin{aligned} f_i^{(0)} = {} & \rho w_i \left[1 + 3\mathbf{e}_i \cdot \mathbf{q} + \tfrac{9}{2}(\mathbf{e}_i \cdot \mathbf{q})^2 - \tfrac{3}{2}q^2\right] + \\ & \rho w_i \left[(a-b)\left\{1 + 3(\mathbf{e}_i \cdot \mathbf{r}) + a\left\{\tfrac{9}{2}(\mathbf{e}_i \cdot \mathbf{r})^2 - \tfrac{3}{2}r^2\right\} + c\right\}\right], \end{aligned} \tag{15}$$

where $w_0 = 4/9$, $w_1 = w_2 = w_3 = w_4 = 1/9$, and $w_5 = w_6 = w_7 = w_8 = 1/36$. The relaxation time τ in (11) is related to the kinematic viscosity v_c by

$$\tau = \frac{6v_c + 1}{2}. \tag{16}$$

A Chapman-Enskog procedure can be applied to determine the macroscopic behavior of the model represented by (15). It can be shown that the macroscopic behavior matches the Navier-Stokes equations. The continuity equation (1) is recovered to $0(\delta^2)$ and $0(M_\alpha^3)$ where $\delta = c_s\tau/L$ and $M_\alpha = \sqrt{3}U/C_s$; L is the macroscopic length and Re is the Reynolds number $= UL/v_c$. The magnetic field equations (4) are solved by a kinetic wave particle split (KWPS) algorithm [6] following the procedure described by Huba and Lyon [5].

4 Computation of Hartmann-Poiseuille Flow

To demonstrate the method, the computation is performed for fully developed Hartmann-Poiseuille flow in a channel. The flow parameters are Hartmann number $H = 54.17$, Mach number $M = 0.19$, Reynolds number $\mathrm{Re} = 2.2 \times 10^6$, Magnetic Reynolds number $\mathrm{Re}\,m = 0.041$. The magnetic field B_0 is applied normal to the plate. The analytical solution for incompressible MHD flow is given as

$$
\begin{aligned}
V_x &= \tfrac{gL}{B_0\rho}\sqrt{\tfrac{\mu}{v_c}}\text{Coth }(H)\left[1-\tfrac{\cosh(Hy/L)}{\cosh(H)}\right], \\
B_x &= \tfrac{gL}{B_0\rho}\left(\tfrac{\sinh(Hy/L)}{\sinh(H)}-\tfrac{y}{L}\right).
\end{aligned} \tag{17}
$$

In (17), $g = \frac{dP}{dx}$ and μ is the electrical resistivity of the medium. Computations are performed on a 80 × 40 lattice. For the given Mach number and speed of sound, the characteristic velocity U is determined. Kinematic viscosity v_c is calculated from the given Reynolds number; the relaxation time τ is determined from (16). Starting from the initial density, velocity and magnetic fields, the equilibrium distribution function is obtained from (15). f_i is initialized as $f_i^{(0)}$. For each time step, the updating of the particle distribution is split into two substeps–collision and advection. The collision process at position $\mathbf{x}$ occurs according to the R.H.S. of LBGK Equation (11). The resulting particle distribution at $\mathbf{x}$, which is a sum of the original distribution and the collision term, is then advected to the nearest neighbor with velocity $\mathbf{e}_i$. Then ρ and $\mathbf{q}$ are computed from (12) using the updated distribution function. These values of ρ and $\mathbf{q}$ are then employed in solving the magnetic field equations on the lattice using the kinetic wave/particle splitting finite-difference method [6]. The new values of the magnetic field are then used in the calculation of the particle distribution function at new time step given by (15). This procedure is repeated until a convergence criterion is satisfied. Figures 2 and 3 show the comparison of the computed solution with the analytical solution. Excellent agreement is obtained.

5 Conclusion

A Lattice-Boltzmann-BGK (LBGK) formulation has been developed for low Mach number MHD flows. The method has been successfully applied to compute the Hartmann-Poiseuille flow. This preliminary investigation indicates the potential of the LBGK method for achieving high efficiency and accuracy on a lattice comparable to a finite-difference grid.

References

1. CHEN, S. and DOOLEN, G.D., Lattice Boltzmann Method for Fluid Flows, *Ann. Rev. Fluid Mechanics*, Vol. 30, pp. 329-364, 1998.
2. CHEN, H., CHEN, S. and MATTHAEUS, W.H., Recovery of the Navier-Stokes Equations Using a Lattice-Gas Boltzmann Method, *Phys. Rev. A*, Vol. 45, pp. R 5339-5342, 1992.
3. QIAN, Y.H., D'HUMIERES, D. and LALLEMAND, P., Lattice BGK models for Navier-Stokes Equations, *Europhys. Lett.*, Vol. 17, pp. 479-484, 1992.
4. HUBA, J.D. and LYON, J.G., A New 3D MHD Algorithm: The Distribution Function Method, *J. Plasma Physics*, Vol. 61, pp. 391-405, 1999.
5. MARTINEZ, D.O., MATTHAEUS, W.H. and CHEN, S., Lattice Boltzmann Magnetohydrodynamics, *Phys. Plasmas*, Vol. 1, pp. 1850-1867, 1994.

6. REKSOPRODJO, H. and AGARWAL, R.K., A Kinetic Scheme for Equations of Ideal Magnetohydrodynamics with a Bi-Temperature Model, AIAA Paper No. 2000-0448, 2000.

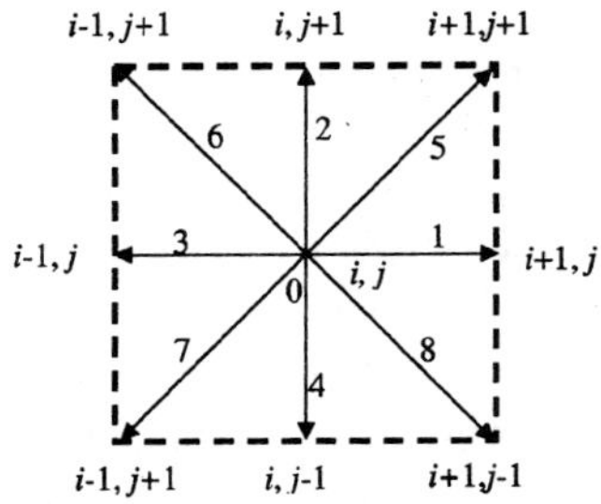

Fig. 1: Nine-Velocity Square Lattice

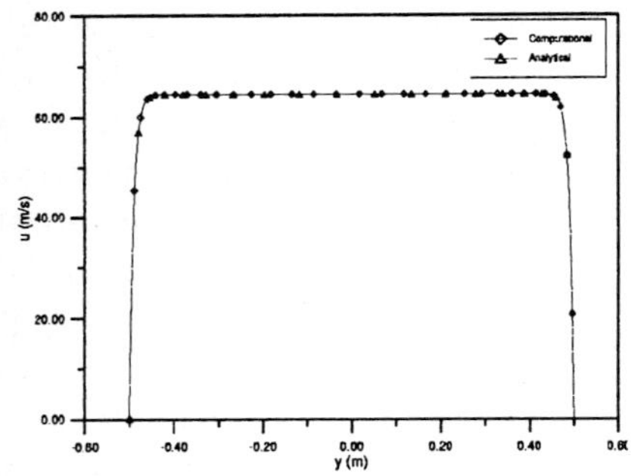

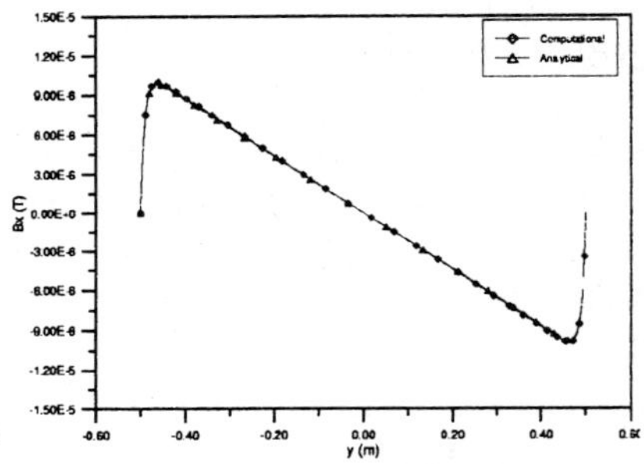

Fig. 2: V_x velocity along the y -axis. Fig. 3: Magnetic field B_x along the y -axis.

Convergence Acceleration in Lattice Boltzmann Method

Nobuyuki Satofuka and Mitsuru Ishikura

Kyoto Institute of Technology, Kyoto 606-8585, Japan

Abstract. Numerical simulations using the lattice Boltzmann (LB) method are presented for the two-dimensional channel flow with/without sudden expansion. Nine-velocity square lattice model is used. In order to improve computational efficiency of the LB method, nonuniform mesh is used, and residual averaging and mutigrid method are also incorporated into the LB method. By using these techniques, convergence to a steady state was considerably accelerated.

1 Introduction

The Navier-Stokes equations can be derived as the macroscopic behavior of hard sphere particles with a Maxwellian velocity distribution function whose evolution is governed by the Boltzmann equation. However, much simpler kinetic models can result in the same macroscopic behavior. This idea led Frisch, Hasslacher, and Pomeau [1] to the invention of a novel technique called lattice gas automata (LGA). Although the LGA method has provided a fast and efficient way for solving partial differential equations, there exist some fundamental problems in this method in simulating realistic fluid flows obeying Navier-Stokes equations. The lattice gas fluid momentum equations cannot be reduced to the Navier-Stokes equations because of two fundamental problems. The first is the non-Galilean invariance property due to the density dependence of the convection coefficient. Second the pressure has explicit and unphysical velocity dependence. To avoid some of these inherent problems, several lattice Boltzmann (LB) models have been proposed. The main feature of LB method is to replace the particle occupation variables (Boolean variables) by the single-particle distribution functions (real variables) $f_i = \langle n_i \rangle$, where $\langle\ \rangle$ denotes a local ensemble average, in the evolution equation, i.e.,the lattice Boltzmann equation. The LB method as a numerical scheme was first proposed by McNamara and Zanetti [2]. The LB model proposed by Chen et al [3], and Qian et al [4]. abandons Fermi-Dirac statistic and applies the single relaxation time approximation first introduced by Bhatnager, Gross, and Krook in 1954 [5], to greatly simplify the collision operator. This model is called the lattice BGK (LBGK) model. In this paper the LBGK method [6] is used to simulate two-dimensional channel flow with/without sudden expansion. In order to improve computational efficiency, use of nonuniform mesh and incorporation of residual averaging and multigrid are proposed.

2 Nine-velocity Square Lattice Model

A square lattice with unit spacing is used on which each node has eight nearest neighbors connected by eight links as shown in Fig. 1. Particles can only reside on the nodes and move to their nearest neighbors along these links in the unit time. Hence, there are two types of moving particles. Particles of type 1 move along the axes with speed $|\mathbf{e}_{1i}| = 1$ and particle of type 2 move along the diagonal directions with speed $|\mathbf{e}_{2i}| = \sqrt{2}$. Rest particles with speed zero are also allowed at each node.

The occupation of the three types of particles is represented by the single-particle distribution function, $f_{\sigma i}(\mathbf{x}, t)$, where subscripts σ and i indicate the type of particle and the velocity direction, respectively. When $\sigma = 0$, there is only f_{01}. The distribution function, $f_{\sigma i}$, is the probability of finding a particle at node $\mathbf{x}$ and time t with velocity $\mathbf{e}_{\sigma i}$. The particle distribution function satisfies the lattice Boltzmann equation

$$f_{\sigma i}(\mathbf{x} + \mathbf{e}_{\sigma i}, t + 1) - f_{\sigma i}(\mathbf{x}, t) = \Omega_{\sigma i} \tag{1}$$

where $\Omega_{\sigma i}$ is the collision operator representing the rate of change of the particle distribution due to collisions. According to Bhatnagar, Gross, and Krook (BGK) , the collision operator is simplified using the single time relaxation approximation. Hence, the lattice Boltzmann BGK (LBGK) equation (in lattice unit) is

$$f_{\sigma i}(\mathbf{x} + \mathbf{e}_{\sigma i}, t + 1) - f_{\sigma i}(\mathbf{x}, t) = -\frac{1}{\tau}\left[f_{\sigma i}(\mathbf{x}, t) - f_{\sigma i}^{(0)}(\mathbf{x}, t)\right] \tag{2}$$

where $f_{\sigma i}^{(0)}(\mathbf{x}, t)$ is the equilibrium distribution at $\mathbf{x}, t$ and τ is the single relaxation time which controls the rate of approach to equilibrium. The density per node, ρ, and the macroscopic velocity, $\mathbf{u}$, are defined in terms of the particle

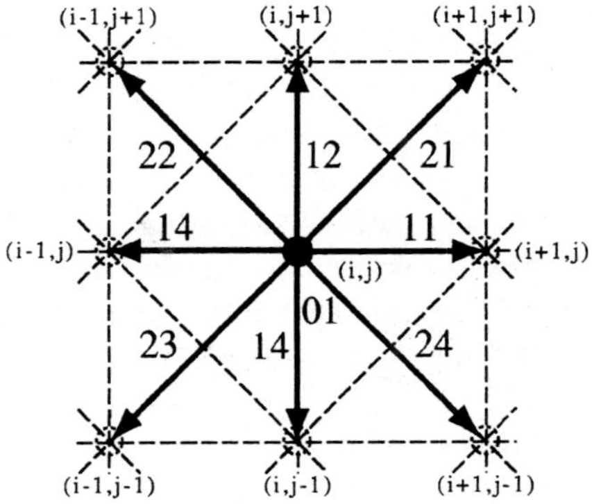

Fig. 1. Square lattice model.

distribution functions by

$$\rho = \sum_{\sigma}\sum_{i} f_{\sigma i}, \quad \rho\mathbf{u} = \sum_{\sigma}\sum_{i} f_{\sigma i}\mathbf{e}_{\sigma i}. \tag{3}$$

A suitable equilibrium distribution can be chosen in the following form for particles of each type

$$\begin{aligned} f_{01}^{(0)} &= \rho\alpha - \frac{3}{2}\rho\mathbf{u}^2, \\ f_{1i}^{(0)} &= \rho\beta + \frac{1}{3}\rho(\mathbf{e}_{1i}\cdot\mathbf{u}) + \frac{1}{2}\rho(\mathbf{e}_{1i}\cdot\mathbf{u})^2 - \frac{1}{6}\rho\mathbf{u}^2, \\ f_{2i}^{(0)} &= \rho\frac{(1-4\beta-\alpha)}{4} + \frac{1}{12}\rho(\mathbf{e}_{2i}\cdot\mathbf{u}) + \frac{1}{8}\rho(\mathbf{e}_{2i}\cdot\mathbf{u})^2 - \frac{1}{24}\rho\mathbf{u}^2. \end{aligned} \tag{4}$$

The relaxation time is related to the viscosity by

$$\tau = \frac{6\nu + 1}{2} \tag{5}$$

where ν is the kinematic viscosity measured in lattice units. Hou et al [6] used the value of $\alpha = 4/9$ and $\beta = 1/9$.

Having chosen the appropriate lattice size and the characteristic velocity, ν can be calculated for a given Re number and then the relaxation time can be determined by using Eq. 5. Starting from an initial density and velocity fields, the equilibrium distribution function can be obtained using Eq. 4, and $f_{\sigma i}(\mathbf{x},t)$ can be initialized as $f_{\sigma i}^{(0)}(\mathbf{x},t)$. For each time step, the updating of the particle distribution can be split into two substeps: collision and advection. The collision process at position $\mathbf{x}$ occurs according to the right-hand side of the lattice Boltzmann equation given as Eq. 2. The resulting particle distribution at $\mathbf{x}$, which is the sum of the original distribution and the collision term, is then advected to the nearest neighbor of $\mathbf{x} + \mathbf{e}_{\sigma i}$, for particle velocity . Then $\rho, \mathbf{u}$ can be computed from the updated using Eq. 3. The updating procedure can be terminated for steady state problems when a certain criterion is satisfied. The method can also be used for transient problems.

3 Two-dimensional Channel with Sudden Expansion

A two-dimensional channel with an expansion ratio of 1:2 and an aspect ratio of 1:4 for each side of expansion is studied here. At a moderate value of Reynolds number such as $Re = 10$, flows in the channel is two-dimensional. The Reynolds number for the system is defined as

$$Re = \frac{H\ U_0}{\nu} \tag{6}$$

where H is the height of the entry section, U_0 is the maximum inlet velocity, and ν is the kinematic viscosity. The Reynolds number is chosen to be 10 for this simulation.

At the entrance (upstream), a parabolic profiles of the horizontal component of velocity, u, with a maximum $U_0 = 0.1$, is enforced, and the vertical component of velocity, v, is set to zero. At the exit (downstream), a constant pressure boundary conditions, $p = 1.0$, is enforced. At the walls a noslip boundary condition is applied. To impose pressure boundary conditions explicitly, it is better to transform the density distribution function $f_{\sigma i}$ to the pressure distribution function $p_{\sigma i}$ using $p_{\sigma i} = \rho c_s^2 f_{\sigma i}$, then Eq. 2 becomes

$$p_{\sigma i}(\mathbf{x} + \mathbf{e}_{\sigma i}, t+1) - p_{\sigma i}(\mathbf{x}, t) = -\frac{1}{\tau}\left[p_{\sigma i}(\mathbf{x}, t) - p_{\sigma i}^{(0)}(\mathbf{x}, t)\right]. \qquad (7)$$

The average density, ρ_0, is set to be 1.0 and the initial value of the velocity field is set to be zero in the interior of the channel.

Two types of computational mesh are used in the simulation: uniform mesh shown in the lower half of Fig. 2, and nonuniform rectangular mesh shown in the upper half of Fig. 2. A nonuniform mesh is certainly desirable in many practical applications. What inhibits the use of a nonuniform mesh in conventional LB methods is coupling of the discretization of physical and momentum spaces. The pressure distribution have to move from one lattice site to another in a single time step to warrant the followed collision process. In the present approach, the computational mesh is uncoupled from the discretization of momentum space and it can have an arbitrary shape. Collisions still take place on the grid points of the computational mesh. After a collision, the pressure distributions at the lattice sites now may not be exactly determined, they can always be calculated using interpolation. After interpolation collision and advec-

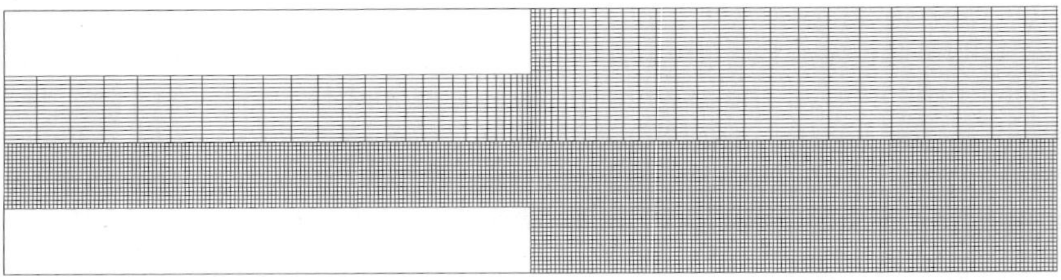

Fig. 2. Nnuniform mesh (upper) and uniform mesh (lower).

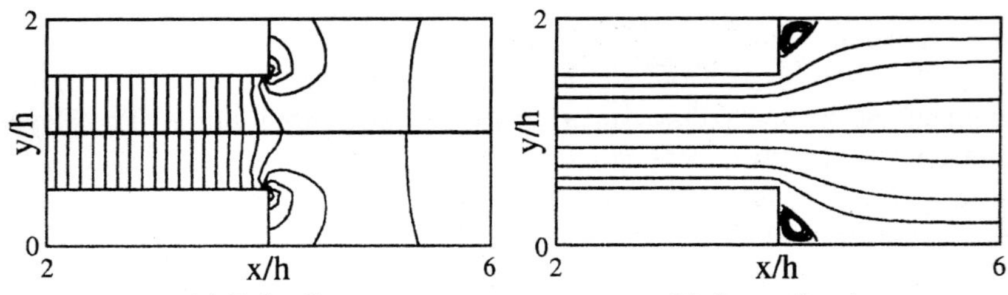

(a) Isobar line (b) Stream function

Fig. 3. Results of simulations.

tion steps are repeated. The uniform rectangular mesh of $128 \times 33 + 129 \times 65$ and the nonuniform one of $27 \times 33 + 28 \times 65$ are used in the simulation. Fig. 3 shows the isobar line (Fig. 3(a)) and the contour line of the stream function (Fig. 3(b)) for flow in the symmetric channel with sudden expansion. In these figures, upper half corresponds to that simulated by using the nonuniform mesh while lower half indicates one for uniform mesh. It is apparent that the agreement is quite excellent. Computational cost of the simulation with the nonuniform mesh is less than 1/5 of that using the uniform mesh.

4 Convergence Acceleration

In traditional numerical method, residual averaging and multigrid method are usually used to accelerate the convergence to a steady state. However these techniques have not been applied in the LB Method. Therefor we tried to simulate two-dimensional channel flow problem using these techniques.

In residual averaging, residual $R_{\sigma i}$ is defined as

$$R_{\sigma i} = p_{\sigma i}(\mathbf{x}, t+1) - p_{\sigma i}(\mathbf{x}, t) \tag{8}$$

and distribution function of next time step, $p_{\sigma i}(\mathbf{x}, t+1)$ is rewritten as

$$p_{\sigma i}(\mathbf{x}, t+1) = p_{\sigma i}(\mathbf{x}, t) + \bar{R}_{\sigma i} \tag{9}$$

where $\bar{R}_{\sigma i}$ is residual which is averaged for each particle. For example,if $p_{\sigma i} = p_{11}$, $\bar{R}_{11}$ is calculated as

$$\bar{R}_{11} = \frac{1}{10}\begin{bmatrix} R_{11}(i-1, j+1) & R_{11}(i, j+1) \\ 2R_{11}(i-1, j) & 4R_{11}(i, j) \\ R_{11}(i-1, j-1) & R_{11}(i, j-1) \end{bmatrix}. \tag{10}$$

In multigrid method, following relations are defined.

$$p^{(0)}_{\sigma i2h} = I^{2h}_{h} p_{\sigma ih} \quad , \quad I^{2h}_{h} = \frac{1}{16}\begin{bmatrix} 1 & 2 & 1 \\ 2 & 4 & 2 \\ 1 & 2 & 1 \end{bmatrix} \tag{11}$$

where subscripts h,$2h$ are grid level expressed by lattice size, while $p^{(0)}_{\sigma i2h}$ is an initial value at coarse grid. Using distribution functions $p_{\sigma i2h}$ at coarse grid which are calculated by Eq.7, distribution functions at fine grid are modified as

$$p^{new}_{h} = p_{h} + I^{h}_{2h}(p_{\sigma i2h} - p^{(0)}_{\sigma i2h}). \tag{12}$$

Fig.4 shows comparison of convergence history of L2-residual. Table 1 shows comparison of CPU times. In these cases computations were carried out with $Re = 10$ on $65 \times 33, 129 \times 65, 257 \times 129$ lattice. The number of time steps using residual averaging is 14% less than that of the traditional LB method. The result using multigrid method with residual averaging is about 70% of that without convergence acceleration. Saving of CPU time using residual averaging is only marginal. CPU time using multigrid method with residual averaging is about 1/2 of that without multigrid. These cases are computed on SGI ONYX R 10000.

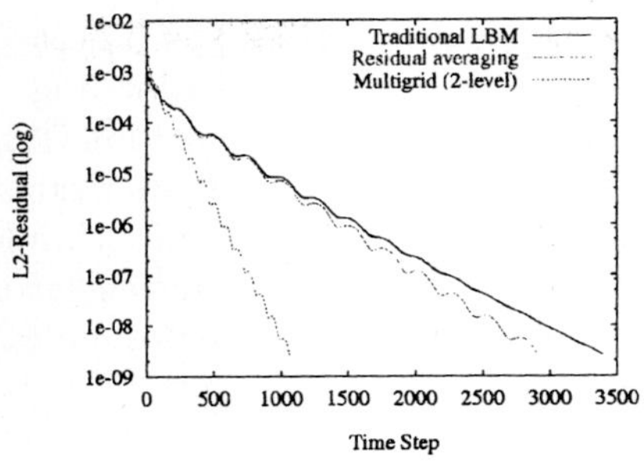

Fig. 4. Convergence history of L2-residual(65 × 33)

Table 1. Comparison of CPU time

lattice size	65×33	129×65	257×129
traditional	57.5	733	10564
resi. ave.	56.5	743	10864
multi.(2-level)	24.7	436	5940
traditional	1	1	1
resi. ave.	0.98	1.01	1.03
multi.(2-level)	0.43	0.59	0.56

5 Conclusion

Numerical simulations of two-dimensional channel flow with/without sudden expansion are presented. By using nonuniform mesh for two-dimensional channel with sudden expansion, the computational time was reduced to 20% of that with uniform mesh. By incorporating residual averaging and multigrid into the lattice Bolzmann method, convergence to steady state was considerably accelerated.

Acknowledgemens

This study was supported in part by the Research for Future Program (97P01101) from Japan Society for the Promotion of Science.

References

1. U. Frisch, B. Hasslacher, Y. Pomeau: Lattice-Gas Automata for the Navier-Stokes Equation, *Phys. Rev. Lett. 56*, pp. 1505-1508, (1986).
2. G. McNamara, G. Zanetti: Use of the Boltzmann Equation to Simulate Lattice-Gas Automata, *Phys. Rev. Lett. 61*, pp. 2332-2335, (1988).
3. H. Chen, W. H. Matthaeus: Recovery of the Navier-Stokes Equation Using a Lattice-Gas Boltzmann Method, *Phys. Rev. A. 45*, pp. 5539-5542, (1992).
4. Y. H. Qian, D. D'Humieres, P. Lallemand: Lattice BGK Models for Navier-Stokes Equation, *Europhys. Lett. 17*, pp. 479-484, (1992).
5. P. L. Bhatnagar, E. P. Gross, M. Krook: A model for Collision Processes in Gases. 1. Small amplitude Processes in Charged and Neutral one-component Systems, *Phys. Rev. 94*, pp. 511-525, (1954).
6. S. Hou, Q. Zou, S. Chen, G. Doolen, A. C. Cogley: Simulation of Cavity Flow by the Lattice Boltzmann Method, *J. Comput. Phys. 118*, pp. 329-347, (1995).

Flow Prediction by Lattice-Boltzmann Methods

Olga Filippova and Dieter Hänel

Institute of Combustion and Gasdynamics, University of Duisburg, D-47048 Duisburg, Germany

Abstract. The present paper is concerned with applications of lattice-Boltzmann methods to aerodynamic flow problems. Lattice Boltzmann methods are based on gas-kinetic representation of fluid dynamics and enable the simulation of macroscopic, continuum flows in the low Mach number limit. With recent extensions of the solution concept, as local grid refinement and boundary fitting, the method is able to deal with strongly anisotropic flow problems over curvilinear geometries. The solution concept will be explained in detail and demonstrated by applications to 2D and 3D flows around cylinders and airfoils.

1 Introduction

The Lattice-Boltzmann method, used here in the form of the lattice-BGK (LBGK) method [1,2], is a discrete solution method of the Boltzmann equation with the simplified collision term by Bhatnagar, Gross, Krook (BGK). This method has reached now a rather high level for very different, complex physical applications. Nevertheless in its basic form it can be considered as the new approach for numerical simulations of incompressible flows (with the restriction that it achieves this state in low-frequency limit).

Although the ability of LBGK scheme to simulate incompressible flows in low-frequency limit was recognized several years ago, the original scheme has needed essential algorithmic developments to deal with high-Reynolds number flows. The difficulties appearing on this way are connected with indirect numerical simulation of the Navier-Stokes equations with the LBGK scheme. The basic simple and efficient gas-kinetic algorithm is indissolubly connected with Cartesian-like "molecular" lattice. Therefore the LBGK scheme in its original form is not able to compete with highly developed conventional CFD methods in simulation of high Reynolds number flows ($Re \sim O(10^3)$) in complex geometry and its main field of application was seen in easy and efficient simulation of small Reynolds number flows ($Re \sim O(10)$) in complex random geometry (as porous media).

To circumvent this limitation, a number of variants have been proposed [3,4], most of which were based on a combination of Lattice Boltzmann scheme with consolidated finite volume or finite difference techniques. This merge has considerably extended the range of application of the Lattice Boltzmann method at a fairly reasonable cost in terms of computational complexity. Nonetheless, these merges can accommodate only relatively smooth variations of the flow field because large deformations of the non-uniform mesh may result in numerical instabilities. Many flows of engineering interest, however, exhibit violent excursions

over highly localized regions (as boundary layers) which require a correspondingly highly clustered mesh.
The popular approach in conventional CFD methods for solution of the flows with localized zones of high gradients is provided by *locally embedded* grids, namely grids in which the lattice/time spacing is refined or coarsened locally. This kind of refinement concept for LBGK methods was developed in [5]. It preserves the second order spatial accuracy of the LBGK scheme, improves locally its time accuracy and enables in addition some acceleration strategies.

The other serious problem of the LBGK method was the treatment of arbitrary curvilinear geometries without impairing of second order spatial accuracy. Widely used "bouncing-back" conditions enable to consider arbitrary, even very complex and irregular geometry, but in the "stair" manner, decreasing the accuracy of the scheme near the boundary to the first order. To overcome this drawback a new boundary fitting formulation, which preserves the second order accuracy of the outer scheme for arbitrary curvilinear geometry, was developed by the authors [6,7] In its application to isolated boundaries in low-frequency limit it doesn't carry instability in LBGK scheme if the resolution of the boundary layer is reasonable, whereas being connected with the other kind of boundary formulations (as, for instance, at the edges of inlet/outlet of the channel) it can provide instability. This instability is evoked by the basic feature of this concept as it is the approximate position of the boundary describing with boundary-fitting formulas. Boundary fitting concept enables the approximation of different types of boundary conditions, as they are no-slip and slip conditions, injection and suction of the fluid from the curvilinear surfaces and wall function formulations for turbulent flows.

With these essential algorithmic extensions the LBGK scheme is able to deal efficiently with strongly anisotropic flow problems over curvilinear geometries. The solution concept is demonstrated in applications to 2D and 3D high-Reynolds number flows around cylinders and airfoils.

2 The LBGK scheme

In Lattice BGK schemes [1,2] the molecular phase space is represented by a Cartesian-like lattice, where the step size $\delta_x = C\delta_t$ corresponds to the "mean free path" of a "molecule" of speed C along link i. The lattice BGK formulation of fluid dynamics reads as follows:

$$f_i(t+\delta_t, \boldsymbol{r}+\boldsymbol{C}_i\delta_t) = f_i(t,\boldsymbol{r}) + \omega[f_i^{\mathrm{eq}}(t,\boldsymbol{r}) - f_i(t,\boldsymbol{r})]$$

Here f_i is a set of discrete populations representing the number density of "molecules" at position $\boldsymbol{r}$ at time t moving along the direction identified by the discrete speed $\boldsymbol{C}_i$. Once the discrete populations are known, fluid pressure and speed are given as moments of the discrete distribution functions at node $(t, \boldsymbol{r})$

$$P = \frac{\rho_0 C^2}{3} \sum_i f_i, \quad \boldsymbol{U} = \sum_i f_i \boldsymbol{C}_i$$

The expression for equilibrium distribution function for simulation of incompressible flows with constant density $\rho_0 = 1$ was proposed in [8]

$$f_i^{\mathrm{eq}} = t_p [\frac{P}{\rho_0 C_s^2} + \frac{U_\alpha C_{i\alpha}}{C_s^2} + \frac{U_\alpha U_\beta}{2C_s^2} (\frac{C_{i\alpha} C_{i\beta}}{{C_s}^2} - \delta_{\alpha\beta})] \tag{1}$$

where index p represents the square modulus of the corresponding "molecular" velocity $\boldsymbol{C}_i$. For 9-speed LBGK models in 2D $p = 0, 1, 2$ and $t_0 = 4/9, t_1 = 1/9, t_2 = 1/36$ and for 15-speed LBGK models in 3D $p = 0, 1, 3$ and $t_0 = 2/9, t_1 = 1/9, t_2 = 1/72$.

3 Numerical Results

Several numerical examples described below have shown that the LBGK concept with the proposed algorithmic extensions is the simple, accurate and efficient tool for dealing with aerodynamic flow problems at high Reynolds numbers.

The simulation of three-dimensional organized coherent structures considered as the onset of turbulent motions in the wake flow behind blunt bodies is one of the main topics of the modern CFD. Lattice BGK scheme with boundary-fitting and local grid refinement has shown its ability to deal with such kind of problems. The spontaneous development of 3-D instability in the wake flow behind the circular cylinder in the channel at $Re = 300$ is presented in Fig. 1, where lateral modes of vortex shedding are shown. Prismatic patch of the finer grid with refinement ratio 3 is applied around the cylinder to resolve accurately boundary layer.

Grid refinement in 3D can be done uniformly in all directions or with the stretching in one direction. The refinement with stretching is especially suitable for consideration of 3D instability of the initially 2D flows (as in the previous example). In this case only several layers of the fine cubic LBGK lattice with $z = const$ are considered instead of the connected lattice and LBGK scheme on these layers is extended by the interpolation of the distribution functions in z-direction as it was proposed in general in [4]. The changes in the relaxation parameters, time-stepping on the fine grid and rescaling of the distribution functions on the interfaces between two grids are performed according to local 2nd order grid refinement concept [5].

2D vortex shedding from the trailing edge of airfoils in cascade for $Re = 7500$ and angle of attack 16° is shown in Fig. 2 for 10 different instants in cycle. Patch with refinement ratio 6 is applied around an airfoil in periodical cell to resolve accurately thin boundary layers. Left column in Fig. 2 shows instantaneous isobars whereas right column shows instantaneous isolines of vorticity in the vicinity of airfoil at the same time.

An extension on turbulent flows at high Re numbers is achieved by coupling a finite-difference solution of the $k-\epsilon$ turbulence model with the LBGK solution. The logarithmic law of wall is used in the boundary-fitting formulation [6].

This allows to solve efficiently turbulent flows around arbitrary curvilinear geometries on the simplest Cartesian grids with second order accuracy in space. LBGK scheme with local grid refinement can be essentially accelerated in the simulation of steady-state problems by the reduction of the amount of time steps on the fine grid [9]. Numerical results are in the good agreement with experimental data. Details one can find in [10]. C_P values on the surface of NACA 4412 profile obtained experimentally [11] and numerically with different models for high-Reynolds number flows are presented in Fig. 3.

Summarizing, the application of the developed LBGK method for several aerodynamical problems has shown the good accuracy and high efficiency of the new numerical approach avoiding expensive grid generation.

4 Figures

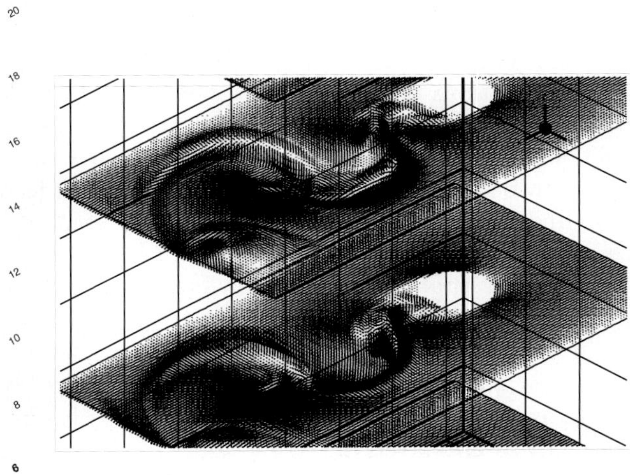

Fig. 1. 3D instability in the wake behind the circular cylinder in the rectangular channel at $Re = 300$. Velocity field in two different cuts $z = const$, v_z is increased 10 times

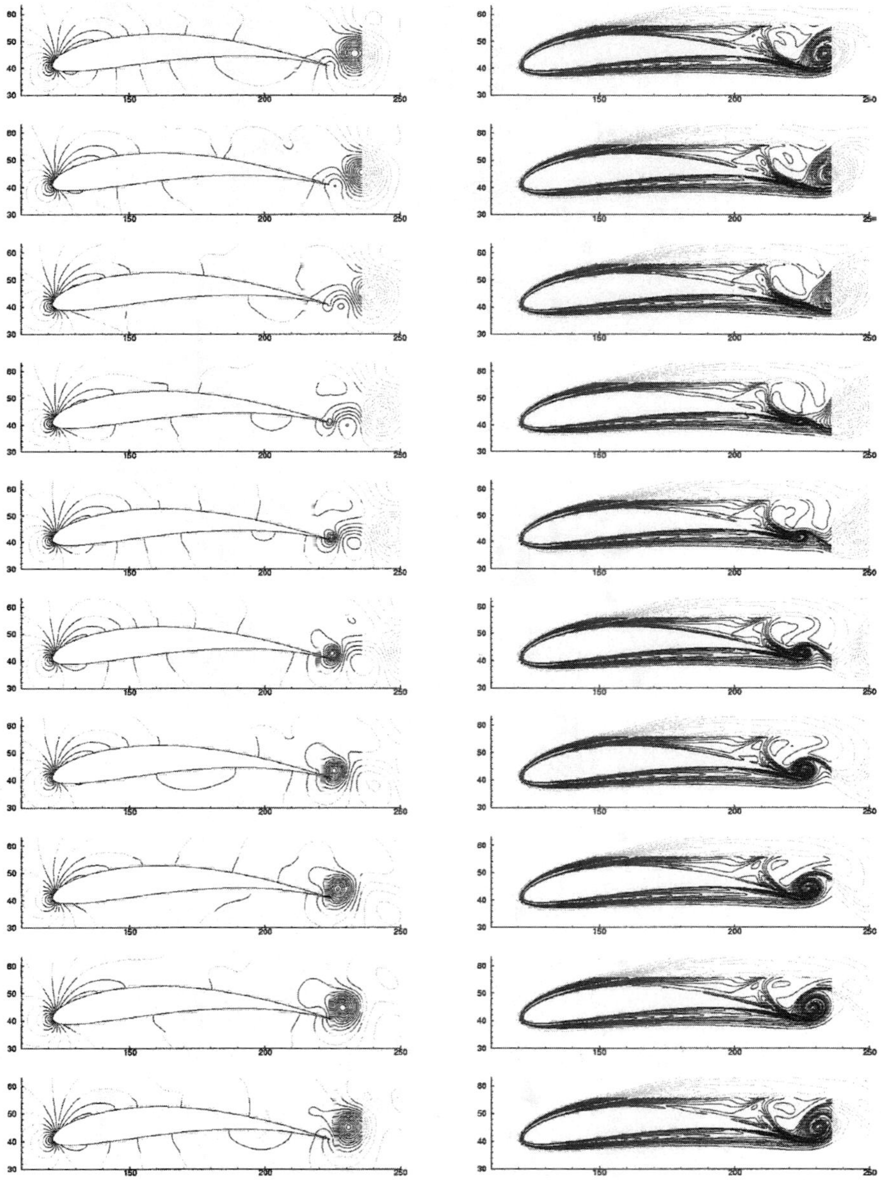

Fig. 2. Instantaneous isobars (left column) and isolines of vorticity (right column) for laminar incompressible flow around an airfoil in cascade ($Re = 7500$, angle of attack 16°) at 10 different instants in cycle. Solid lines - data from the embedded grid (refinement ratio 6), dotted lines - data from the basic coarse grid

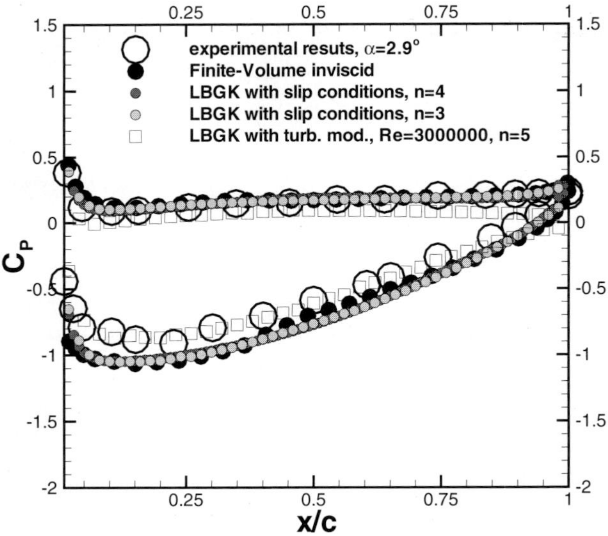

Fig. 3. C_P values for high-Reynolds number flow around an isolated NACA 4412 profile ($\alpha = 2.9^o$) obtained experimentally and numerically. Open circles - experimental data ($Re \sim 3 \cdot 10^6$) [11] solid black circles - numerical results obtained with Finite-Volume solver for inviscid flows, solid light circles - numerical results obtained with multi-scale LBGK scheme with slip conditions on the surface of the airfoil, squares - numerical results obtained with multi-scale LBGK scheme with turbulence modelling ($Re = 3 \cdot 10^6$) [10]

References

1. Y.H. Qian, d'Humieres and P. Lallemand, *Europys. Lett.*, **17**, 479 (1992)
2. S. Chen, Z. Wang, X. Shan and G. D. Doolen, *J. Stat. Phys.*, **81**, 35 (1992)
3. F. Nannelli and S. Succi, *J. Stat. Phys.*, **68**, N 3/4, 401 (1992)
4. X. He, L.-S. Luo and M. Dembo, *J. Comp. Phys.*, **129**, 357 (1996)
5. O. Filippova and D. Hänel, *J. Comp. Phys.*, **147**, 219 (1998)
6. O. Filippova and D. Hänel, *Int. J. Mod. Phys. C*, **9**, N8, 1271 (1998)
7. O. Filippova and D. Hänel, *Computers & Fluids*, **26**, 697 (1997)
8. Q. Zou, S. Hou, S. Chen and G. Doolen, *J. Stat. Phys.*, **81**, 35 (1995)
9. O. Filippova and D. Hänel, submitted in *J. Comp. Phys.*, (1999)
10. O. Filippova, S. Succi, F. Mazzocco, C. Arrighetti, G. Bella and D. Hänel, *submitted in J. Comp. Physics* (2000)
11. R. M. Pinkerton, NACA Rept. N. 563, (1936)

Numerical Simulation of Fluid Motions in a Rotating Field by the Lattice Boltzmann Method

Michihisa Tsutahara, Masaaki Okabayashi, and Takeshi Kataoka

Graduate School of Science and Technology, Kobe University, Rokkodai, Nada, Kobe 657-8501, Japan

Abstract. Rotating fluids are simulated numerically by the lattice Boltzmann method. The 3D15V incompressible lattice Boltzmann model is applied to three-dimensional fluid flows. In a rotating coordinate system, the effect of Coriolis force is incorporated into the local equilibrium distribution function. The averaging method is used in the calculation. The Taylor column is detected by the simulation. The Ekman spiral and the spin-down are simulated and results agree well with theoretical ones.

1 Introduction

In recent years, the lattice Boltzmann method (LBM) has attracted much attention in the computational fluid dynamics [1]. In the LBM, the fluid is regarded as gathering of many particles. The dynamics of the particles evolving in a lattice space consists of two steps: (1) particles at the same lattice site collide according to the collision rules which conserve mass and momentum at each site; (2) after colliding, the particles advect to the next lattice sites in the direction of their velocities. The LBM has the following advantages: (1) simple algorithm; (2) high parallelism; (3) easy treatment of boundary condition. It has been applied to various kinds of flow field with complicated geometry [2][3], such as two-phase fluid flows, flows through porous media, magnetohydrodynamics, and so on. Its efficiency has been confirmed. However, the LBM has not been used very much in a system where the effect of a body force, such as gravity or Coriolis force, is included. In this system, it is necessary to average two distribution functions after the collision stage and after the translating stage because of the excess velocity distributions. The present study deals with simulating incompressible viscous fluid flows in a rotating coordinate system by the LBM.

2 Formulation

In this paper, we use the so-called 3D15V incompressible lattice Boltzmann model. In Fig.1, we consider a cubic lattice in three dimensions and choose the set of vectors associated with the moving directions of particles as follows: $(\pm 2, 0, 0), (0, \pm 2, 0), (0, 0, \pm 2)$ and $(\pm 1, \pm 1, \pm 1)$. The evolution equation for the particle distribution function in space $\mathbf{r}$ and time t is

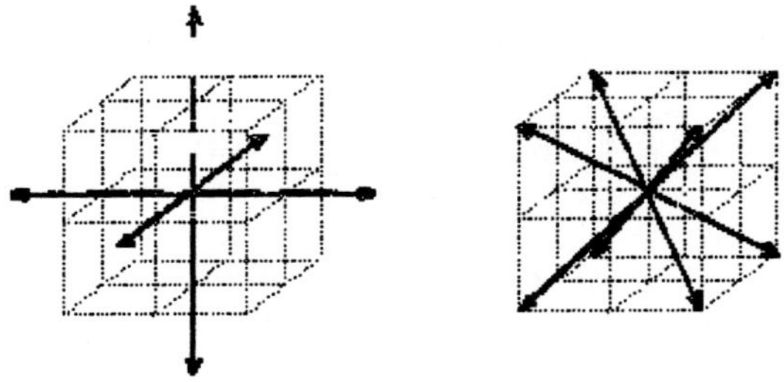

Fig. 1. Velocity vectors of moving particles

$$f_i\left(t+1, r+c_i\right)=f_i(t, r)-\frac{1}{\phi}\left[f_i(t, r)-f_i^{(0)}(t, r)\right] \tag{1}$$

where f_i is the distribution function of particle i, $\mathbf{c}_i$ its velocity, and ϕ the relaxation time. The fluid density and the fluid velocity $\mathbf{u}$ are defined by

$$\rho=\sum_i f_i\ ,\ \mathbf{u}=\frac{1}{\rho} \sum_i f_i \mathbf{c}_i \tag{2}$$

We use the local equilibrium distribution functions $f_i^{(0)}$ proposed in [4]:

$$f_i^{(0)}=A_n+B_n \mathbf{c}_i \cdot \mathbf{u}+C_n\left(\mathbf{c}_i \cdot \mathbf{u}\right)^2+D_n u^2 \tag{3}$$

Subscripts $n=0, n=1$ and $n=2$ indicate rest particles, moving particles with speed 2 and with speed $\sqrt{3}$, respectively. Coefficients A_n, B_n, C_n and D_n are given as follows:

$$\begin{aligned}
& A_0=\frac{\rho}{23}, B_0=0, C_0=0, D_0=-\frac{7 \rho}{24} \\
& A_1=\frac{\rho}{23}, B_1=\frac{\rho}{24}, C_1=\frac{\rho}{32}, D_1=\frac{\rho}{48} \\
& A_2=\frac{2 \rho}{23},\ B_2=\frac{\rho}{12},\ C_2=\frac{\rho}{16},\ D_2=-\frac{\rho}{24}
\end{aligned} \tag{4}$$

The effect of the body force is introduced into the local equilibrium distribution functions. According to the force $\mathbf{F}=\rho\mathbf{a}$ ($\mathbf{a}$: the acceleration), $\mathbf{u}$ in Eq.(3) is replaced by $\mathbf{u}+\phi\mathbf{a}$ at each time step [5]. The impulsive force acts on particles at the collision stage. So that the velocity field obtained by the distribution functions right after the collision stage has excessive velocity in the direction of the body force. After the translating stage, the excessively large pressure gradient will be relaxed. Therefore the correct velocity distribution is obtained by averaging two distribution functions after the collision stage and after the translating stage. According to the multiple scale Chapman-Enskog procedure, one obtains the continuity equation,

$$\frac{\partial \rho}{\partial t} + \frac{\partial}{\partial r_\alpha}(\rho u_\alpha) = 0 \tag{5}$$

and the Navier-Stokes equations,

$$\frac{\partial \rho u_\alpha}{\partial t} + u_\beta \frac{\partial}{\partial r_\beta}(\rho u_\alpha) = -\frac{\partial p}{\partial r_\alpha} + F_\alpha + \nu \frac{\partial^2}{\partial r_\beta^2}(\rho u_\alpha) \tag{6}$$

where p is the pressure, F_α the body force, and ν the kinematic viscosity given as

$$\nu = \frac{2\phi - 1}{3} \tag{7}$$

3 Results

In a rotating coordinate system, Coriolis force acts on all the particles. The rotating axis is in the z direction with angular velocity ω. The Rossby number Ro and the Ekman number Ek are defined as

$$Ro = \frac{U}{\omega H}, \quad Ek = \frac{\nu}{\omega H^2} \tag{8}$$

where U is the reference velocity, ω the angular velocity of the rotating coordinate, and H the characteristic flow length scale.

Firstly, we simulate flow around a cubic body located in a square channel. The number of lattices is 180(x *direction*) $\times$ 50(y *direction*) $\times$ 30(z *direction*) for the flow domain and $10 \times 25 \times 10$ for the body, the center of which is located at $(x, y, z) = (15, 25, 15)$. The left column in Fig. 2 shows the velocity vectors of flow fields for $\omega = 0.01$, after 50000time steps, in which Ro is 0.038 and Ek is 0.038. In this figure, (a), (b) and (c) are the sectional views of the flow field in horizontal planes at $z = 15$,including the body, at $z = 5$, below the body, and in the vertical plane at $y = 15$, respectively. The right column in the same figure shows those for the case of non-rotating case $\omega = 0$. Comparing these two cases, we can detect the existence of the Taylor column under and over the cubic body in the case of rotating case, that is, the flow pattern does not change in the in the horizontal planes whether the plane includes the body or not.

Secondly, we calculate a flow inside an Ekman boundary layer which appears on the top and/or on the bottom of rotating fluids. The number of lattices is $30 \times 30 \times 30$ for the flow domain, in which Ro is 0.033 and Ek is 0.014. In Fig. 3, we plot the fluid velocity components (u, v) along the line $x = 15, y = 15$. At the stationary state, a spiral velocity distribution along the z direction (the Ekman spiral) is detected in the Ekman layer. Calculated result agrees with theoretical one obtained by an analysis under the boundary layer approximation.

Finally, we simulate a vorticity decaying process called the spin-down. The number of lattices is $3 \times 100 \times 100$ for the flow domain. In the initial stage, a fluid is rotating rigidly with dimensionless vorticity 1. Suddenly the wall is stopped. Figure 4 shows the velocity vectors on the planes of $z = 50$ at the dimensionless

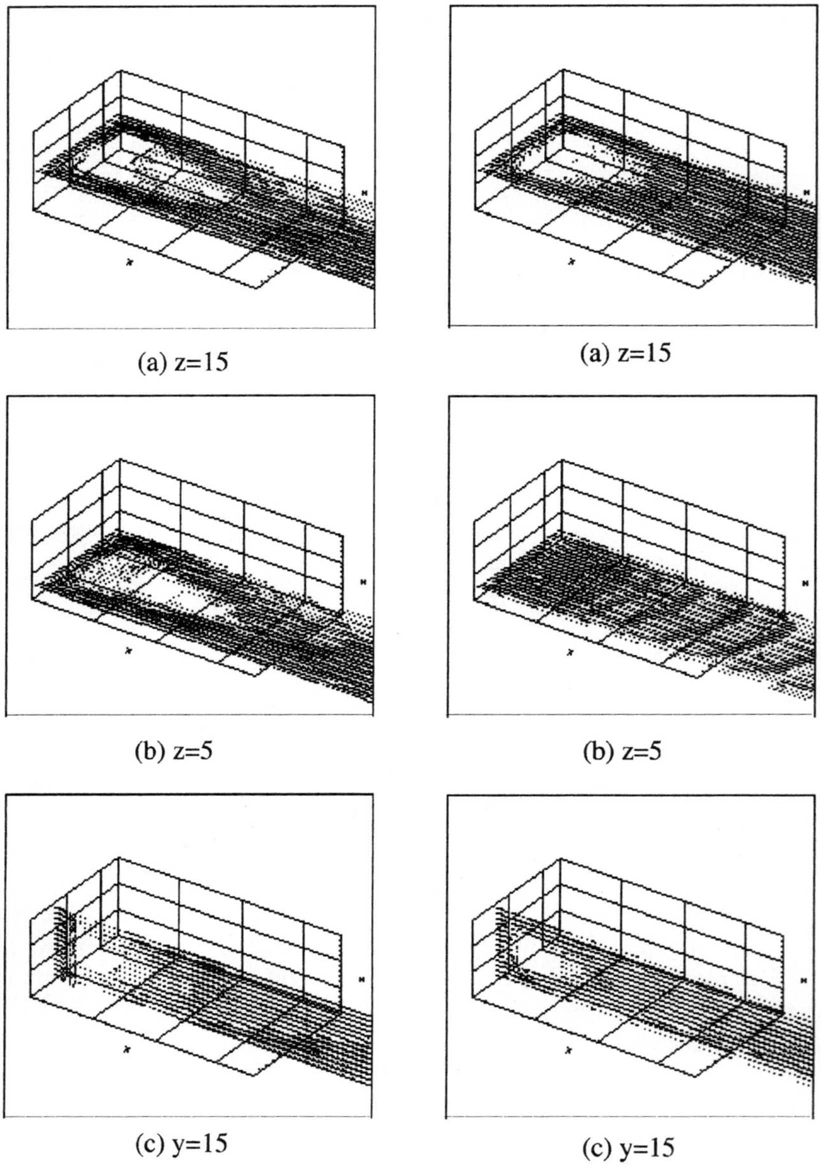

(a) z=15 (a) z=15

(b) z=5 (b) z=5

(c) y=15 (c) y=15

Fig. 2. Velocity vectors on the planes of $z = 15, z = 5$ and $y = 15$ ($\omega = 0.01$: right column, $\omega = 0$: left column)

time $T(= t/(1/\omega)) = 0, 0.3, 3$ and 30. Figure 5 shows the time dependence of the vorticity profiles and agreement between numerical result and theoretical one is good.

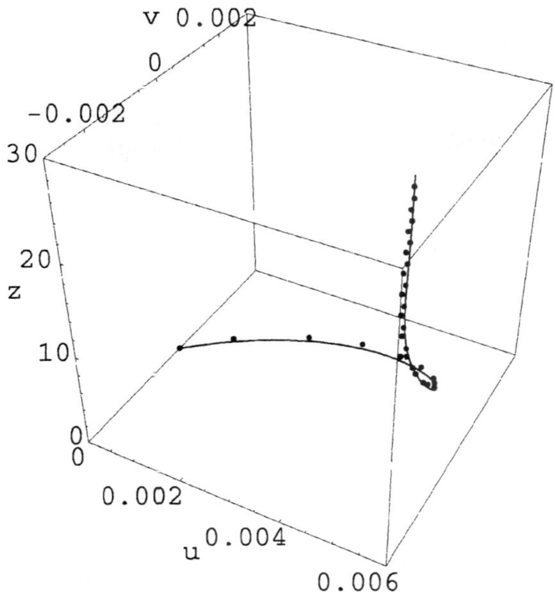

Fig. 3. Plots of the velocity components (u,v) along the line $x = 15, y = 15$ after 10000 time steps. $Ro = 0.033$ and $Ek = 0.014$

4 Conclusions

In this paper, we have simulated an incompressible viscous fluid flow in a rotating system by the LBM. The Taylor column, the Ekman spiral and the spin-down, all of which are characteristic phenomena of rotating fluids, are calculated. The existence of the Taylor column is detected by the calculation. Results of the Ekman spiral and the spin-down agree well with theoretical ones. We have confirmed good applicability of the LBM for simulating flows in a rotating field.

References

1. McNamara G and Zanetti G. Use of the Boltzmann equation to simulate lattice-gas automata. Phys Rev Lett, 1988, 61: 2332-2335
2. Rothman DH and Keller JM. Immiscible cellular-automaton fluids. J Stat Phys, 1988, 52:1119-1124
3. Chen H and Matthaeus WH. New cellular automaton for magnetohydrodynamics. 1987, 58: 1845-1850
4. Chen S, Wang Z, Shan X, Doolen G. Lattice Boltzmann Computational Fluid Dynamics in Three Dimensions. J Stat Phys, 1992, 68: 379-400
5. Shan X. Simulation of Rayleigh-Benard convection using a lattice Boltzmann method, Phys Rev E, 1997, 55, 2780-2787

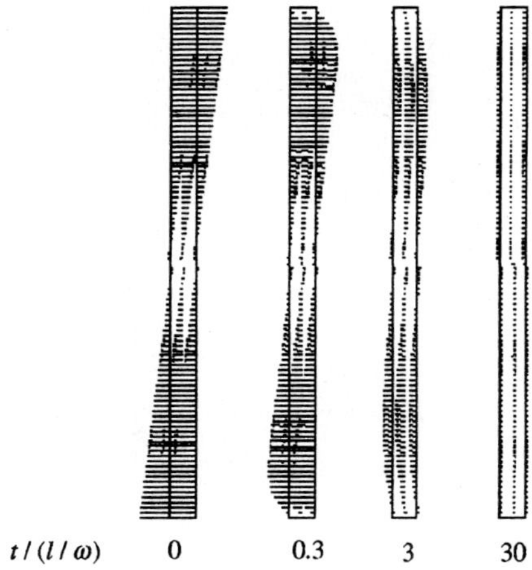

Fig. 4. Velocity vectors of components (u, v)

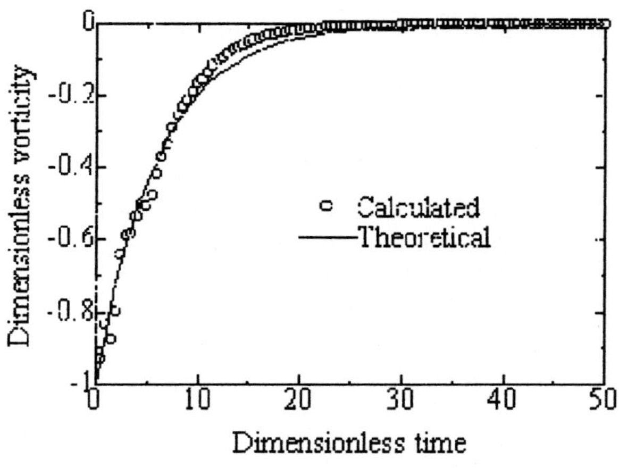

Fig. 5. Time dependence of the dimensionless vorticity. $Ro = 0.02$ and $Ek = 0.027$

Part XV

Aeroacoustics

DNS of Sound Generated by Cylinder Wakes

Nozomu Hatakeyama, Hiroyuki Hosoya, and Osamu Inoue

Institute of Fluid Science, Tohoku University, 2-1-1 Katahira, Aoba-ku,
Sendai 980-8577, Japan

Abstract. Acoustic sounds generated by flows around a circular cylinder are simulated by direct numerical simulations. The two-dimensional, unsteady, compressible Navier-Stokes equations are solved by a finite difference method. The results show that the acoustic sounds are produced and propagate in the radial direction at the same frequency as the vortex shedding. In a far field, the peaks of the sound pressure decay in inverse proportion to the square-root of the distance from the center of the cylinder.

1 Introduction

Computational aeroacoustics is now in progress, and direct numerical simulations (DNS) without solid boundaries have been successfully performed [1,2]. In these DNS, the compressible Navier-Stokes equations were solved by using highly accurate schemes both in space and time so as to capture the sound pressure which is much smaller than the static pressure. Further, non-uniform grid systems were applied in order to realize the vast computational region for an acoustic far field, and special boundary conditions were adopted to remove numerical reflections from the computational boundaries. Following these procedures, acoustic sounds generated by a flow around a circular cylinder are simulated in the present study.

The flow around a cylinder has been studied both experimentally and numerically by a lot of investigators [3–5] for a long time because it is one of the fundamental and abundant phenomena of the fluid mechanics. In spite of its simplicity, there exist a lot of unsettled problems yet. The one of them is the generation mechanism of the acoustic sound by the flow around the cylinder. This has been studied experimentally [6] and numerically [7] to some extent. The numerical analyses have been done in the conventional way that a vorticity-dominant near field is simulated first and then an acoustic far field is obtained by use of approximated equations derived from the acoustic analogy [8].

However, to the authors' knowledge, DNS of acoustic fields with solid boundaries have not yet been reported so far. The purposes of this paper are to compute the acoustic sound in the far field generated by two-dimensional cylinder wakes, by solving directly the compressible Navier-Stokes equations without any theoretical aids of the acoustic analogy, and to increase our understanding of the sound generation and propagation mechanisms.

2 Numerical Procedure

A schematic diagram of the flow around a two-dimensional circular cylinder is presented in Fig. 1. In the Cartesian coordinates (x, y), the uniform flow of the velocity U_∞ parallel to the x direction is considered. Normalized by the static sound velocity c_∞, the streamwise velocity is prescribed by the Mach number $M = U_\infty/c_\infty$. The cylinder of the diameter D is fixed at the origin. The polar coordinates (r, θ) are also used, where the azimuthal angle θ is defined from upstream in the clockwise direction. The Reynolds number is defined as $Re = U_\infty D/\nu$ where ν is the kinematic viscosity. Flow quantities are non-dimensionalized by D, c_∞ and ρ_∞ where ρ_∞ is the ambient density. The physical parameters prescribed are the Prandtl number $Pr = 0.75$ and the ratio of specific heats $\gamma = 1.4$.

The two-dimensional, unsteady, compressible Navier-Stokes equations are solved by a finite difference method. A sixth-order-accurate compact Padé scheme (fourth-order-accurate at the boundary) [9] is used for spatial derivatives, and the fourth-order-accurate Runge-Kutta scheme for time integration. Adiabatic and no-slip conditions are adopted on the cylinder surface, and non-reflecting boundary conditions [10] on the outer boundary so that the numerical wave-reflections from the boundary are removed. Initial flows are given as potential flows except for the boundary layer on the cylinder wall. Within the layer of the thickness estimated by $\delta \sim D/\sqrt{Re}$, a tangent-hyperbolic filter is applied to the potential flow in order to satisfy the no-slip condition on the cylinder. For the early development of the cylinder wakes, the Taylor-type vortex [2] is superposed to the flow at the time $t = 25$ as the perturbation.

Non-uniform O-grid systems are applied in order to simulate the entire field from near to acoustic far fields. The computational domain is divided into three regions of different grid sizes in the radial direction. The first region $[D/2 = 0.5 \le r \le r_\mathrm{v}]$ nearest to the cylinder surface is spaced with the fine grid width increasing from Δr_v, to make a sufficient resolution in the shear layer. Note that the thickness of the layer, δ, varies with respect to the Reynolds number Re as mentioned above. The second is the wake and sound region $[r_\mathrm{v} \le r \le r_\mathrm{s}]$ of the spacing Δr_s. The third buffer region $[r_\mathrm{s} \le r \le r_\mathrm{b}]$, whose spacing increases to the large Δr_b, is necessary for vortices to decay and not to reflect numerically from the outer boundary [5]. The spacings between these sections are smoothly connected by using tangent-hyperbolic curves. The maximum stretching rates of the grid are 4% to the sound region and 9% to the buffer region. The azimuthal spacing is prescribed to an uniform fine $\Delta\theta$ so as to resolve at $r = r_\mathrm{s}$. The resulting test section is $[0.5 \le r \le r_\mathrm{s}] \times [0 \le \theta \le 2\pi]$.

A typical grid system in the case of $Re = 150$ is constructed as follows. The thickness of the boundary layer is estimated as $\delta \sim 0.08$, thus $\Delta r_\mathrm{v} = 0.005$ and $r_\mathrm{v} = 1$ are chosen so that the number of the grid points within the boundary layer is 16. The sound region is prescribed by $r_\mathrm{s} = 100$ with the spacing $\Delta r_\mathrm{s} = 0.2$, and the buffer region $r_\mathrm{b} = 1500$ with $\Delta r_\mathrm{b} = 20$. The azimuthal spacing is prescribed to $\Delta\theta = 0.72°$. The number of the grid points results in 871 (in the radial direction) × 503 (in the azimuthal direction). The time step Δt is 0.0025. The examined Mach numbers range $M = 0.05$–0.3.

3 Results and Discussion

Typical examples at times when the Kármán vortex street is fully developed are shown in Figs. 2–5 in the case of $Re = 150$ and $M = 0.2$. The convergence of the computational results has been confirmed by comparing to the results with a grid system of double resolution in each direction. The influences of the initial perturbation has also been confirmed to be negligible for $t = 500$–1500, because the vortex superposed at $t = 25$ is decayed to go into the buffer region by the time $t = 500$.

First, forces on the cylinder surface, regarded as main sound sources by theory [8], are presented in Fig. 2. Time histories of drag coefficient C_{D} and lift coefficient C_{L} are in Fig. 2 (**a**). They are defined as the ratio of the forces to the kinematic energy at infinity, $\rho_\infty U_\infty^2/2 = M^2/2$. The amplitude of C_{L} is 0.52 whereas that of C_{D} is 0.026. The mean value of the drag coefficient, $\bar{C}_{\mathrm{D}}$, is about 1.34. These results agree with incompressible DNS results [4], but $\bar{C}_{\mathrm{D}}$ is found to increase gradually with the Mach number in the present compressible DNS. Mean and fluctuating surface pressure as a function of the azimuthal angle θ is presented in Fig. 2 (**b**). Pressure coefficient C_p is the time averaged pressure on the cylinder surface normalized by the value at the stagnation point $\theta = 0°$. The vertical lines in Fig. 2 (**b**) connect the largest and smallest instanteneous values. Note that the pressure fluctuations around $\theta = 60°$–$150°$ are large and appear to be bounded by envelopes having the same shapes as the distribution of mean pressure.

Next, the acoustic sounds in the far field are compared to the forces on the cylinder. The sound pressure Δp is defined as $\Delta p = p - p_\infty$, where $p_\infty = \rho_\infty c_\infty^2/\gamma \approx 0.714286$ denotes the ambient pressure. Time history of Δp at the point $r = 100$ (the boundary of the test section) and $\theta = 90°$ (the vertical direction to the uniform flow) is shown in Fig. 3 (**a**). By comparing with the lift coefficient C_{L} in Fig. 2 (**a**), we can readily see that Δp oscillates at the same frequency as the vortex shedding frequency. In fact, the Fourier spectra of Δp and C_{L} are coincident as shown in Fig. 3 (**b**). The Strouhal number is defined by $St = f/M$ where f is the frequency of the periodic vortex shedding. The Strouhal number is evaluated as $St = 0.185$ which is very close to experimental values for $Re = 150$ [3].

Figure 4 shows the acoustic pressure field at $t = 500$, where the contour level $\Delta p_{\mathrm{step}} = 5 \times 10^{-5}$. The black lines indicate the positive values and the grey lines the negative ones. As seen from Fig. 4, rarefaction waves with negative Δp and compression waves with positive Δp are generated alternately around the cylinder at the origin, and propagate toward downstream and upstream, respectively. Behind the cylinder, vortices become weaker with increasing downstream distance, and the influence of the far wake on the sound filed appears only in the sound propagating toward downstream around $\theta \sim 180°$.

Figure 5 shows distributions and decays of the sound pressure plotted along the vertical direction ($\theta = 90°$). In Fig. 5 (**a**), the distributions of Δp are plotted against the radial distance r from the origin at the three different times $t = 500, 505$ and 510. Each peak of the waves is found to propagate and decay. The propagation speed of the waves is equal to the speed of sound c_∞ in the far field, in agreement with the linear acoustic theory. About the other directions, the propagation speeds vary as $c_\theta = c_\infty(1 - M\cos\theta)$ by the mean flow of the medium. In Fig. 5 (**b**), the decays of the positive and the negative peaks are plotted separately. The decaying curves are converged to the lines proportional to $r^{-1/2}$ in the far field about $r > 30$, which is again in accordance with the theory. These results suggest that the sounds generated from the cylinder at low Reynolds numbers are precisely captured by DNS if both the flow dynamics in the near field and the wave propagations in the far field are computed with high-accuracy.

For the other Mach number ranging $M = 0.05$–0.3 and another Reynolds number $Re = 1000$, the results are similar to those described above.

4 Concluding Remarks

By using the DNS of the two-dimensional, unsteady, compressible Navier-Stokes equations, the acoustic sounds generated from the flow around the circular cylinder are successfully computed. The sound frequency is the same as the vortex shedding frequency of the Kármán vortex street. The rarefaction waves and the compression waves are alternately generated and propagate toward downstream and upstream, respectively. The sound pressure also decays proportional to $r^{-1/2}$ in the far acoustic field, which agrees with the theoretical prediction.

The authors thank Mr. Sakari Onuma, Institute of Fluid Science (IFS), Tohoku University, for his technical assistance. Computations were performed with NEC SX-5 at IFS, Tohoku University.

References

1. T. Colonius, S.K. Lele, P. Moin: J. Fluid Mech. **330**, 375 (1997)
2. O. Inoue, Y. Hattori: J. Fluid Mech. **380**, 81 (1999)
3. C.H.K. Williamson: J. Fluid Mech. **328**, 345 (1996)
4. H. Persillon, M. Braza: J. Fluid Mech. **365**, 23 (1998)
5. O. Inoue, T. Yamazaki: Fluid Dyn. Res. **25**, 1 (1999)
6. B. Etkin, G.K. Korbacher, R.T. Keefe: J. Acoust. Soc. Am. **29**, 30 (1957)
7. J.C. Hardin, S.L. Lamkin: AIAA J. **22**, 51 (1984)
8. N. Curle: Proc. Roy. Soc. A **231**, 505 (1955)
9. S.K. Lele: J. Comput. Phys. **103**, 16 (1992)
10. T.J. Poinsot, S.K. Lele: J. Comput. Phys. **101**, 104 (1992)

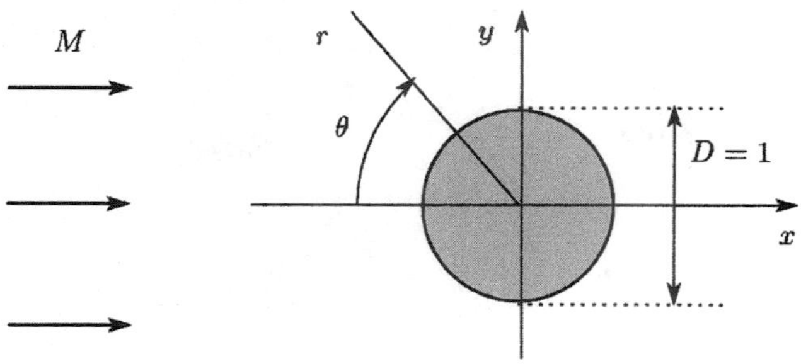

Fig. 1. Schematic diagram of flow

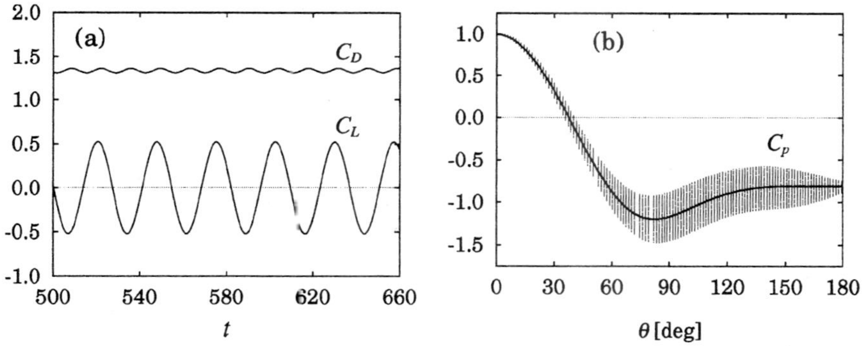

Fig. 2. (**a**) Time history of drag and lift coefficients, C_D and C_L. (**b**) Time average and fluctuations of pressure coefficient C_p. Vertical lines connect largest and smallest instanteneous values at each angle. $Re = 150, M = 0.2$

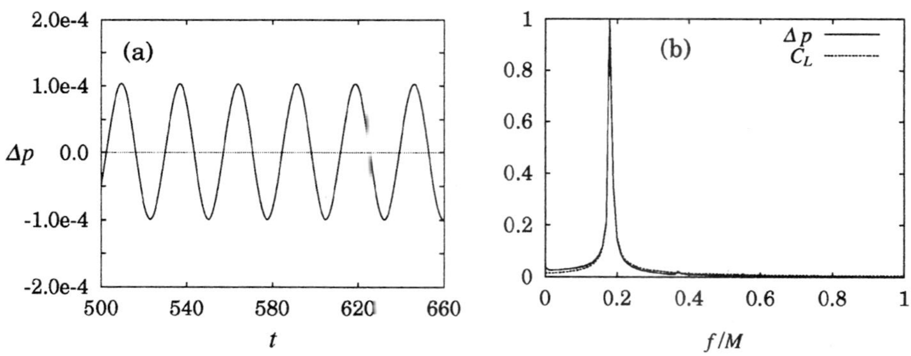

Fig. 3. (**a**) Time history of sound pressure at $r = 100, \theta = 90°$. (**b**) Amplitude spectra of sound pressure and lift coefficient, normalized by each peak value. $Re = 150, M = 0.2$

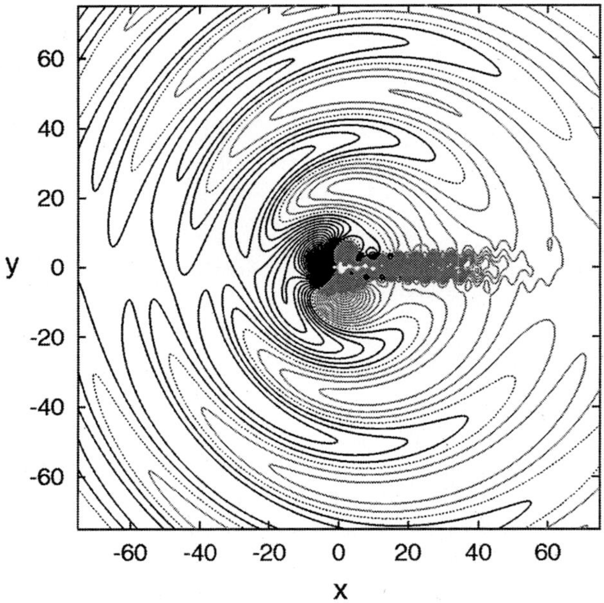

Fig. 4. Contours of sound pressure at $t = 500$ for $Re = 150$, $M = 0.2$. Contour level is $\Delta p_{\mathrm{step}} = 5 \times 10^{-5}$. *Black lines*: positive, *grey*: negative, *dotted*: 0

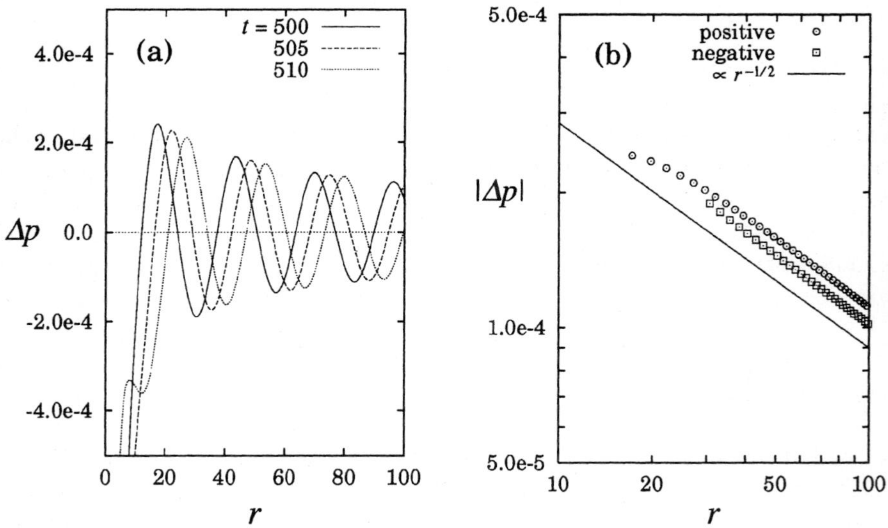

Fig. 5. (**a**) Radial distributions of sound pressure and (**b**) decays of sound pressure with positive and negative peaks displayed separately, at $\theta = 90°$. $Re = 150, M = 0.2$

Numerical Simulation of Aeroacoustic Field in a 2D Cascade Involving a Downstream Moving Grid Using the Space-Time CE/SE Method

X.Y. Wang[1], S.C. Chang[2], and P. Jorgenson[3]

[1]Taitech Inc., NASA Glenn Research Center, Cleveland, Ohio
[2,3]NASA Glenn Research Center, Cleveland, Ohio
[1]email: *wangxy@turbot.grc.nasa.gov*
[2]email: *sin-chung.chang@grc.nasa.gov*
[3]email: *jorgenson@grc.nasa.gov*

Abstract. The space-time conservation element and solution element (CE/SE) method is used to solve an aeroacoustic benchmark problem regarding turbomachinery noise. It concerns the aeroacoustic field generated by the interaction of a convected gust with a 2D cascade of flat-plate airfoils with a downstream moving grid. This tests the accuracy of a numerical method and the ability to model the acoustic wave and the gust across a sliding interface typical of those used in rotor stator interaction problems. The 2D nonlinear Euler Equations are solved and the converged numerical solutions are presented and compared with the corresponding analytical solution. The comparison shows that the CE/SE method is capable of producing accurate solutions in a simple manner.

1 Introduction

The method of space-time conservation element and solution element (abbreviated as the CE/SE method) is an innovative numerical method for solving conservation laws. It is designed to be a numerical method in the field of computational fluid dynamics (CFD). Computational aeroacoustics (CAA) is one of its applications. This method is distinguished from other methods by its very conceptual basis — flux conservation in space and time. Simplicity, generality and accuracy are weighted in the development of this method. Its salient properties are summarized briefly as follows. First, both local and global flux conservations are enforced in space and time instead of in space only. Second, all the dependent variables and their spatial derivatives are considered as individual unknowns to be solved for simultaneously at each grid point. Third, every CE/SE scheme starts from a non-dissipative scheme and numerical dissipation is fully controllable, which result in very low numerical dissipation. Fourth, the flux-based specification of the CE/SE schemes give rise in a natural fashion to an extremely simple yet highly effective non-reflecting boundary condition which is an important issue in CAA. This can be contrasted to the complexity of non-reflecting boundary conditions necessary for traditional numerical methods. A detailed description of this method and the accompanied analysis can be found in [1–4].

A variety of numerical tests have been performed previously to illustrate the accuracy of this method. Applications of this method to CAA problems reveal that the results are comparable to that of a 4th-order compact difference scheme even though the current solver is only 2nd-order accurate. Results show that the present solver can handle both continuous and discontinuous flows very well [5].

In this paper, the 2-D CE/SE Euler solver is applied to study turbomachinery noise. An aeroacoustics benchmark problem involving the interaction of a vortical gust with a cascade of flat-plate airfoils with a downstream moving grid is solved. The numerical solutions are compared with the exact solutions [6] to show the accuracy of the CE/SE method.

2 Application to Gust-Airfoil Problem

Consider the cascade of flat-plate airfoils shown in Fig. 1. The vertical dotted line located at $x = 2$ represents the interface between the stationary and moving zones. The solid and open circles are the grid points that are staggered in time by half time step. No grid point is located at the airfoil trailing edges to avoid the singular flow behavior there.

The mean axial flow is uniform with inflow velocity, U_∞, and static density, ρ_∞. The inflow Mach number, M_∞, is 0.5. The length of each plate is c, and the gap-to-chord ratio, g/c, is 1.0. Flow variables are non-dimensionalized by using U_∞ as the velocity scale, c as the length scale, c/U_∞ as the time scale, ρ_∞ as the density scale and $\rho_\infty U_\infty^2$ as the pressure scale. Thus the mean flow is described in dimensionless variables as

$$\rho = 1, \quad u = 1, \quad v = 0, \quad p = 4.0/1.4 \tag{1}$$

The vortical gust carried by the mean flow has x and y velocity components given by

$$u' = -(v_G k_y/k_x)\cos(k_x x + k_y y - \omega t) \tag{2}$$

and

$$v' = v_G \cos(k_x x + k_y y - \omega t) \tag{3}$$

respectively, where $k_x = k_y = \omega = 5\pi/2$, in which the corresponding period, $T = 2\pi/\omega$, of the gust wave is 4/5. It is assumed that $\rho' = p' = 0$ which enables the gust to satisfy the linearized Euler equations. The gust is propagating from the inlet along the mean flow direction and will interact with airfoils to generate acoustic waves. The entire flow field is simulated using the full Euler equations. To avoid the effect of nonlinear terms, $v_G = 10^{-5}$ is used in the current computation.

2.1 Boundary and Initial Conditions

Having the gust superposed on the mean flow, the flow properties at the inlet are described as:

$$(u_1)^n_{j,k} = \rho \tag{4}$$

$$(u_2)^n_{j,k} = \rho(u + u') \tag{5}$$

$$(u_3)^n_{j,k} = \rho(v + v') \tag{6}$$

$$(u_4)^n_{j,k} = \frac{p}{\gamma - 1} + \frac{\rho}{2}[(u + u')^2 + (v + v')^2] \tag{7}$$

The inlet values of $(u_{mx})^n_{j,k}$ and $(u_{my})^n_{j,k}$ are obtained by taking derivatives of $(u_m)^n_{j,k}$. This is one method for imposing the non-reflecting BC in the CE/SE method [7]. An alternative non-reflecting scheme is used at the outlet which is described as

$$(u_m)^{n+1}_{j,k} = (u_m)^{n+1/2}_{j,k}, \quad (u_{mx})^{n+1}_{j,k} = 0,$$
$$(u_{my})^{n+1}_{j,k} = (u_{my})^{n+1/2}_{j,k} \tag{8}$$

Such a way to impose the non-reflecting boundary conditions can let the flux "stream" out of the spatial domain smoothly without reflection. However, the numerical solutions so obtained at the inlet and outlet are not the correct physical solution. To accurately compute the required sound intensity $\overline{p^2} = \frac{1}{T}\int_{t_0}^{t_0+T} p'^2 dt$ at the inlet($x = -2$) and the outlet($x = 3$), a buffer zone is added at either of those sections. Thus the actual computational domain is $-2 - x_b \leq x \leq 3 + x_b$, where $x_b = 4/5$. The periodic boundary condition is imposed on the upper and lower boundaries, and the reflecting boundary conditions described in [2] are used on the airfoil surfaces.

At $t = 0$, the time-marching variables $(u_m)^n_{j,k}$, $(u_{mx})^n_{j,k}$, and $(u_{my})^n_{j,k}$ in the entire domain are defined in the same way as that for imposing the non-reflecting BCs at the inlet.

2.2 Numerical Results and Discussions

Numerical computations are performed first for the case with a downstream moving grid. The moving speed is 5, and the moving distance is Δy per Δt such that no interpolation or extrapolation is involved near the interface. But the grid movement has to be detected in the computer code. In the computation, $\epsilon = 0.3$ and $\alpha = 0$ are used with $CFL = 0.8$ and a 401x241 uniform grid in the region $-2 \leq x \leq 3$. In all plots shown here, p' is non-dimensionized by dividing it by v_G.

The pressure difference across the flat plate and the sound intensity $\overline{p^2}$ at the inlet and outlet are plotted with the analytical solution in Figs. 2 and 3, respectively. It shows a good agreement in Fig. 2, while there are some discrepancies in Fig. 3, which might be caused by the non-reflecting BCs. Contours of the scattered pressure p' are plotted in Fig. 4 to show the disturbed patterns.

Since the flow in the after part is not rotating, the acoustic field should be identical with that involving a stationary downstream grid. To demonstrate this, the pressure distribution at $y = 0.5, 1.5, 2.5, 3.5$ and the pressure difference across the four flat-plate airfoils at $t = 38T$ are plotted in Fig. 5 and compared with those obtained with a stationary downstream grid under the same computational conditions. It can be seen that the two solutions are identical. Numerical experiments of the case with a stationary downstream grid are referred to [8] and will not be described here.

3 Conclusions

The CE/SE method has been successfully applied to solve a turbomachinery noise problem with a moving downstream grid. The close agreement between computed results and the analytical solution has been achieved. It demonstrates the simplicity and accuracy of the CE/SE method.

4 Acknowledgment

This work was supported by NASA Glenn Research Center through Contract NAS3-97186.

References

[1] S.C. Chang: The method of space-time conservation element and solution element – A new approach for solving the Navier-Stokes and Euler equations. *J. Computational Physics*, **119**, 1995, pp. 295–324.

[2] S.C. Chang, X.Y. Wang, and C.Y. Chow: The space-time conservation element and solution element method – A new high-resolution and genuinely multidimensional paradigm for solving conservation laws. *J. Computational Physics*, **156**, 1999, pp. 89-136.

[3] X.Y. Wang and S.C. Chang: A 2D non-splitting unstructured-triangular-mesh Euler solver based on the method of the space-time conservation element and solution element. Vol. 8, No. 2, pp. 326-340, 1999, *Computational Fluid Dynamics JOURNAL*.

[4] X.Y. Wang and S.C. Chang: A 3D Non-splitting structured/unstructured Euler solver based on the method of space-time conservation element and solution element. *AIAA Paper 98-3278*, Norfolk, June 28–July 1, 1999.

[5] C.Y. Loh, S.C. Chang, J.R. Scott, and S.T. Yu: The space-time conservation element – A new numerical scheme for computational aeroacoustics. *AIAA Paper 96-0276*, January, 1996.

[6] K.C. Hall: Exact solution to category 3 problems – turbomachinery noise. in the proceedings of Second Computational Aeroacoustics (CAA) Workshop on Benchmark Problems, p.41, June, 1997.

[7] S.C. Chang, A. Himansu, C.Y. Loh, X.Y. Wang, S.T. Yu, and P. Jorgenson: Robust and simple non-reflecting boundary conditions for the space-time conservation element and solution element method. *AIAA 97-2077*, June, 1997.

[8] X.Y. Wang, C.Y. Chow, and S.C. Chang: Numerical simulation of gust generated aeroacoustics in a cascade using the space-time conservation element and solution element method. *AIAA Paper 98-0178*, January, 1998.

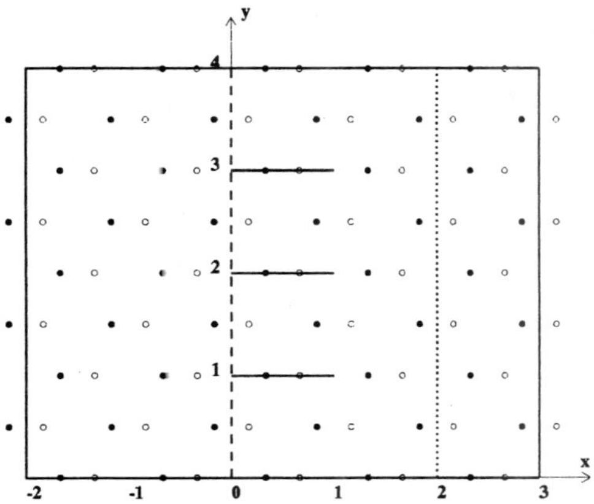

Fig. 1. Schematic sketch of computational domain and mesh used in the computation.

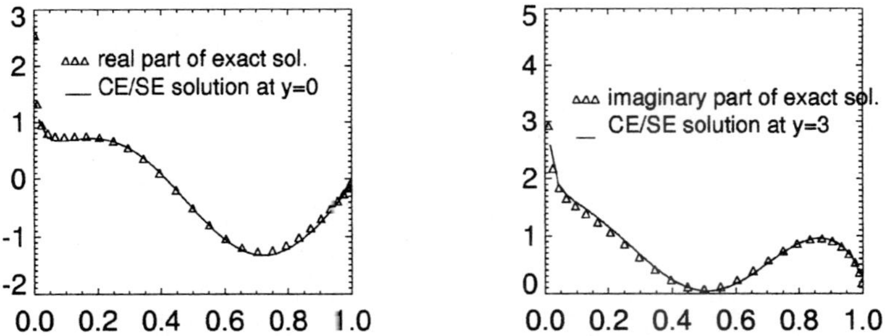

Fig. 2. Pressure difference across the airfoil surface compared with the exact solution.

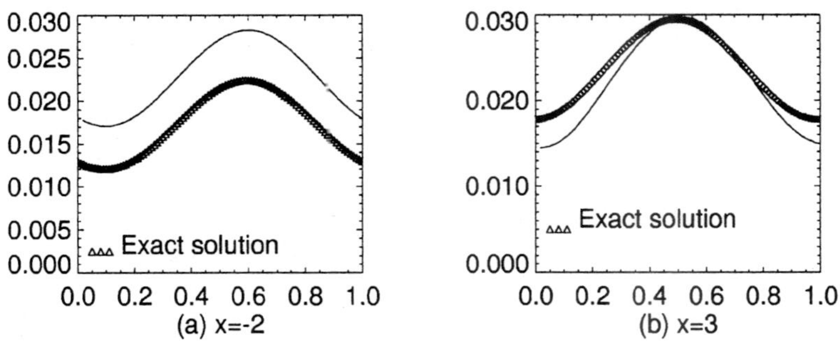

Fig. 3. The sound intensity at $x = -2$ and $x = 3$ compared with the exact solution.

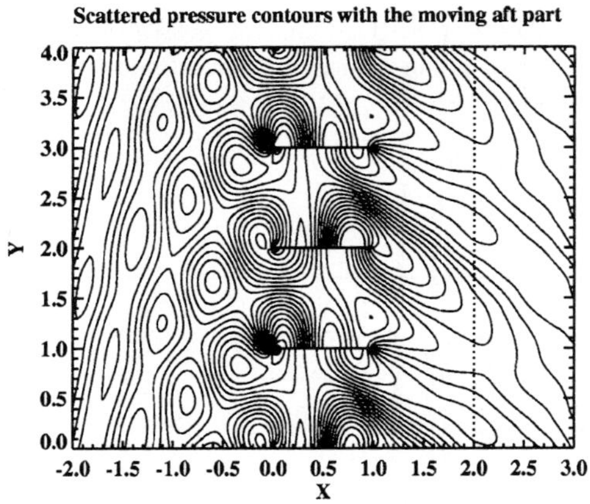

Fig. 4. Computed pressure contours for the case with a downstream moving grid based on a 401x241 grid ($\epsilon = 0.3, \alpha = 0$).

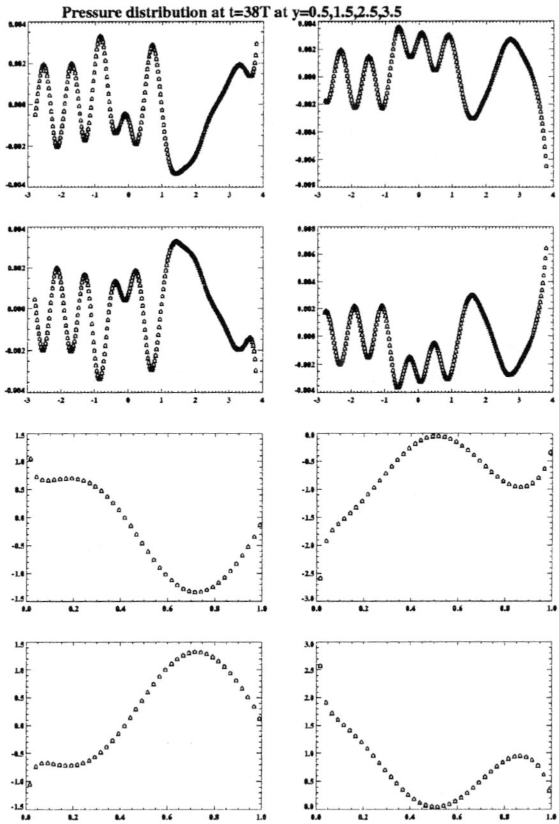

Fig. 5. Comparison of the numerical solutions for the cascade which has a downstream moving grid with those for that which has a stationary downstream grid(the lower four plots are the pressure difference across the flat plates).

An Accurate Method for Computing Propagation of Sound Waves in Non-Uniform Moving Fluid

Igor Men'shov and Yoshiaki Nakamura

Nagoya University, Department of Aerospace Engineering
Furo-cho, Chikusa-ku, Nagoya 464-8603, Japan

Abstract. The present paper aims to extend the Godunov method, a well known numerical approach for computing compressible flows, to aeroacoustics. The main building block in the developed method is the exact solution to the Variational Riemann Problem (VRP). The VRP is to define the first variation of the solution to the inital-value Riemann problem for gas dynamics equations, when the initial data undergo small variations. In the paper the VRP solution is first constructed, which then applied to accurately approximate the numerical acoustic flux in a a finite-volume Godunov-type method for acoustics equations.

1 Introduction

The present paper is devoted to extend the philosophy of the Godunov method for compressible gas dynamics [1] to acoustic problems related to sound wave propagation in non-uniform moving fluids. In accordance with this method, an instantaneous acoustic field at a given time moment is first represented by a piecewise constant distribution, where the distribution of disturbances is considered to be constant within each computational cell. The subsequent development of the approximated acoustic field for a small time period is then treated as the result of the interaction between these cell disturbances.

To realize this approach, a powerful tool like the Riemann Problem (RP) is required, which describes the interaction between two uniform flows. To describe an elementary mechanism of disturbances interaction, we introduce the variational Riemann problem (VRP) that can be read as the problem on the first variation of the solution to the RP with respect to small variations in the initial data. In fact, the VRP treats the interaction between two uniformly distributed disturbance fields, and its solution yields the values of the resultant disturbances. Therefore, this solution can be applied as an elementary constructing block to build a finite volume method for computational aeroacoustics in the same way as in the Godunov method for fluid dynamics.

The present method is based on assumptions of the Linearized Euler Equations (LEEs) model [2,3]: splitting the flow field into two parts, and neglecting non-linear and viscous effects. However, it differs from the latter in the linearization procedure. In our approach the Euler equations are first discretized in space, and then linearized with respect to conservative variables of the base flow. Therefore, the discretized equations for disturbances have the form of conservation laws. Moreover, we obtain the exact solution to the VRP, and apply it to linearize the Godunov numerical flux-function and derive the acoustic flux exactly.

2 Variational Riemann problem

The Variational Riemann Problem (VRP) can be put forward as follows. Let the exact solution $\mathbf{Q}^R = \mathbf{Q}^R(\lambda, \mathbf{Q}_l, \mathbf{Q}_r)$ be given for initial vectors $\mathbf{Q}_l$ and $\mathbf{Q}_r$, where λ is the similarity parameter, $\lambda = x/t$, and $\mathbf{Q}$ represents the vector of conservative gas dynamics parameters. Considering small variations to these vectors such as

$$\mathbf{Q}_l \to \mathbf{Q}_l + \delta\mathbf{Q}_l, \;\; \mathbf{Q}_r \to \mathbf{Q}_r + \delta\mathbf{Q}_r \tag{1}$$

the first variation of the solution is introduced, which can be written as

$$\delta\mathbf{Q}^R = M_l \delta\mathbf{Q}_l + M_r \delta\mathbf{Q}_r \tag{2}$$

by means of the variation matrices (VM) M_l and M_r given by

$$M_i = M_i(\lambda, \mathbf{Q}^R) = \frac{\partial \mathbf{Q}^R}{\partial \mathbf{Q}_i}, \;\; i = l, r \tag{3}$$

The VRP addressed here is how to obtain these VMs for arbitrary initial data of the baseline RP. It is evident that both M_l and M_r are piecewise analytical matrix-functions with the same domains of analyticity as the base non-varied solution $\mathbf{Q}^R(\lambda)$. These domains are defined by a set of waves originating from the initial discontinuity, or in other words, by the wave pattern produced in the non-varied RP.

Moreover, VMs are obviously constant with respect to λ in each analyticity domain except for the centered expansion wave where they vary with λ. In unperturbed zones, we have $M_l = I$ and $M_r = O$ on the left-hand side and $M_l = O$ and $M_r = I$ on the right-hand side, where I is the unit-matrix and O is the null-matrix. The VM for the expansion wave can be obtained in an explicit form by varying Riemann invariants and integrating the resultant system of differential equations [4,6].

In contact zones adjacent to the contact discontinuity, the VMs are constant with λ, and should be defined by considering conjugate relations at the rear characteristic of the expansion wave, at the shock wave, or at the weak discontinuity. Analysing these relations for all possible cases, one can assert that the variational vector $\delta\mathbf{Q}$ in each contact zone is determined for all the cases in the form with one arbitrary constant c_i, where $i = l, r$, as follows:

$$\delta\mathbf{Q} = N^i \mathbf{Q}_i + c_i \mathbf{m}^i \tag{4}$$

Here the matrix N^i and the vector $\mathbf{m}^i$ can be explicitly expressed by using the solution to the base RP. The corresponding rather tedious formulas are not given here, but can be found in [4,6].

To complete the solution, it is necessary to determine two constants c_l and c_r in eq.(4). There are just two conjugate relations for this. These link the variations in pressure and velocity at the contact discontinuity as $[\delta u] = 0, \ [\delta p] = 0$.

With these relations, the constants c_l and c_r are determined in a linear form with respect to the initial variations, $\delta \mathbf{Q}_i, \ i = l, r$, and the variational matrices are explicitly defined throughout the flow by eq.(4). The resultant formulas for these matrices can be represented in a simple compact form (see the references [4,6]).

3 LEE Godunov-type model for acoustic fields

The Linearized Euler Equations (ELL) model for the propagation of acoustic perturbations is obtained under the assumption [2,3,7] that the viscous perturbation fluxes can be neglected, and non-linear inviscid acoustic terms, which are mostly responsible for sound generation processes, can be modeled by a volume source term S for the production of disturbances. This term is commonly derived by using some semi-empirical models constructed on the base of available experimental data. With the above assumptions, the discrete model of the evolution of small flow perturbations against the background of a mean flow takes the form:

$$\omega_i \frac{d\hat{\mathbf{q}}_i}{dt} + \sum_\sigma s_\sigma T_\sigma^{-1} \hat{\mathbf{f}}_\sigma(\overline{\mathbf{q}}, \hat{\mathbf{q}}) = \omega_i \mathbf{S}_i \tag{5}$$

where the hat denotes perturbed values, the overline denotes mean flow parameters, s_σ is the cell interface area, and T_σ is the interface transforming matrix.

The key point of the present approach is how to approximate the acoustic flux $\hat{\mathbf{f}}_\sigma$. We treat this flux in view of the Godunov method as the resultant flux at the cell interface arising from the interaction of perturbations in two neighbouring cells

$$\hat{\mathbf{f}}_\sigma = A(\overline{\mathbf{Q}}_\sigma^R) \hat{\mathbf{Q}}_\sigma^R \tag{6}$$

where A is the Jacobian of the inviscid flux computed at the exact solution to the base RP , and $\hat{\mathbf{Q}}_\sigma^R$ is the variation of this solution that is due to the cell perturbations. Here the vector $\mathbf{Q}$ is related to the vector $\mathbf{q}$ by the interface-based transforming matrix T as $\mathbf{Q} = \mathbf{T}\mathbf{q}$.

The solution to the VRP $\hat{\mathbf{Q}}_\sigma^R$ is expressed in terms of variational matrices derived in the previous section:

$$\hat{\mathbf{Q}}_\sigma^R = M_i \hat{\mathbf{Q}}_i^\sigma + M_{\sigma(i)} \hat{\mathbf{Q}}_{\sigma(i)}^\sigma \tag{7}$$

where $\sigma(i)$ means the neighbour cell, and the superscript σ denotes that the vectors of perturbations are evaluated at the center of the interface. The first order scheme is attained, if the distribution of perturbations is assumed to be constant within the cell. To enhance the accuracy of the method, the disturbance vector of primitive variables is assumed to be a linear function in space within each computational cell [8], and a MUSCL-type approach [9] is applied to calculate the interface values. The time integration of eqs.(5-7) is carried out by the low storage 3rd order Runge-Kutta scheme [5].

The proposed approach makes it also possible to efficiently treat boundary acoustic fluxes. Various types of outer boundary conditions such as inflow perturbations, non-reflected outflow perturbations, and moving wall can be imposed by means of the solution of appropriate VRPs. Details cab be found in [5].

4 Numerical results

The present method has been already validated by checking the acoustic field generated by a monopole source and a dipole source in gas at rest. Those model problems are the acoustic field generated by a sphere executing small pulsations in the radial direction, which models the monopole, and translatory oscillations, which simulates the dipole source. Details of this validation can be found in [5]. Excellent agreement was obtained between the computed and theoretically predicted [10] sound intensities for both short- and long-wave perturbations.

In this section we demonstrate an application of the proposed method to compute propagation of acoustic waves in non-uniform gas flows. It is assumed in these calculations that the base flow is given and not affected by propagating disturbances. Acoustic waves are initiated in the flow by means of an oscillating wall that models a speaker, or by imposing wave-type disturbances at an inlet boundary of the computational domain.

First, the propagation of a plane simple harmonic sound wave across a normally located shock wave is considered. Although this problem is one- dimensional, the calculation is performed using a two- dimensional grid. The computational domain represents a rectangle of $H \times 2H$ sides, where H is the height

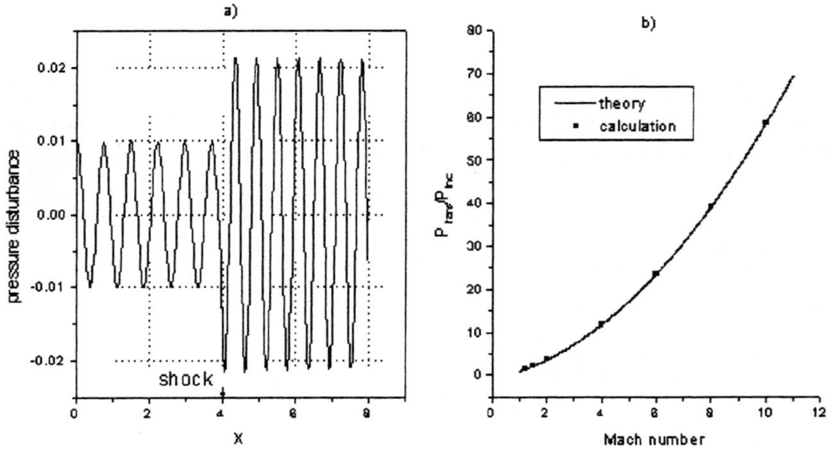

Fig. 1. Passage of a sound wave across the shock of the base flow, $Str = 67$; a) sound pressure along the streamline; b) the amplification factor vs the shock Mach number.

given by H=4 in non-dimensional units. The base flow corresponds to the upstream and downstream of a stationary plane shock wave of a Mach number $M_{sh} = 1.2$, which is located at the middle of the domain. The sound wave is inserted in the upstream at the left boundary by setting the vector of disturbances as $\hat{\mathbf{q}} = \hat{\mathbf{q}}_0 exp(-I\omega t)$. The non-reflecting boundary conditions are used for other boundaries of the computational domain. The grid used in this calculation consists of 320 uniformly distributed cells in the streamwise direction and 10 cells in the normal direction.

Figure 1a) shows results of this calculation for $\omega = 8\pi$, where the computed acoustic pressure is displayed along the streamwise direction. Since the upstream has a supersonic velocity , no sound wave can be reflected from the shock wave, and the waveform of the incident wave in the upstream remains exactly that inserted at the left boundary. The pressure amplitude in the transmitted wave is greater than that in the incident wave. The ratio of the pressure in the incident wave to that in the transmitted wave is shown in Fig. 1b) for several values of the upstream Mach number. The theoretically obtained relation between this amplification factor and the Mach number [10] is also given in this figure. Excellent agreement between the numerical and theoretical data is clearly seen.

Next, the simulation of the propagation of two-dimensional sound waves across a shock wave is considered. In this calculation the base flow and the computational domain are similar to those used to compute the above problem. However in this case, acoustic waves are generated by two symmetrically located sources. These are modeled by two short rigid plates, speakers, located in the upstream, which execute small harmonic oscillations in the transversal direction to the base flow. Simulation parameters for this problem include a uniform grid with 320 cells distributed along the streamwise direction and 160 cells in the transversal direction. All but the speaker boundary are treated by transparent or non-reflecting boundary conditions [5]. The moving wall boundary conditions are used for the speaker boundaries.

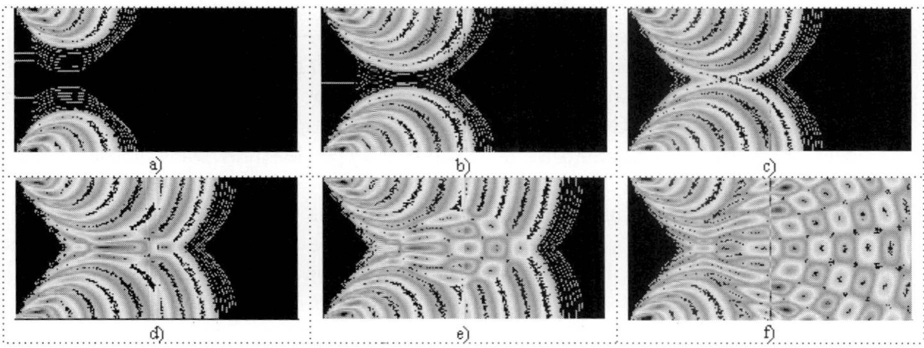

Fig. 2. Computed pressure disturbance in the base flow with a plane shock of $M_{sh} = 1.2$ shown for several time moments; the Strouhal number, $Str = \omega H / c_{base}$, is 39

The results of this calculation are shown in Fig. 2, where the computed sound pressure is presented for several instants after the speakers start generating sound waves of a frequency $\omega = 4\pi$. At the initial stage of this process the sound waves propagate towards each other, and are drifted to the shock wave by the base flow. This is displayed in Figs. 1a), and 1b). Before interacting, the waves pass through the shock wave so that the transmitted waves have the same wave pattern as the incident waves, as can be seen in Fig. 1c). As the leading waves reach the centerline of the computational domain, the interference process occurs between these incident waves, which is shown in Figs. 1d) and 1e). This modifies the wave pattern behind the base flow shock wave, which takes a certain regular cellular structure, as shown in Fig. 1f).

5 Conclusions

A Godunov-type method has been proposed for computing propagation of acoustic disturbances in non-uniform flows. This method can be considered as a direct extension of the Godunov second order scheme of computational fluid dynamics to acoustic equations. A key point of the method is to use the exact solution to the Variational Riemann problem to approximate the acoustic numerical flux. Validation tests have been carried out to examine whether the proposed approach can deal with wave-type solutions. Calculations have been fulfilled for problems with exact analytical solutions: acoustic fields generated by monopole and dipole sources, and the passage of sound waves across a shock in the base flow. The numerical and analytical solutions have been found to be in excellent quantitative agreement. An application of the proposed method to calculate two-dimensional acoustic fields in a shocked base flow has been also demonstrated.

References

1. S.K. Godunov: Mat. Sb. **47**, 3 (1959)
2. R.R. Mankbadi, R. Hixon, S.-H. Shin, L.A. Povinelly: AIAA Journal **36**, 3 (1998)
3. K. Viswanathan, L.N. Sankar: AIAA Journal **33**, 12 (1995)
4. I. Men'shov and Y. Nakamura: 'Variation Matrices in the Riemann Problem with Application to Implicit Godunovs Method'. *AIAA Paper 2000-0920* (2000)
5. I. Men'shov and Y. Nakamura: 'A Godunov-Type Method for Computational Acoustics'. *AIAA Paper 2000-2329* (2000)
6. I. Men'shov and Y. Nakamura: Computers and Fluids **29**, 6 (2000)
7. A. Antonov, T. Kozubskaya: 'Acoustic Noise Simulation for Supersonic Viscous Compressible Gas Flow'. In: *Computational Fluid Dynamics 98, Proceedings of the 4th European CFD Conference, Athens, Greece, 1998*, pp. 48,53
8. T.J. Barth and P.O. Frederickson: 'Higher Order Solution of the Euler Equations on Unstructured Grids Using Quadratic Reconstruction'. *AIAA Paper, 90-0013* (1990)
9. B. van Leer: J.Comput.Phys. **14**(1979)
10. L.D. Landau and E.M. Lifshitz: *Fluid Mechanics.* (Pergamon Press 1982)

Computation of Feedback Aeroacoustic System by the CE/SE Method

Ching Y. Loh[1], Xiao Y. Wang[1], Sin-Chung Chang[2], and Philip C.E. Jorgenson[2]

[1] Taitech Inc.
[2] NASA Glenn Research Center, Cleveland, Ohio 44135, USA
e-mail: Ching.Y.Loh@grc.nasa.gov

Abstract. It is well known that due to vortex shedding in high speed flow over cutouts, cavities, and gaps, intense noise may be generated. Strong tonal oscillations occur in a feedback cycle in which the vortices shed from the upstream edge of the cavity convect downstream and impinge on the cavity lip, generating acoustic waves that propagate upstream to excite new vortices. Numerical simulation of such a complicated process requires a scheme that can : (a) resolve acoustic waves with low dispersion and numerical dissipation, (b) handle nonlinear and discontinuous waves (e.g. shocks), and (c) have an effective (near field) non-reflecting boundary condition (NRBC). The new space time conservation element and solution element method, or CE/SE for short, is a numerical method that meets the above requirements [1-4]. A detailed description of the 2-D CE/SE Euler scheme can be found in [1,2], only a brief sketch is given here.

1 The 2-D CE/SE Euler Scheme

Consider a dimensionless conservation form of the 2-D unsteady Euler equations:

$$\mathbf{U}_t + \mathbf{F}_x + \mathbf{G}_y = 0, \tag{1}$$

where the conservative flow variable vector $\mathbf{U}$ and the flux vectors, $\mathbf{F}$ and $\mathbf{G}$, are given in the usual way. In contrast to other schemes, $\mathbf{U}$ and its spatial derivatives $\mathbf{U_x}$ and $\mathbf{U_y}$ are considered as unknowns. Also, the fluxes $\mathbf{F}$ and $\mathbf{G}$ are conveniently written in terms of the components of $\mathbf{U}$. The integral form of Eq. (1) in the space-time 3-D Euclidean Space E_3 is to be solved:

$$\oint_{S(V)} \mathbf{H_m} \cdot \mathrm{d}\mathbf{S} = 0, \qquad \mathbf{m} = \mathbf{1}, \mathbf{2}, \mathbf{3}, \mathbf{4}, \tag{2}$$

where $S(V)$ denotes the surface around a volume V in E_3 and $\mathbf{H_m} = (\mathbf{F_m}, \mathbf{G_m}, \mathbf{U_m})$. The time t coordinate points in the direction out from the paper surface. The CE/SE method is naturally adapted to an unstructured grid. The unstructured version of the CE/SE scheme can be briefly depicted by Fig. 1. Here, D, E, F are the triangle centers where the flow data at the previous time step are given. In the space-time E_3 space, (2) is applied to the hexagon cylinder $ADBECF$. $\mathbf{U}$ at the hexagon center at the new time level is first evaluated and then by Taylor expansion, $\mathbf{U}$ at the center O of triangle ABC can be obtained. Surface fluxes along the line segment planes AD, DB, BE, EC, CF, FA (called solution elements or

SE's) can be obtained again by Taylor expansions from the 3 centers D, E, and F. In the process, (2) is applied to 3 quadrilateral cylinders $OADB$,$OBEC$ and $OCFA$, which are termed conservation elements ($CE's$).

In the CE/SE scheme, the above flux conservation relation (2) in space-time is the *only* mechanism that transfers information between node points. The conservation element CE is the finite volume to which (2) is to be applied. Discontinuities are allowed to occur in the interior of a conservation element. A solution element SE associated with a grid node (*e.g.*, D, E, F in Fig. 1) is here a set of interface planes in E_3 that passes through this node (*e.g.* $DA, DB, EB, EC, etc.$). The solution variables $\mathbf{U}$, $\mathbf{U_x}$, and $\mathbf{U_y}$ are calculated at this node. Within a given solution element $SE(j,n)$, where j,n are the node index,and time step respectively, the flow variables are not only considered continuous but are also approximated by linear Taylor expansions. Each $S(CE)$ is made up by surface segments belonging to *two* neighboring SE's. The number of equations derived from these flux conservation laws matches the number of unknowns (here 12 scalar unknowns). All the unknowns are solved for based on these relations. No extrapolations (interpolations) across a stencil of cells are needed or allowed.

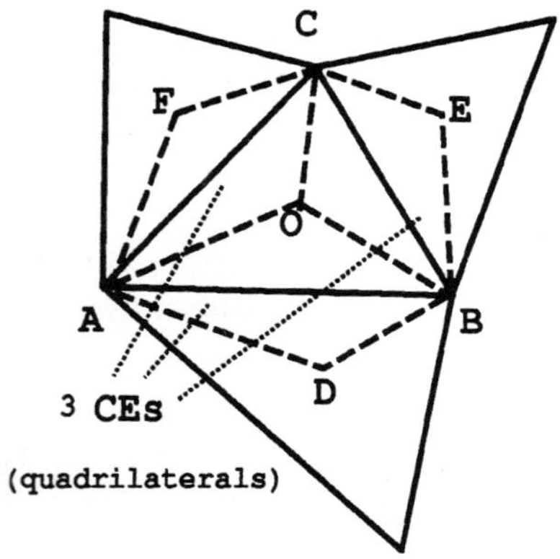

Fig. 1: basic CE/SE grid structure

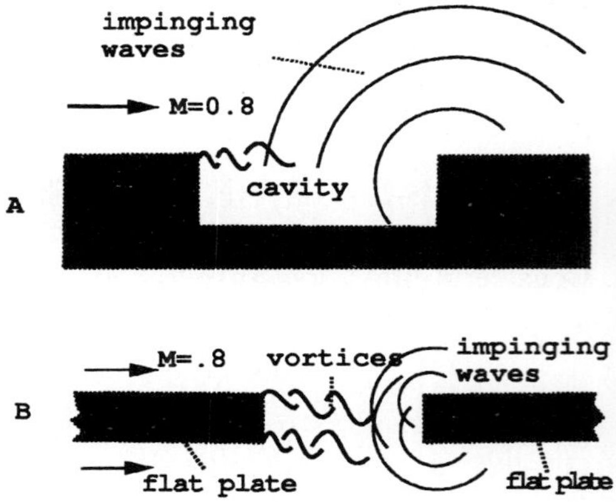

Fig. 2: Sketch of (A) cavity noise, (B) gap noise

For full details of the implementation of the 2-D CE/SE Euler method, leading to the a, $a-\epsilon$ and weighted $a-\epsilon$ schemes, the reader is referred to [3]. The weighted $a-\epsilon$ scheme is used here. ϵ denotes an adjustable parameter that controls the numerical dissipation, and the weighted scheme is intended for cases where discontinuities may be present in the inviscid flow field.

2 The CE/SE Non-Reflecting Boundary Conditions (NRBC)

In the CE/SE scheme, the idea of characteristics stemming from differential equation theory does not properly apply, since we are solving integral equations instead of differential equations. Rather, the following basic criterion is adopted: *In a CE located at the numerical domain boundary, a non-reflecting boundary condition is equivalent to letting the incoming flux from the interior domain to the boundary CE smoothly exit to the exterior of the domain.*

As a matter of fact, the 2-D non-reflecting boundary condition has been proven successful even at the near field boundary. There are various versions of the NRBC. The following are the typical ones employed in our computations. For a grid node (j, n) lying at the domain boundary, the first one, labeled type I, requires that $(\mathbf{U_x})_\mathbf{j}^\mathbf{n} = (\mathbf{U_y})_\mathbf{j}^\mathbf{n} = \mathbf{0}$, while $\mathbf{U_j^n}$ is kept fixed at the initially given steady boundary value. Type II, for cases where there is a substantial gradient in, say, the y direction, requires that $(\mathbf{U_x})_\mathbf{j}^\mathbf{n} = \mathbf{0}, \mathbf{U_j^n} = \mathbf{U_{j'}^{n-1}}, \quad (\mathbf{U_y})_\mathbf{j}^\mathbf{n} = (\mathbf{U_y})_\mathbf{j'}^\mathbf{n-1}$, where j' denotes the nearest interior node to the boundary node j. The proposed non-reflecting boundary conditions above are all simple, truly multi-dimensional and quite effective. Our experiences show that, in general, the reflection amounts to about 1% or even lower.

3 Numerical results for the gap noise problem

As typical aeroacoustic feedback systems, Figure 2 illustrates the geometric configurations which model the gap and cavity noise problems. Both produce sustainable feedback cycle oscillations as a result of vortex shedding (receptivity) at the upstream edge and acoustic feedback.

Figure 3 shows snapshots of the generation and pairing of vortices and the generation of (nonlinear) acoustic waves by vortex- flat plate impingement in the gap noise problem. The noise generated between the rudder and the stablizer of an aeroplane is a typical example of gap noise. In the current computation, there are 50720 triangle elements for the unstructured mesh in the computational domain. These triangles are actually obtained by dividing a rectangular structured mesh cell into 4 pieces (e.g. Figure 4). The rectangular cell keeps a uniform size of $\Delta x = 0.04$ and $\Delta y = 0.01$ around the area of the gap but grows larger near the boundaries. The mean flow follows the x-direction with Mach number $M = 0.8$. Fig. 3 is taken after 10000 time steps are run. At the inlet, the $M = 0.8$ flow is imposed. At the top and bottom the Type I NRBC is used while at the outflow, the Type II NRBC is specified. The domain shown in Fig. 3 is exactly the the computational domain, no buffer zone is used but still, the CE/SE NRBC is very effective.

4 Numerical results for the subsonic cavity noise problem

In the second problem, a $M = 0.8$ subsonic flow passes over a cavity of aspect ratio 6.5. As in the first problem, tonal oscillations occur in a feedback cycle in

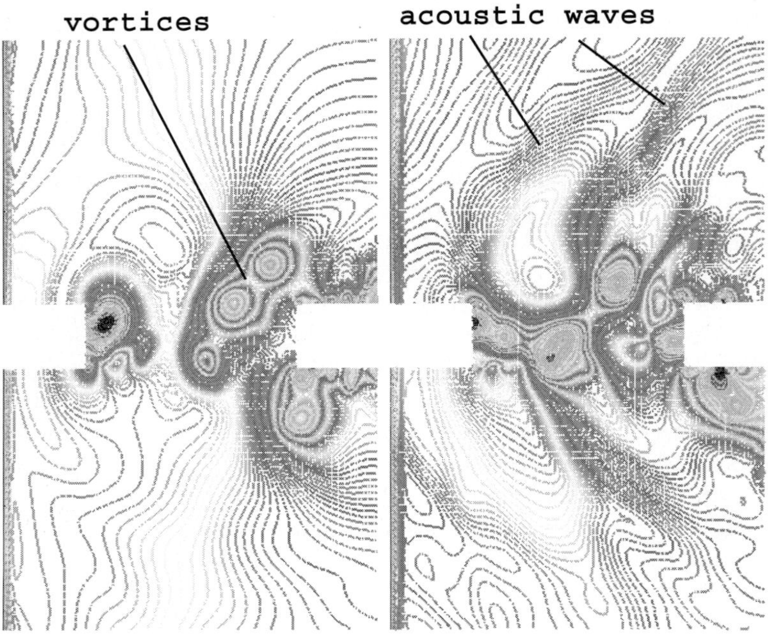

Fig.3: rudder noise; (a) early stage, with vortex shedding and pairing; (b) generation of acoustic waves.

which the vortices shed from the upstream lip of the cavity convect downstream and impinge on the other lip, generating acoustic waves that in turn propagate upstream to excite new vortices.

In the computational domain, 42000 triangular elements are used in the unstructured grid, which is again made from structured rectangular cells (Figure 4). Figure 5 depicts with isobars where the acoustic waves are generated and propagate in a series of snapshots (1-12) in the near field of the cavity. Each snapshot is 3.6 (720 steps) unit apart in time. The boundary conditions are similar to the first example. No visible reflections are observed at the non-reflecting boundaries. At the inflow boundary, upstream propagating waves are well absorbed within a few (4-6) cells, without contaminating the interior domain. From an animation of the solution, the near field acoustic wave structure is quite complicated.

5 Concluding Remarks

Two typical vortex induced, self-excited oscillation aeroacoustic systems are successfully simulated. Through these 'difficult' problems, the capability of the new CE/SE scheme is demonstrated. Specifically,

1. the (2nd order) CE/SE scheme is robust, efficient and yields high resolution, low dispersion results similar to those of higher-order schemes;

Fig. 4 Unstructured grid used in the cavity noise problem, totally 42000 trianglular elements.

2. the novel non-reflecting boundary conditions based on flux balance is simple, genuinely multi-dimensional, and easy to implement.

References

[1] Chang, S. C., "The Method of Space-Time Conservation Element and Solution Element—A New Approach for Solving the Navier-Stokes and Euler Equations," Journal of Computational Physics. v. 119, 295-324 (1995).

[2] Chang, S.-C., Wang, X.-Y. and Chow, C.-Y., "The Space-Time Conservation Element and Solution Element Method: A New High Resolution and Genuinely Multidimensional Paradigm for Solving Conservation Laws ", J. Comp. Phys. v.156, 89-136 (1999).

[3] Loh, Ching Y., Hultgren, Lennart S. and Sin-Chung Chang, "Computing Waves in Compressible Flow Using the Space-Time Conservation Element Solution Element Method," AIAA Paper 98-0369 (1998).

[4] Loh, C. Y., Hultgren, L. S., Chang, S.-C. and Jorgenson, P. C. E., "Vortex Dynamics Simulation in Aeroacoustics by the Space-Time Conservation Element Solution Element Method," AIAA Paper 99-0359 (1999).

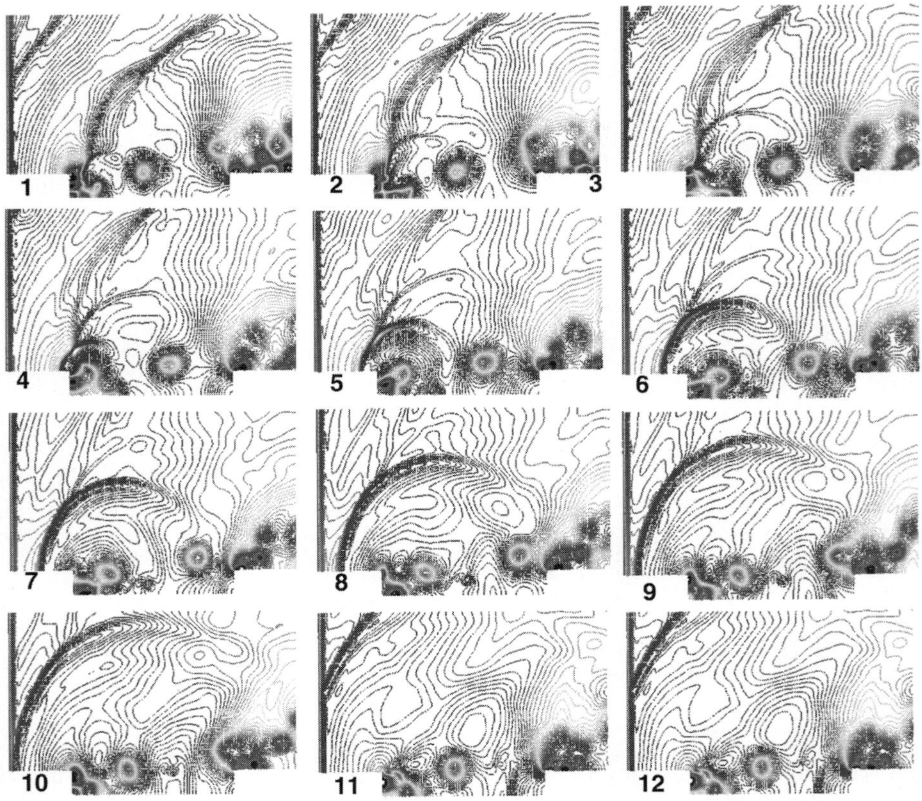

Fig. 5 Isobar snapshots, showing generation of near field non-linear acoustic waves around the cavity.

Numerical Studies of the Separated Flow Around a 2-D Airfoil at Moderate Angle of Attack by Acoustic Excitation

Satoshi Suzuki[1] and Katsuya Ishii[1]

Department of Computational Science and Engineering, Nagoya University, Furo-cho, Chikusa-ku, Nagoya, 464-8603 JAPAN

Abstract. The effects of acoustic excitation on the separated flows around an airfoil with the angle of attack $\alpha = 14°$ at Reynolds number $Re = 1.0 \times 10^5$ are investigated in two and three dimensional numerical simulations. In 2-D simulation, the effects of the frequency and amplitudes of the acoustic wave are investigated. The results show that the acoustic wave with appropriate frequency and with relatively large amplitude dynamically suppresses the leading edge separation. In 3-D simulation, the most effective frequency of the acoustic wave is chosen. The acoustic excitation makes two-dimensional vortical structures in the flow over the suction surface. These phenomena are related to increasing vorticity in the laminar boundary layer and a change of vortical structures near a separation point.

1 Introduction

Several experiments have demonstrated the phenomena of reducing "laminar separation" over two-dimensional airfoils by the acoustic excitation. The response of the separated flow over an airfoil to an acoustic excitation depends on the Reynolds number based on the chord of airfoil, the angle of attack and the frequency of the acoustic wave[1–3]. Asai et al. [3] examined experimentally the response of an airfoil wake to acoustic forcing, in which they used the NACA-0012 airfoil and chord-Reynolds number Re was 2.3×10^4. They showed that at a angle of attack $\alpha = 10°$, the flow is highly receptive to acoustic forcing over a wide range of frequencies. On the other hand, at a stall angle $\alpha = 16°$, only frequencies around the natural shedding frequency can control the vortex shedding. They discussed these phenomena by the instability theory of the wake.

Recently, we performed 2-D simulations of the flow around a NACA-0012 airfoil[4,5]. In [5] we investigated the effect of the excitation frequency at several angles of attack. The flow is changed due to acoustic forcing in a wide range of frequencies at $\alpha = 12°$ [4]. That frequency range becomes narrower with large angles of attack, which is qualitatively in agreement with the experimental results. When $\alpha > 12°$, the flow changes mostly periodically under excitation. However, in the case of $\alpha = 12°$, the flow often experiences irregular changes. In this paper, we study the flow around an airfoil at $\alpha = 14°$ with acoustic excitation, which is a typical case of $\alpha > 12°$. We perform 3-D simulation to compare with the results of 2-D one. We also discuss the effect of amplitude of the acoustic excitation.

2 Numerical Simulation

We simulate the flows past the two and three dimensional NACA-0012 airfoil with acoustic excitation by using the accurate finite difference scheme.

2.1 Governing Equations

The governing equations are the three-dimensional compressible Navier-Stokes equations in a conservation law form expressed in a body-fitted coordinate system :

$$\partial_t \hat{q} + \partial_\xi \hat{M} + \partial_\eta \hat{N} + \partial_\zeta \hat{L} = 0 \ , \tag{1}$$

$$\hat{q}_1 = \frac{1}{J}\rho \ , \hat{q}_2 = \frac{1}{J}\rho u \ , \hat{q}_3 = \frac{1}{J}\rho v \ ,$$

$$\hat{q}_4 = \frac{1}{J}\rho w \ , \hat{q}_5 = \frac{1}{J}e \tag{2}$$

where

$$\hat{M} = \hat{E} - \frac{1}{Re}\hat{R} \ , \ \hat{N} = \hat{F} - \frac{1}{Re}\hat{S} \ , \ \hat{L} = \hat{G} - \frac{1}{Re}\hat{T} \tag{3}$$

with the Reynolds number Re, the time variable t, the streamwise, normal directional and spanwise coordinates ξ, η and ζ respectively. Here ρ is the density, e the internal energy density, p the pressure and u, v and w are the velocity components in the x, y and z direction respectively. The transformation Jacobian is $J = \mid \partial(\xi, \eta, \zeta)/\partial(x, y, z) \mid$. We nondimensionalize the equations by using the density ρ_∞, the acoustic velocity c_∞ at infinity and the chord length l. $\hat{E}$, $\hat{F}$ and $\hat{G}$ are components of the flux vectors. $\hat{R}$, $\hat{S}$ and $\hat{T}$ are components of the viscous flux vector. No turbulence model is used in the present calculation.

2.2 Numerical Algorithm

We use the block penta-diagonal matrix scheme[6,7]. This scheme is essentially second order accurate in time by using the Adams-Bashforth method in the delta form expression. Higher order spatial accuracy for the convective term is achieved by the forth-order accurate five-point central difference and the viscous term is replaced by the second-order accurate three-point difference expression. Forth-order artificial dissipation is added to both implicit and explicit terms for numerical stability.

2.3 Grid System and Boundary Conditions

We use a C-type grid system with $541(\xi) \times 151(\eta)$ points shown in Figure 1 for 2-D simulation, and one with $541(\xi) \times 151(\eta) \times 41(\zeta)$ points for 3-D simulation (wingspan is 1). In these grid systems, 401 points are distributed in ξ direction on the airfoil. Acoustic waves are emitted from a plain speaker on the bottom boundary. The Thompson boundary conditions[8,9] for incoming or outgoing waves are used on the outer boundary conditions.

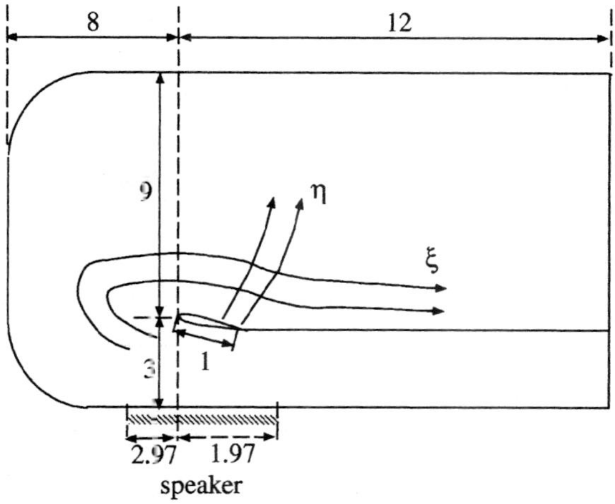

Fig. 1. Computational region

3 Results

The angle of attack is 14° in 2-D and 3-D simulations. In 2-D simulation, frequencies and amplitudes of the acoustic wave are changed. Figure 2 shows the average values of the lift coefficient C_L for the frequencies of emitted acoustic waves at $\alpha = 14°, 15°$ and $16°$. The largest value of C_L becomes smaller as α is large. Figure 3 shows the time histories of C_L excited by the acoustic waves with different amplitude. The relative amplitudes of the acoustic wave to that at infinity δp are chosen from 0.001 to 0.007. In all cases, the frequency of the acoustic wave is 0.195. Figure shows that when the amplitudes of the acoustic waves are large, the amplitudes of C_L oscillation are large. For all cases of $\delta p > 0.004$, the average of C_L is about 0.77. Figure 4 shows the vorticity distribution over the airfoil. The strength of the vorticity in the laminar boundary layer near the leading edge($\xi = 271$) is increased by the excitation. The peak value of the vorticity distribution is changed periodically about between -800 and -1000 with excitation ($\delta p = 0.005, f = 0.195$), while that is mostly constant in the case without excitation. Figure 5 shows the vorticity contour around the airfoil in 2-D simulations without excitation and with excitation respectively. The vorticity contour in the cross-section of $\zeta = const.$ in 3-D simulation is not quite different from Figure 5. With excitation, the slender vortex region around the leading edge attaches to the airfoil and solitary vortices are shed from a half chord. Figure 6 shows top view of spanwise component of vorticity on the airfoil without and with excitation respectively. The leading edge is the left hand side of the figure. Without excitation, prominent spanwise structure is not observed. On the other hand, with excitation, tube-like structures are observed in the spanwise direction.

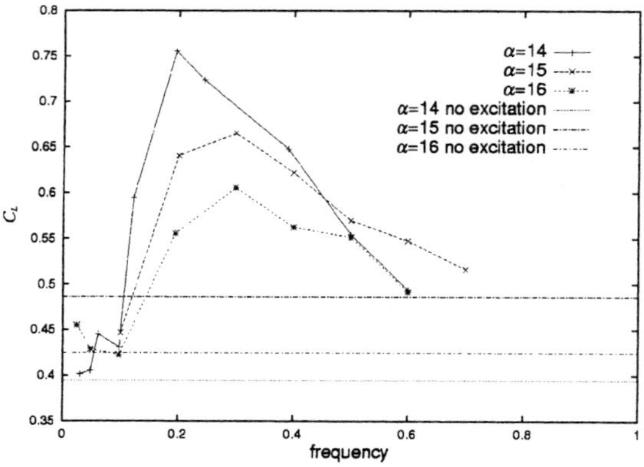

Fig. 2. Average value of C_L vs. excitation frequencies

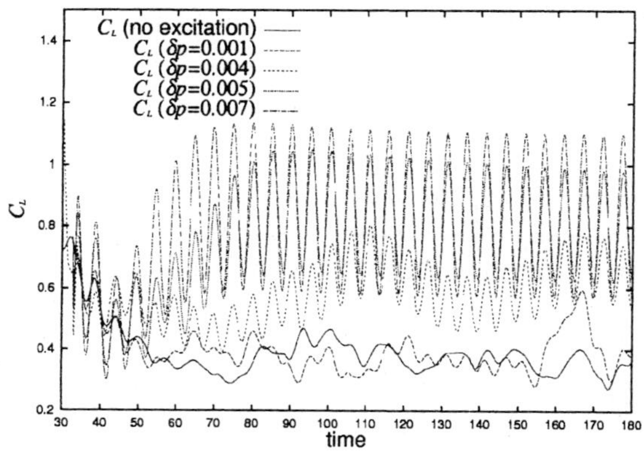

Fig. 3. Time histories of the lift coefficients (δp is the relative amplitude of acoustic wave)

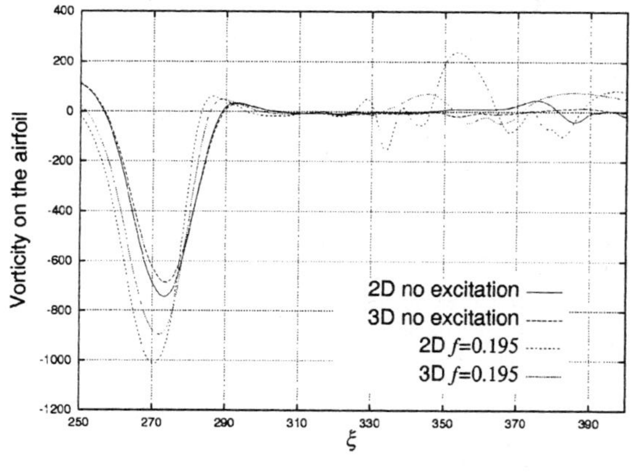

Fig. 4. Vorticity on the airfoil

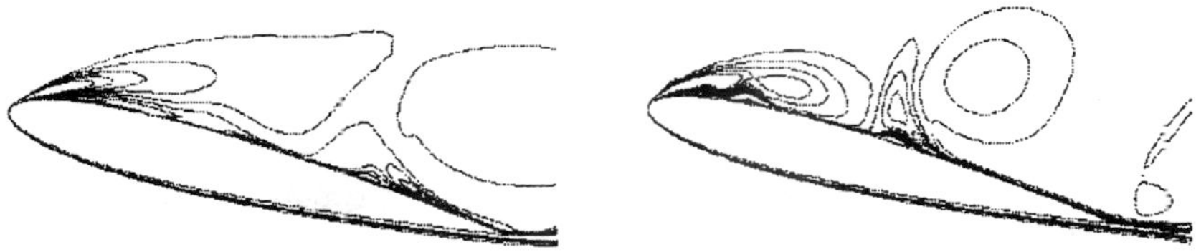

Fig. 5. Vorticity contour: 2-D, without excitation (left) with excitation (right)

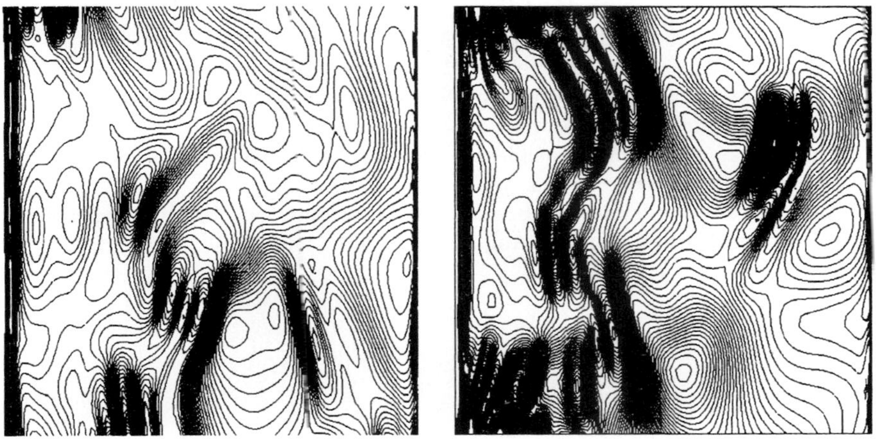

Fig. 6. Top view of spanwise component of vorticity on the airfoil without excitation (left) with excitation (right)

4 Concluding Remarks

The effects of the amplitude of acoustic wave on the flow past the NACA-0012 airfoil are obtained successfully. The 3-D simulations show that the acoustic excitation produces two dimensional structures in the flow field. In 2-D and 3-D simulations, the acoustic excitation strengthen vorticity in the laminar boundary layer. These show that the flow near the laminar separation points is important in the acoustic control problem for airfoils.

5 Acknowledgments

All computations were carried out on the Fujitsu VPP-5000 located at Nagoya University Computation Center and VPP-7000 at RIKEN Computer Center. This work was supported by a Grand-in-Aid for Scientific Research from the Ministry of Education, Sports and Culture.

References

1. K. B. M. Q. Zaman and D. J. McKinzie: 'Control of Laminar Separation Over Airfoils by Acoustic Excitation', AIAA J., Vol. 29, pp. 1075–1083 (1991)
2. M. Nishioka, M. Asai, and S. Yoshida: 'Control of Flow Separation by Acoustic Excitation', AIAA J., Vol. 28, pp. 1909–1915 (1990)
3. M. Asai, T. Seno, M. Nishioka: 'Response of Airfoil Near-Wake to External Forcing', J. Japan Soc. Aeronau. and Space Sci. , Vol. 45, pp. 441–416 (1997)
4. S. Suzuki, K. Ishii: 'Numerical Analysis of the separation on a 2-D airfoil by acoustic excitation' , Computational Fluid Dynamics Journal, vol.9 (2000 to be appeared)
5. S. Suzuki, K. Ishii: 'Numerical Simulation of the Acoustic Control of Separation over 2D Airfoils' 6-th AIAA/CEAS Aeroacoustics Conference (2000)
6. Y. Shida, K. Kuwahara, K. Ono and H. Takami: AIAA J., Vol. 25, p. 408 (1987)
7. K. Ishii, S. Adachi and S. Akishita: 'Numerical Study on Acoustic Emission from Flow Past an Airfoil', J. Phys. Soc. Japan, Vol. 66, No. 7, July, pp. 1995–2001 (1997)
8. Thompson, K. W.: 'Time-Dependent Boundary Conditions for Hyperbolic Systems', J. Comput. Phys., Vol. 68, Jan., pp. 1–24 (1987)
9. Thompson, K. W.: 'Time-Dependent Boundary Conditions for Hyperbolic Systems II', J. Comput. Phys., Vol. 89, Aug., pp. 439–461 (1990)

Part XVI

Adjoint Method

Quality and Level of the Airfoil Shapes Obtained by using Genetic Algorithms and Evolutional Strategies - Meso-evolution Certified by using the Degeneracy Structure Method

Ken Naitoh[1], Egon Krause[2], Guido Dietz[2], and Kunio Kuwahara[3]

[1] Yamagata University, 4-3-16 Jyonan Yonezawa Yamagata, Japan
[2] Aerodynamisches Institute der RWTH Aachen, Germany
[3] Institute of Space and Astronautical Science, Kanagawa, Japan

Abstract. The theoretical method is first introduced for clarifying the quality and level of the optimized results, obtained by using genetic algorithms (GAs) and evolutional strategies (ESs). We apply this method to the problems minimizing the drag force of airfoil. Drag forces are obtained by the numerical calculation code of two-dimensional flows including an instability theory on the transition to turbulence in the boundary layer. It should be stressed that GAs and ESs are relatively easy to get the best-adapted solution for subsonic flow problems. Finally, we will discuss the robustness of the optimized airfoil shapes for small disturbances such as the attachment of insect on airfoil.

1 Introduction

Recently, a lot of optimizations for flows and structures are tried by using the stochastic nonlinear algorithms such as evolutional strategies (ESs), genetic algorithms (GAs), and evolutionary programmings (EPs) [1] [2]. Why are these nonlinear optimization algorithms used? This is because the nonlinear optimization algorithms may find artificial macroevolution, which can not be found by using our brains and by using linear optimization algorithms such as gradient method. However, the applications of nonlinear optimization algorithms to engineering targets have big problems. One problem is whether people can really obtain the best solution for each problem. The other is how the arbitrary parameters in GAs and ESs, such as crossover and mutation rates, can be optimized for each problem. We may be able to obtain much more wonderful solutions, when these parameters are optimized.

In the present report, the theoretical method [3] [4] [5] is first introduced for clarifying the quality and level of the optimized results obtained by GAs and ESs. We apply this method in order to clarify the quality and level of optimized airfoil shapes, which are obtained by the numerical calculation code of two-dimensional flows including an instability theory on the transition to turbulence in the boundary layer and also by GAs and ESs.

Then, we can conclude that the optimized airfoil shape in subsonic regimes will be the best-adapted solution, because the problems in subsonic regimes have dull frequency distributions in the degeneracy structure [3] [4] [5]. It should be stressed that GAs and ESs are often difficult to get the best-adapted solution for supersonic flow problems. Finally, we will show the stability of the optimized airfoil shapes for small disturbances such as the attachment of dust on airfoil.

2 Degeneracy Structure Method

A mathematical hyperspace is proposed for classifying the problems [10]. However, this hyperspace can not be used for actual engineering problems, because this space are extremely complex.

Here, the macroscopic space is introduced to clarify the essential characteristics underlying the optimization problems, when the GAs are employed. This space is spanned by the frequency, the Hamming distance (HD) [3] [10], and fitness value [1](Fig. 1). The curved surface, that is, the frequency distribution for all of the possible strings plotted against HD and fitness value characterizes the essential differences between the engineering problems, when GAs are applied. We call this curved surface in this mathematical space as the degeneracy structure [3]. (The Hamming distance (HD) is defined as the number of the mutated loci from the best-adapted species. The species (strings) having the higher fitness values are expected to emerge widely in GAs.)

GAs consist of the strings having many types of one-dimensional structures and fitness values. For example, we will consider the binary string system with five loci. Thirty-two types of structures such as "11111" and "10101" are possible. Some of the strings having different one-dimensional structures often have a same HD and a same fitness value. These strings are degenerate. Thus, the frequency of strings has some peaks in the two-dimensional surface spanned by the HD and the fitness value. It should be emphasized that redundant strings must not be counted, which have identical string structure and fitness value [3].

We will not be able to generate this frequency distribution for all of the possible strings, when strings are longer than those having fifty loci, because of the limitation of today's computer power. Therefore, we can use the "initial" degeneracy structure, which is evaluated with a limited number of the strings produced by random number generaters in the first generation of GAs. This is because the "initial" degeneracy structure is similar to that obtained by all of the possible strings [3]. Then, the Hamming distance is defined as that from the elite, that is, the best string in the first generation.

It should be emphasized that the loci having no influence on fitness value, introns, must be eliminated before generating the degeneracy structure [5].

Some people will feel that this three-dimensional space is also too complex to be applied to engineering problems. Thus, let us introduce a simplified degeneracy structure in order to characterize the engineering problems optimized with GAs, which is the frequency distribution plotted against only the fitness value.

We know that the dull frequency distributions on the simplified degeneracy structure in the problems such as knapsack problem bring the best-adapted solutions, while GAs can not give us the best-adapted solution for the sharp frequency distributions in the problems of the type of TSP problem (Fig.2) [3]. (The frequency in Fig. 2(a) is homogeneously distributed along the fitness value , although the frequency in Fig. 2(b) is relatively concentrated around the best-adapted species.)

When we use ESs, the HD can be replaced by the structure distance between phenotypes.

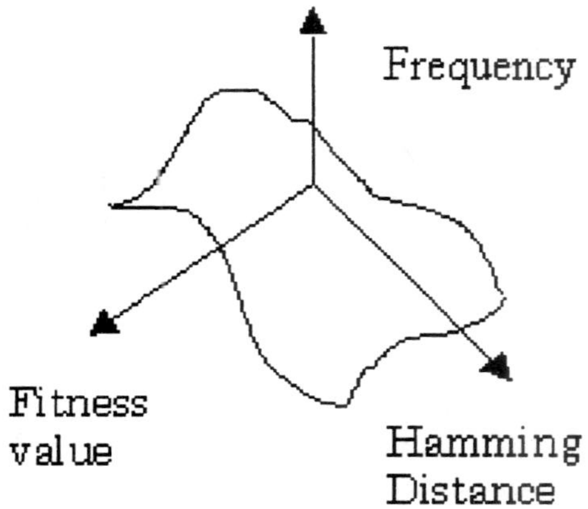

Fig. 1. Degeneracy structure for clarifying the characteristics underlying problems

3 Airfoil Flow Problems

We apply the above-mentioned degeneracy structure method for clarifying the quality and level of the airfoil shapes, optimized by a simple evolutional strategy (ES) [7] [8], which eliminates crossover operator in GAs.

(Problem A) The flow field around the airfoils and the drag forces are obtained by the numerical calculation code of two-dimensional subsonic flows including the instability theory on the transition to turbulence in the boundary layer [7] [8] [9]. Next, we will consider to minimize the drag force by controlling the location where the transition to turbulence occurs on airfoil surfaces. Control of the transition location is performed by varying the two-dimensional shape of upper airfoil surface. The conventional airfoil produced previously [7] [8] is employed as the base one. Mach and Reynolds numbers are about 0.40 and 1,660,000.

(Problem B) The supersonic flow fields around an airfoil are calculated by the numerical calculation code of compressible flows [6].

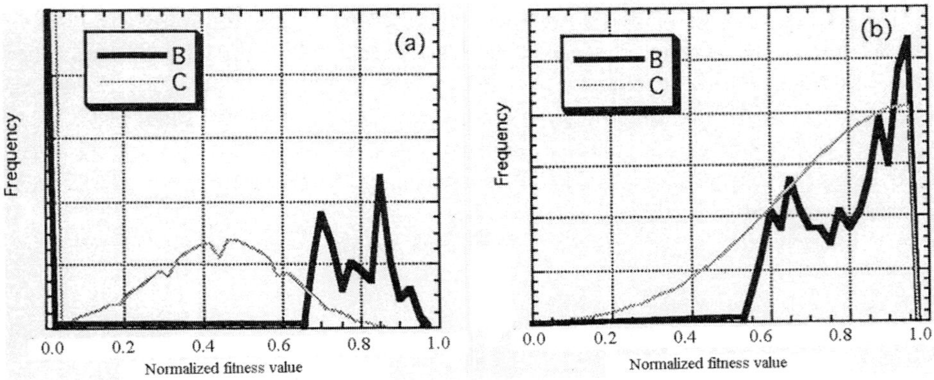

Fig. 2. Simplified Degeneracy structure. (a) Knapsack problem (b) Problem of the type of TSP problem. (The best-adapted species A has the fitness value of 1.0. Line B implies the group of species having the Hamming distance less than 4, which have the string structures close to the best-adapted solution. Line C shows the other species.)

4 Quality and Level of "Optimized" Airfoils

Then, we can conclude that the airfoil "optimized" for Problem A will be the best-adapted solution, because the problem in subsonic regimes has the dull frequency distribution in the simplified degeneracy structure (Fig. 3) [6].

In this case, the drag force is reduced with the order of ten percent. It should be emphasized that this improvement of ten percent is achieved for the base airfoil produced by the former specialists. Figure 4 shows the new airfoil optimized by using a ES for Problem A. Emphasis is also placed on the fact that our brains can not propose this type of uneven geometry of airfoils. In this sense, this airfoil optimized is that of a "meso-evolution" between macroevolution and microevolution, which has not been found by GAs and ESs for flow problems.

Next, we can see that GAs and ESs are often difficult to get the best-adapted solution for supersonic flow problems such as Problem B, since the frequency distribution plotted against the fitness value is relatively sharp (Fig. 5) [6].(Sharp frequency distributions along fitness value will bring an "adapted" solution, that is, local minimum.)

Finally, we will show the robustness of the "optimized" airfoils for small disturbances such as the attachment of dust or insect on airfoil.

It has been shown that the best-adapted species can emerge for the type of Problem A. However, the best-adapted species may lose its advantage in cases where even a small disturbance occurs in the airfoil shapes. This will happen in the cases where the problem has a sharp peak of frequency along the HD direction around the best-adapted species [3]. These sensitivities, $\partial(Frequency)/\partial(HD)$ (the gradient of frequency along HD) and $\partial(Fitness)/\partial(HD)$ (the gradient of fitness value along HD), should be included in the fitness value of each string.

The degeneracy structure method will give us much more benefits, when this method is applied to the complex optimizations such as multiobjective ones. This is because this method becomes essentially complete for the GAs and ESs having complex problems having longer strings and larger populations. The degeneracy structure method also gives us the way to optimize the arbitrary constants in GAs and ESs for each problem [3]. We may produce artificial macroevolutions by using GAs and also by this degeneracy structure method.

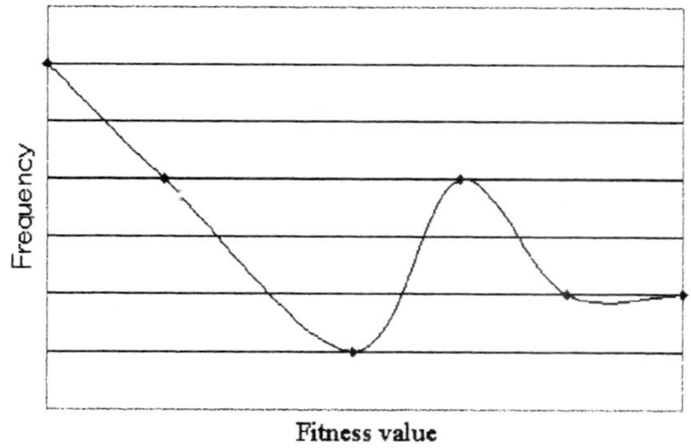

Fig. 3. Frequency distribution plotted against the fitness value in the subsonic flow around a airfoil

Acknowledgement

The present work was partially done, while the first author have worked in Nissan Research Center.

References

1. D.E. Goldberg: 'Genetic algorithms', (Addison-Wesley publishing company, 1994)
2. E. Schoneberg, F. Heinzmann. and S. Feddersen: 'Genetische algorithmen und evolutionsstrategien', (Addison-Wesley publishing company, 1994)
3. K. Naitoh: 'Macroscopic kinetic equation for a genetic algorithm', Japan J. of Industrial and Applied Mathematics **Vol.15, No.1**, 87 (1998)
4. K. Naitoh: 'Four-group equation of genetic algorithm', JSME International Journal, Series C, **Vol.38, No.2** (1995)
5. K. Naitoh: 'Introns for accelerating quasi-macroevolution', JSME International Journal, Series C, **Vol.41, No.3** (1998)
6. K. Naitoh: 'Short report of airfoil optimization by using evolutional strategy', Internal report, RWTH Aachen (1994)
7. G. Dietz: 'Diplomarbeit', Aerodynamisches Institute RWTH, Aachen, (1992)
8. P. Guntermann: 'Entwicklung eines Profilmodells mit variabler Geometrie zur Untersuchung des Transitionsverhaltens in kompressibler Unterschallstromung', Doctoral thesis, Aerodynamisches Institute RWTH, Aachen, (1992)

9. M. Drela and M.B. Giles: 'Viscous-Inviscid Analysis of Transonic and Low Reynolds Number Airfoils, AIAA Journal, Vol.25, No.10, (1987)
10. M. Eigen: 'Steps toward life', (Oxford university press, 1992)

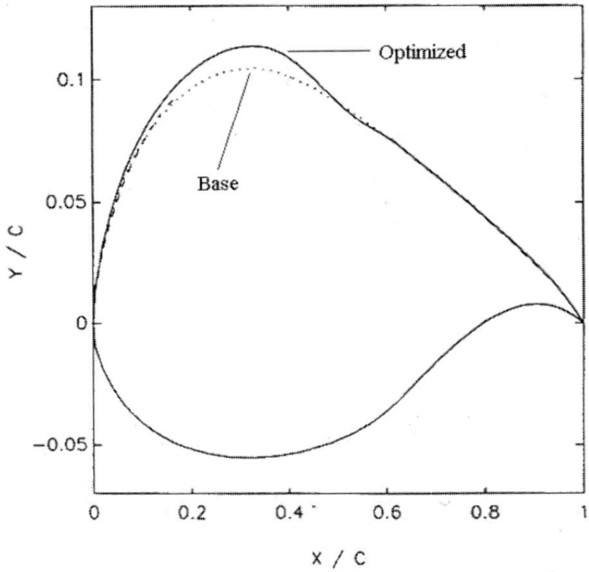

Fig. 4. Airfoil shape optimized for Problem A

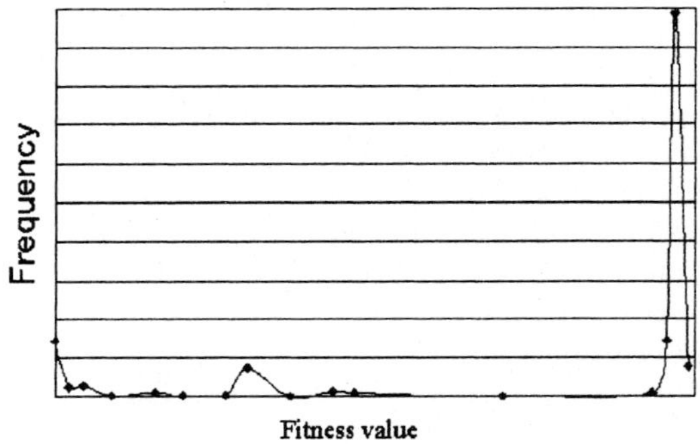

Fig. 5. Frequency distribution plotted against the fitness value in the supersonic flow around a airfoil

Application of the Adjoint Method to the Localization of Sources of Thermocapillary Instabilities

O. Bouizi[1], E. Chénier[2], C. Delcarte[1]

[1] Laboratoire d'Informatique pour la Mécanique et les Sciences de l'Ingénieur (LIMSI), Université Paris-Sud, 91405-Orsay. FRANCE
[2] LMSNPT, Université de Marne-la-Vallée, 77454 Marne-la-Vallée Cedex 02, France.
e-mail: delcarte@@limsi.fr

Key Words: thermocapillary convection, adjoint method, instabilities.

Abstract. Axisymmetric thermocapillary convection is studied in a laterally heated liquidi bridge. The explicite treatment of the viscous stress singularity, localized at the junction of the free and solid surfaces, turns out to significantly influence the flow structure: multiple solutions are exhibited in zero gravity. In order to find the localization of the sources of this thermocapillary instability, a method based on the numerical solution of the system of adjoint equations is used.

1 Introduction

The adjoint equation method offers an efficient way of calculating sensitivities, with respect to parameter changes, of linear functionals defined on the solution of partial differential equations. This method was initiated by Hill [7], who pointed out the role of adjoint systems in the local receptivity problem for boundary layers, and developed for nonlocal Görtler instabilities by Luchini and Bottaro [8]. The adjoint method is nowadays used in various applications such as aerodynamic shape optimization, optimal control of flows, and calculations of fastest growing spatial modes in Tollmien-Schlichting waves.

2 Formulation of the problem

In the present work we present an application of the adjoint method to a problem of floating zone crystal growth in a zero gravity environment [3]. Floating zone crystal growth is a containerless technique to purify crystals and alloys. A rod of polycristalline material, usually Silicon, is passed through a ring heater, producing a zone of molten material held by capillarity. A schematic view of this process is shown in Fig. 1. For simplicity, the ring heater is defined, in the axial direction, as follows: $Q(z) = (1 - z^2)^2$, $-1 \leq z \leq 1$. The thermocapillary convection is described by the Navier-Stokes equations in the Boussinesq

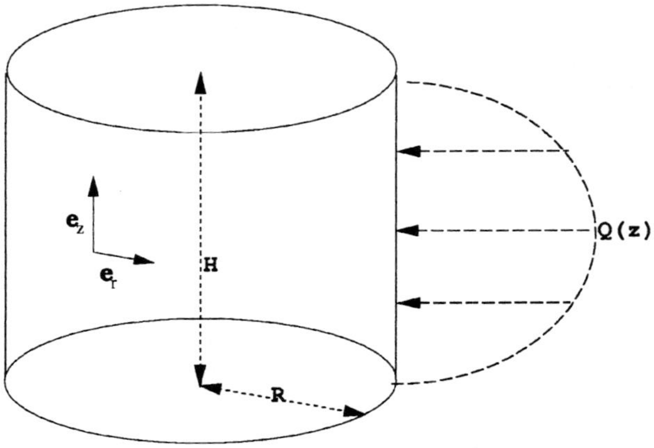

Fig. 1. Floating zone crystal growth: sketch of the configuration

and small capillary number approximations and we express these equations in dimensionless form, using coordinates (r, z), as follows:

$$\begin{aligned}
\frac{\partial u}{\partial t} + \mathbf{u}.\nabla u &= -\frac{\partial p}{\partial r} + Pr\big(\Delta u - \frac{u}{r^2}\big), \\
\frac{\partial w}{\partial t} + \mathbf{u}.\nabla w &= -\frac{\partial p}{\partial z} + Pr\Delta w) \\
\frac{\partial \theta}{\partial t} + \mathbf{u}.\nabla \theta &= \Delta \theta \\
\frac{1}{r}\frac{\partial (ru)}{\partial r} + \frac{\partial w}{\partial z} &= 0.
\end{aligned} \tag{1}$$

where $\nabla = \mathbf{e}_r\frac{\partial}{\partial r} + \mathbf{e}_z\frac{\partial}{\partial z}$, $\Delta = \frac{1}{r}\frac{\partial}{\partial r}(r\frac{\partial}{\partial r}) + \frac{\partial^2}{\partial z^2}$. $\mathbf{u} = u\mathbf{e}_r + w\mathbf{e}_z$ is the velocity field, θ is the temperature and p is the pressure. Pr is the Prandtl number ν/κ, ratio between the momentum and thermal diffusivities, in this study $Pr = 0.01$; the aspect ratio $A = H/R$ is fixed to 2. The boundary conditions are:

$$\bullet\, z = \pm 1 \begin{cases} \mathbf{u} = \mathbf{0} \text{ (no-slip conditions)} \\ \theta = 0 \text{ (solidification temperature)} \end{cases}$$

$$\bullet\, r = 1 \quad \begin{cases} u = 0 \text{ (undeformable free surface)} \\ \frac{\partial w}{\partial r} + Ma\frac{\partial \theta}{\partial z} = 0 \text{ (stress condition)} \\ \frac{\partial \theta}{\partial r} = Q(z) \text{ (heat flux)} \end{cases} \tag{2}$$

The parameter Ma is the Marangoni number, ratio between the characteristic thermocapillary ($V = |\gamma|\delta\theta/\mu$) and thermal diffusion (κ/R) velocities. The velocity and temperature fields are spatially expanded using Chebyshev series with Gauss-Radau and Gauss-Lobatto collocation points in the r and z directions respectively. A second order finite difference time-stepping scheme is chosen, with an implicit treatment of the diffusion terms, the others being explicitly evaluated. A projection diffusion method [1] is employed to uncouple the velocity and pressure fields. The chosen heat flux $Q(z)$, fulfils the regularity requirement:

[1] We refer to [1] for a full general description of the projection diffusion method

$Q(z = \pm 1) = 0$. For the stress condition, a small scale has to be considered in the vicinity of the fusion/solidification fronts where the radial fluxes of axial momentum go from the imposed stress condition to zero. The way to regularize this condition is to introduce a function $f_n(z)$ such that $f_n(\pm 1) = 0$; the boundary condition now reads : $\frac{\partial w}{\partial r} + Ma\frac{\partial \theta}{\partial z} f_n(z) = 0$ with $f_n(z) = (1 - z^{2n})^2$, n being an integer (fixed to 13 in this study). A filtering scale δ can be defined as the distance from the solid surfaces to the points where $f_n(z) = 0.9$. This scale decreases as $1/n$. Results show both quantitative (on bifurcation thresholds) [5], [4] and qualitative (on symmetry propreties) effects of the filter. Namely, for the first time, multiple solutions are exhibited in zero gravity through a pitchfork bifurcation at $Ma = 110$ [3]; Fig 2 (left) shows temperature contours and stream function of one of the two asymmetrical steady flows. We want to localize the spatial origin of this transition.

3 The linearized Navier-Stokes equations

We are concerned with the temporal and spatial development of an 'infinitesimal' disturbance of the flow (u, w, θ, p) about a steady state (U, W, T, P).

In this case the linearized Navier-Stokes equations read:

$$\begin{aligned}
\frac{\partial u}{\partial t} + U\frac{\partial u}{\partial r} + W\frac{\partial u}{\partial z} + u\frac{\partial U}{\partial r} + w\frac{\partial U}{\partial z} &= -\frac{\partial p}{\partial r} + Pr(\Delta - \frac{1}{r^2})u, \\
\frac{\partial w}{\partial t} + U\frac{\partial w}{\partial r} + W\frac{\partial w}{\partial z} + u\frac{\partial W}{\partial r} + w\frac{\partial W}{\partial z} &= -\frac{\partial p}{\partial z} + Pr\Delta w, \\
\frac{\partial \theta}{\partial t} + U\frac{\partial \theta}{\partial r} + W\frac{\partial \theta}{\partial z} + u\frac{\partial T}{\partial r} + w\frac{\partial T}{\partial z} &= \Delta\theta \\
\frac{1}{r}\frac{\partial (ru)}{\partial r} + \frac{\partial w}{\partial z} &= 0
\end{aligned} \tag{3}$$

The boundary conditions are:

$$\bullet z = \pm 1 \begin{cases} u = 0 \\ \theta = 0 \end{cases} \qquad \bullet r = 1 \begin{cases} u = 0 \\ \frac{\partial w}{\partial r} + Ma\frac{\partial \theta}{\partial z} f_n(z) = 0 \\ \frac{\partial \theta}{\partial r} = 0 \end{cases} \tag{4}$$

Temperature and stream function contours of the most unstable eigenmode of the jacobian matrix of (3) are shown in Fig. 2 (center and right). This mode presents a symmetry $u(r,z) = -u(r,-z)$, $w(r,z) = w(r,-z)$, $\theta(r,z) = -\theta(r,-z)$ which breaks the original symmetry of the steady flow: $u(r,z) = u(r,-z)$, $w(r,z) = -w(r,-z)$, $\theta(r,z) = \theta(r,-z)$. We believe, as [6], that the spatial location of the source of this symmetry-breaking bifurcation can be obtained from the maximum amplitude of the most unstable eigenmode of the linearized adjoint problem.

4 The adjoint equations

The adjoint problem of (3) is the following:

Find a vector function $\hat{\mathbf{u}} = (\hat{u}, \hat{w})$ and a scalar function $\hat{\theta}$ satisfying the Navier-Stokes equations in the weak form (see, e.g., [11])

$$\begin{aligned}
\int\int\int\Big[& \Big(\frac{\partial u}{\partial t} + U\frac{\partial u}{\partial r} + W\frac{\partial u}{\partial z} + u\frac{\partial U}{\partial r} + w\frac{\partial U}{\partial z} + \frac{\partial p}{\partial r} - Pr\big(\Delta u - \frac{u}{r^2}\big)\Big)\hat{u} \\
+ & \Big(\frac{\partial w}{\partial t} + U\frac{\partial w}{\partial r} + W\frac{\partial w}{\partial z} + u\frac{\partial W}{\partial r} + w\frac{\partial W}{\partial z} + \frac{\partial p}{\partial z} - Pr\Delta w\Big)\hat{w} \\
+ & \Big(\frac{\partial \theta}{\partial t} + U\frac{\partial \theta}{\partial r} + W\frac{\partial \theta}{\partial z} + u\frac{\partial T}{\partial r} + w\frac{\partial T}{\partial z} - \Delta\theta\Big)\hat{\theta}\Big] rdrdzdt \\
& = \int\int \hat{p}\boldsymbol{\nabla}.\mathbf{u}\, rdrdzdt.
\end{aligned}$$

for some $\hat{p}(r,z)$. By successive integrations by parts, we obtain :

$$\begin{aligned}
\int\int\int\Big[& \Big(\frac{\partial \hat{u}}{\partial t} + U\frac{\partial \hat{u}}{\partial r} + W\frac{\partial \hat{u}}{\partial z} - \hat{u}\frac{\partial U}{\partial r} - \hat{w}\frac{\partial W}{\partial r} - \hat{\theta}\frac{\partial T}{\partial r} - \frac{\partial \hat{p}}{\partial r} + Pr\big(\Delta\hat{u} - \frac{\hat{u}}{r^2}\big)\Big)u \\
+ & \Big(\frac{\partial \hat{w}}{\partial t} + U\frac{\partial \hat{w}}{\partial r} + W\frac{\partial \hat{w}}{\partial z} - \hat{u}\frac{\partial U}{\partial z} - \hat{w}\frac{\partial W}{\partial z} - \hat{\theta}\frac{\partial T}{\partial z} - \frac{\partial \hat{p}}{\partial z} + Pr\Delta\hat{w}\Big)w \\
+ & \Big(\frac{\partial \hat{\theta}}{\partial t} + U\frac{\partial \hat{\theta}}{\partial r} + W\frac{\partial \hat{\theta}}{\partial z} + \Delta\hat{\theta}\Big)\theta + p\boldsymbol{\nabla}.\hat{\mathbf{u}}\Big] rdrdzdt \\
& = \int\int\int\Big(\frac{\partial \rho}{\partial t} + \boldsymbol{\nabla}.\mathbf{J}\Big) rdrdzdt,
\end{aligned}$$

where $\rho = u\hat{u} + w\hat{w} + \theta\hat{\theta}$ and $\mathbf{J} = (J_r, J_z)$ with

$$\begin{aligned}
J_r = {} & U\rho - u\hat{p} + \hat{u}p + Pr\big(u\frac{\partial \hat{u}}{\partial r} - \hat{u}\frac{\partial u}{\partial r}\big) \\
& + Pr\big(w\frac{\partial \hat{w}}{\partial r} - \hat{w}\frac{\partial w}{\partial r}\big) + \theta\frac{\partial \hat{\theta}}{\partial r} - \hat{\theta}\frac{\partial \theta}{\partial r} \\
J_z = {} & W\rho - w\hat{p} + \hat{w}p + Pr\big(u\frac{\partial \hat{u}}{\partial z} - \hat{u}\frac{\partial u}{\partial z}\big) \\
& + Pr\big(w\frac{\partial \hat{w}}{\partial z} - \hat{w}\frac{\partial w}{\partial z}\big) + \theta\frac{\partial \hat{\theta}}{\partial z} - \hat{\theta}\frac{\partial \theta}{\partial z}.
\end{aligned}$$

One thus obtains that $(\hat{u}, \hat{w}, \hat{\theta}, \hat{p})$ represent a solution of the adjoint system

$$\begin{aligned}
\frac{\partial \hat{u}}{\partial t} + U\frac{\partial \hat{u}}{\partial r} + W\frac{\partial \hat{u}}{\partial z} - \hat{u}\frac{\partial U}{\partial r} - \hat{w}\frac{\partial W}{\partial r} - \hat{\theta}\frac{\partial T}{\partial r} &= \frac{\partial \hat{p}}{\partial r} - Pr\big(\Delta - \frac{1}{r^2}\big)\hat{u}, \\
\frac{\partial \hat{w}}{\partial t} + U\frac{\partial \hat{w}}{\partial r} + W\frac{\partial \hat{w}}{\partial z} - \hat{u}\frac{\partial U}{\partial z} - \hat{w}\frac{\partial W}{\partial z} - \hat{\theta}\frac{\partial T}{\partial z} &= \frac{\partial \hat{p}}{\partial z} - Pr\Delta\hat{w}, \\
\frac{\partial \hat{\theta}}{\partial t} + U\frac{\partial \hat{\theta}}{\partial r} + W\frac{\partial \hat{\theta}}{\partial z} &= -\Delta\hat{\theta}, \\
\frac{1}{r}\frac{\partial (r\hat{u})}{\partial r} + \frac{\partial \hat{w}}{\partial z} &= 0
\end{aligned} \tag{5}$$

with boundary conditions:

$$\bullet z = \pm 1 \begin{cases} \hat{\mathbf{u}} = 0 \\ \hat{\theta} = 0 \end{cases} \qquad \bullet r = 1 \begin{cases} \hat{u} = 0 \\ \frac{\partial \hat{w}}{\partial r} + Ma^* \frac{\partial \hat{\theta}}{\partial z} f_n(z) = 0 \\ \frac{\partial \hat{\theta}}{\partial r} = 0 \end{cases} \tag{6}$$

Ma^* is chosen so that:

$$\int\int\int \left(\frac{\partial \rho}{\partial t} + \nabla . \mathbf{J} \right) r dr dz dt = 0$$

Using equations (3) and (4) we obtain

$$Ma^* = Ma \frac{\int_{-1}^{+1} \left[\hat{w} \frac{\partial \theta}{\partial z} \right]_{r=1} f_n(z) dz}{\int_{-1}^{+1} \left[w \frac{\partial \hat{\theta}}{\partial z} \right]_{r=1} f_n(z) dz}.$$

The numerical computation of (5) must be performed by marching backwards in time [2], [9].

The value of $\hat{\theta}$ on the free surface is shown in Fig 3 (left): the maximum is located near $z = 0$. The value of $\frac{\partial \hat{\theta}}{\partial r}$ is about 10^{-20} at $r = 1$. Therefore, the response to a perturbation $q(z)$ of the heat flux, located at $z = 0$, should be maximum. A plot of $\hat{w}$ as a function of z is shown in Fig 3 (right): it is asymmetrical, with a maximum of 0.004 at $z = 0.5$; $\frac{\partial \hat{w}}{\partial r}$ fluctuates between $\pm 7.10^{-9}$ on the free surface. So the most sensitive region for a change of the stress boundary condition would be $z = 0.5$. An extended account of this work with additional results will be presented elsewhere.

References

1. A. Batoul, H. Khallouf and G. Labrosse, *C. R. Acad. Sci. Paris*, 319, Série II, 1455-1461 (1994).
2. T. R. Bewley, R. Temam and M. Ziane, A general framework for robust control in fluid mechanics, *Physica D*, 138, 360-392 (2000).
3. E. Chénier, C. Delcarte and G. Labrosse, Solutions multiples thermocapillaires en zone flottante, *Eur. Phys. J. AP*, 2, 93-97 (1998).
4. E. Chénier, C. Delcarte and G. Labrosse, Existence of physical small scales in buoyant-capillary convection flow, *Lecture Notes in Physics, Proc. of the 16th Int. Conf. on Numerical Methods in Fluid Dynamics, Arcachon, France*, Springer-Verlag, Berlin, 135-140 (1998).
5. E. Chénier, C. Delcarte and G. Labrosse, Stability of the axisymmetric buoyant-capillary flows in a laterally heated liquid bridge, *Physics Fluids*, 11, 527-541 (1999).
6. E. Gadoin and P. Le Quéré, Characterization of unstable modes in partitioned cavities, *Proc. of the 11th Int. Heat Transfer Conf., Kyongju, Korea*, 3, 429-434 (1998).
7. D. C. Hill, Adjoint systems and role in the receptivity problem for boundary layers, *J. Fluid Mech.*, 292, 183-204 (1995).

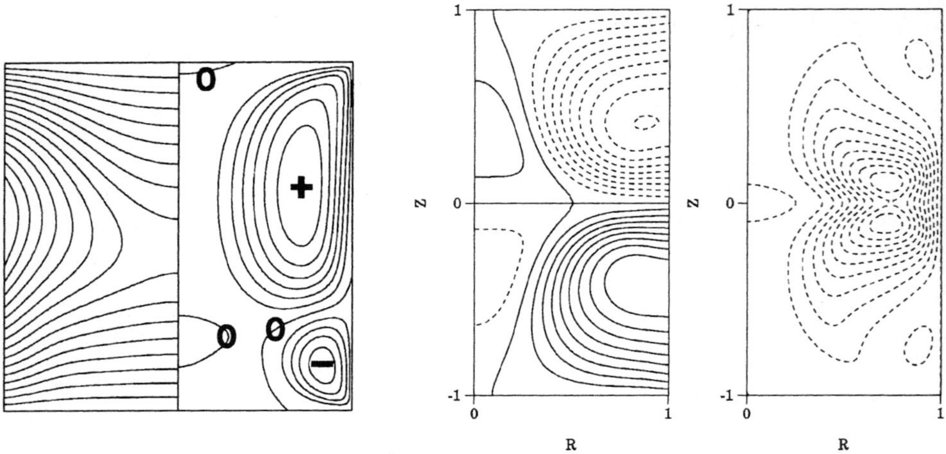

Fig. 2. Temperature and stream function contours of the steady flow (left) and of the most unstable mode (right) at $Ma = 110$.

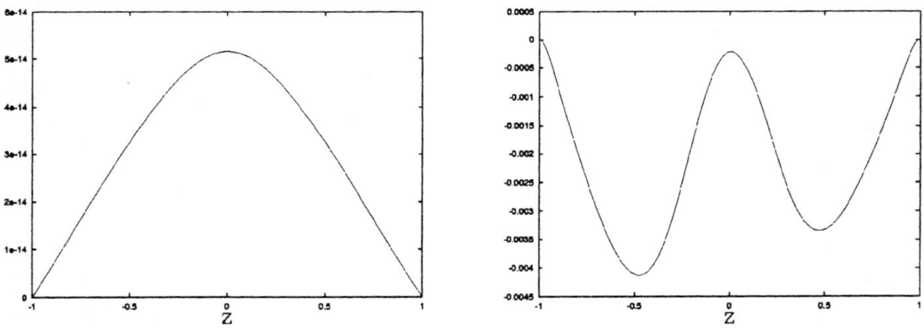

Fig. 3. Temperature (left) and axial velocity (right) of the most unstable adjoint eigenmode on the free surface. .

8. P. LUCHINI AND A. BOTTARO, Görtler vortices: a backward-in-time approach to the receptivity problem, *J. Fluid Mech.*, 363, 1-23 (1998).
9. P. LUCHINI, Reynolds-number-independent instability of the boundary layer over a flat surface: optimal perturbations, *J. Fluid Mech.*, 404, 289-309 (2000).
10. H. SALWEN AND C. E. GROSCH, The continuous spectrum of the Orr-Sommerfeld equations. Part 2. Eigenfunction expansions, *J. Fluid Mech.*, 104, 445-465 (1981).
11. R. TEMAM, *Navier-Stokes equations*, North-Holland, Amsterdam (1979).

Aerodynamic Optimization of Supersonic Transport Wing Using Unstructured Adjoint Method

Hyoung-Jin Kim[1], Daisuke Sasaki[1], Shigeru Obayashi[1], and Kazuhiro Nakahashi[1]

Tohoku University, Sendai, 980-8579 JAPAN

1 Introduction

In the application of gradient-based methods to practical aerodynamic design problems, one of the major concerns is an accurate and efficient calculation of sensitivity derivatives of an aerodynamic objective function. Sensitivity derivatives can be evaluated robustly and efficiently by using a sensitivity analysis code based either on a direct method or on an adjoint method. An adjoint method is preferable in aerodynamic designs because it is more economical when the number of design variables is larger than the total number of an objective function and constraints.

For complex aerodynamic configurations, the unstructured grid approach has several advantages over the structured grid approach. This approach can treat complex geometry with greater efficiency and less effort. It also has a greater flexibility in the adaptive grid refinement/unrefinement; thus the total number of grid points can be saved. Previous works on structured/unstructured sensitivity analysis methods can be found in references. [1-4]

In this study, direct and adjoint sensitivity codes have been developed from a 3-D unstructured Euler solver based on a cell-vertex finite volume method. With the resultant adjoint code, designed are supersonic transport (SST) wings with wing-body-nacelle and wing-body configuration. sensitivities of interior nodes are neglected except those for design variables associated with nacelle translation in order to reduce required computational time for the mesh sensitivity calculation.

2 Sensitivity Analysis

2.1 Direct Method

The discrete residual vector of the nonlinear flow equations is null for a converged flow field solution of steady problems, which can be written symbolically as

$$R[Q, X, \beta] = 0, \tag{1}$$

where Q is a flow variable vector, X grid position vector, β vector of design variables. Direct sensitivity equations can be obtained by direct differentiation of the above equation with respect to β as follows.

$$\frac{dR}{d\beta} = \left[\frac{\partial R}{\partial Q}\right]\frac{dQ}{d\beta} + C,\ where\, C = \left[\frac{\partial R}{\partial X}\right]\frac{dX}{d\beta} + \frac{\partial R}{\partial \beta} \tag{2}$$

The vector C does not depend on the $dQ/d\beta$, and thus, is constant throuthout the solution process. In order to find the solution of Eq.(2) iteratively, a pseudo time term is added to obtain an incremental form, and the LU-SGS scheme is used as was for the flow analysis. The total derivative of the objective function F is given as follows.

$$\frac{dF}{d\beta} = \left[\frac{\partial F}{\partial Q}\right]^T\frac{dQ}{d\beta} + \left[\frac{\partial F}{\partial X}\right]^T\frac{dX}{d\beta} + \frac{\partial F}{\partial \beta} \tag{3}$$

2.2 Adjoint Method

One can introduce adjoint variables and combine Eqs. (2) and (3). Coefficients of the flow variable sensitivity vector $dQ/d\beta$ in the combined equation form the following adjoint equation.

$$\left[\frac{\partial R}{\partial Q}\right]^T \Lambda + \frac{\partial F}{\partial Q} = 0 \tag{4}$$

If one finds the adjoint variable vector Λ which satisfies the above adjoint equation, one can obtain the sensitivity derivative of F with respect to β without any information about the flow variable sensitivity vector $dQ/d\beta$. This makes the computational cost for the sensitivity analysis independent of the number of design variables. Equation (3) eventually becomes to the following form,

$$\frac{dF}{d\beta} = \left[\frac{\partial F}{\partial Q}\right]^T\frac{dQ}{d\beta} + \frac{\partial F}{\partial \beta} + \Lambda^T C \tag{5}$$

The adjoint equation (4) is also solved by the LU-SGS method with a pseudo time term added.

3 Design Methodology

3.1 Design Objective

The objective of the present design study is defined as follows to minimize drag while maintaining a specified lift $C_L{}^*$.

$$F = C_D - \frac{\frac{\partial C_D}{\partial \alpha}}{\frac{\partial C_L}{\partial \alpha}}(C_L - C_L{}^*)$$

3.2 Design Parameters and Grid Modification Method

The wing section geometry is modified adding a linear combination of Hicks and Henne shape functions. We used five design sections along an SST wing span and defined 20 Hicks- Henne design variables and one twist angle per a design section resulting in 105 design variables. For the case with an engine nacelle, the height of diverter is also considered as a design parameter in addition to the 105 design variables. With the new geometry of design sections, vertical coordinates of wing surface node points are linearly interpolated.

For the movement of the grid points with the perturbed surface grid, we used the elliptic partial differential equation method [6] for the grid displacement δ x is used. Required computational time to obtain converged solution was same with that of a few iterations of the Euler solver.

3.3 Grid Sensitivity

The elliptic equation method for the interior grid movement is differentiated to be applied to the grid sensitivity calculation for the vector C in Eq.(5) with respect to each geometric design variable. Since this requires almost the same computational cost with the grid movement procedure, the total computational burden would be a substantial amount if the number of design variables becomes large; say, more than one hundred.

Figure 1 compares the derivatives of the objective function obtained with and without the interior grid sensitivity information for the wing-body-nacelle configuration. Derivatives with respect to the design variables show little difference between the two values except for design variables having indices of 21 - 30, which are defined on the lower surface of the second design section and cause the nacelle to be translated vertically. The nacelle inlet and outlet have sharp edges, and a translation of a body with any singularity such as sharp edges requires interior grid sensitivity information.[2]

In this study, interior grid sensitivities for the ten design variables (21-30) of the design case with nacelle are calculated by the elliptic equation method, while for other design variables, only the surface grid sensitivities are defined.

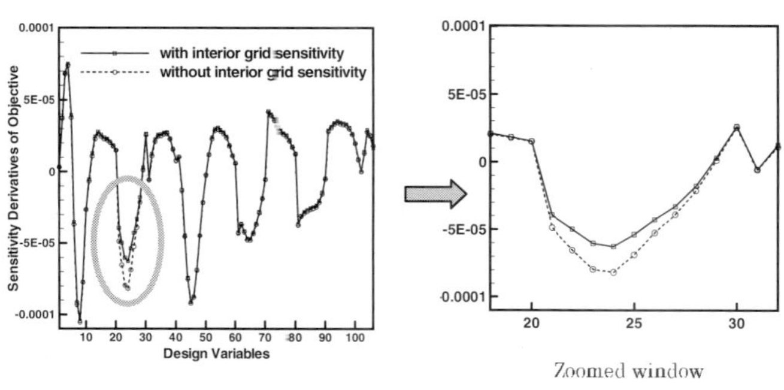

Fig. 1. Comparison of sensitivity derivatives with/without interior grid sensitivity

This simplification approach reduses the computational time for the vector C in Eq.(5) by 75 percent. For the wing-body configuration case, no body-translation is considered, and therefore, interior grid sensitivities for all the design variables can be ignored.

4 Design Results

4.1 Design I; wing-body-nacelle configuration

The present design method is applied to two experimental supersonic transports. One is wing-body-nacelle and the other wing-body configuration, both of which are under development by National Aerospace Laboratory (NAL) of Japan as basic studies for the next generation supersonic transport.[5] Design conditions are a free-stream Mach number of 2.0 and CL of 0.100. The numberof nodes and cells for the adopted volume grid are about 270,000 and 1,500,000, respectively.

As mentioned earlier, total number of design variables is 106. Constraints are imposed so that wing section thickness values at front (5%chord), rear (80%chord) spar position and maximum thickness position (50%chord) should be larger than those of the initial geometry. In addition, the diverter leading edge height is also constrained to be larger than the initial value to prevent boundary layer suction.

The SQP optimization [7] was run for five iterations to minimize the objective function with the geometric constraints. However, no further performance improvement was made after design iteration three. As can be seen in Table 1 the drag coefficient was reduced by 16 counts retaining the lift coefficient as the specified value and satisfying imposed thickness constraints by the design.

Figure 2 shows the surface pressure contours on the wing lower surface. It can be noted that the strength of the impinging shock wave on the wing lower surface generated by the diverter leading edge is greatly reduced through the design procedure.

The leading-edge height of the diverter remained the same as the initial value, since the gradient of the objective function with respect to the height is

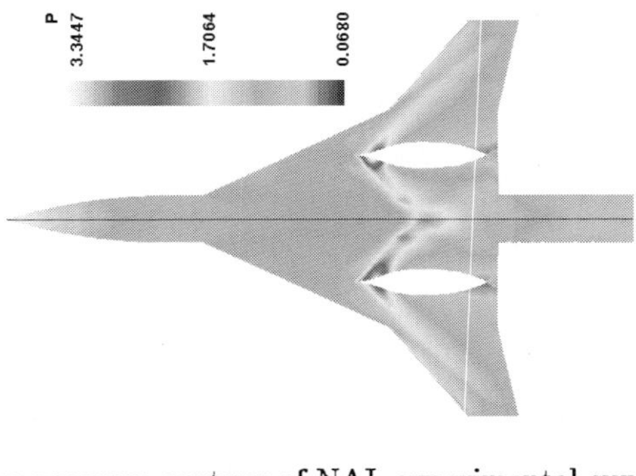

Fig. 2. Lower surface pressure contour of NAL experimental supersonic transport with nacelles, lower:initial, upper: design

positive throughout the design iterations. This is quite natural in a sense that the diverter height increment will increase the aircraft volume and also the pressure drag accordingly.

4.2 Design II; wing-body configuration

Design conditions are the same as previous; free-stream Mach number of 2.0 and CL of 0.100. The initial wing geometry has been designed by an inverse design method with the natural laminar flow (NLF) concept, and shows very good aerodynamic performance at the design condition.[5] The number of nodes and cells for the adopted volume grid are about 260,000 and 1,390,000, respectively. In the present optimization, the same design variables are employed as the design example I except the diverter leading edge height. Therefore, totally 105 design variables are used for the wing section shape modification and twist angles variation. The same thickness constraints are also imposed as example I.

Table 1 summarizes the design results. The SQP optimizer was run for fifteen iterations to obtain a drag coefficient reduced by only one count from 0.006349 to 0.006242 retaining the lift coefficient as the specified value and satisfying imposed thickness constraints.

Figure 3 compares wing section shapes and pressure distributions at a wing section. Pressure distributions show that the suction peak at the leading edge has been increased by the design with other features of pressure distributions being almost the same although the section shapes have been changed remarkably. This implies that the initial shape is already near an optimum, and performance improvement is hard to be made from the initial one. The increased suction peaks of the design wing, therefore, does not meet the NLF concept employed as a design philosophy for the initial wing. Instead, the present design utilizes the leading-edge suction by taking the advantage of the inviscid computation.

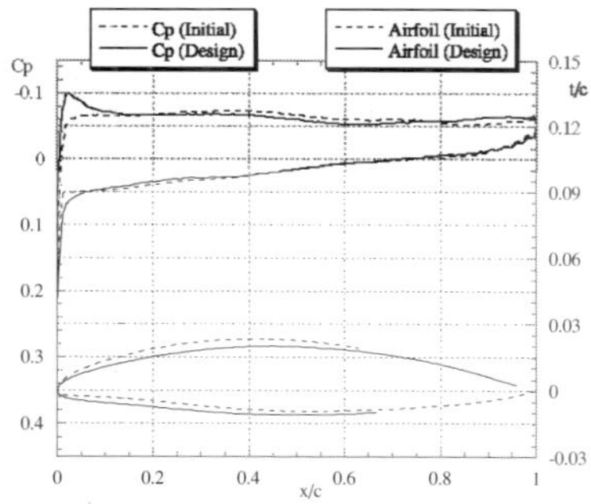

Fig. 3. Wing section shapes and pressure distributions for design example II ($\eta = 0.3$)

Table 1. Design results for SST wing-body-nacelle configuration

	example I (wing-body-nacelle)			example II (wing-body)		
	Initial	Design	$\Delta(\%)$	Initial	Design	$\Delta(\%)$
C_L	0.10017	0.10020	+0.03	0.10002	0.09993	-0.088
C_D	0.020513	0.018918	+7.78	0.006349	0.006242	1.675
L/D	4.883	5.297	+8.48	15.75	16.01	1.614

5 Conclusion Remarks

An aerodynamic design optimization system is developed using the unstructured Euler solver and the discrete adjoint method. For an efficient calculation of terms related with the grid sensitivities, grid sensitivities of interior node points are ignored except those for the design variables associated with nacelle translation. The present method is successfully applied to design the SST wing-body-nacelle and wing-body configurations. For the wing-body-nacelle configuration, the impinging shock wave from the diverter on the wing lower surface has been greatly reduced by five iterations of the SQP optimizer. On the other hand, the initial shape of the wing-body configuration was near optimum itself, and only one count drag reduction was made by fifteen design iterations.

References

1. Newman III, J.C.,Taylor III, A. C., Barnwell, R.W.,Newman, P.A. , and Hou, G.J.-W.: ' Overview of Sensitivity Analysis and Shape Optimization fo r Complex Aerodynamic Configurations' J.Aircraft, **36**, 1 (1999)
2. Anderon, W. K., and Venkatakrishnan, V.: 'Aerodynamic Design Optimization on Unstructured Grids with a Continuous Adjoint Formulation,' AIAA 97-0643, (1997)
3. Reuther, J. J., Jameson, A., Alonso, J. J., Rimlinger, M. J., and Saunders, D.: 'Constrained Multipoint Aerodynamic Shape Optimization Using an Adjoint Formulation and Parallel Computers, Part 1,' J. Aircraft, **36**, 1 (1999) pp51-60
4. Kim,H. J.,Kim, C., Rho, O.H. and Lee, K.: 'Aerodynamic Sensitivity Analysis for Navier-Stokes Equations', AIAA-99-0402,Jan. (1999)
5. Iwamiya, T.: 'NAL SST Project and Aerodynamic Design of Experimental Aircraft,' Proc. Computational Fluid Dynamics '98, Vol.2, ECCOMAS 98, John Wiley & Sons, Ltd. (1998) pp.580-585
6. Crumpton, P.I. and Giles, M. B.: 'Implicit Time Accurate Solutions on Unstructured Dynamic Grids,' AIAA 95-1671, June, (1995)
7. Vanderplaats, G. N.: 'ADS - A Fortran Program For Automated Design Synthesis version 3.00; Users' Guide,' Engineering Design Optimization, INC.,(1987)

Part XVII

Vortex Method

Three-Dimensional Vortex Method for Near-Wall Boundary-Layer Flow Structure

Zinedine Khatir and Peter W. Carpenter

Fluid Dynamics Research Centre, School of Engineering
University of Warwick, Coventry CV4 7AL, UK
E: es2184@eng.warwick.ac.uk, T: 00 44 (0) 2476 523086, F: 00 44 (0) 2476 518922

1 Introduction

The development of vortex motions in fluid flow is responsible for some of the most fascinating aspects of fluid mechanics, such as mixing, shearing, transport, and instability. Such issues arise in a variety of flow regimes, ranging from fundamental mathematical questions in laminar and turbulent flow to sophisticated engineering settings and devices. The discovery of coherent structure in turbulence has encouraged the hope that the study of vortex dynamics will lead to better models and understanding of turbulent flow, thereby solving or at least making less mysterious one of the great unresolved problem of classical physics. A comprehensive review on structures in turbulent flow is available in [1].

The hairpin vortex is postulated to be the basic flow structure of turbulent boundary-layer flows. It is believed that the central features of near-wall flow can be explained in terms of how hairpin vortices interact with the shear flow, with each other, and with the surface layer near the wall.

The first numerical model for studying turbulent structures in three-dimensional shear flows is due to Hama [2] with his 'localised-induction approximation' (LIA). Hama's original motivation was to attempt to extend the vortex modelling from free-shear flows to boundary-layer flows. This LIA vortex calculation has been used sucessfully by various authors [3]–[5]. In a later paper, Moin *et al* [6] analysed numerically the deformation of hairpin-shaped vortex filament under self-induction and in the presence of shear using the Biot-Savart law. More recently, Hon & Walker [7] used the LIA to develop a numerical procedure to investigate the evolution of hairpin vortices in a shear flow.

All previous aforementioned studies neglected the interaction between the hairpin vortices and the background vorticity. Clearly, as mentioned in the paragraph above, it is desirable to study the interaction of hairpin vortices with the shear flow in the near-wall regions. Thus, we have developed a novel computational approach capable of faithfully capturing the physical process and which simulates disturbances in a boundary-layer flow.

2 Computational Model

Our computational model, following [8], is derived itself from the classical approach of hydrodynamic stability wherein a flow field ($\mathbf{u}$) is decomposed into its

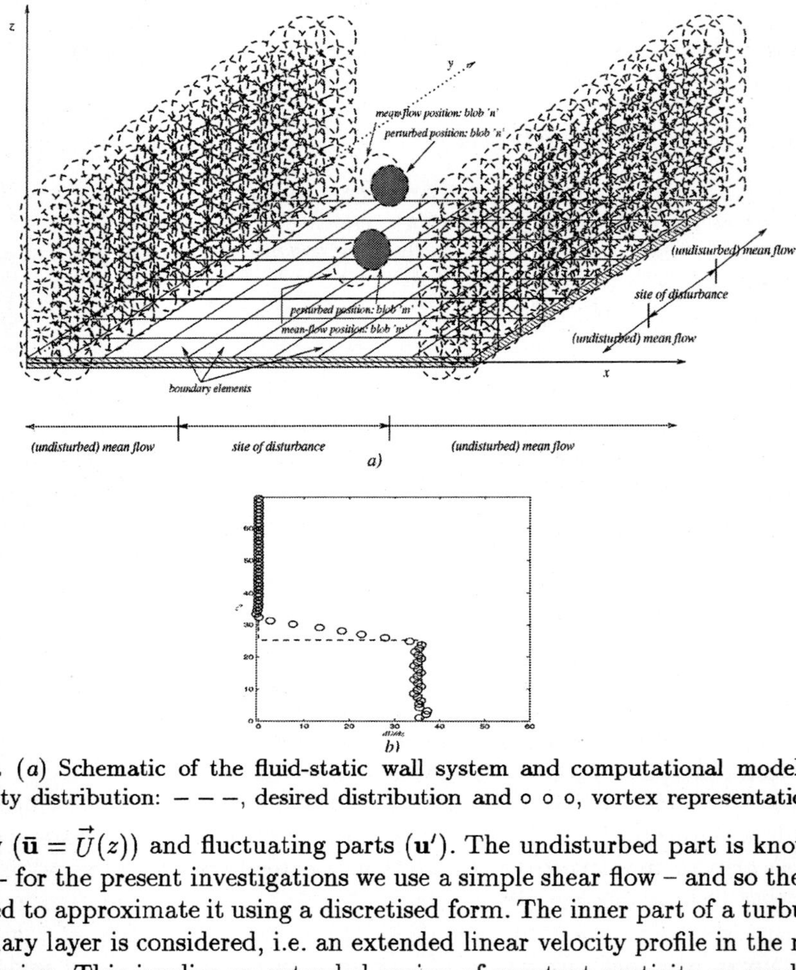

Fig. 1. (*a*) Schematic of the fluid-static wall system and computational model; (*b*) Vorticity distribution: – – –, desired distribution and o o o, vortex representation.

steady ($\bar{\mathbf{u}} = \vec{U}(z)$) and fluctuating parts ($\mathbf{u}'$). The undisturbed part is known *a priori* - for the present investigations we use a simple shear flow – and so there is no need to approximate it using a discretised form. The inner part of a turbulent boundary layer is considered, i.e. an extended linear velocity profile in the near-wall region. This implies an extended region of constant vorticity as modelled in Fig. 1*b*. Our computational efforts are thus focused on a spatially localised site of disturbance to evaluate the perturbation, $\mathbf{u}'$. Apart from computational savings, this approach ensures that upstream and downstream conditions are satisfied exactly because the known analytical solution prevails in regions away from the perturbation. In effect, the inflow and outflow boundaries of our computational domain are located infinitely far upstream and downstream. This obviates disturbance-reflection problems associated with a finite computational 'box' and is a particularly important advantage when both spatially and temporally growing disturbances may be present in the system and when disturbance-energy propagation is possible in both upstream and downstream directions.

The perturbation velocity is decomposed into its rotational and irrotational parts, $\mathbf{u}' = \mathbf{u}'_\omega + \mathbf{u}'_p$; these are modelled, respectively, using discrete-vortex and boundary-element methods. The rotational perturbation is evaluated by summing the differences between the velocity induced by each blob in its perturbed and undisturbed positions as suggested by Fig. 1*a*. We remark that this differencing approach conserves vorticity exactly and ensures that the computation remains localised, because away from the region of disturbance the disturbed and undisturbed vortex blob positions are coincident. The irrotational perturbation is obtained by enforcing the no-flux condition at the wall for the total fluid velocity. The vortex-blobs are convected using the sum of the velocity perturbations and the undisturbed flow. Viscous effects on the perturbations have not yet been modelled. However, viscous effects are included exactly in the undisturbed flow and so our results are Reynolds-number dependent.

As the disturbance spatially spreads, the computational region is extended automatically by increasing the number of vortex and boundary elements.

3 Mathematical Formulation

3.1 Equations of Motion

The equation of motion of incompressible inviscid fluid within a three-dimensional wall-bounded domain D and with boundary surface ∂D are expressed in the vorticity transport form as

$$\frac{D\boldsymbol{\omega}}{Dt} = (\boldsymbol{\omega}.\nabla)\,\mathbf{u}\ ,\quad \nabla\cdot\mathbf{u} = 0\ , \tag{1a}$$

$$\boldsymbol{\omega} = \nabla\times\mathbf{u}\ ,\quad \mathbf{u}\cdot\hat{\mathbf{n}} = 0\ , \tag{1b}$$

where $\boldsymbol{\omega} = (\omega_x, \omega_y, \omega_z)$ is the vorticity as defined in (1b) to be the curl of the fluid velocity, $\frac{D}{Dt}$ is the the material derivative defined as follows $\frac{DQ}{Dt} = \frac{\partial Q}{\partial t} + (\mathbf{u}\cdot\nabla)Q$ and $\nabla\cdot\mathbf{u} = 0$ states that the flow is incompressible. The impermeability condition, in (1b), on ∂D, where $\hat{\mathbf{n}}$ is the outward unit vector normal to the boundary, allows the fluid to move along the solid boundary but not to penetrate it and (x, y, z) are the coordinates in the streamwise, spanwise and wall-normal directions respectively.

3.2 Discrete Vortex Method

The vorticity is decomposed into $\boldsymbol{\omega} = \tilde{\omega} + \boldsymbol{\omega}'$ with $\tilde{\omega} = \tilde{\omega}_y\vec{\jmath}$ and $\boldsymbol{\omega}' = ({\omega_x}', {\omega_y}', {\omega_z}')$, thus initially $\boldsymbol{\omega}' = 0$.

In the equations above, the velocity (and its gradients) are made up of several contributions. They are decomposed into a rotational vortical field, $\mathbf{u}'_\omega(\nabla\mathbf{u}'_\omega)$, and potential component, $\mathbf{u}'_p(\nabla\mathbf{u}'_p)$, according to the Helmholtz decomposition rule:

$$\mathbf{u} = \underbrace{U(z)\,\vec{\imath}}_{meanflow} + \underbrace{\mathbf{u}_\omega{}'}_{Rotational\ disturbance} + \underbrace{\mathbf{u}_p{}'}_{Irrot.\ disturbance} \tag{2}$$

with $\mathbf{u}'_\omega = (u'_\omega, v'_\omega, w'_\omega)$ and with $\mathbf{u}'_p = (u'_p, v'_p, w'_p)$.

The rotational velocity perturbation is calculated using the discrete vortices, the rotational disturbance takes the form:

$$\mathbf{u}'_\omega = (u'_\omega = u_\omega - u^*_\omega, v'_\omega, w'_\omega) \tag{3}$$

where $u^*_\omega \vec{\imath}$ is the velocity induced by the undisturbed vortex elements. In a perfect approximation $u^*_\omega = U(z)$ and it is determined as the initial approximation to the required mean flow field.

The rotational velocity field is formally given by

$$\mathbf{u}_\omega(\mathbf{x},t) = U(z)\ \vec{\imath} + \mathbf{u}'_\omega = \int_D \mathbf{K}(\mathbf{x}-\mathbf{z})\ \boldsymbol{\omega}(\mathbf{z})\ d\mathbf{z}$$

where $\mathbf{K}$ is the Biot-Savart kernel. Under our discrete-vortex approximation this takes the form

$$\mathbf{u}_\omega(\mathbf{x},t) = \sum_{n=1}^{\infty} \mathbf{K}_\delta(\vec{r_n})\ \boldsymbol{\omega}_n \tag{4}$$

where $\vec{r_n} = (x-X_n, y-Y_n, z-Z_n)$ and r_n is its magnitude with (X_n, Y_n, Z_n) the Lagrangian coordinates of the centre of the nth vortex blob. The desinguralised kernel $\mathbf{K}_\delta(\vec{r_n})$ is easily computed from

$$\mathbf{K}_\delta(\vec{r_n}) = f_\delta(r_n)\mathbf{K}(\vec{r_n}), \tag{5}$$

where $f_\delta(r) = f(r/\delta)$ with δ the cutoff parameter and $f(r)$ is the *shape factor* defined as $f(r) = 4\pi \int_0^r s^2\phi(s)\,ds$ and the cutoff function $\phi(\mathbf{x})$ is given by $\phi(\mathbf{x}) = \frac{1}{4\pi r^2} f'(r)$ (e.g. see Hald [9]).

Thus, the $\mathbf{K}(\vec{r_n})$ in (5) is precisely given by:

$$\mathbf{K}_\delta(\vec{r_n}) = \frac{f_\delta(r_n)}{4\pi {r_n}^3} \begin{pmatrix} 0 & z-Z_n & -(y-Y_n) \\ -(z-Z_n) & 0 & x-X_n \\ y-Y_n & -(x-X_n) & 0 \end{pmatrix}. \tag{6}$$

Following [11] the irrotational $\mathbf{u}'_p$ is found by writing $\mathbf{u}'_p = \nabla\phi'$ and solving $\nabla^2\phi' = 0$ using a boundary-element method based on the boundary elements seen in Fig. 1*a* and subject to the kinematic boundary condition:

$$\left(\nabla\phi' + U(z)\ \vec{\imath} + \mathbf{u}'_\omega\right)\cdot\hat{\mathbf{n}} = 0\,. \tag{7}$$

Finally, we have the following closed system of equations for the evolution of the perturbation vorticity and the evolution of the N perturbed vortex elements

$$\begin{cases} \frac{d\mathbf{x}}{dt} = U(z)\ \vec{\imath} + \mathbf{u}'_\omega + \mathbf{u}'_p \\ \frac{d\boldsymbol{\omega}}{dt}(t) = (\boldsymbol{\omega}(t)\cdot\nabla)\left\{U(z)\ \vec{\imath} + \mathbf{u}'_\omega + \mathbf{u}'_p\right\}. \end{cases} \tag{8}$$

4 Numerical Results and Discussion

Numerical results of disturbances in the inner part of a boundary layer flow past a static wall are presented. The simulation is based on experimental conditions given on p. 754 of [12] for the case $dP/dx = 0$. Consider the following shear flow:

$$U(z) = \begin{cases} z\,U_\infty/\delta_e & \text{for } z \le \delta_e \\ U_\infty & \text{for } z > \delta_e\,. \end{cases} \tag{9}$$

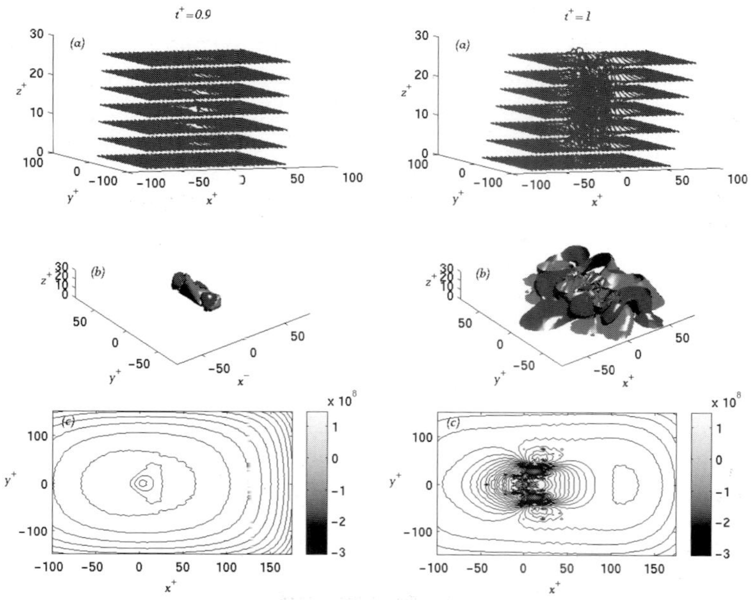

Fig. 2. State of the disturbance at $t^+ = 0.9$ and $t^+ = 1$ initiated by a small 'bump': (a) Positions of the vortex filaments; (b) Isosurface of the spanwise vorticity perturbation ω'_y ($\omega'_y = 0.75$ (light gray) and $\omega'_y = -0.75$ (dark gray)); (c) Contour plot of the total pressure on the wall boundary surface.

The numerical parameters are $U_\infty = 0.15\, m/s$, $\nu = 10^{-6}\, m^2/s$ and the boundary layer thickness $\delta_l = 80\, mm$. The friction velocity is $V_* = 6\, mm/s$. The velocity gradient at the wall is $\lambda_w = 36\, s^{-1}$ and $\delta_e = 4.2\, mm$. The shape factor is chosen to be $f(\mathbf{x}) = (1 - 2e^{-r^3} + e^{-r^3/2})$ so that the mollified Kernel $\mathbf{K}_\delta$ is as $\mathbf{K}_\delta(\mathbf{x}) = (1 - 2e^{-r^3/\delta^3} + e^{-r^3/2\delta^3})\, \mathbf{K}(\mathbf{x})$ with $\delta = 4.32 \times 10^{-4}$. The vorticity field is modelled by means of 7 layers of overlapping discrete vortex elements (Fig. 1b). After the disturbance has been initiated by a small 'bump', the disturbance has spread both upstream and downstream directions and a 'lift-up' has been initiated. Hairpin vortices like structures are being produced (Fig. 2a), consistent with experimental results [13]. These hairpin vortices are found to be inclined at 45° to the flow direction (Fig. 2a), thus in good agreement with experiments [12,13]. It is felt that hairpin vortices play a dominant role in the production of turbulence. The growth of the enstrophy as demonstrated in Fig. 2b with the creation of vortical structures supports the idea [1,12]. The footprints of the total pressure field on the wall in Fig. 2c suggest that those vortical structures induce a complex unsteady three-dimensional effect. Streaklike structures appear to be emerging at $t^+ = 1$ in Fig. 2b which strengthens the belief that those structures, as a result of the interaction with the background shear flow, are a fundamental aspect of turbulent flow fields. Previous studies [5-7] neglected the interaction between the hairpin vortices and the background vorticity. In fact, as shown in

Fig. 2 this interaction is very strong. This emphasises the importance of including these effects of such interactions. Our method fully accounts for this interaction and show good agreement with experimental observations [13-15].

5 Conclusion

A numerical method capable of simulating three-dimensional near-wall boundary-layer flow has been developed. The versatility and simplicity of the method allows the investigation of a wide range of realistic and very complex flow-structure interactions.

References

1. ROBINSON, S.K. 1991. Coherent motions in the turbulent boundary layer, *Ann. Rev. Fluid Mech.* **23**, pp. 601–639.
2. HAMA, F.R. 1962. Progressive Deformation of a Curved Filament by its Own Induction, *Phys. Fluids* **5**, pp. 1156–1162.
3. DHANAK, M.R. & DE BERNARDINI, B. 1981. The evolution of an elliptic vortex ring, *J. Fluid Mech.* **109**, pp. 189–216.
4. DHANAK, M.R. 1981. Interaction between vortex filament and an approaching rigid sphere, *J. Fluid Mech.* **110**, pp. 129–147.
5. AREF, H. & FLINCHEM, E. P. 1984. Dynamics of a vortex filament in a shear flow, *J. Fluid Mech.* **148**, pp. 477–497.
6. MOIN, P. & LEONARD, A. & KIM, J. 1986. Evolution of a curved of a filament into a vortex ring, *Phys. Fluids* **29**(4), pp. 955–963.
7. HON, T,-L & WALKER, J. D. A. 1991. Evolution of hairpin vortices in a shear flow, *Computers and Fluids* **20**(3), pp. 343–358.
8. LUCEY, A. D. & CAFOLLA, G. J. & CARPENTER, P. W. 1997. Numerical simulation of a boundary-layer flow interacting with a passive compliant boundary, *Lecture Notes in Physics* **490**, pp. 406–411.
9. HALD, O. H. 1991. Convergence of vortex methods, In K. Gustafsson and J. Sethian, editors, *Vortex Methods and Vortex Motion*, SIAM, Philadelphia, pp. 741–773.
10. BEALE, J. T. & MAJDA, A. 1985. Higher order accurate vortex methods with explicit velocity kernels, *J. Comput. Phys.* **58**, pp. 188–208.
11. CHORIN, A. J. 1973. Numerical study of slightly viscous flow, *J. Fluid Mech.*, **57**(4), pp. 785–796.
12. KLINE, S. J. & REYNOLDS, W. C. & SCHRAUB, F. A. & RUNSTADLER, P. W. 1967. The structure of turbulent boundary layers, *J. Fluid Mech.*, **30**, pp. 741–773.
13. HEAD, M. R. & BANDYOPHADYAY, P. 1981. New aspects of turbulent boundary layer structure, *J. Fluid Mech.*, **107**, pp. 297-338.
14. ACALAR, M. S. & SMITH, C. R. 1987. A study of hairpin vortices in a laminar boundary layer, Part 1. *J. Fluid Mech.* **175**, pp. 1–41.
15. ACALAR, M. S. & SMITH, C. R. 1987. A study of hairpin vortices in a laminar boundary layer, Part 2. *J. Fluid Mech.* **175**, pp. 43–83.

Vortex Breakdown in Confined Swirling Flow: Numerical Simulation and PIV Measurement

Ying Zheng Liu[1], Han Ping Chen[1], and Hide S.Koyama[2]

[1] School of Power and Energy Engineering
Shanghai Jiao Tong University, Shanghai 200030, P.R.China
[2] Department of Mechanical Engineering
Tokyo Denki University, Tokyo 101-8457, Japan

Abstract. For vortex breakdown of swirling flow confined in cylindrical container with a rotating upper endwall, the whole meridional velocity field at different Re for H/R=2.0 is investigated by both numerical simulation and PIV measurement. To increase the convergence rate and calculation precision of numerical simulation, the modified SIMPLE based scheme with explicit expression of the periphery velocity component and multi staggered grid is used here. With the consideration into the whole meridional velocity field, the numerical results agree well with the experimental measurements via PIV.

1 Introduction

Swirling flows in cylindrical container that are induced from rotating boundaries are important to the designer of fluid mixing devices and semiconductor material processing devices. With simple geometry and boundary conditions, it is suitable to establish high precision controlled apparatus, and evaluate numerical scheme by comparison with experimental results. Although a large quantity of numerical and experimental investigation has been done to gain insight into vortex dynamics of such swirling flows, the mechanism of some complicated vortex phenomena, such as vortex breakdown, is still not generally acknowledged.

Ever since Escudier [1,2] found single, double and triple vortex breakdowns in systematically organized flow visualization studies of swirling flow induced from a rotating endwall inside the closed cylindrical container, such type of setup has been widely employed experimentally and numerically by many investigators. Much numerical work [3–5] has been done on vortex breakdown in confined swirling flow, and PIV (Particle Image Velocimetry) technique was applied to a few experimental investigation [6]. The flow visualizations of Escudier demonstrated that the existence of bubble-type vortex breakdowns at the axis of the container was mainly dependent on $Ar = H/R$ and $Re = \Omega R^2/\nu$, H being the height of the container, R the radius, Ω the angular speed of the rotating endwall and ν the kinematic viscosity of the working fluid. Based on the axisymmetric Navier-Stokes equation, extensive numerical work [3–5] was undertaken to check if the numerical solution could reproduce the observed vortex characteristics and to analyze physical mechanism of the vortex breakdown. As companion numerical work, the systematical comparison between numerical simulation, visualizations and LDA measurements was performed by the authors [7]. Most of the numerical simulation results aforementioned were consistent with the experimental observations on the formation of breakdown bubble(s), but the comparison on the whole fluid motion is still missing. The objective of this paper is to present investigation of the whole meridional flow field by both numerical simulation and experimental quantitative measurements via PIV.

2 Mathematical model

The closed cylindrical container with radius $R = 73mm$ and height $H = 146mm$ ($Ar = 2.0$) is as shown in Fig.1. It is filled with an incompressible homogeneous Newtonian fluid at constant temperature, so the density ρ and the viscosity ν have the constant values in the calculation field. The top endwall is rotating at constant angular velocity. The bottom endwall and the cylindrical sidewall are at rest. On the assumption that the flow is axisymmetric, the Navier-Stokes equations and the continuity equation are written as follows, in cylindrical coordinates (z, r, θ) with corresponding velocity components (u, v, w).

$$\frac{\partial(\rho u)}{\partial t} + \frac{\partial}{\partial z}\left(\rho u^2 - \mu\frac{\partial u}{\partial z}\right) + \frac{1}{r}\frac{\partial}{\partial r}\left(\rho r u v - \mu r\frac{\partial u}{\partial r}\right) = -\frac{\partial p}{\partial z} \quad (1)$$

$$\frac{\partial(\rho v)}{\partial t} + \frac{\partial}{\partial z}\left(\rho u v - \mu\frac{\partial v}{\partial z}\right) + \frac{1}{r}\frac{\partial}{\partial r}\left(\rho r v v - \mu r\frac{\partial v}{\partial r}\right)$$

$$= -\frac{\partial p}{\partial r} - \mu\frac{\nu}{r^2} + \rho\frac{w^2}{r} \quad (2)$$

$$\frac{\partial(\rho r w)}{\partial t} + \frac{\partial}{\partial z}\left(\rho r u w - \mu\frac{\partial(rw)}{\partial z}\right) + \frac{1}{r}\frac{\partial}{\partial r}\left(\rho r^2 v w - \mu r\frac{\partial(rw)}{\partial r}\right)$$

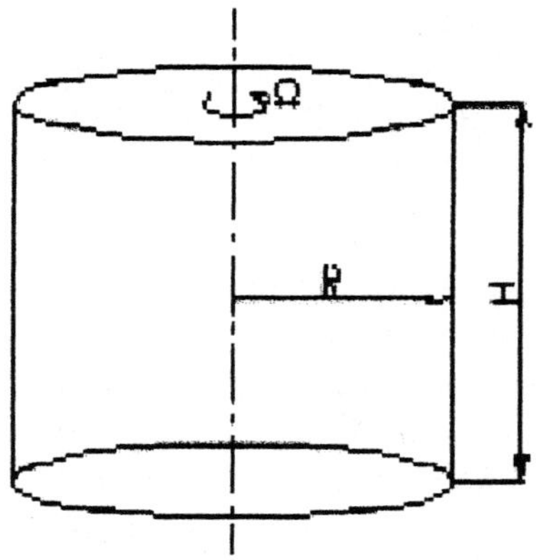

Fig. 1. Mathematical model

$$= -w\left(\frac{\rho v}{r} + \frac{\mu}{r^2}\right) \tag{3}$$

$$\frac{\partial u}{\partial z} + \frac{1}{r}\frac{\partial (vr)}{\partial r} = 0 \tag{4}$$

At the sidewall and endwalls, the no-slip condition is imposed on the velocity field. The initial and boundary conditions as follows:

$$\begin{aligned}
&z = 0, 0 \le r \le R : u = 0, v = 0, w = 0\\
&z = H, 0 \le r \le R : u = 0, v = 0, w = r\Omega\\
&r = 0, 0 \le z \le H : \frac{\partial u}{\partial r}, v = 0, w = 0\\
&r = R, 0 \le z \le H : u = 0, v = 0, w = 0
\end{aligned}$$

The governing equations and the boundary conditions for the flow in the working fluid are solved numerically by the finite control volume method. The diffusive-convective term is discretized with hybrid differencing scheme. The staggered grid system is used to calculate the velocity field. The pressure and the velocity are linked by the modified SIMPLE based pressure correction equation [8] with explicit expression of the periphery velocity component. The change of the periphery velocity component in iterations is considered as the solution to the strong couple between the periphery velocity component and the flow field. For the swirling flow induced from rotating boundary, such modification can increase the convergence rate.

To enhance the convergence rate for fine grid distribution and cut down the truncation errors, the multi staggered grid algorithm is used here. It is FMG-FAS scheme in V cycle. The variables in question are restricted and prolonged between different levels by bilinear interpolation, and the residuals are transferred by summation. The internal coarse grid is assembled by four fine grids, but six for control volumes of u and v velocity components near the boundary. All of the equations in question are treated in the same way, including the pressure correction equation. And all of the variables in numerical simulation are treated as double precision.

The mean cell residual of each equation is used as the convergence criterion of the numerical simulation.

$$\begin{aligned}
&S_m = \sqrt{\Sigma S^2_{(i,j)}}/n_p, S_u = \sqrt{\Sigma S^2_{(i,j)}}/n_u\\
&S_v = \sqrt{\Sigma S^2_{(i,j)}}/n_v, S_w = \sqrt{\Sigma S^2_{(i,j)}}/n_w\\
&S_m, S_u, S_v, S_w < 10^{-10}
\end{aligned}$$

The finest uniform grid used in the present investigation is 128×256, which results in a grid independent solution. And the matrix is solved by means of the tridiagonal matrix algorithm (TDMA). The relaxation factors are 0.5 for each momentum equation and 0.8 for pressure correction equation.

3 Experimental apparatus

A high-quality precision-controlled turntable facility with the same boundary condition as the numerical simulation was fabricated as shown in Fig.2.

A rectangular tank was placed outside of the cylindrical container. To reduce the curvature effects of the glass sidewall, silicone oil (the working fluid) was filled in both the container and the tank. The rotating rate of the motor was under control of the frequency convertor, which made the experimental apparatus encompass a broad range of Re. The output power precision of the convertor is $0.01Hz$. Diameter of the rotating endwall and inner diameter of the cylindrical container was $145.00 \pm 0.01mm$ and $146.34 \pm 0.02mm$ respectively. The height of the container is $146.0 \pm 0.1mm$. Seeding particles were provided by polyamide (38A2111 PSD-20) with a mean diameter of $20\mu m$.

The PIV (Dantec PIV2100 processor) instrument with 200 mj YAG laser (8 ns pulse) and Kodak CCD 10 Plus (15 Hz, 1 M pixels, cross correlation) camera were used for measurement. A laser sheet ($532.0nm$) was arranged to illuminate particle distribution in the meridional plane. The recording CCD camera was placed to be perpendicular to the illuminated region. The PIV 2100 processor processed image maps from the CCD camera into velocity vector maps by means of cross correlation (FlowMap3.11). The maximum out-of-plane velocity was given by $v = \Omega R$. The thickness of the laser sheet and the time between double exposures must be adjusted to accommodate the maximum periphery displacement of most particles during the double successive laser sheet sweeps.

4 Results and conclusions

For the steady swirling flows with and without breakdown, the whole fluid motion in the meridional plane is made clear in Fig.3 via both numerical simulation (right halves) and PIV measurements (left halves).

Fig. 2. Photograpg of experimental apparatus

When Reynolds number is low, it is known that the overall swirling flow inside the container is typical of large-scale vortex flow. The Ekman layer is developed by the rotating endwall through friction and is thrown outwards due to the centrifugal force. Then the continuity of the incompressible fluid flow deflects the boundary layer downwards. With the aid of Stewardson layer on the sidewall, the fluid is transported to the bottom endwall where Ekman layer is also formed for the same reason.The periphery velocity is diminished to a great extent by the stationary bottom endwall, and the centrifugal force is reduced. So the radial pressure gradient in the region is dominant, which generates the radial flow towards the axis. At last, the fluid goes up along the axis to the rotating endwall, and compensates those carried and ejected centrifugally. Therefore, the annual type circulation is formed. When Re is increased to be 1.38×10^3 (a), the conspicuous phenomenon is that the main flow from the bottom endwall turns away from the axis at $Ar = 1/3$ or so, therefore a low velocity region is formed. When $Re = 1.48 \times 10^3$ (b), the main flow is stagnated at the axis with the appearance of a small recirculation zone from the numerical results. It is bubble-type vortex breakdown phenomenon. With the increase of Re (c, d, e, f), the breakdown bubble grows and moves down to the bottom endwall. However, the velocity distribution of the whole meridional field changes little.

For the experimental results via PIV, the velocity vectors near both endwalls and sidewall are not good, since the signal noise is high for cross correlation of the flow field near the solid boundary. With the consideration into the whole field, the numerical simulation agrees well with the experimental results on the whole field fluid motion, location and size of the breakdown bubble and the fluid motion near the breakdown bubble. It is obvious that the fluid flow near the sidewall keeps at the same status with the increase of Reynolds number.

References

1. M.P. Escudier: Experiments in Fluids. **3**, 66 (1984)
2. M.P. Escudier: Progress of Aerospace Science. **25**, 189 (1988)
3. J.M. Lopez: J.Fluid.Mech. **221**, 533 (1990)
4. H.J. Lugt: J.Fluid.Mech. **179**, 179 (1987)
5. N. Tsitverblit: JSME Fluid Dynamics Research. **11**, 19 (1993)
6. P. Buchhave, M.L. Jakobsen et al.: 'PIV: Measurement of the Early Transition of a Rotating Flow in a Closed Cylinder'. In: *4th International Conference on Laser Anemometry, Advance and Applications, Cleverland Ohio,August 5–9, 1991.*
7. Y.Z. Liu, H.P. Chen et al: 'Numerical and Experimental Investigation on Rotating Flow with Vortex Breakdown'. In: *ASME00 Fluids Engineering Summer Conference, Boston, June 11–15, 2000.*
8. Y.Z Liu, H.P. Chen: 'A Rotating Flow Calculation Method Based on the Pressure Linking Equation Including the Rotating Velocity Correction'. In: *Internatioal Conference of Fluxes and Structures in Fluids, Sancta Petersburg, June 10–12, 1999.*

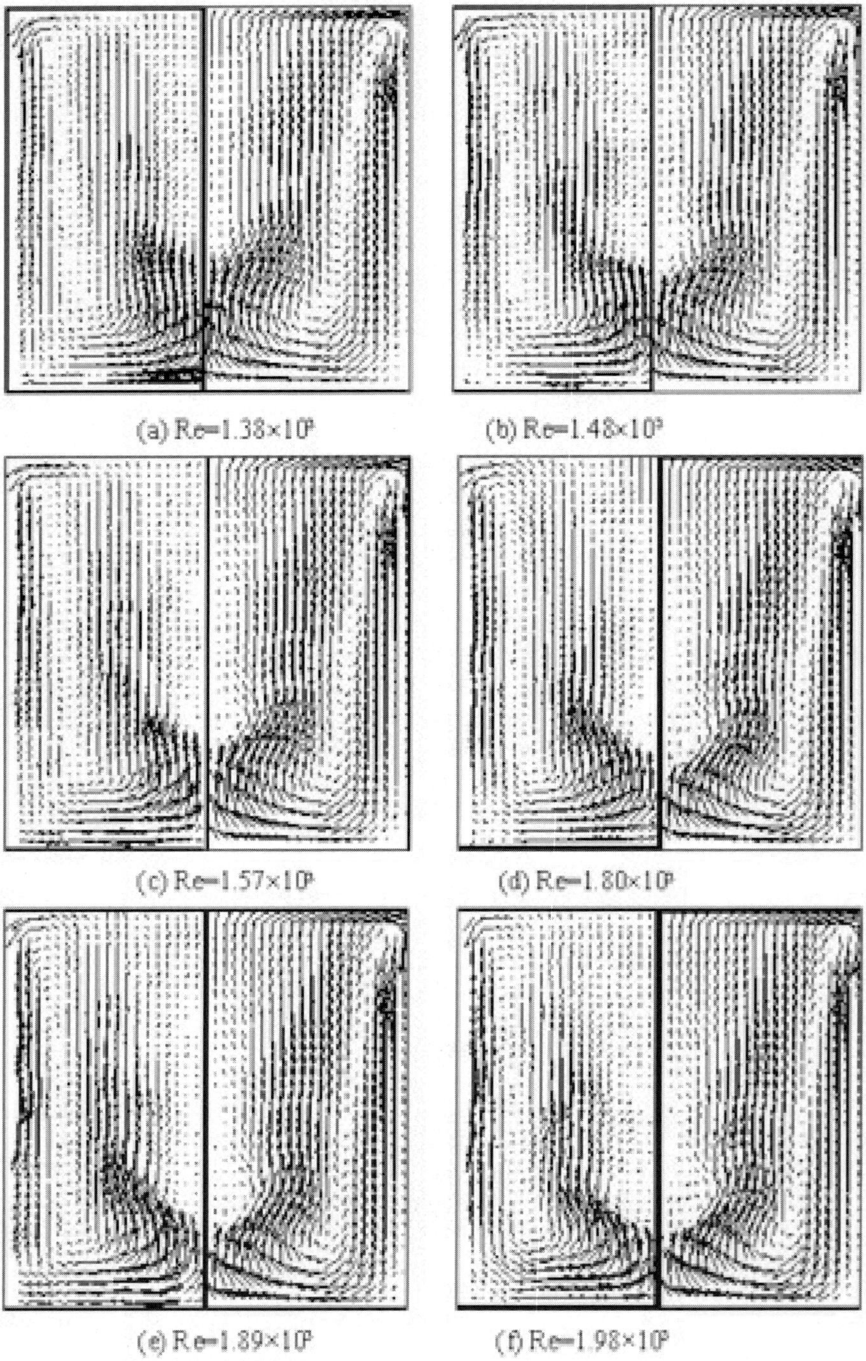

Fig. 3. Meridional velocity distribution: PIV (left halves) and CFD (right halves)

Fluid–Structure Interaction under the Vortex Lock-in Regime

Jun-Ho Hong and Norio Arai

Tokyo Noko University, Koganei, Tokyo 184-8588, Japan

Abstract. In this paper, the fluid–structure interaction under the lock-in regime is investigated the flow past a freely vibrating circular cylinder by 3-D numerical simulations. When the lock-in regime is realized, the three kinds of frequencies, those are the vortex shedding frequency, the body oscillation frequency and the body natural frequency considered the virtual mass effect, are coincided each other and the vortex street is redistributed to be switched over from one row to two.

1 Introduction

The problem of a bluff body oscillating in a free stream has received a great deal of attention. If a bluff body is placed in a flow, it experiences a fluctuating forces those are in the transverse direction (lift force) and the streamwise direction (drag force) by the asymmetric formation of vortices, which can cause a structure to vibrate. It is called the flow-induced vibration. One of the fundamental features of the flow–induced vibration is the ability of the structure oscillating with its natural frequency to synchronize in the vortex shedding frequency. The 'lock-in' regime might cause some serious damages of the body such as to turn down the fatigue strength.

A lot of experiments and numerical studies have been carried out on this problem. Early experimental studies have concentrated on the bluff bodies forced to oscillate in transverse direction to the flow. Griffin [1] have investigated the fluid–structure interaction problem on the freely vibrating circular bodies experimentally. These experimental studies of vortex–induced vibration have established two basic facts. Vortex-induced vibration is a self limiting process, in which the vibration amplitude is bounded approximately within a cylinder diameter. The other is that the phase difference of the lift force relative to the cylinder motion strongly depends on the vibration amplitude. However, the interaction is a strong nonlinear one and it is so difficult to analyze it experimentally.

In this article the problem on the free vibrating elastic body is analyzed by the 3 dimensional numerical simulations, in which the features of the lock-in regime and the wake structure are made clear.

2 Governing equation

In this analysis an incompressible viscous unsteady flow is calculated. The governing equations are the continuity equation and the Navier–Stokes equation, which are formulated by using the MAC method to calculate both the pressure P and the velocity $\boldsymbol{u}$. They are calculated by using the body–fitted coordinate (ξ, η, ζ, τ). The discretization is carried out by using the third order upwind scheme [2].

3 Elastic body

It is assumed that the circular cylinder is mounted as a spring–damper–mass system (Fig.1). The motion of the cylinder, y, can be described by the equation,

$$\frac{d^2y}{dt^2} + \zeta\omega_n\frac{dy}{dt} + \omega_n^2 y = \frac{F(t)}{m}, \tag{1}$$

where $\zeta = c/(2\sqrt{mk})$, $\omega_n = \sqrt{k/m} = 2\pi f_n$. Also, c is the coefficient of viscous damping and k is the rigidity of the spring. m is the mass of cylinder, ζ is a damping ratio, ω_n is a angular natural(eigen) frequency and f_n is the natural(eigen) frequency of the cylinder. In all simulations zero damping ($\zeta = 0$) is assumed since this would produce the vibration amplitude considerably. $F(t)$ is constructed by the total force of the lift and the effect of the virtual mass. The circular cylinder is allowed to vibrate only in transverse direction with an one–degree–of–freedom. Equation (1) can be solved easily using the Runge–Kutta method, once the force coefficient is known from the flow field calculation.

4 Calculation condition

In this calculation, we set the segment length of the circular cylinder $D = 1$ and the Reynolds number $Re = 1.0 \times 10^3$–1.0×10^4. The grid points are about 200,000 ($97 \times 100 \times 20$) covering a $100D$ circular region around the cylinder. The dimensionless time step is 2.5×10^{-3} . The calculation is started by the impulsive start. On the body surface, the moving condition is applied [3]. The Sommerfeld radiation condition is applied in the far field [4].

5 Results and discussion

5.1 Drag coefficient, lift coefficient, and body displacement

The time histories of drag coefficient C_d, lift coefficient C_l and the displacement of circular cylinder Y/D are shown in Fig.2. The results of C_d and C_l in 3-D cases show the smaller and more variable amplitude than the 2-D cases [5]. Although the vortex has diffused to the spanwise direction to be weaker, Y/D, doesn't become so small since the amplitude of vortex–induced vibrations are strongly associated with the phase difference between fluid force C_f and Y/D rather than the strength of C_l.

At the low velocity, C_l shows a quite big fluctuation same as the 2-D case. But, in this velocity range, the vortex shedding frequency is different from the natural frequency of the body and the vibration state doesn't rush into the lock-in regime, therefore Y/D is remaining in a small amplitude state (Fig.2b).

When the flow velocity rises and approaches the critical reduced velocity U_{r-cr} (Fig.2c), the cylinder vibration shows a big amplitude, while the average of lift force becomes smaller than the lower velocity case and C_d shows a big fluctuation. In this case, $U_{r-cr} = 1/St_s = 4.83$ because $St_s = 0.207$ where St_s is the Strouhal number of stationary cylinder. These phenomena agree with the 2-D case but the pattern of C_l around U_{r-cr} shows a little difference. In the range of lock-in regime, the amplitude of Y/D is rather big when the amplitude of C_l is small, consequently, the importance of the phase difference between Y/D and the fluid force is shown clearly.

In the high velocity range, all vibrations are suppressed to make the average values so small (Fig.2d). Then the decreasing mechanism is different from the case of 2-D. The wavelength of C_l curve becomes do not agree with the Y/D curve. In other words, the vortex shedding frequency and the vibration frequency of body becomes do not coincide each other as shown later. Such a phenomenon is not able to be observed in 2-D case.

5.2 Vortex shedding frequency

Figure 3 shows the vortex shedding frequency. The straight line represents the vortex shedding frequency of the stationary body, f_s, and hence the reduced velocity at which $f_s/\tilde{f}_n = 1$ is equal to the inverse of the Strouhal number of the stationary body. Here, the effects by the virtual mass are considered on the eigen frequency of cylinder so that the effective eigen frequency $\tilde{f}_n$ is used to arrange the vortex shedding frequency. The peaks of the power spectrum are shown more than one in the range of lock-in regime because of the complex pattern of C_l.

The vortex shedding frequency of oscillating cylinder f_v increases along the straight line of f_s with the increase of flow velocity in the low speed range. Near the U_{r-cr}, the lock-in regime realized while $f_v/\tilde{f}_n$ is equal to one. Here, the three kinds of frequencies (effective eigen frequency, vortex shedding frequency, and body oscillation frequency) coincide each other ($\tilde{f}_n = f_v = f_b$). When the flow velocity enter into the high speed range, f_v and f_b are separated ($U_\infty/\tilde{f}_n D = 15.8$). While f_v meet a straight line again, f_b keeps the condition which coincide to $\tilde{f}_n$. This phenomenon do not agree with the case of 2-D.

5.3 Isobaric surface

Figure 4 shows the isobaric surfaces. Figure 4b is very similar to the one shown for the case of stationary cylinder (Fig.4a). The width of the wake is very small that seems as a straight line. This vortices pattern influences a small fluctuation force on the body in transverse direction.

In the case of the moderate speed, the lock-in regime is realized, the body oscillation becomes so large and the vortex street forms two parallel rows (Fig.4c). Namely, the vortices pattern is switched over from one row to two rows and the vortex spacing becomes wider obviously in the transverse direction. This spreading generally leads to the increasing of the transverse direction force and the amplitude of the body vibrations and to twist the vortices columns.

In the case of the higher speed (Fig.4d), the rearrangement is carried out to be one row again and the width becomes small.

6 Conclusions

In order to catch the fluid–structure interaction in the range of the lock-in regime, 3 dimensional numerical simulation is carried out on the freely oscillating circular cylinder. The results are summarized as follows.

1. Under the lock-in regime, the three kind of frequencies, those are the vortex shedding frequency f_v , the body oscillation frequency f_b and the body natural (eigen) frequency of the system considered virtual mass effect $\tilde{f_n}$, are coincided each other. Without the lock-in regime, however, f_b shows a different frequency from f_v and f_s while f_v coincides the vortex frequency from the stationary body f_s .
2. When the lock-in regime is realized, the amplitude of vibrations increase remarkably and the vortex street is redistributed to be switched over from one row to two. And the self limiting process is controlled by the phase difference between the body oscillation and the fluid forces (lift and the force due to the virtual mass effect).

References

1. O. M. Griffin: ASME Journal of Pressure Vessel Technology, **102**, pp. 158–166(1980)
2. T. Kawamura, H. Kuwahara: AIAA, **84**–0340 (1984)
3. H. Izumi: Transactions of the JSME, B–**60**–572, pp. 1218–1225 (1994)
4. T. Yoshida, T. Watanabe et. al: Transactions of the JSME, B–**59**–565, pp. 2799–2806 (1993)
5. N. Arai, J.H. Hong et. al: 'Osillation of elastic body'. In: *ECCOMAS 2000, Barcelona, Sep. 11–14* (2000)

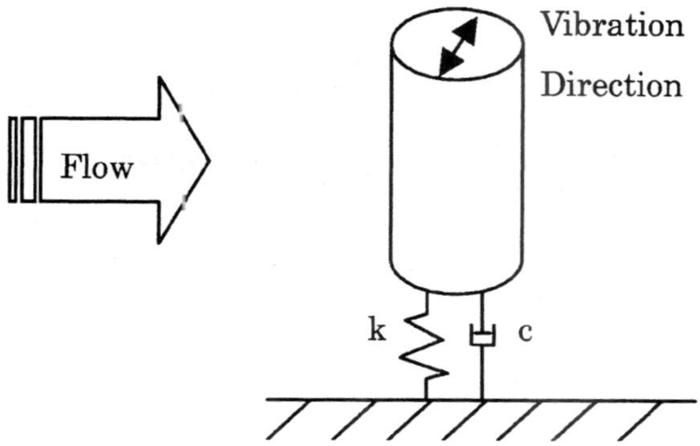

Fig. 1. Schematic diagram of the vibrating system.

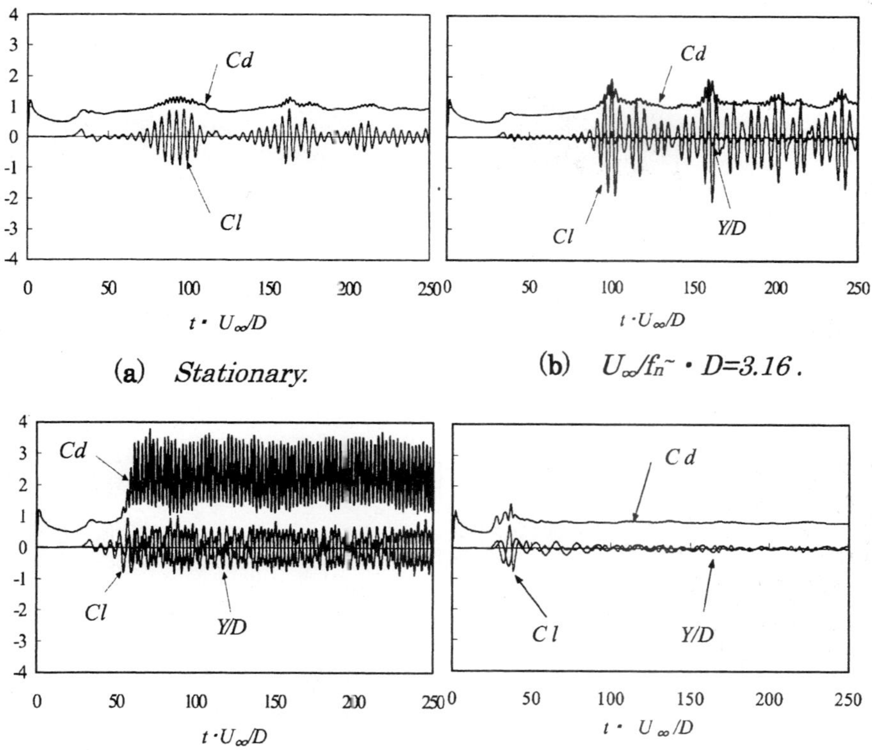

(a) *Stationary.*

(b) $U_\infty/f_n^{\sim} \cdot D=3.16$.

(c) $U_\infty/f_n^{\sim} \cdot D=4.89$.

(d) $U_\infty/f_n^{\sim} \cdot D=15.8$.

Fig. 2. Time histories of C_d, C_l and Y/D.

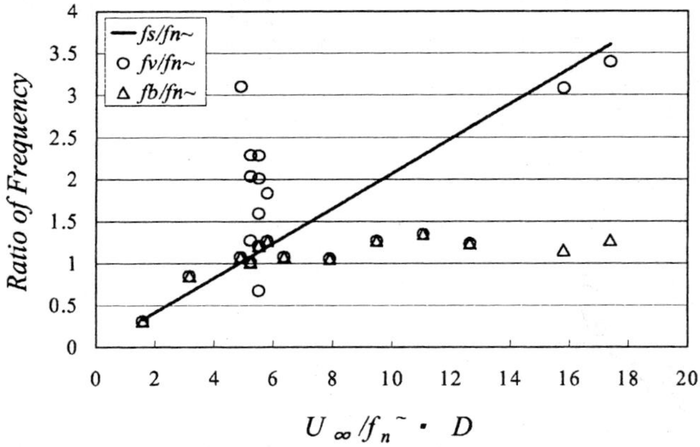

Fig. 3. Variation of the vortex shedding frequency.

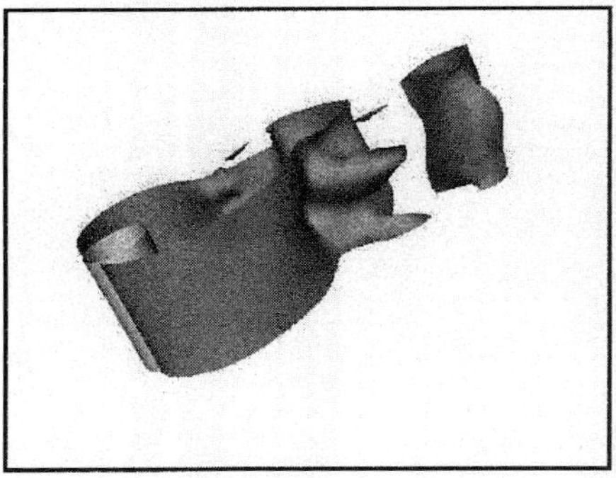

(a) *Stationary.*

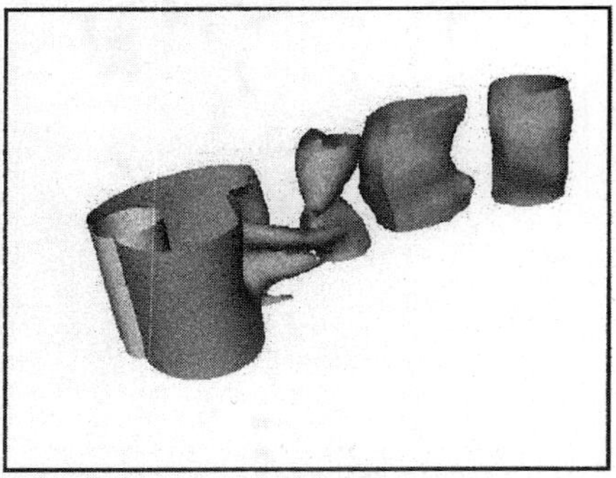

(b) $U_\infty/f_n^{\sim} \cdot D = 3.16$.

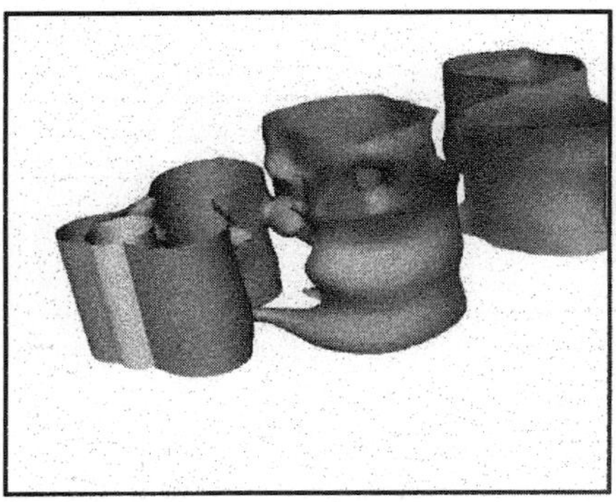

(c) $U_\infty/f_n^{\sim} \cdot D = 5.5$.

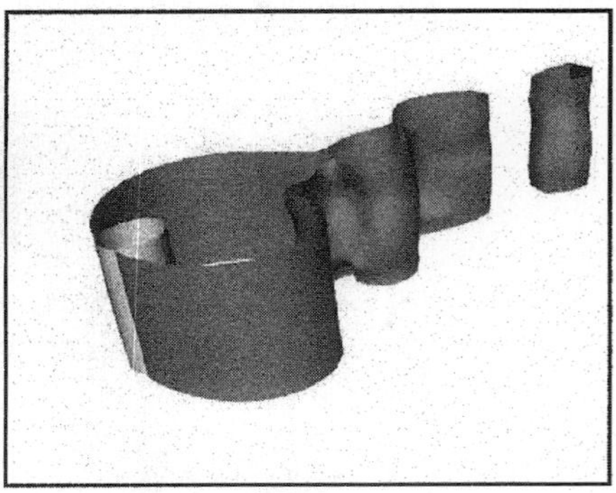

(d) $U_\infty/f_n^{\sim} \cdot D = 15.8$.

Fig. 4. Instantaneous isobaric surface.

Computation of Active Control of Flow past a Circular Cylinder

K. Srinivas

Department of Aeronautical Engineering, The University of Sydney, NSW 2006, Australia. Email: ragh@aero.usyd.edu.au

Abstract. Present investigation is aimed at the possibility of computing active control of flow past a cylinder ie., to develop a feedback loop to limit drag. This phase of the work concentrates on control by imposing rotational oscillations upon the cylinder. A code based on SIMPLEC methodology with a $k-\epsilon$ model of turbulence is employed to compute unsteady flow past a cylinder. In accordance with experiments it is found that for higher amplitudes of rotation and for certain frequencies there is a pronounced reduction in drag. At the same time drag shoots up to large values near about the natural frequency of vortex shedding.

1 Introduction

There have been many studies addressing the issue of control of a flow to regulate features like drag, lift, efficiency or noise. Two powerful methods of control - Passive Control and Active Control have emerged as a result. A good account of these along with a comparison of their merits can be found in Pack and Joslin [4]. Active Control, of interest here, relies generally on the possibility of delaying or accelerating transition, reducing drag etc. and at present is the one much interest is focussed upon.

The purpose of this research is to investigate the feasibility of developing a feedback loop to control the flow past a given body, say, a cylinder. It is known that drag and lift phenomena for this flow are influenced by vortex shedding behind the cylinder. Typically, the envisaged loop will comprise of measuring some signal downstream of the cylinder. This signal is then fed back to the upstream in order to manipulate conditions therein so that vortex shedding is controlled. Such a goal is far from being achieved. This paper describes only the initial effort in this direction.

There have been quite a few experimental studies of active control of flow past a cylinder. It appears that drag control can be achieved by imposing rotational oscillations about the cylinder, injecting mass into the boundary layer or by subjecting the cylinder to cross flow oscillations. Tokumaru and Dimotakis [8] have demonstrated in their exploratory work that considerable control in drag can be achieved by forcing a rotational oscillation about the cylinder. A substantial reduction in c_d was noticed when the forcing Strouhal number was 0.8 at a Reynolds number of 1.5×10^4. Fujisawa et al. [1] have carried out this study near the natural frequency of vortex shedding and noted that drag shoots to a peak value at a Strouhal number of 0.17 followed by a sudden decrease.

Work by Warui and Fujisawa [9] concentrates on control by inducing cross flow oscillations.

Regarding computational work the flow behind a bluff body has been studied extensively [3][6] employing very sophisticated tools and models of turbulence like LES and DNS. Reference [4] gives a detailed account of the computational work at NASA langley Research Centre in the area of control by oscillatory blowing upon an aerofoil. It is interesting to note that there have been at least two studies trying to compute the active control of cylinder flow. Work by Park et al. [5] considers control by blowing and suction, the extent of which is determined by a feedback sensor. Complete suppression of vortex shedding has been achieved for a Reynolds number of 60 while at higher Reynolds numbers control is not realised. Gunzbuger and Lee [2] employ injection and suction through orifices on the surface of the cylinder. The amount of fluid to be injected was determined by a feedback law based on the pressure levels. This control was found to be effective in reducing the rms value of lift.

The present study concentrates on the effect of rotational oscillations upon drag and lift about a circular cylinder. A preliminary account of this investigation dealing with laminar flow computation was presented in Srinivas et al. [7]. The present study is more detailed and extends it to turbulent flows.

Today turbulent flows have been computed with advanced models and methods. But the present application for practical use will call for a simple and fast code. Accordingly we still use only a $k - \epsilon$ model of turbulence. The question that arises naturally is whether a simple model can capture all the complicated effects. To some extent, the present study becomes an investigation into this question.

Section 2 of the paper gives the computational details while section 3 presents and discusses the results obtained.

2 Computational Details

Two-dimensional, Reynolds averaged Navier-Stokes equations are solved in their unsteady form using a code based on SIMPLEC procedure. The equations employ Cartesian components of velocity and non-orthogonal coordinates. A collocated Finite Volume method is used to discretise the equations. The resulting equations are solved using a segregated approach. Temporal integration is carried out by a three level fully implicit scheme of second order accuracy. Details of the approach could be found in Ref. [10].

An O grid is used for computations. It consists typically of 100 equally spaced points along the cylinder and 60 points placed non-uniformly in the radial direction. The grid spacing close to the cylinder is 0.001m, the cylinder diameter being 0.1m. The outer boundary of computations is placed at a distance of about 20 cylinder diameters.

Runs were carried out at Reynolds numbers of 15,000 and 20,000 to enable comparison with the previous experimental investigations. Free stream boundary conditions were imposed at the outer boundary while rotational conditions were forced on the cylinder surface.

Computations were first performed for a stationary cylinder. It is known that it is difficult to obtain vortex shedding for a stationary cylinder if no-slip conditions are imposed upon the surface from the beginning of the computations. It requires a big perturbation to be forced upon the flow. Accordingly asymmetric, fluctuating components of velocity were imposed upon the cylinder surface and subsequently removed. This did enable vortex shedding. Computations continued for a real time of 4 seconds when the vortex shedding had stabilised. The resulting flow field served as the starting condition for the subsequent computations when rotational oscillations were imposed.

The rotational oscillations at a prescribed frequency and amplitude were then imposed upon the cylinder surface as per:

$$v_\theta = \omega_{max} sin(2\pi ft) \tag{1}$$

Where v_θ is the tangential velocity along the cylinder, ω_{max} is the amplitude and f the frequency of oscillations. Usually ω_{max} and f are non-dimensionalised as:

$$v_r = \frac{R\omega_{max}}{U_\infty} \text{ and } S_f = \frac{fD}{U_\infty} \tag{2}$$

Where Sf is the forcing Strouhal number and D the cylinder diameter.

The run was continued till the c_l and c_d behaviour stabilised at their new levels.

It is obvious that the time step size δt plays a crucial role in these computations. Several runs were made to arrive at the value of δt that was large enough to reduce the CPU time requirements; at the same time it should be small enough such that reducing it any further did not alter the results significantly. The results reported here are those obtained with a δt of 0.0005s requiring 2000 time steps to cover a real time of 1 second.

At every time step, inner iterations were performed till the residue decayed by three orders of magnitude. This required as many as 400 iterations before the vortex shedding stabilised. Subsequent calculations required only about 50 iterations.

Pressure distribution around the cylinder was integrated at every time step to obtain unsteady rms values of Cl and mean values of Cd.

Computations were carried out on a 300 MHZ PC and it required about 15 hours for a computation covering a real time of 2.5 seconds.

3 Results and Discussion

3.1 Stationary Cylinder

As is well known frequency of vortex shedding behind a bluff body depends upon the Reynolds number of the flow. As such it becomes necessary of any computational code to predict this dependence properly. Accordingly runs were conducted for a stationary cylinder for various Reynolds numbers ranging from 20 to 20,000. It was found that hardly any shedding occurred below a Reynolds number of 80. Fig. 1 shows the vortex shedding frequency (expressed as Strouhal number) plotted as a function of Reynolds number. A good agreement with previous investigators for low Reynolds Numbers [5][2] and with experiments at the higher Reynolds number [1] is evident.

Fig. 1 also gives the time dependence of c_d and c_l for the stationary cylinder. Clearly, the drag or the lift behaviour stabilises in less than two real seconds of computations. The figure also shows the contributions to c_d and c_l from viscosity and pressure separately. As expected viscosity plays only a minor role in determining c_d or c_l.

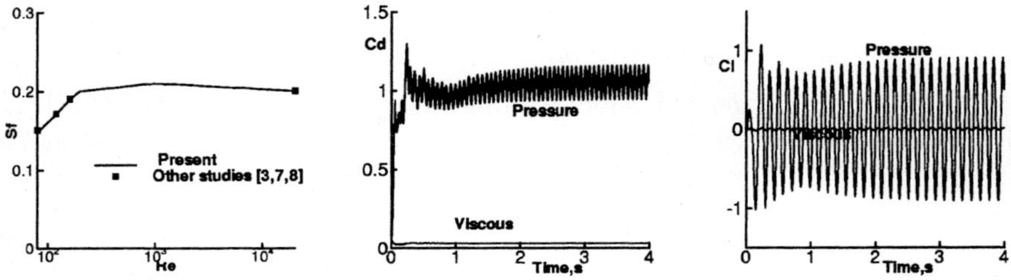

Fig. 1. Effect of Reynolds number upon vortex shedding frequency, and time history of c_d and c_l for stationary cylinder flow

3.2 Effect of Rotational Oscillations

Rotational oscillations with various non-dimensional amplitudes v_r and at different frequencies ie., different forcing Strouhal numbers, S_f were imposed upon the cylinder starting from the stationary solution discussed in the previous section.

The behaviour of mean c_d with the forcing Strouhal number for amplitudes of $v_r = 0.38$ and 2 is shown in Fig. 2 along with corresponding results from experimental studies from others. For both the amplitudes, the present computations do capture the correct trends. Typically drag rises sharply from its stationary value when the forcing frequency is low and decreases at higher frequencies. But quantitative differences do exist between the results. For example, c_d for the stationary cylinder agrees more closely with experiments from Fujisawa et al [1] than with that from Tokumaru and Dimotakis [8]. It is believed that the procedure used by Fujisawa et al is more refined consequently the present prediction seems a very acceptable one. At lower amplitude, ie., $v_r = 0.38$, computed c_d values seem to be shifted to the right compared to the experiments. On the other hand, at the higher amplitude there is an over prediction.

The coefficient of lift, c_l follows similar trends.

Fig. 3 shows the time histories of c_d and c_l for typical values of S_f for the high amplitude case, ie. $v_r = 2$, where the effect of oscillatory rotation is pronounced.

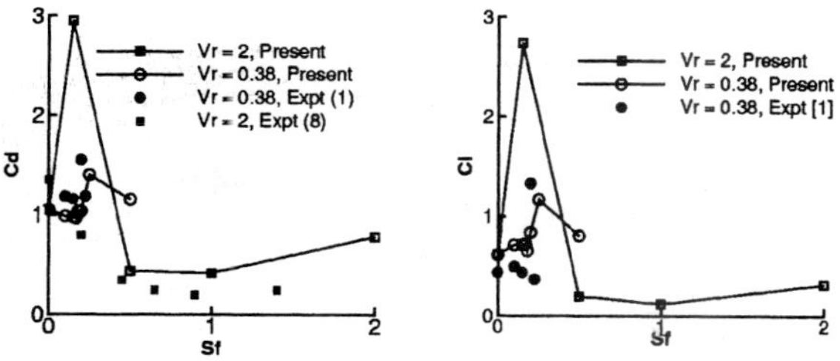

Fig. 2. c_d and c_l as a function of forcing Strouhal number

It may be mentioned that it required v_r to be nearer to or more than unity to notice any clear effect of oscillations.

Contours of c_p shown in Fig. 4 further reveal the effect of rotation. At the lower frequency, one sees a large asymmetry in the contours in the flow direction contributing to the drag. Fig. 5 shows the velocity vectors with instantaneous stream lines. These indicate a highly disturbed flow at the smaller frequency. At the higher frequency the flow looks more symmetrical than the stationary cylinder flow.

Though the study seems promising that the effects of rotational oscillation upon a cylinder can be captured reasonably, there are other questions to be answered. The role of turbulence parameters, the model itself may need a further study. The work is continuing using a C type grid at present. The task of establishing the feedback loop to control vortex shedding is also to be taken up shortly.

4 Acknowledgments

Thanks are due to Dr. C P Mellen, University of Karlsruhe, Germany for many helpful suggestions.

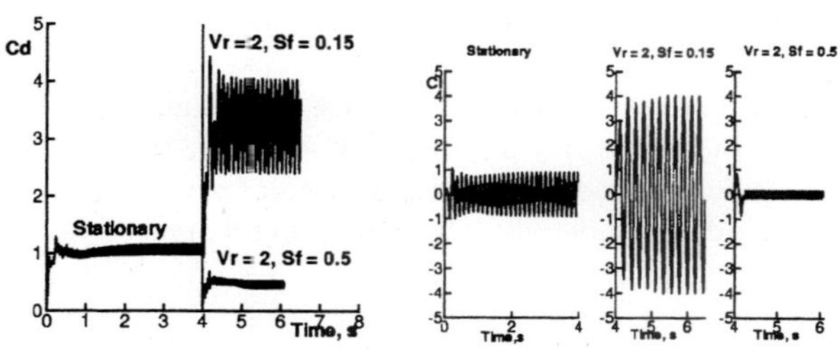

Fig. 3. Time history of c_d and c_l with rotational oscillations

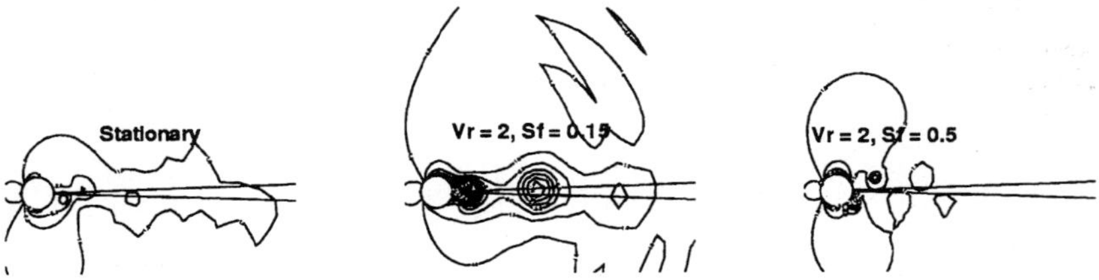

Fig. 4. c_p contours for flow with rotational oscillations

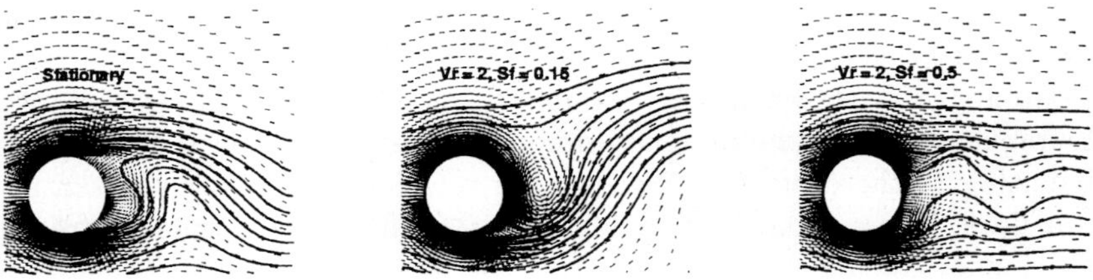

Fig. 5. Velocity vectors and instantaneous streamlines

References

1. N. Fujisawa, K. Ikemoto, and K. Nagoya. Vortex shedding resonance from a rotationally oscillating cylinder. *Journal of Fluid and Structures*, 12:1041–1053, 1998.
2. M. D. Gunzbuger and H. C. Lee. Feedback control of Karman vortex shedding. In *Transactions ASME*, volume 63, pages 828–835, 1996.
3. S. Murakami, S. Iizuka, A. Mochida, and Y. Tominaga. LES analysis of turbulent flows past a square cylinder using various SGS models. *Direct and Large Eddy Simulations*, 2:385–395, 1997.
4. L. G. Pack and R. D. Joslin. Overview of active flow control at NASA Langley Research Centre. Private Communication.
5. D. S. Park, D. M. Ladd, and E. W. Hendricks. Feedback control of von Karman vortex shedding behind a circular cylinder at low reynolds numbers. *Physics of Fluids*, 6(7), 1994.
6. W. Rodi. On the simulation of turbulent flow past bluff bodies. *Journal of Wind Engineering and Industrial Aerodynamics*, 46-47:3–19, 1993.
7. K. Srinivas, N. Fujisawa, and H. W. Aung. Effect of rotational oscillation upon vortex shedding behind a cylinder. In *8th International Symposium on Computational Fluid Dynamics*, September 1999.
8. P. T. Tokumaru and P. E. Dimotakis. Rotary oscillation control of a cylinder wake. *Journal of Fluid Mechanics*, 224:77–90, 1991.
9. H. M. Warui and N. Fujisawa. Feedback control of vortex shedding from a circular cylinder by cross flow oscillations. *Experiments in Fluids*, 21:49–56, 1996.
10. J. Zhu. A low diffusive and oscillation free convection scheme. *Communications in Applied Numerical Methods*, 7:225–232, 1991.

Numerical Investigations of Vortex-Shedding Flowmeters

Ernst von Lavante and Stephan Perpéet

Institute of Turbomachinery, University of Essen, D-45127 Essen, Germany

Abstract. In the present investigation, the problem of accurate determination of mass flows by means of vortex-shedding flowmeter is studied for the case of disturbed inflow. To this end, the flow about bluff bodies of different designs, used in typical vortex-shedding flowmeter, was numerically simulated using a solver of the unsteady, compressible Navier-Stokes equations in three dimensions. The computations were carried out for several types of disturbed inflow conditions, including bends and flow with rotation oriented in the axial direction. The resulting flow fields were analyzed using various methods, including visualization, evaluation of several flow features and DFT of properly chosen variables. Recommendations regarding the operational conditions of the vortex-shedding flowmeters are made.

1 Introduction

Many aerodynamic problems require a volume- or mass-flow data for its quantitative solution. Therefore, a number of methods for flow rate measurement have been developed. One relatively simple flow measurement device is the vortex-shedding flowmeter, in which the mass flow is determined by observing the relationship between vortex-shedding frequency from a bluff body attached inside a channel, and the corresponding mean velocity about it. The bluff body causes the production of a system of periodic vortices, whose frequency can be correlated with the mean flow velocity and, therefore, the mass flow. This procedure assumes a regular and well defined vortex structure as well as shedding mechanism, resulting mostly in linear dependency of the mass flow on the shedding frequency over a wide range of Reynolds numbers. However, it has been observed that certain types of disturbancies in the inflow result in rather irregular pressure signatures of the vortex system or even shift of its charachteristic frequencies, leading to unreliable mass flow data.

2 Numerical Algorithm

The numerical algorithm employed here uses three-dimensional, time-dependent full Navier-Stokes equations describing the conservation of mass, momentum and energy:

$$\frac{\partial Q}{\partial t} + \frac{\partial (F - F_v)}{\partial x} + \frac{\partial (G - G_v)}{\partial y} + \frac{\partial (H - H_v)}{\partial z} = 0. \tag{1}$$

The program is based on the finite-volume method, using a cell-centered formulation on block-structured grids. The spatial discretization is carried out with help of an approximate Riemann-solver. The formulation of the numerical fluxes at the cell interfaces is based on Roe's Flux Difference Splitting [4], as implemented by El-Miligui et al. [2]. The reconstruction of the cell-centered variables to the cell-interface locations for the computation of convective terms was done using a monotone interpolation of the primitve variables $Q = (\rho, u, v, w, p)^T$. The interpolation slope was limited by an appropriate limiter, according to the previously published MUSCL type procedure (see, for example, [2]):

$$Q_R^n = Q_{i+1}^{n-1} - \frac{1}{4} L_{i+1} \left[(1+\kappa) \Delta_{i+1}^- + (1-\kappa) \Delta_{i+1}^+ \right]^{n-1} , \tag{2}$$

using a special quadratic version of the van Albada limiter developed by the present authors:

$$L_i = \frac{2\, {\Delta_i^-}^2\, {\Delta_i^+}^2 + \varepsilon}{{\Delta_i^-}^4 + {\Delta_i^+}^4 + \varepsilon} , \qquad \varepsilon \approx 1 \cdot 10^{-5} . \tag{3}$$

Variation of the accuracy factor κ and switching the limiter on or off results in various schemes from simple first order accurate up to third order formulation. In the present work, the second order accurate upwind biased Fromm scheme ($\kappa = 0$) and the above form of van Albada limiter has been used. The viscous fluxes F_v, G_v and H_v were centrally differenced. The method has been shown to be accurate and effective in simulation of low Mach number viscous flows [6]. Starting with a constant initialization of the scalar flow variables and an appropriate turbulent velocity profile, the integration in time is carried out by a modified explicit Runge-Kutta time stepping and, optionally, an implicit Symmetric-Gauss-Seidel (SGS) scheme, making *CFL* numbers of up to 30 possible.

2.1 Boundary Conditions

For the inlet and outlet planes of the simulated pipes, subsonic quasy-one-dimensional non-reflecting boundary conditions were implemented. These were based on the Riemann invariants normal to the pipe cross-section.

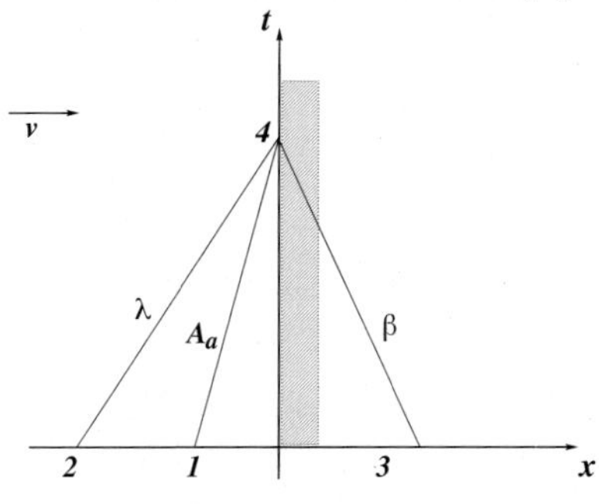

Fig. 1. Characteristics at the free stream boundary

The characteristics in direction normal to the boundary are:

$$\lambda = c + \frac{\gamma - 1}{2} u_n \qquad \text{corresponding to the direction} \qquad \frac{dn}{dt} = u_n + c$$

$$\text{and} \qquad \beta = -c + \frac{\gamma - 1}{2} u_n \qquad \frac{dn}{dt} = u_n - c$$

$$A_a = c \left(\frac{p}{p_{ref}} \right)^{\frac{1-\gamma}{2\gamma}} \qquad \frac{dn}{dt} = u_n \tag{4}$$

with u_n the velocity in direction normal to the boundary, n the normal coordinate, c the local speed of sound and γ the specific heat ratio. A_a is the speed of sound corresponding to some reference pressure p_{ref}. Similarly to enthalpy, A_a represents a state variable and remains constant along the particle path $1 - 4$. For isentropic flow A_a remains always constant. λ and β are the corresponding Riemann invariants. The characteristics for point (4) in Fig. 1 are:

$$A_{a4} = A_{a1} + u_1 \frac{A_{a2} - A_{a1}}{c_2 - u_1}, \quad \lambda_4 = \lambda_2 + \left(\frac{\lambda + \beta}{2} \right)_2 \frac{A_{a4} - A_{a2}}{A_{a2}},$$

$$\beta_4 = \beta_3 + \left(\frac{\lambda + \beta}{2} \right)_3 \frac{A_{a4} - A_{a3}}{A_{a3}}. \tag{5}$$

With the help of A_a, λ, β known at the point (4), the conservative variables can be determined. The above boundary conditions made the pressure waves and other disturbances run out of the domain without any noticeable reflection. Interestingly, the above boundary conditions allowed the normal velocity to become intermittently negative without effecting the stability of the scheme, making it well suited for the current application.
The interzonal boundary conditions enabled data exchange among the blocks of the computational grid and were accomplished by overlapping grids; at the walls, the no-slip solid viscous wall boundary conditions were used.

2.2 Turbulence Model

The highly complex geometrical and physical configurations are making the choise of an appropriate turbulence model very difficult. In the vortical, periodically separating flow with strong local rates of acceleration, any isotropic models are not adequate. The present authors decided, therefore, the use of the concept of large eddy simulation (LES). Several subgrid scale models were tested, including the simple Smagorinsky [5] and the non-linear version according to Germano et al. [1]. Since, however, their influence was negligible, the computations were mostly carried out without any SGS-model, relying only on the numerical damping inherent to the scheme.

3 Results

The bluff body inside the vortex-shedding flowmeter causes separation of periodic vortices from the upper and lower edge of a bluff body, similar to the well known von Kármán vortex street. In the present effort to optimize the bluff body, many different shapes were testet. As a result, an "optimum" shape was found and extensively tested [8] in 2-D, 3-D and experimentally. The resulting plot of the frequency of vortex passage through an ultrasound barrier placed downstream of the bluff body is displayed in Fig. 2 as a function of the time averaged mean velocity in the pipe. Its highly linear behavior in the given velocity range is obvious. Somewhat surprising is the good agreement between the 2-D and 3-D numerical results. The right side of Fig. 2 shows the shape of the optimized bluff body.

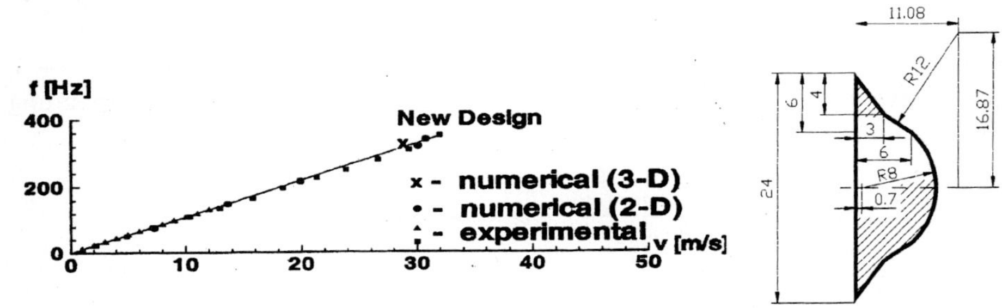

Fig. 2. Frequency vs. pipe mean velocity and optimized bluff body shape

As long as the inflow remains undisturbed, the shedding frequency in the wake of the bluff body is rather constant for most shapes. There are many studies concerning the optimization of the bluff body shapes, investigated under the condition of undisturbed inflow. We decided to examine the development of the vortex-shedding frequency in a more practical view. Flow meters are normally part of a pipe system and therefore the inflow is often nonuniform or even unsteady. Fig. 3 shows a section of the pipe, with a single and a double bend, featuring the protruding bluff body downstream of the bend. The characteristic double vortex system in the wake of the bend can to be seen.

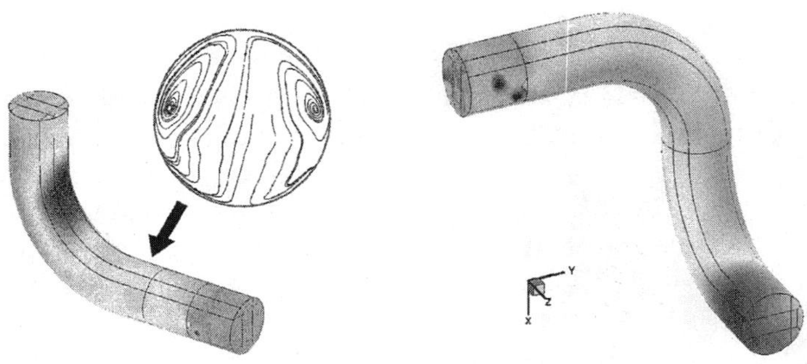

Fig. 3. Pressure Contours of a single (left) and a double bend (right)

Fig. 4 (a) shows the result of DFT analysis of the flow behind the bluff body. The result is shown after introducing a single bend upstream of the meter, displaying a very strong peak at the meter's natural vortex-shedding frequency. In this case, the dominating frequency remains unchanged. Fig. 4 (b) shows the DFT analysis of flow past a double bend. Again, a very strong peak at the meter's frequency is noticeable.

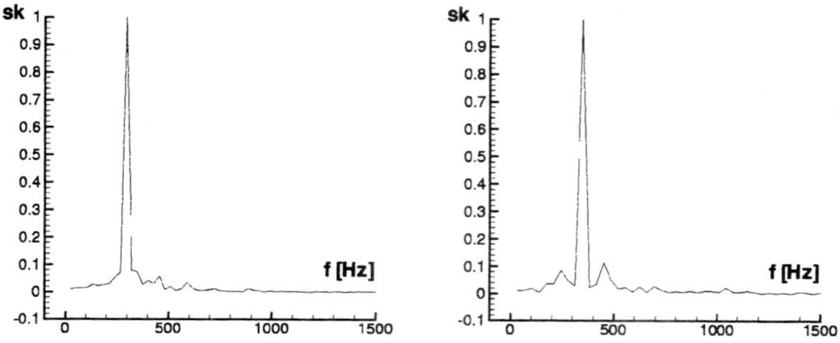

Fig. 4. Shedding frequency for flow downstream a single bend (left) and with a double bend upstream the meter (right)

The signal in Fig. 4 was obtained by observing the modulation of the radial velocity component, as given by the present ultrasound barrier.

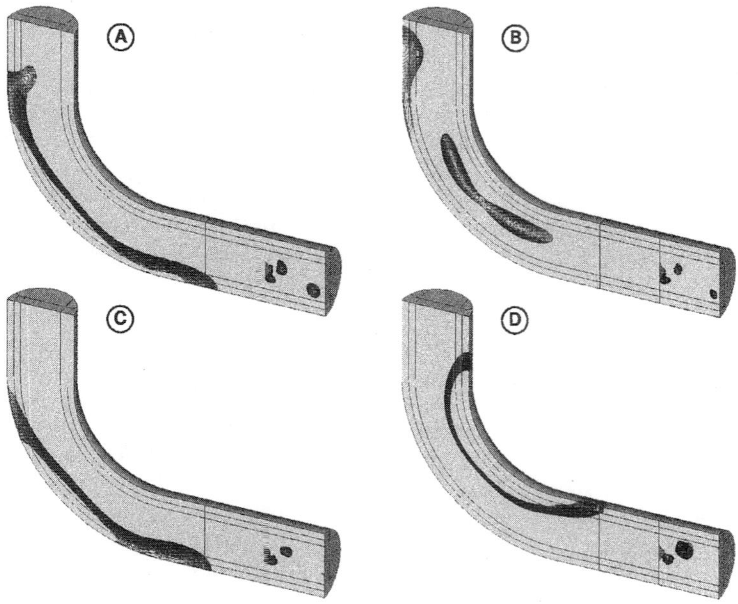

Fig. 5. Highly unsteady flow in the single 90°-bend

However, the pressure field in the single 90°-bend was highly unsteady, with a wave that was alternativly reflected between the lower and upper wall while rotating around the pipe axis and slowly moving upstream (Fig. 5, to the top). This is consistant with behavior reported by Rütten et al., Aachen.
Table 1 summarizes the results. The Strouhal number remains almost constant for all introduced bends. Due to the large tangential velocities in magnitude of the pipe mean velocity the deviation in case of the swirl is noticeable higher.

Table 1. Results

	Swirl	90°-Bend horizontal	90°-Bend vertical	Double Bend	Standard Tube	Exp.
Meshpoints	414300	621400	621400	621400	1.07 Mio.	
Reynolds number	183000	180000	173000	213000	191000	175000
Frequency [Hz]	287	300	295	348	307	280
v [m/s]	27.7	27.29	26.18	32.18	28.9	26.48
Strouhal number	0.249	0.264	0.27	0.26	0.255	0.254

4 Conclusions

Several numerical investigations of unsteady and disturbed pipe flows are presented. The simulations were useful in determining practical demands concerning the installation of the meter. The DFT analysis showed that the vortex-shedding flowmeter will still work properly under disturbed flow conditions.

References

1. Germano, M., Piomelli, U., Moin, P., Cabot, W., *A dynamic subgrid-scale eddy viscosity model*, Phy. Fluids, A 3(7):1760-1765, July 1991.
2. El-Milligui, A., Cannizzaro, F., Melson, N. D., von Lavante, E., *A three dimensional multigrid multistage time stepping scheme for Navier-Stokes equations*, Proceedings of the 9th GAMM Conference on Numerical Methods in Fluid Mechanics, Lausanne, 1991.
3. Mickan, B., Wendt, G., Kramer, R., Dopheide, D., *Systematic investigation of pipe flows and installation effects using laser doppler anemometry*, Flow Measurement and Instrumentation, Vol. 7, No. 3/4, 1996.
4. Roe, P. L., *Characteristic Based Schemes for the Euler Equations*, A Numerical Review of Fluid Mechanics, pp. 337-365, 1986.
5. Smagorinsky, J.S., *General circulation experiments with primitive equations*, Monthly Weather Review, Vol. 91, 1963.
6. von Lavante, E., Yao, J., *Simulation of flow in exhaust manifold of an reciprocating engine*, AIAA 24th Fluid Dynamics Conference, 1983.
7. J. Yao and E. von Lavante, Proceedings of 7th International Symposium on CFD, pp. 689-694, Sept. 15-19, 1997, Beijing, PR China.
8. Hans, V., Windorfer, H., von Lavante, E. and Perpeet, S., *Experimental and numerical optimization of acoustic signals associated with ultrasound measurement of vortex frequencies*, Proceedings of FLOMEKO '98, 1998.

Part XVIII

Two-phase Flows

Flux Schemes for the Numerical Simulation of Two Phase Flows

A. Kumbaro, I. Toumi, and V. Seignole

CEA Saclay, DRN/DMT/SYSCO, F-91191 Gif-sur-Yvette Cedex, FRANCE

Abstract. This paper discusses the development of a new method for the discretization of non linear non conservative systems of partial differential equations occurring in the numerical simulation of two- phase flows. This method is able to treat general non-conservative systems with realistic state equations and to calculate both single-phase and two-phase flows and is flexible enough to be applied to any mesh type, structured or non structured. A classical numerical benchmark is presented to illustrate the method.

1 Introduction

The finding of a unique model for the multiphase flows remains a problem of fundamental research. Various averaged models based on conservation of mass, momentum and energy of the mixture or of each phase are used in the two-phase flow computer tools developed in the nuclear industry. But if the predictions given by a model are correct in a domain of application extrapolations are not generally recommended. Or, if a numerical method is supposed to be suitable to simulate the two-fluid two-phase flow model for example, the same method should be fitted in the case of one phase flow model.

Actually, in the Saclay and Grenoble CEA centres we are developing the two-phase flow module of the TRIO_U code which aims to be a 3D thermal-hydraulic platform. To this end, we are interested in the numerical methods which can be applied to any model, from one phase flow model to two-fluid two-phase flow model and which can treat various constitutive models for the transfer terms. Furthermore, the numerical method must combine robustness and accuracy, respect the fundamental conservation physical laws and handle all types of meshes.

In this paper, we will describe the application of a multidimensional finite volume method implemented in the Trio_U code to the averaged one pressure two-fluid model. This method combines the generalized Roe formulation with the flux scheme. On one hand, the Roe scheme has been shown to drastically reduce numerical diffusion when compared to the numerical methods used in the current generation of thermal-hydraulic codes. On the other hand, the flux scheme allows to treat general non-conservative systems with realistic state equations.

The paper is organized as follows. The next Section briefly describe the standard two-phase flow model implemented in the Trio_U code. The Section 3 is devoted to the presentation of our method. Then Section 4 gives a numerical illustration. Finally, some further developments are indicated.

2 Two-Fluid Two-Phase Flow Model

A standard two-fluid two-phase flow model with additional transport equations forms the basis of our physical model. In 3D, the mass, momentum and energy balance equations reads :

$$\frac{\partial}{\partial t}\alpha_k \rho_k + \nabla \cdot (\alpha_k \rho_k \mathbf{u}_k) = \Gamma_k, \quad \text{with} \quad \sum_{k=v,\ell} \Gamma_k = 0 \tag{1}$$

$$\begin{aligned} \frac{\partial}{\partial t}\alpha_k \rho_k \mathbf{u}_k + \nabla \cdot (\alpha_k \rho_k \mathbf{u}_k \otimes \mathbf{u}_k) + \alpha_k \nabla p - \nabla \cdot (\alpha_k (\overline{\overline{\tau_k}} + \overline{\overline{\tau_k}}^{Re})) \\ = \alpha_k \rho_k \mathbf{f}_{ext} + \mathbf{F}_k^w + \mathbf{F}_k^i + \Gamma_k \mathbf{u}_i \end{aligned} \tag{2}$$

with $\sum_{k=v,\ell} \mathbf{F}_k^i = 0$.

$$\begin{aligned} \frac{\partial}{\partial t}(\alpha_k \rho_k E_k) + p\frac{\partial}{\partial t}\alpha_k + \nabla \cdot (\alpha_k \rho_k H_k \mathbf{u}_k) - \nabla(\alpha_k (q_k + q_k^{Re})) \\ = \alpha_k \rho_k \mathbf{f}^{ext} \mathbf{u}_k + Q_v^w + Q_k^i + \mathbf{F}_k^i \cdot \mathbf{u}_i + \Gamma_k H^i \end{aligned} \tag{3}$$

with $\sum_{k=v,\ell} Q_k^i = 0$.

The model is closed by the Equations of state for each phase,

$$\rho_k = \rho_k(p, h_k), \quad \text{for} \quad k = v, \ell$$

The two-fluid model can be coupled to an additional transport equation for interfacial area concentration [1], [7] and [9] :

$$\frac{\partial a_i}{\partial t} + \mathrm{div}(a_i \mathbf{U}^i) = \Phi_{a_i}. \tag{5}$$

We remark that the two-fluid two-phase flow model is a non conservative one. Furthermore, the basic two-fluid model is well-known to be non-hyperbolic and there is quite an extensive literature on the subject, for instance [10]. It is also well known that the inclusion of regularizing terms such as interface pressure correction terms or virtual mass force terms [11] which contain partial derivatives, makes the system hyperbolic. In order to make the system well posed, we consider the interface pressure term of the form :

$$(p - p_i)\nabla \alpha_k, \quad k = v, \ell \tag{6}$$

3 The Numerical Method

3.1 Condensed form

We can then write the system (1)-(4) in the following condensed form:

$$\frac{\partial \mathbf{v}}{\partial t} + \nabla \cdot \mathbf{f}(\mathbf{v}) + \sum_{j=1}^{d} C_j(\mathbf{v}) \frac{\partial \mathbf{v}}{\partial x_j} = S(\mathbf{v})$$

where $\mathbf{v}$ is the vector of conservative variables and $\mathbf{F}(\mathbf{v})$ is the flux function:

$$\mathbf{v} = (\alpha_v \rho_v, \alpha_l \rho_l, \alpha_v \rho_v \mathbf{u}_v, \alpha_l \rho_l \mathbf{u}_l, \alpha_v \rho_v E_v, \alpha_l \rho_l E_l)^T \tag{8}$$

$$f_\nu(\mathbf{v}) = (\alpha_v \rho_v (\mathbf{u}_v \cdot \nu), \alpha_l \rho_l (\mathbf{u}_l \cdot \nu), \alpha_v \rho_v (\mathbf{u}_v \cdot \nu)\mathbf{u}_v + \alpha_v p \nu, \alpha_l \rho_l (\mathbf{u}_l \cdot \nu)\mathbf{u}_l + \alpha_l p \nu,$$
$$\alpha_v \rho_v H_v (\mathbf{u}_v \cdot \nu), \alpha_l \rho_l H_l (\mathbf{u}_l \cdot \nu))^T \tag{9}$$

with $\mathbf{u}_k \in \mathbf{R}^d$ (d is space dimension), $\mathbf{v} \in \mathbf{R}^m$, $m = 2(d+2)$ and ν a unit vector in $\mathbf{R}^d$. The vector $C_j(\mathbf{v}) \frac{\partial \mathbf{v}}{\partial x_j}$ forms the non conservative part of the system and depends on the differential transfer terms of the model (the choice of the flux function is unique). The source term $S(\mathbf{v})$ contains only algebraic relations.

On the other hand, the system (1)-(4) can also be written as :

$$\frac{\partial \mathbf{w}}{\partial t} + \sum_{j=1}^{d} B_j(\mathbf{w}) \frac{\partial \mathbf{w}}{\partial x_j} = \tilde{S}(\mathbf{w}), \tag{10}$$

with the vector of "primitive" state parameters $\mathbf{w} = (\alpha_v, p, u_v, u_l, h_v, h_l)^T$, where $B_\nu(\mathbf{w}) = \sum_{j=1}^{d} \nu_j B_j(\mathbf{w})$ contains the advection and the transfer parts.

We introduce the Jacobian matrices :

$$J_\nu(\mathbf{v}) = \left(\frac{\partial f_{\nu\ell}(\mathbf{v})}{\partial \mathbf{v}_k} \right)_{1 \le \ell, k \le m} \quad \text{and} \quad T_\nu(\mathbf{w}) = \left(\frac{\partial f_{\nu\ell}(\mathbf{w})}{\partial \mathbf{w}_k} \right)_{1 \le k, \ell \le m} \tag{11}$$

The quasi-linear equations for the vector of conservative variables and for the flux are given by :

$$\frac{\partial \mathbf{v}}{\partial t} + A_\nu(\mathbf{v}) \frac{\partial \mathbf{v}}{\partial \nu} = S(\mathbf{v}) \tag{12}$$

$$\frac{\partial f(\mathbf{v})}{\partial t} + \tilde{A}_\nu(\mathbf{v}) \frac{\partial f_\nu(\mathbf{v})}{\partial \nu} = J_\nu(\mathbf{v}) S(\mathbf{v}) \tag{13}$$

where $\tilde{A}_\nu = (\mathrm{Id} + E_\nu(\mathbf{v})) J_\nu(\mathbf{v})$ and $E_\nu(\mathbf{v})$ is the non conservative matrix obtained in the following manner :

$$E_\nu(\mathbf{v})\frac{\partial f_\nu(\mathbf{v})}{\partial \nu} = C_\nu(\mathbf{v})\frac{\partial \mathbf{v}}{\partial \nu} = D_\nu(\mathbf{w})\frac{\partial \mathbf{w}}{\partial \nu} = D_\nu(\mathbf{w})T_\nu^{-1}\frac{\partial f_\nu(\mathbf{v})}{\partial \nu} \tag{14}$$

Every differential transfer term will give its contribution to the non conservative matrix. The matrices $\tilde{A}_\nu$ and B_ν are similar : $\tilde{A}_\nu = T_\nu B_\nu T_\nu^{-1}$.

3.2 Eigenelements of $\tilde{A}$

In order to determine the eigenvalues of the system, we must find the roots of a polynomial of degree six $P(\lambda) = \det(B - \lambda \mathrm{Id})$. A straightforward computation of this determinant leads to the following polynomial :

$$P(\lambda) = (\lambda - \mathbf{u}_v \cdot \nu)^3(\lambda - \mathbf{u}_l \cdot \nu)^3 P_B(\lambda).$$

To get the acoustic eigenvalues (which are real) we use an iterative algorithm initializing it by approximate eigenvalues given by a perturbation method [11] :

$$\lambda_k^{q+1} = \lambda_k^q - \frac{P_B}{P_B'}(\lambda_k^q),\ k = 1,2,\ \text{with } \lambda_k^0 = \frac{\alpha_v \rho_l \mathbf{u}_v + \alpha_l \rho_v \mathbf{u}_l}{\alpha_l \rho_v + \alpha_v \rho_l} \cdot \nu \pm a_m$$

where a_m is the mixture sound velocity. The void waves are easy to compute using the relations for the sum s and product p of the roots of $P_B(\lambda) = \lambda_{3,4}^2 - s\lambda_{3,4} + p = 0$.

We get the eigenvectors $\mathbf{r}$ and $\mathbf{l}$ of the matrix B_ν by a straightforward computation. Finally, the right and left eigenvectors of $\tilde{A}_\nu$ are obtained respectively by $\mathbf{R} = T_\nu \mathbf{r}$ and $\mathbf{L} = \mathbf{l}T_\nu^{-1}$.

3.3 Roe-type scheme

The explicit multidimensional flux scheme [3]-[7] for the approximation of system (7) is the following :

$$\begin{aligned}\mathbf{v}_k^{n+1} = \mathbf{v}_k^n + (\Delta t_n)S(\mathbf{v}_k^n) - \frac{\Delta t_n}{|K|}\sum_{K\cap L_j} \Phi_{K,L_j}^n(\mathbf{v}_K^n, \mathbf{v}_{L_j}^n; K, L_j) \\ + E_\nu(\mathbf{v}_k^n)(\Phi_{K,L_j}^n(\mathbf{v}_K^n, V_{L_j}^n; K, L_j) - |K\cap L_j| f_\nu(\mathbf{v}_k^n))\end{aligned}$$

with $E_\nu(\mathbf{v}_k^n) = D_\nu(\mathbf{v}_K^n)T_\nu^{-1}(\mathbf{v}_K^n)$.

We distinguish the following main features of our method :

1. The numerical flux at the interface $K \cap L$ with ν the unit external normal, $\Phi_{K,L}^n = |K\cap L| g(\mathbf{v}_K^n, \mathbf{v}_L^n)$, is given as a linear combination of $f_\nu(\mathbf{v}_K^n)$ and $f_\nu(\mathbf{v}_L^n)$:

$$g(\mathbf{v}_K^n, \mathbf{v}_L^n) = \frac{f_\nu(\mathbf{v}_K^n) + f_\nu(\mathbf{v}_L^n)}{2} - \mathrm{sgn}(\tilde{A}_\nu(\mathbf{v}_K^n, \mathbf{v}_L^n))\frac{f_\nu(\mathbf{v}_L^n) - f_\nu(\mathbf{v}_K^n)}{2} \tag{15}$$

This is a flux scheme introduced in [2]. We remark that the numerical flux satisfies the same quasi-linear equation as the flux function du to the relation $\text{sgn}(\tilde{A}_\nu)\tilde{A}_\nu = \tilde{A}_\nu \text{sgn}(\tilde{A}_\nu)$:

$$\frac{\partial g}{\partial t} + \tilde{A}_\nu(\mathbf{v}_K^n, \mathbf{v}_L^n)\frac{\partial g}{\partial \nu} = 0 \tag{16}$$

2. The linearized state at which the matrix $\tilde{A}_\nu$ is evaluated is defined using the averaged values obtained during the computation of $A_\nu^{ROE}(\mathbf{v}_K^n, \mathbf{v}_L^n)$ [11].
3. The flux scheme does not used explicitly the hyperbolic character of the convection operator. If $\tilde{A}(\mathbf{v})$ has complex eigenvalues, they occur by conjugate pairs. For this reason, the method allows to determine whether the hyperbolicity of this operator is necessary to have stable and convergent numerical computations. In [5] we have shown that the presence of the differential transfer terms in the non conservative matrix E is sufficient to make the problem well posed. In this case our method is still able to solve the problem without changing the eigenstructure of the basic model, i.e. using complex eigenvalues.

4 Numerical Results

In order to illustrate the capabilities of the numerical method to capture shocks and to deal with unstructured meshes in a 3D geometry we present here the results obtained with a classical benchmark, the water faucet problem [12] using unstructured three-dimensional meshes. This test case contains some important features at least from the numerical point of view. Moreover, an analytical solution to compare with is available [12]. The solution consists in a shock wave that travels under the gravity effect. Two cases are realised with the presence or not of the interfacial pressure term in the eigenstructure of the system. Figure 1 shows that the solution is stable even for a fine mesh and that the results are the same in both cases. The Figure 2 presents the pressure distribution obtained for a convergent - divergent nozzle.

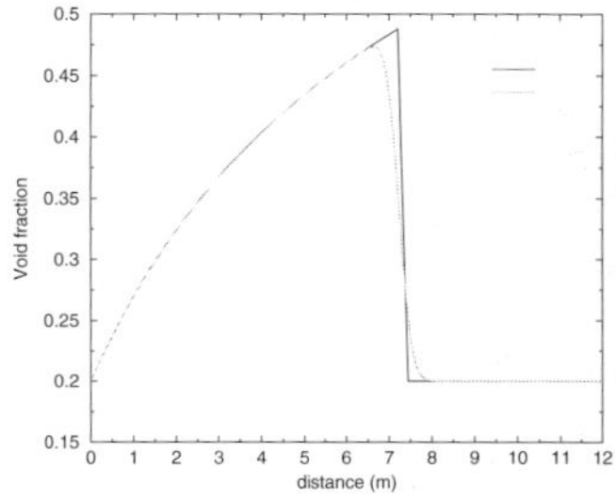

Fig 1. Faucet problem

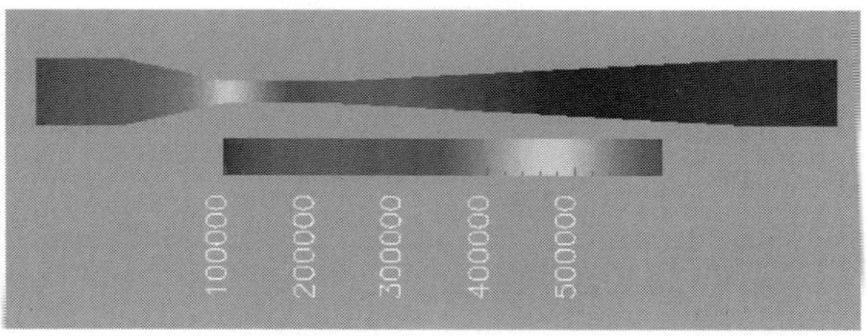

Fig 2. Nozzle flow

5 Conclusions and Further Developments

A new multidimensional numerical method for the two-fluid two-phase flow models has been implemented in the two-phase flow module of the Trio_U code. It associates advanced numerical techniques such as the flux scheme and the approximate Riemann solutions to upwind the flux. The numerical method allows easy various transfer term to be changed, added and integrated transparently to the existing model but the library of the physical closure laws in the code needs to be complemented. The scheme presented is an explicit one but an implicit version is under development. Further developments concern the terms related with dissipation phenomena, the high-order spatial accuracy by variable reconstruction as well as modelling of the interfacial area transport.

References

1. M. Ishii, Thermo-Fluid Dynamic Theory of Two-Phase Flow, Collection de la Direction des Etudes et Recherches d' Électricité de France, Eyrolles, Paris, 1975.
2. J.M. Ghidaglia, Flux Schemes for Solving Non-linear Systems of Conservation Laws, in Proc. of the Meeting in Honour of P.L. Roe, J.J. Chattot and M. Hafez, Eds., Arcachon, July. 1998.
3. J.M. Ghidaglia, A. Kumbaro and G. Le Coq, 1996. Une approche volumes-finis à flux caractéristiques pour la résolution numérique des systèmes hyperboliques de lois de conservation, C.R. Acad. Sc. Paris, 322, I, 981-988.
4. Ghidaglia, J.-M, Kumbaro, A., Le Coq, G, On the Numerical Solution to Two Fluid Models via a Cell Center Finite Volume Method, to appear.
5. Kumbaro, A., Seignole, Ghidaglia, J.-M, Flux Schemes for the Two-Fluid Models of the TRIO_U Code AMIF-ESF Workshop - Computing Methods for Two-Phase Flow, Aussois, jan. 12-14, 2000.
6. A.Kumbaro, I. Toumi, J. Cortes, "Two Perturbation Methods to Upwind the Jacobian Matrix of Two-Fluid Flow Models", Sec. International Symp. On Finite Volumes for Complex Applications, July, 19-23, Duisbourg, Germany, 1999.
7. C. Morel, Modélisation Multi-Dimensionnelle des Ecoulements Gaz Liquide. Application à la Simulation des Ecoulements à Bulles Ascendants en Conduite Verticale. Ph.D. Thesis, Ecole Centrale de Paris, 1998
8. Ransom, V.H., Hicks, D.L., Hyperbolic Two-Pressure Models for Two-Phase Flow, J. Comput. Physics, 53, 14. 1984.
9. H. Städtke, A. Blahak and B. Worth, Modelling of Transport of Interfacial Area Concentration in Two-Phase Flow Systems, Proc. NURETH-8, The 8th Int. Topical Meeting on Nuclear Reactor Thermal-Hydraulics, September 30-October 4, 1997, Kyoto, Japan
10. H.B. Stewart, B. Wendroff, Two-Phase Flows: Models and methods, J. Comput. Phys., 56 , 362-409, 1984
11. Toumi, I., An Upwind Numerical Method for Two-Fluid Two-Phase Flow Models, Nucl. Eng. Design, 123, 147-168, 1996.
12. G. Hewitt, J.M. Delhaye and M. Zuber "Numerical benchmark tests in Multiphase Science and Technology, vol. 3, Hemisphere Washington DC/New York, 1987

High Resolution KFVS Methods for Multi–Fluid Dynamics

Hua–Mo Wu and Hua–Zhong Tang

State Key Laboratory of Scientific and Engineering Computing, Institute of Computational Mathematics and Scientific/Engineering Computing, Chinese Academy of Sciences, Beijing 100080, P.R. China

Abstract. This paper concerns the extension of high resolution kinetic flux–vector splitting (KFVS) finite volume methods. A suitable modification of Maxwellian is given to recover the Euler equations for real gases with an additional conservative equation in the new method. Based on it and the well–known connection between Euler equations and Boltzmann equations, a high resolution KFVS scheme is presented to solve multifluid flows, without any Riemann solvers and adding any nonconservative corrections. The numerical solutions shown are oscillation–free near material fronts, and have correct shock speeds.

1 Preliminaries

Some conservative Euler solvers for gas mixtures have been shown to produce numerical errors near contact discontinuities, if the temperature and the ratio of specific heats are not constant here. Some remedies have also been proposed to avoid these errors, such as applications of nonconservative corrections to conservative Euler solvers [1,2,3,4]. The development of gas–kinetic methods for compressible flow simulations, such as KFVS methods and BGK–type schemes, has attracted much attention in the past few years. They have also provided robust and accurate numerical solutions for various unsteady compressible flows [6,7,8,9]. Recently, Xu [8] extended BGK–based scheme to multicomponent flow calculations by considering independent Maxwellian for each component, and gave the accurate solutions The purpose of this paper is to consider a local equilibrium state different from one in [8], and study high resolution KFVS finite volume methods for solving numerically multicomponent flows.

Consider two–dimensional Euler equations governing two–component compressible nonreacting flows,

$$\frac{\partial U}{\partial t} + \frac{\partial F_1(U)}{\partial x} + \frac{\partial F_2(U)}{\partial y} = 0, \tag{1}$$

where

$$\begin{aligned} U &= [\rho, \rho u_1, \rho u_2, \rho\Phi, E]^T, \\ F_i(U) &= [\rho u_i, \rho u_1 u_i + \delta_{1,i} p, \rho u_2 u_i + \delta_{2,i} p, \rho u_i \Phi, u_i(E+p)]^T, \end{aligned} \tag{2}$$

and ρ, (u_1, u_2), p, and $E = \rho e + \frac{1}{2}\rho u^2$ are the density, the velocity vector, the pressure, and the total energy, respectively. $\Phi = \Phi(\gamma)$ is a smooth function of the ratio of specific heats γ, or the mass fractions α, or the level set function. To close the above system, we need an additional equation, $p = p(\rho, e)$.

It is well–known whenever the particle velocity distribution is Maxwellian, the Euler equations of motion for an ideal gas can be derived from Boltzmann equation

$$f_t + v_1 f_x + v_2 f_y = J(f, f), \tag{3}$$

where f is particle velocity distribution function, (v_1, v_2) is the particle velocity vector, the right hand side term $J(f, f)$ represents a collision term, which vanishes in the Euler limit.

In order to recover system (1) with the general equation of state, we should introduce the following local equilibrium state [6]

$$g = \rho \left(\frac{\lambda_1}{\pi}\right)^{1.5} \left(\frac{\lambda_2}{\pi}\right)^{K/2} e^{-\lambda_1[(v-u)^2+(\phi-\Phi)^2]-\lambda_2\xi^2}, \tag{4}$$

where ξ is the internal variable with K degrees of freedom, $K = 2(2-\gamma)/(\gamma-1)$ in two–dimensional case, and $u^2 = u_1^2 + u_2^2$.

Taking the moments of g in terms of moment vector $\Psi = [1, v_1, v_2, \phi, \frac{1}{2}(v^2 + \xi^2)]^T$ gives

$$\begin{bmatrix} \rho \\ \rho u_1 \\ \rho u_2 \\ \rho\Phi \\ \frac{1}{2}\rho u^2 + \frac{\rho}{2\lambda_1} + \frac{\rho K}{4\lambda_2} \end{bmatrix} = \int_{R^{K+3}} \Psi g \; dv d\xi, \tag{5}$$

and

$$\begin{bmatrix} \rho u_i \\ \rho u_1 u_i + \frac{\rho}{2\lambda_1}\delta_{1,i} \\ \rho u_2 u_i + \frac{\rho}{2\lambda_1}\delta_{2,i} \\ \rho u_i \Phi \\ \frac{1}{2}u_i(\rho u^2 + \frac{2\rho}{\lambda_1} + \frac{\rho K}{2\lambda_2}) \end{bmatrix} = \int_{R^{K+3}} (v_1, v_2)\Psi g \; dv d\xi. \tag{6}$$

Thus if $\lambda_1 = \rho/2p$, $\lambda_2 = \rho K/4(\rho e - p)$, the Euler equations (1) can be cast in the moment form as follows

$$\left\langle \Psi, \frac{\partial g}{\partial t} + v_1 \frac{\partial g}{\partial x} + v_2 \frac{\partial g}{\partial y} \right\rangle = 0, \tag{7}$$

where $<\Psi, g>= \int_{R^{K+3}} \Psi g \; dv d\phi d\xi$.

2 Numerical Methods

Consider an arbitrary quadrilateral A_{ij} as a finite control volume (See Fig.1), and suppose it's bounding surface ∂A_{ij} consists of four surface elements s_l,

$l = 1,2,3,4$, with outward unit normal vector n_l. Integrating Eq.(7) and using the divergence theorem give

$$\frac{\partial}{\partial t}\int_{A_{ij}} \langle \Psi, g\rangle \, dV = -\sum_{l=1}^{4}\int_{R^{K+3}} v_{n_l}\Psi_{s_l} g_{s_l}|s_l| \, d\Xi, \tag{8}$$

where $d\Xi = dv_{n_l}dv_{t_l}d\phi d\xi$, and Ψ_{s_l} and g_{s_l} represent Ψ and g written in terms of the normal and tangential molecular velocity components v_{n_l} and v_{t_l} on each surface s_l, i.e.,

$$\Psi_{s_l} = \left[1,\ v_{n_l}\cos(\alpha_l) - v_{t_l}\sin\alpha_l,\ v_{n_l}\sin(\alpha_l) + v_{t_l}\cos\alpha_l,\ \phi,\ \frac{v^2+\xi^2}{2}\right]^T \equiv \Psi,$$

$$g_{s_l} = \rho\left(\frac{\lambda_1}{\pi}\right)^{1.5}\left(\frac{\lambda_2}{\pi}\right)^{K/2} e^{-\lambda_1[(v_{n_l}-u_{n_l})^2+(v_{t_l}-u_{t_l})^2+(\phi-\Phi)^2]-\lambda_2\xi^2} \equiv g,$$

where α_l is the angle of n_l with the x–axis (measured in the anti–clockwise direction).

Let U_{ij} denote the field vector at the centroid of the volume A_{ij}, then Eq.(8) can be approximated as

$$|A_{ij}|\left(\frac{\partial U}{\partial t}\right)_{ij} = -\sum_{l=1}^{4}(F_{n_l})_{s_l}|s_l|. \tag{9}$$

According to the sign of the molecular velocity component v_{n_l}, the macroscopic flux F_{n_l} can be split into two parts, i.e.

$$F_{n_l} = F_{n_l}^{+} + F_{n_l}^{-} \equiv \int_{R^{K+2}}\int_{v_{n_l}\geq 0} v_{n_l}\Psi_{s_l}g_{s_l}\, d\Xi + \int_{R^{K+2}}\int_{v_{n_l}\leq 0} v_{n_l}\Psi_{s_l}g_{s_l}\, d\Xi,$$

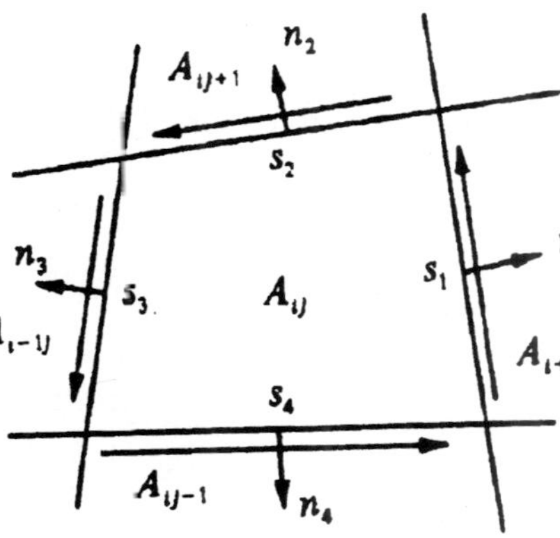

Fig. 1. Finite control volume $A_{i,j}$.

Based on this flux vector splitting, a stable approximate of the Euler equations (1) can be given as

$$U_{ij}^{n+1} = U_{ij}^{n} - \frac{\Delta t}{|A_{ij}|}\sum_{l=1}^{4}(F_{n_l}(U^n))_{s_l}|s_l|, \tag{10}$$

where values of $F_{n_l}(U^n)$ on surface s_l are taken as

$$(F_{n_1}(U^n))_{s_1} = F_{n_1}^{+}(U_{A_{ij}}^{n}) + F_{n_1}^{-}(U_{A_{i+1j}}^{n}),$$
$$(F_{n_2}(U^n))_{s_2} = F_{n_2}^{+}(U_{A_{ij}}^{n}) + F_{n_2}^{-}(U_{A_{ij+1}}^{n}),$$

etc. The split flux vectors $F_{n_l}^{\pm}$ are of explicit forms

$$\begin{aligned} F_{n_l}^{\pm,1}(U) &= \rho < v_{n_l}^{1} >_{\pm}, \\ F_{n_l}^{\pm,2}(U) &= \rho < v_{n_l}^{2} >_{\pm} \cos(\alpha_l) - \rho < v_{n_l}^{1} >_{\pm} < v_{t_l}^{1} > \sin(\alpha_l), \\ F_{n_l}^{\pm,3}(U) &= \rho < v_{n_l}^{2} >_{\pm} \sin(\alpha_l) + \rho < v_{n_l}^{1} >_{\pm} < v_{t_l}^{1} > \cos(\alpha_l), \\ F_{n_l}^{\pm,4}(U) &= \tfrac{1}{2}\left(\rho < v_{n_l}^{3} >_{\pm} + < v_{n_l}^{1} >_{\pm} (< v_{t_l}^{2} > + < \xi^2 >)\right), \end{aligned} \tag{11}$$

where $< \cdots >_{\pm}$ and $< \cdots >$ can be found in [6,7,9].

The scheme (10) is only of first order accuracy. In order to reduce the numerical dissipation, the following form of initial reconstruction

$$\overline{U}_{ij}(x,y) = U_{ij} + (U_x)_{ij}(x - x_{ij}) + (U_y)_{ij}(y - y_{ij}), \quad (x,y) \in A_{ij}, \tag{12}$$

with approximate slopes

$$(U_k)_{ij} = (\mathrm{sgn}(s_{ij}^{+,k}) + \mathrm{sgn}(s_{ij}^{-,k}))\frac{|s_{ij}^{+,k} s_{ij}^{-,k}|}{|s_{ij}^{+,k}| + |s_{ij}^{-,k}|}, \quad k = x \text{ or } y, \tag{13}$$

is used to approximate the cell averaged conservative variables U_{ij} at each time level, where

$$s_{ij}^{\pm,k} = \Delta^{\pm,k} U_{ij} / \Delta^{\pm,k} k_{ij}, \tag{14}$$

and $\Delta^{+,k}$ and $\Delta^{-,k}$ are forward and backward difference operators in k–direction, respectively. Based on the linear reconstruction (12), a high resolution MUSCL–type KFVS scheme can be obtained

$$U_{ij}^{n+1} = U_{ij}^{n} - \frac{\Delta t}{|A_{ij}|}\sum_{l=1}^{4}\left(F_{n_l}^{+}(\overline{U}) + F_{n_l}^{-}(\overline{U})\right)_{s_l}|s_l|, \tag{15}$$

where numerical flux $F_{n_l}^{\pm}(\overline{U})$ on surface s_l can be explicitly given as, for example,

$$\left(F_{n_1}^{+}(\overline{U})\right)_{s_1} = F_{n_1}^{+}(\overline{U}^{0}), \quad \left(F_{n_1}^{-}(\overline{U})\right)_{s_1} = F_{n_1}^{-}(\overline{U}^{1}),$$

with

$$\begin{aligned} \overline{U}^{0} &= U_{ij}^{n} + (U_x)_{ij}(x_{S_1} - x_{ij}) + (U_y)_{ij}(y_{S_1} - y_{ij}), \\ \overline{U}^{1} &= U_{i+1j}^{n} + (U_x)_{i+1j}(x_{S_1} - x_{i+1j}) + (U_y)_{i+1j}(y_{S_1} - y_{i+1j}). \end{aligned}$$

Remark: (i) The KFVS finite volume schemes (10) and (15) do not depend on the particular expression of the equation of state, besides without any Riemann solvers. (ii) Combing the other expansions with different form of nonlinear limiters can also be considered in Eqs.(12), (13) and (14). (iii) The above idea can be easy to be extended to three–dimensional case.

3 Numerical Results

This section will present several numerical experiments for multicomponent flow calculations to demonstrate the performance of MUSCL–type KFVS finite volume methods with third order TVD Runge–Kutta time discretization for different forms of Ψ, the correctness and robustness.

We perform calculations of the Kelvin–Helmholtz instability [5]. As initial conditions, constant pressure and temperature are taken above and below the interface, with zero vertical velocity. Above the interface, the gas moves towards the left with the horizontal velocity $-u_0$,; below the interface, the horizontal velocity is u_0. The initial shape of the interface is assumed to be a sine perturbation. Again we have periodic boundaries in the horizontal direction and walls at the bottom and top. We will use a level set function for Ψ and uniform 128×256 grids in current study, and refer the readers to [5] for details.

Firstly we consider the air–air case. Set the density and the sound speed to 1 everywhere, $u_0 = 0.25$. Amplitude of the initial perturbation $a = 0.1$. In Fig.3, we show that zero contours of Ψ for the computations at time 1, 2, 3, 4, 5, 6, 7, 8. The current results show rollup of a vortex structure as it progresses, and are better than ones in [6], which is shown small oscillations.

Secondly we consider the air–helium ($\gamma = 1.63$) case. The gas below the interface is helium. As initial data, its sound speed is set to 1. We also set the density to 1 everywhere, $u_0 = 0.5$, and the amplitude of the initial perturbation $a = 0.1$. In Fig.4, shown is zero contours of Ψ for the computations at time 1, 2, 3, 4, 5, 6, 7, 8.

Acknowledgments. This research was supported by National Natural Science Foundation of China, Grant No. 19901031, and Laboratory of Computational Physics of Beijing Institute of Applied Physics and Computational Mathematics.

References

1. R. Abgrall: J. Comput. Phys. **125**, 150(1996)
2. P. Jenny, B. Muller, H. Thomann: J. Comput. Phys. **132**, 91(1997)
3. S. Karni: SIAM J. Sci. Statist Comput. **17**, 1019(1996)
4. B. Larrouturou: J. Comput. Phys. **95**, 59(1991)
5. W. Mulder, S. Osher, J. Sethian: J. Comput. Phys. **100**, 209(1992)
6. H.Z. Tang, H.M. Wu, accepted by Computers & Fluids, 2000
7. H.Z. Tang, H.M. Wu: Chinese J. Numer. Math. & Appl. **22**, 375(1999)
8. K. Xu: J. Comput. Phys. **134**, 122(1997)
9. K. Xu, *Gas–kinetic Schemes for Unsteady Compressible Flow Simulations*, In *VKI Fluid Dynamics Lecture Series, 1998–03, 1998*

Fig. 2. Kelvin–Helmholtz instability for air–air case. Shown are zero contours of Ψ for the computations at $t = 1, 2, 3, 4, 5, 6, 7, 8$. $\Psi = 1/(\gamma - 1)$.

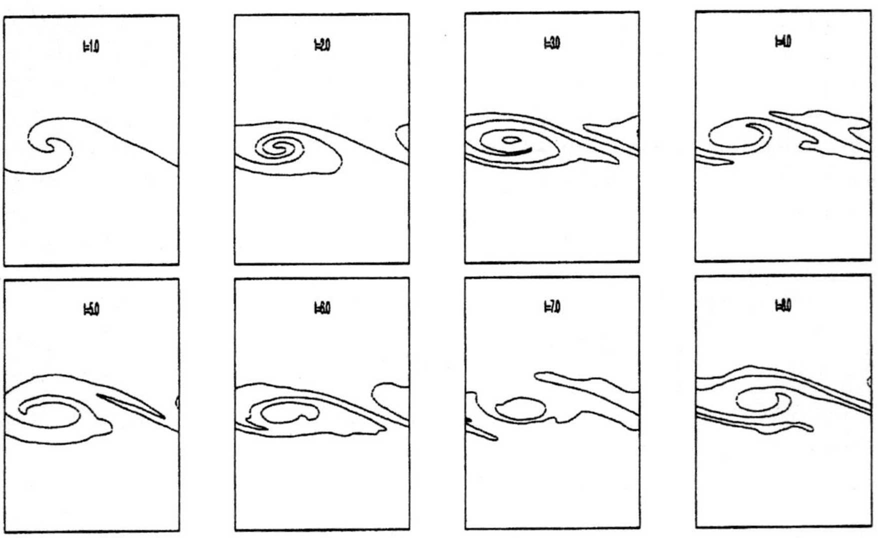

Fig. 3. Kelvin–Helmholtz instability for air–helium case. Shown are zero contours of Ψ for the computations at $t = 1, 2, 3, 4, 5, 6, 7, 8$. $\Psi = 1/(\gamma - 1)$.

Numerical Simulation of Two-Phase Flows with Stiff Source Terms

H. Städtke, G. Franchello, B. Worth

Commisison of the European Communities
Joint Reserach Centre - Ispra Establishment
I-21020 Ispra (Varese)
e-mail: herbert.staedtke@jrc.it

1 Introduction

The numerical simulation of gas-liquid two-phase flow processes remains a challenging task, caused mainly by the inhomogeneous nature of the two-phase mixture and the resulting strong thermal and mechanical non-equilibrium conditions. The approach described herein starts from an extended "two-fluid" representation of the two-phase flow where both phases are treated as interpenetrating continua with source terms representing the interfacial transport processes for mass, momentum and energy. Such a model results from space/time or ensemble averaging of the basic Navier-Stokes equations.

A crucial new aspect of the present approach is the use of a fully hyperbolic model of two-phase flow where all the eigenvalues and the corresponding set of linearly-independent eigenvectors can be expressed as algebraic functions of the major dependent flow parameters. This allows the application of generalized Godunov-type numerical schemes such as Approximate Riemann Solvers and Flux Vector Splitting techniques which make explicit use of the hyperbolicity of the flow equations.

2 Two-Fluid Model of Two-Phase Flow

The development of the two fluid model is largely related to the pioneering work of Ishii [4], Boure [1], and Delhaye and Achard [2]. With the index $i = g$ (gas or vapour), and $i = l$ (liquid), the separated balance equations for mass, momentum and enrgy can be written as

$$\frac{\partial}{\partial t}(\alpha_i \varrho_i) + \nabla \cdot (\alpha_i \varrho_i \, \boldsymbol{v}_i) = \sigma_i^M , \tag{1}$$

$$\frac{\partial}{\partial t}(\alpha_i \varrho_i \, \boldsymbol{v}_i) + \nabla \cdot (\alpha_i \varrho_i \, \boldsymbol{v}_i \, \boldsymbol{v}_i) + \alpha_i \nabla p_i - \nabla \cdot (\alpha_i \bar{\mathbf{T}}_i) = \boldsymbol{f}_i^{int} + \boldsymbol{f}_i^{ext} + \sigma_i^M \, \boldsymbol{v}_i^{ex} , \tag{2}$$

$$\frac{\partial}{\partial t}\left[\alpha_i \varrho_i \left(u_i + \frac{v_i^2}{2}\right)\right] + \nabla \cdot \left[\alpha_i \varrho_i \, \boldsymbol{v}_i \left(h_i + \frac{v_i^2}{2}\right)\right] \nabla (\alpha_i \, \boldsymbol{q}_i) - \nabla (\alpha_i \bar{\mathbf{T}}_i \cdot \boldsymbol{v}_i)$$
$$= \sigma_i^M \left(h + \frac{v_i^2}{2}\right) + \sigma_i^Q + Q_i^{ext} + \boldsymbol{f}^{ext} \cdot \boldsymbol{v}_i + \boldsymbol{f}^{int} \cdot \boldsymbol{v}_i^{fi} . \tag{3}$$

The source terms on the right-hand side of (1) to (3) describe the mass, momentum and energy transfer processes at the interface with the condition:

$$\sum_{i=g,l} \sigma_i^M = 0\,, \qquad \sum_{i=g,l} \boldsymbol{f}_i^{int} = 0\,, \qquad \sum_{i=g,l} \sigma_i^Q = 0\,. \tag{4}$$

The interfacial forces, introduced in the momentum and energy equationas, (2) and (3), can be split into two parts: the interfacial friction force $\boldsymbol{f}_i^{fr}$ and the non-viscous forces $\boldsymbol{f}_i^{nv}$ including virtual mass effects [3].

$$\boldsymbol{f}_i^{int} = \boldsymbol{f}_i^{fr} + \boldsymbol{f}_i^{nv}. \tag{5}$$

Apart from the non-viscous term for the interfacial forces $\boldsymbol{f}_i^{nv}$ which inludes time and space derivative terms, all the other interfacial source terms are assumed to be algebraic functions of state parameters and as such do not enter in the characteristic analysis of the system of governing equations.

The balance equations (1) to (3) represent a system of partial differential equations for the state vector $\boldsymbol{U}$ which can be combined in a compact matrix form as

$$\frac{\partial \boldsymbol{U}}{\partial t} + \mathbf{G} \cdot (\nabla \boldsymbol{U}) = \boldsymbol{C}\,, \tag{6}$$

with $\boldsymbol{C}$ as the vector of source terms. With the appropriate definition for the "non-viscous" part of the interfacial forces, only real eigenvalues appear in the coefficient matrix for the convective part of the flow equations $\mathbf{G}$ together with a corresponding complete set of independent eigenvectors as discussed in more detail in [6,7]. flow parameters.

The eigenvalues (the characteristic velocities of various wave propagation processes) obtained are in ξ-direction:

$$\left.\begin{array}{lll} \lambda_{1,2} = \boldsymbol{v}_g \cdot \boldsymbol{n}_\xi\,, & \boldsymbol{v}_l \cdot \boldsymbol{n}_\xi & \text{void waves}\,, \\ \lambda_{3,4} = \boldsymbol{v} \cdot \boldsymbol{n}_\xi \pm a & & \text{pressure waves}\,, \\ \lambda_{5,6} = \boldsymbol{v}_g \cdot \boldsymbol{n}_\xi\,, & \boldsymbol{v}_l \cdot \boldsymbol{n}_\xi & \text{shear waves}\,, \\ \lambda_{7,8} = \boldsymbol{v}_g \cdot \boldsymbol{n}_\xi\,, & \boldsymbol{v}_l \cdot \boldsymbol{n}_\xi & \text{entropy waves}\,. \end{array}\right\} \tag{7}$$

More details on the complete eigenspectrum of the hyperbolic two-phase flow model including the definition of the average two-phase flow velocity, $\boldsymbol{v}$, and the mixture sound velocity, a, can be found in [6].

Since the convective part of the flow equations (6) is hyperbolic, the coefficient matrix $\mathbf{G}$ can be diagonalized

$$\mathbf{\Lambda} = \mathbf{T}^{-1}\,\mathbf{G}\,\mathbf{T} \tag{8}$$

to yield the diagonal matrix $\mathbf{\Lambda}$ with eigenvalues λ_k along the principal diagonal, where the columns of the transformation matrix $\mathbf{T}$ are the right eigenvectors of the coefficient matrix $\mathbf{G}$. With respect to the individual eigenvalues, the coefficient matrix $\mathbf{G}$ can be split into elementary parts as

$$\mathbf{G} = \sum_{k=1}^{8} \mathbf{G}_k \qquad \text{with} \qquad \mathbf{G}_k = \mathbf{T}\,\mathbf{\Lambda}_k\,\mathbf{T}^{-1}\,, \tag{9}$$

where the diagonal matrix $\mathbf{\Lambda}_k$ includes only the k_{th} eigenvalues.

3 Numerical Method

The governing flow equations can be transformed into a quasi conservative form

$$\frac{\partial \boldsymbol{V}}{\partial t} + \nabla \cdot \boldsymbol{F} + \mathbf{H}^{nc} \nabla \cdot \boldsymbol{F} = \boldsymbol{D} \tag{10}$$

with the state vector of conserved variables $\boldsymbol{V}$ and the corresponding flux vector $\boldsymbol{F}$. The new source term vector $\boldsymbol{D}$ for the conserved variables includes the volume source terms for mass, momentum, and energy as already introduced in the general balance equations (1) to (3).

The space and time derivative terms in the separated momentum equations result in a "non-conservative" part of the coefficient matrix $\mathbf{H}^{nc}$ in (10) defined as

$$\mathbf{H}^{nc} = (\mathbf{J}\,\mathbf{G} - \mathbf{K})\,\mathbf{K}^{-1}, \quad \text{with} \quad \mathbf{J} = \frac{\partial \boldsymbol{V}}{\partial \boldsymbol{U}} \quad \text{and} \quad \mathbf{K} = \frac{\partial \boldsymbol{F}}{\partial \boldsymbol{U}}. \tag{11}$$

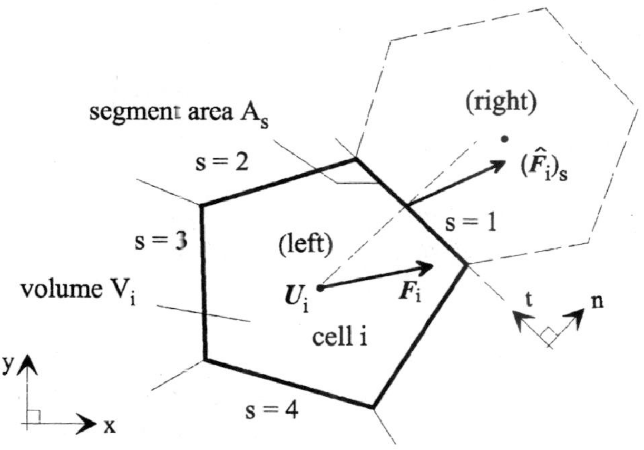

Fig. 1. Finite volume discretization

For the numerical solution scheme, the governing equations (10) are transformed into a finite volume approximation for arbitrary polygon-shaped computational cells i with volume V_i, boundary segment area A_s (Fig. 1), as

$$\boldsymbol{V}_i^{n+1} = \boldsymbol{V}_i^{n} - \frac{\Delta t}{V_i} \sum_s A_s \left(\hat{\boldsymbol{F}}_i\right)_s^{n+1}$$

$$-\frac{\Delta t}{V_i} \sum_s A_s \left(\mathbf{H}_i^{nc}\right)_s^{n+1} \left(\hat{\boldsymbol{F}}_i\right)_s^{n+1} + \boldsymbol{D}_i^{n+1} \Delta t, \tag{12}$$

where the intercell fluxes $\left(\hat{\boldsymbol{F}}_i\right)_s$, the non-conservative part of the coefficient matrix $(\mathbf{H}_i^{nc})$ and the source term vector $\boldsymbol{D}_i$ are evaluated implicitly at the new time level $n+1$.

The numerical fluxes at the cell interfaces are calculated from a series of linearized, quasi one-dimensional Riemann problems as schematically shown in in Fig. 2.

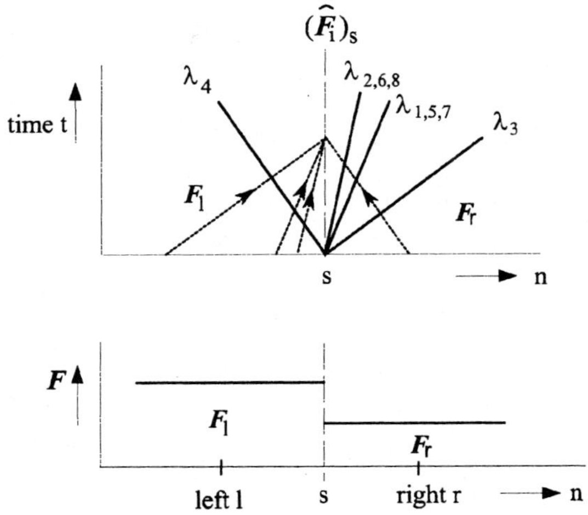

Fig. 2. Linearized Riemann problem for two-phase flow

The "Godunov fluxes" at the cell interfaces, $\hat{\mathbf{F}}_s$, are then calculated as:

$$\hat{\boldsymbol{F}}_s = \sum_{k,\lambda_k \geq 0} \left(\tilde{\mathbf{R}}_s\right)_k) \boldsymbol{F}_l + \sum_{k,\lambda_k<0} \left(\tilde{\mathbf{R}}_s\right)_k \boldsymbol{F}_r \,. \tag{13}$$

The "weighting" factors for the left and right fluxes in equation (13) are the sums of the split coefficient matrices for fluxes ordered with respect to the signs of the corresponding eigenvalues λ_k. These matrices can be obtained from the basis system of equations (6) by a parameter transformation using fluxes as independent parameters

$$\frac{\partial \boldsymbol{F}}{\partial t} + \mathbf{R} \cdot (\nabla \boldsymbol{F}) = \mathbf{K}\, \boldsymbol{C} \,, \tag{14}$$

with the new coefficient matrix

$$\mathbf{R} = \sum_k \mathbf{R}_k \,, \quad \text{with} \quad \mathbf{R}_k = \mathbf{K}\, \mathbf{G}_k\, \mathbf{K}^{-1} \quad \text{and} \quad \tilde{\mathbf{R}}_k = \frac{1}{\lambda_k} \mathbf{R}_k \,. \tag{15}$$

For the time integration of (12) an implicit Newton-Raphson iteration is used where the new-time values for fluxes and source terms are estimated by a first-order Taylor expansion. A nearly second-order accuracy is obtained by a linear reconstruction of the solution in all computational cells following the Monotonic Upstream Scheme for Conservation Laws (MUSCL) approach of Van Leer [8]. More details with respect to the numerical method applied can be found in [6,7].

4 Numerical Example

As an example, the prediction of the start-up of a bubble column is given in Fig. 3 showing the formation and rising of a bubble cloud and its interaction with the two-phase mixture level. The transient is governed by buoyancy and interfacial friction forces, the displacement of the (heavy) liquid phase in front of the bubble cloud and the resulting re-circulation pattern in the liquid phase.

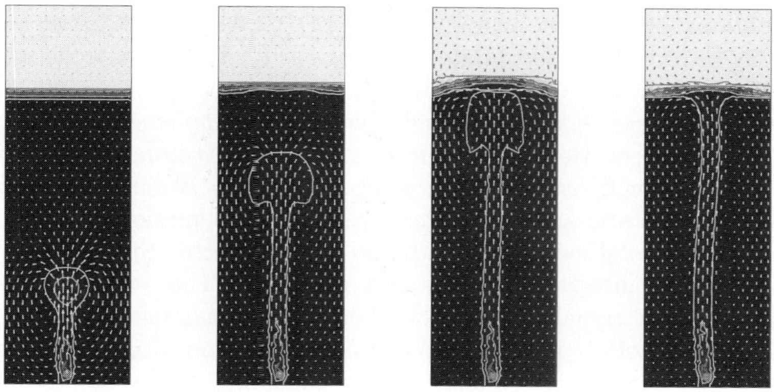

t = 10 s t = 20 s t = 25 s t = 50 s

Fig. 3. Start-up of bubble column with central gas sparger, overlay of volumetric concentration of gas (light) and liquid (dark) with mass flow density vector fields for gas and liquid

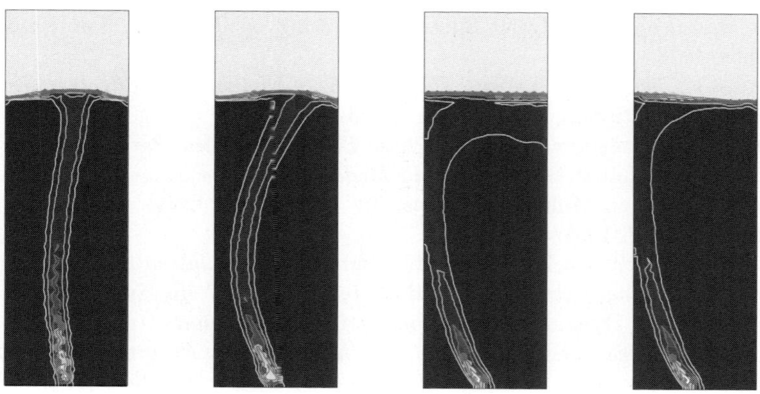

t = 150 s t = 200 s t = 225 s t = 250 s

Fig. 4. Long time behaviour of flat bubble column, spectral plots of volumetric concentration of gas (light) and liquid (dark)

For the long term behaviour (shown in Fig. 4), distinct instabilities occurred with the movement of the bubble plume towards the right (or left) side of the column which is also experimentally confirmed.

5 Summary

Based on a two-fluid approach, a new model for multi-dimensional inhomogeneous non-equilibrium two-phase flow has been developed using separate balance equations for mass, momentum and energy (entropy) for the two (coupled) phases. By introducing appropriate formulations for the non-viscous interfacial forces, a hyperbolic system of equations has been derived, characterized by the existence of only real eigenvalues and a complete complementary set of independent eigenvectors. Both the eigenvalues and eigenvectors can be expressed algebraically as functions of the major dependent flow parameters permitting a complete characteristic analysis of the governing flow equations.

The new model allows the application of Flux Vector Splitting techniques for the numerical integration of the flow equations. This technique combines preservation of the signal propagation along the characteristic wave directions with the property of accurate conservation of mass, momentum and energy.

Results of several test demonstrate the capability of the new models and numerical strategies for the prediction of inhomogeneous, non-equilibrium two-phase flow processes.

References

1. A. Boure, *On a Unified Presentation of the Non-Equilibrium Two-Phase Flow Models*, Proc. of ASME Symp., New York, 1975
2. J. M. Delhaye, J. L. Achard, *On the Averaging Operators Introduced in Two-Phase Flow Modelling*, Proc. CSNI Specialist Meeting on Transient Two-Phase Flow, Toronto, 1976
3. D. Drew, I. Cheng, R. T. Lahey, *The Analysis of Virtual Mass Effects in Two-phase Flow*, Int. J.Multiphase Flow **5**, No. 4, 233-242, 1979
4. M. Ishii: *Thermodynamics of Two-Phase Flow,*, (Eyrolles, Paris, 1976)
5. H. Städtke, R. Holtbecker, *Hyperbolic Model for Inhomogeneous Two-phase Flows*, Int. Conference on Multiphase Flows, '91 TSUKUBA, Univ. of Tsukuba, Japan, September 24-27, 1991
6. H. Städtke, G. Franchello, B. Worth, *Numerical Simulation of Gas-Liquid Two-phase Flows Using Flux Vector Splitting Techniques*, 8th Int. Symposium on Computational Fluid Dynamics, September 5-10, 1999, Germany
7. H. Städtke, G. Franchello, B. Worth, *On the Hyperbolic Nature of Two-phaseFlow Equations: Characteristic Analysis and Related Numerical Methods*, Intern. Conference on Godunov Methods - Theory and Applications, October 18-22, 1999, Oxford, United Kingdom
8. B. Van Leer, *Towards the Ultimate Conservative Difference Scheme, Second Order Sequel to Godunov's Method*, J. Computational Phys. **32**, 1979

Numerical Simulation on the Propagation of an Internal Wave in Multifluid

Hideyuki Oka[1], Tetuya Kawamura[2], and Katsuya Ishii[3]

[1] Ship Research Institute, 6-38-1 Shinkawa, Mitaka, Tokyo 181-0004, Japan
[2] Ochanomizu University, 2-1-1 Ootsuka, Bunkyo-ku, Tokyo 112-8610, Japan
[3] Nagoya University, Furo-cho, Chikusa-ku, Nagoya 464-8603, Japan

Abstract. We numerically investigate the propagation of surface and internal waves which are caused by a gravity current in a container. A level set approach is used to capture interfaces without explicit tracking. To model the complex boundary of the container, a virtual boundary method is applied to the level set formulation. In this study, it is shown that multifluid flows in a deformed container can be simulated through the combination of the above two methods.

1 Introduction

Recently, there is a need for research on the effective use of ocean space and the preservation of global environment, such as the development of a floating airport and the disposal of fossil fuel CO_2 in the deep ocean basins. Hence, it is important to study the safety of various kinds of offshore structures against the wave motion which is caused by an earthquake, a seaquake and so on. When a seaquake takes place, an underwater landslide from a shelf edge may be caused in the ocean. Consequently, a particle-driven flow can occur along the bottom of the sea, and then the internal wave between seawater and particle-laden seawater and the surface wave between the air and seawater are thought to be produced by the turbidity current.

In this study, we make a three-dimensional simulation on the propagation of surface and internal waves on the assumption that the air, the seawater and the particle-laden seawater are incompressible Newtonian fluids, respectively. A level set approach [1],[2] is used to treat free boundaries. We extend our three-dimensional code developed for two-phase flow computations [3],[4] so that it can be applied to three-fluid free boundary problems. In order to model the ocean bed, we try to apply a virtual boundary method [5],[6] in conjunction with the present level set formulation. This method represents a body within a flow field via a forcing term added to the governing equation. Hence, a flow about a body can be computed in a Cartesian computational domain on a regular grid in the course of the solution of the overall fluid equations with the body's effect imposed by a force field.

In this paper, we show that the combination of the level set approach and the virtual boundary method is an effective technique to simulate multifluid flows with solid boundaries of a complex shape.

2 Level set formulation

In this paper, we consider a simple situation of an undersea landslide. Figure 1 shows a vertical section of the schematic diagram of the initial flow configuration. The computational domain is discretized by a uniform, three-dimensional, Cartesian, stationary grid. A staggered grid system is used, in which the fluid velocity $\boldsymbol{u}$, the density ρ and the viscosity μ are located at cell centers whereas the pressure p is evaluated at the vertices of the cells.

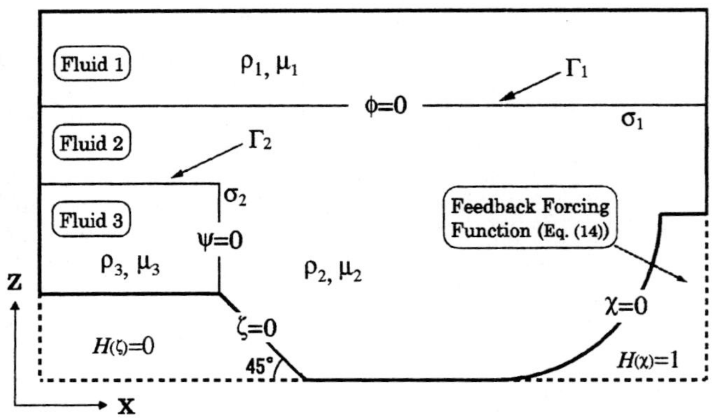

Fig. 1. Schematic diagram of initial flow configuration

In this study, we assume that the three phases are immiscible, incompressible Newtonian fluids which are free of impurities and surfactants, and that the flow is isothermal without phase change. Since the fluid motion of three phases is solved as one field for the entire computational domain, the governing equations are the incompressible Navier-Stokes equations which are written in dimensionless form as follows:

$$\frac{\partial \boldsymbol{u}}{\partial t} + (\boldsymbol{u} \cdot \nabla)\boldsymbol{u} = \frac{1}{\rho}\left(-\nabla p + \frac{1}{Re}\nabla \cdot (2\mu D)\right) + \boldsymbol{g}_{\mathbf{u}} + \frac{1}{\rho We}\kappa\delta(d)\boldsymbol{n} \, , \quad (1)$$

$$\nabla \cdot \boldsymbol{u} = 0 \, , \quad (2)$$

where D is the rate of deformation tensor, and $\boldsymbol{g}_{\mathbf{u}}$ is a dimensionless unit gravitational force. The last term in the right hand side of (1) represents the surface tension force. Here κ is the curvature of the interface, $\delta(\boldsymbol{x})$ is the Dirac delta function, d is the distance to the interface, and $\boldsymbol{n}$ is the normal to the interface. Thus, the effect of surface tension is to act as an additional forcing term in the direction normal to the fluid interface. The dimensionless parameters, Re and We, represent the Reynolds number and the Weber number, respectively, which are defined as follows:

$$Re = \frac{\rho_2 H\sqrt{gH}}{\mu_2} \, , \quad We = \frac{\rho_2 g H^2}{\sigma} \, . \quad (3)$$

Here σ is the surface tension coefficient. In this paper the reference length H is the depth from the interface Γ_1, and the phase speed of a long wave $\sqrt{gH}$ is used as the reference velocity.

A level set function is introduced so that the interface between fluids can be captured without explicit tracking. The level set functions, ϕ and ψ, are smooth distance functions (i.e., $|\nabla\phi| = 1,\ |\nabla\psi| = 1$), and obey the following equations, respectively:

$$\frac{\partial\phi}{\partial t} + \boldsymbol{u}\cdot\nabla\phi = 0\,, \quad \frac{\partial\psi}{\partial t} + \boldsymbol{u}\cdot\nabla\psi = 0\,. \tag{4}$$

In the present simulation, the level set functions ϕ and ψ are defined as additional cell-centered variables which are equal to zero at the interfaces Γ_1 and Γ_2, respectively (i.e., $\Gamma_1 = \{\boldsymbol{x}\in\Omega\,|\,\phi(\boldsymbol{x},t)=0\},\ \Gamma_2 = \{\boldsymbol{x}\in\Omega\,|\,\psi(\boldsymbol{x},t)=0\}$). The first level set function ϕ is chosen such that $\phi<0$ corresponds to the region occupied by fluid 1; $\phi>0$ corresponds to the other regions. For the second level set function, we set $\psi<0$ in the region occupied by fluid 3, and $\phi>0$ in the other regions. Since ϕ and ψ are chosen to be the signed normal distance from Γ_1 and Γ_2, respectively, we can recast the density ρ and the surface tension $\boldsymbol{F}_{st}$ in the level set framework as follows:

$$\rho(\phi,\psi) = \begin{cases} \xi_1 + (1-\xi_1)H(\phi) \text{ if } & \phi\le\epsilon\,, \\ \xi_2 + (1-\xi_2)H(\psi) \text{ if } & \psi\le\epsilon\,, \\ 1 & \text{otherwise}\,. \end{cases} \tag{5}$$

$$\boldsymbol{F}_{\mathrm{st}} = \frac{1}{We_1}\kappa(\phi)\delta(\phi)\nabla\phi + \frac{1}{We_2}\kappa(\psi)\delta(\psi)\nabla\psi\,, \tag{6}$$

where $H(x)$ is the smoothed Heaviside function, ϵ is a half thickness of the interfaces, ξ_1 and ξ_2 are the density ratios ρ_1/ρ_2 and ρ_3/ρ_2, respectively. The viscosity $\mu(\phi,\psi)$ is also written in the same form as $\rho(\phi,\psi)$ using the viscosity ratios $\eta_1(=\mu_1/\mu_2)$ and $\eta_2(=\mu_3/\mu_2)$. For the Weber numbers, their subscripts 1 and 2 correspond to those of the interfaces Γ_1 and Γ_2, respectively.

In the present numerical procedure, we use the second-order projection method for variable density problems [7]. Firstly, one can rewrite (1) in the following form:

$$\boldsymbol{W}_{\mathrm{d}} + \frac{\nabla p}{\rho} = \boldsymbol{W}\,, \tag{7}$$

where $\boldsymbol{W}_{\mathrm{d}}$ represents $\partial\boldsymbol{u}/\partial t$ and $\boldsymbol{W}$ represents

$$-(\boldsymbol{u}\cdot\nabla)\boldsymbol{u} + \boldsymbol{g}_{\mathbf{u}} + \frac{1}{\rho}\left(-\nabla p + \frac{1}{Re}\nabla\cdot(2\mu D) + \boldsymbol{F}_{\mathrm{st}}\right)\,. \tag{8}$$

After taking the divergence of both sides of (7), we use the continuity equation in order to set $\nabla\cdot\boldsymbol{W}_{\mathrm{d}} = 0$. Thus, we obtain the equation for the pressure field:

$$\nabla\cdot\left(\frac{\nabla p}{\rho}\right) = \nabla\cdot\boldsymbol{W}\,. \tag{9}$$

In (9) we solve the pressure with the condition that

$$\frac{\nabla p}{\rho} \cdot \boldsymbol{n} = \boldsymbol{W} \cdot \boldsymbol{n} \quad \boldsymbol{x} \in \partial \Omega \ , \tag{10}$$

where $\partial \Omega$ represents the boundary. Once the pressure field p is determined from (9), we can then update $\boldsymbol{W}_{\mathrm{d}}$ as

$$\boldsymbol{W}_{\mathrm{d}} = \boldsymbol{W} - \frac{\nabla p}{\rho} \ . \tag{11}$$

The procedure of decomposing $\boldsymbol{W}$ into the two parts $\boldsymbol{W}_{\mathrm{d}}$ and $\nabla p/\rho$ is called a projection since the set of divergence free vector fields and the set of gradient fields are orthogonal complements of each other when we use a density weighted norm [7].

We refer the reader to Oka & Ishii [3] for a detailed description of the three-dimensional level set formulation including reinitialization procedures for the level set function.

3 Modeling the boundary of general geometry

The future goal of the present study is to investigate the motion of surface and internal waves which are induced by a turbidity current along the bottom of the sea. Thus, we need to take account of complex boundaries of the container so as to model the ocean floor. At first we consider the way of handling the inclined plane boundary of a level of 45° in the Cartesian grid system. It is relatively easy to treat the velocity boundary condition at that angle. However, in solving (9) for the pressure, it is difficult to explicitly enforce the boundary condition (10) in numerical calculation. Therefore, according to Aleinov et al. [8], instead of solving (9) just in Ω with the boundary conditions specified by (10), we solve

$$\nabla \cdot \left(H(\zeta) \frac{\nabla p}{\rho} \right) = \nabla \cdot \left(H(\zeta) \boldsymbol{W} \right) \ , \tag{12}$$

where ζ is the third level set function which is negative within solid bodies. The zero level set of ζ represents the boundary $\partial \Omega$. The weak solutions of (12) satisfy the boundary conditions (10). Thus, though the computational domain can be a rectangular domain, the inclined plane boundary can be taken into account.

In addition to the above inclined plane boundary, we also consider the curved surface of a cylindrical body with the cross section of one-quarter of a disk as a typical case of a complex boundary. It is generally difficult to treat the velocity boundary condition accurately at such a boundary. Hence, in order to cope with this difficulty, we apply a virtual boundary method [5],[6] in conjunction with the forth level set function χ which is defined to be positive within the solid body. The virtual boundary method employs the following forcing term governed by a feedback loop:

$$\boldsymbol{F}(\boldsymbol{x}_{\mathrm{s}}, t) = \alpha \int_0^t \boldsymbol{U}(\boldsymbol{x}_{\mathrm{s}}, \bar{t}) \, d\bar{t} + \beta \boldsymbol{U}(\boldsymbol{x}_{\mathrm{s}}, t) \ . \tag{13}$$

Here, $\boldsymbol{F}$ is the external force imposed at the body's surface points defined by $\boldsymbol{x}_s$, and $\boldsymbol{U}$ is the fluid velocity at these surface points. The quantities α and β are negative constants. This feedback function is added to the momentum equations, and effectively brings the fluid velocity to zero at the desired points in the flow which define the no-slip boundary. However, Saiki & Biringen [6] showed that some of their results converged to an incorrect solution with the forcing term imposed only on the boundary. They suggested that the forcing function should be applied at the boundary and interior points of the body in order to converge to a correct solution. In this study, since the slip conditions are imposed on all boundaries, the forcing function should be modified so that it may model the full slip boundary. Therefore, we use the feedback forcing function modified as follows:

$$(\boldsymbol{F}(\boldsymbol{x},t)\cdot\boldsymbol{n}_s)\,\boldsymbol{n}_s\,H(\chi)\ ,\quad \boldsymbol{n}_s=\frac{\nabla\chi}{|\nabla\chi|}\ , \tag{14}$$

where $\boldsymbol{n}_s$ is the unit normal vector at each contour surface of χ.

4 Numerical result

We shows the numerical result under the initial conditions depicted in Fig. 1. Each fluid is quiescent at $t = 0$. The computational domain is a rectangular box with full slip boundaries. The dimensionless length of the three sides of the domain is $2L \times 2L \times L$ in the x, y and z directions, respectively, where we have $L = 1.25$ in this paper. The number of the grid points is $48\times48\times24$. The dimensionless parameters, Re, We_1, We_2, ξ_1, ξ_2, η_1 and η_2, are set equal to 500, 1000, 100, 1/1000, 5, 1/100 and 2, respectively. According to Saiki & Biringen [6], the negative constants α and β in the feedback forcing term (13) are set equal to -4×10^5 and -6×10^2 in this simulation. Figure 2 shows the time evolution of the two interfaces, Γ_1 and Γ_2, and the velocity field on the vertical center plane ($y = L$). At the initial stage, fluid 3 starts to flow down along the inclined plane due to the effect of gravity. After fluid 3 reaches the curved boundary, it runs upward along the surface of the body which is modeled by the forcing function (14). Then, it flows away from the curved boundary, producing an internal wave on Γ_2. We can also observe wave motion on Γ_1 throughout this simulation. Hence, it is found that the interface Γ_1 is considerably influenced by the gravity current of fluid 3. In addition, the numerical result shows that the velocity field along the slope and the curved surface is obtained accurately even in a regular Cartesian grid by using the two techniques described in the previous section.

5 Conclusion

In this paper, we have proposed a new numerical technique to simulate the multifluid flow field with complex boundary geometries by combining the virtual boundary method [5],[6] with the level set approach [1],[2]. It has been shown that the gravity current and the subsequent propagation of waves on the interfaces are well simulated through the combination of the two methods.

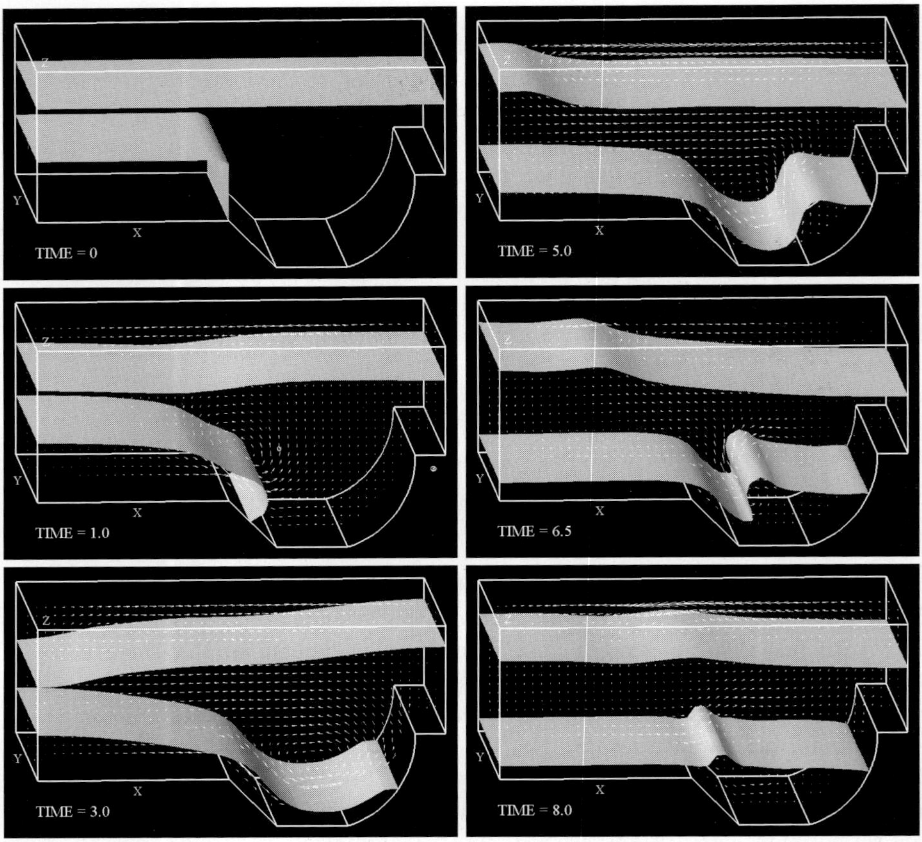

Fig. 2. Time evolution of two interfaces, Γ_1 and Γ_2, and the velocity vectors on the vertical center plane in a container of complex shape. Here Re=500, We_1=1000 and We_2=100. The other parameters, ξ_1, ξ_2, η_1 and η_2 are 1/1000, 5, 1/100 and 2, respectively. Time increases from top to bottom and left to right: t=0, 1, 3, 5, 6.5, 8

References

1. M. Sussman, P. Smereka, S. Osher: J. Comput. Phys. **114**, 146 (1994)
2. M. Sussman, P. Smereka: J. Fluid Mech. **341**, 269 (1997)
3. H. Oka, K.Ishii: J. Phys. Soc. Jpn. **68**-3, 823 (1999)
4. H. Oka, K.Ishii: J. Phys. Soc. Jpn. **69**-2, 392 (2000)
5. D. Goldstein, R. Handler, L. Sirovich: J. Comput. Phys. **105**, 354 (1993)
6. E. M. Saiki, S. Biringen: J. Comput. Phys. **123**, 450 (1996)
7. J. B. Bell, L. Marcus: J. Comput. Phys. **101**, 334 (1992)
8. I. D. Aleinov, E. G. Puckett, M. Sussman: 'Formation of Droplets in Microscale Jetting Devices'. In: *Proceedings of the 3rd ASME·JSME Joint Fluids Engineering Conference FEDSM'99 at San Francisco, California, USA, July 18–23, 1999*, FED-**248**, F-209-06, FEDSM99-7106 on CD-ROM

Conservative Interface-Capturing Approaches for Multi-component Flows

Yang-Yao Niu

Chung-Hua University, Hsin-Chu,30067, Taiwan,ROC. E-mail:yniu@chu.edu.tw

Abstract. In this study, the upwind schemes AUSMDV are presented for computing the dynamics of multi-component flows with stiffened perfect gas. Numerical model is not only demonstrated to maintain pressure equilibrium over contact discontinuities using conservative pressure update, but also avoid spurious oscillations near material interfaces. The benchmark tests involve a two-component shock tube, a shock-contact interaction and a two-dimensional square bubble problem.

1 Introdction

Conservative Euler solvers have been found to produce numerical oscillations near contact discontinuities in multi-component flow problems. As explained in [1-2], the averaging procedure in the conservative formulation causing numerical mixing, often fails to maintain pressure equilibrium among fluid composition cross material interfaces. In order to overcome numerical difficulties from the multi-component calculations, great efforts have been developed toward many directions as reviewed in [3]. Recently, Shyue [3] extended the work of Abgrall [1] from polytropic gases in one dimension to a stiffened gas with the so-called "stiffened" gas equation of state for approximating materials including compressible liquids and solids. Shyue modified the Roe type Riemann solver by assuming velocity and pressure equilibrium in each grid cell. It shows that monotone profiles of flow solutions near material front and pressure equilibrium are achieved. However, the reconstruction of the numerical flux is expressed in a sophisticated way. The derivation of eigensystem will become a tedious work for solving multi-dimensional complicated multi-fluid problems. Unfortunately, another kind of efficient and simple-implementing flux vector splitting such as van Leer [4] shares a serious disadvantage, that is, large dissipation on contact discontinuities. Also as claimed by Abgrall [2], the reconstruction of Riemann solver for the multi-component flows does not work for the van Leer flux splitting, however, one remedy may be to apply the hybrid splitting of Coquel and Liou [5]. In this study, to consider the multi-component flow equations, the volume fraction model proposed by Shyue [3] is regarded as a model equation here. The used system equations are strictly hyperbolic. We first use the hybrid type flux splitting HUS [5] and AUSMD/V [6] to apply for the multi-component flow problems. Numerical solutions of the system equation can be evolved by a predictor-corrector approach with a constant time step. The interface conservative variable can be determined through MUSCL [7] approach with a minmod type limiter.

2 Numerical Method

To consider the multi-component flow equations, the volume fraction model proposed by Shyue [3] is regarded as a model equation here. The transport equation for each volume fraction in the framework of Shyue is expressed in quasi-conservative form. However, in this study, we combine the continuity equation with the non-conservative evolution equation of volume fraction to be a conservation equation. Therefore, the extended Euler equations in conservation form in one dimension is solved. The system equation used is a strictly hyperbolic. That allows us to apply existing hyperbolic solvers as approximate Riemann solvers or flux vector splitting method, which have successfully solved the single phase problems, to simulate the multi-component model. Most previous studies[1-3] modified approximate Riemann solves to keep velocity and pressure equilibrium across the interface for solving the multi-fluid flows. In this study, the hybrid type flux splitting AUSMD/V is applied for the multi-component flow problems without the tedious derivation of eigensystem as usually done in the Riemann solvers.

Numerical solutions of the system equation can be evolved by a predictor-corrector approach with a constant time step. The interface conservative variables can be determined through MUSCL approach [15] with the Minmod limiter.

2.1 Numerical Flux

The interface numerical flux of AUSMD/V is simply defined as

$$F_{\frac{1}{2}} = \frac{1}{2}(\rho M)_{\frac{1}{2}}[\Phi_L + \Phi_R] - \frac{1}{2}|(\rho M)_{\frac{1}{2}}|\Delta_{\frac{1}{2}}\Phi + P_{\frac{1}{2}}, \tag{1}$$

where the interface mass flux is also written as

$$(\rho M)_{\frac{1}{2}} = \rho_L M_L^+ + \rho_R M_R^-, \tag{2}$$

and $\Delta_{\frac{1}{2}}(\bullet) = (\bullet)_R - (\bullet)_L$, $\Phi = [1, u, v, H]$, and $P = [0, p, 0, 0]$.

Here (M_L^+, M_R^-) are defined as

$$M_L^+ = \alpha_L\left(\frac{(M+1)^2}{4} - \frac{M+|M|}{2}\right)_L + \frac{M_L + |M_L|}{2}; \tag{3}$$

$$M_R^- = \alpha_R\left(-\frac{(M-1)^2}{4} - \frac{M-|M|}{2}\right)_R + \frac{M_R - |M_R|}{2}, \tag{4}$$

where the parameter α is determined such that the numerical dissipation for the contact discontinuity vanishes and can have several choices. In this paper, a weighting function based only on the ratio of pressure to ratio.

The following split pressure is used by

$$p^{\pm} = \begin{cases} p(M \pm 1)^2(2 \mp M)/4, & \text{if } |M| > 1; \\ p(M \pm |M|)/2M, & \text{otherwise.} \end{cases} \tag{5}$$

The interface Mach number and pressure are simply a sum of the respective "left" and "right" components,

$$M_{\frac{1}{2}} = M_L^+ + M_R^-, \tag{6}$$

and

$$P_{\frac{1}{2}} = p_L^+ + p_R^-. \tag{7}$$

The difference between the family of AUSMD and AUSMV is only in the treatment of the term $(\rho M^2)_{\frac{1}{2}}$ in the momentum equation. The momentum flux of AUSMV is

$$(\rho M^2)_{AUSMV} = M_L^+(\rho M)_L + M_R^-(\rho M)_R, \tag{8}$$

while that of AUSMD requires the same form as other equations in Eq. (1), i.e.,

$$(\rho M^2)_{AUSMD} =$$

$$\frac{1}{2}[(\rho M)_{\frac{1}{2}}(M_L + M_L) - |(\rho M)_{\frac{1}{2}}|(M_R - M_L)]. \tag{9}$$

A combination of momentum fluxes between AUSMD and AUSMV in terms of a weighted function of pressure gradient is called AUSMDV and expressed as

$$(\rho M^2)_{AUSMDV} = (\frac{1}{2} + S)(\rho M^2)_{AUSMV} +$$

$$(\frac{1}{2} - S)(\rho M^2)_{AUSMD}, \tag{10}$$

where S is a switching function

$$S = \frac{1}{2}\min(1, 10\frac{|p_R - p_L|}{\min(p_R, p_L)}). \tag{11}$$

These numerical fluxes AUSMD/V associated with the above velocity splitting both automatically result in a vanished mass flux without producing excessive numerical dissipation while solving a stationary contact discontinuity. Also the ideal of the common sound of speed in the velocity and pressure splitting enable numerical fluxes for the conserved variables reproducing those of the exact Riemann solution for the contact discontinuities. This is a reason why AUSMD/V allows exact capturing of stationary discontinuities in the single-component gas dynamic problems. Therefore, we investigate their accuracy in solving multi-component flow problems here.

3 Results and Discussions

3.1 Two-component shock tube

In the first test case, the works of Abgrall [2] for a Sod's shock tube problem involving two pure fluids separated by an interface is solved. We consider a shock tube filled on its left part with low-density fluid and on its right part with high-density fluid under different the ratio of specific heat. Velocity and pressure equilibrium are assumed along the whole tube. The solution is plotted at time 0.1. Mixture pressure, mixture density, velocities and void fraction contain exact solutions which are represented as solid lines in each plot. The computed solutions are performed on constant spacing meshes of 100 cells. CFL No.=.5 is chosen to keep numerical stability. The prediction of mixture density, mixture pressure, gas void fraction and gas velocity solved by AUSMDV. Only the results by AUSMDV is demonstrated in Figure 1, respectively. The comparisons show that the agreement between exact solutions and computed results are satisfactory. It is seen that correct behaviors near interfaces are captured in the density and void fraction profiles. The computed shock waves are very well located. It is also shown that velocity and pressure equilibrium are retained in the computations.

3.2 Shock refraction problem

Following the works of Shyue [3], a shock refraction problem describing a gas flow with air and helium separated by an interface, which is hit by the incident shock wave and set into motion , is solved. As what shyue mentioned, after hitting by the incident shock wave, a shock wave transmitted across the interface becomes unstable physically. Thus any erroneous numerical behavior is inclined to result in non-physical dynamics. The results of mixture pressure, mixture density, velocities and gas void fraction containing exact solutions are depicted in Figures 2. The computed solutions are performed on constant spacing meshes of 400 cells. CFL No.=.5 is chosen to keep numerical stability. Here only the solution on 400 cells is shown. The computed volume fraction is well predicted as same as exact data. Based on the same grid size, the AUSMDV achieves quite consistent solutions with the exact profiles in the profile of the contact discontinuity for the velocity and pressure distributions. Besides, the non-physical solutions near interfaces are completely reduced from the computed solutions by AUSMDV scheme.

3.3 Advection of a square bubble

In a two-dimensional case, a evolution of a square bubble in a liquid suggested in Shyue [3] is solved. The computational domain is a square of unit one length. The initial square bubble whose initial dimensions are 0.2x0.2 is located between $0.1 < x < 0.3$ and $0.1 < y < 0.3$ as shown in Figure 3. Velocity and pressure equilibrium are assumed over the whole computational domain. The solutions performed by AUSMD are plotted at time 0.36. The computed solutions are performed on constant spacing meshes of 100x100 cells. CFL No.=0.5 is chosen to keep numerical stability. The contour plot of the density describing a gas bubble along the line x=y is shown in Figure 3. It is seen that the interface between the gas and liquid are slightly smeared, but the shape of a square shape is preserved. The cross-section plots of mixture density and mixture pressure are also predicted in Figure 3, respectively. It is shown that correct behaviors and smooth profiles near interfaces are captured in the computed density and pressure results. It is also seen that pressure equilibrium is retained in the computation.

3.4 Conclusion

This study suggested that the computation of the dynamics of multi-component flow with a stiffened perfect gas is performed on the extended Euler equation with a volume fraction model using the AUSMDV scheme. Numerical results demonstrate that the solution of AUSMDV demonstrates excellent accuracy compared with exact solutions and exhibits oscillation-free material interfaces in all the test cases. The above procedure has provided an alternative to solve the multi-component flow problems and demonstrates the simplicity and accuracy.

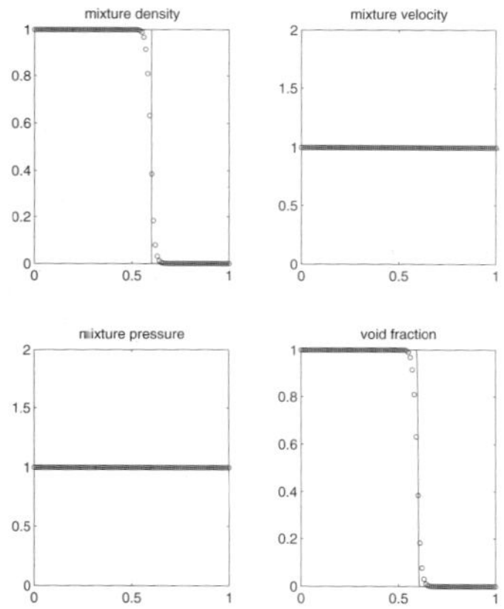

Fig. 1. Numerical solutions in the two-component shock tube by AUSMDV (o— computed solutions on 100 cells, — exact solutions)

References

1. R. Abgrall, Generalization of Roe's Riemann solver to mixtures of Perfect Gas, Rech. Aerospat. 6, 1, (1988)
2. R.Abgrall, How to prevent pressure oscillations in multi-component flow calculations: A quasi conservative approach, Journal of Computational Physics. 125, 150, (1996)
3. Shyue, K.M.., "An efficient Shock-Capturing Algorithm for Compressible Multi-component problems" Journal of Computational Physics, Vol 142, pp.208-242, (1998)
4. Van Leer, B. "Flux Vector Splitting for the Euless Equations", Lecture Notes in Physics Vol.170,pp 507-508, (1982)
5. Coquel, F and Liou, M. S., "Field by Field Hybrid Upwind Splitting Methods" AIAA Paper 93-3302-CP,(1933)

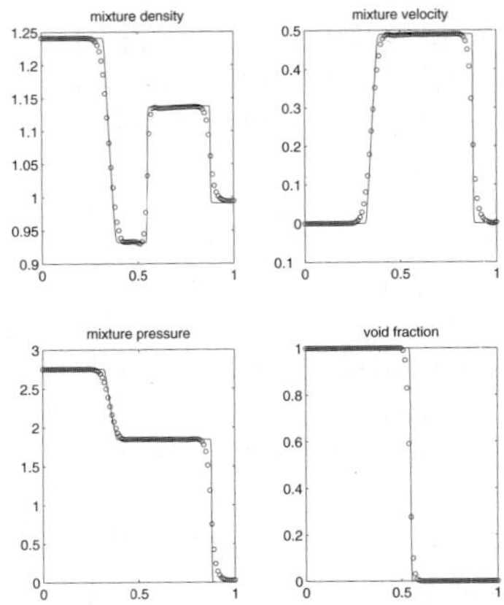

Fig. 2. Numerical solutions in the shock contact interaction problem by AUSMDV (o— computed solutions on 400 cells, —exact solutions)

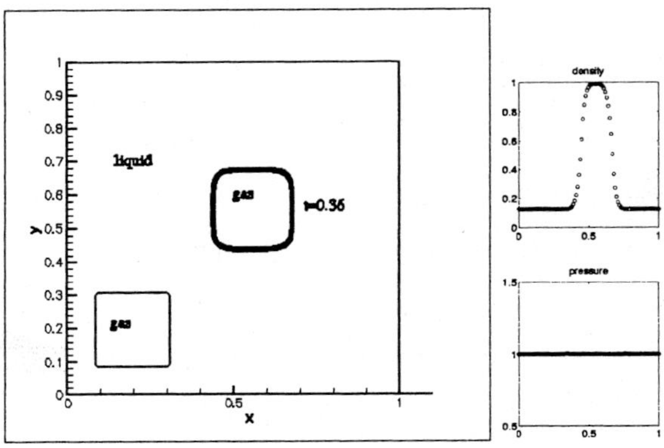

Fig. 3. A density contour plot of a square bubble adevtion (on the left hand side); the mixture pressure and velocity distribution by AUSMDV (on the right hand side)

6. Liou, M. S. and Wada, Y., "A quest towards ultimate numerical flux schemes," "Computational Fluid Dynamics Review 1995" (Eds. M. Hafez and K. Oshima), John Wiley and Sons, pp 251-278, (1995)
7. Van Leer, B. "Towards the ultimate conservative difference, V: A second order sequel to Godunov's method", Journal of Computational Physics, Vol 32, pp.179-186, (1979)
8. Jean-Marc Moschetta and D. I. Pullin "A Robust Low Diffusive Kinetic Scheme for the Navier-Stokes/Euless Equations", Journal of Computational Physics, Vol. 133, pp.193-204, (1997)

Numerical Analysis of Cavitating Flow through a 2-D Decelerating Cascade

Byeong Rog Shin[1], Yuka Iga[2], and Toshiaki Ikohagi[1]

[1] Institute of Fluid Science, Tohoku University, Katahira, Sendai 980-8577, Japan
[2] Graduate School of Engineering, Tohoku University, Sendai 980-8579, Japan

Abstract. A Numerical method for solving the cavity flow is proposed and applied to simulate unsteady cavitating flows through a two-dimensional decelerating cascade. A finite-difference Runge-Kutta method combined with MUSCL TVD scheme is used, and a mesoscopic model which is based on a model of compressible gas-liquid two-phase media is applied to solve above cavitating flows. The present method permits us to treat simply the whole gas-liquid two-phase flow field including wave propagation and large interface deformation in spite of the use of a fixed mesh system. Numerical results such as detailed observations of the unsteady cavity flows and the comparisons of predicted results with available experimental data are provided.

1 Introduction

Cavitation is a phenomenon generally observed in high-speed fluid flow around an obstacle such as impellers of the fluid machinery. It is well known that this cavity flow causes the noise, vibration and damage in hydraulic machine systems [1], so that it has an unfavorable effect on the performance and eventually brings the low efficiency. Recently, it pays attention to the cavitation phenomena which occur in turbopump and inducer of the liquid rocket for reusable launch vehicles in particular. The existence of the unsteady cavity flow is to be a very important factor in the estimation of the performance. Also, the improvement of state of the art on prediction of cavitation is a common subject in development of design of the high-speed fluid devices.

Therefore, in order to clarify the behavior of cavity flow, some efforts to propose gas-liquid two-phase flow models for numerical simulations have been made [2–6]. However, up to date, because the cavity flows have strong unsteady flow phenomena including phase changes, fluid transients, vortex shedding and turbulence, the numerical method to solve these flows is not established yet. In the above investigations, they have had little comprehensive applications to the transient flow range from subcavitation state to supercavitation state. Lately, the present authors have proposed a mathematical cavity flow model [7,8] based on a mesoscopic model taking account of the compressibility of the gas-liquid two-phase media. With this model and TVD-MacCormack scheme [9], the mechanism of developed cavitation has been investigated by Shin and Ikohagi [10] through the application to a couple of cavitating flows around a hydrofoil.

The purpose of this paper is to verify an applicability of the above mathematical model to the wide range of cavity flow through cascade, and to make clear the unsteady behavior of some kinds of cavitation. In this study, the previous TVD-MacCormack method is extended to a high-order MUSCL TVD [11] solution method. As the numerical examples, an unsteady cavity flows through a 2-D decelerating cascade are simulated. And then, detailed cavity flow behavior including the developed cavitation and growing and shedding of the cavity for the above flow fields are investigated, and pressure distributions are compared with available data.

2 Mathematical Model and Governing Equations

Let us consider a local volumetric element in gas-liquid two-phase media, in which numerous particles of infinitesimal bubbles or droplets are assumed to distribute homogeneously. Then, the two-phase media must be modeled mesoscopically by the infinite set of these elements with each different void fraction. That is, cavitation bubbles can be expressed by arranging homogenized elements according to their shapes. Therefore, our mathematical modeling for gas-liquid two-phase media is essentially different from the conventional ones.

Under the our model concept, the pressure is determined by the equation of state for gas-liquid two-phase media, that is written as follows:

$$\rho = \frac{p(p+p_c)}{K(1-Y)p(T+T_0)+RY(p+p_c)T} \tag{1}$$

where, ρ, p, Y, and T are the mixture density, pressure, quality and the temperature, respectively. R is the gas constant and K, P_c and T_0 represent the liquid constant, pressure constant and the temperature constant for water, respectively. This equation is derived from the local equilibrium assumption, and corresponds to the following equations of state for pure liquid (Y=0) [12] and ideal gas (Y=1), respectively.

$$p + p_c = \rho_\ell K(T+T_o);\ \text{for } Y=0, \qquad p = \rho_g RT;\ \text{for } Y=1 \tag{2}$$

where, subscripts ℓ and g mean liquid phase and gas phase. Therefore, the apparent compressibility is considered, and the sound speed c for the isothermal condition becomes $c^2 = (c_1/c_2)(p(p+p_c)/\rho)$, where, $c_1 = Y\{R(p+p_c)-Kp\}+Kp$ and $c_2 = Y\{R(p+p_c)^2-Kp^2\} - \rho KRY(1-Y)p_cT_0 + Kp^2$. And, the relation between the local void fraction α and the quality Y is given as $\rho(1-Y) = (1-\alpha)\rho_\ell$ and $\rho Y = \alpha\rho_g$, where

$$\alpha = \frac{RY(p+p_c)T}{K(1-Y)p(T+T_0)+RY(p+p_c)T} \tag{3}$$

Based on above our modeling concept mentioned above, under the isothermal condition and neglecting the surface tension, the 2-D governing equations for the mixture mass, momentum and the gas-phase mass can be written in the curvilinear coordinates (ξ, η) as follows:

$$\frac{\partial \hat{\boldsymbol{Q}}}{\partial t} + \hat{\boldsymbol{L}}(\hat{\boldsymbol{Q}}) \equiv \frac{\partial \hat{\boldsymbol{Q}}}{\partial t} + \frac{\partial(\hat{\boldsymbol{E}} - \hat{\boldsymbol{E}}_v)}{\partial \xi} + \frac{\partial(\hat{\boldsymbol{F}} - \hat{\boldsymbol{F}}_v)}{\partial \eta} = 0 \tag{4}$$

where $\hat{\boldsymbol{Q}}$ is the unknown variable vector, $\hat{\boldsymbol{E}}$, $\hat{\boldsymbol{F}}$ are the flux vectors and $\hat{\boldsymbol{E}}_v$, $\hat{\boldsymbol{F}}_v$ are the viscous terms presented by the stress tensor τ.

$$\hat{\boldsymbol{Q}} = J\begin{pmatrix}\rho\\ \rho u\\ \rho v\\ \rho Y\end{pmatrix},\qquad \hat{\boldsymbol{E}} = J\begin{pmatrix}\rho U\\ \rho u U + \xi_x p\\ \rho v U + \xi_y p\\ \rho U Y\end{pmatrix},\qquad \hat{\boldsymbol{F}} = J\begin{pmatrix}\rho V\\ \rho u V + \eta_x p\\ \rho v V + \eta_y p\\ \rho V Y\end{pmatrix},$$

$$\hat{\boldsymbol{E}}_v = J\begin{pmatrix}0\\ \xi_x\tau_{xx} + \xi_y\tau_{xy}\\ \xi_x\tau_{yx} + \xi_y\tau_{yy}\\ 0\end{pmatrix},\qquad \hat{\boldsymbol{F}}_v = J\begin{pmatrix}0\\ \eta_x\tau_{xx} + \eta_y\tau_{xy}\\ \eta_x\tau_{yx} + \eta_y\tau_{yy}\\ 0\end{pmatrix}.$$

The Jacobian J of the transformation between Cartesian coordinates x_i and curvilinear coordinates ξ_i is defined by $J = x_\xi y_\eta - x_\eta y_\xi$. u_i is the physical velocity and U_i is the contravariant velocity.

The mixture density ρ and the mixture viscosity [13] μ can be expressed by $\rho = (1-\alpha)\rho_\ell + \alpha\rho_g$, and $\mu = (1-\alpha)(1+2.5\alpha)\mu_\ell + \alpha\mu_g$.

3 Numerical Methods

In this paper, the governing equations (4) are numerically integrated by using the finite-difference third-order Runge-Kutta explicit method. And then the Roe's flux difference splitting (FDS) method [14] with the third-order MUSCL TVD scheme [11] is applied to enhance the numerical stability, especially for the existence of steep gradient in density as well as pressure near the gas-liquid interface. Therefore, the derivative of the flux vector, for instance, $\hat{\boldsymbol{E}}$ with respect to ξ at point i can be written with the numerical flux as $(\partial\hat{\boldsymbol{E}}/\partial\xi) = (\hat{\boldsymbol{E}}_{i+1/2} - \hat{\boldsymbol{E}}_{i-1/2})/\Delta\xi$ and then, the approximate Riemann solver based on the Roe's FDS is applied. Hence, the numerical flux $\hat{\boldsymbol{E}}_{i+1/2}$ is written as,

$$\hat{\boldsymbol{E}}_{i+1/2} = (1/2)\{\hat{\boldsymbol{E}}(\hat{\boldsymbol{Q}}^L_{i+1/2}) + \hat{\boldsymbol{E}}(\hat{\boldsymbol{Q}}^R_{i+1/2}) - |\hat{\boldsymbol{A}}_{i+1/2}|(\hat{\boldsymbol{Q}}^R_{i+1/2} - \hat{\boldsymbol{Q}}^L_{i+1/2})\} \quad (5)$$

where, the Jacobian matrix $|\hat{\boldsymbol{A}}_{i+1/2}|$ of $\hat{\boldsymbol{E}}$ is estimated introducing the Roe's averaging. $\hat{\boldsymbol{Q}}^{L,R}_{i+1/2} = J\boldsymbol{Q}^{L,R}_{i+1/2}$ and

$$\boldsymbol{Q}^L_{i+1/2} = \boldsymbol{Q}_i + (1/4)\{(1-\kappa)D^+\boldsymbol{Q}_{i-1/2} + (1+\kappa)D^-\boldsymbol{Q}_{i+1/2}\} \quad (6)$$

$$\boldsymbol{Q}^R_{i+1/2} = \boldsymbol{Q}_{i+1} - (1/4)\{(1-\kappa)D^-\boldsymbol{Q}_{i+3/2} + (1+\kappa)D^+\boldsymbol{Q}_{i+1/2}\} \quad (7)$$

Here, the flux-limited values of DQ and the mmod function are determined by $D^+\boldsymbol{Q}_{i-1/2}$=mmod$(\delta\boldsymbol{Q}_{i-1/2}, b\delta\boldsymbol{Q}_{i+1/2})$, $D^-\boldsymbol{Q}_{i+1/2}$=mmod$(\delta\boldsymbol{Q}_{i+1/2}, b\delta\boldsymbol{Q}_{i-1/2})$, $\delta\boldsymbol{Q}_{i+1/2} = \boldsymbol{Q}_{i+1} - \boldsymbol{Q}_i$, and mmod$(x,y) = \text{sign}(x)\max[0, \min\{|x|, y\,\text{sign}(x)\}]$. And, the linear combination parameter κ is determined by the range of $-1 \le \kappa \le 1$, and it has an effect on the accuracy, that is Eqs.(6) and (7) has a third-order accuracy at $\kappa = 1/3$. The slope of the flux in the mmod function is controlled by the limiter b. The range of b, $1 \leq b \leq (3-\kappa)/(1-\kappa)$, is determined by the condition of TVD stable.

The governing equations (4) are numerically integrated using the third-order TVD Runge-Kutta explicit method written as

$$\hat{\boldsymbol{Q}}^{(1)} = \hat{\boldsymbol{Q}}^{n} - \Delta t \hat{\boldsymbol{L}}(\hat{\boldsymbol{Q}}^{n}) \tag{8}$$

$$\hat{\boldsymbol{Q}}^{(2)} = (3/4)\hat{\boldsymbol{Q}}^{n} + (1/4)\{\hat{\boldsymbol{Q}}^{(1)} - \Delta t \hat{\boldsymbol{L}}(\hat{\boldsymbol{Q}}^{(1)})\} \tag{9}$$

$$\hat{\boldsymbol{Q}}^{n+1} = (1/3)\hat{\boldsymbol{Q}}^{n} + (2/3)\{\hat{\boldsymbol{Q}}^{(2)} - \Delta t \hat{\boldsymbol{L}}(\hat{\boldsymbol{Q}}^{(2)})\} \tag{10}$$

In this study, we consider the gas-liquid two-phase media to be air-water mixture, so that the constants p_c, K and T_0 for water were estimated as 1944.61MPa, 472.27J/KgK and 3837K, respectively.

4 Numerical Results

Some cavitating flows around a hydrofoil in decelerating cascade are calculated at some different cavitation numbers. The blade profile is Clark Y 11.7% hydrofoil. The up- and down-stream boundaries of the computational domain are located at each 5 times chord length c distances from the leading edge and trailing edge of the hydrofoil. The pitch-chord ratio is 0.9. The stagger angle is 30deg. The cavitation number σ is defined by $\sigma = 2(p_0 - p_v)/\rho_0 u_0^2$ using mean values at one chord length upstream from the leading edge and the vapor pressure p_v. The Reynolds number Re based on the inlet mean velocity is about 2×10^5 and 7×10^5. A body fitted H-type computational grid having 321×121 grid points is used. An ordinary compressible flow boundary condition is imposed.

Figure 1 shows time-averaged pressure distributions for several cavitating flows. As a whole, the comparison of predictions for noncavity flow with experiments using wind tunnel [15] is favorable even though running time for averaging

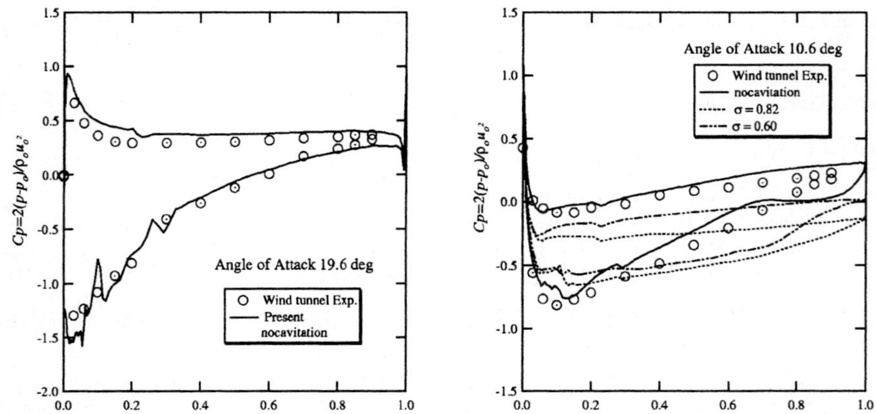

Fig. 1. Comparison of pressure distribution

is not so enough and noncavity flow close to incompressible flow was solved by the present two-phase flow model based on compressible flow. With low cavitation number, relatively low pressure is distributed on suction surface. In particular at σ=0.6 and Re=7×10^5, it shows $Cp \approx -\sigma$ at 10%∼60% chord region.

Figure 2 shows time evolutions of instantaneous void fraction around the hydrofoil at σ=0.6 and angle of attack α of 10.6deg. The behavior of this cavitating flow indicates a kind of transient cavitation state. The occurrence and the development of the sheet cavitation are similar to the subcavitating state at σ=0.82 not shown here. However, the fluctuation on the cavity length and thickness becomes much stronger. The internal structure of the cavity on the void fraction is clear. It can be seen that the cavitation occurred near 10% chord on the suction surface is developing and shedding toward downstream with time. After Fig.2(f), a wavy cloud cavitation was peeled off from the suction surface and then the sheet cavitation was formed again.

A time-averaged void fraction and pressure contours at σ=0.6 and α=10.6deg seen in Fig.1 are shown in Fig.3. The cavity flow pattern indicates aforementioned transient cavitation state. Also, a typical pressure distribution of cascade is shown in this figure, where the region surrounded by inner contour line on suction surface is that representing near vapor pressure.

5 Conclusions

In order to verify an applicability of our two-phase flow model to wide range of cavitations, and to make clear the unsteady behavior of cavity flows, 2-D cavitating flows through a decelerating cascade were numerically simulated. As the result, it showed a successive application to cavitating cascade flows, and a good prediction of pressure distributions in comparison with experiments. From investigation of the void fraction and velocity field in the cavity, complex cavity flow behaviors of the cavitation occurrence and its growing and shedding were understood. Also they give us a lot of information to make out the anomalous phenomena due to unsteady cavity flows in hydromachinery such as turbopumps.

References

1. C.E. Brennen: *Hydrodynamics of Pumps* (Oxford University Press, Oxford 1994)
2. A. Kubota et al.: J. Fluid Mech. **240** (1992), p. 59
3. J.L. Reboud, Y. Delannoy: 'Two-Phase Flow Modeling of Unsteady Cavitation'. In: *Proc. 2nd International Symposium on Cavitation, Tokyo, 1994*, pp. 39–44
4. Y. Chen, S.D. Heister: Computers & Fluids **24** (1995), p. 799
5. M. Deshpande et al.: ASME J. of Fluids Engng. **119** (1997), p. 420
6. S. Clerc: J. Comp. Phys. **161** (2000), p. 354
7. K. Okuda, T. Ikohagi: Trans JSME B **62** (1996), p. 3792
8. B.R. Shin,T. Ikohagi: 'A Numerical Study of Unsteady Cavitating Flows'. In: *Proc. 3rd International Symposium on Cavitation, Grenoble, 1998*, Vol.2, pp. 301–306
9. H.C. Yee: 'Upwind and Symmetric Shock-Capturing Schemes'. NASA TM-89469 (1987)

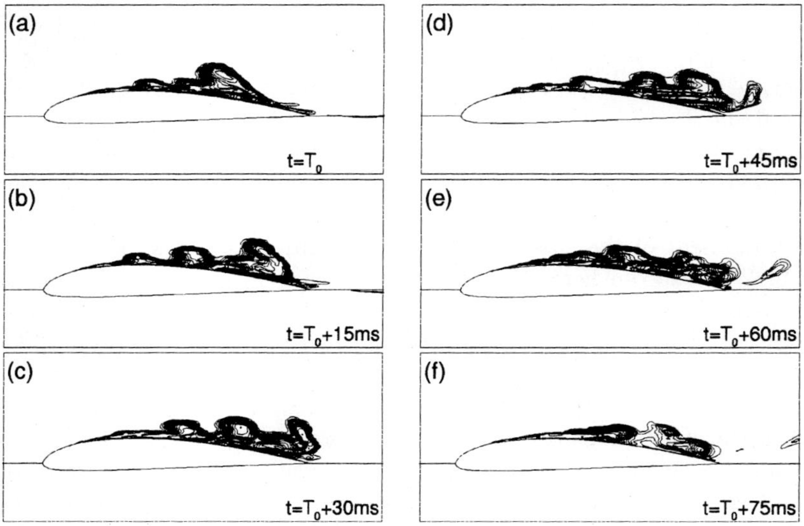

Fig. 2. Time evolution of void fraction at $\sigma = 0.6$ and $\alpha = 10.6°$

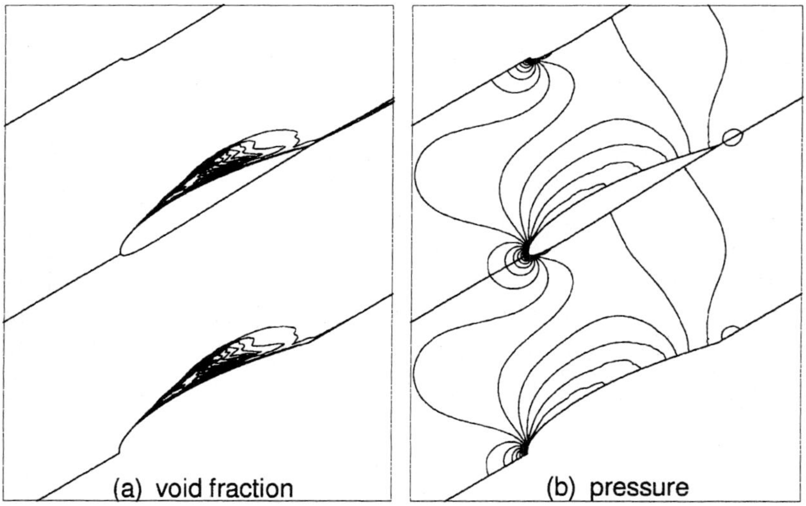

Fig. 3. Time-averaged void fraction and pressure contours at $\sigma = 0.6$

10. B.R. Shin, T. Ikohagi: 'Numerical Analysis of Unsteady Cavity Flows around a Hydrofoil'. ASME FEDSM99-7215 (1999)
11. B. van Leer: J. Comp. Phys. **32** (1979), p. 101
12. H.T. Chen, R. Collins: J. Comp. Phys. **7** (1971), p. 89
13. D.R.H. Beattie, P.B. Whally: Int. J. Multiphase Flow **8** (1982), p. 83
14. P.L. Roe: J. Comp. Phys. **43** (1981), p. 357
15. H. Murai, S. Ibuki: Mem. Inst. High Speed Mech., Tohoku Univ. **47** (1981), p. 117

Part XIX

Fluid Structure Interaction

A Coupled CFD/CSD Methodology for Simulating Structural Response to Airblast and Fragment

Joseph D. Baum[1], Hong Luo[1], Eric L. Mestreau[1], Rainald Löhner[2], Daniele Pelessone[3], and Charles Charman[3]

[1] Center for Applied Computational Sciences, ATG, SAIC, McLean, VA 22102, USA
[2] CSI, George Mason University, Fairfax, VA 22030, USA
[3] General Atomics, San Diego, CA 92121, USA

1 Introduction

Several classes of important engineering problems require the concurrent application of CFD and CSD techniques. Currently, attempts to model these problems are solved either iteratively, requiring several cycles of CFD run followed by CSD run, or by assuming that the CFD and CSD solutions can be decoupled. The various efforts to develop a fluid/structure coupling can be classified according to the complexity level of the approximations used for each of the domains. These range from simple 6 DOF integration to finite elements with complex models for elasto-plastic materials with rupture laws and contact. Similarly, the fluid dynamics approximations range from the potential flow (irotational, inviscid, isentropic flows) to the full Navier-Stokes set of equations. The present research interests focus on non-linear applications, in particular, structures that experience severe deformations due to blast loads. Hence, the fluid applies either the Euler or Reynolds-Averaged Navier-Stokes equations, while elasto-plastic materials with rupture criteria are used for the structural modeling.

The flow solver used is FEFLO98, a 3-D adaptive, unstructured, edge-based hydro- solver based on the Finite-Element Method Flux-Corrected Transport (FEM-FCT) concept [1]. It solves the Arbitrary Lagrangean-Eulerian (ALE) formulation of the Euler and Reynolds-Averaged turbulent, Navier-Stokes equations. Detonation is modeled using the JWL equation of state. The structural dynamics solver is DYNA3D [2], a well-established unstructured, explicit finite element code. DYNA3D is well suited for modeling large deformations, and provides a good base for non-linear materials with elasto-plastic compartmental laws with rupture. Two approaches can be used to model fluid/structure interactions. The so-called 'tight coupling' approach requires solving both CFD & CSD as one coupled set of equations, and would require the complete rewrite of both solvers. The second approach, termed 'loose coupling', decouples the CFD and CSD sets of equations and uses projection methods to transfer interface information between the domains. We adopted the latter method as it enables using pre-existing, well-established codes. The two modules are managed by a 'controller' code, which transfers information between the solvers [3]. This code handles non-matching meshes at the interface and incorporates conservative interpolation schemes and fast techniques for neighbor search. It also manages the time synchronization between the CFD and CSD solvers via a leap-frog approach.

2 Numerical Results

Weapon Fragmentation Study. The coupled technology has been applied to the simulation of the detonation and fragmentation of an experimental weapon. The bomb hangs tip-down at the center of a reinforced concrete room, as if it would have penetrated through the roof. The thick-walled steel weapon is top (i.e., base) ignited. The detonation front propagates from the base to the tip at the C-J detonation velocity, as prescribed in the detonation model.

The structural response (case expansion) is modeled using GA-DYNA [4]. Several CSD meshes of this weapon were tested, using either 8-node hexahedral elements or brick-like parallelepipedal elements, and varying the number of elements from 748 to 8228. The results presented here were obtained with 748 brick elements with a single element across the thickness of the casing. The fragment size distribution for the present simulation is prescribed based on arena test results. A more accurate procedure is described below.

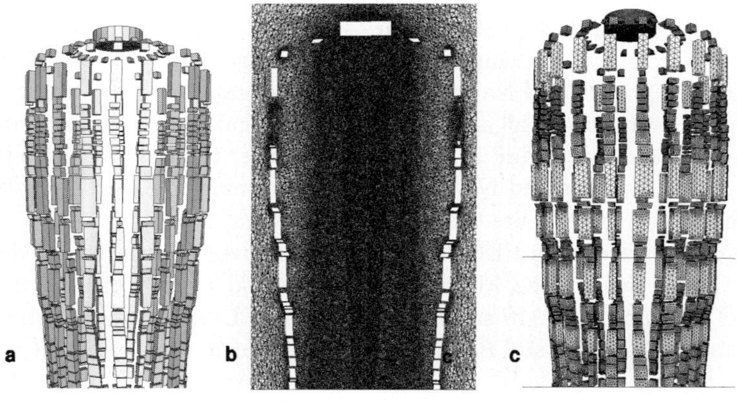

Fig. 1. (**a**) CSD mesh, (**b**) CFD mesh on a plane cut, and (**c**) CFD mesh on the surface of the weapon, $t = 0.550$ ms

After ignition, as the detonation wave propagates from the base to the tip, the high- pressure detonation products force the case to expand. The structural elements fail once the element average strain exceeds 100%. The strain criterion for failure is computed at the center of each element. Each failing fragment is then treated as a separate rigid body for which the trajectory is computed using a 6 DOF integrator linked to the contact algorithms. Once bricks fail, fluid elements are introduced into the narrow gaps separating the fragments.

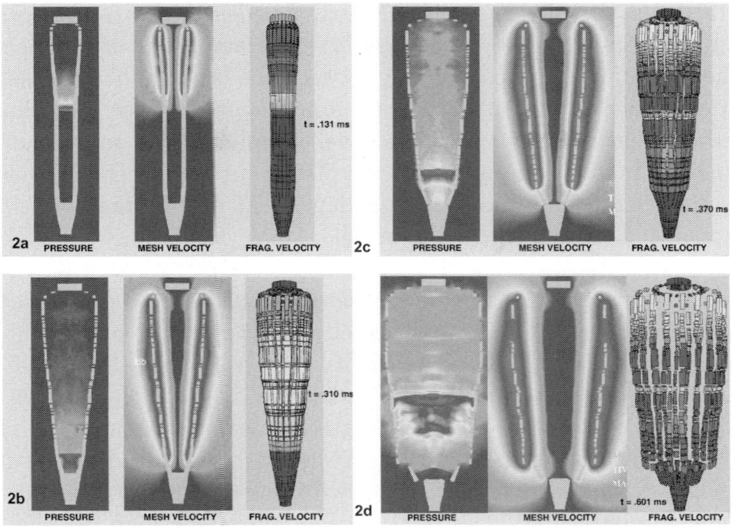

Fig. 2. Propagation of detonation wave and case fragmentation. Results show pressure, mesh velocity and fragment velocity at times 131 ?s, 310 ?s, 370 ?s, and 600 ?s, respectively

The gaps are of the order of a millimeter, which would result in unacceptable small fluid elements and small integration time step. The gap size was increased by shrinking the fragments uniformly around the center of gravity. The topology change due to the breakup requires remeshing of at least part of the domain, a CPU intensive process that is allowed to occur only every 5-8 μs. Thus, the CSD code maintains a list of failed elements, and shrinks them only when allowed.

One important aspect of this class of simulations is the large size disparity between the critical length scales. After fragment break-up, the gap between fragments is several **millimeters**. In contrast, the average fragment length is about ten **centimeters**, and the room length is of the order of ten **meters**. This forced us to attach to each flying fragment entities called grid sources, which enforce local pre-specified element size, ensuring a uniform, high-resolution mesh about each fragment, and thereby reduce the number of local and global remeshings. Figures 1a through 1c show the CSD surface mesh, the CFD mesh on a planar cut through the weapon, and the CFD surface mesh, respectively, at 550 μs. The results show the finely-resolved mesh within the initial HE zone, resulting from the application of a grid source placed along the centerline of the weapon. This fine mesh ensures sufficient resolution of the detonation front, and hence, the correct pressure rise and detonation propagation velocity. On the weapon we form a much finer CFD mesh than the CSD mesh, with rapid increase of mesh size with distance from the centerline. The range-of-influence of the centerline-placed source was specified to produce a fine-resolution mesh not just within the HE zone, but also around the complete volume in which fragmentation occurs. Thus, it guarantees that when the CSD elements fail and fragments are formed, there will be a sufficiently large number of CFD elements within the newly formed gaps opening between the breaking fragments (Fig 1b). As the fragments exit the fine-resolution zone imposed by the centerline source, CFD mesh resolution is reduced to the level specified by the sources attached to the fragment. Finally, examination of the planar cut results (Fig 1b) shows that the mesh size dictated by the sources is dependent on the fragment size as well as the face size. Sources attached to smaller faces yield a finer mesh than those attached to larger faces.

Figures 2a through 2d show a sequence of snapshots. At each time, the panel shows the pressure and CFD mesh velocity contours on a planar cut through the weapon, and the CSD fragment velocity contours. Figure 2a at 131 μs shows detonation wave propagation down (from base to tip) as a planar front, and the radial expansion of the case. The first fragment break-up occurs at 94 μs, for the upper row attached to the heavy base. While the base itself does not fragment, as it does not expand significantly, the row of elements below fails due to shear, not tension. Similarly, the layer of fragments above the nose cone fails due to shear (Fig 2c at 370 μs). Detonation was completed at about $t = 263$ μs and the shock reflected upward (Fig 2b). The relatively small spacing between the expanding fragments ensures that the high-pressure detonation products would be fairly contained within the expanding fragments for an extended period (Fig 2d at t=0.6 ms).

During case expansion, the internal face velocity significantly exceeds the external face velocity, resulting in case thinning. On average, while the CG of the element experiences a 100% strain (break-up criteria), the internal face expands about $145 - 160\%$, compared to about $70 - 80\%$ for the external face. This indicates that during a significant portion of the expansion period, the internal face velocity is about twice the external face velocity. The fragments achieve their terminal velocity within about $120 - 150$ μs after detonation front passage. This is significantly slower then the acceleration period of $60 - 90$ μs for a **serrated** weapon [5]. The final mass- averaged fragment velocity obtained using a strain break-up value of 1.0, was 787.6 m/sec. The experimental measured value was 752 m/sec. To examine the role of the break-up strain value on the final velocity we conducted three more simulations, at break-up strain values of 0.1 ($V_f = 675.53$ m/sec), 0.5 ($V_f = 741.8$ m/sec), and infinity (no break-up, $V_f = 859.7$ m/sec). Hence, the experimental data indicates a break-up strain value of about 0.6, a value that corresponds to an internal expansion of about 90%, and external expansion of about 50%.

A total of five simulations were conducted under this study. The initial mesh for each included about 8.7 million elements, the final one about 19.2 million elements. Once remeshing parallelization was completed, CPU time per simulation was cut to less than a day on a SGI Origin 2000, using 16 processors.

Blast Impact on a Reinforced Concrete Wall. As the next step in the CFD/CSD coupling development effort, we applied the coupled methodology to the simulation of airblast interaction with reinforced concrete wall. The model included two rooms, but only the connecting wall was modeled with the CSD code. All other structural components were treated as rigid. The CFD domain consisted of 9,387 boundary points, 55,281 points and 296,751 elements. The wall included 81,101 nodes, 69,048 solid hexahedron elements in the concrete and 598 beam points in the steel rebars. While the CFD solution was non-adapted, three levels of mesh adaptation [6] were used in the CSD model. The standard DYNA3D element erosion model was used to eliminate failed elements.

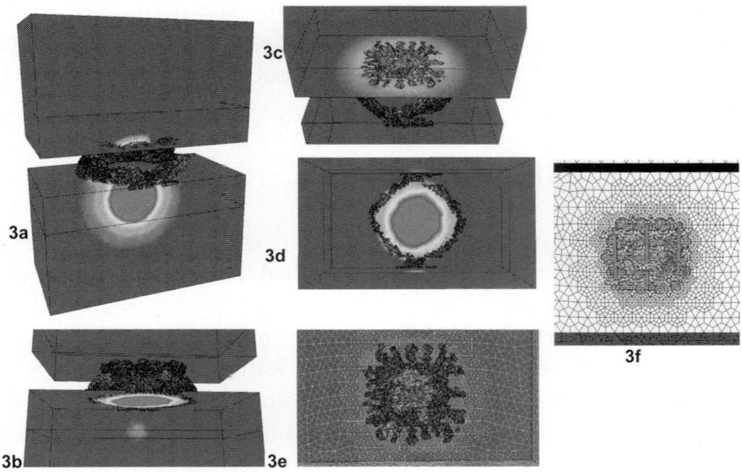

Fig. 3. Results after 400 integration steps: (**a**) Pressure contours on the plane of symmetry superimposed on the CSD mesh velocity; (**b**) A view from below, the blast chamber; (**c**) Top view; (**d**) The adapted CSD mesh showing crown damage in the blast chamber; (**e**) CSD mesh adaptation to the spalation web in the second chamber; and (**f**) The exposed rebars are visible through the damaged and removed concrete

Several new schemes were employed here: 1) A new algorithm to strictly enforce a minimum separation between fragments. Current available contact algorithms are based on penalty methods, where repulsive forces are applied when collision is detected, with the force proportional to the penetration distance. These methods do not ensure a penetration-free contact. We developed a new contact algorithm based purely on geometrical information [4], enforcing a strict separation distance; 2) A new crack propagation model [6] that takes advantage of the CSD H-refinement scheme. As the crack propagates through the material, mesh adaptation is used to ensure the accuracy of the stress wave propagation, and the accurate agglomeration of the elements into discrete fragments. The new model was validated against data for two tests and should reduce the need for expensive arena test data.; 3) The adaptation procedure ensures that each fragment contains several elements. As the elements fail and fragments are formed, each is treated by GA-DYNA as an independent body, with the appropriate volume, mass, momentum and energy. GA-DYNA then keeps track of fragment-to-fragment and fragment-to-wall interactions through a recently-parallellized procedure. GA-DYNA transfers the information to FEFLO98, which treats every fragment as a sphere, allowing for accurate momentum and energy exchange; and 4) A new model that allows rebar data to be interpolated from enclosing elements. The previous model required all nodes to be on the rebar itself.

Figure 3 shows several snapshots taken after 400 time steps. The results show the typical damaged concrete pattern: a crown in the blast room and spallation web on the opposite side. Notice the complex connectivity through the concrete that allows the high pressure to propagate from the peripheral crown to the centered spall zone. Three levels of mesh adaptation were used, enabling accurate prediction of the spallation, crack propagation, element failure and fragment formation.

3 Summary and Conclusions

A recently developed loose-coupling algorithm that combines state-of-the-art CFD and CSD methodologies has been applied to the simulation of weapon detonation and fragmentation, and reinforced concrete wall response to blast loading. Several new routines were developed to enable accurate communication between the codes, especially during case (or wall) fragmentation. The results demonstrate the ability of the coupled methodology to handle the many controlling complex physical phenomena and yield results that are in good agreement with experimental data. While other techniques model weapon fragmentation, the present approach also yields, in addition to the fragment size and velocity distribution, an accurate description of the airblast environment.

Acknowledgements. This research effort was supported by the Defense Threat Reduction Agency. Dr. Michael E. Giltrud, served as the contract technical monitor.

References

1. R. Löhner, K. Morgan, J. Peraire and M. Vahdati - Finite Element Flux- Corrected Transport (FEM-FCT) for the Euler and Navier-Stokes Equations; Int. J. Num. Meth. Fluids 7, 1093-1109 (1987).
2. R.G. Whirley and J.O. Hallquist - DYNA3D, A Nonlinear Explicit, Three- Dimensional Finite Element Code for Solid and Structural Mechanics - User Manual; UCRL-MA-107254 (1991), also Comp. Meth. Appl. Mech. Eng. 33, 725- 757 (1982).
3. J.R. Cebral and R. Löhner - Conservative Load Transfer for Fluid-Structure- Thermal Simulations; Proc. 4th WCCM, Buenos Aires, Argentina, July (1998).
4. D. Pelessone and C.M. Charman - A General Formulation of a Contact Algorithm with Node/Face and Edge/Edge Contacts; 1998 ASME Pressure Vessels and Piping Conference, San Diego, Ca, July (1998).
5. J.D. Baum, H. Luo, R. Löhner, D. Pelessone and C. Charman - A Coupled Fluid/Structure Modeling of Shock Interaction with a Truck; AIAA-96-0795 (1996).
6. D. Pelessone, C.M. Charman, R. Löhner and J.D. Baum A new Crack Propagation Algorithm for modeling Weapon Fragmentation; in preparation (2000).

KFMG based Euler Formulation for Computational Aeroelasticity

R. Krishnamurthy[1], B.S. Sarma[1], and S.M. Deshpande[2]

[1] Defence Research & Developement Laboratory, Hyderabad, India
[2] CFD Centre, Indian Institute of Science, Bangalore, India

Abstract. A two-dimensional computational aeroelastic code called KFMG-CA, based on Kinetic Flux Vector Splitting scheme on moving grids with suitable structural model, has been developed to study the unsteady and aeroelastic behaviour of 2-D airfoils and applied for NACA 64A006 airfoil undergoing pitch and plunge motions and the aeroelastic response characteristics are presented.

1 Introduction

Aerospace vehicles, generally, experience time-dependent aerodynamic forces due to flow unsteadiness and these unsteady forces play an important role in determining the structural behaviour of flight vehicles and at times leading to aeroelastic instability. Computation of unsteady aerodynamic forces on oscillating bodies and their interaction with structural dynamics constitute the rapidly progressing interesting field of computational aeroelasticity. Ballhaus and Goorjian [1] have developed LTRAN2 code to obtain unsteady aerodynamic forces for airfoils and Guruswamy[2] has extended the formulations to predict aeroelastic response. Batina[3] has developed an upwind Euler solver based on van Leer flux vector splitting scheme, to operate on dynamic meshes for predicting the unsteady flow around oscillating airfoils. Wu[4] has carried out Euler/N-S based unsteady and aeroelastic computations for an airfoil in transonic flow. In the present work, an Euler solver based on Deshpande's Kinetic Flux Vector Splitting (KFVS) scheme, which was proved to be robust in handling complex geometries in different flow regimes, has been further developed for moving grid applications. Because of the robustness of KFVS based Euler formulations, it is very tempting to extend this method to moving grids so that problems in unsteady aerodynamics can also be solved with this extended kinetic method called KFMG. Also, a two d.o.f. structural model involving pitch and plunge motions is developed. The aeroelastic code has been developed by integrating this KFMG based Euler solver with this structural model. The steady and unsteady computations for NACA0012 and NACA 64A006 airfoils oscillating in transonic flow and the aeroelastic computations for NACA 64A006 airfoil are presented here.

2 Kinetic Flux vector splitting scheme on Moving Grids (KFMG)

Different upwind schemes are in use to solve Euler equations and in the present study, Kinetic Flux Vector Splitting (KFVS) scheme on stationary grids developed by moving grids. The KFVS solver essentially addresses Boltzmann equations of Kinetic theory of gases and is transformed into Euler equations through a moment method strategy. Extensive studies are carried out using 2-D and 3-D formulations of KFVS scheme in different flow regimes [6,7]. The 1-D KFMG solver based on KFVS scheme developed by Kulkarni and Deshpande[8] revealed that the KFVS scheme has inherent simplicity, especially when applied to moving grid problems. The simplicity stems for the fact that the molecular velocity term present in the governing equations at the Boltzmann level is replaced by the term involving the difference between the molecular velocity and grid velocity [9]. The above aspect has been exploited in deriving the fluxes for moving grids. The details are presented in Ref.[10]. It is interesting to note that, the split fluxes $G^{\pm}_{k,new}$ corresponding to moving grids can be cast in terms of the split fluxes $G^{\pm}_{k,old}$ corresponding to stationary grids and the same can be seen from. We begin now the search for periodic solutions to Hamiltonian systems. All this will be done in the convex case; that is, we shall study the boundary-value problem.

$$G^{\pm}_{k,new}(\rho,\bar{u},\beta)=\begin{bmatrix} G^{(1)\pm}_{k,old}(\rho,\bar{u}_{nk},\beta) \\ G^{(2)\pm}_{k,old}(\rho,\bar{u}_{nk},\beta)+w_1G^{(1)\pm}_{k,old}(\rho,\bar{u}_{nk},\beta) \\ G^{(3)\pm}_{k,old}(\rho,\bar{u}_{nk},\beta)+w_2G^{(1)\pm}_{k,old}(\rho,\bar{u}_{nk},\beta) \\ G^{(4)\pm}_{k,old}(\rho,\bar{u}_{nk},\beta)+\frac{w_1^2+w_2^2}{2}G^{(1)\pm}_{k,old}(\rho,\bar{u}_{nk},\beta) \\ +\,w_1G^{(2)\pm}_{k,old}(\rho,\bar{u}_{nk},\beta)+w_2G^{(3)\pm}_{k,old}(\rho,\bar{u}_{nk},\beta) \end{bmatrix}$$

where, w_1 and w_2 represent the grid velocities. The grid velocity w_1 and w_2 at each grid point during the (n+1) time step are calculated using the location of the point in n and (n+1) time steps.

The wall boundary conditions are treated as a local 1-D piston undergoing expansion and compression as prescribed by Deshpande and Kulkarni[8]. Considering that the velocity normal to the moving boundary alone will contribute towards fluxes, the flux normal to the moving boundary has been derived.

3 Dynamic Mesh Algorithm

Between two successive time steps, the body changes its orientation in case of oscillation and hence the grid has to be moved or regenerated around the deformed body. Regeneration of grid at each time step is not only laborious but also expensive with respect to CPU time of the computer. Hence, a moving grid strategy given by J.T.Batina[3] based on 'spring analogy' is implemented to move the grid so as to get a good quality grid around the oscillating body.

The mesh is moved to conform to the instantaneous position of the body by modeling each edge of each quadrilateral by a spring. The spring stiffness for a given edge i-j is taken to be inversely proportional to the length of the edge. Grid points on the outer boundary are held fixed, and the instantaneous locations of the points on the inner boundary (body) are prescribed. At each time step, the static equilibrium equations in the x and y directions are solved iteratively at each interior node i of the grid for the displacements δ_{x_i} and δ_{y_i} . The spring analogy method is coded to obtain the grid around the new orientation of the oscillating airfoil at every time level.

4 Structural Dynamics Model

A two degrees of freedom structural dynamics model involving pitch and plunge motions of an airfoil has been developed. The definitions of variables and sign conventions have been taken from Ref.[4]. It is assumed that the airfoil is rigid and the amplitude of oscillation is small. Considering the inertia forces, damping forces, elastic forces and aerodynamic forces, the equations of motion of the system can be written as,

$$[M] \begin{bmatrix} \xi'' \\ \alpha'' \end{bmatrix} + [C] \begin{bmatrix} \xi' \\ \alpha' \end{bmatrix} + [K] \begin{bmatrix} \xi \\ \alpha \end{bmatrix} = \frac{4}{\pi \mu k_c^2} \begin{bmatrix} -C_L \\ 2C_M \end{bmatrix}$$

where [M],[C] and [K] are the mass, damping and stiffness matrices respectively, k_c is the reduced frquency, μ is the airfoil-to-air mass ratio, ξ is the non dimensionalised plunging displacement and α is the pitch angle. The structural equations of motion are solved using linear acceleration method. The KFMG based Euler solver has been coupled with structural dynamics model to integrate both aerodynamic and structural equations simultaneously for obtaining the aeroelastic response.

5 Results and Discussions

The capabilities of the dynamic mesh algorithm are demonstrated by considering the grid around NACA 0012 airfoil. The grid is generated by elliptic grid generator ELGRID- TM based on Thomas-Middelcoff procedure [11]. In this example[10], the airfoil is pitched about the quarter chord with an amplitude of 10° and the mesh at a pitch angle of -10° and $+10^o$. The mesh moves smoothly as the airfoil pitches, and the procedure is completely general in the sense that it can treat any realistic oscillating airfoil motions. The formulations of KFVS based Euler solver on dynamically moving meshes, are demonstrated by performing the calculations on NACA 0012 airfoil with grid size 101 x 41. The outer boundary has been chosen as a circle with radius equal to 20 times the chord of the airfoil. Initially, steady state calculations were performed on the airfoil at a free stream Mach number of $M_\infty = 0.85$ and angle of attack of $\alpha = 1^o$. Reconstruction of fluxes for obtaining space accuracy and fourth order

Runge-Kutta for time accuracy have been implemented in the solver. The results obtained from the above solver for $M_\infty = 0.755$ and $\alpha = 0.016^o$ have been taken as the input to perform the unsteady calculations for the airfoil pitching harmonically about the quarter chord with an amplitude of $\alpha_o = 2.51^o$ and a reduced frequency based on semi-chord of k = 0.0814 at $M_\infty = 0.755$ and mean angle of $\alpha_M = 0.016^o$. The instantaneous angle of attack has been calculated based on the following expression:

$$\alpha(t) = \alpha_o + \alpha_M \, sin(\omega t)$$

Unsteady pressure oscillations were obtained for the pitching NACA 0012 airfoil using global time step and fourth order Runge-Kutta (RK-4) method for time accuracy and second order scheme via reconstruction of fluxes for space accuracy. Calculated lift and moment coefficients during third cycle of motion compare well with experimental data available in literature.

Similar calculations were performed on NACA 64A006 airfoil. Initially, steady state analysis has been performed for this airfoil for different free stream Mach numbers M_∞ and $\alpha = 0^o$. The steady pressure distribution along the chord of the airfoil, corresponding to a typical free stream Mach number of 0.85 is shown Fig.1 along with experimental pressure coefficients. The results obtained from this solver for $M_\infty = 0.85$

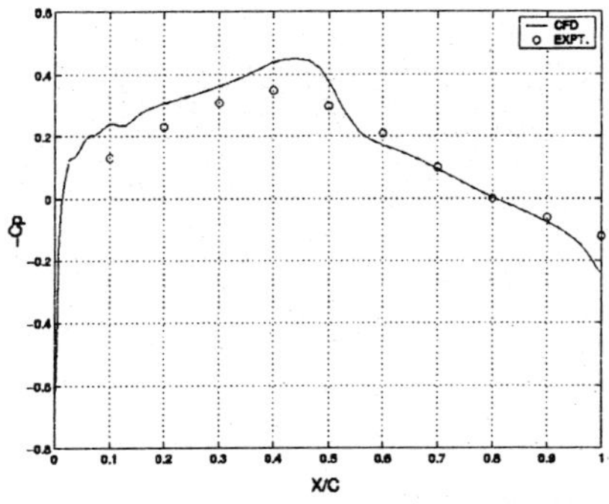

Fig. 1. Pressure distribution along the chord

The results obtained from this solver for $M_\infty = 0.85$ and $\alpha = 0^o$ have been taken as the input data to perform unsteady calculations for the airfoil which oscillates harmonically about the quarter chord point with an amplitude of $\alpha_o = 0.01$ rad. and a reduced frequency based on chord of k = 0.1 at $M_\infty = 0.85$ and the mean angle of attack $\alpha_M = 0^o$. The time evolution of the lift coefficient is plotted with non-dimensional time along with comparison in Fig.2.

For this airfoil also, the grid size of 101 x 41 has been chosen with outer boundary as a circle with radius equal to 20 times the chord of the airfoil, to perform steady and unsteady calculations. For both NACA 0012 and NACA 64A006 airfoils, the steady and unsteady aerodynamic coefficients compare well with corresponding results available in literature.

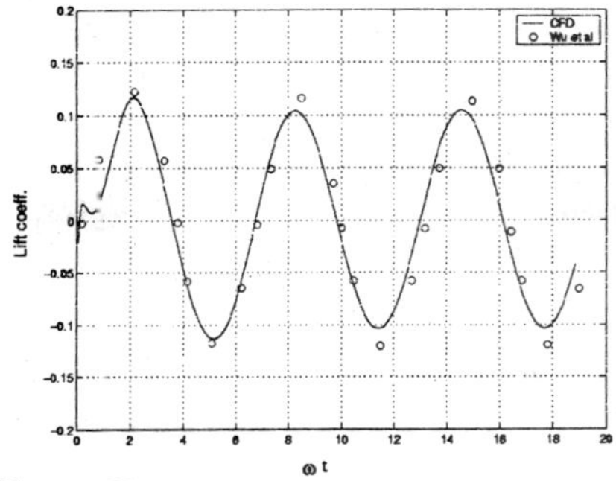

Fig. 2. Unsteady lift coefficient with time

At the end of the third cycle of forced sinusoidal oscillations, the airfoil was released and allowed to follow pitch and plunging motions dictated by the structural dynamics equations. The structural parameters used in the present study are: $\frac{\omega_h}{\omega_\alpha} = 0.2$, $a_h = -0.5$, $x_\alpha = 0.25$, $U^* = 5.5$, $k_c = 0.1$, and the same parameters have been used by Wu et al[4] in a similar study. The value of airfoil-to-air mass ratio(μ) has been parametrically varied from 110 to 150, during this phase of the calculations to study the effects of the airfoil-to-air mass ratio on the flutter characteristics. As shown in Fig. 3, it was found that variations in the airfoil-to-air mass ratio can lead to stable damped oscillations, neutrally stable oscillations, or divergent (flutter) oscillations.

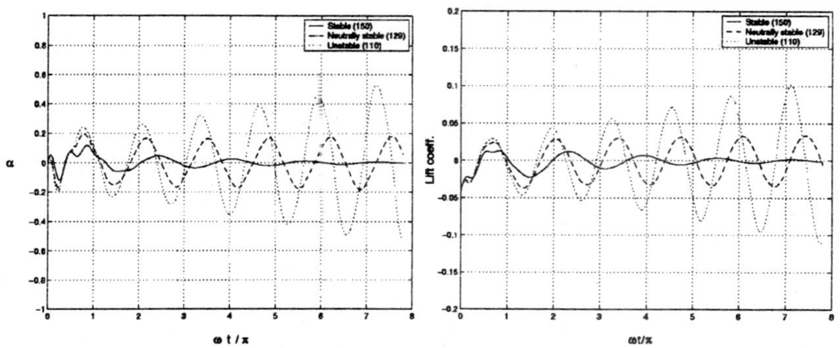

(a) Pitch angle with time (b) Lift coefficient with time

Fig. 3. Aeroelastic response characteristics

6 Conclusions

A two-dimensional Euler solver based on Kinetic Flux Vector Splitting on moving grids (KFMG) integrated with a structural model to predict unsteady flows and aeroelastic response characteristics is developed. The solver has been validated against NACA 0012 airfoil oscillating in transonic flow and unsteady aerodynamic coefficients predicted. Also, the solver has been validated for NACA 64A006 airfoil, by performing both unsteady and aeroelastic calculations, for transonic free stream Mach number. It has been found that, the results compare very well with that available in literature. The accuracy has been improved by implementing fourth order Runge-Kutta method for time accuracy and method of reconstruction of fluxes for space accuracy. Thus, the effectiveness of Kinetic scheme based Euler computations on moving grid to perform computational aeroelastic studies is successfully demonstrated.

Acknowledgements

The first two authors express their gratitude to Director, DRDL for giving encouragement and support during the course of this work. The first author thanks Head, CFD Division,DRDL for the useful discussions while pursuing this work.

References

1. W.F. Ballhaus, P.M. Goorjian: *Implicit Finite Difference Computations of Unsteady Transonic Flows about Airfoils'*. In : AIAA Journal, vol.15, 1977.
2. P.Guru Guruswamy : *'Unsteady Aerodynamic and Aeroelastic Calculations for Wings using Euler Equations'*. In : AIAA Journal vol.28, March 1990.
3. J.T.Batina : *'Unsteady Euler Airfoil Solutions Using Unstructured Dynamic Meshes'*. In : AIAA Journal, Vol.28 Aug. 1990.
4. J.Wu, K.R.V.Kaza, and L.N.Sankar : *'Technique for the Prediction of Airfoil Flutter Characteristics in Separated Flow'*. In : J. of aircraft, vol.26, 1989.
5. S.M.Deshpande : *'Kinetic Theory Based New Upwind Methods for Inviscid Compressible Flows'*. In : AIAA paper 86-0275.
6. J.S.Mathur and S.M.Deshpande : *'Reconstruction on Unstructured Grids Using Kinetic Flux Vector Splitting Method'*. In : NAL Report PD CF 9601
7. S.M.Deshpande, S.Sekar, M.Nagarathinam, R.Krishnamurthy, P.K.Sinha and P.S.Kulkarni : *'Three-Dimensional Upwind Euler Solver using Kinetic Flux Vector Splitting Method (KFVS)'*. In : Lecture Notes in Physics No. 414, Springer-verlag 1992.
8. P.S.Kulkarni and S.M.Deshpande : *'KFVS on Moving Grid for Unsteady Aerodynamics'*. In : Proceedings of Third ECCOMAS Conference, Paris, 1996.
9. R.Krishnamurthy and S.M.Deshpande : *'Differential and Integral Forms of Euler Equations on Moving Grids'*. In : Report No. 97, FM 1, Indian Institute of Science, Bangalore, India (1997)
10. R.Krishnamurthy,B.S.Sarma and S.M.Deshpande : *'KFMG : Kinetic Flux Vector Splitting Scheme on Moving Grids'*. In : Proceedings of Third Asian Conference on Computational Fluid Dynamics,1998
11. R.Krishnamurthy, K.Anandhanarayanan and M.Nagarathinam : *'Elliptic Grid Generation Around Flight Vehicle Configurations'*. In : Proceedings of fourth ECCOMAS Conference, Athens, Greece, 1998.

A Unified Wall Boundary Treatment for Viscous and Inviscid Flows in the CE/SE Method

S.-C. Chang[1], Z.-C. Zhang[2], S.T. John Yu[2], P.C.E. Jorgenson[1]

[1] NASA Glenn Research Center, Cleveland, OH 44135
[2] Mechanical Engineering Department, Wayne State University, Detroit, MI 48202

Abstract. In the setting of the CE/SE method, a new and unified wall boundary treatment for the Navier-Stokes and Euler Equations is proposed. In essence, the shear stress exerted on the fluid by a wall is modeled as a source term as a part of local space-time flux conservation in the vicinity of a wall boundary. When the fluid is inviscid, the source term vanishes and the boundary condition reduces to the usual "slip" condition. On the other hand, when the fluid is viscous, the source-term effect is consistent with the traditional no-slip condition. Numerical results show that the new treatment is robust, efficient and accurate for viscous and inviscid flows.

1 Introduction

In conventional CFD, the slip condition is applied along the wall for inviscid flows, while no-slip condition is applied for viscous flows. In this paper, a unified wall boundary treatment for inviscid and viscous flows is proposed in the setting of the CE/SE method [1-2]. In essence, the shear stress exerted on the fluid by a wall is modeled as a source term as a part of local space-time flux conservation in the vicinity of a wall boundary. When the fluid is inviscid, the source term vanishes and the boundary condition reduces to the usual "slip" condition. On the other hand, when the fluid is viscous, the source-term effect is consistent with the traditional no-slip condition. The accuracy, efficiency and robustness of the new boundary treatment is demonstrated using numerical examples in which reflection of an oblique shock on a wall is calculated for both viscous and inviscid flows.

2 The CE/SE Scheme for 2D Viscous Flows

The 2D Navier-Stokes Equations in conservation form can be written as

$$\frac{\partial u_m}{\partial t} + \frac{\partial f_{im}}{\partial x} + \frac{\partial g_{im}}{\partial y} - \frac{\partial f_{vm}}{\partial x} - \frac{\partial g_{vm}}{\partial y} = 0, \quad m = 1,2,3,4 \tag{1}$$

The column matrices formed by u_m, f_{im}, g_{im}, f_{vm}, g_{vm}, $m = 1,2,3,4$, respectively, are:

$$U = \begin{bmatrix} \rho \\ \rho u \\ \rho v \\ E_t \end{bmatrix} \quad F_i = \begin{bmatrix} \rho u \\ \rho u^2 + p \\ \rho uv \\ (E_t + p)\, u \end{bmatrix} \quad G_i = \begin{bmatrix} \rho v \\ \rho uv \\ \rho v^2 + p \\ (E_t + p)\, v \end{bmatrix}$$

$$F_v = \begin{bmatrix} 0 \\ \tau_{xx} \\ \tau_{xy} \\ \tau_{xx}u + \tau_{xy}v - q_x \end{bmatrix} \quad G_v = \begin{bmatrix} 0 \\ \tau_{xy} \\ \tau_{yy} \\ \tau_{xy}u + \tau_{yy}v - q_y \end{bmatrix} \tag{2}$$

The inviscid fluxes f_{im}, g_{im} are functions of u_m, while the viscous fluxes f_{vm} and g_{vm} are functions of $u_m, \partial u_m/\partial x$, and $\partial u_m/\partial y$. Let $x_1 = x, x_2 = y$ and $x_3 = t$ be the coordinates of a Euclidean space E_3. Then the integral form of Eq. (1) is

$$\oint_{S(V)} \boldsymbol{h_m} \cdot d\boldsymbol{s} = 0, \quad m = 1,2,3,4 \tag{3}$$

where (i) $\boldsymbol{h_m}$=(f_{im}-f_{vm}, g_{im}-g_{vm}, u_m), (ii) $S(V)$ is the boundary of a region V in E_3, and (iii) $d\boldsymbol{s} = d\sigma\boldsymbol{n}$ with $d\sigma$ and $\boldsymbol{n}$, respectively, being the area and outward unit normal of a surface element on $S(V)$. Note that $\boldsymbol{h_m}$ can be decomposed into the inviscid and viscous parts, i.e.,

$$\boldsymbol{h}_m = \boldsymbol{h}_{im} - \boldsymbol{h}_{vm} \tag{4}$$

where $\boldsymbol{h}_{im} = (f_{im}, g_{im}, u_m)$, and $\boldsymbol{h}_{vm} = (f_{vm}, g_{vm}, 0)$.

In the 2D CE/SE method [2], the space-time mesh used (Fig.1(a)) is constructed from a spatial triangle mesh (Fig.1(b)). The unknowns are stored at the space-time mesh points (marked by squares and circles in Fig.1(a)) with their spatial projections being the centroids of the triangles depicted in Fig.1(b). Each space-time mesh point is associated with three conservation elements (CEs) and one solution element (SE). As an example, at point G', the associated CEs are cylinders $E'F'G'D'EFGD$ (CE$^{(1)}$), $A'B'G'F'ABGF$ (CE$^{(2)}$) and $C'D'G'B'CD GB$ (CE$^{(3)}$). The SE is the union of four planes $A'B'C'D'E'F'$, $GG''B''B$, $GG''D''D$ and $GG''F''F$, and their immediate neighborhood. As explained in [2], inside each SE (such as that associated with point G'), u_m, f_{im}, g_{im} are approximated by first-order Taylor's expansions in x, y and t with the expansion coefficients being the functions of the independent unknowns $(u_m)_{G'}$, $(u_{mx})_{G'}$ and $(u_{my})_{G'}$. By imposing numerical analogues of Eq. (3) over CEs and using other conditions, a family of CE/SE Euler solvers, referred to as the Euler a-ε-α-β schemes, is constructed in [2]. Since the current Navier-Stokes solver is an extension of the Euler a-ε-α-β scheme with additional treatments of the viscous terms, only these additional treatments will be discussed in this paper.

To proceed, note that the third (time) component of each vanishes. Thus, in calculating the total viscous flux leaving a CE (such as CE$^{(2)}$) through its boundary, we only need to evaluate the viscous fluxes leaving the CE through its four lateral surfaces $ABA'B'$, $BGB'G'$, $GFG'F'$ and $FAF'A'$. To expain how they are evaluated, as an example, we consider the surface $FAF'A'$ (Fig.1(c)). Let $\Delta\boldsymbol{S} = (S_x, S_y, 0)$ be the unit outward normal vector of $FAF'A'$ multiplied by its area. Then

$$\int_{FAF'A'} \boldsymbol{h}_{vm} \cdot d\boldsymbol{s} \approx S_x f_{vm}\left((u_m)_Q, (u_{mx})_Q, (u_{my})_Q\right) +$$

$$S_y g_{vm}\left((u_m)_Q, (u_{mx})_Q, (u_{my})_Q\right) \tag{5}$$

where Q denotes the centroid of $FAF'A'$. Because $FAF'A'$ belongs to the SE of point A, we assume that

$$(u_m)_Q \approx (u_m)_A + (u_{mx})_A(x_Q - x_A) + (u_{my})_A(y_Q - y_A) + (u_{mt})_A \Delta t/4 \tag{6}$$

Note that, as explained in [2], $(u_{mt})_A$ is a function of the independent unknowns $(u_m)_A$, $(u_{mx})_A$ and $(u_{my})_A$. Furthermore, it is assumed that

$$(u_{mx})_Q \approx (u_{mx})_A, \quad (u_{my})_Q \approx (u_{my})_A \tag{7}$$

Since the above scheme is constructed using a triangular mesh, it is compatible with unstructured meshes. In addition, the above scheme has been extended to become 3D NS solvers, which however is out of the scope of this paper.

3 Unified Wall Boundary Treatment

For simplicity, consider the horizontal wall depicted in Fig. 2(a). No grid point is placed on the wall. Instead, a ghost point E, which is the mirror image of point G with respect to the wall, is introduced. The flow variables u_m and their spatial derivatives, u_{mx} and $u_{my} (m = 1,2,3,4)$, at point E are obtained from those of point G by assuming that, at any time t, the flow fields below and above $\overline{DF}$ are the mirror images of each other, i.e., we have

$$\begin{cases} (u_m)_E = (u_m)_G, & m = 1,2,4 \\ (u_m)_E = -(u_m)_G, & m = 3 \end{cases} \qquad \begin{cases} (u_{mx})_E = (u_{mx})_G, & m = 1,2,4 \\ (u_{mx})_E = -(u_{mx})_G, & m = 3 \end{cases}$$

$$\begin{cases} (u_{my})_E = -(u_{my})_G, & m = 1,2,4 \\ (u_{my})_E = (u_{my})_G, & m = 3 \end{cases} \tag{8}$$

Using the above conditions and the concepts of the dual space-time mesh discussed in [2], one can calculate the flow variables and their spatial derivatives at G for the next time step. The calculation involves an integration over three conservation elements $\mathrm{CE}^{(i)}, i = 1,2,3$ (Fig. 2(b)). However, a special treatment for the calculation involving $\mathrm{CE}^{(1)}$ is needed due to the wall boundary lying across $\mathrm{CE}^{(1)}$.

Let (i) the viscosity μ be a constant; (ii) the wall be an insulated wall; (iii) S be the area of the rectangle $DFF'D'$; (iv) Q_+ and Q_-, respectively, be the points immediately above and below the centoid Q of $DFF'D'$. Then, because (i) $u = v = 0$ at the wall; and (ii) the numerical solution is linear in x, y, and t within a SE, the mass, momentum and energy fluxes entering into the fluid in the triangular cylinder $GDFG'D'F'$ from the wall form the row matrix

$$g_{w+} = S\left(0, -\frac{1}{R_{eL}}\frac{\partial u}{\partial y}, p - \frac{4}{3R_{eL}}\frac{\partial v}{\partial y}, 0\right)_{Q+} \tag{9}$$

On the other hand, the same four fluxes entering into the fluid in the triangular cylinder $EFDE'F'D'$ from the wall form the row matrix

$$g_{w_-} = -S\left(0,\ -\frac{1}{R_{eL}}\frac{\partial u}{\partial y},\ p - \frac{4}{3R_{eL}}\frac{\partial v}{\partial y},\ 0\right)_{Q_-} \tag{10}$$

Using the mirror image conditions, we have

$$\left(\frac{\partial u}{\partial y}\right)_{Q_-} = -\left(\frac{\partial u}{\partial y}\right)_{Q_+},\quad \left(p - \frac{4}{3R_{eL}}\frac{\partial v}{\partial y}\right)_{Q_-} = \left(p - \frac{4}{3R_{eL}}\frac{\partial v}{\partial y}\right)_{Q_+} \tag{11}$$

By using Eqs. (9-11), it is concluded that the mass, momentum, and energy fluxes entering into the fluid in $GDEFG'D'E'F'$, i.e., $CE^{(1)}$, from the wall form the row matrix

$$g_w = g_{w_+} + g_{w_-} = -2S\cdot\left(0,\ \frac{1}{R_{eL}}\frac{\partial u}{\partial y},\ 0,\ 0\right)_{Q_+} \tag{12}$$

In the current treatment, the surviving x-momentum flux in Eq. (12) is treated as a source term in flux balance calculation involving conservation element $CE^{(1)}$. To calculate the flux g_w, we need to calculate $\partial u/\partial y$. For simple laminar flows with enough mesh resolution of the boundary layer, because $u = 0$ at point Q_+, and $u \cong (u_G + u_{G'})/2$ at the midpoint M of $\overline{GG'}$, we assume

$$\left(\frac{\partial u}{\partial y}\right)_{Q_+} \approx \frac{(u_G + u_{G'})/2 - 0}{y_M - y_{Q_+}} \tag{13}$$

Note that, $g_w = 0$ for inviscid flows, and the current treatment becomes the usual "slip" condition. As such, the present boundary treatment is suitable for inviscid as well as viscous flows.

4 Numerical Results

Figure 3 is a schematic of a standard test problem for Navier Stokes solvers. When the shock is strong and the incident shock angle is large enough, boundary layer separation occurs at the shock impinging point. It is assumed that $M_\infty = 2.0, R_e = 2.96\times10^5$ and $\beta = 32.6^o$. The computational domain is $[-1, 1.7]\times[0, 1.2]$. Clustered cells near the solid wall are employed to resolve the boundary layer. Here 28800 isosceles triangles are used.

The new solid boundary treatment is applied along the wall. In the left and upper boundaries, the flow conditions are fixed according to the incoming flows. Non-reflective boundary condition is used in the outlet boundary. The numerical results are shown in Fig. 4 for pressure contours, pressure distribution and skin friction distribution along the wall. The negative skin friction coefficient represents the recirculating flow region. The numerical results, including the separation location and length, compare favorably with the experimental data

[3]. The second example is an inviscid shock reflection problem [4]. This problem has exact Euler solution. We apply the same wall boundary condition treatment with $1/R_{eL} \longrightarrow 0$. A uniform mesh of 9600 conservation elements is used. The numerical results are shown in Fig.5. These results agree very well with the analytical solutions.

5 Concluding Remarks

A unified wall boundary treatment for viscous and inviscid flows has been described. In this treatment, the wall boundary condition for inviscid flows becomes a special case of that for viscous flows. Numerical results show that this new boundary treatment is robust, efficient, and very accurate for viscous as well as inviscid flows. This new treatment could also be easily extended to model complex flow physics along wall, e.g., flow permeation through a porous wall.

References

[1] S.C. Chang: "The Method of Space-Time Conservation Element and Solution Element - A New Approach for Solving the Navier Stokes and Euler Equations," J. Comp. Phys., **119** , pp. 295-324 (1995)

[2] S.C. Chang, X.Y. Wang and C.Y. Chow: "The Space-Time Conservation Element and Solution Element Method: A New High-Resolution and Genuinely Multidimensional Paradigm for Solving Conservation Laws," J. Comp. Phys., **156**, pp.89-136 (1999)

[3] R.J. Hakkinen, I. Greber, L. Trilling and S.S. Abarbanel: "The Interaction of an Oblique Shock Wave with a Laminar Boundary Layer," NASA Memo 2-18-59W (1959)

[4] Yee, H.C., Warming, R.F. and Harten, A., "Implicit Total Variation Diminishing(TVD) Schemes for Steady-State Calculations," AIAA paper 83-1902 (1983)

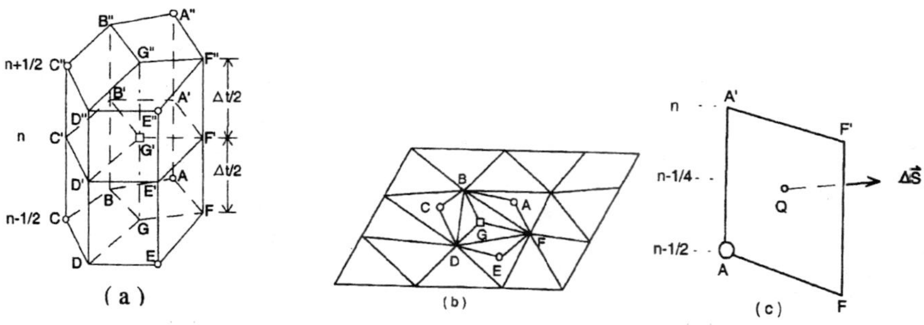

Fig. 1. A schematic of the CE/SE scheme: (a) a space-time mesh; (b) a triangular spatial mesh; (c) calculation of viscous fluxes.

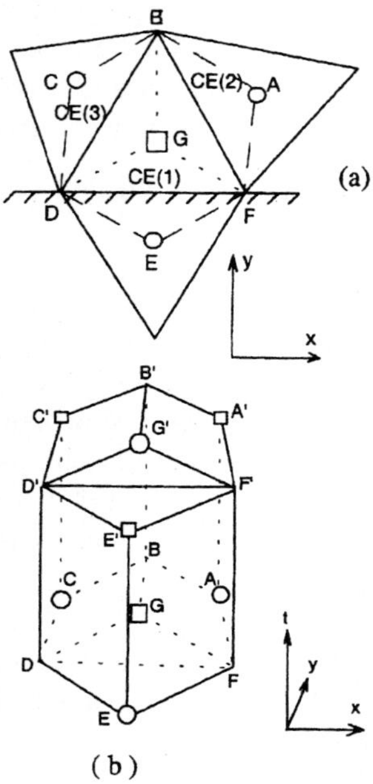

Fig. 2. A schematic for wall boundary treatment: (a) a spatial mesh and (b) CEs near the wall boundary.

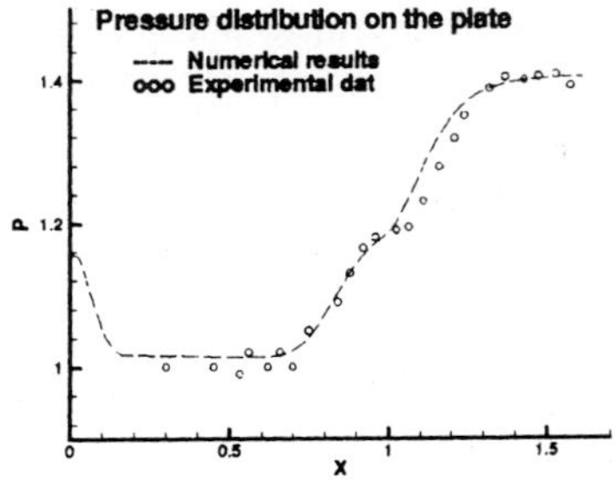

(b) pressure distribution along the wall.

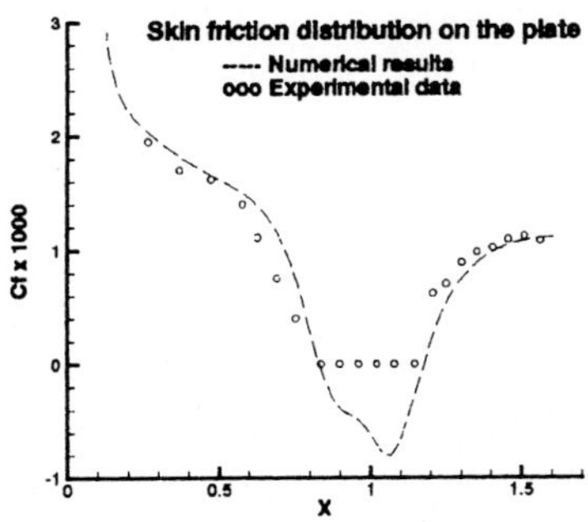

(c) skin friction along the wall.

Fig. 4. Numerical and experimental results of the shock/boundary layer interaction.

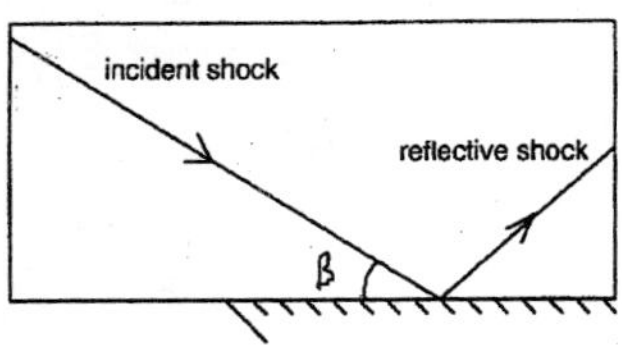

Fig. 3. Shock/boundary layer interaction problem.

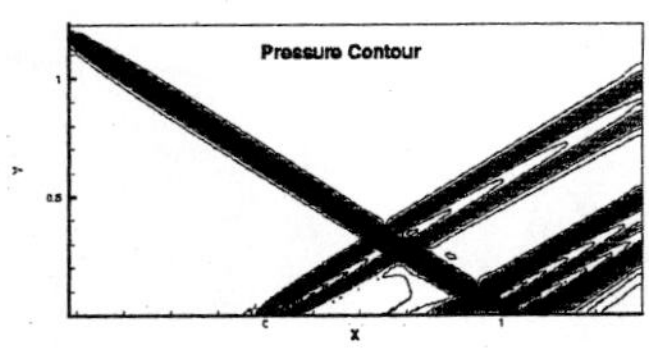

(a) pressure contours.

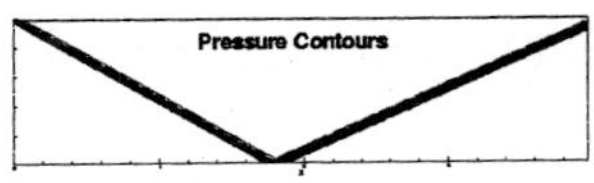

(a) pressure contours.

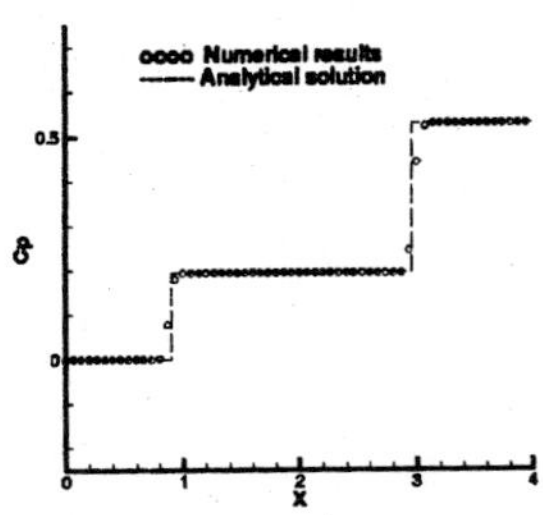

(b) pressure coefficients at $y = 0.5$.

Fig. 5. Euler solutions of the oblique shock reflection problem.

Part XX

Optimization and Design

An Improved Aerodynamic Inverse Design Method for Complex SST Configuration Combined with 3D NS Solver

Kisa Matsushima[1] and Tshiyuki Iwamiya[2]

[1] FUJITSU Ltd., 1-9-3 Nakase Mihama-ku Chiba 261-8588, JAPAN
[2] National Aerospace Laboratory, Tokyo, JAPAN

1 Introduction

A next generation Supersonic Transport (SST) is under research in Japan. The National Aerospace Laboratory (NAL) has been developing several experimental models of the SST since 1995. The project aims to reduce drag by aerodynamic shape design of a wing. The aerodynamic design of the first model which has no propulsion system was over in 1998. The full configuration model with propulsion system is currently being designed with the research for a better design method.

To design the desirable shape of wings for those models, we have developed a design method utilizing CFD [1]. It is an inverse design method based on Takanashi's residual correction concept [2] using an inverse problem solver and Navier-Stokes (N-S) simulation. The method successfully worked for the first SST experimental model which is the wing-fuselage configuration [3]. The designed wing showed desirable performance in N-S flows. The next challenge is to design a wing for a full configuration model of the SST. In order to handle the complicated flow physics attributed to the complex configuration, the inverse design method should be improved. In this article, the improvement of the method and the application of the improved method to the complex airplane are presented.

2 Aerodynamic Design Method

The design procedure for supersonic wings is iterative method. Fig.1 illustrates the procedure. The method determines the wing section's geometry which realizes a specified target pressure distribution at all span stations of a wing. First a baseline shape is to be guessed. Then the flow field around the wing is analyzed by flow simulation to get the current Cp distribution on the wing surface. Next the inverse problem is solved to obtain the geometrical correction value Δf corresponding to the difference between target and current pressure distributions ΔCp. Next, the flow analysis is conducted again to see if the current shape realizes target pressure distribution. If the difference between target and current pressure distributions is negligible, the design is completed. Otherwise, the next step is once again to solve the inverse problem and repeat the design loop. There are two primary parts; one is a flow analysis, and the other is a design part where the inverse problem is solved. Both parts may be completely

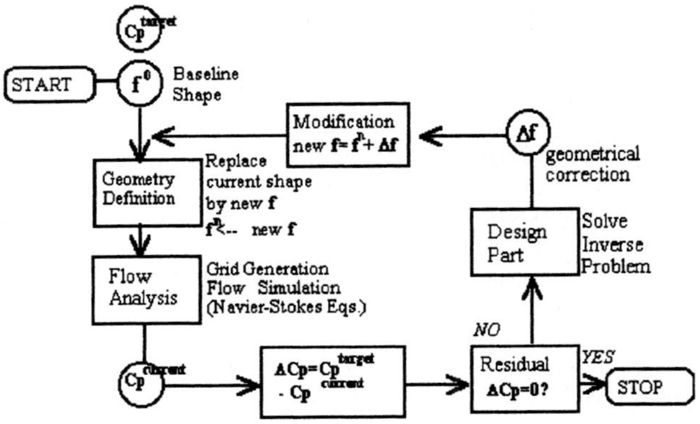

Fig. 1. Design method of residual-correction concept

independent from each other in terms of their algorithms and equations. With the method, the accuracy of a design result depends on the analysis part. The flow analysis of the present design method is done by a Navier-Stokes simulation code [4]. So, designed geometry by the method are valid for Navier-Stokes flows. The efficiency of the design method depends on the design part of the inverse problem. Thus, the simple mathematical model should be selected for the formulation of it. On the other hand, the model should be as exact as possible, which would conflict with the simplicity. The formulation will be discussed later.

3 Complex configuration and flow phenomena

As the first step of the design process, N-S simulation [4] for the flow about the baseline shape of the full configuration SST model is conducted at $M_\infty = 2.0$. The results are shown in Fig. 2. It presents pressure distributions on the complex geometry of the lower surface of the SST model with propulsion system. The locations of three span-stations are also indicated. The right-hand side plot shows the pressure distribution on the lower surface of the wing as well as the location of diverters. From those figures, we learn that the combination of a wing and an engine-nacelle makes complicated flow phenomena. Interaction between them is so strong that it violates the premise of the existing residual-correction method we developed in 1997-1998.

With the inverse problem of the method, it is assumed that every physical phenomenon is mild in the flowfield, which means that entropy does not obviously change there. In order to satisfy this assumption, the boundary condition (ΔCp) for the inverse problem solver is to be reviewed. We also assume that the change of flow physics due to the correction of the wing geometry is small so that we can neglect interaction between the wing and the other components

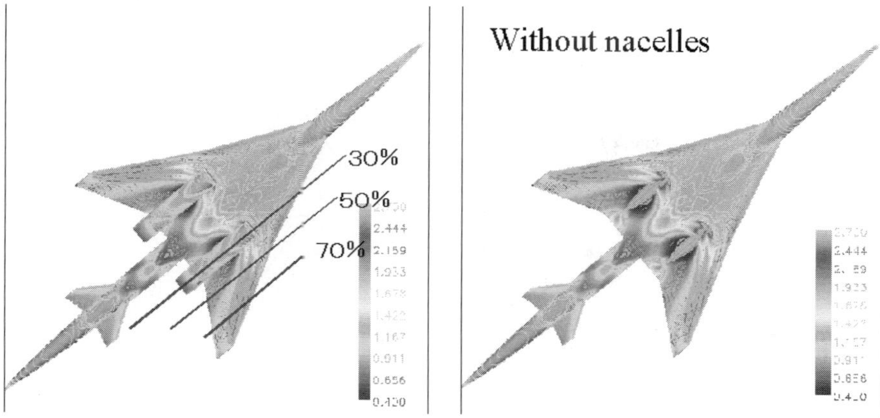

Fig. 2. Pressure distribution on the lower surface

when solving the inverse problem. But, for the design of this configuration case, it is desirable to take the effect into account.

4 Improvements of the Design Method

4.1 Modification on Boundary Condition for the Inverse Problem

The inverse problem is formulated using Green's theorem [3], then the functions appearing in the formulation should be C^2 continuous. To hold this requirement with the complicated physics of the wing-nacelle interaction, the smoothing of profile shape of the current Cp distribution is performed. In addition to it, the relaxation is made with ΔCp at several span-stations.

4.2 Inverse Problem Including Interacting Effect

The inverse problem is re-formulated so as to consider the interacting effect. The supersonic flowfield considered here is indicated in Fig. 3. The x-axis is streamwise, the y-axis is spanwise A wing and the upper portion of a nacelle are located at the $z = 0$ and $z = b$ plane; $\tau_\pm$ and $\tau_{b\pm}$ denote upper(+) or lower(-) surface of the wing and the nacelle portion, respectively. For the formulation, the small perturbation theory is applied as the previous one. The basic equation of the improved version of the inverse problem is

$$\Delta\phi(x,y,z) = -\frac{1}{2\pi}\int\int_{\tau_+}\frac{1}{r_c}\Big[\Delta\phi_\zeta(\xi,\eta,+0) - \Delta\phi_\zeta(\xi,\eta,-0)\Big]d\xi d\eta$$

$$+\frac{1}{2\pi}\int\int_{\tau_+}\frac{z(x-\xi)}{[(y-\eta)^2+z^2]r_c}\big(\Delta\phi_\xi(\xi,\eta,+0) - \Delta\phi_\xi(\xi,\eta,-0)\big)d\xi d\eta$$

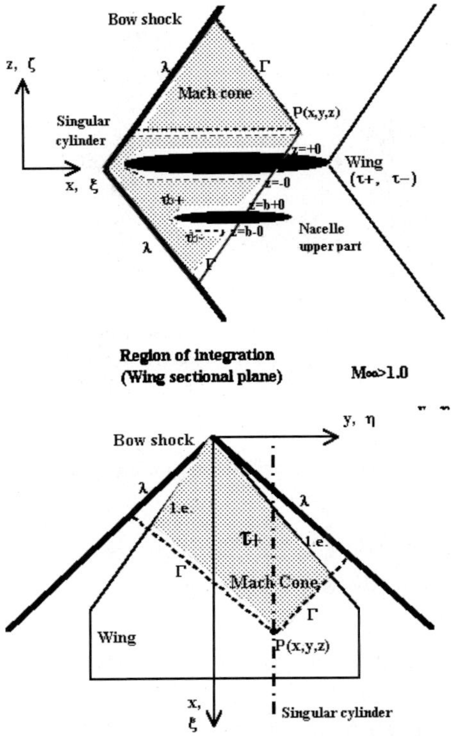

Fig. 3. Coordinates of supersonic flowfield about wing-nacelle system

$$-\frac{1}{2\pi}\int\int_{\tau_{b+}}\frac{1}{r_b}\Big[\Delta\phi_\zeta(\xi,\eta,b+0)-\Delta\phi_\zeta(\xi,\eta,b-0)\Big]d\xi d\eta$$

$$+\frac{1}{2\pi}\int\int_{\tau_{b+}}\frac{z_b(x-\xi)}{[(y-\eta)^2+z_b^2]r_b}\big(\Delta\phi_\xi(\xi,\eta,b+0)-\Delta\phi_\xi(\xi,\eta,b-0)\big)d\xi d\eta \qquad (1)$$

where

$$r_c=\sqrt{(x-\xi)^2-(y-\eta)^2-z^2},\quad r_b=\sqrt{(x-\xi)^2-(y-\eta)^2-z_b^2},\quad z_b=z-b$$

ϕ is the small perturbation potential of the flowfield. Δ means the difference between two flowfields such as a target and a beseline ones. The third and fourth integral terms in Eq. (1) are representing the interacting effect by the nacelle.

According to the thin wing theory, $\Delta\phi_\xi$ is related to the pressure difference and $\Delta\phi_\zeta$ implies the geometrical correction of the wing shape. We need to perform more calculus to lead the eventual equations to relate the pressure difference to the geometrical correction directly.

5 Design using the Improved Method

The method modified by the improvements proposed in Section 4 has been applied to the wing design of the full configuration model. We have been trying two strategies. One is to modify the boundary condition for the inverse problem. It is called ‘strategy-1‘. The other is to use the improved inverse problem solver together with the ‘strategy-1‘. It is called ‘strategy-2‘. The initial states of the surface Cp distributions are shown as the diamond lines in Fig. 4. The + lines indicate the target Cp. The design process at two span stations is presented here,

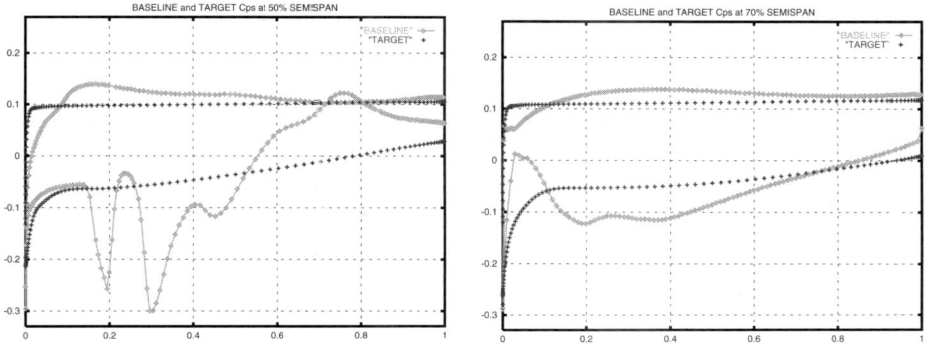

Fig. 4. Initial states at 50% and 70% semi-spans

which are 50% and 70% semi-span stations. At the 50% semi-span station, the interacting effect by the nacelle is strong, while it is weak at the 70% semi-span station. Figures 5 and 6 show the design results at the 50% and 70% semi-span stations, respectively. In each figure, the left-hand side plot is the resulting Cp distribution (square lines) at the fourth design cycle of the design method using strategy-1. The target Cps (+ lines) are also shown in every plot. The right-hand side plot is the resulting Cp distribution (square lines) at the first cycle of the method using strategy-2. Both strategies work. Especially, the improvement with the strategy-2 is distinguished.

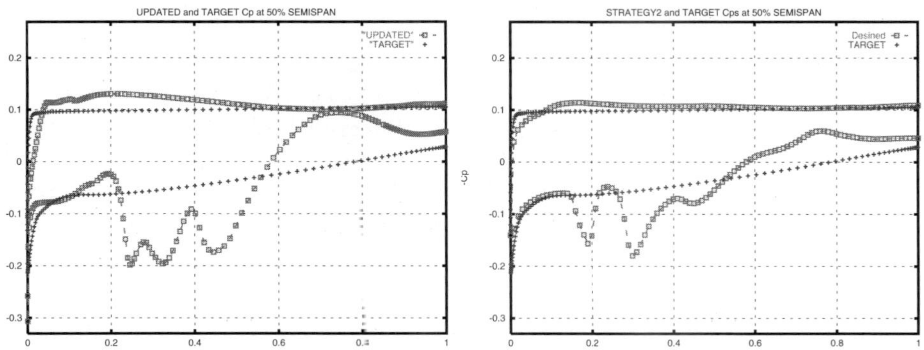

Fig. 5. Design results of 4th design cycle using strategy-1 or 1st cycle using strategy-2 at 50% semi-span

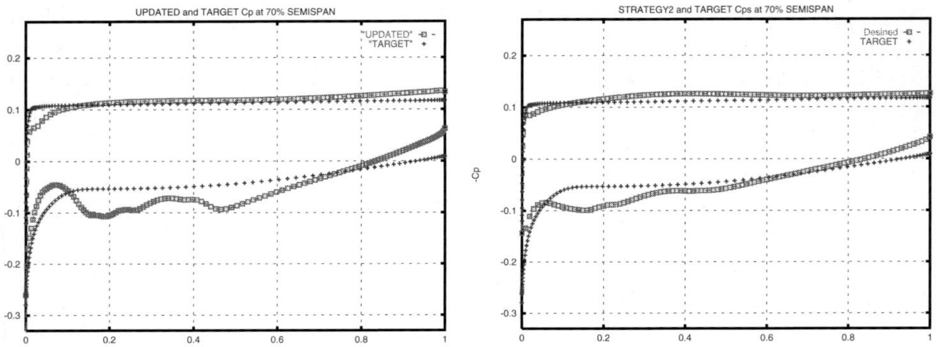

Fig. 6. Design results of 4th design cycle using strategy-1 or 1st cycle using strategy-2 at 70% semi-span

6 Conclusion

An inverse design method which was formerly devised for a wing alone/wing-fuselage system has been applied to wing design for full configuration model of Japanese SST. For this complex case, the method should be improved to handle the interaction between a wing and an engine-nacelle. Two strategies are proposed for the inverse problem solver. One is concerning the boundary condition for it and the other is respecting to the mathematical model of the inverse problem. The preliminary design results have shown that the improved method using the strategies is promising.

References

1. S. Jeong, K. Matsushima et al.: 'Inverse Design Method for Wings of Supersonic Transport'. In: *36 th AIAA Aerospace Sciences Meeting, January, 1998*, AIAA Paper 98–0602, 1998
2. S. Takanashi: J. Aircraft **22**, 8 (1985)
3. K. Matsushima, S. Jeong et al.: ' Supersonic Inverse Design Method and Its Application to Japanese SST '. In: *Lecture Note in Physics 515, 16th ICNMFD in Arcachon, France, July, 1998*, ed. by C.-H. Bruneau (Springer, 1998) pp. 74–79
4. R. Takaki, T. Iwamiya, A. Aoki : ' CFD Analysis Applied to the Supersonic Research Airplan '. In: *International CFD workshop for SST Design in Tokyo, Japan, March, 1998*, ed. by H. Aiso et al. (NAL, 1998) pp. 68–72

Non-Generational Multiobjective Evolution Strategy for Aerofoil Design and Optimisation Problems in CFD

E.J. Whitney and K. Srinivas

Department of Aeronautical Engineering, The University of Sydney, NSW 2006, Australia. Email: eric@aero.usyd.edu.au, ragh@aero.usyd.edu.au

Abstract. This paper presents a new method for the multiobjective design and optimisation of two dimensional aerofoils. The method utilises a modified Evolution Strategy (ES) called a Non-Generational MultiObjective Evolution Strategy (NGMOES). Fitness evaluations are performed by a linear vortex panel method with integral boundary layer operating in parallel. The result of an example run is presented at the end of this paper.

1 Introduction

After having established its capability to compute complex fluid flow, CFD is playing an increasing role in multidisciplinary programs such as aerofoil shape optimisation. In this work, we explore the possibility of utilising simple CFD codes to perform shape optimisation using evolution strategies. This paper describes the optimisation carried out for a typical aerofoil design problem.

Multiobjective problems in aerodynamics can arise when the designer wishes to satisfy a number of conditions which can not be combined in a straightforward manner, or when the outcome of two or more objectives conflict in their final shape. The method presented is applied as a design tool for aerofoil design and optimisation problems. The approach is in contrast to both inverse methods where the flow variables about an aerofoil are specified, and to perturbation methods where an existing shape is refined using classical deterministic optimisation methods (such as conjugate gradient). While significantly slower in function evaluation terms, the method can be started without any preconceived notion of the aerofoil shape, and will find a set of optimum solutions that satisfy a multiple objective problem. Similar methods using Genetic Algorithms (GAs) have been employed, including those by Fanjoy and Crossley [4], in multiobjectives by Obayashi et al. [7] and in parallel by Jones et al. [6].

The problems tackled so far are in the domain of conventional incompressible low-subsonic aerodynamics. Consequently, a conventional linear vortex panel method with one dimensional integral boundary layer is used to evaluate the performance of each potential solution. Because the simplified panel method evaluates very quickly, the optimisation time can be quite short. The panel method includes a simplistic boundary layer shape factor separation model which gives a first order estimate as to the approximate maximum lift.

A method for the design and optimisation two-dimensional aerofoil sections using Non-Generational MultiObjective Evolution trategies is presented. A technique based on the natural concept of survival of the fittest, evolution strategies work from one population of solutions to the next, and attempt to avoid local sub-optima in the search for the global optimum solution to a stated problem. Population based methods have been very successful in the solution of multi-modal and non-analytic problems.

For this work, the standard evolution strategy has been extended to multiobjective problems by the inclusion of a generation overlap scheme, where only some of the population is replaced at each step, and a Pareto-tournament selection, where individual solutions compete for the right to mate in small sub-populations or tournaments. These two extensions have exhibited excellent convergence on multiobjective test cases where there are many local sub-optima and where the set of solutions is discontinuous.

A typical run is presented for a helicopter rotor aerofoil. The objectives represent drag minimisation at two different c_l values, with constraints on the value of c_m, t/c, $c_{l_{min}}$ and $c_{l_{max}}$ achieved. The method is shown to be capable of converging to a good set of solutions to the problem presented.

2 NGMOES Algorithm

The Non-Generational MultiObjective Evolution Strategy is an extension of the standard evolution strategy (ES) developed by Rechenberg and Schwefel [1,8].

Multiobjective optimisation allows a set of solutions to be evolved answering a problem which has two or more objectives which cannot be combined in a rational way. Based on the idea of Pareto optimality (Pareto ranking), solutions are compared to other solutions, using the notion of *Pareto dominance*. For a minimization problem, a solution $(\boldsymbol{x}^*, \boldsymbol{y}^*)$ is Pareto optimal if it is non-dominated, or if there is no other strategy $(\boldsymbol{x}, \boldsymbol{y}) \in \bar{A} \times \bar{B}$ such that $f_A(\boldsymbol{x}, \boldsymbol{y}) \leq f_A(\boldsymbol{x}^*, \boldsymbol{y}^*)$ and $f_B(\boldsymbol{x}, \boldsymbol{y}) \leq f_B(\boldsymbol{x}^*, \boldsymbol{y}^*)$

The NGMOES algorithm implements this concept when selecting individuals for the next parent population, by using a Pareto tournament selection, a method based on that presented by Horn et al. [5]. Pareto tournament selection takes a small subset of the main population, ranks it and selects the non-dominated individual to be a parent. If there are two or more non-dominated individuals, the best is discriminated by niching, or if no niching is used, it is selected at random. Niching operates by determining a function of the Euclidean distance (in fitness space) to the individuals already selected, and then choosing the individual most distant from them. In this way, a level of diversity is maintained in the population.

The generational scheme, based on one presented by Valenzuela-Rendón and Uresti-Charre [9], allows Pareto optimal sets to be evolved gradually, without needing to replace the entire population at each generation. At each generation offspring individuals are combined with the parent individuals and the entire set is re-ranked, with the poorest individuals allowed to die. A kind of “soft elitism”, the set evolves gradually, without discarding already good solutions which can be used for further comparison.

The NGMOES method has displayed excellent performance on problems such as those presented Deb [2].

3 Fitness Evaluation

Fitness evaluations for the NGMOES are performed in parallel across the Local Area Network. The message passing model PVM (Parallel Virtual Machine) is used to enable communication between machines acting as solver and the machine running the NGMOES.

The code is parallelised using the master - slave model where the evolution computation is performed entirely on one machine (the master), and the aerodynamic work is distributed across the network. One solver (a slave) runs on each machine including the master machine if desired. Remote solvers are assigned new jobs as soon as they become free, minimising the solution time for an entire generation. Faster solvers do more work, therefore the heterogeneous network is self load-balancing.

The genotype (the optimised vector of floating point values) is converted to the phenotype (two dimensional aerofoil) via a mean line and thickness distribution model. The mean line is a Beziér curve with a specified number of control points, as is the thickness distribution. The y/c coordinates of the control points form the design variables, and the x/c coordinates are fixed. The surface points are produced by offsetting the thickness distribution perpendicularly about each side of the mean line. This representation is used (rather than an upper / lower surface model) because it is believed to be closer to the problem at hand, reducing correlation between otherwise unrelated design variables.

The panel code used is a simple two dimensional linear vorticity method, using an integral boundary layer with a simple transition and separation model. This code runs very quickly and thus reduces optimisation times for problems where a more sophisticated method is not needed.

Solutions are rejected as candidates if they return insufficient results, as a byproduct of aerodynamic separation. Because linear interpolation of returned data is used for the fitness function, this ensures a more truthful fitness result.

4 Results and Discussion

A typical result is given for a helicopter aerofoil section design problem. A new aerofoil is designed to replace the old NACA 8-H-12 aerofoil. For this problem, the stated objectives are:

- *Objective One:* Minimise c_d at a c_l of 0.2.
- *Objective Two:* Minimise c_d at a c_l of 1.0.
- *Constraints:*
  - $c_{l_{min}}$ is below -0.1.

- $c_{m\alpha=0}$ is between -0.005 and +0.005.
- $c_{l_{max}}$ is above 1.6.
- $(t/c)_{max}$ is between 12% and 13%
- $(x/c)_{max-thickness}$ is between 15% and 35%.

The Reynolds number was 2×10^6 and 131 panelling nodes were used. Six points on the mean line and ten points on the thickness distribution. The angles of attack analysed for each were $-4°$ to $17°$ in $0.5°$ steps. The NGMOES used a population size of 750, a replacement size of 200 and ran for 1200 generations. The tournament size increased linearly from 2 to 15 over the 1200 generations.

The constraints were added equally to each objective to ensure even distribution of penalties. Both objectives and penalties are scaled by the standard order of magnitude of the property to ensure even results. Both objectives and penalties are linear functions.

The algorithm generated 408168 trial solutions, and took approximately four hours running on a heterogenous network of six machines with speeds ranging from 166Mhz to 400Mhz, with the master running at 300Mhz.

A selection of six aerofoils taken from the Pareto optimal front are shown in Fig. 1.

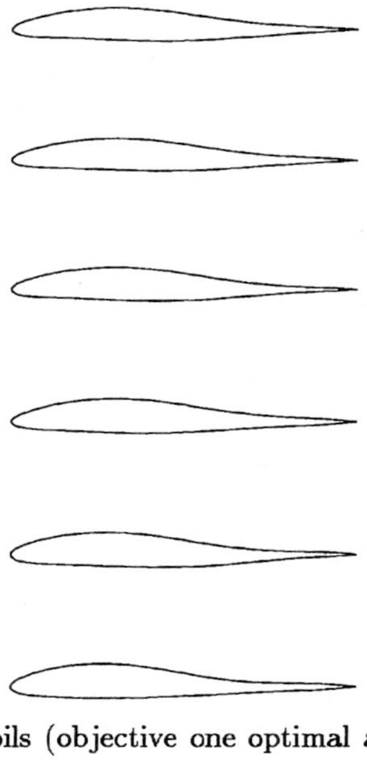

Fig. 1. Pareto optimal aerofoils (objective one optimal at top, objective two optimal at bottom).

The final population (including pareto optimal front) is shown in Fig. 2. The objective one and objective two values are the respective drag coefficients multipled by 10.

A compromise aerofoil shown in Fig. 3 was taken from the Pareto set (aerofoil 4 of 11). The evolved aerofoil (12.82% thick) and NACA 8-H-12 are compared

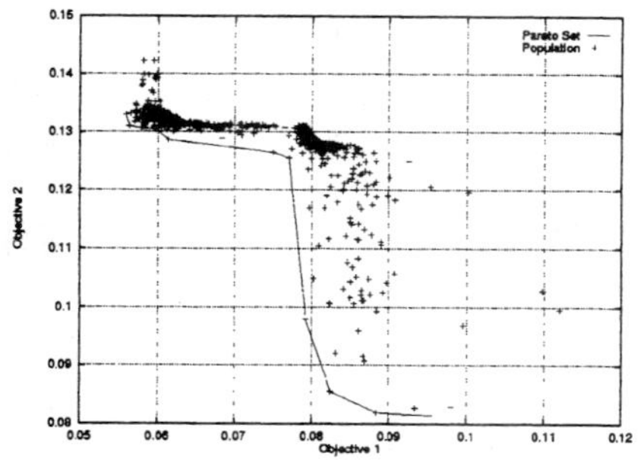

Fig. 2. Final population, and Pareto front indicated by connected points.

quantitatively using MSIS, the two dimensional incompressible Euler solver with coupled boundary layer, written by M. Drela [3]. The aerodynamic properties are shown in Fig. 4 for the smooth condition and Fig. 5 for the rough condition (forced transition at 5% x/c).

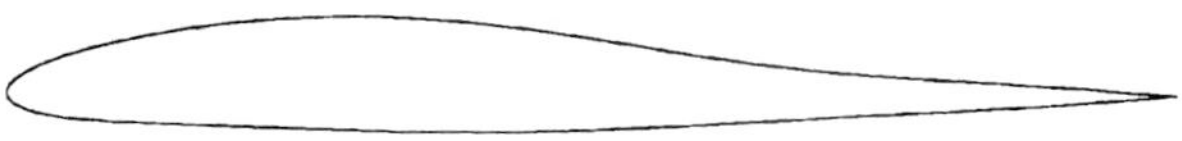

Fig. 3. Selected aerofoil.

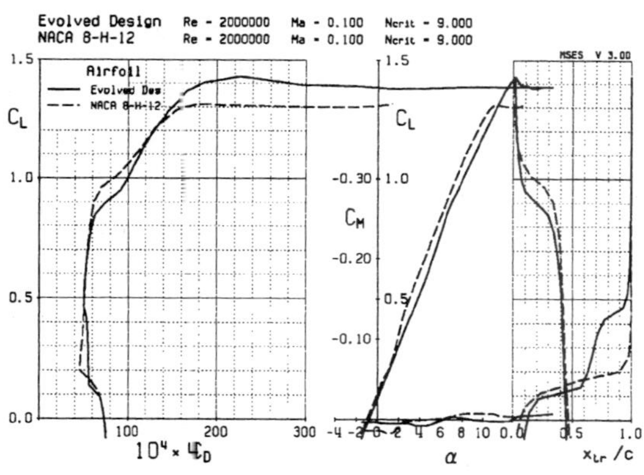

Fig. 4. Selected aerofoil, smooth condition.

The drag properties at normal c_d values compare very well, and are superior at high c_l values. While the predicted $c_{l_{max}}$ does not quite satisfy the give objective, it is superior to the 8-H-12 value. The pitching moments remain practically zero throught the c_l range and are predicted lower than the NACA section. In the rough condition, the c_d properties are similar, except at high c_l values, where they are superior.

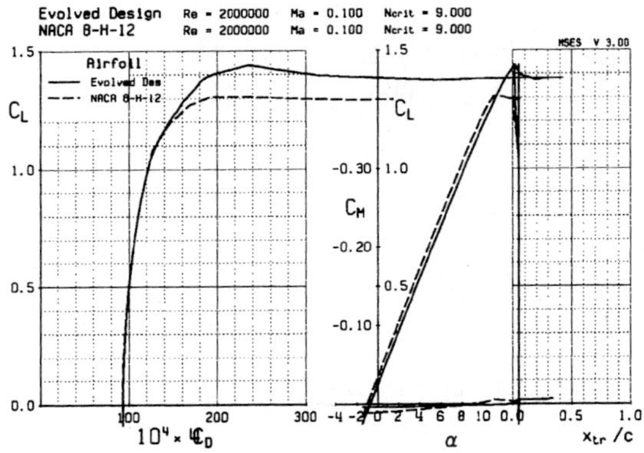

Fig. 5. Selected aerofoil, rough condition.

Further research is being undertaken in the area of Hierarchical Genetic Algorithms (HGAs) for acceleration of the method presented.

References

1. T. Bäck and H.-P. Schwefel. Evolution strategies I: Variants and their computational implementation. In G. Winter, J. Périaux, M. Galán, and P. Cuesta, editors, *Genetic Algorithms in Engineering and Computer Science, Proc. First Short Course EUROGEN-95*, pages 111–126. Wiley, Chichester, Las Palmas de Gran Canaria, Spain, December 4–8, 1995.
2. K. Deb. Multi-objective genetic algorithms: Problem difficulties and construction of test problems. Technical report, Department of Computer Science / LS11, University of Dortmund, 1998.
3. M. Drela. *A User's Guide to MSES 2.95*. MIT Computational Aerospace Sciences Laboratory, September 1996.
4. D. W. Fanjoy and W. A. Crossley. Aerodynamic shape design for rotor airfoils via genetic algorithm. *Journal of the American Helicopter Society*, 43(3):263–270, July 1998.
5. J. Horn, N. Nafpliotis, and D. E. Goldberg. A niched pareto genetic algorithm for multiobjective optimization. Technical report, IlliGAL, 1993.
6. B. R. Jones, W. A. Crossley, and A. S. Lyrintzis. Aerodynamic and aeroacoustic optimization of airfoils via a parallel genetic algorithm. In *7th AIAA/USAF/NASA/ISSMO Symposium on Multidisciplinary Analysis and Optimisation*, 1998.
7. S. Obayashi, T. Tsukahara, and T. Nakamure. Cascade airfoil design by multiobjective genetic algorithms. In *2nd International Conference on Genetic Algorithms in Engineering Systems: Innovations and Applications.*
8. H.-P. Schwefel and T. Bäck. Evolution strategies II: Theoretical aspects. In G. Winter, J. Périaux, M. Galán, and P. Cuesta, editors, *Genetic Algorithms in Engineering and Computer Science, Proc. First Short Course EUROGEN-95*, pages 127–140. Wiley, Chichester, Las Palmas de Gran Canaria, Spain, December 4–8, 1995.
9. M. Valenzuela-Rendón and E. Uresti-Charre. A non-generational genetic algorithm for multiobjective optimization. In T. Bäck, editor, *Seventh International Conference on Genetic Algorithms*, pages 658–665. San Mateo, California, 1997.

CFD Design Study for Modified HOPE-X Aerodynamic Characteristics

Yukimitsu Yamamoto

National Aerospace Laboratory, Tokyo, Japan

Abstract. HOPE-X (H-II Orbiting Plane Experiment) is scheduled for launch in the first decade of 21st century. For several years, systematic aerodynamic study has been conducted by the wind tunnel tests and CFD simulations[1]. Recently, in order to reduce vehicle weight and get high performance, configuration of HOPE-X is re-designed from tip-fin type to two-vertical fin type geometry. For this new aerodynamic configuration design, CFD has been mainly used and Navier-Stokes calculations have been performed from low($M_\infty = 0.4$) to hypersonic speed ($M_\infty = 26$), using NAL NWT (Numerical Wind Tunnel) system[2]. Through these effort of the parametric CFD study, new basic design configuration, HOPE 11 model, is determined. In the present study, this re-designed HOPE 11 aerodynamic characteristics are further investigated in detail from launching to re-entry phase along its flight trajectory. Comparisons with wind tunnel tests are also made.

1 Introduction

HOPE-X (H-II Orbiting Plane Experiment) is scheduled for launch by H-IIA rocket in the first decade of 21st century. HOPE-X is a vertical take off and horizontal landing unmanned re-entry flight demonstrator vehicle. HOPE-11 model, which is the recent re designed model to reduce the dry weight, has twin vertical fin on the leeward fuselage, because previous tip fin model needs a heavy wing structure weight. At the same time, the fuselage length was shortened by 1.5 meters and total length became 13.38 m. The wing reference area was reduced to 50.22 m^2 [3]. Through the re-design of aerodynamic configuration, the vehicle dry mass becomes lower about 1.0 ton than the previous tip fin model. In this paper, numerical aerodynamic study of newly re-designed HOPE 11 model is performed further from low to hypersonic speeds.

First, transonic aerodynamic characteristics are investigated in detail. Systematic analysis are made from subsonic ($M_\infty = 0.4$) to supersonic ($M_\infty = 3.0$) speeds by multi-block Navie-Stokes code. 90 parametric computations have been performed for HOPE 11 model including base flow regions. Sting interference effects of 2.7 % scale wind tunnel model are also investigated.

In this transonic flow study, computations of High Speed Flight Demonstrator II [4] are also made from $M_\infty = 0.8$ to 1.2. Detailed flowfield around ADS (Air Data Sensor) boom extended from the nose are investigated. High Speed Flight Demonstrator vehicle is a 25 % scale model of HOPE-11 and flight experiments will be made in 2002 to confirm transonic aerodynamics.

Second, hypersonic aerodynamic characteristics are investigated. 150 perfect gas computations have been made. At launching phase, study about turbulent heating are carried out at low hypersonic speeds. TPS on the leeward fuselage is not so strong and it is important to predict accurate turbulent heating at launching phase of low angles of attack flight. Comparisons are made with turbulent heat transfer measurements by AEDC hypersonic wind tunnel.

Real gas effects on aerodynamic characteristics, are evaluated, in detail, using 7 species chemically non-equilibrium one temperature N-S code.

2 Numerical method

CFD code developed at NAL for transonic flow analysis is based on Reynolds averaged Navier-Stokes equations with Baldwin-Lomax turbulence model. Numerical approach is the finite volume method using Chakravarthy's TVD scheme[5]. In the hypersonic flow analysis, upwind TVD flux-split method is used for perfect gas flow analysis. Real gas computations are mainly conducted by flux-difference splitting method using 7 species, one temperature, chemically non-equilibrium code[6].

In Fig.1, surface grids of HOPE11 model are shown with detailed dimensions. For this model, 100 points are distributed streamwise from the nose to the fuselage base and 96 points in the circumferential direction.

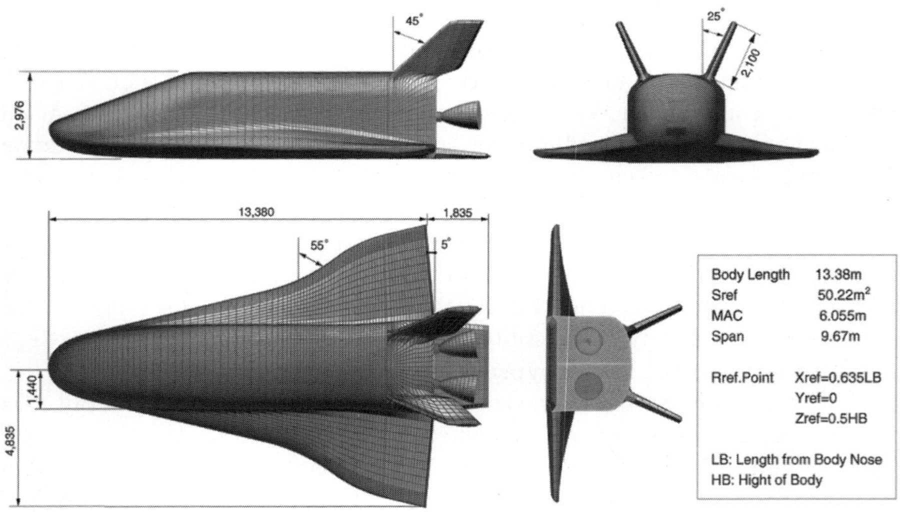

Fig. 1. HOPE-X Basic Configuration

3 Numerical Results

3.1 Transonic Aerodynamic Analysis of HOPE-11 Model with Base Flow Effects

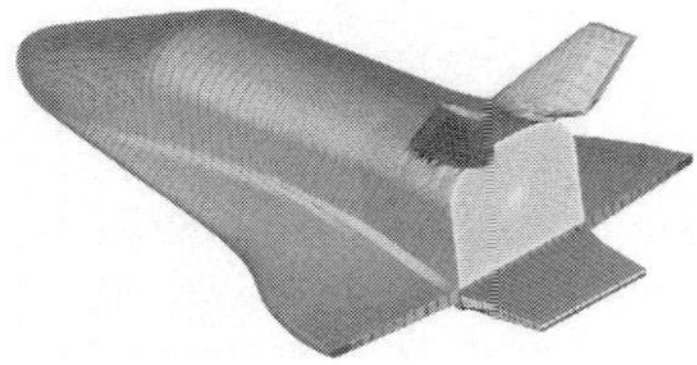

Fig.2

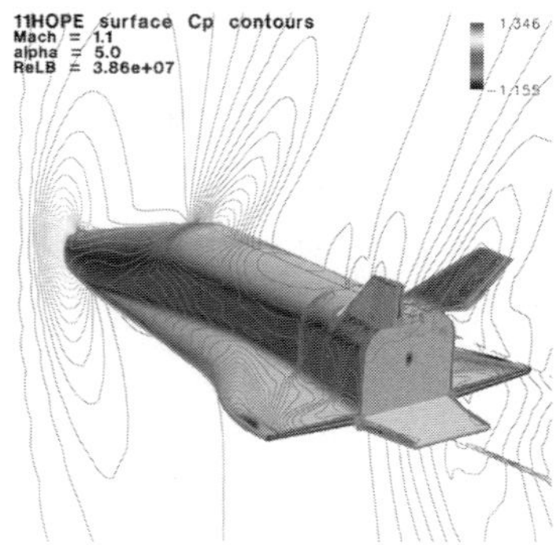

Fig.3

For accurate predictions of transonic aerodynamic performance, base flow effects are exactly estimated by generating multi-block grids. Surface grid are presented in Fig.2. Computations are made at Mach number from 0.4 to 3.0, angles of attack are taken from 0 to 20 degrees with and without side slip effects.

Typical pressure contours are shown in Fig3 at $M_\infty = 1.1$ and $\alpha = 5°$. Sting interference effects are investigated for 2.7% scale HOPE-11 model. Pressure contours are presented in Fig.4. Interference effects on the drag coefficients are shown in Fig.5, where drag with sting is higher than the other cases. Flight model without sting has due to large base pressure effects and lower drag coefficients.

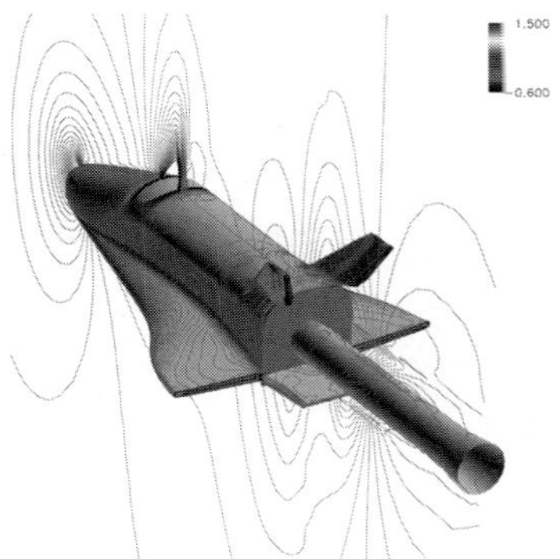

Fig.4

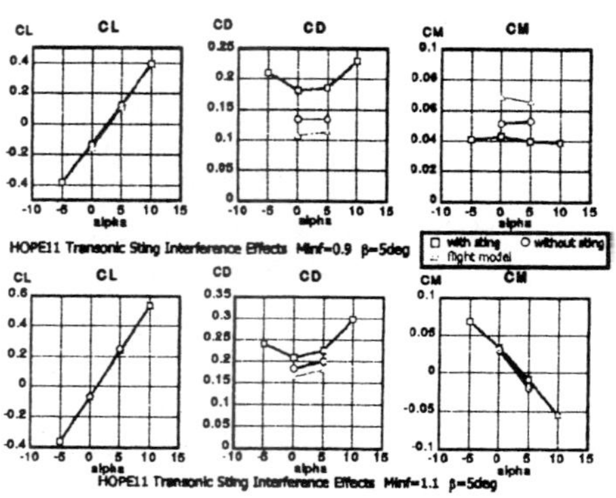

Fig.5

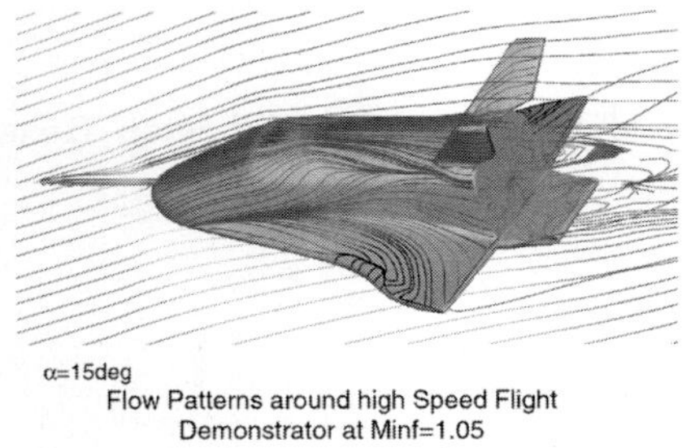

Fig.6

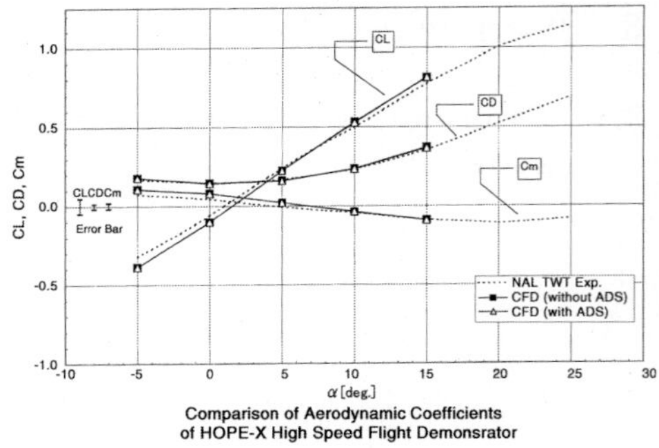

Fig.7

A high-speed flight demonstrator program using 25% scale model of HOPE-11 is planned. For the pre-flight evaluations, analysis of ADS (Air Data Sensor) flowfields are conducted. Figure 6 shows streamline patterns of flight model with ADS. Comparisons with aerodynamic coefficients are presented in Fig.7. It is known that CFD results agrees well with transonic wind tunnel data at NAL and ADS has no influences on total aerodynamic properties. More detailed flow analysis around ADS are also made.

3.2 Hypersonic Aerodynamics and Real Gas Flow Analysis of HOPE-11 Model

It is important to evaluate turbulent heating at launching phase of HOPE-11 flight, because dynamic pressure is high in this flight range and TPS on the leeward fuselage, which is not so thermally strong, may suffer from turbulent heating at low angles of attack. Turbulent heating tests by AEDC HWT are made at $M_\infty = 5.98$. Trip device is attached on the nose surface and turbulent heating is investigated, using 4% scale model.

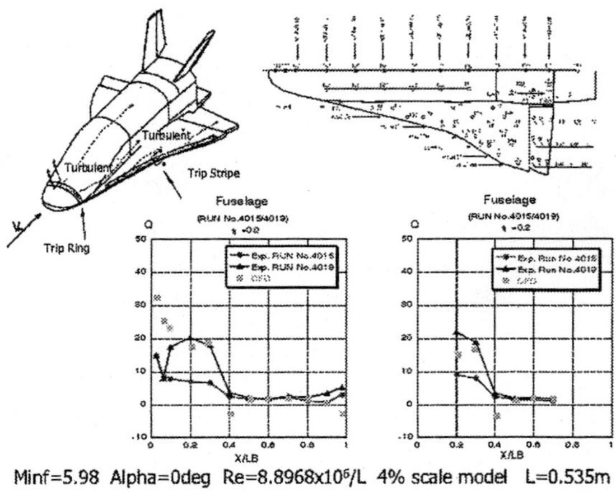

Minf=5.98 Alpha=0deg Re=8.8968x10^6/L 4% scale model L=0.535m

Fig.8

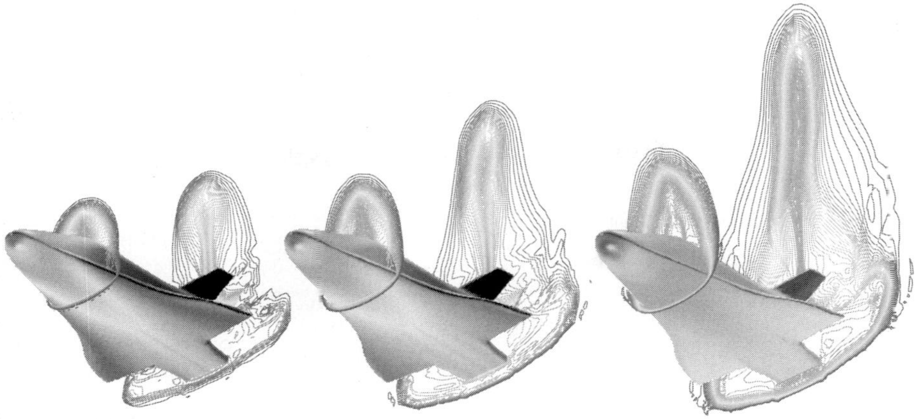

N_2 Mass Fraction
HOPE11 REAL GAS FLOW ANALYSIS
Fig.9

In Fig8, sketch of trip device arrangement and comparisons of heat transfer distributions are plotted on the leeward fuselage. CFD results, using Baldwin Lomax turbulence model, predict turbulent heating very accurately.

Real gas effects have been an important influence on the aerodynamic heating and pitching moment characteristics. These effects are investigated by using chemically non-equilibrium Navie-Stokes code. Computations are made at the same conditions of perfect gas flow analysis.

Mass fraction contours of molecular nitrogen are shown in Fig.9 at $M_\infty = 26$, $\alpha = 30$, 40, and 50 deg. On the nose, mass fraction of N_2 becomes low due to the dissociation caused by high temperature flow behind the bow shock wave. Downstream of this nose region, recombinations of atomic nitrogen occurs and mass fraction of N_2 recovers to free stream value. It is noted that the dissociations of N_2 are not caused so much on the wing leading edge, suffering from high heating due to recompressions and shock-shock interactions.

In order to evaluate aerodynamic characteristics, perfect, frozen and real gas computations are made. In Fig.10, shock wave stand-off distance is depicted at the body end cross section. The differences appears clearly, depending on the each flow models. Shock wave distance is shortened most for the real gas computation. Aerodynamic characteristics such as center of pressure Xcp and pitching moment coefficient Cm are influenced by gas model assumptions. In Fig.11, change of Xcp and Cm, are plotted with Mach numbers at $\alpha = 30$, 40 and 50 deg. Under Mach number of 10, differences is small. However, with the increase of Mach numbers, pitch up phenomena becomes conspicuous by real gas effects. Frozen model has an influence, about half of the total change of pitching moment Cm and center of pressure Xcp at $\alpha = 30$ and 40 deg. However, it is remarked that, at $\alpha = 50$ deg, only chemical reaction effects becomes dominant in aerodynamic change of HOPE 11 model.

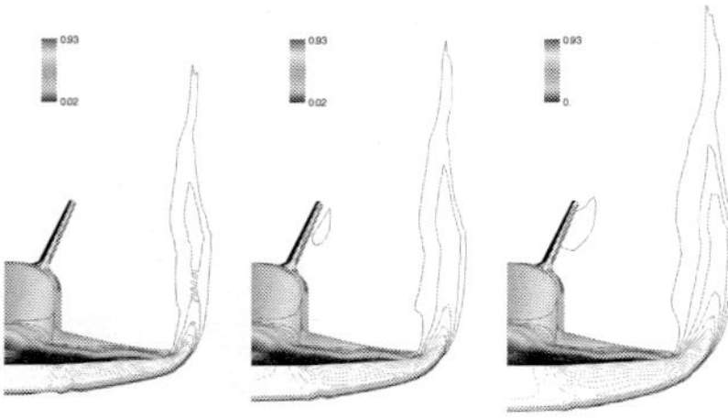

(a) Real Gas (b) Frozen Gas (c) Perfect Gas

Pressure Contours at the Body End Cross Section
(M_∞=26, α=40°)

Fig.10

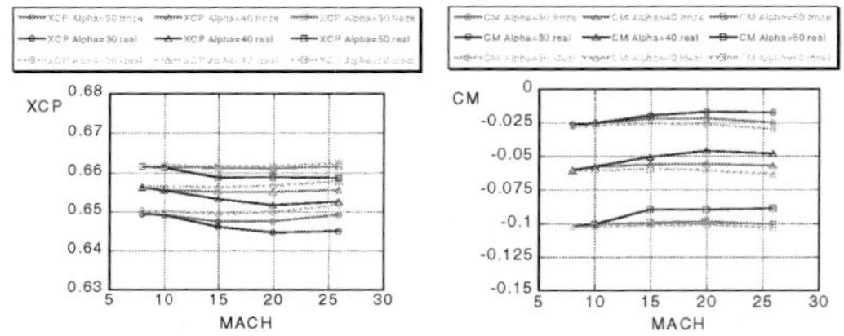

(a) Xcp (b) Cm

Change of Center of Pressure and Pitching Moment of HOPE-11
(Effects of Mach Numbers and Angle of Attack)

Fig.11

4 Conclusions

CFD study of aerodynamic characteristics of HOPE-11 basic re-designed model has been performed. In the present analysis, parametric study is carried out for 340 flow conditions from subsonic to hypersonic speeds along the planned HOPE-X flight trajectory, including launching phase. Real gas effects are studied, by a series of computations using three different gas models. Present CFD simulations contributed greatly to the design work of HOPE-X and the concept of NWT (Numerical Wind Tunnel) at NAL has been almost realized.

References

1. Y.Yamamoto, "Numerical Study of Hypersonic Aerodynamics for Atmospheric Re-entry Problems." ,Proc., of 21st ISTS Symposium, Vol.1, 1998, pp873-880
2. Y.Yamamoto and R.Ito, "Parametric CFD Study for Modified HOPE-X Configurations." Priprint paper, ISTS 2000-e-04, 22nd ISTS Symposium, 2000, Morioka, Japan.
3. S.Ishimoto, et, al., "Modification of HOPE-X Aerodynamics Configuration." Priprint paper, ISTS 2000-e-42, 22nd ISTS Symposium, 2000, Morioka, Japan.
4. M.Yanagihara, et, al., " Simulation Analysis of the HOPE-X Demonstrator.", AIAA Paper99-4875, Nov., 1999
5. Y.Yamamoto and R.Ito, "Multi-block CFD Analysis of Complex Transonic Flow Fields around HOPE-X Configurations." Priprint paper, ISTS 2000-e-02, 22nd ISTS Symposium, 2000, Morioka, Japan.
6. M.Yanagihara, et, al., " Numerical Analysis of Hypersonic Aerodynamics for Atmospheric Re-Entry Problems of HOPE and HYFLEX.", AIAA Paper98-0277, January 12-15, 1998, Reno, Nevada

Transonic Airfoil Optimization Using a Gradient Method

Yasuyoshi Horibata

College of Engineering, Hosei University, 3-7-2, Kajino-cho, Koganei, Tokyo 184-8584, Japan

Abstract. Transonic cascade design is described based on the Navier-Stokes equations. An objective function is defied, and the airfoil shape is optimized by minimizing the objective function using a gradient method. At each search of the method, the gradient is calculated by using the implicit function theorem. An optimization example is presented for a compressor cascade.

1 Introduction

This paper deals with transonic airfoil optimization based on the Navier-Stokes equations. The airfoil is determined by using n_D design variables $\boldsymbol{x}_D$. The objective function $F(\boldsymbol{x}_D)$ is defined which showes an aerodynamic characteristic of the airfoil. The airfoil is optimized by finding the design variables such that the objective function attains its minimum value. The gradient method is used to minimize the objective function. At each search, the gradient of the objective function is required with respect to the design variables. The gradient is calculated using the impliclit function method. The calculation method uses information from a solver for the Navier-Stokes equations, and need not derive the adjoint equations [2,3].

2 Formulation

The airfoil optimization is formulated as a nonlinear optimization problem:

$$\textit{minimize} \quad F(\boldsymbol{x}_D)$$

subject to

$$c_j(\boldsymbol{x}_D) = 0, \quad j = 1, \ldots, m_e$$
$$c_j(\boldsymbol{x}_D) \geq 0, \quad j = m_e + 1, \ldots, m$$

The sequential quadratic programming (SQP) procedure [4] is used to solve this nonlinear optimization problem. The SQP procedure iterates solving a quadratic programming subproblem. At each search, it requires the objective function $F(\boldsymbol{x}_D)$ and its gradient $\boldsymbol{g}$ with respect to the design variables $\boldsymbol{x}_D$.

3 Design Variables

The airfoil shape is generated by a camber line, thickness distribution, and a stagger angle, as shown in Figure 1. Thus, the shape parameters consist of the stagger angle and parameters which determin the camber line and thicness distribution. Let $\boldsymbol{x}_s$ denote the shape parameters. On the other hand, the characterisics of inlet flow should be unchanged during the optimization. For transonic flow simulation, this paper specifies the inlet flow angle, total pressure, total temperature, and outlet static pressure as the boundary conditions. As a result, inlet Mach number is varied as the airfoil shape is varied during optimization. Hence, the outlet static pressure is chosen to be a design variable, and is adjusted such that the inlet Mach number remains unchanged during optimization. Accordingly, the design variables $\boldsymbol{x}_D$ consist of the shape parameters $\boldsymbol{x}_s$ and the outlet static pressure x_p:

$$\boldsymbol{x}_D = \begin{pmatrix} \boldsymbol{x}_s \\ x_p \end{pmatrix}$$

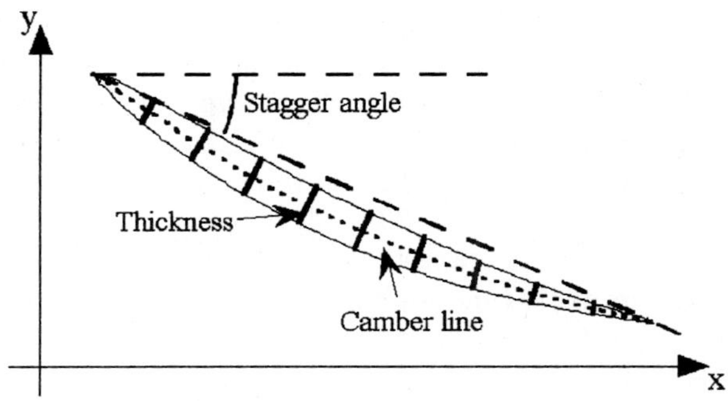

Fig. 1. Airfoil shape generated using a camber line (dotted line), thickness distribution (thick solid line), and a stagger angle.

4 Gradient

The objective function depends on the shape parameters $\boldsymbol{x}_s$ and the outlet static pressure x_p. Hence, the gradient of the objective function with respect to the design variables is written as

$$\boldsymbol{g} = \begin{pmatrix} \boldsymbol{g}_s \\ g_p \end{pmatrix}$$

where $\boldsymbol{g}_s$ and g_p are the gradients of the objective function with respect to the shape parameters $\boldsymbol{x}_s$ and the outlet static pressure x_p, respectively.

4.1 Gradient with respect to the shape parameter

Let $\boldsymbol{x}_g$ and $\boldsymbol{x}_f$ denote the grid coordinates used in flow simulation and flow variables defined at the grid points, respectively. The objective function is evaluated from $\boldsymbol{x}_g$ and $\boldsymbol{x}_f$. Accordingly, the objective function can be regarded as a function of $\boldsymbol{x}_g$ and $\boldsymbol{x}_f$, too. Hence, using the implicit function theorem, $\boldsymbol{g}_s$ is written in terms of the Jacobian matrix J_f as [1]

$$\boldsymbol{g}_s = \boldsymbol{h} - J_s^T (J_f^{-1})^T \nabla_f F \tag{1}$$

where

$$\begin{aligned} h_i &= \sum_j \frac{\partial F(\boldsymbol{x}_f, \boldsymbol{x}_g)}{\partial x_g^j} \frac{\partial x_g^j}{\partial x_s^i} \\ (\nabla_f F)_i &= \frac{\partial F(\boldsymbol{x}_f, \boldsymbol{x}_g)}{\partial x_f^i} \\ J_s &= (\boldsymbol{d}_1 \boldsymbol{d}_2 \ldots \boldsymbol{d}_{n_D - 1}) \end{aligned}$$

Here, $\boldsymbol{d}_i$ is the vector defined from the residual vector $\boldsymbol{R}(\boldsymbol{x}_f, \boldsymbol{x}_g, x_p)$ by

$$\boldsymbol{d}_i = \sum_k \frac{\partial \boldsymbol{R}}{\partial x_g^k} \frac{\partial x_g^k}{\partial x_s^i}$$

4.2 Gradient with respect to the outlet static pressre

Similarly, the gradient with respect to the outlet static pressure x_p is rewritten as [1]

$$g_p = -\boldsymbol{d}_p^T (J_f^{-1})^T \nabla_f F \tag{2}$$

where

$$\boldsymbol{d}_p = \partial \boldsymbol{R} / \partial x_p$$

4.3 System of Linear Equations

The vector $(J_f^{-1})^T \nabla_f F$ on the right-hand sides of equations (1) and (2) is the solution of the system of linear equations

$$J_f^T \boldsymbol{x} = \nabla_f F \tag{3}$$

Equation (3) is solvede by the iterative method [1]. Using the solution, both $\boldsymbol{g}_s$ and g_p are obtained from equations (1) and (2), respectively. The gradient calculation requires only as much CPU time as one flow simulation does.

5 Numerical Results

The developed optimization method is useed for a compressor cascade design. NACA 65 wing is chosen to be the initial airfoil shape.

The objective function is taken to be the sum of the total pressure loss, outlet flow angle deviation, and inlet Mach number deviation:

$$F(\boldsymbol{x}_D) = w_1(P_{ti} - P_{t0}) + w_2(\theta_o - \bar{\theta}_o)^2 + w_3(M_{in} - \bar{M}_{in})^2$$

Here P_{ti} and P_{t0} are the inlet and outlet total pressure, respectively, θ_o and M_{in} are the outlet flow angle and inlet Mach number, respectively, and $\bar{\theta}_o$ and $\bar{M}_{in}$ are their specified values, respectively. The coefficients w_1, w_2, and w_3 are weights.

For a mechanical reason, the maximum value of the thickness should be spesified, and the airfoil area should be more than the specified value. Hence, the following constraints are imposed.

$$(1 - 0.005)\bar{t}_{max} \leq t_{max} \leq (1 + 0.005)\bar{t}_{max}$$

$$\bar{S} \leq S$$

Here, t_{max} and $\bar{t}_{max}$ are the maximum thickness and its specified value, respectively, S and $\bar{S}$ are the airfoil area and its specified minimum value, respectively.

The total pressure loss decreases about 27 % after 60 searches. At that time, the outlet flow angle deviation remains around 0.02 degrees. The inlet Mach number, airfoil area, and maximum thickness hardly vary during the search iteration. Figures 2 and 3 show camber lines and thickness distribution for the initial shape and the shape after the 60 searches, respectively. Figure 4 shows the corresponding pressure coefficient distributions on the initial and optimized shapes. In Figure 5, total pressure loss coefficient and outlet flow angle deviation are plotted against the inlet Mach number for the initial and optimized shapes. In Figure 6, total pressure loss coefficient and outlet flow angle deviation are plotted against the inlet flow angle for the initial and optimized shapes. Both figures show that the optimized shape is superior to the initial one.

6 Conclusions

Based on the Navier-Stokes equations, optimization of a cascade airfoil shape is described. An objective function is minimized using the sequential quadratic programming procedure. The gradient of the objective function is calculated using the implicit function theorem. Thus, CPU time required by the optimization is greatly reduced. A numerical result is presented for a compressor cascade. The total pressure loss is decreased by about 27 % after 60 searches while keeping the outlet flow angle, inlet Mach number, maximum airfoil thickness, and airfoil area to be their specified values.

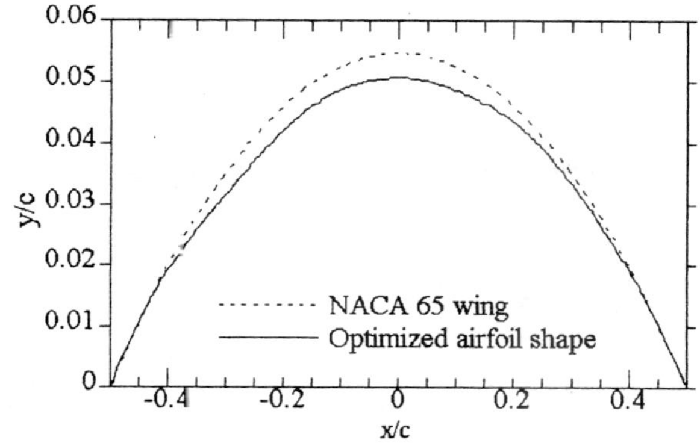

Fig. 2. Camber lines for NACA 65 wing (dotted line) and optimized airfoil shape (solid line).

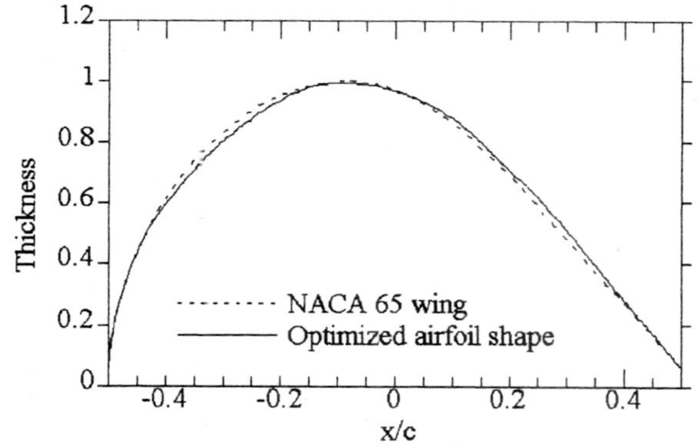

Fig. 3. Thickness distribution for NACA 65 wing (dotted line) and optimized airfoil shape (solid line).

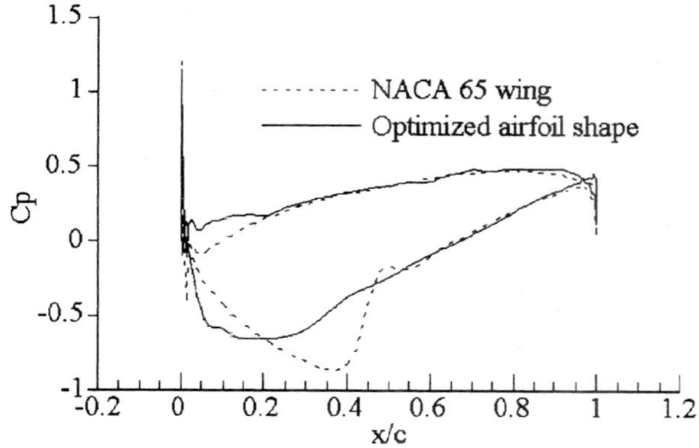

Fig. 4. Pressure coefficient distribution on NACA 65 wing (dotted line) and optimized airfoil shape (solid line).

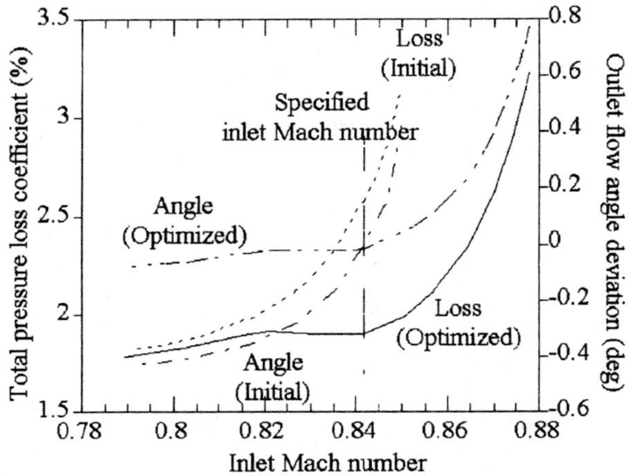

Fig. 5. Total pressure loss coefficient and outlet flow angle deviation vs. inlet Mach number.

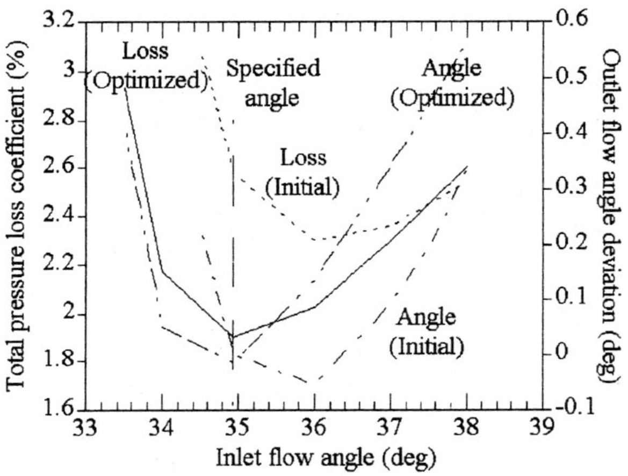

Fig. 6. Total pressure loss coefficient and outlet flow angle deviation vs. inlet flow angle.

References

1. Y. Horibata: AIAA Paper 2000-0786 (2000)
2. A. Jameson, N. A. Pierce, L. Martinelli: AIAA Paper 97-0101 (1997)
3. S. Nadarajah, A. Jameson: AIAA Paper 2000-0667 (2000)
4. K. Schittkowski: Annals of Operations Research **5**, 485 (1986).

Part XXI

Moving Boundaries

A Moving-Mesh Finite-Volume Scheme for Compressible Flows

Kenichi Matsuno, Kiyotaka Mihara, and Nobuyuki Satofuka

Kyoto Institute of Technology, Kyoto 606-8585, JAPAN

1 Introduction

This paper presents a new scheme for calculating compressible viscous flows with traveling wall boundary on moving mesh system. For the moving-mesh system, it is necessary for the scheme to satisfy the geometric conservation laws[1]. To satisfy the geometric conservation laws, a finite-volume formulation in the complete space-time (x, y, t) domain is adopted in the scheme. This treatment makes it possible that the scheme completely satisfies the physical and geometrical conservation laws for solving unsteady flow equation on moving-mesh system. The resultant fully implicit scheme is solved iteratively at every time step. This "inner" iteration is performed in an explicit manner through the efficient and highly stable Rational Runge-Kutta scheme, so that the algorithm of the scheme can be treated explicitly notwithstanding the implicit formulation. This approach of the inner iteration is similar to so-called pseudo-time approach. This paper also gives some discussion and numerical interpretation between the inner iteration strategy and the pseudo-time approach. The paper gives an application of the present scheme to compressible Navier-Stokes flows with fluid/body-motion coupled interaction.

2 Moving-Mesh Finite-Volume Scheme

2.1 Geverning Equation

For an axisymmetric compressible flow, the non-dimensionalized Navier-Stokes equation may be expressed as [2]

$$\frac{\partial}{\partial t}Q + \frac{\partial}{\partial x}\left(E - \frac{1}{Re}E_V\right) + \frac{\partial}{\partial y}\left(F - \frac{1}{Re}F_V\right) + \left(Z - \frac{1}{Re}Z_V\right) = 0. \tag{1}$$

where, x and y are axial and radial coordinates respectively. Here, Q is a vecor of conservative variables, E and F are inviscid fluxes, E_V and F_V are viscous fluxes, Z and Z_V are vectors appearing in axisymmetric cylindrical coordinate system, and Re is the Reynolds number.

Equation(1) can be represented using extended divergent operator $\tilde{\nabla}$ and tensor $\tilde{\boldsymbol{F}}$ as

$$\tilde{\nabla}\tilde{\boldsymbol{F}} + \boldsymbol{Z} = 0. \tag{2}$$

Here,

$$\tilde{\nabla} = \left(\frac{\partial}{\partial x}, \frac{\partial}{\partial y}, \frac{\partial}{\partial t} \right), \quad \tilde{\boldsymbol{F}} = \left(E - \frac{1}{Re} E_V, F - \frac{1}{Re} F_V, Q \right), \tag{3}$$

and $\boldsymbol{Z} = Z - Z_V / Re$.

2.2 Moving-Mesh Finite-Volume Scheme

Since the flow is non-stationary and body-wall boundaries move and change their relative positions, the body-fitted mesh system must dynamically change its shape according to the movement of the wall boundaries. The present flow solution method is a finite-volume method and we construct a control volume in the moving-mesh system. In this paper, we assume that both the number of grid points and grid topology do not change regardless of any flow and wall-boundary-position changes.

Now let (ξ, η) be general curvilinear coordinate system. The coordinate system is assumed to be time-dependent. Thus $\xi = \xi(x, y, t)$ and $\eta = \eta(x, y, t)$.

The present method is a vertex-centered finite-volume method and we define flow variables at the vertex (i, j, n), where i, j, and n are indices of ξ, η, and t directions respectively. Now let the flow and grid system at the time level $n+1$ $(t = t^{n+1})$ be unknown, assuming that the flow variables and grid at the time-level up to n $(t = t^n)$ are completely known.

Figure 1 shows a control volume, which is used for the construction of the present moving-mesh finite-volume scheme. The control volume stands in space-

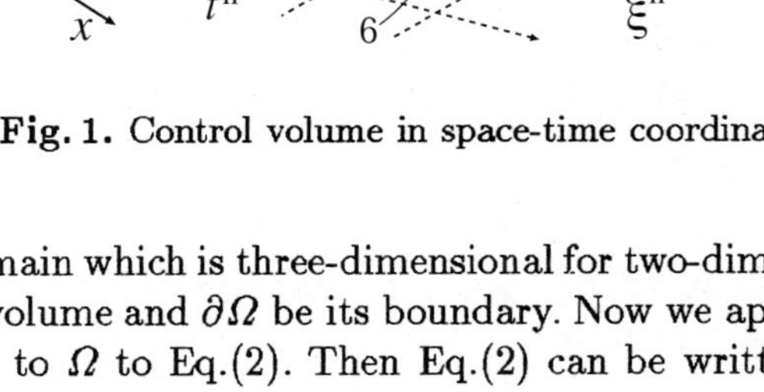

Fig. 1. Control volume in space-time coordinate

time, (x, y, t), domain which is three-dimensional for two-dimensional flows. Now let Ω be control volume and $\partial\Omega$ be its boundary. Now we apply volume integration with respect to Ω to Eq.(2). Then Eq.(2) can be written in integral form using the Gauss theorem:

$$\int_{\Omega} \left(\tilde{\nabla} \tilde{\boldsymbol{F}} + \boldsymbol{Z} \right) dV = \int_{\partial\Omega} \tilde{\boldsymbol{F}} \cdot \tilde{\mathbf{n}} dS + \int_{\Omega} \boldsymbol{Z} dV = 0. \tag{4}$$

where, $\tilde{\mathbf{n}} = (n_x, n_y, n_t)$ is an outward normal vector to the surface of the control volume (cell) and the length of the vector equals to the area of the surface.

With temporal and spatial second-order accuracy, discretizes form of Eq.(4) is

$$\sum_{\ell=1}^{6} \tilde{\boldsymbol{F}}_\ell \cdot \tilde{\mathbf{n}}_\ell + V_{i,j}^{n+1/2} \boldsymbol{Z}_{i,j}^{n+1/2} = 0. \tag{5}$$

Noting that the upper (surface 5 shown in Fig.1) and lower (6 in Fig.1) surfaces of the control volume have only n_t component, equation(5) can be written as,

$$Q_{i,j}^{n+1} (n_t)_5 + Q_{i,j}^{n} (n_t)_6 + \sum_{\ell=1}^{4} \tilde{\boldsymbol{F}}_\ell^{n+1/2} \cdot \tilde{\mathbf{n}}_\ell + V_{i,j}^{n+1/2} \boldsymbol{Z}_{i,j}^{n+1/2} = 0. \tag{6}$$

where $\tilde{\boldsymbol{F}}^{n+1/2} = (\tilde{\boldsymbol{F}}^{n+1} + \tilde{\boldsymbol{F}}^{n})/2$, and $\boldsymbol{Z}^{n+1/2} = (\boldsymbol{Z}^{n+1} + \boldsymbol{Z}^{n})/2$. Here $(n_t)_5$ and $-(n_t)_6$ respectively coincide with the upper- and lower-surface areas of the control volume and $V_{i,j}^{n+1/2}$ is the volume of the control volume in (x, y, t)-domain. Vector $\tilde{\mathbf{n}}_l$, $(l = 1, 2, 3, 4)$ is normal to the side surface of the control volume.

We use the Roe flux difference scheme[3] to estimate the inviscid fluxes E and F. To get high-order accurate flux estimation, we apply MUSCL interpolation[3] technique with *minmod*-limiter to calculate the flux. The viscous flux terms, E_V and F_V, are evaluated by Galerkin-like discretization which results the formula of central-difference method. For $Q_\ell^{n+1/2}$ $(= (Q_\ell^{n+1} + Q_\ell^{n})/2\ )$ is estimated using the Roe-averaged state which is used to evaluate the inviscid fluxes in $\tilde{\boldsymbol{F}}_\ell^{n+1/2}$.

Equation (6) is nonlinear and we have to introduce subiteration (inner iteration) procedure or linearization to solve it. Now we define operator $\mathcal{R}$ as

$$\begin{aligned} \mathcal{R}\left(Q^{n+1}, Q^{n}\right) = & \frac{Q^{n+1} + Q^{n} \frac{(n_t)_6}{(n_t)_5}}{\Delta t} \\ & + \frac{1}{\Delta t \cdot (n_t)_5} \left[\sum_{\ell=1}^{4} \tilde{\boldsymbol{F}}_\ell^{n+1/2} \cdot \tilde{\mathbf{n}}_\ell + V_{i,j}^{n+1/2} \boldsymbol{Z}_{i,j}^{n+1/2} \right]. \end{aligned} \tag{7}$$

Solution at t^{n+1} $(= t^n + \Delta t\ )$ must satisfy $\mathcal{R}\left(Q^{n+1}, Q^{n}\right) = 0$, which is equivalent to Eq.(6). Applying the Newton iteration method to solve $\mathcal{R}\left(Q^{n+1}, Q^{n}\right) = 0$ we get following iteration with δQ as an unknown and ν as iteration index,

$$\frac{\partial}{\partial Q}[\mathcal{R}\left(Q^{n+1<\nu>}, Q^{n}\right)]\delta Q = -\mathcal{R}\left(Q^{n+1<\nu>}, Q^{n}\right) \tag{8}$$

Simplest approximation to the Jacobian matrix of left-hand side of Eq.(8) is, with considering that the Jacobian matrix has the dimension of $1/\Delta t$,

$$\frac{I}{\Delta \tau} \sim \frac{\partial}{\partial Q}[\mathcal{R}\left(Q^{n+1<\nu>}, Q^{n}\right)], \tag{9}$$

Here $\Delta \tau \sim \Delta t$ and I is a unit matrix. Thus Eq.(8) is approximated as $\delta Q/\Delta \tau = -\mathcal{R}\left(Q^{n+1<\nu>}, Q^{n}\right)$, which can be represented in differential form as

$$\frac{dQ^{n+1}}{d\tau} = -\mathcal{R}\left(Q^{n+1}, Q^{n}\right) \tag{10}$$

It is the same form as pseudo-time subiteration approach[4]. A pseudo-time step $\Delta\tau$ has strong relation to Δt, but is not necessary to be constant for the whole of the region. Here we put $\Delta\tau \simeq \Delta t$. An efficient as well as robust scheme should be applied to get converged solution of $dQ^{n+1}/d\tau \to 0$. In this paper, we employ a Rational Runge-Kutta (abbr. RRK) scheme[5] to solve Eq.(10). The RRK scheme advances the pseudo-time step $\Delta\tau$, and $Q^{n+1<\nu+1>}$ is obtained. Iteration is continued until converged solution is obtained, $\nu = 1, 2, \ldots, \nu_{fin}$. Here, ν_{fin} is determined such that the convergence condition, $\mid \mathcal{R}\left(Q^{n+1<\nu_{fin}>}, Q^n\right) \mid< \epsilon$, is satisfied. Converged solution $Q^{n+1} = Q^{n+1<\nu_{fin(Q)}>}$ is obtaind with ϵ of order of 10^{-6} in this paper.

2.3 Equation of Body Motion

When the body motion is generated by the force due to flow and the fluid motion is also affected by the body motion, we have to solve Newton's equation of body motion as well as the fluid dynamics equation simultaneously at each time step. In this paper, we assume that the body moves only in x-direction without deformation of the body shape. Therefore the axisymmeric assumption is also applied to the present study. Thus Newton's equation of body motion is described in vector form of first order equation as,

$$\frac{d\mathbf{x}}{dt} = \mathbf{b}, \quad \text{where} \quad \mathbf{x} = \begin{bmatrix} x \\ u_b \end{bmatrix}, \quad \mathbf{b} = \begin{bmatrix} u_b \\ f/m \end{bmatrix} \tag{11}$$

Here, m is mass of the body and $u_b = dx/dt$ is a speed of the body and used as boundary condition for flow equation. Total force f is calculated by integrating pressure and friction force acting on the body surface. When the acting force is estimated by the value calculated at previous inner iteration step, $< \nu >$, then the following numerical formulation is used to obtain $\mathbf{x}^{n+1<\nu>}$.

$$\frac{\mathbf{x}^{n+1<\nu>} - \mathbf{x}^n}{\Delta t} = \frac{\mathbf{b}^{n+1<\nu>} + \mathbf{b}^n}{2} \tag{12}$$

The obtained new intermediate position and speed of the body, $\mathbf{x}^{n+1<\nu>}$, are returned to the next inner iteration step for flow solution.

2.4 Simultaneous Subiteration

The flow solution is obtained at the position of grid points, while the grid positions are changed related to the location of the body wall. Thus the flow solution and grid positions are both unknown and must be calculated simultaneously. The following procedure is adopted in this paper.

(1) $\nu = 0$; Estimate initial guess: $Q^{n+1<0>} = Q^n$
(2) Solve Eq.(12) and get new position and speed of the body wall.
(3) Generate new grid $(x, y)^{n+1<\nu>}_{i,j}$ according to the new body wall location.
(4) Perform one subiteration for flow solution and get $Q^{n+1<\nu+1>}$.
(5) If the flow solution is converged, then finish subiteration and go to next time step. Otherwise, update the counters $\nu = \nu + 1$ and go to the step (2).

3 Applications

3.1 Verification-GCL

The present finite-volume scheme is constructed on the moving-mesh system. It is essential for the scheme to satisfy "Geometric conservation Law" (GCL condition)[1], which means the scheme can calculate uniform flow on the moving-mesh system without error. To verify it, we apply the present scheme to uniform flow ($\rho_\infty = 1, p_\infty = 1/\gamma, u_\infty = 1, v_\infty = 1$) on the moving-mesh system defined by the equations: $x^n_{i,j} = \bar{x}_i + 0.3 \cdot \Delta x \cos\theta^n_{i,j}$, $y^n_{i,j} = \bar{y}_j + 0.3 \cdot \Delta y \sin\theta^n_{i,j}$, $\theta^n_{i,j} = (\pi/2)(\bar{x}_i + \bar{y}_j)3n/40$, where Δx and Δy are mean grid spacings. The computation with 1000 time steps was performed. Resultant error, defined by $E_{un}(\rho^n) = \{\sum_{i,j}(\rho_\infty - \rho^n_{i,j})^2/(i_{\max} \cdot j_{\max})\}^{1/2}$, was order of 10^{-14}, which shows that the scheme satisfies the GCL condition.

3.2 Gun-Tunnel Problem

The present scheme has been applied to a gun-tunnel-like problem. Figure 2 illustrates the problem and an initial condition. At $t = 0$, the piston, which separates high pressure and low pressure chambers, starts to move due to the force exerted by the fluid. Material density of the piston is assumed twice of the fluid density, hoping quick response of the piston movement to the flow change. Grid system has to be generated at every time step, according to the motion

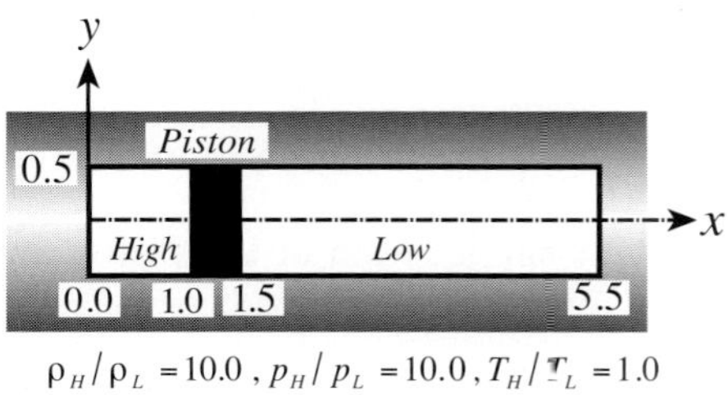

Fig. 2. Gun tunnel problem

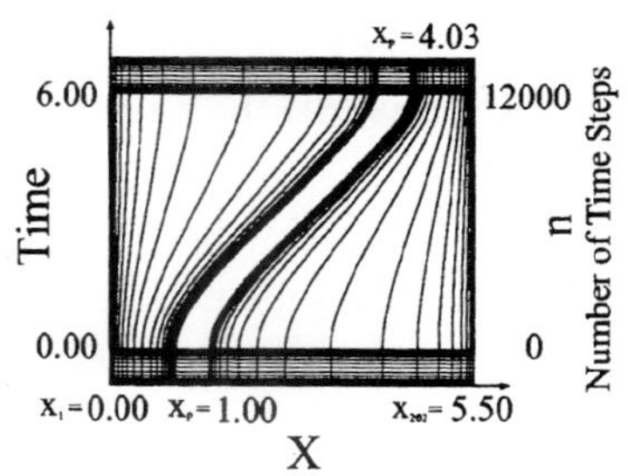

Fig. 3. Trajectories of piston and grid

of the body. The first several grid points adjacent to the wall are fixed to calculate viscous region. The grid points at intermediate region are generated by the simplest proportional transformation from initial grid distribution. Figure 3 illustrates the movement of the grid and piston. In the present problem, we calculated two cases: Case (1) No clearance between the piston (radius r_b) and pipe wall (radius R_p), $r_b/R_p = 1$ and Case (2) Ten percent clearance, $r_b/R_p = 0.90$. Figures 4 and 5 are results at time $t = 4.8$. At $t = 4.8$, the shock wave, reflected at the end wall of the pipe, is traveling toward the piston with interacting with boundary layer. Both cases show typical shock boundary layer interaction phenomena and we can see the typical λ shock. The effect of the clearance between

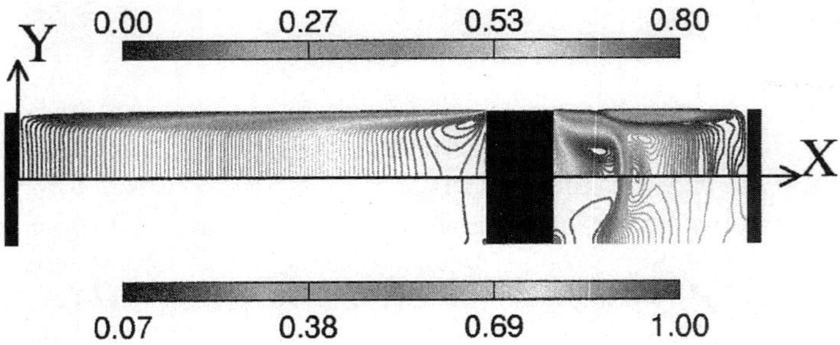

Fig. 4. Mach number(*upper half*, normalized by maximum local Mach number) and pressure(*lower half*, normalized by maximum local pressure) contours at $t = 4.8$, Case (1) No clearance between piston and pipe

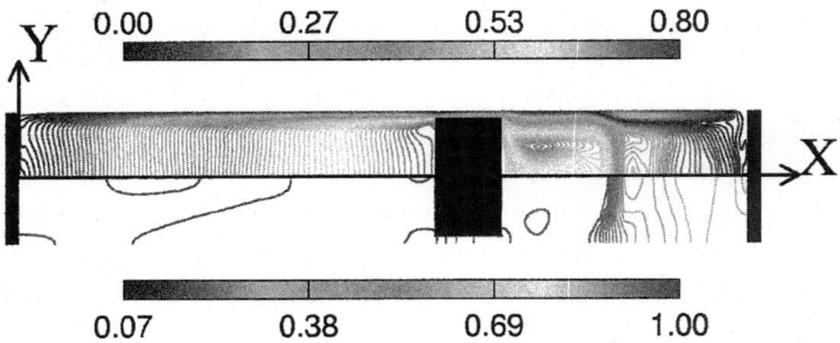

Fig. 5. Mach number(*upper half*, normalized by maximum local Mach number) and pressure(*lower half*, normalized by maximum local pressure) contours at $t = 4.8$, Case (2) 10 percent clearance between piston and pipe

the piston and the pipe can be seen typically at the location of the piston and shock-boundary layer interaction phenomena.

In conclusion, the finite-volume scheme on the moving-mesh system for calculating a system of compressible fluid/body-motion coupled problem with traveling wall boundaries has been presented in this paper. Results of the applications have showed promising features of the scheme.

References

1. H.Zhang et al.: Computers and Fluids, **22**, 9 (1993)
2. K.A. Hoffman et al.: *Fundamental Equations of Fluid Dynamics*, (Engineering Education System, Wichita 1996)
3. C. Hirsch: *Numerical Computation of Internal and External Flows*, (John Wiley & Sons, Chichester 1988)
4. C.L. Rumsey et al.: Computers and Fluids, **25**, 217 (1996).
5. Wanbecq, A.: Computing, **20**, 333 (1978).

A Moving Grid Algorithm Based on the Deformation Method

Zhong Lei[1], Guojun Liao[2], and Gary C. de la Pena[2]

[1] VINAS Co., Ltd., Osaka, Japan 550-0002
[2] Department of Mathematics, University of Texas, Arlington, TX, USA 76019

Abstract. In this paper, a moving grid method based on a deformation method from differential geometry is presented. The node velocity is determined by a monitor function and a vector field which is calculated by a Poisson equation. It is proved that the cell volume is kept to be proportional to the monitor function when the grids are adapted to the solution at each time step. The moving grid method is then applied to a square cavity flow. Results show that the grid indeed follows the monitor function closely. Further research is required on constructing the monitor functions to make the method practical.

1 Grid Generation

To describe the deformation method, let us consider the PDE for $u(\boldsymbol{x},t)$,

$$u_t(\boldsymbol{x},t) = L(u)\,, \tag{1}$$

where u is a scalar or vector variable, L is a differential operator in $\boldsymbol{x}$ only, on a physical domain Ω_p in R^3, for $t > 0$. Assume that we are provided with some positive error estimator $\delta(\boldsymbol{x},t)$. Define a positive monitor (weight) function $f(\boldsymbol{x},t) = C_1/\delta(\boldsymbol{x},t)$, where C_1 is a positive scaling parameter. Note that f is small in regions of large error. Let $\Omega_c = [0,1]^3$ be the computational domain. The deformation method constructs a transformation $\phi : \Omega_c \to \Omega_p$, $\boldsymbol{x} = \phi(\boldsymbol{\xi},t)$ such that

$$\begin{aligned} \det \nabla\phi(\boldsymbol{\xi},t) &= f(\phi(\boldsymbol{\xi},t),t), && t_{n-1} \le t \le t_n, \\ \phi(\boldsymbol{\xi},t_{n-1}) &= \phi_{n-1}(\boldsymbol{\xi}), && \boldsymbol{\xi} \text{ on } \Omega_c, \end{aligned} \tag{2}$$

where $\boldsymbol{\xi}$ is a node of a fixed grid in Ω_c, $\phi_{n-1}(\boldsymbol{\xi})$ represents the coordinates of its image at $t = t_{n-1}$. Note that (2) ensures a direct control of the cell size of the moving grid. Moreover, since the Jacobian determinant of ϕ is positive, ϕ is one-to-one.

The computation of ϕ involves two steps: The first step is to find a vector field $\boldsymbol{v}(\boldsymbol{\xi},t)$ satisfying

$$\operatorname{div} \boldsymbol{v}(\boldsymbol{\xi},t) = -\left(\frac{1}{f}\right)_t, \qquad \boldsymbol{\xi} \in \Omega_c,\ t_{n-1} \le t \le t_n\,. \tag{3}$$

The vector field $\boldsymbol{v}$ can be found by setting $\boldsymbol{v} = \nabla w$, where w satisfies the scalar Poisson equation (for a fixed t)

$$\Delta w(\boldsymbol{\xi},t) = -\frac{\partial}{\partial t}\left(\frac{1}{f(\boldsymbol{\xi},t)}\right), \qquad \boldsymbol{\xi} \in \Omega_c\,. \tag{4}$$

with Neumann boundary condition. The second step is to solve for the new node location $\phi(\boldsymbol{\xi}, t_n)$ at time t_n from the ODE system

$$\frac{\partial}{\partial t}\phi(\boldsymbol{\xi}, t) = \eta(\phi(\boldsymbol{\xi}, t), t), \qquad t_{n-1} \le t \le t_n, \boldsymbol{\xi} \in \Omega_c, \tag{5}$$

$$\phi(\boldsymbol{\xi}, t_{n-1}) = \phi_{n-1}(\boldsymbol{\xi}),$$

where the node velocity $\eta(\boldsymbol{x}, t) = f(\boldsymbol{x}, t)\boldsymbol{v}(\boldsymbol{x}, t)$. The mathematical foundation of the deformation method is provided by the following

Theorem: [4] $\det \nabla\phi(\boldsymbol{\xi}, t) = f(\phi(\boldsymbol{\xi}, t), t)$ holds true for each $\boldsymbol{\xi}$ in Ω_c and each $t > t_0$, provided it is true at $t = t_0$.

In this paper we transform equation (1) through the grid mapping $\boldsymbol{x} = \phi(\boldsymbol{\xi}, t)$ and solve the transformed equation on an orthogonal grid on Ω_c.

Various equidistribution principles can be used to construct the monitor function. A posteriori error estimates (if available), residues, and truncation errors, etc. are redistributed evenly over the whole domain. For instance, in the calculation of Euler flows with shock waves [6], we used

$$f = \frac{C_1}{1 + C_2|\nabla p|}, \tag{6}$$

where p is the pressure, C_2 is a constant for adaptation intensity, C_1 is a normalization parameter. For interface resolution, we can construct f by using the distance d to the interface as follows: Let f be piecewise linear such that

$$f = \begin{cases} 1 & \text{if } |d| > \epsilon_1 \\ \epsilon_2 & \text{if } d = 0, \end{cases} \tag{7}$$

where ϵ_1 and ϵ_2 are small parameters to be chosen by the user. The method has been applied to hyperbolic equations [7] and convection-diffusion problems [4]. Recently, a new moving grid method was formulated based on the level set evolution equations and it acheives the same control over the Jacobian [8]. In this study, the moving grid algorithm described above is applied to incompressible flow problems. For more information visit our website at http://www.uta.edu/math/liao.htm.

2 Governing Equations

For an incompressible flow, the governing equations in Ω_c can be written in conservative form as

$$\frac{\partial u_i}{\partial t} + \frac{1}{J}\frac{\partial}{\partial \xi_j}\left[J\left(U_j - V_j\right)u_i\right] = -\frac{1}{J}\frac{\partial}{\partial \xi_j}\left(J\frac{\partial \xi_j}{\partial x_i}p\right) + \frac{1}{J}\frac{1}{Re}\frac{\partial}{\partial \xi_j}\left(Jq_{jk}\frac{\partial u_i}{\partial \xi_k}\right), \tag{8}$$

$$\frac{\partial \xi_k}{\partial x_i}\frac{\partial}{\partial \xi_k}\left[\frac{1}{J}\frac{\partial}{\partial \xi_j}\left(J\frac{\partial \xi_j}{\partial x_i}p\right)\right] = -\frac{1}{J}\frac{\partial D}{\partial t} + \frac{1}{J}\frac{\partial}{\partial \xi_k}\frac{\partial \xi_k}{\partial x_i}\frac{\partial}{\partial \xi_j}\left[J\left(U_j - V_j\right)u_i\right]. \tag{9}$$

where in the pressure Poisson equation (PPE) (9), the viscous terms have been removed by using the continuity equation. U_i is the contravariant velocity and V_i is determined by the node velocity $\boldsymbol{x}_t$ of the moving grid:

$$U_i = \frac{\partial \xi_i}{\partial x_j} u_j, \; V_i = \frac{\partial \xi_i}{\partial t}. \tag{10}$$

The Jacobian and metrics are defined as

$$J = \frac{\partial(x_1, x_2, x_3)}{\partial(\xi_1, \xi_2, \xi_3)}, \; \frac{\partial \xi_i}{\partial t} = \frac{\partial x_j}{\partial t}\frac{\partial \xi_i}{\partial x_j}, \; q_{jk} = \frac{\partial \xi_j}{\partial x_i}\frac{\partial \xi_k}{\partial x_i}. \tag{11}$$

3 Numerical Method

The marker-and-cell (MAC) method is used with collocated arrangement in this study. The convective terms are discretized using the QUICK scheme to remove numerical wiggles, while the viscous terms are discretized by central difference scheme. The second-order Adams-Bashforth method is used for time integration to solve the momentum equation (8). Both the pressure Poisson equation (PPE) (9) and the deformation Poisson equation (DPE) (4) are discretized by central differences and then solved with the ADI method. An explicit 2nd-order Runge-Kutta method is used to solve (5).

The pseudo code of the moving grid algorithm is as follows

```
Form an initial grid.
Integrate u_i until t = t_0 and find ψ, ω.
Form an adaptive grid at t = t_0 by the static deformation method.
For t = t_n , n = 1, 2, 3, ··· , do:
   Compute RHS(t_n, u_i^n) of (8).
   Compute the new volecity u_i^{n+1} by the time integration.
   Compute f by using the new transient solution u_i^{n+1}, p^{n+1}
   Adapt grids by the deformation method
End do
```

4 Numerical Example

The flow in a square cavity whose top wall is dirven by a lid with uniform velocity has served as a model problem for testing and evaluating numerical techniques. This is a typical complex flow, that a recirculating region and two secondary vortexes exist in a simple geometry. The present study compares results on adapted grids to the benchmark solutions found in Ghia *et al.* [9].

4.1 Steady Flow

A steady solution was first obtained on a fixed uniform 51×51 grid. The result was used to form the monitor function f by

$$f = \begin{cases} 1 & \text{if } |d| > 0.005 \\ 0.1 - 180d & \text{if } -0.005 \le d \le 0 \\ 0.1 + 180d & \text{if } 0 \le d \le 0.005 \end{cases}, \tag{12}$$

where $d = \psi(\psi_{\min} - \psi)$ was selected to refine both the primary and secondary vortexes. Then, adaptive grids were generated by the static version of the deformation method according to f, using the monitor function $g = (1-s) + sf$ from $s = 0$ to $s = 1$. Finally, the problem was solved again on the new grid. The resulting adaptive grid and streamline contours are shown in Fig. 1 using the same contour levels as in [9]. The result agrees well with those presented in [9] using 129×129 grids. The comparison on stream function and vorticity is given in Table 1. The values presented are nodal values.

Table 1. Comparison to Benchmark Solution, Re = 1000

	Uniform Grid (129×129)		Adapted Grid (51×51)
Primary Vortex	BenchMark [9]	MAC	Adaptive MAC
$\psi_{\min}$	-0.117929	-0.1188303	-0.118252
$\omega_{v.c}$	2.04968	2.16926	2.18633
Location (x, y)	$(0.5313, 0.5625)$	$(0.5312, 0.5625)$	$(0.5378, 0.5675)$

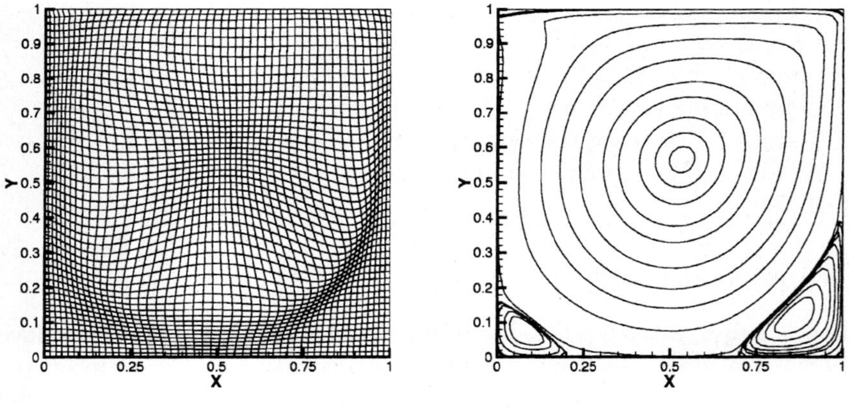

Fig. 1. Steady flow. (**left**) Plot of adapted grid. (**right**) Contour plot of stream function

4.2 Unsteady Flow

For the sake of numerical stability at the early stage, the unsteady solution is first solved on a uniform grid until $t = t_0$. At $t = t_0$ an initial adapted grid is generated according to the monitor function $f_{\text{init}} = (1-s) + sf$ where f is given by

$$f = \begin{cases} 1 & \text{if } |d| > 0.004 \\ 0.344 - 164d & \text{if } -0.004 \le d \le 0 \, . \\ 0.344 + 164d & \text{if } 0 \le d \le 0.004 \end{cases} \tag{13}$$

where $d = \psi(\psi_{\min} - \psi)$ and s goes from 0 to 1. For $t > t_0$ a moving grid is formed using the same f. The moving grid and the streamline contours for selected times are shown in Fig. 2 and Fig. 3. The comparison on stream function and vorticity is given in Table 2. The values presented are nodal values.

Table 2. Comparison to Benchmark Solution, Re = 1000

Primary Vortex	Uniform Grid (129 × 129) BenchMark [9]	Moving Grid (51 × 51) Moving Grid MAC
$\psi_{\min}$	-0.117929	-0.117967
$\omega_{\text{v.c}}$	2.04968	2.19837
Location (x, y)	$(0.5313, 0.5625)$	$(0.5228, 0.5589)$

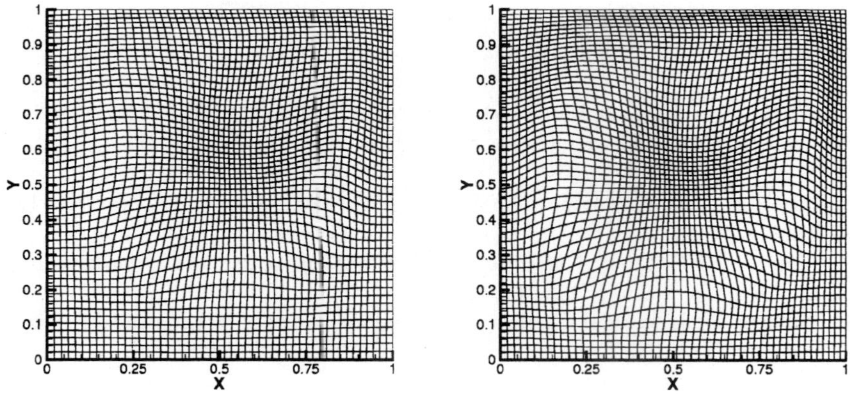

Fig. 2. Unsteady flow. (**left**) Grid plot of moving grid at $t = 15.0$. (**right**) Grid plot of moving grid at $t = 40.0$ ($t_0 = 5, \Delta t = 0.005$)

5 Conclusion

The moving grid is indeed refined in the regions where $\psi \approx 0$ and $\psi \approx \psi_{\min}$ as intended by the monitor function. Further research on proper choice of the

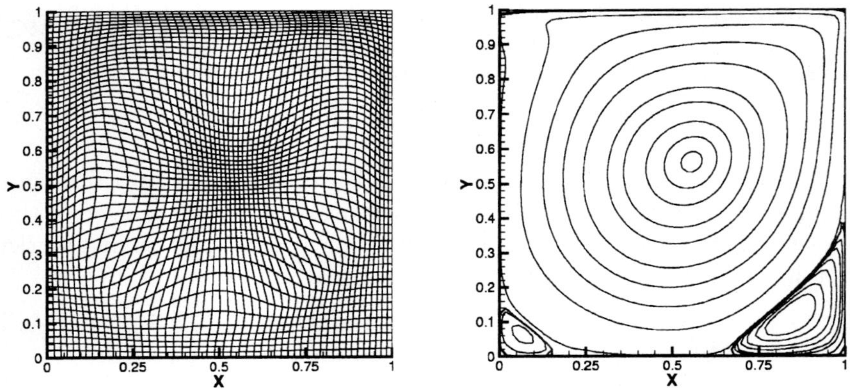

Fig. 3. Unsteady flow. (**left**) Grid plot of moving grid at $t = 60.0$. (**right**) Contour plot of stream function at $t = 60.0$ ($t_0 = 5, \Delta t = 0.005$)

monitor function is necessary. Also, more sophisticated algorithms should be considered for the solution of the deformation ODE system to improve accuracy. Extension to 3D and general domains is straightforward. The deformation method can be used for unstructured grid and in combination with finite-volume and finite-element solvers as well.

References

1. J. Moser: *Volume Elements of a Riemann Manifold*, Trans AMS, **120**, (1965).
2. G. Liao and D. Anderson: *A New Approach to Grid Generation*, Appl. Anal., **44**, (1992).
3. G. Liao and J. Su: *A Moving Grid Method for (1+1) Dimension*, Appl Math Lett., **8**, (1995).
4. B. Semper and G. Liao: *A Moving Grid Finite Element Method using Grid Deformation*, Numer. Meth. PDEs, **11**, (1995).
5. P. Bochev, G. Liao, and G. dela Pena: *Analysis and Computation of Adaptive Moving Grids by Deformation*, Numer. Meth. PDEs, **12**, (1996).
6. F. Liu, S. Ji, and G. Liao: *An Adaptive Grid Method with Cell-Volume Control and its Applications to Euler Flow Calculations*, SIAM J. Sci. Comput., **20**, (1998).
7. G. Liao, G. dela Pena, D.A. Anderson: *A Moving Finite Difference Method for Partial Differential Equations*, Preprint
8. G. Liao, F. Liu, G. dela Pena, D Peng, and S. Osher: *Level-Set-Based Deformation Methods for Adaptive Grids*, J Comp Phys, **159**, (2000).
9. U. Ghia, K.N. Ghia and C.T. Shin: *High-Re Solutions for Incompressible Flow Using the Navier-Stokes Equations and a Multigrid Method* J Comp Phys, **48**, (1982).

A Method for Computing Flows about Complex Configurations with Moving Boundaries

Yoko Takakura[1]

Tokyo Noko University, 2-24-16 Nakacho, Koganei, Tokyo184-8588, Japan

Abstract. A Method was presented for computing flows about complex configurations with moving boundaries by introducing the moving coordinate method together with the cut-cell method and the multi-block method. The present method enables the numerical analysis of flows about moving complex configurations without deforming grids. This method was applied to branch flows for opening process of a high-voltage gas circuit breaker. As results it was shown that the flows are classified into four patterns depending on the relative location of the fixed electrode and the nozzle region of the cylinder, and that the characteristic phenomenon appears at the beginning of movement. It was confirmed that the present method works well for the moving boundary problems about complex configurations.

1 Introduction

There is much difficulty in computing flows when complex configurations are moving. If the complex configurations move away from other bodies, the multi-domain method [1,2] is effective. However, when the moving complex configurations are close or contact to the other bodies which are moving with different velocity or stationary, what is the strategy? In this paper a method is presented to resolve the above-stated difficulty by introducing the moving coordinate method[3] so that the problem on moving complex configurations should be transformed to that on moving simple configurations with stationary complex configurations. The cut-cell method[5] is adopted for the movement of the simple configurations, and the multi-block method[6] for the expression of the complex configurations.

As an application, unsteady flows during opening process of a high-voltage gas circuit breaker are shown. The apparatus is shown in Fig. 1: when it is ordered to interrupt currents, the cylinder and the movable electrode begin to move to the right; then the fluid in the chamber is compressed because of the reducing volume between the moving cylinder and the fixed piston, it flows against the centerline, and it branches to the left and right; the leftward fluid blows through the nozzle onto the fixed electrode to extinguish the arc, and the currents are interrupted. A high voltage of an alternating current, however, might cause regeneration of arc and lowered performances. Therefore CFD analysis is needed to design a high-performance breaker.

2 Numerical Analysis Method

2.1 Governing Equations

In the moving coordinate method[3] coordinate system * moving with velocity $\mathbf{v}_G$ relative to the standard stationary coordinate system is introduced. Let $\mathbf{u} = (u^1, u^2, u^3)$ be the velocity of fluid observed in the standard system, and $\mathbf{u}_* = (u_*^1, u_*^2, u_*^3)$ be the velocity of fluid observed in moving system *. When the Galilean transformation, $\mathbf{u} - \mathbf{v}_G = \mathbf{u}_*$, and the metric relations are used in the well-known conservative form for the Euler equations in general curvilinear coordinate system, then we obtain the governing equation in coordinate system * [4]:

$$\frac{\partial \tilde{\mathbf{q}}}{\partial \tau} + \frac{\partial \tilde{\mathbf{F}}^i}{\partial \xi^i} = \tilde{\mathbf{s}}, \tag{1}$$

$$\tilde{\mathbf{q}} = \frac{1}{J}\begin{bmatrix} \rho \\ \rho u_*^1 \\ \rho u_*^2 \\ \rho u_*^3 \\ e_* \end{bmatrix}, \quad \tilde{\mathbf{F}}^i = \frac{\xi^i_{x^k}}{J}\begin{bmatrix} \rho u_*^k \\ \rho u_*^1 u_*^k + \delta^{1,k} p \\ \rho u_*^2 u_*^k + \delta^{2,k} p \\ \rho u_*^3 u_*^k + \delta^{3,k} p \\ (e_* + p) u_*^k \end{bmatrix}, \quad \tilde{\mathbf{s}} = \frac{1}{J}\begin{bmatrix} 0 \\ -\rho(dv_G^1/dt) \\ -\rho(dv_G^2/dt) \\ -\rho(dv_G^3/dt) \\ -\rho u_*^j (dv_G^j/dt) \end{bmatrix}, \tag{2}$$

where e_* is the total energy observed in moving system * ,

$$e_* = p/(\gamma - 1) + (1/2)\rho(\mathbf{u}_*, \mathbf{u}_*). \tag{3}$$

For the above equation system holds the physical interpretation that the effect of the inertial force due to the acceleration of the moving coordinate system is added to the conservation laws in the stationary system.

The momentum and the total energy are transformed from the standard coordinates to the moving coordinates as:

$$(\rho\mathbf{u})_* = (\rho\mathbf{u}) - \rho\mathbf{v}_G \tag{4}$$

$$e_* = e - \rho(\mathbf{u}, \mathbf{v}_G) + (1/2)\rho(\mathbf{v}_G, \mathbf{v}_G). \tag{5}$$

Simultaneously the followings hold in the transformation from the moving coordinates to the standard coordinates:

$$(\rho\mathbf{u}) = (\rho\mathbf{u})_* + \rho\mathbf{v}_G \tag{6}$$

$$e = e_* + \rho(\mathbf{u}_*, \mathbf{v}_G) + (1/2)\rho(\mathbf{v}_G, \mathbf{v}_G). \tag{7}$$

The density is invariant on the transformation.

In this research, the standard coordinate system is attached to the simple configurations, i.e., the fixed electrode and the fixed piston, and the moving coordinate system * is attached to the complex configurations, i.e., the cylinder and the movable electrode, which are moving to the right. Here equation system (1) with moving coordinate system * including the inertial force $\tilde{\mathbf{s}}$ is solved, where the cylinder and movable electrode are stationary and the fixed electrode and piston are moving to the left. The movement of the simple configurations is carried out by the cut-cell method[5] . To compute flows about the complex configurations the multi-block method[6] is adopted.

2.2 Discretization

The governing equations (1) are non-dimensionalized and discretized by the cell-centered finite volume method with the axsymmetric condition. The spatial discretization is executed by the Chakravarthy-Osher post-process TVD scheme and the time integration is carried out by the first-order explicit method.

2.3 Body-Surface Boundary Condition

As to the cylinder and the movable electrode, since they are stationary in moving system *, the usual slip condition on the stationary surface is imposed.

Let **n** be the unit normal vector to the surface, superscript f be for physical values of the fluid cell adjacent to the surface, and superscript d be for those of the dummy cell inside the body, and then as the slip condition we can use

$$\mathbf{u}_*^d = \mathbf{u}_*^f - 2(\mathbf{u}_*^f, \mathbf{n})\,\mathbf{n}. \tag{8}$$

The 0th-order extrapolation is used for the density and the total energy.
As to the condition of the fixed electrode and the piston which are moving to the left in moving system *, after transforming physical values in the moving system to those in the stationary system, the condition on the stationary surface is imposed, and then physical values should be restored to those in moving system *.

When the velocity transformed to the stationary system, $\mathbf{u} = \mathbf{u}_* + \mathbf{v}_G$, is substituted in eq.(8), then

$$\mathbf{u}_*^d = \mathbf{u}_*^f - 2(\mathbf{u}_*^f + \mathbf{v}_G, \mathbf{n})\,\mathbf{n}. \tag{9}$$

For the total energy transformed to the stationary system, eq.(7), the 0th-order extrapolation is adopted, and then we obtain

$$e_*^d = e_*^f + 2\rho^f(\mathbf{u}_*^f + \mathbf{v}_G, \mathbf{n})(\mathbf{n}, \mathbf{v}_G), \tag{10}$$

where the 0th-order extrapolation is used for the density.

3 Computational Condition

Figure 3 displays the views of grid for the multi-block method (Number of grid points: 201 × 135) about the circuit-breaker model in the moving coordinate system *: (a) and (b) are the same grid but shows different states of closed and open circuits, respectively. The cylinder and the movable electrode move from position (a) up to position (b) with the time history for the location of the cylinder shown in Fig 2. The working fluid is sulfur hexafluoride ($SF_6; \gamma = 1.09$), which is, in initial state, stationary with pressure 0.601[MPa] and temperature 288K].

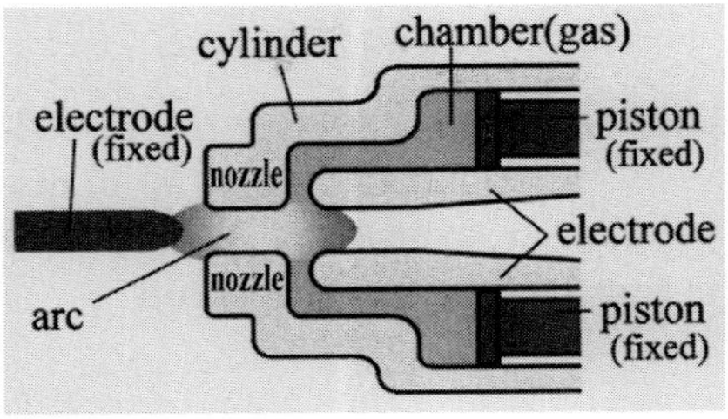

Fig. 1. Apparatus of gas circuit breaker.

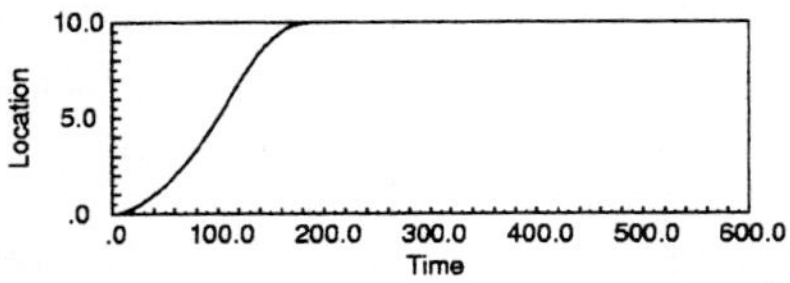

Fig. 2. Time history for location of cylinder.

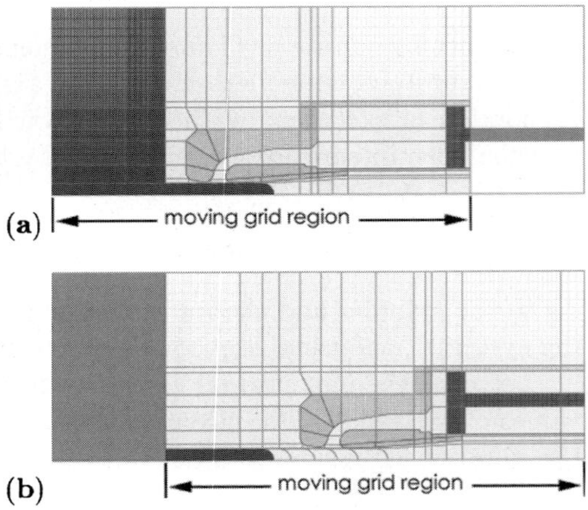

Fig. 3. Views of grid for multi-block method. (**a**) closed position. (**b**) open position.

4 Results and Discussion

Figures 4 (a) $\sim$ (f) show density contours in the opening process. It is observed that flows can be classified into four patterns shown in Figs.4 (b), (c), (d) and (e), depending on the relative location of the fixed electrode and the nozzle part of the cylinder.

In closed state (a), the fluid is stationary. In (b) when the nozzle moves covering the fixed electrode, the fluid spouts out of the slight opening space between the nozzle and the fixed electrode almost horizontally along the electrode. Also in (c) when the nozzle moves covering the tip of fixed electrode, the fluid tends to flow out of the nozzle exit with high speed upwardly along the outer wall of the cylinder, rather than horizontally along the fixed electrode. In (d) when the nozzle passes near the tip of fixed electrode, the fluid gushes obliquely with some angle to the centerline, as if the exit of the nozzle and the tip of the electrode compose another nozzle. In (e) when the nozzle approaches to the maximum moving distance (open state) away from the fixed electrode, the fluid spouts again in parallel to the centerline, generating the structure of shock waves peculiar to the nozzle jet flows. As this nozzle jet flow is supersonic except for the region near the tip of the electrode, the shock wave structure is stable against some change of locations of the electrode. In this stage the chamber pressure reaches the maximum value, and after then it becomes low. Soon in (f), some time after the nozzle stops away from the fixed electrode, the shock-wave structure vanishes.

Figure 4 shows the characteristic phenomenon when the cylinder and the movable electrode began to move: the gas near the outer wall of the cylinder becomes rarefied and the gas near the inner wall is compressed, and they propagate.

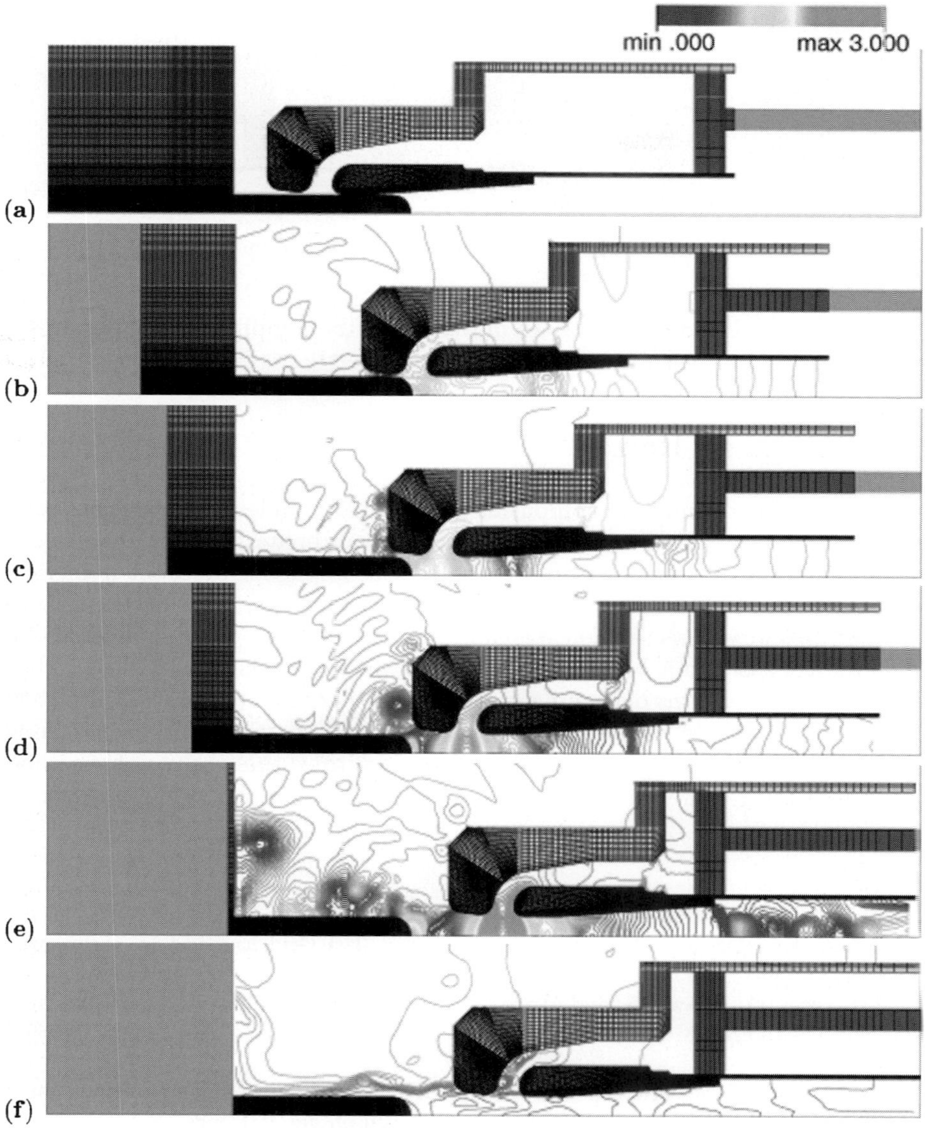

Fig. 4. Patterns of density contours in opening process. (**a**) t=0.0:just before nozzle begins to move. (**b**) t=100.2:when nozzle moves covering fixed electrode. (**c**) t=115.7:when nozzle moves covering tip of fixed electrode. (**d**) t=131.1:when nozzle passes near tip of fixed electrode. (**e**) t=166.5:when nozzle approaches to maximum moving distance. (**f**) t=519.1:after nozzle stops away from fixed electrode.

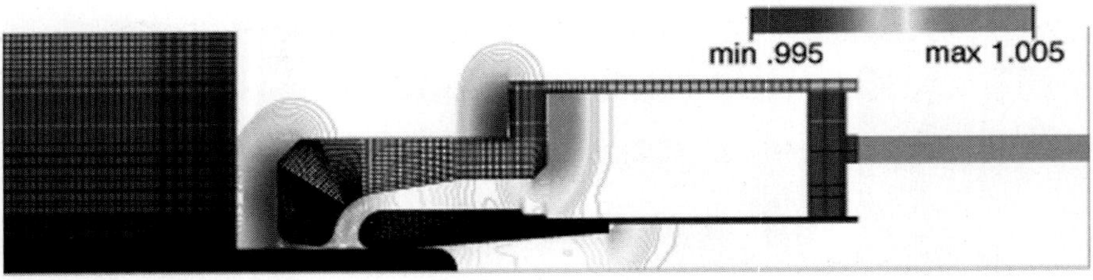

Fig. 5. Characteristic phenomenon when cylinder begins to move: density contours.

This phenomenon indicates that the movement of the complicated configurations is realized by the present method.

5 Concluding Remarks

A Method was presented for computing flows about complex configurations with moving boundaries by combining our moving coordinate method, the cut-cell method and the multi-block method. The present method enables the numerical analysis of flows about complex configurations containing moving boundaries without difficulty. This method was applied to flows in the opening process of a high-voltage gas circuit breaker. As results it was shown that the flows are classified into four patterns depending on the relative location of the fixed electrode and the nozzle region of the cylinder, and that the characteristic phenomenon appears at the beginning of movement. It was confirmed that the present method works well for the moving boundary problems about complex configurations.

References

1. Benek, J.A., Buning, P.G. and Steger, J.L.: 'A 3-D Grid Embedding Technique,' AIAA Paper 85-1523 (1985)
2. Takakura, Y., Ogawa, S. and Wada, Y.: 'Transonic Wind-Tunnel Flows About a Fully Configured Model of Aircraft,' AIAA Journal, Vol. 33, No. 3, pp.557-559 (1995)
3. Takakura,Y., Higashino,F., and Ogawa,T.: 'Unsteady Flow Computations on a Flying Projectile within a Ballistic Range,' Computers & Fluids, Vol.27, No.5-6, pp.645-650 (1998)
4. Takakura,Y.: 'Methods for Computing Flows with Moving Boundaries: Governing Equations, Geometric Conservation Laws, and Applications of Moving-Coordinate Method,' Journal for Japan Society of CFD, Vol.8, No.3, pp.117-129 (2000) in Japanese
5. Yang, G., Causon, D.M., Ingram, D.M., Saunders, R., and Batten, P.: 'A Cartesian Cut Cell Method for Compressible Flows Part B: Moving Body Problems,' The Aeronautical Journal, Vol.101, No.1002, pp.57-65 (1997)
6. Sawada, K. and Takanashi, S.: 'A Numerical Investigation on Wing/Nacelle Interference of USB Configurations,' AIAA Paper 87-0455 (1987)

Development of a New Higher Order Kinetic Boundary Condition and a New Algorithm for Moving Grid Using q-LSKUM

P. Chandra Sekhar[1], A.K. Ghosh[2], and S.M. Deshpande[1]

[1] R & DB Centre of Excellence for CFD, Deptartment of Aerospace Engineering, Indian Institute of Science, Bangalore, India. e-mail: suresh@aero.iisc.ernet.in
[2] CFD group, Aeronautical Development Agency, Bangalore, India. e-mail: tdgakg@ada.ernet.in

Abstract. A new higher order accurate treatment of wall boundary condition has been developed within the framework of q-LSKUM using the specular reflection model of kinetic theory of gases. Using the present q-LSKUM, higher order spatial accuracy for the entire flow domain is achieved. LSKUM being a robust and grid free method, it is very tempting to apply to problems involving moving grid. Advantages inherent to entropy variables have been exploited to develop a new method on moving grids called q-LSKUM-MG.

1 Introduction

The Least Squares Kinetic Upwind Method (LSKUM) based on entropy variables called q-LSKUM [1] has some distinct advantages. The present work considers two further extensions of q-LSKUM, one pertaining to accurate treatment of wall boundary condition for increasing solution accuracy especially in the near field and the other is about modifications in q-LSKUM to tackle unsteady aerodynamics problems invloving moving grid (q-LSKUM-MG). All the earlier applications of LSKUM and q-LSKUM [1],[2] and [3] to flow problems are based on first order accurate treatment of wall boundary condition even though second order state update is used at interior points. It is very tempting to consider the possibility of improving solution accuracy at wall points. We will briefly describe the basic theory of q-LSKUM and then consider its extensions.

2 q-LSKUM

LSKUM belongs to the family of kinetic schemes. The main principle behind these schemes is that the Euler equations are the moments of Boltzmann equation of kinetic theory of gases when the distribution function is a Maxwellian. The 2-D Boltzmann equation in the Euler limit is given by

$$\frac{\partial F}{\partial t} + v_1 \frac{\partial F}{\partial x} + v_2 \frac{\partial F}{\partial y} = 0 \tag{1}$$

where v_1 and v_2 are molecular velocities along x and y directions respectively and F is the Maxwellian. The Maxwellian F is given by

$$F = \frac{\rho}{I_o} \frac{\beta}{\pi} exp[-\beta(v_1 - u_1)^2 - \beta(v_2 - u_2)^2 - \frac{I}{I_o}] \tag{2}$$

where ρ is mass density, u_1 and u_2 are the fluid velocities in x and y directions respectively, $\beta = \frac{1}{2RT}$, T is temperature, R is gas constant, I is the internal energy variable due to non-translational degrees of freedom and I_o is the equilibrium internal energy due to non-translational degrees of freedom. To construct an upwind scheme for the Boltzmann equation(1), splitting is done with respect to the sign of v_1 and v_2 (Courant-Issacson-Rees scheme), that is

$$\frac{\partial F}{\partial t} + \frac{v_1 + |v_1|}{2}\frac{\partial F}{\partial x} + \frac{v_1 - |v_1|}{2}\frac{\partial F}{\partial x} + \frac{v_2 + |v_2|}{2}\frac{\partial F}{\partial y} + \frac{v_2 - |v_2|}{2}\frac{\partial F}{\partial y} = 0 \quad (3)$$

Referring to Fig. 1(a), let the coordinates of the various nodes P_i be denoted by (x_i, y_i) where i=1,2,. . .,n. The discrete approximations of the spatial derivatives F_{xo} and F_{yo} at node P_o in terms of the data at its neighbours are obtained by using weighted least squares expressions.

$$F_{xo} = \frac{(\sum w_i \Delta y_i^2)(\sum w_i \Delta F_i \Delta x_i) - (\sum w_i \Delta x_i \Delta y_i)(\sum w_i \Delta F_i \Delta y_i)}{(\sum w_i \Delta x_i^2)(\sum w_i \Delta y_i^2) - (\sum \Delta x_i \Delta y_i)^2}$$

$$F_{yo} = \frac{(\sum w_i \Delta x_i^2)(\sum w_i \Delta F_i \Delta y_i) - (\sum w_i \Delta x_i \Delta y_i)(\sum w_i \Delta F_i \Delta x_i)}{(\sum w_i \Delta x_i^2)(\sum w_i \Delta y_i^2) - (\sum \Delta x_i \Delta y_i)^2} \quad (4)$$

where $\Delta x_i = x_i - x_o, \Delta y_i = y_i - y_o$ and $\Delta F_i = F_i - F_o$. To get an upwind discrete approximation of F_{xo}, we take the stencil to the left of point of P_o for $v_1 > 0$ and to the right of P_o for $v_1 < 0$. Similarly, to get an upwind discrete approximation of F_{yo}, we take upper and lower stencil of P_o for $v_2 < 0$ and $v_2 > 0$ respectively. Since the choice of stencil is based on upwind principle, these are also called as upwind stencils. This approach is also called as X-Y splitting. It is also possible to split the fluxes quadrantwise and this approach is employed for treating solid wall boundary condition.

The least squares expressions (4) are only first order accurate. Second order accuracy is achieved by using defect-correction procedure employing q-variables. The q-variables (also called entropy variables) were first introduced by Deshpande [4] while studying the symmetric hyperbolic form of Euler equations. The q-variables are related to Maxwellian F and the relation in 2-D is given by

$$\frac{dF}{F} = dq_1 + v_1 dq_2 + v_2 dq_3 + \left(I + \frac{v^2}{2}\right) dq_4 + \frac{3 - 2\gamma}{2 - \gamma}(I - I_o)dq_4 \quad (5)$$

The q-variables in 2-D are also related to primitive variables (ρ, u_1, u_2, β) by

$$\mathbf{q} = \begin{bmatrix} q_1 \\ q_2 \\ q_3 \\ q_4 \end{bmatrix} = \begin{bmatrix} \ln\rho + \frac{1}{\gamma-1}\ln\beta - \beta u^2 \\ 2u_1\beta \\ 2u_2\beta \\ -2\beta \end{bmatrix} \quad (6)$$

where $u^2 = u_1^2 + u_2^2$. The primitive variables can be extracted from $\mathbf{q}$ vector and they can be used to construct the Maxwellian F (2).

Second order accuracy of the discrete approximations of F_{xo} and F_{yo} are obtained by replacing ΔF_i in (4) with $\Delta\tilde{F}_i$ given by

$$\Delta\tilde{F}_i = F(\tilde{q}_i) - F(\tilde{q}_o) \quad (7)$$

where

$$\tilde{q}_i = q_i - \frac{1}{2}\Delta x_i q_{xi} - \frac{1}{2}\Delta y_i q_{yi} \quad (8)$$

$$\tilde{q}_o = q_o - \frac{1}{2}\Delta x_i q_{xo} - \frac{1}{2}\Delta y_i q_{yo} \quad (9)$$

The q-derivatives in (8, 9) are obtained in the following ways:

(A) Using Upwind Stencil (M-I)
Consider an upwind discrete approximation of F_{yo} at P_o (refer Fig. 1(a)) for $v_2 > 0$. The lower stencil (upwind stencil) at P_o has to be used to evaluate this derivative F_{yo}. In order to get higher order accuracy of this derivative F_{yo} at P_o, we need to determine $\tilde{q}_i$ at all nodes in the lower stencil of P_o including P_o, which inturn requires q-derivatives q_{xi} and q_{yi} at these node. These q-derivatives are computed using the least squares expressions (4) and making use of lower stencil at the corresponding P_ith node. Similarly for other derivates of F_{xo} and F_{yo}, corresponding upwind stencil are used with the least squares expressions (4) to evaluate the q-derivatives. This approach was followed by Dauhoo etal [1].
Now consider a point P_2 near the boundary and it has some points on the boundary in its connectivity (see Fig. 1(b)). If the above approach has to be followed, the q-derivatives at point P_3 and P_4 in the connectivity of P_2 cannot be calculated using upwind stencil (lower stencil at P_3 and P_4 are empty). Thus purely upwind stencil (US) method fails for the points near the boundary. Hence, the earlier version of q-LSKUM [1] has second order accuracy for interior points and first order accuracy for points on the boundary and as well as for points close to the boundary. This drawback was overcome by the use of full stencil (FS) to evaluate q-derivatives as discussed below.

(B) Using Full Stencil (M-II)
As discussed above the use of upwind stencil for calculating q-derivatives limits higher order accuracy to interior points only. In the present method, the derivatives F_x and F_y are still calculated using least squares expression (4) and upwind stencil but the q-derivatives q_{xi} and q_{yi} required for defect-correction are calculated using least squares expression (4) and full stencil (all points in the connectivity) at the nodes in the corresponding upwind stencils. The use of full stencil for calculating q-derivative combined with Kinetic Characteristic Boundary Condition (KCBC) [5], make the wall boundary condition higher order accurate. Thus the present approach helps in achieving higher order accuracy for the entire flow domain.

3 q-LSKUM on Moving Grid (q-LSKUM-MG)

LSKUM being a grid free method, can be easily applied to problems involving moving grid. The present formulation of LSKUM-MG [6] is very general, in the sense that if there is no grid movement it will automatically reduce to that for stationary grid. The nice features of q-LSKUM, where solution obtained is higher order accurate for the entire flow domain, combined with that of LSKUM-MG motivated us to apply q-LSKUM to problems involving moving grid (q-LSKUM-MG).

The formulation for moving grid is developed at the Boltzmann level. Upwinding is also introduced at the molecular level based on the sign of molecular velocity relative to the grid velocity. Then moving grid LSKUM at Euler level is obtained by using moment method strategy.

Consider the Boltzmann equation (1). Introducing the grid velocity $\mathbf{w}(w_1, w_2)$, w_1 is along x-direction and w_2 is along y-direction, into the Boltzmann equation (1), it becomes

$$\left(\frac{\partial F}{\partial t} + w_1\frac{\partial F}{\partial x} + w_2\frac{\partial F}{\partial y}\right) + (v_1 - w_1)\frac{\partial F}{\partial x} + (v_2 - w_2)\frac{\partial F}{\partial y} = 0 \tag{10}$$

It can be rewritten as

$$\left(\frac{dF}{dt}\right)_{Mov} + \bar{v}_1\frac{\partial F}{\partial x} + \bar{v}_2\frac{\partial F}{\partial y} = 0 \tag{11}$$

where $\bar{v}_1 = v_1 - w_1$, $\bar{v}_2 = v_2 - w_2$ and the special time derivative is defined as

$$\left(\frac{dF}{dt}\right)_{Mov} = \frac{\partial F}{\partial t} + w_1\frac{\partial F}{\partial x} + w_2\frac{\partial F}{\partial y} \tag{12}$$

The derivative $(\frac{dF}{dt})_{Mov}$ is along the curve $\frac{dx}{dt} = w_1$, $\frac{dy}{dt} = w_2$.

An upwind scheme is obtained by splitting with respect to $\bar{v}_1$ and $\bar{v}_2$ (CIR scheme form).

$$\left(\frac{dF}{dt}\right)_{Mov} + \left(\frac{\bar{v}_1 + |\bar{v}_1|}{2}\right)\frac{\partial F}{\partial x} + \left(\frac{\bar{v}_1 - |\bar{v}_1|}{2}\right)\frac{\partial F}{\partial x} + \left(\frac{\bar{v}_2 + |\bar{v}_2|}{2}\right)\frac{\partial F}{\partial y} + \left(\frac{\bar{v}_2 - |\bar{v}_2|}{2}\right)\frac{\partial F}{\partial y} = 0 \tag{13}$$

The Maxwellian F in terms of $(\bar{v}_1, \bar{v}_2)$ and $(\bar{u}_1, \bar{u}_2)$ is given by

$$F = \frac{\rho}{I_o}\frac{\beta}{\pi} exp[-\beta(\bar{v}_1 - \bar{u}_1)^2 - \beta(\bar{v}_2 - \bar{u}_2)^2 - \frac{I}{I_o}] \tag{14}$$

where $\bar{u}_1 = u_1 - w_1$ and $\bar{u}_2 = u_2 - w_2$. The discrete approximations to the spatial derivatives are calculated using weighted least squares expressions (4) and corresponding upwind stencil. The q-variables are used for defect-correction to get higher order accuracy. The q-variable at node P_i in a frame moving with velocity $\mathbf{w}_j(w_{j1}, w_{j2})$ are given by

$$q_i(\mathbf{w}_j) = \begin{bmatrix} \ln \rho_i + \frac{1}{\gamma - 1}\ln \beta_i - \beta_i(\mathbf{u}_i - \mathbf{w}_j)^2 \\ 2(u_{i_1} - w_{j_1})\beta_i \\ 2(u_{i_2} - w_{j_2})\beta_i \\ -2\beta_i \end{bmatrix} \tag{15}$$

Consider now the state update at a point P_o having grid velocity $\mathbf{w}_o(w_{o1}, w_{o2})$. It may be noted that the q-variables and their derivatives required for defect-correction at node P_o are to be obtained in the frame moving with grid velocity $\mathbf{w}_o$. Using the modified q-variables $\tilde{q}_i$ and $\tilde{q}_o$ defined by

$$\tilde{q}_i(\mathbf{w}_o) = q_i(\mathbf{w}_o) - \frac{1}{2}\Delta x_i q_{x_i}(\mathbf{w}_o) - \frac{1}{2}\Delta y_i q_{y_i}(\mathbf{w}_o) \tag{16}$$

$$\tilde{q}_o(\mathbf{w}_o) = q_o(\mathbf{w}_o) - \frac{1}{2}\Delta x_i q_{x_o}(\mathbf{w}_o) - \frac{1}{2}\Delta y_i q_{y_o}(\mathbf{w}_o) \tag{17}$$

we then get the modified Maxwellian difference

$$\Delta\tilde{F}_i(\mathbf{w}_o) = F(\tilde{q}_i(\mathbf{w}_o)) - F(\tilde{q}_o(\mathbf{w}_o)) \tag{18}$$

The q-derivatives in equations (16, 17) are calculated using least squares expressions with full stencil. The Maxwellian difference ΔF_i in the least squares expressions (4) is replaced with modified Maxwellian difference $\Delta\tilde{F}_i(\mathbf{w}_o)$ to get higher order accuracy.

4 Results

The two methods q-LSKUM with upwind stencil (M-I) and q-LSKUM with full stencil (M-II) are tested for flow past stationary NACA0012 airfoil. Here, the results (C_p plots and Mach contours) for subsonic case (M=0.63, AOA=2 deg.) and for transonic case (M=0.85, AOA=1 deg.) are presented in Fig. 2. The lift coefficient C_L and drag coefficient C_D for these cases are given below:

Method	*Subsonic case*		*Transonic case*	
	C_L	C_D	C_L	C_D
M-I	0.3178	0.019	0.3677	0.0572
M-II	0.3261	0.0009	0.3730	0.0549
AGARD	0.3335	0.00003	0.379	0.0576

It can been seen that for transonic case, the shock is much sharper for M-II and C_L is close to AGARD best value. The results obtained by M-II are in very good agreement with AGARD [7] results for subsonic case. For example, C_L is close to AGARD best value for M-II and further C_D (which is theoretically zero) is again close to zero. By choosing different weights in the least squares expressions (4), solution accuracy can be further improved.

The new method q-LSKUM-MG has been applied to NACA0012 airfoil oscillating sinusodially. The Fig. 3 shows C_l versus angle of attack and some instantaneous C_p plots against AGARD [8] obtained for the test case having M_∞=0.755 and α=0.016^o + 2.51$^o\sin(\omega t)$. These plots are in very good aggreement with AGARD experimental results.

References

1. M.Z. Dauhoo, A.K. Ghosh, V. Ramesh, S.M. Deshpande, *q-LSKUM A New Higher Order Kinetic Method for Euler Equations using Entropy Variables*, ISCFD'99, September 1999, Bremen, Germany.
2. A.K. Ghosh, *Robust Least Squares Kinetic Upwind Method for inviscid compressible flows*, Ph.D. Thesis, Dept. of Aerospace Engg. IISc, Bangalore, June 1996.
3. V. Ramesh, Ph.D. Thesis, Dept. of Aerospace Engg., IISc, Bangalore. (under preparation)
4. S.M. Deshpande, *NASA Technical Paper-2583*, 1986.
5. J.C. Mandal, *Kinetic Upwind Method for Inviscid Compressible Flows*, Ph.D. Thesis, Department of Aerospace Engineering, IISc, 1989.
6. V. Ramesh, S.M. Deshapnde, *Least Squares Kinetic Upwind Method on Moving Grids* , FM report, No. 99 FM1, CFD Centre, Department of Aerospace Engineering, IISc, January 1999.
7. AGARD-AR-211
8. AGARD Report No. 702

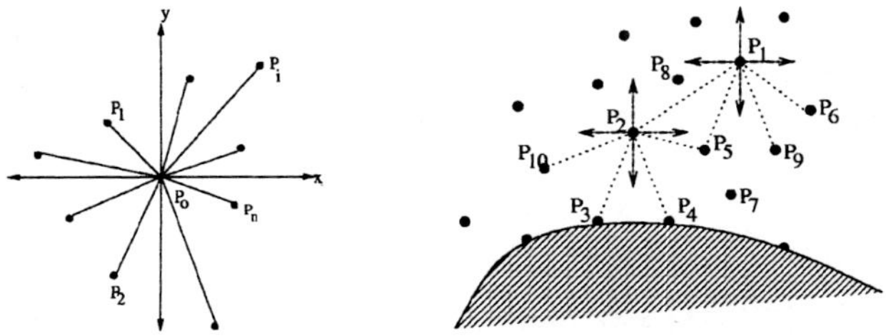

(a) At a point (b) Near boundary

Fig. 1. Figure shows connectivity

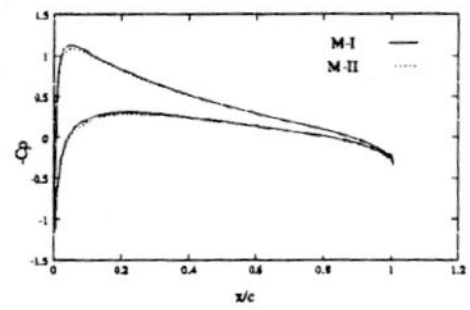

(1a) C_p plot(M=0.63, AOA=2deg.)

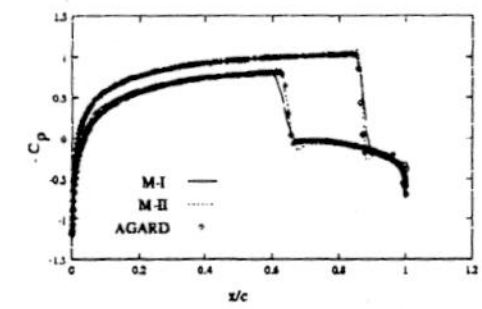

(2a) C_p plot(M=0.85, AOA=1deg.)

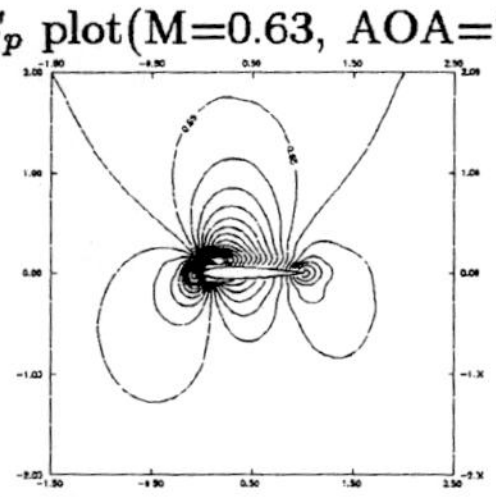

(1b) Mach contour for M-II

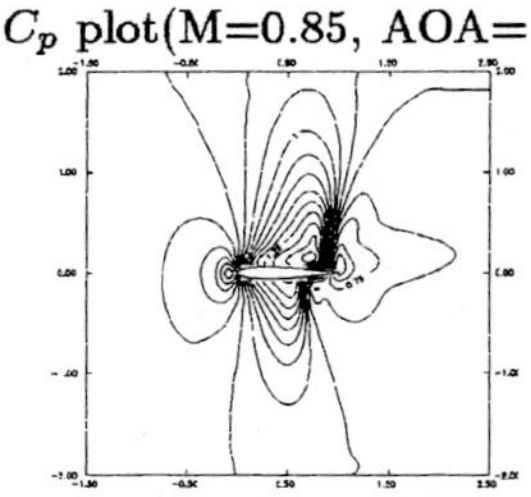

(2b) Mach contour for M-II

Fig. 2. Plots for NACA0012 airfoil using q-LSKUM with upwind stencil (M-I) (with first order boundary treatement) and q-LSKUM with full stencil (M-II) (2nd order treatment for entire domain)

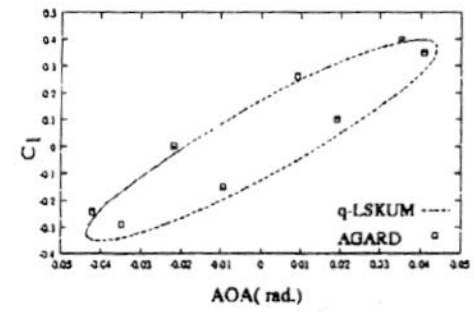

(1a) Lift coefficient versus angle of attack

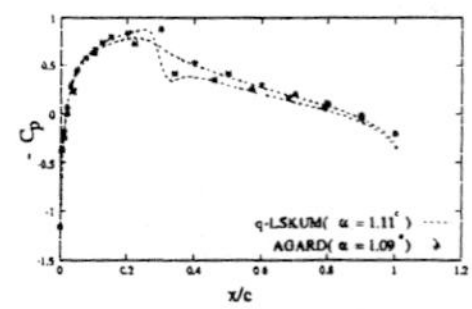

(1b) Cp plot(up)

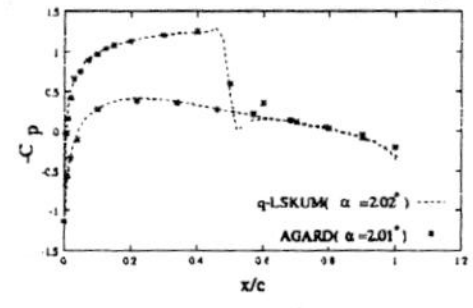

(1c) Cp plot(down)

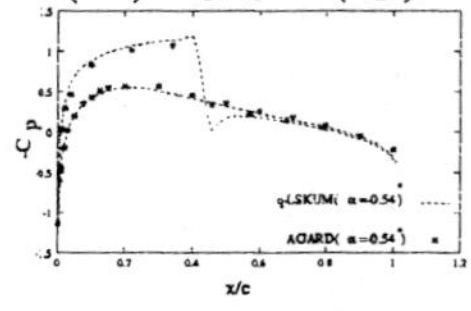

(1d) Cp plot(up)

Fig. 3. q-LSKUM-MG on oscillating NACA0012 airfoil, M=0.755, $\alpha = 0.016^o + 2.51^o \sin(wt)$ and reduced frequency=0.0814

Part XXII

Level Set / VOF

Numerical Simulation of Droplet and Bubble Flows Using Fast Level Set Method

Ruquan Liang and Nobuyuki Satofuka

Kyoto Institute of Technology, Matsugasaki, Sakyo-ku, Kyoto 606-8585, JAPAN

Abstract. The unsteady motions of bubble and droplet in bubble and droplet flows are investigated using a fast level set method combined with a method of lines, and several efficient techniques of level set method are presented. We consider the motions of a droplet falling onto a step in air and two rising bubbles in water.

1 Introduction

Droplet and bubble flows, in which many interesting propagatings of interface are involved, play an important role in many nature and industrial processes such as chemical engineering, petroleum refining, and semi-conductor manufacturing. To fully understand the behaviour of such flows, the basic micro-mechanisms as well as the interaction between different phases need to be well investigated. The approaches available for treating moving interfaces in such investigations include Lagrangian approaches, such as boundary-fitted grid and boundary-integral methods, and Eulerian approaches, such as volume-of-fluid, level set, phase-field, and enthalpy methods. As an Eulerian approach, the level set method has been widely used to captur moving interfaces since its inception in [1]. The merits of this method are well known by now. In this paper the moving interfaces are treated through a fast level set method.

2 Level set method

In briefly, the level set function is a signed normal distance from the interface. Consider a closed moving interface $\Gamma(t)$ propagating with speed $\mathbf{F}$ normal to itself, and $\Omega(t)$ is the region that $\Gamma(t)$ enclosed. We associate with $\Omega(t)$ an auxiliary function $\phi(x,t)$, which is called level set function,

$$\phi(\mathbf{x},t) = \begin{cases} dist(\mathbf{x},\Gamma_{(t)},t) & if\ \mathbf{x}\ outside\ \Gamma_{(t)} \\ -dist(\mathbf{x},\Gamma_{(t)},t) & if\ \mathbf{x}\ inside\ \Gamma_{(t)} \end{cases} \tag{1}$$

$$\Gamma(t) = \{\mathbf{x} \in \Omega : \phi(\mathbf{x},t) = 0\} \tag{2}$$

As one of the nice features of level set method, the unit normal vector of the interface $\mathbf{n}$ and normal speed $\mathbf{F}$ can be simple represented in terms of $\phi(\mathbf{x},t)$,

$$\mathbf{n} = -\frac{\nabla\phi}{|\nabla\phi|} \tag{3}$$

$$\mathbf{F} = \frac{\phi_t}{|\nabla\phi|} \tag{4}$$

We have

$$\phi(\mathbf{x}(\mathbf{t}), t) = c \tag{5}$$

By chain rule, we get

$$\phi_t + \mathbf{u} \cdot \nabla\phi = 0 \tag{6}$$

Moreover, in multiphase cases including gas-liquid-solid three phases, a level vel set of single function can not describe the multiple interfaces with different properties and level set multiple functions are needed. An earlier work on this side was done in [2]. In general, the level set multiple functions are coupled through local and/or global constraints.

3 Reinitialization of level set function

3.1 Reinitialization

The reinitializing problem consists of the stable specification and efficient evaluation of ϕ. In general, even we initialize the level set function ϕ as a signed distance from the interface, while eq.(6) will move the level set $\phi = 0$ at the correct velocity, ϕ will no longer be a distance function and can become irregular at later time. This will necessarily results in the variation of interface thickness in time, making further computation and contour plotting highly inaccurate. Fortunately, moving interface by solving the level set equation differs from solving a general PDE. We can ignore all values of ϕ far from the zero level set and replace the solution ϕ at any time by another function ϕ_0 with the same zero set as ϕ and then take ϕ_0 as the initial data to use. In [3], an iterative procedure was proposed to fulfil above process.

$$\phi_t = sign(\phi_0)(1 - |\nabla\phi|) \tag{7}$$

$$\phi(\mathbf{x}, 0) = \phi_0(\mathbf{x}) \tag{8}$$

3.2 Improvement for reinitialization

By carrying out reinitialization, the level set function will remain a distance function at later time, and this will ensure the interfaces having a finite thickness for all time. Another important issue is mass conservation. For incompressible flows, the total mass is conserved in time. However, even with the above reinitializing procedure, it has been found that considerable amount of total mass is lost in time. To overcome this difficulty, we make an improvement on reinitializing procedure aimed at preserving the total mass in time. We do not make the reinitialization in all grid points. We regard grid points which bound the interface as the nearest points to the interface, and the reinitialization in those points will be neglected. Therefore, it is necessary to pick the grid points bounding the interface out. For grid point (i,j), if any one of the follows,

$$\phi(i,j)\phi(i+1,j) < 0 \tag{9}$$

$$\phi(i,j)\phi(i-1,j) < 0 \tag{10}$$

$$\phi(i,j)\phi(i,j+1) < 0 \tag{11}$$

$$\phi(i,j)\phi(i,j-1) < 0 \tag{12}$$

is satisfied, the grid point (i,j) is regarded as a grid point that bounds the interface.

4 Treatment of artificial interface

Because of the discontinuity for density ρ, numerical instabilities will yield at the interface, especially for large density ratios. In order to overcome the difficulty, a finite thickness 2α is given as the artificial interface, and density ρ become smooth at the interface. As a side effect from the definition of artificial interface 2α, the simulation accuracy decreases when collisions occur. To overcome the problem, we propose a scheme aimed at stripping the half of the artificial interface α outside the real interface when collision occurs. The outline of the scheme is given as follows,

(1) Judge the collision has happened.

(2) Decrease properly the values of level set function in the points where collision has happened, and then update ϕ for one time step.

(3) Reinitialize the level set function.

(4) Judge if the values of ϕ in the points where collision has happened become zero. If not so, go back to step (2).

5 Fast level set method: Narrow tube level set method

The original level set method in [1] updates all the level sets through out all space. This needs considerable computational effort, particularly in three dimensions. In fact, an efficient computational labor should be confine to a narrow tube around the interface. An earlier approach on this side was developed in [4]. In this paper we present our narrow tube method, which pays attention to the balance between work spent updating points in the tube and work spent doing reinitialization. Let $0 < \beta < \gamma$ be two constants which are comparable to Δx whose values will be determined below. For a given open region $\Omega(t)$ with interface $\Gamma(t)$, we define a narrow tube

$$M = \{\mathbf{x} \in \Omega : |\phi(\mathbf{x},t)| < \gamma\} \tag{13}$$

We move the tube with the motion of the interface and reinitialize the level set function only in the tube. Then we also define an extra tube

$$N = \{\mathbf{x} \in \Omega : |\phi(\mathbf{x},t)| < \beta\} \tag{14}$$

to monitor the behavior of the level set function as well as the movement of the interface and trigger the reinitialization only when necessary. When the interface moves and touches the edge of the tube N, the calculation is stopped. A new initial level set function corresponding to the signed distance function and two new tubes M and N are re-built. Moreover, a very thin narrow tube means that computation is done only close to the interface, and reinitialization should be done every time step. A very large narrow tube decreases the effect of fast marching. In our numerical tests, we choose $4\Delta x < \gamma < 8\Delta x$ and $\gamma - \beta = \Delta x$. On the other hand, we introduce two equal-sized $1D$ arrays to store the index of the grid points in the two tubes. We outline the method as follows,

(1) Construct level set function and two narrow tubes M and N.

(2) Update ϕ in tube M for one time step.

(3) Repair the boundary of tube M by interpolation.

(4) Check if the interface touches the edge of tube N. If so, trigger the reinitialization in tube M, and construct two new tubes M and N and then go to step (2). Otherwise, go to step (2).

Moreover, We briefly summarize the present strategy. The Navier-Stokes equations in primitive variable formulations are solved on a staggered grid by the method of lines, and the Poisson equation is iteratively solved by Successive Over Relaxation (SOR) method. A second-order Adams-Bashforth time integration scheme is combined with the QUICK and central finite difference approach for spatial discretization, while CSF model is used to treat the surface tension force on the interface.

6 Numerical results

In this section, we apply the present strategy to several problems to show the power of the strategy. As the first example, we compare the perfomance between our narrow tube level set method and global level set method on the following problem. Four initially circular bubbles rotate in the divergence-free shearing flow given by

$$F(x,y) = \frac{max(1-(1-x^2-y^2)^4,0)}{8(x^2+y^2)}(-y,x) \tag{15}$$

The computation is performed on the domain $D = [-1,1]^2$. Table 1 tabulates the CPU time (in seconds) used in each step for the narrow tube method and global method and their ratios. The computation are performed on an $ONYX$ R8000 workstation. Fig.1 shows the distortion produced by the flow. We can find that the performance in CPU time used is highly good with narrow tube method, especially in grids of large sizes. No distinguishable differences are noticed between the results of narrow tube method and global method.

Next, we present a computation for a droplet falling onto a step, which is what happen in real life. Two level sets are used, namely, φ for the droplet, and ψ for the step. In Fig.2, we compare the results of (2a)$We_{wall} = 1.0$, $We_{gas} = 0.5$ with (2b) $We_{wall} = 20.0$, $We_{gas} = 10.0$. For the two cases, we have $\rho_l/\rho_g = 1000.0$,

Table 1. Comparison of CPU time for Global and Local Level Set Methods

Mesh size	Global	Local	Local/Global
51 × 51	0.0296	0.0154	0.5203
101 × 101	0.1209	0.0623	0.5153
201 × 201	0.5985	0.2847	0.4757
301 × 301	2.3391	0.8195	0.3503
401 × 401	7.2062	1.9739	0.2739

t=0.0 t=8.0(narrow tube method) t=8.0(global method)

Fig. 1. A collection of four initially circular bubbles transported by a divergence-free shearing velocity.

$\mu_l/\mu_g = 100.0$, $Re = 100.0$, $Fr = 1.0$. In case (2a), the surface tension is very strong. The droplet falls onto the second step, and then falls onto the lower wall after the collision happens. In case (2b), the surface tension is not so strong as case (2a). The droplet is separated into two parts after the collision happens. One part falls onto the second step, and then falls onto the lower wall and undergoes oscillation. At the same time, another part remains onto the first step, and forms stable shape of half of ellipse due to the surface tension. Finally, we show a simulation on two rising bubbles to test the narrow tube level set method on fully three-dimensional problems. We have $\rho_g/\rho_l = 1.0/1000.0$, $\mu_g/\mu_l = 1.0/100.0$, $Re = 100.0$, $Fr = 1.0$, $We = 20000.0$. The two bubbles are initially spheres. As the two bubbles rise they deform considerably, and merge into one big bubble. The big bubble collides and extends onto the upper wall at time $t = 9.6$. No distinguishable differences are noticed between the results using our narrow tube method and the results obtained before using global method.

7 Conclusions

The computations of droplet and bubble flows were carried out using a fast level set method. The level set multiple functions enable us to deal with droplet and bubble flows with complex geometry in Cartesian coordinates, while fast level set method gives us the ability to speed up the computations. The improvement for reinitialization and the treatment of artificial interface make the accuracy and the simple implementation of this strategy are very attractive. The numerical

t=0.5 t=1.5 t=5.0

(2a) $We_{wall} = 1.0$, $We_{gas} = 0.5$

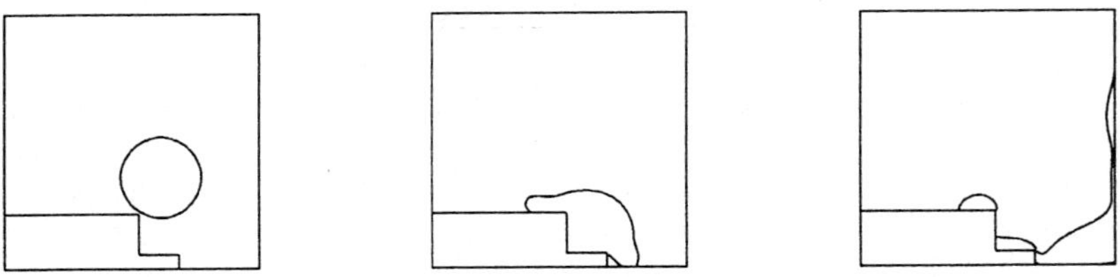

t=0.5 t=1.5 t=5.0

(2b) $We_{wall} = 20.0$, $We_{gas} = 10.0$

Fig. 2. A droplet falls onto a step.

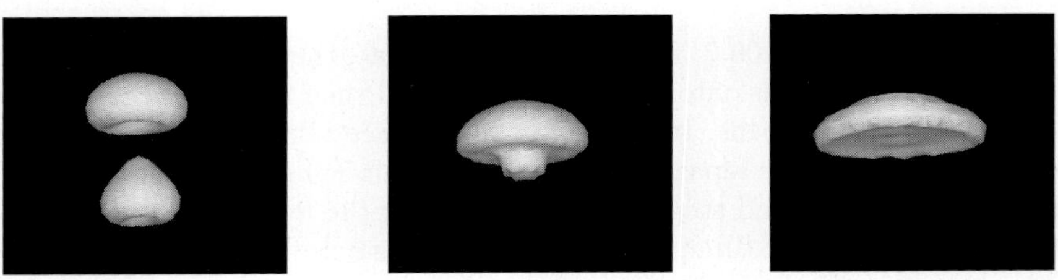

t=1.2 t=5.4 t=9.6

Fig. 3. Two rising bubbles in water.

results reproduce the essential physics of the problems studied, and the mass conservation is satisfactory. We remark that the ultimate goal of level set method is to move the interface, and an adaptive mesh can concentrate computational effort near the interface. We will report the work on this side in a separate paper.

References

1. S. Osher, J.A. Sethian: J. Comput. Phys. 79, 12 (1988)
2. B. Merriman, J. Bence, S. Osher: J. Comput. Phys. 112, 334 (1994)
3. M. Sussman, P. Smereka, S. Osher: J. Comput. Phys. 119, 146 (1994)
4. D. Adalsteinsson, J. A. Sethian: J. Comput. Phys. 118, 269 (1995)
5. R.Q. Liang, N. Satofuka: Jpn. J. JSME. **64-617**, 42 (1998)

A Front Tracking Concept on Arbitrary Grids

F. Bet, D. Hänel, O. Gloth, L. Tran, and R. Vilsmeier

Institute of Combustion and Gasdynamics, University of Duisburg
D-47048 Duisburg, Germany, E-mail: haenel@vug.uni-duisburg.de

Abstract. Level-set concepts are used for front tracking in the frame of Finite-Volume methods on structured and unstructured grids. The tracking method is based on a local transport and normalization of the level set function, combined with a flux separation approach. A tracking concept on structured grids, simplified under the assumptions of free surfaces, is now in intensive use for the prediction of hydrodynamic flows around ships. An extension of this concept to arbitrary grids and different types of discontinuities is under investigation. Both methods are demonstrated with examples for free surface flow around ships on structured grids and for tracking of moving shocks on unstructured grids. Ongoing investigation are concerned with tracking of detonation waves and with interfaces in porous media.

1 Basic relations

The governing equations are the conservation equations for mass, momentum and energy for an inviscid, compressible or incompressible fluid in integral form:

$$\frac{d}{dt}\int_V \mathbf{Q}\, dV + \oint_A \mathbf{H}\cdot\boldsymbol{n}\, dA = 0 \tag{1}$$

A discontinuous solution is given as the weak solution of system (1), which is explored to define the jump conditions and to determine the front speed c. Generally there exist two types of discontinuities. The normal discontinuities, like shocks or detonation waves, move with a speed which is a function of the local jump conditions. The tangential discontinuities, i.e. contact discontinuities or interfaces, are separating two immiscible fluids and are moving with the local fluid velocity.

The front tracking methods are formulated in the Eulerian frame, i.e moving discontinuities are tracked on a fixed grid. The geometry and the movement of a front is described by a level-set function $G(\boldsymbol{r},t)$, [1,2], which is a continuous function over a discontinuity. The discontinuity is defined by a discrete iso-value $G(\boldsymbol{r},t) = G_0 = 0$. The motion of the front with a propagation speed $\boldsymbol{c}$ is described by the convection of the G-function as:

$$\frac{\partial G}{\partial t} + \boldsymbol{c}\cdot\nabla G = 0 \tag{2}$$

For reasons of accuracy the level set function should be a distance function with respect to the front. This requirement is satisfied if the function $G(\boldsymbol{r},t)$ fulfills the normalization condition:

$$|\nabla G| = 1 \tag{3}$$

This work has been supported by the German Research Society (DFG) in the frame of the research program: "Hyperbolic Conservation Laws" (ANUME)

2 Incompressible Flow with Free Surfaces

An essential contribution of resistance of a moving ship stems from the wave drag caused by the formation of surface waves. The wave formation is a dynamic effect, governed by the inviscid equations. The wave forming free surface is a material interface between air and water. If breaking waves and wind forces on the surfaces are not taken into account, then surface tension and viscous stresses are negligible at the interface. These assumptions, used here, which simplify essentially the treatment of the discontinuous interface. In particular the flows in water and air are decoupling, and for the computation of ship flow the air flow is not of interest and can be neglected.
The free surface is defined by the level set function $G(\boldsymbol{r},t) = G_0 = 0$. The kinematic condition of the free surface defines the front speed $\boldsymbol{c}$ of the local fluid velocity, which is used in the transport equation of the level set function Eq. (2):

$$\boldsymbol{c} = \boldsymbol{v}_{water} \quad \text{at} \quad G(\boldsymbol{r},t) = 0 \tag{4}$$

The dynamic condition reduces to equilibrium of pressure forces normal to surface

$$p_{water}(G_0(\boldsymbol{r},t),t) = p_{air} = const. \quad \text{at} \quad G(\boldsymbol{r},t) = 0 \tag{5}$$

The hydrodynamic flow is described by the Euler or Navier-Stokes equations of an incompressible fluid including the gravity force. These equations are formulated in a Finite-Volume manner on a curvilinear, structured grid. To avoid the pressure relaxation by solving a Poisson equation, the method of artificial compressibility is used. The compressibility is introduced by an artificial time derivative of the pressure in the continuity equation, where the artificial speed of sound c is chosen in the order of a typical fluid velocity. In this form, the method is consistent with the steady-state only. With this preconditioning the equations of incompressible fluids behave like the compressible one and are discretized and solved in a similar way. Thus the fluxes are discretized by a modified Roe flux and the system of equations is solved with an explicit multi-stepping scheme.
The method is applied to 2-D and 3-D hydrodynamic flows around ships, [6].

The G-equation Eq. (2) is discretized in the same upwind manner as the conservation equations and is solved by the multi-stepping method at each time step. The front velocity $\boldsymbol{c}_S$ is determined from the local fluid velocity at the position of the surface $G(\boldsymbol{r},t) = 0$. Renormalization is performed after a few time steps using an iterative procedure with a filter function, similar as in [3]. The algorithmic treatment of the surface tracking becomes easy and elegant by the fact, that due to the previous assumptions, the flow in the region of air is decoupled and not of interest for the computations. Thus the grid points above the the surface are used as dummy points, on which the variables from the interesting range in the water are extrapolated, such that the hydrodynamic flow can be computed without interruption at the surface.
The calculation starts from a time level t where all variables are known in the range of water. Then the velocities are extrapolated, usually with second order accuracy, to the dummy points in the "air". The dynamic condition at the surface, Eq. (5), is satisfied by an interpolation of the pressure in the water such

that the pressure on the surface $G = 0$ is equal the prescribed value p_{air}. Then the conservation equations and the level set equations are advanced one time step without interrupting at surface and the procedure starts again.
It has to be mentioned that the accuracy of the treatment of the free surface depends strongly on the interpolation procedure. In the present calculations the wave elevations are assumed to be small compared to the ship length. This restriction is however not crucial for computations of ship flow at usual cruising speeds as a large number of calculations have shown.
A classical example of validation is the flow around a dived airfoil (NACA0012) under an angle of attack of 5° in water flow with the Froude number $Fr = 0.567$. The data are according the experimental studies in [7]. The deformation of the free surface is shown in Fig. 1 for calculation using an adaptive grid (above) and using the present level set method (below). A comparison of the wave height of both methods with the experiment is given in Fig. 2.

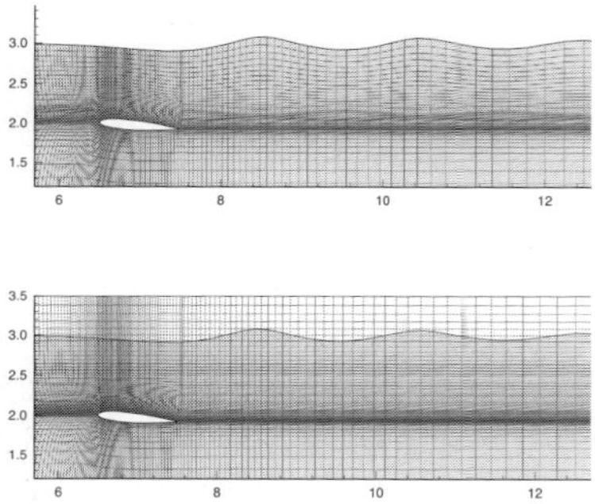

Fig. 1. Grids and free surfaces for a dived NACA0012 airfoil, computed with adaptive grid (above) and with level-set

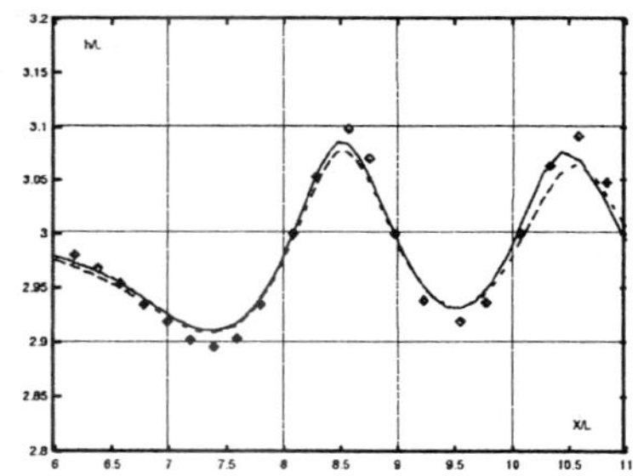

Fig. 2. Comparison of wave heights for dived NACA0012 airfoil, – adaptive grid, - - level-set, o experiment

Comparisons of the wave heights with experiment and the convergence of the drag drag coefficients to the experimental values for 3-D computations of the model ship configurations "series 60 ship" and "Wigley "ship are shown in the Fig 3 and Fig 4.
More complex computations are performed e.g. for inland ships in restricted water.

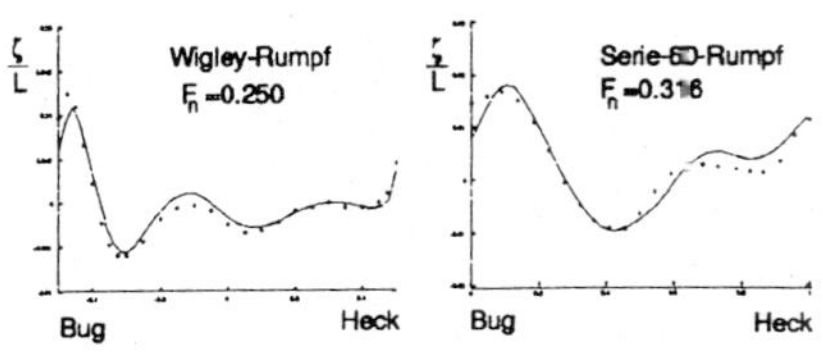

Fig. 3. Comparison of computed and measured wave heights for "series 60" and "Wigley" ships

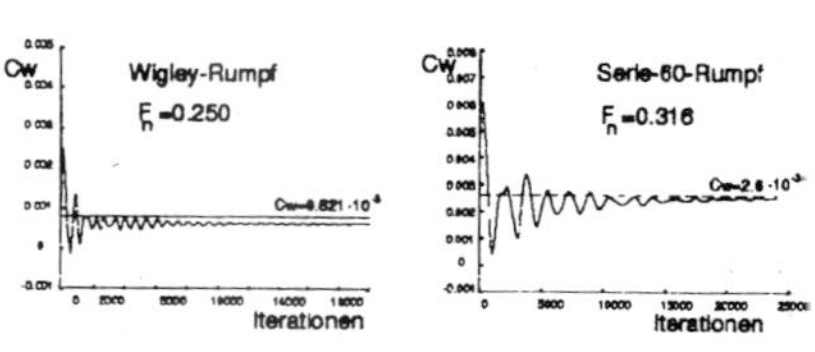

Fig. 4. Convergence of drag coefficients for "series 60" and "Wigley" ships

3 Front tracking on unstructured grids

Aim of this investigation is a generalization of previous tracking concept to arbitrary grids and arbitrary types of discontinuities. The level-set function has shown to be a very flexible tool for the transport and geometrical description of a discontinuity, which can be formulated principally in the same way on unstructured¡grids. However, material interfaces, like free surfaces, are easier to handle because the front velocity is known by the fluid velocity. But for normal discontinuities, like shocks or detonation wave, the numerical determination of the front velocity and their distribution into the scalar field of the level-set function becomes more difficult. The idea of extrapolation of variables to front is adopted from the previous method, which is used here for updating the states left and right of the discontinuity as well for the formulation of cell interface fluxes near the discontinuity. The level set function is formulated locally in a narrow band near the discontinuity. The tracking becomes more flexible but requires special algorithms.

The Finite-Volume solution of the conservation laws Eq. (1) is solved on arbitrary grids employing a nodal finite volume method, Fig. 5. A discrete formulation of the above system reads as follows:

$$\frac{\Delta \mathbf{Q}}{\Delta t}|_{Vd} + \frac{1}{V_{Vd}} \sum_{i=1}^{nv} \mathbf{H}_i \, \boldsymbol{n}_i \, \Delta A_i = 0 \tag{6}$$

where Vd is a control volume with the size V_{Vd} and the sum is carried out over all bounding segments. The Finite-Volume solution is performed with the object-oriented MOUSE package [4], currently in development at the site of the authors.

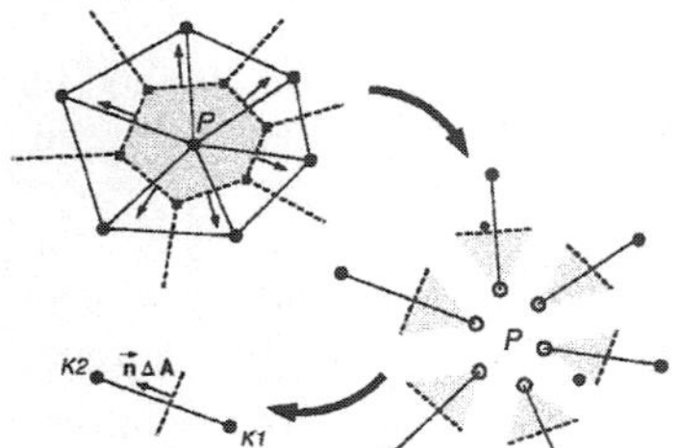

Fig. 5. Control volume

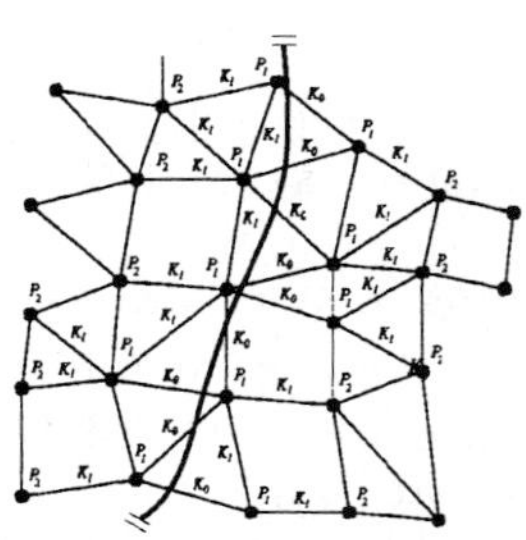

Fig. 6. Neighbourship relations

The level-set solution is performed locally on nodes of the grid in two neighbourship levels only, as sketched in Fig. 6, [5]. These nodes are designed as P_1 and P_2 with the edges K_0 and K_1. The level-set function G is advanced by solving Eq. (2). The integration requires the knowledge of speed c on all these nodes, which however is available for shocks only at the front itself. Therefore the term $c \cdot \nabla G$ is directly evaluated for the edges K_0, where the gradient ∇G can easily be computed on all surrounding nodes employing a Gaussian-type integration on the finite volumes or a least squares method. A directional upwinding is applied in the direction of the edges, combined with a central formulation for the transverse component due to lack of information. Since the discrete values for G are stored at the nodes of the mesh a redistribution of the time derivative

$\frac{\partial G}{\partial t}$, computed at the edges onto the appending nodes has to be performed. As the propagation speed is generally not available at edges K_1 a discretization on these is performed with extrapolated speeds. This residual is however only required, if the corresponding node in the first neighbourship gets overrun.

The normalization of the level-set function according to Eq. (3) G-function normal to the discontinuities. is achieved here with a pointwise iteration process for nodes in the definition range around the discontinuity. The iteration process starts at the nodes of the first neighbourship level P_1, iterating up to convergence, and then proceeds to the next level. Within a level, the pointwise iteration is carried out by a two step method, based on least square approaches. In a first step, the direction of the gradient ∇G^n is computed, in the second step the functional value of G^{n+1} is set to satisfy the slope condition. The least square approach of the first step results in the gradients of G in the second neighbourship and reads:

$$\Phi_{par} = \sum_{j \in E_{par}(i)} \left(\frac{\nabla G_i \cdot \boldsymbol{m}_{ij} - (G_j - G_i)}{|\boldsymbol{m}_{ij}|^{q/2}} \right)^2 \rightarrow \text{minimum}$$

with $\boldsymbol{m}_{ij} = \boldsymbol{X}_j - \boldsymbol{X}_i$ and a distance weighting exponent $q \geq 0$. Step 2 uses again a least square approach, but now projected in the direction of the gradient ∇G_i^n:

$$\Phi_{hyp} = \sum_{j \in E_{hyp}(i)} \left(\frac{C \frac{\nabla G_i^n}{|\nabla G_i^n|} \cdot \boldsymbol{m}_{ij} - (G_j - G_i^{n+1})}{|\boldsymbol{m}_{ij}|^{q/2}} \right)^2 \rightarrow \text{minimum}$$

The result of the second step is a new G-value which is nearly a distance function. Additional corrections have to performed to preserve the exact position of the front after normalization.

Flux separation is used to update the cell interface fluxes on control volumes next to the front. The flux contributions on edges K_0, crossing the front, are computed separately, using variables extrapolated only from the "left" and "right" side, resp. These extrapolated variables define the jump conditions, from which the local front velocity, e.g. with Rankine-Hugoniot conditions, can be determined. As a node of the mesh gets overrun by a front, the variables stored at the corresponding nodes, switch to the other side. At present these new values are obtained by extrapolation from the the surrounding neighbours at the new side.

Note, that this method does not ensure completely conservation, however, different test problems have shown that this tracking concept performs with sufficient accuracy on arbitrary, even rather coarse meshes.

The computational examples, shown here, are concerned with tracking of shock waves. A test of the robustness is the computation of a radial symmetric, focusing shock wave on a triangulated mesh, Fig. 3. The shock profiles remain sharp until the level-set range shrinks to one cell. The density ratio reaches the value of 5.73, approaching the limit of 6 for an infinite Mach number.

The development of a shock to its steady state position for a cylinder at $Ma_\infty = 2$ is shown in Fig. 3. The initial condition is a straight shock, located in front of the cylinder. The fixed grid shown in a), the range of the level-set is given for the initial b), an intermediate c) and for the final state (d). The level $G = 0$ in the final state is shown in e). The position agrees very well with a captured solution.

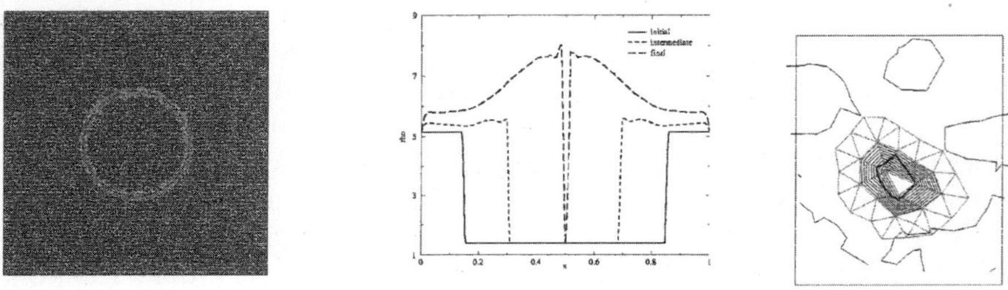

Fig. 7. Shock focusing: Grid (left), density profiles (middle), zoom near focus (right)

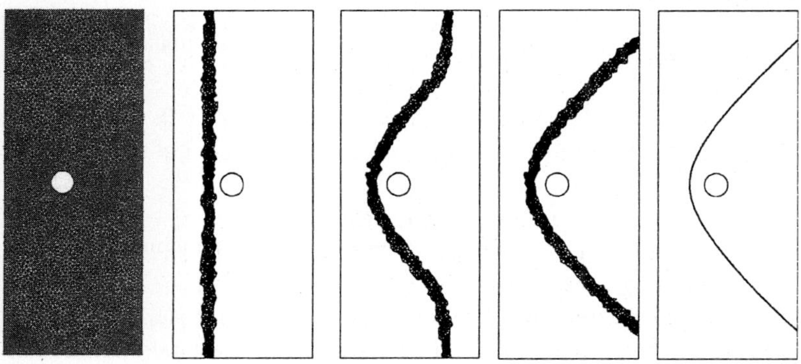

a) b) c) d) e)

Fig. 8. Development of a tracked shock wave for flow with $Ma = 2$ around a cylinder

References

1. A. Kerstein, W. Ashurst, F. Williams: Field Equation for Interface Propagating in an Unsteady Homogeneous Flow Field. *Phys. Rev. A, vol. 37, pp2728-2731*, (1988).
2. J. A. Sethian: Level Set Methods and Fast Marching Methods: Evolving Interfaces in Computational Geometry, Fluid Mechanics, Computer Vision and Materials Science. Cambridge University Press, (1999.)
3. M. Sussman, P. Smereka, S. Osher: A Level Set Approach for Computing Solutions to Incompressible Two-Phase Flow. *J. of Comp. Physics, vol. 114, pp 146-159*, (1994).
4. D. Hänel, A. Dervieux, R. Vilsmeier, O. Gloth, C. Viozat, L. Fournier: Development of Navier-Stokes Solvers on Hybrid Grids. In: Numerical Flow Simulation I, pp 89-111, Vieweg-Verlag, (1999.)
5. L.Tran, R.Vilsmeier, D.Hänel: A local level set method for the treatment of discontinuities on unstructured grids, *FVCA-II 99*, Duisburg, Germany, (1999).
6. F. Bet, D. Hänel, S.D. Sharma: Numerical Simulation of Ship Flow. Proc. of the 22nd Symposium on Naval Hydrodynamics, 9-14 August, Washington (1998).
7. J.H. Duncan: The breaking and non-breaking wave resistance of a two-dimensional airfoil. Journ. of Fluid Mechanics, 126, (1983).

Three Dimensional Computer Simulation of Mould Filling with N Fluids by VOF PLIC and Projection Methods

Jérôme Breil and Jean Paul Caltagirone

Université Bordeaux I, Laboratoire MASTER-ENSCPB,
Av pey Berland, BP 108,
33402 Talence Cedex, France
E-mail: breil@lmaster.u-bordeaux.fr

1 Introduction

The purpose of this paper is to study mould filling using several batches. This process is currently being used by "SNPE Propulsion" to fill the booster rockets of Ariane V. The production cycle of such engines can be broken down into three principal steps [3]. First of all, a stage of malaxation gives us, by mixture of various components, a propellant in thay looks like a very viscous, and homogeneous paste. The loading is then filled by run under the effect of gravity. This filling is carried out in several parts, and for security reasons we cannot prepare the matter necessary to only one mix. It finishes by the cooking of the booster rocket while the propellant reticulates and solidifies itself. Simulations will allow us to validate the quality of the filling but also to foresee the ballistics of the missile. Indeed, several studies have shown the influence of the filling on the thrust curve of missiles.

As shown before, the filling of the loading requires successive castings. However, the dynamic viscosity of the material increases over time, and at the end of the operation, the loading is composed with a different viscosities stacking. Thus, during simulation, it is necessary to track the interface of each fluid independently. Moreover, we solve the Navier-Stokes equations and the transport of the interfaces of the various fluids in 3D because the geometries are relatively complex. The fluid used, propellant, is composed of polymer matrix hardening charged in active particles. Its rheological behavior is Newtonian.

In a first part, we present the numerical methods that we used to solve the Navier-Stokes equations. We especially develop the presentation of a new method: the vectorial projection that enables us to obtain a null divergence speed field. In a second part, we deal with the VOF PLIC method used for the advection of the interface. We develop the method in 2D but the development was also extended to 3D. The last part is devoted to the study of an Ariane V booster rocket filling, taking into account a defect of filling.

2 Flow resolution

The flow is in a first approximation assumed to be incompressible in all the fluids considered and the surface tension is neglected. The Navier-Stokes equations are solved in several stages:

$$\rho\left[\frac{\boldsymbol{V}^*-\boldsymbol{V}^n}{\Delta t}+(\boldsymbol{V}^n.\nabla)\,\boldsymbol{V}^*\right]-dr\nabla(\nabla.\boldsymbol{V}^*)=-\nabla p^n+\boldsymbol{f}+\nabla.\left[\mu(\nabla\boldsymbol{V}^*+\nabla^t\boldsymbol{V}^*)\right] \quad (1a)$$

$$\nabla\,(\nabla.\boldsymbol{V}')=-\nabla\,(\nabla.\boldsymbol{V}^*) \quad (1b)$$

$$\boldsymbol{V}^{n+1}=\boldsymbol{V}^*+\boldsymbol{V}' \quad (1c)$$

$$p^{n+1}=p^n-dr\nabla.\boldsymbol{V}^* \quad (1d)$$

In the first stage (1a), we seek to solve a prediction of the field speed $\boldsymbol{V}^*$ starting from the $\boldsymbol{V}^n$ field using the Augmented Lagrangian method [2]. This method is characterized by the introduction of the Lagrangian parameter, which allows to implicitly impose the incompressibility constraint. Parameter dr, introduced by this method, is a coupling constraint term on the velocity field. It must satisfy the Navier-Stokes equations and the continuity equation. The verification of the two constraints is only possible with an iterative algorithm. However, the convergence of this method is very weak and takes a long time, particularly for the contentment of the incompressibility constraint. The pressure field is obtained as usual in this method (1d).

A correction speed $\boldsymbol{V}'$ is then calculated using $\boldsymbol{V}^*$ by a vectorial projection method (1b) [1]. The boundary conditions on $\boldsymbol{V}'$ are deduced from $\boldsymbol{V}^{n+1}$ and $\boldsymbol{V}^*$. This formulation is equivalent to $\boldsymbol{V}=P\frac{1}{H}\boldsymbol{V}^*$ where H is a sub-space of L^2 with zero divergence and $P\frac{1}{H}$ is the orthogonal projection of L^2 on H.

We obtained $\boldsymbol{V}^{n+1}$ speed by the relation $\boldsymbol{V}^{n+1}=\boldsymbol{V}^*+\boldsymbol{V}'$. The $\boldsymbol{V}^{n+1}$ fields obtained (1c) is null divergence, which is of primary importance for calculating free surface flow. Indeed, to keep the volume constant during the advection of our fluids, we need to have $\nabla.\boldsymbol{V}^{n+1}=0$.

These equations are discretized using the finite volume method on a fixed cartesian grid; the linear systems obtained are solved using a conjugate gradient method, associated with a MILU preconditioning [4].

To validate our method, we used a standard analytical problem: the Green Taylor whirl. Precise solutions were obtained for relatively coarse grids. We show that the correction step does not modify the accuracy of the scheme in space for the velocity and pressure fields.

3 VOF PLIC method

As previously explained, the advection of the interface is carried out using the VOF PLIC method (Piecewise Linear Interface Calculation) [5] [6]. This method can be broken down into three distinct stages. Initially, we use the volumine fraction θ, present on each node, to rebuild the interface using lines in 2D and plans in 3D. To simplify the presentation of the method, we only show in this paper the 2D method (figure 1) and we will only consider the case where AB=BC=h. The 2D and 3D methods have been written in irregular grids, as well. In the first part of the reconstruction, a normal direction to the interface is estimated by a finite-difference formula: $\boldsymbol{m} = \nabla\theta$. The determination of the normal direction and the volumine fraction present in the cell gives us the position of the interface. The most general equation for a straight line in the (x, y) plane with normal $\boldsymbol{m}$ is: $m_x + m_y = r$, in which r is a parameter that needs to be determined. After some symmetrical operations, we can always suppose that $0 < m_x \leq m_y$. The lines of normal direction $\boldsymbol{m}$ which pass by the points B and D give us two critic volumine fractions $\theta_1 = \dfrac{m_x}{2m_y}$ and $\theta_2 = 1 - \theta_1$ (figure 1a). It induces three formulas for r (figures 1 b, c, d):

$$r = \sqrt{2\theta m_x m_y} \text{ if } 0 \leq \theta \leq \theta_1 \tag{2}$$

$$r = (2\theta m_y + m_x) \text{ if } \theta_1 < \theta < \theta_2 \tag{3}$$

$$r = m_x + m_y - \sqrt{2(1-\theta)\, m_x m_y} \text{ if } \theta_2 \leq \theta \leq 1 \tag{4}$$

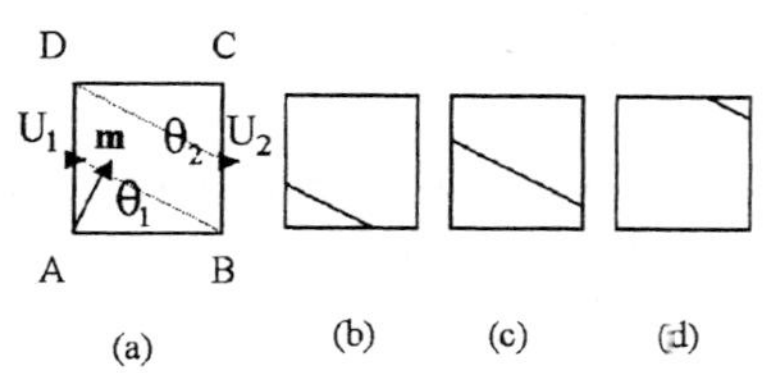

Fig. 1. three types of possible segments in a cell for a normal m given

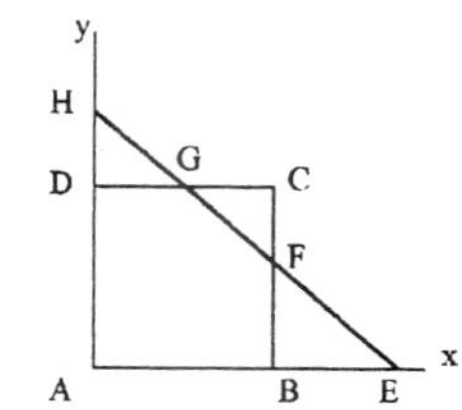

Fig. 2. the area ABFGD gives us the new volume fraction

Now, that the interface is rebuild, we can easily apply the speed field previously calculated on the mesh (figure 1 a) U1 and U2 to a point M(x,y) of the cell. The speed on point M is given by:

$$U = (1-x)\, U_1 + xU_2 \tag{5}$$

Point M moves from its initial position to a new one M'(x',y), where :

$$x' = \left(1 - U_1\frac{\Delta t}{h} + U_2\frac{\Delta t}{h}\right) x + U_1\frac{\Delta t}{h} \tag{6}$$

Using this expression, the new line $m'_x + m_y = r'$ can be calculated, where $m'_x = \dfrac{m}{1 - U_1\frac{\Delta t}{h} + U_2\frac{\Delta t}{h}}$ and $r' = r + m' U_1 \frac{\Delta t}{h}$.

Using the new position of the line, the new value of the volume fraction θ can be evaluated. The method used is an area calculation method. The area is the region delimited by the points ABFGD (figure 2). To obtain this area, it is sufficient to withdraw from triangle AEH the area of triangles BEF and DGH respectively when AE>AB and AH>AD. The area is given by:

$$Area = \frac{r^2}{2m_x m_y} \left(1 - \max\left(r - m_x \mathrm{AB}, 0\right) \left(\frac{r - m_x \mathrm{AB}}{r}\right)^2 - \max\left(r - m_y \mathrm{AD}, 0\right) \left(\frac{r - m_y \mathrm{AD}}{r}\right)^2\right) \quad (7)$$

To determine the new volume fraction present in the cell, it is necessary to calculate the contribution of the cells upstream and downstream to obtain the final voluminal fraction, as shown on figure 3.

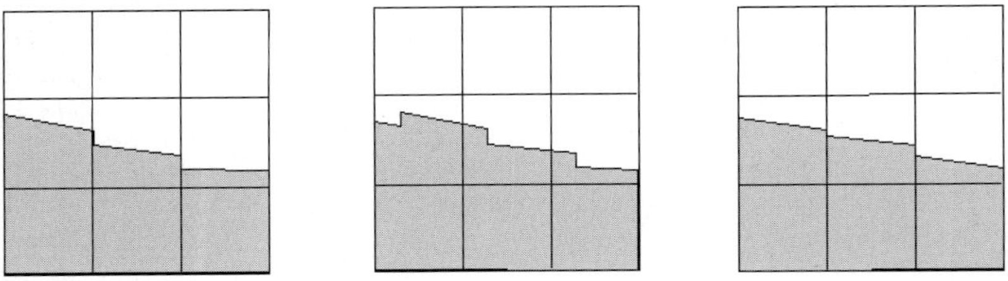

Fig. 3. The three stages of VOF-PLIC method in a positive field speed

The procedure thus programmed for the advection in the X direction is applied to the Y direction. This algorithm using fractional steps simplifies the programming of the method by obviously reducing the number of possible cases in the final calculation of area. Our numerical code is validated by many cases: The Rayleigh-Taylor instabilities, the buckling jets, and cases of pure advection for which analytical speed fields are available.

4 Results

The European space launcher Ariane V is composed by a cryotechnical central stage and by two powder acceleration stages which provide 90 % of the thrust during the first two minutes of flight. The solid propellant motors, ten times bigger than those previously built by European industry, measure 27 meters long for a diameter of 3 meters. Considering their measurements and their mass, close to 270 tons, they are segmented into three parts: a front segment and two big segments, central and rear, of more than 110 tons and each one 10 meters long. The two boosters propel the space launcher out of his fire zone with a thrust of 600 tons each, which explains the importance of a symmetrical thrust.

During the filling, the stratification of the mix will modify the progress of the flame forehead and therefore the thrust curve of the space launcher. In the case of a filling defect due to a cap, in the injector we want to know it's consequences on the combustion. *In situ* test would require a great quantity of propellant and many motors would need to be manufactured, thus proving too expensive and being a general waste this going against certain experimental ethics.

The case we treat in this study is a cylindrical mould 10 meters high and 3 meters in diameter, with a central core of 1.5 meter in diameter. This mould is filled by 3 injectors which are spaced by 0.8 meter of the central axis and separated by 120 degrees. 10 mixes are required to fill this booster rocket and as explained previously, the viscosity of the fluids increases during the filling.

The mix viscosities go from 300 Pa.s to 6000 Pa.s, thus, at the end, the first mix is solid while the last is still fluid. The density of the propellant is $1800kg/m^3$

The speeds used are 5 m/s in the nominal injectors, 2.5 m/s in the blocked injector at the beginning of the filling, 1 m/s in the nominal injectors and 0.5 m/s in the blocked injector at the end. The time step changes during the simulation, because $CFL = U\Delta t/\Delta x = 0.5$ must be maintained for the VOF method. As the geometry becomes too large for a single simulation and as a mix becomes solid during simulation, we perform the total simulation in 4 times on 60*40*40 grids.

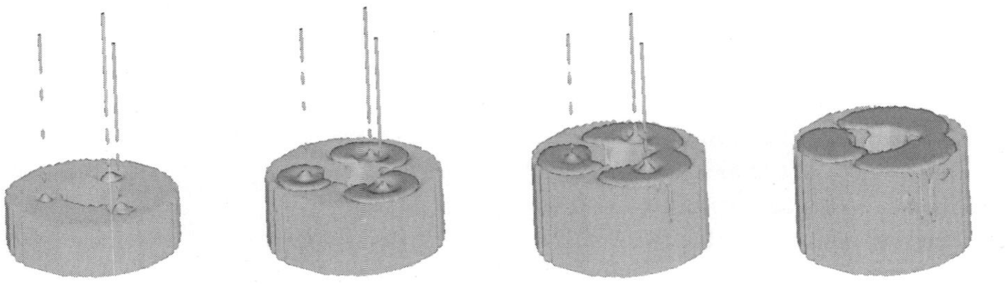

Fig. 4. Filling of the second mix

Figure 4 shows different stages of the second mix filling (the first being initialized). It can be seen that the mix falls in rain. This is due to the acceleration factor which causes constriction and the cut of the paste during the filling. The propellant is injected and when the expected volume is reached, the calculation continues while the final draining takes place. The difference of flow in the injectors appears clearly with the difference of the splash on the first mix which is more important for the nominal injector and on the constriction of the propergol jet which is more important in the blocked injector.

Simulations are then gathered into one file (figure 5) which enable us to visualize the whole loading. Using the transparency view, we can see the good symmetry of the calculation and the influence of the different flows on the stacking. In the middle of the mould, the stack is periodic and the mix comes back up to the surface for an extended period. For the first mix the bottom modifies the stack. Computing time of this loading is about two months. This gives an idea of the complexity of this type of simulation. That's why we can only present the calcul of 8 mixes. The final calculation with the ten mixes is given to the "internal ballistic" specialists who use the result of the filling simulation to perform a surface regression simulating the propellant combustion.

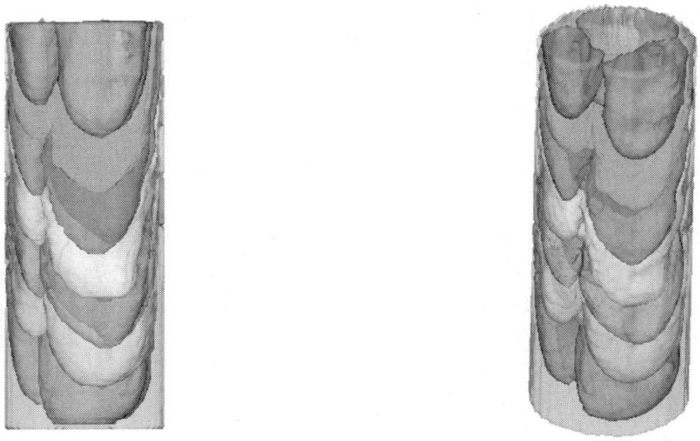

Fig. 5. Booster after the filling of 8 mixes

5 Conclusion and acknowledgments

The code presented is a 3D tool which makes possible to manage several products present in the calculation field, each product having its own properties of density and viscosity. The geometry of the moulds can be easily implemented with the approximation of the Cartesian grid. Ariane V benefits therefore from the numerical progress at all levels, from the manufacturing to the combustion. All these stages are simulated and are closely linked.

We are very grateful to Mr Tauzia, Mrs Marraud and Mr Giraud from "SNPE Propulsion" for the technical support and for allowing us to present these results.

References

1. J. P. Caltagirone, J. Breil: *Sur une méthode de projection vectorielle pour la résolution des équations de Navier-Stokes* (C. R. Acad. Sci. Paris, 1999) t. 327, Série II b, p. 1179–1184
2. M. Fortin, R. Glowinski : *Méthodes de Lagrangien Augmenté; Application à la résolution numérique de problèmes aux limites, Méthodes Mathématiques de l'Informatique* (Dunod, Paris 1982)
3. E. Labat: 'Modélisation des écoulements de propergol solide lors du remplissage de moules'. In: *PHD thesis, Université de Bordeaux I, France, 1994*
4. H. A. Van Der Vorst : *BICGSTAB : A fast and smoothly converging variant of BI-CG for the solution of nonsymetric linear systems* (SIAM Journal of Scientific Statistical Computing, 1982) vol. 13, n?2, p. 631–644.
5. D. L. Youngs : *Time-Dependent Multi-Material Flow with Large Fluid Distorsion* (Numerical Methods for Fluid Dynamics, Reading, UK, Mar. 29-31, 1982) Proceedings of a first Conference, p. 273–285
6. S. Zaleski, J. Li, S. Succi, R. Scardovelli & G. Zanetti : *Direct Numerical Simulation of Flows with Interfaces* (Proceedings of the 2nd International Conference on Multiphase Flow, Kyoto, Apr. 3-7, 1995, v. 2, PT2-1-PT2-12)

Theoretical Aspects of Composite Grid Methods in Computational Fluid Dynamics

Zi-Niu Wu

Tsinghua University, Department of Engineering Mechanics
Beijing 100084, P.R. China

Abstract. The composite grid methods considered in this paper refer to overlapping grid methods and adaptive Cartesian grid method. This paper reviews some recent work concerning the theoretical aspects, such as stability, accuracy, convergence, uniqueness, and conservation, of composite grid methods used in Computational Fluid Dynamics, in order to clarify some hotly debated questions. The most important point to be remarked is that, for composite overlapping grid method, only a nonconservative treatment can stably lead to conservative results. Another important result is that, the adaptive Cartesian grid method is as stable and accurate as the conventional smoothly refined curvilinear grid method, provided they are based on the same difference approximations and that the ratio between the finest mesh and the coarsest mesh is the same.

1 Introduction

One of the major difficulties in Computational Fluid Dynamics is how to treat complex geometry. The most currently used approach to overcome such a difficulty is to use unstructured grids or composite grids such as patched grid, overlapping grid, and adaptive Cartesian grid. There are also other methods such as hybrid Cartesian/unstructured grid method, hybrid structured/unstructured grid method, and gridless method. A particular feature of these methods is that they contain grid interfaces. Taking the composite overlapping grid as example. In this method, the entire domain of computation is filled with subdomains which partly overlap each other. The boundaries of each subdomain need boundary conditions, and in the case of a pure grid boundary (or interface), the boundary conditions (also called interface conditions) are defined through interpolations using state variables (normal interpolation method[3]) or numerical fluxes (flux interpolation method[1]). Such a method is extremely efficient for treating problems with separated geometries, such as the helicopter problems, missile launching, etc. In this paper, we consider the theoretical aspects of the composite grid methods in order to clarify some hotly debated questions.

The theoretical aspects concern the basic behavior of numerical schemes on composite grids and should not be considered as a grid generator problem. They consist of basic problems such as accuracy, stability, convergence, uniqueness, and conservation. The work related to these aspects is discussed in a uniform way in this paper.

Accuracy. The standard question of accuracy can be defined as: giving a pth order accurate numerical scheme in interior points, how to choose the order of accuracy of the interface treatment in order to maintain the global order of accuracy.

Stability. Assuming the interior difference equations to be Cauchy stable, how to ensure stability of the interface treatment?

Convergence. In comparison with a single domain treatment, does a time-dependent composite grid method for steady state computations delay the convergence speed to a steady state?

Uniqueness. The numerical solutions are governed by the difference equations in each subgrid, together with boundary and interface conditions. Are these equations or conditions linearly dependent? If so, then the solutions are non-unique.

Conservation. Conservation has been believed to be very important for problems with discontinuous solutions such as shocks. Assuming that the difference equation in each subgrid is conservative, how to ensure conservation at the grid interface efficiently and in a stable way?

Relations. The last question is: how the accuracy, stability, convergence, uniqueness, and conservation are related to each other?

Though the overlapping grid method is frequently used in incompressible flow problems[7], the present paper mainly addresses results for compressible flows.

2 Statement of the main results

2.1 Accuracy

The standard question of accuracy can be defined as: giving a pth order accurate numerical scheme in interior points, how to choose the order of accuracy of the interface treatment in order to maintain the global order of accuracy? For overlapping grids, the accuracy has been studied by several authors [1,3,5,6,10]. The main conclusions can be summarized as:

1) If there is one or a small number of interfaces, the interface treatment with normal interpolation can be one order less accurate than the interior difference scheme without changing the overall order of accuracy[3,10].

2) If the flux interpolation is used to define interface conditions, then increasing the order of interpolation does not necessarily improve the accuracy[2].

The adaptive Cartesian grid method which contains a series of abruptly refined interfaces, is as accurate as a smooth refinement method having the same total refinement ratio.

2.2 Stability

For overlapping grids, the stability has been studied mainly using the GKS-theory [8,9,13,14]. Assuming the interior difference equations to be Cauchy stable, the following results have been proved:

1) When overlapping grid is used, interface condition defined by flux interpolation is only marginally stable except for one-sided schemes[13]; for interface

condition defined by normal interpolation, the overlapping grid method is stable for dissipative schemes and its stability depends on whether the number of grids in the overlap is odd or even for nondissipative schemes[13];

2) When implicit schemes are used to increase the convergence speed for steady state computation, a time-lagging interface condition should be used in order that the schemes in each subdomain can be solved independently at each time step. This time lagging should be done in a standard way in order to maintain stability[14].

3) For adaptive Cartesian grid method with multiple abruptly refined interfaces, mesh refinement induces a negative or positive numerical dissipation depending on the relative direction of the convection wave. If the wave travels in the direction of decreasing mesh size, then the mesh refinement induces a negative dissipation which narrows the stability domain of a stable difference scheme[17].

2.3 Convergence

When computing steady flows using a time-dependent approach, the convergence speed to a steady state is the main factor to be considered. Since we are interested in the steady state solutions, the time-accuracy of the interface treatment is unimportant. Thus it is convenient to use a time-lagging interface treatment, that is, using values of the previous time step to define the interface condition. This allows us to uncouple the system to be solved and to parallize the algorithm very efficiently[15] provided the time lagging is defined correctly. The time-lagging leads to a loss of convergence speed, or a increase of the number of time iterations necessary to reach a prescribed residual. Increasing the overlapping width reduces the number of required iterations, but increases the CPU time at each iteration. Thus it may have an optimal overlapping width for minimal computational cost. The best convergence occurs at an overlapping width (in terms of the number of mesh points in the overlap)[14,15]:

1) near CFL (Courant-Friedrich-Lewy) for centered dissipative schemes;

2) equal to 2 for upwind schemes.

2.4 Uniqueness

For overlapping grid methods using dissipative difference equations in each subdomain[13,16],

1) the problem has nonunique steady state solution for flux interpolation interface condition;

2) the problem has unique steady state smooth solution for normal interpolation interface condition;

3) the problem has unique steady state discontinuous solution for normal interpolation interface condition and for schemes having enough of dissipation.

For nondissipative difference equations with normal interpolation, the uniqueness of the steady state solutions depends on whether the number of grids in the overlap is odd or even for schemes[13].

2.5 Conservation

Conservation has been believed to be very important for flow computations on overlapping grids[1,2,4,9,12,16]. However, it is shown that a conservative treatment would be only marginally stable[13,9]. The question is: does one absolutely need conservative treatment in order to have conservative solutions? This is quite different to conservation in incompressible flows[11] where only the mass loss is avoided. Moreover, conservation should be considered in the case of direct interaction between the grid interface and a shock wave. In a recent paper[12], the conservation error is analyzed assuming that the solution is smooth at the interface or the Rankine-Hugoniot relation is satisfied. First of all, assuming smooth solutions has no meaning for conservation study since conservation is necessary only for problems with weak solutions. Second, the Rankine-Hugoniot relation is a result of conservation and should not be an assumption for studying conservation. The key point is whether a moving shock can transmit the grid interface. Recently, Wu has proved the following results[16]:

1) When the interface condition is defined by flux interpolation, the shock can get transmitted independently of other conditions;

2) When, inside the shock, the interior difference schemes have a dissipation no smaller than that of the standard Roe scheme, the shock can get transmitted independently of other conditions;

3) When the exact shock speed (properly normalized) is greater than $\frac{1}{2} - \frac{\sqrt{2}}{4}$ but small than $\frac{1}{2} + \frac{\sqrt{2}}{4}$, the shock can get transmitted independently of other conditions.

The above results have been established rigorously for the scalar Burgers' equation and numerically they have been found to remain true for the Euler equations in gas dynamics. After closely examining the dynamic nature of a slowly moving shock wave, we have recently proved the following result (which is submitted to publication):

For the Euler equations in gas dynmaics, a slowly moving shock can transmit the overlapping grid interface with nonconservative normal interpolation, provided that, just inside the shock, the difference scheme has a dissipation no smaller than that of the standard Roe scheme.

3 Conclusions

The unified discussions presented in this paper lead to the following conclusion for composite overlapping grid methods:

(1) the flux interpolation method, which leads to conservative solutions, may have some trouble in practice since it is only marginally stable and may yield nonunique steady state solutions;

(2) the normal interpolation method, which is simpler and more stable than the flux interpolation method, does not provide conservative results according to the classical theory of weak solutions;

(3) thus, it appears that there exists no overlapping grid treatment which enures both stability and conservation.

Luckily, according to the study based on shock/interface interactions, it is now clear that it is not absolutely necessary to use conservative interpolations to define the interface conditions.

One can simply use a scheme which is sufficiently dissipative (TVD, ENO, etc.) and the normal interpolation interface condition in order to have conservative solutions in a stable way.

This is why that correct solutions are often claimed when one just uses a TVD scheme and a simple linear interpolation based on the state variables.

The adaptive Cartesian grid method invloves a series of abruptly refined interfaces. Two opinions are often heared in the past:

1) Opinion 1. The adaptive Cartesian grid method is less accurate and less stable than the usual smooth refinement method, since the former involves abruptly refined interfaces, which easily causes trouble in accuracy and stability.

2) Opinion 2. The adaptive Cartesian grid method is more accurate and more stable than the usual smooth refinement method, since the former involves uniform meshes in each subgrid.

Based on a direct analysis, the present author obtained the following result (submitted to publication):

If the adaptive Cartesian grid method and the smooth refinement method have the same number of mesh points and the same ratio between the finest mesh and coarsest mesh, then both methods are equally stable and equally accurate, provided the two methods are based on the same difference approximation.

Though all the results have been established in a rigorous way, they can also be understood in a heuristical way. First consider the transmission of a shock across a grid interface. The existence of the interface certainly causes an energy loss to the moving numerical shock. Obviously, the shock can transmit the grid interface by overcoming the energy loss in three cases: a) high shock speed (so that the shock itself has a great energy), b) schemes with high dissipation (so that the error created at the interface is efficiently dissipated), c) conservative interpolation (no energy loss at the interface). See the appendix of Wu[16] for a more heuristical explanation. Last consider the existence of optimal overlapping length for implicit schemes with time-lagging interface treatment. For a hyperbolic equation, numerical waves travel a distance no larger than CFL (in terms of number of grid points) at each time iteration. When the overlapping width is equal to CFL, the interaction between the ending points of the overlap vanishes so that the time-lagging treatment becomes as efficient as the single domain one. Thus the optimal overlapping length should be no larger than CFL.

Acknowledgments. Work supported by the National Natural Science Foundations under contract number 19872013.

References

1. M.J. Berger, 'On conservation at grid interfaces', SIAM J. Numer. Anal., **90**, pp.967-983 (1987).

2. Brown D.L., 'An upsplit Godunov method for systems of conservation-laws on curvilinear overlapping grids', Math. Comput. Model, **20**, pp.29-48 (1994).
3. G. Chesshire, D. Henshaw, 'Composite overlapping meshes for the solution of partial differential equations', J. Comput. Phys., **90**, pp.1-64 (1990).
4. G. Chesshire, D. Henshaw, 'A scheme for conservative interpolation on overlapping grids', Siam J Sci Comput., **15**, pp.819-845 (1994).
5. Henshaw D., 'A 4th-order accurate method for the incompressible Navier-Stokes equations on overlapping grids', J Comput. Phys., **113**, pp.13-15 (1994).
6. Jacquotte O.P., Coussement G., 'Structured mesh adaption-space accuracy and interpolation methods', Comput. Meth. Appl. Mech., **101**, pp.397-432 (1992).
7. Liu J, Shyy W, 'Assessment of grid interface treatments for multi-block incompressible viscous flow computation', Comput. Fluids, **25**, pp.719-740 (1996).
8. Olsson F, Petersson NA, 'Stability of interpolation on overlapping grids', Comput. Fluids, **25**, pp.583-605 (1996).
9. E. Part-Enander, B. Sjogreen, 'Conservative and nonconservative interpolation between overlapping grids for finite volume solutions of hyperbolic problems', Comput. Fluids, **23**, pp.551-574 (1994).
10. S.V. Parter, 'On the overlapping grid method for elliptic boundary value problem', SIAM J. Numer. Anal., **36**, pp.819-852 (1999).
11. Shyy W, Liu J, Wright J, 'Pressure-based viscous-flow computation using multiblock overlapped curvilinear grids', Numer. Heat Transfer B-Fluid, **25**, p39 (1994).
12. Tang HS, Zhou T, 'On nonconservative algorithms for grid interfaces', SIAM J. Numer. Anal., **37**, pp.173-193 (1999).
13. Z.N. Wu, 'Uniqueness of steady state solutions for difference equations on overlapping grids', SIAM J. Numer. Anal., **33**, pp.1336-1357 (1996).
14. Z.N. Wu, H. Zou, 'Grid overlapping for implicit parallel computation of compressible flow', J. Comput. Phys., **15**, pp.2-43 (2000).
15. Z.N. Wu, 'Convergence study of an implicit multidomain method for compressible flow', Comput. Fluids, **25**, pp.181-196 (1996).
16. Z.N. Wu, 'Steady and unsteady shock waves on overlapping grids', SIAM J. Sci. Computing, **20**, pp.1851-1874 (1999).
17. Z.N. Wu, 'Anisotropic Cartesian grid method for viscous flow computations', CFD Review (World Sci. Publ.), **1**, pp.93-113 (1998).

Part XXIII

Technical Notes

Numerical Simulation of Particle Motion in Steady Inspiratory Airflow

Tony W.H. Sheu[1], S.K. Wang[2], and S.F. Tsai[2]

Department of Naval Architecture and Ocean Engineering, National Taiwan University,
73 Chou-Shan Rd. Taipei, Taiwan, R. O. C.

Abstract. In this study, we consider equations governing particle trajectories and Navier-Stokes equations for the carrier fluid flow. For accurately modelling such an inhaled particle motion, the solid-gas two-phase model is in use when formulating the inspiratory problem. Under the relatively dilute particle condition in the mixture, equations for two fields are coupled through the interfacial drag in the solid-phase momentum equations. The present study investigates air flow in the trachea that is featured by having a geometric bifurcation in the human central airway. Owing to this geometric complexity, we adcpt finite element analysis for both the gas and solid field equations. In the gas phase, we employ the streamline upwind convection-diffusion finite element model on bi-quadratic elements to simulate the incompressible air flow [1]. As for the simulation of inhaled particle motion, we apply another streamline upwind convection-reaction finite element model on bi-quadratic elements [2]. A feature common to the two employed finite element models is that both of them can provide nodally exact solutions to their respective prototype equations in one dimension. More precisely, the models employed in this paper determine the analytic upwindnig coefficients for the convection-diffusion and the convection-reaction equations for the gas-phase and the solid-phase equations, respectively. In two dimensions, both models introduce physically meaningful artificial damping terms solely in the streamline direction. With these terms added to the formulation, the discrete system is enhanced while solution accuracy is well maintained without deterioration due to numerical diffusion errors. Tests for the investigated problem were conducted and the results are presented in Fig. 1, with an emphasis on the calculation of the target efficiency of the inhaled particles.

Keywords : solid-gas two phases; convection-reaction

References

1. Tony W. H. Sheu, S. F. Tsai, Morten M. T. Wang: 'Discussion of numerical deficiency of applying a partially-weighted upwind finite element model to incompressible Navier-Stokes equations'. Numerical Heat Transfer, Part B: Fundamentals, **32, No.2**, pp. 197-214 (1997)
2. Tony W. H. Sheu, H. Y. Shiah: 'A two-dimensional streamline upwind scheme for the convection-reaction equation'. Int. J. Numer. Meths. in Fluids, (in review)

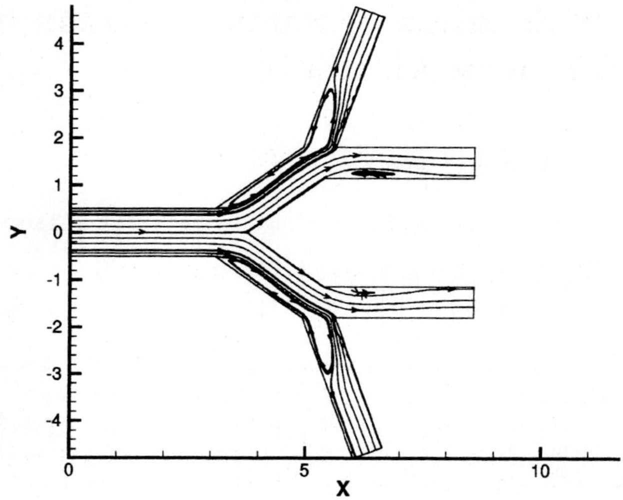

Fig. 1. The streamline plot for showing eddy formation in the air flow.

Extension of Zonal, Spectral Solutions for the Compressible Boundary Layer to the Navier-Stokes Equations

Adriana Nastase

Aerodynamik des Fluges, RWTH–Aachen, 52062 Aachen, Germany

In this paper own zonal, spectral solutions for the partial differential equations (PDE) of the three-dimensional stationary, compressible boundary layer (CBL) given as in [1],[2] for the computation of the flow over flattened, flying configurations (FC) are now extended for the Navier-Stokes layer (NSL). If $\eta = (x_3 - Z(x_1, x_2))/\delta(x_1, x_2)$ is a new coordinate, the spectral forms of the: axial, lateral and vertical velocity's components u_δ, v_δ and w_δ, pressure p, density function $R = ln\rho$ and absolute temperature T (1a-f) and their ten boundary conditions (1g-p), at the NSL-edge ($\eta = 1$), are:

$$u_\delta = u_e \sum_{i=1}^{N} u_i \eta^i \;, \qquad v_\delta = v_e \sum_{i=1}^{N} v_i \eta^i \;, \qquad w_\delta = w_e \sum_{i=1}^{N} w_i \eta^i \;,$$

$$p = p_w + (p_e - p_w) \sum_{i=1}^{N} p_i \eta^i \;, \quad R = R_w + (R_e - R_w) \sum_{i=1}^{N} r_i \eta^i \;,$$

$$T = T_w + (T_e - T_w) \sum_{i=1}^{N} t_i \eta^i \;, \quad w_N = \gamma_{0,N} + \sum_{i=1}^{N-1} \gamma_{i,N} w_i \;,$$

$$u_{N-2} = \alpha_{0,N-2} + \sum_{i=1}^{N-3} \alpha_{i,N-2} u_i \;, \quad v_{N-2} = \alpha_{0,N-2} + \sum_{i=1}^{N-3} \alpha_{i,N-2} v_i \;, \quad \sum_{i=1}^{N} p_i = 1 \;,$$

$$u_{N-1} = \alpha_{0,N-1} + \sum_{i=1}^{N-3} \alpha_{i,N-1} u_i \;, \quad v_{N-1} = \alpha_{0,N-1} + \sum_{i=1}^{N-3} \alpha_{i,N-1} v_i \;, \quad \sum_{i=1}^{N} r_i = 1 \;,$$

$$u_N = \alpha_{0,N} + \sum_{i=1}^{N-3} \alpha_{i,N} u_i \;, \quad v_N = \alpha_{0,N} + \sum_{i=1}^{N-3} \alpha_{i,N} v_i \;, \quad \sum_{i=1}^{N} t_i = 1 \;. \qquad \text{(1a-p)}$$

The spectral forms (1a-f) automatically satisfy the non-slip conditions at FC's wall ($\eta = 0$). Here p_w, R_w, T_w are the given values of p, R and T at wall and u_e, v_e, w_e, p_e, R_e and T_e are the values of u, v, w, p, R and T at NSL's edge, obtained from an inviscid reinforced potential solver used here as outer flow (instead of the parallel undisturbed flow used by Prandtl in his BL) over the modified FC, obtained after the solidification of the NSL, on the FC and u_i, v_i w_i, p_i, r_i, t_i are their free spectral coefficients, which are determined by fulfilling of the NSL's PDE. This extension allows to use the correct asymptotical behaviour along the singular lines like leading edges, the junctions lines wing/fuselage, the ridges on the FC (resulting from the hinge lines of the leading edge flaps), a correct last behaviour at the infinity. For the supersonic flow the additional boundary conditions on characteristic surface can be also correct fulfilled, due to the outer hyperbolical potential flow. If the spectral forms given in (1a-c) are introduced in the impulse equations and seven coefficients, representing their boundary conditions, are eliminated, these equations contain $3N - 7$ variables. By using the collocation method, the impulse equations can be written in $2k + k_1 = 3N - 7$ points and form a quadratical algebraic system (QAS) versus the coefficients u_i, v_i and w_i:

$$\sum_{i=1}^{N-3} u_i \left[\sum_{j=1}^{N-3} (\tilde{A}^{(1)}_{ijk} u_j + \tilde{B}^{(1)}_{ijk} v_j) + \sum_{j=1}^{N-1} \tilde{C}^{(1)}_{ijk} w_j \right] = \tilde{D}^{(1)}_k + \sum_{i=1}^{N-3} (\tilde{A}^{(1)}_{ik} u_i + \tilde{B}^{(1)}_{ik} v_i)$$

$$+ \sum_{i=1}^{N-1} \tilde{C}^{(1)}_{ik} w_i \ , \qquad \sum_{i=1}^{N-3} v_i \left[\sum_{j=1}^{N-3} (\tilde{A}^{(2)}_{ijk} u_j + \tilde{B}^{(2)}_{ijk} v_j) + \sum_{j=1}^{N-1} \tilde{C}^{(2)}_{ijk} w_j \right] = \tilde{D}^{(2)}_k +$$

$$\sum_{i=1}^{N-3} (\tilde{A}^{(2)}_{ik} u_i + \tilde{B}^{(2)}_{ik} v_i) + \sum_{i=1}^{N-1} \tilde{C}^{(2)}_{ik} w_i \ , \qquad \sum_{i=1}^{N-1} w_i \left[\sum_{j=1}^{N-3} (\tilde{A}^{(3)}_{ijk_1} u_j + \tilde{B}^{(3)}_{ijk_1} v_j) + \right.$$

$$\left. \sum_{j=1}^{N-1} \tilde{C}^{(3)}_{ijk_1} w_j \right] = \tilde{D}^{(3)}_{k_1} + \sum_{i=1}^{N-3} (\tilde{A}^{(3)}_{ik_1} u_i + \tilde{B}^{(3)}_{ik_1} v_i) + \sum_{i=1}^{N-1} \tilde{C}^{(3)}_{ik_1} w_i \ . \qquad \text{(2a-c)}$$

Hereby are $k = 1, 2, .., (N - 3)$ and $k_1 = 1, 2, .., (N - 1)$. The coefficients u_i, v_i, w_i are obtained by the iterative solving of this equivalent QAS with slightly variable coefficients. These coefficients depend on the density ρ, viscosity μ and pressure p, inside the NSL, which are here expressed only as functions from the velocity's components inside the NSL as follows. By using the continuity equation, which is non-linear in ρ, but linear in R as in (3a), a linear algebraic equation (3b) versus r_i is obtained, which coefficients depend only on u_i, v_i, w_i:

$$u \frac{\partial R}{\partial x_1} + v \frac{\partial R}{\partial x_2} + w \frac{\partial R}{\partial x_3} = - \left(\frac{\partial u}{\partial x_1} + \frac{\partial v}{\partial x_2} + \frac{\partial w}{\partial x_3} \right) \ , \quad \sum_{i=1}^{N} \nu_{iq} r_i = \nu_q \ . \qquad \text{(3a,b)}$$

An exponential law of the viscosity μ versus T and the physical equation of ideal gas for the pressure p are used as in (4a,b). If μ and p are introduced in the temperature's PDE, the spectral coefficients t_i of T fulfill an algebraic transcendental equation (4c), which coefficients are only function of u_i, v_i, w_i:

$$\mu \equiv \mu_\infty \left[\frac{T}{T_\infty} \right]^{n_1} \ , \quad p = R_g \rho T \ , \quad \sum_{i=1}^{N} h_i t_i + h_0 T^{n_1} = \theta \ . \qquad \text{(4a-c)}$$

The collocation method is used in the equations (3b) and (4c) for the computation of r_i and t_i and the boundary conditions (1p) and (1j) are also fulfilled. This analytical-numerical method is more accurate and reduces the computer time.

References

1. Nastase, A.: The Determination of Hybrid Analytical-Numerical Solutions for the Three-Dimensional Compressible Boundary Layer Equations, ZAMM, 73, (1993).
2. Nastase, A.: A New Spectral Method and Its Aerodynamic Applications, Proc. of the 7th Intern. Symposium on CFD, Beijing, China, 1997.

Analysis and Minimization of Flow Noise in a Centrifugal Pump

M.A. Langthjem and N. Olhoff

Institute of Mechanical Engineering, Aalborg University, Pontoppidanstræde 101, DK-9220 Aalborg East, Denmark

1 Introduction

In a recent review article on noise from centrifugal pumps, Rzentkowski [1] concludes by the words, "At the present stage of knowledge, one can hardly claim that the process of noise generation within centrifugal pumps is understood. This is rather surprising, given the industrial importance of the problem." The present work attempts to give a contribution to a better understanding through a numerical study. Another purpose is to develop a numerical tool for minimization of the noise generation.

The relation between unsteady flow, pressure fluctuations, and noise has been studied experimentally by Chu, Dong and Katz [2], using a pump with a 'two-dimensional' volute, and two-dimensional calculations compare well with measured pressure data. It is found that turbulence can be ignored, but vorticity gives a major contribution to the total pressure variation. A potential flow model thus cannot provide an acceptable description of the flow. A combined vortex panel / discrete vortex method appears to include the flow effects pointed out in Ref. [2] as being important, and exclude those which are not.

2 Numerical method and results

In the present numerical model, the inlet is modeled by a point source and the blades of the impeller are covered with vortex elements with discrete, bound vortices. The casing is covered with panels with distributed vortices. Wake vortices are shed from the trailing edges of the impeller blades and convected downstream with the local induced velocity. After each time increment an equation system for the vortex strengths is solved and the impeller coordinates are updated. Once the vortex strengths are known, the flow velocities can be evaluated at any point.

When the velocity field has been computed, the fluid forces acting on the impeller blades can be evaluated from the unsteady Bernoulli equation. The resulting fluid forces (per unit breadth) are 'lumped' onto the blade control points. These N_f isolated forces act as *acoustic dipoles*. The acoustic pressure is then evaluated by solving the two-dimensional wave equation with N_f dipole terms on the right-hand side. The solution is obtained by considering each harmonic component of the moving forces at a time which gives a Kirchoff-Helmholz type of equation. The boundary term is discretized by using the boundary element method.

Computations have been performed for the 'two-dimensional' laboratory pump of Chu *et al.* [2]. Figures 1*a* and *b* show the positions of the shed vortices at different times. Part *c* shows the velocity vectors. Figure 2 shows (*a*) the acoustic pressure fluctuations at a point near the volute tongue, and (*b*) the corresponding spectrum. The experimental results from Ref. [2] are represented by circles.

Acknowledgements
Financial support from Aalborg University, Grundfos A/S, the Alexander von Humbolt Foundation, and the Danish Ministry of Research is gratefully acknowledged.

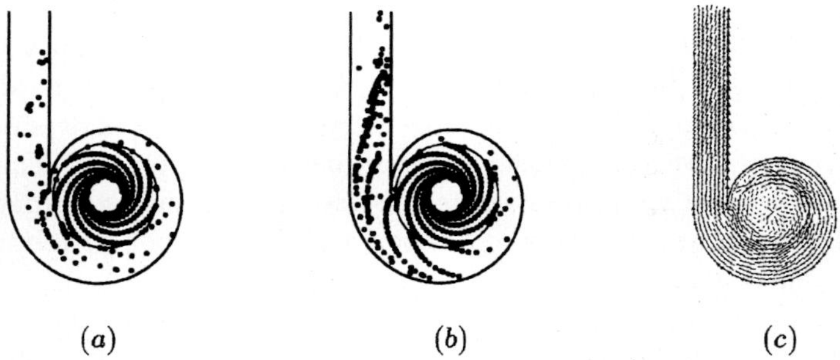

Fig. 1. (*a*) Instantaneous location of the vortices at the nondimensional time $t_* = t/t_{blade} = 19.44$, after impulsive start from rest at $t_* = 0$. t_{blade} is the blade passing time. (*b*) The same at $t_* = 97.22$. (*c*) The velocity field at $t_* = 116.67$.

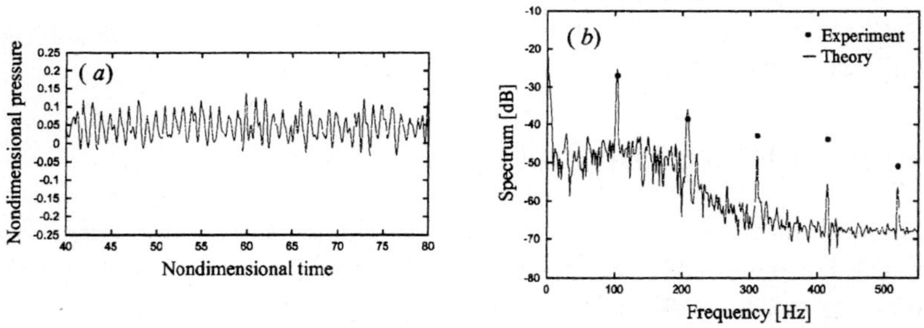

Fig. 2. (*a*) Acoustic pressure fluctuations at a point near the volute tongue. (*b*) The corresponding spectrum.

References

1. G. Rzentkowski: 'Generation and control of pressure pulsations emitted from centrifugal pumps: A review'. In: *PVP-Vol. 328, Flow-Induced Vibration, 1996*, ed. by M.J. Pettigrew (ASME, New York 1996) pp. 439-454.
2. S. Chu, R. Dong, J. Katz: 'Relationship between unsteady flow, pressure fluctuations, and noise in a centrifugal pump - Parts A and B'. Journal of Fluids Engineering **117** (1995)

The Instability of Natural Convection Boundary Layer around a Vertical Heated Flat Plate

Fenggan Zhuang[1] and Jianjun Tao[2]

[1] China Aerospace Science and Technology Corporation Beijing 100830,China
[2] The Institute of Fluid Mechanics, Beijing University of Aeronautics and Astronautics. Beijing, 100083, China

Natural convection widely exists in energy and chemical industries, its simple and important model is boundary layer along a vertical heated flat plate. Considering the coupling - effect of velocity and temperature fields, the instability of the natural convection boundary layer is reduced to an eigenvalue problem of six-order Orr-Sommerfeld equations. In the early theoretical researches Gill&Davey[1] obtained two sets of eigenvalues in their stability analysis of a "buoyancy layer", Hiber&Gebhart[2]pursued this matter and found there was a "loop" in the neutral curve when the Prandtl number was large enough. According to their calculation, there were two sets of eigenvalues as the Pr=0.733, and one of them would be damped downstream (whose amplifying rate was less than zero). In one word, in the past studies[3] only one unstable mode was found, and no further researches have been done, especially on the "loop" region.

A natural convection boundary layer along a vertical isothermal heated flat plate was considered in this paper. Particular emphasis is placed on the calculations around "loop" region. The neutral curves of different Prandtl numbers were calculated and shown in Fig.1. It can be illustrated that (Fig.1(a)) that there is a "loop" in the neutral curve when the Prandtl number increased to Pr=2. But as shown in Fig.1(c) it is difficult to find any "loop" in the neutral curve. This discrepancy can be interpreted clearly in Fig.1(b) that the "loop" in $B-G$ plane is just a twist of the neutral curve in the $B-\alpha_r-G$ space.

Usually, buoyancy driven system can be unstable due to two kinds of instabilities: the first is caused by shear flow, and named as viscosity instability; the other is due to the buoyancy force, named as buoyancy instability. In this paper, we made a first attempt to compare and analyze the contributions of the buoyancy and the viscosity instabilities by computing their neutral curves and amplifying surfaces (composed by isolines of amplifying rate-α_i in the frequency (B)-wave number(α_r)-G space). Sepicailly, bifurcation solutions were found for the buoyancy instability.

For such coupled buoyancy driven system, we take fig.2 as an example to study the topological structure of the amplifying surface. It is illustrated clearly (Fig.2(a)) that, for Pr=2, there is an intersection line in the B-α_r-G space where two sets of eigenvalues have the same Grashof number, wave number and frequency but different amplifying rate. This means that there are bifurcation so-

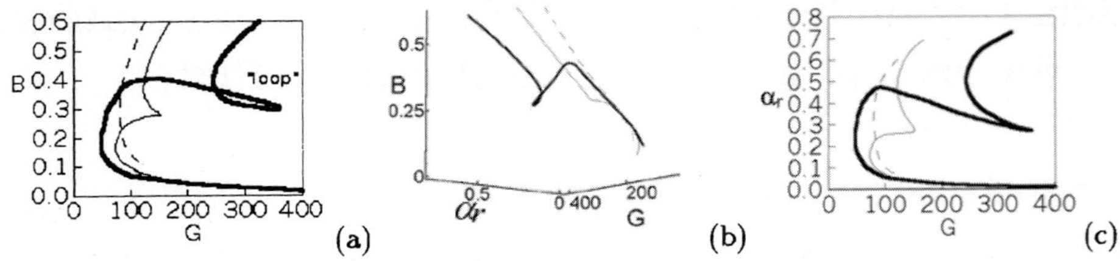

Fig. 1. The neutral curves for natural convection boundary layer at different Prandtl numbers. The dotted line Pr=0.3; the solid line Pr=0.733; the thick solid line Pr=2

lutions in this eigensystem, or in other words, two unstable modes exist along the intersection line. It is a distinct character because in other open incompressible flow systems, such as plume and boundary layer, few multi-unstable modes were found. In addition, the start point of the intersection was found a saddle point in the complex α plane, and the amplifying rates of the two unstable modes presented symmetrical character near the frequency of the saddle point. The criterion for determining the existence of the multi-unstable modes was also discussed.

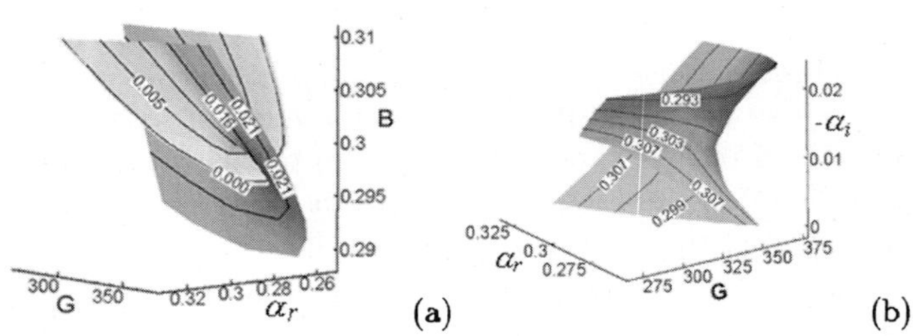

Fig. 2. (a) The amplifying surface in the "loop" region, the amplifying rates were also labeled;(b) The isolines of B in the $(-\alpha_r)-\alpha_r - G$ space. Pr=2

References

1. A.E.Gill and A.Davey: J. Fluid Mech. **35** (1969)
2. C.A.Hiber and B.Gebhart: J.Fluid Mech.**48** (1971)
3. R.A.Brewster and B.Gebhart J.Fluid Mech. **229** (1991)

Semi-implicit and Implicit Solutions of Pulsatile Flow in the Axisymmetric Tube Using a Staggered Grid *

Pavel Burda[1] and Josef Kořenář[2]

[1] Dept. of Math., Czech Techn. Univ., Karlovo n. 13, 121 35 Prague 2, Czech Rep.
[2] Inst. of Hydrodynamics, Czech Ac. Sci., Podbabská 13, 166 12 Prague 6, Czech R.

1 The Problem of Pulsatile Flow

We consider pulsatile flow of incompressible Newtonian fluid in a circular tube in the vicinity of an axisymmetric constriction. The problem is closely related e.g. to predictions of flow properties of blood in the vicinity of stenoses in blood vessels. Mathematically the problem is governed by the Navier-Stokes equation

$$\frac{\partial V}{\partial t} + (V \cdot \nabla)V = -\frac{1}{\rho} grad\, p + \nu \Delta V ,$$

(here V is the velocity vector, p is the pressure, ρ the density, ν is the coefficient of kinematic viscosity, t is time), and by the continuity equation $div\ V = 0$.

The equations are equipped with the pulsating boundary conditions on the inlet and outlet, cf. [1]. Due to axial symmetry, the problem is solved on a 2-dimensional domain Ω, using cylindrical coordinates, cf. [1].

2 Time and Space Discretizations

The semiimplicit scheme described in [1] has been used succesfully to solve the pulsatile flow with lower Reynolds numbers [1], [2].

On the basis of the Crank-Nicolson scheme we come to the implicit scheme, see [4]. In each time step the resulting system of equations is solved iteratively.

Space discretization is done by the MAC method which uses a staggered grid, and which is modified here by using the upwinding technique, cf. [4].

In the approximation of the resulting Poisson equation we utilize the block tridiagonal structure of the matrix to apply a very fast algorithm.

We refine the grid towards the corners caused by nonconvex internal angles of the domain, where the solution has a singularity, cf. [3].

3 Numerical Results for Pulsatile and Oscillatory Flow

We present some results for a constricted tube which represents a model of the part of the aorta where $Re \approx 1700$. We have used here the implicit scheme.

* *This research has been and has been supported partly by the grant No. 201/98/0528 of the GACR agency, partly by the grant No. A 2120801/98 of the ASCR agency, and partly by the research project of the Ministry of Education No. J04/98/210000003.*

On Figure 1 we show the result for pulsatile flow in a certain time of the period. Note that the fluid near the wall flows backwards.

On Figure 2 we show some results for oscillatory flow in a tube extended to a "cavity". We compare our results based on the MAC method, and the results obtained by Z. Skalák who used the finite element method, cf. [5].

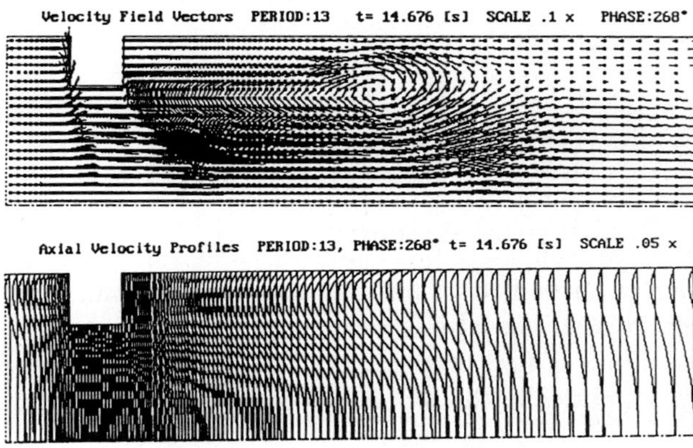

Fig 1: Pulsatile flow in a constricted tube

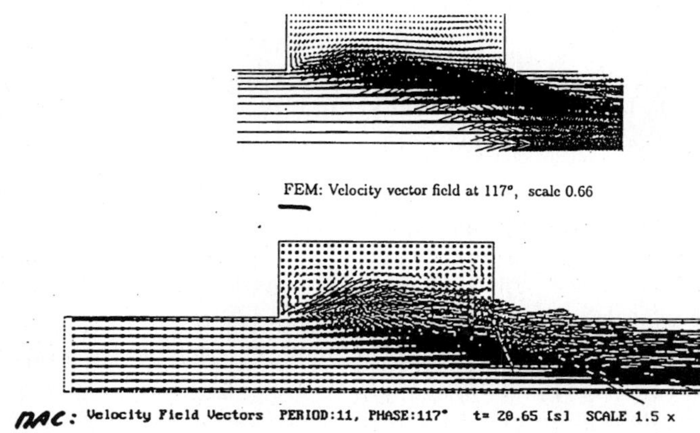

Fig 2: Comparison of FEM solution and MAC solution

References

1. Burda, P., Kořenář, J.: Comm. Inst. Hydrodyn. **19**, 87-110, (1993)
2. Burda, P., Kořenář, J.: ZAMM **74**, T649-T651, (1994)
3. Burda, P.: 'On the F. E. M. for the Navier-Stokes equations in the domains with corner singularities'. In: *Finite Element Methods, Supeconvergence, ...* ed. by M. Křížek, P. Neittaanmäki, R. Stenberg, (Marcel Dekker, New York, 1998), pp. 41-52.
4. Burda, P., Kořenář, J.: 'Second order discretization of pulsatile flow in a constricted tube'. In: *Proc. Conf. Numerical Modelling Prague 1994*, ed. by M. Feistauer, R. Rannacher, K. Kozel, (FMF Prague 1995), pp. 53-58.
5. Skalák, Z.: 'Numerical analysis of the flow in corrugated channels'. In: *ibid.*, pp. 256-263.

Application of the CIP Method to the Simulation of the Laser-Matter Interaction

Takayuki Utsumi[1], Akira Sasaki[1], Tomoaki Kunugi[2], Sadao Fujii[3], and Mikio Akamatsu[4]

[1] Advanced Photon Research Center, Japan Atomic Energy Research Institute 25-1 Miiminami-machi,Neyagawa city, Osaka-fu 572-0019, Japan
[2] Department of Nuclear Engineering, Kyoto University Yoshida, Sakyo, Kyoto, 606-8501, Japan
[3] Kanto Technical Institute, Kawasaki Heavy Industries, Ltd. 118 Futatsuzuka, Noda, Chiba, 278-8585 Japan
[4] Nuclear Systems Department, Kawasaki Heavy Industries, Ltd. 2-6-5 Minami-Suna, Koto-Ku, Tokyo 136-8588 Japan

1 Introduction

The high-peak-power ultrashort pulse laser ($100TW, 20fs, 10Hz$) from Ti:Sapphire laser based on chirped pulse amplification has been successfully developed at Japan Atomic Energy Research Institute and the high-efficiency X-ray sources would be realized in the near future. In the development of the X-ray laser and incoherent X-ray sources, it is essential to improve our understanding of complicated physical processes in high temperature high-density plasmas from both the point of views of experiments and simulation. When the laser pulse strikes a solid target, a hot and dense plasmas in an extreme state of high energy concentration are produced. These complicated processes can be investigated hydrodynamically assuming the system is composed of continuum media or multi-phase/multi-component fluids. In order to simulate these phenomena, it is necessary that sharp interface discontinuities in the density and pressure must be traced accurately in the solid, liquid and gas phases, i.e. compressible and incompressible fluids must be treated simultaneously.

Recently, Cubic Interpolated Propagation (CIP) and CIP-Combined Unified Procedure (C-CUP), were introduced by Yabe et al. [1]. We have extended these methods and developed a hydrodynamic code [2] to apply to the simulation of the laser-matter interaction. Regarding to the material properties, high accuracy and usability are required for numerical hydrodynamic simulations. We have adopted the SESAME [3] and QEOS [4] as the equation of state and TKN model [5] and Drude model for the calculation of the transportation coefficients (thermal conductivity, electrical conductivity, absorption coefficients) of materials.

2 Numerical Results

To verify the code, we carried out the numerical simulation with the condition that a laser pulse incidents on the solid slab and thin foil Aluminum target at normal incidence. The results are shown in the following figures. Size of the analysis system is 1.6μm × 2.4μm, and number of meshes is 80 × 120. The $10^{12} W/cm^2$ laser pulse is irradiated downward from the upper center onto the Al target and the exposure time is $20psec$ with the rectangle time profile. We assumed the targets are surrounded with the transparent ideal air, and the initial pressure and temperature of the system are $0.01Mpa(0.1atm.)$ and $300K$, respectively. The computational time step is $25fsec$. The formation and propagation of laser-produced plasmas are obtained with good numerical stability.

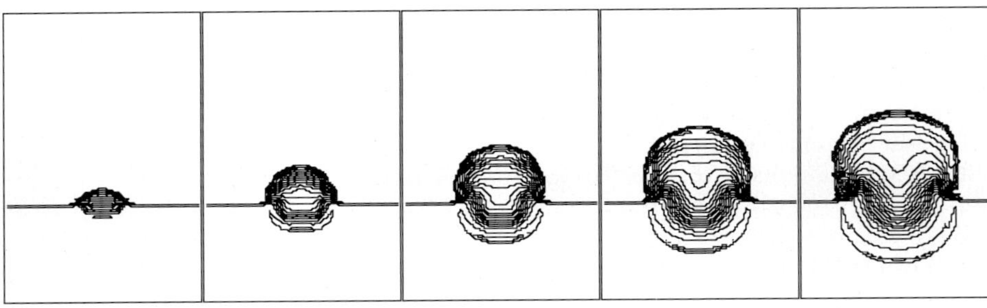

Fig. 1. Temporal changes of density distribution in the case of Al slab target irradiation. The unit is $log_{10}(kg/m^3)$, and the interval of contour is 0.1.

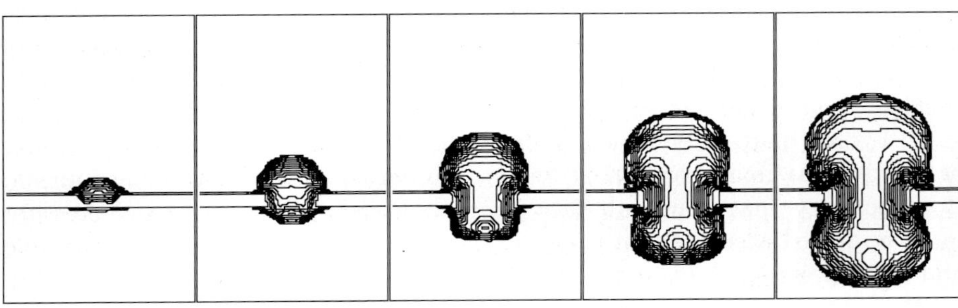

Fig. 2. Temporal changes of density distribution in the case of Al foil target irradiation. The unit is $log_{10}(kg/m^3)$, and the interval of contour is 0.1.

References

1. T. Yabe and T. Aoki: Comput. Phys. Commun. **66**, 219 (1991)
2. T. Utsumi, A. Sasaki, T. Kunugi et al.: CFD journal **8**, 128 (1999)
3. S.H. Kathleen: LA-10160-MS, 1984
4. R. More, K. Warren, D. Young and G. Zimmerman: Phys. of Fluids **31**, 3059 (1988)
5. Y.T. Lee and R.M. More: Phys. of Fluids **27**, 1273 (1984)

Numerical Analysis of Transport Properties in Turbulent Premixed Flames

Shinnosuke Nishiki[1], Tatsuya Hasegawa[1], and Ryutaro Himeno[2]

[1] Nagoya Institute of Technology, Gokiso-cho, Showa-ku, Nagoya, 466-8555 Japan
[2] The Institute of Physical and Chemical Research, 2-1 Hirosawa, Wako-shi, Saitama, 351-0198 Japan

1 Introduction

Numerical simulation becomes to play an important role in design of practical combustors, but useful turbulent combustion models are still required. In this study, DNS of a turbulent premixed flame with single-step irreversible Arrhenius-type reaction is performed in order to evaluate turbulent transport properties.

A premixed flame has the laminar burning velocity of $u_L = 0.600$ m/s, the flame thickness of $\delta = 0.217$ mm and the density ratio of $\rho_u/\rho_b = 7.53$. The temperature dependence of transport coefficients is evaluated by the 0.7 power of temperature, while $Le = 1.0$ and $Pr = 0.75$.

The simulation domain is a box of 8 mm×4 mm×4 mm as shown in Fig. 1. The number of grid points is 512×128×128. The NSCBC proposed by Poinsot and Lele [1] is applied in x direction, while the periodic boundary condition is applied in y, z directions. The 6th-order central finite difference method is used for x direction, and the Fourier spectral collocation method is used for y, z directions. The 3rd-order Runge-Kutta method is used in time integration.

The velocity field of 3D homogeneous isotropic turbulence is initially assigned with a mean inflowing velocity of $U_c = 0.60$ m/s and a velocity caused by expansion of gases. The same homogeneous isotropic turbulence comes into the domain with U_c assuming Taylor?s hypothesis and shifting the phase. The characteristics of the homogeneous isotropic turbulence are $u'/u_L = 0.88$ and $l_t/\delta = 15.9$. Reynolds number based on integral length scale is about $Re_{lt} = 90$.

2 Results and Discussions

A wrinkled flame appears in the turbulent flow as shown in Fig. 2. Streamwise evolution of turbulent fluctuations is shown in Fig. 3. The fluctuations somewhat decay in front of the flame region, but all components increase dramatically in the flame region, especially x component increases more than twice as much as y, z components.

Next the mechanism of turbulence generation in the flame region is analyzed on the basis of the Favre-averaged transport equation of turbulent kinetic energy along streamwise direction. Streamwise evolution of each term is shown in Fig. 4. The mean pressure gradient term (III) and the pressure work term (IV) produce turbulent kinetic energy in the flame region. On the other hand, the mean velocity gradient term (I) and the diffusion and dissipation term (V) decrease turbulent kinetic energy in the flame region. The turbulent diffusion term (II) has a little effect.

$$\frac{D\tilde{k}}{Dt} = \underbrace{-\frac{\overline{\rho u_i'' u_k''}}{\bar{\rho}}\frac{\partial \tilde{u}_i}{\partial x_k}}_{\text{(I)}} \underbrace{-\frac{1}{\bar{\rho}}\frac{\partial \overline{\rho u_i'' u_i'' u_k''}}{2\partial x_k}}_{\text{(II)}} \underbrace{-\frac{\overline{u_i''}}{\bar{\rho}}\frac{\partial \bar{p}}{\partial x_i}}_{\text{(III)}} \underbrace{-\frac{1}{\bar{\rho}}\overline{u_i''\frac{\partial p'}{\partial x_i}}}_{\text{(IV)}} \underbrace{+\frac{1}{Re}\frac{1}{\bar{\rho}}\overline{u_i''\frac{\partial \tau_{ik}}{\partial x_k}}}_{\text{(V)}}$$

3 Conclusions

Database of well-developed premixed turbulent combustion is obtained and obtained results from analyzing the database are as follows: 1) A wrinkled flame appears in the turbulent flow. 2) The turbulent fluctuation of x component increases more than twice as much as y, z components in the flame region. 3) The mean pressure gradient term has an important role on producing of turbulent kinetic energy in the flame region.

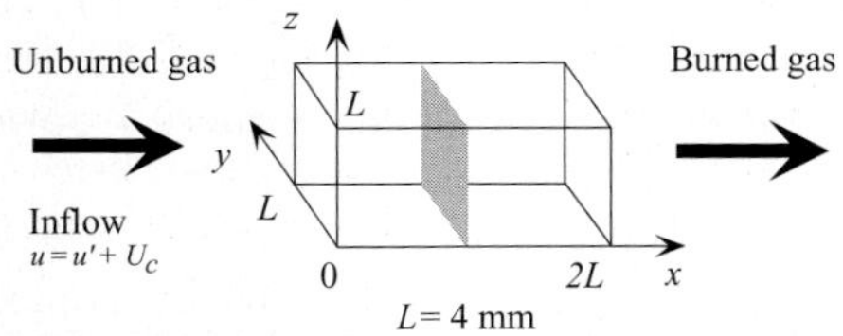

Fig. 1. Simulation domain and coordinate system

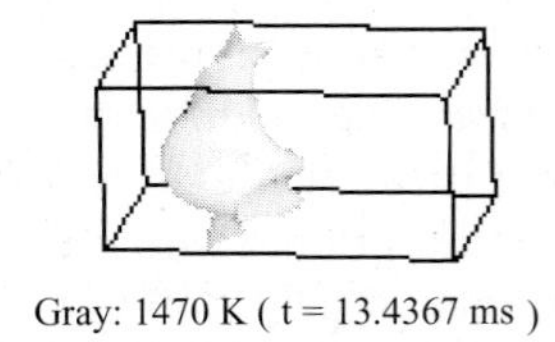

Fig. 2. Contour surface of temperature

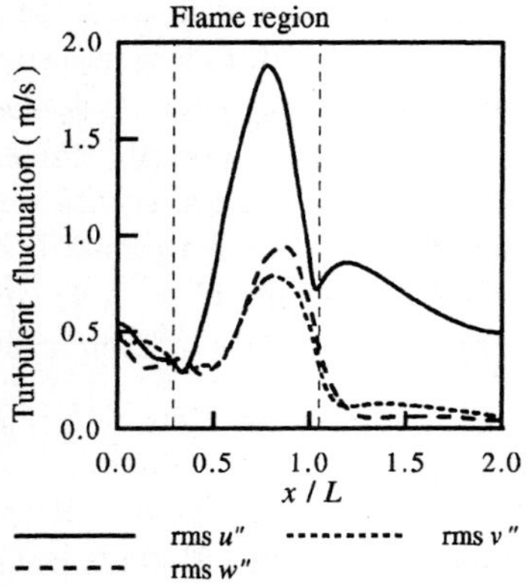

Fig. 3. Streamwise evolution of turbulent fluctuation

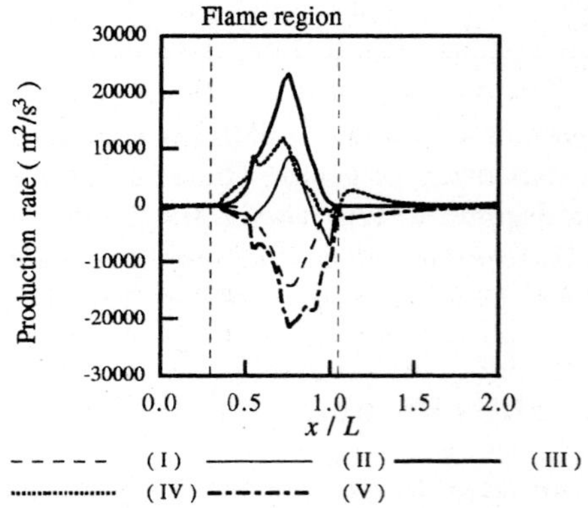

Fig. 4. Streamwise evolution of each term in transport equation of turbulent kinetic energy

References

1. T. Poinsot, S.K. Lele: J. Comp. Phys, **101**, 104(1992)

Lagrangian Block Method

Vincent H. Chu and Wihbel Altai

McGill University, Montreal, Quebec H3A 2K6, Canada

Abstract. Numerical schemes have been developed to minimize numerical diffusion in the simulation of advective process. However, the problem of unphysical oscillations continues to affect the stability of these schemes particularly those that are based on the Eulerian method. Negative undershoot is one of the major difficulty associated with the numerical oscillations. Physical laws are violated and numerical instabilities are initiated when the model parameters, such as the turbulent energy and dissipation rate in the k-ϵ turbulence model, or the tracer mass in the chemistry transport models, become zero and then negative. Procedures have been proposed to insure positive results but the improvement in computational stability is often at the expense of adding numerical diffusion to the simulation. One naturally positive (non-negative) scheme that stands out as a departure from the conventional approach, is the Lagrangian Block Method (LBM). The computation elements of the LBM are blocks and cells. Blocks are advected and diffused by a Lagrangian procedure while cells are fixed with an Eulerian grid. In the original implementation of the method by Egan and Mahoney (1972), advection and diffusion are simulated by separated steps using a hybrid scheme. In the new method introduced by Chu and Altai (1998), by dividing the blocks into 9 portions for re-distribution to the cells, advection and diffusion are simulated in a single step with equal precision. Figures 1 and 2 show the numerical results of the pure advection experiments. The blocks were able to be advected in a uniform velocity by the LBM at high Courant number without numerical diffusion. A small numerical 'distortion' nevertheless was observed in the rotation of the blocks. Unlike the numerical diffusion, the distortion stops after the second rotation (see the concentration profiles in column 2 of Figure 3). It is clear from inspection of the concentration profiles shown in Figure 3, that the numerical distortion is fixed and quickly become independent of time. This distortion effect is to be distinguished from the diffusion effect which cause the mass to spread continuously with time as shown in columns 3 and 4 of the figure. The detail of the LBM and the results of the Large Eddy Simulations of turbulent jets conducted using the LBM are given in Altai and Chu (2000). The LBM is to be distinguished from other semi-Lagrangian methods in that the computation are conducted not only for the mass but also for the dimensions and mass-center location of the blocks in the cells. The blocks are re-defined in every time step by the LBM and in this sense, they are not Lagrangian particles in such method as the Partical-In-Cell.

References

1. Altai, W. and Chu, V. H.: 'Large Eddy Simulation by Lagrangian block method,' *Proc. 1st Int'l Conf. Computational Fluid Dynamics*, Kyoto, Lecture Notes in Physics, Springer-Verlag (2000).

2. Chu, V. H. and Altai, W. 'Generalized second-moment method for advection and diffusion processes,' *Proc. 2nd Intl Symp. on Envir. Hydraulics,* Hong Kong, Balkema Publisher, 357-362 (December, 1998).
3. Egan, B. A. and Mahoney, J. R. 'Numerical modeling of advection and diffusion of urban source pollution,' *J. Appl. Meteorology,* **11**, 312-322 (1972)

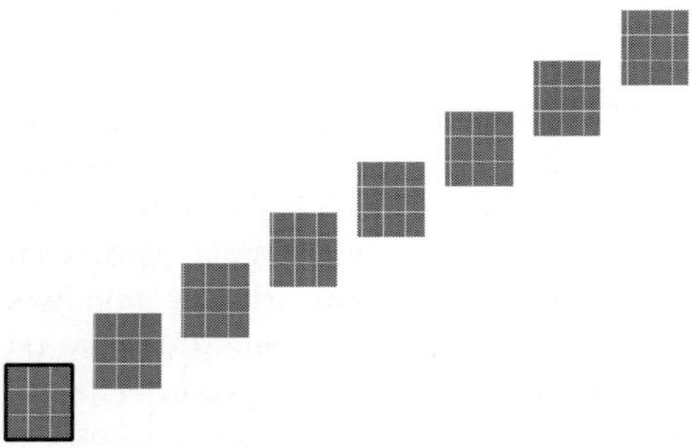

Figure 1: 3x3 blocks of dye move in an oblique direction; time interval 8 Δt, Courant number $Co = 0.93$.

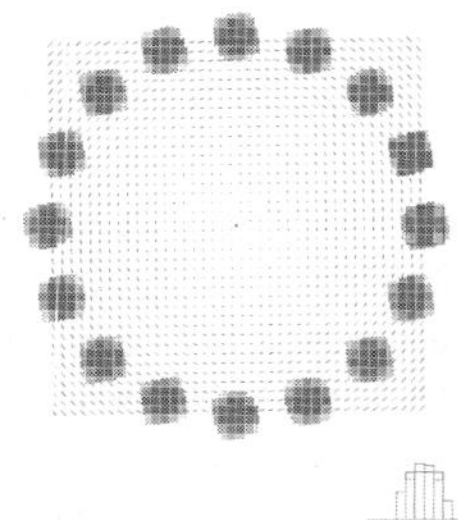

Figure 2: Rotation in a circular velocity field without diffusion; $D = 0$, Péclet number $Pe = \infty$.

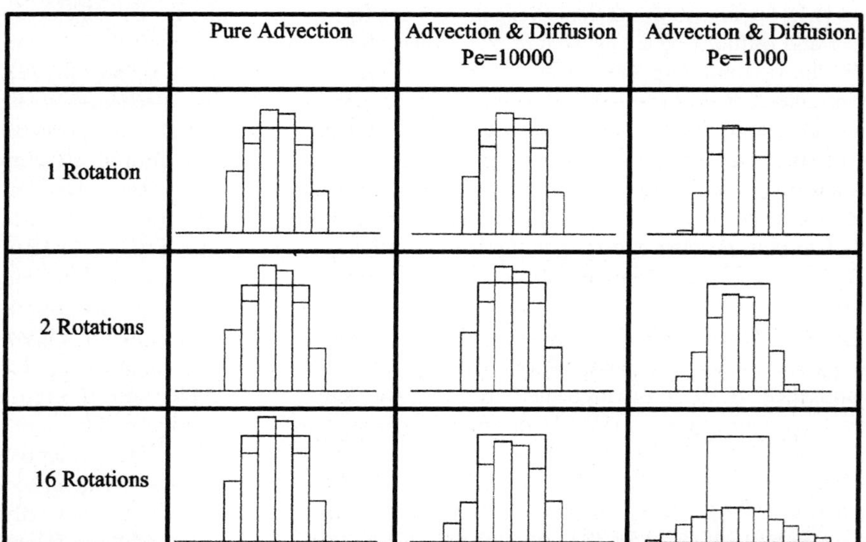

Figure 3: Concentration profiles along the line of symmetry after completing one, two and sixteen rotations.

Grid Generation Based on Harmonic Maps

Reijo Lehtimäki

Helsinki University of Technology, Laboratory of Aerodynamics,
P.O.Box 4400, FIN-02015 HUT, Espoo, Finland

1 Harmonic Grid Generation with Clustering

Harmonic grid generation is attractive due to the existence and uniqueness results for the harmonic map. These results assume the image to have a non-positive Riemannian curvature and a convex boundary, assumptions that are met by the logical domain. In the two-dimensional case, the uniqueness is guaranteed for a flat (Radó [1]) as well as for a curved surface (Schoen and Yau [2]) as the physical domain. Harmonic map is then also differentiable with a non-vanishing Jacobian, in fact, a diffeomorphism. Unfortunately, it seems that the three-dimensional analog is not true [3]. Note that, these results are for the continuous grid and the regularity of the discrete grid is not an obvious implication. Nevertheless, the harmonic method has in practice proven very robust.

Harmonic maps are extremals of the energy functional [4]. Thus, they are solutions of the corresponding Euler equations. In our application the physical space X is a Euclidean space and x^n a Cartesian coordinate system. The metric tensor of that system is an identity matrix and its determinant is one. The Euler equations then reduce to

$$\sum_k \left(\frac{\partial^2 \xi^\lambda}{(\partial x^k)^2} + \Gamma^\lambda_{\alpha\beta}\frac{\partial \xi^\alpha}{\partial x^k}\frac{\partial \xi^\beta}{\partial x^k}\right) = 0 \tag{1}$$

The Christoffel symbols in ξ^n are calculated as $\Gamma^k_{ij} = \bar{q}_{i,j}\cdot\bar{q}^k = \bar{\xi}_{,ij}\cdot q^{kl}\bar{\xi}_{,l}$. Here $\bar{q}_i$ are the covariant and $\bar{q}^i$ the contravariant base vectors, q^{ij} the contravariant components of the metric tensor, and $\bar{\xi}$ the position vector in the logical domain. Comma denotes differentiation with respect to the indices following it.

In Eq. (1), the curvilinear coordinates appear as dependent variables and the Cartesian coordinates as independent. For convenience, the equation is transformed. First-order derivatives are easy since, for a diffeomorphism the Jacobian matrices of the map and its inverse are inverses of eachother. The second-order derivatives can be transformed by setting the Cartesian coordinates equal to themselves mapped to the logical domain and back, and differentiating these identities twice with respect to each of the Cartesian coordinates, using the chain rule of calculus. After some algebra, one obtains the transformed equations

$$\begin{aligned}
(y_\zeta z_\eta - y_\eta z_\zeta)L(x) + (y_\xi z_\zeta - y_\zeta z_\xi)L(y) + (y_\eta z_\xi - y_\xi z_\eta)L(z) - J\Gamma^1_{\alpha\beta}g^{\alpha\beta} &= 0\\
(x_\eta z_\zeta - x_\zeta z_\eta)L(x) + (x_\zeta z_\xi - x_\xi z_\zeta)L(y) + (x_\xi z_\eta - x_\eta z_\xi)L(z) - J\Gamma^2_{\alpha\beta}g^{\alpha\beta} &= 0\\
(x_\zeta y_\eta - x_\eta y_\zeta)L(x) + (x_\xi y_\zeta - x_\zeta y_\xi)L(y) + (x_\eta y_\xi - x_\xi y_\eta)L(z) - J\Gamma^3_{\alpha\beta}g^{\alpha\beta} &= 0
\end{aligned} \tag{2}$$

Here $L(x) = x_{\xi\xi}g^{11} + 2x_{\xi\eta}g^{12} + 2x_{\xi\zeta}g^{13} + x_{\eta\eta}g^{22} + 2x_{\eta\zeta}g^{23} + x_{\zeta\zeta}g^{33}$. The $g^{\alpha\beta}$ are the contravariant components of the metric tensor of the coordinate system in the physical domain induced by the harmonic map. The J is the Jacobian determinant $J = \det(\partial x^i/\partial \xi^j)$. Non-uniform difference approximations formed in the clustered logical domain are used for the derivatives, see the abstract or [5]. This results in a similarly clustered grid in the physical domain.

The middle surface cutting through the blade passage of the NASA 55C fan rotor, consisting of 15 blades (blocks) is shown in Fig. 1 on the left. The grid for the Huygens probe on the right also looks quite nice not regarding the somewhat unsmooth point distribution in the circumferential direction at the center. This is due to the unsmooth point distribution on the boundary and actually demonstrates the ability of the method to carry the boundary point distribution far into the field. The grid is inverted at the corner pointed by the broken arrow. Thus, despite of the beautiful theory in the continuum, the method cannot guarantee a regular discrete grid. Therefore, setting all the Christoffel symbols to zero $\Gamma^k_{ij} = 0$ [5], is perhaps preferable. Equations (2) will then simplify considerably and at least qualitatively the grids obtained with the two methods look quite similar. Note that, clustering the logical domain more to the sides, on the segment that corresponds to the box on the top of the probe, and re-interpolating the logical domain prior to grid generation, is an easy way to obtain a regular grid for the Huygens probe.

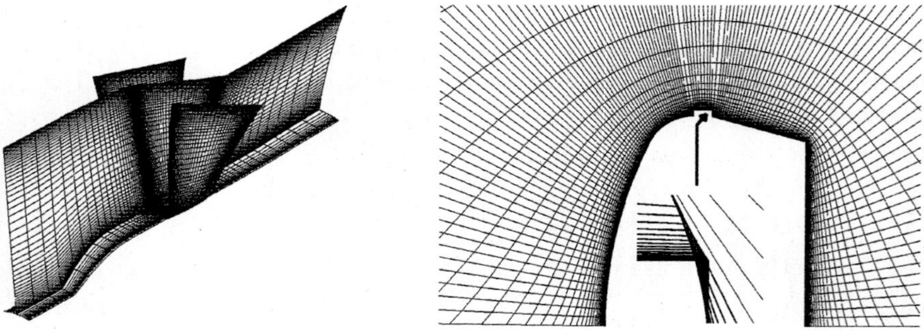

Fig. 1. NASA 55C fan rotor (left) and a close-up of the Huygens probe (right)

References

1. T. Radó: Jahresbericht der Deutchen Mathematiker Vereinigung **35**, 49 (1926)
2. R. Schoen and S.-T. Yau: Inventiones Mathematicae **44**, 265 (1978)
3. G. Liao, T. Pan, and J. Su: Num. Meth. P.D.E. **10**, 21 (1994)
4. J. Eells and L. Lemaire: Bull. London Math. Soc. **10**, 1 (1978)
5. R. Lehtimäki: AIAA Journal **37**, 768 (1999)

Global Stability Analysis of Two-Dimensional Incompressible Flows around an Elliptic Cylinder at Various Incidences

Asei Tezuka and Kojiro Suzuki

The University of Tokyo, Bunkyo, Tokyo 113-8656, JAPAN

Introduction

One of the most famous phenomena of transition from steady flow to unsteady flow is the onset of Kármán vortices. In this study, the global stability analysis was made to investigate the stability of the flowfield around an elliptic cylinder at various incidences from 0 to 90 degrees. The variation of the critical Reynolds number with the incidence was clarified. The code validation is made by analyzing the critical Reynolds number for the circular cylinder.

About the Global Stability Analysis

The incompressible Navier-Stokes equations can be written as: $\mathrm{d}\boldsymbol{u}/\mathrm{d}t = \boldsymbol{f}(\boldsymbol{u})$, $\boldsymbol{u} = (u_1, v_1, \cdots, u_N, v_N)^T$, where N is the number of grid points. Suppose that we have a stationary solution $\boldsymbol{u}_0$ at time t_0. To evaluate the global stability, the linearized equation of the velocity perturbation $\tilde{\boldsymbol{u}}$: $\mathrm{d}\tilde{\boldsymbol{u}}/\mathrm{d}t = (\partial \boldsymbol{f}/\partial \boldsymbol{u})\, \tilde{\boldsymbol{u}} \equiv A\tilde{\boldsymbol{u}}$ must be analyzed. The global stability is investigated by the eigensystem analysis of $A\tilde{\boldsymbol{u}} = \lambda\tilde{\boldsymbol{u}}$. In order to improve the accuracy of eigensystem analysis, we integrate the perturbation equation while T and solve the eigensystem analysis of $B \equiv \exp(AT)$ in place of matrix A. The size of matrix A and B $(2N \times 2N)$ is too large. Using Arnoldi's method, the eigensystem of matrix B are expressed by that of the $M \times M$ $(M \ll N)$ matrix H which correspond to the growth of the initial disturbance distributed over the whole flowfield. The approximate eigenvector of matrix A represents the mode of disturbance. The stability of the mode is evaluated by the real part of the eigenvalue. The period of the oscillation is predicted by the imaginary part of the eigenvalue. Detail of this method is explained in Ref.[1].

The MAC method[2] was used to calculate the Navier-Stokes equations. Four-stage Runge-Kutta method was used for the time integration, and the second order central difference was used to calculate the convection terms. The No-slip boundary conditions, the uniform flow conditions and the outflow conditions are imposed at the cylinder surface, far field and wake, respectively. The number of mesh points is 40×60 in the radial and circumferential directions, respectively.

Results and Discussion

Flow around a Circular Cylinder Whether the flowfield is stable or not is judged by the most unstable mode which has the largest real part of the eigenvalue. The initial disturbance to the velocity vector field was given by a set of random numbers. M is sufficiently large (M=60 in the present study), the dependency of the result on the initial disturbance becomes negligibly small. The critical Reynolds number of Kármán vortices is estimated as 49.35. This result shows a good agreement with several experimental data and the numerical result presented by Chiba[1].

Flow around an Elliptic Cylinder The reference length of the Reynolds number is the major axis of the ellipse. The angle of attack is defined as an angle between flow direction and the major axis of the ellipse. Figure 1a shows the variation of the critical Reynolds number with the angle of attack. When the angle of attack becomes larger, the critical Reynolds number decreases monotonically. The critical Reynolds number is quite sensitive to the angle of attack around 20 $\sim$ 30 degrees. To investigate the effect of the chord length projected normal to the uniform flow (see Fig.1b), we rearrange the result of Fig.1a in which the reference length is the major axis of the ellipse. Figure 1b shows the results when the Reynolds number is based on the projected length. In this case, the critical Reynolds number becomes minimum at angle of attack around 50 degrees. In the curve of Fig.1b, the local maxima are observed at the symmetric configurations, that is, at angles of attack 0 degree and 90 degrees. This means that asymmetry of the main flow may reduce the global stability of the flowfield.

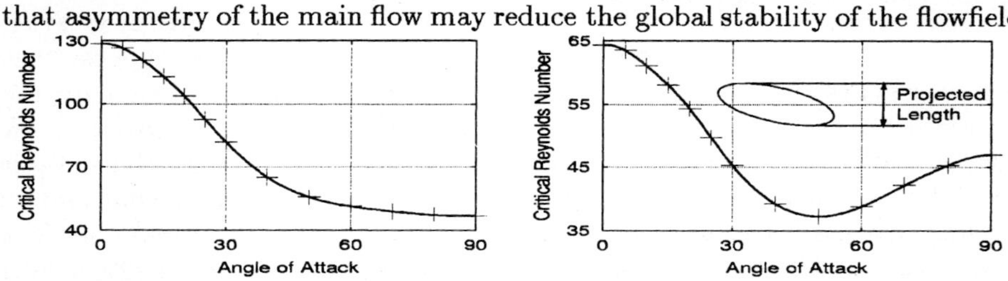

(a) Reference Length: Major Axis (b) Reference Length: Projected Length

Fig. 1. Variation of Critical Reynolds Number with Angle of Attack

Conclusions

- In case of elliptic cylinder, the critical Reynolds number is reduced when the angle of attack increases. The condition that the major axis is normal to incident flow is the most unstable.
- When the projected length is taken as the reference length, the critical Reynolds number becomes minimum at angle of attack about 50 degrees. The global stability is affected by the asymmetry of the main flow.

References

1. Chiba S.: Numerical Study on Global Stability of Incompressible Viscous Flows, Ph.D. thesis, The University of Tokyo, (1997), (In Japanese).
2. Chorin,A.J.: A Numerical Method for Solving Incompressible Viscous Flow Problems, J.Comp.Phys., **2**, 12-26(1967).

Fundamental Study on Propagation of Blade-Vortex Interaction Noise by CFD/FW-H Method

Takashi Aoyama and Shigeru Saito

National Aerospace Laboratory, 7-44-1, Jindaijihigashi-machi, Chofu, Tokyo, Japan

1 Introduction

The blade-vortex interaction (BVI) noise of a helicopter is generated by the sudden change of blade load during the interaction between a blade and previously shed tip vortices. It prevents helicopters from being widely used as a means of inter-city transportation in densely populated area. The intensity and the directivity of the BVI noise is strongly affected by the local intersection angle between the blade and the vortex. However, its effect is not understood well because the interaction is highly complicated. Therefore, the effect of the local intersection angle on the BVI-noise propagation is studied in this paper.

2 Calculation Method

The tip vortex is simply modeled by the Scully model and the BVI noise is calculated by a combined method [1] of a 3D unsteady Euler CFD code and an aeroacoustic code based on the Ffowcs Williams and Hawkings formulation. The effect of the vortex in the CFD code is modeled by the field velocity approach (FVA) [2] in which the effect is taken into account by modifying the grid time metrics to include the velocities induced by the vortex. The time history of acoustic pressure at an observer position is calculated by the aeroacoustic code using the pressure distribution on the blade surface obtained by the CFD code.

3 Results

Figures 1 and 2 show comparisons between the measured and the calculated acoustic pressure histories in Case 1D and Case 1B in Ref. 3, respectively. The measurement was done using the experimental set up shown in Fig. 3, in which the BVI noise was generated by the interaction between the rotating blade and the vortex from the vortex generator. In Fig. 1, the result by FVA is in better agreement with the experimental result compared with that by the surface transpiration approach (STA). In Fig. 2, FVA gives the good result even in the head-on BVI case in which the miss-distance between the blade and the vortex is zero. Figure 4 shows the effect of the local intersection angle (LIA) on the peak radiation direction (PRD) of the BVI noise. The effect is modeled by shifting the location of the line vortex in the x-direction as shown in Fig. 4. The blade is rotating counter-clockwise and intersecting with the line vortex. The blade is not a compact source, so the peak radiation direction is shown from the effective source, which is set to be located at 0.9 rotor radius. The peak radiation direction is dramatically affected by the local intersection angle.

4 Conclusions

The present method predicted the measured BVI noise well and it was indicated that the peak radiation direction was dramatically affected by the local intersection angle.

References

1. Aoyama, T. et al., Calculation of Rotor Blade-Vortex Interaction Noise Using Parallel Super Computer, 22nd European Rotorcraft Forum, No. 81, 1996.
2. Khanna, H, Coupled Wake/CFD Solutions for Rotors in Hover, Master Thesis of the University of Maryland, 1996.
3. Caradonna, F., et al., A Review of Methods for the Prediction of BVI Noise, AHS Technical Specialists' Meeting for Rotorcraft Acoustics and Aerodynamics, 1997.

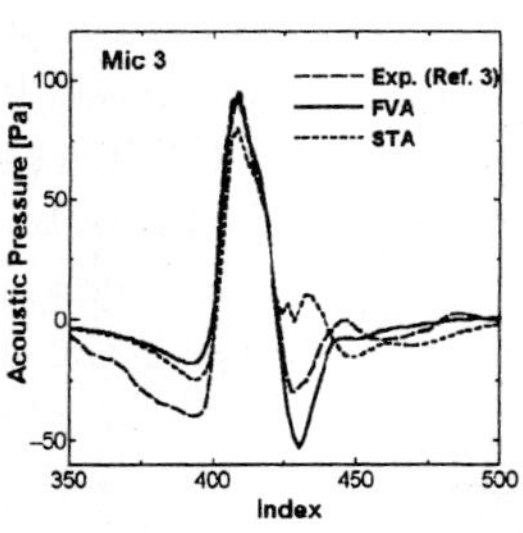

Fig. 1. Noise in parallel BVI.

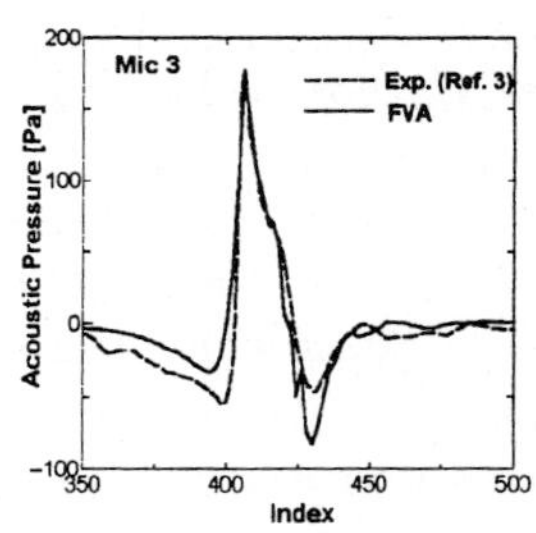

Fig. 2. Noise in parallel head-on BVI.

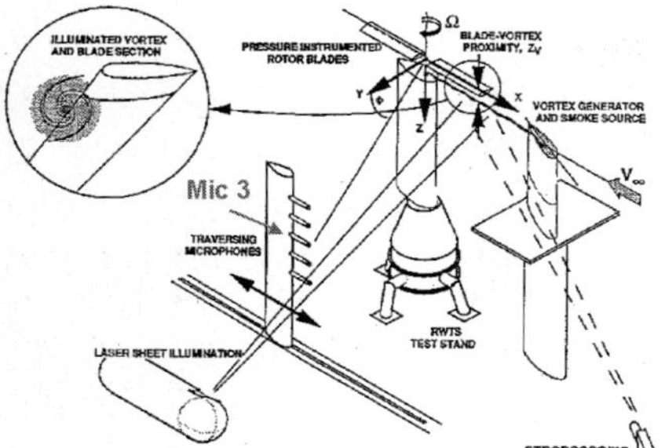

Fig. 3. Experimental set up (Ref. 3).

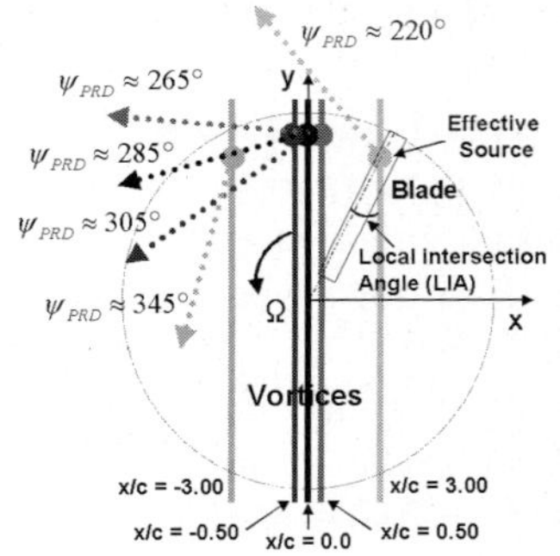

Fig. 4. Effect of LIA on PRD.

Turbulence Modelling of Strongly Swirling Flows

L.K. Yeh and C.A. Lin

Department of Power Mechanical Engineering, National Tsing Hua University, Hsinchu 30013, TAIWAN

1 Introduction

A two-dimensional swirling flow is considerably more complicated than two-dimensional plane flows, for additional strains arise due to the azimuthal motion, requiring the solution for azimuthal momentum. Swirl introduces intense azimuthal streamline curvature and hence curvature-turbulence interaction affects all six independent stress components. Therefore, the natural route is to apply Second-Moment Closures in predicting the Swirling flows. However, this is expensive computationally. One alternative is to adopt a non-linear stress and strain relationship of the Reynolds stresses.This can be achieved by assuming that the Reynolds stresses are taken to be non-linear function of the mean velocity gradients. The present study aims at investigating the capability of variants of non-linear eddy viscosity models on strongly swirling flows.

Within the framework of eddy-viscosity the Reynolds stress is approximated by linear and non-linear stress and strain relations. The stress and strain relationship can be expressed in a general form,

$$-\rho\overline{u_i u_j} = -\frac{2}{3}\delta_{ij}\rho k + \mu_t S_{ij} + F(k, \epsilon, S_{ij}, \Omega_{ij}) \tag{1}$$

where.

$$S_{ij} = \frac{\partial U_i}{\partial x_j} + \frac{\partial U_j}{\partial x_i}, \quad \Omega_{ij} = \frac{\partial U_i}{\partial x_j} - \frac{\partial U_j}{\partial x_i}$$

The function F and the eddy viscosity can be made non-linear functions of the strain rates and k and ϵ. Three non-linear eddy viscosity models (Shih2[1], SSG[2], GL[2]) are adopted in the present study.

2 Results

Experimental data had been obtained by So et al.[3]. This consists of a pipe into which an annular swirling stream is introduced together with a non-swirling central jet. The latter is introduced to inhibit extensive reverse flow (vortex breakdown) along the centre-line. The swirl number, S, $S = \frac{\int_0^{r_o} UWr^2 dr}{r_o \int_0^{r_o} U^2 r dr}$ for this case is 2.25, where r_o is the radius of the pipe and U and W are the axial and tangential velocity, respectively. According to the rule of thumb, a sub-critical

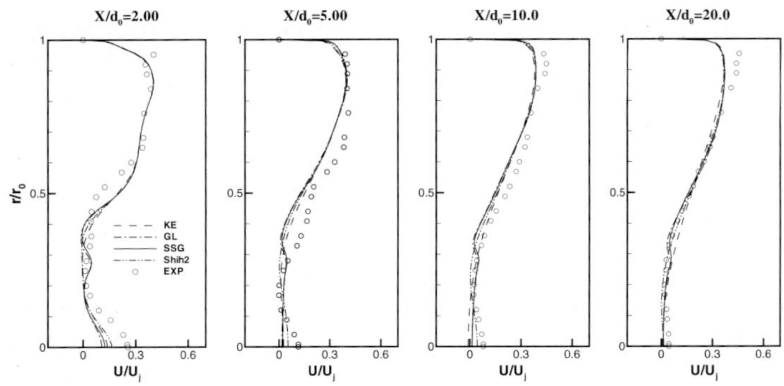

Fig. 1. Axial velocity distributions

state is reached when the maximum swirl velocity to the averaged streamwise velocity exceeds unity. A sub-critical state appears to reflect a strong decay in turbulent mixing, and a corresponding dominance of convective features, making the governing equations nearly hyperbolic in nature. This is confirmed from the measured axial and tangential velocity, shown in Figures 1 and 2. Non-linear model predictions (GL,SSG) and experimental data show the shape of the mean flow profiles to remain similar over the whole length of domain, and this implies that the mixing is weak. The $k - \epsilon$ and Shih2 predictions show, on the contrary, an excessively radial diffusive transport, with a faster decay of the centreline axial velocity and early return of solid body rotation of the swirling motion.

References

1. T. H. Shih, J. Zhu and J. L. Lumely: Computer Methods in Applied Mechanics and Engineering **125**, 287 (1995)
2. T. B. Gatski and C. G. Speziale: Journal of Fluid Mechanics **254**, 59 (1993).
3. R. M. C. So, S. A. Ahmed and H. C. Mongia: NASA CR-3832 (1984)

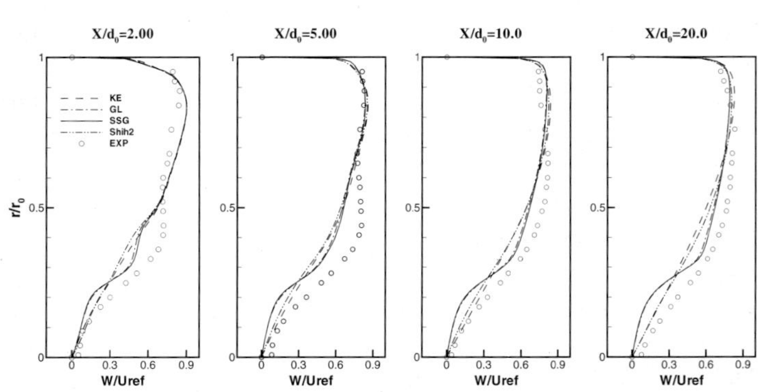

Fig. 2. Tangential velocity distributions

Cytofluid Dynamic Theory for Calculating Two-phase Flows and Bio-chemical Reactions

Ken Naitoh

Yamagata University, 4-3-16 Jyonan Yonezawa Yamagata, Japan

Mesoscopic kinetic theory is proposed, which describes the deformation and atomization processes of liquid droplets, the biological cell divisions, and the unlacing processes of DNA and RNA. The shape of the "parcel" representing liquid droplet, biological cell, and nitrogenous-base surrounded by water molecules for nucleic acids is assumed to be a spheroid deforming with time t. Let $a(t)$ and $b(t)$ denote one-half of the length of the long axis and short axis, respectively. We define the short-to-long-axis ratio of the spheroid as $\gamma(t) = \frac{a(t)}{b(t)}$, the deformation ratio. Then the radius r_d of a parcel when it is a spherical globule can be given by ${r_d}^3 = a(t)b(t)^2$. The surface tension is proportional to r^{-m} with the local radius r and the constant m. The distance Xs is that between the origin and the surface position. These definitions and the potential flow assumption for the parcel lead to the differential equation of parcel deformations [1] [2] [3] [5],

$$\frac{\partial^2 \gamma}{\partial \bar{t}^2} = A(\gamma)[\frac{\partial \gamma}{\partial \bar{t}}]^2 + B(\gamma)\gamma^{\frac{5}{3}-\frac{2}{3}m} + C(\gamma)\frac{\partial^2 \gamma_S}{\partial \bar{t}^2} + D(We), \tag{1}$$

with $\bar{t} = t\sqrt{\frac{8\sigma}{\rho_L {r_d}^{2+m}}}$, $A(\gamma) = \frac{1}{3\gamma}\frac{\gamma^2-2}{\gamma^2-1/2}$, $B(\gamma) = \frac{3}{8}\frac{2\gamma^{2m}-1/\gamma^m-\gamma^m}{\gamma^2-1/2}$, $C(\gamma) = 3\frac{\gamma^{7/3}}{\gamma^2-1/2}$, and $D(We)$, where σ, ρ_L, $\gamma_s(t)[= Xs(t)/r_d]$, and We denote the surface tension coefficient, the density, the dimensionless surface position, and the Weber number.

Based on the proposed theory, previously reported empirical atomization models [4] [6] can be derived along with their arbitrary constants. We show the calculation of the spray formation in a direct-injection engine (Fig. 1(a)) obtained by using this particle model and the compressible turbulence calculation code [7].

This theory also provides the formulation for breakup phenomena when two liquid droplets with spheroid shapes collide at the surface position X_s. The collision of two liquid droplets having about the 2:3 size ratio between the silver and golden ratios is extremely unstable and brings smaller child droplets (Fig.1(b)) [2] [3]. Moreover, it is known that, when a living cell divides, the size ratio of two resulting child cells is around 2:3 [2]. The present model also describes the cell division process.

The size ratio of the double carbon-nitrogen ring in purine bases (A, G) and the single ring in pyrimidine bases (T, C, U) is also around 2:3 (Fig.2(a) [2]). The analysis based on Eq. (1) shows that the bases having the 2:3 size ratio are able to unlace fastest (Fig.2(b)). Therefore, the bases having about the 2:3 size ratio are naturally selected among several molecules. The present theory also clarifies the origin of life between liquid droplets and living cells [5].

References

1. K. Naitoh and Y.Takagi: 'Synthesized Spheroid Particle (SSP) Method for Calculating Spray Phenomena', Trans. of SAEpaper962017 (1996)
2. K. Naitoh: 'Cytofluid dynamic thoery', Proc. of the Annual Meeting of Japan Society of Indust. and Appl. Mathematics, (1998)(to be published in Japan Journal of Industrial and Applied Mathematics)
3. K. Naitoh: 'Cytofluid dynamic theory for Atomization Processes', Gas℘Oil Sci. and Tech., (1999)(Also in IFP Report, 1999 and Proc. of COMODIA98, 1998)
4. K. Naitoh: 'Numerical Simulation of Spray Wall-impingement based on OPT model', Trans of SAEpaper940526, (1994)
5. K. Naitoh: 'Asymmetry model for the Origin of Life', Proc. of the Annual Meeting of Japan Society of Indust. and Appl. Math., (1999)(submitted to the Journal)
6. P.J. O'Rourke and A.A. Amsden: 'The TAB method for Numerical Calculation of Spray Droplet Breakup', Trans of SAEpaper872089, Vol. 54, No.2 (1987)
7. K. Naitoh and K. Kuwahara: 'Large Eddy Simulation and direct simulation of compressible turbulence and reacting flows in engines based on BI-SCALES method', J. of Fluid Dynamics Research 10, (1992)

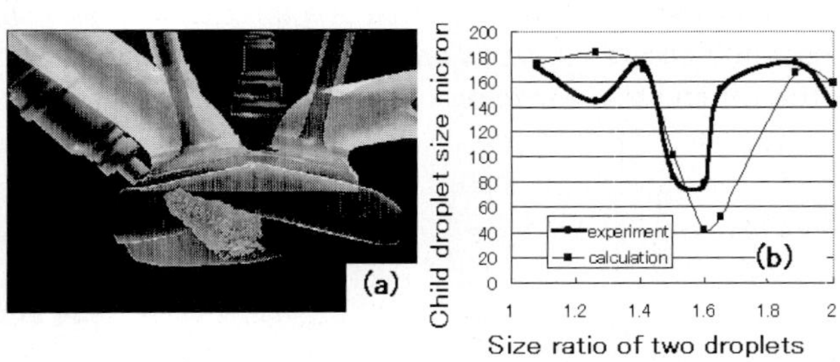

Fig. 1. Atomization processes of liquid droplets (a)Spray in direct-injection gasoline engine (b)Child droplet size for the collisions of two different-sized parent droplets

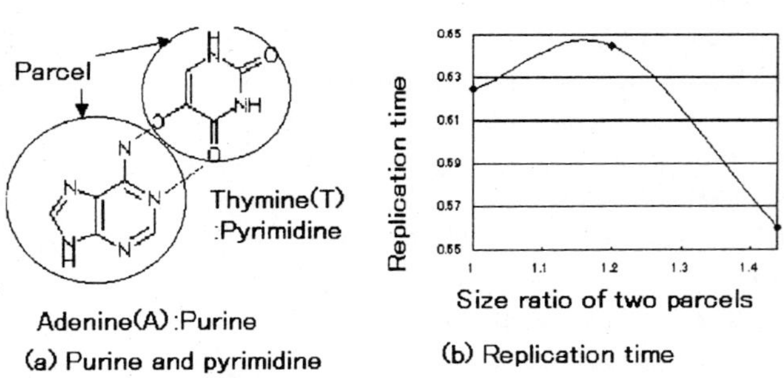

Fig. 2. Calculations for biological systems. (a)Sizes of purine and pyrimidine (b)Time until two different-sized connected parcels unlace [2] [5]

Euler Computations on Arbitrary Grids Using LSKUM

Ramesh V.[1] and Deshpande S.M.[2]

[1] CTFD Division, NAL, Bangalore, India, e-mail : vr@ctfd.cmmacs.ernet.in
[2] Dept. of Aero. Engg.,IISc, Bangalore, India, e-mail : suresh@aero.iisc.ernet.in

In this paper we report the results of computation on arbitrary meshes using the Least Squares Kinetic Upwind Method(LSKUM) in conjunction with a quadtree based preprocessor. LSKUM[1] is a node based Euler solver.The method is based on the fact that moments of the Boltzmann Equation lead to the Euler equation. Consider the 2-D Boltzmann equation $\frac{\partial F}{\partial t} + v_1 \cdot \frac{\partial F}{\partial x} + v_2 \cdot \frac{\partial F}{\partial y} = 0$, where F is the Maxwellian velocity distribution function, v_1 and v_2 are the Cartesian components of the molecular velocity. An upwind update scheme for the above equation is obtained by splitting the molecular velocity into positive and negative parts and using single step Euler for the time derivative. The derivatives F_x and F_y are determined using the least squares approach. Derivatives at any node are evaluated using the information of immediate surrounding nodes, reffered to as the connectivity data. The present method has the capability to work on an arbitrary distribution of points, given the connectivity data associated with them. In the present work we have developed a quadtree[2][3] based preprocessor which gives the connectivity data. Figure 1a shows a NACA 0012 airfoil with an overlapping structured cartesian grid. Figure 1b shows a typical point on the body with its connectivity highlighted along with the background quads used for connectivity generation. Figure 1c shows the density contours for this grid for $M = 1.20$ and $\alpha = 0$. Figure 1d shows the C_p plot. Figure 2a shows yet another application of the method on an arbitrary grid. This consists of an independently distributed set of points on the body combined with embedded blocks of Cartesian grids around the airfoil. The embedded Cartesian grid has three levels of grids. Over this again we have an overlapped polar grid at trailing edge point. The preprocessor then operates on the set of all nodes obtained by simply merging all the grid points to give the connectivity data. Figure 2b shows the connectivity for a point lying in the transition from one grid level to the next one. Figure 2c shows the density contours and Figure 2d shows the C_P plot for $M = 0.85$ and $\alpha = 1.0$. All the results obtained are quite satisfactory and encouraging.

Thus LSKUM solver has been shown to operate on an arbitrary set of points in conjunction with a preprocessor.

References

1. Ghosh, A.K. and Deshpande, S.M, *AIAA paper No. 95-1735.*
2. *VKI - Lecture series 1994-02. Grid generation, Jan 24-28, 1994.*
3. Ramesh, V., *Phd thesis (under preperation) IISc, Bangalore, India.*

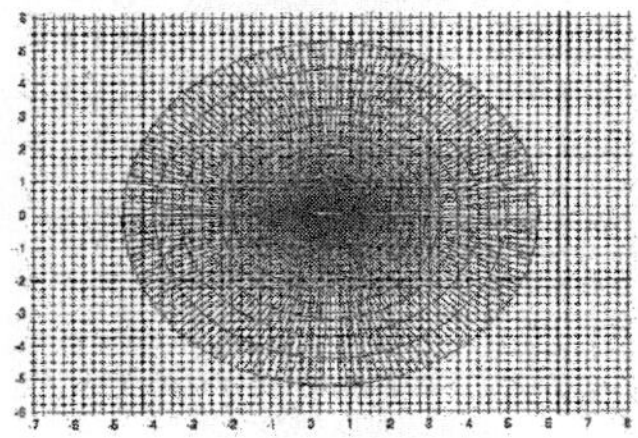

Figure 1a A view of the overlapped Structured-cartesian grid

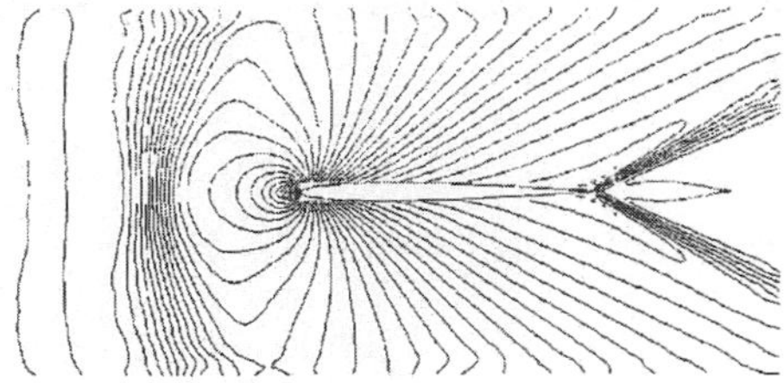

Figure 1c Density Contours NACA0012 : M=1.20 Alpha=0

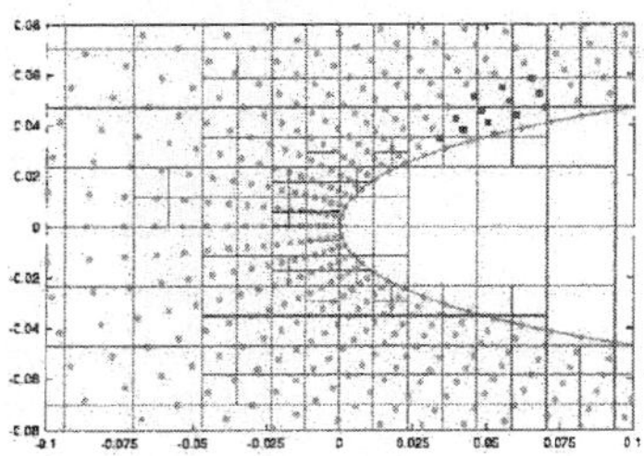

Figure 1b Connectivity for a point on the body

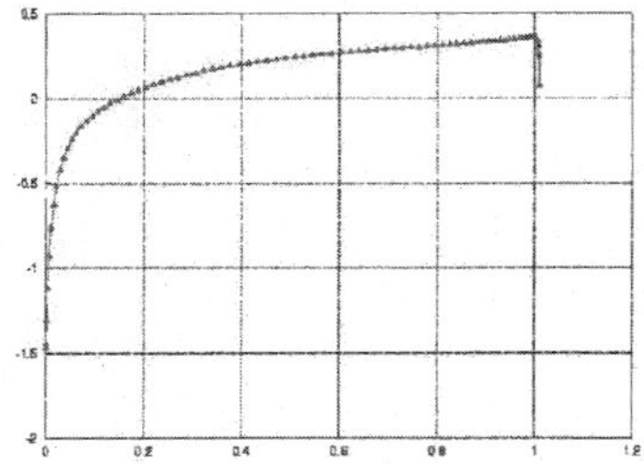

Figure 1d Cp Distribution NACA0012 : M=1.20 Alpha=0

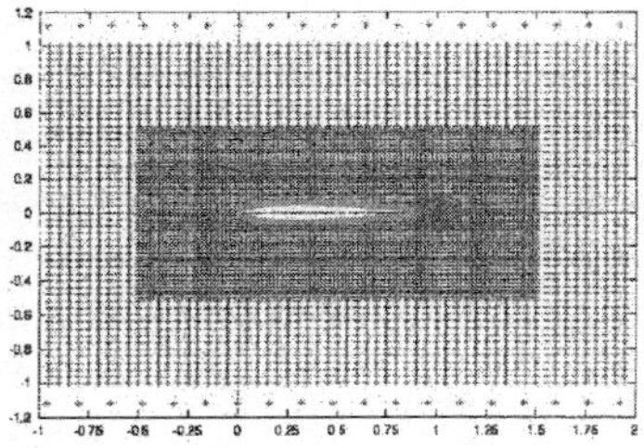

Figure 2a A view of the overlapped polar - Embedded cartesian grid

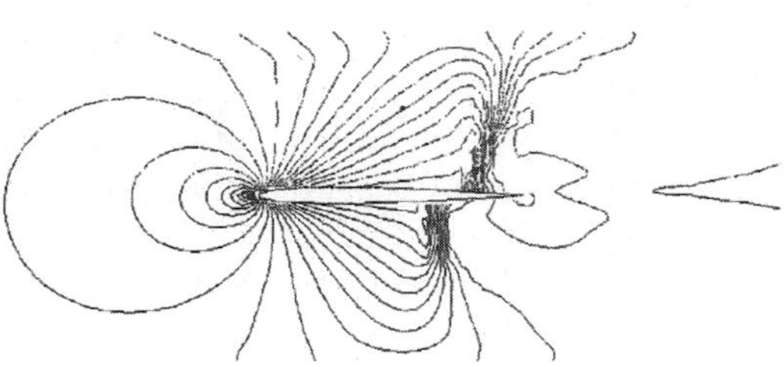

Figure 2c Density Contours NACA0012 : M=0.85 Alpha=1.0

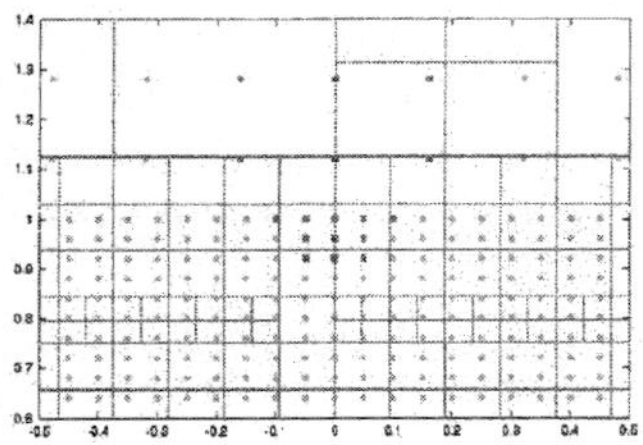

Figure 2b Connectivity for a point in the transition region for the embedded cartesian grid

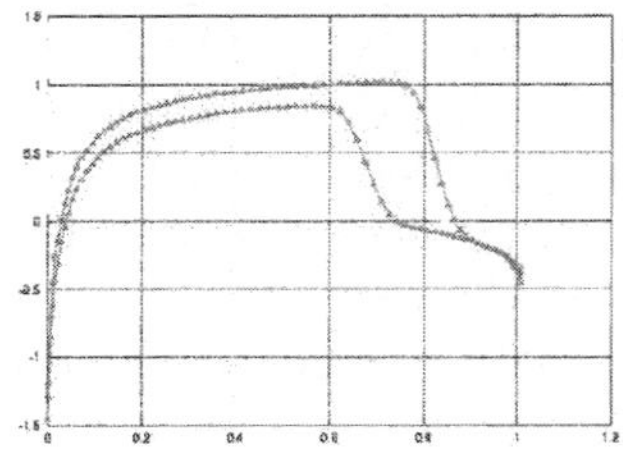

Figure 2d Cp Distribution NACA0012 : M=0.85 Alpha=1.0

Numerical and Analytical Investigation of the Large Scale Flow due to an "Energy Tower"

Moshe Isareli[1], Noa Porrat[1], and Irad Yavneh[1]

Computer Science Department, Technion Institute of Technology, Haifa 32000, Israel

The motivation for this work is the atmospheric flow due to an "energy tower" ([1,8]. The energy tower is a proposed structure intended for exploiting solar energy without the need for collectors. The plan is to sprinkle water at the top of a tall, cylindrical tower. The air near the top cools, its density increases, and it falls down the tower, driving turbines as it flows out at the bottom. The flow field that will evolve at mid-range distance from the tower (tens or hundred of kilometers) might influence the efficiency of the process.There are two questions of particular interest to the designers of the "energy towers". The first is related to the amplification of the angular velocity near the sink at the top of the tower (the "bath tub" effect). The second involves the possibility of intake at the sink (top) of the cold wet air which was ejected at the source (bottom). The above two effects if present can considerably lower the net energy output and consequently the "energy tower" will be uneconomical to build and operate. Previous work considered only the flow inside the tower and its immediate neighbourhood (within one to two diameters, [5,4].

The two main factors which influence the flow field induced by an energy tower at large distances are the stratification of the atmosphere and Coriolis acceleration due to Earth's rotation ([2]). The subject of this work is, therefore, the flow in a rotating stratified fluid, initially at rest, which is forced by a vertically aligned source-sink pair. Stratified flow tends to be predominantly horizontal. On the other hand, rotating systems tend to resist radial flow. The field which is formed in the presence of both rotation and stratification is mainly in the circumferential direction. The character of the flow, and its aspect ratio, are determined by the relative strength of rotation and stratification. In the atmosphere, the effect of stratification is stronger, and the horizontal scale is larger than the vertical. A well-known model for large-scale atmospheric flow, which captures both effects is the quasi-geostrophic model. This model is based on the assumptions of small Rossby number and slow variation of the flow field. This research includes numerical and analytical solutions for the quasi- geostrophic equations driven by a source-sink pair. The tower itself, as an obstacle, does not appear in the model.

The solution was obtained both for the axisymmetric case (with respect to the tower axis) and for the three dimensional flow in the presence of an ambient wind. For the parameter range relevant to an energy tower, the amplitude of the resulting field was found to be small compared to typical winds in the atmosphere. In the present model, some assumptions become valid only a few hours after activating the source-sink pair. The field which is formed immediately after the tower is activated is described by a different balance, which is obtained by a linearization of the equations of motion. The properties of this field were obtained and described for the axisymmetric case. It was found to be composed of a system of waves, propagating from the source and sink points [3]. The typical velocity of this propagation is determined by the strength of stratification and the typical vertical length scale (height of the tower, in our case). After the front moves away the solution coincides with that of the quasi-geostrophic equation. The existence of time independent side winds will cause the advection of the flow field of the "energy tower" and the reduction of its amplitude. For wind velocities larger then 1m/sec. a linearized form becomes a good approximation to the governing equation and the leading balance is the advection of potential vorticity generated at the source or sink. For a smaller wind velocity non linear terms are important and there will be advection of vorticity normal to the wind with a concomitant break of symmetry.

The numerical solution of the approximate three dimensional unsteady quasi-geostrophic model was obtained on a Cartesian grid by a finite difference approximation. The time marching was implicit and the elliptic three dimensional problem at every time step was solved by an efficient multigrid program [7]. Fourth order numerical viscosity was added to the vorticity equation for stabilization, the Jacobian was approximated by the Arakawa scheme. The numerical solution agrees well with the analytical approximations for the "weak" and "strong" wind cases. For intermediate winds there is no analytical solution and the expected "break of symmetry" was found and quantified. The axisymmetric numerical solution breaks down at long times when the source/sink is very concentrated. Following an extensive series of numerical experiments it was concluded that this is not a physical instability [6]. For a "smoothed out" source/sink the symmetric solution persists for long times.

References

1. Carlson, R. Philip, U.S. Patent No. 3,894,393, Power Generation through Controlled Connection (aeroelectric power generation), Assignee-Lockheed Aircraft Corp. Burbank Cal.
2. Gill, A., Atmosphere-Ocean Dynamics (1982).
3. Monismith, S.G., Jeffry B.W., and Yavneh, I., "Anisotropy and coherent vortex structures in planetary turbulence", Science 264 410-413 (1994).
4. Moses, E. J., Toren, M. and Olek, S., Numerical (Computational Fluid Dynamics) Simulation of the Energy Towers, The Israel Electric Corporation Ltd., Strategic Resources Group, Planning, Development & Technology Division, Department of Energy Resources and Thermal Systems Development (1996).
5. Mezhubovski, V., The flow in "Energy Towers", Ph.D. thesis, Technion IIT (1999).
6. Pedlosky, J., Geophisycal Fluid Dynamics, Springer, New York (1979).
7. Yavneh,I., and McWilliams, J.C., "Multigrid solution of stably stratified flows: the quasigeostrophic equations", J. Sci. Comput. 11(1) 47-69 (1996).
8. Zaslavsky, D., Energy from Dry Air or Aeroelectric Power Stations, vol. 1,2, The Faculty of Agricultural Engineering, Technion-Israel Institute of Technology (1987).

Numerical Simulation of Axisymmetric Super– and Hypersonic Separated Flows in Vicinity of Cylinder–Flare Configuration

Igor A. Bedarev and Nataliya N. Fedorova

Institute Theoretical and Applied Mechanics,
Siberian Branch of Russian Academy of Sciences,
Institutskaya, 4/1, 630090 Novosibirsk, Russia

This paper is a continuation of the studies dealt with numerical simulations of supersonic turbulent separated flows. 2–D turbulent supersonic flows in the vicinity of the plane configuration were explored in [1], [2]. The results of numerical modeling of axisymmetric flows in vicinity of a cylinder–flare configuration at various Mach numbers ($M = 3, 5, 7$) are presented in the paper. All the calculations were performed in conditions of the real physical experiments proposed in [3] as the test cases for AGARD Working Group 18 (1997). The Table shows the condition of the experiments investigated in the computations. Here α stands for flare angle, r is the cylinder radius, M_∞ and Re_1 are the freestream Mach and unit Reynolds numbers, P_0 and T_0 are the freestream static pressure and stagnation temperature; δ and C_{f0} are the boundary layer thickness and the skin friction before the interaction, L denotes the laminar case and T is for the turbulent case.

Table 1. Conditions of the experiments investigated in the computations

Case	α, °	r, mm	M_∞	Re_1, m^{-1}	P_0, Pa	T_0, K	δ, mm	$C_{f0} \cdot 10^{-3}$
T	30	25.4	2.85	$1.6 \cdot 10^7$	$1.7 \cdot 10^5$	265	11	1.75
T	35	65.5	5.01	$4.41 \cdot 10^7$	$3.5 \cdot 10^6$	500	2.5	–
T	20	25.4	7.05	$5.66 \cdot 10^6$	$2.45 \cdot 10^6$	890	25	1.22
T	35	25.4	7.05	$5.66 \cdot 10^6$	$2.45 \cdot 10^6$	890	25	1.22

The Favre-averaged Navier- Stokes equations complemented by $k-\omega$ Wilcox's [6] two–equation turbulence models are used as a mathematical model in turbulent case. It's known that some limiter on the minimum of the second model parameter ω (specific dissipation rate of turbulent kinetic energy) is needed in the model to avoid the extremely high turbulence viscosity values. It was shown in [7], that values of this limiter affects on the computational results greatly (see also [2]). Two different methods of inviscid flux approximation, namely, Van Leer [4] and Yanenko [5] were used in computations.

Comparisons of computations with the experiments have been conducted on all the parameters available in experimental database and have shown a good coincidence on the surface pressure and heat transfer distributions, as well as on the velocity and pressures fields. In Figure 1, 2 the numerical and experimental surface pressure and heat transfer distribution are presented for cases $M = 5, 7$ and $\alpha = 35°$. Calculations had been made using Yanenko inviscid flux splitting scheme. This scheme had shown the best resolutions of all the details of the flows in question. It must be mentioned that the Wilcox turbulence model works rather good up to $M = 5$ without any corrections accounted for compressibility effects but for the higher Mach numbers special corrections seems to be necessary.

The work was supported by Russian Foundation for Basic Research (Grants No. 99–01–00565 and 00–01–00891).

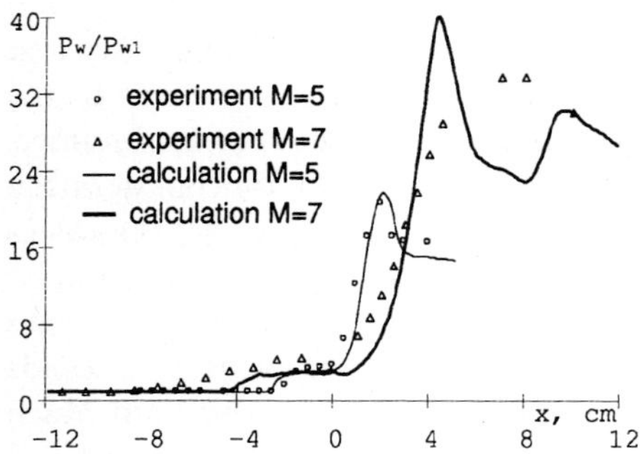

Fig. 1. Numerical and experimental pressure surface distribution ($M = 5, 7$ and $\alpha = 35°$)

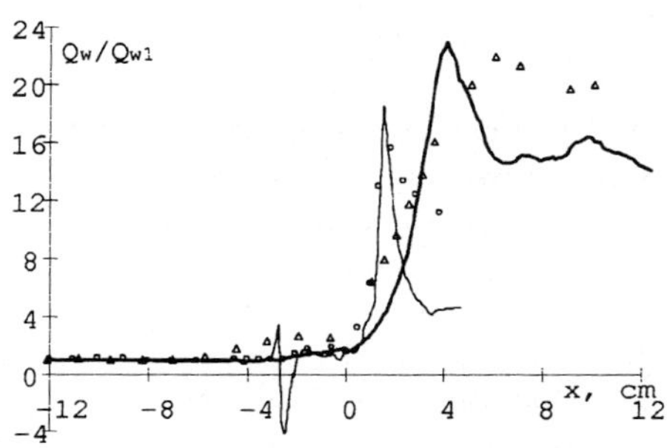

Fig. 2. Numerical and experimental heat transfer ($M = 5, 7$ and $\alpha = 35°$)

References

1. A.V. Borisov, N.N. Fedorova: 'Numerical Simulation of Turbulent Flows near the Forward–Facing Steps', Thermophysics and Aeromechanics', Vol. **3**, No. 1, pp. 69–83 (1996).
2. I.A. Bedarev, A.V. Borisov, N.N. Fedorova: 'Numerical Simulation of the Supersonic Turbulent Separated Flows in Vicinity of the Backward– and Forward– Faced Steps', Computational Fluid Dynamics Journal, (to appear).
3. J.M. Delery and A. G. Panaras: 'Shock Wave / Boundary Layer Interaction In High Mach Number Flows', AGARD–FDP, Working Group 18 Report – Step 1, Subgroup 1 on "Viscous Interaction", Chapter 1.
4. B. Van Leer: 'Flux–vector splitting for the Euler equation', ICASE Technical report No. 82–30 (1982).
5. V.M. Kovenia and N.N. Yanenko: *Method of splitting in the gasdynamic problems* (Novosibirsk, Nauka Publ., 1981) (in Russian).
6. D.C. Wilcox: 'A Half Century Historical Review of the $k - \omega$ Model', AIAA Paper No. 91–0615 (1991).
7. F.R. Menter: 'Influence of Freestream Values on $k - \omega$ Turbulence Model Predictions', AIAA Journ., Vol. **30**, pp. 1657–1659 (1992).

Computational Experiment of a Complex Dusted Plasma Flow

Takehiko Sato and Hideya Nishiyama

Tohoku University, 2-1-1 Katahira, Aoba-ku, Sendai 980-8577, JAPAN

1 Introduction

In this paper, a numerical model taking account of the compressibility and variable transport properties [1] is presented to clarify the RF induction thermal plasma flow impinging on the substrate and the behavior of the injected fine particle in the plasma flow to advance the plasma applications such as plasma spraying and environmental purification. The effects of an applied RF electromagnetic field, particle sizes and particle materials on the in-flight particle characteristics in a plasma flow are clarified by computational experiment.

2 Numerical modeling

The numerical model is derived from the following assumptions. (1) The plasma flow is axisymmetric 2D, continuous and further ideal gas in LTE and optically thin. (2) All injected particles are sphere with same diameter and the internal temperature is uniform in low Bi number. (3) Only drag force is acted on the particle. (4) The thermodynamic and transport properties of the particle both in solid and in liquid phases are respectively as a function of particle temperature. (5) Collision between particles is neglected under the dilute particle-loading condition. (6) There is one-way coupling between the dispersed phase and the carrier phase. (7) The nozzle and the chamber are electrically insulated. (8) The vector potential and the induction electric field has only azimuthal component.

The governing equations are derived for plasma jet, particle and RF induction electromagnetic field. Eulerian approach is adopted for a supersonic plasma flow, which is taken into account the variable thermodynamic and transport properties, Lorentz force and Joule heating. Lagrangian approach is adopted for in-flight particles, which is taken into account the plasma-particle interaction of one way coupling and phase change criterion. The RF induction electromagnetic field is solved by the vector potential equation.

3 Results and discussions

Figure 1 shows the temperature distributions of an impinging plasma jet (a) without RF and (b) with RF electromagnetic field under the conditions of the discharge power 10 kW and the operating pressure 50 kPa. Although the plasma

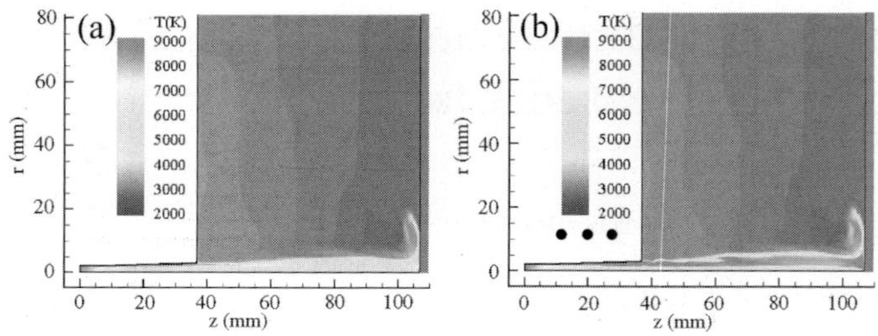

Fig. 1. Plasma temperature distributions (**a**) without RF electromagnetic field, (**b**) with RF electromagnetic field

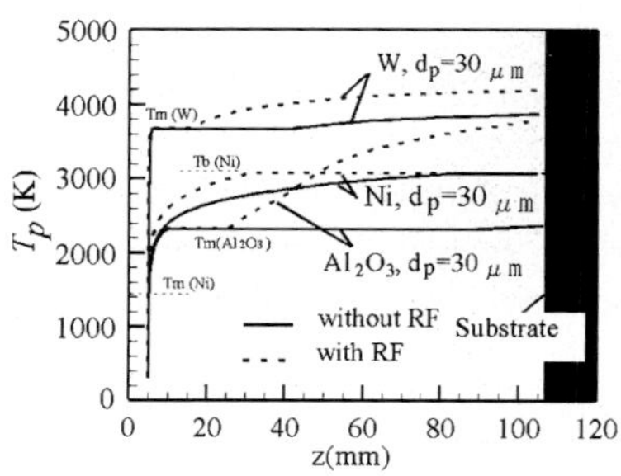

Fig. 2. Axial evolution of particle temperature with and without RF electromagnetic field

velocity is high, the plasma temperature is increased effectively even in the downstream region by applying the RF electromagnetic field due to the production of Joule heating.

Figure 2 shows the axial evolution of particle temperatures for Ni, W, Al_2O_3. The particle temperature increases rapidly up to the melting temperature. The heating rate of the Ni particle is larger than those of the W and Al_2O_3 particles. It is obviously shown that the heating rate increases effectively when the RF electromagnetic field is applied because of the active heat transfer from plasma flow.

4 Conclusions

The important control parameters and the optimum operating conditions of the in-flight particles in a supersonic complex plasma flow are clarified by the computational experiment.

References

1. H. Nishiyama, M. Kuzuhara, O. P. Solonenko and S.Kamiyama, Plasma Chem. and Plasma Process,**19**(3) (1999), pp.363-381

Numerical Study of the Flow around Savonius Rotors under its Operation

Kazuko Miyashita[1] and Tetuya Kawamura[1]

Ochanomizu University, 2-1-1, Ootsuka, Bunkyo-ku, Tokyo, 112-8610, Japan

1 Introduction

The wind force is widely recognized as the environmentally friendly energy and attracts public attention. In order to make an effective windmill, it is very important to analyze the flow field around a windmill. In this study, the Savonius rotor is chosen for the simulation. Fig. 1 is the schematic figure of the Savonius rotor. For the numerical simulation of rotating body such as rotors, it is convenient to use the rotating coordinate system which rotates with the same speed. However, if there is another body which is not rotating or if there are many bodies which rotates at different position and with different speed, it is very difficult to choose one special rotating coordinate system. In these situations, it is natural to use many coordinate systems separately which is suitable for the flow simulation around each body and connect these coordinates adequately. We focus on these points and simulate the flow fields around a windmill by using domain decomposition method in which the whole computation region is divided into several regions and they are connected adequately.

2 Numerical Method

Since the rotational frequency is low enough, the flow around the Savonius rotors is assumed as incompressible. The basic equations are transformed into the rotating coordinate system which rotates the same angular velocity as the rotor when we calculate the flow near the rotors. We use two computational domains. One domain includes rotating rotors and another includes a fixed ground. The computations in the two domains are performed alternatively in every time step. The physical quantities (velocity and pressure) are exchanged through the overlapping region. The numerical method to solve the incompressible Navier-Stokes equation is the standard MAC method [1] with the fractional step method [2]. All the spatial derivatives except the nonlinear term of the Navier-Stokes equation are approximated by the second order central difference. Nonlinear terms are approximated by the third order upwind scheme due to the numerical stability [3]. Euler explicit scheme is employed for the time integration.

3 Results

Typical results obtained by the present study are shown here. The overlap ratio (=S/D, see Fig.2) is chosen to 0.16 and 0.25 and tip speed ratio $\lambda(= R\omega/u_\infty)$ is changed from 0.25 to 1.25. The height of the rotor (= h) in the main calculation is 0.95D. Figure 3 is the time-averaged torque and power coefficients for various tip speed ratio λ. Here S/D=0.16. The torque coefficient decreases almost linearly as the λ increases and it becomes negative around $\lambda = 1$. The power coefficient has its maximum value around $\lambda = 0.75$.

4 Summary

In this study, the flow field around the windmill standing on the fixed ground are computed by using domain decomposition method. As the results, the running performance of the windmill is obtained numerically. Although the Savonius rotor is chosen for the computation, this method can be applied for computations of other kinds of windmill.

References

1. Harlow, F. H. and Welch, J. E.: Phys. Fluids **8**, 2182 (1965)
2. Yanenko, N. N.: *The Method of Fractional Steps*, (Springer-Verla, 1971)
3. Kawamura, T and Kuwahara, K: AIAA Paper **84**, 340 (1984)

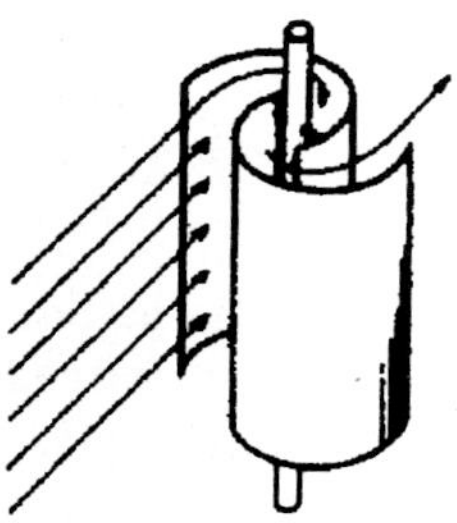

Fig. 1. Savonius rotor

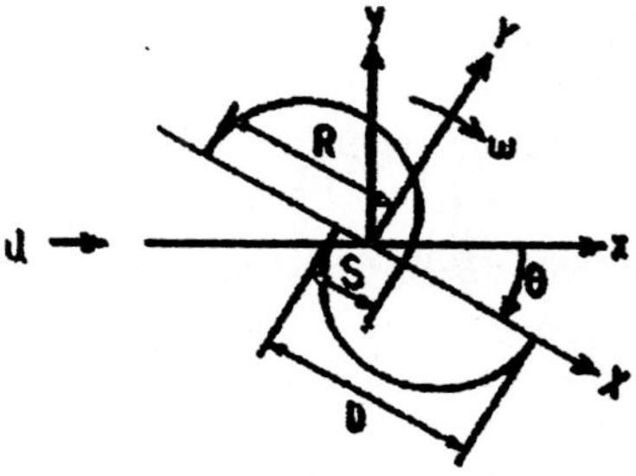

Fig. 2. Rotating coordinate

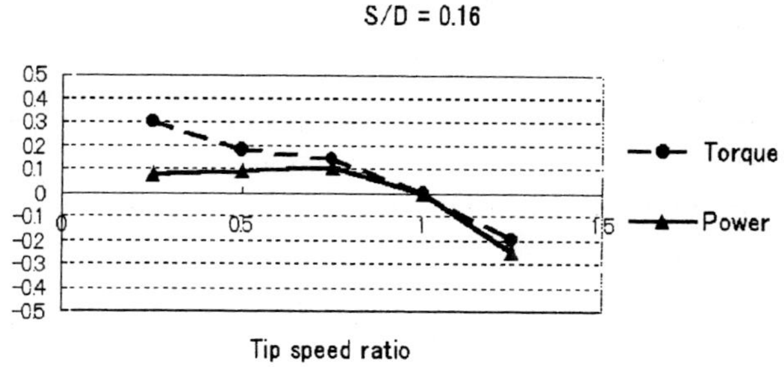

Fig. 3. Time averaged Torque and Power coefficients

Wake Structure around a Decelerated Cylinder

Nobu. Nishikawa[1], T, Shimizu[2], H. Maeda[2], and F Mikami[1]

[1] Faculty of Engineering, Chiba University, Inage-ku, Yayoi 1-33, Chiba, Japan
[2] Graduate Student, Chiba University

1 Introduction

Flows around decelerated circular cylinder are studied with BFC-MAC scheme for a moving frame of reference. The effect of the magnitude of deceleration on the wake length S/d is discussed for the Reynolds number $Re = 36.5$. The curves of wake-length versus "instantaneous Reynolds number", is almost similar for deceleration parameters larger than a critical value. To analyze such a sort of similarity the decomposition of a stream-function to steady part and unsteady part is applied for Psi-Omega formulation.

2 Basic Equation and Numerical Procedure

In the present study, the flow is assumed to be incompressible, laminar, and fluid properties are assumed to be constant. In the present study, we consider a sequential moving mode of a constant velocity period $U(t) = U_0$ and decelerating period, which requires two parameters, Re and α, respectively. With reference velocity U_0 and conventional dimensionless variables [1], the x-wise momentum equations in the moving reference frame are written, as follows. Here, the x–axis is chosen as moving direction.

$$u_t + U^* u_\xi + V^* u_\eta = -\frac{y_\eta p_\xi - y_\xi p_\eta}{J} + \frac{1}{Re}\Delta u - \frac{\alpha}{Re^2} \tag{1}$$

Here, U^*, V^* are contravariant velocity along the body fitted coordinates ξ, η respectively. For BFC grid attached to the moving cylinder with velocity $U(t)$, the outer-boundary velocity is expressed as $U(t) = 1 - \alpha t$. On a C–type grid, at the downstream end of the grid, the flow variables are linearly extrapolated. The implicit Euler expression is applied for time derivatives and then convective terms are approximated by 3-rd order upwind expression. The viscous terms and pressure derivatives are expressed by central differencings. The non-staggered differencing and point Jacobi scheme are applied.

3 Results and Discussion

Here the results for the initial Reynolds number $Re = 36.5$ are mainly discussed. In Fig.1, the wake length S/d values are plotted against the instantaneous Reynolds number $Re_s(t) = U(t)d/\nu$, and the agreement of the present S/d

with the experimental value [2] and the classical numerical results for steady flow summarized in Fig.3 of [3] are satisfactory at small deceleration such as $\alpha = 15$. Such agreement at small α implies that at nearly steady structure is realized at time by time or at each $Re_s(t)$.

"α-similarity" . As already described in the experimental study by [4], the S/d curves by the present calculation, run very closely to each other for the range of α, e.g. $\alpha = 28$, 58, or 139. In the last case the deceleration term is 10% or so of pressure gradient. Thus we call such behavior as "α-similarity".

"Stream-Function Vorticity Formulation". To explain such a similarity for α, the decomposition, such as steady part and unsteady part in Psi-Omega formulation in polar grids is more advantageous for decomposition than the primitive variable scheme. Fig.2, preliminary without decomposition, shows the transient profile of $\Psi - \Psi_{\mathrm{steady}}$ divided by T_{elapse}, where T_{elapse} is measured from the start of deceleration, and Ψ_{steady} is a solution before deceleration. Except at mid-leeside the curves show no difference, which implies a linear dependency upon the time of $\Psi(t)$. Decomposition such as $\Psi(t) = \Psi_{\mathrm{steady}} + T_{\mathrm{elapse}} \Psi_1(t)$ is under development and compared with other types of decomposition. The one with expansion in powers of time by Badr et al.[5] is available. Thus, the α similarity has been attempted to be explained.

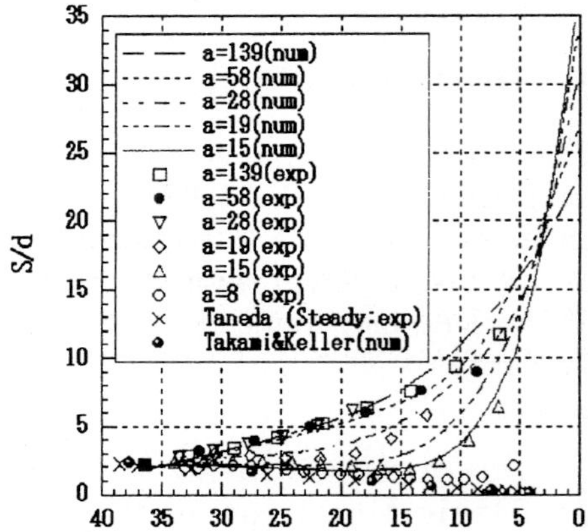

Fig.1 Wake length versus Res(t)

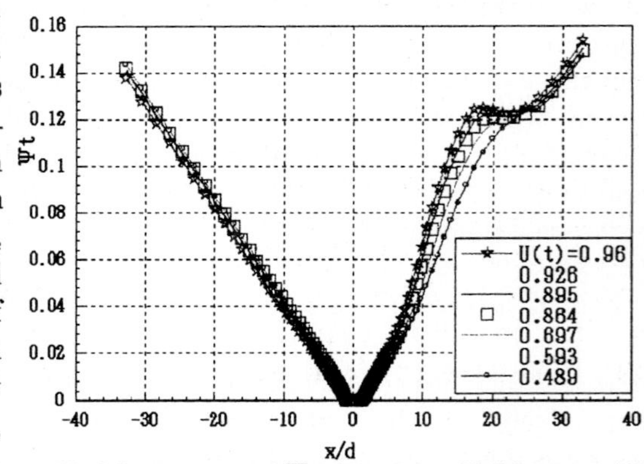

Fig.2 Increase rate of Ψ t=(Ψ — Ψ steady)/Telapse: α=-139, θ=6deg:[X<0],(180-6 :[X>0])deg

References

1. Nishikawa N.,Ishiwada T and Mikami F, Theor. Appl. Mech., **47**, pp223- 232 (1998).
2. Taneda, S, J. Physical Soc. Japan, **11**, p.302(1956)
3. Takami, H and Keller, H, Physics of Fluids Supplement II, pp51-55 (1969).
4. Tatsuno,M. and Taneda, S., Bull. Res. Inst. Appl. Mech, Kyushu Univ, **33**, pp1-33 (1969).
5. Badr, H.M., Dennis S.C.R. and Kocabiyik, S. JFM, **308**, pp97-110 (1996).

Numerical Solution of 3D Flow Fields Utilized for Problems of Flow Through Transonic Stages

Jiří Dobeš[1], Jaroslav Fořt[1], Jiří Fürst[1], Jan Halama[1], and Karel Kozel[1]

Dep. of Technical Mathematics, Faculty of Mechanical Eng.,
TU Prague
Karlovo nám. 13.
121 35 Prague 2, Czech Republic

Abstract. This contribution summarizes results of several numerical methods developed at our department. Presented methods are based on central Lax Wendroff type schemes and upwind schemes with Riemann solver. The results of 2D methods with different types of grid has been discussed for several typical axial and radial cascade geometries, comparisons of 3D results are presented for several geometries of blade row.

1 Governing equations

The 3D inviscid flow is described by the system of Euler equations, which can be written in following form for cartesian coordinates in relative frame of reference:

$$\mathbf{W}_t + \mathbf{F}_x + \mathbf{G}_y + \mathbf{H}_z = \mathbf{Q}, \tag{1}$$

where vector of unknowns $\mathbf{W} = [\rho, \rho w_1, \rho w_2, \rho w_3, e]^T$ consists of density, momentum components and full energy. Next $\mathbf{F}$, $\mathbf{G}$, $\mathbf{H}$ are inviscid fluxes and right hand side $\mathbf{Q}$ consists of Coriolis and centripetal forces. Domain of solution Ω consists of one cascade pitch. The solution $\mathbf{W}$ has to fulfill an integral form of the system (1) and boundary conditions along $\partial\Omega$, i.e. non-permeabilty condition on the walls, periodicity conditions and proper form of conditions on the inlet and outlet cross-sections which describes the radial distribution of flow parameters.

2 Numerical methods and results

The following numerical methods has been developed:
a) simplified TVD Mac Cormack scheme published by Causon;
b) full TVD Mac Cormack scheme;
c) cell vertex scheme based on the Ron-Ho-Ni's scheme;
d) explicit scheme based on modified Roe's Riemann solver;
e) implicit scheme based on Osher's Riemann solver.
First two methods have been developed for structured grid of quadrilateral or hexahedral cell centered finite volumes. Third method has been developed for the same grid type, but cell vertex finite volumes. First two methods use TVD form of damping terms, the third one Jameson's type of artificial viscosity. The

other methods have been developed for general unstructured meshes. Explicit Roe's method is based on 1D approximated Roe's Riemann solver with entropy correction (Harten, Hyman) applied on each cell interface. Higher order method uses MUSCL type of interpolation, van Leer limiter and three stage Runge-Kutta scheme in time direction. Osher's implicit method uses first order approximation in implicit part and higher order approximation on right hand side (piecewise linear reconstruction by weighted ENO). Linear system of equations is solved by restarted GMRES with ILU preconditioning.

Achieved results of 2D flow has been compared with experiments and results of other methods for several typical geometries of axial and radial turbine cascades and compressor axial cascades. Fig. 1 compares results for the turbine rotor cascade with highly twisted blades, typical for the low pressure steam rotor cascades. They are plotted in a form of iso-Mach lines for several cross-sections in span-wise direction.

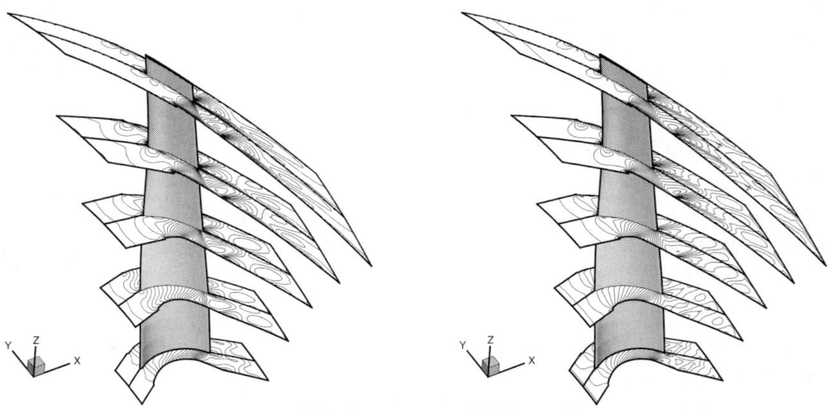

Fig.1: turbine rotor. simplified TVD MC left, cell vertex right

Acknowledgements

This work was supported by the grants of GA CR No. 101/98/K001, 101/98/1229, 101/00/1057 and Research Plane of MSM CR No 212200009.

References

1. D.M. Causon:'High Resolution Finite Volume Schemes and Computational Aerodynamics' In *Notes on Numerical Fluid Mechanics Vol.24, Braunschweig 1989, Vieweg* pp 63-74
2. J. Fořt ,M. Fialová?,J. Fürst ,M. Horák , K. Kozel :' Numerical Solution of Several 2D and 3D Internal Flow Problems.' In *Proceedings of 15th International Conference on Numerical Methods in Fluid Dynamics, Monterey, USA*, eds. P. Kutler, J. Flores, J.-J. Chattot, (Springer-Verlag, Berlin 1997), pp 364-369

On a Generalised Finite Difference Framework and its Order of Accuracy

N. Balakrishnan and Sridar D.

CFD Center, Department of Aerospace Engineering, Indian Institute Science, Bangalore - 560 012

1 Introduction

A new finite difference framework in line with the gridless schemes based on the method of least squares [1,2] for solving conservation laws of gasdynamics is proposed. In the new framework the state update at any given node requires the state in a cluster of grid points in its neighbourhood. Two different possibilities arising out of this new idea called Upwind least squares finite difference method (LSFD-U) and its versatility are demonstrated elsewhere [3–5]. The work presented here can be considered to be a third possibility.

2 The Method

The Euler equations in 2D are given by,

$$U_t + f_x + g_y = 0 \tag{1}$$

Refering to figure 1,'I' is a fictitious interface associated with node 'i', a neighbour of node 'o'. Flux along $\boldsymbol{oi}$ at the fictitious interface $F_{\uparrow I}$ is given by,

$$F_{\uparrow I} = \frac{1}{|\boldsymbol{\Delta r_I}|}(f_I \Delta x_I + g_I \Delta y_I) \tag{2}$$

where $\Delta x_I = x_I - x_O, \Delta y_I = y_I - y_O$. Expanding f_I and g_I in Taylor's series around 'o' we have,

$$\Delta F^e_{\uparrow I} = \frac{1}{|\boldsymbol{\Delta r_I}|}\sum_{q=1}^{1}\left[\frac{\partial^q f}{\partial x^q}\frac{\Delta x_I^{q+1}}{q!} + \sum_{m=0}^{q=1}\left\{\binom{q}{m+1}\frac{\partial^q f}{\partial x^{q-(m+1)}\partial y^{m+1}} + \binom{q}{m}\frac{\partial^q g}{\partial x^{q-m}\partial y^m}\right\}\frac{\Delta x_I^{q-m}\Delta y_I^{m+1}}{q!} + \frac{\partial^q}{\partial y^q}\frac{\Delta y_I^{q+1}}{q!}\right] \tag{3}$$

The interfacial flux F_I and therefore the flux difference $\Delta F_{\uparrow I}$ can be determined using any upwind flux formula [3,4]. The resulting over determined system can be solved by defining an error $E_I = (\Delta F_{\uparrow I} - \Delta F^e_{\uparrow I})|\boldsymbol{\Delta r_I}|$ and minimising $\sum_I E_I^2$ with respect to the derivative of f and g. The derivatives, f_x and g_y thus determined are used in the state update formula. Though this method is a finite difference method, because of its similarity to finite volume method, it has an added advantage that it employs the recent developements in finite volume method such as the method of reconstruction and limiters.

3 Order of Accuracy

The order of accuracy with which the flux derivatives are numerically determined depends on two factors: (1) Accuracy of the estimated interfacial numerical flux. (2) Accuracy of the least squares procedure. The former depends on the reconstruction procedure and the latter depends on the number of terms included in the Taylor's series expansion of $\Delta F_{\uparrow I}$. In this context the following theorm is proven:

If a k-exact reconstruction procedure is employed to determine the directional flux $F_{\uparrow I}$ at the fictitious interface and if terms upto $O(h^l)$ are included in the Taylor's series expansion given by eq(3)., then the first derivatives of the fluxes are determined to $O(h^l)$ for $k \geq l$.

4 Results

The method has been tested out for a wide range of mach numbers [3,4]. (1) Sample results for flow in a channel with a ramp of 10.75° obtained using a grid wherein the grid points are placed randomly are shown in figure 3. (2) The results are obtained using KFVS flux formula and Venkatakrishnan limiter.

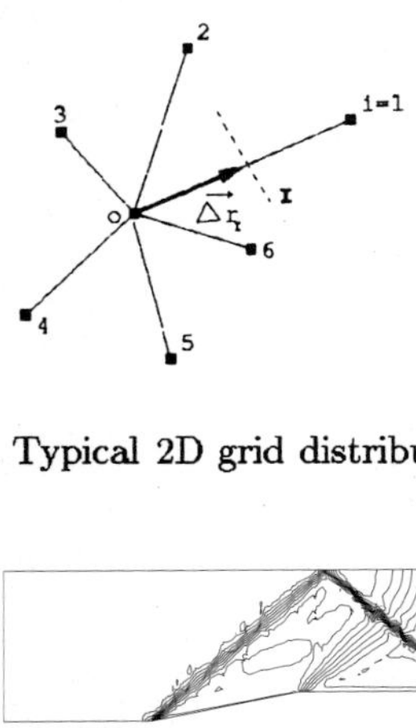

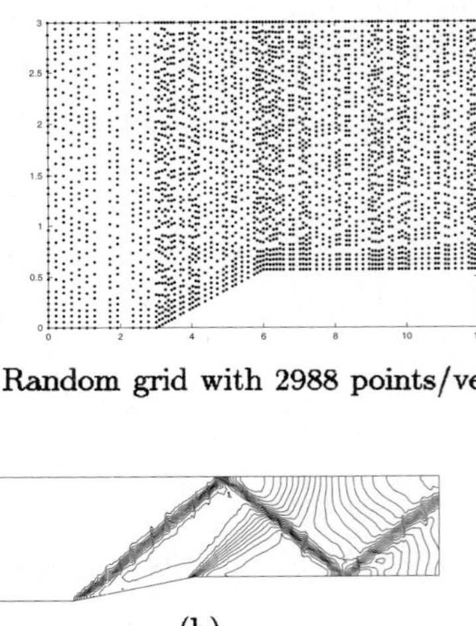

Fig. 1. Typical 2D grid distribution. **Fig. 2.** Random grid with 2988 points/vertices.

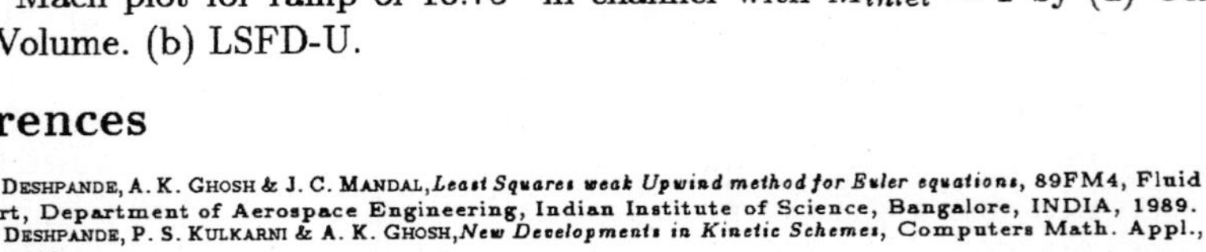

(a) (b)

Fig. 3. Mach plot for ramp of 10.75° in channel with $M_{inlet} = 2$ by (a) Cell-vertex Finite Volume. (b) LSFD-U.

References

1. S. M. Deshpande, A. K. Ghosh & J. C. Mandal, *Least Squares weak Upwind method for Euler equations*, 89FM4, Fluid Mechanics report, Department of Aerospace Engineering, Indian Institute of Science, Bangalore, INDIA, 1989.
2. S. M. Deshpande, P. S. Kulkarni & A. K. Ghosh, *New Developments in Kinetic Schemes*, Computers Math. Appl., Vol 35(2), p75, 1998.
3. N. Balakrishnan & Praveen, C., *A new Upwind Least Squares Finite Difference Scheme (LSFD-U) for Euler equations of Gasdynamics*, finite volume for Complex Applications II-Problems and Perspectives, HERNES Science Publications, Paris, 1999.
4. Praveen, C. & N. Balakrishnan, *A Least Squares based Finite Difference scheme for Compressible Flows*, Proceedings of the 8th Asian Congress of Fluid Mechanics, p533, Dec. 6-10, 1999, Shenzhen, China.
5. N. Balakrishnan, *New Least Squares based Finite Difference Methods*, 99FM9, Fluid Mechanics report, Department of Aerospace Engineering, Indian Institute of Science, Bangalore, INDIA, 1999.

Numerical Simulation of Multi-Dimensional Structure and Mach Reflection of Gaseous Detonation Waves *

Deliang Zhang[1], Wei Xie[1], and Changmin Guo[2]

[1] Institute of Mechanics, Chinese Academy of Science, Beijing 100080, China.
[2] University of Science & Technology of China, Hefei, Anhei, 230026, China.

Abstract. When a detonation wave interacts with a wedge, the Mach reflection is occured. In this paper the DDT process of combustible gases and Mach reflection of detonation waves have been simulated numerically with a new CE/SE method. Simulated results are compared and verified with experiemntal results.

1 Model and Numerical Methods

In numerical simulation a two-step chemical reaction model (ZND model) is adopted for simlicity. In this model there are two steps: induction reaction and exothermic reaction respectively. They are expressed by proceeding parameters: α and β. At first α and β are 1.0, followed by decrease of α to zero, then β

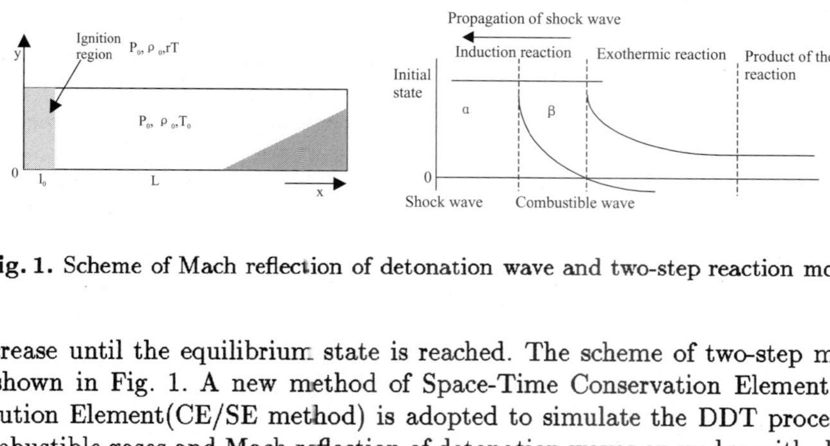

Fig. 1. Scheme of Mach reflection of detonation wave and two-step reaction model

decrease until the equilibrium state is reached. The scheme of two-step model is shown in Fig. 1. A new method of Space-Time Conservation Element and Solution Element(CE/SE method) is adopted to simulate the DDT process of combustible gases and Mach reflection of detonation waves on wedge with chemical reaction. It differes substantially in both concept and methodology from well-established methods: finite differece, finite volume, finite element and spectral method. This is first time to use CE/SE method to simulate the DDT process of combustible gases and Mach reflection of detonation waves with chemical reaction.

* Supported by the National Science Foundation Committee of China (19872068) and the Lab of High Temperature Gasdynamics

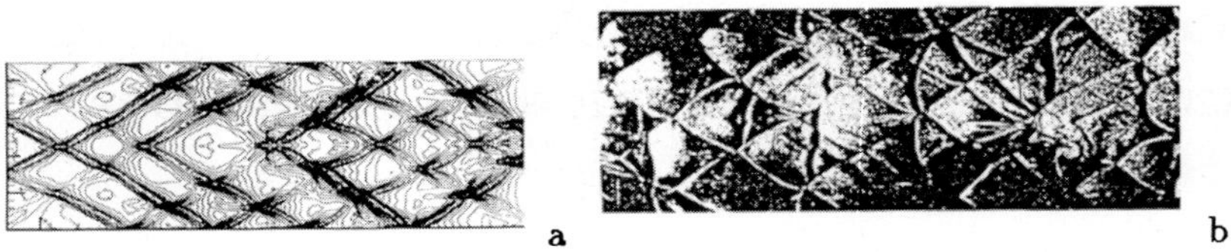

Fig. 2. Detonation cells in $2H_2 + O_2 + Ar$. a) Numerical simuation results; b) Experimental results

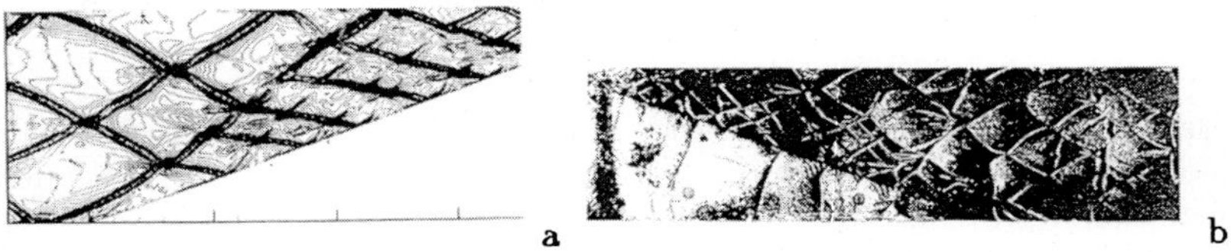

Fig. 3. Cellular structure produced by the Mach reflection of detonation waves in $2H_2 + O_2 + Ar$. a) Numerical simulation of Mach reflection; b) Experimental results of Mach reflection

2 Results

The simulated detonation cells of combustible gases and Mach reflection of detonation waves are shown in Fig. 2 and Fig. 3. The simulated results are compared with expeimental results. Following conclutions are given:

1. the Mach reflection of detonation waves differs totally from the Mach reflection of shock waves. Before and after Mach reflection of detonation waves there is a marked boundary line. Both side of the boundary line the size of cellular steructure is decreased with the increase of wedge angle and initial pressure of combustible gas.

2. the angle χ of the trace of triple point with wedge surface is decreased with the increase of wedge angle. Its change is not obvious with the initial pressure of combustible gases.

3. simulated results show that it is successful to adopt the CE/SE method in the chemical reaction flow of DDT process of combustible gas and Mach reflection of detonation waves. It has following adventages: simplicity of scheme; high resolution and small computational time.

References

1. M.A. Nettleton: *Gaseous Detonation*, (Chapman Hall 1987)
2. J. Meltzer, J.E. Shephered, R. Akbar, A. Sabet: 'Mach Reflection of Detonation Waves'. In: *Progress in Astronautics & Aeronautics*, AIAA, **153**, pp. 78–94, (1991)
3. S. Taki, T. Fujiwara: 'Numerical Simulation on the establishment of Gaseous Detonation'. In: *Progress in Astronautics & Aeronautics*, AIAA, **94**, pp. 186–200, (1984)
4. S.C. Chang: J. of Comput. Phys. **119**, pp. 295–324, (1995)

Robust Feedback Controllers for a Backward Facing Step Flow

Günter Bärwolff[1], Maiko Garwon[2], and Rudibert King[2]

[1] Technical University of Berlin, Dept. of Mathematics, Strasse des 17. Juni 136, MA 6-3, D-10623 Berlin, Germany
[2] Technical University of Berlin, Institute of Process and Plant Technology, Hardenbergstrasse 36a, P2-1, D-10623 Berlin, Germany

1 Flow problem

We consider a backward facing step channel and we investigate the flow for a Reynolds number of 3000 built with the inflow velocity $U_\infty = 2,27\,\frac{m}{s}$ of the block inflow profile and the step height $H = 20\,mm$ for an air flow. Over a spanewise slit in front of the seperation line an acoustic manipulation is realized. Because of the absence of detailed information about the acoustic control, which results in blowing and suction, was simulated by a sine function of the form $U_{jet} = A\ sin(2\pi\, f\, t)$ $(U_1 = U_2 = U_{jet})$. Former calculations with amplitudes of $A = 0.1\,U_\infty$, $A = 0.01\,U_\infty$ and $A = 0.001\,U_\infty$ using a fixed frequency of $f = 50\,Hz$ showed a good agreement with the experimental results [1].

2 Identification and control

Because of the huge amount of computational work for the solution of a mathematical optimization problem for the drag minimization we are looking for computationally "cheap" control strategies. Our interest is focussed on the relation between the amplitude A of the acoustic signal and the recirculation length X_r. X_r is determind as the last intersection with the x-axis of the graph of wall shear stresses (x-direction) averaged in the lateral direction. In the result of the large eddy simulations of the transitional step flow for several parameters constellations different black box models have been identified. In the most simple case a first order model of the form

$$T\,X_r'(t) + X_r(t) = K\,A(t) \tag{1}$$

could be used to describe the dependence of the reattachment length on the amplitude of the harmonic excitation. Figure 1 shows the time response of the recirculation length to a step of the manipulation amplitude based on the LES and the black box model (from $A = 0$, non manipulated flow, to $A = 0.01U_\infty$, manipulated case). A second order approximation gives only a slight improvement of the fit. However, if this approximation is repeated for other Re-numbers different parameters T and K are identified. Therefore a family of models was considered to design a controller. The first black box model was used as a basis for a robust controller design. Different control schemes were tested. Only the very simple P-controller is shown here. It's equation is of the form

$$A(t) = K_p[W(t) - X_r(t)] \quad , \tag{2}$$

where W stands for the setpoint of the recirculation length and k_p is the gain factor of the controller. For a stable closed loop it is necessary to work with the factor $K_p = -0.01\frac{U_\infty}{H}$. If the same controller is to be used for other larger Reynolds numbers, the gain has to be reduced to maintain robust stability. To guarantee zero steady state offset for a constant setpoint a PI-controller can be used which is described by

$$E'(t)K_c\tau_c + E(t)K_c = A'(t)\tau_c \tag{3}$$

with $E(t) = W(t) - X_r(t)$. τ_c, K_c are constant system parameters ([2]). Figure 2 shows the time behavior of the recirculation length and the manipulation amplitude during the work of the P-controller. A recirculation setpoint length $W = 5\,H$ was prescribed. The study of more robust controllers like Smith-predictors to increase performance is under consideration.

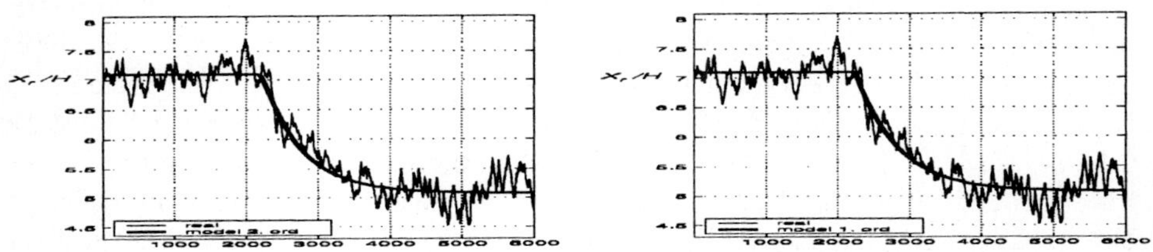

Fig. 1. time response to an amplitude step

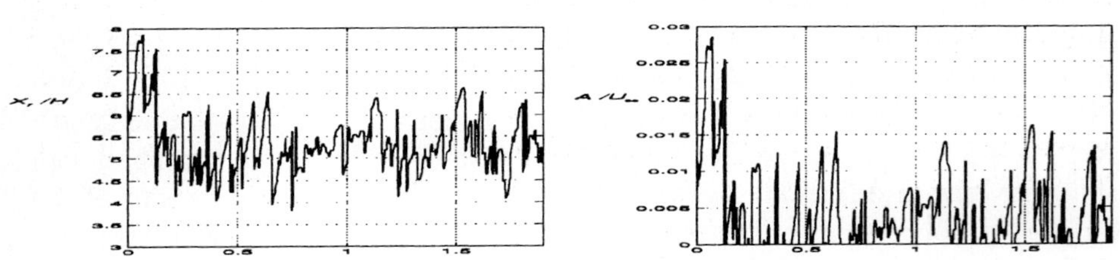

Fig. 2. numerical simulation of the P-controller, $X_r(t)$ and $A(t)$

References

1. A. Huppertz, G. Janke: Ergebnisse der Strömung hinter einer rückwärtsgewandten Stufe durch dreidimensionale Anregung, private communication (1997/1998)
2. M. Garwon: Robuste Regelung einer simulierten Stufenströmung, Diplomarbeit, TU Berlin (2000)

DNS of the Free-Surface Turbulent Flow with Constant Temperature at the Surface

Y. Yamamoto[1], T. Kunugi[1], and A. Serizawa[1]

Department of Nuclear Engineering, Kyoto University
Yoshida, Sakyo, Kyoto, 606-8501, Japan

Abstract. In the present study, Direct Numerical Simulation (DNS) of the free-surface turbulent flow with constant temperature at the surface is performed. Numerical investigations are carried out for two Prandtl numbers 1.0 and 5.0 with the neutral or stable stratification (Ri=27.6), while a Reynolds number of 200, based on the friction velocity and flow depth. As the results, fundamental turbulent statistic data for the influence of Prandtl number change and buoyancy effect are obtained.

1 Numerical Procedures

Governing equations are incompressible Navier-Stokes equations with the Boussinesq approximation,the continuity equation and the energy equation.The integration method is based on a fractional step method [1)], time integration is a second order Adams-Bashforth scheme, and a second order central differencing scheme[2),3)] is adapted for the spatial discretization. As the boundary conditions for fluid motion,[4),5)] free-slip condition at the free-surface, no-slip condition at the bottom wall and the cyclic conditions in the streamwise and spanwise direction are imposed respectively. For the equation of energy, temperatures are constant at the free-surface and the wall. Numerical conditions are tablated in Table 1, where $R_\tau = u_\tau h/\nu$ is Reynolds number, u_τ is the friction velocity, h is the flow depth, $Pr = \nu/\alpha$ is Prandtl number, α is the thermal diffusivity, $Ri = \beta g \Delta T h/u_\tau^2$ is Richardson number, ΔT is the temperature difference between the free-surface and wall. All quantities normalized bythe friction velocity, kinetic viscosity and mean scalar flux at the free-surface, were denoted by the super script $+$.

2 Results and Discussion

Figure 1 shows turbulent intensity profile. Neutral stratified results are good agreements with previous studies.[4),5)] Stable stratified profiles indicate that stable stratification restrains the turbulent motion. Mean scalar profiles and scalar flux and fluctuation profiles are shown in Figs. 2,3. It can be seen that scalar fluctuation is produced by mean velocity gradient near wall and mean scalar gradient near free-surface, but these profiles are distinct from each other. Especially, in the case of $Pr = 5.0$, the turbulent scalar statistics amount to maximums at near free-surface where typical turbulence structures are existent. These may suggest that if the Prandtl number is the higher, the heat transfer is enhanced by turbulent structures at near free-surface.

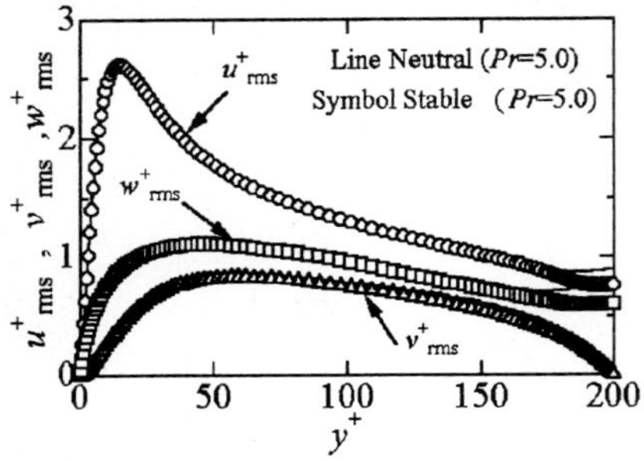

Fig.1 Turbulent intensity profiles

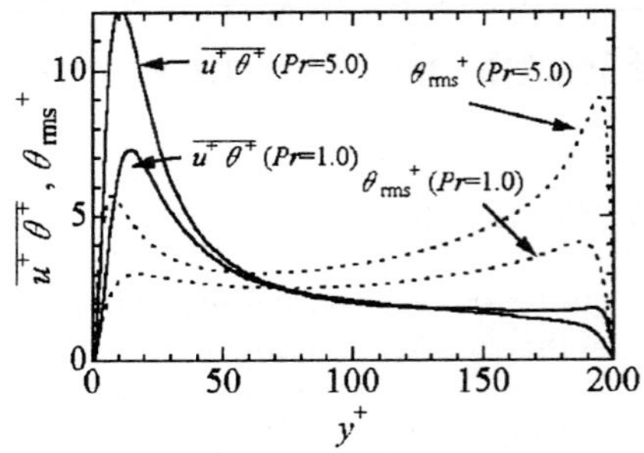

Fig.2 Scalar flux and fluctuation profiles

Table1 Numerical conditions

Re	Grid Number ($x \times y \times z$)	Resolution ($\Delta x^+, \Delta y^+, \Delta z^+$)	*Pr*	*Ri*
200	128×108×128	10.0, 0.5-4.0,5.0	1.0	-
200	256×131×256	5.0,0.5-2.0, 2.5	5.0	-
200	128×108×128	10.0,0.5-4.0,5.0	1.0	27.6
200	256×131×256	5.0, 0.5-2.0, 5.0	5.0	27.6

References

1. A.J. Chorin: Math. Comp. **22**, pp.745-762, (1968)
2. H. Kawamura: *The Recent Developments in Turbulence Research*, (International Academic Publishers 1995), pp.54-60
3. T. Kajishima: Trans.JSME. Ser.B **60, 574**, pp.2058-2063, (1994)(In Japanese)
4. R.A. Handler et al: AIAA J., **31**, pp.1998-2007, (1993)
5. R. Nagaosa and T. Saito: AIChE. J. **43**, (1997)

Finite Spectral Method for Non-Periodic Problems

Jian-Ping Wang[12]

[1] Dept. of Mechanics and Engineering Science, Peking University Beijing 100871, China
[2] Dept. of Environmental and Information Science, Yokkaichi University Yokkaichi, Mie 512-8512, Japan

1 Introduction

Localization of spectral methods has been a significant problem in computational fluid dynamics [1]. Domain decomposition techniques [2] have succeeded in reducing the global effects to each sudomain, but they are still not satisfactory for the needs of flexibility in treating complex phenomenon and geometry. Finite spectral method based on non-periodic Fourier transform is the first attempt of dealing with spectral methods pointwise [3], while more CPU time is required than FFT. In order to compete with finite difference method and finite element method, it is necessary to develop a local spectral method not only being flexible but also efficient.

2 Finite Spectral Method base on Non-Periodic Fourier Integral

2.1 Finite Spectral Approximation

A new finite spectral method was found when approximating a pulse function in Fourier series. The value at each grid point affects its neighbor 2N points as

$$W_N(x) = \frac{1}{2N} \sum_{n=-N}^{N} C_n \exp \frac{\mathrm{i}\pi n x}{l} = \frac{1}{2l} \sum_{n=-N}^{N} C_n \cos \frac{\pi n x}{l} \tag{1}$$

Conversely, an arbitrary function can be constructed by the basis functions $W_N(x)$ of its neighbor 2N points. Non-periodic functions are interpolated by

$$f(x) = \sum_{n=N_1}^{N_2} f(x_j) W_N(x - x_j) \tag{2}$$

It is easy to obtain the formulae of calculating derivatives at each discrete point as

$$f_x(x) = \sum_{n=N_1}^{N_2} f(x_j) W_N^{(1)}(x - x_j) \tag{3}$$

where $W_N^{(1)}(x)$ is the first derivative of $W_N(x)$.

2.2 Schemes for Compressible and Incompressible Flows

Finite spectral NND scheme is proposed for treating discontinuities. Shock tube problems and supersonic viscous flows passing a two-dimensional cylinder were calculated accurately.

The QUICK scheme is introduced to finite spectral method, with which incompressible flows around a cylinder was calculated.

3 Results

Figure 1 (**a**) shows the basis function of finite spectral approximation. Figure 1 (**b**) shows finite spectral interpolation of a non-periodic polynomial.

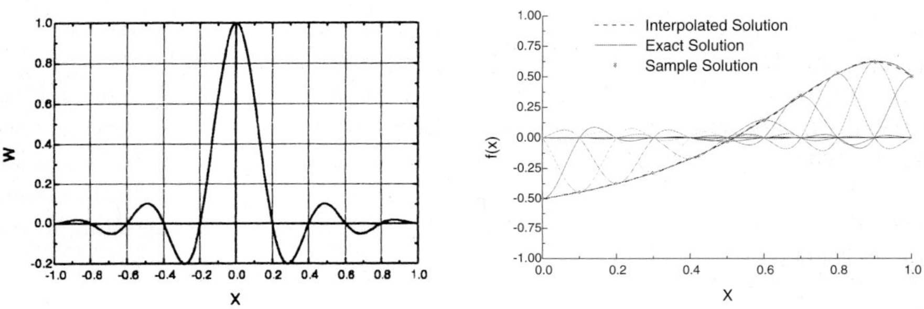

Fig. 1. Finite spectral approximation. (**a**) Expansion of a pulse function in Fourier series. (**b**) Finite spectral interpolation for a non-periodic polynomial

4 Concluding Remarks

The new finite spectral method is a powerful tool in CFD. The characteristics of this method are local property, non-periodicity, orthogonal relation, efficiency and simplicity.

References

1. J. P. Wang: 'Key to Problems in Spectral Methods'. In *Computational Fluid Dynamics Review 1998*. ed. by M. Hafez, K. Oshima (World Scientific, Singapore 1998) pp. 369-378
2. A. T. Patera: J. Comput. Phys. **54**, 468(1984)
3. J. P. Wang: Computers & Fluids **27**, 639(1998)

This work is supported by National Science Foundation of China and National Climbed Program and Ministry of Education of China.

Index

Druck: Strauss Offsetdruck, Mörlenbach
Verarbeitung: Schäffer, Grünstadt